Lecture Notes in Computer Science 14093

Founding Editors

Gerhard Goos
Juris Hartmanis

Editorial Board Members

Elisa Bertino, USA
Wen Gao, China

Bernhard Steffen ⓘ, Germany
Moti Yung ⓘ, USA

Advanced Research in Computing and Software Science
Subline of Lecture Notes in Computer Science

Subline Series Editors

Giorgio Ausiello, *University of Rome 'La Sapienza', Italy*
Vladimiro Sassone, *University of Southampton, UK*

Subline Advisory Board

Susanne Albers, *TU Munich, Germany*
Benjamin C. Pierce, *University of Pennsylvania, USA*
Bernhard Steffen ⓘ, *University of Dortmund, Germany*
Deng Xiaotie, *Peking University, Beijing, China*
Jeannette M. Wing, *Microsoft Research, Redmond, WA, USA*

More information about this series at https://link.springer.com/bookseries/558

Daniël Paulusma · Bernard Ries
Editors

Graph-Theoretic Concepts in Computer Science

49th International Workshop, WG 2023
Fribourg, Switzerland, June 28–30, 2023
Revised Selected Papers

Springer

Editors
Daniël Paulusma (iD)
Durham University
Durham, UK

Bernard Ries (iD)
University of Fribourg
Fribourg, Switzerland

ISSN 0302-9743   ISSN 1611-3349   (electronic)
Lecture Notes in Computer Science
ISBN 978-3-031-43379-5   ISBN 978-3-031-43380-1   (eBook)
https://doi.org/10.1007/978-3-031-43380-1

This Springer imprint is published by the registered company Springer Nature Switzerland AG
The registered company address is: Gewerbestrasse 11, 6330 Cham, Switzerland

Paper in this product is recyclable.

Preface

This volume contains the 33 papers presented at the 49th International Workshop on Graph-Theoretic Concepts in Computer Science (WG 2023). The conference was held in Fribourg, Switzerland from 28 June 2023 to 30 June 2023. About 55 participants from all over the world attended the conference in person, and about 30 participants registered for the conference on-line.

WG has a long-standing tradition. Since 1975, WG has taken place 25 times in Germany, five times in The Netherlands, three times in France, twice in Austria, the Czech Republic, and the UK, and once in Greece, Israel, Italy, Norway, Poland, Slovakia, Spain and Turkey. This was the second time the conference was held in Switzerland.

WG aims to merge theory and practice by demonstrating how concepts from graph theory can be applied to various areas in computer science, or by extracting new graph-theoretic problems from applications. The conference is well balanced with respect to established researchers and junior scientists.

We received 116 submissions, sixteen of which were withdrawn before entering the (single-blind) review process. The Program Committee provided three independent reviews for each submission. After a careful discussion, the Program Committee accepted 33 papers, which yields an acceptance ratio of 33%. Due to the high competition and a limited schedule, there were papers that could not be accepted although they deserved to be.

The program included three inspired invited talks by Flavia Bonomo (University of Buenos Aires, Argentina) on "Generalized List Matrix Partition Problems on Chordal Graphs, Parameterized by Leafage", Eunjung Kim (LAMSADE, Paris-Dauphine University, France) on "Twin-width, Graph Classes and a Bit of Logic" and Nicolas Trotignon, (CNRS, École normale supérieure de Lyon, France) on "Triangle-free Graphs of Large Chromatic Number".

The third WG Test of Time Award, given for a highly influential paper presented at a previous WG conference, was given to Alistair Sinclair and Mark Jerrum for their paper "Approximate Counting, Uniform Generation and Rapidly Mixing Markov Chains" from WG 1987. An excellent fourth invited talk on "35 Years of Counting, Sampling and Mixing" was given by Mark Jerrum (Queen Mary University of London, UK).

The WG 2023 Best Paper award was given to Paul Jungeblut, Samuel Schneider and Torsten Ueckerdt for their paper "Cops and Robber – When Capturing is not Surrounding". The WG 2023 Best Student Paper Award was given to Falko Hegerfeld for his paper "Tight Algorithms for Connectivity Problems Parameterized by Modular-Treewidth", co-authored by Stefan Kratsch. Both awards were sponsored by Springer-Verlag.

We would like to thank the following organisations for their financial support (in alphabetical order): the Canton of Fribourg, the City of Fribourg, Springer-Verlag, the

Swiss National Science Foundation (SNSF) and the University of Fribourg. We would also like to thank all the authors, the members of the Program Committee and additional reviewers, the speakers, session chairs, student helpers and all other participants for their contribution towards making the conference a successful event. In particular, we thank Felicia Lucke for all her excellent work on the Organising Committee.

July 2023   Daniël Paulusma
                                                                 Bernard Ries

Organization

Program Committee

Nick Brettell	Victoria University of Wellington, New Zealand
Yixin Cao	Hong Kong Polytechnic University, China
Clément Dallard	Université d'Orléans, France
Vida Dujmović	McGill University, Canada
David Eppstein	University of California, Irvine, USA
Bruno Escoffier	Sorbonne University, France
Carl Feghali	École Normale Supérieure de Lyon, France
Esther Galby	Hamburg University of Technology, Germany
Danny Hermelin	Ben-Gurion University of the Negev, Israel
Yusuke Kobayashi	Kyoto University, Japan
Stephen Kobourov	University of Arizona, USA
Daniel Král'	Masaryk University, Czech Republic
Michael Lampis	Paris Dauphine University, France
Paloma Lima	IT University of Copenhagen, Denmark
Amer Mouawad	American University of Beirut, Lebanon
Andrea Munaro	University of Parma, Italy
Daniel Paulusma (Chair)	Durham University, UK
Irena Penev	Charles University, Czech Republic
Bernard Ries (Chair)	University of Fribourg, Switzerland
Laura Sanitá	Bocconi University, Italy
Roohani Sharma	Max Planck Institute for Informatics, Germany
Uéverton Souza	Fluminense Federal University, Brazil
Maya Stein	University of Chile, Chile
Csaba Tóth	California State University Northridge, USA
Virginia Vassilevska Williams	Massachusetts Institute of Technology, USA
Shira Zerbib	Iowa State University, USA

Additional Reviewers

Aboulker, Pierre	Akmal, Shyan
Abu-Khzam, Faisal	Alecu, Bogdan
Ackerman, Eyal	Atminas, Aistis
Agrawal, Akanksha	Bampis, Evripidis
Ahmed, Abu Reyan	Bartier, Valentin
Ahn, Jungho	Barát, János
Aigner-Horev, Elad	Baste, Julien
Akhoondian Amiri, Saeed	Belmonte, Rémy

Benedek, Márton
Bentz, Cédric
Bhore, Sujoy
Bonomo, Flavia
Bouquet, Valentin
Boyar, Joan
Bradshaw, Peter
Brandenburg, Franz
Broersma, Hajo
Burgess, Andrea
Cameron, Kathie
Casel, Katrin
Chakraborty, Dibyayan
Chan, Timothy M.
Chaplick, Steven
Chen, Yong
Chepoi, Victor
Chitnis, Rajesh
Choi, Ilkyoo
Curticapean, Radu
Dalirrooyfard, Mina
Dey, Sanjana
Dibek, Cemil
Dourado, Mitre
Dreier, Jan
Duron, Julien
Dvořák, Pavel
Dósa, György
Eiben, Eduard
El Sabeh, Remy
Faria, Luerbio
Fernau, Henning
Figueiredo, Celina
Focke, Jacob
Friggstad, Zachary
Gajarský, Jakub
Garlet Milani, Marcelo
Gaspers, Serge
Geniet, Colin
Ghahremani, Mani
Ghosh, Anirban
Giroudeau, Rodolphe
Golovach, Petr
Gonzalez, Carolina
Haslegrave, John
Hatzel, Meike

Heeger, Klaus
Huang, Shenwei
Huroyan, Vahan
Huszár, Kristóf
Hörsch, Florian
Itzhaki, Yuval
Jacob, Dalu
Jaffke, Lars
Jain, Pallavi
Jana, Satyabrata
Jin, Ce
Kanj, Iyad
Kaul, Matthias
Kellerhals, Leon
Kloks, Ton
Knop, Dušan
Koana, Tomohiro
Korchemna, Viktoriia
Kratochvil, Jan
Krenn, Mario
Krithika, R.
Kryven, Myroslav
Kumar, Ravi
Kwon, O-Joung
Křišťan, Jan Matyáš
La, Hoang
Lagoutte, Aurélie
Lendl, Stefan
Li, Shaohua
Li, Yanjia
Lidicky, Bernard
Liedloff, Mathieu
Lin, Bingkai
Liu, Yunlong
Lucke, Felicia
Maaz, Stephanie
Madaras, Tomas
Maia, Ana Karolinna
Majewski, Konrad
Mann, Felix
Martin, Barnaby
Masařík, Tomáš
Mathew, Rogers
Mathialagan, Surya
Mazzoleni, María Pía
McGinnis, Daniel

Melissinos, Nikolaos
Miyazaki, Shuichi
Mnich, Matthias
Molter, Hendrik
Moore, Benjamin
Mütze, Torsten
Narayanaswamy, N. S.
Newman, Alantha
Nguyen, Tung
Nisse, Nicolas
Okamoto, Yoshio
Okrasa, Karolina
Oliveira, Fabiano
Oostveen, Jelle
Oum, Sang-il
Paesani, Giacomo
Panolan, Fahad
Picouleau, Christophe
Pierron, Théo
Pollner, Tristan
Rajendran, Goutham
Razgon, Igor
Rong, Guozhen
Roy, Sanjukta
Schindl, David
Schirneck, Martin
Schlotter, Ildikó
Sen, Sagnik
Serna, Maria
Siebertz, Sebastian
Sikora, Florian
Simonov, Kirill

Sivaraman, Vaidy
Smith, Siani
Sokołowski, Marek
Sonmez Turan, Meltem
Spence, Richard
Stamoulis, Giannos
Štorgel, Kenny
Sylvester, John
Takazawa, Kenjiro
Tale, Prafullkumar
Telle, Jan Arne
Thilikos, Dimitrios
Tong, Weitian
Tsur, Dekel
Turcotte, Jèrèmie
Ueckerdt, Torsten
van der Poel, Andrew
van Iersel, Leo
van Leeuwen, Erik Jan
Vasilakis, Manolis
Vaxès, Yann
Vialette, Stéphane
Vu, Tung Anh
Weller, Mathias
Wiederrecht, Sebastian
Wulms, Jules
Xefteris, Michalis
Xu, Yinzhan
Yang, Shizhou
Yang, Yongjie
Zerovnik, Janez

Contents

Proportionally Fair Matching
with Multiple Groups

Sayan Bandyapadhyay[1] , Fedor V. Fomin[2] , Tanmay Inamdar[2](✉) ,
and Kirill Simonov[3]

[1] Portland State University, Portland, USA
sayanb@pdx.edu
[2] University of Bergen, Bergen, Norway
{Fedor.Fomin,Tanmay.Inamdar}@uib.no
[3] Hasso Plattner Institute, University of Potsdam, Potsdam, Germany
Kirill.Simonov@hpi.de

Abstract. We study matching problems with the notion of proportional fairness. Proportional fairness is one of the most popular notions of group fairness where every group is represented up to an extent proportional to the final selection size. Matching with proportional fairness or more commonly, proportionally fair matching, was introduced in [Chierichetti et al., AISTATS, 2019]. In this problem, we are given a graph G whose edges are colored with colors from a set C. The task is for given $0 \leq \alpha \leq \beta \leq 1$, to find a maximum (α, β)-balanced matching M in G, that is a matching where for every color $c \in C$ the number of edges in M of color c is between $\alpha|M|$ and $\beta|M|$. Chierichetti et al. initiated the study of this problem with two colors and in the context of bipartite graphs only. However, in many practical applications, the number of colors—although often a small constant—is larger than two. In this work, we make the first step towards understanding the computational complexity of proportionally fair matching with more than two colors. We design exact and approximation algorithms achieving reasonable guarantees on the quality of the matching as well as on the time complexity, and our algorithms work in general graphs. Our algorithms are also supported by suitable hardness bounds.

Keywords: Matching · Fairness · Parameterized Algorithms

1 Introduction

In this paper, we consider the proportionally fair matching problem in general graphs. Matching is one of the most fundamental notions in graph theory whose

Most of this work was done when all four authors were affiliated with the University of Bergen, Norway. The research leading to these results has received funding from the Research Council of Norway via the project BWCA (grant no. 314528), and the European Research Council (ERC) via grant LOPPRE, reference 819416.

D. Paulusma and B. Ries (Eds.): WG 2023, LNCS 14093, pp. 1–15, 2023.
https://doi.org/10.1007/978-3-031-43380-1_1

study can be traced back to the classical theorems of Kőnig [33] and Hall [24]. The first chapter of the book of Lovász and Plummer [37] devoted to matching contains a nice historical overview of the development of the matching problem. The problems of finding a maximum size or a perfect matching are the classical algorithmic problems; an incomplete list of references covering the history of algorithmic improvements on these problems is [18,25,40,41,44], see also the book of Schrijver [47] for a historical overview of matching algorithms. Matchings appear naturally in various applications, e.g., kidney transplant matching [46] or numerous assignment problems like assigning products to customers [45]; students to schools [34]; reviewers to manuscripts [8]; and workers to firms [1]. There are scores of works that study fair versions of matchings [10,20,21,26,28,31,50]. Among these distinct notions of matchings, our work is most relevant to the work on (α, β)-balanced matching of Chierichetti et al. [10]. The notion of (α, β)-balanced matching was formulated in [10] by bringing *proportional* fairness and maximum cardinality matching together. Proportional fairness is based on the concept of disparate impact [19], which in the context of matching is defined as follows. A matching is (α, β)-balanced or proportionally fair if the ratio between the number of edges from each group (a color) and the size of the matching is at least α and at most β.

As a motivating example of proportionally fair matching, consider the product recommendation problem in e-commerce. With the advancement of digital marketing and advertising, nowadays companies are interested in more fine-tuned approaches that help them reach the target groups of customers. These groups may be representative of certain underlying demographic categorizations based on gender, age group, geographic location etc. Thus, the number of groups is often a small constant. In particular, in this contemporary setting, one is interested in finding assignments that involve customers from all target groups and have a balanced impact on all these groups. This assignment problem can be modeled as the proportionally fair matching problem between customers and products. In a realistic situation, one might need to assign many products to a customer and many customers to a product. This can be achieved by computing multiple matchings in an iterative manner while removing the edges from the input graph that are already matched.

In a seminal work, Chierichetti et al. [10] obtained a polynomial-time 3/2-approximation for the size of the matching, when the input graph is bipartite and the number of groups is 2. However, in many real-world situations, like in the above example, it is natural to assume that the number of target groups is more than 2. Unfortunately, the algorithm of [10] strongly exploits the fact that the number of groups $\ell = 2$. It is not clear how to adapt or extend their algorithm when we have more than two groups. The only known algorithm for $\ell > 2$ groups is an $n^{O(\ell)}$-time randomized exact algorithm [10,14], which also works for general graphs. The running time of this algorithm has a "bad" exponential dependence on the number of groups, i.e., the running time is not a *fixed* polynomial in n. Thus, this algorithm quickly becomes impractical if ℓ grows. Our research on proportionally fair matching is driven by the following question. *Do there exist* *efficient algorithms with guaranteed performance for proportionally fair matching when the number of groups ℓ is more than two?*

1.1 Our Results and Contributions

In this work, we obtain several results on the PROPORTIONALLY FAIR MATCHING problem in general graphs with any arbitrary ℓ number of groups.

- First, we show that the problem is extremely hard for any general ℓ number of groups, in the sense that it is not possible to obtain any approximation algorithm in $2^{o(\ell)}n^{O(1)}$ time even on path graphs, unless the Exponential Time Hypothesis (ETH) [27] is false.
- To complement our hardness result, we design a $\frac{1}{4\ell}$-approximation algorithm that runs in $2^{O(\ell)}n^{O(1)}$ time. Our algorithm might violate the lower (α) and upper (β) bounds by at most a multiplicative factor of $(1 + \frac{4\ell}{|OPT|}$ if $|OPT|$ is more than $4\ell^2$, where OPT is any optimum solution. Thus, the violation factor is at most $1 + \frac{1}{\ell}$, and tends to 1 with asymptotic values of $|OPT|$.
- We also consider a restricted case of the problem, referred to as the β-limited case in [10], where we only have the upper bound, i.e., no edges might be present from some groups. In this case, we could improve the approximation factor to $\frac{1}{2\ell}$ and running time to polynomial.
- Lastly, we show that the parameterized version of the problem where one seeks for a proportionally fair matching of size k, can be solved exactly in $2^{O(k)}n^{O(1)}$ time. Thus the problem is fixed-parameter tractable parameterized by k.

All of our algorithms are based on simple schemes. Our approximation algorithms use an iterative peeling scheme that in each iteration, extracts a *rainbow* matching containing at most one edge from every group. The exact algorithm is based on a non-trivial application of the celebrated color-coding scheme [2]. These algorithms appear in Sects. 3, 4, and 5, respectively. The hardness proof is given in Sect. 6.

1.2 Related Work

In recent years, researchers have introduced and studied several different notions of fairness, e.g., disparate impact [19], statistical parity [29,51], individual fairness [15] and group fairness [16]. Kleinberg et al. [32] formulated three notions of fairness and showed that it is theoretically impossible to satisfy them simultaneously. See also [11,12] for similar exposures.

The notion of proportional fairness with multiple protected groups is widely studied in the literature, which is based on disparate impact [19]. Bei et al. [4] studied the *proportional candidate selection problem*, where the goal is to select a subset of candidates with various attributes from a given set while satisfying certain proportional fairness constraints. Goel et al. [22] considered the problem of learning non-discriminatory and proportionally fair classifiers and proposed the *weighted sum of logs* technique. Proportional fairness has also been considered in the context of Federated learning [53]. Additionally, proportional fairness has been studied in the context of numerous optimization problems including voting [17], scheduling [30,38], kidney exchange [5,48], and Traveling Salesman Problem [43].

Several different fair matching problems have been studied in the literature. Huang et al. [26] and Boehmer et al. [7] studied fair b-matching, where matching preferences for each vertex are given as ranks, and the goal is to avoid assigning vertices to the highly ranked vertices as much as possible (see [7,26] for a formal definition). Fair-by-design-matching is studied in [21], where instead of a single matching, a probability distribution over all feasible matchings is computed which guarantees individual fairness. See also [28,31].

Apart from the fair versions of matchings, many constrained versions are also studied [6,49]. [49] studied the Bounded Color Matching (BCM) problem where edges are colored and from each color class, only a given number of edges can be chosen. BCM is a special case of 3-set packing and, hence, admits a 3/4-approximation [49]. We note that the β-limited case of PROPORTIONALLY FAIR MATCHING is a special case of BCM and, thus, a 3/4-approximation follows in this case where the upper bound might be violated by 3/4 factor. One should compare this factor with our violation factor, which asymptotically tends to 1.

Finally, we note that when the input graph is bipartite, a matching has a natural interpretation as an intersection of two matroids. Matroid intersection has a rich literature containing work on both exact [36,39] and approximation algorithms [35]. However, these algorithms do not consider fairness constraints.

2   Preliminaries

For an integer $\ell \geq 1$, let $[\ell] := \{1, 2, \ldots, \ell\}$. Consider any undirected n-vertex graph $G = (V, E)$ such that the edges in E are colored by colors in $C = \{1, \ldots, \ell\}$. The function $\chi : E \to C$ describes the color assignment. For each color $c \in C$, let E_c be the set of edges colored by the color c, i.e., $E_c = \chi^{-1}(c)$. A subset $E' \subseteq E$ is a *matching* in G if no two edges in E' share a common vertex.

Definition 1. (α, β)-***balanced matching.*** *Given* $0 \leq \alpha \leq \beta \leq 1$, *a matching* $M \subseteq E$ *is called* (α, β)-*balanced if for each color* $c \in C$, *we have that* $\alpha \leq \dfrac{|M \cap E_c|}{|M|} \leq \beta$.

Thus a matching is (α, β)-balanced if it contains at least α and at most β fraction of edges from every color. In the PROPORTIONALLY FAIR MATCHING problem, the goal is to find a maximum-sized (α, β)-balanced matching. In the restricted β-limited case of the problem, $\alpha = 0$, i.e., we only have the upper bound.

For $\gamma \leq 1$ and $\Delta \geq 1$, a (γ, Δ)-approximation algorithm for PROPORTIONALLY FAIR MATCHING computes a matching of size at least $\gamma \cdot |\text{OPT}|$, where every color appears in at least α/Δ fraction of the edges and in at most $\beta \cdot \Delta$ fraction. OPT is an optimum (α, β)-balanced matching.

A matching is called a *rainbow matching* if all of its edges have distinct colors. We will need the following result due to Gupta et al. [23].

Theorem 1 (Theorem 2 in [23]). *For some integer $k > 0$, suppose there is a rainbow matching in G of size k. There is a $2^k \cdot n^{O(1)}$ time algorithm that computes a rainbow matching of size k.*

3   A $(\frac{1}{4\ell}, 1 + \frac{4\ell}{|OPT|})$-Approximation for PROPORTIONALLY FAIR MATCHING

In this section, we design an approximation algorithm for PROPORTIONALLY FAIR MATCHING. Let OPT be an optimum (α, β)-balanced matching (which, we assume, exists), $\text{OPT}_c = \text{OPT} \cap E_c$. We design two algorithms: one for the case when $\alpha > 0$ and the other for the complementary β-limited case. In this section, we slightly abuse the notation, and use OPT (resp. OPT_c for some color $c \in C$) to refer to $|\text{OPT}|$ (resp. $|\text{OPT}_c|$). The intended meaning should be clear from the context; however we will disambiguate in case there is a possibility of confusion.

First, we consider the $\alpha > 0$ case. Immediately, we have the following observation.

Observation 1. *For any color $c \in C$, OPT contains at least one edge of color c and, hence, G contains a rainbow matching of size ℓ.*

Our algorithm runs in *rounds*. In the following, we define a round. The input in each round is a subgraph $G' = (V', E')$ of G.

Round. Initially $M = \emptyset$. For every color $1 \le c \le \ell$, do the following in an iterative manner. If there is no edge of color c in G', go to the next color or terminate and return (G', M) if $c = \ell$. Otherwise, pick any edge e of color c from G' and add e to the already computed matching M. Remove all the edges (including e) from G' that share a common vertex with e. Repeat the process for the next color with the current (or updated) graph G' or terminate and return (G', M) if $c = \ell$. Thus in each round, we find a rainbow matching in a greedy manner.

Next, we describe our algorithm. The most challenging part of our algorithm is to ensure that the final matching computed is (α, β)-balanced modulo a small factor, i.e., we need to ensure both the lower and the upper bounds are within a small factor for each color. Note that just the above greedy way of picking edges might not even ensure that at least one edge is selected from each color. We use the algorithm of [23] in the beginning to overcome this barrier. However, the rest of our algorithm is extremely simple.

The Algorithm. We assume that we know the size of OPT. We describe later how to remove this assumption. Apply the algorithm in Theorem 1 on G to compute a rainbow matching M' of size ℓ. If $\text{OPT} \le 4\ell^2$, return $M := M'$ as the solution and terminate. Otherwise, remove all the edges of M' and the edges adjacent to them from G to obtain the graph G_0. Initialize M to M'. Greedily pick matched edges in rounds using the **Round** procedure and add them to M until exactly $\lceil \text{OPT}/(4\ell) \rceil$ edges are picked in total. In particular, the graph G_0 is the input to the 1-st round and G_1 is the output graph of the 1-st round. G_1 is the input to the 2-nd round and G_2 is the output graph of the 2-nd round, and so on. Note that it might be the case that the last round is not completed fully if the size of M is reached to $\lceil \text{OPT}/(4\ell) \rceil$ before the completion of the round.

Note that the above algorithm is oblivious to α and β in the sense that it never uses these values. Nevertheless, we prove that the computed matching is (α, β)-balanced modulo a small factor. Now we analyze our algorithm.

3.1 The Analysis

Let $M_c = M \cap E_c$. Also, let c^* be a color $c \in C$ such that $|\text{OPT}_c|$ is the minimum at $c = c^*$. The proof of the following observation is fairly straightforward and can be found in the full version [3].

Observation 2. $\alpha \leq 1/\ell \leq \beta$.

First we consider the case when $\text{OPT} \leq 4\ell^2$. In this case the returned matching M is a rainbow matching of size exactly ℓ. The existence of such a matching follows by Observation 1. Thus, we immediately obtain a 4ℓ-approximation. As $|M_c|/|M| = 1/\ell$ in this case, by Observation 2, $\alpha \leq |M_c|/|M| \leq \beta$. Thus we obtain the desired result. In the rest of the proof, we analyze the case when $\text{OPT} > 4\ell^2$. We start with the following lemma.

Lemma 1. *The algorithm successfully computes a matching of size exactly* $\lceil \text{OPT}/(4\ell) \rceil$. *Moreover, for each color c with $\text{OPT}_c > 4\ell$ and round $i \in [1, \lceil \text{OPT}_c/(4\ell) \rceil - 1]$, G_{i-1} contains an edge of color c.*

Proof. Note that by Observation 1, the algorithm in Theorem 1 successfully computes a rainbow matching M' of size ℓ. Now consider any color c such that $\text{OPT}_c \leq 4\ell$. For such a color, M already contains at least $1 \geq \lceil \text{OPT}_c/(4\ell) \rceil$ edge. Now consider any other color c with $|\text{OPT}_c| > 4\ell$. Consider the rainbow matching M' computed in the beginning. As $|M'| = \ell$, the edges of M' can be adjacent to at most 2ℓ edges from OPT, since it is a matching. In particular, the edges of M' can be adjacent to at most 2ℓ edges from the set OPT_c. Hence, G_0 contains at least $\text{OPT}_c - 2\ell$ edges of the set OPT_c. Now consider the execution of round $i \geq 1$. At most ℓ edges are chosen in this round. Hence, these edges can be adjacent to at most 2ℓ edges of OPT_c. It follows that at most 2ℓ fewer edges of the set OPT_c are contained in G_i compared to G_{i-1}. As G_0 has at least $\text{OPT}_c - 2\ell$ edges from the set OPT_c of color c and $\text{OPT}_c > 4\ell$, for each of the first $\lceil (\text{OPT}_c - 2\ell)/(2\ell) \rceil = \lceil \text{OPT}_c/(2\ell) \rceil - 1$ rounds, the algorithm will be able to pick an edge of color c. Thus from such a color c with $\text{OPT}_c > 4\ell$, it can safely pick at least $\lceil \text{OPT}_c/(2\ell) \rceil \geq \lceil \text{OPT}_c/(4\ell) \rceil$ edges in total. Now, as $\text{OPT} = \sum_c \text{OPT}_c$, $\sum_{c \in C} \lceil \text{OPT}_c/(4\ell) \rceil \geq \lceil \text{OPT}/(4\ell) \rceil$. It follows that the algorithm can pick at least $\lceil \text{OPT}/(4\ell) \rceil$ edges. As we stop the algorithm as soon as the size of M reaches to $\lceil \text{OPT}/(4\ell) \rceil$, the lemma follows.

Note that the claimed approximation factor trivially follows from the above lemma. Next, we show that M is (α, β)-balanced modulo a small factor that asymptotically tends to 1 with the size of OPT.

Lemma 2. *For each color* $c \in C$, $|M_c| \geq |\text{OPT}_{c^*}|/(4\ell)$.

Proof. If $\text{OPT}_{c^*} \leq 4\ell$, $|M_c| \geq 1 \geq \text{OPT}_{c^*}/(4\ell)$. So, assume that $\text{OPT}_{c^*} > 4\ell$. Now suppose $|M_c| < \text{OPT}_{c^*}/(4\ell)$ for some c. By Lemma 1, for each of the first $\lceil \text{OPT}_c/(4\ell) \rceil - 1 \geq \lceil \text{OPT}_{c^*}/(4\ell) \rceil - 1$ rounds, G_{i-1} contains an edge of color c. It follows that the algorithm was forcibly terminated in some round $i \leq (\text{OPT}_{c^*}/(4\ell)) - 1$. Thus, the number of edges chosen from each color $c' \neq c$ is at most $\text{OPT}_{c^*}/(4\ell)$. Hence,

$$|M| = \sum_{c' \neq c} |M_{c'}| + |M_c| < (\ell - 1) \cdot (\text{OPT}_{c^*}/(4\ell)) + (\text{OPT}_{c^*}/(4\ell)) \leq \lceil \text{OPT}/(4\ell) \rceil.$$

This contradicts Lemma 1, which states that we select exactly $\lceil \text{OPT}/(4\ell) \rceil$ edges.

Corollary 1. *For each color* $c \in C$, $(|M_c|/|M|) \geq \frac{\alpha}{1 + \frac{4\ell}{\text{OPT}}}$.

Proof. By Lemma 2, $|M_c| \geq \text{OPT}_{c^*}/(4\ell)$.

$$\frac{|M_c|}{|M|} \geq \frac{(\text{OPT}_{c^*}/(4\ell))}{\lceil \text{OPT}/(4\ell) \rceil} \geq \frac{(\text{OPT}_{c^*}/(4\ell))}{(\text{OPT}/(4\ell)) + 1} = \frac{(\text{OPT}_{c^*})/(\text{OPT})}{1 + \frac{4\ell}{\text{OPT}}} \geq \frac{\alpha}{1 + \frac{4\ell}{\text{OPT}}}.$$

The last inequality follows as OPT satisfies the lower bound for all colors.

Now we turn to proving the upper bound. Let $\alpha^* = \text{OPT}_{c^*}/\text{OPT}$.

Lemma 3. *For each color* $c \in C$, $|M_c| \leq \frac{\beta}{\alpha^*} \cdot (\text{OPT}_{c^*}/(4\ell)) + 1$.

Proof. Suppose for some $c \in C$, $|M_c| > \frac{\beta}{\alpha^*} \cdot (\text{OPT}_{c^*}/(4\ell)) + 1$. Then the number of rounds is strictly greater than $\frac{\beta}{\alpha^*} \cdot (\text{OPT}_{c^*}/(4\ell))$. Now, for any c', $\text{OPT}_{c'} \geq \alpha^* \cdot \text{OPT}$ and $\text{OPT}_{c'} \leq \beta \cdot \text{OPT}$. Thus, by the definition of α^*, $\frac{\beta}{\alpha^*} \cdot \text{OPT}_{c^*} \geq \text{OPT}_{c'}$. It follows that, for each c', the number of rounds is strictly greater than $\text{OPT}_{c'}/(4\ell)$. Hence, for each $c' \in C$, more than $(\text{OPT}_{c'}/(4\ell)) + 1$ edges have been chosen. Thus, the total number of edges chosen is strictly larger than

$$\sum_{c' \in C} ((\text{OPT}_{c'}/(4\ell)) + 1) \geq \lceil \text{OPT}/(4\ell) \rceil.$$

This contradicts Lemma 1, which states that we select exactly $\lceil \text{OPT}/(4\ell) \rceil$ edges.

Corollary 2. *For each color* $c \in C$, $(|M_c|/|M|) \leq \beta \cdot (1 + \frac{4\ell}{\text{OPT}})$.

Proof. By Lemma 3,

$$\frac{|M_c|}{|M|} \leq \frac{(\beta/\alpha^*) \cdot (\text{OPT}_{c^*}/(4\ell)) + 1}{\lceil \text{OPT}/(4\ell) \rceil} \leq \frac{(\beta/\alpha^*) \cdot (\text{OPT}_{c^*}/(4\ell)) + (\beta/\alpha^*)}{\text{OPT}/(4\ell)}$$

$$= \frac{\beta}{\alpha^*} \cdot \frac{\text{OPT}_{c^*}}{\text{OPT}} \cdot \left(1 + \frac{4\ell}{\text{OPT}}\right)$$

$$= \frac{\beta}{\alpha^*} \cdot \alpha^* \left(1 + \frac{4\ell}{\text{OPT}}\right) = \beta \cdot \left(1 + \frac{4\ell}{\text{OPT}}\right).$$

The second inequality follows, as $\alpha^* \leq \beta$ or $\beta/\alpha^* \geq 1$.

Now let us remove the assumption that we know the size of an optimal solution. Note that $\ell \leq \text{OPT} \leq n$. We probe all values between ℓ and n, and for each such value T run our algorithm. For each matching M returned by the algorithm, we check whether M is $(\frac{\alpha}{(1+4\ell/T)}, \beta \cdot (1 + \frac{4\ell}{T}))$-balanced. If this is the case, then we keep this solution. Otherwise, we discard the solution. Finally, we select a solution of the largest size among the ones not discarded. By the above analysis, with $T = \text{OPT}$, the matching returned satisfies the desired lower and upper bounds, and has size exactly $\lceil \text{OPT}/(4\ell) \rceil$. Finally, the running time of our algorithm is dominated by $2^\ell n^{O(1)}$ time to compute a rainbow matching algorithm, as stated in Theorem 1.

Theorem 2. *There is a $2^\ell \cdot n^{O(1)}$ time $(\frac{1}{4\ell}, 1 + \frac{4\ell}{\text{OPT}})$-approximation algorithm for* PROPORTIONALLY FAIR MATCHING *with $\alpha > 0$.*

4   A Polynomial-Time Approximation in the β-Limited Case

In the β-limited case, again we make use of the **Round** procedure. But, the algorithm is slightly different. Most importantly, we do not apply the algorithm in Theorem 1 in the beginning. Thus, our algorithm runs in polynomial time.

The Algorithm. Assume that we know the size of OPT. If $\text{OPT} \leq 2\ell$, we pick any edge and return it as the solution. Otherwise, we just greedily pick matched edges in rounds using the **Round** procedure with the following two cautions. If for a color, at least $\beta \cdot \text{OPT}/(2\ell)$ edges have already been chosen, do not choose any more edge of that color. If at least $\frac{\text{OPT}}{2\ell} - 1$ edges have already been chosen, terminate.

Now we analyze the algorithm. First note that if $\text{OPT} \leq 2\ell$, the returned matching has only one edge. The upper bound is trivially satisfied and also we obtain a 2ℓ-approximation. Henceforth, we assume that $\text{OPT} > 2\ell$. Before showing the correctness and analysis of the approximation factor, we show the upper bound for each color. Again let M be the computed matching and $M_c = M \cap E_c$. The proof of the following lemma can be found in the full version [3].

Lemma 4. *Algorithm always returns a matching of size at least $(\text{OPT}/2\ell) - 1$.*

Assuming this we have the following proposition.

Proposition 1. *For each color $c \in C$, $|M_c|/|M| \leq \beta \cdot (1 + \frac{2\ell}{|\text{OPT}|})$.*

Proof. By Lemma 4 and the threshold put on each color in the algorithm, $\frac{|M_c|}{|M|} \leq \frac{\beta \cdot \text{OPT}/(2\ell)}{(\text{OPT}/(2\ell)) - 1} \leq \beta \cdot (1 + \frac{2\ell}{\text{OPT}})$ The last inequality follows, as $\text{OPT} > 2\ell$.

Theorem 3. *There is a polynomial time algorithm for* PROPORTIONALLY FAIR MATCHING *in the β-limited case that returns a matching of size at least $(\text{OPT}/2\ell) - 1$ where every color appears in at most $\beta \cdot (1 + 2\ell/\text{OPT})$ fraction of the edges.*

5   An Exact Algorithm for PROPORTIONALLY FAIR MATCHING

Theorem 4. *There is a $2^{O(k)}n^{O(1)}$-time algorithm that either finds a solution of size k for a* PROPORTIONALLY FAIR MATCHING *instance, or determines that none exists.*

Proof. We present two different algorithms using the well-known technique of color coding: one for the case $\alpha = 0$ (β-limited case), and one for the case $\alpha > 0$. *β-limited case.* We aim to reduce the problem to finding a rainbow matching of size k, which we then solve via Theorem 1. The graph G remains the same, however the coloring is going to be different. Namely, for each of the original colors $c \in C$ we color the edges in E_c uniformly and independently at random from a set of k' new colors, where $k' = \lfloor \beta k \rfloor$. Thus, the new instance I' is colored in $\ell \cdot k'$ colors. We use the algorithm of Theorem 1 to find a rainbow matching of size k in the colored graph in I'. Clearly, if a rainbow matching M of size k is found, then the same matching M is a β-limited matching of size k in the original coloring. This holds since by construction for any original color $c \in C$, there are k' new colors in the edge set E_c, and therefore no more than k' edges in $|M \cap E_c|$.

In the other direction, we show that if there exists a β-limited matching M of size k with respect to the original coloring, then with good probability M is a rainbow matching of size k in the new coloring. Assume the original colors c_1, $\ldots$, c_t, for some $1 \le t \le \ell$, have non-empty intersection with M, and for each $j \in [t]$ denote $k_j = |M \cap E_{c_j}|$. Observe that $\sum_{j=1}^{t} k_j = k$, and for each $j \in [t]$, $1 \le k_j \le k'$. The proof of the following claim can be found in the full version [3].

Claim. There exists some $\delta > 0$ such that for each $j \in [t]$:

$$\Pr\left[M \cap \left(\bigcup_{i=1}^{j} E_{c_i}\right) \text{ is a rainbow matching in } I'\right] \ge \exp\left(-\delta \sum_{i=1}^{j} k_i\right), \quad (1)$$

Applying (1) with $j = t$, we obtain that M is a rainbow matching with probability at least $2^{-\delta k}$. By repeating the reduction above $2^{O(k)}$ times independently, we achieve that the algorithm succeeds with constant probability.

The case $\alpha > 0$. We observe that in this case, if a matching is fair it necessarily contains at least one edge from each of the groups. Thus, if the number of groups ℓ is greater than k, we immediately conclude there cannot be a fair matching of size k. Otherwise, we guess how the desired k edges are partitioned between the ℓ groups $C = \{c_1, \ldots, c_\ell\}$. That is, we guess the numbers k_j for $j \in [\ell]$ such that $\sum_{j=1}^{\ell} k_j = k$, and $\alpha k \le k_j \le \beta k$ for each $j \in [\ell]$. From now on, the algorithm is very similar to the β-limited case. For each group c_j, we color the edges of E_{c_j} from a set of k_j colors uniformly and independently at random, where the colors used for each E_{c_j} are non-overlapping. Now we use the algorithm of Theorem 1 to find a rainbow matching of size k. If there is a rainbow matching M of size k, the same matching is a fair matching of size k for the original instance, since in each E_{c_j} exactly k_j edges are chosen, which is at least αk and at most βk. In the other direction, if there is a fair matching M of size k in the original instance, by (1) the matching M is a rainbow matching in the new instance with probability at least $2^{-\delta k}$. Again, by repeating the coloring subprocess independently $2^{O(k)}$ times, we achieve a constant probability of success. Since there are $2^{O(k)}$ options for partitioning k edges into at most $\ell \leq k$ groups, the running time of the whole algorithm is $2^{O(k)}n^{O(1)}$.

Finally, we note that the coloring part in both cases can be derandomized in the standard fashion by using perfect hash families [42], leading to a completely deterministic algorithm.

6   Hardness of Approximation for PROPORTIONALLY FAIR MATCHING

In this section, we show an inapproximability result for PROPORTIONALLY FAIR MATCHING under the Exponential Time Hypothesis (ETH) [27]. ETH states that $2^{\Omega(n)}$ time is needed to solve any generic 3SAT instance with n variables. For our purpose, we need the following restricted version of 3SAT.

3SAT-3
INPUT: Set of clauses $T = \{C_1, \ldots, C_m\}$ in variables $x_1, \ldots, x_n$, each clause being the disjunction of 3 or 2 literals, where a literal is a variable x_i or its negation $\bar{x}_i$. Additionally, each variable appears 3 times.
QUESTION: Is there a truth assignment that simultaneously satisfies all the clauses?

3SAT-3 is known to be NP-hard [52]. We need the following stronger lower bound for 3SAT-3 proved in [13].

Proposition 2 ([13]). *Under ETH, 3SAT-3 cannot be solved in $2^{o(n)}$ time.*

We reduce 3SAT-3 to PROPORTIONALLY FAIR MATCHING which rules out any approximation for the latter problem in $2^{o(\ell)}n^{O(1)}$ time. Our reduction is as follows. For each clause C_i, we have a color i. Also, we have $n-1$ additional colors $m+1, \ldots, m+n-1$. Thus, the set of colors $C = \{1, \ldots, m+n-1\}$. For each variable x_i, we construct a gadget, which is a 3-path (a path with 3 edges). Note that x_i can either appear twice in its normal form or in its negated form, as it appears 3 times in total. Let C_{i_1}, C_{i_2} and C_{i_3} be the clauses where x_i appears. Also, suppose it appears in C_{i_1} and C_{i_3} in one form, and in C_{i_2} in the other form. We construct a 3-path P_i for x_i where the j-th edge has color i_j for $1 \leq j \leq 3$. Additionally, we construct $n-1$ 3-paths $Q_{i,i+1}$ for $1 \leq i \leq n-1$. All edges of $Q_{i,i+1}$ is of color $m+i$. Finally, we glue together all the paths in the following way to obtain a single path. For each $1 \leq i \leq n-1$, we glue $Q_{i,i+1}$ in between P_i and P_{i+1} by identifying the last vertex of P_i with the first vertex of $Q_{i,i+1}$ and the last vertex of $Q_{i,i+1}$ with the first vertex of P_{i+1}. Thus we obtain a path P with exactly $3(n+n-1) = 6n-3$ edges. Finally, we set $\alpha = \beta = 1/(m+n-1)$.

Lemma 5. *There is a satisfying assignment for the clauses in 3SAT-3 if and only if there is an (α, β)-balanced matching of size at least $m + n - 1$.*

Proof. Suppose there is a satisfying assignment for all the clauses. For each clause C_j, consider a variable, say x_i, that satisfies C_j. Then there is an edge of color j on P_i. Add this edge to a set M. Thus, after this step, M contains exactly one edge of color j for $1 \le j \le m$. Also, note that for each path P_i, if the middle edge is chosen, then no other edge from P_i can be chosen. This is true, as the variable x_i can either satisfy the clauses where it appears in its normal form or the clauses where it appears in its negated form, but not both types of clauses. Hence, M is a matching. Finally, for each path $Q_{i,i+1}$, we add its middle edge to M. Note that M still remains a matching. Moreover, M contains exactly one edge of color j for $1 \le j \le m + n - 1$. As $\alpha = \beta = 1/(m + n - 1)$, M is an (α, β)-balanced matching.

Now suppose there is an (α, β)-balanced matching M of size at least $m+n-1$. First, we show that $|M| = m+n-1$. Note that if $|M| > m+n-1$, then the only possibility is that $|M| = 2(m+n-1)$, as $\alpha = \beta$ and at most 2 edges of color j can be picked in any matching for $m+1 \le j \le m+n-1$. Suppose $|M| = 2(m+n-1)$. Then from each $Q_{i,i+1}$, M contains the first and the third edge. This implies, from each P_t, $1 \le t \le n$, we can pick at most one edge. Thus, total number of edges in M is at most $2(n-1) + n$. It follows that $2m + 2n - 2 \le 2n - 2 + n$ or $n \ge 2m$. Now, in 3SAT-3 the total number of literals is $3n$ and at most $3m$, as each variable appears 3 times and each clause contains at most 3 literals. This implies $n \le m$, and we obtain a contradiction. Thus, $|M| = m + n - 1$. Now, consider any P_i. In the first case, the first and third edges of P_i are corresponding to literal x_i and, hence, the middle edge is corresponding to the literal $\bar{x}_i$. If the middle edge is in M, assign 0 to x_i, otherwise, assign 1 to x_i. In the other case, if the middle edge is in M, assign 1 to x_i, otherwise, assign 0 to x_i. We claim that the constructed assignment satisfies all the clauses. Consider any clause C_j. Let $e \in P_i$ be the edge in M of color j for $1 \le j \le m$. Note that e can be the middle edge in P_i or not. In any case, if e is corresponding to $\bar{x}_i$, we assigned 0 to x_i, and if e is corresponding to x_i, we assigned 1 to x_i. Thus, in either case, C_j is satisfied. This completes the proof of the lemma. $\qquad\square$

Note that for a 3SAT-3 instance the total numbers of literals is $3n$. As each clause contains at least 2 literals, $m \le 3n/2$. Now, for the instances constructed in the above proof, the number of colors $\ell = m+n-1 \le 3n/2+n-1 = 5n/2-1$. Thus, the above lemma along with Proposition 2 show that it is not possible to decide whether there is an (α, β)-balanced matching of a given size in time $2^{o(\ell)}n^{O(1)}$. Using this, we also show that even no $2^{o(\ell)}n^{O(1)}$ time approximation algorithm is possible. Suppose there is a $2^{o(\ell)}n^{O(1)}$ time γ-approximation algorithm, where $\gamma < 1$. For our constructed path instances, we apply this algorithm to obtain a matching. Note that the γ-approximate solution M must contain at least one edge of every color, as $\alpha = \beta$. By the proof in the above lemma, $|M|$ is exactly $m + n - 1$. Hence, using this algorithm, we can decide in $2^{o(\ell)}n^{O(1)}$ time whether there is an (α, β)-balanced matching of size $m + n - 1$. But, this is a contradiction, which leads to the following theorem.

Theorem 5. *For any $\gamma > 1$, under ETH, there is no $2^{o(\ell)}n^{O(1)}$ time γ-approxim- ation algorithm for* PROPORTIONALLY FAIR MATCHING, *even on paths.*

7 Conclusions

In this paper, we study the notion of proportional fairness in the context of matchings in graphs, which has been studied by Chierichetti et al. [9]. We obtained approximation and exact algorithms for the proportionally fair matching problem. We also complement these results by showing hardness results. It would be interesting to obtain a $o(\ell)$- or a true $O(\ell)$-approximation for PROPORTIONALLY FAIR MATCHING improving our result. As evident from our hardness result, there is a lower bound of $2^{\Omega(\ell)}n^{O(1)}$ on the running time of such an algorithm.

References

1. Ahmadi, S., Ahmed, F., Dickerson, J.P., Fuge, M., Khuller, S.: An algorithm for multi-attribute diverse matching. In: Bessiere, C. (ed.) Proceedings of the Twenty-Ninth International Joint Conference on Artificial Intelligence, IJCAI 2020, pp. 3–9. ijcai.org (2020)
2. Alon, N., Yuster, R., Zwick, U.: Color-coding. J. ACM (JACM) **42**(4), 844–856 (1995)
3. Bandyapadhyay, S., Fomin, F.V., Inamdar, T., Simonov, K.: Proportionally fair matching with multiple groups. CoRR abs/2301.03862 (2023). https://doi.org/10.48550/arXiv.2301.03862
4. Bei, X., Liu, S., Poon, C.K., Wang, H.: Candidate selections with proportional fairness constraints. In: Seghrouchni, A.E.F., Sukthankar, G., An, B., Yorke-Smith, N. (eds.) Proceedings of the 19th International Conference on Autonomous Agents and Multiagent Systems, AAMAS 2020, Auckland, New Zealand, 9–13 May 2020, pp. 150–158. International Foundation for Autonomous Agents and Multi-agent Systems (2020). https://doi.org/10.5555/3398761.3398784, https://dl.acm.org/doi/10.5555/3398761.3398784
5. Benedek, M., Biró, P., Kern, W., Paulusma, D.: Computing balanced solutions for large international kidney exchange schemes. In: Faliszewski, P., Mascardi, V., Pelachaud, C., Taylor, M.E. (eds.) 21st International Conference on Autonomous Agents and Multiagent Systems, AAMAS 2022, Auckland, New Zealand, 9–13 May 2022, pp. 82–90. International Foundation for Autonomous Agents and Multiagent Systems (IFAAMAS) (2022). https://doi.org/10.5555/3535850.3535861, https://www.ifaamas.org/Proceedings/aamas2022/pdfs/p82.pdf
6. Berger, A., Bonifaci, V., Grandoni, F., Schäfer, G.: Budgeted matching and budgeted matroid intersection via the gasoline puzzle. Math. Program. **128**(1–2), 355–372 (2011)
7. Boehmer, N., Koana, T.: The complexity of finding fair many-to-one matchings. In: Bojanczyk, M., Merelli, E., Woodruff, D.P. (eds.) 49th International Colloquium on Automata, Languages, and Programming, ICALP 2022, 4–8 July 2022, Paris, France. LIPIcs, vol. 229, pp. 27:1–27:18. Schloss Dagstuhl - Leibniz-Zentrum für Informatik (2022). https://doi.org/10.4230/LIPIcs.ICALP.2022.27

8. Charlin, L., Zemel, R.: The toronto paper matching system: an automated paper-reviewer assignment system (2013)
9. Chierichetti, F., Kumar, R., Lattanzi, S., Vassilvitskii, S.: Fair clustering through fairlets. In: Advances in Neural Information Processing Systems, pp. 5029–5037 (2017)
10. Chierichetti, F., Kumar, R., Lattanzi, S., Vassilvitskii, S.: Matroids, matchings, and fairness. In: Chaudhuri, K., Sugiyama, M. (eds.) The 22nd International Conference on Artificial Intelligence and Statistics, AISTATS 2019, 16–18 April 2019, Naha, Okinawa, Japan. Proceedings of Machine Learning Research, vol. 89, pp. 2212–2220. PMLR (2019)
11. Chouldechova, A.: Fair prediction with disparate impact: a study of bias in recidivism prediction instruments. Big Data **5**(2), 153–163 (2017)
12. Corbett-Davies, S., Pierson, E., Feller, A., Goel, S., Huq, A.: Algorithmic decision making and the cost of fairness. In: Proceedings of the 23rd ACM SIGKDD International Conference on Knowledge Discovery and Data Mining, Halifax, NS, Canada, 13–17 August 2017, pp. 797–806. ACM (2017)
13. Cygan, M., Marx, D., Pilipczuk, M., Pilipczuk, M.: Hitting forbidden subgraphs in graphs of bounded treewidth. Inf. Comput. **256**, 62–82 (2017)
14. Czabarka, E., Szekely, L.A., Toroczkai, Z., Walker, S.: An algebraic monte-carlo algorithm for the bipartite partition adjacency matrix realization problem. arXiv preprint arXiv:1708.08242 (2017)
15. Dwork, C., Hardt, M., Pitassi, T., Reingold, O., Zemel, R.: Fairness through awareness. In: Proceedings of the 3rd Innovations in Theoretical Computer Science Conference, pp. 214–226 (2012)
16. Dwork, C., Ilvento, C.: Group fairness under composition. In: Proceedings of the 2018 Conference on Fairness, Accountability, and Transparency (FAT* 2018) (2018)
17. Ebadian, S., Kahng, A., Peters, D., Shah, N.: Optimized distortion and proportional fairness in voting. In: Proceedings of the 23rd ACM Conference on Economics and Computation, pp. 563–600 (2022)
18. Edmonds, J.: Paths, trees, and flowers. Can. J. Math. **17**(3), 449–467 (1965)
19. Feldman, M., Friedler, S.A., Moeller, J., Scheidegger, C., Venkatasubramanian, S.: Certifying and removing disparate impact. In: proceedings of the 21th ACM SIGKDD International Conference on Knowledge Discovery and Data Mining, pp. 259–268 (2015)
20. Freeman, R., Micha, E., Shah, N.: Two-sided matching meets fair division. In: Zhou, Z. (ed.) Proceedings of the Thirtieth International Joint Conference on Artificial Intelligence, IJCAI 2021, Virtual Event / Montreal, Canada, 19–27 August 2021, pp. 203–209. ijcai.org (2021). https://doi.org/10.24963/ijcai.2021/29
21. García-Soriano, D., Bonchi, F.: Fair-by-design matching. Data Min. Knowl. Disc. **34**(5), 1291–1335 (2020). https://doi.org/10.1007/s10618-020-00675-y
22. Goel, N., Yaghini, M., Faltings, B.: Non-discriminatory machine learning through convex fairness criteria. In: Furman, J., Marchant, G.E., Price, H., Rossi, F. (eds.) Proceedings of the 2018 AAAI/ACM Conference on AI, Ethics, and Society, AIES 2018, New Orleans, LA, USA, 02–03 February 2018, p. 116. ACM (2018). https://doi.org/10.1145/3278721.3278722
23. Gupta, S., Roy, S., Saurabh, S., Zehavi, M.: Parameterized algorithms and kernels for rainbow matching. Algorithmica **81**(4), 1684–1698 (2019)
24. Hall, P.: On representatives of subsets. J. London Math. Soc. **10**, 26–30 (1935)
25. Hopcroft, J.E., Karp, R.M.: An $n^{5/2}$ algorithm for maximum matchings in bipartite graphs. SIAM J. Comput. **2**, 225–231 (1973)

26. Huang, C., Kavitha, T., Mehlhorn, K., Michail, D.: Fair matchings and related problems. Algorithmica **74**(3), 1184–1203 (2016)
27. Impagliazzo, R., Paturi, R.: On the complexity of k-sat. J. Comput. Syst. Sci. **62**(2), 367–375 (2001)
28. Kamada, Y., Kojima, F.: Fair matching under constraints: theory and applications (2020)
29. Kamishima, T., Akaho, S., Sakuma, J.: Fairness-aware learning through regularization approach. In: Spiliopoulou, M., et al. (eds.) Data Mining Workshops (ICDMW), 2011 IEEE 11th International Conference on, Vancouver, BC, Canada, 11 December 2011, pp. 643–650. IEEE Computer Society (2011)
30. Kesavan, D., Periyathambi, E., Chokkalingam, A.: A proportional fair scheduling strategy using multiobjective gradient-based African buffalo optimization algorithm for effective resource allocation and interference minimization. Int. J. Commun Syst **35**(1), e5003 (2022)
31. Klaus, B., Klijn, F.: Procedurally fair and stable matching. Econ. Theory **27**(2), 431–447 (2006)
32. Kleinberg, J., Mullainathan, S., Raghavan, M.: Inherent trade-offs in the fair determination of risk scores. In: 8th Innovations in Theoretical Computer Science Conference (ITCS 2017). Schloss Dagstuhl-Leibniz-Zentrum fuer Informatik (2017)
33. Kőnig, D.: Über Graphen und ihre Anwendung auf Determinantentheorie und Mengenlehre. Math. Ann. **77**(4), 453–465 (1916)
34. Kurata, R., Hamada, N., Iwasaki, A., Yokoo, M.: Controlled school choice with soft bounds and overlapping types. J. Artif. Intell. Res. **58**, 153–184 (2017)
35. Linhares, A., Olver, N., Swamy, C., Zenklusen, R.: Approximate multi-matroid intersection via iterative refinement. Math. Program. **183**(1), 397–418 (2020). https://doi.org/10.1007/s10107-020-01524-y
36. Lokshtanov, D., Misra, P., Panolan, F., Saurabh, S.: Deterministic truncation of linear matroids. ACM Trans. Algorithms **14**(2), 14:1-14:20 (2018). https://doi.org/10.1145/3170444
37. Lovász, L., Plummer, M.D.: Matching Theory. AMS (2009)
38. Lu, Y.: The optimization of automated container terminal scheduling based on proportional fair priority. Math. Probl. Eng. 2022 (2022)
39. Marx, D.: A parameterized view on matroid optimization problems. Theoret. Comput. Sci. **410**(44), 4471–4479 (2009)
40. Mądry, A.: Navigating central path with electrical flows: from flows to matchings, and back. In: FOCS 2013, pp. 253–262. IEEE Computer Society (2013)
41. Mucha, M., Sankowski, P.: Maximum matchings via Gaussian elimination. In: FOCS 2004, pp. 248–255. IEEE Computer Society (2004)
42. Naor, M., Schulman, L.J., Srinivasan, A.: Splitters and near-optimal derandomization. In: Proceedings of the 36th Annual Symposium on Foundations of Computer Science (FOCS 1995), pp. 182–191. IEEE (1995)
43. Nguyen, M.H., Baiou, M., Nguyen, V.H., Vo, T.Q.T.: Nash fairness solutions for balanced tsp. In: 10th International Network Optimization Conference (INOC) (2022)
44. Rabin, M.O., Vazirani, V.V.: Maximum matchings in general graphs through randomization. J. Algorithms **10**(4), 557–567 (1989)
45. Ristoski, P., Petrovski, P., Mika, P., Paulheim, H.: A machine learning approach for product matching and categorization. Semant. web **9**(5), 707–728 (2018)
46. Roth, A.E., Sönmez, T., Ünver, M.U.: Efficient kidney exchange: coincidence of wants in markets with compatibility-based preferences. Am. Econ. Rev. **97**(3), 828–851 (2007)

47. Schrijver, A.: Combinatorial Optimization. Polyhedra and Efficiency, vol. A. Springer-Verlag, Berlin (2003)
48. St-Arnaud, W., Carvalho, M., Farnadi, G.: Adaptation, comparison and practical implementation of fairness schemes in kidney exchange programs. arXiv preprint arXiv:2207.00241 (2022)
49. Stamoulis, G.: Approximation algorithms for bounded color matchings via convex decompositions. In: Csuhaj-Varjú, E., Dietzfelbinger, M., Ésik, Z. (eds.) MFCS 2014. LNCS, vol. 8635, pp. 625–636. Springer, Heidelberg (2014). https://doi.org/10.1007/978-3-662-44465-8_53
50. Sun, Z., Todo, T., Walsh, T.: Fair pairwise exchange among groups. In: IJCAI, pp. 419–425 (2021)
51. Thanh, B.L., Ruggieri, S., Turini, F.: k-NN as an implementation of situation testing for discrimination discovery and prevention. In: Apté, C., Ghosh, J., Smyth, P. (eds.) Proceedings of the 17th ACM SIGKDD International Conference on Knowledge Discovery and Data Mining, San Diego, CA, USA, 21–24 August 2011, pp. 502–510. ACM (2011)
52. Yannakakis, M.: Node- and edge-deletion np-complete problems. In: Lipton, R.J., Burkhard, W.A., Savitch, W.J., Friedman, E.P., Aho, A.V. (eds.) Proceedings of the 10th Annual ACM Symposium on Theory of Computing, 1–3 May 1978, San Diego, California, USA, pp. 253–264. ACM (1978)
53. Zhang, G., Malekmohammadi, S., Chen, X., Yu, Y.: Equality is not equity: Proportional fairness in federated learning. arXiv preprint arXiv:2202.01666 (2022)

Reconstructing Graphs from Connected Triples

Paul Bastide[1]([✉]), Linda Cook[2], Jeff Erickson[3], Carla Groenland[4],
Marc van Kreveld[4], Isja Mannens[4], and Jordi L. Vermeulen[4]

[1] LaBRI - Bordeaux University, Bordeaux, France
paul.bastide@ens-rennes.fr
[2] Institute for Basic Science, Discrete Math Group, Daejeon, Republic of Korea
linda.cook@ibs.re.kr
[3] University of Illinois, Urbana-Champaign, Champaign, USA
jeffe@illinois.edu
[4] Utrecht University, Utrecht, The Netherlands
{c.e.groenland,m.j.vankreveld,i.m.e.mannens}@uu.nl

Abstract. We introduce a new model of indeterminacy in graphs:
instead of specifying all the edges of the graph, the input contains all
triples of vertices that form a connected subgraph. In general, different
(labelled) graphs may have the same set of connected triples, making
unique reconstruction of the original graph from the triples impossible.
We identify some families of graphs (including triangle-free graphs) for
which all graphs have a different set of connected triples. We also give
algorithms that reconstruct a graph from a set of triples, and for testing
if this reconstruction is unique. Finally, we study a possible extension of
the model in which the subsets of size k that induce a connected graph
are given for larger (fixed) values of k.

Keywords: Algorithms · Graph reconstruction · Indeterminacy ·
Uncertainty · Connected Subgraphs

1 Introduction

Imagine that we get information about a graph, but not its complete structure
by a list of edges. Does this information uniquely determine the graph? In this
paper we explore the case where the input consists of all triples of vertices whose
induced subgraph is connected. In other words, we know for each given triple of
vertices that two or three of the possible edges are present, but we do not know
which ones. We may be able to deduce the graph fully from all given triples.

LC is supported by the Institute for Basic Science (IBS-R029-C1) and CG by Marie-
Skłodowska Curie grant GRAPHCOSY (number 101063180). MvK and JV are supported by the Netherlands Organisation for Scientific Research (NWO) under project
no. 612.001.651.

D. Paulusma and B. Ries (Eds.): WG 2023, LNCS 14093, pp. 16–29, 2023.
https://doi.org/10.1007/978-3-031-43380-1_2

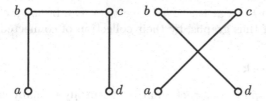

Fig. 1. Two different labelled trees that give the same set of connected triples.

As a simple example, assume we are given the (unordered, labelled) triples abc, bcd, and cde. Then the only (connected) graph that matches this specification by triples is the path a—b—c—d—e. On the other hand, if we are given all possible triples on a set of four vertices a, b, c, d except for abc, then there are several graphs possible. We must have the edges ad, bd, and cd, and zero or one of the edges ab, bc, and ca. See Fig. 1 for another example.

This model of indeterminacy of a graph does not use probability and is perhaps the simplest combinatorial model of partial information. Normally a graph is determined by pairs of vertices which are the edges; now we are given triples of vertices with indeterminacy on the edges between them. As such, we believe this model is interesting to study.

As illustrated in our previous example, there are cases where reconstruction of the graph from the set T of triples is unique and there are cases where multiple (labelled) graphs may have the same set of triples. There are also cases where T is not consistent with any graph, such as $T = \{abc, cde\}$. Can we characterize these cases, and what can we say if we have additional information, for example, when we know that we are reconstructing a tree or a triangle-free graph?

1.1 Our Results

After preliminaries in Sect. 2, we provide two relatively straightforward, general algorithms for reconstruction in Sect. 3. One runs in $O(n^3)$ time when the triples use n vertex labels, and the other runs in $O(n \cdot |T|)$ time when there are $|T|$ triples in the input. These algorithms return a graph that is consistent with the given triples, if one exists, and decide on uniqueness.

Then, we give an $O(|T|)$ time algorithm to reconstruct trees on at least five vertices, provided that the unknown graph is known to be a tree, in Sect. 4. In fact, all triangle-free graphs can be reconstructed, provided we know that the unknown graph is triangle-free. We give an algorithm running in expected $O(|T|)$ time for this in Sect. 5. Moreover, we show that 2-connected outerplanar graphs and triangulated planar graphs can be uniquely reconstructed.

In Sect. 6 we study a natural extension of the model where we are given the connected k-sets of a graph for some fixed $k \geq 4$, rather than the connected triples. We show the largest value of k such that each n-vertex tree is distinguished from other trees by its set of connected k-sets is $\lceil n/2 \rceil$. A similar threshold is shown for the random graph. Finally, we show that graphs with girth strictly larger than k, on at least $2k - 1$ vertices, can be uniquely reconstructed, among the class of thus graphs, by their collection of connected k-sets.

1.2 Related Work

The problem of graph reconstruction arises naturally in many cases where some unknown graph is observed indirectly. For instance, we may have some (noisy) measurement of the graph structure, or only have access to an oracle that answers specific types of queries. Much previous research has been done for specific cases, such as reconstructing metric graphs from a density function [10], road networks from a set of trajectories [1], graphs using a shortest path or distance oracle [19], labelled graphs from all r-neighbourhoods [25], or reconstructing phylogenetic trees [7]. A lot of research has been devoted to the *graph reconstruction conjecture* [21, 29], which states that it is possible to reconstruct any graph on at least three vertices (up to isomorphism) from the multiset of all (unlabeled) subgraphs obtained through the removal of one vertex. This conjecture is open even for planar graphs and triangle-free graphs, but has been proved for outerplanar graphs [15] and maximal planar graphs [22]. We refer the reader to one of the many surveys (e.g. [5, 17, 23, 28]) for further background. Related to our study of the random graph in Sect. 6 is a result from Cameron and Martins [8] from 1993, which implies that for each graph H, with high probability the random graph $G \sim G(n, \frac{1}{2})$ can be reconstructed from the set of (labelled) subsets that induce a copy of H (up to complementation if H is self-complementary).

Many types of uncertainty in graphs have been studied. Fuzzy graphs [27] are a generalisation of fuzzy sets to relations between elements of such sets. In a fuzzy set, membership of an element is not binary, but a value between zero and one. Fuzzy graphs extend this notion to the edges, which now also have a degree of membership in the set of edges. Uncertain graphs are similar to fuzzy graphs in that each edge has a number between zero and one associated with it, although here this number is a probability of the edge existing. Much work has been done on investigating how the usual graph-theoretic concepts can be generalised or extended to fuzzy and uncertain graphs [20, 24].

2 Preliminaries

All graphs in this paper are assumed to be connected, finite, and simple. Let G be an unknown graph with n vertices and let T be the set of all triples of vertices that induce a connected subgraph in G. Since the graph is connected, we can recover the vertex set V of G easily from T. We will use $\overline{T}$ to denote the complement of this set T, i.e. $\overline{T}$ is the set of all triples of vertices for which the induced subgraph is not connected. Note that $|T \cup \overline{T}| = \binom{n}{3} \in \Theta(n^3)$.

Observe that both the presence and absence of a triple gives important information: in the former case, at most one of the three possible edges is absent, whereas in the latter case, at most one of these edges is present.

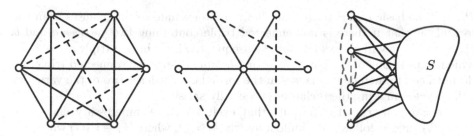

Fig. 2. Three classes of ambiguous triples: a complete graph minus any independent set of edges, a star graph plus any (partial) matching of the leaves, and a path of length four in which all vertices are fully adjacent to some set S. In this last case, we cannot tell the difference between the red and green path. (Color figure online)

It is possible that graphs that are not the same (as labelled graph) or even not isomorphic yield the same set of triples, for example, a path on three vertices and a triangle. We also give examples of larger graphs that cannot be distinguished from their set of connected triples in Fig. 2.

We will make use of (LSD) radix string sorting (as described in e.g. [9]) to sort a collection of t sets of cardinality k in time $\mathcal{O}(tk)$ in several of the algorithms presented in this paper.

3   Algorithm for Finding Consistent Graphs from Triples

Given a set of triples T, we can find a graph G consistent with those triples by solving a 2-SAT formula. The main observation here is that the presence of a triple abc means that at least two of the edges ab, ac and bc must exist, whereas the absence of a triple means at most one of the edges can exist. We can then construct a 2-SAT formula where each variable corresponds to an edge of the graph, and truth represents presence of that edge. For each triple $abc \in T$, we add clauses $(ab \lor ac)$, $(ab \lor bc)$ and $(ac \lor bc)$ to the formula. For each triple $abc \in \overline{T}$, we add clauses $(\neg ab \lor \neg ac)$, $(\neg ab \lor \neg bc)$ and $(\neg ac \lor \neg bc)$. A graph consistent with the set of triples can then be found by solving the resulting 2-SAT formula and taking our set of edges to be the set of true variables in the satisfying assignment. If the formula cannot be satisfied, no graph consistent with T exists.

We can solve the 2-SAT formula in linear time with respect to the length of the formula [2,11]. We add a constant number of clauses for each element of T and $\overline{T}$, so our formula has length $O(|T \cup \overline{T}|)$. As $|T \cup \overline{T}| = \binom{n}{3}$, this gives us an $O(n^3)$ time algorithm to reconstruct a graph with n vertices. However, we prefer an algorithm that depends on the size of T, instead of also on the size of $\overline{T}$. We can eliminate the dependency on the size of $\overline{T}$ by observing that some clauses can be excluded from the formula because the variables cannot be true.

Lemma 1. *We can find a graph G consistent with T in $O(n \cdot |T|)$ time, or output that no consistent graph exists.*

Proof. The basic observation that allows us to exclude certain clauses from the formula is that if there is no connected triple containing two vertices a and b, the variable ab will always be false. Consequently, if we have a triple $abc \in \overline{T}$ for which at most one of the pairs ab, ac and bc appear in some connected triple, we do not need to include its clauses in the formula, as at least two of the variables will be false, making these clauses necessarily satisfied.

We can construct the formula that excludes these unnecessary clauses in $O(n \cdot |T|)$ time as follows. We build a matrix $M(i, j)$, where $i, j \in V(G)$ with each entry containing a list of all vertices with which i and j appear in a connected triple, i.e. $M(i, j) = \{x \mid ijx \in T\}$. This matrix can be constructed in $O(n^2 + |T|)$ time, by first setting every entry to $\emptyset$ (this takes n^2 time) and then running through T, adding every triple abc to the entries $M(a, b)$, $M(b, c)$ and $M(a, c)$. We also sort each list in linear time using e.g. radix sort. As the total length of all lists is $O(|T|)$, this takes $O(n^2 + |T|)$ time in total.

Using this matrix, we can decide which clauses corresponding to triples from $\overline{T}$ to include as follows. For all pairs of vertices (a, b) that appear in some connected triple (i.e. $M(a, b) \neq \emptyset$), we find all x such that $abx \in \overline{T}$. As $M(a, b)$ is sorted, we can find all x in $O(n)$ time by simply recording the missing elements of the list $M(a, b)$. We then check if $M(a, x)$ and $M(b, x)$ are empty. If either one is not, we include the clause associated with $abx \in \overline{T}$ in our formula. Otherwise, we can safely ignore this clause, as it is necessarily satisfied by the variables for ax and bx being false.

Our algorithm takes $O(n)$-time for each non-empty element of $M(i, j)$, of which there are $O(|T|)$, plus $O(n^2)$ time to traverse the matrix. The total time to construct the formula is $O(n^2 + n \cdot |T|)$. As $|T| \in \Omega(n)$ for connected graphs, this simplifies to $O(n \cdot |T|)$ time. The resulting formula also has $O(n \cdot |T|)$ length, and can be solved in time linear in that length.   □

Observe that this is only an improvement on the naive $O(n^3)$ approach if $|T| \in o(n^2)$. We also note that we can test the uniqueness of the reconstruction in the same time using Feder's approach for enumerating 2-SAT solutions [12].

4   Unique Reconstruction of Trees

In this section, we prove the following result.

Theorem 1. *Let T be a set of triples, and let it be known that the underlying graph $G = (V, E)$ is a tree. If $n \geq 5$, then G can be uniquely reconstructed in $O(|T|)$ time.*

Let us briefly examine trees with three or four vertices. A tree with three vertices is always a path and it will always have one triple with all three vertices. We do not know which of the three edges is absent. A tree with four vertices is either a path or a star. The path has two triples and the star has three triples. For the star, the centre is the one vertex that appears in all three triples, and hence the reconstruction is unique. For the path, we will know that the graph is a path, but we will not know in what order the middle two vertices appear (see Fig. 1).

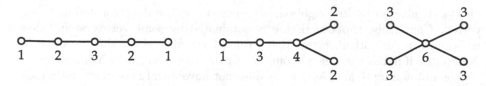

Fig. 3. All trees on five vertices, and the number of triples each vertex occurs in.

Next we consider trees with at least five vertices. We first show that we can recognise all leaves and their neighbours from the triples. In the following, we say that a vertex v *dominates* a vertex u if v appears in all the triples that u appears in. If u is a leaf, then it is dominated by its unique neighbour v. It is possible that u is also dominated by a neighbour of v, but in this case uvw will be the only triple containing u, and v will be dominated by w. Moreover, if $|V| \geq 4$, there exists a triple vwx for some vertex x different from u. This can be used to recognise the leaves as long as $|V| \geq 4$. Moreover, if $|V| \geq 5$, we can also identify the neighbour v of each leaf u as either the unique vertex that dominate u or when u is dominated by two vertices v and w, there exits $x, y \in V$ such that wxy is a triple and w dominates v. We can use this to prove that any tree can be reconstructed from its triples, provided that we know that the result must be a tree and $|V| \geq 5$, since we can iteratively recognise and remove vertices of degree 1, while recording where to 'glue them back at the end' until at most four vertices remain. We can complete the reconstruction via some closer examination of the connected triples in the original tree that contain the remaining vertices.

In order to derive an optimal, $O(|T|)$ time reconstruction algorithm, we will use a further characterisation of vertices of a tree using the triples. The main idea is that we can recognise not only leaves, but also other vertices where we can reduce the tree. If a vertex v has degree 2 in a tree, then there are two nodes w, w' such that every triple with v also contains w or w' (or both). The converse is not true for two reasons: if v is a leaf, it also has the stated property, and if v has degree 3 where at least one neighbour is a leaf, then it has this property as well. This brings us to the following characterisation.

Lemma 2. *A vertex v of a tree G of at least five vertices with triple set T is:*

(i) *a leaf if and only if v is dominated by some vertex w and does not dominate any vertex itself;*

(ii) *if v is not a leaf, then v is (a) a node of degree 2, or (b) a node of degree 3 with at least one leaf neighbour, if and only if there are two nodes w_1, w_2 such that all triples with v also contain w_1 or w_2.*

Moreover, both characterisations can be checked in time $O(|T_v|)$, when the set of triples T_v that include v is given for all $v \in V(G)$.

Proof. A leaf v can necessarily only appear in triples with its adjacent vertex w, as it is not adjacent to any other vertices by definition. A leaf is therefore always dominated by its neighbour w. Since $|V| \geq 5$, v does not dominate any vertex. Conversely, suppose that v is dominated by some vertex w and does not dominate any other vertex. It is straightforward to check that v cannot be dominated if it has three neighbours or if it has two non-leaf neighbours. Since v does not dominate any vertices, it does not have a leaf neighbour. So v must be a leaf itself. This proves (i).

The second characterisation can be seen as follows. If v has degree at least 4, then no w_1, w_2 as in (ii) exist, which is easily verified by looking at the triples with v and its neighbours only. Furthermore, if v has degree 3 and none of its neighbours are leaves, then again there are no such w_1, w_2. On the other hand, in case (a) the two neighbours can be taken as w_1, w_2 and in case (b) w_1, w_2 can be chosen to be two neighbours of v so that the unique neighbour of v that is not in $\{w_1, w_2\}$ is a leaf.

For testing (i), take any triple $vab \in T_v$, and test both a and b separately if they are the sought w. For testing (ii), take any triple $vab \in T_v$. If characterisation (ii) holds, then w_1 must be a or b. We try both as follows: For $w \in \{a, b\}$ we remove all triples with w from T_v. Then $w = w_1$ if and only if all of the remaining triples of T_v all contain some $w_2 \neq v$. We test this by looking at some remaining triple in $vcd \in T_v$ and testing whether either c or d is contained in every other remaining triple. In total, we get four options to test for w_1 and w_2; each option is easily checked in $O(|T_v|)$ time.   □

The vertices of V partition into V', V'', and V''', where V' contains the leaves, V'' contains the vertices that are not leaves but satisfy the second condition of the lemma, and $V''' = V \setminus (V' \cup V'')$. Note that more than half of the vertices of G are in $V' \cup V''$. We next turn to the proof of Theorem 1. We will assume that there is a total order on the vertex set of G and that the connected triples uvw are stored in an ordered tuple with $u < v < w$. For each triple uvw in T, we generate vwu and wuv as well. We collect the triples with the same first vertex to generate T_v for all $v \in V$. We will begin by showing how we recognize whether each vertex is in V', V'' or V'''.

Then, for all $v \in V$, we use T_v to test if v is dominated by some vertex u. We can find the vertices that dominate v, in time $O(|T_v|)$, by examining the first triple vwx and noting that only w and x can dominate v. We then check every other triple in T_v for the presence of w and x. By labeling dominated/dominating vertices as we go we can, in time $O(|T|)$, find all leaves as vertices that are dominated by some vertex and don't dominate any vertex themselves, i.e. all vertices that satisfy condition (i) of Lemma 2. We then check T_v for the remaining vertices, to see if condition (ii) is satisfied and find the vertices w_1 and w_2 in a similar fashion. We again start with the first triple vwx and check if there is some triple vyz that does not contain w or x. If not, then $w1 = w$ and $w_2 = x$. Otherwise we have four candidates w, x, y, z for w_1 and w_2 and we check, for every pair, whether there is a triple that contains neither of them. This can be done in $O(|T_v|)$ time.

After this, we remove all triples containing a leaf from T. Let G' be the graph obtained by removing all leaves and incident edges. Then the new triple set is the set of connected triples for this graph G', and we can recover G from G'.

For all vertices in V'' note that they can no longer be vertices of degree 3 in G', but they may have become leaves. We test this and consider the subset $W \subseteq V''$ of vertices that have not become leaves.

The subgraph of G' induced on W consists of a disjoint union of paths. Let $v \in W$. Then v has exactly two neighbours in $V(G')$ and they are the two vertices w_1, w_2 satisfying the second condition of the Lemma 2. As mentioned above, we can find these two vertices w_1, w_2 for each $v \in W$ in time $O(|T_v|)$. In particular, we know all path components of $G'[W]$, as well as the unique vertices in $V(G') \setminus W$ that the endpoints of any such path are adjacent to. Suppose that v_1—v_2—$\ldots$—v_ℓ is one of the path components of $G[W]$. Let $x_1, x_2 \in V(G') \setminus W'$ such that x_1 is the other neighbour of v_1, and x_2 is the other neighbour of v_k (in G'). We record the edges $x_1 v_1$ and $x_2 v_k$, as well as the edges and vertices in the path $v_1, \ldots, v_\ell$. Then we replace each triple $u x_1 v_1$ by $u x_1 x_2$ and each triple $v_k x_2 u$ by $x_1 x_2 u$. Afterwards, we discard all triples that contain any of $v_1, \ldots, v_\ell$. Let G'' be the graph obtained by deleting $v_1, \ldots, v_\ell$ and adding the edge $\{x_1, x_2\}$. The resulting triple set is the triple set for G'', and we can recover G from G''. We repeat this for all path components. Note that G'' is a tree, if and only if G' is a tree and thus we maintain throughout that the stored triple set corresponds to a tree, and that we can reconstruct the original tree G from knowing this tree and the additional information that we record.

Finally, we also remove all leaves in $V'' \setminus W$ by discarding more triples, similar to the first leaf removal. This process takes time linear in $|T|$, and reduces the number of vertices occurring in T to half or less. We recurse the process on the remaining tree until it has size five, at which point we can uniquely identify the structure of the tree by simply looking at the number of triples each vertex occurs in (see Fig. 3). We may not remove all vertices of V' or V'' if the remaining tree would be smaller than five vertices; in that case, we can simply leave some leaf or not contract the paths in W completely. A standard recurrence shows that the total time used is $O(|T|)$. This finishes the proof of Theorem 1.

We note that if the tree contains no leaves that are siblings, then we do not need to know that the graph is a tree for unique reconstruction.

5   Further Reconstructible Graph Classes

In this section, we give larger classes of graphs for which the graphs that are determined by their set of connected triples.

Theorem 2. *There is an algorithm that reconstructs a graph G on $n \geq 5$ vertices that is known to be triangle-free from its set T of triples in deterministic $O(|T| \log(|T|))$ time or randomized $O(|T|)$ expected time.*

Proof. Let T be the given list of connected triples. For every triple abc we create three ordered copies abc, acb, bca. We then sort the list in lexicographical order in $O(|T|)$ time using radix sort. For every potential edge ab that appears as the first two vertices of some triple we test whether it is an edge as follows.

If we find two triples abc and abd for some $c, d \in V(G)$, we search for the triples acd and bcd. If $ab \notin E(G)$, then we must have that $bc, ac, bd, ad \in E(G)$, and thus both acd and bcd are connected. Therefore, if either triple is not in the list, we know that ab is an edge. Otherwise, we find that $\{a, b, c, d\}$ induces a C_4. We then search for another vertex e that appears in a triple with any of a, b, c, d and reconstruct the labeling of the C_4 as follows. Suppose $G[\{a, b, c, d\}]$ induces a C_4 and we try to retrieve the exact order of the vertices in the cycle. Assume w.l.o.g. that a has the remaining vertex e as a neighbour, then since G is triangle-free, e is not adjacent to b and d. This means that bde is known to be disconnected, whereas abe, ade are connected. If e is not a neighbour of c, then ace is not connected. Hence, if any of a, b, c, d has a private neighbour (one not adjacent to other vertices in the cycle), then we get the labeling of our C_4 (and find that e is a private neighbour of a). If e is adjacent to c besides a, then we know the following two vertex sets also induce C_4's: $abce, acde$. We know e is adjacent to two out of $\{a, b, c\}, \{a, d, c\}$ but not to b and d. So we find e is adjacent to a and c and also have found our labeling.

Suppose abc is the unique triple we found in our list that begins with ab. We check for triples starting in ac or bc. Since one of the three vertices involved must be adjacent to some other vertex d, one of these two potential edges must appear in at least two triples. We can then use the previous methods to reconstruct some subgraph containing a, b and c. The result will tell us whether ab is an edge or not.

Note that we can search our list for a specific triple or a triple starting with a specific pair of vertices, in time $O(\log(|T|))$ using binary search. This means that the above checks can be done in $O(\log(|T|))$ time. By handling potential edges in the order in which they appear in the list, we only need to run through the list once, and thus we obtain a runtime of $O(|T|\log(|T|))$.

Using a data structure for "perfect-hashing" like the one described by Fredman, Komlós and Szemerédi [13], we can query the required triples in $O(1)$ time and thus reconstruct the graph in $O(|T|)$ time. Given a list S of n distinct items from a set of m items, [13] describes an algorithm to create a data-structure that can store n items, and allow for membership in S to be queried in constant time for all m and n. The construction of the data-structure is a randomized process that takes $O(n)$ time in expectation. In our case S is the list of triples and our universe is $V(G) \times V(G) \times V(G)$ so the process takes time $O(T)$. □

We also prove the following two results in the full version [3].

Theorem 3. *We can reconstruct any graph on $n \geq 6$ vertices that is known to be 2-connected and outerplanar from its list of connected triples.*

Our approach is similar to the one for trees: we show that we can identify a vertex of degree two, and remove it from the graph by 'merging' it with one of its neighbours.

A *triangulated planar graph*, also called a *maximal planar graph*, is a planar graph where every face (including the outer face) is a triangle.

Theorem 4. *Let T be a set of triples, and let it be known that the underlying graph $G = (V, E)$ is planar and triangulated. Then G can be uniquely reconstructed from T if $n \geq 7$.*

To show this result, we first show that unique reconstruction of such graphs is possible if they do not contain any separating triangles: A *separating triangle* is a triangle in the graph whose removal would result in the graph being disconnected. We then show we can reduce the problem of reconstructing a triangulated planar graph that contains a separating triangle to reconstructing triangulated planar graphs that do *not* contain a separating triangle.

6   Reconstruction from Connected k-Sets

For $k \geq 2$ and a graph $G = (V, E)$, we define the *connected k-sets* of G as the set $\{X \subseteq V \mid |X| = k$ and $G[X]$ is connected$\}$. We will denote the set of neighbours of a vertex v by $N(v)$.

Observation 1. *For $k' \geq k \geq 2$, the connected k'-sets of a graph are determined by the connected k-sets.*

Indeed, a $(k + 1)$-set $X = \{x_1, \ldots, x_{k+1}\} \subseteq V$ induces a connected subgraph of G if and only if for some $y, z \in X$, both $G[X \setminus \{y\}]$ and $G[X \setminus \{z\}]$ are connected.

Given a class $\mathcal{C}$ of graphs, we can consider the function $k(n)$, where for any integer $n \geq 1$, we define $k(n)$ to be the largest integer $k \geq 2$ such that all (labelled) n-vertex graphs in $\mathcal{C}$ have a different collection of connected k-sets. By Observation 1, asking for the largest such k is a sensible question: reconstruction becomes more difficult as k increases. We will always assume that we only have to differentiate the graph from other graphs in the graph class, and remark that often the *recognition problem* (is $G \in \mathcal{C}$?) cannot be solved even from the connected triples.

First, we give an analogue of Theorem 1. The proof is given in the full version [3].

Theorem 5. *If it is known that the input graph is a tree, then the threshold for reconstructing trees is at $\lceil n/2 \rceil$: we can reconstruct an n-vertex tree from the connected k-sets if $k \leq \lceil n/2 \rceil$ and we cannot reconstruct the order of the vertices in an n-vertex path if $k \geq \lceil n/2 \rceil + 1$.*

In the full version [3], using the theorem above, we give examples showing that for every $k \geq 2$, there are infinitely many graphs that are determined by their connected k-sets but not by their connected $(k + 1)$-sets.

We next show that a threshold near $n/2$ that we saw above for trees, holds for almost every n-vertex graph. The Erdős-Renyi random graph $G \sim G(n, \frac{1}{2})$ has n vertices and each edge is present with probability $\frac{1}{2}$, independently of the other edges. This yields the uniform distribution over the collection of (labelled) graphs on n vertices. If something holds for the random graph with high probability (that is, with a probability that tends to 1 as $n \to \infty$), then we say that it holds for *almost every graph*. The random graph is also interesting since it is often use to deduce the existence of extremal graphs for many problems. More information can be found in e.g. [4,14,18].

We say an n-vertex graph $G = (V, E)$ is *random-like* if the following three properties hold (with log of base 2).

1. For every vertex $v \in V$,

$$n/2 - 3\sqrt{n \log n} \leq |N(v)| \leq n/2 + 3\sqrt{n \log n}.$$

2. For every pair of distinct vertices $v, w \in V$,

$$n/4 - 3\sqrt{n \log n} \leq |N(v) \cap (V \setminus N(w))| \leq n/4 + 3\sqrt{n \log n}.$$

3. There are no disjoint subsets $A, B \subseteq V$ with $|A|, |B| \geq 2 \log n$ such that there are no edges between a vertex in A and a vertex in B.

Lemma 3. *For $G \sim G(n, \frac{1}{2})$, with high probability G is random-like.*

The claimed properties of the random graph are well-known, nonetheless we added a proof in full version for the convenience of the reader [3].

Theorem 6. *For all sufficiently large n, any n-vertex graph G that is random-like can be reconstructed from the set of connected k-sets for $2 \leq k \leq \frac{1}{2}n - 4\sqrt{n \log n}$ in time $O(n^{k+1})$. On the other hand, $G[S]$ is connected for all subsets S of size at least $\frac{1}{2}n + 4\sqrt{n \log n}$.*

In particular, for almost every graph (combining Lemma 3 and Theorem 6), the connectivity of k-tuples for $k \geq \frac{1}{2}n + 4\sqrt{n \log n}$ gives no information whatsoever, whereas for $k \leq \frac{1}{2}n - 4\sqrt{n \log n}$ it completely determines the graph.

Proof (of Theorem 6). Let K be the set of connected k-sets.

We first prove the second part of the statement. Let $k \geq \frac{1}{2}n + 4\sqrt{n \log n}$ be an integer and let S be a subset of V of size at least k. Consider two vertices $u, v \in S$. We will prove that u and v are in the same connected component of $G[S]$. By the first random-like property, there are at most $\frac{1}{2}n + 3\sqrt{n \log n}$ vertices non-adjacent to u, which implies that u has at least $\sqrt{n \log n} - 1 \geq 2 \log n$ neighbours in S. Note that, for the same reason, this is also true for v. Therefore, we can apply the third random-like property on $A = N(u) \cap S$ and $B = N(v) \cap S$ to ensure that there is an edge between the two sets. We conclude that there must exist a path from u to v.

Let us now prove the first part of the statement. Let $2 \leq k \leq \lfloor \frac{1}{2}n - 4\sqrt{n \log n} \rfloor$. Let $u \in V(G)$. We claim that the set of vertices $V \setminus N[v]$ that are not adjacent or equal to u, is the largest set S such that $G[S]$ is connected and $G[S \cup \{u\}]$ is not. Note that the two conditions directly imply that $S \subseteq V \setminus N[u]$. To prove equality, it is sufficient to prove that $G[V \setminus N[u]]$ is connected. Consider two vertices $v, w \in G[V \setminus N[u]]$. By the second random-like property, the sets $A = N(v) \cap (V \setminus N[u])$ and $B = N(w) \cap (V \setminus N[u])$ have size at least $n/4 - 3\sqrt{n \log n}$. By the third property of random-like, there is therefore an edge between A and B.

This proves that there is a path between v and w using vertices in $V \setminus N[u]$, and so $G[V \setminus N[v]]$ is connected.

For each vertex u, the set of vertices it is not adjacent to can now be found by finding the largest set S such that $G[S]$ is connected and $G[S \cup \{u\}]$ is not. Since any such S is a subset of $V \setminus N[u]$, there is a unique maximal (and unique maximum) such S. We now give the $O(n^{k+1})$ time algorithm for this.

We begin by constructing a data structure like the deterministic version described by Fredman et al. [13], which allows us to query the required k-sets in $O(1)$. This takes deterministic time of $O(|K| \log |K|) = O(n^k k \log n) = O(n^{k+1})$. For a vertex v, we give an algorithm to reconstruct the neighbourhood of v in time $O(n^{k+1})$.

1. We first run over the subsets S of size k until we find one for which $G[S]$ is connected but $G[S \cup \{v\}]$ is not. This can be done in time $O(kn^k)$: if $G[S]$ is connected, then $G[S \cup \{v\}]$ is disconnected if and only if $G[S \setminus \{s\} \cup \{v\}]$ is disconnected for all $s \in S$.
2. For each vertex $w \in V \setminus (S \cup \{v\})$, we check whether there is a subset $U \subseteq S$ of size $k-1$ for which $U \cup \{w\}$ is connected, and whether for each subset $U' \subseteq S$ of size $k-2$, $U' \cup \{w, v\}$ is not connected. If both are true for the vertex w, then $G[S \cup \{w\}]$ is connected and $G[S \cup \{w, v\}]$ is not connected, so we add w to S and repeat this step.
3. If no vertex can be added anymore, we stop and output $V \setminus (S \cup \{v\})$ as the set of neighbours of v.

We repeat step 2 at most n times, and each time we try at most n vertices as potential w and run over subsets of size at most $k-1$. Hence, this part runs in time $O(n^{k+1})$. We repeat the algorithm above n times (once per vertex) in order to reconstruct all edges. □

We prove the following analogue to Theorem 2 in the full version [3].

Theorem 7. *Let $k \geq 4$ be an integer. Every graph on at least $2k - 1$ vertices that is known to have no cycles of length at most k is determined by its connected k-sets.*

7   Conclusion

We have presented a new model of uncertainty in graphs, in which we only receive all triples of vertices that form a connected induced subgraph. In a way, this is the simplest model of combinatorial indeterminacy in graphs. We have studied some basic properties of this model, and provided an algorithm for finding a graph consistent with the given indeterminacies. We also proved that trees, triangle-free graphs and various other families of graphs are determined by the connected triples, although we need to know the family the sought graph belongs to. In order to obtain a full characterisation, it is natural to put conditions on the way a triangle may connect to the rest of the graph, for instance, it is not too difficult to recognise that a, b, c induces a triangle if at least two of a, b, c have private neighbour, whereas it is impossible to distinguish whether a, b, c induce a triangle or a path if all three vertices have the same neighbours outside of the triangle. We leave this open for future work.

Similar to what has been done for graph reconstruction (see e.g. [23]), another natural direction is to loosen the objective of reconstruction, and to see if rather than determining the (labelled) graph, we can recover some graph property such as the number of edges or the diameter. A natural question is also how many connected triples are required (when given a 'subcollection', similar to [6,16,26], or when we may perform adaptive queries, as in [19]).

We gave various results in an extension of our model to larger k-sets, including trees and random graphs. There are several other logical extensions to the concept of reconstructing a graph from connected triples. We could define a (k, ℓ)-representation T to contain all k-sets that are connected and contain at least ℓ edges. The definition of connected triples would then be a $(3, 2)$-representation. Note that in this case some vertices may not appear in T, or T might even be empty altogether (e.g. for trees when $\ell \geq k$). Another natural extension would be to specify the edge count for each k-set, but this gives too much information even when $k = n-2$: the existence of any edge $\{u, v\}$ can be determined from the number of edges among vertices in the four sets $V, V \setminus \{u\}, V \setminus \{v\}$ and $V \setminus \{u, v\}$.

Some interesting algorithmic questions remain open as well. In particular, we presented an efficient algorithm to specify whether a collection of connected k-sets, for $k = 3$ uniquely determines a graph, but do not know how to solve this efficiently for larger values of k. Is the following decision problem solvable in polynomial time: given a graph G and an integer k, is G determined by its collection of k-tuples? We note that when k equals 4, membership in coNP is clear (just give another graph with the same connected 4-sets) whereas even NP-membership is unclear.

Finally, a natural question is whether the running time of $O(n \cdot |T|)$ for finding a consistent graph with a set of connected triples (Lemma 1) can be improved to $O(|T|)$ (or expected time $O(|T|)$).

References

1. Ahmed, M., Wenk, C.: Constructing street networks from GPS trajectories. In: Epstein, L., Ferragina, P. (eds.) ESA 2012. LNCS, vol. 7501, pp. 60–71. Springer, Heidelberg (2012). https://doi.org/10.1007/978-3-642-33090-2_7
2. Aspvall, B., Plass, M.F., Tarjan, R.E.: A linear-time algorithm for testing the truth of certain quantified Boolean formulas. Inf. Process. Lett. **8**(3), 121–123 (1979)
3. Bastide, P., et al.: Reconstructing graphs from connected triples (2023)
4. Bollobás, B.: Random graphs. In: Modern Graph Theory. GTM, vol. 184, pp. 215–252. Springer, New York (1998). https://doi.org/10.1007/978-1-4612-0619-4_7
5. Bondy, J.A., Hemminger, R.L.: Graph reconstruction - a survey. J. Graph Theory **1**(3), 227–268 (1977)
6. Bowler, A., Brown, P., Fenner, T.: Families of pairs of graphs with a large number of common cards. J. Graph Theory **63**(2), 146–163 (2010)

7. Brandes, U., Cornelsen, S.: Phylogenetic graph models beyond trees. Discrete Appl. Math. **157**(10), 2361–2369 (2009)
8. Cameron, P.J., Martins, C.: A theorem on reconstruction of random graphs. Comb. Probab. Comput. **2**(1), 1–9 (1993)
9. Cormen, T.H., Leiserson, C.E., Rivest, R.L., Stein, C.: Introduction to Algorithms, pp. 197–200. MIT press (2022)
10. Dey, T.K., Wang, J., Wang, Y.: Graph reconstruction by discrete Morse theory. In: Proceedings of the 34th International Symposium on Computational Geometry. Leibniz International Proceedings in Informatics (LIPIcs), vol. 99, pp. 31:1–31:15 (2018)
11. Even, S., Itai, A., Shamir, A.: On the complexity of timetable and multicommodity flow problems. SIAM J. Comput. **5**(4), 691–703 (1976)
12. Feder, T.: Network flow and 2-satisfiability. Algorithmica **11**(3), 291–319 (1994)
13. Fredman, M.L., Komlós, J., Szemerédi, E.: Storing a sparse table with 0(1) worst case access time. J. ACM **31**(3), 538–544 (1984)
14. Frieze, A., Karonski, M.: Introduction to Random Graphs. Cambridge University Press, New York (2015)
15. Giles, W.B.: The reconstruction of outerplanar graphs. J. Comb. Theory Ser. B **16**(3), 215–226 (1974)
16. Groenland, C., Guggiari, H., Scott, A.: Size reconstructibility of graphs. J. Graph Theory **96**(2), 326–337 (2021)
17. Harary, F.: A survey of the reconstruction conjecture. In: Bari, R.A., Harary, F. (eds.) Graphs and Combinatorics. LNCS, vol. 406, pp. 18–28. Springer, Berlin (1974). https://doi.org/10.1007/BFb0066431
18. Janson, S., Rucinski, A., Luczak, T.: Random Graphs. John Wiley & Sons, Hoboken (2011)
19. Kannan, S., Mathieu, C., Zhou, H.: Graph reconstruction and verification. ACM Trans. Algorithms **14**(4), 1–30 (2018)
20. Kassiano, V., Gounaris, A., Papadopoulos, A.N., Tsichlas, K.: Mining uncertain graphs: an overview. In: Sellis, T., Oikonomou, K. (eds.) ALGOCLOUD 2016. LNCS, vol. 10230, pp. 87–116. Springer, Cham (2017). https://doi.org/10.1007/978-3-319-57045-7_6
21. Kelly, P.J.: On Isometric Transformations. Ph.D. thesis, University of Wisconsin (1942)
22. Lauri, J.: The reconstruction of maximal planar graphs. J. Combi. Theory Ser. B **30**(2), 196–214 (1981)
23. Lauri, J., Scapellato, R.: Topics in Graph Automorphisms and Reconstruction. Cambridge University Press, Cambridge (2016)
24. Mordeson, J.N., Peng, C.S.: Operations on fuzzy graphs. Inf. Sci. **79**(3), 159–170 (1994)
25. Mossel, E., Ross, N.: Shotgun assembly of labeled graphs. IEEE Trans. Netw. Sci. Eng. **6**(2), 145–157 (2017)
26. Myrvold, W.: The degree sequence is reconstructible from $n - 1$ cards. Discrete Math. **102**(2), 187–196 (1992)
27. Rosenfeld, A.: Fuzzy graphs. In: Zadeh, L.A., Fu, K.S., Tanaka, K., Shimura, M. (eds.) Fuzzy Sets and their Applications to Cognitive and Decision Processes, pp. 77–95. Elsevier (1975)
28. Tutte, W.: All the king's horses. A guide to reconstruction. Graph Theory Relat. Top., 15–33 (1979)
29. Ulam, S.M.: A Collection of Mathematical Problems, Interscience Tracts in Pure and Applied Mathematics, vol. 8. Interscience Publishers (1960)

Parameterized Complexity of Vertex Splitting to Pathwidth at Most 1

Jakob Baumann⬥, Matthias Pfretzschner(✉)⬥, and Ignaz Rutter⬥

Universität Passau, 94032 Passau, Germany
{baumannjak,pfretzschner,rutter}@fim.uni-passau.de

Abstract. Motivated by the planarization of 2-layered straight-line drawings, we consider the problem of modifying a graph such that the resulting graph has pathwidth at most 1. The problem PATHWIDTH-ONE VERTEX EXPLOSION (POVE) asks whether such a graph can be obtained using at most k vertex explosions, where a *vertex explosion* replaces a vertex v by $\deg(v)$ degree-1 vertices, each incident to exactly one edge that was originally incident to v. For POVE, we give an FPT algorithm with running time $O(4^k \cdot m)$ and an $O(k^2)$ kernel, thereby improving over the $O(k^6)$-kernel by Ahmed et al. [2] in a more general setting. Similarly, a *vertex split* replaces a vertex v by two distinct vertices v_1 and v_2 and distributes the edges originally incident to v arbitrarily to v_1 and v_2. Analogously to POVE, we define the problem variant PATHWIDTH-ONE VERTEX SPLITTING (POVS) that uses the split operation instead of vertex explosions. Here we obtain a linear kernel and an algorithm with running time $O((6k+12)^k \cdot m)$. This answers an open question by Ahmed et al. [2].

Keywords: Vertex Splitting · Vertex Explosion · Pathwidth 1

1 Introduction

Crossings are one of the main aspects that negatively affect the readability of drawings [20]. It is therefore natural to try and modify a given graph in such a way that it can be drawn without crossings while preserving as much of the information as possible. We consider three different operations.

A *deletion operation* simply removes a vertex from the graph. A *vertex explosion* replaces a vertex v by $\deg(v)$ degree-1 vertices, each incident to exactly one edge that was originally incident to v. Finally, a *vertex split* replaces a vertex v by two distinct vertices v_1 and v_2 and distributes the edges originally incident to v arbitrarily to v_1 and v_2.

Nöllenburg et al. [18] have recently studied the vertex splitting problem, which is known to be NP-complete [11]. In particular, they gave a non-uniform FPT-algorithm for deciding whether a given graph can be planarized with at most k splits. We observe that, since degree-1 vertices can always be inserted into a planar drawing, the vertex explosion model and the vertex deletion model are

© The Author(s), under exclusive license to Springer Nature Switzerland AG 2023
D. Paulusma and B. Ries (Eds.): WG 2023, LNCS 14093, pp. 30–43, 2023.
https://doi.org/10.1007/978-3-031-43380-1_3

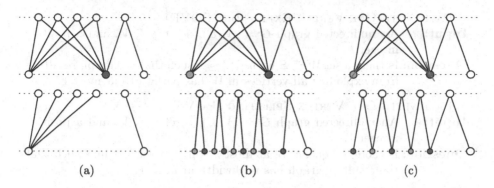

Fig. 1. Given the shown bipartite graph, a crossing-free 2-layered drawing can be obtained using one vertex deletion (a), two vertex explosions (b), or three vertex splits (c).

equivalent for obtaining planar graphs. Note that this is not necessarily the case for other target graph classes (see, for example, Fig. 1). The problem of deleting vertices to obtain a planar graph is also known as VERTEX PLANARIZATION and has been studied extensively in the literature [13, 15–17]. In particular, Jansen et al. [13] gave an FPT-algorithm with running time $O(2^{O(k \log k)} \cdot n)$.

Ahmed et al. [2] investigated the problem of splitting the vertices of a bipartite graph so that it admits a 2-layered drawing without crossings. They assume that the input graph is bipartite and only the vertices of one of the two sets in the bipartition may be split. Under this condition, they give an $O(k^6)$-kernel for the vertex explosion model, which results in an $O(2^{O(k^6)} m)$-time algorithm. They ask whether similar results can be obtained in the vertex splitting model. Figure 1 illustrates the three operations in the context of 2-layered drawings[1].

We note that a graph admits a 2-layer drawing without crossings if and only if it has pathwidth at most 1, i.e., it is a disjoint union of caterpillars [3, 9]. Motivated by this, we more generally consider the problem of turning a graph $G = (V, E)$ into a graph of pathwidth at most 1 by the above operations. In order to model the restriction of Ahmed et al. [2] that only one side of their bipartite input graph may be split, we further assume that we are given a subset $S \subseteq V$, to which we may apply modification operations as part of the input. We define that the new vertices resulting from an operation are also included in S.

More formally, we consider the following problems, all of which have been shown to be NP-hard [1, 19].

---

[1] In this context, minimizing the number of vertex explosions is equivalent to minimizing the number of vertices that are split, since it is always best to split a vertex as often as possible.

PATHWIDTH-ONE VERTEX EXPLOSION (POVE)

Input: An undirected graph $G = (V, E)$, a set $S \subseteq V$, and a positive integer k.

Question: Is there a set $W \subseteq S$ with $|W| \leq k$ such that the graph resulting from exploding all vertices in W has pathwidth at most 1?

PATHWIDTH-ONE VERTEX SPLITTING (POVS)

Input: An undirected graph $G = (V, E)$, a set $S \subseteq V$, and a positive integer k.

Question: Is there a sequence of at most k splits on vertices in S such that the resulting graph has pathwidth at most 1?

We note that the analogous problem with the deletion operation has been studied extensively [8,19,23]. Here, a branching algorithm with running time $O(3.888^k \cdot n^{O(1)})$ [23] and a quadratic kernel [8] are known. Our results are as follows.

First, in Sect. 3, we show that POVE admits a kernel of size $O(k^2)$ and an algorithm with running time $O(4^k m)$, thereby improving over the results of Ahmed et al. [2] in a more general setting.

Second, in Sect. 4, we show that POVS has a kernel of size $16k$ and it admits an algorithm with running time $O((6k+12)^k \cdot m)$. This answers the open question of Ahmed et al. [2].

Finally, in Sect. 5, we consider the problem Π VERTEX SPLITTING(Π-VS), the generalized version of the splitting problem where the goal is to obtain a graph of a specific graph class Π using at most k split operations. Eppstein et al. [10] recently studied the similar problem of deciding whether a given graph G is k-splittable, i.e., whether it can be turned into a graph of Π by splitting every vertex of G at most k times. For graph classes Π that can be expressed in monadic second-order graph logic (MSO_2, see [7]), they gave an FPT algorithm parameterized by the solution size k and the treewidth of the input graph. We adapt their algorithm for the problem Π-VS, resulting in an FPT algorithm parameterized by the solution size k for MSO_2-definable graph classes Π of bounded treewidth. Using a similar algorithm, we obtain the same result for the problem variant using vertex explosions.

2   Preliminaries

A parameterized problem L with parameter k is *non-uniformly fixed-parameter tractable* if, for every value of k, there exists an algorithm that decides L in time $f(k) \cdot n^{O(1)}$ for some computable function f. If there is a single algorithm that satisfies this property for all values of k, then L is *(uniformly) fixed-parameter tractable*.

Given a graph G, we let n and m denote the number of vertices and edges of G, respectively. Since we can determine the subgraph of G that contains no isolated vertices in $O(m)$ time, we assume, without loss of generality, that $n \in O(m)$. For a vertex $v \in V(G)$, we let $N(v) := \{u \in V(G) \mid \mathrm{adj}(v, u)\}$ and $N[v] := N(v) \cup \{v\}$ denote the open and closed neighborhood of v in G, respectively.

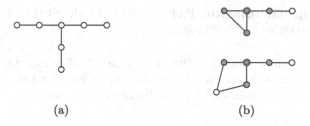

(a)   (b)

Fig. 2. (a) The graph T_2. (b) Two graphs that do not contain T_2 as a subgraph, but both contain N_2 (marked in orange) as a substructure. (Color figure online)

We refer to vertices of degree 1 as *pendant* vertices. For a vertex v of G, we let $\deg^*(v) := |\{u \in N(v) \mid \deg(u) > 1\}|$ denote the degree of v ignoring its pendant neighbors. If $\deg^*(v) = d$, we refer to v as a vertex of *degree* * d. A graph is a *caterpillar* (respectively a *pseudo-caterpillar*), if it consists of a simple path (a simple cycle) with an arbitrary number of adjacent pendant vertices. The path (the cycle) is called the *spine* of the (pseudo-)caterpillar.

Philip et al. [19] mainly characterized the graphs of pathwidth at most 1 as the graphs containing no cycles and no T_2 (three simple paths of length 2 that all share an endpoint; see Fig. 2a) as a subgraph. We additionally use slightly different sets of forbidden substructures. An N_2 *substructure* consists of a *root* vertex r adjacent to three distinct vertices of degree at least 2. Note that every T_2 contains an N_2 substructure, however, the existence of an N_2 substructure does not generally imply the existence of a T_2 subgraph; see Fig. 2b. In the following proposition, we state the different characterizations for graphs of pathwidth at most 1 that we use in this work.

Proposition 1 ($\star^2$). *For a graph G, the following statements are equivalent.*

a) G has pathwidth at most 1
b) every connected component of G is a caterpillar
c) G is acyclic and contains no T_2 subgraph
d) G is acyclic and contains no N_2 substructure
e) G contains no N_2 substructure and no connected component that is a pseudo-caterpillar.

We define the *potential* of $v \in V(G)$ as $\mu(v) := \max(\deg^*(v) - 2, 0)$. The *global potential* $\mu(G) := \sum_{v \in V(G)} \mu(v)$ is defined as the sum of the potentials of all vertices in G. Observe that $\mu(G) = 0$ if and only if G contains no N_2 substructure. The global potential thus indicates how far away we are from eliminating all N_2 substructures from the instance.

Recall that, for the problems POVE and POVS, the set $S \subseteq V(G)$ marks the vertices of G that may be chosen for the respective operations. We say that a set $W \subseteq S$ is a *pathwidth-one explosion set* (POES) of G, if the graph resulting from exploding all vertices in W has pathwidth at most 1.

---

² The proofs of results marked with a star can be found in the full version [4].

3 FPT Algorithms for PATHWIDTH-ONE VERTEX EXPLOSION

In this section, we first show that POVE can be solved in time $O(4^k \cdot m)$ using bounded search trees. Subsequently, we develop a kernelization algorithm for POVE that yields a quadratic kernel in linear time.

3.1 Branching Algorithm

We start by giving a simple branching algorithm for POVE, similar to the algorithm by Philip et al. [19] for the deletion variant of the problem. For an N_2 substructure X, observe that exploding vertices not contained in X cannot eliminate X, because the degrees of the vertices in X remain the same due to the new degree-1 vertices resulting from the explosion. To obtain a graph of pathwidth at most 1, it is therefore always necessary to explode one of the four vertices of every N_2 substructure by Proposition 1. Our branching rule thus first picks an arbitrary N_2 substructure from the instance and then branches on which of the four vertices of the N_2 substructure belongs to the POES. Recall that S denotes the set of vertices of the input graph that can be exploded.

Branching Rule 1. *Let r be the root of an N_2 substructure contained in G and let x, y, and z denote the three neighbors of r in N_2. For every vertex $v \in \{r, x, y, z\} \cap S$, create a branch for the instance $(G', S \setminus \{v\}, k-1)$, where G' is obtained from G by exploding v.*
If $\{r, x, y, z\} \cap S = \emptyset$, reduce to a trivial no-instance instead.

Note that an N_2 substructure can be found in $O(m)$ time by checking, for every vertex v in G, whether v has at least three neighbors of degree at least 2. Also note that vertex explosions do not increase the number of edges of the graph. Since Branching Rule 1 creates at most four new branches, each of which reduces the parameter k by 1, exhaustively applying the rule takes $O(4^k \cdot m)$ time. By Proposition 1, it subsequently only remains to eliminate connected components that are a pseudo-caterpillar. Since a pseudo-caterpillar can (only) be turned into a caterpillar by exploding a vertex of its spine, the remaining instance can be solved in linear time.

Theorem 1. *The problem* PATHWIDTH-ONE VERTEX EXPLOSION *can be solved in time $O(4^k \cdot m)$.*

3.2 Quadratic Kernel

We now turn to our kernelization algorithm for POVE. In this section, we develop a kernel of quadratic size, which can be computed in linear time.

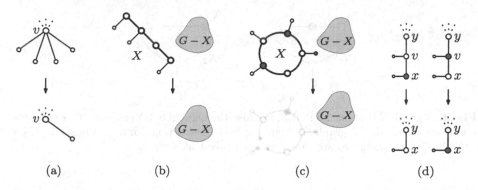

Fig. 3. Examples for Reduction Rules 1 (a), 2 (b), 3 (c), and 4 (d). The vertices of S are marked in green (Color figure online).

We adopt our first two reduction rules from the kernelization of the deletion variant by Philip et al. [19] and show that these rules are also safe for the explosion variant. The first rule reduces the number of pendant neighbors of each vertex to at most one; see Fig. 3a.

Reduction Rule 1. ($\star$). *If G contains a vertex v with at least two pendant neighbors, remove all pendant neighbors of v except one to obtain the graph G' and reduce the instance to $(G', S \cap V(G'), k)$.*

Since a caterpillar has pathwidth at most 1 by Proposition 1, we can safely remove any connected component of G that forms a caterpillar; see Fig. 3b for an example.

Reduction Rule 2. *If G contains a connected component X that is a caterpillar, remove X from G and reduce the instance to $(G - X, S \setminus V(X), k)$.*

If G contains a connected component that is a pseudo-caterpillar, then exploding an arbitrary vertex of its spine yields a caterpillar. If the spine contains no vertex of S, the spine is a cycle that cannot be broken by a vertex explosion. However, by Proposition 1, acyclicity is a necessary condition for a graph of pathwidth at most 1. Hence we get the following reduction rule; see Fig. 3c for an illustration.

Reduction Rule 3. *Let X denote a connected component of G that is a pseudo-caterpillar. If the spine of X contains a vertex of S, remove X from G and reduce the instance to $(G - X, S \setminus V(X), k - 1)$. Otherwise reduce to a trivial no-instance.*

Recall that the degree* of a vertex is the number of its non-pendant neighbors. Our next goal is to shorten paths of degree*-2 vertices to at most two vertices. If we have a path x, y, z of degree*-2 vertices, we refer to y as a *2-enclosed* vertex. Note that exploding a 2-enclosed vertex y cannot eliminate any

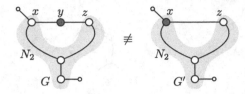

Fig. 4. A graph G that has no POES, because the highlighted N_2 substructure contains no vertex of S. For the graph G' resulting from contracting y into x, the set $\{x\}$ is a POES. The two instances are therefore not equivalent.

N_2 substructures from the instance. By Proposition 1, vertex y can thus only be part of an optimal solution if exploding y breaks cycles. If we want to shorten the chain x, y, z by contracting y into one of its neighbors, we therefore need to ensure that the shortened chain contains a vertex of S if and only if the original chain contained a vertex of S. If $y \in S$, we cannot simply add one of its neighbors, say x, to S in the reduced instance, because exploding x may additionally remove an N_2 substructure; see Fig. 4 for an example. While shortening paths of degree*-2 vertices to at most three vertices is simple, shortening them to length at most 2 (i.e., eliminating all 2-enclosed vertices) is therefore more involved. In the following, we briefly sketch how this can be achieved in linear time. For the specific reduction rules and the corresponding correctness proofs, we refer to the full version of the paper [4].

Lemma 1 ($\star$). *Given an instance of* POVE*, an equivalent instance without 2-enclosed vertices can be computed in $O(m)$ time.*

Sketch of Proof. Given a 2-enclosed vertex y, we show that we can decide greedily whether y is contained in an optimal solution or not. This means that we can either immediately explode y, or we can safely contract it into one of its degree*-2 neighbors. Since y is 2-enclosed, y is not contained in any N_2 substructures and we thus only have to consider cycles containing y. If there exists a cycle C in G with $C \cap S = \{y\}$ (i.e., y is the only splittable vertex of C), then we can immediately explode y. Otherwise, every cycle containing y contains at least one additional vertex of S. In this case, we can show that there exists a minimum POES of G that does not contain y, thus we can remove y from S and contract it into one of its neighbors, thereby preserving all cycles of the instance. To achieve linear running time, we can show that the set of 2-enclosed vertices that should be exploded can be computed globally using a specialized spanning tree. □

To simplify the instance even further, the following reduction rule removes all degree*-2 vertices v that are adjacent to a vertex x of degree* 1; see Fig. 3d for an illustration. Roughly speaking, since v cannot be contained in a cycle and x substitutes v in all N_2 substructures v is contained in, all forbidden substructures are preserved.

Reduction Rule 4 ($\star$). *Let v be a degree*-2 vertex of G with non-pendant neighbors x and y, such that x has degree* 1. Remove v from G and add a new edge xy. If $v \in S$, reduce to $(G - v + xy, (S \setminus \{v\}) \cup \{x\}, k)$. Otherwise reduce to $(G - v + xy, S \setminus \{x\}, k)$.*

Recall that the global potential $\mu(G)$ indicates how far away we are from our goal of eliminating all N_2 substructures from G. With the following lemma, we show that our reduction rules ensure that the number of vertices in the graph G is bounded linearly in the global potential of G.

Lemma 2. *After exhaustively applying Reduction Rules 1–4 and Lemma 1, it holds that $|V(G)| \leq 8 \cdot \mu(G)$.*

Proof. Reduction Rule 2 ensures that G contains no vertices of degree* 0. For $i \in \{1, 2\}$, let V_i denote the set of non-pendant degree*-i vertices of G and let V_3 denote the set of vertices with degree* at least 3. Recall that we defined the global potential as

$$\mu(G) = \sum_{v \in V(G)} \mu(v) = \sum_{v \in V(G)} \max(0, \deg^*(v) - 2).$$

Since all vertices of V_1 and V_2 have degree* at most 2, their potential is 0 and we get

$$\mu(G) = \sum_{v \in V_3} (\deg^*(v) - 2) = \sum_{v \in V_3} \deg^*(v) - 2 \cdot |V_3|.$$

Note that $|V_3| \leq \mu(G)$, because each vertex of degree* at least 3 contributes at least 1 to the global potential. We therefore get

$$\sum_{v \in V_3} \deg^*(v) \leq 3 \cdot \mu(G). \tag{1}$$

By Lemma 1, every vertex in $v \in V_2$ is adjacent to a vertex of $V_1 \cup V_3$, since otherwise, v would be 2-enclosed. However, Reduction Rule 4 additionally ensures that vertices of V_2 cannot be adjacent to vertices of V_1, thus every vertex of V_2 must be adjacent to a vertex of V_3. Note that two adjacent vertices of V_1 would form a caterpillar, which is prohibited by Reduction Rule 2. Therefore, every vertex of V_1 is also adjacent to a vertex of V_3.

Overall, every vertex of V_1 and V_2 is thus adjacent to a vertex of V_3. Note that every vertex $v \in V_1$ must additionally have a pendant neighbor, because otherwise, v itself would be a pendant vertex. Hence every vertex of V_1 and V_2 has degree at least 2 and thus contributes to the degree* of its neighbor in V_3. We therefore have $|V_1| + |V_2| \leq \sum_{v \in V_3} \deg^*(v)$, hence $|V_1| + |V_2| \leq 3 \cdot \mu(G)$ by Eq. 1. Recall that $|V_3| \leq \mu(G)$, thus $|V_1| + |V_2| + |V_3| \leq 4 \cdot \mu(G)$. By Reduction Rule 1, each of these vertices can have at most one pendant neighbor and thus $|V(G)| \leq 8 \cdot \mu(G)$.

With Lemma 2, it now only remains to find an upper bound for the global potential $\mu(G)$. We do this using the following two reduction rules.

Reduction Rule 5. *Let v be a vertex of G with potential $\mu(v) > k$. If $v \in S$, explode v to obtain the graph G' and reduce the instance to $(G', S \setminus \{v\}, k-1)$. Otherwise reduce to a trivial no-instance.*

Proof of Safeness. Since exploding a vertex $u \in V(G) \setminus \{v\}$ decreases $\mu(v)$ by at most one, after exploding at most k vertices in $V(G) \setminus \{v\}$ we still have $\mu(v) > 0$. Because $\mu(v) > 0$ implies that G contains an N_2 substructure, it is therefore always necessary to explode vertex v by Proposition 1. □

Reduction Rule 6. *If $\mu(G) > 2k^2 + 2k$, reduce to a trivial no-instance.*

Proof of Safeness. By Reduction Rule 5 we have $\mu(v) \leq k$ and consequently $\deg^*(v) \leq k + 2$ for all $v \in V(G)$. Hence exploding a vertex v decreases the potential of v by at most k and the potential of each of its non-pendant neighbors by at most 1. Overall, k vertex explosions can therefore only decrease the global potential $\mu(G)$ by at most $k \cdot (2k + 2)$. □

Because Reduction Rule 6 gives us an upper bound for the global potential $\mu(G)$, we can now use Lemma 2 to obtain the kernel.

Theorem 2 ($\star$). *The problem* PATHWIDTH-ONE VERTEX EXPLOSION *admits a kernel of size $16k^2 + 16k$. It can be computed in time $O(m)$.*

4   FPT Algorithms for PATHWIDTH-ONE VERTEX SPLITTING

In this section, we briefly outline how the results from Sect. 3 can be adapted for the split operation. For detailed proofs, we refer to the full version [4].

4.1   Linear Kernel

One can prove that Reduction Rules 1–4 and Lemma 1 we used for POVE are also safe for the problem POVS. Since only these are needed to establish the upper bound of $|V(G)| \leq 8 \cdot \mu(G)$ in Lemma 2, the lemma also applies for POVS.

The main difference to the kernelization of POVE lies in the way the global potential changes due to splits. While a vertex explosion can decrease the global potential linearly in k, we can show that a single vertex split decreases $\mu(G)$ by at most 2. If $\mu(G) > 2k$, we can thus again reduce to a trivial no-instance. Using Lemma 2 with $\mu(G) \leq 2k$, we obtain the following result.

Theorem 3 ($\star$). *The problem* PATHWIDTH-ONE VERTEX SPLITTING *admits a kernel of size $16k$. It can be computed in time $O(m)$.*

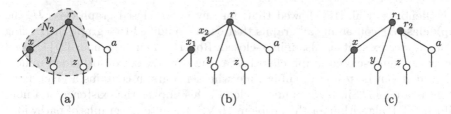

Fig. 5. (a) An N_2 substructure $\{r, x, y, z\}$. (b)-(c) Two possible branches eliminating the N_2 substructure. The former splits off edge rx at x, the latter splits off the edges rz and ra at r.

4.2 Branching Algorithm

As in Sect. 3.1, our branching algorithm for POVS eliminates every N_2 substructure of G by branching on which of its four vertices should be split. In this case, however, we need to additionally consider the possible ways to split a single vertex. The following lemma helps us limit the number of suitable splits.

Lemma 3 (⋆). *For every instance of* POVS, *there exists a minimum sequence of splits such that every split operation splits off at most two edges.*

Theorem 4 (⋆). *The problem* POVS *can be solved in time* $O((6k + 12)^k \cdot m)$.

Sketch of Proof. From the kernelization, we use Reduction Rule 1 reducing pendant vertices, and the above rule that yields the bound $\mu(G) \leq 2k$. Together, these two rules ensure that each vertex has degree at most $2k + 3$. We now branch on the way of splitting an N_2 substructure with root r and neighbors $\{x, y, z\}$ as above (see Fig. 5). If we split r, then, by Lemma 3, we may assume that we split off one of the neighbors $\{x, y, z\}$, together with at most one other neighbor of r; these are $3 \cdot (2k + 3)$ choices. If we split a vertex $v \in \{x, y, z\}$, then it is necessary that we only split off the edge rv at v, thus there is only one possibility for each of them. Overall, we thus find a branching vector of size $6k + 12$. □

5 FPT Algorithms for Splitting and Exploding to MSO₂-Definable Graph Classes of Bounded Treewidth

While Sect. 4 focused on the problem of obtaining graphs of pathwidth at most 1 using at most k vertex splits on the input graph, we now consider the problem of splitting vertices to obtain other graph classes. With the following problem, we generalize the problem POVS.

Π Vertex Splitting(Π-VS)
    Input: An undirected graph $G = (V, E)$, a set $S \subseteq V$, and a positive integer k.
    Question: Is there a sequence of at most k splits on vertices in S such that the resulting graph is contained in Π?

Nöllenburg et al. [18] showed that, for any minor-closed graph class Π, the graph class Π_k containing all graphs that can be modified to a graph in Π using at most k vertex splits is also minor-closed. Robertson and Seymour [21] showed that every minor-closed graph class has a constant-size set of forbidden minors and that it can be tested in cubic time whether a graph contains a given fixed graph as a minor. Since Π_k is minor-closed, this implies the existence of a non-uniform FPT-algorithm for the problem Π-VS. Because the graphs of pathwidth at most 1 form a minor-closed graph class, this includes the problem POVS.

Proposition 2 ([18]). *For every minor-closed graph class Π, the problem Π-VS is non-uniformly FPT parameterized by the solution size k.*

We say that a graph class Π is *MSO$_2$-definable*, if there exists an MSO$_2$ (monadic second-order graph logic, see [7]) formula φ such that $G \models \varphi$ if and only if $G \in \Pi$. In the following, we show that the problem Π-VS is uniformly FPT parameterized by k if Π is MSO$_2$-definable and has bounded treewidth. Since every minor-closed graph class is MSO$_2$-definable, this improves the result from Proposition 2 for graph classes of bounded treewidth.

Eppstein et al. [10] showed that the problem of deciding whether a given graph G can be turned into a graph of class Π by splitting each vertex of G at most k times can be expressed as an MSO$_2$ formula on G, if Π itself is MSO$_2$-definable. Using Courcelle's Theorem [6], this yields an FPT-algorithm parameterized by k and the treewidth of the input graph. Their algorithm exploits the fact that the split operations create at most k copies of each vertex in the graph. Since the same also applies for the problem Π-VS, where we may apply at most k splits overall, their algorithm can be straightforwardly adapted for Π-VS, thereby implying the following result.

Corollary 1. *For every MSO$_2$-definable graph class Π, the problem Π-VS is FPT parameterized by the solution size k and the treewidth of the input graph.*

For a graph class Π of bounded treewidth, we let $\text{tw}(\Pi)$ denote the maximum treewidth among all graphs in Π. With the following lemma, we show that, if the target graph class Π has bounded treewidth, then every yes-instance of Π-VS must also have bounded treewidth.

Proposition 3. *For a graph class Π of bounded treewidth, let $\mathcal{I} = (G, S, k)$ be an instance of Π-VS. If $\text{tw}(G) > k + \text{tw}(\Pi)$, then $\mathcal{I}$ is a no-instance.*

Proof. We first show that a single split operation can reduce the treewidth of G by at most 1. Assume, for the sake of contradiction, that we can obtain a graph G' of treewidth less than $\text{tw}(G) - 1$ by splitting a single vertex v of G into vertices v_1 and v_2 of G'. Let $\mathcal{T}$ denote a minimum tree decomposition of G'. Remove all occurences of v_1 and v_2 in $\mathcal{T}$ and add v to every bag of $\mathcal{T}$. Observe that the result is a tree decomposition of size less than $\text{tw}(G)$ for G, a contradiction. A single split operation thus decreases the treewidth of the graph by at most 1. Since every graph $G' \in \Pi$ has $\text{tw}(G') \leq \text{tw}(\Pi)$, it is thus impossible to obtain a graph of Π with at most k vertex splits if $\text{tw}(G) > k + \text{tw}(\Pi)$.   □

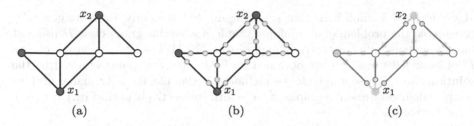

Fig. 6. (a) An instance $(G, S, 2)$ of Π-VE. (b) The corresponding auxiliary graph $G^{\times}$ obtained by subdividing each edge in G twice. (c) The graph obtained by exploding $\{x_1, x_2\}$ in G is the highlighted minor of $G^{\times}$. Since Π is MSO$_2$-definable, one can express Π-VE using an MSO$_2$ formula on $G^{\times}$.

Given a graph class Π of bounded treewidth, we first determine in time $f(k + \mathrm{tw}(\Pi)) \cdot n$ whether the treewidth of G is greater than $k + \mathrm{tw}(\Pi)$ [5]. If this is the case, then we can immediately report a no-instance by Proposition 3. Otherwise, we know that $\mathrm{tw}(G) \leq k + \mathrm{tw}(\Pi)$. Since $\mathrm{tw}(\Pi)$ is a constant, we have $\mathrm{tw}(G) \in O(k)$, and thus Corollary 1 yields the following result.

Theorem 5. *For every MSO$_2$-definable graph class Π of bounded treewidth, the problem Π-VS is FPT parameterized by the solution size k.*

Vertex Explosion. We now briefly sketch how these results extend to the problem variant Π VERTEX EXPLOSION(Π-VE) using vertex explosions instead of vertex splits. In this case, for minor-closed graph classes Π, the set of yes-instances of Π-VE is not minor-closed in general, thus the non-uniform FPT algorithm used to obtain Proposition 2 does not work for Π-VE. Additionally, the FPT-algorithm by Eppstein et al. [10] for MSO$_2$-definable graph classes cannot be straightforwardly adapted for Π-VE, since the number of new vertices resulting from explosions is not bounded by a function in k. However, using the approach illustrated in Fig. 6, we obtain the following results.

Lemma 4 ($\star$). *For every MSO$_2$-definable graph class Π, the problem Π-VE is FPT parameterized by the treewidth of the input graph.*

Theorem 6 ($\star$). *For every MSO$_2$-definable graph class Π of bounded treewidth, the problem Π-VE is FPT parameterized by the solution size k.*

We remark that, for arbitrary graph classes Π, the question whether a graph of Π can be obtained by applying arbitrarily many vertex splits to at most k vertices in the input graph is not equivalent to Π-VE.

6   Conclusion

In this work, we studied the problems PATHWIDTH-ONE VERTEX EXPLOSION and PATHWIDTH-ONE VERTEX SPLITTING, obtaining an efficient branching algorithm and a small kernel for each variant. Subsequently, we more generally considered the problem of obtaining a graph of a specific graph class Π using at most k vertex splits (respectively explosions). For MSO_2-definable graph classes Π of bounded treewidth, we obtained an FPT algorithm parameterized by the solution size k. These graph classes include, for example, the outerplanar graphs, the pseudoforests, and the graphs of treewidth (respectively pathwidth) at most c for some constant c.

Instead of splitting vertices to obtain a graph of pathwidth at most 1, one can also consider obtaining graphs of treewidth at most 1, i.e., forests. Since, in this context, the degree-1 vertices resulting from an explosion can simply be reduced, the explosion model is equivalent to the problem FEEDBACK VERTEX SET, a well-studied NP-complete [14] problem that admits a quadratic kernel [22]. In the full version of this paper [4], we show that the problem of splitting vertices of a graph to obtain a forest is equivalent to the problem FEEDBACK EDGE SET, which asks whether a given graph can be made acyclic using at most k edge deletions; a problem that can be solved by computing an arbitrary spanning forest of the graph. Firbas [12] independently obtained the same result.

References

1. Ahmed, R., et al.: Splitting vertices in 2-layer graph drawings. IEEE Comput. Graph. Appl. **43**(3), 24–35 (2023). https://doi.org/10.1109/MCG.2023.3264244
2. Ahmed, R., Kobourov, S.G., Kryven, M.: An FPT algorithm for bipartite vertex splitting. In: Angelini, P., von Hanxleden, R. (eds.) Graph Drawing and Network Visualization - 30th International Symposium, GD 2022. LNCS, vol. 13764, pp. 261–268. Springer, Cham (2022). https://doi.org/10.1007/978-3-031-22203-0_19
3. Arnborg, S., Proskurowski, A., Seese, D.: Monadic second order logic, tree automata and forbidden minors. In: Börger, E., Kleine Büning, H., Richter, M.M., Schönfeld, W. (eds.) CSL 1990. LNCS, vol. 533, pp. 1–16. Springer, Heidelberg (1991). https://doi.org/10.1007/3-540-54487-9_49
4. Baumann, J., Pfretzschner, M., Rutter, I.: Parameterized complexity of vertex splitting to pathwidth at most 1. CoRR abs/2302.14725 (2023). https://doi.org/10.48550/arXiv.2302.14725
5. Bodlaender, H.L.: A linear time algorithm for finding tree-decompositions of small treewidth. In: Kosaraju, S.R., Johnson, D.S., Aggarwal, A. (eds.) Proceedings of the Twenty-Fifth Annual ACM Symposium on Theory of Computing, pp. 226–234. ACM (1993). https://doi.org/10.1145/167088.167161
6. Courcelle, B.: The monadic second-order logic of graphs. i. recognizable sets of finite graphs. Inf. Comput. **85**(1), 12–75 (1990). https://doi.org/10.1016/0890-5401(90)90043-H
7. Cygan, M.: Parameterized Algorithms. Springer, Cham (2015). https://doi.org/10.1007/978-3-319-21275-3
8. Cygan, M., Pilipczuk, M., Pilipczuk, M., Wojtaszczyk, J.O.: An improved FPT algorithm and a quadratic kernel for pathwidth one vertex deletion. Algorithmica **64**(1), 170–188 (2012). https://doi.org/10.1007/s00453-011-9578-2
9. Eades, P., McKay, B.D., Wormald, N.C.: On an edge crossing problem. In: Proceedings of the 9th Australian Computer Science Conference, vol. 327, p. 334 (1986)

10. Eppstein, D., et al.: On the planar split thickness of graphs. Algorithmica **80**(3), 977–994 (2017). https://doi.org/10.1007/s00453-017-0328-y
11. Faria, L., de Figueiredo, C.M.H., Mendonça, C.F.X.: Splitting number is NP-complete. In: Hromkovič, J., Sýkora, O. (eds.) WG 1998. LNCS, vol. 1517, pp. 285–297. Springer, Heidelberg (1998). https://doi.org/10.1007/10692760_23
12. Firbas, A.: Establishing Hereditary Graph Properties via Vertex Splitting. Diploma thesis, Technische Universität Wien (2023). https://doi.org/10.34726/hss.2023.103864
13. Jansen, B.M.P., Lokshtanov, D., Saurabh, S.: A near-optimal planarization algorithm. In: Chekuri, C. (ed.) Proceedings of the Twenty-Fifth Annual ACM-SIAM Symposium on Discrete Algorithms (SODA), pp. 1802–1811. SIAM (2014). https://doi.org/10.1137/1.9781611973402.130
14. Karp, R.M.: Reducibility among combinatorial problems. In: Miller, R.E., Thatcher, J.W. (eds.) Proceedings of a Symposium on the Complexity of Computer Computations, pp. 85–103. The IBM Research Symposia Series, Plenum Press, New York (1972). https://doi.org/10.1007/978-1-4684-2001-2_9
15. Kawarabayashi, K.: Planarity allowing few error vertices in linear time. In: 50th Annual IEEE Symposium on Foundations of Computer Science, FOCS 2009, pp. 639–648. IEEE Computer Society (2009). https://doi.org/10.1109/FOCS.2009.45
16. Lewis, J.M., Yannakakis, M.: The node-deletion problem for hereditary properties is NP-complete. J. Comput. Syst. Sci. **20**(2), 219–230 (1980). https://doi.org/10.1016/0022-0000(80)90060-4
17. Marx, D., Schlotter, I.: Obtaining a planar graph by vertex deletion. In: Brandstädt, A., Kratsch, D., Müller, H. (eds.) WG 2007. LNCS, vol. 4769, pp. 292–303. Springer, Heidelberg (2007). https://doi.org/10.1007/978-3-540-74839-7_28
18. Nöllenburg, M., Sorge, M., Terziadis, S., Villedieu, A., Wu, H., Wulms, J.: Planarizing graphs and their drawings by vertex splitting. In: Angelini, P., von Hanxleden, R. (eds.) GD 2022. LNCS, vol. 13764, pp. 232–246. Springer, Cham (2022). https://doi.org/10.1007/978-3-031-22203-0_17
19. Philip, G., Raman, V., Villanger, Y.: A quartic kernel for pathwidth-one vertex deletion. In: Thilikos, D.M. (ed.) WG 2010. LNCS, vol. 6410, pp. 196–207. Springer, Heidelberg (2010). https://doi.org/10.1007/978-3-642-16926-7_19
20. Purchase, H.C., Cohen, R.F., James, M.: Validating graph drawing aesthetics. In: Brandenburg, F.J. (ed.) GD 1995. LNCS, vol. 1027, pp. 435–446. Springer, Heidelberg (1996). https://doi.org/10.1007/BFb0021827
21. Robertson, N., Seymour, P.D.: Graph minors. XIII. The disjoint paths problem. J. Comb. Theory, Ser. B **63**(1), 65–110 (1995). https://doi.org/10.1006/jctb.1995.1006
22. Thomassé, S.: A $4k^2$ kernel for feedback vertex set. ACM Trans. Algorithms **6**(2), 32:1–32:8 (2010). https://doi.org/10.1145/1721837.1721848
23. Tsur, D.: Faster algorithm for pathwidth one vertex deletion. Theor. Comput. Sci. **921**, 63–74 (2022). https://doi.org/10.1016/j.tcs.2022.04.001

Odd Chromatic Number of Graph Classes

Rémy Belmonte[1], Ararat Harutyunyan[2], Noleen Köhler[2(✉)],
and Nikolaos Melissinos[3]

[1] Université Gustave Eiffel, CNRS, LIGM, 77454 Marne-la-Vallée, France
`remy.belmonte@u-pem.fr`
[2] Université Paris-Dauphine, PSL University, CNRS UMR7243, LAMSADE,
Paris, France
`ararat.harutyunyan@lamsade.dauphine.fr, noleen.kohler@dauphine.psl.eu`
[3] Department of Theoretical Computer Science, Faculty of Information Technology,
Czech Technical University in Prague, Prague, Czech Republic

Abstract. A graph is called *odd* (respectively, *even*) if every vertex has odd (respectively, even) degree. Gallai proved that every graph can be partitioned into two even induced subgraphs, or into an odd and an even induced subgraph. We refer to a partition into odd subgraphs as an *odd colouring* of G. Scott [Graphs and Combinatorics, 2001] proved that a graph admits an odd colouring if and only if it has an even number of vertices. We say that a graph G is k-odd colourable if it can be partitioned into at most k odd induced subgraphs. We initiate the systematic study of odd colouring and odd chromatic number of graph classes. In particular, we consider for a number of classes whether they have bounded odd chromatic number. Our main results are that interval graphs, graphs of bounded modular-width and graphs of bounded maximum degree all have bounded odd chromatic number.

Keywords: Graph classes · Vertex partition problem · Odd colouring · Colouring variant · Upper bounds

1 Introduction

A graph is called *odd* (respectively even) if all its degrees are odd (respectively even). Gallai proved the following theorem (see [8], Problem 5.17 for a proof).

Theorem 1. *For every graph G, there exist:*

- *a partition (V_1, V_2) of $V(G)$ such that $G[V_1]$ and $G[V_2]$ are both even;*
- *a partition (V_1', V_2') of $V(G)$ such that $G[V_1']$ is odd and $G[V_2']$ is even.*

Ararat Harutyunyan is supported by the grant from French National Research Agency under JCJC program (DAGDigDec: ANR-21-CE48-0012).
Noleen Köhler is supported by the grant from French National Research Agency under JCJC program (ASSK: ANR-18-CE40-0025-01).
Nikolaos Melissinos is partially supported by the CTU Global postdoc fellowship program.

© The Author(s), under exclusive license to Springer Nature Switzerland AG 2023
D. Paulusma and B. Ries (Eds.): WG 2023, LNCS 14093, pp. 44–58, 2023.
https://doi.org/10.1007/978-3-031-43380-1_4

This theorem has two main consequences. The first one is that every graph contains an induced even subgraph with at least $|V(G)|/2$ vertices. The second is that every graph can be *even coloured* with at most two colours, i.e., partitioned into two (possibly empty) sets of vertices, each of which induces an even subgraph of G. In both cases, it is natural to wonder whether similar results hold true when considering odd subgraphs.

The first question, known as the *odd subgraph conjecture* and mentioned already by Caro [3] as part of the graph theory folklore, asks whether there exists a constant $c > 0$ such that every graph G contains an odd subgraph with at least $|V(G)|/c$ vertices. In a recent breakthrough paper, Ferber and Krivelevich proved that the conjecture is true.

Theorem 2 ([5]). *Every graph G with no isolated vertices has an odd induced subgraph of size at least $|V(G)|/10000$.*

The second question is whether every graph can be partitioned into a bounded number of odd induced subgraphs. We refer to such a partition as an *odd colouring*, and the minimum number of parts required to odd colour a given graph G, denoted by $\chi_{odd}(G)$, as its *odd chromatic number*. This can be seen as a variant of proper (vertex) colouring, where one seeks to partition the vertices of a graph into odd subgraphs instead of independent sets. An immediate observation is that in order to be odd colourable, a graph must have all its connected components be of even order, as an immediate consequence of the handshake lemma. Scott [11] proved that this necessary condition is also sufficient. Therefore, graphs can generally be assumed to have all their connected components of even order, unless otherwise specified.

Motivated by this result, it is natural to ask how many colours are necessary to partition a graph into odd induced subgraphs. As Scott showed [11], there exist graphs with arbitrarily large odd chormatic number. On the computational side, Belmonte and Sau [2] proved that the problem of deciding whether a graph is k-odd colourable is solvable in polynomial time when $k \leq 2$, and NP-complete otherwise, similarly to the case of proper colouring. They also show that the k-odd colouring problem can be solved in time $2^{O(k \cdot rw)} \cdot n^{O(1)}$, where k is the number of colours and rw is the rank-width of the input graphs. They then ask whether the problem can be solved in FPT time parameterized by rank-width alone, i.e., whether the dependency on k is necessary. A positive answer would provide a stark contrast with proper colouring, for which the best algorithms run in time $n^{2^{O(rw)^2}}$ (see, e.g., [7]), while Fomin et al. [6] proved that there is no algorithm that runs in time $n^{2^{o(rw)}}$, unless the ETH fails.[1]

On the combinatorial side, Scott showed that there exist graphs that require $\Theta(\sqrt{n})$ colours. In particular, the *subdivided clique*, i.e., the graph obtained from a complete graph on n vertices by subdividing[2] every edge once requires

---

[1] While Fomin et al. proved the lower bound for clique-width, it also holds for rank-width, since rank-width is always at most clique-width.

[2] Subdividing an edge uv consists in removing uv, adding a new vertex w, and making it adjacent to exactly u and v.

exactly n colours, as the vertices obtained by subdividing the edges force their two neighbours to be given distinct colours. More generally, and by the same argument, given any graph G, the graph H obtained from G by subdividing every edge once has $\chi_{\text{odd}}(H) = \chi(G)$, and H is odd colourable if and only if $|V(H)| = |V(G)|+|E(G)|$ is even. Note that a subdivided clique is odd colourable if and only if the subdivided complete graph K_n satisfies $n \in \{k : k \equiv 0 \vee k \equiv 3 \pmod 4\}$. Surprisingly, Scott also showed that only a sublinear number of colours is necessary to odd colour a graph, i.e., every graph of even order G has $\chi_{\text{odd}}(G) \leq cn(\log\log n)^{-1/2}$. As Scott observed, this bound is quite weak, and he instead conjectures that the lower bound obtained from the subdivided clique is essentially tight:

Conjecture 1 (Scott, 2001) . Every graph G of even order has $\chi_{\text{odd}}(G) \leq (1 + o(1))c\sqrt{n}$.

One way of seeing Conjecture 1 is to consider that subdivided cliques appear to be essentially the graphs that require most colours to be odd coloured. More specifically, consider the family $\mathcal{B}$ of graphs G' obtained from a graph G by adding, for every pair of vertices $u, v \in V(G)$, a vertex w_{uv} and edges uw_{uv} and vw_{uv}, and G' has even order. Note that subdivided cliques of even order are exactly those graphs in $\mathcal{B}$ where graph G is edgeless, and that the graphs in $\mathcal{B}$ have $\chi_{\text{odd}}(G') = |V(G)| \in \Theta(\sqrt{|V(G')|})$. A question closely related to Conjecture 1 is whether if a class of graphs $\mathcal{G}$ does not contain arbitrarily large graphs of $\mathcal{B}$ as induced subgraphs, then $\mathcal{G}$ has odd chromatic number $\mathcal{O}(\sqrt{n})$, i.e., they satisfy Conjecture 1. This question was already answered positively for some graph classes. In fact, the bounds provided were constant. It was shown in [2] that every cograph can be odd coloured using at most three colours, and that graphs of treewidth at most k can be odd coloured using at most $k + 1$ colours. In fact, those results can easily be extended to all graphs admitting a join, and H-minor free graphs, respectively. Using a similar argument, Aashtab et al. [1] showed that planar graphs are 4-odd colourable, and this is tight due to subdivided K_4 being planar and 4-odd colourable, as explained above. They also proved that subcubic graphs are 4-odd colourable, which is again tight due to subdivided K_4, and conjecture that this result can be generalized to all graphs, i.e., $\chi_{\text{odd}}(G) \leq \Delta + 1$, where Δ denotes the maximum degree of G. Observe that none of those graph classes contain arbitrarily large graphs from $\mathcal{B}$ as induced subgraphs. On the negative side, bipartite graphs and split graphs contain arbitrarily large graphs from $\mathcal{B}$, and therefore the bound of Conjecture 1 is best possible. In fact, Scott specifically asked whether the conjecture holds for the specific case of bipartite graphs.

Our Contribution. Motivated by these first isolated results and Conjecture 1, we initiate the systematic study of the odd chromatic number in graph classes, and determine which have bounded odd chromatic number. We focus on graph classes that do not contain large graphs from $\mathcal{B}$ as induced subgraphs. Our main results are that graphs of bounded maximum degree, interval graphs and graphs of bounded modular width all have bounded odd chromatic number.

In Sect. 3, we prove that every graph G of even order and maximum degree Δ has $\chi_{\text{odd}}(G) \leq 2\Delta - 1$, extending the result of Aashtab et al. on subcubic graphs to graphs of bounded degree. We actually prove a more general result, which provides additional corollaries for graphs of large girth. In particular, we obtain that planar graphs of girth 11 are 3-odd colourable. We also obtain that graphs of girth at least 7 are $\mathcal{O}(\sqrt{n})$-odd colourable. While this bound is not constant, it is of particular interest as subdivided cliques have girth exactly 6.

In Sect. 4 we prove that every graph with all connected components of even order satisfies $\chi_{\text{odd}}(G) \leq 3 \cdot mw(G)$, where $mw(G)$ denotes the modular-width of G. This significantly generalizes the cographs result from [2] and provides an important step towards proving that graphs of bounded rank-width have bounded odd chromatic number, which in turn would imply that the ODD CHROMATIC NUMBER is FPT when parameterized by rank-width alone.

Finally, we prove in Sect. 5 that every interval graph with all components of even order is 6-odd colourable. Additionally, every proper interval graph with all components of even order is 3-odd colourable, and this bound is tight.

We would also like to point out that all our proofs are constructive and furthermore a (not necessarily) optimal odd-colouring with the number of colours matching the upper bound can be computed in polynomial time. In particular, the proof provided in [8] of Theorem 1, upon which we rely heavily is constructive, and both partitions can easily be computed in polynomial time. An overview of known results and open cases is provided in Fig. 1 below.

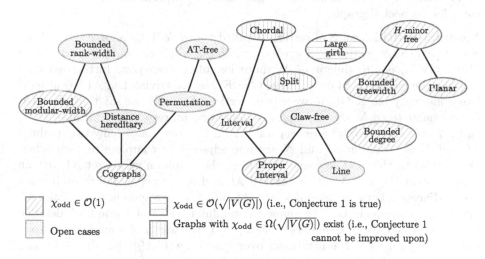

Fig. 1. Overview of known and open cases.

2   Preliminaries

For a positive integer i, we denote by $[i]$ the set of integers j such that $1 \leq j \leq i$. A partition of a set X is a tuple $\mathcal{P} = (P_1, \ldots, P_k)$ of subsets of X such that $X =$

$\bigcup_{i \in [k]} P_i$ and $P_i \cap P_j = \emptyset$, i.e., we allow parts to be empty. Let $\mathcal{P} = (P_1, \ldots, P_k)$ be a partition of X and $Y \subseteq X$. We let $\mathcal{P}|_Y$ be the partition of Y obtained from $(P_1 \cap Y, \ldots, P_k \cap Y)$ by removing all empty parts. A partition $(Q_1, \ldots, Q_\ell)$ of X is a *coarsening* of a partition $(P_1, \ldots, P_k)$ of X if for every P_i and every Q_j either $P_i \cap Q_j = \emptyset$ or $P_i \cap Q_j = P_i$, i.e., every Q_j is the union of P_i's.

Every graph in this paper is simple, undirected and finite. We use standard graph-theoretic notation, and refer the reader to [4] for any undefined notation. For a graph G we denote the set of vertices of G by $V(G)$ and the edge set by $E(G)$. Let G be a graph and $S \subseteq V(G)$. We denote an edge between u and v by uv. The *order* of G is $|V(G)|$. The *degree* (respectively, *open neighborhood*) of a vertex $v \in V(G)$ is denoted by $d_G(v)$ (respectively, $N_G(v)$). We denote the subgraph induced by S by $G[S]$. $G \setminus S = G[V(G) \setminus S]$. The *maximum* degree of any vertex of G is denoted by Δ. We denote paths and cycles by tuples of vertices. The *girth* of G is the length of a shortest cycle of G. Given two vertices u and v lying in the same connected component of G, we say an edge e *separates* u and v if they lie in different connected components of $G \setminus \{e\}$.

A graph is called odd (even, respectively) if every vertex has odd (respectively, even) degree. A partition $(V_1, \ldots, V_k)$ of $V(G)$ is a *k-odd colouring*[3] of G if $G[V_i]$ induces an odd subgraphs of G for every $i \in [k]$. We say a graph is k-odd colourable if it admits a k-odd colouring. The *odd chromatic number* of G, denoted by $\chi_{\text{odd}}(G)$, is the smallest integer k such that G is k-odd colourable. The empty graph (i.e., $V(G) = \emptyset$) is considered to be both even and odd. Since every connected component can be odd coloured separately, we only need to consider connected graphs.

Modular-width. A set S of vertices is called a *module* if, for all $u, v \in S, N(u) \cap S = N(v) \cap S$. A partition $\mathcal{M} = (M_1, \ldots, M_k)$ of $V(G)$ is a module partition of G if every M_i is a module in G. Without loss of generality, we further ask that any module partition $\mathcal{M}$ of G, unless $G = K_1$, is non-trivial, i.e., $\mathcal{M}$ has at least two non-empty parts. Given two sets of vertices X and Y, we say that X and Y are *complete to each other* (*completely non-adjacent*, respectively) if $uv \in E(G)$ ($uv \notin E(G)$, respectively) for every $u \in X, v \in Y$. Note that for any two modules M and N in G, either M and N are non-adjacent or complete to each other. We let $G_{\mathcal{M}}$ be the module graph of $\mathcal{M}$, i.e., the graph on vertex set $\mathcal{M}$ with an edge between M_i and M_j if and only if M_i and M_j are complete to each other (non-adjacency between modules M_i, M_j in $G_{\mathcal{M}}$ corresponds to M_i and M_j being non-adjacent in G). We define the modular width of a graph G, denoted by $\text{mw}(G)$, recursively as follows. $\text{mw}(K_1) = 1$, the width of a module partition $(M_1, \ldots, M_k)$ of G is the maximum over k and $\text{mw}(G[M_i])$ for all $i \in [k]$ and $\text{mw}(G)$ is the minimum width of any module partitions of G.

---

[3] This definition of odd colouring is not to be confused with the one introduced by Petrusevski and Skrekovski [10], which is a specific type of proper colouring.

3 Graphs of Bounded Degree and Graphs of Large Girth

In this section, we study Scott's conjecture (Conjecture 1) as well as the conjecture made by Aashtab et al. [1] which states that $\chi_{odd}(G) \leq \Delta + 1$ for any graph G. We settle Conjecture 1 for graphs of girth at least 7, and prove that $\chi_{odd}(G) \leq 2\Delta - 1$ for any graph G, thus obtaining a weaker version of the conjecture of Aashtab et al. To this end, we prove the following more general theorem, which implies both of the aforementioned results.

Theorem 3. *Let $\mathcal{H}$ be a class of graphs such that:*

- *$K_2 \in \mathcal{H}$*
- *$\mathcal{H}$ is closed under vertex deletion and*
- *there is a $k \geq 2$ such that any connected graph $G \in \mathcal{H}$ satisfies at least one of the following properties:*
  (I) G has two pendant vertices u, v such that $N_G(u) = N_G(v)$ or
  (II) G has two adjacent vertices u, v such that $d_G(u) + d_G(v) \leq k$.

Then every graph $G \in \mathcal{H}$ with all components of even order has $\chi_{odd}(G) \leq k-1$.

Proof. First notice that $\mathcal{H}$ is well defined as K_2 has the desired properties. The proof is by induction on the number of vertices. Let $|V(G)| = 2n$.

For $n = 1$, since G is connected, we have that $G = K_2$ which is odd. Therefore, $\chi_{odd}(G) = 1 \leq k - 1$ (recall that $k \geq 2$). Let G be a graph of order $2n$. Notice that we only need to consider the case where G is connected as, otherwise, we can apply the inductive hypothesis to each of the components of G. Assume first that G has two pendant vertices u, v such that $N_G(u) = N_G(v) = \{w\}$. Then, since $G \setminus \{u, v\}$ is connected and belongs to $\mathcal{H}$, by induction, there is an odd colouring of $G \setminus \{u, v\}$ that uses at most $k - 1$ colours. Let $(V_1, \ldots, V_{k-1})$ be a partition of $V(G) \setminus \{u, v\}$ such that $G[V_i]$ is odd for all $i \in [k - 1]$. We may assume that $w \in V_1$. We give a partition $V'_1, \ldots, V'_{k-1}$ of $V(G)$ by setting $V'_1 = V_1 \cup \{u, v\}$ and $V'_i = V_i$ for all $i \in [k] \setminus \{1\}$. Notice that for all $i \in [k - 1]$, $G[V''_i]$ is odd. Therefore, $\chi_{odd}(G) \leq k - 1$.

Thus, we assume that G has an edge $uv \in E(G)$ such that $d_G(u) + d_G(v) \leq k$. We may assume that $k \geq 3$ for otherwise the theorem follows. We consider two cases; $G \setminus \{u, v\}$ is connected and $G \setminus \{u, v\}$ is disconnected.

Assume that $G \setminus \{u, v\}$ is connected. Since $G \setminus \{u, v\}$ has $|V(G) \setminus \{u, v\}| = 2n - 2$ and belongs to $\mathcal{H}$, by induction, there is an odd colouring of it that uses at most $k - 1$ colours. Let $(V_1, \ldots, V_{k-1})$ be a partition of $V(G) \setminus \{u, v\}$, such that $G[V_i]$ is odd of all $i \in [k - 1]$. We give a partition of G into $k - 1$ odd graphs as follows. Since $|N_G(\{u, v\})| \leq k - 2$, there exists $\ell \in [k - 1]$ such that $V_\ell \cap N_G(\{u, v\}) = \emptyset$. We define a partition $(U_1, \ldots, U_{k-1})$ of $V(G)$ as follows. For all $i \in [k - 1]$, if $i \neq \ell$, we define $U_i = V_i$, otherwise we set $U_i = V_i \cup \{u, v\}$. Notice that for all $i \neq \ell$, $G[U_i]$ is odd since $U_i = V_i$. Also, since $N_{G[U_\ell]}[v] = N_{G[U_\ell]}[u] = \{u, v\}$ and $G[V_\ell]$ is odd, we conclude that $G[U_\ell]$ is odd. Thus, $\chi_{odd}(G) \leq k - 1$.

Now, we consider the case where $G \setminus \{u, v\}$ is disconnected. First, we assume that there is at least one component in $G \setminus \{u, v\}$ of even order. Let U be the set of vertices of this component. By induction, $\chi_{\text{odd}}(G[U]) \leq k - 1$ and $\chi_{\text{odd}}(G \setminus U) \leq k - 1$. Furthermore, $|N_G(\{u, v\}) \cap U| \leq k - 3$ because $G \setminus \{u, v\}$ has at least two components. Let $(U_1, \ldots, U_{k-1})$ be a partition of U such that $G[U_i]$ is odd for all $i \in [k - 1]$. Also, let $(V_1, \ldots, V_{k-1})$ be a partition of $V(G) \setminus U$ such that $G[V_i]$ is odd for all $i \in [k - 1]$. We may assume that $V_i \cap \{u, v\} = \emptyset$ for all $i \in [k - 3]$. Since $|N_G(\{u, v\}) \cap U| \leq k - 3$, there are at least two indices $l, l' \in [k - 1]$ such that $U_l \cap N_G(\{u, v\}) = U_{l'} \cap N_G(\{u, v\}) = \emptyset$. We may assume that $l = k - 2$ and $l' = k - 1$. We define a partition $(V_1', \ldots, V_{k-1}')$ of $V(G)$ as follows. For all $i \in [k - 1]$ we define $V_i' = U_i \cup V_i$. We claim that $G[V_i']$ is odd for all $i \in [k - 1]$. To show the claim, we consider two cases; either $V_i' \cap \{u, v\} = \emptyset$ or not. If $V_i' \cap \{u, v\} = \emptyset$, since the only vertices in $V(G) \setminus U$ that can have neighbours in U are v and u we have that $G[V_i']$ is odd. Indeed, this holds because $U_i \cap N_G(V_i) = \emptyset$ and both $G[U_i]$ and $G[V_i]$ are odd. If $V_i' \cap \{u, v\} \neq \emptyset$, then $i = k - 2$ or $i = k - 1$. In both cases, we know that $U_i \cap N_G(V_i) = \emptyset$ because the only vertices in $V(G) \setminus U$ that may have neighbours in U are v and u and we have assumed that u, v do not have neighbours in $U_{k-2} \cup U_{k-1}$. So, $G[V_i']$ is odd because $U_i \cap N_G(V_i) = \emptyset$ and both $G[U_i]$ and $G[V_i]$ are odd.

Thus, we can assume that all components of $G \setminus \{u, v\}$ are of odd order. Let $\ell > 0$ be the number of components, denoted by $V_1, \ldots, V_\ell$, of $G \setminus \{u, v\}$ and note that ℓ must be even. We consider two cases, either for all $i \in [\ell]$, one of $G[V_i \cup \{u\}]$ or $G[V_i \cup \{v\}]$ is disconnected, or there is at least one $i \in [\ell]$ such that both $G[V_i \cup \{u\}]$ and $G[V_i \cup \{v\}]$ are connected.

In the first case, for each V_i, $i \in [\ell]$ we call w_i the vertex in $\{u, v\}$ such that $G[V_i \cup \{w_i\}]$ is connected. Note that w_i is uniquely determined, i.e., only one of u and v can be w_i for each $i \in [\ell]$. Now, by induction, for all $i \in [\ell]$, $G[V_i \cup \{w_i\}]$ has $\chi_{\text{odd}}(G[V_i \cup \{w_i\}]) \leq k - 1$. Let, for each $i \in [\ell]$, $(V_1^i, \ldots, V_{k-1}^i)$ denote a partition of $V_i \cup \{w_i\}$ such that $G[V_j^i]$ be odd, for all $j \in [k - 1]$. Furthermore, we may assume that for each $i \in [\ell]$, if $v \in V_i \cup \{w_i\}$, then $v \in V_{k-2}^i$. Also, we can assume that for each $i \in [\ell]$, if $u \in V_i \cup \{w_i\}$, then $u \in V_{k-1}^i$. Finally, let $I = \{i \in [\ell] \mid w_i = u\}$ and $J = \{i \in [\ell] \mid w_i = v\}$.

We consider two cases. If $|I|$ is odd, then $|J|$ is odd since $\ell = |I| + |J|$ is even. Then, we claim that for the partition $(U_1, \ldots, U_{k-1})$ of $V(G)$ where $U_i = \bigcup_{j \in [\ell]} V_i^j$ it holds that $G[U_i]$ is odd for all $i \in [k - 1]$. First notice that $(U_1, \ldots, U_{k-1})$ is indeed a partition of $V(G)$. Indeed, the only vertices that may belong in more than one set are u and v. However, v belongs only to some sets V_{k-2}^i, and hence it is no set U_i except U_{k-2}. Similarly, u belongs to no set U_i except U_{k-1}. Therefore, it remains to show that $G[U_i]$ is odd for all $i \in [k - 1]$. We will show that for any $i \in [k - 1]$ and for any $x \in U_i$, $|N_G(x) \cap U_i|$ is odd. Let $x \in U_i \setminus \{u, v\}$, for some $i \in [k - 1]$. Then we know that $N_G(x) \cap U_i = N_G(x) \cap V_i^j$ for some $j \in [\ell]$. Since $G[V_i^j]$ is odd for all $i \in [k - 1]$ and $j \in [\ell]$ we have that $|N_G(x) \cap U_i| = |N_G(x) \cap V_i^j|$ is odd. Therefore, we only need to consider u and v. Notice that $v \in U_{k-2} = \bigcup_{j \in [\ell]} V_{k-2}^j$ (respectively, $u \in U_{k-1} = \bigcup_{j \in [\ell]} V_{k-1}^j$). Also, v (respectively, u) is included in V_{k-2}^j (respectively, V_{k-1}^j) only if $j \in I$ (respectively, $j \in J$). Since $G[V_{k-2}^j]$ (respectively, $G[V_{k-1}^j]$) is odd for any $j \in [\ell]$ we have that $|N(v) \cap V_{k-2}^j|$ (respectively, $|N(u) \cap V_{k-1}^j|$) is odd for any

$j \in I$ (respectively, $j \in J$). Finally, since $|I|$ and $|J|$ are odd, we have that $|N_G(v) \cap U_{k-2}| = \sum_{j \in I} |N(v) \cap V_{k-2}^j|$ and $|N_G(u) \cap U_{k-1}| = \sum_{j \in I} |N(u) \cap V_{k-1}^j|$ are both odd. Therefore, for any $i \in [k-1]$, $G[U_i]$ is odd and $\chi_{\text{odd}}(G) \leq k-1$.

Now, suppose that both $|I|$ and $|J|$ are even. We consider the partition $(U_1, \ldots, U_{k-1})$ of $V(G)$ where, for all $i \in [k-3]$ $U_i = \bigcup_{j \in [\ell]} V_i^j$, $U_{k-2} = \bigcup_{j \in J} V_{k-2}^j \cup \bigcup_{j \in I} V_{k-1}^j$ and $U_{k-1} = \bigcup_{j \in I} V_{k-2}^j \cup \bigcup_{j \in J} V_{k-1}^j$. We claim that for this partition it holds that $G[U_i]$ is odd for all $i \in [k-1]$. First notice that $(U_1, \ldots, U_{k-1})$ is indeed a partition of $V(G)$. Indeed, this is clear for all vertices except for v and u. However, v only belongs to sets of type V_{k-2}^i for $i \in I$, and u only belongs to sets of type V_{k-1}^i for $i \in J$. Therefore, u or v belong to no set U_i except U_{k-1}. We will show that for any $i \in [k-1]$ and $x \in U_i$, $|N_G(x) \cap U_i|$ is odd. Let $x \in U_i \setminus \{u, v\}$, for some $i \in [k-1]$. Then we know that $N_G(x) \cap U_i = N_G(x) \cap V_i^j$ for some $j \in [\ell]$. Since $G[V_i^j]$ is odd for all $i \in [k-1]$ and $j \in [\ell]$ we have that $|N_G(x) \cap U_i| = |N_G(x) \cap V_i^j|$ is odd. Therefore, we only need to consider v and u. Note that $u, v \in U_{k-1}$. Since both $|I|$ and $|J|$ are even and $U_{k-1} = \bigcup_{j \in I} V_{k-2}^j \cup \bigcup_{j \in J} V_{k-1}^j$, we have that $|N_G(v) \cap U_{k-1} \setminus \{u\}|$ and $|N_G(u) \cap U_{k-1} \setminus \{v\}|$ are both even. Finally, since $uv \in E(G)$ we have that $|N_G(v) \cap U_{k-1}|$ and $|N_G(u) \cap U_{k-1}|$ are both odd. Hence, $\chi_{\text{odd}}(G) \leq k-1$.

Now we consider the case where there is at least one $i \in [\ell]$ where both $G[V_i \cup \{v\}]$ and $G[V_i \cup \{u\}]$ are connected. We define the following sets I and J. For each $i \in [\ell]$, (i) $i \in J$, if $G[V_i \cup \{v\}]$ is disconnected, and (ii) $i \in I$, if $G[V_i \cup \{u\}]$ is disconnected. Finally, for the rest of the indices, $i \in [\ell]$, which are not in $I \cup J$, it holds that both $G[V_i \cup \{v\}]$ and $G[V_i \cup \{u\}]$ are connected. Call this set of indices X and note that by assumption $|X| \geq 1$. Since $|I| + |J| + |X|$ is even, it is easy to see that there is a partition of X into two sets X_1 and X_2 such that both $I' := I \cup X_1$ and $J' := J \cup X_2$ have odd size. Let $V_I = \bigcup_{i \in I'} V_i$ and $V_J = \bigcup_{i \in J'} V_i$. Now, by induction, we have that $\chi_{\text{odd}}(G[V_I \cup \{v\}]) \leq k-1$ and $\chi_{\text{odd}}(G[V_J \cup \{u\}]) \leq k-1$. Assume that $(V_1^I, \ldots, V_{k-1}^I)$ is a partition of V_I and $(V_1^J, \ldots, V_{k-1}^J)$ is a partition of V_J such that for any $i \in [k-1]$, $G[V_i^I]$ and $G[V_i^J]$ are odd. Without loss of generality, we may assume that $v \in V_1^I$ and $u \in V_{k-1}^J$. Since $|X| \geq 1$, note that both $d_G(u)$ and $d_G(v)$ are at least two, which implies that $d_G(u) \leq k-2$ and $d_G(v) \leq k-2$. Therefore, there exists $i_0 \in [k-2]$ such that $N_G(v) \cap V_{i_0}^I = \emptyset$ and $j_0 \in [k-1] \setminus \{1\}$ such that $N_G(v) \cap V_{j_0}^J = \emptyset$. We reorder the sets V_i^J, $i \in [k-2]$, so that $i_0 = 1$ and we reorder the sets V_i^I, $i \in [k-1] \setminus \{1\}$ so that $j_0 = k-1$. Note that this reordering does not change the fact that $v \in V_1^I$ and $u \in V_{k-1}^J$. Consider the partition $(U_1, \ldots, U_{k-1})$ of $V(G)$, where $U_i = V_i^I \cup V_i^J$. We claim that for all $i \in [k-1]$, $G[U_i]$ is odd. Note that for any $x \in U_i$, we have $N_G(x) \cap U_i = N_G(x) \cap V_i^I$ or $N_G(x) \cap U_i = N_G(x) \cap V_i^J$. Since for any $i \in [k-1]$, $G[V_i^I]$ and $G[V_i^J]$ are odd we conclude that $G[U_i]$ is odd for any $i \in [k-1]$. $\square$

Notice that the class of graphs G of maximum degree Δ satisfies the requirements of Theorem 3. Indeed, this class is closed under vertex deletions and any connected graph in the class has least two adjacent vertices u, v such that $d_G(u) + d_G(v) \leq 2\Delta$. Therefore, the following corollary holds.

Corollary 1. *For every graph G with all components of even order,* $\chi_{\mathrm{odd}}(G) \leq 2\Delta - 1$.

Next, we prove Conjecture 1 for graphs of girth at least seven.

Corollary 2. *For every graph G with all components of even order of girth at least 7,* $\chi_{\mathrm{odd}}(G) \leq \frac{3\sqrt{|V(G)|}}{2} + 1$. $\hspace{2cm} (*)^4$.

One may wonder if graphs of sufficiently large girth have bounded odd chromatic number. In fact, this is far from being true, which we show in the next.

Proposition 1. *For every integer g and k, there is a graph G such that every component of G has even order, G is of girth at least g and* $\chi_{\mathrm{odd}}(G) \geq k$. $\hspace{1cm} (*)$

Next, we obtain the following result for sparse planar graphs.

Corollary 3. *For every planar graph G with all components of even order of girth at least 11,* $\chi_{\mathrm{odd}}(G) \leq 3$. $\hspace{3cm} (*)$

The upper bound in Corollary 3 is tight as C_{14}, the cycle of length 14, has $\chi_{\mathrm{odd}}(C_{14}) = 3$.

4   Graphs of Bounded Modular-Width

In this section we consider graphs of bounded modular-width and show that we can upper bound the odd chromatic number by the modular-width of a graph.

Theorem 4. *For every graph G with all components of even order,* $\chi_{\mathrm{odd}}(G) \leq 3\,\mathrm{mw}(G)$.

In order to prove Theorem 4 we show that every graph G is 3-colourable for which we have a module partition $\mathcal{M}$ such that the module graph $G_{\mathcal{M}}$ exhibits a particular structure, i.e., is either a star Lemma 1 or a special type of tree Lemma 2. The following is an easy consequence of Theorem 1 which will be useful to colour modules and gain control over the parity of parts in case of modules of even size.

Remark 1. For every non-empty graph G of even order, there exists a partition (V_1, V_2, V_3) of $V(G)$ with $|V_2|, |V_3|$ being odd such that $V[G_1]$ is odd and $G[V_2]$, $G[V_3]$ are even. This can be derived from Theorem 1 by taking an arbitrary vertex $v \in V(G)$, setting $V_3 := \{v\}$ and then using the existence of a partition (V_1, V_2) of $V(G) \setminus \{v\}$ such that $G[V_1]$ is odd and $G[V_2]$ is even.

Lemma 1. *For every connected graph G of even order with a module partition* $\mathcal{M} = \{M_1, \ldots, M_k\}$ *such that* $G_{\mathcal{M}}$ *is a star,* $\chi_{\mathrm{odd}}(G) \leq 3$.

---

[4] For every result which is marked by $(*)$ the proof can be found in the full version of the paper.

Proof of A. ssume that in $G_\mathcal{M}$ the vertices $M_2, \ldots, M_k$ have degree 1. We refer to M_1 as the centre and to $M_2, \ldots, M_k$ as leaves of $G_\mathcal{M}$. We further assume that $|M_2|, \ldots, |M_\ell|$ are odd and $|M_{\ell+1}|, \ldots, |M_k|$ are even for some $\ell \in [k]$. We use the following two claims.

Claim 1. *If $W \subseteq V(G)$ with $G[W \cap M_i]$ is odd for every $i \in [k]$, then $G[W]$ is odd.*

Proof. First observe that the degree of any vertex $v \in W \cap M_1$ in $G[W]$ is $d_{G[W \cap M_1]}(v) + \sum_{i=2}^{k} |W \cap M_i|$. Since $d_{G[W \cap M_1]}(v)$ is odd and $|W \cap M_i|$ is even for every $i \in \{2, \ldots, k\}$ (which follows from $G[W \cap M_i]$ being odd by the handshake lemma) we get that $d_{G[W]}(v)$ is odd. For every $i \in \{2, \ldots, k\}$ the degree of any vertex $v \in W \cap M_i$ in $G[W]$ is $d_{G[W \cap M_i]}(v) + |W \cap M_1|$ which is odd (again, because $|W \cap M_1|$ must be even). Hence $G[W]$ is odd.   ◇

Claim 2. *If $W \subseteq V(G)$ such that $G[W \cap M_i]$ is even for every $i \in [k]$, $|W \cap M_1|$ is odd and $|\{i \in \{2, \ldots, k\} : |W \cap M_i| \text{ is odd}\}|$ is odd, then $G[W]$ is odd.*

Proof. Since $G_\mathcal{M}$ is a star and M_1 its centre we get that the degree of any vertex $v \in W \cap M_i$ for any $i \in \{2, \ldots, k\}$ is $d_{G[W \cap M_i]}(v) + |W \cap M_1|$. Since $|W \cap M_1|$ is odd and $d_{G[W \cap M_i]}(v)$ is even we get that every $v \in W \cap M_i$ for every $i \in \{2, \ldots, k\}$ has odd degree in $G[W]$. Moreover, the degree of $v \in W \cap M_1$ is $d_{G[W \cap M_1]}(v) + \sum_{i=2}^{k} |W \cap M_i|$. Since $d_{G[W \cap M_1]}(v)$ is even and $|\{i \in \{2, \ldots, k\} : |W \cap M_i| \text{ is odd}\}|$ is odd $d_{G[W]}(v)$ is odd. We conclude that $G[W]$ is odd.   ◇

First consider the case that $|M_1|$ is odd. Since G is of even order this implies that there must be an odd number of leaves of $G_\mathcal{M}$ of odd size and hence ℓ is even. Using Theorem 1 we let (W_1^i, W_2^i) be a partition of M_i such that $G[W_1^i]$ is odd and $G[W_2^i]$ is even for every $i \in [k]$. Note that since $G[W_1^i]$ is odd $|W_1^i|$ has to be even and hence $|W_2^i|$ is odd if and only if $i \in [\ell]$. We define $V_1 := \bigcup_{i \in [k]} W_1^i$ and $V_2 := \bigcup_{i \in [k]} W_2^i$. Note that (V_1, V_2) is a partition of G. Furthermore, $G[V_1]$ is odd by Claim 1 and $G[V_2]$ is odd by Claim 2. For an illustration see Fig. 2.

Now consider the case that $|M_1|$ is even. We first consider the special case that $\ell = 1$, i.e., there is no $i \in [k]$ such that $|M_i|$ is odd. In this case we let (W_1^i, W_2^i, W_3^i) be a partition of M_i for $i \in \{1, 2\}$ such that $G[W_1^i]$ is odd, $G[W_2^i]$, $G[W_3^i]$ are even and $|W_2^i|$, $|W_3^i|$ are odd which exists due to Remark 1. For $i \in \{3, \ldots, k\}$ we let (W_1^i, W_2^i) be a partition of M_i such that $G[W_1^i]$ is odd and $G[W_2^i]$ is even which exists by Theorem 1. We define $V_1 := \bigcup_{i \in [k]} W_1^i$, $V_2 := \bigcup_{i \in [k]} W_2^i$ and $V_3 := W_3^1 \cup W_3^2$. As before we observe that (V_1, V_2, V_3) is a partition of $V(G)$, $G[V_1]$ is odd by Claim 1 and $G[V_2], G[V_3]$ are even by Claim 2. For an illustration see Fig. 2.

Lastly, consider the case that $|M_1|$ is even and $\ell > 1$. By Remark 1 there is a partition (W_1^1, W_2^1, W_3^1) of M_1 such that $G[W_1^1]$ is odd, $G[W_2^1]$, $G[W_3^1]$ are even and $|W_2^1|$, $|W_3^1|$ are odd. For $i \in \{2, \ldots, k\}$ we let (W_1^i, W_2^i) be a partition of M_i such that $G[W_1^i]$ is odd and $G[W_2^i]$ is even which exists by Theorem 1.

We define $V_1 := \bigcup_{i \in [k]} W_1^i$, $V_2 := W_2^1 \cup \bigcup_{i=3}^k W_2^i$ and $V_3 := W_3^1 \cup W_2^2$. Note that (V_1, V_2, V_3) is a partition of $V(G)$. Furthermore, $G[V_1]$ is odd by Claim 1 and $G[V_3]$ is odd by Claim 2. Additionally, since $|M_1|$ is even there is an even number of $i \in \{2, \ldots, k\}$ such that $|M_i|$ is odd. Since for each $i \in \{2, \ldots, k\}$ for which $|M_i|$ is odd, $|W_1^i|$ must be odd, we get that $|\{i \in \{2, \ldots, k\} : |V_1 \cap M_i| \text{ is odd}\}|$ is odd (note that $V_1 \cap M_2 = \emptyset$ because $W_2^2 \subseteq V_3$). Hence we can use Claim 2 to conclude that $G[V_2]$ is odd. For an illustration see Fig. 2.   □

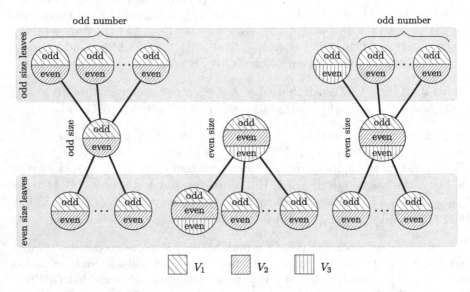

Fig. 2. Schematic illustration of the three cases in the proof of Lemma 1. Depicted is the module graph $G_{\mathcal{M}}$ along with a partition of the modules into sets V_1, V_2 and V_3 such that $G[V_i]$ is odd for $i \in [3]$.

Let G be a connected graph of even order with module partition $\mathcal{M} = (M_1, \ldots, M_k)$ such that $G_{\mathcal{M}}$ is a tree. For an edge e of $G_{\mathcal{M}}$ we let X_e and Y_e be the two components of the graph obtained from $G_{\mathcal{M}}$ by removing e. We say that the tree $G_{\mathcal{M}}$ is colour propagating if the following properties hold.

(i) $|\mathcal{M}| \geq 3$.
(ii) Every non-leaf module has size one.
(iii) $|\bigcup_{M \in V(X_e)} M|$ is odd for every $e \in E(G_{\mathcal{M}})$ not incident to any leaf of $G_{\mathcal{M}}$.

Lemma 2. *For every connected graph G of even order with a module partition $\mathcal{M} = (M_1, \ldots, M_k)$ such that $G_{\mathcal{M}}$ is a colour propagating tree, $\chi_{\text{odd}}(G) \leq 2$.*

Proof. To find an odd colouring (V_1, V_2) of G, we first let (W_1^i, W_2^i) be a partition of M_i such that $G[W_1^i]$ is odd and $G[W_2^i]$ is even for every $i \in [k]$. The partitions

(W_1^i, W_2^i) exist due to Theorem 1. Note that (ii) implies that for every module M_i which is not a leaf $|W_2^i| = 1$ and $W_1^i = \emptyset$. We define $V_1 := \bigcup_{i \in [k]} W_1^i$ and $V_2 := \bigcup_{i \in [k]} W_2^i$.

To argue that (V_1, V_2) is an odd colouring of G first consider any $v \in V(G)$ such that $v \in M_i$ for some leaf M_i of $G_{\mathcal{M}}$. Condition (i) implies that $G_{\mathcal{M}}$ must have at least three vertices and hence the neighbour M_j of M_i cannot be a leaf due to $G_{\mathcal{M}}$ being a tree. Hence $|M_j| = 1$ by (ii). Hence, if $v \in W_1^i$, then $d_{G[V_1]}(v) = d_{G[W_1^i]}(v)$ since $W_1^i = \emptyset$ and therefore $d_{G[V_1]}(v)$ is odd. Further, if $v \in W_2^i$, then $d_{G[V_2]}(v) = d_{G[W_2^i]}(v) + 1$ since $|W_2^j| = 1$ and hence $d_{G[V_2]}(v)$ is odd. Hence the degree of any vertex $v \in M_i$ is odd in $G[V_1]$, $G[V_2]$ respectively.

Now consider any vertex $v \in V(G)$ such that $M_i = \{v\}$ for some non-leaf M_i of $G_{\mathcal{M}}$. Let $M_{i_1}, \ldots, M_{i_\ell}$ be the neighbours of M_i in $G_{\mathcal{M}}$. Let e_j be the edge $M_i M_{i_j} \in E(G)$ for every $j \in [\ell]$. Without loss of generality, assume that $M_i \notin V(X_{e_j})$ for every $j \in [\ell]$. By (iii) we have that $|\bigcup_{M \in V(X_{e_j})} M|$ is odd whenever M_{i_j} is not a leaf in $G_{\mathcal{M}}$. Hence, by (ii), $|X_{e_j}| \equiv |M_{i_j}| \pmod 2$ for every $j \in [\ell]$ for which M_{i_j} is not a leaf in $G_{\mathcal{M}}$. On the other hand, as a consequence of the handshake lemma we get that $|W_2^{i_j}|$ is odd if and only if $|M_{i_j}|$ is odd. Hence the following holds for the parity of the degree of v in $G[V_2]$.

$$d_{G[V_2]}(v) = |\{j \in [m] : d_{G_{\mathcal{M}}}(M_{i_j}) \geq 2\}| + \bigcup_{\substack{j \in [m] \\ d_{G_{\mathcal{M}}}(M_{i_j}) = 1}} |W_2^{i_j}| \equiv |V(G) \setminus M_i| \pmod 2.$$

Since G has even order, $d_{G[V_2]}(v)$ is odd and (V_1, V_2) is an odd colouring of G. □

We now show that, given a graph G with module partition $\mathcal{M}$, we can decompose the graph in such a way that the module graph of any part of the decomposition is either a star or a colour propagating tree. Here we consider the module graph with respect to the module partition $\mathcal{M}$ restricted to the part of the decomposition we are considering. To obtain the decomposition we use a spanning tree $G_{\mathcal{M}}$ and inductively find a non-separating star, i.e., a star whose removal does not disconnect the graph, or a colour propagating tree. In order to handle parity during this process we might separate a module into two parts.

Lemma 3. *For every connected graph G of even order and module partition $\mathcal{M} = (M_1, \ldots, M_k)$ there is a partition $\widehat{\mathcal{M}}$ of $V(G)$ with at most $2k$ many parts such that there is a coarsening $\mathcal{P}$ of $\widehat{\mathcal{M}}$ with the following properties. $|P|$ is even for every part P of $\mathcal{P}$. Furthermore, for every part P of $\mathcal{P}$ we have that $\widehat{\mathcal{M}}|_P$ is a module partition of $G[P]$ and $G[P]_{\widehat{\mathcal{M}}|_P}$ is either a star (with at least two vertices) or a colour propagating tree.* (∗)

Proof 1. Without loss of generality assume that G is connected. Furthermore, let $k := \text{mw}(G)$ and $\mathcal{M} = (M_1, \ldots, M_k)$ be a module partition of G. Let $\widehat{\mathcal{M}}$ be a partition of $V(G)$ with at most $2k$ parts and $\mathcal{P}$ be a coarsening of $\widehat{\mathcal{M}}$ as in Lemma 3. First observe that $\widehat{\mathcal{M}}|_P$ must contain at least two parts for every part P of $\mathcal{P}$ as $\widehat{\mathcal{M}}|_P$ is a module partition of $G[P]$. Since $\widehat{\mathcal{M}}$ has at most $2k$ parts and

$\mathcal{P}$ is a coarsening of $\widehat{\mathcal{P}}$ this implies that $\mathcal{P}$ has at most k parts. Since $G[P]_{\widehat{\mathcal{M}}|_P}$ is either a star or a colour propagating tree we get that $\chi_{\mathrm{odd}}(G[P]) \leq 3$ for every part P of $\mathcal{P}$ by Lemma 1 and Lemma 2. Using a partition (W_1^P, W_2^P, W_3^P) of $G[P]$ such that $G[W_i^P]$ is odd for every $i \in [3]$ for every part P we obtain a global partition of G into at most $3k$ parts such that each part induces an odd subgraph. □

Since deciding whether a graph is k-odd colourable can be solved in time $2^{O(k\,\mathrm{rw}(G))}$ [2, Theorem 6] and $\mathrm{rw}(G) \leq \mathrm{cw}(G) \leq \mathrm{mw}(G)$, where $\mathrm{cw}(G)$ denotes the clique-width of G and $\mathrm{rw}(G)$ rank-width, we obtain the following as a corollary.

Corollary 4. *Given a graph G and a module partition of G of width m the problem of deciding whether G can be odd coloured with at most k colours can be solved in time $2^{O(m^2)}$.*

5 Interval Graphs

In this section we study the odd chromatic number of interval graphs and provide an upper bound in the general case as well as a tight upper bound in the case of proper interval graphs. We use the following lemma in both proofs.

Lemma 4. *Let G be a connected interval graph and $P = (p_1, \ldots, p_k)$ a maximal induced path in G with the following property.*

(Π) $\ell_{p_1} = \min\{\ell_v : v \in V(G)\}$ and for every $i \in [k-1]$ we have that $r_{p_{i+1}} \geq r_v$ for every $v \in N_G(p_i)$.

Then every $v \in V(G)$ is adjacent to at least one vertex on P. (∗)

To prove that the odd chromatic number of proper interval graphs is bounded by three we essentially partition the graph into maximal even sized cliques greedily in a left to right fashion.

Theorem 5. *For every proper interval graph G with all components of even order, $\chi_{\mathrm{odd}}(G) \leq 3$ and this bound is tight.*

Proof. We assume that G is connected. Fix an interval representation of G and denote the interval representing vertex $v \in V(G)$ by $I_v = [\ell_v, r_v]$ where $\ell_v, r_v \in \mathbb{R}$. Let $P = (p_1, \ldots, p_k)$ be a maximal induced path in G as in Lemma 4. For every vertex $v \in V(G) \setminus \{p_1, \ldots, p_k\}$ let $i_v \in [k]$ be the index such that p_{i_v} is the first neighbour of v on P. Note that this is well defined by Lemma 4. For $i \in [k]$ we let Y_i be the set with the following properties.

(Π)1_i $\{v \in V(G) : i_v = i\} \subseteq Y_i \subseteq \{v \in V(G) : i_v = i\} \cup \{p_i, p_{i+1}\}$.

(Π)2_i $p_i \in Y_i$ if and only if $\left| \{p_1, \ldots, p_{i-1}\} \cup \bigcup_{j \in [i-1]} \{v \in V(G) : i_v = j\} \right|$ is even.

(Π)3_i $p_{i+1} \in Y_i$ if and only if $\left| \{p_1, \ldots, p_i\} \cup \bigcup_{j \in [i]} \{v \in V(G) : i_v = j\} \right|$ is odd.

First observe that $(Y_1, \ldots, Y_k)$ is a partition of $V(G)$ as $(\Pi 2)_i$ and $(\Pi 3)_i$ imply that every p_i is in exactly one set Y_i. Furthermore, $|Y_i|$ is even for every $i \in [k]$ since $(\Pi 1)_i$ and $(\Pi 3)_i$ imply that $\left| Y_i \cup \{p_1, \ldots, p_i\} \cup \bigcup_{j \in [i-1]} \{v \in V(G) : i_v = j\} \right|$ is even and $(\Pi 2)_i$ implies that $\left| (\{p_1, \ldots, p_i\} \cup \bigcup_{j \in [i-1]} \{v \in V(G) : i_v = j\}) \setminus Y_i \right|$ is even. Since $v \in V(G) \setminus \{p_1, \ldots, p_k\}$ is not adjacent to p_{i_v-1} we get that $\ell_v \in I_{p_{i_v}}$. Since G is a proper interval graph this implies that $r_{p_{i_v}} \leq r_v$ and hence v is adjacent to p_{i_v+1}. Hence $(\Pi 1)_i$ implies that $G[Y_i]$ must be a clique since $Y_i \cap \{p_1, \ldots, p_k\} \subseteq \{p_i, p_{i+1}\}$ for every $i \in [k]$. Furthermore, $N_G(Y_i)$ and Y_{i+3} are disjoint since $r_v \leq r_{p_{i+1}}$ for every $v \in Y_i$ by property (Π) and $r_{p_{i+1}} < \ell_{p_{i+3}} \leq r_w$ for every $w \in Y_{i+3}$ since P is induced. Hence we can define an odd-colouring (V_1, V_2, V_3) of G in the following way. We let $V_j := \bigcup_{i \equiv j \pmod 3} Y_i$ for $j \in [3]$. Note that since $N_G(Y_i) \cap Y_{i+3}$ we get that $d_{G[Y_i]}(v) = d_{G[V_j]}(v)$ for $i \equiv j \pmod 3$ which is odd (as Y_i is a clique of even size). Hence $G[V_j]$ is odd for every $j \in [3]$.

To see that the bound is tight consider the graph G consisting of K_4 with two pendant vertices u, w adjacent to different vertices of K_4. Clearly, G is a proper interval graph and further $\chi_{\text{odd}}(G) = 3$. □

We use a similar setup (i.e., a path P covering all vertices of the graph G) as in the proof of Theorem 5 to show our general upper bound for interval graphs. The major difference is that we are not guaranteed that sets of the form $\{p_i\} \cup \{v \in V(G) : i_v = i\}$ are cliques. To nevertheless find an odd colouring with few colours of such sets we use an odd/even colouring as in Theorem 1 of $\{v \in V(G) : i_v = i\}$ and the universality of p_i. Hence this introduces a factor of two on the number of colours. Furthermore, this approach prohibits us from moving the p_i around as in the proof of Theorem 5. As a consequence we get that the intervals of vertices contained in a set Y_i span a larger area of the real line than in the proof of Theorem 5. This makes the analysis more technical.

Theorem 6. *For every interval graph G with all components of even order,* $\chi_{\text{odd}}(G) \leq 6$. $\hspace{2cm}$ $(*)$

Note that we currently are unaware whether the bound from Theorem 6 is tight or even whether there is an interval graph G with $\chi_{\text{odd}}(G) > 3$.

6   Conclusion

We initiated the systematic study of odd colouring on graph classes. Motivated by Conjecture 1, we considered graph classes that do not contain large graphs from a given family as induced subgraphs. Put together, these results provide evidence that Conjecture 1 is indeed correct. Answering it remains a major open problem, even for the specific case of bipartite graphs.

Several other interesting classes remain to consider, most notably line graphs and claw-free graphs. Note that odd colouring a line graph $L(G)$ corresponds to colouring the edges of G in such a way that each colour class induces a bipartite graph where every vertex in one part of the bipartition has odd degree, and every vertex in the other colour part has even degree. This is not to be confused with the notion of odd k-edge colouring, which is a (not necessarily proper) edge colouring with at most k colours such that each nonempty colour class induces a graph in which every vertex is of odd degree. It is known that all simple graphs can be odd 4-edge coloured, and every loopless multigraph can be odd 6-edge coloured (see e.g., [9]). While (vertex) odd colouring line graphs is not directly related to odd edge colouring, this result leads us to believe that line graphs have bounded odd chromatic number.

Finally, determining whether Theorem 4 can be extended to graphs of bounded rank-width remains open. We also believe that the bounds in Theorem 6 and Corollary 1 are not tight and can be further improved. In particular, we believe that the following conjecture, first stated in [1], is true:

Conjecture 2 (Aashtab et al., 2023). Every graph G of even order has $\chi_{\mathrm{odd}}(G) \leq \Delta + 1$.

References

1. Aashtab, A., Akbari, S., Ghanbari, M., Shidani, A.: Vertex partitioning of graphs into odd induced subgraphs. Discuss. Math. Graph Theory **43**(2), 385–399 (2023)
2. Belmonte, R., Sau, I.: On the complexity of finding large odd induced subgraphs and odd colorings. Algorithmica **83**(8), 2351–2373 (2021)
3. Caro, Y.: On induced subgraphs with odd degrees. Discret. Math. **132**(1–3), 23–28 (1994)
4. Diestel, R.: Graph Theory, 4th Edn., vol. 173 of Graduate texts in mathematics. Springer, Cham (2012)
5. Ferber, A., Krivelevich, M.: Every graph contains a linearly sized induced subgraph with all degrees odd. Adv. Math. **406**, 108534 (2022)
6. Fomin, F.V., Golovach, P.A., Lokshtanov, D., Saurabh, S., Zehavi, M.: Clique-width III: Hamiltonian cycle and the odd case of graph coloring. ACM Trans. Algorithms **15**(1), 9:1-9:27 (2019)
7. Ganian, R., Hliněný, P., Obdržálek, J.: A unified approach to polynomial algorithms on graphs of bounded (bi-)rank-width. Eur. J. Comb. **34**(3), 680–701 (2013)
8. Lovász, L.: Combinatorial Problems and Exercises. North-Holland (1993)
9. Petrusevski, M.: Odd 4-edge-colorability of graphs. J. Graph Theory **87**(4), 460–474 (2018)
10. Petrusevski, M., Skrekovski, R.: Colorings with neighborhood parity condition (2021)
11. Scott, A.D.: On induced subgraphs with all degrees odd. Graphs Comb. **17**(3), 539–553 (2001)

Deciding the Erdős-Pósa Property
in 3-Connected Digraphs

Julien Bensmail[1], Victor Campos[2], Ana Karolinna Maia[2], Nicolas Nisse[1(✉)],
and Ana Silva[2]

[1] Université Côte d'Azur, CNRS, Inria, I3S, Sophia Antipolis, France
{julien.bensmail,nicolas.nisse}@inria.fr
[2] ParGO, Universidade Federal do Ceará, Fortaleza, Brazil

Abstract. A (di)graph H has the Erdős-Pósa (EP) property for (butterfly) minors if there exists a function $f : \mathbb{N} \to \mathbb{N}$ such that, for any $k \in \mathbb{N}$ and any (di)graph G, either G contains at least k pairwise vertex-disjoint copies of H as (butterfly) minor, or there exists a subset T of at most $f(k)$ vertices such that H is not a (butterfly) minor of $G - T$. It is a well known result of Robertson and Seymour that an undirected graph has the EP property if and only if it is planar. This result was transposed to digraphs by Amiri, Kawarabayashi, Kreutzer and Wollan, who proved that a strong digraph has the EP property for butterfly minors if, and only if, it is a butterfly minor of a cylindrical grid. Contrary to the undirected case where a graph is planar if, and only if, it is the minor of some grid, not all planar digraphs are butterfly minors of a cylindrical grid. In this work, we characterize the planar digraphs that have a butterfly model in a cylindrical grid. In particular, this leads to a linear-time algorithm that decides whether a weakly 3-connected strong digraph has the EP property.

Keywords: Erdős-Pósa property · Planar digraphs · Butterfly minor

1 Introduction

A classical result by Erdős and Pósa [5] states that there is a function $f : \mathbb{N} \to \mathbb{N}$ such that, for every k, every graph G contains either k pairwise vertex-disjoint cycles or a set T of at most $f(k)$ vertices such that $G - T$ is acyclic. The generalization of Erdős and Pósa's result for digraphs and directed cycles was conjectured by Younger [13] and proved by Reed et al. [7].

(Partially) supported by: FUNCAP MLC-0191-00056.01.00/22 and PNE-0112-00061.01.00/16, and CNPq 303803/2020-7, the CAPES-Cofecub project Ma 1004/23, by the project UCA JEDI (ANR-15-IDEX-01) and EUR DS4H Investments in the Future (ANR-17-EURE-004), the ANR Digraphs, the ANR Multimod and the Inria Associated Team CANOE. The full versions of omitted or sketched proofs can be found in [3].

D. Paulusma and B. Ries (Eds.): WG 2023, LNCS 14093, pp. 59–71, 2023.
https://doi.org/10.1007/978-3-031-43380-1_5

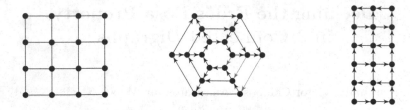

Fig. 1. The (4×4)-grid (left), the (3×6)-cylindrical grid $C_{3,6}$ (middle), and the directed wall (right) obtained from $C_{3,6}$ by removing the three red arcs. (Color figure online)

We say that H is a *minor* of G if H is obtained from a subgraph of G by a sequence of edge contractions. If H is a digraph and we restrict the contractions in the previous definition to *butterfly contractions* [6], we get the definition of a butterfly minor. We say that a graph H has the *Erdős-Pósa (EP) property for minors* if there is a function $f : \mathbb{N} \to \mathbb{N}$ such that, for every k, every graph G contains either k pairwise vertex-disjoint copies of H as a minor or a set T of at most $f(k)$ vertices such that H is not a minor of $G - T$. By changing graph into digraph and minor into butterfly minor, the previous definition can be adapted into the *EP property for butterfly minors in digraphs*. In this view, if H is the undirected graph with a unique vertex and a unique loop on it and D is the digraph obtained from H by orienting its loop edge, then Erdős and Pósa proved that H has the EP property for minors while Reed et al. proved that D has the EP property for butterfly minors.

The results of Erdős and Pósa and Reed et al. were generalized by Robertson and Seymour [8] for undirected graphs and by Amiri et al. [1] for digraphs. Robertson and Seymour [8] proved that an undirected graph G has the EP property for minors if, and only if, G is planar. Amiri et al. [1] proved that a strong digraph D has the EP property for butterfly minors if, and only if, D is a butterfly minor of a cylindrical grid (see Fig. 1). The results of Robertson and Seymour [8] and Amiri et al. [1] are similar since an undirected graph is planar if, and only if, it is a minor of some grid [9]. Contrary to the undirected case, not all planar digraphs are butterfly minors of a cylindrical grid. In this paper, we provide a structural characterization of planar digraphs that are butterfly minors of a cylindrical grid. In particular, such characterization leads to a linear-time algorithm that decides whether a weakly 3-connected strong digraph has the EP property for butterfly minors.

Although planarity is a necessary condition for a digraph to be a butterfly minor of a cylindrical grid, it is not sufficient. For example, the two planar digraphs of Fig. 2 are not butterfly minors of any cylindrical grid. To see this, first note that they are planar, weakly 3-connected, and have essentially a unique (up to the outerface) embedding in the plane, according to Whitney's Theorem [12]. Note also that, in a cylindrical grid, any embedding is such that there is a point in the plane around which all directed cycles go, and in the same direction. We refer to this as being concentric and with same orientation. Now, in the

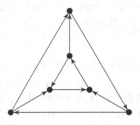 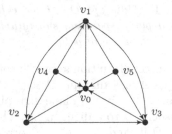

Fig. 2. Two planar digraphs L (left) and R (right) which are not butterfly minors of any cylindrical grid.

digraph L of Fig. 2, the matching between the two directed triangles forces that, in any planar embedding, either the two triangles are not concentric or they have opposite orientations. On the other hand, the digraph R of Fig. 2 is acyclic but it is not a minor of any cylindrical grid. To see why, note that, if R was a butterfly minor of a cylindrical grid, then, because R is acyclic, it would also be a butterfly minor of a *directed wall*, which is the digraph obtained by cutting a cylindrical grid along "parallel" arcs (see Fig. 1). Note that, in an embedding of a directed wall similar to the one given in Fig. 1, no arc goes down. This means that some relative positions of the vertices of R in a directed wall are forced. Namely, the two sources v_4 and v_5 of R must be below each of their out-neighbors, vertex v_1 must be below its three out-neighbors, and the universal sink v_0 must be above every other vertex. It can then be checked that these positions must lead to some crossing arcs. This second example shows that sources and sinks may play an important role in the fact that a planar digraph may or may not be a butterfly minor of a cylindrical grid. In a way, our main result tells that the above two examples fully characterize the reasons why a planar digraph cannot be a butterfly minor of a cylindrical grid.

To formally state our main result, we need a few definitions. Given a digraph $D = (V, A)$ and $\emptyset \neq X \subset V(D)$, the set of arcs between X and $V \backslash X$ is denoted by $(X, V \backslash X)$. We say that $(X, V \backslash X)$ is a *dicut* if there are no arcs from $V \backslash X$ to X. A *dijoin path* P of D is a directed path in D whose arc-set intersects the arc-set of every dicut of D. A *plane digraph* is a planar digraph together with a planar embedding. Recall also that, given a plane digraph H, H^* denotes its *dual*. That is, the dual digraph H^* of H (with a fixed planar embedding) is the digraph that has a vertex for each face of the embedding of H and H^* has an arc $e^* = \{u, v\}$ for each two faces u and v in the embedding of H that are separated from each other by an arc $e \in E$. Moreover, each dual arc e^* is oriented by a $90°$ clockwise turn from the corresponding primal arc e. For instance, if a face of a plane digraph H is "surrounded" by a directed cycle oriented clockwise (resp., counter-clockwise), then the corresponding vertex of H^* is a source (resp., a sink).

We can now state our main result.

Theorem 1. *A digraph D is a butterfly minor of a cylindrical grid if, and only if, D has a plane spanning supergraph H with neither sources nor sinks such that H* admits a dijoin path.*

To get further intuition about Theorem 1, consider the definition of a *feedback arc set* $F \subseteq A$ of a digraph $D = (V, A)$, which is any subset of arcs such that $D - F$ is acyclic. Given a plane digraph D, it is known that every directed cycle of D is associated to a dicut in D^*. This implies that a set of arcs is a feedback arc set of D if, and only if, the corresponding set of its dual edges intersects the arc-set of every dicut of D^* [2]. Therefore, the fact that D^* admits a dijoin path means that such path intersects the arcs of a feedback arc set of D "in the same direction", i.e., intersects the drawing of each directed cycle of D, with each intersection occurring in the same orientation. This is equivalent to being concentric and with the same orientation. This condition (D^* admits a dijoin path) allows avoiding the kind of planar digraphs as exemplified by the digraph L in Fig. 2. In turn, the difficulties exemplified in digraph R in Fig. 2 are dealt with by the existence of a supergraph H with neither sources nor sinks.

Structure of the Paper and Algorithmic Applications. We first prove that if D is a plane digraph with neither sources nor sinks such that D^* has a dijoin path, then D is a butterfly minor of a cylindrical grid (Theorem 4). Observe that this gives us the sufficiency part of Theorem 1. We then show that if D is a butterfly minor of a cylindrical grid, then D has a planar embedding such that D^* admits a dijoin path (Theorem 5). Observe that D might still have sources and sinks, so the remainder of the proof consists in adding arcs to D in order kill all sources and sinks (Lemma 2).

Theorems 4 and 5 have the following important corollary:

Corollary 1. *Any digraph D without sources or sinks is a butterfly minor of a cylindrical grid if and only if D admits a planar embedding s.t. D* has a dijoin path.*

Note that the planar digraph R in Fig. 2 is acyclic. So, whatever be its planar embedding, the dual is strongly connected, i.e., R^* has no dicuts. Therefore, every planar embedding of R is such that R^* has a trivial dijoin path (the empty path). Therefore, unfortunately, there is no hope that the condition on sources and sinks can be removed from Corollary 1.

Note that any *strongly connected digraph* (or *strong*) D satisfies the conditions of Corollary 1. Together with the result of Amiri et al. [1], this implies that:

Corollary 2. *Any strong digraph D has the EP property for butterfly minors if, and only if, D admits a planar embedding such that D* has a dijoin path.*

By Whitney's Theorem [12], any weakly 3-connected planar digraph D has a unique (up to the outerface) planar embedding (computable in linear time). Since deciding whether the dual of a plane digraph admits a dijoin path can be done in linear time, then our result has the following algorithmic application:

Corollary 3. *Deciding whether a weakly 3-connected strong digraph has the EP property for butterfly minors can be done in linear time.*

Section 2 is devoted to defining the main notions and to present previously known results used in this paper. Section 3 is devoted to digraphs with neither sources nor sinks. Section 4 is devoted to obtaining the supergraph with neither sinks nor sources.

2   Preliminaries

Planar Digraphs and Duality. In this section, we present a number of simple known facts concerning planar graphs and their duals. The interested reader can find formal definitions and proofs for such facts in most books on graph theory (e.g., [11]).

Given a digraph $D = (V, A)$ and $e \in A$, let $D \backslash e = (V, A \backslash \{e\})$ and let D/e be the digraph obtained from D after contracting the arc e.

Observation 1. *Let $D = (V, A)$ be a plane digraph, and $e \in A$ be any arc of D. Then, $(D \backslash e)^* = D^*/e^*$ and $(D/e)^* = D^* \backslash e^*$.*

A *dicut* of a digraph $D = (V, A)$ is a partition $(X, V \backslash X)$ of the vertex-set such that X is a non empty proper subset of V and there are no arcs from $V \backslash X$ to X. The arc-set of $(X, V \backslash X)$ is the set of arcs from X to $(X, V \backslash X)$. A *dijoin* $X \subseteq A(D)$ of D is a set of arcs intersecting all dicuts' arc-sets of D. A *dijoin path* (resp., *dijoin walk*) of D is a dijoin inducing a directed path (resp., directed walk) in D. That is, a dijoin path/walk P of D is a directed path/walk whose arc-set intersects the arc-set of every dicut of D.

Observation 2. *A digraph D admits a dijoin path if, and only if, the decomposition of D into strongly connected components has a single source component and a single sink component.*

Observation 3. *Let $D = (V, A)$ be a digraph with a dijoin path P, and $e \in A \backslash A(P)$. Then, P is a dijoin path of $D \backslash e$.*

Observation 4. *Let $D = (V, A)$ be a digraph with a dijoin path P, and $e \in A$. Let P' be obtained from P by contracting e if $e \in A(P)$, and $P' = P$ otherwise. Then, P' is a dijoin walk of D/e.*

Observation 5. *Let $D = (V, A)$ be a digraph with a dijoin path P, and $v \in V$ be an isolated vertex. Then, P is a dijoin path of $D \backslash v$.*

Observation 6. *Every digraph with a dijoin walk admits a dijoin path.*

Butterfly Models and Cylindrical Grids. We now present the formal definition of butterfly models. Let G and H be two digraphs. A (*butterfly*) *model* of G in H is a function $\eta : V(G) \cup A(G) \to \mathcal{S}(H)$, where $\mathcal{S}(H)$ denotes the set of all subdigraphs of H, such that:

– for every $v \in V(G)$, $\eta(v)$ is a subdigraph of H being the orientation of some tree such that $V(\eta(v))$ can be partitioned into $(\{r_v\}, I_v, O_v)$ where
  • $\eta(v)[O_v \cup \{r_v\}]$ is an out-arborescence rooted in r_v (thus in which all non-root vertices have in-degree 1), called the *out-tree* of v,
  • $\eta(v)[I_v \cup \{r_v\}]$ is an in-arborescence rooted in r_v (thus in which all non-root vertices have out-degree 1), called the *in-tree* of v;
– for every two distinct $u, v \in V(G)$, $\eta(u)$ and $\eta(v)$ are vertex-disjoint;
– for every $(x, y) \in A(G)$, $\eta(xy)$ is a directed path of H from the out-tree of x to the in-tree of y, with internal vertices disjoint from every vertex of $\eta(u)$ for every $u \in V(G)$, and from every internal vertex of $\eta(uv)$ for every $(u, v) \in A(G) \setminus \{(x, y)\}$.

Throughout this work, given a model of G in H, we will refer to the arcs $e \in A(H) \cap \bigcup_{f \in A(G)} A(\eta(f))$ as the **blue arcs** of the model, and to the arcs $e \in A(H) \cap \bigcup_{v \in V(G)} A(\eta(v))$ as the **black arcs**. A vertex of H incident to at least one black arc will be referred to as a **black vertex**.

A model of G in H is *minimal* if, for every $v \in V(G)$ and for every leaf w of $\eta(v)$, w is incident to some blue arc. Note that, up to removing the leaves that do not satisfy this property from $\eta(v)$, we can always assume to be working on a minimal model.

Butterfly contracting an arc $(u, v) \in A(D)$ of some digraph D consists in contracting the arc (u, v) if $d^-(v) = 1$ or $d^+(u) = 1$. A digraph G is a *butterfly minor* of some digraph H if G can be obtained from H by deleting arcs, deleting vertices, and butterfly contracting arcs. Note that if G is a butterfly minor of H, then G can be obtained by first removing some arcs, then removing isolated vertices, and finally performing butterfly contractions.

Observation 7 [1]. *A digraph G is a butterfly minor of some digraph H if, and only if, G has a butterfly model in H.*

We now deal with cylindrical grids. Let $n, m \in \mathbb{N}^*$. The *cyclindrical grid* $C_{n,2m}$ can be seen as a set of n concentric directed cycles having the same direction and linked through $2m$ directed paths that alternate directions (see Fig. 3). Formally, $C_{n,2m}$ is the digraph with vertex-set $\{(i, j) \mid 0 \leq i < n, 0 \leq j < 2m\}$, and with the following arc-set. For every $0 \leq i < n$ and $0 \leq j < 2m$, we have $((i, j), (i, j+1 \mod m)) \in A(C_{n,2m})$, and the directed cycle induced by $\{(i, j) \mid 0 \leq j < m\}$ is called the ith *column* of $C_{n,2m}$. For every $0 \leq i < n-1$ and $0 \leq j < m$, we have $((i, 2j), (i+1, 2j)) \in A(C_{n,2m})$ and $((i, 2j+1), (i-1, 2j+1)) \in A(C_{n,2m})$. Moreover, for every $0 \leq j < 2m$, the directed path induced by $\{(i, j) \mid 0 \leq i < n\}$ is called the jth *row* of $C_{n,2m}$.

Throughout this work, we consider that any $C_{n,2m}$ is embedded in the plane so that its first column coincides with the outerface (see Fig. 3). Hence, we may naturally refer to left/right and top/bottom such that the first (last) column is the leftmost (rightmost) and the first (last) row is the bottommost (topmost). The arcs of a column are referred to as *vertical arcs*. Note that all vertical arcs are going up. The arcs of a row are the *horizontal arcs*. Moreover, the arcs of even (resp., odd) rows are horizontal to the right (resp., to the left).

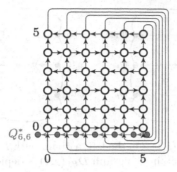

Fig. 3. A planar embedding of the cylindrical grid $C_{6,6}$. The red directed path $Q^*_{6,6}$ is the dijoin path defined and used in Sect. 4. (Color figure online)

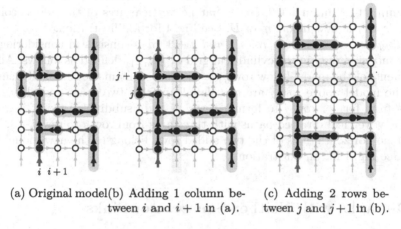

(a) Original model (b) Adding 1 column between i and $i+1$ in (a).   (c) Adding 2 rows between j and $j+1$ in (b).

Fig. 4. Green rows and columns are added. Blue arcs belong to the images of some arcs of G by η. Grey subtrees (with black vertices and arcs) are the images of some vertices of G by η. (Color figure online)

Since $C_{n,2m}$ is strong, we get that $C^*_{n,2m}$ is a DAG. Moreover, $C^*_{n,2m}$ has a unique sink t^*, corresponding to the outerface of the given embedding of $C_{n,2m}$, and a unique source s^*, corresponding to the face of $C_{n,2m}$ bounded by the last column of $C_{n,2m}$. Note that if P^* is any directed path from s^* to t^* in $C^*_{n,2m}$, then P^* is a dijoin path, i.e., it intersects all dicuts of $C^*_{n,2m}$ (or, equivalently, P^* "crosses" all directed cycles of $C_{n,2m}$).

Let η be a butterfly model of a digraph G in $C_{n,2m}$. We will deal with η through a few operations. Due to lack of space, we only present them informally.

- *Adding one column between columns i and $i+1$ in η* consists in considering the new model η' of G in the cylindrical grid $C_{n+1,2m}$ obtained as follows. Roughly, the left part of the model (between columns 0 to i) does not change, one new column is added (with abscissa $i+1$), and the right part of the model (between former columns $i+1$ to n) is translated by one column to the right.

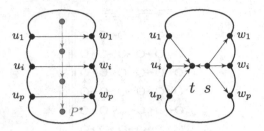

Fig. 5. Construction of $D_{P^*}(s,t)$. On the left, a dijoin path P^* is represented by dashed red arcs. On the right, the obtained digraph $D_{P^*}(s,t)$ is depicted. (Color figure online)

The horizontal arcs of the model that were going from former columns i to $i+1$ are subdivided once, i.e., they are now directed paths that go from column i to column $i+2$. Note that no vertical arcs of the added column belong to the new model η' of D. See Fig. 4 for an illustration.

- *Adding two rows between rows j and $j+1$ in η consists in considering the new model η' of G in the cylindrical grid $C_{n,2(m+1)}$ defined as follows.* All the elements of the model below row j, or in row j, remain the same, all elements of the model above row j are translated up from two rows, and all vertical arcs from former row j to former row $j+1$ are subdivided twice, i.e., they are now vertical directed paths with three arcs from row j to row $j+3$. Note that no horizontal arcs of the two added rows belong to the new model η' of D. See Fig. 4 for an illustration.

3   Digraphs with Neither Sources nor Sinks

Let D be a plane digraph such that D^* has a dijoin path P^* with arcs $(e_1^*, \cdots, e_p^*)$. Let $D_{P^*}(s,t)$ be obtained from D as follows (see Fig. 5 to follow the construction). For every $i \in \{1, \cdots, p\}$, let $e_i = (u_i, w_i)$ be the arc of D corresponding to e_i^*. Subdivide e_i into three arcs (u_i, t_i), (t_i, s_i), and (s_i, w_i). Then, remove (t_i, s_i), and, for every $i \in \{1, \cdots, p\}$, identify the vertices $t_1, \cdots, t_p$ into one vertex t, and the vertices $s_1, \cdots, s_p$ into one vertex s. Finally, add an arc from s to t. Note that $V(D_{P^*}(s,t)) = V(D) \cup \{s,t\}$ and, for every $v \in V(D)$, the in-degree (resp., out-degree) of v in D is the same as in $D_{P^*}(s,t)$. Since P^* is a dijoin path of D^*, the set $\{e_i\}_{i \leq m}$ is a feedback arc set of D [2]. Therefore:

Observation 8. *Let D be a plane digraph such that D^* has a dijoin path P^*. If D has neither sources nor sinks, then $D_{P^*}(s,t)$ is a planar DAG having s as unique source and t as unique sink.*

A *visibility representation* of a graph G is a mapping of $V(G)$ into non-intersecting horizontal segments[1] $\{h_u\}_{u \in V(G)}$, together with a mapping $\{t_e\}_{e \in E(G)}$ of the edges into vertical segments such that for every $uv \in E(G)$,

---

[1] Here, segment means line segment in the plane.

we get that t_{uv} has endpoints in h_u and h_v, and t_{uv} does not cross h_w for every $w \neq u, v$.

Theorem 2 ([4]). *Every planar graph admits a visibility representation.*

Here, we apply the approach presented in [10] to our context in order to obtain a butterfly model of a planar digraph D into a cylindrical grid, if one exists. For this, we slightly adapt their definitions to our purposes.

We consider a visibility representation $(\{h_u\}_{u \in V}, \{t_e\}_{e \in A})$ of $D = (V, A)$ to be drawn on the plane, and, given two horizontal (vertical) segments s_1, s_2, we write $s_1 \leq s_2$ if the y-coordinate (x-coordinate) of s_1 is smaller than the one of s_2. Now, given a DAG $D = (V, A)$, we say that a visibility representation $(\{h_u\}_{u \in V}, \{t_e\}_{e \in A})$ of D is *increasing* if $h_u \leq h_v$ for every arc $(u, v) \in A$ (in other words, the arcs are all directed upwards).

In [10], in order to construct a visibility representation, the authors show that they can obtain an orientation D of a graph G that is acyclic, has exactly one source s and exactly one sink t, and $(s, t) \in A(D)$ (they call such a digraph a *PERT-digraph*). After they obtain this orientation, they use a total order $(v_1, \ldots, v_n)$ of $V(G)$ that meets the orientation D, and then construct a visibility representation such that $s_1 < s_2 < \ldots < s_n$, where s_i denotes the y-coordinate of h_{v_i}, for every $i \in \{1, \ldots, n\}$. Observe that, because the order meets the orientation, we get that this is an increasing visibility representation. Their representation also has the property that the x-coordinate of arc (s, t) is smaller than the x-coordinate of every other edge of G. In short, even though they use a different terminology, the results presented in [10] actually show that the theorem below holds. The interested reader can check this is true by observing, in their algorithm W-VISIBILITY, that after they obtain the desired orientation D (line 2), they only work on D itself; also, the increasing order over the y-coordinates is ensured in line 5.1 of their algorithm.

Theorem 3 ([10]). *Let D be a planar DAG with unique source s and unique sink t, and such that $(s, t) \in A(D)$. Then, D admits an increasing visibility representation such that each horizontal segment has a distinct y-coordinate, and the x-coordinate of the segment of (s, t) is smaller than the x-coordinate of the segment of every other arc of D.*

Theorem 4. *Let $D = (V, A)$ be a digraph without sources or sinks. If D has a planar embedding such that D^* admits a dijoin path P^*, then D has a butterfly model in $C_{n, 2m}$ for some $n, m \in \mathbb{N}^*$.*

Sketch of the Proof. By Observation 8, $D_{P^*}(s, t)$ is a DAG with a unique source s, a unique sink t, and $(s, t) \in A(D_{P^*}(s, t))$. By Theorem 3, there exists an increasing visibility representation of $D_{P^*}(s, t)$. Let $V(D_{P^*}(s, t)) = \{s = v_1, \ldots, t = v_n\}$ be ordered increasingly according to their y-coordinates on the representation and suppose, without loss of generality, that the y-coordinate of $h_{v_1} = h_s$ is 0 and the difference between the y-coordinates of h_{v_i} and $h_{v_{i-1}}$ is 2 (their value on the constructed increasing visibility representation are all different, so we just need to adjust it). Observe that, in this case, the y-coordinate of h_{v_i} is $2i - 2$. We will build a model of D in some cylindrical grid as follows.

For each $v_i \in V(D)$ (hence $i \notin \{1, n\}$), let h'_{v_i} be the segment equivalent to h_{v_i}, but in the upper row. In other words, h'_{v_i} has y-coordinate $2i - 1$, and leftmost and rightmost x-coordinates equal to the ones of h_{v_i}. The idea of the proof is to relate v_i with the path formed by the union of paths associated to h'_{v_i} and h_{v_i} in the cylindrical grid. Since all arcs of D point upwards, we get that the subpath associated to h'_{v_i} (i.e., in row $2i - 1$) corresponds to the out-tree of v_i, while the subpath associated to h_{v_i} (i.e., in row $2i - 2$) corresponds to the in-tree of v_i. $\qquad\square$

Note that Theorem 4 allows us to prove the "if" part of Theorem 1.

Theorem 5. *If a digraph $D = (V, A)$ has a butterfly model in $C_{n,2m}$ for some $n, m \in \mathbb{N}^*$, then D has a planar embedding such that D^* admits a dijoin path.*

Proof. Consider the planar embedding of $C_{n,2m}$ such that the outerface contains its first column (see Fig. 3) and let P^* be any directed path from the single source of $C^*_{n,2m}$ to its single sink. Note that P^* is a dijoin path of $C^*_{n,2m}$.

By Observation 7, D is a butterfly minor of $C_{n,2m}$. Let $s_1, \cdots, s_q$ be the sequence of operations allowing to get D from $C_{n,2m}$ where these operations are ordered in such a way that first arcs are removed, then isolated vertices are removed and, finally, butterfly contractions are performed. For every $0 \leq i \leq q$, let G_i be the digraph obtained after the ith operation (so $G_0 = C_{n,2m}$ and $G_q = D$). We show, by induction on $0 \leq i \leq q$, how to obtain a directed path P_i which is a dijoin path of G_i^*. In particular, it holds for $i = 0$ by taking $P_0 = P^*$.

Let $i \geq 1$. If s_i consists in removing an arc e_i of G_{i-1} then, if $e_i^* \in A(P_{i-1})$, let $P_i' = P_{i-1}/e_i^*$, and let $P_i' = P_{i-1}$ otherwise. By Observations 1 and 4, P_i' is a dijoin walk of G_i^* and, by Observation 6, G_i^* admits a dijoin path P_i. If s_i consists in removing an isolated vertex, then, by Observation 5, $P_i = P_{i-1}$ is a dijoin path of G_i^*. And if s_i is a butterfly contraction of the arc $e_i \in A(G_{i-1})$, where $e_i^* \notin A(P_{i-1})$, then Observations 1 and 3 ensure us that $P_i = P_{i-1}$ is a dijoin path of G_i^*.

Finally, let us consider the case when s_i consists in butterfly contracting an arc $e_i = (u, v) \in A(G_{i-1})$ such that $e_i^* \in A(P_{i-1})$. Let us assume that $d^-(v) = 1$ (the case when $d^+(u) = 1$ is symmetric). Observe that $d^+(v) > 0$ as otherwise e_i^* would be a loop, contradicting that P_{i-1} is a directed path. Then, let $\{f_1, \cdots, f_q\}$ be the set of out-arcs of v ordered clockwise in the embedding of D in the plane. Then, let P_i' be the directed walk obtained by replacing e_i^* in P_{i-1} by the directed walk consisting of the arcs $f_1^*, f_2^*, \cdots, f_q^*$. Note that P_i' is a dijoin walk in G_i^*. Indeed, consider the set of arcs K of a dicut of G_i^*. If K is also a dicut in G_{i-1}^*, then $e_i^* \notin K$ and P_i' intersects K since P_{i-1} is a dijoin path and so intersects K. Otherwise, $e_i^* \in K$ which implies that $\{f_1^*, f_2^*, \cdots, f_q^*\} \cap K \neq \emptyset$, and so P_i' intersects K. Finally, by Observation 6, G_i^* admits a dijoin path P_i. $\qquad\square$

Theorems 4 and 5 prove the following corollary which corresponds to Theorem 1 in the case of digraphs with neither sources nor sinks (and in particular, Corollary 3 is a special case of the following corollary).

Corollary 4. *A digraph D without sources or sinks is a butterfly minor of a cylindrical grid if, and only if, D admits a planar embedding such that D^* has a dijoin path. Moreover, if D is weakly 3-connected, then this can be decided in linear time.*

Proof. Due to the weakly 3-connectivity of D, and by Whitney's Theorem, D has a unique (up to the outerface) planar embedding (and a unique dual). Given such an embedding, checking the existence of a dijoin path can be done in linear time by Observation 2.   □

Recall that a strong digraph D has the EP property for butterfly minors if and only if D is a butterfly minor of a cylindrical grid [1]. Together with our result, we get:

Corollary 3. *Deciding whether a weakly 3-connected strong digraph D has the EP property for butterfly minors can be done in linear time.*

4 Digraphs with Sources and Sinks

We have seen that if D is a butterfly minor of a cylindrical grid, then D has a planar embedding such that D^* admits a dijoin path (Theorem 5). As we want to show that this holds for a *spanning supergraph with neither sources nor sinks*, it remains to "kill" sources and sinks in D. This is done in this section.

Given a cylindrical grid $C_{n,2m}$ with the canonical planar embedding described previously, let $Q^*_{n,2m}$ be the directed path of the dual $C^*_{n,2m}$ whose arcs correspond exactly to all arcs of $C_{n,2m}$ that go from the last (topmost) row to the first (bottommost) row. Note that $Q^*_{n,2m}$ is a dijoin path of $C^*_{n,2m}$ (see Fig. 3). The next lemma states that if a digraph D has a model in a cylindrical grid, then it is possible to get a model such that $Q^*_{n,2m}$ only crosses blue arcs of this model.

Lemma 1. *If a digraph D has a butterfly model η in $C_{n,2m}$, then D has a butterfly model in $C_{n',2m'}$ for some $n' \geq n$ and $m' \geq m$ such that no black arcs of this model are dual of an arc of $Q^*_{n',2m'}$.*

Lemma 2. *If a digraph $D = (V, A)$ has a butterfly model in $C_{n,2m}$, then D has a spanning supergraph with neither sources nor sinks that has a butterfly model in $C_{n',2m'}$ for some $n' \geq n$ and $m' \geq m$.*

Sketch of the Proof. If D has no sources nor sinks, then we are done, so suppose otherwise. In what follows, given a source s in D (resp., a sink t), we describe a process that builds a model for an in-arc that we add to s (resp., an out-arc that we add to t), so that the obtained supergraph has also a model in a cylindrical grid and has one less source (resp., sink). By iteratively applying such process, we get the desired conclusions.

Let us consider a butterfly model η of D with a dijoin path P^* as in Lemma 1, and suppose that D has a source s (the case of a sink is symmetric). We will add a new arc (z, s) for some $z \in V(D)$, and a directed path Q (finishing in the in-tree of the model of s) for modelling this arc in the existing model η. The difficulty is to find the vertex of an out-tree in which we can start Q from. First let us add two columns between any two consecutive columns and let us add two rows between any two consecutive rows. We also add one column to the left and one column to the right of the cylindrical grid.

Let r_s be the root of $\eta(s)$. Note that, by assuming η to be minimal, we get that the in-tree of the model of s in η is reduced to its root. Let a_1 be the vertex below r_s and b_1 the vertex below a_1. Since we have just added two rows between any two rows, and because the in-tree of s is reduced to r_s, we get that a_1, b_1 are not part of the model of any vertex nor arc. Let Q initially contain just the arc (a_1, r_s) (this will actually be the last arc of Q). Let us assume that Q has been built up to some vertex a_h, i.e., $Q = (a_h, a_{h-1}, \cdots, a_1, r_s)$, and additionally assume that the vertex b_h below a_h is not part of the model of any vertex nor arc (this is the case for $h = 1$). Let w be the vertex below b_h.

- If w is not part of the model of any vertex nor arc, then let $a_{h+1} = b_h$ and let $b_{h+1} = w$, and we continue to build Q.
- If w is in the out-tree of some vertex, then add $(w, b_h), (b_h, a_h)$ to the end of Q to be done.
- If w is part of the in-tree of some vertex a (and not of its out-tree, i.e., w is not the root of the model of a), then assume that the row of a_h goes to the right (the other case is symmetric). Let x be the left neighbor of a_h, y be the vertex below x (and to the left of b_h), and z the vertex below y (and to the left of w). Note that, since w is part of the model and b_h is not, then the rows of b_h and a_h are rows that have been added just before starting the process. In particular, this implies that either both x and y belong to the model of some $e \in V(D) \cup A(D)$, or neither x nor y is part of any model. We can then prove that the former case is not possible because it would contradict the fact that w is part of the in-tree (and not of the out-tree) of the model of a. In the latter case, we set $a_{h+1} = x$ and $b_{h+1} = y$ and go on.
- If w is part of the model of some arc $e = (u, v) \in A(D)$. We apply similar arguments and omit the proof because of space constraints.

The above process is not ensured to finish because if might happen that vertices a_h and b_h are already on the outerface of the model η. The next two cases allow to ensure that our process will actually terminate. For this purpose, we use the dijoin path P^*.

- If (b_h, a_h) crosses the dijoin path P^* and P^* does not cross any blue arc, then let us consider the closest row under P^* that contains a vertex of the model. W.l.o.g., let us assume that this row goes to the right and let x be the rightmost vertex of this row that is part of the model of some vertex v^* of D. By minimality of the model and because x has no out-neigbour in the model of any vertex or arc, then, x must be the root of $\eta(v^*)$ (which is actually an in-tree). Now, we add to $\eta(v^*)$: the up-going arc (x, y) (y being the up-neighbor of x), and the horizontal directed path Y starting from y to the leftmost vertex of this row (then y becomes the new root of $\eta(v^*)$ and the path Y will be considered as its out-tree). To conclude this case, add at the beginning of Q, the directed path from the path Y of $\eta(v^*)$ (added in previous paragraph) to b_h.

– If (b_h, a_h) crosses P^* which crosses some blue arc, then we apply similar arguments and omit the proof because of space constraints. ☐

Further Work. An interesting question is whether there exists a structural condition on the sources and sinks of a digraph D that corresponds to being a butterfly minor of a cylindrical grid (avoiding to invoke a supergraph without sources or sinks). This may help to answer the question of the computational complexity of deciding if a strong digraph D has the EP property when D is not weakly 3-connected. Since the class of digraphs that are butterfly minors of a cylindrical grid is closed under taking butterfly minors, it would also be interesting to characterize the minimal forbidden butterfly minors for this class.

References

1. Amiri, A., Kawarabayashi, K., Kreutzer, S., Wollan, P.: The Erdos-Pósa property for directed graphs. arXiv preprint arXiv:1603.02504 (2016)
2. Bang-Jensen, J., Gutin, G.: Digraphs: Theory, Algorithms and Applications, 2nd edn. Springer, London (2008). https://doi.org/10.1007/978-1-84800-998-1
3. Bensmail, J., Campos, V., Maia, A.K., Nisse, N., Silva, A.: Deciding the Erdös-Pósa property in 3-connected digraphs (2023). www.inria.hal.science/hal-04084227
4. Duchet, P., Hamidoune, Y., Las Vergnas, M., Meyniel, H.: Representing a planar graph by vertical lines joining different levels. Discrete Math. **46**, 319–321 (1983)
5. Erdős, P., Pósa, L.: On independent circuits contained in a graph. Can. J. Math. **17**, 347–352 (1965)
6. Johnson, T., Robertson, N., Seymour, P., Thomas, R.: Directed tree-width. J. Comb. Theory Ser. B **82**(1), 138–154 (2001)
7. Reed, B., Robertson, N., Seymour, P., Thomas, R.: Packing directed circuits. Combinatorica **16**(4), 535–554 (1996). https://doi.org/10.1007/BF01271272
8. Robertson, N., Seymour, P.: Graph minors. V. Excluding a planar graph. J. Comb. Theory Ser. B **41**, 92–114 (1986)
9. Robertson, N., Seymour, P.: Graph minors. VII. Disjoint paths on a surface. J. Comb. Theory Ser. B **45**(2), 212–254 (1988)
10. Tamassia, R., Tollis, I.G.: A unified approach to visibility representations of planar graphs. Discrete Comput. Geom. **1**, 321–341 (1986). https://doi.org/10.1007/BF02187705
11. West, D.B.: Introduction to Graph Theory, 2 edn. Prentice Hall (2000)
12. Whitney, H.: 2-isomorphic graphs. Am. J. Math. **55**(1), 245–254 (1933)
13. Younger, D.: Graphs with interlinked directed circuits. In: Proceedings of the Mid-West Symposium on Circuit Theory, vol. 2 (1973)

New Width Parameters for Independent Set: One-Sided-Mim-Width and Neighbor-Depth

Benjamin Bergougnoux[1] (ID), Tuukka Korhonen[2]([✉]) (ID), and Igor Razgon[3]

[1] University of Warsaw, Warsaw, Poland
benjamin.bergougnoux@mimuw.edu.pl
[2] University of Bergen, Bergen, Norway
tuukka.korhonen@uib.no
[3] Birkbeck University of London, London, UK
i.razgon@bbk.ac.uk

Abstract. We study the tractability of the maximum independent set problem from the viewpoint of graph width parameters, with the goal of defining a width parameter that is as general as possible and allows to solve independent set in polynomial-time on graphs where the parameter is bounded. We introduce two new graph width parameters: one-sided maximum induced matching-width (o-mim-width) and neighbor-depth. O-mim-width is a graph parameter that is more general than the known parameters mim-width and tree-independence number, and we show that independent set and feedback vertex set can be solved in polynomial-time given a decomposition with bounded o-mim-width. O-mim-width is the first width parameter that gives a common generalization of chordal graphs and graphs of bounded clique-width in terms of tractability of these problems.

The parameter o-mim-width, as well as the related parameters mim-width and sim-width, have the limitation that no algorithms are known to compute bounded-width decompositions in polynomial-time. To partially resolve this limitation, we introduce the parameter neighbor-depth. We show that given a graph of neighbor-depth k, independent set can be solved in time $n^{O(k)}$ even without knowing a corresponding decomposition. We also show that neighbor-depth is bounded by a polylogarithmic function on the number of vertices on large classes of graphs, including graphs of bounded o-mim-width, and more generally graphs of bounded sim-width, giving a quasipolynomial-time algorithm for independent set on these graph classes. This resolves an open problem asked by Kang, Kwon, Strømme, and Telle [TCS 2017].

Keywords: Graph width parameters · Mim-width · Sim-width · Independent set

Due to space limits, most of technicals details are omitted or just sketched. The full version of the paper is available on arXiv [2]. Tuukka Korhonen was supported by the Research Council of Norway via the project BWCA (grant no. 314528).

D. Paulusma and B. Ries (Eds.): WG 2023, LNCS 14093, pp. 72–85, 2023.
https://doi.org/10.1007/978-3-031-43380-1_6

1   Introduction

Graph width parameters have been successful tools for dealing with the intractability of NP-hard problems over the last decades. While tree-width [25] is the most prominent width parameter due to its numerous algorithmic and structural properties, only sparse graphs can have bounded tree-width. To capture the tractability of many NP-hard problems on well-structured dense graphs, several graph width parameters, including clique-width [7], mim-width [26], Boolean-width [6], tree-independence number [9,27], minor-matching hypertree width [27], and sim-width [20] have been defined. A graph parameter can be considered to be more general than another parameter if it is bounded whenever the other parameter is bounded. For a particular graph problem, it is natural to look for the most general width parameter so that the problem is tractable on graphs where this parameter is bounded. In this paper, we focus on the maximum independent set problem (INDEPENDENT SET).

Let us recall the standard definitions on branch decompositions. Let V be a finite set and $\mathbf{f} : 2^V \to \mathbb{Z}_{\geq 0}$ a symmetric set function, i.e., for all $X \subseteq V$ it holds that $\mathbf{f}(X) = \mathbf{f}(V \setminus X)$. A branch decomposition of $\mathbf{f}$ is a pair (T, δ), where T is a cubic tree and δ is a bijection mapping the elements of V to the leaves of T. Each edge e of T naturally induces a partition (X_e, Y_e) of the leaves of T into two non-empty sets, which gives a partition $(\delta^{-1}(X_e), \delta^{-1}(Y_e))$ of V. We say that the width of the edge e is $\mathbf{f}(e) = \mathbf{f}(\delta^{-1}(X_e)) = \mathbf{f}(\delta^{-1}(Y_e))$, the width of the branch decomposition (T, δ) is the maximum width of its edges, and the branchwidth of the function $\mathbf{f}$ is the minimum width of a branch decomposition of $\mathbf{f}$. When G is a graph and $\mathbf{f} : 2^{V(G)} \to \mathbb{Z}_{\geq 0}$ is a symmetric set function on $V(G)$, we say that the $\mathbf{f}$-*width* of G is the branchwidth of $\mathbf{f}$.

Vatshelle [26] defined the maximum induced matching-width (mim-width) of a graph to be the mim-width where $\mathrm{mim}(A)$ for a set of vertices A is defined to be the size of a maximum induced matching in the bipartite graph $G[A, \overline{A}]$ given by edges between A and $\overline{A}$, where $\overline{A} = V(G) \setminus A$. He showed that given a graph together with a branch decomposition of mim-width k, any locally checkable vertex subset and vertex partitioning problem (LC-VSVP), including INDEPENDENT SET, DOMINATING SET, and GRAPH COLORING with a constant number of colors, can be solved in time $n^{\mathcal{O}(k)}$. Mim-width has gained a lot of attention recently [1,4,5,17–19,22]. While mim-width is more general than clique-width and bounded mim-width captures many graph classes with unbounded clique-width (e.g. interval graphs), there are many interesting graph classes with unbounded mim-width where INDEPENDENT SET is known to be solvable in polynomial-time. Most notably, chordal graphs, and even their subclass split graphs, have unbounded mim-width, but it is a classical result of Gavril [15] that INDEPENDENT SET can be solved in polynomial-time on them. More generally, all width parameters in a general class of parameters that contains mim-width and was studied by Eiben, Ganian, Hamm, Jaffke, and Kwon [11] are unbounded on split graphs.

With the goal of providing a generalization of mim-width that is bounded on chordal graphs, Kang, Kwon, Strømme, and Telle [20] defined the parameter *special induced matching-width* (sim-width). Sim-width of a graph G is the sim-width where $\mathtt{sim}(A)$ for a set of vertices A is defined to be the maximum size of an induced matching in G whose every edge has one endpoint in A and another in $\overline{A}$. The key difference of $\mathtt{mim}$ and $\mathtt{sim}$ is that $\mathtt{mim}$ ignores the edges in $G[A]$ and $G[\overline{A}]$ when determining if the matching is induced, while $\mathtt{sim}$ takes them into account, and therefore the sim-width of a graph is always at most its mim-width. Chordal graphs have sim-width at most one [20]. However, it is not known if INDEPENDENT SET can be solved in polynomial-time on graphs of bounded sim-width, and indeed Kang, Kwon, Strømme, and Telle asked as an open question if INDEPENDENT SET is NP-complete on graphs of bounded sim-width [20].

In this paper, we introduce a width parameter that for the INDEPENDENT SET problem, captures the best of both worlds of mim-width and sim-width. Our parameter is inspired by a parameter introduced by Razgon [24] for classifying the OBDD size of monotone 2-CNFs. For a set of vertices A, let $E(A)$ denote the edges of the induced subgraph $G[A]$. For a set $A \subseteq V(G)$, we define the upper-induced matching number $\mathtt{umim}(A)$ of A to be the maximum size of an induced matching in $G - E(\overline{A})$ whose every edge has one endpoint in A and another in $\overline{A}$. Then, we define the *one-sided maximum induced matching-width* (o-mim-width) of a graph to be the omim-width where $\mathtt{omim}(A) = \min(\mathtt{umim}(A), \mathtt{umim}(\overline{A}))$. In particular, o-mim-width is like sim-width, but we ignore the edges on one side of the cut when determining if a matching is induced. Clearly, the o-mim-width of a graph is between its mim-width and sim-width. Our first result is that the polynomial-time solvability of INDEPENDENT SET on graphs of bounded mim-width generalizes to bounded o-mim-width. Moreover, we show that the interest of o-mim-width is not limited to INDEPENDENT SET by proving that the FEEDBACK VERTEX SET problem is also solvable in polynomial time on graphs of bounded o-mim-width.

Theorem 1. *Given an n-vertex graph together with a branch decomposition of o-mim-width k,* INDEPENDENT SET *and* FEEDBACK VERTEX SET *can be solved in time $n^{\mathcal{O}(k)}$.*

We also show that o-mim-width is bounded on chordal graphs. In fact, we show a stronger result that o-mim-width of any graph is at most its *tree-independence number* (tree-α), which is a graph width parameter defined by Dallard, Milanič, and Štorgel [9] and independently by Yolov [27], and is known to be at most one on chordal graphs.

Theorem 2. *Any graph with tree-independence number k has o-mim-width at most k.*

We do not know if there is a polynomial-time algorithm to compute a branch decomposition of bounded o-mim-width if one exists, and the corresponding question is notoriously open also for both mim-width and sim-width. Because of this, it is also open whether INDEPENDENT SET can be solved in polynomial-time on graphs of bounded mim-width, and more generally on graphs of bounded o-mim-width.

In our second contribution we partially resolve the issue of not having algorithms for computing branch decompositions with bounded mim-width, o-mim-width, or sim-width. We introduce a graph parameter *neighbor-depth*.

Definition 3. *The neighbor-depth (nd) of a graph G is defined recursively as follows:*

1. $\mathbf{nd}(G) = 0$ *if and only if $V(G) = \emptyset$,*
2. *if G is not connected, then $\mathbf{nd}(G)$ is the maximum value of $\mathbf{nd}(G[C])$ where $C \subseteq V(G)$ is a connected component of G,*
3. *if $V(G)$ is non-empty and G is connected, then $\mathbf{nd}(G) \leq k$ if and only if there exists a vertex $v \in V(G)$ such that $\mathbf{nd}(G \setminus N[v]) \leq k-1$ and $\mathbf{nd}(G \setminus \{v\}) \leq k$.*

In the case (3) of Definition 3, we call the vertex v the pivot-vertex *witnessing $\mathbf{nd}(G) \leq k$.*

By induction, the neighbor-depth of all graphs is well-defined. We show that neighbor-depth can be computed in $n^{\mathcal{O}(k)}$ time and also INDEPENDENT SET can be solved in time $n^{\mathcal{O}(k)}$ on graphs of neighbor-depth k.

Theorem 4. *There is an algorithm that given a graph G of neighbor-depth k, determines its neighbor-depth and solves INDEPENDENT SET in time $n^{\mathcal{O}(k)}$.*

We show that graphs of bounded sim-width have neighbor-depth bounded by a polylogarithmic function on the number of vertices.

Theorem 5. *Any n-vertex graph of sim-width k has neighbor-depth $\mathcal{O}(k \log^2 n)$.*

Theorems 4 and 5 combined show that INDEPENDENT SET can be solved in time $n^{\mathcal{O}(k \log^2 n)}$ on graphs of sim-width k, which in particular is quasipolynomial time for fixed k. This resolves, under the mild assumption that NP $\not\subseteq$ QP, the question of Kang, Kwon, Strømme, and Telle, who asked if INDEPENDENT SET is NP-complete on graphs of bounded sim-width [20, Question 2].

Neighbor-depth characterizes branching algorithms for INDEPENDENT SET in the following sense. We say that an independent set branching tree of a graph G is a binary tree whose every node is labeled with an induced subgraph of G, so that (1) the root is labeled with G, (2) every leaf is labeled with the empty graph, and (3) if a non-leaf node is labeled with a graph $G[X]$, then either (a) its children are labeled with the graphs $G[L]$ and $G[R]$ where (L, R) is a partition of X with no edges between L and R, or (b) its children are labeled with the graphs $G[X \setminus N[v]]$ and $G[X \setminus \{v\}]$ for some vertex $v \in X$. Note that such a tree corresponds naturally to a branching approach for INDEPENDENT SET, where we branch on a single vertex and solve connected components independently of each other. Let $\beta(G)$ denote the smallest number of nodes in an independent set branching tree of a graph G. Neighbor-depth gives both lower- and upper-bounds for $\beta(G)$.

Theorem 6. *For all graphs G, it holds that $2^{\mathbf{nd}(G)} \leq \beta(G) \leq n^{\mathcal{O}(\mathbf{nd}(G))}$.*

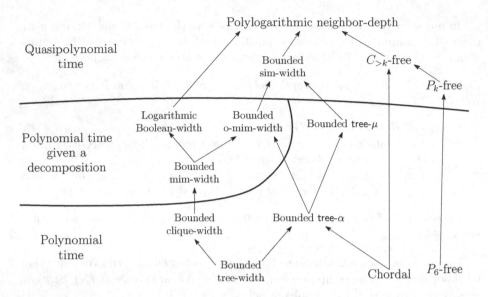

Fig. 1. Hierarchy of some graph classes with polylogarithmically bounded neighbor-depth, divided vertically on whether the best known algorithm for INDEPENDENT SET on the class is polynomial time, polynomial time given a decomposition (and quasipolynomial without a decomposition), or quasipolynomial time.

By observing that some known algorithms for INDEPENDENT SET in fact construct independent set branching trees implicitly, we obtain upper bounds for neighbor-depth on some graph classes purely by combining the running times of such algorithms with Theorem 6. In particular, for an integer k, we say that a graph is $C_{>k}$-free if it does not contain induced cycles of length more than k. Gartland, Lokshtanov, Pilipczuk, Pilipczuk and Rzazewski [14] showed that INDEPENDENT SET can be solved in time $n^{\mathcal{O}(\log^3 n)}$ on $C_{>k}$-free graphs for any fixed k, generalizing a result of Gartland and Lokshtanov on P_k-free graphs [13]. By observing that their algorithm is a branching algorithm that (implicitly) constructs an independent set branching tree, it follows from Theorem 6 that the neighbor-depth of $C_{>k}$-free graphs is bounded by a polylogarithmic function on the number of vertices.

Proposition 7. *For every fixed integer k, $C_{>k}$-free graphs with n vertices have neighbor-depth at most $\mathcal{O}(\log^4 n)$.*

Along the same lines as Proposition 7, a polylogarithmic upper bound for neighbor-depth can be also given for graphs with bounded induced cycle packing number, using the quasipolynomial algorithm of Bonamy, Bonnet, Déprés, Esperet, Geniet, Hilaire, Thomassé, and Wesolek [3].

In Fig. 1 we show the hierarchy of inclusions between some of the graph classes discussed in this paper, and the known algorithmic results for INDEPENDENT SET on those classes. All the inclusions shown are proper, and all the inclusions between these classes appear in the figure. Some of the inclusions are proven in Sects. 3 and 4, and some of the non-inclusions in the full version of the paper [2]. Note that bounded Boolean-width is equivalent to bounded clique-width [26]. The polynomial-time algorithm for INDEPENDENT SET on P_6-free graphs is from [16], the definition of tree-μ and polynomial-time algorithm for INDEPENDENT SET on graphs of bounded tree-μ is from [27], and the definition of Boolean-width and a polynomial-time algorithm for INDEPENDENT SET on graphs of logarithmic Boolean-width is from [6]. The inclusion of logarithmic Boolean-width in polylogarithmic neighbor-depth follows from Theorem 5 and the fact the sim-width of a graph is at most its Boolean-width. Polynomial-time algorithm for INDEPENDENT SET on graphs of bounded clique-width follows from [8,23].

Organization of this Paper. We prove the part of Theorem 1 on INDEPENDENT SET and Theorem 2 in Sect. 3. Theorem 5 is proved in Sect. 4. Proofs omitted in this version of the paper due to space constraints are provided in the full version in [2].

2   Preliminaries

The size of a set V is denoted by $|V|$ and its power set is denoted by 2^V. We let $\max(\emptyset) := -\infty$. Our graph terminology is standard and we refer to [10].

The subgraph of G induced by a subset X of its vertex set is denoted by $G[X]$. We also use the notation $G \setminus X = G[V(G) \setminus X]$. For two disjoint subsets of vertices X and Y of $V(G)$, we denote by $G[X,Y]$ the bipartite graph with vertex set $X \cup Y$ and edge set $\{xy \in E(G) : x \in X$ and $y \in Y\}$. Given two disjoint set of vertices X, Y, we denote by $E(X)$ the set of edges of $G[X]$ and by $E(X,Y)$ the set of edges of $G[X,Y]$. For a set of edges E' of G, we denote by $G - E'$ the graph with vertex set $V(G)$ and edge set $E(G) \setminus E'$.

An *independent set* is a set of vertices that induces an edgeless graph. Given a graph G with a weight function $w : V(G) \to \mathbb{Z}_{\geq 0}$, the problem INDEPENDENT SET asks for an independent set of maximum weight, where the weight of a set $X \subseteq V(G)$ is $\sum_{x \in X} w(x)$. A *feedback vertex set* is the complement of a set of vertices inducing a forest (i.e. acyclic graph). The problem FEEDBACK VERTEX SET asks for a feedback vertex set of minimum weight.

A *matching* in a graph G is a set $M \subseteq E(G)$ of edges having no common endpoint. We denote by $V(M)$ the set of vertices incident to M. An *induced matching* is a matching M such that $G[V(M)]$ does not contain any other edges than M. Given two disjoint subsets A, B of $V(G)$, we say that a matching M is a (A,B)-matching if every edge of M has one endpoint in A and the other in B.

Width Parameters. We refer to the introduction for the definitions of branch-decomposition and f-width, we recall below the definitions of mim-width, sim-width and o-mim-width.

- The maximum induced matching-width (mim-width) [26] of a graph G is the mim-width where $\mathtt{mim}(A)$ is the size of a maximum induced matching of the graph $G[A, \overline{A}]$.
- The special induced matching-width (sim-width) [20] of a graph G is the sim-width where $\mathtt{sim}(A)$ is the size of maximum induced $(A, \overline{A})$-matching in the graph G.
- Given a graph G and $A \subseteq V(G)$, the *upper-mim-width* $\mathtt{umim}(A)$ of A is the size of maximum induced $(A, \overline{A})$-matching in the graph $G - E(\overline{A})$. The one-sided maximum induced matching-width (o-mim-width) of G is the omim-width where $\mathtt{omim}(A) := \min(\mathtt{umim}(A), \mathtt{umim}(\overline{A}))$.

The following is a standard lemma that f-width at most k implies balanced cuts with f-width at most k.

Lemma 8. *Let G be a graph, $X \subseteq V(G)$ a set of vertices with $|X| \geq 2$, and $\mathtt{f} : 2^{V(G)} \to \mathbb{Z}_{\geq 0}$ a symmetric set function. If the $\mathtt{f}$-width of G is at most k, then there exists a bipartition $(A, \overline{A})$ of $V(G)$ with $\mathtt{f}(A) \leq k$, $|X \cap A| \leq \frac{2}{3}|X|$, and $|X \cap \overline{A}| \leq \frac{2}{3}|X|$.*

A tree decomposition of a graph G is a pair $(T, \mathtt{bag})$, where T is a tree and $\mathtt{bag} : V(T) \to 2^{V(G)}$ is a function from the nodes of T to subsets of vertices of G called *bags*, satisfying that (1) for every edge $uv \in E(G)$ there exists a node $t \in V(T)$ so that $\{u, v\} \subseteq \mathtt{bag}(t)$, and (2) for every vertex $v \in V(G)$, the set of nodes $\{t \in V(T) \colon v \in \mathtt{bag}(t)\}$ induces a non-empty and connected subtree of T. The width of a tree decomposition is the maximum size of $\mathtt{bag}(t)$ minus one, and the treewidth of a graph is the minimum width of a tree decomposition of the graph.

For a set of vertices $X \subseteq V(G)$, we denote by $\alpha(X)$ the maximum size of an independent set in $G[X]$. The independence number of a tree decomposition $(T, \mathtt{bag})$ is the maximum of $\alpha(\mathtt{bag}(t))$ over $t \in V(T)$ and it is denoted by $\alpha(T, \mathtt{bag})$. The tree-independence number of a graph (tree-α) is the minimum independence number of a tree decomposition of the graph [9,27].

For a set of vertices $X \subseteq V(G)$, we denote by $\mu(X)$ the maximum size of an induced matching in G so that for each edge of the matching, at least one of the endpoints of the edge is in X. For a tree decomposition $(T, \mathtt{bag})$, we denote by $\mu(T, \mathtt{bag})$ the maximum of $\mu(\mathtt{bag}(t))$ over $t \in V(T)$. Yolov [27] defined the minor-matching hypertree width (tree-μ) of a graph to be the minimum $\mu(T, \mathtt{bag})$ of a tree decomposition $(T, \mathtt{bag})$ of G.

3   O-Mim-Width

In this section, we prove the part of Theorem 1 on INDEPENDENT SET and Theorem 2. We start with some intermediary results. The following reveals an important property of cuts of bounded upper-mim-width. Razgon proved a similar statement in [24]. To simplify the statements of this section, we fix an n-vertex graph G with a weight function $w : V(G) \to \mathbb{Z}_{\geq 0}$.

Lemma 9. *Let $A \subseteq V(G)$. For every $X \subseteq A$ that is the union of t independent sets, there exists $X' \subseteq X$ of size at most $t \cdot \mathrm{umim}(A)$ such that $N(X) \setminus A = N(X') \setminus A$. In particular, we have $|\{N(X) \setminus A : X \in \mathsf{IS}(A)\}| \leq n^{\mathrm{umim}(A)}$ where $\mathsf{IS}(A)$ is the set of independent sets of $G[A]$.*

Proof. It is sufficient to prove the lemma for $t = 1$, since if X is the union of t independent sets $X_1, \ldots, X_t$, then the case $t = 1$ implies that, for each $i \in [1, t]$, there exits $X_i' \subseteq X_i$ such that $N(X_i) \setminus A = N(X_i') \setminus A$ and $|X_i'| \leq \mathrm{umim}(A)$. It follows that $X' = X_1' \cup \cdots \cup X_t' \subseteq X$, $N(X) \setminus A = N(X') \setminus A$ and $|X'| \leq t \cdot \mathrm{umim}(A)$.

Let X be an independent set of $G[A]$. If for every vertex $x \in X$, there exists a vertex $y_x \in \overline{A}$ such that $N(y_x) \cap X = \{x\}$, then $\{xy_x : x \in X\}$ is an induced $(A, \overline{A})$-matching in $G - E(\overline{A})$. We deduce that either $|X| \leq \mathrm{umim}(A)$ or there exists a vertex $x \in X$ such that $N(X) \setminus A = N(X \setminus \{x\}) \setminus A$. Thus, we can recursively remove vertices from X to find a set $X' \subseteq X$ of size at most $\mathrm{umim}(A)$ and such that $N(X) \setminus A = N(X') \setminus A$. In particular, the latter implies that $\{N(X) \setminus A : X \in \mathsf{IS}(A)\} = \{N(X) \setminus A : X \in \mathsf{IS}(A) \wedge |X| \leq \mathrm{umim}(A)\}$. We conclude that $|\{N(X) \setminus A : X \in \mathsf{IS}(A)\}| \leq n^{\mathrm{umim}(A)}$. $\square$

To solve INDEPENDENT SET and FEEDBACK VERTEX SET, we use the general toolkit developed in [1] with a simplified notation adapted to our two problems. This general toolkit is based on the following notion of representativity between sets of partial solutions. In the following, the collection $\mathcal{S}$ represents the set of solutions, in our setting $\mathcal{S}$ consists of either all the independent sets or all the set of vertices inducing a forest.

Definition 10. *Given $\mathcal{S} \subseteq 2^{V(G)}$, for every $\mathcal{A} \subseteq 2^{V(G)}$ and $Y \subseteq V(G)$, we define $\mathrm{best}_\mathcal{S}(\mathcal{A}, Y) := \max\{w(X) : X \in \mathcal{A} \wedge X \cup Y \in \mathcal{S}\}$. Given $A \subseteq V(G)$ and $\mathcal{A}, \mathcal{B} \subseteq 2^A$, we say that $\mathcal{B}$ $(\mathcal{S}, A)$-represents $\mathcal{A}$ if for every $Y \subseteq \overline{A}$, we have $\mathrm{best}_\mathcal{S}(\mathcal{A}, Y) = \mathrm{best}_\mathcal{S}(\mathcal{B}, Y)$.*

Observe that if there is no $X \in \mathcal{B}$ such that $X \cup Y \in \mathcal{S}$, then $\mathrm{best}_\mathcal{S}(\mathcal{B}, Y) = \max(\emptyset) = -\infty$. It is easy to see that the relation "$(\mathcal{S}, A)$-represents" is an equivalence relation.

The following is an application of Theorem 4.1 from [1]. It proves that a routine for computing small representative sets can be used to design a dynamic programming algorithm.

Theorem 11 ([1])**.** *Let $\mathcal{S} \subseteq 2^{V(G)}$. Assume that there exists a constant c and an algorithm that, given $A \subseteq V(G)$ and $\mathcal{A} \subseteq 2^A$, computes in time $|\mathcal{A}| n^{\mathcal{O}(\mathrm{omim}(A))}$ a subset $\mathcal{B}$ of $\mathcal{A}$ such that $|\mathcal{B}| \leq n^{c \cdot \mathrm{omim}(A)}$ and $\mathcal{B}$ $(\mathcal{S}, A)$-represents $\mathcal{A}$. Then, there exists an algorithm, that given a branch decomposition $\mathcal{L}$ of G, computes in time $n^{\mathcal{O}(\mathrm{omim}(\mathcal{L}))}$ a set of size at most $n^{c \cdot \mathrm{omim}(A)}$ that contains an element in $\mathcal{S}$ of maximum weight.*

The following lemma provides a routine to compute small representative sets for INDEPENDENT SET. We denote by $\mathcal{I}$ the set of all independent sets of G.

Lemma 12. *Let $k = \mathrm{omim}(A)$. Given a collection $\mathcal{A} \subseteq 2^A$, we can compute in time $|\mathcal{A}| n^{\mathcal{O}(k)}$ a subset $\mathcal{B}$ of $\mathcal{A}$ such that $\mathcal{B}$ $(\mathcal{I}, A)$-represents $\mathcal{A}$ and $|\mathcal{B}| \leq n^k$.*

Proof. Let $\mathcal{A} \subseteq 2^A$. We compute $\mathcal{B}$ from the empty set as follows:

- If $\mathtt{umim}(A) = k$, then, for every $Y \in \{N(X) \setminus A : X$ is an independent in $\mathcal{A}\}$, we add to $\mathcal{B}$ an independent set $X \in \mathcal{A}$ of maximum weight such that $Y = N(X) \setminus A$.
- If $\mathtt{umim}(A) > k$, then, for each subset $Y \subseteq \overline{A}$ with $|Y| \leq k$, we add to $\mathcal{B}$ a set $X \in \mathcal{A}$ of maximum weight such that $X \cup Y$ is an independent set (if such X exists).

It remains to prove the runtime. First, we prove that $|\mathcal{B}| \leq n^k$. This is straightforward when $\mathtt{umim}(A) > k$. When $\mathtt{umim}(A) = k$, Lemma 9 implies that $|\{N(X) \setminus A : X$ is an independent in $\mathcal{A}\}| \leq n^k$ and thus, we have $|\mathcal{B}| \leq n^k$.

Next, we prove that $\mathcal{B}$ $(\mathcal{I}, A)$-represents $\mathcal{A}$, i.e. for every $Y \subseteq \overline{A}$, we have that $\mathtt{best}_{\mathcal{I}}(\mathcal{A}, Y) = \mathtt{best}_{\mathcal{I}}(\mathcal{B}, Y)$. Let $Y \subseteq \overline{A}$. As $\mathcal{B}$ is subset of $\mathcal{A}$, we have $\mathtt{best}_{\mathcal{I}}(\mathcal{B}, Y) \leq \mathtt{best}_{\mathcal{I}}(\mathcal{A}, Y)$. In particular, if there is no $X \in \mathcal{A}$ such that $X \cup Y$ is an independent set, then we have $\mathtt{best}_{\mathcal{I}}(\mathcal{A}, Y) = \mathtt{best}_{\mathcal{I}}(\mathcal{B}, Y) = -\infty$.

Suppose from now that $\mathtt{best}_{\mathcal{I}}(\mathcal{A}, Y) \neq -\infty$ and let $X \in \mathcal{A}$ such that $X \cup Y$ is an independent set and $w(X) = \mathtt{best}_{\mathcal{I}}(\mathcal{A}, Y)$. We distinguish the following cases:

- If $\mathtt{umim}(A) = k$, then, by construction, there exists an independent set $W \in \mathcal{B}$ such that $N(X) \setminus A = N(W) \setminus A$ and $w(X) \leq w(W)$. As $X \cup Y$ is an independent set, we deduce that $N(X) \cap Y = N(W) \cap Y = \emptyset$ and thus $W \cup Y$ is an independent set.
- If $\mathtt{umim}(A) > k$, then $\mathtt{umim}(\overline{A}) = k$ as $\mathtt{omim}(A) = \min(\mathtt{umim}(A), \mathtt{umim}(\overline{A})) = k$. By Lemma 9, there exists an independent set $Y' \subseteq Y$ of size at most k such that $N(Y) \setminus \overline{A} = N(Y') \setminus \overline{A}$. As $Y' \subseteq Y$, we know that $X \cup Y'$ is an independent set. Thus, by construction there exists a set $W \in \mathcal{B}$ such that $W \cup Y'$ is an independent set and $w(X) \leq w(W)$. Since $N(Y) \setminus A = N(Y') \setminus A$, we deduce that $W \cup Y$ is an independent set.

In both cases, there exists $W \in \mathcal{B}$ such that $W \cup Y$ is an independent set and $w(X) \leq w(W) \leq \mathtt{best}_{\mathcal{I}}(\mathcal{B}, Y)$. Since $\mathtt{best}_{\mathcal{I}}(\mathcal{B}, Y) \leq \mathtt{best}_{\mathcal{I}}(\mathcal{A}, Y) = w(X)$, it follows that $w(X) = \mathtt{best}_{\mathcal{I}}(\mathcal{A}, Y) = \mathtt{best}_{\mathcal{I}}(\mathcal{B}, Y)$. As this holds for every $Y \subseteq \overline{A}$, we conclude that $\mathcal{B}$ $(\mathcal{I}, A)$-represents $\mathcal{A}$.

It remains to prove the running time. Computing $\mathtt{omim}(A) = k$ and checking whether $\mathtt{umim}(A) = k$ can be done by looking at every set of $k + 1$ edges and check whether one of these sets is an induced $(A, \overline{A})$-matching in $G - E(\overline{A})$ and in $G - E(A)$. This can be done in time $\mathcal{O}(\binom{n^2}{k+1} n^2) = n^{\mathcal{O}(k)}$ time. When $\mathtt{umim}(A) > k$, it is clear that computing $\mathcal{B}$ can be done in time $|\mathcal{A}| n^{\mathcal{O}(k)}$. This is also possible when $\mathtt{umim}(A) = k$ as Lemma 9 implies that $|\{N(X) \setminus A : X$ is an independent set in $\mathcal{A}\}| \leq n^k$. □

We obtain the following by using Theorem 11 with the routine of Lemma 12.

Theorem 13. *Given an n-vertex graph with a branch decomposition of o-mim-width k, we can solve* INDEPENDENT SET *in time $n^{\mathcal{O}(k)}$.*

We show that the o-mim-width of a graph is upper bounded by its tree-independence number.

We say that a branch decomposition is *on a set* $V(G)$ if it is a branch decomposition of some function $\mathbf{f} : 2^{V(G)} \to \mathbb{Z}_{\geq 0}$. Next we give a general lemma for turning tree decompositions of G into branch decompositions on $V(G)$.

Lemma 14. *Let* (T, bag) *be a tree decomposition of a graph* G. *There exists a branch decomposition* (T', δ) *on the set* $V(G)$ *so that for every bipartition* $(A, \overline{A})$ *of* $V(G)$ *given by an edge of* (T', δ), *there exists a bag of* (T, bag) *that contains either* $N(A)$ *or* $N(\overline{A})$.

Then we restate Theorem 2 and prove it using Lemma 14.

Theorem 2. *Any graph with tree-independence number* k *has o-mim-width at most* k.

Proof. Let G be a graph with tree-independence number k and (T, bag) a tree decomposition of G with independence number $\alpha(T, \mathsf{bag}) = k$. By applying Lemma 14 we turn (T, bag) into a branch decomposition on $V(G)$ so that for every partition $(A, \overline{A})$ of $V(G)$ given by the decomposition, either $N(A)$ or $N(\overline{A})$ has independence number at most k. Now, if $N(A)$ has independence number at most k, then $\mathtt{umim}(\overline{A}) \leq k$, and if $N(\overline{A})$ has independence number at most k, then $\mathtt{umim}(A) \leq k$, so we have that $\mathtt{omim}(A) \leq k$, and therefore the o-mim-width of the branch decomposition is at most k. □

With similar arguments, we also prove the following.

Theorem 15. *Any graph with minor-matching hypertreewidth* k *has sim-width at most* k.

4   Neighbor-Depth of Graphs of Bounded Sim-Width

In this section we show that graphs of bounded sim-width have poly-logarithmic neighbor-depth, i.e., Theorem 5. The idea of the proof will be that given a cut of bounded sim-width, we can delete a constant fraction of the edges going over the cut by deleting the closed neighborhood of a single vertex. This allows to first fix a balanced cut according to an optimal decomposition for sim-width, and then delete the edges going over the cut in logarithmic depth.

We say that a vertex $v \in V(G)$ *neighbor-controls* an edge $e \in E(G)$ if e is incident to a vertex in $N[v]$. In other words, v neighbor-controls e if $e \notin E(G \setminus N[v])$.

Lemma 16. *Let* G *be a graph and* $A \subseteq V(G)$ *so that* $\mathtt{sim}(A) \leq k$. *There exists a vertex* $v \in V(G)$ *that neighbor-controls at least* $|E(A, \overline{A})|/2k$ *edges in* $E(A, \overline{A})$.

Proof. Suppose the contradiction, i.e., that all vertices of G neighbor-control less than $|E(A, \overline{A})|/2k$ edges in $E(A, \overline{A})$. Let $M \subseteq E(A, \overline{A})$ be a maximum induced $(A, \overline{A})$-matching, having size at most $|M| \leq \mathtt{sim}(A) \leq k$, and let $V(M)$ denote the set of vertices incident to M. Now, an edge in $E(A, \overline{A})$ cannot be added to M if and only if one of its endpoints is in $N[V(M)]$. In particular, an edge in $E(A, \overline{A})$ cannot be added to M if and only if there is a vertex in $V(M)$ that neighbor-controls it. However, by our assumption, the vertices in $V(M)$ neighbor-control strictly less than

$$|V(M)| \cdot |E(A, \overline{A})|/2k = |E(A, \overline{A})|$$

edges of $E(A, \overline{A})$, so there exists an edge in $E(A, \overline{A})$ that is not neighbor-controlled by $V(M)$, and therefore we contradict the maximality of M. $\qquad \square$

Now, the idea will be to argue that because sim-width is at most k, there exists a balanced cut $(A, \overline{A})$ with $\mathtt{sim}(A) \le k$, and then select the vertex v given by Lemma 16 as the pivot-vertex. Here, we need to be careful to persistently target the same cut until the graph is disconnected along it.

Theorem 5. *Any n-vertex graph of sim-width k has neighbor-depth $\mathcal{O}(k \log^2 n)$*

Proof. For integers $n \ge 2$ and $k, t \ge 0$, we denote by $\mathtt{nd}(n, k, t)$ the maximum neighbor-depth of a graph that

1. has at most n vertices,
2. has sim-width at most k, and
3. has a cut $(A, \overline{A})$ with $\mathtt{sim}(A) \le k$, $|E(A, \overline{A})| \le t$, $|A| \le 2n/3$, and $|\overline{A}| \le 2n/3$.

We observe that if a graph G satisfies all of the conditions 1–3, then any induced subgraph of G also satisfies the conditions. In particular, note that n can be larger than $|V(G)|$, and in the condition 3, the cut should be balanced with respect to n but not necessarily with respect to $|V(G)|$.

We will prove by induction that

$$\mathtt{nd}(n, k, t) \le 1 + 4k(\log_{3/2}(n) \cdot \log(n^2 + 1) + \log(t + 1)). \qquad (1)$$

This will then prove the statement, because by Lemma 8 any graph with n vertices and sim-width k satisfies the conditions with $t = n^2$.

First, when $n \le 2$ this holds because any graph with at most two vertices has neighbor-depth at most one. We then assume that $n \ge 3$ and that Eq. (1) holds for smaller values of n and first consider the case $t = 0$.

Let G be a graph that satisfies the conditions 1–3 with $t = 0$. Because $t = 0$, each connected component of G has at most $2n/3$ vertices, and therefore satisfies the conditions with $n' = 2n/3$, $k' = k$, and $t' = (2n/3)^2$. Therefore, by induction each component of G has neighbor-depth at most $\mathtt{nd}(2n/3, k, (2n/3)^2)$. Because the neighbor-depth of G is the maximum neighbor-depth over its components, we get that

$$\begin{aligned}
\mathtt{nd}(G) \le\; & \mathtt{nd}(2n/3, k, (2n/3)^2) \\
\le\; & 1 + 4k(\log_{3/2}(2n/3) \cdot \log((2n/3)^2 + 1) + \log((2n/3)^2 + 1)) \\
\le\; & 1 + 4k((\log_{3/2}(n) - 1) \cdot \log((2n/3)^2 + 1) + \log((2n/3)^2 + 1)) \\
\le\; & 1 + 4k(\log_{3/2}(n) \cdot \log((2n/3)^2 + 1)) \le 1 + 4k(\log_{3/2}(n) \cdot \log(n^2 + 1)),
\end{aligned}$$

which proves that Eq. (1) holds when $t = 0$.

We then consider the case when $t \geq 1$. Assume that Eq. (1) does not hold and let G be a counterexample that is minimal under induced subgraphs. Note that this implies that G is connected, and every proper induced subgraph G' of G has neighbor-depth at most $1 + 4k(\log_{3/2}(n) \cdot \log(n^2 + 1) + \log(t + 1))$. We can also assume that $t = |E(A, \overline{A})|$.

Now, by Lemma 16 there exists a vertex $v \in V(G)$ that neighbor-controls at least $t/2k$ edges in $E(A, \overline{A})$. We will select v as the pivot-vertex. By the minimality of G, we have that $\mathbf{nd}(G \setminus \{v\}) \leq 1 + 4k(\log_{3/2}(n) \cdot \log(n^2 + 1) + \log(t + 1))$, so it suffices to prove that $\mathbf{nd}(G \setminus N[v]) \leq 1 + 4k(\log_{3/2}(n) \cdot \log(n^2 + 1) + \log(t + 1)) - 1$. Because v neighbor-controls at least $t/2k$ edges in $E(A, \overline{A})$, the graph $G \setminus N[v]$ satisfies the conditions with $n' = n$, $k' = k$, and $t' = t - t/2k$. We denote

$$\alpha = \frac{t' + 1}{t + 1} = 1 - \frac{t/2k}{t + 1} \leq 1 - \frac{t/2k}{2t} \leq 1 - \frac{1}{4k}.$$

Now we have that

$$\mathbf{nd}(G) \leq \mathbf{nd}(n, k, t - t/2k) + 1 \leq 2 + 4k(\log_{3/2}(n) \cdot \log(n^2 + 1) + \log(\alpha \cdot (t + 1)))$$
$$\leq 2 + 4k(\log_{3/2}(n) \cdot \log(n^2 + 1) + \log(\alpha) + \log(t + 1))$$
$$\leq 2 + 4k \log(\alpha) + 4k(\log_{3/2}(n) \cdot \log(n^2 + 1) + \log(t + 1))$$
$$\leq 2 - 4k \cdot \frac{1}{4k} + 4k(\log_{3/2}(n) \cdot \log(n^2 + 1) + \log(t + 1))$$
$$\leq 1 + 4k(\log_{3/2}(n) \cdot \log(n^2 + 1) + \log(t + 1)),$$

which proves that Eq. (1) holds when $t \geq 1$, and therefore completes the proof. $\square$

5 Conclusion

We conclude with some open problems. First, as already discussed, it is still open if independent set can be solved in polynomial-time on graphs of bounded mim-width, because it is not known how to construct a decomposition of bounded mim-width if one exists. It would be very interesting to resolve this problem by either giving an algorithm for computing decompositions of bounded mim-width, or by defining an alternative width parameter that is more general than mim-width and allows to solve INDEPENDENT SET in polynomial-time when the parameter is bounded.

The class of graphs of polylogarithmic neighbor-depth generalizes several classes where INDEPENDENT SET can be solved in (quasi)polynomial time. Another interesting class where INDEPENDENT SET can be solved in polynomial-time and which, to our knowledge, could have polylogarithmic neighbor-depth is the class of graphs with polynomial number of minimal separators [12]. It would be interesting to show that this class has polylogarithmic neighbor-depth. More generally, Korhonen [21] studied a specific model of dynamic programming algorithms for INDEPENDENT SET, in particular, tropical circuits for independent set, and it appears plausible that all graphs with polynomial size tropical circuits for independent set could have polylogarithmic neighbor-depth.

References

1. Bergougnoux, B., Dreier, J., Jaffke, L.: A logic-based algorithmic meta-theorem for mim-width. In: Proceedings of the 2023 Annual ACM-SIAM Symposium on Discrete Algorithms (SODA), pp. 3282–3304. SIAM (2023). https://doi.org/10.1137/1.9781611977554.ch125
2. Bergougnoux, B., Korhonen, T., Razgon, I.: New width parameters for independent set: one-sided-mim-width and neighbor-depth. CoRR abs/2302.10643 (2023). https://doi.org/10.48550/arXiv.2302.10643
3. Bonamy, M., et al.: Sparse graphs with bounded induced cycle packing number have logarithmic treewidth. In: Bansal, N., Nagarajan, V. (eds.) Proceedings of the 2023 ACM-SIAM Symposium on Discrete Algorithms, SODA 2023, Florence, Italy, 22–25 January 2023, pp. 3006–3028. SIAM (2023). https://doi.org/10.1137/1.9781611977554.ch116
4. Brettell, N., Horsfield, J., Munaro, A., Paesani, G., Paulusma, D.: Bounding the mim-width of hereditary graph classes. J. Graph Theory **99**(1), 117–151 (2022). https://doi.org/10.1002/jgt.22730
5. Brettell, N., Horsfield, J., Munaro, A., Paulusma, D.: List k-colouring P_t-free graphs: a mim-width perspective. Inf. Process. Lett. **173**, 106168 (2022). https://doi.org/10.1016/j.ipl.2021.106168
6. Bui-Xuan, B., Telle, J.A., Vatshelle, M.: Boolean-width of graphs. Theor. Comput. Sci. **412**(39), 5187–5204 (2011). https://doi.org/10.1016/j.tcs.2011.05.022
7. Courcelle, B., Engelfriet, J., Rozenberg, G.: Handle-rewriting hypergraph grammars. J. Comput. Syst. Sci. **46**(2), 218–270 (1993). https://doi.org/10.1016/0022-0000(93)90004-G
8. Courcelle, B., Makowsky, J.A., Rotics, U.: Linear time solvable optimization problems on graphs of bounded clique-width. Theory Comput. Syst. **33**(2), 125–150 (2000). https://doi.org/10.1007/s002249910009
9. Dallard, C., Milanič, M., Štorgel, K.: Treewidth versus clique number. II. Tree-independence number. CoRR abs/2111.04543 (2022). https://doi.org/10.48550/arXiv.2111.04543
10. Diestel, R.: Graph Theory. Graduate Texts in Mathematics, vol. 173, 4th edn. Springer, Cham (2012)
11. Eiben, E., Ganian, R., Hamm, T., Jaffke, L., Kwon, O.: A unifying framework for characterizing and computing width measures. In: Braverman, M. (ed.) 13th Innovations in Theoretical Computer Science Conference, ITCS 2022, Berkeley, CA, USA, 31 January–3 February 2022. LIPIcs, vol. 215, pp. 63:1–63:23. Schloss Dagstuhl - Leibniz-Zentrum für Informatik (2022). https://doi.org/10.4230/LIPIcs.ITCS.2022.63
12. Fomin, F.V., Todinca, I., Villanger, Y.: Large induced subgraphs via triangulations and CMSO. SIAM J. Comput. **44**(1), 54–87 (2015). https://doi.org/10.1137/140964801
13. Gartland, P., Lokshtanov, D.: Independent set on P_k-free graphs in quasi-polynomial time. In: Irani, S. (ed.) 61st IEEE Annual Symposium on Foundations of Computer Science, FOCS 2020, Durham, NC, USA, 16–19 November 2020, pp. 613–624. IEEE (2020). https://doi.org/10.1109/FOCS46700.2020.00063

14. Gartland, P., Lokshtanov, D., Pilipczuk, M., Pilipczuk, M., Rzazewski, P.: Finding large induced sparse subgraphs in $C_{>t}$-free graphs in quasipolynomial time. In: Khuller, S., Williams, V.V. (eds.) STOC 2021: 53rd Annual ACM SIGACT Symposium on Theory of Computing, Virtual Event, Italy, 21–25 June 2021, pp. 330–341. ACM (2021). https://doi.org/10.1145/3406325.3451034

15. Gavril, F.: Algorithms for minimum coloring, maximum clique, minimum covering by cliques, and maximum independent set of a chordal graph. SIAM J. Comput. 1(2), 180–187 (1972). https://doi.org/10.1137/0201013

16. Grzesik, A., Klimosová, T., Pilipczuk, M., Pilipczuk, M.: Polynomial-time algorithm for maximum weight independent set on P_6-free graphs. ACM Trans. Algorithms 18(1), 4:1–4:57 (2022). https://doi.org/10.1145/3414473

17. Jaffke, L., Kwon, O., Strømme, T.J.F., Telle, J.A.: Mim-width III. Graph powers and generalized distance domination problems. Theor. Comput. Sci. 796, 216–236 (2019). https://doi.org/10.1016/j.tcs.2019.09.012

18. Jaffke, L., Kwon, O., Telle, J.A.: Mim-width I. Induced path problems. Discret. Appl. Math. 278, 153–168 (2020). https://doi.org/10.1016/j.dam.2019.06.026

19. Jaffke, L., Kwon, O., Telle, J.A.: Mim-width II. The feedback vertex set problem. Algorithmica 82(1), 118–145 (2020). https://doi.org/10.1007/s00453-019-00607-3

20. Kang, D.Y., Kwon, O., Strømme, T.J.F., Telle, J.A.: A width parameter useful for chordal and co-comparability graphs. Theor. Comput. Sci. 704, 1–17 (2017). https://doi.org/10.1016/j.tcs.2017.09.006

21. Korhonen, T.: Lower bounds on dynamic programming for maximum weight independent set. In: 48th International Colloquium on Automata, Languages, and Programming, ICALP 2021, Glasgow, Scotland, 12–16 July 2021 (Virtual Conference), pp. 87:1–87:14 (2021). https://doi.org/10.4230/LIPIcs.ICALP.2021.87

22. Munaro, A., Yang, S.: On algorithmic applications of sim-width and mim-width of (H_1, H_2)-free graphs. CoRR abs/2205.15160 (2022). https://doi.org/10.48550/arXiv.2205.15160

23. Oum, S., Seymour, P.D.: Approximating clique-width and branch-width. J. Comb. Theory Ser. B 96(4), 514–528 (2006). https://doi.org/10.1016/j.jctb.2005.10.006

24. Razgon, I.: Classification of OBDD size for monotone 2-CNFs. In: Golovach, P.A., Zehavi, M. (eds.) 16th International Symposium on Parameterized and Exact Computation, IPEC 2021, Lisbon, Portugal, 8–10 September 2021. LIPIcs, vol. 214, pp. 25:1–25:15. Schloss Dagstuhl - Leibniz-Zentrum für Informatik (2021). https://doi.org/10.4230/LIPIcs.IPEC.2021.25

25. Robertson, N., Seymour, P.D.: Graph minors. III. Planar tree-width. J. Comb. Theory Ser. B 36(1), 49–64 (1984). https://doi.org/10.1016/0095-8956(84)90013-3

26. Vatshelle, M.: New width parameters of graphs. Ph.D. thesis, University of Bergen, Norway (2012). www.hdl.handle.net/1956/6166

27. Yolov, N.: Minor-matching hypertree width. In: Czumaj, A. (ed.) Proceedings of the Twenty-Ninth Annual ACM-SIAM Symposium on Discrete Algorithms, SODA 2018, New Orleans, LA, USA, 7–10 January 2018, pp. 219–233. SIAM (2018). https://doi.org/10.1137/1.9781611975031.16

Nonplanar Graph Drawings with k Vertices per Face

Carla Binucci[1]([✉])[iD], Giuseppe Di Battista[2][iD], Walter Didimo[1][iD],
Seok-Hee Hong[3][iD], Michael Kaufmann[4][iD], Giuseppe Liotta[1][iD], Pat Morin[5][iD],
and Alessandra Tappini[1][iD]

[1] Università degli Studi di Perugia, Perugia, Italy
{carla.binucci,walter.didimo,giuseppe.liotta,alessandra.tappini}@unipg.it
[2] Università degli Studi Roma Tre, Rome, Italy
giuseppe.dibattista@uniroma3.it
[3] University of Sydney, Camperdown, Australia
seokhee.hong@sydney.edu.au
[4] University of Tübingen, Tübingen, Germany
mk@informatik.uni-tuebingen.de
[5] Carleton University, Ottawa, Canada
morin@scs.carleton.ca

Abstract. The study of nonplanar graph drawings with forbidden or desired crossing configurations has a long tradition in geometric graph theory, and received an increasing attention in the last two decades, under the name of *beyond-planar graph drawing*. In this context, we introduce a new hierarchy of graph families, called k^+-*real face graphs*. For any integer $k \geq 1$, a graph G is a k^+-real face graph if it admits a drawing Γ in the plane such that the boundary of each face (formed by vertices, crossings, and edges) contains at least k vertices of G. We give tight upper bounds on the maximum number of edges of k^+-real face graphs. In particular, we show that 1^+-real face and 2^+-real face graphs with n vertices have at most $5n - 10$ and $4n - 8$ edges, respectively. Also, if all vertices are constrained to be on the boundary of the external face, then 1^+-real face and 2^+-real face graphs have at most $3n - 6$ and $2.5n - 4$ edges, respectively. We also study relationships between k^+-real face graphs and beyond-planar graph families with hereditary property.

Keywords: beyond-planar graph drawing · k^+-real face graphs · edge density

1 Introduction

The study of nonplanar graph drawings with forbidden substructures has a long tradition in geometric graph theory (see, e.g., [31]). In the last two decades, this

Research started at the Summer Workshop on Graph Drawing (SWGD) 2022, and partially supported by: (*i*) MIUR, grant 20174LF3T8 "AHeAD: efficient Algorithms for HArnessing networked Data"; (*ii*) Dipartimento di Ingegneria - Università degli Studi di Perugia, Ricerca di Base, grants RICBA21LG and RICBA22CB.

topic, often recognized as *beyond-planar graph drawing*, has become increasingly popular. This growth in interest is due in part to human cognitive experiments aimed at estimating the impact of crossing configurations on graph visualization readability. See [22,25,27] for recent surveys or books on the subject.

With a widely-accepted terminology, a *beyond-planar graph family* is a type of nonplanar graphs that can be drawn in the plane by avoiding some given edge crossing configurations or by guaranteeing some properties about edge crossings. For example, for a given positive integer k, the family of *k-planar graphs* consists of those graphs that can be drawn in the plane with at most k crossings per edge [28,34], while *k-quasi planar graphs* are graphs that can be drawn in the plane without k mutually crossing edges [1,4,6,33,35]. Again, *right-angle-crossing graphs* (also known as *RAC graphs*) are those graphs that admit a straight-line drawing in which any two crossing edges form angles of 90° at their crossing point [20,21]; generalizations and variants of RAC drawings have also been proposed (see, e.g., [3,18,24]). Refer to [22] for other notable beyond-planar graph families. Given a family $\mathcal{F}$ of beyond-planar graphs, one of the most relevant problems is establishing the maximum *edge density* for the elements of $\mathcal{F}$, i.e., the maximum number of edges that an n-vertex graph in $\mathcal{F}$ can have with respect to n. This is a classical *Turán-type* problem with a long tradition in extremal graph theory [9,15,29] and represents one of the core research topics in the literature on graph drawing beyond planarity. For example, it is known that n-vertex 1-planar graphs and 2-planar graphs have at most $4n - 8$ edges and $5n - 10$ edges, respectively, and both these bounds are tight, in the sense that there are graphs in these families that can actually achieve them [34]. A graph of $\mathcal{F}$ that is maximally dense (i.e., whose number of edges is the maximum possible over its number of vertices) is usually called an *optimal* graph of $\mathcal{F}$.

A similar research direction investigates how different beyond-planar graph families relate to each other in terms of inclusion, partly exploiting edge density results [22]. For instance, it has been shown that the family of simple k-planar graphs is a subset of $(k + 1)$-quasi planar graphs for any $k \geq 2$ [7].

Contribution. In this paper, we propose a new hierarchy of graph families, which we call k^+-*real face graphs*, for any positive integer k. Namely, consider any drawing Γ of a graph G in the plane, where edge crossing points are regarded as dummy vertices. The drawing Γ divides the plane into connected regions, called *faces*: if no two edges of G cross in Γ, the boundary of each face consists of vertices (and edges) of G; otherwise, the boundaries of some faces contain dummy vertices. We say that G is a k^+-*real face graph* if it admits a drawing such that each face boundary contains at least k real vertices, i.e., k vertices of G (k^+ stands for "k or more"). By definition, for any $k \geq 1$, the family of $(k + 1)^+$-real face graphs is included in the family of k^+-real face graphs. From a theoretical perspective, studying k^+-real face graphs generalizes to nonplanar graphs the study of planar graphs that admit a crossing-free drawing whose face sizes are above a desired threshold [5,23,30]. Also, finding k^+-real face graphs can be regarded as a generalization to nonplanar graphs of the classical guarding planar graph problem [16], where the vertices that cover the face set

Table 1. Summary of density results in this paper; n denotes the number of vertices.

Graph Family	Crossings $(\chi \leq)$	Edges $(m \leq)$	Ref.
k^+-real face graphs $(k \geq 3)$	$\frac{2-k}{k} \cdot m + n - 2$	$\frac{k}{k-2}(n-2)$	Lemma 1, Theorem 1
2^+-real face graphs	$n - 2$	$4n - 8$	Lemma 1, Theorem 2
1^+-real face graphs	$m + n - 2$	$5n - 10$	Lemma 1, Theorem 3
outer k^+-real face graphs $(k \geq 3)$	$\frac{2-k}{k} \cdot m + \frac{k-1}{k} \cdot n - 1$	$\frac{k-1}{k-2} \cdot n - \frac{k}{k-2}$	Lemma 2, Theorem 4
outer 2^+-real face graphs	$\frac{1}{2}n - 1$	$2.5n - 4$	Lemma 2, Theorem 5
outer 1^+-real face graphs	$m - 1$	$3n - 6$	Lemma 2, Theorem 6

are the (real) vertices of G. From a more practical perspective, the interest in k^+-real face graphs is motivated by the intuition that faces that mostly consist of crossing points could make the graph layout less readable; indeed, the number of real vertices per face can be regarded as a measure of how much the drawing is far from being planar; in particular, one would avoid, when possible, faces formed only by crossing points. Our results can be summarized as follows:

- We provide tight upper bounds on the edge density of k^+-real face graphs, for all values of k, both in the general case (Sect. 3) and in the constrained scenario in which all vertices of the graph are forced to stay on the external face (Sect. 4); see Table 1 for a summary of our results. The constrained scenario can be regarded as a generalization of the study of outerplanar graphs to our graphs, which we call *outer k^+-real face graphs*. We note that similar constraints have been previously studied in other families of beyond-planar graphs (see, e.g., [10,12,13,19,26]).
- We establish inclusion relationships between k^+-real face graphs and families of beyond-planar graphs with hereditary property, such as h-planar and h-quasi planar graphs (Sect. 5). In particular, we show that, for any positive integer k, the family of k^+-real face graphs is not included in any beyond-planar graph family with hereditary property. However, this is not always the case if we restrict our attention to optimal graphs.

For space reasons, some proofs have been omitted or sketched.

2   Basic Definitions

Let G be a graph. We assume that G is connected and simple, meaning that it contains neither multiple edges nor self-loops (if G is not connected we can treat each connected component of G independently). We denote by $V(G)$ and $E(G)$ the set of vertices and the set of edges of G, respectively. A *drawing* Γ of G maps each vertex $v \in V(G)$ to a distinct point in the plane and each edge $uv \in E(G)$ to a simple Jordan arc between the points corresponding to u and v. We assume that Γ is a *simple* drawing, that is: (i) adjacent edges do not intersect, except at their common endpoint; (ii) two independent (i.e., non-adjacent) edges intersect at most in one of their interior points, called a *crossing point*; and (iii) no three edges intersect at a common crossing point.

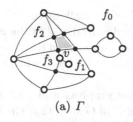

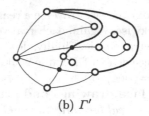

(a) Γ (b) Γ'

Fig. 1. (a) A nonplanar drawing Γ of a graph G with 5 crossings. White circles are the real-vertices of Γ (i.e., the vertices of G) and black circles are the crossing-vertices of Γ. Graph G has $n = 10$ vertices and $m = 14$ edges. Drawing Γ has $\nu = 15$ vertices, $\mu = 24$ edges, and $\varphi = 11$ faces. Face f_0 is the external face. The shaded face is a 0-real face. Face f_2 is a 2-real triangle and f_3 is a 2-real quadrilateral. The boundary of f_1 is not a simple cycle as vertex v is traversed twice while walking along its boundary. It follows that $\deg_\Gamma^r(f_1) = 4$, $\deg_\Gamma^c(f_1) = 3$, and $\deg_\Gamma(f_1) = 7$. (b) A 2^+-real face drawing Γ' of G, obtained from the previous one by rerouting two edges of G (the thicker ones).

Refer to Fig. 1 for an illustration of the next definitions. Let Γ be a drawing of G. A *vertex* of Γ is either a point corresponding to a vertex of G, called a *real-vertex*, or a point corresponding to a crossing point, called a *crossing-vertex*. Observe that a crossing-vertex has degree 4. We remark that in the literature a plane graph obtained by replacing crossing points with dummy vertices is often referred to as a *planarization* [17].

We denote by $V(\Gamma)$ the set of vertices of Γ. An *edge* of Γ is a curve connecting two vertices of Γ; an edge of Γ whose endpoints are both real-vertices coincides with an edge of G. We denote by $E(\Gamma)$ the set of edges of Γ. Drawing Γ subdivides the plane into topologically connected regions, called *faces*. The boundary of each face consists of a circular sequence of vertices and edges of Γ. We denote by $F(\Gamma)$ the set of faces of Γ. Exactly one face in $F(\Gamma)$ corresponds to an infinite region of the plane, called the *external face* (or *outer face*) of Γ; the other faces are the *internal faces* of Γ. When the boundary of a face f of Γ contains a vertex v (or an edge e), we also say that f contains v (or e).

From now on, we denote by $n = |V(G)|$ and $m = |E(G)|$ the number of vertices and the number of edges of G, respectively. For a drawing Γ of G, we denote by $\nu = |V(\Gamma)|$, $\mu = |E(\Gamma)|$, and $\varphi = |F(\Gamma)|$ the number of vertices, edges, and faces of Γ, respectively. Also, we denote by $\chi = |V(\Gamma) \setminus V(G)| = \nu - n$ the number of crossing-vertices of Γ.

- **Degree of vertices and faces.** For a vertex $v \in V(G)$, denote by $\deg_G(v)$ the *degree of v in G*, i.e., the number of edges incident to v. Analogously, for a vertex $v \in V(\Gamma)$, denote by $\deg_\Gamma(v)$ the *degree of v in Γ*. For a face $f \in F(\Gamma)$, denote by $\deg_\Gamma(f)$ the *degree of f*, i.e., the number of times we traverse vertices (either real- or crossing-vertices) while walking on the boundary of f clockwise. Each vertex contributes to $\deg_\Gamma(f)$ the number of times we traverse it (possibly more than once if the boundary of f is not a simple cycle). Also, denote by $\deg_\Gamma^r(f)$ the *real-vertex degree of f*, i.e., the number of times we traverse a real-vertex of Γ while walking on the boundary of f clockwise. Again, each real-vertex contributes to $\deg_\Gamma^r(f)$ the number of times we traverse it. Finally, $\deg_\Gamma^c(f)$ denotes the number of times we traverse a crossing-vertex of Γ while walking on the boundary of f clockwise. Clearly, $\deg_\Gamma(f) = \deg_\Gamma^r(f) + \deg_\Gamma^c(f)$.

- k^+-**real face drawings and graphs.** Given a graph G and a positive integer k, a k^+-*real face drawing* of G is a drawing Γ of G such that the boundary of each face of Γ has at least k real-vertices. If G admits a k^+-real face drawing, we say that G is a k^+-*real face graph*. An *outer k^+-real face drawing* of G is a k^+-real face drawing Γ of G such that all its real-vertices are on the boundary of the outer face. If G admits an outer k^+-real face drawing we say that G is an *outer k^+-real face graph*. We say that a face $f \in F(\Gamma)$ is an *h-real face*, where h is a non-negative integer, if $\deg_\Gamma^r(f) = h$. An h-real face of degree d is called an *h-real d-gon*. An h-real 3-gon is also called an *h-real triangle*, and an h-real 4-gon is also called an *h-real quadrilateral*. We say that an edge $e = uv \in E(\Gamma)$ is an *h-real edge* ($h \in \{0,1,2\}$) if $|\{u,v\} \cap V(G)| = h$, i.e., e contains h real-vertices.

3   Density of k^+-Real Face Graphs

In this section, we prove tight upper bounds on the number of edges that a k^+-real face graph can have. We start by proving the following upper bound on the number χ of crossing-vertices in a k^+-real face drawing.

Lemma 1. *Let Γ be a k^+-real face drawing of a graph G. We have:*

$$\chi \leq \frac{2-k}{k} \cdot m + n - 2 \tag{1}$$

Proof. By hypothesis, each face $f \in F(\Gamma)$ contains at least k real-vertices (i.e., at least k vertices of G). Since each real-vertex $v \in V(G)$ can belong to at most $\deg_G(v)$ faces of Γ and since $\sum_{v \in V(G)} \deg_G(v) = 2m$, we have that the number φ of faces of Γ is such that $\varphi \leq \frac{2m}{k}$. Also, the number of edges μ of Γ is such that $\mu = m + 2\chi$. Hence, by Euler's formula applied to Γ, we have $\varphi = \mu + 2 - \nu = m + 2\chi + 2 - n - \chi$, and hence $\varphi = m + \chi + 2 - n$. It follows that $\chi = \varphi - m + n - 2 \leq \frac{2m}{k} - m + n - 2 = \frac{2-k}{k} \cdot m + n - 2$. □

3.1 k^+-Real Face Graphs, with $k \geq 2$

We first consider the case $k \geq 3$ and then the case $k = 2$.

Theorem 1. *Let k be a positive integer such that $k \geq 3$. If G is a k^+-real face graph with n vertices and m edges, then $m \leq \frac{k}{k-2}(n-2)$, and this bound is tight. Also, the optimal n-vertex k^+-real face drawings are exactly the n-vertex planar drawings in which each face is a simple k-gon.*

Proof (Sketch). Let Γ be any k^+-real face drawing of G. When $k \geq 3$, the term $\frac{2-k}{k}$ is negative and, equivalently, $\frac{k}{k-2}$ is positive. Since the number χ of crossing-vertices of Γ cannot be negative, i.e., $\chi \geq 0$, by Eq. (1) of Lemma 1 we have that the number of edges m must satisfy the inequality $m \leq \frac{k}{k-2}(n-2)$.

For the tightness of the bound, just consider the family of planar embedded graphs such that each face has k vertices. Any n-vertex graph in this family has $m = \frac{k}{k-2}(n-2)$ edges. Also, one can prove that every k^+-real face drawing with $\frac{k}{k-2}(n-2)$ edges is planar and all its faces have degree k. $\qquad\square$

Theorem 2. *If G is a 2^+-real face graph with n vertices and m edges, then $m \leq 4n - 8$, and this bound is tight. Also, the optimal n-vertex 2^+-real face graphs are exactly the optimal 1-planar graphs.*

Proof (Sketch). Let Γ be any 2^+-real face drawing of G. By Eq. (1) of Lemma 1, with $k = 2$, we get $\chi \leq n - 2$. Since $\mu \leq 3\nu - 6$, and since $\mu = m + 2\chi$ and $\nu = n + \chi$, we have $m \leq \chi + 3n - 6$, and therefore $m \leq n - 2 + 3n - 6 = 4n - 8$. This proves that $4n - 8$ is an upper bound on the number of edges of G.

About the tightness of the bound, consider the family of 1-planar graphs, i.e., graphs that admit a drawing Γ with at most one crossing per edge. Each face of Γ has at least two real-vertices (see also [36]), thus Γ is a 2^+-real face drawing. In particular, for $n = 8$ and for every $n \geq 12$, there exists an optimal 1-planar graph with n vertices and $4n - 8$ edges [14,34]. Also, it can be proven that every optimal 2^+-real face drawing Γ of G is also a 1-planar drawing of G. $\square$

3.2 1^+-Real Face Graphs

To prove an upper bound on the number of edges in 1^+-real face graphs, we use *discharging* techniques. See for example [2,4,24] for other papers that use similar approaches. Following [4], we consider a *charging function* ch : $F(\Gamma) \to \mathbb{R}$ such that, for each $f \in F(\Gamma)$, we set:

$$\text{ch}(f) = \deg_\Gamma(f) + \deg_\Gamma^r(f) - 4 = 2\deg_\Gamma^r(f) + \deg_\Gamma^c(f) - 4 \qquad (2)$$

The value ch(f) is called the *initial charge* of f. By using Euler's formula, it is not difficult to prove that the following relationship holds (for details, see [4]):

$$\sum_{f \in F(\Gamma)} \text{ch}(f) = 4n - 8 \qquad (3)$$

The idea of a discharging technique is to derive from the initial charging function ch a new function ch$'$ that satisfies the next two properties (see also [4]):
C1. ch$'(f) \geq \alpha \deg_\Gamma^r(f)$, for some real number $\alpha > 0$;
C2. $\sum_{f \in F(\Gamma)} \text{ch}'(f) \leq \sum_{f \in F(\Gamma)} \text{ch}(f)$
If $\alpha > 0$ is a real number for which a charging function ch$'$ satisfies C1 and C2, by Eq. (3) we have: $4n - 8 = \sum_{f \in F(\Gamma)} \text{ch}(f) \geq \sum_{f \in F(\Gamma)} \text{ch}'(f) \geq$

$\alpha \sum_{f \in F(\Gamma)} \deg_\Gamma^r(f)$. Also, since $\sum_{f \in F(\Gamma)} \deg_\Gamma^r(f) = \sum_{v \in V(G)} \deg_G(v) = 2m$, we get the following:

$$m \le \frac{2}{\alpha}(n-2) \tag{4}$$

Thus, Eq. (4) can be exploited to prove upper bounds on the edge-density of a graph for specific values of α, whenever we find a charging function ch' that satisfies C1 and C2. We are now ready to present the main result of this section.

Theorem 3. *Let G be a 1^+-real face graph with n vertices and m edges. We have that $m \le 5n - 10$, and this bound is tight.*

Proof (Sketch). Let Γ be a 1^+-real face drawing of G. We first augment Γ and G as follows. If some face f of Γ contains a pair u and v of real-vertices but does not contain an edge uv on its boundary, then we augment Γ (and G) with an edge uv drawn in the interior of f, in such a way that it does not create any crossing. We then repeat this process until every pair of real-vertices in each face f is connected by an edge on the boundary of f. Note that, this augmentation is not unique and may introduce multiple edges in G. However, it does not create any 0-real faces and any faces of degree two in the drawing; also, the drawing remains a 1^+-real face drawing. If Γ' denotes the drawing resulting from the augmentation on Γ, for each face $f \in F(\Gamma')$ we have that: (a) $\deg_{\Gamma'}(f) \ge 3$; and (b) $3 \ge \deg_{\Gamma'}^r(f) \ge 1$. Also, denoted by G' the graph resulting from the augmentation on G, we have $V(G') = V(G)$ and $E(G) \subseteq E(G')$; hence, an upper bound on the number of edges m' of G' is also an upper bound on the number of edges m of G.

Suppose given on Γ' the initial charging function $\text{ch} : F(\Gamma') \to \mathbb{R}$ of Eq. (2). If we are able to define a charging function $\text{ch}' : F(\Gamma') \to \mathbb{R}$ that satisfies C1 and C2 for $\alpha = \frac{2}{5}$, then by Eq. (4) we get $m \le m' \le 5n - 10$, and we are done.

We show how to define ch'. For every face $f \in \Gamma'$, we initially set $\text{ch}'(f) = \text{ch}(f) = 2\deg_{\Gamma'}^r(f) + \deg_{\Gamma'}^c(f) - 4$. With this choice and with $\alpha = \frac{2}{5}$, function ch' satisfies C2. Also, C1 becomes $2\deg_{\Gamma'}^r(f) + \deg_{\Gamma'}^c - 4 \ge \frac{2}{5}\deg_{\Gamma'}^r(f)$, that is, $8\deg_{\Gamma'}^r(f) + 5\deg_{\Gamma'}^c(f) \ge 20$. Hence, since $\deg_{\Gamma'}(f) \ge 3$, C1 is always satisfied for each face f such that either $\deg_{\Gamma'}^r(f) \ge 2$, or $\deg_{\Gamma'}^r(f) = 1$ and $\deg_{\Gamma'}^c(f) \ge 3$. It follows that, the only faces that do not satisfy C1 are the 1-real triangles, i.e., each face t for which $\deg_{\Gamma'}^r(t) = 1$ and $\deg_{\Gamma'}^c(t) = 2$. Indeed, for a 1-real triangle t the initial charge equals 0, thus we need to suitably increase the value of $\text{ch}'(t)$.

For each 1-real triangle t, let f be the face incident to the unique 0-real edge of t; see Fig. 2a. Observe that $\deg_{\Gamma'}(f) \ge 4$. Indeed, if it were $\deg_{\Gamma'}(f) = 3$ then G would contain two parallel edges (which is impossible because G is simple) or there would be two adjacent edges of G that cross in Γ (which is impossible because Γ is simple). Also, since Γ' is a 1^+-real face drawing, we have $\deg_{\Gamma'}^r(f) \ge 1$. We apply a discharging operation, by moving a fraction $\frac{2}{5}$ of charge from f to t across their shared 0-real edge. In this way, we set $\text{ch}'(t) = \frac{2}{5}$ and reduce $\text{ch}'(f)$ by $\frac{2}{5}$. The total charge of Γ' determined by ch' does not change (hence C2 is still satisfied) but now $\text{ch}'(t)$ satisfies C1.

Since for a face f the reduction of $\text{ch}'(f)$ by $\frac{2}{5}$ occurs across a 0-real edge of f, the number of times this happens is at most $\deg_{\Gamma'}^c(f) - 1$. Therefore, after we

(a)   (b)

Fig. 2. Illustration for the proof of Theorem 3: (a) A 1-real triangle t and an adjacent face f that moves a charge of $\frac{2}{5}$ towards t across a 0-real edge. (b) A 1-real quadrilateral f that moves two charges of $\frac{2}{5}$ towards two adjacent 1-real triangles t_1 and t_2; face f recovers a charge of $\frac{1}{5}$ from a 2-real triangle f' that shares a vertex x with f, t_1, and t_2

have applied a discharging operation for each 1-real triangle, the charge $\mathrm{ch}'(f)$ of each face f of degree at least four is such that:

$$\mathrm{ch}'(f) \geq 2\deg_{\Gamma'}^{r}(f) + \deg_{\Gamma'}^{c}(f) - 4 - \frac{2}{5}\deg_{\Gamma'}^{c}(f) + \frac{2}{5} = 2\deg_{\Gamma'}^{r}(f) + \frac{3}{5}\deg_{\Gamma'}^{c}(f) - \frac{18}{5}$$

Hence, f satisfies C1 (i.e., $\mathrm{ch}'(f) \geq \frac{2}{5}\deg_{\Gamma'}^{r}(f)$) if this relation holds:

$$8\deg_{\Gamma'}^{r}(f) + 3\deg_{\Gamma'}^{c}(f) \geq 18 \tag{5}$$

It can be easily verified that the above relation is always satisfied for a face f of degree at least four, except when f is a 1-real quadrilateral (which consists of one real-vertex and 3 crossing-vertices). Indeed, if f is a 1-real quadrilateral it could have moved a fraction $\frac{2}{5}$ of charge towards a 1-real triangle t_1 and a fraction $\frac{2}{5}$ of charge towards another 1-real triangle t_2; see Fig. 2b. Both t_1 and t_2 share a crossing-vertex x with f and with another face f'. In this case $\mathrm{ch}'(f) = \mathrm{ch}(f) - \frac{4}{5} = 1 - \frac{4}{5} = \frac{1}{5} = \frac{2}{5}\deg_{\Gamma'}^{r}(f) - \frac{1}{5}$, thus f has a deficit of $\frac{1}{5}$. Observe that the boundary of f' contains two real-vertices adjacent to x, which are connected by an edge due to the edge augmentation initially performed on Γ. Hence, f' is a 2-real triangle and at this point we have $\mathrm{ch}'(f') = \mathrm{ch}(f') = 1 = \frac{2}{5}\deg_{\Gamma'}^{r}(f') + \frac{1}{5}$. It follows that $\mathrm{ch}'(f')$ has a surplus of $\frac{1}{5}$, and we can move this surplus from f' to f, i.e., we increase $\mathrm{ch}'(f)$ by $\frac{1}{5}$ and decrease $\mathrm{ch}'(f')$ by $\frac{1}{5}$. Since this reduction of $\mathrm{ch}(f')$ can happen at most once for f', both f and f' satisfy C1 at the end of this operation. This completes the proof that $m \leq 5n - 10$.

As for the tightness of the bound, consider any n-vertex optimal 2-planar drawing Γ. Such a drawing has $5n - 10$ edges and it is composed of a planar pentangulation (i.e., every face is a simple cycle of degree five) plus five crossing edges inside each pentagon [11]. A 1^{+}-real face drawing with $n' > n$ vertices and $5n' - 10$ edges is obtained from Γ by adding a vertex inside each pentagon and connecting it to all vertices of the pentagon (see Fig. 3 for an illustration). $\quad\square$

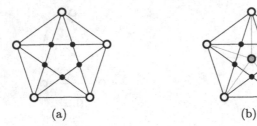

Fig. 3. (a) Pentagonal face of an optimal 2-planar drawing. (b) Augmenting each pentagonal face with a vertex and five edges (in gray) to make the drawing 1^+-real face.

4 Density of Outer k^+-Real Face Graphs

In this section, we provide tight upper bounds on the maximum number of edges that an outer k^+-real face graph can have, depending on k. For an outer k^+-real face drawing Γ of a graph G, we denote by $F_{\text{int}}(\Gamma) \subset F(\Gamma)$ the subset of internal faces of Γ. Additionally, we denote by φ_{int} the number of internal faces of Γ, that is $\varphi_{\text{int}} = |F_{\text{int}}(\Gamma)|$. Notice that $\varphi = \varphi_{\text{int}} + 1$. As for k^+-real face graphs, we first give an upper bound on the number χ of crossing-vertices in an outer k^+-real face drawing (the proof relies on similar arguments).

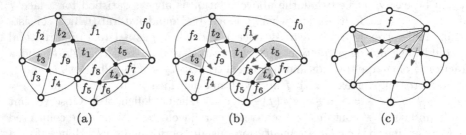

Fig. 4. (a) An edge-maximal outer 1^+-real face drawing of a graph with 7 vertices and 14 edges. Face f_1 is a 2-real 4-gon; $f_2, \ldots, f_7$ are 2-real triangles; f_8 and f_9 are 1-real quadrilaterals; $t_1, \ldots, t_5$ are 1-real triangles. (b) Arrows show a mapping of the 1-real triangles that satisfies Property (c) of Lemma 3. (c) Another example of an edge-maximal outer 1^+-real face drawing with 8 vertices and 16 edges. Face f is a 3-real triangle. The shaded faces are the 1-real triangles; a mapping of these triangles that satisfies Property (c) of Lemma 3 is shown.

Lemma 2. *Let G be a graph and let k be a positive integer. If Γ is an outer k^+-real face drawing of G then the following holds:*

$$\chi \leq \frac{2-k}{k} \cdot m + \frac{k-1}{k} \cdot n - 1 \qquad (6)$$

4.1 Outer k^+-Real Face Graphs, with $k \geq 2$

We first consider the case $k \geq 3$ and then the case $k = 2$. The proof of the next theorem is similar to the proof of Theorem 1 and it has been omitted.

Theorem 4. *Let k be a positive integer such that $k \geq 3$. If G is an outer k^+-real face graph with n vertices and m edges, then $m \leq \frac{k-1}{k-2} \cdot n - \frac{k}{k-2}$, and this bound is tight. Also, the optimal n-vertex outer k^+-real face drawings are exactly the n-vertex outerplanar drawings in which each internal face is a simple k-gon.*

Theorem 5. *Let G be an outer 2^+-real face graph with n vertices and m edges. We have that $m \leq 2.5n - 4$, and this bound is tight. Also, the n-vertex optimal outer 2^+-real face graphs are exactly the optimal outer-1-planar graphs.*

Proof (Sketch). Let Γ be any outer 2^+-real face drawing of G. By Lemma 2, with $k = 2$, we get $\chi \leq \frac{n}{2} - 1$. If we remove from Γ exactly one edge of G per crossing-vertex, we get an outerplanar graph with $m' = m - \chi$ edges and n vertices. Since a maximal outerplanar graph with n vertices has at most $2n - 3$ edges, we have $m - \chi \leq 2n - 3$, and therefore $m \leq 2n - 3 + \chi \leq 2n - 3 + \frac{n}{2} - 1$, that is, $m \leq \frac{5}{2}n - 4 = 2.5n - 4$. This proves that $2.5n - 4$ is an upper bound on the number of edges of G. Also, it can be proven that the bound is tight and that every optimal outer 2^+-real face drawing Γ is also 1-planar; since optimal outer-1-planar graphs have at most $2.5n - 4$ edges [8,19], this implies that optimal outer 2^+-real face graphs are exactly the optimal outer-1-planar graphs. $\square$

4.2 Outer 1^+-Real Face Graphs

As for 1^+-real face graphs, we use discharging techniques to prove an upper bound on the number of edges of outer 1^+-real face graphs. An outer 1^+-real face Γ is *edge-maximal* if the drawing obtained by adding to Γ any new edge between two of its real-vertices is no longer outer 1^+-real face. An example of edge-maximal outer 1^+-real face drawing is illustrated in Fig. 4a; as Theorem 6 will show, this graph is however not optimal, as for any $n \geq 3$ there exist outer 1^+-real face graphs that contain $3n - 6$ edges. Another edge-maximal outer 1^+-real face drawing that is not optimal is shown in Fig. 4c.

We now present a key result about the structure of edge-maximal outer 1^+-real face drawings.

Lemma 3. *Let G be an n-vertex outer 1^+-real face graph, with $n \geq 4$, and let Γ be an edge-maximal outer 1^+-real face drawing of G. The following properties hold:*

a) *The boundary of the external face is a simple cycle that consists of exactly n real-vertices and no crossing-vertices.*
b) *Each internal face of Γ is either a 3-real triangle, or a 2-real d-gon ($d \geq 3$), or a 1-real triangle, or a 1-real quadrilateral.*

c) *We can map each 1-real triangle to exactly one face of Γ that is either a 2-real d-gon, for $d \geq 4$, or a 1-real quadrilateral, in such a way that: (i) at most $(d-3)$ 1-real triangles are mapped to the same 2-real d-gon; and (ii) at most two 1-real triangles are mapped to the same 1-real quadrilateral.*

d) *The number of 3-real triangles plus the number of 2-real d-gons is exactly n, and the number of 1-real quadrilaterals is at most $n - 4$.*

Theorem 6. *If G is an outer 1^+-real face graph with n vertices and m edges, then $m \leq 3n - 6$, and this bound is tight.*

Proof (Sketch). To prove the upper bound, it is enough to concentrate on edge-maximal outer 1^+-real face drawings of G. Let Γ be such a drawing. If G has three vertices, then Γ is a 3-cycle and the statement trivially holds. Assume that $n \geq 4$. We exploit a discharging technique as for Theorem 3. In this case, we want to show the existence of a charging function ch' that satisfies C1 and C2 for $\alpha = \frac{2}{3}$. If such a function exists then, by Eq. (4), we get $m \leq 3n-6$. For each face $f \in F(\Gamma)$, initially set $ch'(f) = ch(f)$, where $ch(f)$ is the charging function of Eq. (2). Denote by f_0 the external face of Γ. Based on Properties (a) and (b) of Lemma 3, $\deg_\Gamma(f_0) = \deg_\Gamma^r(f_0) = n$ and each internal face of Γ is either a 3-real triangle, or a 2-real d-gon, or a 1-real triangle, or a 1-real quadrilateral. At this point, we have:

- $ch'(f_0) = 2\deg_\Gamma^r(f_0) + \deg_\Gamma^c(f_0) - 4 = 2n - 4$; the charge excess of f_0 with respect to $\frac{2}{3}\deg_\Gamma^r(f_0)$ is $\frac{4}{3}n - 4$;
- If f is a 3-real triangle, then $ch'(f) = 2$; it has no charge excess/deficit;
- If f is a 2-real d-gon, $ch'(f) = d-2$; hence, if $d = 3$ (i.e., f is a 2-real triangle) f has a charge deficit of $\frac{1}{3}$, while if $d \geq 4$ it has an excess of $d - \frac{10}{3}$;
- If f is a 1-real triangle then $ch(f) = 0$ and f has a charge deficit of $\frac{2}{3}$;
- If f is a 1-real quadrilateral then $ch(f) = 1$ and f has a charge excess of $\frac{1}{3}$.

We modify ch' by moving charges from faces with an excess to faces with a deficit, in such a way that C1 is satisfied. Based on the above analysis, the only faces with a deficit are the 2-real triangles and the 1-real triangles. We map each 1-real triangle to either a 2-real d-gon (with $d \geq 4$) or to a 1-real quadrilateral, so that the mapping satisfies Property (c) of Lemma 3. This mapping tells for each face with a deficit from which face it will receive charges; see Fig. 4. Property (d) of Lemma 3 is used to prove that the new charging function satisfies C1.   □

5   Inclusion Relationships

In this section, we study inclusion relationships between the families of k^+-real face graphs and other beyond-planar graph families. As already observed, k^+-real face graphs form a hierarchy of families, that is, for each integer $k \geq 1$, the family of $(k + 1)^+$-real face graphs is properly included in the family of k^+-real face graphs.

Note that the hierarchy of k-planar graphs has the opposite behavior, i.e., each k-planar graph is also a $(k+1)$-planar graph (for any $k \geq 1$). The results of Sect. 3 provide insights about inclusion relationships between the hierarchies of k-planar graphs and of k^+-real face graphs, for $k \in \{1, 2\}$. Namely, Theorem 2 shows that 1-planar graphs are 2^+-real face

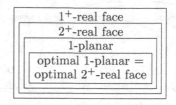

Fig. 5. Inclusion relationships between k^+-real face graphs and k-planar graphs, for $k = 1, 2$.

graphs and that the families of optimal 1-planar graphs and optimal 2^+-real face graphs coincide. These relationships are summarized in Fig. 5.

We now show a more general result about the relationship between k^+-real face graphs and any other beyond-planar graph family with hereditary property. This result (Theorem 7) excludes that, for any fixed positive integer k, there exists some beyond-planar graph families with hereditary property that contain all k^+-real face graphs. In the following, we formalize this concept.

Let $\mathcal{F}$ be a family of beyond-planar graphs. We say that $\mathcal{F}$ has the *hereditary property* if any subgraph of a graph in $\mathcal{F}$ also belongs to $\mathcal{F}$. Most of the beyond-planar graph families studied in the literature (see, e.g., [22]) have the hereditary property. On the contrary, the family of k^+-real face graphs (for any $k \geq 1$) does not necessarily satisfy this property, as removing vertices from a k^+-real face graph makes it impossible in some cases to guarantee at least k real-vertices per face; for example, if we remove the central vertex in the drawing of Fig. 3b, the drawing is no longer a 1^+-real face drawing. Nonetheless, it is immediate to see that if we remove from a k^+-real face graph any subset of edges but no vertices, the resulting subgraph is still a k^+-real face graph.

Lemma 4. *For any integer $k > 0$ and for any family $\mathcal{F}$ of beyond-planar graphs with hereditary property, there exists a k^+-real face graph not belonging to $\mathcal{F}$.*

Proof. Let G be any (connected) graph such that $G \notin \mathcal{F}$. Consider any drawing Γ of G. If Γ is already a k^+-real face drawing, we are done. Otherwise, we augment Γ into a new drawing by suitably adding new vertices and edges; refer to Fig. 6 for an example. We first consider the set of 0-real faces of Γ (i.e., faces whose boundary contains only crossings). If this set is not empty, there must be a 0-real face f that is adjacent to a face f' containing a real-vertex v. Add to Γ a new real-vertex u in the interior of f and connect u to v with an edge that crosses exactly one edge shared by f and f'. In this way, the set of 0-real faces is decreased by one element. Iterate this procedure until there is no more 0-real faces in the drawing. Now, consider every face f of Γ that contains $1 \leq h < k$ real-vertices (if any). Arbitrarily select a real-vertex v of f, and attach to v a chain of $k - h$ vertices in the interior of f. This creates a new face f' in place of f, which contains k real-vertices. Once all those faces have been processed, the underlying graph G' of the resulting drawing is a (connected) k^+-real face graph. Also, since $G \subseteq G'$ and $\mathcal{F}$ has the hereditary property, we have that $G' \notin \mathcal{F}$. $\square$

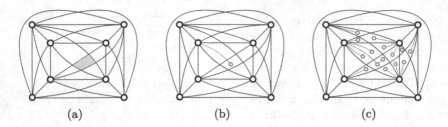

Fig. 6. (a) An initial drawing of a graph that is not 1-planar; it has one 0-real face (shaded). (b) An augmentation that removes the 0-real face; the new elements are red. (c) A further augmentation that makes the drawing 2^+-real face. (Color figure online)

Lemma 4 immediately implies the following.

Theorem 7. *For any positive integer k, the family of k^+-real face graphs is not included in any beyond-planar graph family with hereditary property.*

A consequence of Theorem 7 is that, for any integer $k > 0$, the family $\mathcal{K}$ of k^+-real face graphs is incomparable with any beyond-planar graph family $\mathcal{F}$ with hereditary property whose edge density is higher than the edge density of $\mathcal{K}$. For instance, each family of k^+-real face graphs is incomparable with the families of h-planar graphs for $h \geq 3$. Indeed, Theorem 7 proves the existence of a 1^+-real face graph that is not h-planar; on the other hand, since the maximum number of edges of an h-planar graph, for $h \geq 3$, can be higher than $5n-10$ [32,34], there exist h-planar graphs that are not 1^+-real face graphs. Similarly, each family of k^+-real face graphs is incomparable with the family of h-quasi planar graphs, for every $h \geq 3$, as 3-quasi planar graphs can have up to $6.5n - 20$ edges [1].

6   Open Problems

We conclude with two open research questions.

OP(1) The maximum edge density of 2-planar graphs is the same as the one of 1^+-real face graphs, and Theorem 6 implies that there are 1^+-real face graphs that are not 2-planar. An open question is whether there exist 2-planar graphs that are not 1^+-real face graphs. Note that, every 2-planar drawing of an optimal 2-planar graph G is not 1^+-real face, as it contains 0-real faces. However, one cannot exclude in principle that G admits a 1^+-real face drawing that is not 2-planar.

OP(2) Another interesting research direction is to establish the complexity of testing whether a graph is k^+-real face or outer k^+-real face for a given k. In particular, are these problems NP-hard?

Acknowledgments. We thank Vida Dujmović for valuable discussion.

References

1. Ackerman, E.: On the maximum number of edges in topological graphs with no four pairwise crossing edges. Discret. Comput. Geom. **41**(3), 365–375 (2009). https://doi.org/10.1007/s00454-009-9143-9
2. Ackerman, E.: On topological graphs with at most four crossings per edge. Comput. Geom. **85** (2019). https://doi.org/10.1016/j.comgeo.2019.101574
3. Ackerman, E., Fulek, R., Tóth, C.D.: On the size of graphs that admit polyline drawings with few bends and crossing angles. In: Brandes, U., Cornelsen, S. (eds.) GD 2010. LNCS, vol. 6502, pp. 1–12. Springer, Heidelberg (2011). https://doi.org/10.1007/978-3-642-18469-7_1
4. Ackerman, E., Tardos, G.: On the maximum number of edges in quasi-planar graphs. J. Comb. Theory Ser. A **114**(3), 563–571 (2007). https://doi.org/10.1016/j.jcta.2006.08.002
5. Ali, P., Dankelmann, P., Mukwembi, S.: The radius of k-connected planar graphs with bounded faces. Discret. Math. **312**(24), 3636–3642 (2012). https://doi.org/10.1016/j.disc.2012.08.019
6. Alon, N., Erdős, P.: Disjoint edges in geometric graphs. Discret. Comput. Geom. **4**, 287–290 (1989). https://doi.org/10.1007/BF02187731
7. Angelini, P., et al.: Simple k-planar graphs are simple (k+1)-quasiplanar. J. Comb. Theory Ser. B **142**, 1–35 (2020). https://doi.org/10.1016/j.jctb.2019.08.006
8. Auer, C., et al.: Outer 1-planar graphs. Algorithmica **74**(4), 1293–1320 (2016). https://doi.org/10.1007/s00453-015-0002-1
9. Avital, S., Hanani, H.: Graphs. Gilyonot Lematematika **3**, 2–8 (1966)
10. Bekos, M.A., Cornelsen, S., Grilli, L., Hong, S., Kaufmann, M.: On the recognition of fan-planar and maximal outer-fan-planar graphs. Algorithmica **79**(2), 401–427 (2017). https://doi.org/10.1007/s00453-016-0200-5
11. Bekos, M.A., Kaufmann, M., Raftopoulou, C.N.: On optimal 2- and 3-planar graphs. In: SoCG. LIPIcs, vol. 77, pp. 16:1–16:16. Schloss Dagstuhl - Leibniz-Zentrum für Informatik (2017). https://doi.org/10.4230/LIPIcs.SoCG.2017.16
12. Binucci, C., et al.: Algorithms and characterizations for 2-layer fan-planarity: from caterpillar to stegosaurus. J. Graph Algorithms Appl. **21**(1), 81–102 (2017). https://doi.org/10.7155/jgaa.00398
13. Binucci, C., et al.: Fan-planarity: properties and complexity. Theor. Comput. Sci. **589**, 76–86 (2015). https://doi.org/10.1016/j.tcs.2015.04.020
14. Bodendiek, R., Schumacher, H., Wagner, K.: Über 1-optimale graphen. Math. Nachr. **117**, 323–339 (1984)
15. Bollobás, B.: Extremal Graph Theory. Academic Press, New York (1978)
16. Bose, P., Kirkpatrick, D.G., Li, Z.: Worst-case-optimal algorithms for guarding planar graphs and polyhedral surfaces. Comput. Geom. **26**(3), 209–219 (2003). https://doi.org/10.1016/S0925-7721(03)00027-0
17. Di Battista, G., Eades, P., Tamassia, R., Tollis, I.G.: Graph Drawing: Algorithms for the Visualization of Graphs. Prentice-Hall, Hoboken (1999)
18. Di Giacomo, E., Didimo, W., Liotta, G., Meijer, H.: Area, curve complexity, and crossing resolution of non-planar graph drawings. Theory Comput. Syst. **49**(3), 565–575 (2011). https://doi.org/10.1007/s00224-010-9275-6
19. Didimo, W.: Density of straight-line 1-planar graph drawings. Inf. Process. Lett. **113**(7), 236–240 (2013). https://doi.org/10.1016/j.ipl.2013.01.013
20. Didimo, W.: Right angle crossing drawings of graphs. In: Hong, S.-H., Tokuyama, T. (eds.) Beyond Planar Graphs, pp. 149–169. Springer, Singapore (2020). https://doi.org/10.1007/978-981-15-6533-5_9

21. Didimo, W., Eades, P., Liotta, G.: Drawing graphs with right angle crossings. Theor. Comput. Sci. **412**(39), 5156–5166 (2011). https://doi.org/10.1016/j.tcs.2011.05.025

22. Didimo, W., Liotta, G., Montecchiani, F.: A survey on graph drawing beyond planarity. ACM Comput. Surv. **52**(1), 4:1–4:37 (2019). https://doi.org/10.1145/3301281

23. Du Preez, B.: Plane graphs with large faces and small diameter. Australas. J. Comb. **80**(3), 401–418 (2021)

24. Dujmovic, V., Gudmundsson, J., Morin, P., Wolle, T.: Notes on large angle crossing graphs. Chicago J. Theor. Comput. Sci. **2011** (2011)

25. Hong, S.-H.: Beyond planar graphs: introduction. In: Hong, S.-H., Tokuyama, T. (eds.) Beyond Planar Graphs, pp. 1–9. Springer, Singapore (2020). https://doi.org/10.1007/978-981-15-6533-5_1

26. Hong, S., Eades, P., Katoh, N., Liotta, G., Schweitzer, P., Suzuki, Y.: A linear-time algorithm for testing outer-1-planarity. Algorithmica **72**(4), 1033–1054 (2015). https://doi.org/10.1007/s00453-014-9890-8

27. Hong, S., Kaufmann, M., Kobourov, S.G., Pach, J.: Beyond-planar graphs: algorithmics and combinatorics (Dagstuhl Seminar 16452). Dagstuhl Rep. **6**(11), 35–62 (2016). https://doi.org/10.4230/DagRep.6.11.35

28. Kobourov, S.G., Liotta, G., Montecchiani, F.: An annotated bibliography on 1-planarity. Comput. Sci. Rev. **25**, 49–67 (2017). https://doi.org/10.1016/j.cosrev.2017.06.002

29. Kupitz, Y.S.: Extremal problems in combinatorial geometry. Lecture notes series, Matematisk institut, Aarhus universitet (1979)

30. Lan, Y., Shi, Y., Song, Z.: Extremal h-free planar graphs. Electron. J. Comb. **26**(2), 2 (2019). https://doi.org/10.37236/8255

31. Pach, J.: Geometric graph theory. In: Handbook of Discrete and Computational Geometry, 2nd edn., pp. 219–238. Chapman and Hall/CRC (2004). https://doi.org/10.1201/9781420035315.ch10

32. Pach, J., Radoicic, R., Tardos, G., Tóth, G.: Improving the crossing lemma by finding more crossings in sparse graphs. Discret. Computat. Geom. **36**(4), 527–552 (2006). https://doi.org/10.1007/s00454-006-1264-9

33. Pach, J., Törocsik, J.: Some geometric applications of Dilworth's theorem. Discret. Comput. Geom. **12**, 1–7 (1994). https://doi.org/10.1007/BF02574361

34. Pach, J., Tóth, G.: Graphs drawn with few crossings per edge. Combinatorica **17**(3), 427–439 (1997). https://doi.org/10.1007/BF01215922

35. Suk, A., Walczak, B.: New bounds on the maximum number of edges in k-quasi-planar graphs. Comput. Geom. **50**, 24–33 (2015). https://doi.org/10.1016/j.comgeo.2015.06.001

36. Suzuki, Y.: 1-planar graphs. In: Hong, S.-H., Tokuyama, T. (eds.) Beyond Planar Graphs, pp. 47–68. Springer, Singapore (2020). https://doi.org/10.1007/978-981-15-6533-5_4

Computational Complexity of Covering Colored Mixed Multigraphs with Degree Partition Equivalence Classes of Size at Most Two (Extended Abstract)

Jan Bok[1,3]([✉]) [ID], Jiří Fiala[2] [ID], Nikola Jedličková[2] [ID], Jan Kratochvíl[2]([✉]) [ID], and Michaela Seifrtová[2] [ID]

[1] Computer Science Institute, Faculty of Mathematics and Physics, Charles University, Prague, Czech Republic
[2] Department of Applied Mathematics, Faculty of Mathematics and Physics, Charles University, Prague, Czech Republic
{fiala,jedlickova,honza,mikina}@kam.mff.cuni.cz
[3] Université Clermont Auvergne, CNRS, Clermont Auvergne INP, Mines Saint-Étienne, LIMOS, 63000 Clermont-Ferrand, France
jan.bok@uca.fr

Abstract. The notion of graph covers (also referred to as locally bijective homomorphisms) plays an important role in topological graph theory and has found its computer science applications in models of local computation. For a fixed target graph H, the H-COVER problem asks if an input graph G allows a graph covering projection onto H. Despite the fact that the quest for characterizing the computational complexity of H-COVER had been started more than 30 years ago, only a handful of general results have been known so far.

In this paper, we present a complete characterization of the computational complexity of covering colored graphs for the case that every equivalence class in the degree partition of the target graph has at most two vertices. We prove this result in a very general form. Following the lines of current development of topological graph theory, we study graphs in the most relaxed sense of the definition - the graphs are mixed (they may have both directed and undirected edges), may have multiple edges, loops, and semi-edges. We show that a strong P/NP-co dichotomy holds true in the sense that for each such fixed target graph H, the H-COVER problem is either polynomial time solvable for arbitrary inputs, or NP-complete even for simple input graphs.

1 Introduction

The notion of *graph covers* stems from topology and is viewed as a discretization of the notion of covers of topological spaces. Apart from being used in combinatorics as a tool for constructing large highly symmetric graphs [3–6], this notion has found computer science applications in the theory of local computation [2,13–15,17,30]. In this paper we aim to contribute to the kaleidoscope of

D. Paulusma and B. Ries (Eds.): WG 2023, LNCS 14093, pp. 101–115, 2023.
https://doi.org/10.1007/978-3-031-43380-1_8

results about computational complexity of graph covers. We first briefly comment on the known results and show where our main result is placed among them. The formal definitions of graphs under consideration (Definition 1) and of graph covering projections (Definitions 2 and 3) are presented in Sect. 2, as well as the detailed definition of the so called degree reducing reduction (Definition 4), the concept of the degree partition of a graph (Proposition 1) and identification of several special graphs which play the key role in our characterization in Theorem 2 (Definition 5).

Despite the efforts and attention that graph covers received in the computer science community, their computational complexity is still far from being fully understood. Bodlaender [9] proved that deciding if one graph covers another one is an NP-complete problem, if both graphs are part of the input. Abello et al. [1] considered the variant when the target graph, say H, is fixed, i.e., a parameter of the problem, and the question is if an input graph covers H (this decision problem will be referred to as H-COVER). They showed examples of graphs H for which the problem is polynomial time solvable as well as examples for which it is NP-complete, but most importantly, they were the first to formulate the goal of a complete characterization of the computational complexity of the H-COVER problem, depending on the target graph H. Some of the explicit questions of Abello et al. [1] were answered by Kratochvíl et al. in [25,27], some of the NP-hardness results have been strengthened to planar input graphs by Bílka et al. [8]. A connection to a generalization of the Frequency Assignment Problem has been identified through partial covers [7,18,21]. The computationally even more sophisticated problem of *regular covers* has been treated in [19]. In a recent paper [11], the authors initiated the study of the complexity of H-COVER for graphs that allow multiple edges and loops, and also semi-edges. This is motivated by the recent development of topological graph theory where it has now become standard to consider this more general model of graphs [29,31–34]. The graphs with semi-edges were also introduced and used in mathematical physics, e.g. by Getzler and Karpanov [22]. It should be pointed out right away that considering loops, multiple edges and directed edges was shown necessary already in [26], where it is proved that in order to fully understand the computational complexity of H-COVER for *simple* undirected graphs H (i.e., undirected graphs without multiple edges, loops, and semi-edges), it is necessary and sufficient to understand the complexity of the problem for colored mixed multigraphs of minimum degree greater than 2. All papers from that era restrict their attention to covers of connected graphs. Disconnected target graphs are carefully treated in detail only in [10], where it is argued that the right way to define covers of disconnected graphs is to request that the preimages of all vertices have the same size. Such covers are called *equitable covers* in [10], and in the current paper we adopt this view and require graph covers to be equitable in case of covering disconnected graphs.

Apart from several isolated results (which also include a complete characterization of the complexity of H-COVER for connected simple undirected graphs

H with at most 6 vertices [25]), the following general results have been known about the complexity of H-COVER for infinite classes of graphs:

1. Polynomial time solvability of H-COVER for connected simple undirected graphs H which have at most two vertices in every equivalence class of their degree partitions [25].
2. NP-completeness of H-COVER for regular simple undirected graphs H of valency at least three [20,27].
3. Complete characterization of the complexity of H-COVER for undirected (multi)graphs H (without semi-edges) on at most three vertices [28].
4. Complete characterization of the complexity of H-COVER for colored mixed (multi)graphs H on at most two vertices [26] (for graphs without semi-edges) and [11] (with semi-edges allowed).

It turns out that so far all the known NP-hard instances of H-COVER remain NP-hard for simple graphs on input. This has led Bok et al. to formulating the following conjecture.

Strong Dichotomy Conjecture for Graph Covers [12]. *For every graph H, the H-COVER problem is either polynomial time solvable for arbitrary input graphs, or it is NP-complete for simple graphs as input.*

The main result of our paper is a complete characterization of the computational complexity of H-COVER for graphs H, each of whose equivalence classes of the degree partition has at most 2 vertices. This provides a common generalization of results 1 and 4.

Theorem 1. *The H-COVER problem satisfies Strong Dichotomy for graphs H such that each equivalence class of the degree partition has at most 2 vertices.*

The actual characterization is somewhat technical and it follows from Theorem 2 in Sect. 2, presented after the formal definitions of all the notions and special graphs that are needed for it. The characterization goes much farther beyond the motivating results from [25,26]. The main novel points are the following:

- For simple graphs H, the H-COVER problem is always polynomial time solvable (if H has all equivalence classes of size at most 2), while for general graphs, already graphs with 2 vertices may define NP-complete cases (and even graphs with 1 vertex when semi-edges are allowed).
- For simple graphs H, the polynomial time algorithm is based on 2-SATISFIABILITY, while in case of general graphs, our polynomial time algorithm is a blend of 2-SATISFIABILITY and PERFECT MATCHING algorithms; this is somewhat surprising, since these two approaches are known to be incompatible in some other situations.
- The NP-complete cases are proved for simple input graphs, which is in line with the Strong Dichotomy Conjecture as stated in [12] (in contrast to many previous results which allowed multiple edges and loops in the input graphs).

2   Preliminaries

2.1   Definitions

Throughout the paper we will be working with the most general notion of a *graph* which allows multiple edges, loops, directed edges and also semi-edges and whose elements – both edges and vertices – are colored. A semi-edge is a pendant edge, incident to just one vertex (and adding just 1 to the degree of this vertex, unlike the loop, which adds 2 to the degree). In figures, semi-edges are depicted as lines with one loose end, the other one being the vertex incident to the semi-edge. To avoid any possible confusion, we present a formal definition.

Definition 1. *A* graph *is a quadruple* $G = (V, \Lambda, \iota, c)$, *where* V *is a (finite) set of* vertices, $\Lambda = \overline{E} \cup \overrightarrow{E} \cup \overline{L} \cup \overrightarrow{L} \cup S$ *is the set of* edges *of* G, $\iota : \Lambda \longrightarrow \binom{V}{2} \cup (V \times V) \cup V$ *is the* incidence mapping *of edges, and* $c : V \cup E \longrightarrow C$ *is a* coloring *of the vertices and edges. The edges of* $\overline{E}$ *are called* normal undirected edges *and they satisfy* $\iota(e) \in \binom{V}{2}$, *the edges of* $\overrightarrow{E}$ *are* normal directed edges *(and* $\iota(e) \in (V \times V) \setminus \{(u, u) : u \in V\}$), *the edges of* $\overline{L}$ ($\overrightarrow{L}$) *are* undirected (directed, respectively) loops *(and we have* $\iota(e) \in V$ *in both cases), and finally the edges of* S *are called* semi-edges *(and again* $\iota(e) \in V$).

The vertex set and edge set of a graph G will be denoted by $V(G)$ and $\Lambda(G)$, respectively, and a similar notation will be used for $\overline{E}(G), \overrightarrow{E}(G), \overline{L}(G), \overrightarrow{L}(G)$ and $S(G)$. Since we can distinguish vertices from edges, and directed edges from the undirected ones, we assume without loss of generality that colors of vertices, of directed edges and of undirected ones are different. However, we allow directed loops and directed normal edges to have the same color, as well as undirected normal edges, undirected loops and semi-edges. Edges with the same incidence function are called *parallel*. A graph is called *simple* if it has no parallel edges, no pair of opposite directed normal edges, no loops and no semi-edges. When talking about a disjoint union of graphs, we assume that the graphs are vertex (and therefore also edge) disjoint. The following definition presents the main notion of the paper.

Definition 2. *Let* G *and* H *be connected graphs colored by the same sets of colors. A* covering projection *from* G *to* H *is a pair of color-preserving mappings* $f_V : V(G) \longrightarrow V(H)$, $f_E : \Lambda(G) \longrightarrow \Lambda(H)$ *such that*

– *the preimage of an undirected normal edge of* H *incident with vertices* $u, v \in V(H)$ *is a perfect matching in* G *spanning* $f^{-1}(u) \cup f^{-1}(v)$, *each edge of the matching being incident with one vertex of* $f^{-1}(u)$ *and with one vertex of* $f^{-1}(v)$;
– *the preimage of a directed normal edge of* H *leading from a vertex* $u \in V(H)$ *to a vertex* $v \in V(H)$ *is a perfect matching in* G *spanning* $f^{-1}(u) \cup f^{-1}(v)$, *each edge of the matching being oriented from a vertex of* $f^{-1}(u)$ *to a vertex of* $f^{-1}(v)$;

- the preimage of an undirected loop of H incident with a vertex $u \in V(H)$ is a disjoint union of cycles in G spanning $f^{-1}(u)$;
- the preimage of a directed loop of H incident with a vertex $u \in V(H)$ is a disjoint union of directed cycles in G spanning $f^{-1}(u)$; and
- the preimage of a semi-edge of H incident with a vertex $u \in V(H)$ is a disjoint union of semi-edges and normal edges spanning $f^{-1}(u)$ (each vertex of $f^{-1}(u)$ being incident to exactly one semi-edge and no normal edges, or exactly one normal edge and no semi-edges, from the preimage).

We say that G covers H, and write $G \longrightarrow H$, if there exists a covering projection from G to H. Informally speaking, if G covers H via a covering projection (f_V, f_E) and if an agent moves along the edges of G and in every moment sees only the label $f_V(u)$ (or $f_E(e)$) of the vertex (edge) he/she is currently visiting, plus the labels of the incident edges (vertices, respectively), then the agent cannot distinguish whether he/she is moving through the covering graph G or the target graph H. Mind the significant difference between undirected loops and semi-edges. The presence of an undirected loop incident with a vertex, say u, means that there are two ways how to move from u to u along this loop, while for a semi-edge, there is just one way. The same holds true for their preimages in covering projections (undirected cycles, or isolated edges). An example of a graph and a possible cover is depicted in Fig. 1 right.

In [11], a significant role of semi-edges was noted. A color-preserving vertex-mapping $f_V : V(G) \longrightarrow V(H)$ is called *degree-obedient* if for any edge color α, any vertex $u \in V(G)$ and any vertex $x \in V(H)$, the number of edges of color α that lead from u to a vertex from $f_V^{-1}(x)$ in G is the same as the number of edges of color α leading from $f_V(u)$ to x in H, counting those edges that may map onto each other in a covering projection (e.g., if $x = f_V(u)$ and H has ℓ undirected loops and s semi-edges incident with x, and u is incident with k loops, n normal undirected edges with both end-vertices in $f_V^{-1}(x)$ and t semi-edges, then $t \leq s$ and $2k + n + t = 2\ell + s$; analogously for other types of edges). It is proved in [11] that every degree-obedient vertex-mapping extends to a covering projection if H has no semi-edges, and also when G has no semi-edges and is bipartite.

It follows straightforwardly from the definition of graph covering that the preimages of any two vertices have the same size. For disconnected graphs, we add this requirement to the definition.

Definition 3. Let G and H be graphs and let $f = (f_V, f_E): G \longrightarrow H$ be a pair of incidence-compatible color-preserving mappings. Then f is a covering projection of G to H if for each component G_i of G, the restricted mapping $f|_{G_i} : G_i \longrightarrow H$ is a covering projection of G_i onto some component of H, and for every two vertices $u, v \in V(H)$, $|f^{-1}(u)| = |f^{-1}(v)|$.

Another notion we need to recall is that of the *degree partition* of a graph. This is a standard notion for simple undirected graphs, cf. [16], and it can be naturally generalized to graphs in general. A partition of the vertex set of a graph G is *equitable* if every two vertices of the same class of the partition a) have the same color, and b) have the same number of neighbors along edges of the same color in every class (including its own). The *degree partition* of a graph is then the coarsest equitable partition. It can be found in polynomial time, and moreover, a canonical linear ordering of the classes of the degree partition comes out from the algorithm. Let $V(G) = \bigcup_{i=1}^{k} V_i$ be the degree partition of G, in the canonical ordering. The *degree refinement matrix* of G is a $k \times k$ matrix M_G whose entries are vectors indexed by edge colors expressing that every vertex $u \in V_i$ has $M_{i,j,c}$ neighbors in V_j along edges of color c (if $i = j$ and c is a color of directed edges, then every vertex $u \in V_i$ has $M_{i,j,c}$ in-neighbors and $M_{i,j,c}$ out-neighbors in V_i along edges of color c). The following is proved in [26] for graphs without semi-edges, the extension to graphs with semi-edges is straightforward.

Proposition 1. *Let G and H be graphs and let $V(G) = \bigcup_{i=1}^{k} V_i$ and $V(H) = \bigcup_{i=1}^{\ell} W_i$ be the degree partitions of their vertex sets, in the canonical orderings. If G covers H, then $k = \ell$, the degree refinement matrices of G and H are equal, and for any covering projection $f : G \longrightarrow H$, $f(V_i) = W_i$ holds true for every $i = 1, 2, \ldots, k$.*

The classes of the degree partition will be further referred to as *blocks*. Once we have determined the degree partition of a graph, we will re-color the vertices so that vertices in different blocks are distinguished by vertex-colors (representing the membership to blocks), and recolor and de-orient the edges so that edges connecting vertices from different blocks are undirected and so that for any edge color, either all edges of this color belong to the same block, or they are connecting vertices from the same pair of blocks. The degree partition will remain unchanged after such a re-coloring.

A *block graph* of G is a subgraph G' of G whose vertex set is the union of some blocks of G, and such that for every edge color α, G' either contains all edges of color α that G contains, or none. A block graph G' of G is *induced* if G' contains all edges of G on the vertices of $V(G')$. A block graph is *monochromatic* if it contains edges of at most one color. A *uniblock graph* is a block graph whose vertex set is a single block of G. An *interblock graph* of G is a block graph whose vertices belong to two blocks of G, and each of its edges is incident with vertices from both blocks (i.e., with one vertex from each block).

As a local bijection, any graph covering maintains vertex degrees. In particular, vertices of degree one are mapped onto vertices of degree one and once we choose the image of such a vertex, the image of its neighbor is uniquely determined. Applied inductively, this proves the well known fact that the only connected cover of a (rooted) tree is an isomorphic copy of the tree itself. (Note here, that by definition a tree is a connected graph that does not contain cycles, parallel edges, oppositely oriented directed edges, loops, and semi-edges.) As a special case, the only connected cover of a path is the path itself. These observations are the basis of the following degree reducing reduction which has been introduced in [26] for graphs without semi-edges, and generalized to graphs with semi-edges in [10].

Definition 4. (Degree reducing reduction) *Let G be a non-tree graph.*

1. *Determine all vertices that belong to cycles in G or that are incident with semi-edges or that lie on paths connecting aforementioned vertices. Determine all maximal subtrees pending on these vertices. Determine the isomorphism types of these subtrees, introduce a new vertex color for each isomorphism type, delete each subtree and color its root by the color corresponding to the ismomorphism type of the deleted tree. In this way we obtain a graph with minimum degree at least 2 (or a single-vertex graph).*
2. *Determine all maximal paths with at least one end-vertex of degree greater than 2 and all inner vertices being of degree exactly 2. (For this step, a cycle is viewed as a path whose end-vertices are equal.) Determine all color patterns of the sequences of vertex colors, edge colors and edge directions along such paths, and introduce a new color for each such pattern. Replace each such path by a new edge of this color as follows:*
    2.1. *If both end-vertices of the path are of degree greater than 2 and the color pattern, say π, is symmetric, the path gets replaced by an undirected edge (or loop) of color π.*
    2.2. *If both end-vertices of the path are of degree greater than 2 and the color pattern π is asymmetric, the path gets replaced by a directed edge (or loop) of color π.*
    2.3. *If the path ends with a semi-edge (the other end of the path must be a vertex of degree greater than 2), replace it by a semi-edge incident with its end-vertex of degree greater than 2, and color it with color $\pi\alpha$, where π is the color pattern along the path without the ending semi-edge, and α is the color of the semi-edge. In this case, consider the colors corresponding to $\pi\alpha$ (on one sided open paths) and $\pi\alpha\pi^{-1}$ on symmetric paths ending with vertices of degree greater than 2 on both sides, as the same color (this enables a path of color pattern $\pi\alpha\pi^{-1}$ be mapped on the one sided open path in a covering projection).*

Denote the resulting graph by G. Note that G is a path or a cycle (if Step 2 was void) or has minimum degree greater than 2.

Fig. 1. An example of the application of the degree reducing reduction. An example of a graph cover of the reduced graph is depicted in the right.

The reduced graph can be constructed in polynomial time. The usefulness of this reduction is observed in [26] and [10]:

Observation 1. *Given graphs G and H, perform the degree reducing reduction on both of them simultaneously. Then $G \longrightarrow H$ if and only if $G \longrightarrow H$.*

Finally, for a subset $W \subseteq V(G)$, we denote by $G[W]$ the subgraph induced by W. If α is an edge color, then G^α denotes the spanning subgraph of G containing exactly the edges of color α.

2.2   Our Results

In order to describe the results, we introduce the formal notation of certain small graphs. We denote by

- $F(b, c)$ the one-vertex graph with b semi-edges and c loops;
- $FD(c)$ the one-vertex graph with c directed loops;
- $W(k, m, \ell, p, q)$ the two-vertex graph with ℓ parallel undirected edges joining its two vertices and with k (q) semi-edges and m (p) undirected loops incident with one (the other one, respectively) of its vertices;
- $WD(m, \ell, m)$ the directed two-vertex graph with m directed loops incident with each of its vertices, the two vertices being connected by ℓ directed edges in each direction;
- $FF(c)$ the two-vertex graph connected by c parallel undirected edges, with the two vertices being distinguishable to belong to different blocks;
- $FW(b)$ the three-vertex graph with bundles of b parallel edges connecting one vertex to each of the remaining two; and
- $WW(b, c)$ the graph on four vertices obtained from a 4-cycle by replacing the edges of a perfect matching by bundles of b parallel edges, and replacing the edges of the complementary matching by bundles of c parallel edges, the two independent sets of size 2 belonging to different blocks.

Edges of all of these graphs are uncolored (or, equivalently, monochromatic). We shall only consider W graphs having $k + 2m = 2p + q$. See the illustration in Fig. 2 for the graphs defined here.

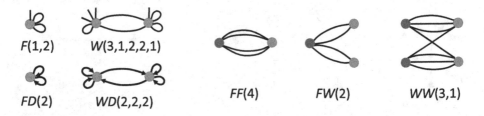

$F(1,2)$ $W(3,1,2,2,1)$

$FD(2)$ $WD(2,2,2)$ $FF(4)$ $FW(2)$ $WW(3,1)$

Fig. 2. Examples of the small graphs we are considering.

Definition 5. *For the convenience of the reader, the maximal harmless monochromatic uniblock and interblock graphs are depicted in Fig. 3. A regular monochromatic uniblock graph with at most two vertices is called*

- *harmless if it is isomorphic to $F(b,0)$, $b \leq 2$, $F(1,c)$, $F(0,c)$, $FD(c)$, $W(2,0,0,0,2)$, $W(2,0,0,1,0)$, $W(0,c,0,c,0)$, $W(1,c,0,c,1)$, $W(0,0,c,0,0)$, $W(1,0,1,0,1)$, $WD(c,0,c)$, $WD(0,c,0)$, $WD(1,1,1)$ (c being an arbitrary nonnegative integer),*
- *harmful if it is isomorphic to $F(b,c)$ such that $b \geq 2$ and $b + c \geq 3$, or to $W(k,m,\ell,p,q)$ such that $\ell \geq 1$ and $k + 2m + \ell = q + 2p + \ell \geq 3$, or to the disjoint union of $F(b,c)$ and $F(b',c')$ such that at least one of them is harmful, or to $WD(c,b,c)$ such that $b \geq 1, c \geq 1$ and $b + c \geq 3$.*

A monochromatic interblock graph is called

- *harmless if it is isomorphic to $FF(c)$ or $WW(0,c)$ (with c being an arbitrary nonnegative integer), or to $FW(0)$, $FW(1)$, or $WW(1,1)$,*
- *dangerous if it is isomorphic to $FW(2)$, and*
- *harmful if it is isomorphic to $FW(c)$ for $c \geq 3$, or to $WW(b,c)$ such that $b \geq 1, c \geq 1$ and $b + c \geq 3$.*

Note that under the assumption that each degree partition equivalence class has size at most two, every monochromatic uniblock graph as well as every monochromatic interblock graph fall in exactly one of the above described categories. The choice of the terminology is explained by the following theorem.

Theorem 2. *Suppose all blocks of a graph H have sizes at most 2. Then the following hold true:*

1. *If all monochromatic uniblock and interblock graphs of H are harmless, then the H-COVER problem is solvable in polynomial time (for arbitrary input graphs).*
2. *If at least one of the monochromatic uniblock or interblock graphs of H is harmful, then the H-COVER problem is NP-complete even for simple input graphs.*
3. *If the minimum degree of H is greater than 2 and H contains a dangerous monochromatic interblock graph, then the H-COVER problem is NP-complete even for simple input graphs.*

Observe that Theorem 2 implies that *H*-COVER is polynomial time solvable if and only if every monochromatic uniblock graph defines a polynomial time solvable instance and the monochromatic interblock graphs are such that either each vertex has at most one neighbor, or each vertex has degree at most two. For the interblock graphs, this is also very close to saying that each monochromatic interblock graph itself defines a polynomial time solvable instance, but not quite. The one and only exception is the graph $FW(2)$. Indeed, $FW(2)$-COVER is polynomial time solvable (since it reduces to $F(2)$-Cover), but with the additional condition that all vertices have degrees greater than 2, the presence of $FW(2)$ in *H* leads to NP-completeness of *H*-COVER (this will be shown in detail in Sect. 4).

3   Proof of Theorem 2 - Polynomial Cases

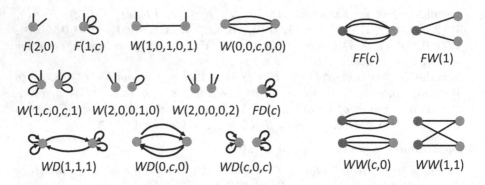

Fig. 3. The maximal harmless monochromatic uniblock (left) and interblock (right) graphs (c is an arbitrary non-negative integer).

Here we sketch an algorithm that proves Part 1 of Theorem 2. It clearly runs in polynomial time, the details can be found in the journal version of the paper.

Algorithm

1. Compute the degree partitions of G (the input graph) and H (the target graph). Reorder the equivalence classes W_i of the degree partition of H so that $W_1, \ldots, W_s$ are singletons and $W_{s+1}, \ldots, W_k$ contain two vertices each, and reorder the degree partition equivalence classes V_i of the input graph G accordingly. Denote further, for every $i = 1, \ldots, s$, by a_i the vertex in W_i, and, for every $i = s + 1, \ldots, k$, by b_i, c_i the vertices of W_i.
2. Check that the degree refinement matrices of G and H are indeed the same.
3. Decide if the edges within $G[V_i]$ can be mapped onto the edges of $H[W_i]$ to form a covering projection, for each $i = 1, 2, \ldots, s$. (This step amounts to checking degrees and the numbers of semi-edges incident with the vertices, as well as checking that monochromatic subgraphs contain perfect matchings in case of semi-edges in the target graph).
4. Preprocess the two-vertex equivalence classes W_i, $i = s + 1, \ldots, k$ when $H[W_i]$ contains semi-edges (this may impose conditions on some vertices of V_i, whether they can map on b_i or c_i).
5. Using 2-SATISFIABILITY, find a degree-obedient vertex mapping from V_i onto W_i for each $i = s + 1, \ldots, k$, which fulfills the conditions observed in Step 4. (For every vertex $u \in V_i$, introduce a variable x_u with the interpretation that x_u is true if u is mapped onto b_i and it is false when x_u is mapped onto c_i. The harmless block graphs are such that either all neighbors of a vertex u must be mapped onto the same vertex, and thus the value of the corresponding variables are all the same (e.g., for $WW(0, c)$), or u has exactly two neighbors which should map onto different vertices (e.g., for $WW(1, 1)$)

meaning that the corresponding variables must get opposite values. All the situations that arise from harmless block graphs can be described by clauses of size 2).

6. Complete the covering projection by defining the mapping on edges in case a degree-obedient vertex mapping was found in Step 5, or conclude that G does not cover H otherwise. (The existence and polynomial time constructability of covering projections from degree-obedient vertex mappings for such instances have been proven in [11]).

4 Proof of Theorem 2 - NP-Hard Cases

The proof is technical and involves several NP-hardness reductions. We will provide an overview of its main steps, the details will appear in the full version of the paper.

Step 1. We first argue that H-COVER restricted to simple input graphs is NP-complete for every harmful uniblock or interblock graph H. These cases have been proved previously in [1,10,11,26,28], however, some of them only for input graphs allowing parallel edges. An extra care was thus needed to strengthen the NP-hardness results for simple input graphs.

Step 2. Covering the dangerous graph $FW(2)$ itself is polynomial time decidable, since red$FW(2) = F(2)$ is harmless. However, if $FW(2)$ is a monochromatic interblock graph of H, all vertices of H have degrees greater than 2, and H does not contain any harmful monochromatic uniblock or interblock graph, then H contains a block graph which is reducible to one of the graphs from Fig. 4 (if subscript k is used in the name of a graph from this figure, it refers to the number of parallel red edges or loops). This can be proven by a straightforward case analysis.

Step 3. For every graph H from Fig. 4, H-COVER is NP-complete for simple input graphs (for those graphs indexed by k, we claim the statement for every $k \geq 3$ in case of H_k and H'_k, for every $k \geq 2$ in case of B'_k, for every $k \geq 1$ in case of $B_k, C'_k, D_k, D'_k, L_k, L'_k, M_k$ and M'_k, and for every $k \geq 0$ in case of C_k). We prove this by reductions from MONOTONE 2-IN-4-SATISFIABILITY which is known to be NP-complete [23]. The 25 graphs can be grouped into a few groups which are handled en bloc by unified reductions tailored on the groups.

Step 4. The last step is to show that H-COVER for simple input graphs polynomially reduces to H'-COVER for simple input graphs, when H is a block graph of H', and H is a harmful graph or one of the graphs from Fig. 4. This is a step which is usually called the garbage collection. We describe a way how to construct a simple graph G' from a simple graph G (an input of H-COVER), so that $G' \longrightarrow H'$ if and only if $G \longrightarrow H$. Note that this is somewhat simpler in case when H is balanced in the sense that in every two-vertex block and for every edge color, both vertices are incident with the same number of semi-edges of this color.

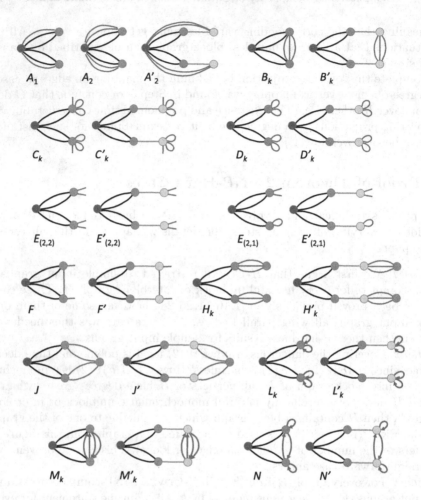

Fig. 4. Block graphs forced by $FW(2)$.

5   Proof of Theorem 1

Suppose H is a connected graph each of whose equivalence classes of the degree partition has at most 2 vertices. The H-COVER problem can be solved in constant or linear time if H is a tree or a cycle or a path (possibly ending with semi-edges). Otherwise, consider the reduced graph H, reduced via the degree reducing reduction of Definition 4. It is important that H also has at most two vertices in each equivalence class of its degree partition. If H is a path or a cycle, then H-COVER is solvable in polynomial time, and so is H-COVER, due to Observation 1.

If H contains a vertex of degree greater than 2, then all vertices of H have degrees greater than 2. If all monochromatic uniblock and interblock graphs of H are harmless, then H-COVER is polynomially solvable for general input graphs, and so is H-COVER, due to Observation 1. If H contains a harmful or a dangerous uniblock or interblock graph, then H-COVER is NP-complete for simple input graphs by Parts 2 and 3 of Theorem 2. If G is a simple graph as an input to the H-COVER problem, the reverse operation to the degree reducing reduction gives a simple graph G such that $G \longrightarrow H$ if and only if $G \longrightarrow H$. Hence H-COVER is also NP-complete for simple input graphs.

6 Concluding Remarks

The polynomial algorithm of Sect. 3 combines two approaches - finding perfect matchings and solving 2-SATISFIABILITY. It is well known that these two problems are polynomial time solvable. It may be somewhat surprising that so is their combination, e.g., in comparison with the so called compatible 2-factor problem [24], whose instances solvable in polynomial time are of two types, one solved by a reduction to perfect matching, the other one solved by 2-SATISFIABILITY, but if restrictions of both types are present in the same instance, the problem becomes NP-complete.

Note further that the polynomiality of the polynomial time solvable case does not depend on the target graph being fixed. If H is a graph with at most 2 vertices in each block of the degree partition, and all monochromatic block and interblock graphs are harmless, then the algorithm described in Sect. 3 remains polynomial time even if H is part of the input.

We believe that the method developed above has a much wider potential and we conjecture the following:

Conjecture. *Let H be a block graph of a graph H'. Then H-COVER for simple input graphs polynomially reduces to H'-COVER for simple input graphs.*

And of course, the ultimate goal is to prove (or disprove) the Strong Dichotomy Conjecture for graph covers parameterized by the target graph, ideally with a complete catalog of the polynomially solvable cases.

Acknowledgments. All co-authors were supported by research grant GAčR 20-15576S of the Czech Science Foundation, Nikola Jedličková was further supported by SVV–2020–260578 and GAUK 370122. Jan Bok was partially financed by the ANR project GRALMECO (ANR-21-CE48-0004) and the French government IDEX-ISITE initiative 16-IDEX-0001 (CAP 20-25).

References

1. Abello, J., Fellows, M.R., Stillwell, J.C.: On the complexity and combinatorics of covering finite complexes. Aust. J. Comb. **4**, 103–112 (1991)
2. Angluin, D.: Local and global properties in networks of processors. In: Proceedings of the Twelfth Annual ACM Symposium on Theory of Computing, STOC 1980, pp. 82–93. Association for Computing Machinery, New York (1980)
3. Biggs, N.: Algebraic Graph Theory. Cambridge University Press, Cambridge (1974)

4. Biggs, N.: Covering biplanes. In: The Theory and Applications of Graphs, Fourth International Conference, Kalamazoo, pp. 73–79. Wiley (1981)
5. Biggs, N.: Constructing 5-arc transitive cubic graphs. J. Lond. Math. Soc. **II**(26), 193–200 (1982)
6. Biggs, N.: Homological coverings of graphs. J. Lond. Math. Soc. **II**(30), 1–14 (1984)
7. Bílka, O., Lidický, B., Tesař, M.: Locally injective homomorphism to the simple weight graphs. In: Ogihara, M., Tarui, J. (eds.) TAMC 2011. LNCS, vol. 6648, pp. 471–482. Springer, Heidelberg (2011). https://doi.org/10.1007/978-3-642-20877-5_46
8. Bílka, O., Jirásek, J., Klavík, P., Tancer, M., Volec, J.: On the complexity of planar covering of small graphs. In: Kolman, P., Kratochvíl, J. (eds.) WG 2011. LNCS, vol. 6986, pp. 83–94. Springer, Heidelberg (2011). https://doi.org/10.1007/978-3-642-25870-1_9
9. Bodlaender, H.L.: The classification of coverings of processor networks. J. Parallel Distrib. Comput. **6**, 166–182 (1989)
10. Bok, J., Fiala, J., Jedličková, N., Kratochvíl, J., Seifrtová, M.: Computational complexity of covering disconnected multigraphs. In: Bampis, E., Pagourtzis, A. (eds.) FCT 2021. LNCS, vol. 12867, pp. 85–99. Springer, Cham (2021). https://doi.org/10.1007/978-3-030-86593-1_6
11. Bok, J., Fiala, J., Hliněný, P., Jedličková, N., Kratochvíl, J.: Computational complexity of covering multigraphs with semi-edges: small cases. In: Bonchi, F., Puglisi, S.J. (eds.) 46th International Symposium on Mathematical Foundations of Computer Science, MFCS 2021, Tallinn, Estonia, 23–27 August 2021. LIPIcs, vol. 202, pp. 21:1–21:15. Schloss Dagstuhl - Leibniz-Zentrum für Informatik (2021)
12. Bok, J., Fiala, J., Jedličková, N., Kratochvíl, J., Rzążewski, P.: List covering of regular multigraphs. In: Bazgan, C., Fernau, H. (eds.) IWOCA 2022. LNCS, vol. 13270, pp. 228–242. Springer, Cham (2022). https://doi.org/10.1007/978-3-031-06678-8_17
13. Chalopin, J.: Local computations on closed unlabelled edges: the election problem and the naming problem. In: Vojtáš, P., Bieliková, M., Charron-Bost, B., Sýkora, O. (eds.) SOFSEM 2005. LNCS, vol. 3381, pp. 82–91. Springer, Heidelberg (2005). https://doi.org/10.1007/978-3-540-30577-4_11
14. Chalopin, J., Métivier, Y., Zielonka, W.: Local computations in graphs: the case of cellular edge local computations. Fund. Inform. **74**(1), 85–114 (2006)
15. Chalopin, J., Paulusma, D.: Graph labelings derived from models in distributed computing: a complete complexity classification. Networks **58**(3), 207–231 (2011)
16. Corneil, D.G., Gotlieb, C.C.: An efficient algorithm for graph isomorphism. J. Assoc. Comput. Mach. **17**, 51–64 (1970)
17. Courcelle, B., Métivier, Y.: Coverings and minors: applications to local computations in graphs. Eur. J. Comb. **15**, 127–138 (1994)
18. Fiala, J., Kratochvíl, J.: Complexity of partial covers of graphs. In: Eades, P., Takaoka, T. (eds.) ISAAC 2001. LNCS, vol. 2223, pp. 537–549. Springer, Heidelberg (2001). https://doi.org/10.1007/3-540-45678-3_46
19. Fiala, J., Klavík, P., Kratochvíl, J., Nedela, R.: Algorithmic aspects of regular graph covers with applications to planar graphs. CoRR abs/1402.3774 (2014)
20. Fiala, J., Kratochvíl, J.: Locally constrained graph homomorphisms – structure, complexity, and applications. Comput. Sci. Rev. **2**(2), 97–111 (2008)
21. Fiala, J., Kratochvíl, J., Pór, A.: On the computational complexity of partial covers of theta graphs. Electron. Notes Discret. Math. **19**, 79–85 (2005)
22. Getzler, E., Kapranov, M.M.: Modular operads. Compos. Math. **110**(1), 65–125 (1998)

23. Kratochvíl, J.: Complexity of hypergraph coloring and Seidel's switching. In: Bod-laender, H.L. (ed.) WG 2003. LNCS, vol. 2880, pp. 297–308. Springer, Heidelberg (2003). https://doi.org/10.1007/978-3-540-39890-5_26

24. Kratochvíl, J., Poljak, S.: Compatible 2-factors. Discret. Appl. Math. **36**(3), 253–266 (1992)

25. Kratochvíl, J., Proskurowski, A., Telle, J.A.: Complexity of graph covering prob-lems. In: Mayr, E.W., Schmidt, G., Tinhofer, G. (eds.) WG 1994. LNCS, vol. 903, pp. 93–105. Springer, Heidelberg (1995). https://doi.org/10.1007/3-540-59071-4_40

26. Kratochvíl, J., Proskurowski, A., Telle, J.A.: Complexity of colored graph covers I. Colored directed multigraphs. In: Möhring, R.H. (ed.) WG 1997. LNCS, vol. 1335, pp. 242–257. Springer, Heidelberg (1997). https://doi.org/10.1007/BFb0024502

27. Kratochvíl, J., Proskurowski, A., Telle, J.A.: Covering regular graphs. J. Comb. Theory Ser. B **71**(1), 1–16 (1997)

28. Kratochvíl, J., Telle, J.A., Tesař, M.: Computational complexity of covering three-vertex multigraphs. Theor. Comput. Sci. **609**, 104–117 (2016)

29. Kwak, J.H., Nedela, R.: Graphs and their coverings. Lecture Notes Ser. **17**, 118 (2007)

30. Litovsky, I., Métivier, Y., Zielonka, W.: The power and the limitations of local computations on graphs. In: Mayr, E.W. (ed.) WG 1992. LNCS, vol. 657, pp. 333–345. Springer, Heidelberg (1993). https://doi.org/10.1007/3-540-56402-0_58

31. Malnič, A., Marušič, D., Potočnik, P.: Elementary abelian covers of graphs. J. Algebraic Combin. **20**(1), 71–97 (2004)

32. Malnič, A., Nedela, R., Škoviera, M.: Lifting graph automorphisms by voltage assignments. Eur. J. Comb. **21**(7), 927–947 (2000)

33. Mednykh, A.D., Nedela, R.: Harmonic Morphisms of Graphs: Part I: Graph Cov-erings. Vydavatelstvo Univerzity Mateja Bela v Banskej Bystrici, 1st edn. (2015)

34. Nedela, R., Škoviera, M.: Regular embeddings of canonical double coverings of graphs. J. Comb. Theory Ser. B **67**(2), 249–277 (1996)

Cutting Barnette Graphs Perfectly is Hard

Édouard Bonnet, Dibyayan Chakraborty$^{(\boxtimes)}$, and Julien Duron

Univ Lyon, CNRS, ENS de Lyon, Université Claude Bernard Lyon 1,
LIP UMR5668, Lyon, France
{edouard.bonnet,dibyayan.chakraborty,julien.duron}@ens-lyon.fr

Abstract. A *perfect matching cut* is a perfect matching that is also a cutset, or equivalently a perfect matching containing an even number of edges on every cycle. The corresponding algorithmic problem, PERFECT MATCHING CUT, is known to be NP-complete in subcubic bipartite graphs [Le & Telle, TCS '22] but its complexity was open in planar graphs and in cubic graphs. We settle both questions at once by showing that PERFECT MATCHING CUT is NP-complete in 3-connected cubic bipartite planar graphs or *Barnette graphs*. Prior to our work, among problems whose input is solely an undirected graph, only DISTANCE-2 4-COLORING was known NP-complete in Barnette graphs. Notably, HAMILTONIAN CYCLE would only join this private club if Barnette's conjecture were refuted.

1 Introduction

Deciding if an input graph admits a perfect matching, i.e., a subset of its edges touching each of its vertices exactly once, notoriously is a tractable task. There is indeed a vast literature, starting arguably in 1947 with Tutte's characterization via determinants [38], of polynomial-time algorithms deciding PERFECT MATCHING (or returning actual solutions) and its optimization generalization MAXIMUM MATCHING.

In this paper, we are interested in another containment of a spanning set of disjoint edges –perfect matching– than as a subgraph. As containing such a set of edges as an induced subgraph is a trivial property[1] (only shared by graphs that are themselves disjoint unions of edges), the meaningful other containment is as a *semi-induced subgraph*. By that we mean that we look for a bipartition of the vertex set or *cut* such that the edges of the perfect matching are "induced" in the corresponding cutset (i.e., the edges going from one side of the bipartition to the other), while we do not set any requirement on the presence or absence of edges within each side of the bipartition.

---

[1] Note however that the induced variant of MAXIMUM MATCHING is an interesting problem that happens to be NP-complete [36].

D. Paulusma and B. Ries (Eds.): WG 2023, LNCS 14093, pp. 116–129, 2023.
https://doi.org/10.1007/978-3-031-43380-1_9

This problem was in fact introduced as the PERFECT MATCHING CUT (PMC for short) problem[2] by Heggernes and Telle who show that it is NP-complete [18]. As the name PERFECT MATCHING CUT suggests, we indeed look for a perfect matching that is also a cutset. Le and Telle further show that PMC remains NP-complete in subcubic bipartite graphs of arbitrarily large girth, whereas it is polynomial-time solvable in a superclass of chordal graphs, and in graphs without a particular subdivided claw as an induced subgraph [26]. An in-depth study of the complexity of PMC when forbidding a single induced subgraph or a finite set of subgraphs has been carried out [14, 28].

We look at Le and Telle's hardness constructions and wonder what other properties could make PMC tractable (aside from chordality, and forbidding a finite list of subgraphs or induced subgraphs). A simpler reduction for bipartite graphs is first presented. Let us briefly sketch their reduction (without thinking about its correctness) from MONOTONE NOT-ALL-EQUAL 3-SAT, where given a negation-free 3-CNF formula, one seeks a truth assignment that sets in each clause a variable to true and a variable to false. Every variable is represented by an edge, and each 3-clause, by a (3-dimensional) cube with three anchor points at three pairwise non-adjacent vertices of the cube. One endpoint of the variable gadget is linked to the anchor points corresponding to this variable among the clause gadgets. Note that this construction creates three vertices of degree 4 in each clause gadget, and vertices of possibly large degree in the variable gadgets. Le and Telle then reduce the maximum degree to at most 3, by appropriately subdividing the cubes and tweaking the anchor points, and replacing the variable gadgets by cycles.

Notably the edge subdivision of the clause gadgets creates degree 2-vertices, which are not easy to "pad" with a third neighbor (even more so while keeping the construction bipartite). And indeed, prior to our work, the complexity of PMC in cubic graphs was open. Let us observe that on cubic graphs, the problem becomes equivalent to partitioning the vertex set into two sets each inducing a disjoint union of (independent) cycles. The close relative, MATCHING CUT, where one looks for a mere matching that is also a cutset, while NP-complete in general [5], is polynomial-time solvable in *subcubic* graphs [2,33]. The complexity of MATCHING CUT has further been examined in subclasses of planar graphs [2,35], when forbidding some (induced) subgraphs [13,14,28,29], on graphs of bounded diameter [25,29], and on graphs of large minimum degree [4]. MATCHING CUT has also been investigated with respect to parameterized complexity, exact exponential algorithms [21,24], and enumeration [16].

It was also open if PMC is tractable on planar graphs. Note that Bouquet and Picouleau [3] show that a related problem, DISCONNECTED PERFECT MATCHING, where one looks for a perfect matching that contains a cutset, is NP-

---

[2] The authors consider the framework of (k, σ, ρ)-partition problem, where k is a positive integer, and σ, ρ are sets of non-negative integers, and one looks for a vertex-partition into k parts such that each vertex of each part has a number of neighbors in its own part in σ, and a number of other neighbors in ρ; hence, PMC is then the $(2, \mathrm{N}, \{1\})$-partition problem.

complete on planar graphs of maximum degree 4, on planar graphs of girth 5, and on 5-regular bipartite graphs [3]. They incidentally call this related problem PERFECT MATCHING CUT but subsequent references [14,26] use the name DISCONNECTED PERFECT MATCHING to avoid confusion. We will observe that PMC is equivalent to asking for a perfect matching containing an even number of edges from every cycle of the input graph (See Lemma 1 and 2). The sum of even numbers being even, it is in fact sufficient that the perfect matching contains an even number of edges from every element of a cycle basis. There is a canonical cycle basis for planar graphs: the bounded faces. This gives rise to the following neat reformulation of PMC in planar graphs: is there a perfect matching containing an even number of edges along each face?

While MATCHING CUT is known to be NP-complete on planar graphs [2, 35], it could have gone differently for PMC for the following "reasons." NOT-ALL-EQUAL 3-SAT, which appears as the *right* starting point to reduce to PMC, is tractable on planar instances [32]. In planar graphs, perfect matchings are *simpler* than arbitrary matchings in that they alone [39] can be counted efficiently [20,37]. Let us finally observe that MAXIMUM CUT can be solved in polynomial time in planar graphs [17].

In fact, we show that the reformulations for cubic and planar graphs cannot help algorithmically, by simultaneously settling the complexity of PMC in cubic and in planar graphs, with the following stronger statement.

Theorem 1. PERFECT MATCHING CUT *is NP-hard in 3-connected cubic bipartite planar graphs.*

Not very many problems are known to be NP-complete in cubic bipartite planar graphs. Of the seven problems defined on mere undirected graphs from Karp's list of 21 NP-complete problems [19], only HAMILTONIAN PATH is known to remain NP-complete in this class, while the other six problems admit a polynomial-time algorithm. Restricting ourselves to problems where the input is purely an undirected graph[3], besides HAMILTONIAN PATH/CYCLE [1,34], MINIMUM INDEPENDENT DOMINATING SET was also shown NP-complete in cubic bipartite planar graphs [27], as well as P_3-PACKING [23] (hence, an equivalent problem phrased in terms of disjoint dominating and 2-dominating sets [31]), and DISTANCE-2 4-COLORING [11]. To our knowledge, MINIMUM DOMINATING SET is only known NP-complete in *subcubic* bipartite planar graphs [15,22].

It is interesting to note that the reductions for HAMILTONIAN PATH, HAMILTONIAN CYCLE, MINIMUM INDEPENDENT DOMINATING SET, and P_3-PACKING all produce cubic bipartite planar graphs that are *not* 3-connected. Notoriously, lifting the NP-hardness of HAMILTONIAN CYCLE to the 3-connected case would require to disprove Barnette's conjecture[4] (and that would be indeed suffi-

---

[3] Among problems with edge orientations, vertex or edge weights, or prescribed subsets of vertices or edges, the list is significantly longer, and also includes MINIMUM WEIGHTED EDGE COLORING [7], LIST EDGE COLORING and PRECOLORING EXTENSION [30], k-IN-A-TREE [8], etc.

[4] Which precisely states that every polyhedral (that is, 3-connected planar) cubic bipartite graphs admits a hamiltonian cycle.

cient [12]). Note that hamiltonicity in cubic graphs is equivalent to the existence of a perfect matching that is *not* an edge cut (i.e., whose removal is not disconnecting the graph). We wonder whether there is something inherently simpler about *3-connected* cubic bipartite planar graphs, which would go beyond hamiltonicity (assuming that Barnette's conjecture is true).

Let us call *Barnette* a 3-connected cubic bipartite planar graph. It appears that, prior to our work, DISTANCE-2 4-COLORING was the only *vanilla* graph problem shown NP-complete in Barnette graphs [11]. Arguing that DISTANCE-2 4-COLORING is a problem on *squares* of Barnette graphs more than it is on Barnette graphs, a case can be made for PERFECT MATCHING CUT to be the first natural problem proven NP-complete in Barnette graphs.

Outline of the Proof. We reduce the NP-complete problem MONOTONE NOT-ALL-EQUAL 3-SAT with exactly 4 occurrences of each variable [6] to PMC. Observe that flipping the value of every variable of a satisfying assignment results in another satisfying assignment. We thus see a solution to MONOTONE NOT-ALL-EQUAL 3-SAT simply as a bipartition of the set of variables.

As we already mentioned, NOT-ALL-EQUAL 3-SAT restricted to planar instances (i.e., where the variable-clause incidence graph is planar) is in P. We thus have to design *crossing* gadgets in addition to *variable* and *clause* gadgets. Naturally our gadgets are bipartite graphs with vertices of degree 3, except for some special *anchors*, vertices of degree 2 with one incident edge leaving the gadget.

The variable gadget is designed so that there is a unique way a perfect matching cut can intersect it. It might seem odd that no "binary choice" happens within it. The role of this gadget is only to serve as a baseline for which side of the bipartition the variable lands in, while the "truth assignments" take place in the clause gadgets. (Actually the same happens with Le and Telle's first reduction [26], where the variable gadget is a single edge, which has to be in any solution).

Our variable gadget consists of 36 vertices, including 8 anchor points; see Fig. 1. (We will later explain why we have 8 anchor points and not simply 4, that is, one for each occurrence of the variable.) Note that in all the figures, we adopt the following convention:

- black edges cannot (or can no longer) be part of a perfect matching cut,
- red edges are in every perfect matching cut,
- each blue edge e is such that at least one perfect matching cut within its gadget includes e, and at least one excludes e, and
- brown edges are blue edges that were indeed chosen in the solution.

Let us recall that PMC consists of finding a perfect matching containing an even number of edges from each cycle. Thus we look for a perfect matching M such that every path (or walk) between v and w contains a number of edges of M whose parity only depends on v and w. If this parity is even v and w are on the *same side*, and if it is odd, v and w are on *opposite sides*. The 8 anchor points of each variable gadget are forced on the same side. This is the *side of the variable*.

At the core of the clause gadget is a subdivided cube of blue edges; see Fig. 2. There are three vertices (u_1, u_8, u_{14} on the picture) of the subdivided cube that are forced on the same side as the corresponding three variables. Three perfect matching cuts are available in the clause gadget, each separating (i.e., putting on opposite sides) a different vertex of $\{u_1, u_8, u_{14}\}$ from the other two. Note that this is exactly the semantics of a not-all-equal 3-clause. We in fact need two copies of the subdivided cube, partly to increase the degree of some subdivided vertices, partly for the same reason we duplicated the anchor vertices in the variable gadgets. (The latter will be explained when we present the crossing gadgets.) Increasing the degree of all the subdivided vertices complicate further the gadget and create two odd faces. Fortunately these two odd faces have a common neighboring even face. We can thus "fix" the parity of the two odd faces by plugging the sub-gadget D_j in the even face. We eventually need a total of 112 vertices, including 6 anchor points.

Let us now describe the crossing gadgets. Basically we want to replace every intersection point of two edges by a 4-vertex cycle. This indeed propagates black edges (those that cannot be in any solution). The issue is that going through such a crossing gadget flips one's side. As we cannot guarantee that a variable "wire" has the same parity of intersection points towards each clause gadget it is linked to, we duplicate these wires. At a previous intersection point, we now have two parallel wires crossing two other parallel wires, making four crossings. The gadget simply consists of four 4-vertex cycles; see Fig. 3. This explains why we have 8 anchor points (not 4) in each variable gadget, and 6 anchor points (not 3) in each clause gadget.

2   Preliminaries

For a graph G, we denote by $V(G)$ its set of vertices and by $E(G)$ its set of edges. For $U \subseteq V(G)$, the *subgraph of G induced by U*, denoted as $G[U]$, is the graph obtained from G by removing the vertices not in U. We shall use $E_G(U)$ (or $E(U)$ when G is clear) as a shorthand for $E(G[U])$. For $M \subset E(G)$, $G - M$ is the spanning subgraph of G obtained by removing the edges in M (while preserving their endpoints). We may use *k-cycle* as a short-hand for the k-vertex cycle.

Given two disjoint sets $X, Y \subseteq V(G)$ we denote by $E(X, Y)$ the set of edges between X and Y. A set $M \subseteq E(G)$ is a *cutset*[5] of G if there is a proper bipartition $X \uplus Y = V(G)$, called *cut*, such that $M = E(X, Y)$. Note that a cut fully determines a cutset, and among connected graphs a cutset fully determines a cut. When dealing with connected graphs, we may speak of *the* cut of a cutset. For $X \subseteq V(G)$ the set of *outgoing edges of X* is $E(X, V(G) \setminus X)$. For a cutset M of a connected graph G, and $u, v \in V(G)$, we say that u and v are on the

---

[5] We avoid using the term "edge cut" since, for some authors, an edge cut is, more generally, a subset of edges whose deletion increases the number of connected components.

same side (resp. on *opposite sides*) of M if u and v are on the same part (resp. on different parts) of the cut of M.

A *matching* (resp. *perfect matching*) of G is a set $M \subset E(G)$ such that each vertex of G is incident to at most (resp. exactly) one edge of M. A *perfect matching cut* is a perfect matching that is also a cutset. For $M \subseteq E(G)$ and $U \subseteq V(G)$, we say that M is a *perfect matching cut of* $G[U]$ if $M \cap E(U)$ is so.

Due to space constraints, the proof of statements marked with ($\star$) are deferred to the long version [10].

3   Proof of Theorem 1

Before we give our reduction, we start with a handful of useful lemmas and observations, which we will later need.

3.1   Preparatory Lemmas

Lemma 1 ($\star$). *Let G be a graph, and $M \subseteq E(G)$. Then M is a cutset if and only if for every cycle C of G, $|E(C) \cap M|$ is even.*

Lemma 2. *Let G be a plane graph, and $M \subseteq E(G)$. Then M is a cutset if and only if for any facial cycle C of G, $|E(C) \cap M|$ is even.*

Proof. The forward implication is a direct consequence of Lemma 1. The converse comes from the known fact that the bounded faces form a cycle basis; see for instance [9]. If H is a subgraph of G, let $\tilde{H}$ be the vector of $\mathbb{F}_2^{E(G)}$ with 1 entries at the positions corresponding to edges of H. Thus, for any cycle C of G, we have $\tilde{C} = \Sigma_{1 \leqslant i \leqslant k} \tilde{F}_i$ where F_i are facial cycles of G. And $|M \cap E(C)|$ has the same parity as $\Sigma_{1 \leqslant i \leqslant k} |M \cap E(F_i)|$, a sum of even numbers.   □

Lemma 3. *Let M be a perfect matching cut of a cubic graph G. Let C be an induced 4-vertex cycle of G. Then, exactly one of the following holds:*

(a) $E(C) \cap M = \emptyset$ and the four outgoing edges of $V(C)$ belong to M.
(b) $|E(C) \cap M| = 2$, the two edges of $E(C) \cap M$ are disjoint, and none of the outgoing edges of $V(C)$ belongs to M.

Proof. The number of edges of M within $E(C)$ is even by Lemma 2. Thus $|E(C) \cap M| \in \{0, 2\}$, as all four edges of $E(C)$ do not make a matching.

Suppose that $E(C) \cap M = \emptyset$. As M is a perfect matching, for every $v \in V(C)$ there is an edge in M incident to v and not in $E(C)$. As G is cubic, every outgoing edge of $V(C)$ is in M.

Suppose instead that $|E(C) \cap M| = 2$. As M is a matching, the two edges of $E(C) \cap M$ do not share an endpoint. It implies that all the four vertices of C are touched by these two edges. Thus no outgoing edge of $V(C)$ can be in M.   □

Corollary 1 ($\star$). *Let M be a perfect matching of a cubic graph G. Let C_1, C_2 two vertex-disjoint induced 4-vertex cycles of G such that there is an edge between $V(C_1)$ and $V(C_2)$. Then $E(C_1) \cap M \neq \emptyset$ if and only if $E(C_2) \cap M \neq \emptyset$.*

Lemma 4. *Let M be a perfect matching cut of a cubic graph G. If a 6-cycle has three outgoing edges in M, then all six outgoing edges are in M.*

Proof. Let C be a 6-cycle. Since M is a perfect matching cut, $|E(C) \cap M|$ is even. Hence, $|E(C) \cap M|$ is either 0 or 2. If $|E(C) \cap M| = 2$, four vertices of C are touched by $E(C) \cap M$, which rules out that three outgoing edges of $V(C)$ are in M. Thus $E(C) \cap M = \emptyset$ and, G being cubic, all outgoing edges of $V(C)$ are in M.   □

Lemma 5. *Let M a perfect matching cut of a cubic bipartite graph G. Suppose C is a 6-cycle $v_1 v_2 \ldots v_6$ of G, such that $v_2 v_3$, $v_3 v_4$, $v_5 v_6$ and $v_6 v_1$ are in some induced 4-cycles. Then $M \cap E(C) = \emptyset$.*

Proof. By applying Lemma 3 on the 4-cycle containing $v_2 v_3$, and the one containing $v_6 v_1$, it holds that $v_1 v_2 \in M \Leftrightarrow v_3 v_4 \in M \Leftrightarrow v_5 v_6 \in M$. Thus none of these three edges can be in M, because C would have an odd number of edges in M. Symmetrically, no edge among $v_2 v_3$, $v_4 v_5$ and $v_6 v_1$ can be in M. Thus no edge of C is in M.   □

Observation 1. *Let G be a graph and M be a perfect matching cut of G. Let u, v be two vertices of G. Then for any path P between u and v, $|E(P) \cap M|$ is even if and only if u and v are on the same side of M. Note that implies that for any paths P, Q from u to v, $|E(P) \cap M|$ and $|E(Q) \cap M|$ have same parity.*

3.2 Reduction

We will prove Theorem 1 by reduction from the NP-complete MONOTONE NOT-ALL-EQUAL 3SAT-E4 [6]. In MONOTONE NOT-ALL-EQUAL 3SAT-E4, the input is a 3-CNF formula where each variable occurs exactly four times, each clause contains exactly three distinct literals, and no clause contains a negated literal. Here we say that a truth assignment on the variables *satisfies* a clause C if at least one literal of C is true and at least least one literal of C is false. The objective is to decide whether there is a truth assignment that satisfies all clauses. We can safely assume (and we will) that the variable-clause incidence graph $\mathrm{inc}(I)$ of I has no cutvertex among its "variable" vertices; see long version.

Let I be an instance of MONOTONE NOT-ALL-EQUAL 3SAT-E4 with variables $x_1, x_2, \ldots, x_n$ and clauses $m = 4n/3$ clauses $C_1, C_2, \ldots, C_m$. We shall construct, in polynomial time, an equivalent PMC-instance $G(I)$ that is Barnette.

Our reduction consists of three steps. First we construct a cubic graph $H(I)$ by introducing *variable gadgets* and *clause gadgets*. Then we *draw* $H(I)$ on the plane, i.e., we map the vertices of $H(I)$ to a set of points on the plane, and the edges of $H(I)$ to a set of simple curves on the plane. We shall refer to this drawing as $\mathcal{R}$. Note that, this drawing may not be planar, i.e., two simple curves (or analogously the corresponding edges) might intersect at a point which is not their endpoints. Finally, we eliminate the crossing points by introducing *crossing gadgets*. (Recall that if the clause-variable graph of an MONOTONE NOT-ALL-EQUAL 3SAT-E4 instance is planar, then its satisfiability can be tested in polynomial time; hence, we do need crossing gadgets.) The resulting graph $G(I)$ is Barnette, and we shall prove that $G(I)$ has a perfect matching if and only if I is a positive instance of MONOTONE NOT-ALL-EQUAL 3SAT-E4. Below we formally describe the above steps.

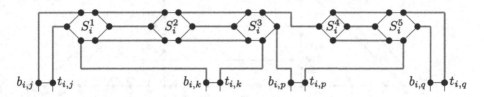

Fig. 1. Variable Gadget $\mathcal{X}_i$ corresponding to the variable x_i appearing in the clauses C_j, C_k, C_p, C_q with $j < k < p < q$.

1. For each variable x_i, let $\mathcal{X}_i$ denote a fresh copy of the graph shown in Fig. 1. Note that the variable x_i appears in exactly four clauses, say, C_j, C_k, C_p, C_q with $j < k < p < q$. The *variable gadget* $\mathcal{X}_i$ contains the special vertices $t_{i,j}$, $b_{i,j}, t_{i,k}, b_{i,k}, t_{i,p}, b_{i,p}, t_{i,q}, b_{i,q}$ as shown in the figure. We recall that red edges are those forced in any perfect matching cut, while black edges cannot be in any solution. An essential part of the proof will consist of justifying the edge colors in our figures.

   For each clause $C_j = (x_a, x_b, x_c)$ with $a < b < c$ let C_j denote a new copy of the graph shown in Fig. 2. The *clause gadget* C_j contains the special vertices $t'_{a,j}$, $b'_{a,j}, t'_{b,j}, b'_{b,j}, t'_{c,j}, b'_{c,j}$, as shown in the figure. Then for each variable x_i that appears in the clause C_j, introduce two new edges $E_{ij} = \{t_{i,j}t'_{i,j}, b_{i,j}b'_{i,j}\}$. Let $H(I)$ denote the graph defined as follows.

$$V(H(I)) = \bigcup_{i=1}^{n} V(\mathcal{X}_i) \cup \bigcup_{j=1}^{m} V(C_j)$$

$$E(H(I)) = \bigcup_{i=1}^{n} E(\mathcal{X}_i) \cup \bigcup_{j=1}^{m} E(C_j) \cup \bigcup_{x_i \in C_j} E_{ij}.$$

   We assign to each edge $e \in E_{i,j}$ its variable as $\mathrm{var}(e) = i$. Note that, for a variable gadget $\mathcal{X}_i$, there are exactly eight edges that have one endpoint in $V(\mathcal{X}_i)$ and the other endpoint not in $V(\mathcal{X}_i)$.

2. In the next step, we generate a drawing $\mathcal{R}$ of $H(I)$ on the plane according to the following procedure.
   (a) For each variable x_i, we embed $\mathcal{X}_i$ as a translate of the variable gadget of Fig. 1 into $[0, 1] \times [2i, 2i + 1]$.
   (b) For each clause C_j, we embed C_j as a translate of the clause gadget of Fig. 2 into $[2, 3] \times [2j, 2j + 1]$.

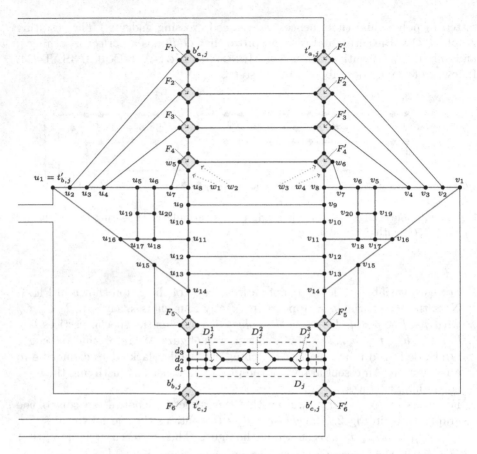

Fig. 2. Clause gadget $C_j = (x_a, x_b, x_c)$ with $a < b < c$. A red edge is selected in any perfect matching cut. A blue edge is selected in some perfect matching cut. A black edge is never selected in any perfect matching cut. (Color figure online)

(c) Two edges incident to vertices in the same variable gadget or same clause gadget do not intersect in $\mathcal{R}$. For two variables $x_i, x_{i'}$ and clauses $C_j, C_{j'}$ with $x_i \in C_j, x_{i'} \in C_{j'}$, exactly one of the following holds:

    i For each pair of edges $(e, e') \in E_{ij} \times E_{i'j'}$, e and e' intersect exactly once in $\mathcal{R}$. When this condition is satisfied, we call $(E_{ij}, E_{i'j'})$ a *crossing quadruple*. Moreover, we ensure that the interior of the subsegment of $e \in E_{ij}$ between its two intersection points with edges of $E_{i'j'}$ is not crossed by any edge;

    ii There is no pair of edges $(e, e') \in E_{ij} \times E_{i'j'}$ such that e and e' intersect in $\mathcal{R}$;

3. For each crossing quadruples $(E_{ij}, E_{i'j'})$ replace the four crossing points shown in Fig. 3a by the crossing gadget shown in Fig. 3b.

Let $G(I)$ denote the resulting graph. We shall need the following definitions.

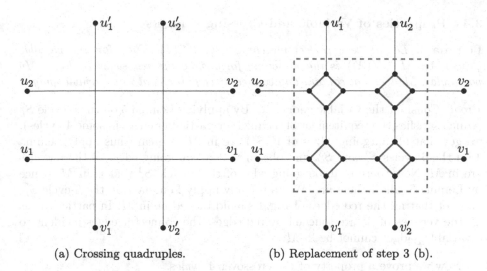

(a) Crossing quadruples.   (b) Replacement of step 3 (b).

Fig. 3. Replacement of a crossing by a crossing gadget.

Definition 1. *Any edge of $G(I)$ whose both endpoints are not contained withing the same gadget (variable, clause, or crossing) is a* connector edge. *Any endpoint of a connector edge is called a* connector vertex. *For a connector edge e incident to a crossing gadget, $var(e)$ is the index of the variable gadget it was originally going to. To each connector edge uv, we associate the variable $var(uv)$ to both u and v, denoted $var(u), var(v)$.*

Now we shall distinguish some 4-cycles of $G(I)$.

Definition 2. *An (induced) 4-cycle C of $G(I)$ is a* crossover *4-cycle if it belongs to some crossing gadget.*

Definition 3. *An induced 4-cycle C of $G(I)$ is* special *if C is identical to F_i or F_i' of some C_j.*

The special 4-cycles of a particular clause gadget C_j are highlighted in Fig. 2. In the next section, we show that $G(I)$ is indeed a 3-connected cubic bipartite planar graph.

3.3 $G(I)$ Is Barnette

Lemma 6 ($\star$). *The graph $G(I)$ is 3-connected.*

Lemma 7 ($\star$). *The graph $G(I)$ is Barnette.*

3.4 Properties of Variable and Crossing Gadgets

Lemma 8. *Let M be a perfect matching cut of $G(I)$. Then for any variable gadget $\mathcal{X}_i$, $M \cap V(\mathcal{X}_i)$ is the matching formed by the red edges in Fig. 1. In particular, M does not contain any connector edge incident to a variable gadget.*

Proof. Consider the variable gadget $\mathcal{X}_i$. By applying Lemma 5 on the 6-cycle S_i^2 (which satisfies the requirement of having four particular edges in some 4-cycles), we get that all outgoing edges of $V(S_i^2)$ are in M. We can thus apply Lemma 4 on the 6-cycles S_i^1 and S_i^3, and obtain that all outgoing edges of these cycles are in M. Now there is an outgoing edge of the 4-cycle S_i^4 that is in M, hence by Lemma 3, all of them are. We can finally apply Lemma 4 on the 6-cycle S_i^5, and get that all the red edges of Fig. 1 should indeed be in M. In particular, as all the vertices of $\mathcal{X}_i$ are touched by red edges, the connector edges incident to a variable gadget cannot be in M. $\qquad\square$

Now we prove a property of the crossover 4-cycles.

Lemma 9 ($\star$). *Let M be a perfect matching cut of $G(I)$ and F be a crossover 4-cycle. Then $|E(F)| = 2$.*

Corollary 2 ($\star$). *For any perfect matching M of $G(I)$, M contains no connector edges.*

3.5 Properties of Clause Gadgets

Observe that D_j is an induced subgraph of the variable gadget $\mathcal{C}_j$.

Lemma 10 ($\star$). *Any perfect matching cut of $G(I)$ contains the edges of D_j drawn in red in Fig. 2.*

Lemma 11 ($\star$). *Let M be a perfect matching cut of $G(I)$ and F be a special 4-cycle of $\mathcal{C}_j$. Then $|E(F) \cap M| = 2$, and no outgoing edge of $V(F)$ is in M.*

Lemma 12 ($\star$). *Let M be a perfect matching cut of $G(I)$ and $\mathcal{C}_j$ be a clause gadget. Let $U_j = \{u_1, \ldots, u_{20}\}$, and $V_j = \{v_1, \ldots, v_{20}\}$. Then no outgoing edge of U_j or of V_j is in M.*

See the definition of L_j^i (and the symmetric R_j^i in V_j) in Fig. 4.

Definition 4. *We say that a perfect matching cut M of $G(I)$ is of type i in $\mathcal{C}_j$ with $i \in \{1, 2, 3\}$, if $M \cap E(U_j \cup V_j) = L_j^i \cup R_j^i$.*

Lemma 13 ($\star$). *Let M be a perfect matching cut of $G(I)$ and $\mathcal{C}_j$ be a clause gadget. Then there exists exactly one integer $i \in \{1, 2, 3\}$ such that M is of type i in $\mathcal{C}_j$.*

As a direct consequence of Lemma 13, we get the following.

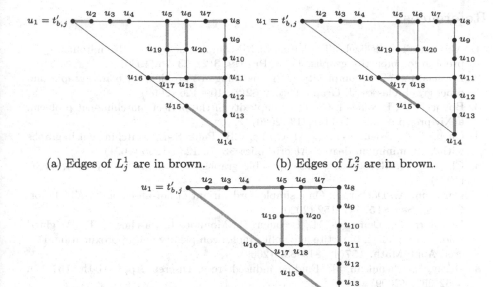

(a) Edges of L_j^1 are in brown.   (b) Edges of L_j^2 are in brown.

(c) Edges of L_j^3 are in brown.

Fig. 4. The three types of perfect matching cuts within a clause gadget.

Lemma 14. *Let M be a perfect matching cut of $G(I)$ and (A, B) be the cut of M. The vertices u_1, u_8, u_{14} of a clause gadget C_j cannot all be on the same side of M. More precisely:*

1. L_j^1 *sets u_1 to one side of M, and u_8, u_{14} to the other;*
2. L_j^2 *sets u_{14} to one side of M, and u_1, u_8 to the other;*
3. L_j^3 *sets u_8 to one side of M, and u_1, u_{14} to the other.*

3.6 Existence of Perfect Matching Cut Implies Satisfiability

Lemma 15 ($\star$). *If $G(I)$ has a perfect matching cut then I is a positive instance.*

3.7 Satisfiability Implies the Existence of a Perfect Matching Cut

Lemma 16 ($\star$). *If I has a satisfying assignment then $G(I)$ has a perfect matching cut.*

We finally get Theorem 1, due to Lemmas 15, 16, 7.

Acknowledgments. We are much indebted to Carl Feghali for introducing us to the topic of (perfect) matching cuts, and presenting us with open problems that led to the current paper. We also wish to thank him and Kristóf Huszár for helpful discussions on an early stage of the project.

References

1. Akiyama, T., Nishizeki, T., Saito, N.: NP-completeness of the Hamiltonian cycle problem for bipartite graphs. J. Inf. Process. **3**(2), 73–76 (1980)
2. Bonsma, P.S.: The complexity of the matching-cut problem for planar graphs and other graph classes. J. Graph Theory **62**(2), 109–126 (2009)
3. Bouquet, V., Picouleau, C.: The complexity of the perfect matching-cut problem. arXiv preprint arXiv:2011.03318 (2020)
4. Chen, C.-Y., Hsieh, S.-Y., Le, H.-O., Le, V.B., Peng, S.-L.: Matching cut in graphs with large minimum degree. Algorithmica **83**(5), 1238–1255 (2021)
5. Chvátal, V.: Recognizing decomposable graphs. J. Graph Theory **8**(1), 51–53 (1984)
6. Darmann, A., Döcker, J.: On a simple hard variant of not-all-equal 3-SAT. Theor. Comput. Sci. **815**, 147–152 (2020)
7. De Werra, D., Demange, M., Escoffier, B., Monnot, J., Paschos, V.T.: Weighted coloring on planar, bipartite and split graphs: complexity and approximation. Discret. Appl. Math. **157**(4), 819–832 (2009)
8. Derhy, N., Picouleau, C.: Finding induced trees. Discret. Appl. Math. **157**(17), 3552–3557 (2009)
9. Diestel, R.: Graph Theory. Graduate Texts in Mathematics, vol. 173, 4th edn. Springer, Heidelberg (2012)
10. Bonnet, É., Chakraborty, D., Duron, J.: Cutting Barnette graphs perfectly is hard. arXiv:2302.11667 (2023)
11. Feder, T., Hell, P., Subi, C.S.: Distance-two colourings of Barnette graphs. Eur. J. Comb. **91**, 103210 (2021)
12. Feder, T., Subi, C.S.: On Barnette's conjecture. Electron. Colloquium Comput. Complex. TR06-015 (2006)
13. Feghali, C.: A note on matching-cut in P_t-free graphs. Inf. Process. Lett. **179**, 106294 (2023)
14. Feghali, C., Lucke, F., Paulusma, D., Ries, B.: New hardness results for (perfect) matching cut and disconnected perfect matching. CoRR, abs/2212.12317 (2022)
15. Garey, M.R., Johnson, D.S.: Computers and Intractability: A Guide to the Theory of NP-Completeness. W. H. Freeman (1979)
16. Golovach, P.A., Komusiewicz, C., Kratsch, D., Le, V.B.: Refined notions of parameterized enumeration kernels with applications to matching cut enumeration. J. Comput. Syst. Sci. **123**, 76–102 (2022)
17. Hadlock, F.: Finding a maximum cut of a planar graph in polynomial time. SIAM J. Comput. **4**(3), 221–225 (1975)
18. Heggernes, P., Telle, J.A.: Partitioning graphs into generalized dominating sets. Nord. J. Comput. **5**(2), 128–142 (1998)
19. Karp, R.M.: Reducibility among combinatorial problems. In: Miller, R.E., Thatcher, J.W. (eds.) Proceedings of a Symposium on the Complexity of Computer Computations. The IBM Research Symposia Series, IBM Thomas J. Watson Research Center, Yorktown Heights, New York, USA, 20–22 March 1972, pp. 85–103. Plenum Press, New York (1972)
20. Kasteleyn, P.: Graph theory and crystal physics. Graph Theory Theor. Phys. 43–110 (1967)
21. Komusiewicz, C., Kratsch, D., Le, V.B.: Matching cut: kernelization, single-exponential time FPT, and exact exponential algorithms. Discret. Appl. Math **283**, 44–58 (2020)

22. Korobitsin, D.V.: On the complexity of domination number determination in monogenic classes of graphs (1992)
23. Kosowski, A., Małafiejski, M., Żyliński, P.: Parallel processing subsystems with redundancy in a distributed environment. In: Wyrzykowski, R., Dongarra, J., Meyer, N., Waśniewski, J. (eds.) PPAM 2005. LNCS, vol. 3911, pp. 1002–1009. Springer, Heidelberg (2006). https://doi.org/10.1007/11752578_121
24. Kratsch, D., Le, V.B.: Algorithms solving the matching cut problem. Theor. Comput. Sci. **609**, 328–335 (2016)
25. Le, H.-O., Le, V.B.: A complexity dichotomy for matching cut in (bipartite) graphs of fixed diameter. Theor. Comput. Sci. **770**, 69–78 (2019)
26. Le, V.B., Telle, J.A.: The perfect matching cut problem revisited. Theor. Comput. Sci. **931**, 117–130 (2022)
27. Loverov, Ya.A., Orlovich, Y.L.: NP-completeness of the independent dominating set problem in the class of cubic planar bipartite graphs. J. Appl. Ind. Math. **14**, 353–368 (2020)
28. Lucke, F., Paulusma, D., Ries, B.: Finding matching cuts in H-free graphs. In: Bae, S.W., Park, H. (eds.) 33rd International Symposium on Algorithms and Computation, ISAAC 2022. LIPIcs, Seoul, Korea, 19–21 December 2022, vol. 248, pp. 22:1–22:16. Schloss Dagstuhl - Leibniz-Zentrum für Informatik (2022)
29. Lucke, F., Paulusma, D., Ries, B.: On the complexity of matching cut for graphs of bounded radius and H-free graphs. Theor. Comput. Sci. **936**, 33–42 (2022)
30. Marx, D.: NP-completeness of list coloring and precoloring extension on the edges of planar graphs. J. Graph Theory **49**(4), 313–324 (2005)
31. Miotk, M., Topp, J., Żyliński, P.: Disjoint dominating and 2-dominating sets in graphs. Discret. Optim. **35**, 100553 (2020)
32. Moret, B.M.E.: Planar NAE3SAT is in P. SIGACT News **19**(2), 51–54 (1988)
33. Moshi, A.M.: Matching cutsets in graphs. J. Graph Theory **13**(5), 527–536 (1989)
34. Munaro, A.: On line graphs of subcubic triangle-free graphs. Discret. Math. **340**(6), 1210–1226 (2017)
35. Patrignani, M., Pizzonia, M.: The complexity of the matching-cut problem. In: Brandstädt, A., Le, V.B. (eds.) WG 2001. LNCS, vol. 2204, pp. 284–295. Springer, Heidelberg (2001). https://doi.org/10.1007/3-540-45477-2_26
36. Stockmeyer, L.J., Vazirani, V.V.: NP-completeness of some generalizations of the maximum matching problem. Inf. Process. Lett. **15**(1), 14–19 (1982)
37. Temperley, H.N.V., Fisher, M.E.: Dimer problem in statistical mechanics-an exact result. Philos. Mag. **6**(68), 1061–1063 (1961)
38. Tutte, W.T.: The factorization of linear graphs. J. Lond. Math. Soc. **1**(2), 107–111 (1947)
39. Vadhan, S.P.: The complexity of counting in sparse, regular, and planar graphs. SIAM J. Comput. **31**(2), 398–427 (2001)

Metric Dimension Parameterized by Treewidth in Chordal Graphs

Nicolas Bousquet, Quentin Deschamps$^{(\boxtimes)}$, and Aline Parreau

Univ. Lyon, Université Lyon 1, CNRS, LIRIS UMR 5205, 69621 Lyon, France
{nicolas.bousquet,quentin.deschamps,aline.parreau}@univ-lyon1.fr

Abstract. The metric dimension has been introduced independently by Harary, Melter [11] and Slater [15] in 1975 to identify vertices of a graph G using its distances to a subset of vertices of G. A *resolving set* X of a graph G is a subset of vertices such that, for every pair (u, v) of vertices of G, there is a vertex x in X such that the distance between x and u and the distance between x and v are distinct. The metric dimension of the graph is the minimum size of a resolving set. Computing the metric dimension of a graph is NP-hard even on split graphs and interval graphs. Bonnet and Purohit [2] proved that the metric dimension problem is W[1]-hard parameterized by treewidth. Li and Pilipczuk strengthened this result by showing that it is NP-hard for graphs of treewidth 24 in [14]. In this article, we prove that metric dimension is FPT parameterized by treewidth in chordal graphs.

1 Introduction

Determining the position of an agent on a network is a central problem. One way to determine its position is to place sensors on nodes of the network and the agents try to determine their positions using their positions with respect to these sensors. More formally, assume that agents know the topology of the graph. Can they, by simply looking at their position with respect to the sensors determine for sure their position in the network? Conversely, where do sensors have to be placed to ensure that any agent at any possible position can easily determine for sure its position? These questions received a considerable attention in the last decades and have been studied in combinatorics under different names such as metric dimension, identifying codes, locating dominating sets...

Let $G = (V, E)$ be a graph and s, u, v be three vertices of G. We say that s *resolves* the pair (u, v) if the distance between s and u is different from the distance between s and v. A *resolving set* of a graph $G = (V, E)$ is a subset S of vertices of G such that any vertex of G is identified by its distances to the vertices of the resolving set. In other words, S is a resolving set if for every pair (u, v) of vertices of G, there is a vertex s of S such that s resolves (u, v). The *metric dimension* of G, denoted by $\dim(G)$, is the smallest size of a resolving set of G.

This notion has been introduced in 1975 by Slater [15] for trees and by Harary and Melter [11] for graphs to simulate the moves of a sonar. The associated

D. Paulusma and B. Ries (Eds.): WG 2023, LNCS 14093, pp. 130–142, 2023.
https://doi.org/10.1007/978-3-031-43380-1_10

decision problem, called the METRIC DIMENSION problem, is defined as follows: given a graph G and an integer k, is the metric dimension of G is at most k?

The METRIC DIMENSION problem is NP-complete [9] even for restricted classes of graphs like planar graphs [4]. Epstein et al. [6] proved that this problem is NP-complete on split graphs, bipartite and co-bipartite graphs. The problem also is NP-complete on interval graphs [8] or sub-cubic graphs [12]. On the positive side, computing the metric dimension is linear on trees [11,15] and polynomial in outer-planar graphs [4].

Parameterized Algorithms. In this paper, we consider the METRIC DIMENSION problem from a parameterized point of view. We say a problem Π is *fixed parameter tractable* (FPT) for a parameter k if any instance of size n and parameter k can be decided in time $f(k) \cdot n^{O(1)}$. Two types of parameters received a considerable attention in the literature: the size of the solution and the "width" of the graph (for various widths, the most classical being the treewidth).

Hartung and Nichterlein proved in [12] that the METRIC DIMENSION problem is W[2]-hard parameterized by the size of the solution. Foucaud et al. proved that it is FPT parameterized by the size of the solution in interval graphs in [8]. This result was extended by Belmonte et al. who proved in [1] that METRIC DIMENSION is FPT parameterized by the size of the solution plus the tree-length of the graph. In particular, it implies that computing the metric dimension for chordal graph is FPT parameterized by the size of the solution.

METRIC DIMENSION is FPT parameterized by the modular width [1]. Using Courcelle's theorem, one can also remark that it is FPT parameterized by the treedepth of the graph as observed in [10]. METRIC DIMENSION has been proven W[1]-hard parameterized by the treewidth by Bonnet and Purohit in [2]. Li and Pilipczuk strengthened this result by showing that it is NP-complete for graphs of treewidth, and even pathwidth, 24 in [14]. While METRIC DIMENSION is polynomial on graphs of treewidth 1 (forests), its complexity is unknown for graphs of treewidth 2 is open (even if it is known to be polynomial for outerplanar graphs). Our main result is the following:

Theorem 1. METRIC DIMENSION *is FPT parameterized by treewidth on chordal graphs. That is,* METRIC DIMENSION *can be decided in time* $O(n^3 + n^2 \cdot f(\omega))$ *on chordal graphs of clique number* ω.

Recall that, on chordal graphs, the treewidth is equal to the size of a maximum clique minus one. Our proof is based on a dynamic programming algorithm. One of the main difficulty to compute the metric dimension is that a pair of vertices might be resolved by a vertex far from them in the graph. This non-locality implies that it is not simple to use classical algorithmic strategies like divide-and-conquer, induction or dynamic programming since a single edge or vertex modification somewhere in the graph might change the whole solution[1].

---

[1] The addition of a single edge in a graph might modify the metric dimension by $\Omega(n)$, see e.g. [7].

The first ingredient of our algorithm consists in proving that, given a chordal graph, if we are using a clique tree of a desirable form and make some simple assumptions on the shape of an optimal solution, we can ensure that resolving a pair of vertices close to a separator implies that we resolve all the pairs of vertices in the graph. Using this lemma, we build a dynamic programming algorithm that computes the minimum size of a resolving set containing a given vertex in FPT-time parameterized by treewdith.

The special type of clique tree used in the paper, inspired from [13], is presented in Sect. 2.1. We then give some properties of resolving sets in chordal graphs in Sect. 2.2. These properties will be needed to prove the correctness and the running time of the algorithm. Then, we present the definition of the extended problem in Sect. 3.1 and the rules of the dynamic programming in Sect. 3.2 where we also prove the correction of the algorithm. We end by an analysis of the complexity of the algorithm in Sect. 4.

Further Work. The function of the treewidth in our algorithm is probably not optimal and we did not try to optimize it to keep the algorithm as simple as possible. A first natural question is the existence of an algorithm running in time $2^\omega \cdot Poly(n)$ for chordal graphs.

We know that Theorem 1 cannot be extended to bounded treewidth graphs since METRIC DIMENSION is NP-hard on graphs of treewidth at most 24 [14]. One can nevertheless wonder if our proof technique can be adapted to design polynomial time algorithms for graphs of treewidth at most 2 on which the complexity status of METRIC DIMENSION is still open.

Our proof crucially relies on the fact that a separator X of a chordal graph is a clique and then the way a vertex in a component of $G \setminus X$ interacting with vertices in another component of $G \setminus X$ is simple. One can wonder if there is a tree decomposition in G where all the bags have diameter at most C, is it true that METRIC DIMENSION is FPT parameterized by the size of the bags plus C. Note that, since METRIC DIMENSION is NP-complete on chordal graphs, the problem is indeed hard parameterized by the diameter of the bags only.

2   Preliminaries

2.1   Nice Clique Trees

Unless otherwise stated, all graphs considered in this paper are undirected, simple, finite and connected. For standard terminology and notations on graphs, we refer the reader to [3]. Let us first define some notations we use throughout the article.

Let $G = (V, E)$ be a graph where V is the set of vertices of G and E the set of edges; we let $n = |V|$. For two vertices x and y in G, we denote by $d(x, y)$ the length of a shortest path between x and y and call it *distance between x and y*. For every $x \in V$ and $U \subseteq V$, the *distance between x and U*, denoted by $d(x, U)$, is the minimum distance between x and a vertex of U. Two vertices x and y are *adjacent* if $xy \in E$. A *clique* is a graph where all the pairs of vertices are adjacent. We denote by ω the size of a maximum clique. Let U be a set of vertices of G. We denote by $G \setminus U$ the subgraph of G induced by the set of vertices $V \setminus U$. We say that U is a *separator* of G if $G \setminus U$ is not connected. If two vertices x and y of $V \setminus U$ belong to two different connected components in $G \setminus U$, we say that U *separates* x and y. If a separator U induces a clique, we say that U is a *clique separator* of G.

Definition 1. *A* tree-decomposition *of a graph G is a pair (X, T) where T is a tree and $X = \{X_i | i \in V(T)\}$ is a collection of subsets (called bags) of $V(G)$ such that:*

- *$\bigcup_{i \in V(T)} X_i = V(G)$.*
- *For each edge $xy \in E(G), x, y \in X_i$ for some $i \in V(T)$.*
- *For each $x \in V(G)$, the set $\{i | x \in X_i\}$ induces a connected sub-tree of T.*

Let G be a graph and (X, T) a tree-decomposition of G. The *width* of the tree-decomposition (X, T) is the biggest size of a bag minus one. The *treewidth* of G is the smallest width of (X, T) amongst all the tree-decompositions (X, T) of G.

Chordal graphs are graphs with no induced cycle of length at least 4. A characterization given by Dirac in [5] ensures chordal graphs are graphs where minimal vertex separators are cliques. Chordal graphs admit tree-decompositions such that all the bags are cliques. We call such a tree-decomposition a *clique tree*.

Our dynamic programming algorithm is performed in a bottom-up way on a clique tree of the graph with more properties than the one given by Definition 1. These properties permit to simplify the analysis of the algorithm. We adapt the decomposition of [13, Lemma 13.1.2] to get this tree-decomposition.

Lemma 2. *Let $G = (V, E)$ be a chordal graph and r a vertex of G. There exists a clique tree (X, T) such that (i) T contains at most $7n$ nodes, (ii) T is rooted in a node that contains only the vertex r, (iii) T contains only four types of nodes, that are:*

- *Leaf nodes, $|X_i| = 1$ which have no child.*
- *Introduce nodes i which have exactly one child j, and that child satisfies $X_i = X_j \cup \{v\}$ for some vertex $v \in V(G) \setminus X_j$.*
- *Forget nodes i which have exactly one child j, and that child satisfies $X_i = X_j \setminus \{v\}$ for some vertex $v \in X_j$.*
- *Join node i which have exactly two children i_1 and i_2, and these children satisfy $X_i = X_{i_1} = X_{i_2}$.*

Moreover, such a clique tree can be found in linear time.

In the following, a clique tree with the properties of Lemma 2 will be called a *nice clique tree* and we will only consider nice clique trees (X, T) of chordal graphs G.

Given a rooted clique tree (T, X) of G, for any node i of T, we define the *subgraph of G rooted in X_i*, denoted by $T(X_i)$, as the subgraph induced by the subset of vertices of G contained in at least one of the bags of the sub-tree of T rooted in i (i.e. in the bag of i or one of its descendants).

2.2 Clique Separators and Resolving Sets

In this section, we give some technical lemmas that will permit to bound by $f(\omega)$ the amount of information we have to remember in the dynamic programming algorithm.

Lemma 3. *Let K be a clique separator of G and G_1 be a connected component of $G \setminus K$. Let G_{ext} be the subgraph of G induced by the vertices of $G_1 \cup K$ and $G_{int} = G \setminus G_{ext}$. Let $x_1, x_2 \in V(G_{int})$ be such that $|d(x_1, K) - d(x_2, K)| \geq 2$. Then, every vertex $s \in V(G_{ext})$ resolves the pair (x_1, x_2).*

Before proving Lemma 5, let us state a technical lemma.

Lemma 4. *Let G be a chordal and T be a nice clique tree of G. Let X, Y be two bags of T such that $X \cap Y = \emptyset$. Assume that there exist $x \in X, y \in Y$ such that $d(x, y) \geq 2$ and let z be a neighbour of x that appears in the bag the closest to Y in T amongst all the bags on the path between X and Y. Then z belongs to a shortest path between x and y.*

Lemma 5. *Let S be a subset of vertices of a chordal graph G. Let X, Y and Z be three bags of a nice tree-decomposition T of G such that Z is on the path P between X and Y in T. Denote by $P = X_1, \ldots Z \ldots X_p$ the bags of P with $X = X_1$ and $Y = X_p$. Let x be a vertex of X and y a vertex of Y with $d(x, Z) \geq 2$ and $d(y, Z) \geq 2$. Assume that any pair of vertices (u, v) with $u \in X_2 \cup \ldots \cup Z$, $v \in Z \cup \ldots \cup X_p$, $d(u, Z) < d(x, Z)$ and $d(v, Z) < d(y, Z)$ is resolved by S. Then the pair (x, y) is resolved by S.*

Proof. Let i_1 be such that $X_{i_1} \cap N[x] \neq \emptyset$ and for every $j > i_1$, $X_j \cap N[x] = \emptyset$ and i_2 be such that $X_{i_2} \cap N[y] \neq \emptyset$ and for $j < i_2$, $X_j \cap N[y] = \emptyset$. Let x' be the only neighbour of x in X_{i_1} and y' be the only neighbour of y in X_{i_2}. They are unique by definition of nice tree-decomposition. Note that $d(x, y) \geq 4$ since $d(x, Z) \geq 2$ and $d(y, Z) \geq 2$. So $N[x]$ is not adjacent to $N[y]$ and then $i_1 < i_2$. By Lemma 4, x' is on a shortest path between x and Z and y' is on a shortest path between y and Z. So $d(x', Z) < d(x, Z)$ and $d(y', Z) < d(y, Z)$. By hypothesis, there is a vertex $s \in S$ resolving the pair (x', y'). Let us prove that s resolves the pair (x, y).

If s belongs to $N[x]$ or to $N[y]$ then s resolves the pair (x, y) since $d(x, y) \geq 4$. So we can assume that $d(s, x) \geq 2$ and $d(s, y) \geq 2$. Let X_s be a bag of T containing s and X'_s be the closest bag to X_s on P between X and Y.

Case 1: $s \in X_{i_1}$ and $s \in X_{i_2}$. Then, $d(s, x') \leq 1$ and $d(s, y') \leq 1$. The vertex s resolves the pair (x', y') so $d(s, x') \neq d(s, y')$ so $s = x'$ or $s = y'$. Assume by symmetry that $s = x'$, then $d(s, x) = 1$ and $d(s, y) \geq 3$ because $d(x, y) \geq 4$. So s resolves the pair (x, y).

Case 2: s belongs to exactly one of X_{i_1} or X_{i_2}. By symmetry assume that $s \in X_{i_1}$. By Lemma 4, y' is on a shortest path between y and s. So $d(s, y) = d(s, y') + 1$. As s belongs to X_{i_1} then $d(x', s) \leq 1$ and $d(x, s) \leq 2$. As $d(y', s) \neq d(x', s)$ we have $d(y', s) \geq 2$, so $d(s, y) \geq 3$. Thus s resolves the pair (x, y).

Case 3: $s \notin X_{i_1}$ and $s \notin X_{i_2}$. First, we consider the case where X'_s is between X_{i_1} and X_{i_2}. Then, $d(s,x) = d(s,x') + 1$ and $d(s,y) = d(s,y') + 1$ by Lemma 4 as X_{i_1} separates y and s and X_{i_2} separates x and s. Thus, s resolves the pair (x,y).

By symmetry, we can now assume that X'_s is between X and X_{i_1}. Since $i_1 < i_2$, X_{i_2} separates s and y. So $d(s,y) = d(s,y') + 1$ by Lemma 4. To conclude we prove that $d(s,x') < d(s,y')$. Let Q be a shortest path between s and y'. The bag X_{i_1} separates s and y' so $Q \cap X_{i_1} \neq \emptyset$. Let $y_1 \in Q \cap X_{i_1}$. By definition of Q, $d(s,y') = d(s,y_1) + d(y_1,y')$. Since $y_1, x' \in X_{i_1}$ and X_{i_1} is a clique, we have that $y_1 \in N[x']$ and so, $y_1 \neq y'$. So $d(y_1,y') \neq 0$. We also have $d(s,x') \leq d(s,y_1) + 1$ because y_1 is a neighbour of x'. As $d(s,x') \neq d(s,y')$, this ensures $d(s,x') < d(s,y')$. So s resolves the pair (x,y) because $d(s,x) \leq d(s,x') + 1 < d(s,y') + 1 = d(s,y)$. $\qquad\square$

The following lemma is essentially rephrasing Lemma 5 to get the result on a set of vertices.

Lemma 6. *Let G be a chordal graph and S be a subset of vertices of G. Let T be a nice clique tree of G. Let X be a bag of T and let $T_1 = (X_1, E_1)$ and $T_2 = (X_2, E_2)$ be two connected components of $T \setminus X$. Assume that any pair of vertices (u,v) of $(X_1 \cup X) \times (X_2 \cup X)$ with $d(u,X) \leq 2$ and $d(v,X) \leq 2$ is resolved by S. Then any pair of vertices (u,v) of (X_1, X_2) with $|d(u,X) - d(v,X)| \leq 1$ is resolved by S.*

3   Algorithm Description

In this section, we fix a vertex v of a chordal graph G and consider a nice clique tree (T,X) rooted in v which exists by Lemma 2. We present an algorithm computing the smallest size of a resolving set of G containing v.

3.1   Extension of the Problem

Our dynamic programming algorithm computes the solution of a generalization of metric dimension which is easier to manipulate when we combine solutions. In this new problem, we will represent some vertices by vectors of distances. We define notations to edit vectors.

Definition 7. *Given a vector $\mathbf{r}$, the notation $\mathbf{r}_i$ refers to the i-th coordinate of $\mathbf{r}$.*

- *Let $\mathbf{r} = (r_1, \ldots, r_k) \in \mathbb{N}^k$ be a vector of size k and $m \in \mathbb{N}$. The vector $\mathbf{r}' = \mathbf{r}|\mathbf{m}$ is the vector of size $k+1$ with $r'_i = r_i$ for $1 \leq i \leq k$ and $r'_{k+1} = m$.*
- *Let $\mathbf{r} = (r_1, \ldots, r_k) \in \mathbb{N}^k$ be a vector of size k. The vector $\mathbf{r}^-$ is the vector of size $k-1$ with $r_i^- = r_i$ for $1 \leq i \leq k-1$.*

Definition 8. *Let i be a node of T and let $X_i = \{v_1, \ldots, v_k\}$ be the bag of i. For a vertex x of G, the distance vector $\mathbf{d}_{\mathbf{X}_i}(\mathbf{x})$ of x to X_i is the vector of size k such that, for $1 \leq j \leq k$, $\mathbf{d}_{\mathbf{X}_i}(\mathbf{x})_j = d(x, v_j)$. We define the set $d_{\leq 2}(X_i)$ as the set of distance vectors of the vertices of $T(X_i)$ at distance at most 2 of X_i in G (i.e. one of the coordinate is at most 2).*

Definition 9. *Let G be a graph and $K = \{v_1, \ldots, v_k\}$ be a clique of G. Let x be a vertex of G. The trace of x on K, denoted by $\mathbf{Tr_K}(x)$, is the vector $\mathbf{r}$ of $\{0,1\}^k \setminus \{1, \ldots, 1\}$ such that for every $1 \leq i \leq k$, $d(x, v_i) = a + \mathbf{r}_i$ where $a = d(x, K)$.*

Let S be a subset of vertices of G. The trace $Tr_K(S)$ of S in K is the set of vectors $\{\mathbf{Tr_K}(x), x \in S\}$.

The trace is well-defined because for a vertex x and a clique K, the distance between x and a vertex of K is either $d(x, K)$ or $d(x, K) + 1$.

Definition 10. *Let $\mathbf{r_1}, \mathbf{r_2}$ and $\mathbf{r_3}$ be three vectors of same size k. We say that $\mathbf{r_3}$ resolves the pair $(\mathbf{r_1}, \mathbf{r_2})$ if*

$$\min_{1 \leq i \leq k} (\mathbf{r_1} + \mathbf{r_3})_i \neq \min_{1 \leq i \leq k} (\mathbf{r_2} + \mathbf{r_3})_i.$$

Lemma 11. *Let K be a clique separator of G and G_1 be a connected component of $G \setminus K$. Let (x, y) be a pair of vertices of $G \setminus G_1$ and let $\mathbf{r}$ be a vector of size $|K|$. If $\mathbf{r}$ resolves the pair $(\mathbf{d_K}(\mathbf{x}), \mathbf{d_K}(\mathbf{y}))$, then any vertex $s \in V(G_1)$ with $\mathbf{Tr_K}(s) = \mathbf{r}$ resolves the pair (x, y).*

Proof. Let s be a vertex of G_1 such that $\mathbf{Tr_K}(s) = \mathbf{r}$. The clique K separates s and x (resp. y) so $d(x, s) = \min_{1 \leq i \leq |K|}(\mathbf{d_K}(\mathbf{x}) + \mathbf{Tr_K}(s))_i + d(K, s)$ (resp. $d(y, s) = \min_{1 \leq i \leq |K|}(\mathbf{d_K}(\mathbf{y}) + \mathbf{Tr_K}(s))_i + d(K, s))$. The vector $\mathbf{r}$ resolves the pair $(\mathbf{d_K}(\mathbf{x}), \mathbf{d_K}(\mathbf{y}))$. So $d(x, s) \neq d(y, s)$ and s resolves the pair (x, y). □

Definition 12. *Let K be a clique separator of G and G_1, G_2 be two (non necessarily distinct) connected components of $G \setminus K$. Let M be a set of vectors and let $x \in V(G_1) \cup K$ and $y \in V(G_2) \cup K$. If a vector $\mathbf{r}$ resolves the pair $(\mathbf{d_K}(\mathbf{x}), \mathbf{d_K}(\mathbf{y}))$, we say that $\mathbf{r}$ resolves the pair (x, y). We say that the pair of vertices (x, y) is resolved by M if there exists a vector $\mathbf{r} \in M$ that resolves the pair (x, y).*

We can now define the generalised problem our dynamic programming algorithm actually solves. We call it the EXTENDED METRIC DIMENSION problem (EMD for short). We first define the instances of this problem.

Definition 13. *Let i be a node of T. An instance for a node i of the EMD problem is a 5-uplet $I = (X_i, S_I, D_{int}(I), D_{ext}(I), D_{pair}(I))$ composed of the bag X_i of i, a subset S_I of X_i and three sets of vectors satisfying*

- *$D_{int}(I) \subseteq \{0,1\}^{|X_i|}$ and $D_{ext}(I) \subseteq \{0,1\}^{|X_i|}$,*
- *$D_{pair}(I) \subseteq \{0,1,2,3\}^{|X_i|} \times \{0,1,2,3\}^{|X_i|}$,*

- $D_{ext}(I) \neq \emptyset$ or $S_I \neq \emptyset$,
- For each pair of vectors $(\mathbf{r_1}, \mathbf{r_2}) \in D_{pair}(I)$, there exist two vertices $x \in T(X_i)$ with $\mathbf{d_{X_i}}(\mathbf{x}) = \mathbf{r_1}$ and $d(x, X_i) \leq 2$ and $y \notin T(X_i)$ with $\mathbf{d_{X_i}}(\mathbf{y}) = \mathbf{r_2}$ and $d(y, X_i) \leq 2$.

Definition 14. A set $S \subseteq T(X_i)$ is a solution for an instance I of the EMD problem if

- **(S1)** Every pair of vertices of $T(X_i)$ is either resolved by a vertex in S or resolved by a vector of $D_{ext}(I)$.
- **(S2)** For each vector $\mathbf{r} \in D_{int}(I)$ there exists a vertex $s \in S$ such that $\mathbf{Tr_{X_i}}(s) = \mathbf{r}$.
- **(S3)** For each pair of vector $(\mathbf{r_1}, \mathbf{r_2}) \in D_{pair}(I)$, for any vertex $x \in T(X_i)$ with $\mathbf{d_{X_i}}(\mathbf{x}) = \mathbf{r_1}$ and any vertex $y \notin T(X_i)$ with $\mathbf{d_{X_i}}(\mathbf{y}) = \mathbf{r_2}$, if $d(x, X_i) \leq 2$ and $d(y, X_i) \leq 2$ the pair (x, y) is resolved by S.
- **(S4)** $S \cap X_i = S_I$.

In the rest of the paper, for shortness, we will refer to an instance of the EMD problem only by an instance.

Definition 15. Let I be an instance. We denote by $\dim(I)$ the minimum size of a set $S \subseteq T(X_i)$ which is a solution of I. If such a set does not exist we define $\dim(I) = +\infty$. We call this value the extended metric dimension of I.

We now explain the meaning of each element of I. Firstly, a solution S must resolve any pair in $T(X_i)$, possibly with a vector of $D_{ext}(I)$ which represents a vertex of $V \setminus T(X_i)$ in the resolving set. Secondly, for all $\mathbf{r}$ in $D_{int}(I)$, we are forced to select a vertex in $T(X_i)$ whose trace is $\mathbf{r}$. This will be useful to combine solutions since it will be a vector of D_{ext} in other instances. The elements in $D_{pair}(I)$ will also be useful for combinations. In some sense $D_{pair}(I)$ is the additional gain of S compared to the main goal to resolve $T(X_i)$. The set S_I constrains the intersection between S and X_i by forcing a precise subset of X_i to be in S.

The following lemma is a consequence of Definition 14. It connects the definition of the extended metric dimension with the metric dimension.

Lemma 16. Let G be a graph, T be a nice tree-decomposition of G and r be the root of T. Let I_0 be the instance $(\{r\}, \{r\}, \emptyset, \emptyset, \emptyset)$, then $\dim(I_0)$ is the smallest size of a resolving set of G containing r.

To ensure that our algorithm works well, we will need to use Lemma 3 in some subgraphs of G. This is possible only if we know that the solution is not included in the subgraph. This corresponds to the condition $D_{ext}(I) \neq \emptyset$ or $S_I \neq \emptyset$ and this is why the algorithm computes the size of a resolving set containing the root of T.

3.2 Dynamic Programming

We explain how we can compute the extended metric dimension of an instance I given the extended metric dimension of the instances on the children of X_i in T. The proof is divided according to the different type of nodes.

Leaf Node. Computing the extended metric dimension of an instance for a leaf node can be done easily with the following lemma:

Lemma 17. *Let I be an instance for a leaf node i and v be the unique vertex of X_i. Then,*

$$
\dim(I) = \begin{cases} 0 & \text{if } S_I = \emptyset, \ D_{int}(I) = \emptyset \text{ and } D_{pair}(I) = \emptyset \\ 1 & \text{if } S_I = \{v\} \text{ and } D_{int}(I) \subseteq \{(\mathbf{0})\} \\ +\infty & \text{otherwise} \end{cases}
$$

Proof. Let I be an instance for i. If $S_I = \emptyset$, only the set $S = \emptyset$ can be a solution for I. This set is a solution only if $D_{int}(I) = \emptyset$ and $D_{pair}(I) = \emptyset$. If $S_I = \{v\}$, only the set $S = \{v\}$ can be a solution for I. This is a solution only if $D_{int}(I)$ is empty or only contains the vector $\mathbf{Tr}_{\mathbf{x}_i}(v)$. $\qquad\square$

In the rest of the section, we treat the three other types of nodes. For each type of nodes we will proceed as follows: define some conditions on the instances on children to be compatible with I, and prove an equality between the extended metric dimension on compatible children instances and the extended metric dimension of the instance of the node.

Join Node. Let I be an instance for a join node i and let i_1 and i_2 be the children of i.

Definition 18. *A pair of instances (I_1, I_2) for (i_1, i_2) is compatible with I if*

- *(J1) $S_{I_1} = S_{I_2} = S_I$,*
- *(J2) $D_{ext}(I_1) \subseteq D_{ext}(I) \cup D_{int}(I_2)$ and $D_{ext}(I_2) \subseteq D_{ext}(I) \cup D_{int}(I_1)$,*
- *(J3) $D_{int}(I) \subseteq D_{int}(I_1) \cup D_{int}(I_2)$,*
- *(J4) Let $C_1 = \{(\mathbf{r}, \mathbf{t}) \in D_{pair}(I_1) \text{ such that } \mathbf{r} \notin d_{\leq 2}(X_{i_1})\}$ and $C_2 = \{(\mathbf{r}, \mathbf{t}) \in D_{pair}(I_2) \text{ such that } \mathbf{r} \notin d_{\leq 2}(X_{i_2})\}$. Let $D_1 = \{(\mathbf{r}, \mathbf{t}) \in d_{\leq 2}(X_{i_1}) \times d_{\leq 2}(G \setminus X_{i_1}) \text{ such that there exists } \mathbf{u} \in D_{int}(I_2) \text{ resolving the pair } (\mathbf{r}, \mathbf{t})\}$ and $D_2 = \{(\mathbf{r}, \mathbf{t}) \in d_{\leq 2}(X_{i_2}) \times d_{\leq 2}(G \setminus X_{i_2}) \text{ such that there exists } \mathbf{u} \in D_{int}(I_1) \text{ resolving the pair } (\mathbf{r}, \mathbf{t})\}$ Then $D_{pair}(I) \subseteq (C_1 \cup D_1 \cup D_{pair}(I_1)) \cap (C_2 \cup D_2 \cup D_{pair}(I_2))$,*
- *(J5) For all $\mathbf{r_1} \in d_{\leq 2}(X_{i_1})$, for all $\mathbf{r_2} \in d_{\leq 2}(X_{i_2})$, $(\mathbf{r_1}, \mathbf{r_2}) \in D_{pair}(I_1)$ or $(\mathbf{r_2}, \mathbf{r_1}) \in D_{pair}(I_2)$ or there exists $\mathbf{t} \in D_{ext}(I)$ such that $\mathbf{t}$ resolves the pair $(\mathbf{r_1}, \mathbf{r_2})$.*

Condition **(J4)** represents how the pairs of vertices of $V(T(X_{i_1})) \times V(T(X_{i_2}))$ can be resolved. A pair $(\mathbf{r}, \mathbf{t})$ is in $(C_1 \cup D_1 \cup D_{pair}(I_1))$ if all the pairs of vertices (x, y) with $x \in V(T(X_{i_1}))$ and $y \in V(T(X_{i_2}))$ are resolved. If $(\mathbf{r}, \mathbf{t})$ is in C_1, no pair (x, y) with $x \in V(T(X_{i_1}))$ and $y \in V(T(X_{i_2}))$ exists, if $(\mathbf{r}, \mathbf{t})$ is in D_1 the pairs of vertices are resolved by a vertex outside of $V(T(X_{i_1}))$ and if $(\mathbf{r}, \mathbf{t})$ is in $D_{pair}(I_1)$ the pairs of vertices are resolved by a vertex of $V(T(X_{i_1}))$. So a pair $(\mathbf{r}, \mathbf{t})$ is resolved if the pair is in $(C_1 \cup D_1 \cup D_{pair}(I_1))$ and in $(C_2 \cup D_2 \cup D_{pair}(I_2))$.

Let $\mathcal{F}_J(I)$ be the set of pairs of instances compatible with I. We want to prove the following lemma:

Lemma 19. *Let I be an instance for a join node i. Then,*

$$\dim(I) = \min_{(I_1, I_2) \in \mathcal{F}_J(I)} (\dim(I_1) + \dim(I_2) - |S_I|).$$

We prove the equality by proving the two inequalities in the next lemmas.

Lemma 20. *Let (I_1, I_2) be a pair of instances for (i_1, i_2) compatible with I with finite values for $\dim(I_1)$ and $\dim(I_2)$. Let $S_1 \subseteq V(T(X_{i_1}))$ be a solution for I_1 and $S_2 \subseteq V(T(X_{i_2}))$ be a solution for I_2. Then $S = S_1 \cup S_2$ is a solution for I. In particular,*

$$\dim(I) \leq \min_{(I_1, I_2) \in \mathcal{F}_J(I)} (\dim(I_1) + \dim(I_2) - |S_I|).$$

Proof. Let us prove that the conditions of Definition 14 are satisfied.
(S1) Let (x, y) be a pair of vertices of $T(X_i)$. Assume first that $x \in V(T(X_{i_1}))$ and $y \in V(T(X_{i_1}))$. Either (x, y) is resolved by a vertex of S_1 and then by a vertex of S or (x, y) is resolved by a vector $\mathbf{r} \in D_{ext}(I_1)$. By condition **(J2)**, $\mathbf{r} \in D_{ext}(I)$ or $\mathbf{r} \in D_{int}(I_2)$. If $\mathbf{r} \in D_{ext}(I)$ then (x, y) is resolved by a vector of $D_{ext}(I_1)$. Otherwise, there exists a vertex $t \in S_2$ such that $\mathbf{Tr}_{X_{i_2}}(t) = \mathbf{r}$. So $t \in S$ and t resolves the pair (x, y). The case $x \in V(T(X_{i_2}))$ and $y \in V(T(X_{i_2}))$ is symmetric. So we can assume that $x \in V(T(X_{i_1}))$ and $y \in V(T(X_{i_2}))$. If $d(x, X_i) \leq 2$ and $d(y, X_i) \leq 2$, the condition **(J5)** ensures that the pair (x, y) is resolved by S or by a vector of $D_{ext}(I)$. Otherwise, either $|d(x, X_i) - d(y, X_i)| \leq 1$ and (x, y) is resolved by Lemma 6 or $|d(x, X_i) - d(y, X_i)| \geq 2$ and (x, y) is resolved by Lemma 3 because $D_{ext}(I) \neq \emptyset$ or $S_I \neq \emptyset$.
(S2) Let $\mathbf{r} \in D_{int}(I)$. By compatibility, the condition **(J3)** ensures that $\mathbf{r} \in D_{int}(I_1)$ or $\mathbf{r} \in D_{int}(I_2)$. As $S = S_1 \cup S_2$, S contains a vertex s such that $\mathbf{Tr}_{X_i}(s) = \mathbf{r}$.
(S3) Let $(\mathbf{r}, \mathbf{t}) \in D_{pair}(I)$ and (x, y) with $x \in V(T(X_i))$ such that $\mathbf{d}_{X_i}(\mathbf{x}) = \mathbf{r}$ and $y \notin T(X_i)$ such that $\mathbf{d}_{X_i}(\mathbf{y}) = \mathbf{t}$. Without loss of generality assume that $x \in V(T(X_{i_1}))$.
   By compatibility, $(\mathbf{r}, \mathbf{t}) \in (C_1 \cup D_1 \cup D_{pair}(I_1)) \cap (C_2 \cup D_2 \cup D_{pair}(I_2))$ so in $C_1 \cup D_1 \cup D_{pair}(I_1)$. If $(\mathbf{r}, \mathbf{t}) \in D_{pair}(I)_1$, then there exists $s \in S_1$ that resolves the pair (x, y) so the pair is resolved by S. If $(\mathbf{r}, \mathbf{t}) \in D_1$, there exists $\mathbf{u} \in D_{int}(I_2)$ such that $\mathbf{u}$ resolves the pair $(\mathbf{r}, \mathbf{t})$. By compatibility, there exists $s \in S_2$ such that $\mathbf{Tr}_{X_i}(s) = \mathbf{u}$. So s resolves the pair (x, y). And $(\mathbf{r}, \mathbf{t}) \notin C_1$ since x belongs to $T(X_{i_1})$ with vector distance $\mathbf{r}$.
(S4) is clear since $X_{i_1} = X_{i_2} = X_i$.
   Thus, $\dim(I) \leq \dim(I_1) + \dim(I_2) - |S_I|$ is true for any pair of compatible instances (I_1, I_2) so $\dim(I) \leq \min_{(I_1, I_2) \in \mathcal{F}_J(I)} (\dim(I_1) + \dim(I_2) - |S_I|)$. $\square$

Lemma 21. *Let I be an instance for a join node i and let i_1 and i_2 be the children of i. Then,*

$$\dim(I) \geq \min_{(I_1, I_2) \in \mathcal{F}_J(I)} (\dim(I_1) + \dim(I_2) - |S_I|).$$

Proof. If $\dim(I) = +\infty$ then the result indeed holds. So we can assume that $\dim(I)$ is finite. Let S be a solution for I of minimal size. Let $S_1 = S \cap T(X_{i_1})$ and $S_2 = S \cap T(X_{i_2})$. We define now two instances I_1 and I_2 for i_1 and i_2. Let $S_{I_1} = S_{I_2} = S_I$, $D_{int}(I_1) = Tr_{X_i}(S_1)$, $D_{int}(I_2) = Tr_{X_i}(S_2)$, $D_{ext}(I_1) = D_{ext}(I) \cup D_{int}(I_2)$ and $D_{ext}(I_2) = D_{ext}(I) \cup D_{int}(I_1)$. To build the sets $D_{pair}(I_1)$ and $D_{pair}(I_2)$ we make the following process that we explain for $D_{pair}(I_1)$. For all pairs of vectors $(\mathbf{r}, \mathbf{t})$ of $(d_{\leq 2}(X_{i_1}), d_{\leq 2}(G \setminus X_{i_1}))$, consider all the pairs of vertices (x, y) with $x \in V(T(X_{i_1}))$, $y \in V(G \setminus T(X_{i_1}))$, $\mathbf{r} \in d_{\leq 2}(X_i)$, $\mathbf{t} \in d_{\leq 2}(G \setminus X_{i_1}))$, $\mathbf{d}_{\mathbf{X}_i}(\mathbf{x}) = \mathbf{r}$ and $\mathbf{d}_{\mathbf{X}_i}(\mathbf{y}) = \mathbf{t}$. If all the pairs are resolved by vertices of S_1 (that is for each pair, there exists a vertex of S_1 that resolves the pair), then add $(\mathbf{r}, \mathbf{t})$ to $D_{pair}(I_1)$.

Checking that (I_1, I_2) is compatible with I, that S_1 is a solution of I_1, and that S_2 is a solution of I_2 is straightforward. It consists of checking conditions of respectively Definition 18 and Definition 14.

Finally we prove the announced inequality. Since S is a minimal solution for I, we have $\dim(I) = |S|$. The sets S_1 and S_2 are solutions for S_1 and S_2 so $\dim(I_1) \leq |S_1|$ and $\dim(I_2) \leq |S_2|$. Since $|S| = |S_1| + |S_2| - |S_I|$, $\dim(I) \geq \dim(I_1) + \dim(I_2) - |S_I|$, giving the result.   □

Lemma 19 is a direct consequence of Lemma 20 and Lemma 21.

Introduce Node. We now consider an instance I for an introduce node i. Let j be the child of i and $v \in V$ be such that $X_i = X_j \cup \{v\}$. Let $X_i = \{v_1, \ldots, v_k\}$ with $v = v_k$. The tree $T(X_i)$ contains one more vertex than its child. The definition of the compatibility is slightly different if we consider the same set as a solution (type 1) or if we add this vertex to the resolving set (type 2).

Definition 22. *An instance I_1 is compatible with I of type 1 (resp. 2) if*

- *(I1) $S_I = S_{I_1}$ (resp. $= S_{I_1} \cup \{v\}$).*
- *(I2) For all $\mathbf{r} \in D_{ext}(I)$, $\mathbf{r}^- \in D_{ext}(I_1)$ (resp. or $\mathbf{r} = (0, \ldots, 0)$).*
- *(I3) For all $\mathbf{r} \in D_{int}(I)$, $\mathbf{r}_k = 1$ and $\mathbf{r}^- \in D_{int}(I_1)$ (resp. or $\mathbf{r} = (1, \ldots, 1, 0)$).*
- *(I4) For all $(\mathbf{r}, \mathbf{t}) \in D_{pair}(I)$, $(\mathbf{r}^-, \mathbf{t}^-) \in D_{pair}(I_1)$.*
- *(I5) If I_1 is of type 1, for all $(\mathbf{r}, \mathbf{t})$ with $\mathbf{t} = (0, \ldots, 0)$, $(\mathbf{r}, \mathbf{t}) \in D_{pair}(I_1)$.*

Lemma 23. *Let I be an instance for an introduce node i. Let $\mathcal{F}_1(I)$ be the set of instances I_1 for i_1 compatible with I of type 1 and $\mathcal{F}_2(I)$ be the set of instances I_2 for i_1 compatible with I of type 2. Then,*

$$\dim(I) = \min \{ \min_{I_1 \in \mathcal{F}_1(I)} \{\dim(I_1)\}; \min_{I_2 \in \mathcal{F}_2(I)} \{\dim(I_2) + 1\}\}.$$

The proof of Lemma 23 consists in proving both inequalities similarly to Lemma 19. One inequality comes from the fact that we can get a solution of I from any compatible instance. The other consists in building a solution for a compatible instance from a minimal solution for I.

Forget Node. The construction for the forget nodes is similar to the one for introduce nodes. The main difference is that a vertex is removed from the bag so we have to keep the information about this vertex. The full construction leads to a similar equality between the extended metric dimension of an instance and the extended metric dimension of the compatible instances of its child.

3.3   Algorithm

Given as input a nice clique tree, the algorithm computes the extended metric dimension bottom up from the leaves. The algorithm computes the extended metric dimension for leaves using Lemma 17, for join nodes using Lemma 19, for introduce nodes using Lemma 23 and forget nodes using a similar lemma. The correction of the algorithm is straightforward by these lemmas.

We denote this algorithm by IMD in the following which takes as input a nice clique tree T and outputs the minimal size of a resolving set of G containing the root of T.

4   Proof of Theorem 1

Let us finally explain how we can compute the metric dimension of G. The following lemma is a consequence of Lemma 16.

Lemma 24. *The metric dimension of G is $\min_{v \in V(G)}\{IMD(T(v))\}$ where $T(v)$ is a nice clique tree of G rooted in v.*

So, n executions of the IMD algorithm with different inputs are enough to compute the metric dimension. Lemma 2 ensures that we can find for any vertex v of G a nice clique tree in linear time, the last part is to compute the complexity of the IMD algorithm.

Lemma 25. *The algorithm for IMD runs in time $O(n(T)^2 + n(T) \cdot f(\omega))$ where $n(T)$ is the number of vertices of the input tree T and $f = O(\omega^2 \cdot 2^{O(4^{2^\omega})})$ is a function that only depends on the size of a maximum clique ω.*

We now have all the ingredients to prove Theorem 1:

Proof. For each vertex v of G, one can compute a nice clique tree of size at most $7n$ according to Lemma 2. Given this clique tree, the IMD algorithm outputs the size of a smallest resolving set containing v by Lemma 16 in time $O(n(T)^2 + n(T) \cdot f(\omega))$ for a computable function f according to Corollary 25. Repeat this for all vertices of G permits to compute the metric dimension of G by Lemma 24 in time $O(n^3 + n^2 \cdot f(\omega))$. $\square$

References

1. Belmonte, R., Fomin, F.V., Golovach, P.A., Ramanujan, M.S.: Metric dimension of bounded tree-length graphs. CoRR abs/1602.02610 (2016)
2. Bonnet, É., Purohit, N.: Metric dimension parameterized by treewidth. Algorithmica **83**(8), 2606–2633 (2021)
3. Chartrand, G., Lesniak, L., Zhang, P.: Graphs and Digraphs, 6th edn. Chapman and Hall/CRC (2015)
4. Díaz, J., Pottonen, O., Serna, M., van Leeuwen, E.J.: On the complexity of metric dimension. In: Epstein, L., Ferragina, P. (eds.) Algorithms - ESA 2012 (2012)
5. Dirac, G.A.: On rigid circuit graphs. Abh. Math. Semin. Univ. Hambg. **25**, 71–76 (1961). https://doi.org/10.1007/BF02992776
6. Epstein, L., Levin, A., Woeginger, G.J.: The (weighted) metric dimension of graphs: hard and easy cases. Algorithmica **72**(4), 1130–1171 (2015)
7. Eroh, L., Feit, P., Kang, C.X., Yi, E.: The effect of vertex or edge deletion on the metric dimension of graphs. J. Comb **6**(4), 433–444 (2015)
8. Foucaud, F., Mertzios, G.B., Naserasr, R., Parreau, A., Valicov, P.: Identification, location-domination and metric dimension on interval and permutation graphs. II. Algorithms and complexity. Algorithmica **78**(3), 914–944 (2017)
9. Garey, J.: A guide to the theory of NP-completeness. J. Algorithms (1979)
10. Gima, T., Hanaka, T., Kiyomi, M., Kobayashi, Y., Otachi, Y.: Exploring the gap between treedepth and vertex cover through vertex integrity. Theor. Comput. Sci. **918**, 60–76 (2022)
11. Harary, F., Melter, R.A.: On the metric dimension of a graph. Ars Combinatoria **2**, 191–195 (1975)
12. Hartung, S., Nichterlein, A.: On the parameterized and approximation hardness of metric dimension. In: 2013 IEEE Conference on Computational Complexity, pp. 266–276. IEEE (2013)
13. Kloks, T.: Treewidth: Computations and Approximations. Springer, Heidelberg (1994)
14. Li, S., Pilipczuk, M.: Hardness of metric dimension in graphs of constant treewidth. Algorithmica **84**(11), 3110–3155 (2022)
15. Slater, P.J.: Leaves of trees. Congressus Numerantium **14** (1975)

Efficient Constructions
for the Győri-Lovász Theorem on Almost
Chordal Graphs

Katrin Casel$^{(\boxtimes)}$ ⓘ, Tobias Friedrich ⓘ, Davis Issac ⓘ,
Aikaterini Niklanovits$^{(\boxtimes)}$ ⓘ, and Ziena Zeif ⓘ

Hasso Plattner Institute, University of Potsdam, 14482 Potsdam, Germany
{Katrin.Casel,Tobias.Friedrich,Davis.Issac,Aikaterini.Niklanovits,
Ziena.Zeif}@hpi.de

Abstract. In the 1970s, Győri and Lovász showed that for a k-connected n-vertex graph, a given set of terminal vertices $t_1, \ldots, t_k$ and natural numbers $n_1, \ldots, n_k$ satisfying $\sum_{i=1}^{k} n_i = n$, a connected vertex partition $S_1, \ldots, S_k$ satisfying $t_i \in S_i$ and $|S_i| = n_i$ exists. However, polynomial time algorithms to actually compute such partitions are known so far only for $k \leq 4$. This motivates us to take a new approach and constrain this problem to particular graph classes instead of restricting the values of k. More precisely, we consider k-connected chordal graphs and a broader class of graphs related to them. For the first class, we give an algorithm with $\mathcal{O}(n^2)$ running time that solves the problem exactly, and for the second, an algorithm with $\mathcal{O}(n^4)$ running time that deviates on at most one vertex from the required vertex partition sizes.

Keywords: Győri-Lovász theorem · chordal graphs · HHD-free graphs

1 Introduction

Partitioning a graph into connected subgraphs is a fundamental task in graph algorithms. Such *connected* partitions occur as desirable structures in many application areas such as image processing [8], road network decomposition [9], and robotics [17].

From a theoretical point of view, the existence of a partition into connected components with certain properties also gives insights into the graph structure. In theory as well as in many applications, one is interested in a connected partition that has a given number of subgraphs of chosen respective sizes. With the simple example of a star-graph, it is observed that not every graph admits a connected partition for any such choice of subgraph sizes. More generally speaking, if there exists a small set of t vertices whose removal disconnects a graph (*separator*), then any connected partition into $k > t$ subgraphs has limited choice of subgraph sizes. Graphs that do not contain such a separator of size less than k are called *k-connected*.

© The Author(s), under exclusive license to Springer Nature Switzerland AG 2023
D. Paulusma and B. Ries (Eds.): WG 2023, LNCS 14093, pp. 143–156, 2023.
https://doi.org/10.1007/978-3-031-43380-1_11

On the other hand, Győri and Lovász independently showed that k-connectivity is not just necessary but also sufficient to enable a connected partitioning into k subgraphs of required sizes, formally stated by the following result.

Győri-Lovász Theorem ([4,7]). Let $k \geq 2$ be an integer, $G = (V, E)$ a k-connected graph, $t_1, \ldots, t_k \in V$ distinct vertices and $n_1, \ldots, n_k \in \mathbb{N}$ such that $\sum_{i=1}^{k} n_i = |V|$. Then G has disjoint connected subgraphs $G_1, \ldots G_k$ such that $|V(G_i)| = n_i$ and $t_i \in V(G_i)$ for all $i \in [k]$.

The caveat of this famous theorem is that the constructive proof of it yields an exponential time algorithm. Despite this result being known since 1976, to this day we only know polynomial constructions for restricted values of k. Specifically, in 1990 Suzuki et al. [15] provided such an algorithm for $k = 2$ and also for $k = 3$ [14]. Moreover in 1994 Wada et al. [16] also provided an extended result for $k = 3$. Nakano et al. [10] gave a linear time algorithm for the case where $k = 4$, G is planar and the given terminals are located on the same face of a plane embedding of G, while in 2016 Hoyer and Thomas [5] provided a polynomial time algorithm for the general case of $k = 4$. And so far, this is where the list ends, thus for $k \geq 5$ it remains open whether there even exists a polynomial time construction.

Towards a construction for general k, we consider restricting the class of k-connected graphs instead of the values of k. More precisely, we consider (generalizations of) *chordal* k-connected graphs. A graph is called *chordal*, if it does not contain an induced cycle of length more than three. The restriction to chordal graphs is known to often yield tractability for otherwise NP-hard problems, for example chromatic number, clique number, independence number, clique covering number and treewidth decomposition [13]. Apart from the interest chordal graphs have from a graph theoretic point of view, their structural properties have also been proven useful in biology when it comes to studying multidomain proteins and network motifs (see e.g. [11,12]).

Our Contribution. To the best of our knowledge, this paper is the first to pursue the route of restricting the Győri-Lovász Theorem to special graph classes in order to develop a polynomial construction for general values of k on a nontrivial subclass of k-connected graphs. We believe that in general considering the structure of the minimal separators of a graph is promising when it comes to developing efficient algorithms for the Győri-Lovász Theorem.

We give a constructive version of the Győri-Lovász Theorem for *chordal* k-connected graphs with a running time in $\mathcal{O}(|V|^2)$. Observe here that this construction works for all values of k. Then we show how this result can be generalized in two directions.

First, we generalize our result to the vertex weighted version of the Győri-Lovász Theorem (as proven independently by Chandran et al. [2], Chen et al. [3] and Hoyer [5]), specifically deriving the following theorem.

Theorem 1. *Let $k \geq 2$ be an integer, $G = (V, E, w)$ a vertex-weighted k-connected chordal graph with $w \colon V \to \mathbb{N}$ and $w_{max} := \max_{u \in V} w(u)$, $t_1, \ldots, t_k \in$*

V *distinct vertices, and* $w_1, \ldots, w_k \in \mathbb{N}$ *with* $w_i \geq w(t_i)$ *for all* $i \in [k]$ *and* $\sum_{i=1}^{k} w_i = w(V)$. *A partition* $S_1, \ldots, S_k$ *of* V, *such that* $G[S_i]$ *is connected,* $t_i \in S_i$ *and* $w_i - w_{max} < w(S_i) < w_i + w_{max}$, *for all* $i \in [k]$, *can be computed in time* $\mathcal{O}(|V|^2)$.

We further use this weighted version to derive an approximate version of the Győri-Lovász Theorem for a larger graph class. Specifically we define I_j^i to contain all graphs that occur from two distinct chordless C_j's that have at least i vertices in common. We focus on I_4^2-free combined with HH-free graphs. More specifically, we consider the subclass of k-connected graphs that contain no hole or house as an induced subgraph (see preliminaries for the definitions of structures such as hole, house etc.) and that does not contain two distinct induced C_4 that share more than one vertex. We call this class of graphs HHI_4^2-free . Note that HHI_4^2-free , apart from being a strict superclass of chordal graphs, is also a subclass of HHD-free graphs (that is house, hole, domino-free graphs), a graph class studied and being used in a similar manner as chordal graphs as it is also a class where the minimum fill-in set is proven to be polynomially time solvable [1] (see also [6] for NP-hard problems solved in polynomial time on HHD-free graphs). Taking advantage of the fact that given an HHI_4^2-free graph, the subgraph formed by its induced C_4 has a treelike structure, we are able to derive the following result.

Theorem 2. *Let* $k \geq 2$ *be an integer,* $G = (V, E, w)$ *a vertex-weighted* k-*connected* HHI_4^2-*free graph with* $w \colon V \to \mathbb{N}$ *and* $w_{max} := \max_{u \in V} w(u)$, $t_1, \ldots, t_k \in V$ *distinct vertices, and* $w_1, \ldots, w_k \in \mathbb{N}$ *with* $w_i \geq w(t_i)$ *for all* $i \in [k]$ *and* $\sum_{i=1}^{k} w_i = w(V)$. *A partition* $S_1, \ldots, S_k$ *of* V, *such that* $G[S_i]$ *is connected,* $t_i \in S_i$ *and* $w_i - 2w_{max} < w(S_i) < w_i + 2w_{max}$, *for all* $i \in [k]$, *can be computed in time* $\mathcal{O}(|V|^4)$.

Notice that the above theorem implies a polynomial time algorithm with an additive error of 1 for the unweighted case.

2   Preliminaries

All graphs mentioned in this paper are undirected, finite and simple. Given a graph G and a vertex $v \in V(G)$ we denote its *open neighborhood* by $N_G(v) := \{u \in V(G) \mid uv \in E(G)\}$ and by $N_G[v]$ its *closed neighborhood*, which is $N(v) \cup \{v\}$. Similarly we denote by $N_G(S) := \bigcup_{v \in S} N_G(v) \setminus S$ the open neighborhood of a vertex set $S \subseteq V(G)$ and by $N_G[S] := N_G(S) \cup S$ its closed neighborhood. We omit the subscript G when the graph we refer to is clear from the context. A vertex $v \in V(G)$ is *universal to a vertex set* $S \subset V(G)$ if $S \subseteq N(v)$. Let G be a graph and $S \subseteq V(G)$. The *induced subgraph* from S, denoted by $G[S]$, is the graph with vertex set S and all edges of $E(G)$ with both endpoints in S.

A graph G is *chordal* if any cycle of G of size at least 4 has a chord (i.e., an edge linking two non-consecutive vertices of the cycle). A vertex $v \in V(G)$ is called *simplicial* if $N[v]$ induces a clique. Based on the existence of simplicial vertices in chordal graphs, the following notion of vertex ordering was given. Given a graph G, an ordering of its vertices $(v_1, \ldots, v_n)$ is called *perfect elimination ordering* (p.e.o.) if v_i is simplicial in $G[\{v_i, v_{i+1}, \ldots, v_n\}]$ for all $i \in [n]$. Given such an ordering $\sigma : V(G) \to \{1, \ldots, n\}$ and a vertex $v \in V(G)$ we call $\sigma(v)$ the *p.e.o. value of* v. Rose et al. [13] proved that a p.e.o. of any chordal graph can be computed in linear time.

Let $e = \{u, v\}$ be an edge of G. We denote by G/e the graph G', that occurs from G by the contraction of e, that is, by removing u and v from G and replacing it by a new vertex z whose neighborhood is $(N(u) \cup N(v)) \setminus \{u, v\}$.

A graph G is *connected* if there exists a path between any pair of distinct vertices. Moreover, a graph is *k-connected* for some $k \in \mathbb{N}$ if after the removal of any set of at most $k - 1$ distinct vertices G remains connected. Given a graph G and a vertex set $S \subseteq V(G)$, we say that S is a *separator* of G if its removal disconnects G. We call S a *minimal separator* of G if the removal of any subset $S' \subseteq V(G)$ with $|S'| < |S|$ results in a connected graph.

We now define some useful subgraphs, see also Fig. 1 for illustrations. An induced chordless cycle of length at least 5 is called a *hole*. The graph that occurs from an induced chordless C_4 where exactly two of its adjacent vertices have a common neighbor is called a *house*. When referring to the induced C_3 part of a house we call it *roof* while the induced C_4 is called *body*. Two induced C_4 sharing exactly one edge form a *domino*. A graph that contains no hole, house or domino as an induced subgraph is called HHD-free. We call a graph that consists of two C_4 sharing a vertex, and an edge that connects the two neighbors of the common vertex in a way that no other C_4 exists a *double house*.

Lastly, let $G = (V, E)$ be a k-connected graph, let $t_1, \ldots, t_k \in V$ be k distinct vertices, and let $n_1, \ldots, n_k$ be natural numbers satisfying $\sum_{i=1}^{k} n_i = |V|$. We call $S_1, \ldots S_k \subseteq V(G)$ a *GL-Partition of* G if $S_1, \ldots S_k$ forms a partition of $V(G)$, such that for all $i \in [k]$ we have that $G[S_i]$ is connected, $t_i \in S_i$ and $|S_i| = n_i$. When there exists an $l \in \mathbb{N}$, such that for such a partition only $n_i - l \leq |S_i| \leq n_i + l$ holds instead of $|S_i| = n_i$, we say that $S_1, \ldots, S_k$ is a *GL-Partition of* G *with deviation* l.

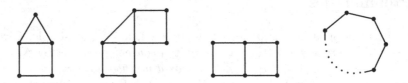

Fig. 1. Specific subgraphs used throughout the paper, from left to right: house, double house, domino and hole example

3   GL-Partition for Chordal Graphs

We present a simple, implementable algorithm with quadratic running time that computes GL-Partitions in chordal graphs. We then show that a slight modification of our algorithm is sufficient to compute a GL-Partition on a vertex weighted graph, thus proving Theorem 1.

3.1   GL-Partition for Unweighted Chordal Graphs

For simplicity, we first prove the restricted version of Theorem 1 to unweighted graphs. We use a p.e.o. to compute a vertex partition, as described formally in Algorithm 1. This algorithm receives as input a k-connected chordal graph $G = (V, E)$, terminal vertices $t_1, \ldots, t_k \in V$, and natural numbers $n_1, \ldots, n_k$ satisfying $\sum_{i=1}^{k} n_i = n$, and outputs connected vertex sets $S_1, \ldots, S_k \subseteq V$ such that $|S_i| = n_i$ and $t_i \in S_i$. In the beginning of the algorithm we initialize each set S_i to contain only the corresponding terminal vertex t_i, and add vertices iteratively to the *non-full sets* (S_i's that have not reached their demanded size). We say a vertex v *is assigned* if it is already part of some S_i and *unassigned* otherwise. At each iteration, the unassigned neighborhood of the union of the previously non-full sets is considered, and the vertex with the minimum p.e.o. value is selected to be added to a non-full set. In case there is more than one non-full set in the neighborhood of this vertex, it is added to the one with lowest priority, where the priority of each set is defined to be the largest p.e.o. value of its vertices so far. The algorithm terminates once all vertices are assigned, in $\mathcal{O}(|V|^2)$ time.

Algorithm 1: ChordalGL

    Input: k-connected chordal graph $G = (V, E)$, terminal vertices $t_1, \ldots, t_k \in V$,
            and natural numbers $n_1, \ldots, n_k$ satisfying $\sum_{i=1}^{k} n_i = n$
    Output: Connected vertex sets $S_1, \ldots, S_k \subseteq V$ such that $|S_i| = n_i$ and $t_i \in S_i$

1   $\sigma \leftarrow$ Compute p.e.o. of G as function $\sigma \colon V \to [|V|]$
2   $S_i \leftarrow \{t_i\}$, for all $i \in [k]$
3   **while** $\bigcup_{i \in [k]} S_i \neq V(G)$ **do**
4      $I \leftarrow \{i \in [k] \mid |S_i| < n_i\}$
5      $V' \leftarrow N(\bigcup_{i \in I} S_i) \backslash \bigcup_{i \in [k]} S_i$
6      $v' \leftarrow arg\,min_{v \in V'}\,\sigma(v)$
7      $J \leftarrow \{i \in I \mid v' \in N(S_i)\}$
8      $j' \leftarrow arg\,min_{j \in J}\,\max(\sigma(S_j))$
9      $S_{j'} \leftarrow S_{j'} \cup \{v'\}$
10 **end**
11 **return** $S_1, \ldots, S_k$

For the correctness of Algorithm 1 it is enough to show that the unassigned neighborhood V' of all non-full sets is not empty in each iteration of the while-loop, since this implies that we enlarge a non-full set (in the algorithm denoted by $S_{j'}$) by one vertex (in the algorithm denoted by v') while maintaining the size of all remaining sets. That is, in each iteration we make progress in the sense that $|\bigcup_{i \in [k]} S_i|$ increases while maintaining the invariant $|S_i| \leq n_i$ for all S_i's. Note that $v' \in N(S_{j'})$ which in turn implies that $G[S_i]$ is always connected for all $i \in [k]$. Finally, by $\sum_{i=1}^{k} n_i = n$ and through the way we update I we ensure that the algorithm (or while-loop) terminates as $\bigcup_{i \in [k]} S_i = V$ only if we have $|S_i| = n_i$ for all S_i's.

Towards proving the required Lemmata for the correctness of Algorithm 1 we make the following observation for the p.e.o. of a graph.

Lemma 1. *Let σ be a p.e.o of a graph $G = (V, E)$ and $P = \{v_1, v_2, \ldots, v_k\}$ a vertex set of G that induces a simple path with endpoints v_1 and v_k. Then $\sigma(v_i) > \min\{\sigma(v_1), \sigma(v_k)\}$ for all $i = 2, \ldots, k-1$.*

Lemma 2. *In each iteration of the while-loop in Algorithm 1 we have $V' \neq \varnothing$.*

Proof. We first define the *z-connecting* neighborhood of a vertex v to be the neighbors of v that are included in some induced path connecting v to z.

We prove that every non-full set S_i contains a vertex in its neighborhood $N(S_i)$ that is unassigned, which implies that $V' \neq \varnothing$. Assume for a contradiction that at some iteration of our algorithm there is an non-full set S_i whose neighborhood is already assigned to other sets. Let v be the vertex of S_i of maximum σ value among its vertices and z be the vertex of maximum σ value among the unassigned vertices. Note that $vz \notin E(G)$. Let $\mathcal{P}$ be the set of all simple induced paths of G with endpoints z and v. Consider now the following cases:

1. If $\sigma(z) > \sigma(v)$, we get from Lemma 1 that every internal vertex of each path in $\mathcal{P}$ has higher σ value than v. Note that no vertex of S_i is an internal vertex of some path in $\mathcal{P}$, since all of them have smaller σ value than v by the selection of v. Denote the z-connecting neighborhood of v by C.

   Let a, b be two vertices in C and assume that $a, b \in S_j$ for some j. Assume also that during our algorithm, a is added to S_j before b. Since all vertices of S_i have smaller σ value than both a and b, and a is added to S_j before b, the moment b is added to S_j, S_i has already been formed. Consider now the iteration that this happens. Since $b \in N(v)$, $G[S_i \cup \{b\}]$ is connected. Moreover since $\sigma(a) > \sigma(v)$ and S_i is not full, b should be added to S_i instead of S_j. As a result each set apart from S_i contains at most one such neighbor of v, and hence $|C| < k$.

   Observe that $G \backslash C$ has no induced path connecting z and v which in turn implies that $G \backslash C$ has no $z - v$ path in general. However, this contradicts the k-connectivity of G.

2. If $\sigma(z) < \sigma(v)$, since z is the unassigned vertex of the highest σ value among all unassigned vertices, and by Lemma 1 all vertices in $\mathcal{P}$ have greater σ value than z, all of its v-connecting neighbors in $\mathcal{P}$ are already assigned in some set. Denote the set of v-connecting neighbors of z by C.

   Assume now that there are two vertices of C, a and b, that are contained in some S_j and assume also without loss of generality that a was added to S_j before b. Note that since $\sigma(z) < \sigma(b)$ at each iteration of our algorithm z is considered before b to be added to some set if the induced graph remains connected. As a result, after a is added to S_j, the induced subgraph $G[S_j \cup \{z\}]$ is connected and hence z should be added to S_j before b.

This means that each set contains at most one v-connecting neighbor of z and therefore $|C| < k$. Since $G\backslash C$ has no induced path connecting z and v, there is no z-v-path in $G\backslash C$, which contradicts the k-connectivity.

Corollary 1. *At each iteration of Algorithm 1, unless all vertices are assigned, the neighborhood of each non-full set contains at least one unassigned vertex.*

In the weighted case we use the above corollary of Lemma 2. In particular, it follows from Corollary 1 that as long as we do not declare a set to be full, we ensure that we are able to extend it by a vertex in its neighborhood that is unassigned. Note that in the weighted case we do not know in advance how many vertices are in each part.

3.2 GL-Partition for Weighted Chordal Graphs

With a slight modification of Algorithm 1 we can compute the weighted version of a GL-Partition . In particular, we prove Theorem 1.

The input of our algorithm differs from the unweighted case by having a positive vertex-weighted graph $G = (V, E, w)$ and instead of demanded sizes $n_1, \ldots, n_k$ we have demanded weights $w_1, \ldots, w_k$ for our desired vertex sets $S_1, \ldots, S_k$, where $\sum_{i=1}^{k} w_i = w(V)$. Note also that $w(S_i)$ is not allowed to deviate more than $w_{max} = \max_{v \in V} w(v)$ from w_i, i.e. $w_i - w_{max} < w(S_i) < w_i + w_{max}$.

Again we set each terminal vertex t_i to a corresponding set S_i, and enlarge iteratively the *non-full weighted sets* (S_i's that are not declared as full). One difference to the previous algorithm is that we declare a set S_i as *full weighted set*, if together with the next vertex to be potentially added its weight would exceed w_i. After that, we decide whether to add the vertex with respect to the currently full weighted sets. Similar to Algorithm 1 we interrupt the while-loop if $S_1, \ldots, S_k$ forms a vertex partition of V and the algorithm terminates. However, to ensure that we get a vertex partition in every case, we break the while-loop when only one non-full weighted set is left and assign all remaining unassigned vertices to it.

Observe that we can make use of Corollary 1, since Algorithm 2 follows the same priorities concerning the p.e.o. as Algorithm 1. Basically, it implies that as long we do not declare a set as full weighted set and there are still unassigned vertices then those sets have unassigned vertices in their neighborhood.

We conclude this section by extending the above algorithms to graphs having distance $k/2$ from being chordal. In particular this corollary is based on the observation that an edge added to a graph does not participate in any of the parts those algorithms output if both of its endpoints are terminal vertices.

Corollary 2. *Let G be a k-connected graph which becomes chordal after adding $k/2$ edges. Given this set of edges, a GL-Partition (also its weighted version) can be computed in polynomial time but without fixed terminals.*

Algorithm 2: WeightedChordalGL

Input: k-connected vertex-weighted chordal graph $G(V, E, w)$, terminal vertices
$t_1, \ldots, t_k \in V$, and positive weights $w_1, \ldots, w_k$ satisfying
$\sum_{i=1}^{k} w_i = w(V)$

Output: Connected vertex sets $S_1, \ldots, S_k \subseteq V$ such that
$w_i - w_{\max} < w(S_i) < w_i + w_{\max}$ and $t_i \in S_i$

1 $\sigma \leftarrow$ Compute p.e.o. of G as function $\sigma: V \rightarrow [|V|]$
2 $S_i \leftarrow \{t_i\}$, for all $i \in [k]$
3 $I \leftarrow \{i \in [k] \mid w(S_i) < w_i\}$
4 **while** $|I| \neq 1$ **and** $\bigcup_{i \in [k]} S_i \neq V(G)$ **do**
5 $\quad$ $V' \leftarrow N(\bigcup_{i \in I} S_i) \backslash \bigcup_{i \in [k]} S_i$
6 $\quad$ $v' \leftarrow arg\,min_{v \in V'} \sigma(v)$
7 $\quad$ $J \leftarrow \{i \in I \mid v' \in N(S_i)\}$
8 $\quad$ $j' \leftarrow arg\,min_{j \in J} \max(\sigma(S_j))$
9 $\quad$ **if** $w(S_{j'}) + w(v') < w_{j'}$ **then**
10 $\quad$ $\quad | \quad S_{j'} \leftarrow S_{j'} \cup \{v'\}$
11 $\quad$ **end**
12 $\quad$ **else**
13 $\quad$ $\quad | \quad I \leftarrow I \backslash \{j'\}$
14 $\quad$ $\quad | \quad$ **if** $\sum_{i \in [k] \backslash I}(w_i - w(S_i)) \geq 0$ **or** $w(S_{j'}) + w(v') = w_{j'}$ **then**
15 $\quad$ $\quad | \quad \quad | \quad S_{j'} \leftarrow S_{j'} \cup \{v'\}$
16 $\quad$ $\quad | \quad$ **end**
17 $\quad$ **end**
18 **end**
19 If $|I| = 1$, assign all vertices $V \backslash \bigcup_{i \in [k]} S_i$ (possibly empty) to S_j with $j \in I$.

---

4 GL-Partition for HHI$_4^2$-free

This section is dedicated to the proof of Theorem 2. The underlying idea for this result is to carefully contract edges to turn a k-connected HHI$_4^2$-free graph into a chordal graph that is still k-connected. Note that we indeed have to be very careful here to find a set of contractions, as we need it to satisfy three seemingly contradicting properties: removing all induced C_4, preserving k-connectivity, and contracting at most one edge adjacent to each vertex. The last property is needed to bound the maximum weight of the vertices in the contracted graph. Further, we have to be careful not to contract terminal vertices.

The computation for the unweighted case of the partition for Theorem 2 is given in Algorithm 3 below, which is later extended to the weighted case as well. Note that we can assume that $n_i \geq 2$ since if $n_i = 1$ for some $i \in [k]$ we simply declare the terminal vertex to be the required set and remove it from G. This gives us a $(k - 1)$-connected graph and $k - 1$ terminal vertices.

Before starting to prove the Lemmata required for the correctness of Algorithm 3 we give a structural insight which is used in almost all proofs of the following Lemmata.

Lemma 3. *Given an HHI_4^2-free graph G and an induced C_4, $C \subseteq V(G)$, then any vertex in $V(G) \backslash C$ that is adjacent to two vertices of C is universal to C. Moreover, the set of vertices that are universal to C induces a clique.*

Algorithm 3: HHI_4^2-free GL

Input: k-connected HHI_4^2-free graph $G(V, E)$, terminal vertices
$t_1, \dots, t_k \in V$, and positive integers $n_1, \dots, n_k \geq 2$ satisfying
$\sum_{i=1}^{k} n_i = n$
Output: Connected vertex sets $S_1, \dots, S_k \subseteq V$ such that
$n_i - 1 \leq |S_i| \leq n_i + 1$ and $t_i \in S_i$

1 Add an edge between each pair of non-adjacent terminals that are part of
  an induced C_4

2 $\mathcal{C} \leftarrow$ Set of all induced C_4 in G.

3 $G' \leftarrow (\bigcup_{C \in \mathcal{C}} V(C), \bigcup_{C \in \mathcal{C}} E(C))$

4 $E' \leftarrow \varnothing$

5 **while** $\mathcal{C} \neq \varnothing$ **do**

6   Select three vertices v_1, v_2, v_3 in G' and the corresponding cycle $C \in \mathcal{C}$
      that satisfies that for all $C' \in \mathcal{C} \backslash \{C\}$ we have $V(C') \cap \{v_1, v_2, v_3\} = \varnothing$.

7   Pick a vertex v from v_1, v_2, v_3 that is not a terminal vertex and add an
      incident edge of v in $G'[\{v_1, v_2, v_3\}]$ to E'.

8   Remove the cycle C from $\mathcal{C}$ and the vertices v_1, v_2, v_3 from G'.

9 **end**

10 Transform G to a weighted graph G'' by contracting each edge of E' in G,
    assigning to each resulting vertex as weight the number of original
    vertices it corresponds to.

11 $S_1, \dots S_k \leftarrow$ Run Algorithm 2 with G'', the given set of terminals
    $t_1, \dots, t_k$, and the size (or weight) demands $n_1, \dots, n_k$ as input.

12 Reverse the edge contraction of E' in the sets $S_1, \dots, S_k$ accordingly.

Lemma 4. *Let G be an HHI_4^2-free graph. If G contains a double house as a subgraph then at least one of the two C_4 in it has a chord.*

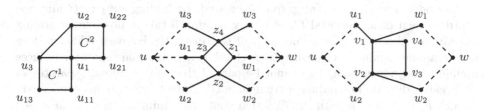

Fig. 2. Illustrations for the vertex namings used in proofs, from left to right: Lemma 4, Lemma 7 and Lemma 8

The following lemma captures the essence of why the algorithm provided in this section cannot be applied also on HHD-free graphs, since it holds for

HHI_4^2-free graphs but not for HHD-free graphs. Think for example of a simple path P of length 5 and a vertex disjoint induced chordless C_4, C. Consider also each vertex of P being universal to C. Observe that this graph is HHD- but not HHI_4^2-free . Every two non adjacent vertices of C together with the endpoints of P create an induced chordless C_4. Adding a chord connecting the two endpoints of P creates a hole and hence the resulting graph is not HHD-free.

Lemma 5. *Let G be an HHI_4^2-free graph and $C = \{v_1, v_2, v_3, v_4\}$ an induced C_4 in G. Then the graph G' created by adding the chord v_1v_3 to G is HHI_4^2-free and has one less induced C_4 than G.*

An essential property of the graph class we work on is being closed under contraction, since our algorithm is based on contracting edges iteratively until the resulting graph becomes chordal. Before proving this property though, although "after an edge contraction a new cycle is created" is intuitively clear, we formally define what it means for a C_4 to be "new".

Definition 1. *Let G be a graph, $uv \in E(G)$ and $G' = G/uv$. Let also w be the vertex of G' that is created by the contraction of uv. We say that an induced cycle C containing w in G' is new if $N_C(w) \nsubseteq N_G(v)$ and $N_C(w) \nsubseteq N_G(u)$.*

Lemma 6. *HHI_4^2-free graphs are closed under contraction of an edge of an induced C_4.*

In order to prove that the contractions of our algorithm do not affect the connectivity, we first study the possible role of vertices on an induced C_4 in minimal separators in HHI_4^2-free graphs.

Lemma 7. *Let G be a k-connected HHI_4^2-free graph for $k \geq 5$. Then no three vertices of an induced C_4 belong in the same minimal separator.*

Proof. Let G be a k-connected HHI_4^2-free graph for $k \geq 5$ and v_1, v_2, v_3, v_4 vertices that induce a C_4, C. Assume that v_1, v_2, v_3 belong in the a same minimal separator S and hence, $(G \backslash \{v_1, v_2, v_3\})$ is only $k - 3$ connected. Let also u and w be two distinct vertices belonging in different connected components of $G \backslash S$.

Consider now the chordal graph G' created, by adding v_2v_4 to C and one chord to each other induced C_4 of G. By Lemma 5 this is possible by adding exactly one chord to each induced C_4 of G - in particular each addition does not create new induced C_4. Since G' is chordal each minimal separator induces a clique, and hence v_1, v_2, v_3 cannot be part of the same minimal separator in G' because they do not induce a triangle in G'. Thus $G' \backslash S$ remains connected.

Let P_1 be a $u - w$ path in $G' \backslash S$ that contains a minimal number of added edges. Let $z_1 z_3 \in E(P_1)$ be one of the added edges, such that z_3 is closer to u on P_1 than z_1. Note that z_1 and z_3 are part of some induced C_4, $C' = \{z_1, z_2, z_3, z_4\}$ in G. Since z_1z_3 cannot be replaced by neither z_1z_2, z_2z_3, nor z_1z_4, z_4z_3 (otherwise we get a path with strictly less added edges than P_1) it follows that $z_2, z_4 \in S$.

We will use the $u - w$ paths through S in G to reach a contradiction. Since S is a minimal $u - w$ separator in G, there are two internally vertex disjoint $u - w$ paths P_2 and P_3, with $P_2 \cap S = \{z_2\}$ and $P_3 \cap S = \{z_4\}$. Let w_2 be the neighbor of z_2 on P_2 closer to w, w_1 the respective neighbor of z_1 on P_1 and w_3 the respective neighbor of z_4 on P_3. Let also u_1, u_2, u_3 be the corresponding neighbors of these paths closer to u. See the illustration in Fig. 2 for these namings, keeping in mind that it could be $w_1 \in \{w_2, w_3\}$ or $u_1 \in \{u_2, u_3\}$ or also $w_1 = w_2 = w_3 = w$ or $u_1 = u_2 = u_3 = u$.

We claim that, in G, z_3 is adjacent to a vertex on $P_2^{[w_2,w]}$ or $P_3^{[w_3,w]}$. Assume otherwise, and assume that $P_2^{[w_2,w]}$, $P_3^{[w_3,w]}$ are induced paths in G (shortcut them otherwise). If $w_2 = w_3$ then notice that $w_2 = w = w_3$. In order for z_3, z_2, z_4, w not to induce a C_4 with three common vertices to C, z_3 has to be adjacent to w which is on $P_2^{[w_2,w]}$. If $w_2 \neq w_3$ then assume without loss of generality that $w_2 \neq w$. In order to not be a hole, there has to be a chord in the cycle build by $P_2^{[w_2,w]}$, $P_3^{[w_3,w]}$ with z_4, z_3, z_2. By assumption, this chord cannot be from z_3, so it has to involve z_4 or z_2. Since $P_2^{[w_2,w]}$ and $P_3^{[w_3,w]}$ are induced and $w_2 \neq w$, either w_2 is adjacent to z_4, or $w_3 \neq w$ is adjacent to z_2. Both cases create a C_4 that has three vertices in common with C, $(w_2, z_2, z_3, z_4,$ and $w_3, z_2, z_3, z_4,$ resp.) and since $z_4 z_2 \notin E(G)$, the added chord for these C_4 has to be $w_2 z_3$, resp. $w_3 z_3$, leading again to z_3 being adjacent to some vertex on $P_2^{[w_2,w]}$ or $P_3^{[w_3,w]}$.

Thus we conclude that z_3 is adjacent to a vertex x on $P_2^{[w_2,w]}$ or $P_3^{[w_3,w]}$ in G. This however allows to create a path from u to w with (at least) one added edge less than P_1 in G (since P_2, P_3 do not contain any added edges). Specifically, if x is on P_2 we get $P_1' = P_1^{[u,z_3]} x P_2^{[x,w]}$ and if $x \in P_3$, $P_1' = P_1^{[u,z_3]} x P_3^{[x,w]}$.

Since C' was an arbitrary cycle we conclude that v_1, v_2, v_3 cannot be part of the same minimal separator in G.

Lemma 8. *Let G be an HHI_4^2-free k-connected graph and $C = \{v_1, v_2, v_3, v_4\}$ be an induced C_4. The graph $G' = G/v_1 v_2$ is still k-connected.*

Now, we finally look specifically at Algorithm 3, and first show that its subroutine creating G'' works correctly.

Lemma 9. *Given an HHI_4^2-free graph G, the vertices selected in line 6 of Algorithm 3 indeed exist as long as an induced C_4 exists.*

Proof. Let G be an HHI_4^2-free graph and $\mathcal{C}$ the set of all induced C_4 in G, consider the bipartite graph T constructed through the following procedure: Its vertices are partitioned into two sets B, and S referred to as *big* and *small* vertices of T, respectively. Each big vertex represents an induced C_4 of $\mathcal{C}$ while each small vertex represents a vertex of G participating in at least two induced C_4. Each small vertex is adjacent to the big vertices which represent a C_4 this vertex participates in. We claim that with this definition T is indeed a tree (actually a forest). Assume now for a contradiction that T contains a cycle and let C be one of the shortest such cycles in T.

First, consider the case that C has length $l \geq 6$. Since T is bipartite, due to its construction, l is even and the vertices of $C = \{s_1, b_1, s_2, b_2, \ldots, s_{l/2}, b_{l/2}\}$ alternate between big and small. We denote by $P_{b_i}^{s_w, s_z}$ a shortest path containing edges from the C_4 represented by the big vertex b_i with endpoints the vertices represented by s_w and s_z. Due to C, the cycle $P_{b_1}^{s_1, s_{l/2}} \ldots P_{b_l/2}^{s_{l/2-1}, s_{l/2}}$ exists in G as a subgraph. Note that since we have assumed that C is a minimal length cycle of T it is also chordless. Hence, in order for a hole not to be an induced subgraph of G at least one chord must exist connecting two vertices corresponding to two small ones of T. This however would create a double house as a subgraph with the two C_4 forming it being the two that correspond to big vertices of T. By Lemma 4 this means that one of the C_4 is not induced, a contradiction to the construction of T. Notice also that in the case where $l = 6$ we directly find a double house and reach a contradiction using the same arguments.

Moreover the assumption that $l = 4$, leads us to a contradiction to the fact that two C_4 have at most one vertex in common. Hence, T is a forest and the vertices mentioned in line 6 are the ones belonging only to a cycle represented by one leaf belonging in B.

Lemma 10. *Given an HHI_4^2-free graph G, lines 1–10 of Algorithm 3 transforms G into a weighted chordal graph G'', with the same connectivity as G and such that each vertex from G is involved in at most one edge contraction to create G''.*

At last, notice that we can easily alter Algorithm 3 to also work for weighted graphs, with the simple change of setting the weights of a vertex in G'' in line 10 to the sum of the weights of the original vertices it was contracted from. With this alteration, we can conclude now the proof of Theorem 2 with the following.

Lemma 11. *Algorithm 3 works correctly and runs in time $\mathcal{O}(|V|^4)$.*

Proof. By Lemma 10, G'' is a chordal graph with maximum vertex weight $2w_{max}$. Further, observe that we did not merge terminal vertices with each other, thus we can properly run Algorithm 2 on it. By the correctness of this algorithm (Theorem 1), we know that $S_1, \ldots, S_k$ in line 11 is a GL-partition for G'' with deviation $2w_{max}$. Since reversing edge-contraction does not disconnect these sets, the unfolded sets $S_1, \ldots, S_k$ are thus also a GL-partition for G with deviation $2w_{max}$; note here that the only edges we added to create G'' are between terminal vertices, which are in separate sets S_i by definition.

The most time consuming part of Algorithm 3 is the preprocessing to transform the input graph into a weighted chordal graph which requires $\mathcal{O}(|V|^4)$ time in order to find all the induced C_4 (note that the induced C_4 are at most $(n-4)/3$ since they induce a tree).

Moreover, as is the case for chordal graphs, we can sacrifice terminals to enlarge the considered graph class.

Corollary 3. *Let G be a k-connected graph which becomes HHI_4^2-free after adding $k/2$ edges. Then, given those edges, a GL-Partition of G with deviation 1 (also its weighted version with deviation $2w_{max} - 1$) can be computed in polynomial time but without fixed terminals.*

References

1. Broersma, H., Dahlhaus, E., Kloks, T.: Algorithms for the treewidth and minimum fill-in of HHD-free graphs. In: International Workshop on Graph-Theoretic Concepts in Computer Science (WG), pp. 109–117 (1997). https://doi.org/10.1007/BFb0024492

2. Chandran, L.S., Cheung, Y.K., Issac, D.: Spanning tree congestion and computation of generalized Györi-Lovász partition. In: International Colloquium on Automata, Languages, and Programming, (ICALP). LIPIcs, vol. 107, pp. 32:1–32:14 (2018). https://doi.org/10.4230/LIPIcs.ICALP.2018.32

3. Chen, J., Kleinberg, R.D., Lovász, L., Rajaraman, R., Sundaram, R., Vetta, A.: (Almost) tight bounds and existence theorems for single-commodity confluent flows. J. ACM **54**(4), 16 (2007). https://doi.org/10.1145/1255443.1255444

4. Győri, E.: On division of graphs to connected subgraphs, combinatorics. In: Colloq. Math. Soc. Janos Bolyai, 1976 (1976)

5. Hoyer, A.: On the independent spanning tree conjectures and related problems. Ph.D. thesis, Georgia Institute of Technology (2019)

6. Jamison, B., Olariu, S.: On the semi-perfect elimination. Adv. Appl. Math. **9**(3), 364–376 (1988)

7. Lovász, L.: A homology theory for spanning tress of a graph. Acta Math. Hungar. **30**(3–4), 241–251 (1977)

8. Lucertini, M., Perl, Y., Simeone, B.: Most uniform path partitioning and its use in image processing. Discrete Appl. Math. **42**(2), 227–256 (1993). https://doi.org/10.1016/0166-218X(93)90048-S

9. Möhring, R.H., Schilling, H., Schütz, B., Wagner, D., Willhalm, T.: Partitioning graphs to speedup Dijkstra's algorithm. ACM J. Exp. Algorithmics **11**, 2–8 (2006). https://doi.org/10.1145/1187436.1216585

10. Nakano, S., Rahman, M.S., Nishizeki, T.: A linear-time algorithm for four-partitioning four-connected planar graphs. Inf. Process. Lett. **62**(6), 315–322 (1997). https://doi.org/10.1016/S0020-0190(97)00083-5

11. Przytycka, T.M.: An important connection between network motifs and parsimony models. In: Apostolico, A., Guerra, C., Istrail, S., Pevzner, P.A., Waterman, M. (eds.) RECOMB 2006. LNCS, vol. 3909, pp. 321–335. Springer, Heidelberg (2006). https://doi.org/10.1007/11732990_27

12. Przytycka, T.M., Davis, G.B., Song, N., Durand, D.: Graph theoretical insights into evolution of multidomain proteins. J. Comput. Biol. **13**(2), 351–363 (2006). https://doi.org/10.1089/cmb.2006.13.351

13. Rose, D.J., Tarjan, R.E., Lueker, G.S.: Algorithmic aspects of vertex elimination on graphs. SIAM J. Comput. **5**(2), 266–283 (1976). https://doi.org/10.1137/0205021

14. Suzuki, H., Takahashi, N., Nishizeki, T., Miyano, H., Ueno, S.: An algorithm for tri-partitioning 3-connected graphs. J. Inf. Process. Soc. Japan **31**(5), 584–592 (1990)

15. Suzuki, H., Takahashi, N., Nishizeki, T.: A linear algorithm for bipartition of biconnected graphs. Inf. Process. Lett. **33**(5), 227–231 (1990). https://doi.org/10.1016/0020-0190(90)90189-5

16. Wada, K., Kawaguchi, K.: Efficient algorithms for tripartitioning triconnected graphs and 3-edge-connected graphs. In: van Leeuwen, J. (ed.) WG 1993. LNCS, vol. 790, pp. 132–143. Springer, Heidelberg (1994). https://doi.org/10.1007/3-540-57899-4_47
17. Zhou, X., Wang, H., Ding, B., Hu, T., Shang, S.: Balanced connected task allocations for multi-robot systems: an exact flow-based integer program and an approximate tree-based genetic algorithm. Expert Syst. Appl. **116**, 10–20 (2019). https://doi.org/10.1016/j.eswa.2018.09.001

Generating Faster Algorithms for d-Path Vertex Cover

Radovan Červený$^{(\boxtimes)}$ and Ondřej Suchý$^{(\boxtimes)}$

Department of Theoretical Computer Science, Faculty of Information Technology,
Czech Technical University in Prague, Prague, Czech Republic
{radovan.cerveny,ondrej.suchy}@fit.cvut.cz

Abstract. Many algorithms which exactly solve hard problems require branching on more or less complex structures in order to do their job. Those who design such algorithms often find themselves doing a meticulous analysis of numerous different cases in order to identify these structures and design suitable branching rules, all done by hand. This process tends to be error prone and often the resulting algorithm may be difficult to implement in practice.

In this work, we aim to automate a part of this process and focus on the simplicity of the resulting implementation.

We showcase our approach on the following problem. For a constant d, the d-PATH VERTEX COVER problem (d-PVC) is as follows: Given an undirected graph and an integer k, find a subset of at most k vertices of the graph, such that their deletion results in a graph not containing a path on d vertices as a subgraph. We develop a fully automated framework to generate parameterized branching algorithms for the problem and obtain algorithms outperforming those previously known for $3 \leq d \leq 8$, e.g., we show that 5-PVC can be solved in $O(2.7^{k} \ast n^{O(1)})$ time.

1 Introduction

The motivation behind this paper is to renew the interest in computer aided design of graph algorithms which was initiated by Gramm et al. [22]. Many parameterized branching algorithms follow roughly the same pattern: 1) perform a meticulous case analysis; 2) based on the analysis, construct branching and reduction rules; 3) argue that once the rules cannot be applied, some specific structure is achieved. Also, depending on how "deeply" you perform the case analysis, you may slightly improve the running time of the algorithm, but bring nothing new to the table.

This paper aims to provide a framework which could help in the first two steps of the pattern at least for some problems. We phrase the framework for a rather general problem which is as follows. For any nonempty finite set of

---

*The authors acknowledge the support of the OP VVV MEYS funded project CZ.02.1.01/0.0/0.0/16_019/0000765 "Research Center for Informatics" and the Grant Agency of the CTU in Prague funded grant No. SGS20/208/OHK3/3T/18.

D. Paulusma and B. Ries (Eds.): WG 2023, LNCS 14093, pp. 157–171, 2023.
https://doi.org/10.1007/978-3-031-43380-1_12

connected graphs $\mathcal{F}$ we define the problem $\mathcal{F}$-SUBGRAPH VERTEX DELETION, $\mathcal{F}$-SVD, where, given a graph $G = (V, E)$ and an integer k, the task is to decide whether there is a subset S of at most k vertices of G such that $G \backslash S$ does not contain any graph from $\mathcal{F}$ as a subgraph (not even as a non-induced one). While we only apply the framework to the problem of d-PVC defined later, the advantage of phrasing the framework for $\mathcal{F}$-SVD is twofold. First, it makes it easier to apply it to other problems. Second, the general notation introduced makes the description less cluttered.

Since the problem is NP-complete for most reasonable choices of $\mathcal{F}$, as follows from the meta-theorem of Lewis and Yannakakis [27], any algorithm solving the problem exactly is expected to have exponential running time. In this paper we aim on the parameterized analysis of the problem, that is, to confine the exponential part of the running time to a specific parameter of the input, presumably much smaller than the input size. In particular, we only use the most standard parameter, which is the desired size of the solution k, also called *the budget*. Algorithms achieving running time $f(k)n^{O(1)}$ are called *parameterized, fixed-parameter tractable*, or *FPT*. See Cygan et al. [12] for a broader introduction to parameterized algorithms.

To understand how parameterized branching algorithms typically work, consider the following simple recursive algorithm for $\mathcal{F}$-SVD. We find in the input graph G an occurrence F' of graph F from $\mathcal{F}$. We know that at least one of the vertices of F' must be in any solution. Hence, for each vertex of F' we try adding it to a prospective solution, decreasing the remaining budget by one, and recursing. The recursion is stopped when the budget is exhausted, or there are no more occurrences of graphs from $\mathcal{F}$ in G, i.e., we found a solution. It is easy to analyze that this algorithm has running time[1] $\mathcal{O}^*(d^k)$, where d is the number of vertices of the largest graph in $\mathcal{F}$. Many parameterized branching problems follow a similar scheme, branching into a constant number of alternatives in each step, for each alternative making a recursive call with the budget (or some other parameter) decreased by some constant.

One can improve upon this trivial algorithm by looking at F' together with its surroundings. Working with this larger graph F'' often allows for more efficient branching as now multiple *overlapping* occurences of graphs from $\mathcal{F}$ may appear in F'' instead of just one. Our framework and that of Gramm et al. [22] rely upon this observation, as they iteratively take larger and larger graphs into consideration—similarly to what a human would do, but on a much larger scale.

The fundamental novelty of our framework in comparison to that of Gramm et al. [22] is that we are able to identify which vertices of the graph F'' under consideration can still have outside neighbors and which do not. We call the latter "red". This way we are able to say that if you find an occurrence of F'' in the input graph, you can be sure that the red vertices do not have neighbors in the input graph apart from those that are in F''.

This additional information allows us to eliminate some branches of the constructed branching rules, rapidly improving their efficiency. It also reduces the

---

[1] The $\mathcal{O}^*()$ notation suppresses all factors polynomial in the input size.

number of graphs we need to consider and also allows us to design better reduction rules to aid our framework.

We apply the general framework to the problem of d-PATH VERTEX COVER (d-PVC). The problem lies in determining a subset S of vertices of a given graph $G = (V, E)$ of at most a given size k such that $G \backslash S$ does not contain a path on d vertices (even not a non-induced one). It was first introduced by Brešar et al. [2], but its NP-completeness for any $d \geq 2$ follows already from the above-mentioned meta-theorem of Lewis and Yannakakis [27]. The 2-PVC problem corresponds to the well known VERTEX COVER problem and the 3-PVC problem is also known as MAXIMUM DISSOCIATION SET or BOUNDED DEGREE-ONE DELETION. The d-PVC problem is motivated by the field of designing secure wireless communication protocols [31] or route planning and speeding up shortest path queries [20].

As mentioned above, d-PVC is directly solvable by a trivial FPT algorithm that runs in $\mathcal{O}^*(d^k)$ time. However, since d-PVC is a special case of d-HITTING SET, it follows from the results of Fomin et al. [17] that for any $d \geq 4$ we have an algorithm solving d-PVC in $\mathcal{O}^*((d - 0.9245)^k)$ time. For $d \geq 6$ algorithms with even better running times are presented in the work of Fernau [15].

In order to find more efficient solutions, the problem has been extensively studied in a setting where d is a small constant. This is in particular the case for the 2-PVC (VERTEX COVER) problem [1,3,6,8,11,13,29,30], where the algorithm of Chen, Kanj, and Xia [10] for a long time held the best known running time of $\mathcal{O}^*(1.2738^k)$, but recently Harris and Narayanaswamy [23] claimed the running time of $\mathcal{O}^*(1.25288^k)$. For 3-PVC, Tu [37] used iterative compression to achieve a running time of $\mathcal{O}^*(2^k)$. This was later improved by Katrenič [24] to $\mathcal{O}^*(1.8127^k)$, by Xiao and Kou [40] to $\mathcal{O}^*(1.7485^k)$ by using a branch-and-reduce approach and finally by Tsur [34] to $\mathcal{O}^*(1.713^k)$. For the 4-PVC problem, Tu and Jin [38] again used iterative compression and achieved a running time of $\mathcal{O}^*(3^k)$ and Tsur [35] gave the current best algorithm that runs in $\mathcal{O}^*(2.619^k)$ time. The authors of this paper developed an $\mathcal{O}^*(4^k)$ algorithm for 5-PVC [4]. For $d = 5$, 6, and 7 Tsur [36] discovered algorithms for d-PVC with running times $\mathcal{O}^*(3.945^k)$, $\mathcal{O}^*(4.947^k)$, and $\mathcal{O}^*(5.951^k)$, respectively.

Using our automated framework, we are able to present algorithms with improved running times for some d-PVC problems when parameterized by the size of the solution k. The results are summarized in Table 1.

Further Related Work. The only other approach to generating algorithms with provable worst-case running time upper bounds we are aware of is limited to algorithms for SAT [14,25,26].

Several moderately exponential exact algorithms are known for 2-PVC and 3-PVC [7,39,41].

Full Version of the Paper. Due to space constraints, we omit most technical details from this extended abstract. We refer the kind reader to the full version of the paper [5].

Table 1. Improved running times of some d-PVC problems.

d-PVC	Previously known	Our result	Our # of rules
2-PVC	$\mathcal{O}^*(1.25288^k)$ [23]	$\mathcal{O}^*(1.3294^k)$	9,345,243
3-PVC	$\mathcal{O}^*(1.713^k)$ [34]	$\mathcal{O}^*(1.708^k)$	1,226,384
4-PVC	$\mathcal{O}^*(2.619^k)$ [35]	$\mathcal{O}^*(2.138^k)$	911,193
5-PVC	$\mathcal{O}^*(3.945^k)$ [36]	$\mathcal{O}^*(2.636^k)$	739,542
6-PVC	$\mathcal{O}^*(4.947^k)$ [36]	$\mathcal{O}^*(3.334^k)$	414,247
7-PVC	$\mathcal{O}^*(5.951^k)$ [36]	$\mathcal{O}^*(3.959^k)$	5,916,297
8-PVC	$\mathcal{O}^*(7.0237^k)$ [15]	$\mathcal{O}^*(5.654^k)$	296,044

2   Fundamental Definitions and Basic Observations

In this paper we are going to assume that vertex sets of all graphs are finite subsets of $\mathbb{N}$, the set of all non-negative integers, i.e., we have a set of all graphs. Furthermore, when adding a graph into a set of graphs, we only add the graph if none of the graphs already in the set is isomorphic to it. Similarly, when forming a set of graphs we only add one representative for each isomorphism class. Finally, when subtracting a graph from a set, we remove from the set all graphs isomorphic to it.

For any nonempty finite set of connected graphs $\mathcal{F}$ we define the problem:

$\mathcal{F}$-SUBGRAPH VERTEX DELETION, $\mathcal{F}$-SVD	
INPUT:	A graph $G = (V, E)$, an integer $k \in \mathbb{N}$
OUTPUT:	A set $S \subseteq V$, such that $|S| \leq k$ and no subgraph of $G \backslash S$ is isomorphic to a graph in $\mathcal{F}$

We call $\mathcal{F}$ of $\mathcal{F}$-SVD a *bump-inducing* set. We call a graph G *bumpy* if it contains some graph from the bump-inducing set $\mathcal{F}$ as a subgraph. We call a vertex subset S a *solution* (for a graph $G = (V, E)$), if the graph $G \backslash S$ is not bumpy. Since $\mathcal{F}$ is finite, checking if G is bumpy is polynomial in the size of G.

Next we define a variant of a supergraph with a restriction that the original graph has to be an induced subgraph of the supergraph.

Definition 1 (expansion, i-expansion, σ, σ_i, σ^*). *Let H be a connected graph. A graph G is an* expansion *of H, if G is connected, $V(H) \subseteq V(G)$ and $G[V(H)] = H$. It is an i-expansion for $i \in \mathbb{N}$ if furthermore $|V(G)| = |V(H)| + i$. For $i \in \mathbb{N}$ let $\sigma_i(H)$ denote the set of all i-expansions of H (note again that we take only one representative for each isomorphism class). As shorthand, we will use $\sigma(H) = \sigma_1(H)$. Let $\sigma^*(H) = \bigcup_{i \in \mathbb{N}} \sigma_i(H)$ denote the set of all expansions of H.*

The following (restricted) variant of a branching rule is the building block of our algorithm.

Definition 2 (Subgraph branching rule). *A subgraph branching rule is a triple* $(H, R, \mathcal{B})$, *where* H *is a connected bumpy graph,* $R \subseteq V(H)$ *is a set of red vertices (representing the vertices supposed not to have neighbors outside H), and* $\mathcal{B} \subseteq (2^{V(H)} \setminus \{\emptyset\})$ *is a non-empty set of branches.*

Definition 3 (An application of a subgraph branching rule). *We say that a subgraph branching rule* $(H, R, \mathcal{B})$ *applies to graph* G, *if* G *contains an induced subgraph* H' *isomorphic to* H *by isomorphism* $\phi : V(H) \to V(H')$ *(witnessing isomorphism) and for every* $r \in R$ *we have* $N_G(\phi(r)) \subseteq V(H')$. *In other words, the vertices of* H' *corresponding to red vertices only have neighbors inside the subgraph* H'. *If the rule applies and the current instance is* (G, k), *then the algorithm makes for each* $B \in \mathcal{B}$ *a recursive call with instance* $(G \setminus \phi(B), k - |B|)$.

Note that we do not allow $\emptyset \in \mathcal{B}$. Therefore the budget gets reduced and we are making progress in every branch.

Definition 4 (Correctness of a subgraph branching rule). *A subgraph branching rule* $(H, R, \mathcal{B})$ *is correct, if for every* G *and every solution* S *for* G *such that* $(H, R, \mathcal{B})$ *applies to* G *and* $\phi : V(H) \to V(H')$ *is the witnessing isomorphism, there exists a solution* S' *for* G *with* $|S'| \leq |S|$ *and a branch* $B \in \mathcal{B}$ *such that* $\phi(B) \subseteq S'$.

Definition 5 (Branching factor of a subgraph branching rule). *For any subgraph branching rule* $(H, R, \mathcal{B})$ *let* $bf((H, R, \mathcal{B}))$ *be the branching factor of the branches in* $\mathcal{B}$, *i.e., the unique positive real solution of the equation:* $1 = \sum_{B \in \mathcal{B}} x^{-|B|}$ *(see [19, Chapter 2.1 and Theorem 2.1] for more information on (computing) branching factors).*

Observation 1. *For any connected bumpy graph* H *and any* $R \subseteq V(H)$ *we can always construct at least one correct subgraph branching rule.*

The following definition formalizes a function that, given a graph H and a set of vertices R, computes a set $\mathcal{B}$ of branches such that $(H, R, \mathcal{B})$ is a correct subgraph branching rule.

Definition 6 (Brancher). *A brancher is a function which assigns to any connected bumpy graph* H *and* $R \subseteq V(H)$ *a correct branching rule* $\tau(H, R) = (H, R, \mathcal{B})$ *for some non-empty set* $\mathcal{B} \subseteq (2^{V(H)} \setminus \{\emptyset\})$. *For a brancher* τ *as a shorthand let* $\tau(H) = \tau(H, \emptyset)$. *For a set of graphs* $\{H_1, H_2, \ldots, H_r\}$ *we will have* $\tau(\{H_1, H_2, \ldots, H_r\}) = \{\tau(H_1), \tau(H_2), \ldots, \tau(H_r)\}$.

The above observation shows that at least one brancher exists.

Definition 7. *For a set of subgraph branching rules* $\mathcal{L} = (\varrho_1, \varrho_2, \ldots, \varrho_r)$ *where* $\varrho_i = (H_i, R_i, \mathcal{B}_i)$ *we will denote* $\Psi(\mathcal{L}) = \max\{|V(H_i)| \mid (H_i, R_i, \mathcal{B}_i) \in \mathcal{L}\}$ *the maximum number of vertices among the graphs of the subgraph branching rules in* $\mathcal{L}$.

The framework makes possible to introduce a number of handmade reduction[2] or branching rules, denoted as $\mathcal{A}$, to help the generating algorithm steer it away from some difficult corner cases. Typically, their purpose is to ensure some substructures no longer appear in the input graph.

Next we define the crucial property of a set of subgraph branching rules which forms a base for the proof of correctness of the generated algorithm.

Definition 8. *A set of subgraph branching rules* $\mathcal{L} = (\varrho_1, \varrho_2, \ldots, \varrho_r)$ *is called exhaustive with respect to* $\mathcal{A}$ *if every rule* ϱ_i *is correct and for every connected bumpy graph* G *to which no handmade rule in* $\mathcal{A}$ *applies and which has at least* $\Psi(\mathcal{L})$ *vertices there is a subgraph branching rule* ϱ_i *in* $\mathcal{L}$ *that applies to* G. *If the set is exhaustive with respect to* $\emptyset$, *that is, even without any handmade rules, we will omit the "with respect to* $\mathcal{A}$*" clause.*

In the process of generating the algorithm, we aim to maintain an exhaustive set of subgraph branching rules at all times.

The following observation identifies our starting set of graphs.

Observation 2. *Let* $\mathcal{F}$ *be the bump-inducing set of some* $\mathcal{F}$-SVD *problem. Let* $f = \max_{H \in \mathcal{F}} |V(H)|$. *Let* $L = \{F_1, F_2, \ldots, F_r\}$ *be the set of all connected bumpy graphs with* f *vertices. Let* τ *be a brancher. Then the set of subgraph branching rules* $\mathcal{L} = \tau(L)$ *is exhaustive.*

3   The Output Algorithm and Its Correctness

Our goal will be to obtain a set $\mathcal{L}$ of subgraph branching rules with good branching factors which is exhaustive with respect to $\mathcal{A}$. This section summarizes how we use the set to design an algorithm for $\mathcal{F}$-SVD once we obtain such a set. We call the algorithm $(\mathcal{A}, \mathcal{L})$-Algorithm for $\mathcal{F}$-SVD and its pseudocode is available in Algorithm 1.

The algorithm first applies some trivial stopping conditions (lines 3 to 5). Then it applies the rules from $\mathcal{A}$ (lines 6 to 7). Next, if every connected component is small, it finds a solution for each of them separately by a brute force (lines 8 to 12). Finally, it takes a component which is large enough and finds a subgraph branching rule from $\mathcal{L}$ that applies to the component and applies it by making the appropriate recursive calls (lines 13 to 18).

The following theorem states that this algorithm is indeed correct.

Theorem 1. *Let* $\mathcal{A}$ *be a list of handmade rules and* $\mathcal{L}$ *be a set of subgraph branching rules. If* $\mathcal{L}$ *is exhaustive with respect to* $\mathcal{A}$, *all rules in* $\mathcal{A}$ *are correct and can be applied in polynomial time, and each branching rule in* $\mathcal{A} \cup \mathcal{L}$ *has branching factor at most* β, *then the* $(\mathcal{A}, \mathcal{L})$-*Algorithm for* $\mathcal{F}$-SVD *is correct and runs in* $\mathcal{O}^*(\beta^k)$ *time.*

---

[2] Roughly speaking, a reduction rule is a polynomial-time procedure that replaces the input instance with another one, preserving the answer.

Implementation Considerations. We want to emphasize, that the effort needed to implement the algorithm does not grow with the number of generated rules in $\mathcal{L}$ as the code that implements the mechanic on line 14 remains the same regardless of the number of rules. Further, the generated list $\mathcal{L}$ is given encoded in a machine-readable format, which further simplifies the implementation.

Algorithm 1. $(\mathcal{A}, \mathcal{L})$-Algorithm for $\mathcal{F}$-SVD

1: Let $\mathcal{A}$ be a list of handmade rules and $\mathcal{L}$ be a set of subgraph branching rules.
2: **function** $SolveRecursively(G, k)$
3:    **if** $k < 0$ **then** Return NO.
4:    **if** G is not bumpy **then** Return YES.
5:    **if** $k = 0$ **then** Return NO.
6:    **if** Some rule from $\mathcal{A}$ can be applied to G **then**
7:       Find the first rule ϱ_A from $\mathcal{A}$ that can be applied to G. Apply ϱ_A to G and return the corresponding answer (might involve recursive calls to $SolveRecursively$).
8:    **if** Each bumpy connected component of G has less than $\Psi(\mathcal{L})$ vertices **then**
9:       Find the optimal solution for each component separately by brute-force.
10:       Let the solutions be $S_1, S_2, \dots, S_c$.
11:       **if** $\sum_{i=1}^{c} |S_i| \leq k$ **then** Return YES.
12:       **else** Return NO.
13:    Let C be the vertices of the bumpy connected component of G with at least $\Psi(\mathcal{L})$ vertices.
14:    Find a branching rule $(H, R, \mathcal{B})$ from the set $\mathcal{L}$ that can be applied to $G[C]$.
15:    Let ϕ be the corresponding isomorphism.
16:    **for** $B \in \mathcal{B}$ **do**
17:       **if** $SolveRecursively(G \backslash \phi(B), k - |B|)$ outputs YES **then** Return YES.
18:    Return NO.

---

4   The Generating Algorithm

In this section we describe the algorithm to generate a suitable list of subgraph branching rules.

For a fixed $\mathcal{F}$-SVD problem the input of the algorithm are the bump-inducing set $\mathcal{F}$, a function $Handled_\mathcal{A}$ which can identify the situations handled by the handmade branching and reduction rules in $\mathcal{A}$, and the target branching factor $\beta \in \mathbb{R}$. We assume that the handmade rules in $\mathcal{A}$ are correct in the context of the given $\mathcal{F}$-SVD problem, they can be applied in polynomial time, and that the branching rules have branching factors at most β. The output of the algorithm is an ordered list of subgraph branching rules $\mathcal{L}$, exhaustive with respect to $\mathcal{A}$, such that every rule in $\mathcal{L}$ has branching factor at most β. The algorithm will be called the $(\mathcal{F}, \mathcal{A}, \beta)$-Algorithm and its output satisfies the assumptions of Theorem 1.

4.1   Overview of the Algorithm

The algorithm maintains an ordered list and a set of connected bumpy graphs named L_{good} and L_{bad}, respectively. The list L_{good} stores graphs that already give rise to good subgraph branching rules, whereas the set L_{bad} represents the substructures for which the algorithm did not find any effective way to tackle them yet. The algorithm starts with L_{good} empty and L_{bad} being the set from Observation 2.

The algorithm works in rounds and in each round it tries to move as many graphs currently in L_{bad} to L_{good}. Firstly, for each graph in L_{bad}, the algorithm "colors red" the vertices that cannot have any outside neighbors. This process is done by the *Color* function introduced below. Secondly, for each now colored graph in L_{bad}, it checks whether the substructure can be handled by some hand-made rule in $\mathcal{A}$. If it does, the graph is moved from L_{bad} to the end of L_{good}. Otherwise, it designs a subgraph branching rule for it with the smallest branching factor it can achieve and if the branching factor of the produced rule is at most β, it again moves the graph from L_{bad} to the end of L_{good}. In one round, the algorithm repeats the above steps as long as possible. Once no graph from L_{bad} can be moved to L_{good} this way, the algorithm replaces all graphs in L_{bad} by all their 1-expansions and starts a next round. This corresponds to deepening the analysis, i.e., considering larger parts of the input graph at once.

4.2 *Color* function

Let H be a connected graph and F be a set of connected graphs. The vertex $v \in V(H)$ will be colored red, i.e., we put v into R, if all 1-expansions of H, where the vertex v has more neighbors in the 1-expansion of H than in H itself, are already also expansions of some graphs in F.

In our algorithm, H is some graph from L_{bad} and F is the list L_{good} (see Fig. 1 for an example).

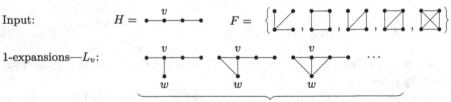

All of them expansions of some graphs in F? $\Rightarrow$ Color v red.

Output:　　　　H with $R = $ •——•——• after considering all of $V(H)$.

Fig. 1. Illustration of the *Color* function.

5　Generating Subgraph Branching Rules

Once we have a graph H together with its red vertices $R \subseteq V(H)$, we want to generate a correct subgraph branching rule for it with as small branching factor as possible.

5.1   Overview of the Approach

We start by brute forcing all the local solutions for the graph H, we keep only those that are inclusion-wise minimal, and we use them to get our initial set of branches $\mathcal{B}_{min}$. It is easy to see, that the resulting subgraph branching rule $(H, R, \mathcal{B}_{min})$ is correct, but not very efficient.

To improve this rule, we employ a function called *DominanceFree* (described below) which uses the red vertices R to filter out some unnecessary branches. Let us label the result as $\mathcal{B}_{df}$.

Finally, we further optimize $\mathcal{B}_{df}$ with the following observation.

Observation 3. *Let $(H, R, \mathcal{B})$ be a correct subgraph branching rule. For any $A \subseteq V(H), A \neq \emptyset$ construct the branches $\mathcal{B}_A = \{B \mid B \in \mathcal{B} \wedge A \not\subseteq B\} \cup \{A\}$. The subgraph branching rule $(H, R, \mathcal{B}_A)$ is correct.*

We greedily improve the branches $\mathcal{B}_{df}$ by repeatedly trying all possible replacements and picking those that minimize the branching factor the most. Let us label the result $\mathcal{B}_{adj}$.

The final generated subgraph branching rule is then $(H, R, \mathcal{B}_{adj})$.

5.2   *DominanceFree* function

The input of the function is a connected bumpy graph H, $R \subseteq V(H)$, and $\mathcal{B}$ such that $(H, R, \mathcal{B})$ is a correct subgraph branching rule.

The point is that if a vertex v has no neighbors outside H (the red vertices), then it might be more beneficial to have a different vertex in the solution instead of v. We call this the dominance between branches.

The basic idea is to take a subset R^* of the red vertices and replace all vertices of the solution in this set by the open neighborhood $N_H(R^*) \backslash R^*$. We only want to do that if this does not increase the size of the solution and if $H[R^*]$ is not bumpy. To increase the power of this notion, we do this in a graph $H' = H \backslash B^{del}$, where B^{del} is a set of vertices shared by both the branches.

Definition 9 (Dominated branch). *Let $(H, R, \mathcal{B})$ be a correct subgraph branching rule. We say that branch $B \in \mathcal{B}$ is dominated by branch $B_d \in \mathcal{B}$ if $B_d \neq B$ and there exists a subset $B^{del} \subsetneq B$ such that for $H' = H \backslash B^{del}$, $R' = R \backslash B^{del}$ there exists a subset $R^* \subseteq R', R^* \neq \emptyset$ such that the following holds:*

1. $H[R^*]$ is not bumpy,
2. $|R^* \cap B| \geq |N_{H'}(R^*) \backslash R^*| \geq 1$,
3. $B_d \subseteq (B \cup N_{H'}(R^*)) \backslash R^*$.

Note that if $N_{H'}(R^*) \backslash R^* = \emptyset$, then $B_d \subseteq B$, a case that cannot appear in $\mathcal{B}_{min}$.

Lemma 1. *If $(H, R, \mathcal{B})$ is a correct subgraph branching rule and branch $B \in \mathcal{B}$ is dominated by branch $B_d \in \mathcal{B}$, then $(H, R, \mathcal{B} \backslash \{B\})$ is a correct subgraph branching rule.*

The purpose of the *DominanceFree* function is to remove branches that are dominated by other branches. However, as there might be cycles of dominance, we have to be a little bit more careful. Consider directed graph $G_{\mathcal{B}} = (\mathcal{B}, E_{\mathcal{B}})$ such that $(B_i, B_j) \in E_{\mathcal{B}}$ if and only if B_i is dominated by B_j. Let $C_1, C_2, \ldots, C_c$ be the strongly connected components of $G_{\mathcal{B}}$. By $rep(C_i)$ we denote an arbitrary, but fixed, branch $B \in C_i$. A component C_i is called a *sink* component if there is no other component $C_j, i \neq j$ such that there exists an edge $(B_i, B_j) \in E_{\mathcal{B}}$ where $B_i \in C_i$ and $B_j \in C_j$. The *DominanceFree* function returns the branches $\mathcal{B}_{df} = \{rep(C_i) \mid i \in \{1, 2, \ldots, c\} \wedge C_i \text{ is a sink component}\}$.

6 Applying $(\mathcal{F}, \mathcal{A}, \beta)$-Algorithm to d-PVC

We are now going to show the specifics of applying the $(\mathcal{F}, \mathcal{A}, \beta)$-Algorithm to the d-PATH VERTEX COVER problem. It is easy to see that d-PVC equals $\mathcal{F}$-SVD for $\mathcal{F} = \{P_d\}$.

6.1 Handmade Rules

For the $(\mathcal{F}, \mathcal{A}, \beta)$-Algorithm to work for interesting values of β, we provide two handmade polynomial time reduction rules to $\mathcal{A}$ that are correct for d-PVC.

Reduction Rule 1 (Red component reduction for d-PVC.) Let (G, k) be an instance of d-PVC. Let $v \in V(G)$ be a vertex such that there are at least two P_d-free connected components C_1, C_2 in $G \backslash v$. If there is a P_d in $G[\{v\} \cup V(C_1) \cup V(C_2)]$, reduce (G, k) to instance $(G \backslash (\{v\} \cup V(C_1) \cup V(C_2)), k-1)$ which corresponds to taking v into a solution. Otherwise, let P_i^1 be the longest path in $G[\{v\} \cup V(C_1)]$ starting in v and let P_j^2 be the longest path in $G[\{v\} \cup V(C_2)]$ starting in v. Assume, without loss of generality, that $i \leq j$. Then, reduce the instance (G, k) to $(G \backslash V(C_1), k)$.

Reduction Rule 2 (Red star reduction for d-PVC, $d \geq 4$) Let (G, k) be the instance of d-PVC. Suppose there exists a subset $C \subseteq V(G)$, $|C| \leq \lfloor \frac{d}{2} \rfloor - 1$ for which there is a subset $L \subseteq V(G)$ such that $\forall v \in L, N(v) = C$ and $|L| \geq 2|C|$. Let $x \in L$. Then reduce instance (G, k) to instance $(G \backslash \{x\}, k)$.

We now discuss how to incorporate these reduction rules into the $(\mathcal{F}, \mathcal{A}, \beta)$-Algorithm. Note that as a part of $\mathcal{A}$, if the rule applies, we would make a call of $SolveRecursively(G \backslash (\{v\} \cup V(C_1) \cup V(C_2)), k-1)$, $SolveRecursively(G \backslash V(C_1), k)$, or $SolveRecursively(G \backslash \{x\}, k)$, respectively, and return the answer obtained. The following two lemmata describe the function $Handled_{\mathcal{A}}$.

Lemma 2. *For the case of the red component reduction rule, let H be a connected bumpy graph and $R \subseteq V(H)$ be its red vertices. If there is a vertex $v \in V(H)$ for which there are at least two d-path free connected components C_1, C_2 in $H \backslash v$ with $V(C_1), V(C_2) \subseteq R$ then the pair H, R is handled by the red component reduction rule, i.e., whenever any subgraph branching rule $(H, R, \mathcal{B})$ would apply to a graph G, the red component reduction rule would also apply to G.*

Lemma 3. *For the case of d-PVC, $d \geq 4$. Let H be a connected bumpy graph and $R \subseteq V(H)$ be its red vertices. If there is a subset $C \subseteq V(H), |C| \leq \lfloor \frac{d}{2} \rfloor - 1$ for which there is a subset $L \subseteq R$ such that $\forall v \in L, N(v) = C$ and $2|C| \leq |L|$, then the pair H, R is handled by the red star reduction rule.*

6.2 Obtained Results

With careful implementation the $(\mathcal{F}, \mathcal{A}, \beta)$-Algorithm together with our hand-made reduction rules is able to achieve the results as summarized in Table 1. Note that $\mathcal{F}$ is fixed to $\{P_d\}$, $\mathcal{A}$ is as described in the previous subsection and the only parameter that varies is β. The question is then whether the algorithm finishes with the given β or not. The table contains, for each d, the least values of β for which our implementation of the algorithm finished. The full source code of the implementation is available at https://github.com/generating-algorithms/generating-dpvc. We also provide a separate repository https://github.com/generating-algorithms/generating-dpvc-data with annotated descriptions of the obtained algorithms. These are basically logs of the successful computation paths taken by the algorithm and are, to some extent, verifiable by hand. Sadly, we were not able to improve the running time of 2-PVC, but we do not know whether it is a limitation of the algorithm itself or a limitation of time, space, and resources.

To better understand the behavior of the generating algorithm, we provide plots of the number of branching rules and time it takes to achieve target branching factor. The runs depicted in the plots were performed on a virtual computer with 255 CPU cores and 128 GB of RAM.

The main point we would like to emphasize is that for the cases of d-PVC, $d \geq 4$, the first algorithms outperforming the state of the art were found in a matter of seconds and minutes.

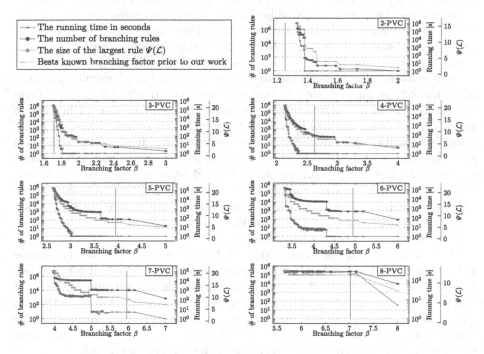

7   Future Research Directions

We provided a framework to generate parameterized branching algorithms tailored for specific vertex deletion problems. In comparison, the framework of Gramm et al. [22] is also suited for problems where the task is to either delete or even add *edges* to the graph. We wonder whether some of our ideas can be translated to the edge setting.

While there are rather few studies on computer generated algorithms with provable worst-case running time upper bounds, there are quite some papers that use computer aided *analysis* of algorithms. In particular, the *Measure & Conquer* approach, introduced by Fomin et al. [18], is popular especially for moderately exponential algorithms [28,32,33,41]. Here the idea is to use simple rules, while measuring the progress not only based on the number of vertices resolved, but also on how favorably the remaining graph is structured, e.g., how many vertices of rather low degree are present. The hope is to capture that some unfavorable branching significantly improves the structure so that a favorable branching appears subsequently. To accomplish this, the analysis of a single rule is often split into many cases, based, e.g., on the degrees of the vertices involved. The computer is then used to optimize the values assigned to favorable structures so as to prove the lowest possible worst-case running time upper bound. Other approaches trying to amortize between the rules with bad branching factors and those with good branching factors include *branching potential* [21] or *labeled search trees* [9]. See also Fernau and Raible [16] for an older survey of the topic.

It may seem interesting to combine the automated generation framework with a computer assisted analysis of the algorithm. However, first, it seems that the computer assisted analysis still requires a non-trivial amount of human intervention, e.g., in design of the measure and cases to be distinguished by the computer. Therefore it seems to be limited to algorithms with few branching rules and does not scale to thousands of rules. Second, the favorable structure we gain, if it can be captured in an automated manner at all, is then exploited in the immediate neighborhood of the finished branching to gain the advantage. Hence, we might possibly as well create a single branching rule encompassing both the structures and "amortize within the rule". Of course, many variants of such a rule would be necessary. This is the approach already prevalent in our framework. However, the sizes of the rules necessary might be beyond the reach of our implementation. The question is whether some transfer of "branching potential" or some other kind of advantage can be explicitly included in the construction of the rules in order to enable this advanced analysis.

Finally, an obvious open question is whether there are, e.g., some handmade rules that would help our algorithm generate a faster algorithm for VERTEX COVER (2-PVC). The fastest known algorithms of Chen, Kanj, and Xia [10] and Harris and Narayanaswamy [23] are rather complex to both analyze (both from the running time and correctness perspective) and implement. We made some experiments with the *struction* and *vertex-domination* rules from [10], but these did not seem to improve the performance of the generating algorithm.

References

1. Balasubramanian, R., Fellows, M.R., Raman, V.: An improved fixed-parameter algorithm for vertex cover. Inf. Process. Lett. **65**(3), 163–168 (1998). https://doi.org/10.1016/S0020-0190(97)00213-5
2. Brešar, B., Kardoš, F., Katrenič, J., Semanišin, G.: Minimum k-path vertex cover. Discret. Appl. Math. **159**(12), 1189–1195 (2011). https://doi.org/10.1016/j.dam.2011.04.008
3. Buss, J.F., Goldsmith, J.: Nondeterminism within P. SIAM J. Comput. 22(3), 560–572 (1993). https://doi.org/10.1137/0222038
4. Červený, R., Suchý, O.: Faster FPT algorithm for 5-path vertex cover. In: Rossmanith, P., Heggernes, P., Katoen, J. (eds.) 44th International Symposium on Mathematical Foundations of Computer Science, MFCS 2019, August 26–30, 2019, Aachen, Germany. LIPIcs, vol. 138, pp. 32:1–32:13. Schloss Dagstuhl - Leibniz-Zentrum für Informatik (2019). https://doi.org/10.4230/LIPIcs.MFCS.2019.32
5. Červený, R., Suchý, O.: Generating faster algorithms for d-path vertex cover (2021). arxiv.org/abs/2111.05896
6. Chandran, L.S., Grandoni, F.: Refined memorization for vertex cover. Inf. Process. Lett. **93**(3), 123–131 (2005). https://doi.org/10.1016/j.ipl.2004.10.003
7. Chang, M., Chen, L., Hung, L., Liu, Y., Rossmanith, P., Sikdar, S.: Moderately exponential time algorithms for the maximum bounded-degree-1 set problem. Discret. Appl. Math. **251**, 114–125 (2018). https://doi.org/10.1016/j.dam.2018.05.032
8. Chen, J., Kanj, I.A., Jia, W.: Vertex cover: Further observations and further improvements. J. Algorithms **41**(2), 280–301 (2001). https://doi.org/10.1006/jagm.2001.1186
9. Chen, J., Kanj, I.A., Xia, G.: Labeled search trees and amortized analysis: improved upper bounds for NP-hard problems. Algorithmica **43**(4), 245–273 (2005). https://doi.org/10.1007/s00453-004-1145-7
10. Chen, J., Kanj, I.A., Xia, G.: Improved upper bounds for vertex cover. Theor. Comput. Sci. **411**(40–42), 3736–3756 (2010). https://doi.org/10.1016/j.tcs.2010.06.026
11. Chen, J., Liu, L., Jia, W.: Improvement on vertex cover for low-degree graphs. Networks **35**(4), 253–259 (2000)
12. Cygan, M., et al.: Parameterized Algorithms. Springer, Cham (2015). https://doi.org/10.1007/978-3-319-21275-3
13. Downey, R.G., Fellows, M.R.: Fixed parameter tractability and completeness. In: Ambos-Spies, K., Homer, S., Schöning, U. (eds.) Complexity Theory: Current Research, Dagstuhl Workshop, February 2–8, 1992, pp. 191–225. Cambridge University Press (1992)
14. Fedin, S.S., Kulikov, A.S.: Automated proofs of upper bounds on the running time of splitting algorithms. In: Downey, R., Fellows, M., Dehne, F. (eds.) IWPEC 2004. LNCS, vol. 3162, pp. 248–259. Springer, Heidelberg (2004). https://doi.org/10.1007/978-3-540-28639-4_22
15. Fernau, H.: Parameterized algorithmics for d-Hitting Set. Int. J. Comput. Math. **87**(14), 3157–3174 (2010). https://doi.org/10.1080/00207160903176868
16. Fernau, H., Raible, D.: Searching trees: an essay. In: Chen, J., Cooper, S.B. (eds.) TAMC 2009. LNCS, vol. 5532, pp. 59–70. Springer, Heidelberg (2009). https://doi.org/10.1007/978-3-642-02017-9_9
17. Fomin, F.V., Gaspers, S., Kratsch, D., Liedloff, M., Saurabh, S.: Iterative compression and exact algorithms. Theor. Comput. Sci. **411**(7–9), 1045–1053 (2010). https://doi.org/10.1016/j.tcs.2009.11.012

18. Fomin, F.V., Grandoni, F., Kratsch, D.: A measure & conquer approach for the analysis of exact algorithms. J. ACM **56**(5), 25:1–25:32 (2009). https://doi.org/10. 1145/1552285.1552286

19. Fomin, F.V., Kratsch, D.: Exact Exponential Algorithms. Springer, Heidelberg (2010). https://doi.org/10.1007/978-3-642-16533-7

20. Funke, S., Nusser, A., Storandt, S.: On k-path covers and their applications. VLDB J. **25**(1), 103–123 (2016). https://doi.org/10.1007/s00778-015-0392-3

21. Gaspers, S.: Exponential Time Algorithms - Structures, Measures, and Bounds. VDM Verlag Dr. Mueller e.K. (2010). https://www.cse.unsw.edu.au/sergeg/ SergeBookETA2010_print.pdf

22. Gramm, J., Guo, J., Hüffner, F., Niedermeier, R.: Automated generation of search tree algorithms for hard graph modification problems. Algorithmica **39**(4), 321–347 (2004). https://doi.org/10.1007/s00453-004-1090-5

23. Harris, D.G., Narayanaswamy, N.S.: A faster algorithm for vertex cover parameterized by solution size. CoRR abs/2205.08022 (2022), https://arxiv.org/abs/2205. 08022

24. Katrenič, J.: A faster FPT algorithm for 3-path vertex cover. Inf. Process. Lett. **116**(4), 273–278 (2016). https://doi.org/10.1016/j.ipl.2015.12.002

25. Kojevnikov, A., Kulikov, A.S.: A new approach to proving upper bounds for MAX-2-SAT. In: Proceedings of the Seventeenth Annual ACM-SIAM Symposium on Discrete Algorithms, SODA 2006, Miami, Florida, USA, January 22–26, 2006, pp. 11–17. ACM Press (2006). http://dl.acm.org/citation.cfm?id=1109557.1109559

26. Kulikov, A.S.: Automated generation of simplification rules for SAT and MAXSAT. In: Bacchus, F., Walsh, T. (eds.) SAT 2005. LNCS, vol. 3569, pp. 430–436. Springer, Heidelberg (2005). https://doi.org/10.1007/11499107_35

27. Lewis, J.M., Yannakakis, M.: The node-deletion problem for hereditary properties is NP-complete. J. Comput. Syst. Sci. **20**(2), 219–230 (1980). https://doi.org/10. 1016/0022-0000(80)90060-4

28. Lokshtanov, Daniel, Saurabh, Saket, Suchý, Ondřej: Solving MULTICUT faster than 2^n. In: Schulz, Andreas S.., Wagner, Dorothea (eds.) ESA 2014. LNCS, vol. 8737, pp. 666–676. Springer, Heidelberg (2014). https://doi.org/10.1007/978-3-662-44777-2_55

29. Niedermeier, R., Rossmanith, P.: Upper bounds for vertex cover further improved. In: Meinel, C., Tison, S. (eds.) STACS 1999. LNCS, vol. 1563, pp. 561–570. Springer, Heidelberg (1999). https://doi.org/10.1007/3-540-49116-3_53

30. Niedermeier, R., Rossmanith, P.: An efficient fixed-parameter algorithm for 3-hitting set. J. Discrete Algorithms **1**(1), 89–102 (2003). https://doi.org/10.1016/ S1570-8667(03)00009-1

31. Novotný, M.: Design and analysis of a generalized canvas protocol. In: Samarati, P., Tunstall, M., Posegga, J., Markantonakis, K., Sauveron, D. (eds.) WISTP 2010. LNCS, vol. 6033, pp. 106–121. Springer, Heidelberg (2010). https://doi.org/10. 1007/978-3-642-12368-9_8

32. van Rooij, J.M.M., Bodlaender, H.L.: Exact algorithms for dominating set. Discret. Appl. Math. **159**(17), 2147–2164 (2011). https://doi.org/10.1016/j.dam.2011.07. 001

33. van Rooij, J.M.M., Bodlaender, H.L.: Exact algorithms for edge domination. Algorithmica **64**(4), 535–563 (2012). https://doi.org/10.1007/s00453-011-9546-x

34. Tsur, D.: Parameterized algorithm for 3-path vertex cover. Theor. Comput. Sci. **783**, 1–8 (2019). https://doi.org/10.1016/j.tcs.2019.03.013

35. Tsur, D.: An $O^*(2.619^k)$ algorithm for 4-path vertex cover. Discret. Appl. Math. **291**, 1–14 (2021). https://doi.org/10.1016/j.dam.2020.11.019

36. Tsur, D.: Faster parameterized algorithms for two vertex deletion problems. Theor. Comput. Sci. 940(Part), 112–123 (2023). https://doi.org/10.1016/j.tcs.2022.10.044
37. Tu, J.: A fixed-parameter algorithm for the vertex cover P_3 problem. Inf. Process. Lett. **115**(2), 96–99 (2015). https://doi.org/10.1016/j.ipl.2014.06.018
38. Tu, J., Jin, Z.: An FPT algorithm for the vertex cover P_4 problem. Discret. Appl. Math. **200**, 186–190 (2016). https://doi.org/10.1016/j.dam.2015.06.032
39. Xiao, M., Kou, S.: Exact algorithms for the maximum dissociation set and minimum 3-path vertex cover problems. Theor. Comput. Sci. **657**, 86–97 (2017). https://doi.org/10.1016/j.tcs.2016.04.043
40. Xiao, M., Kou, S.: Kernelization and parameterized algorithms for 3-path vertex cover. In: Proc. TAMC 2017, pp. 654–668 (2017). https://doi.org/10.1007/978-3-319-55911-7_47
41. Xiao, M., Nagamochi, H.: Exact algorithms for maximum independent set. Inf. Comput. **255**, 126–146 (2017). https://doi.org/10.1016/j.ic.2017.06.001

A New Width Parameter of Graphs Based on Edge Cuts: α-Edge-Crossing Width

Yeonsu Chang[1], O-joung Kwon[1,2(✉)] [ID], and Myounghwan Lee[1]

[1] Department of Mathematics, Hanyang University, Seoul, South Korea
{yeonsu,ojoungkwon,sycuel}@hanyang.ac.kr
[2] Discrete Mathematics Group, Institute for Basic Science (IBS), Daejeon, South Korea

Abstract. We introduce graph width parameters, called α-edge-crossing width and edge-crossing width. These are defined in terms of the number of edges crossing a bag of a tree-cut decomposition. They are motivated by edge-cut width, recently introduced by Brand et al. (WG 2022). We show that edge-crossing width is equivalent to the known parameter tree-partition-width. On the other hand, α-edge-crossing width is a new parameter; tree-cut width and α-edge-crossing width are incomparable, and they both lie between tree-partition-width and edge-cut width.

We provide an algorithm that, for a given n-vertex graph G and integers k and α, in time $2^{O((\alpha+k)\log(\alpha+k))}n^2$ either outputs a tree-cut decomposition certifying that the α-edge-crossing width of G is at most $2\alpha^2 + 5k$ or confirms that the α-edge-crossing width of G is more than k. As applications, for every fixed α, we obtain FPT algorithms for the LIST COLORING and PRECOLORING EXTENSION problems parameterized by α-edge-crossing width. They were known to be W[1]-hard parameterized by tree-partition-width, and FPT parameterized by edge-cut width, and we close the complexity gap between these two parameters.

Keywords: α-edge-crossing width · List Coloring · FPT algorithm

1 Introduction

Tree-width is one of the basic parameters in structural and algorithmic graph theory, which measures how well a graph accommodates a decomposition into a tree-like structure. It has an important role in the graph minor theory developed by Robertson and Seymour [14–16]. For algorithmic aspects, there are various

A full version of the paper is available at https://arxiv.org/abs/2302.04624.
Y. Chang, O. Kwon, and M. Lee are supported by the National Research Foundation of Korea (NRF) grant funded by the Ministry of Science and ICT (No. NRF-2021K2A9A2A11101617 and RS-2023-00211670). O. Kwon is also supported by Institute for Basic Science (IBS-R029-C1).

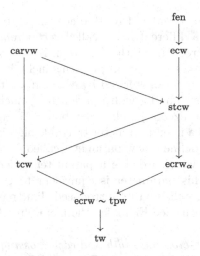

Fig. 1. The hierarchy of the mentioned width parameters. For two width parameters A and B, $A \to B$ means that every graph class of bounded A has bounded B, but there is a graph class of bounded B and unbounded A. Also, $A \sim B$ means that two parameters A and B are equivalent. fen, carvw, ecw, tcw, stcw, ecrw$_\alpha$, ecrw, tpw, and tw denote feedback edge set number, carving-width, edge-cut width, tree-cut width, slim tree-cut width, α-edge-crossing width, edge-crossing width, tree-partition-width, and tree-width, respectively.

fundamental problems that are NP-hard on general graphs, but fixed parameter tractable (FPT) parameterized by tree-width, that is, that can be solved in time $f(k)n^{O(1)}$ on n-vertex graphs of tree-width k for some computable function f. However, various problems are still W[1]-hard parameterized by tree-width. For example, LIST COLORING is W[1]-hard parameterized by tree-width [6].

Recently, edge counterparts of tree-width have been considered. One of such parameters is the *tree-cut width* of a graph introduced by Wollan [17]. Similar to the relationship between tree-width and graph minors, Wollan established a relationship between tree-cut width and weak immersions, and discussed structural properties. Since tree-cut width is a weaker parameter than tree-width, one could expect that some problems that are W[1]-hard parameterized by tree-width, are fixed parameter tractable parameterized by tree-cut width. But still several problems, including LIST COLORING, remain W[1]-hard parameterized by tree-cut width [3,7,8,10,11].

This motivates Brand et al. [2] to consider a more restricted parameter called the *edge-cut width* of a graph. For the edge-cut width of a graph G, a maximal spanning forest F of G is considered as a decomposition tree. For each vertex v of F, the *local feedback edge set* of v is the number of edges $e \in E(G) \backslash E(F)$ where the unique cycle of the graph obtained from F by adding e contains v, and the *edge-cut width* of F is the maximum local feedback edge set plus one over all vertices of G. The edge-cut width of G is the minimum edge-cut width among all maximal spanning forests of G. The edge-cut width with respect to a maximal spanning forest was also considered by Bodlaender [1] to bound the tree-width of certain graphs, with a different name called *vertex remember number*. Brand et al. showed that the tree-cut width of a graph is at most its edge-cut width. Furthermore, they showed that several problems including LIST COLORING are fixed parameter tractable parameterized by edge-cut width.

A natural question is to find a width parameter f such that graph classes of bounded f strictly generalize graph classes of bounded edge-cut width, and also LIST COLORING admits a fixed parameter tractable algorithm parameterized by f. This motivates us to define a new parameter called α-*edge-crossing width*. By relaxing the condition, we also define a parameter called *edge-crossing width*, but it turns out that this parameter is equivalent to tree-partition-width [5]. Recently, Ganian and Korchemna [9] introduced slim tree-cut width which also generalizes edge-cut width. See Fig. 1 for the hierarchy of new parameters and known parameters.

We define the α-*edge-crossing width* and *edge-crossing width* of a graph. For a graph G, a pair $\mathcal{T} = (T, \mathcal{X})$ of a tree T and a collection $\mathcal{X} = \{X_t \subseteq V(G) : t \in V(T)\}$ of disjoint sets of vertices in G, called bags (allowing empty bags), with the property $\bigcup_{t \in V(T)} X_t = V(G)$ is called the *tree-cut decomposition* of G. For a node $t \in V(T)$, let $T_1, T_2, \cdots, T_m$ be the connected components of $T - t$, and let $\mathrm{cross}_\mathcal{T}(t)$ be the number of edges incident with two distinct sets in $\{\bigcup_{t \in V(T_i)} X_t : 1 \leqslant i \leqslant m\}$. We say that such an edge crosses X_t. The *crossing number* of $\mathcal{T}$ is $\max_{t \in V(T)} \mathrm{cross}_\mathcal{T}(t)$, and the *thickness* of $\mathcal{T}$ is $\max_{t \in V(T)} |X_t|$. For a positive integer α, the α-*edge-crossing width* of a graph G, denoted by $\mathrm{ecrw}_\alpha(G)$, is the minimum crossing number over all tree-cut decompositions of G whose thicknesses are at most α. The *edge-crossing width* of $\mathcal{T}$ is the maximum of the crossing number and the thickness of $\mathcal{T}$. The *edge-crossing width* of G, denoted by $\mathrm{ecrw}(G)$, is the minimum *edge-crossing width* over all tree-cut decompositions of G.

It is not difficult to see that the 1-edge-crossing width of a graph is at most its edge-cut width minus one, as we can take the completion of its optimal maximal spanning forest for edge-cut width into a tree as a tree-cut decomposition with small crossing number.

We provide an FPT approximation algorithm for α-edge-crossing width. We adapt an idea for obtaining an FPT approximation algorithm for tree-cut width due to Kim et al. [12].

Theorem 1. *Given an n-vertex graph G and two positive integers α and k, one can in time $2^{\mathcal{O}((\alpha+k)\log(\alpha+k))}n^2$ either*

- *output a tree-cut decomposition of G with thickness at most α and crossing number at most $2\alpha^2 + 5k$, or*
- *correctly report that $\mathrm{ecrw}_\alpha(G) > k$.*

As applications of α-edge-crossing width, we show that LIST COLORING and PRECOLORING EXTENSION are FPT parameterized by α-edge-crossing width. They were known to be W[1]-hard parameterized by tree-cut width (and so by tree-partition-width) [7], and FPT parameterized by edge-cut width [2] and by slim tree-cut width [9]. We close the complexity gap between these parameters.

Theorem 2. *For a fixed positive integer α, the* LIST COLORING *and* PRECOLORING EXTENSION *problems are FPT parameterized by α-edge-crossing width.*

We remark that the EDGE-DISJOINT PATHS problem is one of the W[1]-hard problems parameterized by tree-width that motivates to study width parameters based on edge cuts. Ganian and Ordyniak [10] recently proved that EDGE-DISJOINT PATHS is NP-hard on graphs admitting a vertex cover of size 3. This implies that for every $\alpha \geqslant 3$, this problem is NP-hard on graphs of α-edge-crossing width 0.

This paper is organized as follows. In Sect. 2, we give basic definitions and notations. In Sect. 3, we establish the relationship between width parameters as presented in Fig. 1. We present an FPT approximation algorithm for α-edge-crossing width in Sect. 4 and discuss algorithmic applications in Sect. 5. We conclude and present some open problems in Sect. 6.

Proofs of statements marked with "$\star$" are deferred to the full version.

2   Preliminaries

For a set X and a positive integer n, we call $\binom{X}{n}$ the set of all subsets of X of size exactly n. Let $\mathbb{N}$ be the set of all non-negative integers, and for a positive integer n, let $[n] = \{1, 2, \cdots, n\}$.

For a graph G, we denote by $V(G)$ and $E(G)$ the vertex set and the edge set of G, respectively. Let G be a graph. For a set S of vertices in G, let $G[S]$ denote the subgraph of G induced by S, and let $G - S$ denote the subgraph of G obtained by removing all the vertices in S. For $v \in V(G)$, let $G - v := G - \{v\}$. For an edge e of G, let $G - e$ denote the graph obtained from G by deleting e. The set of neighbors of a vertex v is denoted by $N_G(v)$, and the *degree* of v is the size of $N_G(v)$. For two disjoint sets S_1, S_2 of vertices in G, we denote by $\delta_G(S_1, S_2)$ the set of edges incident with both S_1 and S_2 in G.

An edge e of a connected graph G is a *cut edge* if $G - e$ is disconnected. A connected graph is *2-edge-connected* if it has no cut edges.

For two graphs G and H, we say that G is a *subdivision* of H if G can be obtained from H by subsequently subdividing edges.

A *tree-decomposition* of a graph G is a pair of a tree T and a family of sets $\{B_t\}_{t \in V(T)}$ of vertices in G such that (1) $V(G) = \bigcup_{t \in V(T)} B_t$, (2) for every edge uv of G, there exists a node t of T such that $u, v \in B_t$, and (3) for every vertex v of G, the set $\{t \in V(T) : v \in B_t\}$ induces a subtree of T. The *width* of a tree-decomposition is $\max_{t \in V(T)} |B_t| - 1$, and the *tree-width* of a graph, denoted by $\operatorname{tw}(G)$, is the maximum width over all its tree-decompositions.

A tree-decomposition $(T, \{B_t\}_{t \in V(T)})$ is called *rooted* if T is a rooted tree. A rooted tree-decomposition $(T, \{B_t\}_{t \in V(T)})$ with a root r is called *nice* if (1) for a non-root leaf t of T, $|B_t| = 1$, and (2) if a node t is not a non-root leaf of T, then it is one of the following;

- (Forget node) t has one child t' and $B_t = B_{t'} \backslash \{v\}$ for some $v \in B_{t'}$.
- (Introduce node) t has one child t' and $B_t = B_{t'} \cup \{v\}$ for some $v \in V(G) \backslash B_{t'}$.
- (Join node) t has exactly two children t_1 and t_2 and $B_t = B_{t_1} = B_{t_2}$.

Theorem 3 (Korhonen [13]). *There is an algorithm running in $2^{\mathcal{O}(w)} n$ time, that given an n-vertex graph G and an integer w, either outputs a tree-decomposition of G of width at most $2w + 1$ or reports that the tree-width of G is more than w.*

By applying the following lemma, we can find a nice tree-decomposition.

Lemma 1 (folklore; see Lemma 7.4 in [4]). *Given a tree-decomposition of an n-vertex graph G of width w, one can construct a nice tree-decomposition $(T, \mathcal{B})$ of width w with $|V(T)| = \mathcal{O}(wn)$ in $\mathcal{O}(w^2 \cdot \max(|V(T)|, n))$ time.*

3 Relationships Between Width Parameters

We compare width parameters as presented in Fig. 1 in the full version. Among the relations, we prove the following inequality. This will be mainly used in our approximation algorithm for α-edge-crossing width in Sect. 4.

Lemma 2. *For every graph G and every positive integer α, $\mathrm{tw}(G) \leqslant 5 \, \mathrm{ecrw}(G) - 1$ and $\mathrm{tw}(G) \leqslant 3 \, \mathrm{ecrw}_\alpha(G) + 2\alpha - 1$.*

Proof. We show the first inequality. Let $k = \mathrm{ecrw}(G)$. Let $\mathcal{T} = (T, \{X_t\}_{t \in V(T)})$ be a tree-cut decomposition of G of edge-crossing width $\mathrm{ecrw}(G)$. We consider $\mathcal{T}$ as a rooted tree-cut decomposition with root node r. Let $\sigma : V(G) \to V(T)$ be the function where v is contained in $X_{\sigma(v)}$.

We construct a rooted tree-decomposition $(T, \{B_t\}_{t \in V(T)})$ as follows. For each node t of T, let F_t be the set of edges ab of G satisfying that either

- (type 1) the path between $\sigma(a)$ and $\sigma(b)$ in T contains t as an internal node, or
- (type 2) the path between $\sigma(a)$ and $\sigma(b)$ in T has length at least 1, and contains t as an end node, and the subtree of T rooted at t does not contain the end node of this path other than t.

Let B_t be the union of X_t and the set of vertices of G incident with an edge in F_t.

Since $\mathrm{cross}_T(t) \leqslant k$, the number of the edges of type 1 is at most k. So, because of this type, we put at most $2k$ vertices into B_t. If t is the root node, then there is no edge of type 2. Assume that t is not the root node, and let t' be its parent. For type 2, $\sigma(a) = t$ or $\sigma(b) = t$, and either the vertex in $\{a, b\}$ that is not contained in X_t is contained in $B_{t'}$ or ab crosses $X_{t'}$. Since the number of edges ab of type 2 crossing $X_{t'}$ is at most k, we may add at most k vertices other than $X_t \cup X_{t'}$. As $|X_t \cup X_{t'}| \leqslant 2k$, in total, we have that $|B_t| \leqslant (2k) + (2k) + k = 5k$.

We now verify that $(T, \{B_t\}_{t \in V(T)})$ is a tree-decomposition. Since $X_t \subseteq B_t$ for each $t \in V(T)$, every vertex of G appears in some bag. Let $ab \in E(G)$ and assume that there is no bag of T containing both a and b. If the path between $\sigma(a)$ and $\sigma(b)$ in T contains some node t of T as an internal node, then by the construction, B_t contains both a and b. Assume that there is no node in T that is an internal node of the path between $\sigma(a)$ and $\sigma(b)$ in T. This means that $\sigma(a)$ is adjacent to $\sigma(b)$ in T. By symmetry, we assume that $\sigma(a)$ is the parent of $\sigma(b)$. Then ab is an edge of type 2 for the node $t = \sigma(b)$, and thus, $\{a, b\} \subseteq B_{\sigma(b)}$. Thus, $(T, \{B_t\}_{t \in V(T)})$ satisfies the second condition.

Lastly, to see that $(T, \{B_t\}_{t \in V(T)})$ satisfies the third condition, let $a \in V(G)$. For every vertex $b \in V(G)$ adjacent to a in G, let P_{ab} be the path between $\sigma(a)$ and $\sigma(b)$ in T. We add a to B_x for all $x \in V(P_{ab} - \{a, b\})$. This is the only procedure to add a to some B_x. Since $B_{\sigma(a)}$ contains a, the subtree of T induced by the union of all t where $a \in B_t$ is connected. This implies that $(T, \{B_t\}_{t \in V(T)})$ satisfies the third condition.

It is straightforward to verify the second inequality with the same argument.

We show that stcw $\leqslant$ ecrw$_\alpha$ and ecrw$_\alpha \nleqslant$ stcw. We also show that ecrw$_\alpha \nleqslant$ tcw and tcw $\nleqslant$ ecrw$_\alpha$. For positive integers k and n, let $S_{k,n}$ be the graph obtained from $K_{1,n}$ by replacing each edge with k internally vertex-disjoint paths of length 2.

Lemma 3 (Ganian and Korchemna [9]). *The set* $\{S_{2,n} : n \in \mathbb{N}\}$ *has unbounded slim tree-cut width.*

Lemma 4 ($\star$). *(1)* $\{S_{3,n} : n \in \mathbb{N}\}$ *has 1-edge-crossing width at most 2.*
(2) $\{S_{3,n} : n \in \mathbb{N}\}$ *has unbounded tree-cut width.*

Lemma 5. *For every positive integer α,* ecrw$_\alpha \nleqslant$ stcw *and* ecrw$_\alpha \nleqslant$ tcw.

Proof. Note that $S_{2,n}$ is isomorphic to an induced subgraph of $S_{3,n}$. So, by (1) of Lemma 4, $S_{2,n}$ has 1-edge crossing width at most 2. On the other hand, Lemma 3 shows that $\{S_{2,n} : n \in \mathbb{N}\}$ has unbounded slim tree-cut width. This shows that ecrw$_1 \nleqslant$ stcw. Since ecrw$_1 \leqslant$ ecrw$_\alpha$, we have ecrw$_\alpha \nleqslant$ stcw.

By (1) and (2) of Lemma 4, $\{S_{3,n} : n \in \mathbb{N}\}$ has 1-edge crossing width at most 2, but unbounded tree-cut width. Therefore, ecrw$_1 \nleqslant$ tcw. Since ecrw$_1 \leqslant$ ecrw$_\alpha$, we have ecrw$_\alpha \nleqslant$ tcw.

We now show that tcw $\nleqslant$ ecrw$_\alpha$. For all positive integers k and n, we construct a graph G_k^n as follows. Let $A := \{a_i : i \in [n]\}$, $B = \binom{A}{2}$, and $B_k = \{(W, \ell) : W \in B$ and $\ell \in [k]\}$. Let G_k^n be the graph such that $V(G_k^n) = A \cup B_k$, and for $a \in A$ and $(W, \ell) \in B_k$, a is adjacent to (W, ℓ) in $E(G_k^n)$ if and only if $a \in W$.

Lemma 6. *For every positive integer α,* tcw $\nleqslant$ ecrw$_\alpha$.

Proof. Observe that $\{G_k^{\alpha+1} : k \in \mathbb{N}\}$ has unbounded α-edge-crossing width.
We claim that for every k, $G_k^{\alpha+1}$ has tree-cut width at most $\alpha + 1$. Let (A, B_k) be the bipartition of $G_k^{\alpha+1}$ given by the definition. Let T be a star with center t

and leaves $t_1, \ldots, t_{k\binom{\alpha+1}{2}}$. Let $X_t = A$ and each X_{t_i} consists of a vertex of B_k. Let $\mathcal{T} = (T, \{X_v\}_{v \in V(T)})$. Note that the 3-center of H_t has only vertices of A. Thus, it has at most $\alpha + 1$ vertices. Also, for every edge e of T, $\mathrm{adh}_\mathcal{T}(e) \leqslant 2$. So, $\mathcal{T}$ is a tree-cut decomposition of tree-cut width at most $\alpha + 1$.

To show $\mathrm{stcw} \leqslant \mathrm{ecrw}_\alpha$, we use another equivalent parameter called *super edge-cut width* introduced by Ganian and Korchemna [9]. The *super edge-cut width* $\mathrm{sec}(G)$ of a graph G is defined as the minimum edge-cut width of (H, T) over all supergraphs H of G and maximal spanning forests T of H.

Lemma 7 ($\star$). *For every positive integer α, $\mathrm{sec} \leqslant \mathrm{ecrw}_\alpha$.*

4   An FPT Approximation Algorithm for α-Edge-Crossing Width

We present an FPT approximation algorithm for α-edge-crossing width. We similarly follow the strategy to obtain a 2-approximation algorithm for tree-cut width designed by Kim et al. [12]. We formulate a new problem called CONSTRAINED STAR-CUT DECOMPOSITION, which corresponds to decomposing a large leaf bag in a tree-cut decomposition, and we want to apply this subalgorithm recursively. By Lemma 2, we can assume that a given graph admits a tree-decomposition of bounded width, and we design a dynamic programming to solve CONSTRAINED STAR-CUT DECOMPOSITION on graphs of bounded tree-width.

For a weight function $\gamma : V(G) \to \mathbb{N}$ and a non-empty vertex subset $S \subseteq V(G)$, we define $\gamma(S) := \sum_{v \in S} \gamma(v)$ and $\gamma(\varnothing) := 0$.

CONSTRAINED STAR-CUT DECOMPOSITION

Input : A graph G, two positive integers α, k, and a weight function $\gamma : V(G) \to \mathbb{N}$

Question : Determine whether there is a tree-cut decomposition $\mathcal{T} = (T, \{X_t\}_{t \in V(T)})$ of G such that

- T is a star with center t_c and it has at least one leaf,
- $|X_{t_c}| \leqslant \alpha$ and $\mathrm{cross}_\mathcal{T}(t_c) \leqslant k$,
- for each leaf t of T, $\gamma(X_t) \leqslant \alpha^2 + 2k$ and $|\delta_G(X_t, X_{t_c})| \leqslant \alpha^2 + k$, and
- there is no leaf q of T such that $X_q = V(G)$.

The following lemma explains how we will adapt an algorithm for CONSTRAINT STAR-CUT DECOMPOSITION.

Lemma 8. *Let G be a graph, let α, k be positive integers, and let S be a set of vertices in G. Assume that $|S| \geqslant \alpha + 1$ and $|\delta_G(S, V(G) \backslash S)| \leqslant 2\alpha^2 + 4k$. For each vertex $v \in S$, let $\gamma_S(v) = |\delta_G(\{v\}, V(G) \backslash S)|$.*

If $\mathrm{ecrw}_\alpha(G) \leqslant k$, then $(G[S], \alpha, k, \gamma_S)$ is a Yes-instance of CONSTRAINT STAR-CUT DECOMPOSITION.

Proof. Let $\mathcal{T} = (T, \{X_t\}_{t \in V(T)})$ be a tree-cut decomposition of G of thicknesses at most α and crossing number at most k. For an edge $e = uv$ of T, let $T_{e,u}$ and $T_{e,v}$ be two subtrees of $T - uv$ which contain u and v, respectively.

We want to identify a node t_c of T that will correspond to the central node of the resulting star decomposition. First, we define an extension γ on $V(G)$ of the weight function γ_S on S as $\gamma(v) = \gamma_S(v)$ if $v \in S$ and $\gamma(v) = 0$, otherwise. We orient an edge $e = xy \in E(T)$ from x to y if the edge e satisfies at least one of the following rules; (Rule 1) $S \cap (\bigcup_{t \in V(T_{e,x})} X_t) = \varnothing$ and (Rule 2) $\gamma(\bigcup_{t \in V(T_{e,y})} X_t) > \alpha^2 + 2k$. Note that an edge may have no direction.

Claim ($\star$). Every edge has at most one direction.

So, T has a node whose incident edges have no direction or a direction to the node. Take such a node as a central node t_c. Let $T_1, \ldots, T_m$ be the connected components of $T - t_c$ such that for every $i \in [m]$, $\bigcup_{t \in V(T_i)} X_t \cap S \neq \varnothing$. Note that there is at least one such component, because $|S| > \alpha$ and $|X_{t_c}| \leqslant \alpha$.

Let $(T', \{X'_t\}_{t \in V(T')})$ be a tree-cut decomposition of $G[S]$ such that (1) T' is a star with the central node t_c and leaves $t_1, \ldots, t_m$, (2) $X'_{t_c} = X_{t_c} \cap S$, and (3) for every $i \in [m]$, $X'_{t_i} = \left(\bigcup_{t \in V(T_i)} X_t \right) \cap S$. We claim that $(T', \{X'_t\}_{t \in V(T')})$ satisfies the conditions of answer of CONSTRAINED STAR-CUT DECOMPOSITION. By the construction of decomposition, the first condition holds. Since $X'_{t_c} = X_{t_c} \cap S$ and the crossing number of t_c in $(T, \{X_t\}_{t \in V(T)})$ is at most k, the second condition also holds.

Let t_i be a leaf of T' and let $V_i = \left(\bigcup_{t \in V(T_i)} X_t \right)$. By Rule 2 of the orientation, we have that $\gamma(X'_{t_i}) \leqslant \alpha^2 + 2k$. Let t' be the node in T_i that is adjacent to t_c in T. Since $|X_{t_c}| \leqslant \alpha$ and $|X_{t'}| \leqslant \alpha$, we have $|\delta_G(X_{t_c}, X_{t'})| \leqslant \alpha^2$. Furthermore, because $\mathrm{cross}_T(t') \leqslant k$, we have $|\delta_G(X_{t_c}, V_i \backslash X_{t'})| \leqslant k$. Thus, we have $|\delta_{G[S]}(X'_{t_c}, X'_{t_i})| \leqslant |\delta_G(X_{t_c}, V_i)| \leqslant \alpha^2 + k$.

Lastly, we claim that there is no leaf q of T' such that $X'_q = S$. Suppose that there is a leaf q of T' such that $X'_q = S$. This means that there is no other leaf in T' and $X'_{t_c} = \varnothing$. Let T^* be the connected component of $T - t_c$ for which $\bigcup_{t \in V(T^*)} X_t \cap S = X'_q$. Then for other connected component T^{**} of $T - t_c$, $\bigcup_{t \in V(T^{**})} X_t \cap S = \varnothing$, and therefore the edge of T between T^{**} and the node t_c is oriented towards t_c. Then the edge of T between T^* and t_c should be oriented towards T^*, because $X_{t_c} \cap S = \varnothing$. This contradicts the choice of t_c. This proves the lemma.

We now devise an algorithm for CONSTRAINT STAR-CUT DECOMPOSITION on graphs of bounded tree-width.

Lemma 9. *Let (G, α, k, γ) be an instance of CONSTRAINED STAR-CUT DECOMPOSITION and let $(T, \{B_t\}_{t \in V(T)})$ be a nice tree-decomposition of width at most w. In $2^{\mathcal{O}((k+w)\log(w(\alpha+k)))}|V(T)|$ time, one can either output a solution of (G, α, k, γ), or correctly report that (G, α, k, γ) is a No-instance.*

Proof. Let r be the root of T. We design a dynamic programming to compute a solution of CONSTRAINED STAR-CUT DECOMPOSITION. For each node t of T, let A_t be the union of all bags $B_{t'}$ where t' is a descendant of t in T.

Let $Z \subseteq V(G)$ be a set. A pair $(\mathcal{X}, \mathcal{P})$ of a sequence $\mathcal{X} = (X_0, X_1, \ldots, X_{2k}, Y)$ and a partition $\mathcal{P}$ of Y is *legitimate* with respect to Z if

- $X_0, X_1, \ldots, X_{2k}, Y$ are pairwise disjoint subsets of Z that are possibly empty,
- $(\bigcup_{i \in \{0,1,\ldots,2k\}} X_i) \cup Y = Z$,
- $|X_0| \leqslant \alpha$,
- for each $i \in [2k]$, $|\delta_G(X_i, X_0)| \leqslant \alpha^2 + k$ and $\gamma(X_i) \leqslant \alpha^2 + 2k$,
- $\sum_{\{i,j\} \in \binom{[2k]}{2}} |\delta_G(X_i, X_j)| \leqslant k$,
- for each $i \in [2k]$ and each $P \in \mathcal{P}$, $|\delta_G(X_0, P)| \leqslant \alpha^2 + k$, and $|\delta_G(X_i, P)| = 0$,
- for any distinct sets $P_i, P_j \in \mathcal{P}$, $|\delta_G(P_i, P_j)| = 0$.

Claim $(\star)$. (G, α, k, γ) is a Yes-instance if and only if there is a legitimate pair $(\mathcal{X}, \mathcal{P})$ with respect to $V(G)$ with $\mathcal{X} = (X_0, X_1, \ldots, X_{2k}, Y)$ such that any set of $X_1, \ldots, X_{2k}$ or a set of $\mathcal{P}$ is not the whole set $V(G)$.

For a legitimate pair $(\mathcal{X}, \mathcal{P})$ with respect to Z and $Z' \subseteq Z$, let $\mathcal{X}|_{Z'} = (X_0 \cap Z', X_1 \cap Z', \ldots, X_{2k} \cap Z', Y \cap Z')$ and $\mathcal{P}|_{Z'} = \{P \cap Z' : P \in \mathcal{P}, P \cap Z' \neq \varnothing\}$. We can see that $(\mathcal{X}|_{Z'}, \mathcal{P}|_{Z'})$ is legitimate with respect to Z', because the constraints are the number of vertices in a set, the number of edges between two sets, and the sum of γ-values. Based on this fact, we will recursively store information about all legitimate pairs with respect to A_t for nodes t.

Let $t \in V(T)$. A tuple $(I, \mathcal{Q}, C_1, C_2, D_1, D_2, a, b)$ is a *valid tuple* at t if

- $I : B_t \to \{0, 1, \ldots, 2k, 2k+1\}$,
- $\mathcal{Q}$ is a partition of $I^{-1}(2k+1)$,
- $C_1 : [2k] \to \{0, 1 \ldots, \alpha^2 + 2k\}$,
- $C_2 : \mathcal{Q} \to \{0, 1 \ldots, \alpha^2 + 2k\}$,
- $D_1 : [2k] \to \{0, 1 \ldots, \alpha^2 + k\}$,
- $D_2 : \mathcal{Q} \to \{0, 1 \ldots, \alpha^2 + k\}$, and
- a, b are two integers with $0 \leqslant a \leqslant \alpha$ and $0 \leqslant b \leqslant k$.

A valid tuple $(I, \mathcal{Q}, C_1, C_2, D_1, D_2, a, b)$ at a node t represents a legitimate pair $(\mathcal{X}, \mathcal{P}) = ((X_0, X_1, \ldots, X_{2k}, Y), \mathcal{P})$ with respect to A_t if

- for every $v \in B_t$, $I(v) = i$ if and only if $v \in \begin{cases} X_i & \text{if } 0 \leqslant i \leqslant 2k \\ Y & \text{if } i = 2k+1 \end{cases}$
- $\mathcal{Q} = \mathcal{P}|_{B_t}$,
- for each $i \in [2k]$, $C_1(i) = \gamma(X_i)$,
- for each $Q \in \mathcal{Q}$, $C_2(Q) = \gamma(Q)$,
- for each $i \in [2k]$, $D_1(i) = |\delta_G(X_i \cap A_t, X_0 \cap A_t)|$,
- for each $Q \in \mathcal{Q}$, $D_2(Q) = |\delta_G(Q \cap A_t, X_0 \cap A_t)|$,
- $a = |X_0 \cap A_t|$,
- $b = \sum_{\{i,j\} \subseteq \binom{[2k]}{2}} |\delta_G(X_i, X_j)|$.

We say that a valid tuple is a *record* at t if it represents some legitimate pair with respect to A_t. Let $\mathcal{R}(t)$ be the set of all records at t. It is not difficult to verify that there is a legitimate pair with respect to A_t if and only if there is a record at t. So, $\mathcal{R}(r) \neq \varnothing$ if and only if (G, α, k, γ) is a Yes-instance.

It is known that there is a constant d such that the number of partitions of a set of m elements is at most dm^m. We define a function

$$\zeta(x) = (2k + 2)^x (dx^x)(\alpha^2 + 2k + 1)^{2(x-1)+4k}(\alpha + 1)(k + 1).$$

Observe that if B_t has size q, then the number of all possible valid tuples at t is at most $\zeta(q)$. Note that $\zeta(q) = 2^{\mathcal{O}((q+k)\log(q(\alpha+k)))}$. We describe how to store all records in $\mathcal{R}(t)$ for each node t of T. Due to the space constraint, we only deal with an introduction node.

(Case. t is an introduce node with child t' such that $B_t = B_{t'} \cup \{v\}$.)
We construct a set $\mathcal{R}^*$ from $\mathcal{R}(t')$ as follows. Let

$$\mathcal{J}' = (I', \mathcal{Q}', C_1', C_2', D_1', D_2', a', b') \in \mathcal{R}(t').$$

For every $i \in \{0, 1, \ldots, 2k, 2k + 1\}$ and $Q^* \in \mathcal{Q}' \cup \{\varnothing\}$ when $i = 2k + 1$, we construct a new tuple $\mathcal{J} = (I, \mathcal{Q}, C_1, C_2, D_1, D_2, a, b)$ as follows:

- $I(w) = \begin{cases} I'(w) \text{ if } w \in B_{t'} \\ i \qquad \text{ if } w = v \end{cases}$,

- $\mathcal{Q} = \begin{cases} \mathcal{Q}' & \text{if } 0 \leqslant I(v) \leqslant 2k \\ (\mathcal{Q}'\backslash\{Q^*\}) \cup \{Q^* \cup \{v\}\} & \text{if } I(v) = 2k + 1 \end{cases}$,

- for every $j \in [2k]$, $C_1(j) = C_1'(j) + \gamma(I^{-1}(j))$,

- for every $Q \in \mathcal{Q}$, $C_2(Q) = \begin{cases} C_2'(Q) & \text{if } Q \neq Q^* \cup \{v\} \\ C_2'(Q^*) + \gamma(v) & \text{if } Q = Q^* \cup \{v\} \end{cases}$,

- for every $j \in [2k]$, $D_1(j) = D_1'(j) + |\delta_G(I^{-1}(j_2), X_0 \cap B_{t'})|$,

- for every $Q \in \mathcal{Q}$, $D_2(Q) = \begin{cases} D_2'(Q) & \text{if } Q \neq Q^* \cup \{v\} \\ D_2'(Q^*) + |\delta_G(\{v\}, X_0 \cap B_{t'})| & \text{if } Q = Q^* \cup \{v\} \end{cases}$,

- $a = \begin{cases} a' + 1 \text{ if } I(v) = 0 \\ a' \qquad \text{otherwise} \end{cases}$,

- $b = \begin{cases} b' + |\delta_G(\{v\}, I^{-1}([2k]\backslash\{I(v)\}))| \text{ if } 1 \leqslant I(v) \leqslant 2k \\ b' \qquad\qquad\qquad\qquad\qquad\qquad \text{otherwise} \end{cases}$.

We add this tuple to $\mathcal{R}^*$ whenever it is valid and

- if $1 \leqslant I(v) \leqslant 2k$, then there is no edge between v and $I^{-1}(2k + 1)$ in G,
- if $I(v) = 2k + 1$, then there is no edge between v and $I^{-1}(\{1, \ldots, 2k\})$ in G and there is no edge between v and $I^{-1}(2k + 1)\backslash(Q^* \cup \{v\})$ in G.

We claim that $\mathcal{R}^* = \mathcal{R}(t)$. First we show that $\mathcal{R}^* \subseteq \mathcal{R}(t)$. Let $\mathcal{J}$ be a valid tuple constructed as above from $\mathcal{J}' \in \mathcal{R}(t')$. We have to show that $\mathcal{J}$ represents some legitimate pair with respect to A_t. Since $\mathcal{J}' \in \mathcal{R}(t')$, it represents a legitimate pair $(\mathcal{X}' = (X_0', \ldots, X_{2k}', Y'), \mathcal{P}')$ with respect to $A_{t'}$.

If $0 \leqslant i \leqslant 2k$, then we obtain $\mathcal{X}$ from $\mathcal{X}'$ by replacing X_i' with $X_i' \cup \{v\}$ and set $\mathcal{P} = \mathcal{P}'$. If $i = 2k + 1$, then we obtain $\mathcal{X}$ from $\mathcal{X}'$ by replacing Y' with

$Y' \cup \{v\}$ and adding v to the part $P^* \in \mathcal{P}'$ with $P^* \cap B_t = Q^*$ when $Q^* \in \mathcal{Q}'$ or adding a single part $\{v\}$ when $Q^* = \varnothing$, to obtain a new partition $\mathcal{P}$ of $Y' \cup \{v\}$. Then it is straightforward to verify that $(\mathcal{X}, \mathcal{P})$ is a legitimate pair with respect to A_t and $\mathcal{J}$ represents it. This shows that $\mathcal{J} \in \mathcal{R}(t)$.

To show $\mathcal{R}(t) \subseteq \mathcal{R}^*$, suppose $\mathcal{J} = (I, \mathcal{Q}, C_1, C_2, D_1, D_2, a, b) \in \mathcal{R}(t)$. Then there is a legitimate pair $(\mathcal{X}, \mathcal{P})$ represented by $\mathcal{J}$. Since a pair $(\mathcal{X}|_{B_{t'}}, \mathcal{P}|_{B_{t'}})$ is legitimate, there is a record $\mathcal{J}' \in \mathcal{R}(t')$ which represents the pair. By the construction, $\mathcal{J}$ is computed from $\mathcal{J}'$. Hence $\mathcal{R}(t) \subseteq \mathcal{R}^*$.

We take one record in $\mathcal{R}(t')$ and construct a new tuple as explained above. After then, we check its validity. Note that $|\mathcal{R}(t')| \leqslant \zeta(w)$. Checking the validity takes time $\mathcal{O}(w + k)$. Hence, $\mathcal{R}(t)$ is computed in $2^{\mathcal{O}((w+k)\log(w(\alpha+k)))}$ time.

For other nodes t, we can also compute $\mathcal{R}(t)$ in time $2^{\mathcal{O}((w+k)\log(w(\alpha+k)))}$. Overall, the algorithm runs in $2^{\mathcal{O}((w+k)\log(w(\alpha+k)))}|V(T)|$ time.

Theorem 4. *Given an n-vertex graph G and two positive integers α and k, one can in time $2^{\mathcal{O}((\alpha+k)\log(\alpha+k))}n^2$ either*

- *output a tree-cut decomposition of G with thickness at most α and crossing number at most $2\alpha^2 + 5k$, or*
- *correctly report that $\mathrm{ecrw}_\alpha(G) > k$.*

Proof. We recursively apply the algorithm for CONSTRAINED STAR-CUT DECOMPOSITION as follows. At the beginning, we consider a trivial tree-decomposition with one bag containing all the vertices. In the recursive steps, we assume that we have a tree-cut decomposition $\mathcal{T} = (T, \{X_t\}_{t \in V(T)})$ such that

(i) for every internal node t of T, $|X_t| \leqslant \alpha$ and $\mathrm{cross}_\mathcal{T}(t) \leqslant 2\alpha^2 + 5k$,
(ii) for every leaf node t of T, $|\delta_G(X_t, V(G)\backslash X_t)| \leqslant 2\alpha^2 + 4k$.

If all leaf bags have size at most α, then this decomposition has thickness at most α and crossing number at most $2\alpha^2 + 5k$. Thus, we may assume that there is a leaf bag X_ℓ having at least $\alpha + 1$ vertices.

We apply Theorem 3 for $G[X_\ell]$ with $w = 3k + 2\alpha - 1$. Then in time $2^{\mathcal{O}(k+\alpha)}n$, either we have a tree-decomposition of width at most $2(3k + 2\alpha - 1) + 1 = 6k + 4\alpha - 1$ or we report that $\mathrm{tw}(G) \geqslant \mathrm{tw}(G[X_\ell]) > 3k + 2\alpha - 1$. In the latter case, by Lemma 2, we have $\mathrm{ecrw}_\alpha(G) > k$. Thus, we may assume that we have a tree-decomposition of $G[X_\ell]$ of width at most $6k + 4\alpha - 1$. By applying Lemma 1, we can find a nice tree-decomposition $(F, \{B_t\}_{t \in V(F)})$ of $G[X_\ell]$ of width at most $6k + 4\alpha - 1$ with $|V(F)| = \mathcal{O}((k + \alpha)n)$.

We define γ on X_ℓ so that $\gamma(v) = |\delta_G(\{v\}, V(G)\backslash X_\ell)|$. We run the algorithm in Lemma 9 for the instance $(G[X_\ell], \alpha, k, \gamma)$. Then in time $2^{\mathcal{O}((\alpha+k)\log(\alpha+k))}|V(F)|$, one can either output a solution of $(G[X_\ell], \alpha, k, \gamma)$, or correctly report that $(G[X_\ell], \alpha, k, \gamma)$ is a No-instance. In the latter case, we have $\mathrm{ecrw}_\alpha(G) > k$, by Lemma 8. In the former case, let $\mathcal{T}^* = (T^*, \{Y_t\}_{t \in V(T^*)})$ be the outcome, where q_c is the center of T^* and $q_1, \ldots, q_m$ are the leaves of T^*. Then we modify the tree-cut decomposition $\mathcal{T}$ by replacing X_ℓ with Y_{q_c} and then attaching bags Y_{q_i} to Y_{q_c}, where corresponding nodes are q_c and $q_1, \ldots, q_m$. Let $\mathcal{T}'$ be the resulting tree-cut decomposition.

Observe that $|Y_{q_c}| \leqslant \alpha$ and $\mathrm{cross}_{\mathcal{T}^*}(q_c) \leqslant k$. So, we have $\mathrm{cross}_{\mathcal{T}'}(q_c) \leqslant k + (2\alpha^2 + 4k) = 2\alpha^2 + 5k$. Also, for each $i \in [m]$, we have

$$|\delta_G(Y_{q_i}, V(G) \backslash Y_{q_i})| \leqslant |\delta_G(Y_{q_i}, V(G) \backslash X_\ell)| + |\delta_G(Y_{q_i}, Y_{q_c})| + \mathrm{cross}_{\mathcal{T}^*}(q_c)$$
$$\leqslant (\alpha^2 + 2k) + (\alpha^2 + k) + k = 2\alpha^2 + 4k.$$

Therefore, we obtain a refined tree-cut decomposition with properties (i) and (ii). Note that by the last condition of the solution for CONSTRAINT STAR-CUT DECOMPOSITION, new leaf bags have size less than X_ℓ. Thus, the algorithm will terminate in at most n recursive steps. When this procedure terminates, we either obtain a tree-cut decomposition of G of thickness at most α and crossing number at most $2\alpha^2 + 5k$ or, conclude that $\mathrm{ecrw}_\alpha(G) > k$.

The total running time is $(2^{\mathcal{O}((\alpha+k)\log(\alpha+k))}n) \cdot n = 2^{\mathcal{O}((\alpha+k)\log(\alpha+k))}n^2$.

5   Algorithmic Applications on Coloring Problems

We show that LIST COLORING and PRECOLORING EXTENSION are fixed parameter tractable parameterized by α-edge-crossing width for every fixed α.

A vertex-coloring $f : V(G) \rightarrow \mathbb{N}$ on a graph G is said to be *proper* if $f(u) \neq f(v)$ for all edges $uv \in E(G)$. For a given set $\{L(v) \subseteq \mathbb{N} : v \in V(G)\}$, a coloring $c : V(G) \rightarrow \mathbb{N}$ is called an *L-coloring* if $c(v) \in L(v)$ for all $v \in V(G)$.

LIST COLORING
Input : A graph G and a set of lists $\mathcal{L} = \{L(v) \subseteq \mathbb{N} : v \in V(G)\}$
Question : Does G admit a proper L-coloring $c : V(G) \rightarrow \bigcup \mathcal{L}$?

Assume that a tree-cut decomposition of the input graph G of thickness at most α and crossing number w is given. In the dynamic programming, we need to store colorings on $w + \alpha$ boundaried vertices. Using Lemma 10 the number of colorings to store can be reduced to $g(w + \alpha)$ for some function g. For $V \in \mathbb{N}^q$, $W \in \mathbb{N}^t$, $B \subseteq [q] \times [t]$, we say that (V, W) is *B-compatible* if $V[i] \neq W[j]$ for all $(i, j) \in B$. When we have a vertex partition (X, Y) of a graph G, possible colorings on boundaried vertices in X and Y will be related to vectors V and W.

Lemma 10 ($\star$). *Let q and t be positive integers, and let $B \subseteq [q] \times [t]$. For every set $\mathcal{P}$ of distinct vectors in $\mathbb{N}^q$, there is a subset $\mathcal{P}^*$ of $\mathcal{P}$ of size at most $q! 2^{\frac{q(q+1)}{2}} t^{q-1}(t+1)$ satisfying that for every $W \in \mathbb{N}^t$, if there is $V \in \mathcal{P}$ where (V, W) is B-compatible, then there is $V^* \in \mathcal{P}^*$ where (V^*, W) is B-compatible. Furthermore, such a set $\mathcal{P}^*$ can be computed in time $\mathcal{O}(|\mathcal{P}|2^{q^2}t^{q+2})$.*

Let G be a graph. For disjoint sets S, T of vertices in G and functions $g : S \rightarrow \mathbb{N}$ and $h : T \rightarrow \mathbb{N}$, we say that (S, g) is *compatible* with (T, h) if for every edge vw with $v \in S$ and $w \in T$, $g(v) \neq h(w)$. If g and h are clear from the context, we simply say that S and T are compatible.

Proof. (of Theorem 2). We describe the algorithm for connected graphs. If a given graph is disconnected, then we can apply the algorithm for each component. We assume that G is connected. We may assume that each list $L(v)$ has size at most the degree of v; otherwise, we can freely color v after coloring $G - v$.

Let $\mathrm{ecrw}_\alpha(G) = k$. Using the algorithm in Theorem 4, we obtain a tree-cut decomposition $\mathcal{T} = (T, \mathcal{X} = \{X_t\}_{t \in V(T)})$ of the input graph G of thickness at most α and crossing number $w \leqslant 2\alpha^2 + 5k$. We consider it as a rooted decomposition by choosing a root node r with $X_r \neq \varnothing$.

For every node $t \in V(T)$, we denote by T_t the subtree of T rooted at t, and let $G_t = G[\bigcup_{v \in V(T_t)} X_v]$. For every $t \in V(T)$, let $\partial(t)$ be the graph H where $E(H)$ is the set of edges incident with both $V(G_t)$ and $V(G) \backslash V(G_t)$, and $V(H)$ is the set of vertices in G incident with an edge in $E(H)$. Let $\hat{\partial}(t) := (V(\partial(t)) \cap V(G_t)) \cup X_t$. Note that $|\hat{\partial}(t)| \leqslant \alpha + w$ for any node t of T.

Let $t \in V(T)$. A coloring g on $\hat{\partial}(t)$ is *valid at* t if there is a proper L-coloring f of G_t for which $f|_{\hat{\partial}(t)} = g$. Clearly, the problem is a Yes-instance if and only if there is a valid coloring at the root node.

Let $\zeta = \max\{(\alpha + w + 1)! 2^{\frac{(\alpha+w)(\alpha+w+1)}{2}} (\alpha+w)^{\alpha+w-1}, (w + 2\alpha - 1)^\alpha\}$.

For each node $t \in V(T)$, let $Q[t]$ be the set of all valid colorings at t. We will recursively construct a subset $Q^*[t] \subseteq Q[t]$ of size at most ζ such that

($\square$) for every proper L-coloring h on $G - V(G_t)$, if there is a valid coloring $g \in Q[t]$ compatible with h, then there is $g^* \in Q^*[t]$ compatible with h.

We describe how to construct $Q^*[t]$ depending on whether t is a non-root leaf. This is easy when t is a leaf. Let t_p be the parent of t when t is not the root.

We classify the children of t into two types. Let A_1 be the set of all children p of t such that $\partial(p) \backslash \hat{\partial}(p) \subseteq X_t$, and let A_2 be the set of all other children of t. Note that $|A_2| \leqslant 2w$ because $(T, \mathcal{X})$ has crossing number at most w. Let $C[t]$ be the set of all proper L-colorings on X_t. Clearly, $|C[t]| \leqslant n^\alpha$.

(Step 1.) We first find the set $C'[t]$ of all proper L-colorings f such that for each $x \in A_1$, there exists $g_x \in Q^*[x]$ that is compatible with f. This can be checked by recursively choosing $x \in A_1$, and comparing each coloring in $C[t]$ with a coloring in $Q^*[x]$, and then remaining one that has a compatible coloring in $Q^*[x]$. The whole procedure runs in time $\mathcal{O}(|A_1| \cdot |Q^*[x]| \cdot n^\alpha \cdot (\alpha + w)^2) = \mathcal{O}(\zeta \cdot n^{\alpha+1} \cdot (\alpha + w)^2)$, because each $Q^*[x]$ has the size at most ζ and $|A_1| \leqslant n$.

(Step 2.) Next, we compute the set $I[t]$ of all tuples U in $\prod_{x \in A_2} Q^*[x]$ such that for all distinct $x, y \in A_2$, $U(x)$ and $U(y)$ are compatible, where $U(x)$ denotes the coordinate of U that comes from $Q^*[x]$. Since $|A_2| \leqslant 2w$, we have $|\prod_{x \in A_2} Q^*[x]| \leqslant \zeta^{2w}$. The set $I[t]$ can be computed in time $\mathcal{O}(\zeta^{2w} \cdot w^2 \cdot (\alpha + w)^2)$.

(Step 3.) Lastly, we construct $Q'[t]$ from $I[t]$ and $C'[t]$ as follows. For every $U \in I[t]$ and every $g \in C'[t]$ where $U(x)$ and g are compatible for all $x \in A_2$, we obtain a new function g' on $\hat{\partial}(t)$ such that $g'(v) = (U(x))(v)$ if $v \in \hat{\partial}(x)$ for some $x \in A_2$, and $g'(v) = g(v)$ if $v \in X_t$, and add it to $Q'[t]$. This can be done in time $\mathcal{O}(|I[t]| \cdot |C'[t]| \cdot (\alpha(\alpha + w))^{2w}) = \mathcal{O}(\zeta^{2w} \cdot n^\alpha \cdot (\alpha(\alpha + w))^{2w})$. This stores valid colorings at t and the size of $Q'[t]$ is at most $|I[t]| \times |C'[t]|$. Using Lemma 10, we find a subset $Q^*[t]$ of $Q'[t]$ of size at most ζ. This can be computed in time

$\mathcal{O}(|Q'[t]| \cdot 2^{(\alpha+w)^2} \cdot (\alpha+w)^{\alpha+w+2}) = \mathcal{O}(\zeta^{2w} \cdot n^{\alpha} \cdot 2^{(\alpha+w)^2} \cdot (\alpha+w)^{\alpha+w+2})$. The total running time for this case is $\mathcal{O}(\zeta^{2w} \cdot n^{\alpha+1} \cdot 2^{(\alpha+w)^2} \cdot (\alpha+w)^{\alpha+4w+2})$.

For the correctness, we prove that $(\star)$ $Q^*[t]$ satisfies the property ($\square$).

As $|V(T)| = \mathcal{O}(n)$, the algorithm runs in time $\mathcal{O}(\zeta^{2w} \cdot n^{\alpha+2} \cdot 2^{(\alpha+w)^2} \cdot (\alpha+w)^{\alpha+4w+2}) = 2^{\mathcal{O}((\alpha^2+k)^3)} n^{\alpha+2}$.

PRECOLORING EXTENSION asks whether a precoloring on a vertex set S can be extended to a k-coloring of G. By making a list for a vertex S to the assigned color, and making a list for a neighbor of $v \in S$ which avoids the assigned color of v, we can reduce to LIST COLORING.

6   Conclusion

In this paper, we introduced a width parameter called α-edge-crossing width, which lies between edge-cut width and tree-partition-width, and which is incomparable with tree-cut width. We showed that LIST COLORING and PRECOLORING EXTENSION are FPT parameterized by α-edge-crossing width. It would be interesting to find more problems that are W[1]-hard parameterized by tree-partition-width, but FPT by α-edge-crossing width for any fixed α. There are six more problems that are known to admit FPT algorithms parameterized by slim tree-cut width, but W[1]-hard parameterized by tree-cut width [9], and these problems are candidates for the next research.

We also introduced edge-crossing width, that is equivalent to tree-partition-width. However, our proof is based on the characterization of graphs of bounded tree-partition-width due to Ding and Oporowski [5], and finding an elementary upper bound of tree-partition-width in terms of edge-crossing width is an interesting problem. More specifically, we ask whether there is a constant c such that for every graph G, $\mathrm{tpw}(G) \leqslant c \cdot \mathrm{ecrw}(G)$.

References

1. Bodlaender, H.L.: A partial k-arboretum of graphs with bounded treewidth. Theor. Comput. Sci. **209**(1–2), 1–45 (1998). https://doi.org/10.1016/S0304-3975(97)00228-4
2. Brand, C., Ceylan, E., Hatschka, C., Ganian, R., Korchemna, V.: Edge-cut width: an algorithmically driven analogue of treewidth based on edge cuts (2022). WG2022 accepted. arXiv:2202.13661
3. Bredereck, R., Heeger, K., Knop, D., Niedermeier, R.: Parameterized complexity of stable roommates with ties and incomplete lists through the lens of graph parameters. Inf. Comput. **289**(part A) (2022). Paper No. 104943, 41. https://doi.org/10.1016/j.ic.2022.104943
4. Cygan, M., et al.: Parameterized Algorithms, 1st edn. Springer, Heidelberg (2015). https://doi.org/10.1007/978-3-319-21275-3
5. Ding, G., Oporowski, B.: On tree-partitions of graphs. Discrete Math. **149**(1), 45–58 (1996). https://doi.org/10.1016/0012-365X(94)00337-I. www.sciencedirect.com/science/article/pii/0012365X9400337I

6. Fellows, M.R., et al.: On the complexity of some colorful problems parameterized by treewidth. Inf. Comput. **209**(2), 143–153 (2011). https://doi.org/10.1016/j.ic.2010.11.026

7. Ganian, R., Kim, E.J., Szeider, S.: Algorithmic applications of tree-cut width. SIAM J. Discrete Math. **36**(4), 2635–2666 (2022). https://doi.org/10.1137/20M137478X

8. Ganian, R., Korchemna, V.: The complexity of Bayesian network learning: revisiting the superstructure. In: Ranzato, M., Beygelzimer, A., Dauphin, Y., Liang, P., Vaughan, J.W. (eds.) Advances in Neural Information Processing Systems, vol. 34, pp. 430–442. Curran Associates, Inc. (2021). www.proceedings.neurips.cc/paper/2021/file/040a99f23e8960763e680041c601acab-Paper.pdf

9. Ganian, R., Korchemna, V.: Slim tree-cut width (2022). arXiv:2206.15091

10. Ganian, R., Ordyniak, S.: The power of cut-based parameters for computing edge-disjoint paths. Algorithmica **83**(2), 726–752 (2021). https://doi.org/10.1007/s00453-020-00772-w

11. Gözüpek, D., Özkan, S., Paul, C., Sau, I., Shalom, M.: Parameterized complexity of the MINCCA problem on graphs of bounded decomposability. Theor. Comput. Sci. **690**, 91–103 (2017). https://doi.org/10.1016/j.tcs.2017.06.013

12. Kim, E.J., Oum, S.I., Paul, C., Sau, I., Thilikos, D.M.: An FPT 2-approximation for tree-cut decomposition. Algorithmica **80**(1), 116–135 (2018). https://doi.org/10.1007/s00453-016-0245-5

13. Korhonen, T.: A single-exponential time 2-approximation algorithm for treewidth. In: 2021 IEEE 62nd Annual Symposium on Foundations of Computer Science–FOCS 2021, Los Alamitos, CA, pp. 184–192. IEEE Computer Society (2022)

14. Robertson, N., Seymour, P.D.: Graph minors. V. Excluding a planar graph. J. Comb. Theory Ser. B **41**(1), 92–114 (1986). https://doi.org/10.1016/0095-8956(86)90030-4

15. Robertson, N., Seymour, P.D.: Graph minors. XX. Wagner's conjecture. J. Comb. Theory Ser. B **92**(2), 325–357 (2004). https://doi.org/10.1016/j.jctb.2004.08.001

16. Robertson, N., Seymour, P.: Graph minors. X. Obstructions to tree-decomposition. J. Comb. Theory Ser. B **52**(2), 153–190 (1991)

17. Wollan, P.: The structure of graphs not admitting a fixed immersion. J. Comb. Theory Ser. B **110**, 47–66 (2015). https://doi.org/10.1016/j.jctb.2014.07.003

Snakes and Ladders: A Treewidth Story

Steven Chaplick, Steven Kelk[(✉)], Ruben Meuwese, Matúš Mihalák,
and Georgios Stamoulis

Department of Advanced Computing Sciences, Maastricht University,
Maastricht, The Netherlands
{s.chaplick,steven.kelk,r.meuwese,matus.mihalak,
georgios.stamoulis}@maastrichtuniversity.nl

Abstract. Let G be an undirected graph. We say that G contains a ladder of length k if the $2 \times (k+1)$ grid graph is an induced subgraph of G that is only connected to the rest of G via its four cornerpoints. We prove that if all the ladders contained in G are reduced to length 4, the treewidth remains unchanged (and that this bound is tight). Our result indicates that, when computing the treewidth of a graph, long ladders can simply be reduced, and that minimal forbidden minors for bounded treewidth graphs cannot contain long ladders. Our result also settles an open problem from algorithmic phylogenetics: the common chain reduction rule, used to simplify the comparison of two evolutionary trees, is treewidth-preserving in the display graph of the two trees.

Keywords: Treewidth · Reduction rules · Phylogenetics

1 Introduction

This is a story about treewidth, but it starts in the world of biology. A phylogenetic tree on a set of leaf labels X is a binary tree representing the evolution of X. These are studied extensively in computational biology [16]. Given two such trees a natural aim is to quantify their topological dissimilarity [12]. Many such dissimilarity measures have been devised and they are often NP-hard to compute, stimulating the application of techniques from parameterized complexity [7]. Recently there has been a growing focus on treewidth. This is because, if one takes two phylogenetic trees on X and identifies leaves with the same label, we obtain an auxiliary graph structure known as the *display graph* [6]. Crucially, the treewidth of this graph is often bounded by a function of the dissimilarity measure that we wish to compute [13]. This has led to the use of Courcelle's Theorem within phylogenetics (see e.g. [11,13]) and explicit dynamic programs running over tree decompositions; see [10] and references therein. In [14] the spin-off question was posed: is the treewidth of the display graph actually a meaningful measure of phylogenetic dissimilarity *in itself* - as opposed to purely being a

R. Meuwese was supported by the Dutch Research Council (NWO) KLEIN 1 grant *Deep kernelization for phylogenetic discordance*, project number OCENW.KLEIN.305.

D. Paulusma and B. Ries (Eds.): WG 2023, LNCS 14093, pp. 187–200, 2023.
https://doi.org/10.1007/978-3-031-43380-1_14

route to efficient algorithms? A closely-related question was whether parameter-preserving reduction rules, applied to two phylogenetic trees to shrink them in size, also preserve the treewidth of the display graph? The well-known *sub-tree reduction rule* is certainly treewidth preserving [14]. However, the question remained whether the *common chain reduction rule* [2] is treewidth-preserving. A common chain is, informally, a sequence of leaf labels $x_1, \ldots, x_k$ that has the same order in both trees. Concretely, the question arose [14]: is it possible to truncate a common chain to *constant length* such that the treewidth of the display graph is preserved? Common chains form ladder-like structures in the display graph, i.e., this question is about how far ladders can be reduced in length without causing the treewidth to decrease.

In this article we answer this question affirmatively, and more generally. Namely, we do not restrict ourselves to display graphs, but consider arbitrary graphs. A *ladder L of length* $k \geq 1$ of a graph G is a $2 \times (k+1)$ grid graph such that L induces (only) itself and that L is only connected to the rest of the graph by its four cornerpoints. First, we prove that a ladder L can be reduced to length 4 without causing the treewidth to decrease, and that this is best possible: reducing to length 3 sometimes causes the treewidth to decrease. We also show that if $tw(G) \geq 4$ then reduction to length 3 is safe and, again, best possible. These tight examples are also shown to exist for higher treewidths. Returning to phylogenetics, and thus when G is a display graph, we leverage the extra structure in these graphs to show that common chains can be reduced to 4 leaf labels (and thus the underlying ladder to length 3) without altering the treewidth: this result is thus slightly stronger than on general G.

Our proofs are based on first principles: we directly modify a tree decomposition to get what we need. In doing so we come across the problem that, unless otherwise brought under control, the set of bags that contain a given ladder vertex of G can wind and twist through the tree decomposition in very pathological ways. Getting these *snakes* under control is where much of the hard work and creativity lies, and is the inspiration for the title of this paper.

From a graph-theoretic perspective our results have the following significance. First, it is standard folklore that shortening paths (i.e. suppressing vertices of degree 2) is treewidth-preserving, but there is seemingly little in the literature about shortening recursive structures that are slightly more complex than paths, such as ladders. (Note that Sanders [15] did consider ladders, but only for recognizing graphs of treewidth at most 4, and in such a way that the reduction destroys the ladder topology). Second, our results imply a new safe reduction rule for the computation of treewidth; a survey of other reduction rules for treewidth can be found in [1]. Third, we were unable to find sufficiently precise machinery, characterisations of treewidth or restricted classes of tree decomposition in the literature that would facilitate our results. Perhaps most closely related to our ladders are the more general *protrusions*: low treewidth subgraphs that "hang" from a small boundary [9, Ch. 15-16]. There are general (algorithmic) results [5] wherein one can safely cut out a protrusion and replace it with a graph of parameter-proportional size instead – these are based on a problem having *finite integer index* [4]. Such techniques might plausibly be used to prove that there is *some* constant to which ladders might safely be shortened, but our tight bounds seem out of their reach. Finally, the results imply that minimal forbidden minors for bounded treewidth cannot have long ladders.

Due to space limitations a number of proofs have been deferred to an appendix, which can be found in the arXiv version of this article [8].

2   Preliminaries

We follow [14] for notation. A *tree decomposition* of an undirected graph $G = (V, E)$ is a pair $(\mathcal{B}, \mathbb{T})$ where $\mathcal{B} = \{B_1, \ldots, B_q\}$, $B_i \subseteq V(G)$, is a multiset of *bags* and $\mathbb{T}$ is a tree whose q nodes are in bijection with $\mathcal{B}$, and

(tw1) $\cup_{i=1}^{q} B_i = V(G)$;
(tw2) $\forall e = \{u, v\} \in E(G), \exists B_i \in \mathcal{B}$ s.t. $\{u, v\} \subseteq B_i$;
(tw3) $\forall v \in V(G)$, all the bags B_i that contain v form a connected subtree of $\mathbb{T}$.

The *width* of $(\mathcal{B}, \mathbb{T})$ is equal to $\max_{i=1}^{q} |B_i| - 1$. The *treewidth* of G, denoted $tw(G)$, is the smallest width among all tree decompositions of G. Given a tree decomposition $\mathbb{T}$ of a graph G, we denote by $V(\mathbb{T})$ the (multi)set of its bags and by $E(\mathbb{T})$ the set of its edges. Property (tw3) is also known as *running intersection property*. Without loss of generality, we consider only connected graphs G.

Note that subdividing an edge $\{u, v\}$ of G with a new degree-2 vertex uv does not change the treewidth of G. In the other direction, suppression of degree-2 vertices is also treewidth preserving *unless* it causes the only cycle in a graph to disappear (e.g. if G is a triangle); unlike [14] we will never encounter this boundary case. An equivalent definition of treewidth is based on chordal graphs. Recall that a graph G is chordal if every induced cycle in G has exactly three vertices. The treewidth of G is the minimum, ranging over *all* chordal completions $c(G)$ of G (we add edges until G becomes chordal), of the size of the maximum clique in $c(G)$ minus one. Under this definition, each bag of a tree decomposition of G naturally corresponds to a maximal clique in a chordal completion of G [3].

We say that a graph H is a *minor* of another graph G if H can be obtained from G by deleting edges and vertices and by contracting edges.

A *ladder L of length $k \geq 1$* is a $2 \times (k + 1)$ grid graph. A *square* of L is a set of vertices of L that induce a 4-cycle in L. We call the endpoints of L, i.e., the degree-2 vertices of L, the *cornerpoints* of L. We say that a graph G *contains* L if the following holds (see Fig. 1 for illustration):

1. The subgraph induced by vertices of L is L itself.
2. Only cornerpoints of L can be incident to an edge with an endpoint outside L.

Observe that a ladder of length k is a minor of the ladder of length $(k + 1)$. Treewidth is non-increasing under the action of taking minors, so reducing the length of a ladder in a graph cannot increase the treewidth of the graph.

Suppose G contains a ladder L. We say that L *disconnects* G if L contains a square $\{u, v, w, x\}$ such that the two horizontal edges of the square (following

Fig. 1, these are the edges $\{u, w\}$ and $\{v, x\}$) form an edge cut of the entire graph G. Note that a square of L has this property if and only if all squares of L do. Also, if we reduce the length of a ladder L to obtain a shorter ladder L', L' disconnects G if and only if L does. We recall a number of results from Section 5.2 of [14]; these will form the starting point for our work.

Lemma 1 ([14]). *Suppose G contains a disconnecting ladder L. The ladder L can be increased arbitrarily in length without increasing the treewidth of G.*

For the more general case, the following weaker result is known.

Lemma 2 ([14]). *Suppose G has $tw(G) \geq 3$ and contains a ladder. If the ladder is increased arbitrarily in length, the treewidth of G increases by at most one.*

We now make the following (new) observation; the proof is in the appendix.

Observation 1. *Suppose G contains a ladder L of length 2 or longer. If L is not disconnecting, then $tw(G) \geq 3$.*

We can leverage Observation 1 to reformulate Lemma 2 without the $tw(G) \geq 3$ assumption. However it then only applies to ladders of size at least two.

Lemma 3. *Suppose G contains a ladder L with length at least 2. If L is increased arbitrarily in length, the treewidth of the graph increases by at most one.*

If we start from a sufficiently long ladder, can the ladder be increased in length without increasing the treewidth? Past research has the following partial result.

Theorem 1 ([14]). *Let G be a graph with $tw(G) = k$. There is a value $f(k)$ such that if G contains a ladder of length $f(k)$ or longer, the ladder can be increased in length arbitrarily without altering (in particular: increasing) the treewidth.*

Ideally we would like a single, universal value *that does not depend on k*. In this article we will show that such a single, universal constant does exist.

3   Results

We first consider graphs of treewidth at least 4; we later remove this restriction.

Theorem 2. *Let G be a graph with $tw(G) \geq 4$. If G has a ladder L of length 3 or higher, the ladder can be lengthened arbitrarily without changing the treewidth.*

Proof. Due to Lemma 1 we can assume that L is not disconnecting. Our general strategy is to show that if G contains the ladder L shown in Fig. 1, we can insert an extra 'rung' in the ladder without increasing the treewidth, thus obtaining a ladder with one extra square (see Fig. 2). The extension of the ladder by one square can then be iterated to obtain an arbitrary length ladder.

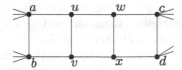

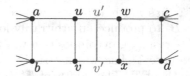

Fig. 1. A ladder L of length 3 with corner points a, b, c, d.

Fig. 2. Inserting a new edge $\{u', v'\}$ into ladder L results in ladder L' of length 4.

Let L be the ladder shown in Fig. 1, and assume that G contains L. Let $(\mathcal{B}, \mathbb{T})$ be a minimum-width tree decomposition for G. We proceed with a case analysis. The cases are cumulative: we will assume that earlier cases do not hold.

Case 1. Suppose that $\mathcal{B}$ contains a bag B such that all four vertices from one of the squares of L are in B. Let $\{u, v, w, x\}$, say, be the square of L contained in bag B, where the position of the vertices is as in Fig. 1. We prolong the ladder as in Fig. 2 and create a valid tree decomposition for the new graph as follows: we introduce a new size-5 bag $B' = \{u', u, v, w, x\}$ which we attach pendant to B in the tree decomposition, and a new size-5 bag $B'' = \{u', v', v, w, x\}$ which we attach pendant to B'. Observe that this is a valid tree decomposition for the new graph. Due to the fact that $tw(G) \geq 4$, the treewidth does not increase, and the statement follows. Note that in this construction B'' contains all four of $\{u', w, v', x\}$, which is a square of the new ladder, so the construction can be applied iteratively many times as desired to produce a ladder of arbitrary length.

Case 2. Suppose that $\mathcal{B}$ contains a bag B such that $|B \cap \{a, u, w, c\}| \geq 2$ and $|B \cap \{b, v, x, d\}| \geq 2$. Let h_1, h_2 be two distinct vertices from $B \cap \{a, u, w, c\}$ and l_1, l_2 be two distinct vertices from $B \cap \{b, v, x, d\}$.

Observe that it is possible to partition the sequence a, u, w, c into two disjoint intervals H_1, H_2, and the sequence b, v, x, d into two disjoint intervals L_1, L_2 such that $h_1 \in H_1$, $h_2 \in H_2$, $l_1 \in L_1$ and $l_2 \in L_2$. If we contract the edges and vertices in each of H_1, H_2, L_1, L_2 we obtain a new graph G' which is a minor of G. Note that G' is similar to G except that the ladder now has two fewer squares – the three original squares have been replaced by a square whose corners correspond to H_1, H_2, L_1, L_2. This square might contain a diagonal but we simply delete this. We have $tw(G') \leq tw(G)$ because treewidth is non-increasing under taking minors. Now, by projecting the contraction operations onto $(\mathcal{B}, \mathbb{T})$ in the usual way[1], we obtain a tree decomposition $(\mathcal{B}', \mathbb{T}')$ for G' such that the width of $\mathbb{T}'$ is less than or equal to the width of $\mathbb{T}$. The bag in $(\mathcal{B}', \mathbb{T}')$ corresponding to B, let us call this B', contains all four vertices H_1, H_2, L_1, L_2. Clearly, $\mathbb{T}'$ is a valid tree decomposition for G'. We distinguish two subcases.

---

[1] In every bag of the decomposition vertices from H_1 all receive the vertex label H_1, and similarly for the other subsets H_2, L_1, L_2.

1. If $\mathbb{T}'$ has width at least 4, we can repeatedly apply the Case 1 transformation to B' to produce an arbitrarily long ladder without raising the width of $\mathbb{T}'$. The resulting decomposition will thus have width no larger than $\mathbb{T}$.

2. Suppose $\mathbb{T}'$ has width strictly less than 4, and thus strictly less than the width of $\mathbb{T}$. The width of $\mathbb{T}'$ is at least 3 because of the bag containing H_1, H_2, L_1, L_2. Case 1 introduces size-5 bags and can thus raise the width of the decomposition by at most 1. Hence we again obtain a decomposition whose width is no larger than $\mathbb{T}$ for a graph with an arbitrarily long ladder.

This concludes Case 2. Moving on, any chordalization of G must add the diagonal $\{w, v\}$ and/or the diagonal $\{u, x\}$. Hence we can assume that there is a bag containing $\{u, w, v\}$ and another bag containing $\{v, w, x\}$. (If the other diagonal is added we can simply flip the labelling of the vertices in the horizontal axis i.e. $a \Leftrightarrow b, u \Leftrightarrow v$ and so on). As Case 1 does not hold we can assume that the bag containing $\{u, w, v\}$ is distinct from the bag containing $\{v, w, x\}$.

For the benefit of later cases we impose extra structure on our choice of minimum-width tree decomposition of G. The *distance* of decomposition $(\mathcal{B}, \mathbb{T})$ is the minimum, ranging over all pairs of bags B_1, B_2 such that B_1 contains $\{u, w, v\}$ and B_2 contains $\{v, w, x\}$, of the length of the path in $\mathbb{T}$ from B_1 to B_2.

We henceforth let $(\mathcal{B}, \mathbb{T})$ be a minimum-width tree decomposition of G such that, ranging over all minimum-width tree decompositions, the distance is minimized. Clearly such a tree decomposition exists.

Let B_1, B_2 be two bags from $\mathcal{B}$ with $\{u, w, v\} \subseteq B_1, \{v, w, x\} \subseteq B_2$ which achieve this minimum distance. Let P be the path of bags from B_1 to B_2, including B_1 and B_2. We assume that P is oriented left to right, with B_1 at the left end and B_2 on the right. As Case 2 does not hold, we obtain the following.

Observation 2. B_1 *does not contain* b, x *or* d, *and* B_2 *does not contain* a, u, c.

Case 3. B_1 **and** B_2 **are adjacent in** P. Although this could be subsumed into a later case it introduces important machinery; we therefore treat it separately.

Subcase 3.1: Suppose $a \in B_1$ (or, completely symmetrically, $d \in B_2$). Note that in this case all the edges in G incident to u are covered by B_1. Hence, we can safely delete u from all bags except B_1. Next, we create a new bag $B^* = \{a, u, w, v\}$ and attach it pendant to B_1, and finally we replace u with x in B_1. It can be easily verified that this is a valid tree decomposition for G and that the width is not increased, so it is still a minimum-width tree decomposition. However, B_1 is now a candidate for Case 2, and we are done. Note that replacing u with x in B_1 is only possible because B_1 is next to B_2 in P.

Subcase 3.2: Suppose Subcase 3.1 does not hold. Then $a \notin B_1$ (and, symmetrically, $d \notin B_2$). Putting all earlier insights together, we see $a, b, x, d \notin B_1$ and $a, u, c, d \notin B_2$. Observe that a, which is not in B_2, is not in any bag to the right of B_2. If it was, then the fact that some bag contains the edge $\{a, u\}$, and the running intersection property, entails that B_2 would contain at least one of a and u, neither of which is permitted. Hence, if a appears in bags other than B_1, they are all in the left part of the decomposition. Completely symmetrically, if d is in bags other than B_2, they are all in the right part of the decomposition. Because of this, b can only appear on the left of the decomposition (because the edge $\{a, b\}$ has to be covered) and c can only be on the right of the decomposition (because of the edge $\{c, d\}$). Summarising, B_1 (respectively, B_2) does not contain a or b (respectively, c or d) and all bags containing a or b (respectively, c or d) are in the left (respectively, right) part of the decomposition. Note that $c \notin B_1$. This is because edge $\{c, d\}$ has to be in some bag, and this must necessarily be to the right of B_2: but then running intersection puts at least one of c, d in B_2, contradiction. Symmetrically, $b \notin B_2$. So $a, b, c, d, x \notin B_1$ and $a, b, c, d, u \notin B_2$.

We now describe a construction that we will use extensively: *reeling in (the snakes) a and b*. Observe that, due to coverage of the edge $\{a, u\}$, and running intersection, there is a simple path of bags p_{ua} starting at B_1 that all contain u such that the endpoint of the path also contains a. The path will necessarily be entirely on the left of the decomposition. Due to coverage of the edge $\{b, v\}$ there is an analogously-defined simple path p_{vb}. (Note that p_{ua} and p_{vb} both exit B_1 via the same bag B'. If they exited via different bags, coverage of the edge $\{a, b\}$ would force at least one of a, b to be in B_1, yielding a contradiction). Now, in the bags along p_{ua}, except B_1, we relabel u to be a, and in the bags along p_{vb}, except B_1, we relabel v to be b. This is no longer necessarily a valid tree decomposition, because coverage of the edges $\{u, a\}$ and $\{v, b\}$ is no longer guaranteed, but we shall address this in due course. Next we delete the vertices u, w, v from all bags on the left of the decomposition, except B_1; they will not be needed. (The only reason that w would be in a bag on the left, would be to meet c, since B_1 and B_2 already cover the edges $\{u, w\}$ and $\{w, x\}$. But then, due to coverage of the edge $\{c, d\}$ and the fact that d only appears on the right of the decomposition, running intersection would put at least one of c, d in B_1, contradiction.) Observe that B' contains $\{a, b\}$. We replace B_1 with 5 copies of itself, and place these bags in a path such that the leftmost copy is adjacent to B', the rightmost copy is adjacent to B_2, and all other bags that were originally adjacent to B_1 can (arbitrarily) be made adjacent to the leftmost copy. In the 5 copied bags we replace $\{u, w, v\}$ respectively with: $\{a, u', b\}$, $\{u', b, v'\}$, $\{u', u, v'\}$, $\{v', u, v\}$ and $\{u, w, v\}$. It can be verified that this is a valid tree decomposition for G', and our construction did not inflate the treewidth - we either deleted vertices from bags or relabelled vertices that were already in bags - so we are done. The operation can easily be telescoped, if desired, to achieve an arbitrarily long ladder.

Case 4. P contains at least one bag other than B_1 and B_2.

Observation 3. *All bags in P contain v, w, by the running intersection property.*

We partition the bags of the decomposition into (i) B_1, (ii) bags *left* of B_1, (iii) B_2, (iv) bags *right* of B_2, (v) all other bags (which we call the *interior*).

Recall that $b, d, x \notin B_1$, $a, c, u \notin B_2$ (because Case 2 does not hold). The proofs of the following two observations are in the appendix.

Observation 4. *No bag in the interior contains u or x. B_1 does not contain x, and no bag on the left contains x. Symmetrically, B_2 does not contain u, and no bag on the right contains u.*

Observation 5. *At least one of the following is true: $a \in B_1$, a is in a bag on the left. Symmetrically, at least one of the following is true: $d \in B_2$, d is in a bag on the right.*

Now, suppose w is somewhere on the left. We will show that then either w can be deleted from the bags on the left, or Case 2 holds. A symmetrical analysis will hold if v is somewhere on the right. Specifically, the only possible reason for w to be on the left would be to cover the edge $\{w, c\}$ – all other edges incident to w are already covered by B_1 and B_2. If no bags on the left contain c, we can simply delete w from all bags on the left. On the other hand, if some bag on the left contains c, then $c \in B_1$, because: $d \notin B_1$, the need to cover the edge $\{c, d\}$, the presence of d on the other 'side' of the decomposition (Observation 5), and running intersection. So we have that $c, u, w, v \in B_1$. This bag already covers all edges incident to w, except possibly the edge $\{w, x\}$. To address this, we replace w everywhere in the tree decomposition with x - this is a legal tree decomposition because some bag contains $\{w, x\}$ - and then add a bag $B' = \{u, w, x, c\}$ pendant to B. This new bag serves to cover all edges incident to w. But B_1 now contains u, v, c, x, so Case 2 applies, and we are done! Hence, we can assume that w is nowhere on the left, and, symmetrically, that v is nowhere on the right. In fact, the above argument can, independently of w, be used to trigger Case 2 whenever $c \in B_1$ or $b \in B_2$. So at this stage of the proof we know: $b, c, d, x \notin B_1$ (and c is not on the left) and $a, b, c, u \notin B_2$ (and b is not on the right).

Subcase 4.1: Suppose $a \notin B_1$. Then, a must only be on the left. It cannot be in the interior (or on the right) because the edge $\{u, a\}$ must be covered, $a \notin B_1$, $u \in B_1$, and u is not in the interior (Observation 4). Because a is on the left, and because some bag must contain the edge $\{a, b\}$, b must also be on the left. In fact b is only on the left. The presence of b both on the left and in the interior (or on the right) would force b into B_1 by running intersection, contradicting the fact that $b \notin B_1$. So a, b are only on the left. We are now in a situation similar to Subcase 3.2. We use the same *reeling in a and b* construction and we are done.

Subcase 4.2: Suppose $a \in B_1$. Note that here u has all its incident edges covered by B_1, so u can be deleted from all other bags. Recall that $b \notin B_1$. Due to edge $\{b, v\}$ some bag must contain both v and b. Suppose there is such a bag on the left. We attach a new bag $\{a, u, w, v\}$ pendant to B_1 and delete u from B_1. We put x in B_1 and to ensure running intersection we replace v with x in all bags anywhere to the right of B_1. This is safe, because in the part of the decomposition right of B_1, v only needs to meet x (and not b, because v meets b on the left). Thus, B_1 now contains $\{a, v, w, x\}$ and Case 2 can be applied.

Hence, we conclude that $\{v, b\}$ is not in a bag on the left. Because of this v can safely be deleted from all bags on the left. That is because any path p_{vb} that starts at B_1 and finishes at a bag containing b must go via the interior. In fact, such a path must avoid B_2, and is thus entirely contained in the interior.

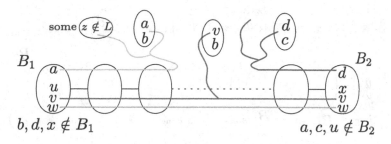

Fig. 3. Path p_{ab} goes via the interior, but it cannot be relabelled to b because it is used by other paths p_{az} to some neighbour z of a that does *not* lie on the ladder.

It avoids B_2 because $a, b \notin B_2$ and $\{v, b\}$ cannot be in a bag to the right: if it was, coverage of edge $\{a, b\}$, the fact that $a \in B_1$ and running intersection would mean that at least one of a and b is in B_2, yielding a contradiction.

The only case that remains is $a \in B_1$, $\{v, b\}$ is not in a bag on the left and thus p_{vb} is in the interior. By symmetry, we assume that $d \in B_2$, $\{w, c\}$ is not in a bag on the right and thus p_{wc} is in the interior. Consider any path p_{ab} starting at B_1, defined in the now familiar way. Note that no bag on the left of B_1 can contain b. This is because $\{v, b\}$ is in a bag in the interior: hence if b was also on the left, b would then by running intersection be in B_1 and we would be in an earlier case. This means that p_{ab} must go via the interior. Suppose the following operation gives a valid tree decomposition: delete u from B_1, attach a new bag $B^* = \{a, u, w, v\}$ pendant to B_1, and relabel all occurrences of a along the path p_{ab} (except in B_1) with b. Then we are done, because we are back in Case 2. A symmetrical situation holds for the path p_{dc}.

Assume therefore that this transformation does not give a valid tree decomposition. This is the most complicated case to deal with. It is depicted in Fig. 3. The issue here is that the path p_{ab} (respectively, p_{dc}) necessarily goes via the interior, but cannot be relabelled with b (respectively, c) because the path is also part of p_{az} (respectively, p_{dz}) where z is some non-ladder vertex that is adjacent to a (respectively d). We deal with this as follows. We argue that some bag in the decomposition *must* contain a, b, v (and possibly other vertices). Suppose this is not the case. By standard chordalization arguments, every chordalization adds at least one diagonal edge to every square of the ladder. If $\{a, b, v\}$ are not together in a bag, then this is because the corresponding chordalization did not add the diagonal $\{a, v\}$ to square $\{a, u, b, v\}$. Hence, the chordalization must have added the diagonal $\{u, b\}$. This would in turn mean that some bag contains $\{a, u, b\}$. Such a bag must be on the interior, because this is the only place that b can be found. However, no bags in the interior contain u – contradiction.

Hence, some bag B' indeed contains $\{a, b, v\}$. Again, because b is only on the interior, B' must be in the interior. There could be multiple such bags, but this does not harm us. Let $B_{v\text{-done}}$ be the rightmost bag on the path P that is part of a path, starting from B_1, from a to some bag B' containing $\{a, b, v\}$. Let $B_{a\text{-done}}$ be the rightmost bag on P that contains a. Note that $B_{v\text{-done}}$ contains

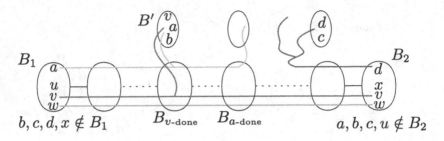

Fig. 4. The bags B', $B_{v\text{-done}}$ and $B_{a\text{-done}}$ illustrated. Note that $B_{a\text{-done}}$ cannot be the penultimate bag on the path P from B_1 to B_2, due to the presence of d in that bag.

a (because of running intersection: $a \in B_1$ and $a \in B'$) and v, w (because it lies on P). We also have $a, v, w \in B_{a\text{-done}}$. By construction, $B_{v\text{-done}}$ is either equal to $B_{a\text{-done}}$ or left of it. This is important because it means that the only reason v might need to be in bags to the right of $B_{a\text{-done}}$ is to reach a bag containing x (i.e. to cover the edge $\{v, x\}$) – all other edges are already covered elsewhere in the decomposition; in particular, edge $\{v, b\}$ is covered by the bag B' containing $\{a, b, v\}$. See Fig. 4 for clarification.

Recall that none of the bags on the path P contain both a and d. (If they did, there would be a bag containing $\{a, d, v, w\}$ and we would be in Case 2, done.) We also know that some path p_{dc} goes via the interior and thus that the penultimate bag on P (i.e. the one before B_2) thus definitely contains d. (To clarify: $c \notin B_2$, $d \in B_2$, the edge $\{c, d\}$ must be covered, and c is only in the interior). Combining these insights tells us that this penultimate bag definitely does *not* contain a, and hence $B_{a\text{-done}}$ is *not* equal to the penultimate bag; it is further left. This fact is crucial. Consider $B_{a\text{-done}}$ and the bag immediately to its right on P. Between these two bags we insert a copy of $B_{a\text{-done}}$, call it B_r, remove a from B_r (i.e. forget it), and add the element x to it instead. Finally, we switch v to x in all bags on P right of B_r, including B_2 itself, and delete v from all bags in the tree decomposition that are anywhere to the right of B_r; there is no point having them there. It requires some careful checking but this is a valid (minimum-width) tree decomposition. Moreover, B_r contains w, v, x. The fact that $B_{a\text{-done}}$ was not the penultimate bag of P, means that the length of the path from B_1 to B_r is strictly less than the length of the path from B_1 to B_2: contradiction on the assumption that these were the closest bags containing $\{u, w, v\}$ and $\{w, v, x\}$ respectively. We are done. ∎

We now deal with the situation when the $tw(G) \geq 4$ assumption is removed.

Lemma 4. *If G has a ladder L of length 5 or longer, the ladder can be increased in length arbitrarily without altering (in particular: increasing) the treewidth. This holds irrespective of the treewidth of G.*

Proof. Let L be a ladder of length 5 or longer. We can assume that L is not disconnecting and $tw(G) \leq 3$. We select the three most central squares and label these as in Fig. 1. These are flanked on both sides by at least one other square. Hence, a, b, c, d each has exactly one neighbour outside the 3 squares, let us call these a', b', c', d' respectively, where $\{a', b'\}$ is an edge and $\{c', d'\}$ is an edge. Now, $tw(G) = 3$ because L is not disconnecting. The only part of the proof of Theorem 2 that does not work for $tw(G) = 3$ is Case 1 and (indirectly) Case 2 because these create size-5 bags. We show that neither case can hold.

Consider Case 1. Let B be a bag containing one of the three most central squares S of the ladder (these are the only squares to which Case 1 is ever applied). A *small* tree decomposition is one where no bag is a subset of another. If a tree decomposition is not small, then by running intersection it must contain two adjacent bags $B^\dagger, B^\ddagger$ such that $B^\dagger \subseteq B^\ddagger$. The two bags can then be safely merged into $B^\ddagger$. By repeating this a small tree decomposition can be obtained without raising the width of the original minimum-width decomposition. Furthermore, some bag B will still exist containing S. If B has five or more vertices we immediately have $tw(G) \geq 4$ and we are done. Otherwise, let B' be any bag adjacent to B; such a bag must exist because G has more than 4 vertices. Due to the smallness of the decomposition we have $B \not\subseteq B'$ and $B' \not\subseteq B$. Hence, $B \cap B' \subset B$ and $B \cap B' \subset B'$. A *separator* is a subset of vertices whose deletion disconnects the graph. Now, $B \cap B'$ is by construction, and the definition of tree decompositions a separator of G. However, due to our use of the three central squares, S is not a separator, and no subset of it is a separator either; the inclusion of a', b', c', d' and the edges $\{a', b'\}$ and $\{c', d'\}$, alongside the fact that L is not disconnecting, ensure this. This yields a contradiction. Hence Case 1 implies $tw(G) \geq 4$ i.e. it cannot happen when $tw(G) = 3$.

We are left with Case 2. This case replaces the three centremost squares with a single square, and deletes any diagonals that this single square might have, to obtain a new graph G'. We have $tw(G') \leq tw(G)$, by minors. Note that $tw(G') \geq 3$ because the shorter ladder in G' (which has length at least 3) is still disconnecting. Hence, $tw(G') = tw(G) = 3$. The decomposition $\mathbb{T}'$ of G' obtained by projecting the contraction operations onto the tree decomposition $\mathbb{T}$ of G, is a valid tree decomposition (as argued in Case 2) with the property that the width of $\mathbb{T}'$ is less than or equal to the width of $\mathbb{T}$. $\mathbb{T}'$ cannot have width less than 3, so it must have width 3. Hence it is a tree decomposition of G' in which all bags have at most four vertices. We then transform $\mathbb{T}'$ into a small tree composition: this does not raise the width of the decomposition, and every bag prior to the transformation either survives or is absorbed into another. Consider the bag B' containing H_1, H_2, L_1, L_2. The presence of a', b', c', d' in G' and the fact that the ladder in G' is not disconnecting, means that H_1, H_2, L_1, L_2 is not a separator for G', and neither is any subset of those four vertices. But the intersection of B' with any neighbouring bag *must* be a separator. Hence B' must contain a fifth vertex, contradiction. So Case 2 cannot happen when $tw(G) = 3$.   ∎

We can, however, still do better. The following proof is in the appendix.

Theorem 3. *If G has a ladder L of length 4 or longer, the ladder can be increased in length arbitrarily without altering (in particular: increasing) the treewidth. This holds irrespective of the treewidth of G.*

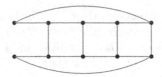

Fig. 5. A graph of treewidth 3 that contains a ladder with 3 squares, shown in red. Increasing the length of the ladder by 1 square increases the treewidth to 4. (Color figure online)

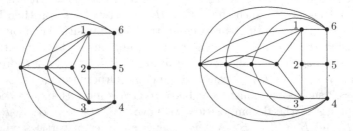

Fig. 6. These graphs have treewidth 4 (left) and 5 (right) and each one has a length 2 ladder, induced by $\{1, 2, 3, 4, 5, 6\}$. In each case increasing the length of the ladder to length 3 increases the treewidth of the graph by one.

Tightness. The constant 4 in the statement of Theorem 3 is equal to the constant obtained for the 'bottleneck' case $tw(G) = 3$. An improved constant 3 for this case is not possible, as Fig. 5 shows. It is natural to ask whether, when $tw(G) \geq 4$, we can start from ladders of length 2, rather than 3. This is also not possible. See Fig. 6. In fact, we have examples up to treewidth 20. These can be found at https://github.com/skelk2001/snakes_and_ladders. We conjecture that this holds for all treewidths, but defer this to future work.

Implications for Phylogenetics. A phylogenetic tree is a binary tree whose leaves are bijectively labelled by a set X. The *display graph* of T_1, T_2 on X is obtained by identifying leaves with the same label. In [14] it was asked whether *common chains* could be truncated to constant length without lowering the treewidth of the display graph. Theorem 3 establishes that the answer is *yes*. In fact, due to the restricted structure of display graphs, we can prove a stronger result: truncation to 4 leaves (i.e. 3 squares) is safe, and this is best possible. Theorem 4 summarizes this. We provide full details in the appendix.

Theorem 4. *Let T_1, T_2 be two unrooted binary phylogenetic trees on the same set of taxa X, where $|X| \geq 4$ and $T_1 \neq T_2$. Then exhaustive application of the subtree reduction and the common chain reduction (where common chains are reduced to 4 leaf labels) does not alter the treewidth of the display graph. This is best possible, because there exist tree pairs where truncation of common chains to length 3 does reduce the treewidth of the display graph (see Fig 7).*

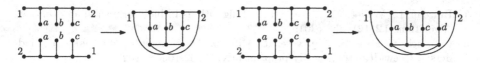

Fig. 7. Lengthening the common chain $\{a, b, c\}$ to $\{a, b, c, d\}$ in these phylogenetic trees causes the treewidth of the display graph to increase. Equivalently: shortening common chains to 3 leaf labels is not guaranteed to preserve treewidth in the display graph.

4 Future Work

A number of interesting open questions remain. First, can the results in this paper be made constructive? That is, given a minimum-width tree decomposition for a graph with a short ladder, but not necessarily the special distance-minimizing one assumed in our proofs, can we manipulate it to obtain a tree decomposition of the same width after the ladder is lengthened? Second, can we prove that for every starting treewidth, not just $tw(G) \leq 20$, extending a ladder of length 2 sometimes causes the treewidth to increase? Third, can our results be (elegantly) generalized to more complex recursive structures than ladders? This requires careful analysis of which parts of our proof are specific to ladders. Finally, if our results are generalized to more general structures, the question arises of how to implement these results as reduction rules: this requires efficient algorithmic recognition of "long" structures in order to produce the "short" variant.

Acknowledgements. We thank Hans Bodlaender and Bart Jansen for insightful feedback. We also thank the members of our department for useful discussions.

References

1. Abu-Khzam, F.N., Lamm, S., Mnich, M., Noe, A., Schulz, C., Strash, D.: Recent advances in practical data reduction. In: Bast, H., Korzen, C., Meyer, U., Penschuck, M. (eds.) Algorithms for Big Data. LNCS, vol. 13201, pp. 97–133. Springer, Cham (2022). https://doi.org/10.1007/978-3-031-21534-6_6
2. Allen, B., Steel, M.: Subtree transfer operations and their induced metrics on evolutionary trees. Ann. Comb. **5**, 1–15 (2001)
3. Blair, J., Peyton, B.: Graph Theory and Sparse Matrix Computation, chap. An Introduction to Chordal Graphs and Clique Trees. In: George, A., Gilbert, J.R., Liu, J.W.H. (eds.) Graph Theory and Sparse Matrix Computation. The IMA Volumes in Mathematics and its Applications, vol. 56, pp. 1–29. Springer, New York (1993). https://doi.org/10.1007/978-1-4613-8369-7_1
4. Bodlaender, H.L., van Antwerpen-de Fluiter, B.: Reduction algorithms for graphs of small treewidth. Inf. Comput. **167**(2), 86–119 (2001)
5. Bodlaender, H.L., Fomin, F.V., Lokshtanov, D., Penninkx, E., Saurabh, S., Thilikos, D.M.: (meta) kernelization. J. ACM **63**(5), 44:1–44:69 (2016)
6. Bryant, D., Lagergren, J.: Compatibility of unrooted phylogenetic trees is FPT. Theoret. Comput. Sci. **351**(3), 296–302 (2006)

7. Bulteau, L., Weller, M.: Parameterized algorithms in bioinformatics: an overview. Algorithms **12**(12), 256 (2019)
8. Chaplick, S., Kelk, S., Meuwese, R., Mihalak, M., Stamoulis, G.: Snakes and ladders: a treewidth story. arXiv preprint arXiv:2302.10662 (2023)
9. Fomin, F.V., Lokshtanov, D., Saurabh, S., Zehavi, M.: Kernelization: Theory of Parameterized Preprocessing. Cambridge University Press, Cambridge (2019)
10. van Iersel, L., Jones, M., Weller, M.: Embedding phylogenetic trees in networks of low treewidth. In: Chechik, S., Navarro, G., Rotenberg, E., Herman, G. (eds.) 30th Annual European Symposium on Algorithms, ESA 2022, September 5–9, 2022, Berlin/Potsdam, Germany. LIPIcs, vol. 244, pp. 69:1–69:14. Schloss Dagstuhl - Leibniz-Zentrum für Informatik (2022). arXiv preprint arXiv:2207.00574
11. Janssen, R., Jones, M., Kelk, S., Stamoulis, G., Wu, T.: Treewidth of display graphs: bounds, brambles and applications. J. Graph Algorithms Appl. **23**(4), 715–743 (2019)
12. John, K.S.: The shape of phylogenetic treespace. Syst. Biol. **66**(1), e83 (2017)
13. Kelk, S., van Iersel, L., Scornavacca, C., Weller, M.: Phylogenetic incongruence through the lens of monadic second order logic. J. Graph Algorithms Appl. **20**(2), 189–215 (2016)
14. Kelk, S., Stamoulis, G., Wu, T.: Treewidth distance on phylogenetic trees. Theoret. Comput. Sci. **731**, 99–117 (2018)
15. Sanders, D.: On linear recognition of tree-width at most four. SIAM J. Discret. Math. **9**(1), 101–117 (1996)
16. Steel, M.: Phylogeny: Discrete and Random Processes in Evolution. SIAM (2016)

Parameterized Results on Acyclic Matchings with Implications for Related Problems

Juhi Chaudhary$^{(\boxtimes)}$ and Meirav Zehavi

Department of Computer Science, Ben-Gurion University of the Negev, Beersheba, Israel
juhic@post.bgu.ac.il, meiravze@bgu.ac.il

Abstract. A matching M in a graph G is an *acyclic matching* if the subgraph of G induced by the endpoints of the edges of M is a forest. Given a graph G and a positive integer ℓ, ACYCLIC MATCHING asks whether G has an acyclic matching of *size* (i.e., the number of edges) at least ℓ. In this paper, we first prove that assuming $\mathsf{W}[1] \not\subseteq \mathsf{FPT}$, there does not exist any FPT-approximation algorithm for ACYCLIC MATCHING that approximates it within a constant factor when parameterized by ℓ. Our reduction is general in the sense that it also asserts FPT-inapproximability for INDUCED MATCHING and UNIQUELY RESTRICTED MATCHING. We also consider three below-guarantee parameters for ACYCLIC MATCHING, viz. $\frac{n}{2} - \ell$, $\mathsf{MM}(\mathsf{G}) - \ell$, and $\mathsf{IS}(\mathsf{G}) - \ell$, where n is the number of vertices in G, $\mathsf{MM}(\mathsf{G})$ is the *matching number* of G, and $\mathsf{IS}(\mathsf{G})$ is the *independence number* of G. We note that the result concerning the below-guarantee parameter $\frac{n}{2} - \ell$ is the most technical part of our paper. Also, we show that ACYCLIC MATCHING does not exhibit a polynomial kernel with respect to vertex cover number (or vertex deletion distance to clique) plus the size of the matching unless $\mathsf{NP} \subseteq \mathsf{coNP/poly}$.

Keywords: Acyclic Matching · Parameterized Algorithms · Kernelization Lower Bounds · Induced Matching · Uniquely Restricted Matching

1 Introduction

Matchings form a central topic in graph theory and combinatorial optimization [31]. In addition to their theoretical fruitfulness, matchings have various practical applications, such as assigning new physicians to hospitals, students to high schools, clients to server clusters, kidney donors to recipients [32], and so on. Moreover, matchings can be associated with the concept of *edge colorings* [3, 6,40], and they are a useful tool for finding optimal solutions or bounds in

The authors are supported by the European Research Council (ERC) project titled PARAPATH (101039913).

competitive optimization games on graphs [2,21]. Matchings have also found applications in Parameterized Complexity and Approximation Algorithms. For example, *Crown Decomposition*, a construction widely used for kernelization, is based on classical matching theorems [12], and matchings are one of the oldest tools in designing approximation algorithms for graph problems (e.g., consider the classical 2-approximation algorithm for VERTEX COVER [1]).

A matching M is a $\mathcal{P}$-*matching* if $G[V_M]$ (the subgraph induced by the endpoints of edges in M) has the property $\mathcal{P}$, where $\mathcal{P}$ is some graph property. The problem of deciding whether a graph admits a $\mathcal{P}$-matching of a given size (number of edges) has been investigated for several graph properties [17,20,22,36,38,39]. If the property $\mathcal{P}$ is that of being a graph, a disjoint union of edges, a forest, or having a unique perfect matching, then a $\mathcal{P}$-matching is a *matching* [33], an *induced matching* [39], an *acyclic matching* [20], and a *uniquely restricted matching* [22], respectively. In this paper, we focus on the parameterized complexity of acyclic matchings, but also discuss implications for the cases of induced matchings and uniquely restricted matchings. In particular, we study the parameterized complexity of these problems.

Problem Definitions and Related Works. Given a graph G and a positive integer ℓ, ACYCLIC MATCHING asks whether G has an acyclic matching of size at least ℓ. Goddard et al. [20] introduced the concept of acyclic matching, and since then, it has gained significant popularity [3,4,18,36,38]. ACYCLIC MATCHING is known to be NP-complete for perfect elimination bipartite graphs [38], star-convex bipartite graphs [36], and dually chordal graphs [36]. On the positive side, ACYCLIC MATCHING is known to be polynomial-time solvable for chordal graphs [4] and bipartite permutation graphs [38]. Fürst and Rautenbach [18] characterized the graphs for which every maximum matching is acyclic and gave linear-time algorithms to compute a maximum acyclic matching in graph classes such as P_4-free graphs and $2P_3$-free graphs. With respect to approximation, Panda and Chaudhary [36] showed that ACYCLIC MATCHING is hard to approximate within factor $n^{1-\epsilon}$ for every $\epsilon > 0$ unless P = NP. Apart from that, Baste et al. [4] showed that finding a maximum cardinality 1-degenerate matching in a graph G is equivalent to finding a maximum acyclic matching in G.

From the viewpoint of Parameterized Complexity, Hajebi and Javadi [24] studied ACYCLIC MATCHING. See Table 1 for a tabular view of known parameterized results concerning acyclic matchings.

Recall that a matching M is an induced matching if $G[V_M]$ is a disjoint union of $K_2's$. In fact, note that every induced matching is an acyclic matching, but the converse need not be true. Given a graph G and $\ell \in \mathbb{N}$, INDUCED MATCHING asks whether G has an induced matching of size at least ℓ. INDUCED MATCHING has been studied extensively due to its various applications and relations with other graph problems [7,11,15,25–27,29,34,37,39,42]. Given a graph G and a positive integer ℓ, INDUCED MATCHING BELOW TRIVIALITY (IMBT) asks whether G has an induced matching of size at least ℓ with the parameter $k = \frac{n}{2} - \ell$, where $n = |V(G)|$. IMBT has been studied in the literature,

Table 1. Overview of parameterized results for ACYCLIC MATCHING. Here, c_4 and tw denote the number of cycles with length four and the treewidth of the input graph, respectively.

	Parameter	Result
1	size of the matching	W[1]-hard on bipartite graphs [24]
2	size of the matching	FPT on line graphs [24]
3	size of the matching	FPT on every proper minor-closed class of graphs [24]
4	size of the matching plus c_4	FPT [24]
5	tw	FPT [9, 24]

albeit under different names: Moser and Thilikos [35] gave an algorithm to solve IMBT in $\mathcal{O}(9^k \cdot n^{\mathcal{O}(1)})$ time. Subsequently, Xiao and Kou [41] developed an algorithm running in $\mathcal{O}(3.1845^k \cdot n^{\mathcal{O}(1)})$ time. INDUCED MATCHING with respect to some new *below guarantee* parameters have also been studied recently [28].

Next, recall that a matching M is a uniquely restricted matching if $G[V_M]$ has exactly one perfect matching. Note by definition that an acyclic matching is always a uniquely restricted matching, but the converse need not be true. Given a graph G and a positive integer ℓ, UNIQUELY RESTRICTED MATCHING asks whether G has a uniquely restricted matching of size at least ℓ. The concept of uniquely restricted matching, motivated by a problem in Linear Algebra, was introduced by Golumbic et al. [22]. Some more results related to uniquely restricted matchings can be found in [5,6,8,17,20,22].

Our Contributions and Methods. Proofs of the results marked with a (∗) are omitted due to lack of space and can be found in the full version (see [10]).

In Sect. 3, we show that it is very unlikely that there exists any FPT-approximation algorithm for ACYCLIC MATCHING that approximates it to a constant factor. Our simple reduction also asserts the FPT-inapproximability of INDUCED MATCHING and UNIQUELY RESTRICTED MATCHING. In particular, we have the following theorem.

Theorem 1. *Assuming* W[1] $\not\subseteq$ FPT*, there is no* FPT *algorithm that approximates any of the following to any constant when the parameter is the size of the matching:* (1) ACYCLIC MATCHING*;* (2) INDUCED MATCHING*;* (3) UNIQUELY RESTRICTED MATCHING.

If $\mathcal{R} \in \{\text{acyclic, induced, uniquely restricted}\}$, then the $\mathcal{R}$ *matching number* of G is the maximum cardinality of an $\mathcal{R}$ matching among all $\mathcal{R}$ matchings in G. We denote by AM(G), IM(G), and URM(G), the *acyclic matching number*, the *induced matching number*, and the *uniquely restricted matching number*, respectively, of G. Also, we denote by IS(G) the *independence number* of G.

In order to prove Theorem 1, we exploit the relationship of IS(G) with the following: AM(G), IM(G), and URM(G). While the relationship of AM(G) and IM(G) with IS(G) is straightforward, extra efforts are needed to state a similar relation between URM(G) and IS(G), which may be of independent interest as well. Also, we note that INDUCED MATCHING is already known to be W[1]-hard, even for bipartite graphs [34]. However, for UNIQUELY RESTRICTED MATCHING, Theorem 1 is the first to establish the W[1]-hardness of the problem.

In Sect. 4, we consider below-guarantee parameters for ACYCLIC MATCHING, i.e., parameterizations of the form $UB - \ell$ for an upper bound UB on the size of any acyclic matching in G. Since the inception of below-guarantee (and above-guarantee) parameters, there has been significant progress in the area concerning graph problems parameterized by such parameters (see a recent survey paper by Gutin and Mnich [23]). It is easy to observe that in a graph G, the size of any acyclic matching is at most $\frac{n}{2}$, where $n = |V(G)|$. This observation yields the definition of ACYCLIC MATCHING BELOW TRIVIALITY, which, given a graph G with $|V(G)| = n$ and a positive integer ℓ, asks whether G has an acyclic matching of size at least ℓ. Here, the parameter is $k = n - 2\ell$.

Note that for ease of exposition, we are using the parameter $n - 2\ell$ instead of $\frac{n}{2} - \ell$. Next, we have the following theorem.

Theorem 2. *There exists a randomized algorithm that solves* AMBT *with success probability at least* $1 - \frac{1}{e}$ *in* $10^k \cdot n^{\mathcal{O}(1)}$ *time.*

The proof of Theorem 2 is the most technical part of our paper. The initial intuition in proving Theorem 2 was to take a cue from the existing literature on similar problems (which are quite a few) like INDUCED MATCHING BELOW TRIVIALITY. However, induced matchings have many nice properties, which do not hold for acyclic matchings in general, and thus make it more difficult to characterize edges or vertices that should necessarily belong to an optimal solution of ACYCLIC MATCHING BELOW TRIVIALITY. However, there is one nice property about ACYCLIC MATCHING, and it is that it is closely related to the FEEDBACK VERTEX SET problem as follows. Given an instance (G, ℓ) of ACYCLIC MATCHING, where $n = |V(G)|$, G has an acyclic matching of size ℓ if and only if there exists a (not necessarily minimal) feedback vertex set, say, X, of size $n - 2\ell$ such that $G - X$ has a perfect matching. We use randomization techniques to find a (specific) feedback vertex set of the input graph and then check whether the remaining graph (which is a forest) has a matching of the desired size or not. Note that the classical randomized algorithm for FEEDBACK VERTEX SET (which is a classroom problem now) [12] cannot be applied "as it is" here. In other words, since our ultimate goal is to find a matching, we cannot get rid of the vertices or edges of the input graph by applying some reduction rules "as they are". Instead, we can do something meaningful if we store the information about everything we delete or modify, and maintain some specific property (called Property R by us) in our graph. We also use our own lemma (Lemma 7) to pick a vertex in our desired feedback vertex set with high probability. Then, using an algorithm for MAX WEIGHT MATCHING (see Sect. 2), we compute an acyclic matching of size at least ℓ, if such a matching exists.

In light of Theorem 2, for AMBT, we ask if ACYCLIC MATCHING is FPT for natural parameters smaller than $\frac{n}{2} - \ell$. Here, an obvious upper bound on AM(G) is the matching number of G. Thus, we consider the below-guarantee parameter

$MM(G) - \ell$, where $MM(G)$ denotes the matching number of G, which yields ACYCLIC MATCHING BELOW MAXIMUM MATCHING (AMBMM). Given a graph G and a positive integer ℓ, AMBMM asks whether G has an acyclic matching of size at least ℓ. Note that the parameter in AMBMM is $k = MM(G) - \ell$.

In [18], Fürst and Rautenbach showed that deciding whether a given bipartite graph of maximum degree at most 4 has a maximum matching that is also an acyclic matching is NP-hard. Thus, for $k = 0$, the AMBMM problem is NP-hard, and we have the following result.

Corollary 1. AMBMM *is* para-NP-hard *even for bipartite graphs of maximum degree at most 4.*

Next, consider the following lemma.

Lemma 1. (*) *If G has an acyclic matching M of size ℓ, then G has an independent set of size at least ℓ. Moreover, given M, the independent set is computable in polynomial time.*

By Lemma 1, for any graph G, $IS(G) \geq AM(G)$, which yields the ACYCLIC MATCHING BELOW INDEPENDENT SET (AMBIS) problem. Given a graph G and a positive integer ℓ, AMBIS asks whether G has an acyclic matching of size at least ℓ. Note that the parameter in AMBIS is $k = IS(G) - \ell$.

In [36], Panda and Chaudhary showed that ACYCLIC MATCHING is hard to approximate within factor $n^{1-\epsilon}$ for any $\epsilon > 0$ unless P $=$ NP by giving a polynomial-time reduction from INDEPENDENT SET. We notice that with a more careful analysis of the proof, the reduction given in [36] can be used to show that AMBIS is NP-hard for $k = 0$. Therefore, we have the following result.

Corollary 2. AMBIS *is* para-NP-hard.

We note that Hajebi and Javadi [24] showed that ACYCLIC MATCHING parameterized by treewidth (tw) is FPT by using Courcelle's theorem. Since $tw(G) \leq vc(G)^1$, this result immediately implies that ACYCLIC MATCHING is FPT with respect to the parameter vc. We complement this result by showing that it is unlikely for ACYCLIC MATCHING to admit a polynomial kernel when parameterized not only by vc, but also when parameterized jointly by vc plus the size of the acyclic matching. In particular, we have the following.

Theorem 3. ACYCLIC MATCHING *does not admit a polynomial kernel when parameterized by vertex cover number plus the size of the matching unless* NP $\subseteq$ coNP/poly.

Parameterization by the size of a *modulator* (a set of vertices in a graph whose deletion results in a graph that belongs to a well-known and easy-to-handle graph class) is another natural choice of investigation. We observe that with only a minor modification in our construction (in the proof of Theorem 3), we derive the following result.

---

[1] We denote by $vc(G)$, the *vertex cover number* of a graph G.

Theorem 4. ACYCLIC MATCHING *does not admit a polynomial kernel when parameterized by the vertex deletion distance to clique plus the size of the matching unless* NP $\subseteq$ coNP/poly.

Due to space constraints, the section concerning negative kernelization results has been deferred to the full version of the paper (see [10]).

2 Preliminaries

For $k \in \mathbb{N}$, let $[k] = \{1, 2, \ldots, k\}$. We consider only simple and undirected graphs unless stated otherwise. For a graph G, let $V(G)$ denote its vertex set, and $E(G)$ denote its edge set. For a graph G, the subgraph of G induced by $S \subseteq V(G)$ is denoted by $G[S]$, where $G[S] = (S, E_S)$ and $E_S = \{xy \in E(G) : x, y \in S\}$. Given a matching M, a vertex $v \in V(G)$ is M-*saturated* if v is incident on an edge of M. Given a graph G and a matching M, we use the notation V_M to denote the set of M-saturated vertices. The *matching number* of G is the maximum cardinality of a matching among all matchings in G, and we denote it by $\mathsf{MM}(G)$. The edges in a matching M are *matched edges*. A matching that saturates all the vertices of a graph is a *perfect matching*. If $uv \in M$, then v is the M-*mate* of u and vice versa. Given a weighted graph G with a weight function $\mathsf{w} : E(G) \to \mathbb{R}$ and a weight $\mathsf{W} \in \mathbb{R}$, MAX WEIGHT MATCHING asks whether G has a matching with weight at least W in G, and can be solved in $\mathcal{O}(m\sqrt{n}\log(N))$ time, where $m = |E(G)|$, $n = |V(G)|$, and the weights are integers lying in $[0, N]$ [14].

The *degree* of a vertex v, denoted by $d_G(v)$, is $|N_G(v)|$. When there is no ambiguity, we do not use the subscript G. The minimum degree of graph G is denoted by $\delta(G)$. We use the notation $\widehat{d}(u, v)$ to represent the distance between two vertices u and v in a graph G. We denote by $\mathsf{IS}(G)$, the *independence number* of a graph G. Given a (multi)graph G and a positive integer ℓ, FEEDBACK VERTEX SET asks whether there exists a feedback vertex set in G of size at most ℓ. A *factor* of a graph G is a spanning subgraph of G (a subgraph with vertex set $V(G)$). A k-*factor* of a graph is a k-regular subgraph of order n. In particular, a 1-*factor* is a perfect matching. Let K_n denote a *complete graph* with n vertices. The size of a clique modulator of minimum size is known as the *vertex deletion distance to a clique*. For a graph G and a set $X \subseteq V(G)$, we use $G - X$ to denote $G[V(G) \backslash X]$.

Standard notions in Parameterized Complexity not explicitly defined here can be found in [12,13]. In the framework of Parameterized Complexity, each instance of a problem Π is associated with a positive integer *parameter* k. A parameterized problem Π is *fixed-parameter tractable* (FPT) if there is an algorithm that, given an instance (I, k) of Π, solves it in time $f(k) \cdot |I|^{\mathcal{O}(1)}$, for some computable function $f(\cdot)$. Due to space constraints, the definition of FPT-*approximation algorithm* has been deferred to the full version (see [10]).

3 FPT-Inapproximation Results

To obtain our result, we need the following proposition.

Proposition 1 ([30]). *Assuming* W[1] $\not\subseteq$ FPT*, there is no* FPT-*algorithm that approximates* CLIQUE *to any constant.*

From Proposition 1, we derive the following corollary.

Corollary 3. *Assuming* W[1] $\not\subseteq$ FPT*, there is no* FPT-*algorithm that approximates* INDEPENDENT SET *to any constant.*

Before presenting our reduction from INDEPENDENT SET, we establish the relationship of IS(G) with the following: AM(G), IM(G), and URM(G), which will be critical for the arguments in the proof of our main theorem in this section.

Relation of IS(G) with AM(G), IM(G) and URM(G).

Lemma 1. (∗) *If G has an acyclic matching M of size ℓ, then G has an independent set of size at least ℓ. Moreover, given M, the independent set is computable in polynomial time.*

Since every induced matching is also an acyclic matching, the next lemma directly follows from Lemma 1.

Lemma 2. *If G has an induced matching M of size ℓ, then G has an independent set of size at least ℓ. Moreover, given M, the independent set is computable in polynomial time.*

Proving a similar lemma for uniquely restricted matching is more complicated. To this end, we need the following notation. Given a graph G, an *even cycle* (i.e., a cycle with an even number of edges) in G is said to be an *alternating cycle* with respect to a matching M if every second edge of the cycle belongs to M. The following proposition characterizes uniquely restricted matchings in terms of alternating cycles.

Proposition 2 ([22]). *Let G be a graph. A matching M in G is uniquely restricted if and only if there is no alternating cycle with respect to M in G.*

Our proof will also identify some bridges based on the following proposition.

Proposition 3 ([19]). *A graph with a unique 1-factor has a bridge that is matched.*

Lemma 3. *If G has a uniquely restricted matching M of size ℓ, then G has an independent set of size at least $\frac{\ell+1}{2}$. Moreover, given M, the independent set is computable in polynomial time.*

Proof. Let M be a uniquely restricted matching in G of size ℓ. By the definition of a uniquely restricted matching, $G[V_M]$ has a unique perfect matching (1-factor). Let H and I be two sets. Initialize $H = G[V_M]$ and $I = \emptyset$. Next, we design an iterative algorithm, say, Algorithm FIND, to compute an independent set in H with the help of Proposition 3. Algorithm FIND does the following:

While H has a connected component of size at least four, go to 1.

1. Pick a bridge, say, e, in H that belongs to M, and go to 2. (The existence of e follows from Proposition 3.)
2. If e is a pendant edge, then remove e along with its endpoints from H, and store the pendant vertex incident on e to I. Else, go to 3.
3. Remove e along with its endpoints from H.

After recursively applying 1–3, Algorithm FIND arbitrarily picks exactly one vertex from each of the remaining connected components and adds them to I.

Now, it remains to show that I is an independent set of H (and therefore also of G) of size at least $\frac{\ell+1}{2}$. For this purpose, we note that Algorithm FIND gives rise to a recursive formula (defined below).

Let R_h denote a lower bound on the maximum size of an independent set in H of a maximum matching of size h. First, observe that if H has a matching of size one, then it is clear that H has an independent set of size at least 1 (we can pick one of the endpoints of the matched edge). Thus, $R_1 = 1$. Now, we define how to compute R_h recursively.

$$R_h = \min\{R_{h-1} + 1, \min_{1 \leq i \leq h-2} R_i + R_{h-i-1}\}. \tag{1}$$

The first term in (1) corresponds to the case where the matched bridge is a pendant edge. On the other hand, the second term in (1) corresponds to the case where the matched bridge is not a pendant edge. In this case, all the connected components have at least one of the matched edges. Next, we claim the following,

$$R_h \geq \frac{h+1}{2}. \tag{2}$$

We prove our claim by applying induction on the maximum size of a matching in H. Recall that $R_1 = 1$. Next, by the induction hypothesis, assume that (2) is true for all $k < h$. Note that since $h - 1 < h$, (2) is true for $k = h - 1$, i.e., $R_{h-1} \geq \frac{h}{2}$. To prove that (2) is true for $k = h$, we first assume that the first term, i.e., $R_{h-1} + 1$ gives the minimum in (1). In this case, $R_h \geq \frac{h}{2} + 1 = \frac{h+2}{2} \geq \frac{h+1}{2}$. Next, assume that the second term gives the minimum in (1) for some $i', 1 \leq i' \leq h - 2$. In this case, note that $i' < h$ and $h - i' - 1 \leq h - 2 < h$, and thus, by the induction hypothesis, (2) holds for both $R_{i'}$ and $R_{h-i'-1}$. Therefore, $R_h \geq \frac{i'+1}{2} + \frac{h-i'}{2} = \frac{h+1}{2}$. □

Remark 1. Throughout this section, let RESTRICTED $\in$ {ACYCLIC, INDUCED, UNIQUELY RESTRICTED} and $\mathcal{R} \in$ {acyclic, induced, uniquely restricted}.

By Lemmas 1–3, we have the following corollary.

Corollary 4. If a graph G has an $\mathcal{R}$ matching M of size ℓ, then G has an independent set of size at least $\frac{\ell+1}{2}$. Moreover, given M, the independent set is computable in polynomial time.

Now, consider the following construction.

Construction. Given a graph G, where $V(G) = \{v_1, \ldots, v_n\}$, we construct a graph $H = \mathsf{reduce}(G)$ as follows. Let G^1 and G^2 be two copies of G. Let $V^1 = \{v_1^1, \ldots, v_n^1\}$ and $V^2 = \{v_1^2, \ldots, v_n^2\}$ denote the vertex sets of G^1 and G^2, respectively. Let $V(H) = V(G^1) \cup V(G^2)$ and $E(H) = E(G^1) \cup E(G^2) \cup \{v_i^1 v_i^2 : i \in [n]\}$. Let us call the edges between G^1 and G^2 *vertical edges*.

Lemma 4. (∗) *Let G and H be as defined above. If H has $\mathcal{R}$ matching of the form $M = \{v_1^1 v_1^2, \ldots, v_p^1 v_p^2\}$, then $I_M = \{v_1, \ldots, v_p\}$ is an independent set of G.*

Hardness of Approximation Proof. To prove Theorem 1, we first suppose that RESTRICTED MATCHING can be approximated within a ratio of $\alpha > 1$, where $\alpha \in \mathbb{R}^+$ is a constant, by some FPT-approximation algorithm, say, Algorithm $\mathcal{A}$. By the definition of $\mathcal{A}$, the following is true:

i) If H does not have an $\mathcal{R}$ matching of size ℓ, then the output of $\mathcal{A}$ is arbitrary (indicating that (H, ℓ) is a No-instance).
ii) If H has an $\mathcal{R}$ matching of size ℓ, then $\mathcal{A}$ returns an $\mathcal{R}$ matching, say, X, such that $|X| \geq \frac{\ell}{\alpha}$ and $|X| \leq \mathsf{opt}(H)$, where $\mathsf{opt}(H)$ denotes the optimal size of an $\mathcal{R}$ matching in H.

Next, we propose an FPT-approximation algorithm, say, Algorithm $\mathcal{B}$, to compute an FPT-approximate solution for INDEPENDENT SET as follows. Given an instance (G, ℓ) of INDEPENDENT SET, Algorithm $\mathcal{B}$ first constructs an instance (H, ℓ) of RESTRICTED MATCHING, where $H = \mathsf{reduce}(G)$. Algorithm $\mathcal{B}$ then solves (H, ℓ) by using Algorithm $\mathcal{A}$. If Algorithm $\mathcal{A}$ returns an $\mathcal{R}$ matching X, then Algorithm $\mathcal{B}$ returns an independent set of size at least $\frac{|X|}{8} (\geq \frac{\ell}{8\alpha})$. Else, the output is arbitrary.

Now, it remains to show that Algorithm $\mathcal{B}$ is an FPT-approximation algorithm for INDEPENDENT SET with an approximation factor of $\beta > 1$, where $\beta \in \mathbb{R}^+$, which we will show with the help of the following two lemmas.

Lemma 5. (∗) *Algorithm $\mathcal{B}$ approximates INDEPENDENT SET within a constant factor $\beta > 1$, where $\beta \in \mathbb{R}^+$.*

Lemma 6. (∗) *Algorithm $\mathcal{B}$ runs in FPT time.*

By Corollary 4 and Lemmas 5 and 6, we have Theorem 1.

4   FPT Algorithm for AMBT

In this section, we prove that AMBT is FPT by giving a randomized algorithm that runs in time $10^k \cdot n^{\mathcal{O}(1)}$, where $n = |V(G)|$ and $k = n - 2\ell$.

First, we define some terminology that is crucial for proceeding further in this section. A graph G *has property* R if $\delta(G) \geq 2$ and no two adjacent vertices of G have degree exactly 2. A path P is a *maximal degree-2 path* in G if: (i) it has at least two vertices, (ii) the degree of each vertex in P (including the endpoints) is exactly 2, and (iii) it is not contained in any other degree-2 path.

If we replace a maximal degree-2 path P with a single vertex, say, v_P, of degree exactly 2 (in G), then we call this operation PATH-REPLACEMENT(P,v_P) (note that the neighbors of v_P are the neighbors of the endpoints of P that do not belong to P). Furthermore, we call the newly introduced vertex (that replaces a maximal degree-2 path in G) *virtual vertex*. Note that if both endpoints of P have a common neighbor, then this gives rise to multiple edges in G. Next, if there exists a cycle, say, C, of length $p \geq 2$ such that the degree of each vertex in C is exactly 2 (in G), then the PATH-REPLACEMENT operation also identifies such cycles and replaces each of them with a virtual vertex having a self-loop, and the corresponding maximal degree-2 path, in this case, consists of all the vertices of C. Therefore, it is required for us to consider AMBT in the more general setting of multigraphs, where the graph obtained after applying the PATH-REPLACEMENT operation may contain multiple edges and self-loops. We also note that multiple edges and self-loops are cycles.

We first present a lemma (Lemma 7) that is crucial to prove Theorem 2.

Lemma 7. (∗) *Let G be a graph on n vertices with the property* R. *Then, for every feedback vertex set X of G, more than $\frac{|E(G)|}{5}$ of the edges of G have at least one endpoint in X.*

Now, consider Algorithm 1.

Observe that the task of Algorithm 1 is first to modify an input graph G to a graph that has property R. By abuse of notation, we call this modified graph G. Since G is non-empty and G has property R, then G definitely has a cycle, and by Lemma 7, with probability at least $\frac{1}{10}$, we pick one vertex, say v, that belongs to a specific feedback vertex set of G. We store this vertex in a set $\widehat{X}$. After removing v from G, we also decrease k by 1. We again repeat the process until either the graph becomes empty or k becomes non-positive while there are still some cycles left in the graph; we return No in the latter case, and the sets A, Z, and $\widehat{X}$ in the former case.

Given an input graph G and a positive integer k, let $\widehat{X}$ be a virtual feedback vertex set returned by Algorithm 1. Note that the set $\widehat{X}$ contains a combination of virtual vertices and the vertices from the set $V(G)$. Let $\widehat{V} \subseteq \widehat{X}$ be the set of virtual vertices. If $v_P \in \widehat{V}$, then there exists some maximal degree-2 path P such that $(v_P, P) \in A$. If all the vertices in P are from $V(G)$, then we say that the set of vertices in P is *safe* for v_P. On the other hand, if the path P contains some virtual vertices, then note that there exist maximal degree-2 paths corresponding to these virtual vertices as well. In this case, we recursively replace the virtual vertices present in P with their corresponding maximal degree-2 paths until we obtain a set that contains vertices from $V(G)$ only, and we say that these vertices are safe for v_P. The process of obtaining a set of safe vertices corresponding to virtual vertices is shown in Fig. 1. Note that, for the graph shown in Fig. 1 (i), if $\widehat{X} = \{v_{P_3}, v_{P_4}, v_{P_6}\}$ is a virtual feedback vertex set returned by Algorithm 1 corresponding to $X = \{i, k, a\}$, then the safe set corresponding to v_{P_3} is $\{i, j\}$, corresponding to v_{P_4} is $\{k, l\}$, and corresponding to v_{P_6} is $\{a, b, c, d, e, f, g, h\}$.

Algorithm 1

Input: A graph G and a positive integer k;
Output: A set $\widehat{X}$ of size at most k, a set Z, and a set A or No;
Initialize $Z \leftarrow \emptyset$, $A \leftarrow \emptyset$, $\widehat{X} \leftarrow \emptyset$;
while $(V(G) \neq \emptyset)$ **do**

   while $(\delta(G) \leq 1)$ **do**
      Pick a vertex $v \in V(G)$ such that $d(v) \leq 1$;
      $Z \leftarrow Z \cup \{v\}$;
      $V(G) \leftarrow V(G) \backslash \{v\}$;

   while (*there exists a maximal degree-2 path P in G*) **do**
      PATH-REPLACEMENT(P,v_P);
      $A \leftarrow A \cup \{(P, v_P)\}$;

   if ($k > 0$ *and G has a cycle*) **then**
      if (*G has a self-loop at some v*) **then**
         $\widehat{X} \leftarrow \widehat{X} \cup \{v\}$;
         $V(G) \leftarrow V(G) \backslash \{v\}$;
         $k \leftarrow k - 1$;

      else
         Pick an edge $e \in E(G)$;
         Pick an endpoint v of e;
         $\widehat{X} \leftarrow \widehat{X} \cup \{v\}$;
         $V(G) \leftarrow V(G) \backslash \{v\}$;
         $k \leftarrow k - 1$;

   else if ($k \leq 0$ *and G has a cycle*) **then**
      return No;

return $\widehat{X}, Z, A$;

Remark 2. Throughout this section, if $\widehat{X}$ is a virtual feedback vertex set returned by Algorithm 1, then let $\widehat{V} \subseteq \widehat{X}$ denote the set of virtual vertices.

Definition 1. *Let $\widehat{X}$ be a virtual feedback vertex set returned by Algorithm 1 if given as input a graph G and a positive integer k. A set $X \subseteq V(G)$ is compatible with $\widehat{X}$ if the following hold: (1) For every $v_P \in \widehat{V}$, X contains at least one vertex from the set of safe vertices corresponding to v_P; (2) For every $v \in \widehat{X} \backslash \widehat{V}$, v belongs to X; (3) $|X| \leq k$.*

Lemma 8. (*) *Given a graph G and a positive integer k, if $\widehat{X}$ is a virtual feedback vertex set returned by Algorithm 1, then any set $X \subseteq V(G)$ compatible with $\widehat{X}$ is a feedback vertex set of G.*

Lemma 9. (*) *Let G be a graph and $k \in \mathbb{N}$. Then, for any feedback vertex set X of G of size at most k, with probability at least 10^{-k}, Algorithm 1 returns a virtual feedback vertex set $\widehat{X}$ such that X is compatible with $\widehat{X}$.*

Remark 3. Let Algorithm $\mathcal{A}$ be any algorithm that solves MAX WEIGHT MATCHING in polynomial time.

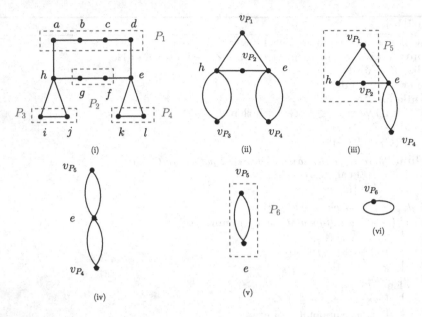

Fig. 1. After applying PATH-REPLACEMENT to paths P_1-P_4 in (i), we obtain the graph in (ii), which has property R. We assume that Algorithm 1 picks v_{P_3} in $\widehat{X}$. After removing v_{P_3}, we obtain the graph shown in (iii) that has a maximal degree-2 path P_5. After applying PATH-REPLACEMENT to P_5, we obtain the graph in (iv). We assume that Algorithm 1 picks v_{P_4} in $\widehat{X}$. After removing v_{P_4}, we obtain the graph shown in (v). Note that the PATH-REPLACEMENT operation identifies the cycle shown in (v) as a maximal degree-2 path and replaces it with v_{P_6} with a self-loop, as shown in (vi). Algorithm 1 then picks v_{P_6} in $\widehat{X}$.

Next, consider Algorithm 2.

Lemma 10. (∗) *Let G, ℓ, $\widehat{X}$, G_W, M_W, and M be as defined in Algorithm 2. If M_W is of weight at least $\ell + |\widehat{V}| \cdot c$, then M is an acyclic matching in G of size at least ℓ.*

Lemma 11. (∗) *Let (G, ℓ) be a Yes-instance of ACYCLIC MATCHING with $n = |V(G)|$. Then, with probability at least $10^{2\ell-n}$, Algorithm 2 returns an acyclic matching M in G of size at least ℓ.*

Lemma 12. (∗) *Let (G, ℓ) be a No-instance of ACYCLIC MATCHING with $n = |V(G)|$. Then, with probability 1, Algorithm 2 returns No.*

We can improve the success probability of Algorithm 1 and thus Algorithm 2, by repeating it, say, t times, and returning a No only if we are not able to find a virtual feedback vertex set of size at most k in each of the repetitions. Clearly, due to Lemma 12, given a No-instance, even after repeating the procedure t times, we will necessarily get No as an answer. However, given a Yes-instance,

---

Algorithm 2

Input: An instance (G, ℓ) of ACYCLIC MATCHING with $n = |V(G)|$;
Output: An acyclic matching M in G of size at least ℓ or No;
Call Algorithm 1 with input $(G, n - 2\ell)$;
if (*Algorithm 1 returns No*) **then**
  ∟ return No;
else if (*Algorithm 1 returns a virtual feedback vertex set* $\widehat{X}$) **then**
  | Initialize $G_W = G$;
  | For every $v_P \in \widehat{V}$, add a vertex w_P to G_W and make it adjacent to every vertex
  | in the safe set corresponding to v_P. Call all edges introduced here *new edges*.
  | Remove all $v \in \widehat{X} \backslash \widehat{V}$ from G_W;
  | Assign weight $c = |E(G)| + 1$ to all new edges of G_W and weight 1 to all the
  | remaining edges of G_W;
  | Call Algorithm $\mathcal{A}$ with input $(G_W, \ell + |\widehat{V}| \cdot c)$;
  | **if** (*Algorithm $\mathcal{A}$ returns a matching M_W of weight at least* $\ell + |\widehat{V}| \cdot c$) **then**
  | | $M = \{e \in M_W : \text{weight of } e \text{ is } 1\}$;
  | ∟ return M;
  | **else**
  | ∟ return No;

---

we return a No only if all t repetitions return an incorrect No, which, by Lemma 11, has probability at most

$$(1 - 10^{-k})^t \leq (e^{-10^{-k}})^t \leq \frac{1}{e^{10^{-k}t}}. \tag{3}$$

Note that we are using the identity $1 + x \leq e^x$ in (3). In order to obtain a constant failure probability, we take $t = 10^k$. By taking $t = 10^k$, the success probability becomes at least $1 - \frac{1}{e}$. Thus, we have the following theorem.

Theorem 1. *There exists a randomized algorithm that solves* AMBT *with success probability at least* $1 - \frac{1}{e}$ *in* $10^k \cdot n^{\mathcal{O}(1)}$ *time.*

5   Conclusion and Future Research

Based on the results given by Moser and Sikdar [34], we note that both ACYCLIC MATCHING and UNIQUELY RESTRICTED MATCHING admit quadratic kernels (with respect to the maximum degree of the input graph) for bounded degree graphs. In fact, for UNIQUELY RESTRICTED MATCHING, the quadratic kernel can be further improved to a linear kernel. The following is true for any maximum uniquely restricted matching.

Property $\widehat{P}$: If M is a maximum uniquely restricted matching in a graph G, then for each vertex $v \in V(G)$, there exists a vertex $u \in V_M$ such that $d(u, v) \leq 1$.

One possible direction for future research is to seek a below-guarantee parameter smaller than the parameter $\frac{n}{2} - \ell$, so that ACYCLIC MATCHING remains

FPT. Also, it would be interesting to see if the running time in Theorem 2 can be substantially improved. Apart from that, we strongly believe that the arguments presented in this work (in Sect. 4) will be useful for other future works concerning problems where one seeks a solution that, among other properties, satisfies that it is itself, or its complement, a feedback vertex set, or, much more generally, an alpha-cover (see [16]). A more detailed conclusion can be found in the full version (see [10]).

References

1. Ausiello, G., Crescenzi, P., Gambosi, G., Kann, V., Marchetti-Spaccamela, A., Protasi, M.: Complexity and Approximation: Combinatorial Optimization Problems and their Approximability properties. Springer, Heidelberg (2012)
2. Bachstein, A., Goddard, W., Lehmacher, C.: The generalized matcher game. Discret. Appl. Math. **284**, 444–453 (2020)
3. Baste, J., Fürst, M., Rautenbach, D.: Approximating maximum acyclic matchings by greedy and local search strategies. In: Kim, D., Uma, R.N., Cai, Z., Lee, D.H. (eds.) COCOON 2020. LNCS, vol. 12273, pp. 542–553. Springer, Cham (2020). https://doi.org/10.1007/978-3-030-58150-3_44
4. Baste, J., Rautenbach, D.: Degenerate matchings and edge colorings. Discret. Appl. Math. **239**, 38–44 (2018)
5. Baste, J., Rautenbach, D., Sau, I.: Uniquely restricted matchings and edge colorings. In: Bodlaender, H.L., Woeginger, G.J. (eds.) WG 2017. LNCS, vol. 10520, pp. 100–112. Springer, Cham (2017). https://doi.org/10.1007/978-3-319-68705-6_8
6. Baste, J., Rautenbach, D., Sau, I.: Upper bounds on the uniquely restricted chromatic index. J. Graph Theory **91**(3), 251–258 (2019)
7. Cameron, K.: Induced matchings. Discret. Appl. Math. **24**(1–3), 97–102 (1989)
8. Chaudhary, J., Panda, B.S.: On the complexity of minimum maximal uniquely restricted matching. Theoret. Comput. Sci. **882**, 15–28 (2021)
9. Chaudhary, J., Zehavi, M.: $\mathcal{P}$-matchings parameterized by treewidth. In: Paulusma, D., Ries, B. (eds.) WG 2023. LNCS, vol. 14093, pp. 217–231. Springer, Cham (2023)
10. Chaudhary, J., Zehavi, M.: Parameterized results on acyclic matchings with implications for related problems. arXiv preprint arXiv:2307.05446v1 (2023)
11. Cooley, O., Draganic, N., Kang, M., Sudakov, B.: Large induced matchings in random graphs. SIAM J. Discret. Math. **35**(1), 267–280 (2021)
12. Cygan, M., et al.: Parameterized Algorithms, vol. 5. Springer, Cham (2015)
13. Downey, R.G., Fellows, M.R.: Fundamentals of Parameterized Complexity, vol. 4. Springer, London (2013)
14. Duan, R., Su, H.-H.: A scaling algorithm for maximum weight matching in bipartite graphs. In Proceedings of the Twenty-third Annual ACM-SIAM Symposium on Discrete Algorithms. SIAM, pp. 1413–1424 (2012)
15. Erman, R., Kowalik, Ł, Krnc, M., Waleń, T.: Improved induced matchings in sparse graphs. Discret. Appl. Math. **158**(18), 1994–2003 (2010)
16. Fomin, F.V., Lokshtanov, D., Misra, N., Saurabh, S.: Planar f-deletion: approximation, kernelization and optimal FPT algorithms. In: 2012 53rd Annual Symposium on Foundations of Computer Science. IEEE, pp. 470–479 (2012)
17. Francis, M.C., Jacob, D., Jana, S.: Uniquely restricted matchings in interval graphs. SIAM J. Discret. Math. **32**(1), 148–172 (2018)

18. Fürst, M., Rautenbach, D.: On some hard and some tractable cases of the maximum acyclic matching problem. Ann. Oper. Res. **279**, 291–300 (2019)
19. Gabow, H.N.: Algorithmic proofs of two relations between connectivity and the 1-factors of a graph. Discret. Math. **26**(1), 33–40 (1979)
20. Goddard, W., Hedetniemi, S.M., Hedetniemi, S.T., Laskar, R.: Generalized subgraph-restricted matchings in graphs. Discret. Math. **293**(1–3), 129–138 (2005)
21. Goddard, W., Henning, M.A.: The matcher game played in graphs. Discret. Appl. Math. **237**, 82–88 (2018)
22. Golumbic, M.C., Hirst, T., Lewenstein, M.: Uniquely restricted matchings. Algorithmica **31**, 139–154 (2001)
23. Gutin, G., Mnich, M.: A survey on graph problems parameterized above and below guaranteed values. arXiv preprint arXiv:2207.12278 (2022)
24. Hajebi, S., Javadi, R.: On the parameterized complexity of the acyclic matching problem. Theoret. Comput. Sci. **958**, 113862 (2023)
25. Kanj, I., Pelsmajer, M.J., Schaefer, M., Xia, G.: On the induced matching problem. J. Comput. Syst. Sci. **77**(6), 1058–1070 (2011)
26. Klemz, B., Rote, G.: Linear-time algorithms for maximum-weight induced matchings and minimum chain covers in convex bipartite graphs. Algorithmica **84**(4), 1064–1080 (2022)
27. Ko, C., Shepherd, F.B.: Bipartite domination and simultaneous matroid covers. SIAM J. Discret. Math. **16**(4), 517–523 (2003)
28. Koana, T.: Induced matching below guarantees: average paves the way for fixed-parameter tractability. In: Proceedings of the 40th International Symposium on Theoretical Aspects of Computer Science (STACS), pp. 39:1–39:21 (2023)
29. Kowalik, L., Luzar, B., Skrekovski, R.: An improved bound on the largest induced forests for triangle-free planar graphs. Discrete Math. Theor. Comput. Sci. **12**(1), 87–100 (2010)
30. Lin, B.: Constant approximating k-clique is w [1]-hard. In: Proceedings of the 53rd Annual ACM SIGACT Symposium on Theory of Computing, pp. 1749–1756 (2021)
31. Lovász, L., Plummer, M.D.: Matching Theory. Annals of Discrete Mathematics, vol. 29. North-Holland, Amsterdam (1986)
32. Manlove, D.: Algorithmics of Matching Under Preferences, vol. 2. World Scientific, Singapore (2013)
33. Micali, S., Vazirani, V.V.: An o($\sqrt{|V|}|e|$) algoithm for finding maximum matching in general graphs. In 21st Annual Symposium on Foundations of Computer Science (sfcs 1980). IEEE, pp. 17–27 (1980)
34. Moser, H., Sikdar, S.: The parameterized complexity of the induced matching problem. Discret. Appl. Math. **157**(4), 715–727 (2009)
35. Moser, H., Thilikos, D.M.: Parameterized complexity of finding regular induced subgraphs. J. Discrete Algorithms **7**(2), 181–190 (2009)
36. Panda, B.S., Chaudhary, J.: Acyclic matching in some subclasses of graphs. Theoret. Comput. Sci. **943**, 36–49 (2023)
37. Panda, B.S., Pandey, A., Chaudhary, J., Dane, P., Kashyap, M.: Maximum weight induced matching in some subclasses of bipartite graphs. J. Comb. Optim. **40**(3), 713–732 (2020)
38. Panda, B.S., Pradhan, D.: Acyclic matchings in subclasses of bipartite graphs. Discrete Math. Algorithms Appl. **4**(04), 1250050 (2012)
39. Stockmeyer, L.J., Vazirani, V.V.: Np-completeness of some generalizations of the maximum matching problem. Inf. Process. Lett. **15**(1), 14–19 (1982)
40. Vizing, V.G.: On an estimate of the chromatic class of a p-graph. Diskret Analiz **3**, 25–30 (1964)

41. Xiao, M., Kou, S.: Parameterized algorithms and kernels for almost induced matching. Theoret. Comput. Sci. **846**, 103–113 (2020)
42. Zito, M.: Induced matchings in regular graphs and trees. In: Widmayer, P., Neyer, G., Eidenbenz, S. (eds.) WG 1999. LNCS, vol. 1665, pp. 89–101. Springer, Heidelberg (1999). https://doi.org/10.1007/3-540-46784-X_10

$\mathcal{P}$-Matchings Parameterized by Treewidth

Juhi Chaudhary$^{(\boxtimes)}$ and Meirav Zehavi

Department of Computer Science, Ben-Gurion University of the Negev,
BeerSheba, Israel
juhic@post.bgu.ac.il, meiravze@bgu.ac.il

Abstract. A *matching* is a subset of edges in a graph G that do not share an endpoint. A matching M is a $\mathcal{P}$-*matching* if the subgraph of G induced by the endpoints of the edges of M satisfies property $\mathcal{P}$. For example, if the property $\mathcal{P}$ is that of being a matching, being acyclic, or being disconnected, then we obtain an *induced matching*, an *acyclic matching*, and a *disconnected matching*, respectively. In this paper, we analyze the problems of the computation of these matchings from the viewpoint of Parameterized Complexity with respect to the parameter *treewidth*.

Keywords: Matching · Treewidth · Parameterized Algorithms · Exponential Time Hypothesis

1 Introduction

Matching in graphs is a central topic of Graph Theory and Combinatorial Optimization [28]. Matchings possess both theoretical significance and practical applications, such as the assignment of new physicians to hospitals, students to high schools, clients to server clusters, kidney donors to recipients [29], and so on. Additionally, the field of competitive optimization games on graphs has witnessed substantial growth in recent years, where matching serves as a valuable tool for determining optimal solutions or bounds in such games [1, 19]. The study of matchings is closely related to the concept of *edge colorings* as well [2, 5, 38], and the minimum number of matchings into which the edge set of a graph G can be partitioned is known as the *chromatic index* of G [38].

Given a graph G, MAXIMUM MATCHING is the problem of finding a matching of maximum size (number of edges) in G. A matching M is said to be a $\mathcal{P}$-*matching* if $G[V_M]$ (the subgraph of G induced by the endpoints of edges in M) has property $\mathcal{P}$, where $\mathcal{P}$ is some graph property. The problem of deciding whether a graph admits a $\mathcal{P}$-matching of a given size has been investigated for many different properties [4, 16, 18, 20, 34, 37]. If the property $\mathcal{P}$ is that of being a graph, a disjoint union of $K_2's$, a forest, a connected graph, a disconnected graph, or having a unique perfect matching, then a $\mathcal{P}$-matching is a *matching* [30], an

---

The authors are supported by the European Research Council (ERC) project titled PARAPATH (101039913).

induced matching [37], an *acyclic matching* [18], a *connected matching*[1] [18], a *disconnected matching*[2] [18], and a *uniquely restricted matching* [20], respectively. Notably, only the optimization problem corresponding to matching [30] and connected matching [18] are polynomial-time solvable for a general graph, while the decision problems corresponding to other above-mentioned variants of matching are NP-complete [18,20,37].

Given a graph G and a positive integer ℓ, the INDUCED MATCHING problem asks whether G has an induced matching of size at least ℓ. The concept of induced matching was introduced by Stockmeyer and Vazirani as the "risk-free" marriage problem in 1982 [37]. Since then, this concept, and the corresponding INDUCED MATCHING problem, have been studied extensively due to their wide range of applications and connections to other graph problems [8,12,23,25,31, 37]. Similarly, the ACYCLIC MATCHING problem considers a graph G and a positive integer ℓ, and asks whether G contains an acyclic matching of size at least ℓ. Goddard et al. [18] introduced the concept of acyclic matching, and since then, it has gained significant popularity in the literature [2,3,17,34]. For a fixed $c \in \mathbb{N}$, a matching M is *c-disconnected* if $G[V_M]$ has at least c connected components. In the c-DISCONNECTED MATCHING problem, given a graph G and a positive integer ℓ, we seek to determine if G contains a c-disconnected matching of size at least ℓ. In the c-DISCONNECTED MATCHING problem, if c is a part of the input, then the problem that arises is known as the DISCONNECTED MATCHING problem. Goddard et al. [18] introduced the concept of disconnected matching and asked about the complexity of determining the maximum size of a matching whose vertex set induces a disconnected graph, which is a restricted version of c-disconnected matching studied in this paper.

The parameter considered in this paper is *treewidth*, a structural parameter that indicates how much a graph resembles a tree. Robertson and Seymour introduced the notion of treewidth in their celebrated work on graph minors [35], and since then, over 4260 papers on google scholar consider treewidth as a parameter in the context of Parameterized Complexity. In practice also, graphs of bounded treewidth appear in many different contexts; for example, many probabilistic networks appear to have small treewidth [7]. Thus, concerning the problems studied in this paper, after the *solution size*, treewidth is one of the most natural parameters. In fact, many of the problems investigated in this paper have already been analyzed with respect to treewidth as the parameter.

Related Work. In what follows, we present a brief survey of algorithmic results concerning the variants of matchings discussed in this paper.

Induced Matching: From the viewpoint of Parameterized Complexity, in [31], Moser and Sikdar showed that INDUCED MATCHING is fixed-parameter tractable (FPT) when parameterized by treewidth by developing an $\mathcal{O}(4^{\text{tw}} \cdot n)$-time dynamic programming algorithm. In the same paper, when the parameter

---

[1] This name is also used for a different problem where we are asked to find a matching M such that every pair of edges in M has a common edge [9].

[2] In this paper, we are using a different (more general) definition for disconnected matching than the one mentioned in [18].

is the size of the matching ℓ, INDUCED MATCHING was shown to be FPT for line graphs, planar graphs, bounded-degree graphs, and graphs of girth at least 6 that include C_4-free graphs[3]. On the other hand, for the same parameter, that is, ℓ, the problem is W[1]-hard for bipartite graphs [31]. Song [36] showed that given a Hamiltonian cycle in a Hamiltonian bipartite graph, INDUCED MATCH-ING is W[1]-hard with respect to ℓ and cannot be solved in time $n^{o(\sqrt{\ell})}$ unless W[1] = FPT, where $n = |V(G)|$. INDUCED MATCHING with respect to *below guarantee* parameterizations have also been studied [26,32,39].

Acyclic Matching: From the viewpoint of Parameterized Complexity, Hajebi and Javadi [22] showed that ACYCLIC MATCHING is FPT when parameterized by treewidth using Courcelle's theorem. Furthermore, they showed that the problem is W[1]-hard on bipartite graphs when parameterized ℓ. However, under the same parameter, the authors showed that the problem is FPT for line graphs, C_4-free graphs, and every proper minor-closed class of graphs. In the same paper, ACYCLIC MATCHING was shown to be FPT when parameterized by the size of the matching plus the number of cycles of length four in the given graph. Some more results concerning acyclic matchings can be found in [2,11,17,34].

c-Disconnected Matching and Disconnected Matching: For every fixed integer $c \geq 2$, c-DISCONNECTED MATCHING is known to be NP-complete even for bounded diameter bipartite graphs [21]. On the other hand, for $c = 1$, c-DISCONNECTED MATCHING is the same as MAXIMUM MATCHING, which is known to be polynomial-time solvable [30]. Regarding disconnected match-ings, DISCONNECTED MATCHING is NP-complete for chordal graphs [21] and polynomial-time solvable for interval graphs [21]. Also, since INDUCED MATCH-ING is a special case of DISCONNECTED MATCHING, we note that DISCON-NECTED MATCHING is NP-hard for every graph class for which INDUCED MATCH-ING is NP-hard [21].

From the viewpoint of Parameterized Complexity, Gomes et al. [21] proved that for graphs with a polynomial number of minimal separators, DISCON-NECTED MATCHING parameterized by the number of connected components, belongs to XP. Furthermore, unless NP $\subseteq$ coNP/poly, DISCONNECTED MATCH-ING does not admit a polynomial kernel when parameterized by vertex cover (vc) plus ℓ nor when parameterized by the vertex deletion distance to clique plus ℓ. In the same paper, the authors also proved that DISCONNECTED MATCHING is FPT when parameterized by treewidth (tw). They used the standard dynamic pro-gramming technique, and the running time of their algorithm is $\mathcal{O}(8^{\mathrm{tw}} \cdot \eta_{\mathrm{tw}+1}^3 \cdot n^2)$, where η_i is the i-th Bell number (the *Bell number* η_i counts the number of dif-ferent ways to partition a set that has exactly i elements).

Our Contribution and Methods. In this paper, we consider the parameter to be the treewidth of the input graph, and as is customary in the field, we suppose that the input also consists of a tree decomposition $\mathcal{T} = (\mathbb{T}, \{\mathcal{B}_x\}_{x \in V(\mathbb{T})})$ of width tw of the input graph. Proofs of the results marked with a (*) are omitted and can be found in the full version (see [10]).

---

[3] Here, C_n denotes a cycle on n vertices.

First, we present a $3^{\text{tw}} \cdot \text{tw}^{\mathcal{O}(1)} \cdot n$ time algorithm for INDUCED MATCHING in graphs with n vertices, improving upon the $\mathcal{O}(4^{\text{tw}} \cdot n)$ time bound by Moser and Sikdar [31]. For this purpose, we use a nice tree decomposition that satisfies the "deferred edge property" (defined in Sect. 2) and the fast subset convolution (see Sect. 2) for the join nodes. Due to space constraints, this section has been deferred to the full version of the paper (see [10]).

Theorem 1. INDUCED MATCHING *can be solved in* $3^{\text{tw}} \cdot \text{tw}^{\mathcal{O}(1)} \cdot n$ *time by a deterministic algorithm.*

In Sect. 3, we present a $6^{\text{tw}} \cdot n^{\mathcal{O}(1)}$ time algorithm for ACYCLIC MATCH-ING in graphs with n vertices, improving the result by Hajebi and Javadi [22], who proved that ACYCLIC MATCHING parameterized by tw is FPT. They used Courcelle's theorem, which is purely theoretical, and thus the hidden parameter dependency in the running time is huge (a tower of exponents). To develop our algorithm, we use the *Cut & Count* method introduced by Cygan et al. [13] in addition to the fast subset convolution. The Cut & Count method allows us to deal with connectivity-type problems through randomization; here, randomization arises from the usage of the *Isolation Lemma* (see Sect. 2).

Theorem 2. ACYCLIC MATCHING *can be solved in* $6^{\text{tw}} \cdot n^{\mathcal{O}(1)}$ *time by a randomized algorithm. The algorithm cannot give false positives and may give false negatives with probability at most* $\frac{1}{3}$.

In Sect. 4, we present a $(3c)^{\text{tw}} \cdot \text{tw}^{\mathcal{O}(1)} \cdot n$ time algorithm for c-DISCONNECTED MATCHING in graphs with n vertices. We use dynamic programming along with the fast subset convolution for the join nodes. This resolves an open question by Gomes et al. [21], who asked whether c-DISCONNECTED MATCHING can be solved in a single exponential time with vertex cover (vc) as the parameter. Since for any graph G, $\text{tw}(G) \leq \text{vc}(G)$, we answer their question in the affirmative.

Theorem 3. *For a fixed positive integer* $c \geq 2$, c-DISCONNECTED MATCHING *can be solved in* $(3c)^{\text{tw}} \cdot \text{tw}^{\mathcal{O}(1)} \cdot n$ *time by a deterministic algorithm.*

In Sect. 5, we present a lower bound for the time complexity of DISCON-NECTED MATCHING, proving that for any choice of a constant c, an $\mathcal{O}(c^{\text{tw}} \cdot n)$-time algorithm for DISCONNECTED MATCHING is unlikely. In fact, we prove that even an $\mathcal{O}(c^{\text{pw}} \cdot n)$-time algorithm is not possible, where pw is the pathwidth of the graph which is bounded from below by the treewidth.

Theorem 4. *Assuming the Exponential Time Hypothesis, there is no* $2^{o(\text{pw} \log \text{pw})} \cdot n^{\mathcal{O}(1)}$-*time algorithm for* DISCONNECTED MATCHING.

2   Preliminaries

Graph-Theoretic Notations and Definitions. For a graph G, let $V(G)$ denote its vertex set and $E(G)$ denote its edge set. Given a matching M, a vertex $v \in V(G)$ is M-*saturated* if v is incident on an edge of M. Given a graph G and a matching M, let V_M denote the set of M-saturated vertices and $G[V_M]$ denote the subgraph of G induced by V_M. A matching that saturates all the vertices of a graph is a *perfect matching*. If $uv \in M$, then v is the M-*mate* of u, and vice versa. Standard graph-theoretic terms not defined here can be found in [15].

A *cut* of a set $X \subseteq V(G)$ is a pair (X_l, X_r) with $X_l \cap X_r = \emptyset$ and $X_l \cup X_r = X$, where X is an arbitrary subset of $V(G)$. When X is immaterial, we do not mention it explicitly. A cut (X_l, X_r) is *consistent* in a subgraph H of G if $u \in X_l$ and $v \in X_r$ implies $uv \notin E(H)$. For a graph G, let $cc(G)$ denote the number of connected components of G. For a graph G, a *coloring* on a set $X \subseteq V(G)$ is a function $f : X \to S$, where S is any set and the elements of S are called *colors*.

Algebraic Definitions. For a set X, let 2^X denote the set of all subsets of X. For a positive integer k, let $[k]$ denote the set $\{1, \ldots, k\}$. In the set $[k] \times [k]$, a *row* is a set $\{i\} \times [k]$ and a *column* is a set $[k] \times \{i\}$ for some $i \in [k]$. For two integers, a and b, we use $a \equiv b$ to indicate that a is even if and only if b is even. If $w : U \to \{1, \ldots, N\}$, then for $S \subseteq U$, $w(S) = \sum_{e \in S} w(e)$. For definitions of *ring* and *semiring*, we refer the readers to any elementary book on abstract algebra. Given an integer $n > 1$, called a *modulus*, two integers a and b are *congruent modulo* n if there is an integer k such that $a - b = kn$. Note that two integers are said to be *congruent modulo* 2 if they have the same parity (that is, either both are odd or both are even). For a set S, we use the notation $|S|_2$ to denote the number of elements in set S congruent modulo 2.

Subset Convolution

Definition 1. *Let B be a finite set and $\mathbb{R}$ be a semiring. Then, the subset convolution of two functions $f, g : 2^B \to \mathbb{R}$ is the function $(f * g) : 2^B \to \mathbb{R}$ such that for every $Y \subseteq B$,* $(f*g)(Y) = \sum_{X \subseteq Y} f(X)g(Y \setminus X).$ $\qquad(1)$

Equivalently, (1) can be written as $(f * g)(Y) = \sum_{\substack{A \cup B = Y \\ A \cap B = \emptyset}} f(A)g(B).$

Given f and g, a direct evaluation of $f * g$ for all $X \subseteq Y$ requires $\Omega(3^n)$ semiring operations, where $n = |B|$. However, we have the following results.

Proposition 5 ([6,13]). *For two functions $f, g : 2^B \to \mathbb{Z}$, where $n = |B|$ and $\mathbb{Z}$ is a ring, given all the 2^n values of f and g in the input, all the 2^n values of the subset convolution of $f * g$ can be computed in $\mathcal{O}(2^n \cdot n^3)$ arithmetic operations.*

Proposition 6 ([13]). *For two functions $f, g : 2^B \to \{-P, \ldots, P\}$, where $n = |B|$, given all the 2^n values of f and g in the input, all the 2^n values of the subset convolution of $f * g$ over the integer max-sum semiring can be computed in time $2^n \cdot n^{\mathcal{O}(1)} \cdot \mathcal{O}(P \log P \log \log P)$.*

Treewidth. Due to space constraints, the definitions of tree decomposition, nice tree decomposition, path decomposition, treewidth, and pathwidth have been deferred to the full version (see [10]). We denote the treewidth and pathwidth of a graph by tw and pw, respectively. Standard notions in Parameterized Complexity not explicitly defined here can be found in [13].

For our problems, we want the standard nice tree decomposition to satisfy an additional property, and that is, among the vertices present in the bag of a join node, no edges have been introduced yet. To achieve this, we use another known type of node, an *introduce edge node*, which is defined as follows:

Introduce Edge Node: x has exactly one child y, and x is labeled with an edge $uv \in E(G)$ such that $u, v \in \mathcal{B}_x$ and $\mathcal{B}_x = \mathcal{B}_y$. We say that uv is *introduced* at x.

The use of introduce edge nodes enables us to add edges one by one in our nice tree decomposition. We additionally require that every edge is introduced exactly once. For every vertex $v \in V(G)$, there exists a unique highest node $t(v)$ such that $v \in \mathcal{B}_{t(v)}$. Further, without loss of generality, assume $t(v)$ is an ancestor of $t(u)$. Our nice tree decomposition will insert the introduce edge bags (introducing edges of the form uv) between $t(u)$ and its parent in an arbitrary order. If a nice tree decomposition having introduce edge nodes satisfies these additional conditions, then we say that it exhibits the *deferred edge* property. Note that, given a tree decomposition of a graph G, where $n = |V(G)|$, a nice tree decomposition satisfying the *deferred edge* property of equal width, at most $\mathcal{O}(\text{tw} \cdot n)$ nodes, and at most $\text{tw} \cdot n$ edges can be found in $\text{tw}^{\mathcal{O}(1)} \cdot n$ time [13,24]. Furthermore, for each node x of the tree decomposition, let V_x be the union of all the bags present in the subtree of $\mathbb{T}$ rooted at x, including $\mathcal{B}_x$. Then, for each node x of the tree decomposition, define the subgraph G_x of G as follows:

$G_x = (V_x, E_x = \{e : e \text{ is introduced in the subtree of } \mathbb{T} \text{ rooted at } x\})$.

Now, consider the following two definitions.

Definition 2 (Valid Coloring). *Given a node x of $\mathbb{T}$, a coloring $d : \mathcal{B}_x \to \{0, 1, 2\}$ is valid on $\mathcal{B}_x$ if there exists a coloring $\widehat{d} : V_x \to \{0, 1, 2\}$ in G_x, called a valid extension of d, such that the following hold: (i) $\widehat{d}$ restricted to $\mathcal{B}_x$ is exactly d, (ii) The subgraph induced by the vertices colored 1 under $\widehat{d}$ has a perfect matching, (iii) Vertices colored 2 under $\widehat{d}$ must all belong to $\mathcal{B}_x$.*

Definition 3 (Correct Coloring). *Given a graph G and a set $X \subseteq V(G)$, two colorings $f_1, f_2 : X \to \{0, 1, 2\}$ are correct for a coloring $f : X \to \{0, 1, 2\}$ if the following conditions hold: (i) $f(v) = 0$ if and only if $f_1(v) = f_2(v) = 0$, (ii) $f(v) = 1$ if and only if $(f_1(v), f_2(v)) \in \{(1, 2), (2, 1)\}$, (iii) $f(v) = 2$ if and only if $f_1(v) = f_2(v) = 2$.*

In Sects. 3 and 4, we use different colors to represent the possible states of a vertex in a bag $\mathcal{B}_x$ of $\mathcal{T}$ with respect to a matching M as follows:

- **white(0):** A vertex colored 0 is not saturated by M.
- **black(1):** A vertex colored 1 is saturated by M, and the edge between the vertex and its M-mate has also been introduced in G_x.

- **gray(2):** A vertex colored 2 is saturated by M, and either its M-mate has not yet been introduced in G_x, or the edge between the vertex and its M-mate has not yet been introduced in G_x.

Definition 4 (Monte Carlo Algorithms with False Negatives). *An algorithm is Monte Carlo with false negatives if when queried about the existence of an object: If it answers yes, then it is true, and if it answers no, then it is correct with probability at least $\frac{2}{3}$ (here, the constant $\frac{2}{3}$ is chosen arbitrarily).*

Cut & Count Method. The Cut & Count method was introduced by Cygan et al. [14]. It is a tool for designing algorithms with a single exponential running time for problems with certain connectivity requirements. The method is mainly divided into the following two parts:

The Cut part: Let S denote the set of feasible solutions. Here, we relax the connectivity requirement by considering a set $\mathcal{R} \supseteq S$ of possibly connected candidate solutions. Furthermore, we consider a set $\mathcal{C}$ of pairs (X, C), where $X \in \mathcal{R}$ and C is a consistent cut of X.

The Count part: Here, we compute the cardinality of $|\mathcal{C}|_2$ using a subprocedure. Non-connected candidate solutions $X \in \mathcal{R} \setminus S$ cancel since they are consistent with an even number of cuts and the connected candidates $x \in S$ remain.

More information on the Cut & Count method can be found in [13,14].

Isolation Lemma. Let U be a universe. A function $\mathsf{w} : U \to \mathbb{Z}$ isolates a set family $\mathcal{F} \subseteq 2^U$ if there is a unique $S' \in \mathcal{F}$ with $\mathsf{w}(S') = \min_{S \in \mathcal{F}} \mathsf{w}(S)$. Let $\mathcal{F} \subseteq 2^U$ be a set family over a universe U with $|\mathcal{F}| > 0$. For each $u \in U$, choose a weight $\mathsf{w}(u) \in \{1, 2, \ldots, N\}$ uniformly and independently at random. Then, *isolation lemma* states that $\mathsf{prob}(\mathsf{w}\ isolates\ \mathcal{F}) \geq 1 - \frac{|U|}{N}$ [33].

3   Algorithm for Acyclic Matching

We use the Cut & Count technique along with a concept called *markers* (see [14]). Given that the ACYCLIC MATCHING problem does not impose an explicit connectivity requirement, we can proceed by selecting the (presumed) forest obtained after choosing the vertices saturated by an acyclic matching M and using the following result:

Proposition 7 ([14]). *A graph G with n vertices and m edges is a forest if and only if it has at most $n - m$ connected components.*

Our solution set contains pairs (X, P), where $X \subseteq V(G)$ is a set of M-saturated vertices and $P \subseteq V(G)$ is a set of marked vertices (markers) such that each connected component in $G[X]$ contains at least one marked vertex. Markers will be helpful in bounding the number of connected components in $G[X]$ by $n' - m'$, where $n' = |X|$ and m' is the number of edges in $G[X]$

(so that Proposition 7 can be applied). Since we will use the Isolation lemma, we will be assigning random weights to the vertices of X. Furthermore, note that two pairs from our solution set with different sets of marked vertices are necessarily considered to be two different solutions. For this reason, we assign random weights both to the vertices of X and vertices of P.

Throughout this section, as the universe, we take the set $U = V(G) \times \{\mathbf{F}, \mathbf{P}\}$, where $V(G) \times \{\mathbf{F}\}$ is used to assign weights to vertices of the chosen forest and $V(G) \times \{\mathbf{P}\}$ is used to assign weights to vertices chosen as markers. Also, throughout this section, we assume that we are given a weight function $\mathsf{w} : U \to \{1, 2, \ldots, N\}$, where $N = 3|U| = 6|V(G)|$.

Let us first consider the Cut part and define the objects we will count.

Definition 5. *Let G be a graph with n vertices and m edges. For integers $0 \leq A \leq n, 0 \leq B \leq m, 0 \leq C \leq n$, and $0 \leq W \leq 2Nn$, we define the following:*

1. $\mathcal{R}_W^{A,B,C} = \{(X, P) : X \subseteq V(G) \wedge |X| = A \wedge G[X]$ *contains exactly B edges $\wedge$ $G[X]$ has a perfect matching $\wedge P \subseteq X \wedge |P| = C \wedge \mathsf{w}(X \times \{\mathbf{F}\}) + \mathsf{w}(P \times \{\mathbf{P}\}) = W\}$.*
2. $\mathcal{S}_W^{A,B,C} = \{(X, P) \in \mathcal{R}_W^{A,B,C} : G[X]$ *is a forest containing at least one marker from the set P in each connected component$\}$.*
3. $\mathcal{C}_W^{A,B,C} = \{((X, P), (X_l, X_r)) : (X, P) \in \mathcal{R}_W^{A,B,C} \wedge P \subseteq X_l \wedge (X_l, X_r)$ *is a consistent cut of $G[X]\}$.*

We call $\mathcal{R} = \bigcup_{A,B,C,W} \mathcal{R}_W^{A,B,C}$ the family of *candidate solutions*, $\mathcal{S} = \bigcup_{A,B,C,W} \mathcal{S}_W^{A,B,C}$ the family of *solutions*, and $\mathcal{C} = \bigcup_{A,B,C,W} \mathcal{C}_W^{A,B,C}$ the family of *cuts*.

Let us now define the count part.

Lemma 8 (*). *Let $G, A, B, C, W, \mathcal{C}_W^{A,B,C}$, and $\mathcal{S}_W^{A,B,C}$ be as defined in Definition 5. Then, for every A, B, C, W satisfying $C \leq A - B$, $|\mathcal{C}_W^{A,B,C}|_2 \equiv |\mathcal{S}_W^{A,B,C}|$.*

Remark 1. Condition $C \leq A - B$ is necessary for Lemma 8 as otherwise (if $A - B < C$), it is not possible to bound the number of connected components in $G[X]$ by $A - B$. As a result, the elements of $\mathcal{S}_W^{A,B,C}$ could not be identified, and Proposition 7 could not be applied.

By Isolation Lemma [33], we have the following lemma.

Lemma 9. *Let G and $\mathcal{S}_W^{A,B,C}$ be as defined in Definition 5. For each $u \in U$, where U is the universe, choose a weight $\mathsf{w}(u) \in \{1, 2, \ldots, 3|U|\}$ uniformly and independently at random. For some A, B, C, W satisfying $C \leq A - B$, if $|\mathcal{S}_W^{A,B,C}| > 0$, then $\mathsf{prob}(\mathsf{w}$ isolates $\mathcal{S}_W^{A,B,C}) \geq \frac{2}{3}$.*

The following observation helps us in proving Theorem 2.

Observation 10 (*). *G admits an acyclic matching of size $\frac{\ell}{2}$ if and only if there exist integers B and W such that the set $\mathcal{S}_W^{\ell, B, \ell - B}$ is nonempty.*

Now we describe a procedure that, given a nice tree decomposition $\mathcal{T}$ with the deferred edge property, a weight function $\mathsf{w} : U \rightarrow \{1, 2, \ldots, N\}$, and integers A, B, C, W as defined in Definition 5 and satisfying $C \leq A - B$, computes $\left| \mathcal{C}_W^{A,B,C} \right|_2$ using dynamic programming. For this purpose, consider the following.

Definition 6. *For every bag $\mathcal{B}_x$ of the tree decomposition $\mathcal{T}$, for every integer $0 \leq a \leq n, 0 \leq b < n, 0 \leq c \leq n, 0 \leq w \leq 12n^2$, for every coloring $d : \mathcal{B}_x \rightarrow \{0, 1, 2\}$, for every coloring $s : \mathcal{B}_x \rightarrow \{0, l, r\}$, we define the following:*

1. $\mathcal{R}_x[a, b, c, d, w] = \{(X, P) : d \text{ is a valid coloring of } \mathcal{B}_x \wedge X \text{ is the set of} \\ \text{vertices colored 1 or 2 under some valid extension } \widehat{d} \text{ of } d \text{ in } G_x \wedge |X| = \\ a \wedge |E_x \cap E(G[X])| = b \wedge P \subseteq X \backslash \mathcal{B}_x \wedge |P| = c \wedge \mathsf{w}(X \times \{\boldsymbol{F}\}) + \mathsf{w}(P \times \{\boldsymbol{P}\}) = w\}.$
2. $\mathcal{C}_x[a, b, c, d, w] = \{((X, P), (X_l, X_r)) : (X, P) \in \mathcal{R}_x[a, b, c, d, w] \wedge P \subseteq X_l \wedge \\ (X, (X_l, X_r)) \text{ is a consistently cut subgraph of } G_x\}.$
3. $\widetilde{\mathcal{A}}_x[a, b, c, d, w, s] = |\{((X, P), (X_l, X_r)) \in \mathcal{C}_x[a, b, c, d, w] : (s(v) = l \Rightarrow v \in \\ X_l) \wedge (s(v) = r \Rightarrow v \in X_r) \wedge (s(v) = 0 \Rightarrow v \notin X)\}|.$

Remark 2. In Definition 6, we assume $b < n$ because otherwise, an induced subgraph containing b edges is definitely not a forest.

The intuition behind Definition 6 is that the set $\mathcal{R}_x[a, b, c, d, w]$ contains all pairs (X, P) that could potentially be extended to a candidate solution from $\mathcal{R}$ (with cardinality and weight restrictions as prescribed by a, b, c, and w), and the set $\mathcal{C}_x[a, b, c, d, w]$ contains all consistently cut subgraphs of G_x that could potentially be extended to elements of $\mathcal{C}$ (with cardinality and weight restrictions as prescribed by a, b, c, and w). The number $\widetilde{\mathcal{A}}_x[a, b, c, d, w, s]$ counts precisely those elements of $\mathcal{C}_x[a, b, c, d, w]$ for which $s(v)$ describes whether for every $v \in \mathcal{B}_x$, v lies in X_l, X_r, or outside X depending on whether $s(v)$ is l, r, or 0, respectively. We have a table $\mathcal{A}$ with an entry $\mathcal{A}_x[a, b, c, d, w, s]$ for each bag $\mathcal{B}_x$ of $\mathbb{T}$, for integers $0 \leq a \leq n, 0 \leq b < n, 0 \leq c \leq n, 0 \leq w \leq 12n^2$, for every coloring $d : \mathcal{B}_x \rightarrow \{0, 1, 2\}$, and for every coloring $s : \mathcal{B}_x \rightarrow \{0, l, r\}$. We say that s and d are compatible if for every $v \in \mathcal{B}_x$, the following hold: $d(v) = 0$ if and only if $s(v) = 0$. Note that we have at most $\mathcal{O}(\mathrm{tw} \cdot n)$ many choices for x, at most n choices for a, b, and c, at most $12n^2$ choices for w, and at most 5^{tw} many compatible choices for d and s. Whenever s is not compatible with d, we do not store the entry $\mathcal{A}_x[a, b, c, d, w, s]$ and assume that the access to such an entry returns 0. Therefore, the size of table $\mathcal{A}$ is bounded by $\mathcal{O}(5^{\mathrm{tw}} \cdot \mathrm{tw} \cdot n)$. We will show how to compute the table $\mathcal{A}$ so that the following will be satisfied.

Lemma 11. *If d is valid and d is compatible with s, then $\mathcal{A}_x[a, b, c, d, w, s]$ stores the value $\widetilde{\mathcal{A}}_x[a, b, c, d, w, s]$. Else, the entry $\mathcal{A}_x[a, b, c, d, w, s]$ stores the value 0.*

By Lemma 8, we seek values $\left| \mathcal{C}_W^{A,B,C} \right|_2$. By Observation 10, Definition 6, and Lemma 11, it suffices to compute values $\mathcal{A}_r[\ell, B, \ell - B, \emptyset, W, \emptyset]$ for all B and W, where r is the root of the decomposition $\mathcal{T}$. (Note that $\mathcal{A}_r[\ell, B, \ell - B, \emptyset, W, \emptyset] = \left| \mathcal{C}_W^{\ell, B, \ell - B} \right|$, and we will calculate the modulo 2 separately.) Further, to achieve the time complexity we aim to achieve, we decide whether to mark a vertex or not in its forget bag. Our algorithm computes $\mathcal{A}_x[a, b, c, d, w, s]$ for all bags $\mathcal{B}_x \in \mathcal{T}$ in a bottom-up manner for all integers $0 \le a \le n, 0 \le b < n, 0 \le c \le n, 0 \le w \le 12n^2$, and for all compatible colorings $d : \mathcal{B}_x \to \{0, 1, 2\}$ and $s : \mathcal{B}_x \to \{0, l, r\}$. Further details are deferred to the full version (see [10]).

We note that, by the naive method, the evaluation of all leaf nodes, introduce vertex and edge nodes, and forget nodes can be done in $5^{\mathrm{tw}} \cdot n^{\mathcal{O}(1)}$ time, but the evaluation of all join nodes altogether can be done in $7^{\mathrm{tw}} \cdot n^{\mathcal{O}(1)}$ time. However, the fast subset convolution can be used to handle join nodes more efficiently, and therefore the evaluation of all join nodes altogether can be done in $6^{\mathrm{tw}} \cdot n^{\mathcal{O}(1)}$ time, and hence we have Theorem 2.

4   Algorithm for c-Disconnected Matching

We first define the following notion.

Definition 7 (Fine Coloring). *Given a node x of $\mathbb{T}$ and a fixed integer $c \ge 2$, a coloring $f : \mathcal{B}_x \to \{0, 1, \ldots, c\}$ is a fine coloring if there exists a coloring $\widehat{f} : V_x \to \{0, 1, \ldots, c\}$ in G_x, a fine extension of f, such that the following hold: (i) $\widehat{f}$ restricted to $\mathcal{B}_x$ is f, (ii) If $uv \in E_x$, $\widehat{f}(u) \ne 0$, $\widehat{f}(v) \ne 0$, then $\widehat{f}(u) = \widehat{f}(v)$.*

Note that point (ii) in Definition 7 implies that whenever two vertices in G_x have an edge between them, then they should get the same color under a fine extension except possibly when either of them is colored 0.

Before we begin the formal description of the algorithm, let us briefly discuss the idea that yields us a single exponential running time for the c-DISCONNECTED MATCHING problem rather than a slightly exponential running time[4] (which is common for most of the naive dynamic programming algorithms for connectivity type problems). We will use Definition 7 to partition the vertices of V_x into color classes (at most c). Note that we do not require in Definition 7 that $G_x[\widehat{f}^{-1}(i)]$ for any $i \in [c]$ is a connected graph. This is the crux of our efficiency. Specifically, this means that we do not keep track of the precise connected components of $G[V_M]$ in G_x for a matching M, yet Definition 7 is sufficient for us.

Now, let us discuss our ideas more formally. We have a table $\mathcal{A}$ with an entry $\mathcal{A}_x[d, f, \widehat{c}]$ for each bag $\mathcal{B}_x$, for every coloring $d : \mathcal{B}_x \to \{0, 1, 2\}$, for every coloring $f : \mathcal{B}_x \to \{0, 1, \ldots, c\}$, and for every set $\widehat{c} \subseteq \{1, 2, \ldots, c\}$. We say that d and f are *compatible* if for every $v \in \mathcal{B}_x$, the following hold: $d(v) = 0$ if and only if $f(v) = 0$. We say that f and $\widehat{c}$ are compatible if for any $v \in \mathcal{B}_x$, $f(v) \in \widehat{c}$. Note that we have at most $\mathcal{O}(\mathrm{tw} \cdot n)$ many choices for x, at most $(2c + 1)^{\mathrm{tw}}$ many choices for compatible d and f, and at most 2^c choices for $\widehat{c}$. Furthermore, whenever f is not compatible with d or $\widehat{c}$, we do not store the entry $\mathcal{A}_x[d, f, \widehat{c}]$ and assume that the access to such an entry returns $-\infty$. Therefore, the size of table $\mathcal{A}$ is bounded by $\mathcal{O}((2c + 1)^{\mathrm{tw}} \cdot \mathrm{tw} \cdot n)$. The following definition specifies the value each entry $\mathcal{A}_x[d, f, \widehat{c}]$ of $\mathcal{A}$ is supposed to store.

---

[4] That is, running time $2^{\mathcal{O}(\mathrm{tw})} \cdot n^{\mathcal{O}(1)}$ rather than $\mathrm{tw}^{\mathcal{O}(\mathrm{tw})} \cdot n^{\mathcal{O}(1)}$..

Definition 8. *If d is valid, f is fine, f is compatible with d and $\widehat{c}$, and there exists a fine extension $\widehat{f}$ of f such that $\widehat{c}$ equals the set of distinct non-zero colors assigned by $\widehat{f}$, then the entry $\mathcal{A}_x[d, f, \widehat{c}]$ stores the maximum number of vertices that are colored 1 or 2 under some valid extension $\widehat{d}$ of d in G_x such that for every $v \in V_x$, $\widehat{d}(v) = 0$ if and only if $\widehat{f}(v) = 0$. Otherwise, the entry $\mathcal{A}_x[d, f, \widehat{c}]$ stores the value $-\infty$.*

Since the root of $\mathbb{T}$ is an empty node, note that the maximum number of vertices saturated by any c-disconnected matching is exactly $\mathcal{A}_r[\emptyset, \emptyset, \{1, 2, \ldots, c\}]$, where r is the root of $\mathbb{T}$. We now provide recursive formulas to compute the entries of table $\mathcal{A}$.

Leaf Node: For a leaf node x, we have that $\mathcal{B}_x = \emptyset$. Hence there is only one possible coloring on $\mathcal{B}_x$, that is, the empty coloring (for both d and f). Since f and G_x are empty, the only compatible choice for $\widehat{c}$ is $\{\}$, and we have $\mathcal{A}_x[\emptyset, \emptyset, \{\}] = 0$.

Introduce Vertex Node: Suppose that x is an introduce vertex node with child node y such that $\mathcal{B}_x = \mathcal{B}_y \cup \{v\}$ for some $v \notin \mathcal{B}_y$. For every coloring $d : \mathcal{B}_x \to \{0, 1, 2\}$, every set $\widehat{c} \subseteq \{1, 2, \ldots, c\}$, and every coloring $f : \mathcal{B}_x \to \{0, 1, \ldots, c\}$ such that f is compatible with d and $\widehat{c}$, we have the following recursive formula:

$$\mathcal{A}_x[d, f, \widehat{c}] = \begin{cases} \mathcal{A}_y[d|_{\mathcal{B}_y}, f|_{\mathcal{B}_y}, \widehat{c}] & \text{if } d(v) = 0, \\ -\infty & \text{if } d(v) = 1, \\ \max\{\mathcal{A}_y[d|_{\mathcal{B}_y}, f|_{\mathcal{B}_y}, \widehat{c} \setminus \{f(v)\}] + 1, \\ \mathcal{A}_y[d|_{\mathcal{B}_y}, f|_{\mathcal{B}_y}, \widehat{c}] + 1\} & \text{if } d(v) = 2. \end{cases}$$

Introduce Edge Node: Suppose that x is an introduce edge node that introduces an edge uv, and let y be the child of x. For every coloring $d : \mathcal{B}_x \to \{0, 1, 2\}$, every set $\widehat{c} \subseteq \{1, 2, \ldots, c\}$, and every coloring $f : \mathcal{B}_x \to \{0, 1, \ldots, c\}$ such that f is compatible with d and $\widehat{c}$, we consider the following cases:

If at least one of $d(u)$ or $d(v)$ is 0, then

$$\mathcal{A}_x[d, f, \widehat{c}] = \mathcal{A}_y[d, f, \widehat{c}].$$

Else, if at least one of $d(u)$ or $d(v)$ is 2, and $f(u) = f(v)$, then

$$\mathcal{A}_x[d, f, \widehat{c}] = \mathcal{A}_y[d, f, \widehat{c}].$$

Else, if both $d(u)$ and $d(v)$ are 1, and $f(u) = f(v)$, then

$$\mathcal{A}_x[d, f, \widehat{c}] = \max\{\mathcal{A}_y[d, f, \widehat{c}], \mathcal{A}_y[d_{\{u,v\} \to 2}, f, \widehat{c}]\}.$$

Else, $\mathcal{A}_x[d, f, \widehat{c}] = -\infty$.

Forget Node: Suppose that x is a forget vertex node with a child y such that $\mathcal{B}_x = \mathcal{B}_y \setminus \{u\}$ for some $u \in \mathcal{B}_y$. For every coloring $d : \mathcal{B}_x \to \{0, 1, 2\}$, every

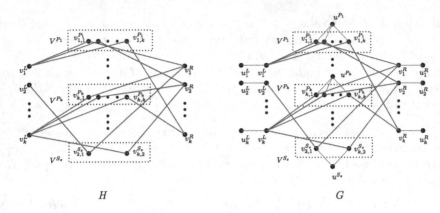

Fig. 1. An illustration of the construction of G from H. Here, we assume that $S_s = \{(2,1),(k,2)\}$.

set $\widehat{c} \subseteq \{1, 2, \ldots, c\}$, and every coloring $f : \mathcal{B}_x \to \{0, 1, \ldots, c\}$ such that f is compatible with d and $\widehat{c}$, we have

$$\mathcal{A}_x[d, f, \widehat{c}] = \max\{\mathcal{A}_y[d_{u \to 0}, f_{u \to 0}, \widehat{c}], \max_{\overline{c} \in \widehat{c}}\{\mathcal{A}_y[d_{u \to 1}, f_{u \to \overline{c}}, \widehat{c}]\}\}.$$

Join Node: Let x be a join node with children y_1 and y_2. For every coloring $d : \mathcal{B}_x \to \{0, 1, 2\}$, every set $\widehat{c} \subseteq \{1, 2, \ldots, c\}$, and for every coloring $f : \mathcal{B}_x \to \{0, 1, \ldots, c\}$ such that f is compatible with d and $\widehat{c}$, we have

$$\mathcal{A}_x[d, f, \widehat{c}] = \max_{d_1, d_2}\{ \max_{\substack{\widehat{c}_{y_1}, \widehat{c}_{y_2} \\ \widehat{c}_{y_1} \cup \widehat{c}_{y_2} = \widehat{c}}} \{\mathcal{A}_{y_1}[d_1, f, \widehat{c}_{y_1}] + \mathcal{A}_{y_2}[d_2, f, \widehat{c}_{y_2}] - |d^{-1}(1)| - |d^{-1}(2)|\}\},$$

where $d_1 : \mathcal{B}_{y_1} \to \{0, 1, 2\}$, $d_2 : \mathcal{B}_{y_2} \to \{0, 1, 2\}$, $\widehat{c}_{y_1}, \widehat{c}_{y_2} \subseteq \{1, 2, \ldots, c\}$ such that f is compatible with d_1, d_2, $\widehat{c}_{y_1}$, and $\widehat{c}_{y_2}$, and d_1, d_2 are correct for d.

Further details are deferred to the full version (see [10]). We note that, by the naive method, the evaluation of all leaf nodes, introduce vertex and edge nodes, and forget nodes can be done in $(2c + 1)^{\mathrm{tw}} \cdot \mathrm{tw}^{\mathcal{O}(1)} \cdot n$ time, but the evaluation of all join nodes altogether can be done in $(4c+2)^{\mathrm{tw}} \cdot \mathrm{tw}^{\mathcal{O}(1)} \cdot n$ time. However, the fast subset convolution can be used to handle join nodes more efficiently, and therefore the evaluation of all join nodes altogether can be done in $(3c)^{\mathrm{tw}} \cdot \mathrm{tw}^{\mathcal{O}(1)} \cdot n$ time, and hence we have Theorem 3.

5   Lower Bound for Disconnected Matching

We give a reduction from $k \times k$ HITTING SET. The input of $k \times k$ HITTING SET consists of a family of sets $S_1, \ldots, S_m \subseteq [k] \times [k]$ such that each set contains at most one element from each row of $[k] \times [k]$, and the question is to determine if there exists a set $\hat{S}$ containing exactly one element from each row such that $\hat{S} \cap S_i \neq \emptyset$ for every $i \in [m]$? Here, the parameter is k.

Proposition 12 ([27]). *Assuming Exponential Time Hypothesis, there is no* $2^{o(k \log k)} \cdot n^{\mathcal{O}(1)}$*-time algorithm for* $k \times k$ HITTING SET.

Our reduction is inspired by the reduction given by Cygan et al. in [14] to prove that there is no $2^{o(\text{pw} \log \text{pw})} \cdot n^{\mathcal{O}(1)}$-time algorithm for MAXIMALLY DISCONNECTED DOMINATING SET unless the Exponential Time Hypothesis fails. Given an instance $(k, S_1, \ldots, S_m)$ of $k \times k$ HITTING SET, we construct an equivalent instance $(G, 3k + m, k)$ of DISCONNECTED MATCHING in polynomial time. First, we define a simple gadget that will be used in our construction.

Definition 9 (Star Gadget). *By adding a* star gadget *to a vertex set* $X \subseteq V(G)$*, we mean the following construction: We introduce a new vertex of degree* $|X|$ *and connect it to all vertices in* X*.*

If we attach a star gadget to multiple vertex disjoint subsets of H, then we have the following lemma.

Lemma 13 (∗). *Let* H *be a graph and let* G *be the graph constructed from* H *by adding star gadgets to vertex disjoint subsets* $X_1, \ldots, X_l$ *of* $V(H)$*. Assume we are given a path decomposition* $\widetilde{\mathcal{P}}$ *of* H *of width* pw *with the following property: For each* X_i*,* $i \in [l]$*, there exists a bag in* $\mathcal{P}$ *that contains* X_i*. Then, in polynomial time, we can construct a path decomposition of* G *of width at most* pw $+ 1$*.*

Now, consider the following construction (see Fig. 1 for an illustration).

Construction $\mathcal{D}$**.** Let $P_i = \{i\} \times [k]$ be a set containing all elements in the i-th row in the set $[k] \times [k]$. We define $\mathcal{S} = \{S_s : s \in [m]\} \cup \{P_i : i \in [k]\}$. Note that for each $X \in \mathcal{S}$, we have $|X| \leq k$, as each S_s, $s \in [m]$ contains at most one element from each row and $|P_i| = k$ for each $i \in [k]$.

First, let us define a graph H. We start by introducing vertices v_i^L for each $i \in [k]$ and vertices v_j^R for each $j \in [k]$. Then, for each set $X \in \mathcal{S}$, we introduce vertices $v_{i,j}^X$ for every $(i,j) \in X$. Let $V^X = \{v_{i,j}^X : (i,j) \in X\}$. We also introduce the edge set $\{v_i^L v_{i,j}^X\} \cup \{v_{i,j}^X v_j^R\}$ for each $X \in \mathcal{S}$ and $i, j \in [k]$. This ends the construction of H. Now, we construct a graph G from the graph H as follows: For each $i \in [k]$ and $j \in [k]$, we attach star gadgets to vertices v_i^L and v_j^R. Furthermore, for each $X \in \mathcal{S}$, we attach star gadgets to X. For each $i \in [k]$ (resp. $j \in [k]$), let u_i^L (resp. u_j^R) denote the unique vertex in the star gadget corresponding to v_i^L (resp. v_j^R). For each $X \in \mathcal{S}$, let u^X denote the unique vertex in the star gadget corresponding to X. Let $E^X = \{v_{i,j}^X u^X : (i,j) \in X\}$.

We now provide a pathwidth bound on G.

Lemma 14 (∗). *Let* H *and* G *be as defined in Construction* $\mathcal{D}$*. Then, the pathwidth of* G *is at most* $3k$*.*

Lemma 15 (∗). *Let* G *be as defined in Construction* $\mathcal{D}$*. If the initial* $k \times k$ HITTING SET *instance is a Yes-instance, then there exists a matching* M *in* G *such that* $|M| = 3k + m$ *and* $G[V_M]$ *has exactly* k *connected components.*

Lemma 16 (∗). *Let G be as defined in Construction $\mathcal{D}$. If there exists a matching M in G such that $|M| \geq 3k + m$ and $G[V_M]$ has at least k connected components, then the initial $k \times k$ HITTING SET instance is a Yes-instance.*

By Proposition 12 and Lemmas 14, 15 and 16, we have Theorem 4.

6   Conclusion

A detailed conclusion (where we briefly discuss the SETH lower bounds) can be found in the full version (see [10]).

References

1. Bachstein, A., Goddard, W., Lehmacher, C.: The generalized matcher game. Discret. Appl. Math. **284**, 444–453 (2020)
2. Baste, J., Fürst, M., Rautenbach, D.: Approximating maximum acyclic matchings by greedy and local search strategies. In: Kim, D., Uma, R.N., Cai, Z., Lee, D.H. (eds.) COCOON 2020. LNCS, vol. 12273, pp. 542–553. Springer, Cham (2020). https://doi.org/10.1007/978-3-030-58150-3_44
3. Baste, J., Rautenbach, D.: Degenerate matchings and edge colorings. Discret. Appl. Math. **239**, 38–44 (2018)
4. Baste, J., Rautenbach, D., Sau, I.: Uniquely restricted matchings and edge colorings. In: Bodlaender, H.L., Woeginger, G.J. (eds.) WG 2017. LNCS, vol. 10520, pp. 100–112. Springer, Cham (2017). https://doi.org/10.1007/978-3-319-68705-6_8
5. Baste, J., Rautenbach, D., Sau, I.: Upper bounds on the uniquely restricted chromatic index. J. Graph Theory **91**(3), 251–258 (2019)
6. Björklund, A., Husfeldt, T., Kaski, P., Koivisto, M.: Fourier meets möbius: fast subset convolution. In: Proceedings of the Thirty-Ninth Annual ACM Symposium on Theory of Computing, pp. 67–74 (2007)
7. Bodlaender, H.L., et al.: A tourist guide through treewidth (1992)
8. Cameron, K.: Induced matchings. Discret. Appl. Math. **24**(1–3), 97–102 (1989)
9. Cameron, K.: Connected matchings. In: Jünger, M., Reinelt, G., Rinaldi, G. (eds.) Combinatorial Optimization—Eureka, You Shrink! LNCS, vol. 2570, pp. 34–38. Springer, Heidelberg (2003). https://doi.org/10.1007/3-540-36478-1_5
10. Chaudhary, J., Zehavi, M.: $\mathcal{P}$-matchings parameterized by treewidth. arXiv preprint arXiv:2307.09333v1 (2023)
11. Chaudhary, J., Zehavi, M.: Parameterized results on acyclic matchings with implications for related problems. In: Paulusma, D., Ries, B. (eds.) WG 2023. LNCS, vol. 14093, pp. 201–216. Springer, Heidelberg (2023)
12. Cooley, O., Draganic, N., Kang, M., Sudakov, B.: Large induced matchings in random graphs. SIAM J. Discret. Math. **35**(1), 267–280 (2021)
13. Cygan, M., et al.: Parameterized Algorithms, vol. 5. Springer, Heidelberg (2015). https://doi.org/10.1007/978-3-319-21275-3
14. Cygan, M., Nederlof, J., Pilipczuk, M., Pilipczuk, M., Van Rooij, J.M., Wojtaszczyk, J.O.: Solving connectivity problems parameterized by treewidth in single exponential time. ACM Trans. Algorithms (TALG) **18**(2), 1–31 (2022)
15. Diestel, R.: Graph Theory. Graduate Text GTM, vol. 173. Springer, Heidelberg (2012)

16. Francis, M.C., Jacob, D., Jana, S.: Uniquely restricted matchings in interval graphs. SIAM J. Discret. Math. **32**(1), 148–172 (2018)
17. Fürst, M., Rautenbach, D.: On some hard and some tractable cases of the maximum acyclic matching problem. Ann. Oper. Res. **279**, 291–300 (2019)
18. Goddard, W., Hedetniemi, S.M., Hedetniemi, S.T., Laskar, R.: Generalized subgraph-restricted matchings in graphs. Discret. Math. **293**(1–3), 129–138 (2005)
19. Goddard, W., Henning, M.A.: The matcher game played in graphs. Discret. Appl. Math. **237**, 82–88 (2018)
20. Golumbic, M.C., Hirst, T., Lewenstein, M.: Uniquely restricted matchings. Algorithmica **31**, 139–154 (2001)
21. Gomes, G.C., Masquio, B.P., Pinto, P.E., dos Santos, V.F., Szwarcfiter, J.L.: Disconnected matchings. Theoret. Comput. Sci. **956**, 113821 (2023)
22. Hajebi, S., Javadi, R.: On the parameterized complexity of the acyclic matching problem. Theoret. Comput. Sci. **958**, 113862 (2023)
23. Klemz, B., Rote, G.: Linear-time algorithms for maximum-weight induced matchings and minimum chain covers in convex bipartite graphs. Algorithmica **84**(4), 1064–1080 (2022)
24. Kloks, T.: Treewidth: Computations and Approximations. Springer, Heidelberg (1994). https://doi.org/10.1007/BFb0045375
25. Ko, C., Shepherd, F.B.: Bipartite domination and simultaneous matroid covers. SIAM J. Discret. Math. **16**(4), 517–523 (2003)
26. Koana, T. Induced matching below guarantees: average paves the way for fixed-parameter tractability. arXiv preprint arXiv:2212.13962 (2022)
27. Lokshtanov, D., Marx, D., Saurabh, S.: Slightly superexponential parameterized problems. SIAM J. Comput. **47**(3), 675–702 (2018)
28. Lovász, L., Plummer, M.D.: Matching Theory. Annals of Discrete Mathematics, vol. 29. North-Holland, Amsterdam (1986)
29. Manlove, D.: Algorithmics of Matching Under Preferences, vol. 2. World Scientific, Singapore (2013)
30. Micali, S., Vazirani, V.V.: An o($\sqrt{|V|}|e|$) algorithm for finding maximum matching in general graphs. In: 21st Annual Symposium on Foundations of Computer Science (SFCS 1980), pp. 17–27. IEEE (1980)
31. Moser, H., Sikdar, S.: The parameterized complexity of the induced matching problem. Discret. Appl. Math. **157**(4), 715–727 (2009)
32. Moser, H., Thilikos, D.M.: Parameterized complexity of finding regular induced subgraphs. J. Discret. Algorithms **7**(2), 181–190 (2009)
33. Mulmuley, K., Vazirani, U.V., Vazirani, V.V.: Matching is as easy as matrix inversion. In Proceedings of the Nineteenth Annual ACM Symposium on Theory of Computing, pp. 345–354 (1987)
34. Panda, B., Chaudhary, J.: Acyclic matching in some subclasses of graphs. Theoret. Comput. Sci. **943**, 36–49 (2023)
35. Robertson, N., Seymour, P.D.: Graph minors. II. Algorithmic aspects of tree-width. J. Algorithms **7**(3), 309–322 (1986)
36. Song, Y.: On the induced matching problem in Hamiltonian bipartite graphs. Geor. Math. J. **28**(6), 957–970 (2021)
37. Stockmeyer, L.J., Vazirani, V.V.: Np-completeness of some generalizations of the maximum matching problem. Inf. Process. Lett. **15**(1), 14–19 (1982)
38. Vizing, V.G.: On an estimate of the chromatic class of a P-graph. Diskret analiz **3**, 25–30 (1964)
39. Xiao, M., Kou, S.: Parameterized algorithms and kernels for almost induced matching. Theoret. Comput. Sci. **846**, 103–113 (2020)

Algorithms and Hardness for Metric Dimension on Digraphs

Antoine Dailly[1]([✉]), Florent Foucaud[1], and Anni Hakanen[1,2]

[1] Université Clermont Auvergne, CNRS, Clermont Auvergne INP, Mines
Saint-Étienne, LIMOS, 63000 Clermont-Ferrand, France
{antoine.dailly,florent.foucaud,anni.hakanen}@uca.fr
[2] Department of Mathematics and Statistics, University of Turku,
20014 Turku, Finland

Abstract. In the METRIC DIMENSION problem, one asks for a
minimum-size set R of vertices such that for any pair of vertices of the
graph, there is a vertex from R whose two distances to the vertices of the
pair are distinct. This problem has mainly been studied on undirected
graphs and has gained a lot of attention in the recent years. We focus
on directed graphs, and show how to solve the problem in linear-time on
digraphs whose underlying undirected graph (ignoring multiple edges)
is a tree. This (nontrivially) extends a previous algorithm for oriented
trees. We then extend the method to unicyclic digraphs (understood
as the digraphs whose underlying undirected multigraph has a unique
cycle). We also give a fixed-parameter-tractable algorithm for digraphs
when parameterized by the directed modular-width, extending a known
result for undirected graphs. Finally, we show that METRIC DIMENSION
is NP-hard even on planar triangle-free acyclic digraphs of maximum
degree 6.

1 Introduction

The metric dimension of a (di)graph G is the smallest size of a set of vertices
that distinguishes all vertices of G by their vectors of distances from the vertices
of the set. This concept was introduced in the 1970s by Harary and Melter [14]
and by Slater [30] independently. Due to its interesting nature and numerous
applications (such as robot navigation [18], detection in sensor networks [30]
or image processing [22], to name a few), it has enjoyed a lot of attention. It
also has been studied in the more general setting of metric spaces [3], and is
generally part of the rich area of identification problems of graphs and other
discrete structures [20].

More formally, let us denote by $\mathrm{dist}(x, y)$ the distance from x to y in a
digraph. Here, the distance $\mathrm{dist}(x, y)$ is taken as the length of a shortest directed

Research funded by the French government IDEX-ISITE initiative 16-IDEX-0001 (CAP
20–25) and by the ANR project GRALMECO (ANR-21-CE48-0004).
A. Hakanen—Research supported by the Jenny and Antti Wihuri Foundation and
partially by Academy of Finland grant number 338797.

D. Paulusma and B. Ries (Eds.): WG 2023, LNCS 14093, pp. 232–245, 2023.
https://doi.org/10.1007/978-3-031-43380-1_17

path from x to y; if no such path exists, $\mathrm{dist}(x, y)$ is infinite, and we say that y is not *reachable* from x. We say that a set S is a *resolving set* of a digraph G if for any pair of distinct vertices v, w from G, there is a vertex x in S with $\mathrm{dist}(x, v) \neq \mathrm{dist}(x, w)$. Furthermore, we require that every vertex of G is reachable from at least one vertex of S. The *metric dimension* of G is the smallest size of a resolving set of G, and a minimum-size resolving set of G is called a *metric basis* of G.[1]

We denote by METRIC DIMENSION the computational version of the problem: given a (di)graph G, determine its metric dimension.

For undirected graphs, METRIC DIMENSION has been extensively studied, and its non-local nature makes it highly nontrivial from an algorithmic point of view. On the hardness side, METRIC DIMENSION was shown to be NP-hard for planar graphs of bounded degree [7], split, bipartite and line graphs [9], unit disk graphs [17], interval and permutation graphs of diameter 2 [11], and graphs of pathwidth 24 [19]. On the positive side, it can easily be solved in linear time on trees [4, 14, 18, 30]. More involved polynomial-time algorithms exist for unicyclic graphs [29] and, more generally, graphs of bounded cyclomatic number [9]; outerplanar graphs [7]; cographs [9]; chain graphs [10]; cactus-block graphs [16]; bipartite distance-hereditary graphs [24]. There are fixed parameter tractable (FPT) algorithms for the undirected graph parameters max leaf number [8], tree-depth [13], modular-width [2] and distance to cluster [12], but FPT algorithms are highly unlikely to exist for parameters solution size [15] and feedback vertex set [12].

Due to the interest for METRIC DIMENSION on undirected graphs, it is natural to ask what can be said in the context of digraphs. The metric dimension of digraphs was first studied in [5] under a somewhat restrictive definition; for our definitions, we follow the recent paper [1], in which the algorithmic aspects of METRIC DIMENSION on digraphs have been addressed. We call *oriented graph* a digraph without directed 2-cycles. A *directed acyclic digraph* (DAG for short) has no directed cycles at all. The *underlying multigraph* of a digraph is the one obtained by ignoring the arc orientations; its *underlying graph* is obtained from it by ignoring multiple edges. In a digraph, a *strongly connected component* is a subgraph where every vertex is reachable from all other vertices. Note that for the METRIC DIMENSION problem, undirected graphs can be seen as a special type of digraphs where each arc has a symmetric arc.

The NP-hardness of METRIC DIMENSION was proven for oriented graphs in [27] and, more recently, for bipartite DAGs of maximum degree 8 and maximum distance 4 [1] (the *maximum distance* being the length of a longest directed path without shortcuts). A linear-time algorithm for METRIC DIMENSION on oriented trees was given in [1].

---

[1] The definition that we use has been called *strong metric dimension* in [1], as opposed to *weak metric dimension*, where one single vertex may be unreachable from any resolving set vertex. The former definition seems more natural to us. However, the term *strong metric dimension* is already used for a different concept, see [25]. Thus, to prevent confusion, we avoid the prefix *strong* in this paper.

Our Results. We generalize the linear-time algorithm for METRIC DIMENSION on oriented trees from [1] to all digraphs whose underlying graph is a tree. In other words, here we allow 2-cycles. This makes a significant difference with oriented trees, and as a result our algorithm is nontrivial. We then extend the used methods to solve METRIC DIMENSION in linear time for unicyclic digraphs (digraphs with a unique cycle). Then, we prove that METRIC DIMENSION can be solved in time $f(t)n^{O(1)}$ for digraphs of order n and modular-width t (a parameter recently introduced for digraphs in [31]). This extends the same result for undirected graphs from [2], and is the first FPT algorithm for METRIC DIMENSION on digraphs. Finally, we complement the hardness result from [1] by showing that METRIC DIMENSION is NP-hard even for planar triangle-free DAGs of maximum degree 6 and maximum distance 4. For results marked with ($*$), we omit the full proof due to space constraints; those can be found in [6].

2   Digraphs Whose Underlying Graph is a Tree

For the sake of convenience, we call *di-tree* a digraph whose underlying graph is a tree. Trees are often the first non-trivial class to study for a graph problem. METRIC DIMENSION is no exception to this, having been studied in the first papers for the undirected [4,14,18,30] and the oriented [1] cases. In the undirected case, a minimum-size resolving set can be found by taking, for each vertex of degree at least 3 spanning k legs, the endpoint of $k-1$ of its legs (a *leg* is an induced path spanning from a vertex of degree at least 3, having its inner vertices of degree 2, and ending in a leaf). In the case of oriented trees, taking all the sources (a *source* is a vertex with no in-neighbour) and $k-1$ vertices in each set of k in-twins yields a metric basis (two vertices are *in-twins* if they have the same in-neighbourhood). Our algorithm, being on di-trees (which include both undirected trees and oriented trees), will reuse those strategies, but we will need to refine them in order to obtain a metric basis. The first refinement is of the notion of in-twins:

Definition 1. *A strongly connected component E of a di-tree is an* escalator *if it satisfies the following conditions:*

1. *its underlying graph is a path with vertices $e_1, \ldots, e_k$ ($k \geq 2$);*
2. *there is a unique vertex $y \notin E$ such that the arc $\overrightarrow{ye_1}$ (resp. $\overrightarrow{ye_k}$) exists;*
3. *there can be any number (possibly, zero) of vertices $z \notin E$ such that the arc $\overrightarrow{e_k z}$ (resp. $\overrightarrow{e_1 z}$) exists; for every $i \in \{1, \ldots, k-1\}$ (resp. $i \in \{2, \ldots, k\}$), no arc $\overrightarrow{e_i z}$ with $z \notin E$ exists.*

Definition 2. *In a di-tree, a set of vertices $A = \{a_1, \ldots, a_k\}$ is a set of* almost-in-twins *if there is a vertex x such that:*

1. *for every $i \in \{1, \ldots, k\}$, the arc $\overrightarrow{xa_i}$ exists and the arc $\overrightarrow{a_i x}$ does not exist;*
2. *for every $i \in \{1, \ldots, k\}$, either a_i is a trivial strongly connected component and $N^-(a_i) = \{x\}$, or a_i is the endpoint of an escalator and $N^-(a_i) = \{x, y\}$ where y is its neighbour in the escalator.*

Note that regular in-twins are also almost-in-twins. The second refinement is the following (for a given vertex x in a strongly connected component with C as an underlying graph, we call $d_C(x)$ the degree of x in C):

Definition 3. *Given the underlying graph C of a strongly connected component of a di-tree and a set D of vertices, we call a set S of vertices inducing a path of order at least 2 in C a special leg if it verifies the four following properties:*

1. *S has a unique vertex v such that $v \in D$ or $d_C(v) \geq 3$;*
2. *S has a unique vertex w such that $d_C(w) = 1$, furthermore $w \notin D$: w is called the endpoint of S;*
3. *all of the other vertices x of S verify $d_C(x) = 2$ and $x \notin D$;*
4. *at least one of the vertices $y \in S \setminus \{w\}$ has an out-arc $\overrightarrow{yz}$ with $z \notin C$.*

Note that several special legs can span from the same vertex, from which regular legs can also span. Algorithm 1, illustrated in Fig. 1, computes a metric basis of a di-tree.

Explanation of Algorithm 1. The algorithm will compute a metric basis $\mathcal{B}$ of a di-tree T in linear-time. The first thing we do is to add every source in T to $\mathcal{B}$ (line 1). Then, for every set of almost-in-twins, we add all of them but one to $\mathcal{B}$ (lines 2–3). Those two first steps, depicted in Fig. 1a, are the ones used to compute the metric basis of an orientation of a tree [1], and as such they are still necessary for managing the non-strongly connected components of the di-tree. Note that we are specifically managing sets of *almost-in-twins*, which include sets of in-twins, since it is necessary to resolve the specific case of escalators. The rest of the algorithm consists in managing the strongly connected components.

For each strongly connected component having C as an underlying graph, we first identify each vertex x of C that has an in-arc coming from **outside** C. Indeed, since x is the "last" vertex of a path coming from outside C, there are vertices of $\mathcal{B}$ "behind" this in-arc (or they can themselves be a vertex in $\mathcal{B}$), which we will call $\mathcal{B}_x$. However, the vertices in $\mathcal{B}_x$ can be "projected" on x since, T being a di-tree, x is on every shortest path from the vertices of $\mathcal{B}$ "behind" the in-arc to the vertices of C. Hence, we will mark x as a **dummy vertex** (lines 5–7, depicted in Figure 1b): we will consider that it is in $\mathcal{B}$ for the rest of this step, and acts as a representative of the set $\mathcal{B}_x$ with respect to C.

We then have to manage some specific cases whenever C is a path (lines 8–17). Indeed, the last two steps of the algorithm do not always work under some conditions. Those specific conditions are highlighted in the proof.

The last two steps are then applied. First, we have to consider the **special legs** defined in Definition 3. The idea behind those special legs is the following: for every out-arc $\overrightarrow{yz}$ with y in the special leg and z outside of C, any vertex in the metric basis "before" the start of the special leg will not distinguish z and the next neighbour of y in the special leg. Hence, we have to add at least one vertex to $\mathcal{B}$ for each special leg, and we choose the endpoint of the special leg (lines 18–19, depicted in Fig. 1c). Finally, we apply the well-known algorithm for

---

Algorithm 1: An algorithm computing the metric basis of a di-tree.

    Input   : A di-tree T.
    Output: A metric basis $\mathcal{B}$ of T.

1 $\mathcal{B} \leftarrow$ Every source of T
2 **foreach** *set I of almost-in-twins* **do**
3   Add $|I| - 1$ vertices of I to $\mathcal{B}$

4 **foreach** *strongly connected component with C as an underlying graph* **do**
5      $D \leftarrow \emptyset$
6   **foreach** *arc $\overrightarrow{uv}$ with $v \in C$ and $u \notin C$* **do**
7   Add v to D
8   **if** *C is a path with endpoints x and y* **then**
9   **if** *there is no vertex in $C \cap D$* **then**
10   **if** *there is no out-arc from C to outside of C* **then**
11   Add x to $\mathcal{B}$
12   **else if** *there is an out-arc from x (resp. y) to outside of C and no other out-arc from C to outside of C* **then**
13   Add y (resp. x) to $\mathcal{B}$
14   **else**
15   Add x and y to $\mathcal{B}$
16   **else if** *there is exactly one vertex w in $C \cap D$, w is neither x nor y, and there is no out-arc from w to outside of C* **then**
17   Add x to $\mathcal{B}$
18   **foreach** *special leg L of C* **do**
19   Add the endpoint of L to $\mathcal{B}$
20   **foreach** *vertex of degree ≥ 3 in C from which span $k \geq 2$ legs of C that do not have a vertex in $\mathcal{B}$ or in D* **do**
21   Add the endpoint of $k - 1$ such legs to $\mathcal{B}$

22 **return** $\mathcal{B}$

---

computing the metric basis of a tree to the remaining parts of C (lines 20–21, depicted in Fig. 1d). The special legs and the legs containing a dummy vertex, being already resolved, are not considered in this part.

Theorem 4 (∗). *Algorithm 1 computes a metric basis of a di-tree in linear time.*

3 Orientations of Unicyclic Graphs

A unicyclic graph U is constituted of a cycle C with vertices $c_1, \ldots, c_n$, and each vertex c_i is the root of a tree T_i (we can have T_i be simply the isolated c_i itself). The metric dimension of an undirected unicyclic graph has been studied in [26,28,29]. In [26], Poisson and Zhang proved bounds for the metric dimension

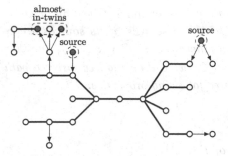

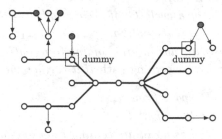

(a) The first step (lines 1-3) is to add every source and manage sets of almost-in-twins.

(b) The second step (lines 5-7) is to mark the dummy vertices of the strongly connected component.

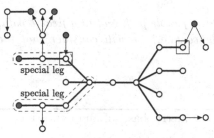

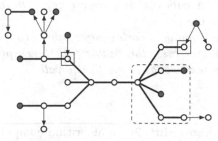

(c) The third step (lines 18-19) is to manage all the special legs.

(d) The fourth and final step (lines 20-21) is to manage the remaining legs with a common ancestor.

Fig. 1. Illustration of Algorithm 1. For the sake of simplicity, there are only two strongly connected components, for which we only represent the underlying graph with bolded edges, so every bolded edge is a 2-cycle. One of the two strongly connected components is a simple path that does not require any action. Vertices in the metric basis are colored in red. (Color figure online)

of a unicyclic graph in terms of the metric dimension of a tree we obtain by removing one edge from the cycle. Sedlar and Škrekovski showed more recently that the metric dimension of a unicyclic graph is one of two values in [28], and then the exact value of the metric dimension based on the structure of the graph in [29]. In this section, we will show that one can compute a metric basis of an orientation of a unicyclic graph in linear time. The algorithm mostly consists in using sources and in-twins, with a few specific edge cases to consider.

In this section, an *induced directed path* $\overrightarrow{P}$ is the orientation of an induced path with only one source and one sink which are its two endpoints. It is said to be *spanning from u* if u is its source endpoint, and its *length* is its number of edges. We also need the following definition:

Definition 5. *Let $\vec{U}$ be the orientation of a unicyclic graph. Given an orientation of a cycle $\vec{C}$ of even length $n = 2k$ with two sources, if its sources are c_i and c_{i+2}, its sinks are c_{i+1} and c_{i+1+k}, and there are, in $\vec{C} \setminus \{c_i, c_{i+2}\}$, neither in-twins nor in-arcs coming from outside of $\vec{C}$, we call an induced directed path $\vec{P}$ an* concerning path *if it verifies the three following properties:*

1. *$\vec{P}$ spans from c_{i+1};*
2. *$\vec{P}$ has length $k - 2$;*
3. *$\vec{P}$ has no in-arc coming from outside of $\vec{P} \cup \vec{C}$;*

Furthermore, if, for every vertex in $\vec{P}$ belonging to a nonempty set I of in-twins, every vertex in I belongs to a concerning path, then, we call $\vec{P}$ an unfixable path.

A path that is a concerning path, but not an unfixable path, will be called a fixable path.

Finally, a vertex might belong both to an unfixable path and to a fixable path; in this case, the fixable path takes precedence (i.e., we will consider that the vertex belongs to the fixable path).

Algorithm 2: An algorithm computing the metric basis of an orientation of a unicyclic graph.

Input : An orientation $\vec{U}$ of a unicyclic graph U.
Output: A metric basis $\mathcal{B}$ of $\vec{U}$.

1 Add to $\mathcal{B}$ every source of $\vec{U}$
2 Apply the **special cases** in Algorithm 3
3 **foreach** *set I of in-twins in $\vec{U}$ that are not already in $\mathcal{B}$* **do**
4   **if** *all the vertices of I are in concerning paths* **then**
5   Add $|I| - 1$ vertices of I to $\mathcal{B}$, prioritizing vertices in unfixable paths
6   **else**
7   Add $|I| - 1$ vertices of I to $\mathcal{B}$, prioritizing vertices in the cycle $\vec{C}$ or in concerning paths, if there are any
8 **return** $\mathcal{B}$

Explanation of Algorithm 2. The algorithm will compute a metric basis $\mathcal{B}$ of an orientation $\vec{U}$ of a unicyclic graph U in linear-time. The first thing we do is to add every source in $\vec{U}$ to $\mathcal{B}$ (line 1). We will also manage the sets of in-twins in $\vec{U}$ (lines 3–7), which we need to do after taking care of some special cases that might influence the choice of in-twins. When we have the choice, we prioritize taking in-twins that are in the cycle to guarantee reachability of vertices in the cycle. Note that those two sets (along with the right priority) are enough in most cases.

Algorithm 3: Special cases of Algorithm 2.

1 **if** *the cycle $\overrightarrow{C}$ has no sink, there is no in-arc coming from outside of C, and no vertex of C is in a set of in-twin* **then**

2 $\quad$ Add c_1 to $\mathcal{B}$

3 **if** *the cycle $\overrightarrow{C}$ has no sink, there is exactly one in-arc $\overrightarrow{uc_i}$ with $u \notin \overrightarrow{C}$, no vertex c_j with $j \neq i$ is an in-twin or has in-arc coming from outside of $\overrightarrow{C}$, and u has an out-neighbour v with $N^-(v) = \{u\}$* **then**

4 $\quad$ Add c_i to $\mathcal{B}$

5 **if** *the cycle $\overrightarrow{C}$ has exactly one source c_i* **then**

6 $\quad$ **if** *the one sink is either c_{i-1} or c_{i+1}, and no vertex c_j with $j \neq i$ is an in-twin or has an in-arc coming from outside of $\overrightarrow{C}$* **then**

7 $\quad\quad$ Add c_{i-1} to $\mathcal{B}$

8 $\quad$ **else if** *the one sink is c_{i+k} with $k > 1$, $|\overrightarrow{C}| \geq 2k$, c_{i+k-1} (resp. c_{i+k+1}) has an out-neighbour v such that $N^-(v) = \{c_{i+k-1}\}$ (resp. $N^-(v) = \{c_{i+k+1}\}$), no vertex in $\{c_{i-1}, c_{i-2}, \ldots, c_{i+k}\}$ (resp. $\{c_{i+1}, c_{i+2}, \ldots, c_{i+k}\}$) has an in-arc, and no vertex in $\{c_{i-2}, c_{i-3}, \ldots, c_{i+k+1}\}$ (resp. $\{c_{i+2}, c_{i+3}, \ldots, c_{i+k-1}\}$) is an in-twin* **then**

9 $\quad\quad$ Add c_{i-1} (resp. c_{i+1}) to $\mathcal{B}$

10 $\quad$ **else if** *the one sink is c_{i+k} with $k > 1$, $|\overrightarrow{C}| = 2k$, c_{i+k-1} has an out-neighbour v_- such that $N^-(v_-) = \{c_{i+k-1}\}$, c_{i+k+1} has an out-neighbour v_+ such that $N^-(v_+) = \{c_{i+k+1}\}$, no vertex in $\overrightarrow{C}$ except c_i has an in-arc, no vertex in $\overrightarrow{C} \setminus \{c_i, c_{i-1}, c_{i+1}\}$ is an in-twin, and c_{i-1} and c_{i+1} are not in a set I of in-twins verifying $|I| \geq 3$* **then**

11 $\quad\quad$ Add c_{i+k} to $\mathcal{B}$

12 **if** *the cycle $\overrightarrow{C}$ has exactly two sources c_i and c_{i+2}, $|\overrightarrow{C}| = 2k$ with $k > 2$, the two sinks are c_{i+1} and c_{i+1+k}, no vertex from $\overrightarrow{C}$ except c_i and c_{i+2} is an in-twin or has an in-arc coming from outside of $\overrightarrow{C}$, there is at least one unfixable path, and there is no fixable path* **then**

13 $\quad$ Add c_{i+1} to $\mathcal{B}$

We then have to manage six specific cases (line 2). **Those special cases are handled in Algorithm 3.** The first two special cases occur when the cycle has no sink. First, if the cycle has no sink, no in-twin, and no arc coming from outside, then, we have to add one vertex of the cycle to $\mathcal{B}$ in order to maintain reachability (lines 1–2). Then, if the cycle has no sink, only one in-arc $\overrightarrow{uc_i}$ is coming from outside of it, and there is a vertex v with $N^-(v) = \{u\}$, then, we have to add either c_i or v to $\mathcal{B}$ in order to resolve them (lines 3–4).

The next three special cases occur when the cycle has one sink. First, if there is only one sink in the cycle, it is an out-neighbour of the source, and no vertex from the cycle apart from the source is an in-twin or has an in-arc coming from outside of the cycle, then we need to add one of the out-neighbours of the source in the cycle to $\mathcal{B}$ in order to resolve them (lines 6–7).

Then, there are two specific cases when the cycle has one sink, both based on the same principle. Both happen when the source is c_i, the sink is c_{i+k}, it has no in-arc, and the cycle contains at least $2k$ vertices. In the fourth special case (lines 8–9), the vertex c_{i+k-1} has an out-neighbour v verifying $N^-(v) = \{c_{i+k-1}\}$. We can see that, if no vertex in the other path from c_i to c_{i+k} (the path going through $c_{i-1}, c_{i-2}, \ldots, c_{i+k+1}$) is in $\mathcal{B}$, then, v and c_{i+k} will not be resolved. Those vertices can be added to $\mathcal{B}$ if they have an in-arc or if they are an in-twin (they will have priority). However, note that c_{i-1} might be an in-twin of c_{i+1}, in which case it should be added to $\mathcal{B}$, resolving the conflict. Hence, if none of $c_{i-1}, c_{i-2}, \ldots, c_{i+k+1}$ has an in-arc or is an in-twin, then, we can add c_{i-1} to $\mathcal{B}$ in order to resolve v and c_{i+k}. Note that, in this case, in comparison to just the sources and the resolution of sets of in-twins, we add one more vertex to $\mathcal{B}$ if c_{i-1} is the only in-twin of c_{i+1}. The same reasoning can be made with the symmetric case.

The fifth special case (lines 10–11) occurs when the cycle contains exactly $2k$ vertices and both c_{i+k-1} and c_{i+k+1} have an out-neighbour (respectively v_- and v_+) with in-degree 1: the pairs of vertices (v_-, c_{i+k}) and (v_+, c_{i+k}) might not be resolved. We can see that any in-arc or in-twin along a path from c_i to c_{i+k} will resolve c_{i+k} and the v pendant on the other path (thus either fully resolving those two pairs, or bringing us back to the previous special case), except if c_{i-1} and c_{i+1} are the only in-twins in the cycle and if they do not have another in-twin. Hence, if no vertex from the cycle except c_i has an in-arc, no vertex from the cycle except c_i, c_{i-1} and c_{i+1} is an in-twin, and c_{i-1} and c_{i+1} do not have another in-twin, then, we need to add at least one more vertex to $\mathcal{B}$ in order to resolve the two pairs of vertices, and adding c_{i+k} does exactly that.

Finally, the sixth special case is more complex (lines 12–13 and consideration in the choice of in-twins). Assume that the cycle $\overrightarrow{C}$ is of even length n, has neither in-twin nor in-arc coming from outside (except the sources), and that there are two sinks in the $\overrightarrow{C}$: one at distance 1 from the sources, and the other at the opposite end of $\overrightarrow{C}$. Now, if the first sink has spanning concerning paths, then, the second cycle and the endpoints of those concerning paths might not be resolved, since they are at the same distance $(\frac{n}{2} - 1)$ of both sources of $\overrightarrow{C}$. Thus, we need to apply a strategy in order to resolve those vertices while trying to not add a supplementary vertex to $\mathcal{B}$. This is done by considering the two kinds of concerning paths, and having a priority in the selection of in-twins.

All the other cases of the cycle are already resolved through the sources and in-twins steps.

Theorem 6 (∗). *Algorithm 2 computes a metric basis of an orientation of a unicyclic graph in linear time.*

4   Modular Width

In a digraph G, a set $X \subseteq V(G)$ is a *module* if every vertex not in X 'sees' all vertices of X in the same way. More precisely, for each $v \in V(G) \setminus X$ one of the following holds: (i) $(v,x),(x,v) \in E(G)$ for all $x \in X$, (ii) $(v,x),(x,v) \notin E(G)$ for all $x \in X$, (iii) $(v,x) \in E(G)$ and $(x,v) \notin E(G)$ for all $x \in X$, (iv) $(v,x) \notin E(G)$ and $(x,v) \in E(G)$ for all $x \in X$. The singleton sets, $\emptyset$, and $V(G)$ are trivially modules of G. We call the singleton sets the *trivial modules* of G.

The graph $G[X]$ where X is a module of G is called a *factor* of G. A family $\mathcal{X} = \{X_1, \ldots, X_s\}$ is a *factorization* of G if $\mathcal{X}$ is a partition of $V(G)$, and each X_i is a module of G. If X and Y are two non-intersecting modules, then the relationship between $x \in X$ and $y \in Y$ is one of (i)-(iv) and always the same no matter which vertices x and y are exactly. Thus, given a factorization $\mathcal{X}$, we can identify each module with a vertex, and connect them to each other according to the arcs between the modules. More formally, we define the *quotient* $G/\mathcal{X}$ with respect to the factorization $\mathcal{X}$ as the graph with the vertex set $\mathcal{X} = \{X_1, \ldots, X_s\}$ and $(X_i, X_j) \in E(G/\mathcal{X})$ if and only if $(x_i, x_j) \in E(G)$ where $x_i \in X_i$ and $x_j \in X_j$. A quotient depicts the connections of the different modules of a factorization to each other while omitting the internal structure of the factors. Each factor itself can be factorized further (as long as it is nontrivial, i.e. not a single vertex). By factorizing the graph G and its factors until no further factorization can be done, we obtain a *modular decomposition* of G. The *width* of a decomposition is the maximum number of sets in a factorization (or equivalently, the maximum number of vertices in a quotient) in the decomposition. The *modular width* of G is defined as the minimum width over all possible modular decompositions of G, and we denote it by $\mathrm{mw}(G)$. An optimal modular decomposition of a digraph can be computed in linear time [21]. METRIC DIMENSION for undirected graphs was shown to be fixed parameter tractable when parameterized by modular width by Belmonte et al. [2]. We will generalize their algorithm to directed graphs.

The following result lists several useful observations.

Proposition 7 (*). *Let* $\mathcal{X} = \{X_1, \ldots, X_s\}$ *be a factorization of* G, *and let* $W \subseteq V(G)$ *be a resolving set of* G.

(i) *For all* $x, y \in X_i$ *and* $z \in X_j$, $i \neq j$, *we have* $\mathrm{dist}_G(x,z) = \mathrm{dist}_G(y,z)$ *and* $\mathrm{dist}_G(z,x) = \mathrm{dist}_G(z,y)$.

(ii) *For all* $x \in X_i$ *and* $y \in X_j$, $i \neq j$, *we have* $\mathrm{dist}_G(x,y) = \mathrm{dist}_{G/\mathcal{X}}(X_i, X_j)$.

(iii) *For all* $x, y \in V(G)$ *we have either* $\mathrm{dist}_G(x,y) \leq \mathrm{mw}(G)$ *or* $\mathrm{dist}_G(x,y) = \infty$.

(iv) *The set* $\{X_i \in \mathcal{X} \mid W \cap X_i \neq \emptyset\}$ *is a resolving set of the quotient* $G/\mathcal{X}$.

(v) *For all distinct* $x, y \in X_i$, *where* $X_i \in \mathcal{X}$ *is nontrivial, we have* $\mathrm{dist}_G(w,x) \neq \mathrm{dist}_G(w,y)$ *for some* $w \in W \cap X_i$.

(vi) *Let* $w_1, w_2 \in X_i$. *If* $\mathrm{dist}_G(w_1,x) \neq \mathrm{dist}_G(w_2,x)$, *then* $x \in X_i$ *and* $\mathrm{dist}_G(w_1,x) \neq \mathrm{dist}_G(w_1,y)$ *or* $\mathrm{dist}_G(w_2,x) \neq \mathrm{dist}_G(w_2,y)$ *for each* $y \notin X_i$.

The basic idea of our algorithm (and that of [2]) is to compute metric bases that satisfy certain conditions for the factors and combine these local solutions into a global solution. We know that nontrivial modules must contain elements of a resolving set, as modules must be resolved locally (Proposition 7 (i)). While combining the local solutions of nontrivial modules, we need to make sure that a vertex $x \in X_i$, where X_i is nontrivial, is resolved from all $y \notin X_i$. If x and y are resolved as described in Proposition 7 (vi), then we need to do nothing special.

However, if $x \in X_i$ is such that $\text{dist}_G(w, x) = d$ for all $w \in W_i$ and a fixed $d \in \{1, \ldots, \text{mw}(G), \infty\}$, there might exist a vertex $y \notin X_i$ such that W_i does not resolve x and y. We call such a vertex x d-*constant* (with respect to W_i). We need to keep track of d-constant vertices and make sure they are resolved when we combine the local solutions. There are at most $\text{mw}(G) + 1$ d-constant vertices in each factor due to Proposition 7 (iii). We need to also make sure vertices in different modules that contain no elements of the solution set are resolved. To do this, we might need to include some vertices from the trivial modules in addition to the vertices we have included from the nontrivial modules.

In the algorithm presented in [2], the problems described above are dealt with by computing values $w(H, p, q)$ for every factor H, where $w(H, p, q)$ is the minimum cardinality of a resolving set of H (with respect to the distance in G) where some vertex is 1-constant iff $p = true$ and some vertex is 2-constant iff $q = true$ (for undirected graphs these are the only two relevant cases). The same values are then computed for the larger graph by combining different solutions of the factors and taking their minimum. Our generalization of this algorithm is along the same lines as the original, however, we have more boolean values to keep track of. One difference to the techniques of the original algorithm is that we do not use the auxiliary graphs Belmonte et al. use. These auxiliary graphs were needed to simulate the distances of the vertices of a factor in G as opposed to only within the factor. In our approach, we simply use the distances in G.

Theorem 8 (*). *The metric dimension of a digraph G with $\text{mw}(G) \le t$ can be computed in time $\mathcal{O}(t^5 2^{t^2} n + n^3 + m)$ where $n = |V(G)|$ and $m = |E(G)|$.*

Proof (sketch). Let us consider one level of an optimal modular decomposition of G. Let H be a factor somewhere in the decomposition, and let $\mathcal{X} = \{X_1, \ldots, X_s\}$ be the factorization of H according to the modular decomposition. For the graph H (and its nontrivial factors $H[X_i]$) we denote by $w(H, \mathbf{p})$ the minimum cardinality of a set $W \subseteq V(H)$ such that

(i) W resolves $V(H)$ in G,
(ii) $\mathbf{p} = (p_1, \ldots, p_{\text{mw}(G)}, p_\infty)$ where $p_d = true$ if and only if H contains a d-constant vertex with respect to W.

If such a set does not exist, then $w(H, \mathbf{p}) = \infty$. In order to compute the values $w(H, \mathbf{p})$, we next introduce the auxiliary values $\omega(\mathbf{p}, I, P)$. The values $w(H[X_i], \mathbf{p})$ are assumed to be known for all $\mathbf{p}$ and nontrivial modules X_i. Let the factorization $\mathcal{X}$ be labeled so that the modules X_i are trivial for $i \in \{1, \ldots, h\}$ and nontrivial for $i \in \{h+1, \ldots, s\}$. Let $I \subseteq \{1, \ldots, h\}$ and $P = (\mathbf{p}^{h+1}, \ldots \mathbf{p}^s)^T$.

We define $\omega(\mathbf{p}, I, P) = |I| + \sum_{i=h+1}^{s} w(H[X_i], \mathbf{p}^i)$ if the conditions (a)–(d) hold. In what follows, a representative of a module X_i is denoted by x_i.

(a) The set $Z = \{X_i \in \mathcal{X} \mid i \in I \cup \{h+1, \ldots, s\}\}$ resolves the quotient $H/\mathcal{X}$ with respect to the distances in G.
(b) For $d \in \{1, \ldots, \text{mw}(G), \infty\}$ and $i \in \{h+1, \ldots, s\}$, if $p_d^i = true$, then for each trivial module $X_j = \{x_j\}$ where $j \notin I$ we have $\text{dist}_G(x_i, x_j) \neq d$ or there exists $X_k \in Z \setminus \{X_i\}$ such that $\text{dist}_G(x_k, x_i) \neq \text{dist}_G(x_k, x_j)$.

(c) For $d_1, d_2 \in \{1, \ldots, \mathrm{mw}(G), \infty\}$ and distinct $i, j \in \{h+1, \ldots, s\}$, if $p_{d_1}^i = p_{d_2}^j = true$, then $\mathrm{dist}_G(x_i, x_j) \neq d_1$, or $\mathrm{dist}_G(x_j, x_i) \neq d_2$, or there exists $X_k \in Z \setminus \{X_i, X_j\}$ such that $\mathrm{dist}_G(x_k, x_i) \neq \mathrm{dist}_G(x_k, x_j)$.

(d) For all $d \in \{1, \ldots, \mathrm{mw}(G), \infty\}$, we have $p_d = true$ (in $\mathbf{p}$) if and only if for some $i \in \{1, \ldots, h\} \setminus I$ we have $\mathrm{dist}_G(x_j, x_i) = d$ for all $X_j \in Z$, or for some $i \in \{h+1, \ldots, s\}$ we have $p_d^i = true$ and $\mathrm{dist}_G(x_j, x_i) = d$ for all $X_j \in Z \setminus \{X_i\}$.

If these conditions cannot be met, then we set $\omega(\mathbf{p}, I, P) = \infty$.

When we know all the values $\omega(\mathbf{p}, I, P)$, we can easily calculate $w(H, \mathbf{p})$ since $w(H, \mathbf{p}) = \min_{I,P} \omega(\mathbf{p}, I, P)$ (proof omitted due to lack of space).

To conclude the proof, we note that $w(G, \mathbf{p})$ is the minimum cardinality of a resolving set that gives some vertices the specific distance combinations according to $\mathbf{p}$. Thus, the metric dimension of G is $\min w(G, \mathbf{p})$ where the minimum is taken over all $\mathbf{p}$ such that $p_\infty = false$. $\qquad\square$

5 NP-Hardness for Restricted DAGs

We now complement the hardness result from [1], which was for bipartite DAGs of maximum degree 8 and maximum distance 4.

Theorem 9 (*). METRIC DIMENSION *is NP-complete, even on planar triangle-free DAGs of maximum degree 6 and maximum distance 4.*

Proof (sketch). We reduce from VERTEX COVER on 2-connected planar cubic graphs, which is known to be NP-complete [23, Theorem 4.1].

Given a 2-connected planar cubic graph G, we construct a DAG G' as follows. First of all, note that by Petersen's theorem, G contains a perfect matching $M \subset E(G)$, that can be constructed in polynomial time. A planar embedding of G can also be constructed in polynomial time, so we fix one. We let $V(G') = V(G) \bigcup_{e=uv \in E(G)} \{a_e, b_e, c_e, d_e^u, d_e^v\} \bigcup_{e=uv \in M} \{f_e, g_e, h_e\}$. For every edge $e = uv$ of G, we add the arcs $\{\overrightarrow{a_e b_e}, \overrightarrow{b_e c_e}, \overrightarrow{c_e d_e^u}, \overrightarrow{c_e d_e^v}, \overrightarrow{u d_e^u}, \overrightarrow{v d_e^v}\}$. For every edge $e = uv$ of the perfect matching M of G, assuming the neighbours of u (in the clockwise cyclic order with respect to the planar embedding of G) are v, x, y and those of v are u, s, t, we arbitrarily fix one side of the edge uv to place the vertices f_e, g_e and h_e (say, on the side that is close to the edges ux and vt). We add the arcs $\{\overrightarrow{f_e g_e}, \overrightarrow{g_e c_e}, \overrightarrow{g_e h_e}, \overrightarrow{h_e u}, \overrightarrow{h_e v}, \overrightarrow{c_e c_{uy}}, \overrightarrow{c_e c_{vs}}, \overrightarrow{h_e c_{ux}}, \overrightarrow{h_e c_{vt}}\}$.

Using the embedding of G, G' can also be drawn in a planar way, it has maximum degree 6 (the vertices of type c_e are of degree 6), has no triangles, and no shortest directed path of length 4.

We claim that G has a vertex cover of size at most k if and only if G' has metric dimension at most $k + |E(G)| + |M|$ (proof omitted due to lack of space). $\qquad\square$

6   Conclusion

METRIC DIMENSION can be solved in polynomial time on outerplanar graphs, using an involved algorithm [7]. Can one generalize our algorithms for trees and unicyclic graphs to solve METRIC DIMENSION for directed (or at least, oriented) outerplanar graphs in polynomial time? Extending our algorithm to cactus graphs already seems nontrivial.

Is METRIC DIMENSION NP-hard on planar bipartite subcubic DAGs?

Also, it would be interesting to see which hardness results known for METRIC DIMENSION of undirected graphs also hold for DAGs, or for oriented graphs.

References

1. Araujo, J., et al.: On finding the best and worst orientations for the metric dimension. Algorithmica 1–41 (2023)
2. Belmonte, R., Fomin, F.V., Golovach, P.A., Ramanujan, M.S.: Metric dimension of bounded tree-length graphs. SIAM J. Discret. Math. **31**(2), 1217–1243 (2017)
3. Blumenthal, L.M.: Theory and Applications of Distance Geometry. Oxford University Press, Oxford (1953)
4. Chartrand, G., Eroh, L., Johnson, M.A., Oellermann, O.R.: Resolvability in graphs and the metric dimension of a graph. Discret. Appl. Math. **105**(1), 99–113 (2000)
5. Chartrand, G., Raines, M., Zhang, P.: The directed distance dimension of oriented graphs. Math. Bohem. **125**, 155–168 (2000)
6. Dailly, A., Foucaud, F., Hakanen, A.: Algorithms and hardness for metric dimension on digraphs. arXiv preprint arXiv:2307.09389 (2023)
7. Díaz, J., Pottonen, O., Serna, M.J., van Leeuwen, E.J.: Complexity of metric dimension on planar graphs. J. Comput. Syst. Sci. **83**(1), 132–158 (2017)
8. Eppstein, D.: Metric dimension parameterized by max leaf number. J. Graph Algorithms Appl. **19**(1), 313–323 (2015)
9. Epstein, L., Levin, A., Woeginger, G.J.: The (weighted) metric dimension of graphs: hard and easy cases. Algorithmica **72**(4), 1130–1171 (2015)
10. Fernau, H., Heggernes, P., van 't Hof, P., Meister, D., Saei, R.: Computing the metric dimension for chain graphs. Inf. Process. Lett. **115**(9), 671–676 (2015)
11. Foucaud, F., Mertzios, G.B., Naserasr, R., Parreau, A., Valicov, P.: Identification, location-domination and metric dimension on interval and permutation graphs. II. Algorithms and complexity. Algorithmica **78**(3), 914–944 (2017)
12. Galby, E., Khazaliya, L., Inerney, F.M., Sharma, R., Tale, P.: Metric dimension parameterized by feedback vertex set and other structural parameters. In: 47th International Symposium on Mathematical Foundations of Computer Science, MFCS 2022, 22–26 August 2022, Vienna, Austria. LIPIcs, vol. 241, pp. 51:1–51:15. Schloss Dagstuhl - Leibniz-Zentrum für Informatik (2022)
13. Gima, T., Hanaka, T., Kiyomi, M., Kobayashi, Y., Otachi, Y.: Exploring the gap between treedepth and vertex cover through vertex integrity. Theoret. Comput. Sci. **918**, 60–76 (2022)
14. Harary, F., Melter, R.A.: On the metric dimension of a graph. Ars Combin. **2**, 191–195 (1976)
15. Hartung, S., Nichterlein, A.: On the parameterized and approximation hardness of metric dimension. In: Proceedings of the 28th Conference on Computational Complexity, CCC 2013, K.lo Alto, California, USA, 5–7 June 2013, pp. 266–276. IEEE Computer Society (2013)

16. Hoffmann, S., Elterman, A., Wanke, E.: A linear time algorithm for metric dimension of cactus block graphs. Theoret. Comput. Sci. **630**, 43–62 (2016)
17. Hoffmann, S., Wanke, E.: METRIC DIMENSION for gabriel unit disk graphs Is NP-complete. In: Bar-Noy, A., Halldórsson, M.M. (eds.) ALGOSENSORS 2012. LNCS, vol. 7718, pp. 90–92. Springer, Heidelberg (2013). https://doi.org/10.1007/978-3-642-36092-3_10
18. Khuller, S., Raghavachari, B., Rosenfeld, A.: Landmarks in graphs. Discret. Appl. Math. **70**(3), 217–229 (1996)
19. Li, S., Pilipczuk, M.: Hardness of metric dimension in graphs of constant treewidth. Algorithmica **84**(11), 3110–3155 (2022)
20. Lobstein, A.: Watching systems, identifying, locating-dominating and discriminating codes in graphs: a bibliography (2022). https://www.lri.fr/~lobstein/debutBIBidetlocdom.pdf
21. McConnell, R.M., de Montgolfier, F.: Linear-time modular decomposition of directed graphs. Discret. Appl. Math. **145**(2), 198–209 (2005)
22. Melter, R.A., Tomescu, I.: Metric bases in digital geometry. Comput. Vision Graph. Image Process. **25**(1), 113–121 (1984)
23. Mohar, B.: Face covers and the genus problem for apex graphs. J. Combin. Theory Ser. B **82**(1), 102–117 (2001)
24. Moscarini, M.: Computing a metric basis of a bipartite distance-hereditary graph. Theoret. Comput. Sci. **900**, 20–24 (2022)
25. Oellermann, O.R., Peters-Fransen, J.: The strong metric dimension of graphs and digraphs. Discret. Appl. Math. **155**(3), 356–364 (2007)
26. Poisson, C., Zhang, P.: The metric dimension of unicyclic graphs. J. Combin. Math. Combin. Comput. **40**, 17–32 (2002)
27. Rajan, B., Rajasingh, I., Cynthia, J.A., Manuel, P.: Metric dimension of directed graphs. Int. J. Comput. Math. **91**(7), 1397–1406 (2014)
28. Sedlar, J., Škrekovski, R.: Bounds on metric dimensions of graphs with edge disjoint cycles. Appl. Math. Comput. **396**, 125908 (2021)
29. Sedlar, J., Škrekovski, R.: Vertex and edge metric dimensions of unicyclic graphs. Discret. Appl. Math. **314**, 81–92 (2022)
30. Slater, P.J.: Leaves of trees. Congressius Numer. **14**, 549–559 (1975)
31. Steiner, R., Wiederrecht, S.: Parameterized algorithms for directed modular width. In: Changat, M., Das, S. (eds.) CALDAM 2020. LNCS, vol. 12016, pp. 415–426. Springer, Cham (2020). https://doi.org/10.1007/978-3-030-39219-2_33

Degreewidth: A New Parameter
for Solving Problems on Tournaments

Tom Davot[1]([✉])(iD), Lucas Isenmann[2](iD), Sanjukta Roy[3](iD),
and Jocelyn Thiebaut[4](iD)

[1] Université de Technologie de Compiègne, CNRS, Heudiasyc, Compiègne, France
tom.davot@hds.utc.fr
[2] Université de Montpellier, Montpellier, France
lucas.isenmann@umontpellier.fr
[3] Pennsylvania State University, State College, USA
sanjukta@psu.edu
[4] Faculty of Information Technology, Czech Technical University in Prague,
Prague, Czech Republic
jocelyn.thiebaut@cvut.cz

Abstract. In the paper, we define a new parameter for tournaments called degreewidth which can be seen as a measure of how far is the tournament from being acyclic. The degreewidth of a tournament T denoted by $\Delta(T)$ is the minimum value k for which we can find an ordering $\langle v_1, \ldots, v_n \rangle$ of the vertices of T such that every vertex is incident to at most k backward arcs (*i.e.* an arc (v_i, v_j) such that $j < i$). Thus, a tournament is acyclic if and only if its degreewidth is zero. Additionally, the class of sparse tournaments defined by Bessy *et al.* [ESA 2017] is exactly the class of tournaments with degreewidth one.

We study computational complexity of finding degreewidth. We show it is NP-hard and complement this result with a 3-approximation algorithm. We provide a $O(n^3)$-time algorithm to decide if a tournament is sparse, where n is its number of vertices.

Finally, we study classical graph problems DOMINATING SET and FEEDBACK VERTEX SET parameterized by degreewidth. We show the former is fixed-parameter tractable whereas the latter is NP-hard even on sparse tournaments. Additionally, we show polynomial time algorithm for FEEDBACK ARC SET on sparse tournaments.

Keywords: Tournaments · NP-hardness · graph-parameter · feedback arc set · approximation algorithm · parameterized algorithms

1 Introduction

A tournament is a directed graph such that there is exactly one arc between each pair of vertices. Tournaments form a very rich subclass of digraphs which

Sanjukta Roy was affiliated to Faculty of Information Technology, Czech Technical University in Prague when majority of this work was done. Jocelyn Thiebaut was supported by the CTU Global postdoc fellowship program.

© The Author(s), under exclusive license to Springer Nature Switzerland AG 2023
D. Paulusma and B. Ries (Eds.): WG 2023, LNCS 14093, pp. 246–260, 2023.
https://doi.org/10.1007/978-3-031-43380-1_18

has been widely studied both from structural and algorithmic point of view [4]. Unlike for complete graphs, a number of classical problems remain difficult in tournaments and therefore interesting to study. These problems include DOM-INATING SET [14], WINNER DETERMINATION [22], or maximum cycle packing problems. For example, DOMINATING SET is W[2]-hard on tournaments with respect to solution size [14]. However, many of these problems become easy on acyclic tournaments (*i.e.* without directed cycle). Therefore, a natural question that arises is whether these problems are easy to solve on tournaments that are close to being acyclic. The phenomenon of a tournament being "close to acyclic" can be captured by minimum size of a *feedback arc set* (fas). A fas is a collection of arcs that, when removed from the digraph (or, equivalently, reversed) makes it acyclic. This parameter has been widely studied, for numerous applications in many fields, such as circuit design [19], or artificial intelligence [5,13]. However, the problem of finding a minimum fas on tournaments (the problem is then called *FAST* for FEEDBACK ARC SET IN TOURNAMENTS), remained opened for over a decade before being proven NP-complete [3,10]. From the approximability point of view, van Zuylen and Williamson [25] provided a 2-approximation of FAST, and Kenyon-Mathieu and Schudy [21] a PTAS algorithm. On the parameterized-complexity side, Feige [15] as well as Karpinski and Schudy [20] independently proved an $2^{O(\sqrt{k})} + n^{O(1)}$ running-time algorithm. Another way to define FAST is to consider the problem of finding an ordering of the vertices $\langle v_1, \ldots, v_n \rangle$ minimising the number of arcs (v_i, v_j) with $j < i$; such arcs are called *backward arcs*. Then, it is easy to see that a tournament is acyclic if and only if it admits an ordering with no backward arcs. Several parameters exploiting an ordering with specific properties have been studied in this sense [18] such as the cutwidth. Given an ordering of vertices, for each prefix of the ordering we associate a cut defined as the set of backward arcs with head in the prefix and tail outside of it. Then cutwidth is the minimum value, among all the orderings, of the maximum size of any possible cut w.r.t the ordering (a formal definition is introduced in next section). It is well-known that computing cutwidth is NP-complete [17], and has an $O(\log^2(n))$-approximation on general graphs [23]. Specifically on tournaments, one can compute an optimal ordering for the cutwidth by sorting the degrees according to the in-degrees [16].

In this paper, we propose a new parameter called *degreewidth* using the concept of backward arcs in an ordering of vertices. Degreewidth of a tournament is the minimum value, among all the orderings, of the maximum number of backward arcs incident to a vertex. Hence, an acyclic tournament is a tournament with degreewidth zero. Furthermore, one can notice that tournaments with degreewidth at most one are the same as the *sparse tournaments* introduced in [8,24]. A tournament is *sparse* if there exists an ordering of vertices such that the backward arcs form a matching. It is known that computing a maximum sized arc-disjoint packing of triangles and computing a maximum sized arc-disjoint packing of cycles can be done in polynomial time [7] on sparse tournaments.

To the best of our knowledge this paper is the first to study the parameter degreewidth. As we will see in the next part, although having similarities with the cutwidth, this new parameter differs in certain aspects. We first study structural and computational aspects of degreewidth. Then, we show how it can be used to solve efficiently some classical problems on tournaments.

Our Contributions and Organization of the Paper. Next section provides the formal definition of degreewidth and some preliminary observations. In Sect. 3, we first study the degreewidth of a special class of tournaments, called regular tournaments, of order $2k+1$ and prove they have degreewidth k. We then prove that it is NP-hard to compute the degreewidth in general tournaments. We finally give a 3-approximation algorithm to compute this parameter which is tight in the sense that it cannot produce better than 3-approximation for a class of tournaments.

Then in Sect. 4, we focus on tournaments with degreewidth one, *i.e.*, the sparse tournaments. Note that it is claimed in [8] that there exists a polynomial-time algorithm for finding such ordering, but the only available algorithm appearing in [24, Lemma 35.1, p.97] seems to be incomplete (see discussion Subsect. 4.2). We first define a special class of tournaments that we call U-tournaments. We prove there are only two possible sparse orderings for such tournaments. Then, we give a polynomial time algorithm to decide if a tournament is sparse by carefully decomposing it into U-tournaments.

Finally, in Sect. 5 we study degreewidth as a parameter for some classical graph problems. First, we show an FPT algorithm for DOMINATING SET w.r.t degreewidth. Then, we focus on tournaments with degreewidth one. We design an algorithm running in time $O(n^3)$ to compute a FEEDBACK ARC SET on tournaments on n vertices with degreewidth one. However, we show that FEEDBACK VERTEX SET remains NP-complete on this class of tournaments.

Due to paucity of space the missing proofs are deferred to full version [12].

2   Preliminaries

2.1   Notations

In the following, all the digraphs are simple, that is without self-loop and multiple arcs sharing the same head and tail, and all cycles are directed cycles. The *underlying graph* of a digraph D is an undirected graph obtained by replacing every arc of D by an edge. Furthermore, we use $[n]$ to denote the set $\{1, 2, \ldots, n\}$.

A tournament is a digraph where there is exactly one arc between each pair of vertices. It can alternatively be seen as an orientation of the complete graph. Let T be a tournament with vertex set $\{v_1, \ldots, v_n\}$. We denote $N^+(v)$ the *out-neighbourhood* of a vertex v, that is the set $\{u \mid (v, u) \in A(T)\}$. Then, T being a tournament, the *in-neighbourhood* of the vertex v denoted $N^-(v)$ corresponds to $V(T) \setminus (N^+(v) \cup \{v\})$. The *out-degree* (resp. *in-degree*) of v denoted $d^+(v)$ (resp. $d^-(v)$) is the size of its out-neighbourhood (resp. in-neighbourhood).

A tournament T of order $2k + 1$ is *regular* if for any vertex v, we have $d^+(v) = d^-(v) = k$. Let X be a subset of $V(T)$. We denote by $T - X$ the subtournament induced by the vertices $V(T) \setminus X$. Furthermore, when X contains only one vertex $\{v\}$ we simply write $T - v$ instead of $T - \{v\}$. We also denote by $T[X]$ the tournament induced by the vertices of X. Finally, we say that $T[X]$ *dominates* T if, for every $x \in X$ and every $y \in V(T) \setminus X$, we have $(x, y) \in A(T)$. For more definitions on directed graphs, please refer to [4].

Given a tournament T, we equip the vertices of T with is a strict total order $\prec_\sigma$. This operation also defines an ordering of the set of vertices denoted by $\sigma := \langle v_1, \ldots, v_n \rangle$ such that $v_i \prec_\sigma v_j$ if and only if $i < j$. Given two distinct vertices u and v, if $u \prec_\sigma v$ we say that u is *before* v in σ; otherwise, u is *after* v in σ. Additionally, an arc (u, v) is said to be *forward* (resp. *backward*) if $u \prec_\sigma v$ (resp. $v \prec_\sigma u$). A topological ordering is an ordering without any backward arcs. A tournament that admits a topological ordering does not contain a cycle. Hence, it is said to be *acyclic*.

A *pattern* $p_1 := \langle v_1, \ldots, v_k \rangle$ is a sequence of vertices that are consecutive in an ordering. Furthermore, considering a second pattern $p_2 := \langle u_1, \ldots, u_{k'} \rangle$ where $\{v_1, \ldots, v_k\}$ and $\{u_1, \ldots, u_{k'}\}$ are disjoint, the pattern $\langle p_1, p_2 \rangle$ is defined by $\langle v_1, \ldots, v_k, u_1, \ldots, u_{k'} \rangle$.

Degreewidth. Given a tournament T, an ordering σ of its vertices $V(T)$ and a vertex $v \in V(T)$, we denote $d_\sigma(v)$ to be the number of backward arcs incident to v in σ, that is $d_\sigma(v) := |\{u \mid u \prec_\sigma v, u \in N^+(v)\} \cup \{u \mid v \prec_\sigma u, u \in N^-(v)\}|$. Then, we define the degreewidth of a tournament with respect to the ordering σ, denoted by $\Delta_\sigma(T) := \max\{d_\sigma(v) \mid v \in V(T)\}$. Note that $\Delta_\sigma(T)$ is also the maximum degree of the underlying graph induced by the backward arcs of σ. Finally, we define the degreewidth $\Delta(T)$ of the tournament T as follows.

Definition 1. *The degreewidth of a tournament T, denoted $\Delta(T)$, is defined as $\Delta(T) := \min_{\sigma \in \Sigma(T)} \Delta_\sigma(T)$, where $\Sigma(T)$ is the set of possible orderings for $V(T)$.*

As mentioned before, this new parameter tries to measure how far a tournament is from being acyclic. Indeed, it is easy to see that a tournament T is acyclic if and only if $\Delta(T) = 0$. Additionally, when degreewidth of a tournament is one, it coincides with the notion of sparse tournaments, introduced in [8].

Remark. The definition of degreewidth naturally extends to directed graphs and we hope it will be an exciting parameter for problems on directed graphs. However, in this article we study this as a parameter for tournaments which is well-studied in various domains [2,9,22]. Moreover, degreewidth also gives a succinct representation of a tournament. Informally, sparse graphs[1] are graphs with a low density of edges. Hence, it may be surprising to talk about sparsity in tournaments. However, if a tournament on n vertices admits an ordering σ where the backward arcs form a matching, then it can be encoded by σ and the set of backward arcs (at most $n/2$). Thus, the size of the encoding for such tournament is $O(n)$, instead of $O(n^2)$. For a tournament with degreewidth k, the same reasoning implies that it can be encoded in $O(kn)$ space.

---

[1] Not to be confused with sparse tournaments that has an arc between every pair of vertices, hence, is not a sparse graph.

2.2   Links to Other Parameters

Feedback Arc/Vertex Set. A *feedback arc set* (fas) is a collection of arcs that, when removed from the digraph (or, equivalently, reversed) makes it acyclic. The size of a minimum fas is considered for measuring how far the digraph is from being acyclic. In this context, degreewidth comes as a promising alternative. Finding a small subset of arcs hitting all substructures (in this case, directed cycles) of a digraph is one of the fundamental problems in graph theory. Note that we can easily bound the degreewidth of a tournament by its minimum fas f.

Observation 1. *For any tournament T, we have $\Delta(T) \leq |f|$.*

Note however that the opposite is not true; it is possible to construct tournaments with small degreewidth but large fas, see Fig. 1(a).

(a) Example of a tournament with degreewidth one but fas (resp. fvs) $\frac{|V(T)|}{3}$.

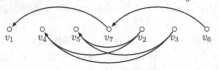

(b) Example of a tournament T with fvs one (v_7) but degreewidth $\frac{|V(T)|-3}{2}$. The topological ordering of $T - v_7$ is $\langle v_1, v_2, v_3, v_4, v_5, v_6 \rangle$.

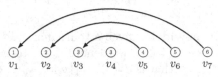

(c) Example of a tournament with degreewidth one but cutwidth $\frac{|V(T)|-1}{2}$. Since the vertices are sorted by increasing in-degrees (values inside the vertices), this is an optimal ordering for the cutwidth.

Fig. 1. Link between degreewidth and other parameters. All the non-depicted arcs are forward.

Similarly, a *feedback vertex set* (fvs) consists of a collection of vertices that, when removed from the digraph makes it acyclic. However, – unlike the feedback arc set – the link between feedback vertex set and degreewidth seems less clear; we can easily construct tournaments with low degreewidth and large fvs (see Fig. 1(a)) as well as large degreewidth and small fvs (see Fig. 1(b)).

Cutwidth. Let us first recall the definition of the cutwidth of a digraph. Given an ordering $\sigma := \langle v_1, \ldots, v_n \rangle$ of the vertices of a digraph D, we say that a prefix of σ is a sequence of consecutive vertices $\langle v_1, \ldots, v_k \rangle$ for some $k \in [n]$. We associate for each prefix of σ a *cut* defined as the set of backward arcs with head in the prefix and tail outside of it. The *width* of the ordering σ is defined as the size of a maximum cut among all the possible prefixes of σ. The cutwidth of D, $ctw(D)$, is the minimum width among all orderings of the vertex set of D.

Intuitively, the difference between the cutwidth and the degreewidth is that the former focuses on the backward arcs going "above" the intervals between the vertices while the latter focuses on the backward arcs coming from and to the vertices themselves. Observe that for any tournament T, the degreewidth is bounded by a function of the cutwidth. Formally, we have the following

Observation 2. *For any tournament T, we have $\Delta(T) \leq 2ctw(T)$.*

Note however that the opposite is not true; it is possible to construct tournaments with small degreewidth but large cutwidth, see Fig. 1(c). We remark that the graph problems that we study parameterized by degreewidth, namely, minimising fas, fvs, and dominating set are FPT w.r.t cutwidth [1,11].

3 Degreewidth

In this section, we present some structural and algorithmical results for the computation of degreewidth. We first introduce the following lemma that provides a lower bound on the degreewith.

Lemma 1. *Let T be a tournament. Then $\Delta(T) \geq \min_{v \in V(T)} d^-(v)$ and $\Delta(T) \geq \min_{v \in V(T)} d^+(v)$.*

3.1 Degreewidth of Regular Tournaments

Theorem 1. *Let T be a regular tournament of order $2k+1$. Then $\Delta(T) = k$. Furthermore, for any ordering σ, by denoting u and v respectively the first and last vertices in σ, we have $d_\sigma(u) = d_\sigma(v) = k$.*

Note that regular tournaments contain many cycles; therefore it is not surprising that their degreewidth is large. This corroborates the idea that this parameter measures how far a tournament is from being acyclic.

3.2 Computational Complexity

We now show that computing the degreewidth of a tournament is NP-hard by defining a reduction from BALANCED 3-SAT(4), proven NP-complete [6] where each clause contains exactly three unique literals and each variable occurs two times positively and two times negatively.

Let φ be a BALANCED 3-SAT(4) formula with m clauses $c_1, \ldots, c_m$ and n variables $x_1, \ldots, x_n$. In the construction, we introduce several regular tournaments of size W or $\frac{W+1}{2} + n + m$, where W is value greater than $n^3 + m^3$. Note that $n + m$ is necessarily odd since $4n = 3m$. By taking a value $W = 3 \mod 4$, we ensure that every regular tournament of size W or $\frac{W+1}{2} + n + m$ has an odd number of vertices.

Construction 1. *Let φ be a BALANCED 3-SAT(4) formula with m clauses $c_1, \ldots, c_m$ clauses and n variables $x_1, \ldots, x_n$ such that n is odd and m is even. Let $W = 3 \mod 4$ be an integer greater than $n^3 + m^3$. We construct a tournament T as follows.*

- *Create two regular tournaments A and D of order $\frac{W+1}{2} + m + n$ such that D dominates A.*
- *Create two regular tournaments B and C of order W such that A dominates $B \cup C$, B dominates C and $B \cup C$ dominates D.*

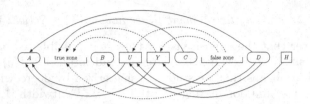

Fig. 2. Example of a nice ordering. A rectangle represents an acyclic tournament, while a rectangle with rounded corners represents a regular tournament. A plain arc between two patterns P and P' represents the fact that there is a backward arc between every pair of vertices $v \in P$ and $v' \in P'$. A dashed arc means some backward arcs may exist between the patterns.

- *Create an acyclic tournament X of order $2n$ with topological ordering $\langle v_1, v_1', \ldots, v_n, v_n' \rangle$ such that $A \cup C$ dominates X and X dominates $B \cup D$.*
- *Create an acyclic tournament Y of order $2m$ with topological ordering $\langle q_1, q_1', \ldots, q_m, q_m' \rangle$ such that $B \cup D$ dominates Y and Y dominates $A \cup C$.*
- *For each clause c_ℓ and each variable x_i of φ,*
  - *if x_i occurs positively in c_ℓ, then $\{v_i, v_i'\}$ dominates $\{q_\ell, q_\ell'\}$,*
  - *if x_i occurs negatively in c_ℓ, then $\{q_\ell, q_\ell'\}$ dominates $\{v_i, v_i'\}$,*
  - *if x_i does not occur in c_ℓ, then introduce the paths (v_i, q_ℓ, v_i') and (v_i', q_ℓ', v_i).*
- *Introduce an acyclic tournament $U = \{u_i^p, \bar{u}_i^p \mid i \leq n, p \leq 2\}$ of order $4n$ such that U dominates $A \cup Y \cup C$ and $B \cup D$ dominates U. For each variable x_i, add the following paths,*
  - *for all variable $x_k \neq x_i$ and all $p \leq 2$, introduce the paths (v_k, u_i^p, v_k') and $(v_k', \bar{u}_i^p, v_k)$,*
  - *introduce the paths (v_i, u_i^1, v_i'), (v_i', u_i^2, v_i), $(v_i, \bar{u}_i^1, v_i')$ and $(v_i', \bar{u}_i^2, v_i)$.*
- *Finally, introduce an acyclic tournament $H = \{h_1, h_2\}$ with topological ordering $\langle h_1, h_2 \rangle$ and such that $A \cup B \cup C \cup X \cup Y \cup D$ dominates H and H dominates U.*

We call a vertex of X a *variable vertex* and a vertex of Y a *clause vertex*. Furthermore, we say that the vertices (v_i, v_i') (resp. (q_ℓ, q_ℓ')) is a *pair of variable vertices* (resp. *pair of clause vertices*).

Definition 2. *Let T be a tournament resulting from Construction 1. An ordering σ of T is nice if:*

- *$\Delta_\sigma(A) = \frac{|A|-1}{2}$, $\Delta_\sigma(B) = \frac{|B|-1}{2}$, $\Delta_\sigma(C) = \frac{|C|-1}{2}$, and $\Delta_\sigma(D) = \frac{|D|-1}{2}$,*
- *σ respects the topological ordering of $U \cup Y$,*
- *$A \prec_\sigma B \prec_\sigma U \prec_\sigma Y \prec_\sigma C \prec_\sigma D \prec_\sigma H$, and*
- *for any variable x_i, either $A \prec_\sigma v_i \prec_\sigma v_i' \prec_\sigma B$ or $C \prec_\sigma v_i \prec_\sigma v_i' \prec_\sigma D$.*

An example of a nice ordering is depicted in Fig. 2. Let σ be a nice ordering, we call the pattern corresponding to the vertices between A and B, the *true zone* and the pattern after the vertices of C the *false zone*. Let (q_ℓ, q_ℓ') be a pair of clause vertices and let (v_i, v_i') be a pair of variable vertices such that

Fig. 3. Example of a tournament where the approximate algorithm can return an ordering σ_{app} (on the left) with degreewidth three while the optimal solution is one in σ_{opt} (on the right). Coloured vertices are the ones incident to the maximum number of backward arcs. all non-depicted arcs are forward arcs.

x_i occurs positively (resp. negatively) in c_ℓ in φ. We say that the pair (v_i, v_i') satisfies (q_ℓ, q_ℓ') if v_i and v_i' both belong to the true zone (resp. false zone). Note that there is no backward arc between $\{q_\ell, q_\ell'\}$ and $\{v_i, v_i'\}$ if and only if (v_i, v_i') satisfies (q_ℓ, q_ℓ'). Notice also that for any pair of variable vertices (v_i, v_i') such that x_i does not appear in c_ℓ and (v_i, v_i') is either in the true zone or in the false zone, then there is exactly two backward arcs between $\{q_\ell, q_\ell'\}$ and $\{v_j, v_j'\}$.

Let φ be an instance of BALANCED 3-SAT(4) and T its tournament resulting from Construction 1. We show that φ is satisfiable if and only if there exists an ordering σ of T such that $\Delta_\sigma(T) < W + 2m + 3n + 4$, which yields the following.

Theorem 2. *Given a tournament T and an integer k, it is NP-complete to compute an ordering σ of T such that $\Delta_\sigma(T) \leq k$.*

3.3 An Approximation Algorithm to Compute Degreewidth

In this subsection, we prove that sorting the vertices by increasing in-degree is a tight 3-approximation algorithm to compute the degreewidth of a tournament. Intuitively, the reasons why it returns a solution not too far from the optimal are twofold. Firstly, observe that the only optimal ordering for acyclic tournaments (*i.e.* with degreewidth 0) is an ordering with increasing in-degrees. Secondly, this strategy also gives an optimal solution for cutwidth in tournaments.

Theorem 3. *Ordering the vertices by increasing order of in-degree is a tight 3-approximation algorithm to compute the degreewidth of a tournament (see Fig. 3).*

4 Results on Sparse Tournaments

In this section, we focus on tournaments with degreewidth one, called sparse tournaments. The main result of this section is that unlike in the general case, it is possible to compute in polynomial time a sparse ordering of a tournament (if it exists). We begin with an observation about sparse orderings (if it exists).

Lemma 2. *Let T be a sparse tournament of order $n > 4$ and σ be an ordering of its vertices. If σ is a sparse ordering, then for any vertex v such that $d^-(v) = i$, the only possible positions of v in σ are $\{i, i + 1, i + 2\} \cap [n]$.*

Note that Lemma 2 gives immediately an exponential running-time algorithm to decide if a tournament is sparse. However, we give in Subsect. 4.2 a polynomial running-time algorithm for this problem. Before that we study a useful subclass of sparse tournaments, we call the U-tournaments.

4.1 U-Tournaments

In this subsection, we study one specific type of tournaments called U-tournaments. Informally, they correspond to the acyclic tournaments where we reversed all the arcs of its Hamiltonian path.

Definition 3. *For any integer $n \geq 1$, we define U_n as the tournament on n vertices with $V(U_n) = \{v_1, v_2, \ldots, v_n\}$ and $A(U_n) = \{(v_{i+1}, v_i) \mid \forall i \in [n-1]\} \cup \{(v_i, v_j) \mid 1 \leq i < n, i+1 < j \leq n\}$. We say that a tournament of order n is a U-tournament if it is isomorphic to U_n.*

Figures 4(a) and 4(d) depict respectively the tournaments U_7 and U_8. This family of tournaments seems somehow strongly related to sparse tournaments and the following results will be useful later for both the polynomial algorithm to decide if a tournament is sparse and the polynomial algorithm for minimum feedback arc set in sparse tournaments. To do so, we prove that each U-tournament of order $n > 4$ has exactly two sparse orderings of its vertices that we formally define.

Definition 4. *Let $P(k) = \langle v_{k+1}, v_k \rangle$ be a pattern of two vertices of U_n for some integer $k \in [n-1]$. For any integer $n \geq 2$, we define the following special orderings of U_n:*

- *if n is even:*
  - *$\Pi(U_n)$ is the ordering given by $\langle v_1, P(2), P(4), \ldots, P(n-2), v_n \rangle$.*
  - *$\Pi_{1,n}(U_n)$ is the ordering given by $\langle P(1), P(3), \ldots, P(n-2), P(n) \rangle$.*
- *if n is odd:*
  - *$\Pi_1(U_n)$ is the ordering given by $\langle P(1), P(3), \ldots, P(n-2), v_n \rangle$.*
  - *$\Pi_n(U_n)$ is the ordering given by $\langle v_1, P(2), P(4), \ldots, P(n-3), P(n-1) \rangle$.*

Figures 4(b) and 4(c) (and Figs. 4(e) and 4(f)) depict the orderings $\Pi_1(U_7)$ and $\Pi_7(U_7)$ (resp. $\Pi(U_8)$ and $\Pi_{1,8}(U_8)$) of the tournament U_7 (resp. U_8). One can notice that these orderings are sparse and the subscript of Π indicates the vertex (or vertices) without a backward arc incident to it in this ordering. In the following, we prove that when $n > 4$ there are no other sparse orderings of U_n. However, note that there are three possible sparse orderings of U_3 (namely, $\Pi_1(U_3)$ and $\Pi_3(U_3)$ defined previously, as well as $\Pi_2(U_3) := \langle v_3, v_2, v_1 \rangle$) and three sparse orderings of U_4 (namely, $\Pi(U_4)$, $\Pi_{1,4}(U_4)$ as defined before, and $\Pi'(U_4) := \langle v_2, v_4, v_1, v_3 \rangle$).

Theorem 4. *For each integer $n > 4$ there are exactly two sparse orderings of U_n. Specifically, if n is even, these two sparse orderings are $\Pi(U_n)$ and $\Pi_{1,n}(U_n)$; otherwise, the two sparse orderings are $\Pi_1(U_n)$ and $\Pi_n(U_n)$.*

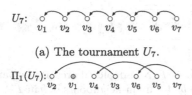

(a) The tournament U_7.

$\Pi_1(U_7)$: <image with v_2 v_1 v_4 v_3 v_6 v_5 v_7>

(b) The sparse ordering $\Pi_1(U_7)$. Note that v_1 is the only vertex not incident to any backward arc.

$\Pi_7(U_7)$: <image with v_1 v_3 v_2 v_5 v_4 v_7 v_6>

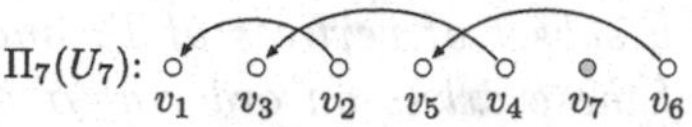

(c) The sparse ordering $\Pi_7(U_7)$. Note that v_7 is the only vertex not incident to any backward arc.

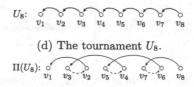

(d) The tournament U_8.

$\Pi(U_8)$: <image with v_1 v_3 v_2 v_5 v_4 v_7 v_6 v_8>

(e) The sparse ordering $\Pi(U_8)$. The dashed forward arcs is a minimum feedback arc set of the tournament. Note that all the vertices are incident to one backward arc.

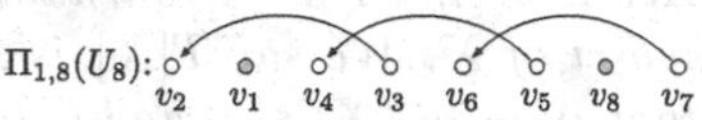

$\Pi_{1,8}(U_8)$: <image with v_2 v_1 v_4 v_3 v_6 v_5 v_8 v_7>

(f) The sparse ordering $\Pi_{1,8}(U_8)$. Note that v_1 and v_8 are the only vertices not incident to any backward arc.

Fig. 4. The tournaments U_7 and U_8 and their sparse orderings. The non-depicted arcs are forward arcs.

4.2 A Polynomial Time Algorithm for Sparse Tournaments

We give here a polynomial algorithm to compute a sparse ordering of a tournament (if any). First of all, let us recall a classical algorithm to compute a topological ordering of a tournament (if any): we look for the vertex v with the smallest in-degree; if v has in-degree one or more, we have a certificate that the tournament is not acyclic. Otherwise, we add v at the beginning of the ordering, and we repeat the reasoning on $T - v$, until $V(T)$ is empty.

The idea of the original "proof" in [24, Lemma 35.1, p.97] was similar: considering the set of vertices X of smallest in-degrees, put X at the beginning of the ordering, and remove X from the tournament. However, potential backward arcs from the remaining vertices of $V \setminus X$ to X may have been forgotten. For example, consider a tournament over 9 vertices consisting of a U_5 (with vertex set $\{v_1, \ldots, v_5\}$) that dominates a U_4 (with vertex set $\{u_1, \ldots, u_4\}$) except for the backward arc (u_4, v_5). It is sparse $(\langle \Pi_5(U_5), \Pi_{1,4}(U_4) \rangle)$ but the algorithm returns the (non-sparse) ordering $\langle \Pi_1(U_5), \Pi_{1,4}(U_4) \rangle$ (v_5 is incident to two backward arcs). The problem is that this algorithm is too "local"; it will always prefer the sparse ordering $\Pi_1(U_{2k+1})$ over $\Pi_{2k+1}(U_{2k+1})$, but it may be necessary to take the latter. Therefore, to correct this, we needed a much more involved algorithm, requiring the study of the U-tournaments and the notion of quasi-domination (see Definition 6). Indeed, unlike the algorithm for the topological ordering, we may have to look more carefully how the vertices with low in-degrees are connected to the rest of the digraph. These correspond to the case where there exists a U-sub-tournament of T which either dominates or "quasi-dominates" (see Definition 6) the tournament T. Because of the latter possibility (where a backward arc (a, b) is forced to appear), we need to look for specific sparse orderings, called M-sparse orderings (where a or b should not be end-

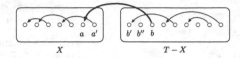

Fig. 5. An example where X (b, a)-quasi-dominates T. Non-depicted arcs are forward. The vertex a' is an out-neighbour of a in X, and b', b'' are in-neighbours of b in $T - X$.

vertices of other backward arcs). As all the sparse orderings for U-tournaments have been described, we can derive a recursive algorithm.

Definition 5. *Let T be a tournament, X be a subset of vertices of T, and M be a subset of X. We say $T[X]$ is M-sparse if there exists an ordering σ of X such that $\Delta_{\sigma(T[X])}(X) \leq 1$ and $d_\sigma(v) = 0$ for all $v \in M$. In that case, σ is said to be an M-sparse ordering of $T[X]$.*

For example, $U_4[\{v_1, v_2, v_3\}]$ is $\{v_2\}$-sparse, because there exists a sparse ordering $\sigma := \langle v_3, v_2, v_1 \rangle$ of $U_4[\{v_1, v_2, v_3\}]$ such that $d_\sigma(v_2) = 0$. We remark that T is sparse if and only if T is $\emptyset$-sparse. In fact, the algorithm described in this section computes a $\emptyset$-sparse ordering of the given tournament (if any).

Definition 6 (see Fig. 5). *Given a tournament T and two of its vertices a and b, we say that a subset of vertices X quasi-dominates T if:*

- *there exists an arc $(b, a) \in A(T)$ such that $a \in X$ and $b \notin X$,*
- *$(u, v) \in A(T)$ for every $(u, v) \in (X \times (V(T) \setminus X)) \setminus \{(a, b)\}$,*
- *$d^-(b) \geq |X| + 1$, and*
- *the vertex a has an out-neighbour in X.*

In this case, we also say X (b, a)-quasi-dominates T.

We can create the algorithm `isUkMsparse` which given $(v_1, \ldots, v_k)$ a U-tournament and M a subset of these vertices, returns a boolean which is True if and only if this tournament is M-sparse. We can also create the algorithm `getUsubtournament` which given T a tournament, and $X = (u_1, \ldots, u_k)$ a list of vertices such that $d^-(u_1) = 1$ and $d^-(u_i) = i - 1$ and $(u_i, u_{i-1}) \in A(T)$ for all $i \in \{2, \ldots, k\}$, returns a U-subtournament dominating or quasi-dominating T. With these two previous algorithms, we can derive Algorithm 3 `isMsparse`.

Algorithm 1: getUsubtournament

Data: T a tournament, and $X = (u_1, \ldots, u_k)$ a list of vertices such that $d^-(u_1) = 1$ and
    $d^-(u_i) = i - 1$ and $(u_i, u_{i-1}) \in A(T)$ for all $i \in \{2, \ldots, k\}$.
Result: A U-subtournament dominating or quasi-dominating T.
1   $w \longleftarrow$ a vertex of $N^-(u_k) \setminus X$;
2   **if** $d^-(w) = d^-(u_k)$ **then return** $X \cup \{w\}$ /* this set dominates T */ ;
3   **else if** $d^-(w) = d^-(u_k) + 1$ **then return** getUsubtournament$(T, X \cup \{w\})$;
4   **else return** X /* this set (w, u_k)-quasi-dominates T */ ;

Algorithm 2: isUkMsparse

Data: $(v_1, \ldots, v_k)$ a U_k tournament, M a subset of the vertices of U_k
Result: True if U_k is M-sparse and False otherwise
1   **if** $k \leq 2$ **then return** *True* ;
2   **else if** $k = 3$ **then return** $|M| \leq 1$;
3   **else if** k *is even* **then return** $|M \setminus \{v_1, v_k\}| = 0$;
4   **else if** k *is odd* **then return** $(v_1 \notin M$ *or* $v_k \notin M)$ *and* $|M \setminus \{v_1, v_k\}| = 0$;

Algorithm 3: isMsparse

Data: T a tournament, M a subset of the vertices of T
Result: True if T is M-sparse and False otherwise

1   **if** $|V(T)| \leq 1$ **then** **return** *True* ;
2   **else if** $\min_{v \in V(T)} d^-(v) \geq 2$ **then** **return** *False* ;
3   **else if** $\min_{v \in V(T)} d^-(v) = 0$ **then**
4      $v \longleftarrow$ the vertex of in-degree 0;
5   **return** isMsparse$(T - v, M \setminus \{v\})$;
6   **else if** $|\{v \in V(T) : d^-(v) = 1\}| = 1$ **then**
7      $v, w \longleftarrow$ two vertices such that $d^-(v) = 1$ and $(w, v) \in A(T)$;
8   **return** $v \notin M$ *and* isMsparse$(T - v, (M \cup \{w\}) \setminus \{v\})$;
9   **else**
10      $v, w \longleftarrow$ two vertices of in-degree 1 such that $(w, v) \in A(T)$;
11      $X \longleftarrow$ getUsubtournament$(T, (v, w))$;
12   **if** X *dominates* T **then**
13   **return** *(*isUkMsparse$(X, M \cap X)$ *and* isMsparse$(T - X, M \setminus X))$;
14   **else**
15         $a, b \longleftarrow$ the vertices such that X (b, a)-quasi-dominates T;
16   **return** *(*isUkMsparse$(X, (M \cup \{a\}) \cap X)$ *and* isMsparse$(T - X, (M \cup \{b\}) \setminus X))$;

Theorem 5. *Algorithm 3 is correct. Hence, it is possible to decide if a tournament T with n vertices is sparse in $\mathcal{O}(n^3)$ by calling* isMsparse*$(T, \emptyset)$.*

Observe that we can easily modify Algorithm 3 to obtain a sparse ordering (if exists). Next corollary follows from the above algorithm.

Corollary 1. *The vertex set of a sparse tournament on n vertices can be decomposed into a sequence $U_{n_1}, U_{n_2}, \ldots, U_{n_\ell}$ for some $\ell \leq n$ such that each $T[U_{n_i}]$ dominates or quasi-dominates $T[\bigcup_{i < j \leq \ell} U_{n_j}]$ and $\sum_{i \in [\ell]} n_i = n$.*

5   Degreewidth as a Parameter

5.1   Dominating Set Parameterized by Degreewidth

A set of vertices X of a directed graph G is a *dominating set (DS)* if for each vertex $v \in V(G) \setminus X$, we have $N^+(v) \cap X \neq \emptyset$. Observe that in graphs where degreewidth is zero, DS is of size one. Similarly, for tournaments with degreewidth equals to one, the DS is of size at most two. That is, we have trivial solutions for DS for acyclic and sparse tournaments. This motivates us to look for FPT algorithm parameterized by degreewidth. In the following, we develop an FPT algorithm for DOMINATING SET using universal families. Before that we observe that size of a dominating is always bounded by the size of degreewidth.

Observation 3. *The size of a minimum dominating set of a tournament T is at most $\Delta(T) + 1$.*

Theorem 6. DOMINATING SET *is FPT in tournaments with respect to degreewidth.*

5.2   FAST and FVST in Sparse Tournaments

A *forbidden pattern* corresponds to the patterns $\Pi(U_{2k})$ for any $k \geq 1$ as well as $\Pi'(U_4) := \langle v_2, v_4, v_1, v_3 \rangle$. An example of the forbidden pattern $\Pi(U_8)$ is depicted in Fig. 4(e). We say a sparse ordering has *forbidden pattern* if a contiguous subsequence of the ordering is a forbidden pattern. Intuitively, the problem of such patterns is that the set of their backward arcs is not a minimum fas. Hopefully, we can use Theorem 4 in such a way that if the pattern $\Pi(U_{2k})$ appears, we can restructure it into $\Pi_{1,2k}(U_{2k})$.

If a sparse ordering does not contain a forbidden pattern then its set of backward arcs is a fas. Hence, we obtain the following result.

Theorem 7. *FAST is solvable in time $O(n^3)$ in sparse tournaments on n vertices.*

For FVST, we show that the problem is difficult to solve on sparse tournaments.

Theorem 8. *FVST is NP-complete on sparse tournaments.*

6   Conclusion

In this paper, we studied a new parameter for tournaments, called degreewidth. We showed that it is NP-hard to decide if degreewidth is at most k, for some natural number k and we proceeded to design a 3-approximation for the degreewidth. One may ask if there is a PTAS for this problem. Then, we investigated sparse tournaments, *i.e.*, tournaments with degreewidth one and developed a polynomial time algorithm to compute a sparse ordering. Is it possible to generalise this result by providing an FPT algorithm to compute the degreewidth? We also showed that FAST can be solved in polynomial time in sparse tournaments, matching with the known result that ARC-DISJOINT TRIANGLES PACKING and ARC-DISJOINT CYCLE PACKING are both polynomial in sparse tournaments [7]. Therefore, the question arise: can this parameter be used to provide an FPT algorithm for FAST in the general case?

Furthermore, we showed an FPT algorithm for DS w.r.t degreewidth. Are there other domination problems e.g., perfect code, partial dominating set, or connected dominating set that is FPT w.r.t degreewidth? Lastly, we also can wonder if this parameter is useful for general digraphs.

Acknowledgements. We would like to thank Frédéric Havet for pointing us a counter-example to the polynomial running-time algorithm in [24, Lemma 35.1, p. 97].

References

1. Alber, J., Bodlaender, H.L., Fernau, H., Kloks, T., Niedermeier, R.: Fixed parameter algorithms for dominating set and related problems on planar graphs. Algorithmica **33**(4), 461–493 (2002)

2. Allesina, S., Levine, J.M.: A competitive network theory of species diversity. Proc. Natl. Acad. Sci. **108**(14), 5638–5642 (2011)
3. Alon, N.: Ranking tournaments. SIAM J. Discret. Math. **20**(1), 137–142 (2006). https://doi.org/10.1137/050623905
4. Bang-Jensen, J., Gutin, G.Z.: Digraphs - Theory, Algorithms and Applications. Springer Monographs in Mathematics, 2nd edn. Springer, Heidelberg (2009). https://doi.org/10.1007/978-1-84800-998-1
5. Bar-Yehuda, R., Geiger, D., Naor, J., Roth, R.M.: Approximation algorithms for the feedback vertex set problem with applications to constraint satisfaction and Bayesian inference. SIAM J. Comput. **27**(4), 942–959 (1998). https://doi.org/10.1137/S0097539796305109
6. Berman, P., Karpinski, M., Scott, A.D.: Approximation hardness of short symmetric instances of MAX-3SAT. Electron. Colloquium Comput. Complex. (049) (2003). http://eccc.hpi-web.de/eccc-reports/2003/TR03-049/index.html
7. Bessy, S., et al.: Packing arc-disjoint cycles in tournaments. In: Rossmanith, P., Heggernes, P., Katoen, J. (eds.) 44th International Symposium on Mathematical Foundations of Computer Science, MFCS 2019, 26–30 August 2019, Aachen, Germany. LIPIcs, vol. 138, pp. 27:1–27:14. Schloss Dagstuhl - Leibniz-Zentrum für Informatik (2019). https://doi.org/10.4230/LIPIcs.MFCS.2019.27
8. Bessy, S., Bougeret, M., Thiebaut, J.: Triangle packing in (sparse) tournaments: approximation and kernelization. In: Pruhs, K., Sohler, C. (eds.) 25th Annual European Symposium on Algorithms, ESA 2017, 4–6 September 2017, Vienna, Austria. LIPIcs, vol. 87, pp. 14:1–14:13. Schloss Dagstuhl - Leibniz-Zentrum für Informatik (2017). https://doi.org/10.4230/LIPIcs.ESA.2017.14
9. Brandt, F., Fischer, F.: PageRank as a weak tournament solution. In: Deng, X., Graham, F.C. (eds.) WINE 2007. LNCS, vol. 4858, pp. 300–305. Springer, Heidelberg (2007). https://doi.org/10.1007/978-3-540-77105-0_30
10. Charbit, P., Thomassé, S., Yeo, A.: The minimum feedback arc set problem is NP-hard for tournaments. Comb. Probab. Comput. **16**(1), 1–4 (2007). https://doi.org/10.1017/S0963548306007887
11. Chen, J., Liu, Y., Lu, S., O'sullivan, B., Razgon, I.: A fixed-parameter algorithm for the directed feedback vertex set problem. In: Proceedings of the Fortieth Annual ACM Symposium on Theory of Computing, pp. 177–186 (2008)
12. Davot, T., Isenmann, L., Roy, S., Thiebaut, J.: DegreeWidth: a new parameter for solving problems on tournaments. CoRR abs/2212.06007 (2022). https://doi.org/10.48550/arXiv.2212.06007
13. Dechter, R.: Enhancement schemes for constraint processing: backjumping, learning, and cutset decomposition. Artif. Intell. **41**(3), 273–312 (1990). https://doi.org/10.1016/0004-3702(90)90046-3
14. Downey, R.G., Fellows, M.R.: Parameterized computational feasibility. In: Clote, P., Remmel, J.B. (eds.) Feasible Mathematics II. Progress in Computer Science and Applied Logic, vol. 13, pp. 219–244. Springer, Boston (1995). https://doi.org/10.1007/978-1-4612-2566-9_7
15. Feige, U.: Faster fast (feedback arc set in tournaments). CoRR abs/0911.5094 (2009). http://arxiv.org/abs/0911.5094
16. Fradkin, A.O.: Forbidden structures and algorithms in graphs and digraphs. Ph.D. thesis, USA (2011). aAI3463323
17. Gavril, F.: Some NP-complete problems on graphs. In: Proceedings of the 11th Conference on Information Sciences and Systems. Johns Hopkins University, Baltimore (1977)

18. Gurski, F., Rehs, C.: Comparing linear width parameters for directed graphs. Theory Comput. Syst. **63**(6), 1358–1387 (2019). https://doi.org/10.1007/s00224-019-09919-x

19. Johnson, D.B.: Finding all the elementary circuits of a directed graph. SIAM J. Comput. **4**(1), 77–84 (1975). https://doi.org/10.1137/0204007

20. Karpinski, M., Schudy, W.: Faster algorithms for feedback arc set tournament, Kemeny rank aggregation and betweenness tournament. In: Cheong, O., Chwa, K.-Y., Park, K. (eds.) ISAAC 2010. LNCS, vol. 6506, pp. 3–14. Springer, Heidelberg (2010). https://doi.org/10.1007/978-3-642-17517-6_3

21. Kenyon-Mathieu, C., Schudy, W.: How to rank with few errors. In: Johnson, D.S., Feige, U. (eds.) Proceedings of the 39th Annual ACM Symposium on Theory of Computing, San Diego, California, USA, 11–13 June 2007, pp. 95–103. ACM (2007). https://doi.org/10.1145/1250790.1250806

22. Laslier, J.F.: Tournament Solutions and Majority Voting, vol. 7. Springer, Heidelberg (1997)

23. Leighton, T., Rao, S.: Multicommodity max-flow min-cut theorems and their use in designing approximation algorithms. J. ACM (JACM) **46**(6), 787–832 (1999)

24. Thiebaut, J.: Algorithmic and structural results on directed cycles in dense digraphs. (Résultats algorithmiques et structurels sur les cycles orientés dans les digraphes denses). Ph.D. thesis, University of Montpellier, France (2019). https://tel.archives-ouvertes.fr/tel-02491420

25. van Zuylen, A., Williamson, D.P.: Deterministic pivoting algorithms for constrained ranking and clustering problems. Math. Oper. Res. **34**(3), 594–620 (2009). https://doi.org/10.1287/moor.1090.0385

Approximating Bin Packing with Conflict Graphs via Maximization Techniques

Ilan Doron-Arad and Hadas Shachnai[✉]

Computer Science Department, Technion, Haifa 3200003, Israel
{idoron-arad,hadas}@cs.technion.ac.il

Abstract. We give a comprehensive study of *bin packing with conflicts* (BPC). The input is a set I of items, sizes $s : I \rightarrow [0, 1]$, and a conflict graph $G = (I, E)$. The goal is to find a partition of I into a minimum number of independent sets, each of total size at most 1. Being a generalization of the notoriously hard graph coloring problem, BPC has been studied mostly on polynomially colorable conflict graphs. An intriguing open question is whether BPC on such graphs admits the same best known approximation guarantees as classic bin packing.

We answer this question negatively, by showing that (in contrast to bin packing) there is no asymptotic polynomial-time approximation scheme (APTAS) for BPC already on seemingly easy graph classes, such as *bipartite* and *split* graphs. We complement this result with improved approximation guarantees for BPC on several prominent graph classes. Most notably, we derive an asymptotic 1.391-approximation for bipartite graphs, a 2.445-approximation for perfect graphs, and a $\left(1 + \frac{2}{e}\right)$-approximation for split graphs. To this end, we introduce a generic framework relying on a novel interpretation of BPC allowing us to solve the problem via *maximization* techniques. Our framework may find use in tackling BPC on other graph classes arising in applications.

1 Introduction

We study the *bin packing with conflicts (BPC)* problem. We are given a set I of n items, sizes $s : I \rightarrow [0, 1]$, and a conflict graph $G = (I, E)$ on the items. A *packing* is a partition $(A_1, \ldots, A_t)$ of I into independent sets called *bins*, such that for all $b \in \{1, \ldots, t\}$ it holds that $s(A_b) = \sum_{\ell \in A_b} s(\ell) \leq 1$. The goal is to find a packing in a minimum number of bins. Let $\mathcal{I} = (I, s, E)$ denote a BPC instance. We note that BPC is a generalization of *bin packing (BP)* (where $E = \emptyset$) as well as the graph coloring problem (where $s(\ell) = 0 \ \forall \ell \in I$).[1] BPC captures many real-world scenarios such as resource clustering in parallel computing [2], examination scheduling [21], database storage [16], and product delivery [3]. As the special case of graph coloring cannot be approximated within a ratio better than $n^{1-\varepsilon}$ [28], most of the research work on BPC has focused on families of

---

[1] See the formal definitions of *graph coloring* and *independent sets* in Sect. 2.

A full version of the paper is available in [6].

conflict graphs which can be optimally colored in polynomial time [4,5,8,15–17,22,23].

Let OPT = OPT($\mathcal{I}$) be the value of an optimal solution for an instance $\mathcal{I}$ of a minimization problem $\mathcal{P}$. As in the bin packing problem, we distinguish between *absolute* and *asymptotic* approximation. For $\alpha \geq 1$, we say that $\mathcal{A}$ is an absolute α-approximation algorithm for $\mathcal{P}$ if for any instance $\mathcal{I}$ of $\mathcal{P}$ we have $\mathcal{A}(\mathcal{I})/\text{OPT}(\mathcal{I}) \leq \alpha$, where $\mathcal{A}(\mathcal{I})$ is the value of the solution returned by $\mathcal{A}$. Algorithm $\mathcal{A}$ is an *asymptotic* α-approximation algorithm for $\mathcal{P}$ if for any instance $\mathcal{I}$ it holds that $\mathcal{A}(\mathcal{I}) \leq \alpha\text{OPT}(\mathcal{I}) + o(\text{OPT}(\mathcal{I}))$. An APTAS is a family of algorithms $\{\mathcal{A}_\varepsilon\}$ such that, for every $\varepsilon > 0$, $\mathcal{A}_\varepsilon$ is a polynomial time asymptotic $(1+\varepsilon)$-approximation algorithm for $\mathcal{P}$. An *asymptotic fully polynomial-time approximation scheme (AFPTAS)* is an APTAS $\{\mathcal{A}_\varepsilon\}$ such that $\mathcal{A}_\varepsilon(\mathcal{I})$ runs in time poly($|\mathcal{I}|, \frac{1}{\varepsilon}$), where $|\mathcal{I}|$ is the encoding length of the instance $\mathcal{I}$.

It is well known that, unless P=NP, BP cannot be approximated within ratio better than $\frac{3}{2}$ [10]. This ratio is achieved by First-Fit Decreasing (FFD) [26].[2] Also, BP admits an AFPTAS [19], and an additive approximation algorithm which packs any instance $\mathcal{I}$ in at most OPT($\mathcal{I}$) + $O(\log(\text{OPT}(\mathcal{I})))$ bins [14]. Despite the wide interest in BPC on polynomially colorable graphs, the intriguing question whether BPC on such graphs admits the same best known approximation guarantees as classic bin packing remained open.

Table 1. Known results for Bin Packing with Conflict Graphs

	Absolute		Asymptotic	
	Lower Bound	Upper Bound	Lower Bound	Upper Bound
General graphs	$n^{1-\varepsilon}$ [28]	$O\left(\frac{n(\log\log n)^2}{(\log n)^3}\right)$ [13]	$n^{1-\varepsilon}$ [28]	$O\left(\frac{n(\log\log n)^2}{(\log n)^3}\right)$ [13]
Perfect graphs	·	**2.445** (2.5 [8])	**c > 1**	**2.445** (2.5 [8])
Chordal graphs	·	$\frac{7}{3}$ [8]	**c > 1**	$\frac{7}{3}$ [8]
Cluster graphs	·	2 [1]		1 [5]
Cluster complement	·	**3/2**	**3/2**	**3/2**
Split graphs	·	**1 + 2/e** (2 [15])	**c > 1**	**1 + 2/e** (2 [15])
Bipartite graphs	·	$\frac{5}{3}$ [15]	**c > 1**	**1.391** ($\frac{5}{3}$ [15])
Partial k-trees	·	$2 + \varepsilon$ [17]		1 [16]
Trees	·	$\frac{5}{3}$ [15]		·
No conflicts	$\frac{3}{2}$ [10]	$\frac{3}{2}$ [26]		1 [25]

We answer this question negatively, by showing that (in contrast to bin packing) there is no APTAS for BPC even on seemingly easy graph classes, such as *bipartite* and *split* graphs. We complement this result with improved approximation guarantees for BPC on several prominent graph classes. For BPC on bipartite graphs, we obtain an asymptotic 1.391-approximation. We further derive improved bounds of 2.445 for perfect graphs, $\left(1 + \frac{2}{e}\right)$ for split graphs, and $\frac{5}{3}$ for

---

[2] We give a detailed description of Algorithm FFD in [6].

bipartite graphs.[3] Finally, we obtain a tight $\frac{3}{2}$-asymptotic lower bound and an absolute $\frac{3}{2}$-upper bound for graphs that are the complements of cluster graphs (we call these graphs below *complete multi-partite*).

Table 1 summarizes the known results for BPC on various classes of graphs. New bounds given in this paper are shown in boldface. Entries that are marked with · follow by inference, either by using containment of graph classes (trees are partial k-trees), or since the hardness of BPC on all considered graph classes follows from the hardness of classic BP. Empty entries for lower bounds follow from tight upper bounds.

1.1 Related Work

The BPC problem was introduced by Jansen and Öhring [17]. They presented a general algorithm that initially finds a coloring of the conflict graph, and then packs each color class separately using the First-Fit Decreasing algorithm. This approach yields a 2.7-approximation for BPC on perfect graph. The paper [17] includes also a 2.5-approximation for subclasses of perfect graphs on which the corresponding *precoloring extension problem* can be solved in polynomial time (e.g., interval and chordal graphs). The authors present also a $(2 + \varepsilon)$-approximation algorithm for BPC on cographs and partial k-trees.

Epstein and Levin [8] present better algorithms for BPC on perfect graphs (2.5-approximation), graphs on which the precoloring extension problem can be solved in polynomial time ($\frac{7}{3}$-approximation), and bipartite graphs ($\frac{7}{4}$-approximation). Their techniques include matching between *large* items and a sophisticated use of new item *weights*. Recently, Huang et al. [15] provided fresh insights to previous algorithms, leading to $\frac{5}{3}$-approximation for BPC on bipartite graphs and a 2-approximation on split graphs.

Jansen [16] presented an AFPTAS for BPC on *d-inductive* conflict graphs, where $d \geq 1$ is some constant. This graph family includes trees, grid graphs, planar graphs, and graphs with constant treewidth. For a survey of *exact* algorithms for BPC see, e.g., [15].

1.2 Techniques

There are several known approaches for tackling BPC instances. One celebrated technique introduced by Jansen and Öhring [17] relies on finding initially a minimum coloring of the given conflict graph, and then packing each color class using a bin packing heuristic, such as First-Fit Decreasing. A notable generalization of this approach is the sophisticated integration of *precoloring extension* [8,17], which completes an initial partial coloring of the conflict graph, with no increase to the number of color classes. Another elegant technique is a matching-based algorithm, applied by Epstein and Levin [8] and by Huang et al. [15].

---

[3] Recently, Huang et al. [15] obtained a $\frac{5}{3}$-approximation for bipartite graphs, simultaneously and independently of our work. We note that the techniques of [15] are different than ours, and their algorithm is more efficient in terms of running time.

The best known algorithms (prior to this work), e.g., for perfect graphs [8] and split graphs [15] are based on the above techniques. While the analyses of these algorithms are tight, the approximation guarantees do not match the existing lower bounds for BPC on these graph classes; thus, obtaining improved approximations requires new techniques.

In this paper we present a novel point of view of BPC involving the solution of a maximization problem as a subroutine. We first find an *initial packing* of a subset $S \subseteq I$ of items, which serves as a baseline packing with *high potential* for adding items (from $I \setminus S$) without increasing the number of bins used. The remaining items are then assigned to extra bins using a simple heuristic. Thus, given a BPC instance, our framework consists of the following main steps.

1. Find an initial packing $\mathcal{A} = (A_1, \ldots, A_m)$ of high potential for $S \subseteq I$.
2. Maximize the total size of items in $\mathcal{A}$ by adding items in $I \setminus S$.
3. Assign the remaining (unpacked) items to extra bins using a greedy approach respecting the conflict graph constraints.

The above generic framework reduces BPC to cleverly finding an initial packing of high potential, and then efficiently approximating the corresponding maximization problem, while exploiting structural properties of the given conflict graph. One may view classic approaches for solving BP (e.g., [20]), as an application of this technique: find an initial packing of high potential containing the *large* items; then add the *small* items using First-Fit. In this setting, the tricky part is to find an initial high potential packing, while adding the small items is trivial. However, in the presence of a conflict graph, solving the induced maximization problem is much more challenging.

Interestingly, we are able to obtain initial packings of high potential for BPC on several conflict graph classes. To solve the maximization problem, we first derive efficient approximation for maximizing the total size of items within a *single* bin. Our algorithm is based on finding a maximum weight independent set of *bounded* total size in the graph, combined with enumeration over items of large sizes. Using the single bin algorithm, the maximization problem is solved via application of the *separable assignment problem (SAP)* [9] framework, adapted to our setting. Combined with a hybrid of several techniques (to efficiently handle different types of instances) this leads to improved bounds for BPC on perfect, split, and bipartite graphs (see Sects. 3, 4, and the full version of the paper [6]). Our framework may find use in tackling BPC on other graph classes arising in applications.

1.3   Organization

In Sect. 2 we give some definitions and preliminary results. Section 3 presents an approximation algorithm for BPC on perfect graphs and an asymptotic approximation on bipartite graphs. In Sect. 4 we give an algorithm for split graphs. We present our hardness results in Sect. 5 and conclude in Sect. 6. Due to space constraints, some of our results and proofs are given in the full version of the paper [6].

2 Preliminaries

For any $k \in \mathbb{R}$, let $[k] = \{1, 2, \ldots, \lfloor k \rfloor\}$. Also, for a function $f : A \to \mathbb{R}_{\geq 0}$ and a subset of elements $C \subseteq A$, we define $f(C) = \sum_{e \in C} f(e)$.

2.1 Coloring and Independent Sets

Given a graph $G = (V, E)$, an *independent set* in G is a subset of vertices $S \subseteq V$ such that for all $u, v \in S$ it holds that $(u, v) \notin E$. Let $\mathsf{IS}(G)$ be the collection of all independent sets in G. Given weight function $w : V \to \mathbb{R}_{\geq 0}$, a *maximum independent set w.r.t. w* is an independent set $S \in \mathsf{IS}(G)$ such that $w(S)$ is maximized. A *coloring* of G is a partition $(V_1, \ldots, V_t)$ of V such that $\forall i \in [t] : V_i \in \mathsf{IS}(G)$; we call each subset of vertices V_i *color class i*. Let $\chi(G)$ be the minimum number of colors required for a coloring of G. A graph G is *perfect* if for every induced subgraph G' of G the cardinality of the maximum clique of G' is equal to $\chi(G')$; note that G' is also a perfect graph. The following well known result is due to [12].

Lemma 2.1. *Given a perfect graph $G = (V, E)$, a minimum coloring of G and a maximum weight independent set of G can be computed in polynomial time.*

2.2 Bin Packing with Conflicts

Given a BPC instance $\mathcal{I}$, let $G_{\mathcal{I}} = (I, E)$ denote the conflict graph of $\mathcal{I}$. A *packing* of a subset of items $S \subseteq I$ is a partition $\mathcal{B} = (B_1, \ldots, B_t)$ of S such that, for all $i \in [t]$, B_i is an independent set in $G_{\mathcal{I}}$, and $s(B_i) \leq 1$. Let $\#\mathcal{B}$ be the number of bins (i.e., entries) in $\mathcal{B}$.

In this paper we consider BPC on several well studied classes of perfect graphs and the acronym BPC refers from now on to perfect conflict graphs. For *bin packing with bipartite conflicts (BPB)*, where the conflict graph is bipartite, we assume a bipartition of V is known and given by X_V and Y_V. Recall that $G = (V, E)$ is a split graph if there is a partition K, S of V into a clique and an independent set, respectively. We call this variant of BPC *bin packing with split graph conflicts (BPS)*.

The following notation will be useful while enhancing a partial packing by new items. For two packings $\mathcal{B} = (B_1, \ldots, B_t)$ and $\mathcal{C} = (C_1, \ldots, C_r)$, let $\mathcal{B} \oplus \mathcal{C} = (B_1, \ldots, B_t, C_1, \ldots, C_r)$ be the *concatenation* of $\mathcal{B}$ and $\mathcal{C}$; also, for $t = r$ let $\mathcal{B} + \mathcal{C} = (B_1 \cup C_1, \ldots, B_t \cup C_t)$ be the *union* of the two packings; note that the latter is not necessarily a packing. We denote by $\mathsf{items}(\mathcal{B}) = \bigcup_{i \in [t]} B_i$ the set of items in the packing $\mathcal{B}$. Finally, let $\mathcal{I} = (I, s, E)$ be a BPC instance and $T \subseteq I$ a subset of items. Define the BPC instances $\mathcal{I} \cap T = (T, s, E_T)$ and $\mathcal{I} \setminus T = (I \setminus T, s, E_{I \setminus T})$ where for all $X \in \{T, I \setminus T\}$ $E_X = \{(u, v) \in E \mid u, v \in X\}$.

2.3 Bin Packing Algorithms

We use $\mathcal{I} = (I, s)$ to denote a BP instance, where I is a set of n items for some $n \geq 1$, and $s : I \to [0, 1]$ is the size function. Let $L_{\mathcal{I}} = \{\ell \in I \mid s(\ell) > \frac{1}{2}\}$ be the set of *large* items, $M_{\mathcal{I}} = \{\ell \in I \mid \frac{1}{3} < s(\ell) \leq \frac{1}{2}\}$ the set of *medium* items, and $S_{\mathcal{I}} = \{\ell \in I \mid s(\ell) \leq \frac{1}{3}\}$ the set of *small* items. Our algorithms use as building blocks also algorithms for BP. The results in the next two lemmas are tailored for our purposes. We give the detailed proofs in [6].[4]

Lemma 2.2. *Given a BP instance $\mathcal{I} = (I, s)$, there is a polynomial-time algorithm First-Fit Decreasing (FFD) which returns a packing $\mathcal{B} = (B_1, \ldots, B_t)$ of $\mathcal{I}$ where $\#\mathcal{B} \leq (1 + 2 \cdot \max_{\ell \in I} s(\ell)) \cdot s(I) + 1$. Moreover, it also holds that $\#\mathcal{B} \leq |L_{\mathcal{I}}| + \frac{3}{2} \cdot s(M_{\mathcal{I}}) + \frac{4}{3} \cdot s(S_{\mathcal{I}}) + 1$.*

Lemma 2.3. *Given a BP instance $\mathcal{I} = (I, s)$, there is a polynomial-time algorithm AsymptoticBP which returns a packing $\mathcal{B} = (B_1, \ldots, B_t)$ of $\mathcal{I}$ such that $t = \mathrm{OPT}(\mathcal{I}) + o(\mathrm{OPT}(\mathcal{I}))$. Moreover, if $\mathrm{OPT}(\mathcal{I}) \geq 100$ then $t \leq 1.02 \cdot \mathrm{OPT}(\mathcal{I})$.*

3   Approximations for Perfect and Bipartite Graphs

In this section we consider the bin packing problem with a perfect or bipartite conflict graph. Previous works (e.g., [8,17]) showed the usefulness of the approach based on finding first a minimum coloring of the given conflict graph, and then packing each color class as a separate bin packing instance (using, e.g., algorithm FFD). Indeed, this approach yields efficient approximations for BPC; however, it does reach a certain limit. To enhance the performance of this *coloring based* approach, we design several subroutines. Combined, they cover the problematic cases and lead to improved approximation guarantees (see Table 1).

Our first subroutine is the coloring based approach, with a simple modification to improve the asymptotic performance. For each color class $C_i, i = 1, \ldots, k$ in a minimum coloring of the given conflict graph, we find a packing of C_i using FFD, and another packing using AsymptoticBP (see Lemma 2.3). We choose the packing which has smaller number of bins. Finally, the returned packing is the concatenation of the packings of all color classes. The pseudocode of Algorithm Color_Sets is given in Algorithm 1.

Algorithm 1. Color_Sets($\mathcal{I} = (I, s, E)$)

1: Compute a minimum coloring $\mathcal{C} = (C_1, \ldots, C_k)$ of $G_{\mathcal{I}}$.
2: Initialize an empty packing $\mathcal{B} \leftarrow ()$.
3: **for** $i \in [k]$ **do**
4:      Compute $\mathcal{A}_1 \leftarrow \mathsf{FFD}((C_i, s))$ and $\mathcal{A}_2 \leftarrow \mathsf{AsymptoticBP}((C_i, s))$.
5:      $\mathcal{B} \leftarrow \mathcal{B} \oplus \arg\min_{\mathcal{A} \in \{\mathcal{A}_1, \mathcal{A}_2\}} \#\mathcal{A}$.
6: **end for**
7: Return $\mathcal{B}$.

For the remainder of this section, fix a BPC instance $\mathcal{I} = (I, s, E)$. The performance guarantees of Algorithm Color_Sets are stated in the next lemma.

---

[4] For more details on algorithms FFD and AsymptoticBP see, e.g., [27].

Lemma 3.1. *Given a BPC instance $\mathcal{I} = (I, s, E)$, Algorithm Color_Sets returns in polynomial time in $|\mathcal{I}|$ a packing $\mathcal{B}$ of $\mathcal{I}$ such that $\#\mathcal{B} \le \chi(G_{\mathcal{I}}) + |L_{\mathcal{I}}| + \frac{3}{2} \cdot s(M_{\mathcal{I}}) + \frac{4}{3} \cdot s(S_{\mathcal{I}})$. Moreover, if $\mathcal{I}$ is a BPB instance then $\#\mathcal{B} \le \frac{3}{2} \cdot |L_{\mathcal{I}}| + \frac{4}{3} \cdot (\text{OPT}(\mathcal{I}) - |L_{\mathcal{I}}|) + o(\text{OPT}(\mathcal{I}))$.*

Note that the bounds may not be tight for instances with many large items. Specifically, if $|L_{\mathcal{I}}| \approx \text{OPT}(\mathcal{I})$ then a variant of Algorithm Color_Sets was shown to yield a packing of at least $2.5 \cdot \text{OPT}(\mathcal{I})$ bins [8]. To overcome this, we use an approach based on the simple yet crucial observation that there can be at most one large item in a bin. Therefore, we view the large items as *bins* and assign items to these bins to maximize the total size packed in bins including large items. We formalize the problem initially on a single bin.

Definition 3.2. *In the* bounded independent set problem (BIS) *we are given a graph $G = (V, E)$, a weight function $w : V \to \mathbb{R}_{\ge 0}$, and a budget $\beta \in \mathbb{R}_{\ge 0}$. The goal is to find an independent set $S \subseteq V$ in G such that $w(S)$ is maximized and $w(S) \le \beta$. Let $\mathcal{I} = (V, E, w, \beta)$ be a BIS instance.*

Towards solving BIS, we need the following definitions. For $\alpha \in (0, 1]$, $\mathcal{A}$ is an α-approximation algorithm for a maximization problem $\mathcal{P}$ if, for any instance $\mathcal{I}$ of $\mathcal{P}$, $\mathcal{A}$ outputs a solution of value at least $\alpha \cdot OPT(\mathcal{I})$. A *polynomial-time approximation scheme (PTAS)* for $\mathcal{P}$ is a family of algorithms $\{A_\varepsilon\}$ such that, for any $\varepsilon > 0$, A_ε is a polynomial-time $(1 - \varepsilon)$-approximation algorithm for $\mathcal{P}$. A *fully PTAS (FPTAS)* is a PTAS $\{A_\varepsilon\}$ where, for all $\varepsilon > 0$, A_ε is polynomial also in $\frac{1}{\varepsilon}$. We now describe a PTAS for BIS. Fix a BIS instance $\mathcal{I} = (V, E, w, \beta)$ and $\varepsilon > 0$. As there can be at most ε^{-1} items with weight at least $\varepsilon \cdot \beta$ in some optimal solution OPT for $\mathcal{I}$, we can *guess* this set F of items via enumeration. Then, to add smaller items to F, we define a residual graph G_F of items with weights at most $\varepsilon \cdot \beta$ which are not adjacent to any item in F. Formally, define $G_F = (V_F, E_F)$, where

$$V_F = \{v \in V \backslash F \mid w(v) \le \varepsilon \cdot \beta, \forall u \in F : (v, u) \notin E\}, \quad E_F = \{(u, v) \in E \mid u, v \in V_F\}$$

Now, we find a maximum weight independent set S in G_F. Note that this can be done in polynomial time for perfect and bipartite graphs. If $w(F \cup S) \le \beta$ then we have an optimal solution; otherwise, we discard iteratively items from S until the remaining items form a feasible solution for $\mathcal{I}$. Since we discard only items with relatively small weights, we lose only an ε-fraction of the weight relative to the optimum. The pseudocode for the scheme is given in Algorithm 2.

Lemma 3.3. *Algorithm 2 is a PTAS for BIS.*

We note that by a result of [7], unless P=NP, BIS does not admit an *efficient* PTAS, even on bipartite graphs.[5] Thus, our PTAS for this problem is of an independent interest.

---

[5] An *efficient PTAS* is a PTAS $\{A_\varepsilon\}$ where, for all $\varepsilon > 0$, the running time of A_ε is given by $f(1/\varepsilon)$ times a polynomial of the input size.

Algorithm 2. PTAS$((V, E, w, \beta), \varepsilon)$

1: Initialize $A \leftarrow \emptyset$.
2: **for** all independent sets $F \subseteq V$ in (V, E) s.t. $|F| \leq \varepsilon^{-1}, w(F) \leq \beta$ **do**
3:　　Define the residual graph $G_F = (V_F, E_F)$.
4:　　Find a maximum independent set S of G_F w.r.t. w.
5:　　**while** $w(F \cup S) > \beta$ **do**
6:　　　　Choose arbitrary $z \in S$.
7:　　　　Update $S \leftarrow S \setminus \{z\}$.
8:　　**end while**
9:　　**if** $w(A) < w(F \cup S)$ **then**
10:　　　　Update $A \leftarrow F \cup S$.
11:　　**end if**
12: **end for**
13: Return A.

We now define our maximization problem for multiple bins. We solve a slightly generalized problem in which we have an initial partial packing in t bins. Our goal is to add to these bins (from unpacked items) a subset of items of maximum total size. Formally,

Definition 3.4. *Given a BPC instance* $\mathcal{I} = (I, s, E)$, $S \subseteq I$, *and a packing* $\mathcal{B} = (B_1, \ldots, B_t)$ *of* S, *define the* maximization problem *of* $\mathcal{I}$ *and* $\mathcal{B}$ *as the problem of finding a packing* $\mathcal{B} + \mathcal{C}$ *of* $S \cup T$, *where* $T \subseteq I \setminus S$ *and* $\mathcal{C} = (C_1, \ldots, C_t)$ *is a packing of* T, *such that* $s(T)$ *is maximized.*

Our solution for BIS is used to obtain a $(1 - \frac{1}{e} - \varepsilon)$-approximation for the maximization problem described in Definition 3.4. This is done using the approach of [9] for the more general *separable assignment problem (SAP)*.

Lemma 3.5. *Given a BPC instance* $\mathcal{I} = (I, s, E)$, $S \subseteq I$, *a packing* $\mathcal{B} = (B_1, \ldots, B_t)$ *of* S, *and a constant* $\varepsilon > 0$, *there is an algorithm* MaxSize *which returns in time polynomial in* $|\mathcal{I}|$ *a* $(1 - \frac{1}{e} - \varepsilon)$-*approximation for the maximization problem of* $\mathcal{I}$ *and* $\mathcal{B}$. *Moreover, given an FPTAS for BIS on the graph* (I, E), *the weight function* s, *and the budget* $\beta = 1$, MaxSize *is a* $(1 - \frac{1}{e})$-*approximation algorithm for the maximization problem of* $\mathcal{I}$ *and* $\mathcal{B}$.

We use the above to obtain a feasible solution for the instance. This is done via a reduction to the maximization problem of the instance with a singleton packing of the large items and packing the remaining items in extra bins. Specifically, in the subroutine MaxSolve, we initially put each item in $L_{\mathcal{I}}$ in a separate bin. Then, additional items from $S_{\mathcal{I}}$ and $M_{\mathcal{I}}$ are added to the bins using Algorithm MaxSize (defined in Lemma 3.5). The remaining items are packed using Algorithm Color_Sets. The pseudocode of the subroutine MaxSolve is given in Algorithm 3.

The proof of Lemma 3.6 uses Lemmas 3.1, 3.3, and 3.5.

Lemma 3.6. *Given a BPC instance* $\mathcal{I} = (I, s, E)$ *and an* $\varepsilon > 0$, *Algorithm* MaxSolve *returns in polynomial time in* $|\mathcal{I}|$ *a packing* $\mathcal{C}$ *of* $\mathcal{I}$ *such that there are* $0 \leq x \leq s(M_{\mathcal{I}})$ *and* $0 \leq y \leq s(S_{\mathcal{I}})$ *such that the following holds.*

Algorithm 3. MaxSolve($\mathcal{I} = (I, s, E), \varepsilon$)

1: Define $T \leftarrow (\{\ell\} \mid \ell \in L_{\mathcal{I}})$.
2: $\mathcal{A} \leftarrow$ MaxSize($\mathcal{I}, L_{\mathcal{I}}, T, \varepsilon$).
3: $\mathcal{B} \leftarrow$ Color_Sets($\mathcal{I} \setminus$ items($\mathcal{A}$)).
4: Return $\mathcal{A} \oplus \mathcal{B}$.

---

1. $x + y \leq \mathrm{OPT}(\mathcal{I}) - |L_{\mathcal{I}}| + \left(\frac{1}{e} + \varepsilon\right) \cdot \frac{|L_{\mathcal{I}}|}{2}$.
2. $\#\mathcal{C} \leq \chi(G_{\mathcal{I}}) + |L_{\mathcal{I}}| + \frac{3}{2} \cdot x + \frac{4}{3} \cdot y$.

Lemma 3.6 improves significantly the performance of Algorithm Color_Sets for instances with many large items. However, Algorithm MaxSize may prefer small over medium items; the latter items will be packed by Algorithm Color_Sets (see Algorithm 3). The packing of these medium items may harm the approximation guarantee. Thus, to tackle instances with many medium items, we use a reduction to a maximum matching problem for packing the large and medium items in at most $\mathrm{OPT}(\mathcal{I})$ bins.[6] Then, the remaining items can be packed using Algorithm Color_Sets. The graph used for the following subroutine Matching contains all large and medium items; there is an edge between any two items which can be assigned to the same bin in a packing of the instance $\mathcal{I}$. Formally,

Definition 3.7. *Given a BPC instance* $\mathcal{I} = (I, s, E)$, *the auxiliary graph of* $\mathcal{I}$ *is* $H_{\mathcal{I}} = (L_{\mathcal{I}} \cup M_{\mathcal{I}}, E_H)$, *where* $E_H = \{(u, v) \mid u, v \in L_{\mathcal{I}} \cup M_{\mathcal{I}}, s(\{u, v\}) \leq 1, (u, v) \notin E\}$.

Algorithm Matching finds a maximum matching in $H_{\mathcal{I}}$ and outputs a packing of the large and medium items where pairs of items taken to the matching are packed together, and the remaining items are packed in extra bins using Algorithm Color_Sets. The pseudocode of the subroutine Matching is given in Algorithm 4.

Algorithm 4. Matching($\mathcal{I} = (I, s, E)$)

1: Find a maximum matching $\mathcal{M}$ in $H_{\mathcal{I}}$.
2: $\mathcal{B} \leftarrow (\{u, v\} \mid (u, v) \in \mathcal{M}) \oplus (\{v\} \mid v \in M_{\mathcal{I}} \cup L_{\mathcal{I}}, \forall u \in M_{\mathcal{I}} \cup L_{\mathcal{I}} : (u, v) \notin \mathcal{M})$.
3: Return $\mathcal{B} \oplus$ Color_Sets($\mathcal{I} \setminus (M_{\mathcal{I}} \cup L_{\mathcal{I}})$).

The proof of Lemma 3.8 follows by noting that the cardinality of a maximum matching in $H_{\mathcal{I}}$ in addition to the number of unmatched vertices in $L_{\mathcal{I}} \cup M_{\mathcal{I}}$ is at most $\mathrm{OPT}(\mathcal{I})$.

Lemma 3.8. *Given a BPC instance* $\mathcal{I} = (I, s, E)$, *Algorithm* Matching *returns in polynomial time in* $|\mathcal{I}|$ *a packing* $\mathcal{A}$ *of* $\mathcal{I}$ *such that* $\#\mathcal{A} \leq \mathrm{OPT}(\mathcal{I}) + \chi(G_{\mathcal{I}}) + \frac{4}{3} \cdot s(S_{\mathcal{I}})$.

---

[6] We note that a maximum matching based technique for BPC is used also in [8,15].

We now have the required components for the approximation algorithm for BPC and the asymptotic approximation for BPB. Our algorithm, ApproxBPC, applies all of the above subroutines and returns the packing which uses the smallest number of bins. We use $\varepsilon = 0.0001$ for the error parameter in MaxSolve. The pseudocode of ApproxBPC is given in Algorithm 5.

Algorithm 5. ApproxBPC($\mathcal{I}$)
___
1: Let $\varepsilon = 0.0001$.
2: Compute $\mathcal{A}_1 \leftarrow$ Color_Sets($\mathcal{I}$), $\mathcal{A}_2 \leftarrow$ MaxSolve($\mathcal{I}, \varepsilon$), $\mathcal{A}_3 \leftarrow$ Matching($\mathcal{I}$).
3: Return $\arg\min_{\mathcal{A} \in \{\mathcal{A}_1, \mathcal{A}_2, \mathcal{A}_3\}} \#\mathcal{A}$.

We give below the main result of this section. The proof follows by the argument that the subroutines Color_Sets, MaxSolve, and Matching handle together most of the difficult cases. Specifically, if the instance contains many large items, then MaxSolve produces the best approximation. If there are many large and medium items, then Matching improves the approximation guarantee. Finally, for any other case, our analysis of the Color_Sets algorithm gives us the desired ratio. We summarize with the next result.

Theorem 3.9. *Algorithm 5 is a 2.445-approximation for BPC and an asymptotic 1.391-approximation for BPB.*

4   Split Graphs

In this section we enhance the use of maximization techniques for BPC to obtain an absolute approximation algorithm for BPS. In particular, we improve upon the recent result of Huang et al. [15]. We use as a subroutine the maximization technique as outlined in Lemma 3.5. Specifically, we start by obtaining an FPTAS for the BIS problem on split graphs. For the following, fix a BPS instance $\mathcal{I} = (I, s, E)$. It is well known (see, e.g., [11]) that a partition of the vertices of a split graph into a clique and an independent set can be found in polynomial time. Thus, for simplicity we assume that such a partition of the split graph G is known and given by K_G, S_G. We note that an FPTAS for the BIS problem on split graphs follows from a result of Pferschy and Schauer [24] for *knapsack with conflicts*, since split graphs are a subclass of chordal graphs. We give a simpler FPTAS for our problem in [6].

Lemma 4.1. *There is an algorithm* FPTAS-BIS *that is an FPTAS for the BIS problem on split graphs.*

Our next goal is to find a suitable initial packing $\mathcal{B}$ to which we apply MaxSize. Clearly, the vertices $K_{G_\mathcal{I}}$ must be assigned to different bins. Therefore, our initial packing contains the vertices of $K_{G_\mathcal{I}}$ distributed to $|K_{G_\mathcal{I}}|$ bins as $\{\{v\} \mid v \in K_{G_\mathcal{I}}\}$. In addition, let $\alpha \in \{0, 1, \ldots, \lceil 2 \cdot s(I) \rceil + 1\}$ be a *guess* of $\mathrm{OPT}(\mathcal{I}) - |K_{G_\mathcal{I}}|$;

then, $(\emptyset)_{i\in[\alpha]}$ is a packing of α bins that do not contain items. Together, the two above packings form the initial packing $\mathcal{B}_\alpha$. Our algorithm uses MaxSize to add items to the existing bins of $\mathcal{B}_\alpha$ and packs the remaining items using FFD. Note that we do not need an error parameter ε, since we use MaxSize with an FPTAS (see Lemma 3.5). For simplicity, we assume that $\mathrm{OPT}(\mathcal{I}) \geq 2$ (else we can trivially pack the instance in a single bin) and omit the case where $\mathrm{OPT}(\mathcal{I}) = 1$ from the pseudocode. We give the pseudocode of our algorithm for BPS in Algorithm 6.

Algorithm 6. Split-Approx($\mathcal{I} = (I, s, E)$)

1: **for** $\alpha \in \{0, 1, \ldots, \lceil 2 \cdot s(I) \rceil + 1\}$ **do**
2:     Define $\mathcal{B}_\alpha = \{\{v\} \mid v \in K_{G_\mathcal{I}}\} \oplus (\emptyset)_{i\in[\alpha]}$
3:     $\mathcal{A}_\alpha \leftarrow$ MaxSize($\mathcal{I}, K_{G_\mathcal{I}}, \mathcal{B}_\alpha$).
4:     $\mathcal{A}_\alpha^* \leftarrow \mathcal{A}_\alpha \oplus$ FFD($\mathcal{I} \setminus$ items($\mathcal{A}_\alpha$)).
5: **end for**
6: Return $\arg\min_{\alpha\in\{0,1,\ldots,\lceil 2\cdot s(I)\rceil+1\}} \#\mathcal{A}_\alpha^*$.

By Lemmas 4.1 and 3.5 we have a $\left(1 - \frac{1}{e}\right)$-approximation for the maximization problem of the BPS instance $\mathcal{I}$ and an initial partial packing $\mathcal{B}$. Hence, for a correct guess $\alpha = \mathrm{OPT}(\mathcal{I}) - |K_{G_\mathcal{I}}|$, the remaining items to be packed by FFD are of total size at most $\frac{s(I)}{e}$ and can be packed in $\frac{2\cdot\mathrm{OPT}(\mathcal{I})}{e}$ bins. Thus, we have

Theorem 4.2. *Algorithm 6 is a $\left(1 + \frac{2}{e}\right)$-approximation for* BPS.

5   Asymptotic Hardness for Bipartite and Split Graphs

In this section we show that there is no APTAS for BPB and BPS, unless $P = NP$. We use a reduction from the *Bounded 3-dimensional matching (B3DM)* problem, that is known to be MAX SNP-complete [18].

For the remainder of this section, let $c > 2$ be some constant. A B3DM instance is a four-tuple $\mathcal{J} = (X, Y, Z, T)$, where X, Y, Z are three disjoint finite sets and $T \subseteq X \times Y \times Z$; also, for each $u \in X \cup Y \cup Z$ there are at most c triples in T to which u belongs. A *solution* for $\mathcal{J}$ is $M \subseteq T$ such that for all $u \in X \cup Y \cup Z$ it holds that u appears in at most one triple of M. The objective is to find a solution M of maximum cardinality. Let $\mathrm{OPT}(\mathcal{J})$ be the value of an optimal solution for $\mathcal{J}$. We use in our reduction a *restricted* instance of B3DM defined as follows.

Definition 5.1. *For $k \in \mathbb{N}$, a B3DM instance $\mathcal{J}$ is k-restricted if $\mathrm{OPT}(\mathcal{J}) \geq k$.*

In the next lemma we show the hardness of k-restricted B3DM. Intuitively, since B3DM instances $\mathcal{J}$ with $\mathrm{OPT}(\mathcal{J}) \leq k$ are polynomially solvable for a fixed k (e.g., by exhaustive enumeration), it follows that restricted-B3DM must be hard to approximate, by the hardness result of Kann [18].

Lemma 5.2. *There is a constant $\alpha < 1$ such that for any $k \in \mathbb{N}$ there is no α-approximation for the k-restricted B3DM problem unless P=NP.*

We give below the main idea of our reduction, showing the asymptotic hardness of BPB and BPS. A more formal description and the proof of Lemma 5.2 are given in [6]. For a sufficiently large $n \in \mathbb{N}$, let $\mathcal{J} = (X, Y, Z, T)$ be an n-restricted instance of B3DM, and let the components of $\mathcal{J}$, together with appropriate indexing, be $U = X \cup Y \cup Z$ and T, where

$$X = \{x_1, \ldots, x_{\tilde{x}}\}, Y = \{y_1, \ldots, y_{\tilde{y}}\}, Z = \{z_1, \ldots, z_{\tilde{z}}\}, T = \{t_1, \ldots, t_{\tilde{t}}\}.$$

We outline our reduction for BPB and later show how it can be modified to yield the hardness result for BPS. Given an n-restricted B3DM instance, we construct a sequence of BPB instances. Each BPB instance contains an item for each element $u \in U$, and an item for each triple $t \in T$. There is an edge (u, t) if $u \in U$ and $t \in T$, and u does not appear in t, i.e., we forbid packing an element u in the same bin with a triple not containing u, for any $u \in U$. Since we do not know the exact value of $\text{OPT}(\mathcal{J})$, we define a family of instances with different number of *filler items*; these items are packed in the optimum of our constructed BPB instance together with elements not taken to the solution for $\mathcal{J}$.

Specifically, for a *guess* $i \in \{n, n+1, \ldots, |T|\}$ of $\text{OPT}(\mathcal{J})$, we define a BPB instance $\mathcal{I}_i = (I_i, s, E)$. The set of items in $\mathcal{I}_i$ is $I_i = U \cup P_i \cup T \cup Q_i$, where P_i, Q_i are a set of $\tilde{t} - i$ (filler) items and a set of $\tilde{x} + \tilde{y} + \tilde{z} - 3 \cdot i$ (filler) items, respectively, such that $P_i \cap U = \emptyset$ and $Q_i \cap U = \emptyset$. The bipartite (conflict) graph of $\mathcal{I}_i$ is $G_i = (I_i, E)$, where $E = E_X \cup E_Y \cup E_Z$ is defined as follows.

$$E_X = \{(x, t) \mid x \in X, t = (x', y, z) \in T, x \neq x'\}$$
$$E_Y = \{(y, t) \mid y \in Y, t = (x, y', z) \in T, y \neq y'\}$$
$$E_Z = \{(z, t) \mid z \in Z, t = (x, y, z') \in T, z \neq z'\}$$

Finally, define the sizes of items in $\mathcal{I}_i$ to be

$$\forall u \in U, p \in P_i, q \in Q_i, t \in T: \quad s(u) = 0.15, s(p) = 0.45, s(q) = 0.85, s(t) = 0.55.$$

By the above, the only way to pack three items from $x, y, z \in U$ with a triple $t \in T$ is if $(x, y, z) = t$; also, $s(\{x, y, z, t\}) = 1$. For an illustration of the reduction see Fig. 1.

Given a packing $(A_1, \ldots, A_q)$ for the BPB instance $\mathcal{I}_i$, we consider all *useful bins* A_b in the packing, i.e., $A_b = \{x, y, z, t\}$, where $x \in X, y \in Y, z \in Z$ and $t = (x, y, z)$. The triple t from bin A_b is taken to our solution for the original n-restricted B3DM instance $\mathcal{J}$. Note that taking all triples as described above forms a feasible solution for $\mathcal{J}$, since each element is packed only once. Thus, our goal becomes to find a packing for the reduced BPB instance with a maximum number of useful bins. Indeed, since $s(A_b) = 1$ for any useful bin A_b, finding a packing with many useful bins coincides with an efficient approximation for BPB.

For the optimal guess $i^* = \text{OPT}(\mathcal{J})$, it is not hard to see that the optimum for the BPB instance $\mathcal{I}_{i^*}$ satisfies $s(I_{i^*}) = \text{OPT}(\mathcal{I}_{i^*})$; that is, all bins in the optimum

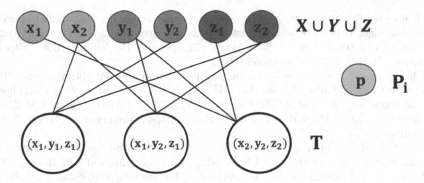

Fig. 1. An illustration of the BPB instance $\mathcal{I}_i = (I_i, s, E)$, where $i = \text{OPT}(\mathcal{J}) = 2$. The optimal solution for $\mathcal{I}_i$ contains the bins $\{x_1, y_1, z_1, (x_1, y_1, z_1)\}, \{x_2, y_2, z_2, (x_2, y_2, z_2)\}$, and $\{p, (x_1, y_2, z_1)\}$; this corresponds to an optimal solution $(x_1, y_1, z_1), (x_2, y_2, z_2)$ for the original B3DM instance. Note that in this example $Q_i = \emptyset$.

are *fully* packed. For a sufficiently large n, and assuming there is an APTAS for BPB, we can find a packing of $\mathcal{I}_{i^*}$ with a large number of bins that are fully packed. A majority of these bins are useful, giving an efficient approximation for the original B3DM instance. A similar reduction to BPS is obtained by adding to the bipartite conflict graph of the BPB instance an edge between any pair of vertices in T; thus, we have a *split* conflict graph. We summarize the above discussion in the next result (the proof is given in [6]).

Theorem 5.3. *There is no* APTAS *for* BPB *and* BPS, *unless* P=NP.

6 Discussion

In this work we presented the first theoretical evidence that BPC on polynomially colorable graphs is harder than classic bin packing, even in the special cases of bipartite and split graphs. Furthermore, we introduced a new generic framework for tackling BPC instances, based on a reduction to a maximization problem. Using this framework, we improve the state-of-the-art approximations for BPC on several well studied graph classes.

We note that better bounds for the maximization problems solved within our framework will imply improved approximation guarantees for BPC on perfect, bipartite, and split graphs. It would be interesting to apply our techniques to improve the known results for other graph classes, such as chordal graphs or partial k-trees.

References

1. Adany, R., et al.: All-or-nothing generalized assignment with application to scheduling advertising campaigns. ACM Trans. Algorithms **12**(3), 38:1–38:25 (2016)

2. Beaumont, O., Bonichon, N., Duchon, P., Larchevêque, H.: Distributed approximation algorithm for resource clustering. In: Shvartsman, A.A., Felber, P. (eds.) SIROCCO 2008. LNCS, vol. 5058, pp. 61–73. Springer, Heidelberg (2008). https://doi.org/10.1007/978-3-540-69355-0_7

3. Christofides, N.: The vehicle routing problem. Combinatorial optimization (1979)

4. Doron-Arad, I., Kulik, A., Shachnai, H.: An APTAS for bin packing with clique-graph conflicts. In: Lubiw, A., Salavatipour, M. (eds.) WADS 2021. LNCS, vol. 12808, pp. 286–299. Springer, Cham (2021). https://doi.org/10.1007/978-3-030-83508-8_21

5. Doron-Arad, I., Kulik, A., Shachnai, H.: An AFPTAS for bin packing with partition matroid via a new method for LP rounding. In: Proceedings of APPROX (2023)

6. Doron-Arad, I., Shachnai, H.: Approximating bin packing with conflict graphs via maximization techniques. arXiv preprint arXiv:2302.10613 (2023)

7. Doron-Arad, I., Shachnai, H.: Tight bounds for budgeted maximum weight independent set in bipartite and perfect graphs. arXiv preprint arXiv:2307.08592 (2023)

8. Epstein, L., Levin, A.: On bin packing with conflicts. SIAM J. Optim. **19**(3), 1270–1298 (2008)

9. Fleischer, L., Goemans, M.X., Mirrokni, V.S., Sviridenko, M.: Tight approximation algorithms for maximum separable assignment problems. Math. Oper. Res. **36**(3), 416–431 (2011)

10. Garey, M.R., Johnson, D.S.: Computers and intractability. A Guide to the (1979)

11. Golumbic, M.C.: Algorithmic Graph Theory and Perfect Graphs. Elsevier, Amsterdam (2004)

12. Grötschel, M., Lovász, L., Schrijver, A.: Geometric Algorithms and Combinatorial Optimization, vol. 2. Springer, Berlin (2012)

13. Halldórsson, M.M.: A still better performance guarantee for approximate graph coloring. Inf. Process. Lett. **45**(1), 19–23 (1993)

14. Hoberg, R., Rothvoß, T.: A logarithmic additive integrality gap for bin packing. In: Proceedings of the Twenty-Eighth Annual ACM-SIAM Symposium on Discrete Algorithms, pp. 2616–2625. SIAM (2017)

15. Huang, Z., Zhang, A., Dósa, G., Chen, Y., Xiong, C.: Improved approximation algorithms for bin packing with conflicts. Int. J. Found. Comput. Sci. 1–16 (2023)

16. Jansen, K.: An approximation scheme for bin packing with conflicts. J. Comb. Optim. **3**(4), 363–377 (1999)

17. Jansen, K., Öhring, S.R.: Approximation algorithms for time constrained scheduling. Inf. Comput. **132**(2), 85–108 (1997)

18. Kann, V.: Maximum bounded 3-dimensional matching is MAX SNP-complete. Inf. Process. Lett. **37**(1), 27–35 (1991)

19. Karmarkar, N., Karp, R.M.: An efficient approximation scheme for the one-dimensional bin-packing problem. In: 23rd Annual Symposium on Foundations of Computer Science, pp. 312–320. IEEE (1982)

20. de La Vega, W.F., Lueker, G.S.: Bin packing can be solved within $1+\varepsilon$ in linear time. Combinatorica **1**(4), 349–355 (1981)

21. Laporte, G., Desroches, S.: Examination timetabling by computer. Comput. Oper. Res. **11**(4), 351–360 (1984)

22. McCloskey, B., Shankar, A.: Approaches to bin packing with clique-graph conflicts. University of California, Computer Science Division (2005)

23. Oh, Y., Son, S.: On a constrained bin-packing problem. Technical Report CS-95-14 (1995)

24. Pferschy, U., Schauer, J.: The knapsack problem with conflict graphs. J. Graph Algorithms Appl. **13**(2), 233–249 (2009)

25. Rothvoß, T.: Approximating bin packing within O(log OPT * log log OPT) bins. In: 54th Annual IEEE Symposium on Foundations of Computer Science, pp. 20–29. IEEE Computer Society (2013)
26. Simchi-Levi, D.: New worst-case results for the bin-packing problem. Naval Res. Logist. (NRL) **41**(4), 579–585 (1994)
27. Vazirani, V.V.: Approximation Algorithms. Springer, Berlin, Heidelberg (2001)
28. Zuckerman, D.: Linear degree extractors and the inapproximability of max clique and chromatic number. In: Proceedings of the Thirty-Eighth Annual ACM Symposium on Theory of Computing, pp. 681–690 (2006)

α_i-Metric Graphs: Radius, Diameter and all Eccentricities

Feodor F. Dragan[1] and Guillaume Ducoffe[2(✉)]

[1] Computer Science Department, Kent State University, Kent, USA
`dragan@cs.kent.edu`
[2] National Institute for Research and Development in Informatics and University of Bucharest, Bucharest, Romania
`guillaume.ducoffe@ici.ro`

Abstract. We extend known results on chordal graphs and distance-hereditary graphs to much larger graph classes by using only a common metric property of these graphs. Specifically, a graph is called α_i-metric ($i \in \mathcal{N}$) if it satisfies the following α_i-metric property for every vertices u, w, v and x: if a shortest path between u and w and a shortest path between x and v share a terminal edge vw, then $d(u,x) \geq d(u,v) + d(v,x) - i$. Roughly, gluing together any two shortest paths along a common terminal edge may not necessarily result in a shortest path but yields a "near-shortest" path with defect at most i. It is known that α_0-metric graphs are exactly ptolemaic graphs, and that chordal graphs and distance-hereditary graphs are α_i-metric for $i = 1$ and $i = 2$, respectively. We show that an additive $O(i)$-approximation of the radius, of the diameter, and in fact of all vertex eccentricities of an α_i-metric graph can be computed in total linear time. Our strongest results are obtained for α_1-metric graphs, for which we prove that a central vertex can be computed in subquadratic time, and even better in linear time for so-called (α_1, Δ)-metric graphs (a superclass of chordal graphs and of plane triangulations with inner vertices of degree at least 7). The latter answers a question raised in (Dragan, *IPL*, 2020). Our algorithms follow from new results on centers and metric intervals of α_i-metric graphs. In particular, we prove that the diameter of the center is at most $3i + 2$ (at most 3, if $i = 1$). The latter partly answers a question raised in (Yushmanov & Chepoi, *Mathematical Problems in Cybernetics*, 1991).

Keywords: metric graph classes · chordal graphs · α_i-metric · radius · diameter · vertex eccentricity · eccentricity approximating trees · approximation algorithms

1 Introduction

Euclidean spaces have the following nice property: if the geodesic between u and w contains v, and the geodesic between v and x contains w, then their

This work was supported by a grant of the Romanian Ministry of Research, Innovation and Digitalization, CCCDI - UEFISCDI, proect number PN-III-P2-2.1-PED-2021-2142, within PNCDI III.

union must be the geodesic between u and x. In 1991, Chepoi and Yushmanov introduced α_i-metric properties ($i \in \mathcal{N}$), as a way to quantify by how much a graph is close to satisfy this above requirement [39] (see also [10, 11] for earlier use of α_1-metric property). All graphs $G = (V, E)$ occurring in this paper are connected, finite, unweighted, undirected, loopless and without multiple edges. The *length* of a path between two vertices u and v is the number of edges in the path. The *distance* $d_G(u, v)$ is the length of a shortest path connecting u and v in G. The *interval* $I_G(u, v)$ between u and v consists of all vertices on shortest (u, v)-paths, that is, it consists of all vertices (metrically) between u and v: $I_G(u, v) = \{x \in V : d_G(u, x) + d_G(x, v) = d_G(u, v)\}$. Let also $I_G^o(u, v) = I_G(u, v) \setminus \{u, v\}$. If no confusion arises, we will omit subindex G.

α_i**-metric property:** if $v \in I(u, w)$ and $w \in I(v, x)$ are adjacent, then
$$d(u, x) \geq d(u, v) + d(v, x) - i = d(u, v) + 1 + d(w, x) - i.$$

Roughly, gluing together any two shortest paths along a common terminal edge may not necessarily result in a shortest path (unlike in the Euclidean space) but yields a "near-shortest" path with defect at most i. A graph is called α_i-metric if it satisfies the α_i-metric property. α_i-Metric graphs were investigated in [10, 11, 39]. In particular, it is known that α_0-metric graphs are exactly the distance-hereditary chordal graphs, also known as ptolemaic graphs [32]. Furthermore, α_1-metric graphs contain all chordal graphs [10] and all plane triangulations with inner vertices of degree at least 7 [25]. α_2-Metric graphs contain all distance-hereditary graphs [39] and, even more strongly, all HHD-free graphs [13]. Evidently, every graph is an α_i-metric graph for some i. Chepoi and Yushmanov in [39] also provided a characterization of all α_1-metric graphs: They are exactly the graphs where all disks are convex and the graph W_6^{++} from Fig. 1 is forbidden as an isometric subgraph (see [39] or Theorem 5). This nice characterization was heavily used in [3] in order to characterize δ-hyperbolic graphs with $\delta \leq 1/2$.

Let the *eccentricity* of a vertex v in G be defined as $e_G(v) = \max_{u \in V} d_G(u, v)$. The *diameter* and the *radius* of a graph are defined as $diam(G) = \max_{u \in V} e_G(u)$ and $rad(G) = \min_{u \in V} e_G(u)$, respectively. Let the *center* of a graph G be defined as $C(G) = \{u \in V : e_G(u) = rad(G)\}$. Each vertex from $C(G)$ is called a *central* vertex. In this paper, we investigate the radius, diameter, and all eccentricities computation problems in α_i-metric graphs. Understanding the eccentricity function of a graph and being able to efficiently compute or estimate the diameter, the radius, and all vertex eccentricities is of great importance. For example, in the analysis of complex networks, the eccentricity of a vertex is used to measure its importance: the *eccentricity centrality index* of v [33] is defined as $\frac{1}{e(v)}$. Furthermore, the problem of finding a central vertex is one of the most famous facility location problems. In [39], the following nice relation between the diameter and the radius of an α_i-metric graph G was established: $diam(G) \geq 2rad(G) - i - 1$. Recall that for every graph G, $diam(G) \leq 2rad(G)$ holds. Authors of [39] also raised a question[1] whether the diameter of the center of an α_i-metric graph can

---

[1] It is conjectured in [39] that $diam(C(G)) \leq i + 2$ for every α_i-metric graph G.

be bounded by a linear function of i. It is known that the diameters of the centers of chordal graphs or of distance-hereditary graphs are at most 3 [11,39].

Related Work. There is a naive algorithm which runs a BFS from each vertex in order to compute all eccentricities. It has running time $O(nm)$ on an n-vertex m-edge graph. Interestingly, this is conditionally optimal for general graphs as well as for some restricted families of graphs [1,5,15,36] since, under plausible complexity assumptions, neither the diameter nor the radius can be computed in truly subquadratic time (i.e., in $O(n^a m^b)$ time, for some positive a, b such that $a + b < 2$). In a quest to break this quadratic barrier, there has been a long line of work presenting more efficient algorithms for computing the diameter and/or the radius, or even better all eccentricities, on some special graph classes. For example, linear-time algorithms are known for computing all eccentricities of interval graphs [26,34]. Extensions of these results to several superclasses of interval graphs are also known [6,7,16,17,21,28–31]. Chordal graphs are another well-known generalization of interval graphs, for which the diameter can unlikely be computed in subquadratic time [5]. For all that, there is an elegant linear-time algorithm for computing the radius and a central vertex of a chordal graph [12]. Until this work there has been little insight about how to extend this nice result to larger graph classes (a notable exception being the work in [13]). This intriguing question is partly addressed in our paper.

Since the existence of subquadratic time algorithm for exact diameter or radius computation is unlikely, even for simple families of graphs, a large volume of work was also devoted to approximation algorithms [1,2,9,36,38]. Authors of [9] additionally address a more challenging question of obtaining an additive c-approximation for the diameter, i.e., an estimate D such that $diam(G) - c \leq D \leq diam(G)$. A simple $\tilde{O}(mn^{1-\epsilon})$ time algorithm achieves an additive n^ϵ-approximation and, for any $\epsilon > 0$, getting an additive n^ϵ-approximation algorithm for the diameter running in $O(n^{2-\epsilon'})$ time for any $\epsilon' > 2\epsilon$ would falsify the Strong Exponential Time Hypothesis (SETH). However, much better additive approximations can be achieved for graphs with bounded (metric) parameters. For example, a vertex furthest from an arbitrary vertex has eccentricity at least $diam(G) - 2$ for chordal graphs [12] and at least $diam(G) - \lfloor k/2 \rfloor$ for k-chordal graphs [8]. Later, those results were generalized to all δ-hyperbolic graphs [14,15,23,24]. In [20], we also introduce a natural generalization of α_i-metric and hyperbolic graphs, which we call a (λ, μ)-*bow metric*: namely, if two shortest paths $P(u, w)$ and $P(v, x)$ share a common shortest subpath $P(v, w)$ of length more than λ (that is, they overlap by more than λ), then the distance between u and x is at least $d(u, v) + d(v, w) + d(w, x) - \mu$. δ-Hyperbolic graphs are $(\delta, 2\delta)$-bow metric and α_i-metric graphs are $(0, i)$-bow metric. (α_1, Δ)-Metric graphs form an important subclass of α_1-metric graphs and contain all chordal graphs and all plane triangulations with inner vertices of degree at least 7. In [25], it was shown that every (α_1, Δ)-metric graph admits an eccentricity 2-approximating spanning tree, i.e., a spanning tree T such that $e_T(v) - e_G(v) \leq 2$ for every vertex v. Finding similar results for general α_1-metric graphs was left as an open problem in [25].

Our Contribution. We prove several new results on intervals, eccentricity and centers in α_i-metric graphs, and their algorithmic applications, thus answering open questions in the literature [18,25,39]. To list our contributions, we need to introduce on our way some additional notations and terminology.

Section 2 is devoted to general α_i-metric graphs ($i \geq 0$). The set $S_k(u,v) = \{x \in I(u,v) : d(u,x) = k\}$ is called a *slice* of the interval $I(u,v)$ where $0 \leq k \leq d(u,v)$. An interval $I(u,v)$ is said to be λ-thin if $d(x,y) \leq \lambda$ for all $x, y \in S_k(u,v)$, $0 < k < d(u,v)$. The smallest integer λ for which all intervals of G are λ-thin is called the *interval thinness* of G. The disk of radius r and center v is defined as $\{u \in V : d(u,v) \leq r\}$, and denoted by $D(v,r)$. In particular, $N[v] = D(v,1)$ and $N(v) = N[v] \setminus \{v\}$ denote the closed and open neighbourhoods of a vertex v, respectively. More generally, for any vertex-subset S and a vertex u, we define $d(u,S) = \min_{v \in S} d(u,v)$, $D(S,r) = \bigcup_{v \in S} D(v,r)$, $N[S] = D(S,1)$ and $N(S) = N[S] \setminus S$. We say that a set of vertices $S \subseteq V$ of a graph $G = (V, E)$ is d^k-convex if for every two vertices $x, y \in S$ with $d(x,y) \geq k \geq 0$, the entire interval $I(x,y)$ is in S. For $k \leq 2$, this definition coincides with the usual definition of convex sets in graphs [4,10,37]. We show first that, in α_i-metric graphs G, the intervals are $(i+1)$-thin, and the disks (and, hence, the centers $C(G)$) are d^{2i-1}-convex. The main result of Sect. 2.1 states that the diameter of the center $C(G)$ of G is at most $3i + 2$, thus answering a question raised in [39].

Let $F_G(v)$ be the set of all vertices of G that are most distant from v. A pair x, y is called a pair of mutually distant vertices if $x \in F_G(y), y \in F_G(x)$. In Sect. 2.2, we show that an additive $O(i)$-approximation of the radius and of the diameter of an α_i-metric graph G with m edges can be computed in $O(m)$ time. For that, we carefully analyze the eccentricities of most distant vertices from an arbitrary vertex and of mutually distant vertices. In Sect. 2.3, we present three approximation algorithms for all eccentricities, with various trade-offs between their running time and the quality of their approximation. Hence, an additive $O(i)$-approximation of all vertex eccentricities of an α_i-metric graph G with m edges can be computed in $O(m)$ time.

Section 3 is devoted to α_1-metric graphs. The eccentricity function $e(v)$ of a graph G is said to be *unimodal*, if for every non-central vertex v of G there is a neighbor $u \in N(v)$ such that $e(u) < e(v)$ (that is, every local minimum of the eccentricity function is a global minimum). We show in Sect. 3.1 that the eccentricity function on α_1-metric graphs is almost unimodal and we characterize non-central vertices that violate the unimodality (that is, do not have a neighbor with smaller eccentricity). Such behavior of the eccentricity function was observed earlier in chordal graphs [25], in distance-hereditary graphs [22] and in all (α_1, Δ)-metric graphs [25]. In Sect. 3.2, we show that the diameter of $C(G)$ is at most 3. This generalizes known results for chordal graphs [11] and for (α_1, Δ)-metric graphs [25]. Finally, based on these results we present in Sect. 3.3 a local-search algorithm for finding a central vertex of an arbitrary α_1-metric graph in subquadratic time. Our algorithm even achieves linear runtime on (α_1, Δ)-metric graphs, thus answering an open question from [25].

All omitted proofs can be found in our technical report [19].

2   General Case of α_i-Metric Graphs for Arbitrary $i \geq 0$

First we present two important lemmas for what follows.

Lemma 1. *Let G be an α_i-metric graph, and let u, v, x, y be vertices such that $x \in I(u,v)$, $d(u,x) = d(u,y)$, and $d(v,y) \leq d(v,x) + k$. Then, $d(x,y) \leq k + i + 2$.*

Lemma 2. *If G is an α_i-metric graph, then its interval thinness is at most $i+1$.*

2.1   Centers of α_i-Metric Graphs

We provide an answer to a question raised in [39] whether the diameter of the center of an α_i-metric graph can be bounded by a linear function of i. For that, we show first that every disk must be d^{2i-1}-convex.

Lemma 3. *Every disk of an α_i-metric graph G is d^{2i-1}-convex. In particular, the center $C(G)$ of an α_i-metric graph G is d^{2i-1}-convex.*

Next auxiliary lemma is crucial in obtaining many results of this section.

Lemma 4. *Let G be an α_i-metric graph. For any $x, y, v \in V$ and any integer $k \in \{0, \ldots, d(x,y)\}$, there is a vertex $c \in S_k(x,y)$ such that $d(v,c) \leq \max\{d(v,x), d(v,y)\} - \min\{d(x,c), d(y,c)\} + i$ and $d(v,c) \leq \max\{d(v,x), d(v,y)\} + i/2$. For an arbitrary vertex $z \in I(x,y)$, we have $d(z,v) \leq \max\{d(x,v), d(y,v)\} - \min\{d(x,z), d(y,z)\} + 2i + 1$ and $d(z,v) \leq \max\{d(x,v), d(y,v)\} + 3i/2 + 1$. Furthermore, $e(z) \leq \max\{e(x), e(y)\} - \min\{d(x,z), d(y,z)\} + 2i + 1$ and $e(z) \leq \max\{e(x), e(y)\} + 3i/2 + 1$ when $v \in F(z)$.*

Using Lemma 4, one can easily prove that the diameter of the center $C(G)$ of an α_i-metric graph G is at most $4i + 3$. Below we improve the bound.

Theorem 1. *If G is an α_i-metric graph, then $diam(C(G)) \leq 3i + 2$.*

Proof. Let $r = rad(G)$. Suppose by contradiction $diam(C(G)) > 3i + 2$. Since $C(G)$ is d^{2i-1}-convex, there exist $x, y \in C(G)$ such that $d(x,y) = 3i + 3$ and $I(x,y) \subseteq C(G)$. Furthermore, for every $u \in V$ such that $\max\{d(u,x), d(u,y)\} < r$, $I(x,y) \subseteq D(u, r-1)$ because the latter disk is also d^{2i-1}-convex. Therefore, for every $z \in I(x,y)$, $F(z) \subseteq F(x) \cup F(y)$. Let ab be an edge on a shortest xy-path such that $d(a,x) < d(b,x)$. Assume $F(b) \not\subseteq F(a)$. Let $v \in F(b) \setminus F(a)$. Since G is α_i-metric, $d(v,y) \geq d(v,b) + d(b,y) - i = r + (d(b,y) - i)$. Therefore, $d(b,y) \leq i$. In the same way, if $F(a) \not\subseteq F(b)$, then $d(a,x) \leq i$. By induction, $F(z) \subseteq F(x)$ ($F(z) \subseteq F(y)$, respectively) for every $z \in I(x,y)$ such that $d(y,z) \geq i + 1$ ($d(x,z) \geq i + 1$, respectively). In particular, if $i + 1 \leq t \leq d(x,y) - i - 1$, then $F(z) \subseteq F(x) \cap F(y)$ for every $z \in S_t(x,y)$. Note that the above properties are also true for every $x', y' \in C(G)$ with $d(x',y') \geq 2i - 1$, as d^{2i-1}-convexity argument can still be used.

Let $c \in I(x, y)$ be such that $F(c) \subseteq F(x) \cap F(y)$ and $k := |F(c)|$ is minimized. We claim that $k < |F(x) \cap F(y)|$. Indeed, let $v \in F(x) \cap F(y)$ be arbitrary. By Lemma 4, some vertex $c_v \in S_{i+1}(x, y)$ satisfies $d(c_v, v) \leq r - 1$. Then, $F(c_v) \subseteq (F(x) \cap F(y)) \setminus \{v\}$, and $k \leq |F(c_v)| \leq |F(x) \cap F(y)| - 1$ by minimality of c. Let $y_c \in I(x, y)$ be such that $F(y_c) \cap F(x) \cap F(y) \subseteq F(c)$ and $d(x, y_c)$ is maximized. We have $y_c \neq y$ because $F(x) \cap F(y) \not\subseteq F(c)$. Therefore, the maximality of $d(x, y_c)$ implies the existence of some $v \in (F(x) \cap F(y)) \setminus F(c)$ such that $d(v, y_c) = r - 1$. Since G is α_i-metric, $d(v, y) \geq d(v, y_c) + d(y_c, y) - i = r + (d(y_c, y) - i - 1)$. As a result, $d(y_c, y) \leq i + 1$. Then, for every $z \in S_{i+1}(x, y_c)$, since we have $d(z, y_c) = d(x, y) - i - 1 - d(y_c, y) \geq d(x, y) - 2i - 2 = i + 1$, $F(z) \subseteq F(x) \cap F(y) \cap F(y_c) \subseteq F(c)$. By minimality of k, $F(z) = F(c)$. However, let $v \in F(c)$ be arbitrary. By Lemma 4, there exists some $c' \in S_{i+1}(x, y_c)$ such that $d(c', v) \leq r - 1$, thus contradicting $F(c') = F(c)$. $\square$

2.2 Approximating Radii and Diameters of α_i-Metric Graphs

In this subsection, we show that a vertex with eccentricity at most $rad(G) + O(i)$ and a vertex with eccentricity at least $diam(G) - O(i)$ of an α_i-metric graph G can be found in parameterized linear time. We summarize algorithmic results of this section in the following theorem.

Theorem 2. *There is a linear $(O(m))$ time algorithm which finds vertices v and c of an m-edge α_i-metric graph G such that $e(v) \geq diam(G) - 5i - 2$, $e(c) \leq rad(G) + 4i + (i+1)/2 + 2$ and $C(G) \subseteq D(c, 4i + (i+1)/2 + 2)$. Furthermore, there is an almost linear $(O(im))$ time algorithm which finds vertices v and c of G such that $e(v) \geq diam(G) - 3i - 2$, $e(c) \leq rad(G) + 2i + 1$ and $C(G) \subseteq D(c, 4i + 3)$.*

Our algorithms are derived from the following new properties of an α_i-metric graph G, whose proofs are omitted due to lack of space.

- Let x, y be a pair of mutually distant vertices. Then, $d(x, y) \geq 2rad(G) - 4i - 3$ and $d(x, y) \geq diam(G) - 3i - 2$. Furthermore, any middle vertex z of a shortest path between x and y satisfies $e(z) \leq \lceil d(x, y)/2 \rceil + 2i + 1 \leq rad(G) + 2i + 1$, and $C(G) \subseteq D(z, 4i + 3)$. There is also a vertex c in $S_{\lfloor d(x,y)/2 \rfloor}(x, y)$ with $e(c) \leq rad(G) + i$.
- Let $v \in V$, $x \in F(v)$ and $y \in F(x)$. Then, $e(x) = d(x, y) \geq 2rad(G) - 2i - diam(C(G)) \geq 2rad(G) - 5i - 2 \geq diam(G) - 5i - 2$. Furthermore, any middle vertex z of a shortest path between x and y satisfies $e(z) \leq \min\{rad(G) + 4i + (i+1)/2 + 2, \lceil d(x, y)/2 \rceil + 7i + 3\}$ and $C(G) \subseteq D(z, 4i + (i+1)/2 + 2)$.

In particular, using at most $O(i)$ breadth-first-searches, one can generate a sequence of vertices $v := v_0, x := v_1, y := v_2, v_3, \ldots v_k$ with $k \leq 5i + 4$ such that each v_i is most distant from v_{i-1} and v_k, v_{k-1} are mutually distant vertices (because the initial value $d(x, y) \geq diam(G) - 5i - 2$ can be improved at most $5i + 2$ times). Therefore, a pair of mutually distant vertices of an α_i-metric graph can be computed in $O(im)$ total time, thus proving Theorem 2.

For every vertex $v \in V \setminus C(G)$ of a graph G we can define a parameter $loc(v) = \min\{d(v,x) : x \in V, e(x) < e(v)\}$ and call it the *locality* of v. It shows how far from v a vertex with a smaller eccentricity than that one of v exists. In α_i-metric graphs, the locality of each vertex is at most $i + 1$.

Lemma 5. *Let G be an α_i-metric graph. Then, for every vertex v, $loc(v) \leq i+1$.*

In α_i-metric graphs, the difference between the eccentricity of a vertex v and the radius of G shows how far vertex v can be from the center $C(G)$ of G.

Lemma 6. *Let G be an α_i-metric graph and k be a positive integer. Then, for every vertex v of G with $e(v) \leq rad(G) + k$, $d(v, C(G)) \leq k + i$.*

Proof. Let x be a vertex from $C(G)$ closest to v. Consider a neighbor z of x on an arbitrary shortest path from x to v. Necessarily, $e(z) = e(x) + 1 = rad(G) + 1$. Consider a vertex $u \in F(z)$. We have $d(u,x) = rad(G)$ and $x \in I(z,u)$, $z \in I(x,v)$. By the α_i-metric property, $d(v,u) \geq d(v,x) + d(x,u) - i = d(v,x) - i + rad(G)$. As $e(v) \geq d(v,u)$ and $e(v) \leq rad(G) + k$, we get $rad(G) + k \geq e(v) \geq d(v,x) - i + rad(G)$, i.e., $d(v,x) \leq i + k$.   □

As an immediate corollary of Lemma 6 we get.

Corollary 1. *Let G be an α_i-metric graph. Then, for every vertex v of G, $d(v, C(G)) + rad(G) \geq e(v) \geq d(v, C(G)) + rad(G) - i$.*

So, in α_i-metric graphs, to approximate the eccentricity of a vertex v up-to an additive one-sided error i, one needs to know only $rad(G)$ and the distance from v to the center $C(G)$ of G.

2.3 Approximating all Eccentricities in α_i-Metric Graphs

In this subsection, we show that the eccentricities of all vertices of an α_i-metric graph G can be approximated with an additive one-sided error at most $O(i)$ in (almost) linear total time. The following first result is derived from the interesting property that the distances from any vertex v to two mutually distant vertices give a very good estimation on the eccentricity of v.

Theorem 3. *Let G be an α_i-metric graph with m-edges. There is an algorithm which in total almost linear ($O(im)$) time outputs for every vertex $v \in V$ an estimate $\hat{e}(v)$ of its eccentricity $e(v)$ such that $e(v) - 3i - 2 \leq \hat{e}(v) \leq e(v)$.*

A spanning tree T of a graph G is called an *eccentricity k-approximating spanning tree* if for every vertex v of G $e_T(v) \leq e_G(v) + k$ holds [35]. All $(\alpha_1, \triangle)$-metric graphs (including chordal graphs and the underlying graphs of 7-systolic complexes) admit eccentricity 2-approximating spanning trees [25]. An eccentricity 2-approximating spanning tree of a chordal graph can be computed in linear time [18]. An eccentricity k-approximating spanning tree with minimum k can be found in $O(nm)$ time for any n-vertex, m-edge graph G [27]. It is also known [15,23] that if G is a δ-hyperbolic graph, then G admits an eccentricity $(4\delta + 1)$-approximating spanning tree constructible in $O(\delta m)$ time and an eccentricity (6δ)-approximating spanning tree constructible in $O(m)$ time.

Lemma 7. *Let G be an α_i-metric graph with m edges. If c is a middle vertex of any shortest path between a pair x, y of mutually distant vertices of G and T is a $BFS(c)$-tree of G, then, for every vertex v of G, $e_G(v) \leq e_T(v) \leq e_G(v) + 4i + 2$. That is, G admits an eccentricity $(4i + 2)$-approximating spanning tree constructible in $O(im)$ time.*

Lemma 8. *Let G be an α_i-metric graph with m edges, and let $z \in V$, $x \in F(z)$ and $y \in F(x)$. If c is a middle vertex of any shortest path between x and y and T is a $BFS(c)$-tree of G, then, for every vertex v of G, $e_G(v) \leq e_T(v) \leq e_G(v) + 9i + 5$. That is, G admits an eccentricity $(9i + 5)$-approximating spanning tree constructible in $O(m)$ time.*

It is a folklore by now that the eccentricities of all vertices in any tree $T = (V, U)$ can be computed in $O(|V|)$ total time. Consequently, by Lemma 7 and Lemma 8, we get the following additive approximations for the vertex eccentricities in α_i-metric graphs.

Theorem 4. *Let G be an α_i-metric graph with m edges. There is an algorithm which in total linear $(O(m))$ time outputs for every vertex $v \in V$ an estimate $\hat{e}(v)$ of its eccentricity $e(v)$ such that $e(v) \leq \hat{e}(v) \leq e(v) + 9i + 5$. Furthermore, there is an algorithm which in total almost linear $(O(im))$ time outputs for every vertex $v \in V$ an estimate $\hat{e}(v)$ of its eccentricity $e(v)$ such that $e(v) \leq \hat{e}(v) \leq e(v) + 4i + 2$.*

3   Graphs with α_1-Metric

Now we concentrate on α_1-metric graphs, which contain all chordal graphs and all plane triangulations with inner vertices of degree at least 7 (see [10, 11, 25, 39]). For them we get much sharper bounds. First we recall some known results.

Theorem 5 ([39]). *G is an α_1-metric graph if and only if all disks $D(v, k)$ $(v \in V, k \geq 1)$ of G are convex and G does not contain the graph W_6^{++} from Fig. 1 as an isometric subgraph.*

Fig. 1. Forbidden isometric subgraph W_6^{++}.

Lemma 9 ([37]). *All disks $D(v, k)$ $(v \in V, k \geq 1)$ of a graph G are convex if and only if for every vertices $x, y, z \in V$ and $v \in I(x, y)$, $d(v, z) \leq \max\{d(x, z), d(y, z)\}$.*

Letting z to be from $F(v)$, we get:

Corollary 2. *If all disks $D(v, k)$ $(v \in V, k \geq 1)$ of a graph G are convex then for every vertices $x, y \in V$ and $v \in I(x, y)$, $e(v) \leq \max\{e(x), e(y)\}$.*

Lemma 10 ([25]). *Let G be an α_1-metric graph and x be its arbitrary vertex with $e(x) \geq rad(G) + 1$. Then, for every vertex $z \in F(x)$ and every neighbor v of x in $I(x, z)$, $e(v) \leq e(x)$ holds.*

3.1   The Eccentricity Function on α_1-Metric Graphs is Almost Unimodal

We prove the following theorem.

Theorem 6. *Let G be an α_1-metric graph and v be an arbitrary vertex of G. If*

(i) $e(v) > rad(G) + 1$ or
(ii) $e(v) = rad(G) + 1$ and $diam(G) < 2\, rad(G) - 1$,

then there must exist a neighbor w of v with $e(w) < e(v)$.

Theorem 6 says that if a vertex v with $loc(v) > 1$ exists in an α_1-metric graph G then $diam(G) \geq 2rad(G) - 1$, $e(v) = rad(G) + 1$ and $d(v, C(G)) = 2$. Two α_1-metric graphs depicted in Fig. 2 show that this result is sharp.

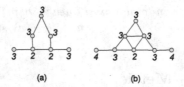

(a)   (b)

Fig. 2. Sharpness of the result of Theorem 6. (a) An α_1-metric graph G with $diam(G) = 2rad(G) - 1$ and a vertex (topmost) with locality 2. (b) A chordal graph (and hence an α_1-metric graph) G with $diam(G) = 2rad(G)$ and a vertex (topmost) with locality 2. The number next to each vertex indicates its eccentricity.

We formulate three interesting corollaries of Theorem 6.

Corollary 3. *Let G be an α_1-metric graph. Then,*

(i) if $diam(G) < 2rad(G) - 1$ (i.e., $diam(G) = 2rad(G) - 2$) then every local minimum of the eccentricity function on G is a global minimum.
(ii) if $diam(G) \geq 2rad(G) - 1$ then every local minimum of the eccentricity function on G is a global minimum or is at distance 2 from a global minimum.

Corollary 4. *For every α_1-metric graph G and any vertex v, the following formula is true: $d(v, C(G)) + rad(G) \geq e(v) \geq d(v, C(G)) + rad(G) - \epsilon$, where $\epsilon \leq 1$, if $diam(G) \geq 2rad(G) - 1$, and $\epsilon = 0$, otherwise.*

A path $(v = v_0, \ldots, v_k = x)$ of a graph G from a vertex v to a vertex x is called *strictly decreasing* (with respect to the eccentricity function) if for every i $(0 \leq i \leq k-1)$, $e(v_i) > e(v_{i+1})$. It is called *decreasing* if for every i $(0 \leq i \leq k-1)$, $e(v_i) \geq e(v_{i+1})$. An edge $ab \in E$ of a graph G is called *horizontal* (with respect to the eccentricity function) if $e(a) = e(b)$.

Corollary 5. *Let G be an α_1-metric graph and v be an arbitrary vertex. Then, there is a shortest path $P(v, x)$ from v to a closest vertex x in $C(G)$ such that:*

(i) if $diam(G) < 2rad(G) - 1$ (i.e., $diam(G) = 2rad(G) - 2$) then $P(v, x)$ is strictly decreasing;

(ii) if $diam(G) \geq 2rad(G) - 1$ then $P(v, x)$ is decreasing and can have only one horizontal edge, with an end-vertex adjacent to x.

3.2 Diameters of Centers of α_1-Metric Graphs

In this section, we provide sharp bounds on the diameter and the radius of the center of an α_1-metric graph. Previously, it was known that the diameter (the radius) of the center of a chordal graph is at most 3 (at most 2, respectively) [11]. To prove our result, we will need a few technical lemmas.

Lemma 11. *Let G be an α_1-metric graph. Then, for every shortest path $P = (x_1, x_2, x_3, x_4, x_5)$ and a vertex u of G with $d(u, x_i) = k$ for all $i \in \{1, \ldots, 5\}$, there exist vertices t, w, s such that $d(t, u) = d(s, u) = k - 1$, $k - 2 \leq d(w, u) \leq k - 1$, and t is adjacent to x_1, x_2, w and s is adjacent to x_4, x_5, w.*

Lemma 12. *Let G be an α_1-metric graph. Then, for every shortest path $P = (x_1, x_2, x_3, x_4, x_5)$ and a vertex u of G with $d(u, x_i) = k$ for all $i \in \{1, \ldots, 5\}$, there exists a shortest path $Q = (y_1, y_2, y_3)$ such that $d(u, y_i) = k - 1$, for each $i \in \{1, \ldots, 3\}$, and $N(y_1) \cap P = \{x_1, x_2\}$, $N(y_2) \cap P = \{x_2, x_3, x_4\}$ and $N(y_3) \cap P = \{x_4, x_5\}$.*

Theorem 7. *Let G be an α_1-metric graph. For every pair of vertices s, t of G with $d(s, t) \geq 4$ there exists a vertex $c \in I^\circ(s, t)$ such that $e(c) < \max\{e(s), e(t)\}$.*

Proof. It is sufficient to prove the statement for vertices s, t with $d(s, t) = 4$. We know, by Corollary 2, that $e(c) \leq \max\{e(s), e(t)\}$ for every $c \in I(s, t)$. Assume, by way of contradiction, that there is no vertex $c \in I^\circ(s, t)$ such that $e(c) < \max\{e(s), e(t)\}$. Let, without loss of generality, $e(s) \leq e(t)$. Then, for every $c \in I^\circ(s, t)$, $e(c) = e(t)$. Consider a vertex $c \in S_1(s, t)$. If $e(c) > e(s)$, then $e(c) = e(s) + 1$. Consider a vertex z from $F(c)$. Necessarily, $z \in F(s)$. Applying the α_1-metric property to $c \in I(s, t)$, $s \in I(c, z)$, we get $e(c) = e(t) \geq d(t, z) \geq d(c, t) + d(s, z) = 3 + e(s) = 2 + e(c)$, which is impossible. So, $e(s) = e(c) = e(t)$ for every $c \in I^\circ(s, t)$. Consider an arbitrary shortest path $P = (s = x_1, x_2, x_3, x_4, x_5 = t)$ connecting vertices s and t. We claim that for any vertex $u \in F(x_3)$ all vertices of P are at distance $k := d(u, x_3) = e(x_3)$ from u. As $e(x_i) = e(x_3)$, we know that $d(u, x_i) \leq k$ ($1 \leq i \leq 5$). Assume $d(u, x_i) = k - 1$, $d(u, x_{i+1}) = k$, and $i \leq 2$. Then, the α_1-metric property applied to $x_i \in I(u, x_{i+1})$ and $x_{i+1} \in I(x_i, x_{i+3})$ gives $d(x_{i+3}, u) \geq k - 1 + 2 = k + 1$, which is a contradiction with $d(u, x_{i+3}) \leq k$. So, $d(u, x_1) = d(u, x_2) = k$. By symmetry, also $d(u, x_4) = d(u, x_5) = k$. Hence, by Lemma 12, for the path $P = (x_1, x_2, x_3, x_4, x_5)$, there exists a shortest path $Q = (y_1, y_2, y_3)$ such that $d(u, y_i) = k - 1$, for each $i \in \{1, \ldots, 3\}$, and $N(y_1) \cap P = \{x_1, x_2\}$, $N(y_2) \cap P =$

$\{x_2, x_3, x_4\}$ and $N(y_3) \cap P = \{x_4, x_5\}$. As $y_i \in I^o(x_1, x_5) = I^o(s, t)$ for each $i \in \{1, \dots, 3\}$, we have $e(y_i) = e(x_3) = k$.

All the above holds for every shortest path $P = (s = x_1, x_2, x_3, x_4, x_5 = t)$ connecting vertices s and t. Now, assume that P is chosen in such a way that, among all vertices in $S_2(s, t)$, the vertex x_3 has the minimum number of furthest vertices, i.e., $|F(x_3)|$ is as small as possible. As y_2 also belongs to $S_2(s, t)$ and has u at distance $k - 1$, by the choice of x_3, there must exist a vertex $u' \in F(y_2)$ which is at distance $k - 1$ from x_3. Applying the previous arguments to the path $P' := (s = x_1, x_2, y_2, x_4, x_5 = t)$, we will have $d(x_i, u') = d(y_2, u') = k$ for $i = 1, 2, 4, 5$ and, by Lemma 12, get two more vertices v and w at distance $k - 1$ from u' such that $vx_1, vx_2, wx_4, wx_5 \in E$ and $vy_2, wy_2 \notin E$. By convexity of disk $D(u', k - 1)$, also $vx_3, wx_3 \in E$. Now consider the disk $D(x_2, 2)$. Since y_3, w are in the disk and x_5 is not, vertices w and y_3 must be adjacent. But then vertices y_2, x_3, w, y_3 form an induced cycle C_4, which is forbidden because disks in G must be convex.

Thus, a vertex $c \in I^o(s, t)$ with $e(c) < \max\{e(s), e(t)\}$ must exist. □

Corollary 6. *Let G be an α_1-metric graph. Then, $diam(C(G)) \leq 3$ and $rad(C(G)) \leq 2$.*

Corollary 6 generalizes an old result on chordal graphs [11]. Finally, note that results of Theorem 7 and Corollary 6 are sharp.

3.3 Finding a Central Vertex of an α_1-Metric Graph

We present a local-search algorithm for computing a central vertex of an arbitrary α_1-metric graph in subquadratic time (Theorem 8). Our algorithm even achieves linear runtime on an important subclass of α_1-metric graphs, namely, (α_1, Δ)-metric graphs (Theorem 9), thus answering an open question from [25] where this subclass was introduced. The (α_1, Δ)-metric graphs are exactly the α_1-metric graphs that further satisfy the so-called triangle condition: for every vertices u, v, w such that u and v are adjacent, and $d(u, w) = d(v, w) = k$, there must exist some common neighbour $x \in N(u) \cap N(v)$ such that $d(x, w) = k - 1$. Chordal graphs, and plane triangulations with inner vertices of degree at least 7, are (α_1, Δ)-metric graphs (see [10,11,25,39]).

We first introduce the required new notations and terminology for this part. In what follows, let $\text{proj}(v, A) = \{a \in A : d(v, a) = d(v, A)\}$ denote the metric projection of a vertex v to a vertex subset A. For every k such that $0 \leq k \leq d(v, A)$, we define $S_k(A, v) = \bigcup\{S_k(a, v) : a \in \text{proj}(v, A)\}$. A *distance-$k$ gate* of v with respect to A is a vertex v^* such that $v^* \in \bigcap\{I(a, v) : a \in \text{proj}(v, A)\}$ and $d(v^*, A) \leq k$. If $k = 1$, then following [12] we simply call it a gate. Note that every vertex v such that $d(v, A) \leq k$ is its own distance-k gate. A cornerstone of our main algorithms is that, in α_1-metric graphs, for every *closed neighbourhood* (for every *clique*, resp.), every vertex has a gate (a distance-two gate, resp.). Proofs are omitted due to lack of space.

The problem of computing gates has already attracted some attention, *e.g.*, see [12]. We use this routine in the design of our main algorithms.

Lemma 13 ([30]). *Let A be an arbitrary subset of vertices in some graph G with m edges. In total $O(m)$ time, we can map every vertex $v \notin A$ to some vertex $v^* \in D(v, d(v, A) - 1) \cap N(A)$ such that $|N(v^*) \cap A|$ is maximized. Furthermore, if v has a gate with respect to A, then v^* is a gate of v.*

The efficient computation of distance-two gates is more challenging. We present a subquadratic-time procedure that only works in our special setting.

Lemma 14. *Let K be a clique in some α_1-metric graph G with m edges. In total $O(m^{1.41})$ time, we can map every vertex $v \notin K$ to some distance-two gate v^* with respect to K. Furthermore, in doing so we can also map v^* to some independent set $J_K(v^*) \subseteq D(v^*, 1)$ such that $\text{proj}(v^*, K)$ is the disjoint union of neighbour-sets $N(w) \cap K$, for every $w \in J_K(v^*)$.*

Then, we turn our attention to the following subproblem: being given a vertex x in an α_1-metric graph G, either compute a neighbour y such that $e(y) < e(x)$, or assert that x is a local minimum for the eccentricity function (but not necessarily a central vertex). Our analysis of the next algorithms essentially follows from the results of Sect. 3.1. We first present the following special case, for which we obtain a better runtime than for the more general Lemma 16.

Lemma 15. *Let x be an arbitrary vertex in an α_1-metric graph G with m edges. If $e(x) \geq rad(G) + 2$, then $\bigcap \{N(x) \cap I(x, z) : z \in F(x)\} \neq \emptyset$, and every neighbour y in this subset satisfies $e(y) < e(x)$. In particular, there is an $O(m)$-time algorithm that either outputs a $y \in N(x)$ such that $e(y) < e(x)$, or asserts that $e(x) \leq rad(G) + 1$.*

Note that Lemma 15 relies on the existence of gates for every vertex with respect to $D(x, 1)$, and that it uses Lemma 13 as a subroutine. We can strengthen Lemma 15 as follows, at the expenses of a higher runtime.

Lemma 16. *Let x be an arbitrary vertex in an α_1-metric graph G with m edges. There is an $O(m^{1.41})$-time algorithm that either outputs a $y \in N(x)$ such that $e(y) < e(x)$, or asserts that x is a local minimum for the eccentricity function. If G is (α_1, Δ)-metric, then its runtime can be lowered down to $O(m)$.*

The improved runtime for (α_1, Δ)-metric graphs comes from the property that for every clique in such a graph, every vertex has a gate [25], and that gates are easier to compute than distance-two gates.

Theorem 8. *If G is an α_1-metric graph with m edges, then a vertex x_0 such that $e(x_0) \leq rad(G) + 1$ can be computed in $O(m)$ time. Furthermore, a central vertex can be computed in $O(m^{1.71})$ time.*

Let us sketch our algorithm for general α_1-metric graphs. By Theorem 2, we can compute in $O(m)$ time a vertex x_0 such that $e(x_0) \leq rad(G) + 3$. We repeatedly apply Lemma 15 until we can further assert that $e(x_0) \leq rad(G) + 1$ (and, hence, by Theorem 6, $d(x_0, C(G)) \leq 2$). Since there are at most two calls to this local-search procedure, the runtime is in $O(m)$. Now, if x_0 has small enough degree ($\leq m^{.29}$), then we can apply Lemma 16 to every vertex of $D(x_0, 1)$ in order to compute a minimum eccentricity vertex within $D(x_0, 2)$, which must be central. Otherwise, we further restrict our search for a central vertex to $X_1 := D(x_0, 5) \cap D(z_0, e(x_0) - 1)$, for some arbitrary $z_0 \in F(x_0)$, which must be a superset of $C(G)$ provided that x_0 is non-central. Starting from an arbitrary vertex of X_1, if we apply Lemma 15 at most five times, then we can either extract some $x_1 \in X_1$ such that $e(x_1) \leq e(x_0)$, or assert that x_0 is central. If $e(x_1) < e(x_0)$, then x_1 is central. Otherwise, we repeat our above procedure for x_1. Doing so, we compute a decreasing chain of subsets $X_0 = V \supset X_1 \supset \ldots \supset X_i \supset \ldots X_T$, and vertices $x_0, x_1, \ldots, x_i, \ldots, x_T$ such that $x_i \in X_i \setminus X_{i+1}$ for every i, $0 \leq i \leq T$. We continue until we compute a vertex of smaller eccentricity than x_0, or we reach $X_{T+1} = \emptyset$ (in which case, x_0 is central).

To lower the runtime to $O(m)$ for the (α_1, Δ)-metric graphs, we use a different approach that is based on additional properties of these graphs. Unfortunately, these properties crucially depend on the triangle condition.

Theorem 9. *If G is an (α_1, Δ)-metric graph with m edges, then a central vertex can be computed in $O(m)$ time.*

Roughly, the algorithm starts from a vertex x which is a local minimum for the eccentricity function of G. We run a core procedure which either outputs two adjacent vertices $u, v \in D(x, 1)$ such that $e(u) = e(v) = e(x)$ and $F(u), F(v)$ are not comparable by inclusion, or outputs a central vertex. In the former case, we can either assert that x is central, or extract a central vertex from $S_{e(x)-1}(y, z)$ for some arbitrary $y \in F(u) \setminus F(v)$, $z \in F(v) \setminus F(u)$. Indeed, we can apply Lemma 16 to the latter slice because $S_{e(x)-1}(y, z) \subseteq D(w, 1)$ for some w [25].

References

1. Abboud, A., Vassilevska Williams, V., Wang, J.: Approximation and fixed parameter subquadratic algorithms for radius and diameter in sparse graphs. In: SODA, pp. 377–391. SIAM (2016)
2. Backurs, A., Roditty, L., Segal, G., Vassilevska Williams, V., Wein, N.: Towards tight approximation bounds for graph diameter and eccentricities. In: STOC 2018, pp. 267–280 (2018)
3. Bandelt, H.-J., Chepoi, V.: 1-hyperbolic graphs. SIAM J. Discr. Math. **16**, 323–334 (2003)
4. Boltyanskii, V.G., Soltan, P.S.: Combinatorial geometry of various classes of convex sets [in Russian]. Štiinţa, Kishinev (1978)
5. Borassi, M., Crescenzi, P., Habib, M.: Into the square: on the complexity of some quadratic-time solvable problems. Electron. Notes TCS **322**, 51–67 (2016)
6. Brandstädt, A., Chepoi, V., Dragan, F.F.: The algorithmic use of hypertree structure and maximum neighbourhood orderings. DAM **82**, 43–77 (1998)
7. Corneil, D., Dragan, F.F., Habib, M., Paul, C.: Diameter determination on restricted graph families. DAM **113**, 143–166 (2001)

8. Corneil, D.G., Dragan, F.F., Köhler, E.: On the power of BFS to determine a graph's diameter. Networks **42**, 209–222 (2003)
9. Chechik, S., Larkin, D.H., Roditty, L., Schoenebeck, G., Tarjan, R.E., Vassilevska Williams, V.: Better approximation algorithms for the graph diameter. In: SODA 2014, pp. 1041–1052 (2014)
10. Chepoi, V.: Some d-convexity properties in triangulated graphs. In: Mathematical Research, vol. 87, pp. 164–177. Ştiinţa, Chişinău (1986). (Russian)
11. Chepoi, V.: Centers of triangulated graphs. Math. Notes **43**, 143–151 (1988)
12. Chepoi, V., Dragan, F.: A linear-time algorithm for finding a central vertex of a chordal graph. In: van Leeuwen, J. (ed.) ESA 1994. LNCS, vol. 855, pp. 159–170. Springer, Heidelberg (1994). https://doi.org/10.1007/BFb0049406
13. Chepoi, V., Dragan, F.F.: Finding a central vertex in an HHD-free graph. DAM **131**(1), 93–111 (2003)
14. Chepoi, V.D., Dragan, F.F., Estellon, B., Habib, M., Vaxès, Y.: Diameters, centers, and approximating trees of δ-hyperbolic geodesic spaces and graphs. In: Proceedings of the 24th Annual ACM Symposium on Computational Geometry (SoCG 2008), 9–11 June 2008, College Park, Maryland, USA, pp. 59–68 (2008)
15. Chepoi, V., Dragan, F.F., Habib, M., Vaxès, Y., Alrasheed, H.: Fast approximation of eccentricities and distances in hyperbolic graphs. J. Graph Algorithms Appl. **23**, 393–433 (2019)
16. Dragan, F.F.: Centers of graphs and the Helly property (in Russian). Ph.D. thesis, Moldava State University, Chişinău (1989)
17. Dragan, F.F.: HT-graphs: centers, connected R-domination and Steiner trees. Comput. Sci. J. Moldova (Kishinev) **1**, 64–83 (1993)
18. Dragan, F.F.: An eccentricity 2-approximating spanning tree of a chordal graph is computable in linear time. Inf. Process. Lett. **154**, 105873 (2020)
19. Dragan, F.F., Ducoffe, G.: α_i-metric graphs: radius, diameter and all eccentricities. CoRR, abs/2305.02545 (2023)
20. Dragan, F.F., Ducoffe, G.: α_i-metric graphs: hyperbolicity. In: Preparation (2022–2023)
21. Dragan, F.F., Ducoffe, G., Guarnera, H.M.: Fast deterministic algorithms for computing all eccentricities in (hyperbolic) Helly graphs. In: Lubiw, A., Salavatipour, M. (eds.) WADS 2021. LNCS, vol. 12808, pp. 300–314. Springer, Cham (2021). https://doi.org/10.1007/978-3-030-83508-8_22
22. Dragan, F.F., Guarnera, H.M.: Eccentricity function in distance-hereditary graphs. Theor. Comput. Sci. **833**, 26–40 (2020)
23. Dragan, F.F., Guarnera, H.M.: Eccentricity terrain of δ-hyperbolic graphs. J. Comput. Syst. Sci. **112**, 50–65 (2020)
24. Dragan, F.F., Habib, M., Viennot, L.: Revisiting radius, diameter, and all eccentricity computation in graphs through certificates. CoRR, abs/1803.04660 (2018)
25. Dragan, F.F., Köhler, E., Alrasheed, H.: Eccentricity approximating trees. Discret. Appl. Math. **232**, 142–156 (2017)
26. Dragan, F.F., Nicolai, F., Brandstädt, A.: LexBFS-orderings and powers of graphs. In: d'Amore, F., Franciosa, P.G., Marchetti-Spaccamela, A. (eds.) WG 1996. LNCS, vol. 1197, pp. 166–180. Springer, Heidelberg (1997). https://doi.org/10.1007/3-540-62559-3_15
27. Ducoffe, G.: Easy computation of eccentricity approximating trees. DAM **260**, 267–271 (2019)
28. Ducoffe, G.: Around the diameter of AT-free graphs. JGT **99**(4), 594–614 (2022)

29. Ducoffe, G.: Beyond Helly graphs: the diameter problem on absolute retracts. In: Kowalik, Ł, Pilipczuk, M., Rzążewski, P. (eds.) WG 2021. LNCS, vol. 12911, pp. 321–335. Springer, Cham (2021). https://doi.org/10.1007/978-3-030-86838-3_25

30. Ducoffe, G.: Distance problems within Helly graphs and k-Helly graphs. Theor. Comput. Sci. **946**, 113690 (2023)

31. Ducoffe, G., Dragan, F.F.: A story of diameter, radius, and (almost) Helly property. Networks **77**, 435–453 (2021)

32. Howorka, E.: A characterization of distance-hereditary graphs. Quart. J. Math. Oxford Ser. **28**, 417–420 (1977)

33. Koschützki, D., Lehmann, K.A., Peeters, L., Richter, S., Tenfelde-Podehl, D., Zlotowski, O.: Centrality indices. In: Brandes, U., Erlebach, T. (eds.) Network Analysis. LNCS, vol. 3418, pp. 16–61. Springer, Heidelberg (2005). https://doi.org/10.1007/978-3-540-31955-9_3

34. Olariu, S.: A simple linear-time algorithm for computing the center of an interval graph. Int. J. Comput. Math. **34**, 121–128 (1990)

35. Prisner, E.: Eccentricity-approximating trees in chordal graphs. Discret. Math. **220**, 263–269 (2000)

36. Roditty, L., Vassilevska Williams, V.: Fast approximation algorithms for the diameter and radius of sparse graphs. In: STOC, pp. 515–524. ACM (2013)

37. Soltan, V.P., Chepoi, V.D.: Conditions for invariance of set diameters under d-convexification in a graph. Cybernetics **19**, 750–756 (1983). (Russian, English transl.)

38. Weimann, O., Yuster, R.: Approximating the diameter of planar graphs in near linear time. ACM Trans. Algorithms **12**(1), 12:1–12:13 (2016)

39. Yushmanov, S.V., Chepoi, V.: A general method of investigation of metric graph properties related to the eccentricity. In: Mathematical Problems in Cybernetics, vol. 3, pp. 217–232. Nauka, Moscow (1991). (Russian)

Maximum Edge Colouring Problem
On Graphs That Exclude a Fixed Minor

Zdeněk Dvořák[1] and Abhiruk Lahiri[2(✉)]

[1] Charles University, 11800 Prague, Czech Republic
`rakdver@iuuk.mff.cuni.cz`
[2] Heinrich Heine University, 40225 Düsseldorf, Germany
`abhiruk@hhu.de`

Abstract. The maximum edge colouring problem considers the maximum colour assignment to edges of a graph under the condition that every vertex has at most a fixed number of distinct coloured edges incident on it. If that fixed number is q we call the colouring a maximum edge q-colouring. The problem models a non-overlapping frequency channel assignment question on wireless networks. The problem has also been studied from a purely combinatorial perspective in the graph theory literature.

We study the question when the input graph is sparse. We show the problem remains NP-hard on 1-apex graphs. We also show that there exists PTAS for the problem on minor-free graphs. The PTAS is based on a recently developed Baker game technique for proper minor-closed classes, thus avoiding the need to use any involved structural results. This further pushes the Baker game technique beyond the problems expressible in the first-order logic.

Keywords: Polynomial-time approximation scheme · Edge colouring · Minor-free graphs

1 Introduction

For a graph $G = (V, E)$, an *edge q-colouring* of G is a mapping $f: E(G) \to \mathbb{Z}^+$ such that the number of distinct colours incident on any vertex $v \in V(G)$ is bounded by q, and the *spread* of f is the total number of distinct colours it uses. The *maximum edge q-chromatic number* $\overline{\chi}'_q(G)$ of G is the maximum spread of an edge q-colouring of G.

A more general notion has been studied in the combinatorics and graph theory communities in the context of extremal problems, called *anti-Ramsey number*. For given graphs G and H, the *anti-Ramsey number* $ar(G, H)$ denotes the maximum number of colours that can be assigned to edges of G so that there does not exist any subgraph isomorphic to H which is *rainbow*, i.e., all the edges

Supported by project 22-17398S (Flows and cycles in graphs on surfaces) of Czech Science Foundation of the subgraph receive distinct colours under the colouring. The maximum edge q-chromatic number of G is clearly equal to $\mathrm{ar}(G, K_{1,q+1})$, where $K_{1,q+1}$ is a star with $q+1$ edges.

The notion of anti-Ramsey number was introduced by Erdős and Simonovits in 1973 [14]. The initial studies focused on determining tight bounds for $\mathrm{ar}(G, H)$. A lot of research has been done on the case when $G = K_n$, the complete graph, and H is a specific type of a graph (a path, a complete graph, ...) [26,29,31]. For a comprehensive overview of known results in this area, we refer interested readers to [17]. Bounds on $\mathrm{ar}(K_n, H)$ where H is a star graph are reported in [21,24]. Gorgol and Lazuka computed the exact value of $\mathrm{ar}(K_n, H)$ when H is $K_{1,4}$ with an edge added to it [18]. For general graph G, Montellano-Ballesteros studied $\mathrm{ar}(G, K_{1,q})$ and reported an upper bound [25].

The algorithmic aspects of this problem started gaining attention from researchers around fifteen years ago, due to its application to wireless networks [27]. At that time there was a great interest to increase the capacity of *wireless mesh networks* (which are commonly called *wireless broadband* nowadays). The solution that became the industry standard is to use multiple channels and transceivers with the ability to simultaneously communicate with many neighbours using multiple radios over the channels [27]. Wireless networks based on the IEEE 802.11a/b/g and 802.16 standards are examples of such systems. But, there is a physical bottleneck in deploying this solution. Enabling every wireless node to have multiple radios can possibly create an interface and thus reduce reliability. To circumvent that, there is a limit on the number of channels simultaneously used by any wireless node. In the IEEE 802.11 b/g standard and IEEE 802.11a standard, the numbers of permittable simultaneous channels are 3 and 12, respectively [34].

If we model a wireless network as a graph where each wireless node corresponds to a vertex of the graph, then the problem can be formulated as a maximum edge colouring problem. The nonoverlapping channels can be associated with distinct colours. On each vertex of the graph, the number of distinctly coloured edges allowed to be incident on it captures the limit on the number of channels that can be used simultaneously at each wireless node. The question of how many channels can be used simultaneously by a given network translates into the number of colours that can be used in a maximum edge colouring.

Devising an efficient algorithm for the maximum edge q-colouring problem is not an easy task. In [1], the problem is reported NP-hard for every $q \geq 2$. The authors further showed that the problem is hard to approximate within a factor of $(1 + \frac{1}{q})$ for every $q \geq 2$, assuming the *unique games conjecture* [2]. A simple 2-approximation algorithm for the maximum edge 2-colouring problem is reported in [15]. The same algorithm from [15] has an approximation ratio of 5/3 with the additional assumption that the graph has a perfect matching [2]. It is also known that the approximation ratio can be improved to 8/5 if the input graph is assumed to be triangle-free [7]. An almost tight analysis of the algorithm is known for the maximum edge q-colouring problem ($q \geq 3$) when the input graph satisfies certain degree constraints [6]. The $q = 2$ case is also known to be fixed-parameter tractable [19].

In spite of several negative theoretical results, the wireless network question continued drawing the attention of researchers due to its relevance in applications. There are several studies focusing on improving approximation under further assumptions on constraints that are meaningful in practical applications [23,30,33,34]. This motivates us to study the more general question on a graph class that captures the essence of wireless mesh networks. Typically, disk graphs and unit disk graphs are well-accepted abstract models for wireless networks. But they can capture more complex networks than what a real-life network looks like [30]. By definition, both unit disk graphs and disk graphs can have arbitrary size cliques. In a practical arrangement of a wireless mesh network, it is quite unlikely to place too many wireless routers in a small area. In other words, a real-life wireless mesh network can be expected to be fairly sparse and avoid large cliques. In this paper, we focus on a popular special case of sparse networks, those avoiding a fixed graph as a minor. In particular, this includes the graphs that can be made planar by deletion of a bounded number k of vertices (the k-apex graphs).

From a purely theoretical perspective, the graphs avoiding a fixed minor are interesting on their own merit. Famously, they admit the structural decomposition devised by Robertson and Seymour [28], but also have many interesting properties that can be shown directly, such as the existence of sublinear separators [3] and admitting layered decomposition into pieces of bounded weak diameter [22]. They have been also intensively studied from the algorithmic perspective, including the PTAS design. Several techniques for this purpose have been developed over the last few decades. The *bidimensionality technique* bounds the treewidth of the graph in terms of the size of the optimal solution and uses the balanced separators to obtain the approximation factor [10,16]. A completely different approach based on local search is known for unweighted problems [5,20]. Dvořák used thin systems of overlays [12] and a generalization of Baker's layering approach [4,13] to obtain PTASes for a wide class of optimization problems expressible in the first-order logic and its variations.

1.1 Our results

Our contribution is twofold. First, we show that the maximum edge q-colouring problem is NP-hard on 1-apex graphs. Our approach is similar in spirit to the approximation hardness reduction for the problem on general graphs [1].

Secondly, we show that there exists a PTAS for the maximum edge q-colouring problem for graphs avoiding a fixed minor. The result uses the *Baker game* approach devised in [13], avoiding the use of involved structural results. The technique was developed to strengthen and simplify the results of [9] giving PTASes for monotone optimization problems expressible in the first-order logic. Our work demonstrates the wider applicability of this technique to problems not falling into this framework.

2 Preliminaries

A graph H is a *minor* of a graph G if a graph isomorphic to H can be obtained from a subgraph of G by a series of edge contractions. We say that G is H-*minor-free* if G does not contain H as a minor. A graph is called *planar* if it can be drawn in the plane without crossings. A graph G is a k-*apex graph* if there exists a set $A \subseteq V(G)$ of size at most k such that $G - A$ is planar. The k-apex graphs are one of the standard examples of graphs avoiding a fixed minor; indeed, they are K_{k+5}-minor-free.

Given a function f assigning colours to edges of a graph G and a vertex $v \in V(G)$, we write $f(v)$ to denote the set $\{f(e) : e \text{ is adjacent to } v\}$, and $f(G) = \{f(e) : e \in E(G)\}$. Recall that f is an edge q-colouring of G if and only if $|f(v)| \leq q$ for every $v \in V(G)$, and the maximum edge q-chromatic number of G is

$$\overline{\chi}_q'(G) = \max\{|f(G)| : f \text{ is an edge } q\text{-colouring of } G\}.$$

A *matching* in a graph G is a set of edges of G where no two are incident with the same vertex. A matching M is *maximal* if it is not a proper subset of any other matching. Note that a maximal matching is not necessarily the largest possible. Let $|G|$ denote $|V(G)| + |E(G)|$. For all other definitions related to graphs not defined in this article, we refer readers to any standard graph theory textbook, such as [11].

3 PTAS for Minor-Free Graphs

Roughly speaking, we employ a divide-and-conquer approach to approximate $\overline{\chi}_q'(G)$, splitting G into vertex disjoint parts $G_1, \ldots, G_m$ in a suitable way, solving the problem for each part recursively, and combining the solutions. An issue that we need to overcome is that it may be impossible to compose the edge q-colourings, e.g., if an edge (v_1, v_2) joins distinct parts and disjoint sets of q colours are used on the neighbourhoods of v_1 and v_2 already. To overcome this issue, we reserve the colour 0 to be used to join the "boundary" vertices. This motivates the following definition.

For a set S of vertices of a graph G, an edge q-colouring f is S-*composable* if $|f(v) \setminus \{0\}| \leq q-1$ for every $v \in S$. Let $\overline{\chi}_q'(G, S)$ denote the maximum number of non-zero colours that can be used by an S-composable edge q-colouring of G. Let us remark that G has an S-composable edge q-colouring using any non-negative number $k' \leq \overline{\chi}_q'(G, S)$ of non-zero colours, as all edges of any colour $c \neq 0$ can be recoloured to 0.

Observation 1 *For any graph G, we have $\overline{\chi}_q'(G) = \overline{\chi}_q'(G, \emptyset)$, and $\overline{\chi}_q'(G, S) \leq \overline{\chi}_q'(G)$ for any $S \subseteq V(G)$.*

We need the following approximation for $\overline{\chi}_q'(G, S)$ in terms of the size of a maximal matching, analogous to one for edge 2-colouring given in [15]. Let us remark that the S-composable edge q-colouring problem is easy to solve for $q = 1$, since we have to use colour 0 on all edges of each component intersecting S and we can use a distinct colour for all edges of any other component. Consequently, in all further claims, we assume $q \geq 2$.

Observation 2 *For any graph G, any $S \subseteq V(G)$, any maximal matching M in G, and any $q \geq 2$,*

$$|M| \leq \overline{\chi}'_q(G, S) \leq \overline{\chi}'_q(G) \leq 2q|M|.$$

Proof. We can assign to each edge of M a distinct positive colour and to all other edges (if any) the colour 0, obtaining an S-composable edge 2-colouring using $|M|$ non-zero colours. On the other hand, consider the set X of vertices incident with the edges of M. By the maximality of M, the set X is a vertex cover of G, i.e., each edge of G is incident with a vertex of X, and thus at most $q|X| = 2q|M|$ colours can be used by any edge q-colouring of G.

In particular, as we show next, the lower bound implies that the S-composable edge q-colouring problem is fixed-parameter tractable when parameterized by the value of the solution (a similar observation on the maximum edge 2-colouring is reported in [19]).

Observation 3 *There exists an algorithm that, given a graph G, a set $S \subseteq V(G)$, and integers $q \geq 2$ and s, in time $O_{q,s}(|G|)$ returns an S-composable edge q-colouring of G using at least $\min(\overline{\chi}'_q(G, S), s)$ colours.*

Proof. We can in linear time find a maximal matching M in G. If $|M| \geq s$, we return the colouring that gives each edge of M a distinct non-zero colour and all other edges colour 0. Otherwise, the set X of vertices incident with M is a vertex cover of G of size at most $2s - 2$, and thus G has treewidth at most $2s - 2$. Note also that for any s', there exists a formula $\varphi_{s',q}$ in monadic second-order logic such that $G, S, E_0, \ldots, E_{s'} \models \varphi_{s',q}$ if and only if $E_0, \ldots, E_{s'}$ is a partition of the edges of G with all parts except possibly for E_0 non-empty such that the function f defined by letting $f(e) = i$ for each $i \in \{0, \ldots, s'\}$ and $e \in E_i$ is an S-composable edge q-colouring of G. Therefore, we can find an S-composable edge q-colouring of G with the maximum number $s' \leq s$ of non-zero colours using Courcelle's theorem [8] in time $O_{q,s}(|G|)$.

A *layering* of a graph G is a function $\lambda \colon V(G) \to \mathbb{Z}^+$ such that $|\lambda(u) - \lambda(v)| \leq 1$ for every edge $(u, v) \in E(G)$. In other words, the graph is partitioned into layers $\lambda^{-1}(i)$ for $i \in \mathbb{Z}^+$ such that edges of G only appear within the layers and between the consecutive layers. Baker [4] gave a number of PTASes for planar graphs based on the fact that in a layering of a connected planar graph according to the distance from a fixed vertex, the union of a constant number of consecutive layers induces a subgraph of bounded treewidth. This is not the case for graphs avoiding a fixed minor in general, however, a weaker statement expressed in terms of Baker game holds. We are going to describe that result in more detail in the following subsection. Here, let us state the key observation that makes layering useful for approximating the edge q-chromatic number.

For integers $r \geq 2$ and m such that $0 \leq m \leq r-1$, the (λ, r, m)-stratification of a graph G is the pair (G', S') such that

- G' is obtained from G by deleting all edges uv such that $\lambda(u) \equiv m \pmod{r}$ and $\lambda(v) \equiv m+1 \pmod{r}$, and
- S' is the set of vertices of G incident with the edges of $E(G) \setminus E(G')$.

Lemma 1. *Let G be a graph, S a subset of its vertices, and $q, r \geq 2$ integers. Let λ be a layering of G. For $m \in \{0, \ldots, r-1\}$, let (G_m, S_m) be the (λ, r, m)-stratification of G.*

- $\overline{\chi}'_q(G_m, S \cup S_m) \leq \overline{\chi}'_q(G, S)$ *for every $m \in \{0, \ldots, r-1\}$.*
- *There exists $m \in \{0, \ldots, r-1\}$ such that $\overline{\chi}'_q(G_m, S \cup S_m) \geq \left(1 - \frac{6q}{r}\right)\overline{\chi}'_q(G, S)$.*

Proof. Given an $(S \cup S_m)$-composable edge q-colouring of G_m, we can assign the colour 0 to all edges of $E(G) \setminus E(G_m)$ and obtain an S-composable edge q-colouring of G using the same number of non-zero colours, which implies that $\overline{\chi}'_q(G_m, S \cup S_m) \leq \overline{\chi}'_q(G, S)$.

Conversely, consider an S-composable edge q-colouring f of G using $k = \overline{\chi}'_q(G, S)$ non-zero colours. For $m \in \{0, \ldots, r-1\}$, let B_m be the bipartite graph with vertex set S_m and edge set $E(G) \setminus E(G_m)$ and let M_m be a maximal matching in B_m. Let $\mathcal{P}$ be a partition of the set $\{0, \ldots, r-1\}$ into at most three disjoint parts such that none of the parts contains two integers that are consecutive modulo r. For each $P \in \mathcal{P}$, let $M_P = \bigcup_{m \in P} M_m$, and observe that M_P is a matching in G. By Observation 2, it follows that $k \geq |M_P|$, and thus

$$3k \geq |\mathcal{P}|k \geq \sum_{P \in \mathcal{P}} |M_P| = \sum_{m=0}^{r-1} |M_m|.$$

Hence, we can fix $m \in \{0, \ldots, r-1\}$ such that $|M_m| \leq \frac{3}{r}k$. By Observation 2, any edge q-colouring of B_m, and in particular the restriction of f to the edges of B_m, uses at most $2q|M_m| \leq \frac{6q}{r}k$ distinct colours.

Let f' be the edge q-colouring of G obtained from f by recolouring all edges whose colour appears on the edges of B_m to colour 0. Clearly f' uses at least $\left(1 - \frac{6q}{r}\right)k$ non-zero colours. Moreover, each vertex $v \in S_m$ is now incident with an edge of colour 0, and thus $|f'(v) \setminus \{0\}| \leq q-1$. Therefore, the restriction of f' to $E(G_m)$ is an $(S \cup S_m)$-composable edge q-colouring, implying that

$$\overline{\chi}'_q(G_m, S \cup S_m) \geq \left(1 - \frac{6q}{r}\right)k = \left(1 - \frac{6q}{r}\right)\overline{\chi}'_q(G, S).$$

Hence, if $r \gg q$, then a good approximation of $\overline{\chi}'_q(G_m, S \cup S_m)$ for all $m \in \{0, \ldots, r-1\}$ gives a good approximation for $\overline{\chi}'_q(G, S)$. We will also need a similar observation for vertex deletion; here we only get an additive approximation in general, but as long as the edge q-chromatic number is large enough, this suffices (and if it is not, we can determine it exactly using Observation 3).

Lemma 2. *Let G be a graph, S a set of its vertices, and v a vertex of G. Let $S' = (S \setminus \{v\}) \cup N(v)$. For any integer $q \geq 2$, we have*

$$\overline{\chi}'_q(G, S) \geq \overline{\chi}'_q(G - v, S') \geq \overline{\chi}'_q(G, S) - q,$$

and in particular if $\varepsilon > 0$ and $\overline{\chi}'_q(G, S) \geq q/\varepsilon$, then

$$\overline{\chi}'_q(G - v, S') \geq (1 - \varepsilon)\overline{\chi}'_q(G, S).$$

Proof. Any S'-composable edge q-colouring of $G - v$ extends to an S-composable edge q-colouring of G by giving all edges incident on v colour 0, implying that $\overline{\chi}'_q(G, S) \geq \overline{\chi}'_q(G - v, S')$. Conversely, any S-composable edge q-colouring of G can be turned into an S'-composable edge q-colouring of $G - v$ by recolouring all edges whose colour appears on the neighbourhood of v to 0 and restricting it to the edges of $G - v$. This loses at most q non-zero colours (those appearing on the neighborhood of v), and thus $\overline{\chi}'_q(G - v, S') \geq \overline{\chi}'_q(G, S) - q$.   □

3.1   Baker game

For an infinite sequence $\mathbf{r} = r_1, r_2, \ldots$ and an integer $s \geq 0$, let $\mathrm{tail}(\mathbf{r})$ denote the sequence $r_2, r_3, \ldots$ and let $\mathrm{head}(\mathbf{r}) = r_1$. *Baker game* is played by two players Destroyer and Preserver on a pair $(G, \mathbf{r})$, where G is a graph and $\mathbf{r}$ is a sequence of positive integers. The game stops when $V(G) = \emptyset$, and Destroyer's objective is to minimise the number of rounds required to make the graph empty. In each round of the game, either

- Destroyer chooses a vertex $v \in V(G)$, Preserver does nothing and the game moves to the state $(G \setminus \{v\}, \mathrm{tail}(\mathbf{r}))$, or
- Destroyer selects a layering λ of G, Preserver selects an interval I of $\mathrm{head}(\mathbf{r})$ consecutive integers and the game moves to the state $(G[\lambda^{-1}(I)], \mathrm{tail}(\mathbf{r}))$. In other words, Preserver selects $\mathrm{head}(\mathbf{r})$ consecutive layers and the rest of the graph is deleted.

Destroyer *wins* in k rounds on the state $(G, \mathbf{r})$ if regardless of Preserver's strategy, the game stops after at most k rounds. As we mentioned earlier Destroyer's objective is to minimise the number of rounds of this game and it is known that they will succeed if the game is played on a graph that forbids a fixed minor (the upper bound on the number of rounds depends only on the sequence $\mathbf{r}$ and the forbidden minor, not on G).

Theorem 1 (Dvořák [13]). *For every graph F and every sequence $\mathbf{r} = r_1, r_2, \ldots$ of positive integers, there exists a positive integer k such that for every graph G avoiding F as a minor, Destroyer wins Baker game from the state $(G, \mathbf{r})$ in at most k rounds. Moreover, letting $n = |V(G)|$, there exists an algorithm that preprocesses G in time $O_F(n^2)$ and then in each round determines a move for Destroyer (leading to winning in at most k rounds in total) in time $O_{F,\mathbf{r}}(n)$.*

Let us now give the algorithm for approximating the edge q-chromatic number on graphs for which we can quickly win Baker game.

Lemma 3. *There exists an algorithm that, given*

- *a graph G, a set $S \subseteq V(G)$, an integer $q \geq 2$, and*
- *a sequence $\mathbf{r} = r_1, r_2, \ldots$ of positive integers such that* Destroyer *wins Baker game from the state $(G, \mathbf{r})$ in at most k rounds, and in each state that arises in the game is able to determine the move that achieves this in time T,*

returns an S-composable edge q-colouring of G using at least $\left(\prod_{i=1}^{k}\left(1 - \frac{6q}{r_i}\right)\right) \cdot \overline{\chi}'_q(G, S)$ non-zero colours, in time $O_{\mathbf{r},k,q}(|G|T)$.

Proof. First, we run the algorithm from Observation 3 with $s = \lceil r_1/3 \rceil$. If the obtained colouring uses less than s non-zero colours, it is optimal and we return it. Otherwise, we know that $\overline{\chi}'_q(G, S) \geq s$. In particular, $E(G) \neq \emptyset$, and thus Destroyer have not won the game yet.

Let $R = \left(\prod_{i=2}^{k}\left(1 - \frac{6q}{r_i}\right)\right)$. Let us now consider two cases depending on Destroyer's move from the state $(G, \mathbf{r})$.

- Suppose that Destroyer decides to delete a vertex $v \in V(G)$. We apply the algorithm recursively for the graph $G - v$, set $S' = (S \setminus \{v\}) \cup N(v)$, and the sequence $\mathrm{tail}(\mathbf{r})$, obtaining an S'-composable edge q-colouring f of $G - v$ using at least $R \cdot \overline{\chi}'_q(G - v, S')$ non-zero colours. By Lemma 2 with $\varepsilon = \frac{q}{s}$, we conclude that f uses at least

$$R \cdot \overline{\chi}'_q(G - v, S') \geq R(1 - \varepsilon)\overline{\chi}'_q(G, S) \geq R\left(1 - \frac{6q}{r_1}\right)\overline{\chi}'_q(G, S)$$

non-zero colours. We turn f into an S-composable edge q-colouring of G by giving all edges incident on v colour 0 and return it.
- Suppose that Destroyer chooses a layering λ. We now recurse into several subgraphs, each corresponding to a valid move of Preserver. For each $m \in \{0, \ldots, r_1 - 1\}$, let (G_m, S_m) be the (λ, r_1, m)-stratification of G_m. Note that G_m is divided into parts $G_{m,1}, \ldots, G_{m,t_m}$, each contained in the union of r_1 consecutive layers of λ. For each $m \in \{0, \ldots, r_1 - 1\}$ and each $i \in \{1, \ldots, t_m\}$, we apply the algorithm recursively for the graph $G_{m,i}$, set $S_{m,i} = (S_m \cup S) \cap V(G_{m,i})$, and the sequence $\mathrm{tail}(\mathbf{r})$, obtaining an $S_{m,i}$-composable edge q-colouring $f_{m,i}$ of $G_{m,i}$ using at least $R \cdot \overline{\chi}'_q(G_{m,i}, S_{m,i})$ non-zero colours. Let f_m be the union of the colourings $f_{m,i}$ for $i \in \{1, \ldots, t_m\}$ and observe that f_m is an $(S \cup S_m)$-composable edge q-colouring of G_m using at least $R \cdot \overline{\chi}'_q(G_m, S \cup S_m)$ non-zero colours. We choose $m \in \{0, \ldots, r_1 - 1\}$ such that f_m uses the largest number of non-zero colours, extend it to an S-composable edge q-colouring of G by giving all edges of $E(G) \setminus E(G_m)$ colour 0, and return it. By Lemma 1, the colouring uses at least

$$R \cdot \overline{\chi}'_q(G_m, S \cup S_m) \geq R\left(1 - \frac{6q}{r_1}\right)\overline{\chi}'_q(G, S)$$

non-zero colours, as required.

For the time complexity, note that each vertex and edge of G belongs to at most $\prod_{i=1}^{d} r_i$ subgraphs processed at depth d of the recursion, and since the depth of the recursion is bounded by k, the sum of the sizes of the processed subgraphs is $O_{r,k,q}(|G|)$. Excluding the recursion and time needed to select Destroyer's moves, the actions described above can be performed in linear time. Consequently, the total time complexity is $O_{r,k,q}(|G|T)$.

Our main result is then just a simple combination of this lemma with Theorem 1.

Theorem 2. *There exists an algorithm that given an F-minor-free graph G and integers $q, p \geq 2$, returns in time $O_{F,p,q}(|G|^2)$ an edge q-colouring of G using at least $(1 - 1/p)\overline{\chi}'_q(G)$ colours.*

Proof. Let $\mathbf{r}$ be the infinite sequence such that $r_i = 10pqi^2$ for each positive integer i, and let k be the number of rounds in which Destroyer wins Baker game from the state $(G', \mathbf{r})$ for any F-minor-free graph G', using the strategy given by Theorem 1. Note that

$$R = \prod_{i=1}^{k}\left(1 - \frac{6q}{r_i}\right) \geq 1 - \sum_{i=1}^{\infty} \frac{6q}{r_i}$$

$$= 1 - \frac{3}{5p}\sum_{i=1}^{\infty}\frac{1}{i^2} = 1 - \frac{3}{5p}\cdot\frac{\pi^2}{6} \geq 1 - \frac{1}{p}.$$

Let $n = |G|$. After running the preprocessing algorithm from Theorem 1, we apply the algorithm from Lemma 3 with $S = \emptyset$ and $T = O_{F,\mathbf{r}}(n) = O_{F,p,q}(n)$, obtaining an edge q-colouring of G using at least $R \cdot \overline{\chi}'_q(G, \emptyset) = R \cdot \overline{\chi}'_q(G) \geq (1 - 1/p)\overline{\chi}'_q(G)$ colours, in time $O_{F,p,q}(n^2)$. □

4   Hardness on 1-apex graphs

In this section, we study the complexity of the maximum edge 2-colouring problem on 1-apex graphs. We present a reduction from PLANAR ($\leq 3, 3$)-SAT which is known to be NP-hard [32].

The *incidence graph* $G(\varphi)$ of a Boolean formula φ in conjunctive normal form is the bipartite graph whose vertices are the variables appearing in φ and the clauses of φ, and each variable is adjacent exactly to the clauses in which it appears. A Boolean formula φ in conjunctive normal form is called PLANAR ($\leq 3, 3$)-SAT if

- each clause of φ contains at most three distinct literals,
- each variable of φ appears in exactly three clauses,
- the incidence graph $G(\varphi)$ is planar.

In PLANAR $(\leq 3, 3)$-SAT problem, we ask whether such a formula φ has a satisfying assignment.

We follow the strategy used in [1], using an intermediate *maximum edge* $1, 2$-*colouring* problem. The instance of this problem consists of a graph G, a function $g \colon V(G) \to \{1, 2\}$, and a number t. An *edge g-colouring* of G is an edge colouring f such that $|f(v)| \leq g(v)$ for each $v \in V(G)$. The objective is to decide whether there exists an edge g-colouring of G using at least t distinct colours. We show the maximum edge $\{1, 2\}$-colouring problem is NP-hard on 1-apex graphs by establishing a reduction from PLANAR $(\leq 3, 3)$-SAT problem. We then use this result to show that the maximum edge q-colouring problem on planar graphs is NP-hard when $q \geq 2$. Let us start by establishing the intermediate result.

Lemma 4. *The maximum edge $\{1, 2\}$-colouring problem is NP-hard even when restricted on the class of 1-apex graphs.*

Proof. Consider a given PLANAR $(\leq 3, 3)$-SAT formula φ with m clauses and n variables and a plane drawing of its incidence graph $G(\varphi)$. Let the clauses of φ be $c_1, \ldots, c_m$ and the variables $x_1, \ldots, x_n$; we use the same symbols for the corresponding vertices of $G(\varphi)$.

Let H be a graph obtained from $G(\varphi)$ as follows. For all $j \in \{1, 2, \ldots, n\}$, if the clauses in which x_j appears are $c_{\ell_{j,1}}$, $c_{\ell_{j,2}}$, and $c_{\ell_{j,3}}$, split x_j to three vertices $x_{j,1}$, $x_{j,2}$, and $x_{j,3}$, where $x_{j,a}$ is adjacent to $c_{\ell_{j,a}}$ for $a \in \{1, 2, 3\}$. For $1 \leq a < b \leq 3$, add a vertex $n_{j,a,b}$ and if x_j appears positively in $c_{\ell_{j,a}}$ and negatively in $c_{\ell_{j,b}}$ or vice versa, make it adjacent to $x_{j,a}$ and $x_{j,b}$ (otherwise leave it as an isolated vertex). Finally, we add a new vertex u adjacent to c_i for $i \in \{1, \ldots, m\}$ and to $n_{j,a,b}$ for $j \in \{1, \ldots, n\}$ and $1 \leq a < b \leq 3$. Clearly, H is a 1-apex graph, since $H - u$ is planar.

Let us define the function $g \colon V(H) \to \{1, 2\}$ as follows:

- $g(u) = 1$,
- $g(c_i) = 2$ for all $i \in \{1, 2, \ldots, m\}$,
- $g(x_{j,a}) = 1$ for all $j \in \{1, 2, \ldots, n\}$ and $a \in \{1, 2, 3\}$, and
- $g(n_{j,a,b}) = 2$ for all $j \in \{1, 2, \ldots, n\}$ and $1 \leq a < b \leq 3$.

First, we show if there exists a satisfying assignment for the formula φ, then H has an edge g-colouring using $n+1$ colours. For $i \in \{1, \ldots, n\}$, choose a vertex $x_{j,a}$ adjacent to c_i such that the (positive or negative) literal of c_i containing the variable x_j is true in the assignment, and give colour i to the edge $(c_i, x_{j,a})$ and all other edges incident on $x_{j,a}$ (if any). All other edges receive colour 0.

Clearly, u is only incident with edges of colour 0, for each $j \in \{1, \ldots, n\}$ and $a \in \{1, \ldots, 3\}$ all edges incident on $x_{j,a}$ have the same colour, and for each $i \in \{1, \ldots, m\}$, the edges incident on c_i have colours 0 and i. Finally, consider a vertex $n_{j,a,b}$ for some $j \in \{1, \ldots, n\}$ and $1 \leq a < b \leq 3$ adjacent to $x_{j,a}$ and $x_{j,b}$. By the construction of H, the variable x_j appears positively in $c_{\ell_{j,a}}$ and negatively in $c_{\ell_{j,b}}$ or vice versa, and thus at most one of the corresponding literals is true in the assignment. Hence, $n_{j,a,b}$ is incident with edges of colour 0 and of at most one of the colours $\ell_{j,a}$ and $\ell_{j,b}$.

Conversely, suppose that there exists an edge g-colouring f of H using at least $m+1$ distinct colours, and let us argue that there exists a satisfying assignment for φ. Since $g(u) = 1$, we can without loss of generality assume that each edge incident with u has colour 0. If a colour $c \neq 0$ is used to colour the edge $(n_{j,a,b}, x_{j,k})$ for some $j \in \{1, \ldots, n\}$, $1 \leq a < b \leq 3$, and $k \in \{a, b\}$, then since $g(x_{j,k}) = 1$, this colour is also used on the edge $(x_{j,k}, c_{\ell_{j,k}})$. Hence, every non-zero colour appears on an edge incident with a clause. Since each clause is also joined to u by an edge of colour 0, it can be only incident with edges of one other colour. Since f uses at least $m + 1$ colours, we can without loss of generality assume that for $i \in \{1, \ldots, m\}$, there exists an edge $(c_i, x_{j,a})$ for some $j \in \{1, \ldots, n\}$ and $a \in \{1, \ldots, 3\}$ of colour i. Assign to x_j the truth value that makes the literal of c_i in which it appears true.

We only need to argue that this rule does not cause us to assign to x_j both values true and false. If that were the case, then there would exist $1 \leq a < b \leq 3$ such that the variable x_j appears positively in clause $c_{\ell_{j,a}}$ and negatively in clause $c_{\ell_{j,b}}$ or vice versa, the edge corresponding to the variable $x_{j,a}$ has colour $\ell_{j,a}$ and the edge corresponding to the variable $x_{j,b}$ has colour $\ell_{j,b}$. However, since $g(x_{j,a}) = g(x_{j,b}) = 1$, this would imply that $n_{j,a,b}$ is incident with the edge $(n_{j,a,b}, x_{j,a})$ of colour $\ell_{j,a}$, the edge $(n_{j,a,b}, x_{j,b})$ of colour $\ell_{j,b}$, and the edge $(n_{j,a,b}, u)$ of colour 0, which is a contradiction.

Therefore, we described how to transform in polynomial time a PLANAR $(\leq 3, 3)$-SAT instance φ to an equivalent instance $H, g, t = m+1$ of the maximum edge $\{1, 2\}$-colouring problem.   □

Now we are ready to prove the main theorem of this section. The proof strategy is similar to the APX-hardness proof in [1]. We include the details for completeness.

Theorem 3. *For an arbitrary integer $q > 2$ the maximum edge q-colouring problem is NP-hard even when the input instance is restricted to 1-apex graphs.*

Proof. We construct a reduction from the maximum edge $\{1, 2\}$-colouring problem on 1-apex graphs. Let G, g, t be an instance of this problem, and let $n = |V(G)|$ and $r = |\{v \in V(G) : g(v) = 1\}|$. We create a graph G' from G by adding for each vertex $v \in V(G)$ exactly $q - g(v)$ pendant vertices adjacent to v. Clearly, G' is an 1-apex graph. We show that G has an edge g-colouring using at least t distinct colours if and only if G' has an edge q-colouring using at least $t + r + (q - 2)n$ colours.

In one direction, given an edge g-colouring of G using at least t colours, we colour each of the added pendant edges using a new colour, obtaining an edge q-colouring of G using at least $t + r + (q - 2)n$ colours.

Conversely, let f be an edge q-colouring of G' using at least $t + r + (q - 2)n$ colours. Process the vertices $v \in V(G)$ one by one, performing for each of them the following operation: For each added pendant vertex u adjacent to v in order, let c' be the colour of the edge (u, v), delete u, and if v is incident with an edge e of colour $c \neq c'$, then recolour all remaining edges of colour c' to c. Note that the number of eliminated colours is bounded by the number $r + (q - 2)n$ of pendant vertices, and thus the resulting colouring still uses at least t colours. Moreover, at each vertex $v \in V(G)$, we either end up with all edges incident on v having the same colour or we eliminated one colour from the neighbourhood of v for each adjacent pendant vertex; in the latter case, since $|f(v)| \leq q$ and v is adjacent to $q - g(v)$ pendant vertices, at most $g(v)$ colours remain on the edges incident on v. Hence, we indeed obtain an edge g-colouring of G using at least t colours. $\square$

5   Future directions

We conclude with some possible directions for future research. The maximum edge 2-colouring problem on 1-apex graphs is NP-hard. But the complexity of the problem is unknown when the input is restricted to planar graphs. We consider this an interesting question left unanswered. The best-known approximation ratio is known to be 2, without any restriction on the input instances. Whereas, a lower bound of $\left(\frac{1+q}{q}\right)$, for $q \geq 2$ is known assuming unique games conjecture. There are not many new results reported in the last decade that bridge this gap. We think, even a $(2 - \varepsilon)$ algorithm, for any $\varepsilon > 0$, will be a huge progress towards that direction. The Baker game technique can yield PTASes for monotone optimization problems beyond problems expressible in the first-order logic. Clearly, the technique can't be extended to the entire class of problems expressible in the monadic second-order logic. It will be interesting to characterise the problems expressible in the monadic second-order logic where the Baker game yield PTASes.

Acknowledgement. The second author likes to thank Benjamin Moore, Jatin Batra, Sandip Banerjee and Siddharth Gupta for helpful discussions on this project. He also likes to thank the organisers of Homonolo for providing a nice and stimulating research environment.

References

1. Adamaszek, A., Popa, A.: Approximation and hardness results for the maximum edge q-coloring problem. In: Cheong, O., Chwa, K.-Y., Park, K. (eds.) ISAAC 2010, Part II. LNCS, vol. 6507, pp. 132–143. Springer, Heidelberg (2010). https://doi.org/10.1007/978-3-642-17514-5_12
2. Adamaszek, A., Popa, A.: Approximation and hardness results for the maximum edge Q-coloring problem. J. Discret. Algorithms **38-41**, 1–8 (2016). https://doi.org/10.1016/j.jda.2016.09.003
3. Alon, N., Seymour, P.D., Thomas, R.: A separator theorem for graphs with an excluded minor and its applications. In: Proceedings of the 22nd Annual ACM Symposium on Theory of Computing, 13–17 May 1990, Baltimore, Maryland, USA, pp. 293–299. ACM (1990). https://doi.org/10.1145/100216.100254
4. Baker, B.S.: Approximation algorithms for NP-complete problems on planar graphs. J. ACM **41**(1), 153–180 (1994). https://doi.org/10.1145/174644.174650
5. Cabello, S., Gajser, D.: Simple PTAS's for families of graphs excluding a minor. Discret. Appl. Math. **189**, 41–48 (2015). https://doi.org/10.1016/j.dam.2015.03.004

6. Chandran, L.S., Hashim, T., Jacob, D., Mathew, R., Rajendraprasad, D., Singh, N.: New bounds on the anti-Ramsey numbers of star graphs. CoRR abs/1810.00624 (2018). https://arxiv.org/abs/1810.00624

7. Chandran, L.S., Lahiri, A., Singh, N.: Improved approximation for maximum edge colouring problem. Discrete Appl. Math. **319**, 42–52 (2022). https://doi.org/10.1016/j.dam.2021.05.017

8. Courcelle, B.: The monadic second-order logic of graphs. I. Recognizable sets of finite graphs. Inf. Comput. **85**(1), 12–75 (1990). https://doi.org/10.1016/0890-5401(90)90043-H

9. Dawar, A., Grohe, M., Kreutzer, S., Schweikardt, N.: Approximation schemes for first-order definable optimisation problems. In: 21th IEEE Symposium on Logic in Computer Science (LICS 2006), 12–15 August 2006, Seattle, WA, USA, Proceedings, pp. 411–420. IEEE Computer Society (2006). https://doi.org/10.1109/LICS.2006.13

10. Demaine, E.D., Hajiaghayi, M.T.: Bidimensionality: new connections between FPT algorithms and PTASs. In: Proceedings of the Sixteenth Annual ACM-SIAM Symposium on Discrete Algorithms, SODA 2005, Vancouver, British Columbia, Canada, 23–25 January 2005, pp. 590–601. SIAM (2005). https://dl.acm.org/citation.cfm?id=1070432.1070514

11. Diestel, R.: Graph Theory. Graduate Texts in Mathematics, vol. 173, 4th edn. Springer, Heidelberg (2012)

12. Dvořák, Z.: Thin graph classes and polynomial-time approximation schemes. In: Proceedings of the Twenty-Ninth Annual ACM-SIAM Symposium on Discrete Algorithms, SODA 2018, New Orleans, LA, USA, 7–10 January 2018, pp. 1685–1701. SIAM (2018). https://doi.org/10.1137/1.9781611975031.110

13. Dvořák, Z.: Baker game and polynomial-time approximation schemes. In: Proceedings of the 2020 ACM-SIAM Symposium on Discrete Algorithms, SODA 2020, Salt Lake City, UT, USA, 5–8 January 2020, pp. 2227–2240. SIAM (2020). https://doi.org/10.1137/1.9781611975994.137

14. Erdős, P., Simonovits, M., Sós, V.T.: Anti-Ramsey theorems. Infinite Finite Sets (Colloquium, Keszthely, 1973; dedicated to P. Erdős on his 60th birthday) **10**(II), 633–643 (1975)

15. Feng, W., Zhang, L., Wang, H.: Approximation algorithm for maximum edge coloring. Theor. Comput. Sci. **410**(11), 1022–1029 (2009). https://doi.org/10.1016/j.tcs.2008.10.035

16. Fomin, F.V., Lokshtanov, D., Raman, V., Saurabh, S.: Bidimensionality and EPTAS. In: Proceedings of the Twenty-Second Annual ACM-SIAM Symposium on Discrete Algorithms, SODA 2011, San Francisco, California, USA, 23–25 January 2011, pp. 748–759. SIAM (2011). https://doi.org/10.1137/1.9781611973082.59

17. Fujita, S., Magnant, C., Ozeki, K.: Rainbow generalizations of Ramsey theory: a survey. Graphs Combin. **26**(1), 1–30 (2010). https://doi.org/10.1007/s00373-010-0891-3

18. Gorgol, I., Lazuka, E.: Rainbow numbers for small stars with one edge added. Discuss. Math. Graph Theory **30**(4), 555–562 (2010). https://doi.org/10.7151/dmgt.1513

19. Goyal, P., Kamat, V., Misra, N.: On the parameterized complexity of the maximum edge 2-coloring problem. In: Mathematical Foundations of Computer Science 2013–38th International Symposium, MFCS 2013, Klosterneuburg, Austria, 26–30 August 2013, pp. 492–503 (2013). https://doi.org/10.1007/978-3-642-40313-2_44

20. Har-Peled, S., Quanrud, K.: Approximation algorithms for polynomial-expansion and low-density graphs. SIAM J. Comput. **46**(6), 1712–1744 (2017). https://doi.org/10.1137/16M1079336

21. Jiang, T.: Edge-colorings with no large polychromatic stars. Graphs Combin. **18**(2), 303–308 (2002). https://doi.org/10.1007/s003730200022

22. Klein, P.N., Plotkin, S.A., Rao, S.: Excluded minors, network decomposition, and multicommodity flow. In: Proceedings of the Twenty-Fifth Annual ACM Symposium on Theory of Computing, 16–18 May 1993, San Diego, CA, USA, pp. 682–690. ACM (1993). https://doi.org/10.1145/167088.167261

23. Kodialam, M.S., Nandagopal, T.: Characterizing the capacity region in multi-radio multi-channel wireless mesh networks. In: Proceedings of the 11th Annual International Conference on Mobile Computing and Networking, MOBICOM 2005, Cologne, Germany, 28 August–2 September 2005, pp. 73–87. ACM (2005), https://doi.org/10.1145/1080829.1080837

24. Manoussakis, Y., Spyratos, M., Tuza, Z., Voigt, M.: Minimal colorings for properly colored subgraphs. Graphs Combin. **12**(1), 345–360 (1996). https://doi.org/10.1007/BF01858468

25. Montellano-Ballesteros, J.J.: On totally multicolored stars. J. Graph Theory **51**(3), 225–243 (2006). https://doi.org/10.1002/jgt.20140

26. Montellano-Ballesteros, J.J., Neumann-Lara, V.: An anti-Ramsey theorem. Combinatorica **22**(3), 445–449 (2002). https://doi.org/10.1007/s004930200023

27. Raniwala, A., Chiueh, T.: Architecture and algorithms for an IEEE 802.11-based multi-channel wireless mesh network. In: INFOCOM 2005. 24th Annual Joint Conference of the IEEE Computer and Communications Societies, 13–17 March 2005, Miami, FL, USA, pp. 2223–2234 (2005). https://doi.org/10.1109/INFCOM.2005.1498497

28. Robertson, N., Seymour, P.D.: Graph minors. XVI. Excluding a non-planar graph. J. Combin. Theory Ser. B **89**(1), 43–76 (2003). https://doi.org/10.1016/S0095-8956(03)00042-X

29. Schiermeyer, I.: Rainbow numbers for matchings and complete graphs. Discret. Math. **286**(1–2), 157–162 (2004). https://doi.org/10.1016/j.disc.2003.11.057

30. Sen, A., Murthy, S., Ganguly, S., Bhatnagar, S.: An interference-aware channel assignment scheme for wireless mesh networks. In: Proceedings of IEEE International Conference on Communications, ICC 2007, Glasgow, Scotland, UK, 24–28 June 2007, pp. 3471–3476. IEEE (2007). https://doi.org/10.1109/ICC.2007.574

31. Simonovits, M., Sós, V.: On restricted colourings of K_n. Combinatorica **4**(1), 101–110 (1984). https://doi.org/10.1007/BF02579162

32. Tovey, C.A.: A simplified NP-complete satisfiability problem. Discret. Appl. Math. **8**(1), 85–89 (1984). https://doi.org/10.1016/0166-218X(84)90081-7

33. Wan, P., Al-dhelaan, F., Jia, X., Wang, B., Xing, G.: Maximizing network capacity of MPR-capable wireless networks. In: 2015 IEEE Conference on Computer Communications, INFOCOM 2015, Kowloon, Hong Kong, 26 April–1 May 2015, pp. 1805–1813. IEEE (2015). https://doi.org/10.1109/INFOCOM.2015.7218562

34. Wan, P., Cheng, Y., Wang, Z., Yao, F.F.: Multiflows in multi-channel multi-radio multihop wireless networks. In: INFOCOM 2011. 30th IEEE International Conference on Computer Communications, 10–15 April 2011, Shanghai, China, pp. 846–854. IEEE (2011). https://doi.org/10.1109/INFCOM.2011.5935308

Bounds on Functionality and Symmetric Difference – Two Intriguing Graph Parameters

Pavel Dvořák[1,2]($\boxtimes$), Lukáš Folwarczný[1,3], Michal Opler[4], Pavel Pudlák[3], Robert Šámal[1], and Tung Anh Vu[1]

[1] Faculty of Mathematics and Physics, Charles University, Prague, Czech Republic
{koblich,samal}@iuuk.mff.cuni.cz, tung@kam.mff.cuni.cz
[2] Tata Institute of Fundamental Research, Mumbai, India
[3] Institute of Mathematics, Czech Academy of Sciences, Prague, Czech Republic
{folwarczny,pudlak}@math.cas.cz
[4] Faculty of Information Technology, Czech Technical University,
Prague, Czech Republic
michal.opler@fit.cvut.cz

Abstract. [Alecu et al.: Graph functionality, JCTB2021] define functionality, a graph parameter that generalizes graph degeneracy. They research the relation of functionality to many other graph parameters (tree-width, clique-width, VC-dimension, etc.). Extending their research, we prove a logarithmic lower bound for functionality of random graph $G(n, p)$ for large range of p. Previously known graphs have functionality logarithmic in number of vertices. We show that for every graph G on n vertices we have $\mathrm{fun}(G) \leq O(\sqrt{n \log n})$ and we give a nearly matching $\Omega(\sqrt{n})$-lower bound provided by projective planes.

Further, we study a related graph parameter *symmetric difference*, the minimum of $|N(u) \Delta N(v)|$ over all pairs of vertices of the "worst possible" induced subgraph. It was observed by Alecu et al. that $\mathrm{fun}(G) \leq \mathrm{sd}(G) + 1$ for every graph G. We compare fun and sd for the class INT of interval graphs and CA of circular-arc graphs. We let INT_n denote the n-vertex interval graphs, similarly for CA_n.

Alecu et al. ask, whether fun(INT) is bounded. Dallard et al. answer this positively in a recent preprint. On the other hand, we show that $\Omega(\sqrt[4]{n}) \leq \mathrm{sd}(\mathrm{INT}_n) \leq O(\sqrt[3]{n})$. For the related class CA we show that $\mathrm{sd}(\mathrm{CA}_n) = \Theta(\sqrt{n})$.

We propose a follow-up question: is fun(CA) bounded?

P. Dvořák—Supported by Czech Science Foundation GAČR grant #22-14872O.
L. Folwarczný—Supported by Czech Science Foundation GAČR grant 19-27871X.
M. Opler—Supported by Czech Science Foundation GAČR grant 22-19557S.
P. Pudlák—Supported by Czech Science Foundation GAČR grant 19-27871X.
R. Šámal—Partially supported by grant 22-17398S of the Czech Science Foundation. This project has received funding from the European Research Council (ERC) under the European Union's Horizon 2020 research and innovation programme (grant agreement No 810115).
T. A. Vu—Supported by Czech Science Foundation GAČR grant 22-22997S.

D. Paulusma and B. Ries (Eds.): WG 2023, LNCS 14093, pp. 305–318, 2023.
https://doi.org/10.1007/978-3-031-43380-1_22

Keywords: Functionality · Finite Projective Plane · Symmetric Difference · Intersection Graphs

1   Introduction

Let $G = (V, E)$ be a graph and $v \in V$ be a vertex. The set of neighbors of v in G is denoted by $N_G(v)$, we omit the subscript if the graph is clear from the context. An adjacency matrix A_G of G is 0-1 matrix such that its rows and columns are indexed by vertices of G and $A[u, v] = 1$ if and only if u and v are connected by an edge. Now, we define the functionality and symmetric difference of a graph – two principle notion of this papers – as introduced by Alecu et al. [1], and implicitly also by Atminas et al. [3].

A vertex v of a graph $G = (V, E)$ is a *function* of vertices $u_1, \ldots, u_k \in V$ (different from v) if there exists a boolean function f of k variables such that for any vertex $w \in V \setminus \{v, u_1, \ldots, u_k\}$ it holds that $A[v, w] = f(A[v, u_1], \ldots, A[v, u_k])$. Informally, we can determine if v and w are connected from the adjacencies of v with the u_i's. The *functionality* $\text{fun}_G(v)$ of a vertex v in G is the minimum k such that v is a function of k vertices of G. We drop the subscript and write just $\text{fun}(v)$ if the graph G is clear from the context. Then, the *functionality* $\text{fun}(G)$ of a graph G is defined as

$$\text{fun}(G) = \max_{H \subseteq G} \min_{v \in V(H)} \text{fun}_H(v),$$

where the maximum is taken over all induced subgraphs H of G.

It is observed in [3] that if $\text{fun}(G) \leq k$ then we can encode G using $n(2^k + (k + 1) \log n)$ bits, where n is the number of vertices of G. Thus, if every graph G in some graph class $\mathcal{G}$ has bounded functionality then $\mathcal{G}$ contains at most $2^{O(n \log n)}$ graphs on n vertices. Such classes are said to be of factorial growth [4] and include diverse classes of practical importance (interval graphs, line graphs, forests, planar graphs, more generally all proper minor-closed classes). Thus, Alecu et al. [1] introduce functionality as a tool to study graph classes of factorial growth, and the related Implicit graph conjecture (although this conjecture was recently disproved [6]). This was also our original motivation. Moreover, functionality is a natural generalization of the graph degeneracy, as the degree of a vertex v is a trivial upper bound for the functionality of v. Thus, it deserves a study for its own sake.

Alecu et al. [1] research the relation of functionality to many other graph parameters: in particular they provide a linear upper bound in terms of clique-width and a lower bound in terms of some function of VC-dimension. They also give a lower bound for the functionality of the hypercube that is linear in the dimension (i.e., logarithmic in the number of vertices).

Another parameter related to functionality is the so-called symmetric difference. Given two vertices u, v of G, let $\text{sd}_G(u, v)$ (or just $\text{sd}(u, v)$ when the graph is clear from the context) be the number of vertices different from u and v that are adjacent to exactly one of u and v. The *symmetric difference* sd(G) of a graph G is defined as

$$\max_{H \subseteq G} \min_{u,v \in V(H)} sd_H(u,v),$$

where the maximum is again taken over all induced subgraphs. We may view sd(u, v) as the size of the set $(N(u) \Delta N(v)) \setminus \{u, v\}$, which explains the term "symmetric difference". It is noted by Alecu et al. [1] that fun(G) $\leq$ sd(G) + 1. However, there is no lower bound in terms of sd as there are graphs of constant functionality and polynomial symmetric difference – for example the interval graphs. This was shown by Theorem 3.2 of Dallard et al. [5] and by our Theorem 7, while Corollary 5.3 of [5] shows that sd(INT) is unbounded, without providing explicit lower bounds.

The intersection graph of a family of sets F is a graph $G = (V, E)$ where $V = \{v_1, \ldots, v_n\}$ and two vertices v_i and v_j are connected if and only if the corresponding sets S_i and S_j of F intersect. An interval graph is an intersection graph of n intervals on a real line. A circular arc graph is an intersection graph of n arcs of a circle. We let INT_n denote the family of all intersection graphs with n vertices, INT the family of all interval graphs. In the same vein, we define CA_n and CA for circular arc graphs.

1.1   Our Results

In this paper, we show several lower and upper bounds for functionality and symmetric difference of various graph classes. As far as we know, there were only logarithmic lower bounds for functionality [1,3,5]. Thus, it would have been possible that the functionality is at most logarithmic (similarly to the VC-dimension [7]). However, we show that the functionality of the incidence graph of a finite projective plane of order k is exactly $k + 1$, i.e., roughly $\sqrt{n}$. We complement this result with an almost matching upper bound that functionality of any graph is at most $O(\sqrt{n \log n})$. Further, we show for $\frac{3 \log^2 n}{n} \leq p \leq 1 - \frac{3 \log^2 n}{n}$ that a random graph $G(n, p)$ has at least logarithmic functionality with probability $1 - o(1)$. Note that if $p \leq o\left(\frac{\log n}{n}\right)$ then fun($G(n, p)$) $\leq o(\log n)$ with high probability because the minimum degree of $G(n, p)$ for such p is $o(\log n)$ with high probability and fun(G) is bounded by the minimum degree of the graph G. Similarly for $p \geq 1 - o\left(\frac{\log n}{n}\right)$, as the functionality of a graph G and its complement $\bar{G}$ is the same, and complement of $G(n, p)$ is $G(n, 1 - p)$. Thus, our range of p is almost optimal – up to a logarithmic factor. Further, we prove that any vertex of $G(n, \frac{1}{2})$ is a function of at most $O(\log n)$ vertices with high probability. Unfortunately, it does not imply that functionality of $G(n, \frac{1}{2})$ is logarithmic as it still can contain an induced subgraph of higher functionality. Overall, it suggests that graphs with polynomial functionality are quite rare and should be very structured (like the case of finite projective plane).

Further, we study symmetric difference of interval and circular arc graphs. We show that symmetric difference of circular arc graph is $\Theta(\sqrt{n})$, i.e., we prove that any circular arc graph has symmetric difference at most $O(\sqrt{n})$ and we present a circular arc graph of symmetric difference $\Omega(\sqrt{n})$. Recently, it was shown that interval graphs have bounded functionality and unbounded symmetric difference [5]. Even though it was not explicitly mentioned, the construction given by Dallard et al. [5] leads to the lower bound $\Omega(\sqrt[4]{n})$ for the symmetric difference of interval graphs. We independently came up with a different construction leading to the same lower bound, however analysis of our construction is simpler than the one from the previous work. For interval graphs, we also present the upper bound $O(\sqrt[3]{n})$ for symmetric difference. Thus, we are leaving a gap between the lower and upper bound. However, we show the symmetric difference of interval graphs is polynomial and strictly smaller than symmetric difference of circular arc graphs.

2   Functionality

2.1   Finite Projective Planes

Recall that a finite projective plane is a pair $(X, \mathcal{L})$, where X is a finite set and $\mathcal{L} \subseteq 2^X$, satisfying the following axioms [8]:

1. For every $p \neq q \in X$, there is exactly one subset of $\mathcal{L}$ containing p and q.
2. For every $L \neq M \in \mathcal{L}$, we have $|L \cap M| = 1$.
3. There exists a subset $Y \subseteq X$ of size 4 such that $|L \cap Y| \leq 2$ for every $L \in \mathcal{L}$.

Elements of X are called points and elements of $\mathcal{L}$ are called lines. We note that for every k which is a power of a prime, a finite projective plane with the following properties can be constructed. Each line contains exactly $k + 1$ points. Each point is incident to exactly $k + 1$ lines. The total number of points is $k^2 + k + 1$. The total number of lines is also $k^2 + k + 1$. The number k is called the order of the finite projective plane.

The incidence graph of a finite projective plane is a bipartite graph with one part X and the second part $\mathcal{L}$. In this graph, $x \in X$ is adjacent to $\ell \in \mathcal{L}$ iff x is incident to ℓ. The following theorem shows that the incidence graph of a finite projective plane has functionality approximately $\sqrt{n}$, where n is the number of vertices. Alecu et al. [1] have shown that there exists a function f such that for every graph G we have $\mathrm{vc}(G) \leq f(\mathrm{fun}(G))$. Our result complements this inequality by showing that the functionality of a graph cannot be upper bounded by its VC-dimension as it is known that the VC-dimension of a finite projective plane of any order is 2 [2].

Theorem 1. *Consider a finite projective plane of order k and its incidence graph G. Then, $\mathrm{fun}(G) = k + 1$. Moreover, for any proper induced subgraph $G' \subset G$ we have $\mathrm{fun}(G') \leq k$.*

Proof. First, note that G is a $(k + 1)$-regular graph, thus $\mathrm{fun}(G) \leq k + 1$. Since G is connected, every proper subgraph $G' \subset G$ contains a vertex of degree at most k. Thus, $\mathrm{fun}(G') \leq k$.

It remains to prove that the functionality of every vertex $v \in V(G)$ is at least $k + 1$. Let ℓ be a line of the projective plane. Because of the point-line duality of finite projective planes [8], it is enough to show that the functionality of ℓ is at least $k+1$. Let $S = \{p_1, \ldots, p_a, \ell_1, \ldots, \ell_b\}$ be a set of points p_i and lines ℓ_j (distinct from ℓ) such that $|S| = a + b \le k$. We will prove that there exist vertices u and w satisfying:

1. The vertices u and w are not in S and they are not adjacent to any vertices in S. They are also different from ℓ.
2. The vertex u is adjacent to ℓ, i.e., u is a point incident to ℓ.
3. The vertex w is not adjacent to ℓ, i.e., w is a point not incident to ℓ or it is a line.

Once we prove the existence of these two vertices, we are done as their existence implies that ℓ is not a function of S.

There are $k+1$ points on each line and every two lines intersect in one point. It follows that there is a point q on ℓ which is distinct from each p_i and it is not incident with any ℓ_i. Thus, the vertex q is not adjacent to any vertex in S and it is adjacent to ℓ. We set $u = q$.

Let $L = \{\ell, \ell_1, \ldots, \ell_b\}$. We will prove the existence of the vertex w by considering two cases. The first case is $b = k$, i.e., the set S consists of k lines and no points. We may consider any line $\ell' \notin L$ as the vertex w. The vertex w is not adjacent to any vertex in S as it contains only lines and it is not adjacent to the line ℓ either. Thus, the vertex w satisfies the sought properties.

The second case is $b \le k - 1$. Let P be the set of points $p_1, \ldots, p_a$ together with all the points that are incident to some line from L. We will show that $|P| \le k^2 + k$, which implies existence of a vertex $w \in V(G) \setminus P$ satisfying the sought properties. First, note that if $b = 0$ (i.e., S contains only points), then $|P| \le k + 1 + a = 2k + 1$. Next, we suppose that $b \ge 1$ which means that L contains at least two lines. Recall that each pair of line intersects in a point. Thus, there are at most $(b+1)(k+1) - 1$ points incident to the lines in L (the -1 comes from the fact that we have at least two lines). Moreover, we have a points $p_1, \ldots, p_a$. Thus, $|P| \le (b+1)(k+1) - 1 + a$. Since $a + b \le k$, we have $a \le k - b$. Therefore,

$$|P| \le (b+1)(k+1) - 1 + k - b = bk + 2k \le k^2 + k.$$

This concludes the second case and also the whole proof.   □

2.2   Upper Bound for General Graphs

Theorem 2. *If G is a graph with n vertices, then* $\mathrm{fun}(G) \le \sqrt{c \cdot n \ln n}$ *for any $c > 3$, if n is big enough.*

Proof. We show that there is $v \in V(G)$ such that $\mathrm{fun}_G(v) \le d(n) := \sqrt{c \cdot n \ln n}$. As $d(n)$ is increasing, this suffices to show existence of such vertex also in a subgraph of G. We will write $d = d(n)$.

First, suppose that $|\mathrm{sd}(u,v)| < d$ for some u, v of G. Since $\mathrm{sd}(u,v) = (N(u) \triangle N(v)) \setminus \{u, v\}$, the vertex v is a function of the set $\mathrm{sd}(u,v) \cup \{u\}$. Next, suppose all sets $\mathrm{sd}(u, v)$ have at least d vertices. In this case we choose an arbitrary vertex $v \in V(G)$. We also choose a random set $S \subseteq V(G)$ by independently putting each vertex of G different from v to S with probability $p = d/n$. Suppose v is not a function of S. Then, there exists $u_1 \in N(v) \setminus S$ and $u_2 \notin N(v) \cup S \cup \{v\}$ such that neighbors of u_1 and u_2 in S are the same, that happens if and only if $\mathrm{sd}(u_1, u_2) \cap S = \emptyset$. We bound the probability of this event by the union bound

$$\Pr\left[\exists u_1 \in N(v) \setminus S, u_2 \notin N(v) \cup S \cup \{v\} : S \cap \mathrm{sd}(u_1, u_2) = \emptyset\right]$$
$$\leq \sum_{u_1, u_2} (1 - p)^{|\mathrm{sd}(u_1, u_2)| - 1} \leq n^2 (1 - p)^{d-1}.$$

The -1 in the exponent is caused by $v \in \mathrm{sd}(u_1, u_2)$, but v cannot be chosen to belong to S. Thus, the probability that v is not a function of S is at most $n^2 e^{-p(d-1)}$, which is strictly smaller than $\frac{1}{n}$ whenever $c > 3$ and n is big enough. Clearly, the expected size of S is $p(n-1) = \frac{n-1}{n}d$. Thus by Markov inequality, $\Pr[|S| > d] \leq \frac{n-1}{n} = 1 - \frac{1}{n}$. This means that with the positive probability $|S| \leq d$ and v is function of S and we can conclude that $\mathrm{fun}_G(v) \leq d$. $\square$

2.3   Random Graphs

In this section, we prove our results about functionality of random graphs.

Theorem 3. *The functionality of the random graph $G = G(n, p)$ is $\Omega(\log n)$ with probability at least $1 - \frac{1}{n^{\log n}}$ for any $\frac{3 \log^2 n}{n} \leq p \leq 1 - \frac{3 \log^2 n}{n}$.*

Proof. Note that $\mathrm{fun}(G) = \mathrm{fun}(\bar{G})$, where $\bar{G}$ is the complement of G. Since $G(n, 1-p)$ is the complement of $G(n, p)$, we can suppose that $p \leq \frac{1}{2}$. We will prove that functionality of G is larger than $k = \frac{1}{2} \log n$ (w.h.p.). Let $v, u_1, \ldots, u_k$ be vertices of G. We will show that v is function of $u_1, \ldots, u_k$ only with small probability.

First suppose that $\frac{3 \log^2 n}{\sqrt{n}} \leq p \leq \frac{1}{2}$. We divide the rest of vertices $V' = V(G) \setminus \{v, u_1, \ldots, u_k\}$ into buckets according to the adjacency to $u_1, \ldots, u_k$, i.e., for each subset $S \subseteq \{u_1, \ldots, u_k\}$ we create a bucket B_S consisting of vertices that are adjacent to all vertices in S and are not adjacent to any vertex in $\{u_1, \ldots, u_k\} \setminus S$. There are at most 2^k buckets and therefore, there is a bucket B containing at least $\frac{|V'|}{2^k} \geq \frac{\sqrt{n}}{2}$ vertices. However by definition of functionality, that means v is connected to all vertices in B or to none of them. This event occurs only with a probability

$$p^{|B|} + (1 - p)^{|B|} \leq 2 \cdot (1 - p)^{\frac{\sqrt{n}}{2}} \leq 2 \cdot e^{-p \cdot \frac{\sqrt{n}}{2}} \leq 2^{-\frac{3}{2} \cdot \log^2 n}.$$

We have $n \cdot \binom{n-1}{k} < n^{\frac{\log n}{2}}$ possibilities how to choose $v, u_1, \ldots, u_k$. Thus by the union bound, we have that $\mathrm{fun}(G) \leq k$ with probability at most $\frac{1}{n^{\log n}}$.

Now, suppose that $\frac{3 \log^2 n}{n} \leq p \leq \frac{3 \log^2 n}{\sqrt{n}}$. Let E' be edges between $u_1, \ldots, u_k$ and $V' = V \setminus \{v, u_1, \ldots, u_k\}$. We bound the expected size of E' as follows.

$$\mathbb{E}[|E'|] = p \cdot k \cdot |V'| \leq \frac{3}{2} \sqrt{n} \cdot \log^3 n$$

By the Chernoff bound, we have that probability that $|E'| \geq 3\sqrt{n} \cdot \log^3 n = \ell$ is at most $e^{-\sqrt{n}\log^3 n}$. Thus with high probability, there are at least $|V'| - \ell \geq \frac{2n}{3}$ vertices in V' such that there is no edge between them and $u_1, \ldots, u_k$. Denote such vertices as B_0. Now, we proceed similarly as in the previous case. Again, the vertex v has to be connected to all vertices in B_0 or none of them, which occurs only with the following probability.

$$p^{|B_0|} + (1-p)^{|B_0|} \leq 2 \cdot (1-p)^{\frac{2n}{3}} \leq 2 \cdot e^{-p \cdot \frac{2n}{3}} \leq 2^{-2 \cdot \log^2 n}.$$

Thus, the probability that v is function of $u_1, \ldots, u_k$ is at most

$$e^{-\sqrt{n}\log^3 n} + 2^{-2 \cdot \log^2 n} \leq 2^{-\frac{3}{2} \cdot \log^2 n}.$$

By the same union bound as before, we have that $\mathrm{fun}(G) \leq k$ with probability at most $\frac{1}{n^{\log n}}$. $\qquad\square$

Proposition 1. *Each vertex of the random graphs $G = G(n, \frac{1}{2})$ is a function of at most $3 \log n$ vertices with probability at least $1 - \frac{1}{n}$.*

Proof. Let $k = 3 \log n$ and V' be a set of k vertices. We will show that with high probability there are no two vertices $v_1, v_2 \notin V'$ such that $N(v_1) \cap V' = N(v_2) \cap V'$, i.e., they have the same neighborhood in V'. Let $v_1, v_2 \notin V'$, then

$$\Pr[v_1, v_2 : N(v_1) \cap V' = N(v_2) \cap V'] = 2^{-k}.$$

By the union bound, we have:

$$\Pr[\exists v_1, v_2 \notin V' : N(v_1) \cap V' = N(v_2) \cap V'] \leq \frac{n^2}{2^k} = \frac{1}{n}. \qquad (1)$$

Thus, each vertex outside of V' has a unique neighborhood to V'. Therefore, V' determines adjacency for any vertex not in V'. $\qquad\square$

3 Symmetric Difference

In this section, we will prove our lower and upper bounds for symmetric difference of interval and circular arc graphs.

3.1 Circular Arc Graphs

In this section, we prove that symmetric difference of circular arc graphs[1] is $\Theta(\sqrt{n})$. More formally, we prove the following two theorems.

Theorem 4. *Any circular arc graph $G \in \mathrm{CA}_n$ has symmetric difference at most $O(\sqrt{n})$.*

---

[1] Intersection graphs of arcs of a cycle, see Sect. 1 for the proper definition.

Theorem 5. *There is a circular arc graph $G \in \mathrm{CA}_n$ of symmetric difference at least $\Omega(\sqrt{n})$.*

First, we prove the upper bound, i.e., that every circular arc graph has symmetric difference at most $O(\sqrt{n})$. We use the following notations for arcs. Let a, b be two points of a circle. Then, an arc $r = [a, b]$ is an arc beginning in a and going in clockwise direction to b. We call a as the starting point of r and b as the ending point of a.

Proof (Proof of Theorem 4). Let circle arc graph $G = (V, E)$ be an intersection graph of a set of arcs $R = \{r_1, \ldots, r_n\}$ of a circle C. Without loss of generality, we can suppose that the circumference of C is $2n$, endpoints of all arcs $r_i \in R$ are integer points and are different for all arcs. Thus, each integer point $\{0, \ldots, 2n - 1\}$ of C is an endpoint of exactly one arc in R.

Consider an arc $r = [a, b] \in R$. We again suppose that the points a and b are in clockwise order and the arc r starts in a and goes clockwise to b. We represent the arc r as a point (a, b) in the plane $\mathbb{R}^2$. Note that all these points are in the square S with corners in the points $(0, 0)$ and $(2n - 1, 2n - 1)$ (some points maybe on the border of S). We divide the square S into subsquares of size $k \times k$ for $k = \frac{2n-1}{\lfloor \sqrt{n} \rfloor - 1}$. Note that we have strictly less than n such subsquares and $k = \Theta(\sqrt{n})$. Thus, there is at least one subsquare that contains two points representing arcs, say $r = [a, b]$ and $r' = [a', b']$. It follows that $|a - a'|, |b - b'| \leq k$. Suppose that $a' > a$ and $b' > b$, other cases are analogous. Then, each arc counted in $\mathrm{sd}(r, r')$ has to start or end in an integer point from the interval $[a, a' - 1]$ or $[b, b' - 1]$. Since there are at most $2k$ integer points in these two intervals and each integer point of C is an endpoint of exactly one arc of R, we conclude that $\mathrm{sd}(r, r') \leq 2k \leq O(\sqrt{n})$. $\qquad\square$

Now, we give a construction of a circular arc graph of symmetric difference at least $\Omega(\sqrt{n})$. Let n be a square of an integer, i.e., $n = d^2$ for some $d \in \mathbb{N}$. We consider a circle C of circumference n and a set P of integer points of C, i.e., $P = \{0, \ldots, n - 1\}$ ordered in clockwise direction. The length $|r|$ of an arc $r = [a, b]$ is equal to $b - a \pmod{n}$. We say the arc $[a, b]$ is *integral* if both a and b are integers.

We will represent each point $p \in P$ as two integer indices $0 \leq i, j < d = \sqrt{n}$ such that $p = i \cdot d + j$. Note that each point p has a unique such representation and we denote it as $(i, j)_d$. Let R be a set of arcs $[(i, j)_d, (j, i)_d]$ for all possible $i \neq j$ such that length of each arc in R is at most $\frac{n}{4}$. Note that we require that $i \neq j$, thus we do not consider zero-length arcs consisting only of a point of a form $(i, i)_d$. See Fig. 1 for an illustration. Let G be the intersection graph of arcs in R. We we will prove that $\mathrm{sd}(G) \geq \Omega(\sqrt{n})$. First, we prove two auxiliary lemmas, Lemma 1 and 2. Lemma 1 asserts that each sufficiently long arc of C contains a lot of starting and ending points of arcs in R. Lemma 2 states that for any two arcs r and r' we will find a long arc s that is a subarc of only one of the arcs r and r' (say r). Thus by Lemma 1, the arc r intersects many arcs that go "away" from the arc r' and that is enough to imply Theorem 5.

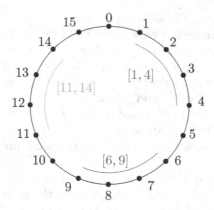

Fig. 1. An example of the circular arc graph lower bound construction for $n = 16$ and $d = 4$. This graph contains three arcs corresponding to points $1 = (0,1)_d$, $6 = (1,2)_d$, and $11 = (2,3)_d$. For an example, the arc corresponding to point $3 = (0,3)_d$ is omitted since $12 - 3 = 9 > 4 = \frac{n}{4}$ where $12 = (3,0)_d$.

Lemma 1. *Let s be an integral arc of C of length at least $d-1$. Then, it contains at least $\frac{d}{5}$ integer points such that they are starting points of arcs in R. Similarly, it contains at least $\frac{d}{5}$ integer points such that they are ending points of arcs in R.*

Proof. Since s is integral and its length is at least $d - 1$, it contains at least d consecutive integer points. Thus, there is an integer j such that s contains all points of the set $\{(j,t)_d, \ldots, (j, d-1)_d, (j+1, 0)_d, \ldots, (j+1, t-1)_d\}$ for some $t \in [d]$. For any $\ell \in [d]$, we define $j_\ell = j$ if $\ell \geq t$ and $j_\ell = j + 1$ otherwise. It follows that s contains the point $(j_\ell, \ell)_d$ for every $\ell \in [d]$. Consider a subset of these points $S = \{(j_k, k)_d \mid 1 \leq (k - j \mod d) \leq \frac{d}{5}\}$. Observe that $|S| = \frac{d}{5}$. We claim that each point of S is a starting point of an arc in R. Let r be an arc $[(j_k, k)_d, (k, j_k)_d]$ where $1 \leq (k - j \mod d) \leq \frac{d}{5}$.

$$|r| = k \cdot d + j_k - j_k \cdot d - k \mod n$$

$$\leq (k - j_k) \cdot d + d \leq \frac{d}{5} \cdot d + d \leq \frac{n}{4} \qquad \text{for sufficiently large } n$$

Thus, $r \in R$. Analogously, the set $E = \{(j_k, k)_d \mid 1 \leq (j+1-k \mod d) \leq \frac{d}{5}\}$ of size $\frac{d}{5}$ contains ending points of arcs in R. $\qquad\square$

Lemma 2. *Let $r = [(i,j)_d, (j,i)_d]$ and $r' = [(i',j')_d, (j',i')_d]$ be two arcs in R. Then, at least one of the arcs r and r' contains an integral subarc s of length $d - 2$ that is disjoint from the other arc.*

Proof. If the arcs r and r' are disjoint then the existence of s is trivial as each arc in R has length at least $d - 1$ (the arcs in R of length exactly $d - 1$ are those of the form $[(k, k+1)_d, (k+1, k)_d]$).

Thus, suppose that r and r' intersect and without loss of generality suppose that the ending point of r lies inside r', i.e. we read in clockwise order the points

$(i', j')_d$, $(j, i)_d$ and $(j', i')_d$. Consider an arc $s' = [(j, i)_d, (j', i')_d]$. Note that arc s' is a subarc of r' and intersects r only in the point $(j, i)_d$. Thus, if $|s'| \geq d - 1$, the arc r' would contain the sought integral subarc s. We have $|s'| = (j' - j) \cdot d + i' - i$ (mod n). Since $0 \leq i, i', j, j' \leq d - 1$, it holds that $|(j' - j) \cdot d + i' - i| \leq d^2 - 1 < n$. Thus, if $|s'| \leq d - 2$ then $j = j'$ or $j' - j = 1$ (mod n) and $i - i' \geq 2$ (note that this holds even when $j = n - 1$ and $j' = 0$).

First, suppose that $j' - j = 1$ (mod n) (and $i - i' \geq 2$). Then, the start point of r also belongs to r' and the arc $s' = [(i', j')_d, (i, j)_d]$ has length at least $d - 1$ and again it is a subarc of r' and intersects the arc r only in the single point $(i, j)_d$.

The last case is when $j' = j$. This implies that $i \neq i'$ as all endpoints are unique. Moreover, since the point $(j, i)_d$ precedes the point $(j', i')_d$ in clockwise order, we get that $i < i'$. Then, the arc $s' = [(i, j)_d, (i', j')_d] = [(i, j)_d, (i', j)_d]$ has length at least d and is a subarc of only the arc r with the exception of its ending point $(i', j')_d$.   □

Now, we are ready to prove Theorem 5.

Proof (Proof of Theorem 5). Consider two arcs $r = [(i, j)_d, (j, i)_d]$ and $r' = [(i', j')_d, (j', i')_d]$ in R. Recall that all arcs in R have length at most $\frac{n}{4}$. Thus, we can suppose that the arc $s' = [(j', i')_d, (i, j)_d]$ has length at least $\frac{n}{4}$ and is disjoint from r and r', except for the endpoints $(j', i')_d$ and $(i, j)_d$. Note that, this implies that the endpoints $(j, i)_d$ and $(j', i')_d$ of r and r', respectively, are in clockwise order. By Lemma 2, we suppose that r contains a subarc s of length $d - 2$ that is disjoint with r' (the other case when s is a subarc of r' is analogous). Let L be a set of points in s such that they are ending points of arcs in R (distinct from r). By Lemma 1, we have that $|L| \geq \frac{d}{5} - 1$.

Let $t = [a, b]$ be an arc with the ending point b in L. The arc t intersects the arc r as the ending point b is a point of $s \subseteq r$. Since the arc s' has length at least $\frac{n}{4}$ and $t \in R$, the starting point a has to be a point of s' and thus, the arc t is disjoint from r'. Therefore, $\mathrm{sd}(r, r') \geq |L| \geq \Omega(\sqrt{n})$.   □

3.2   Interval Graphs

In this section, we will prove that symmetric difference of interval graphs[2] is still fixed power of n but strictly less than the symmetric difference of circular arc graphs. In particular, we will prove the following two theorems.

Theorem 6. *Any interval graph $G \in \mathrm{INT}_n$ has symmetric difference at most $O(\sqrt[3]{n})$.*

Theorem 7. *There is an interval graph $G \in \mathrm{INT}_n$ of symmetric difference at least $\Omega(\sqrt[4]{n})$.*

---

[2] Intersection graphs of intervals of the real line, see Sect. 1 for the proper definition.

Note that the existence of interval graphs of arbitrarily high symmetric difference is proved in Corollary 5.3 of Dallard et al. [5]. While they do not provide explicit bounds, their proof gives the same $\Omega(\sqrt[4]{n})$ bound as ours. However, our proof of the lower bound is self-contained and we believe it to be simpler. We start with a proof of the upper bound.

Proof (Proof of Theorem 6). Let $G = (V, E)$ be an intersection graph of intervals $R = \{r_1, \ldots, r_n\}$ with $r_i = [a_i, b_i]$ and let $V = \{1, \ldots, n\}$. Without loss of generality, we suppose for clarity that all a_i's and b_i's are different points. The intervals are numbered in the order given by their starting points, i.e., for every two indices $i, j \in [n]$ we have $i < j$ if and only if $a_i < a_j$. For an interval r_i, we define two sets:

1. $A_i = \{j \mid a_i < b_j\}$, i.e., it contains the indices of intervals in R that end after the interval r_i starts.
2. $B_i = \{j \mid a_j < b_i\}$, i.e., it contains the indices of intervals in R that start before the interval r_i ends.

Note that $N(i) = A_i \cap B_i$. Moreover, $\mathrm{sd}(i, j) \leq |N(i) \Delta N(j)| \leq |A_i \Delta A_j| + |B_i \Delta B_j|$. Note that for each pair i, j, it holds that $A_i \subseteq A_j$ or $A_j \subseteq A_i$, thus $A_i \Delta A_j$ is $A_i \setminus A_j$ or $A_j \setminus A_i$ and analogously with B_i and B_j. We will prove that there are two intervals r_i and r_j such that $|A_i \Delta A_j| + |B_i \Delta B_j| \leq O(\sqrt[3]{n})$.

Let $d \in \mathbb{N}$ be a parameter. We will find two vertices $i, j \in V$ such that $\mathrm{sd}(i, j) \leq O(\max\{d, d + \frac{n}{d^2}\})$. Thus, if we set $d = \sqrt[3]{n}$ we would get $\mathrm{sd}(i, j) \leq O(\sqrt[3]{n})$. Since any subgraph of an interval graph is again an interval graph, the upper bound for $\mathrm{sd}(G)$ will follow. Let $D_\ell = A_\ell \setminus A_{\ell+1}$. In other words, the set D_ℓ contains intervals of R that end after the interval r_ℓ starts but before the interval $r_{\ell+1}$ starts. Note that $B_i = \{1, \ldots, \ell\}$ where ℓ is the unique index such that $i \in D_\ell$. Let k be the largest index such that $\sum_{\ell \leq k} |D_\ell| \leq d + 2$. Note that for any two indices $i, j \leq k$, it holds that $|A_i \Delta A_j| \leq \sum_{\ell \leq k} |D_\ell|$.

First, we prove that if $k \leq d^2$, then $\mathrm{sd}(i, j) \leq 2d + 2$. Suppose that actually $\sum_{\ell \leq k} |D_\ell| \leq d$. Then, $|D_{k+1}| \geq 3$. Let $i, j \in D_{k+1}$ such that $i, j \leq k$. Then, $B_i = B_j$ and $|A_i \Delta A_j| \leq d$, therefore $\mathrm{sd}(i, j) \leq d$.

From now, we suppose that $d \leq \sum_{\ell \leq k} |D_\ell| \leq d + 2$. Let $p \leq k$ be an index such that $|D_p| \geq 2$ (if such p exists) and $i, j \in D_p = A_p \setminus A_{p+1}$. Thus, $B_i = B_j$. Since $i, j \leq p \leq k$, then $|A_i \Delta A_j| \leq d + 2$ and $\mathrm{sd}(i, j) \leq d + 2$. Note that so far, we did not use the assumption that $k \leq d^2$.

Now, suppose that for all $\ell \leq k$ it holds that $|D_\ell| \leq 1$. Since $k \leq d^2$ there exists two indices $p < q \leq k$ such that $D_p, D_q \neq \emptyset$ and $q - p \leq d$ (there are at least d indices $\ell \leq k$ such that $|D_\ell| = 1$). Let $i \in D_p$ and $j \in D_q$. Since $B_i = \{1, \ldots, p\}, B_j = \{1, \ldots, q\}$, we have $|B_j \setminus B_i| \leq d$. Further, since $i, j \leq k$, we have $|A_i \Delta A_j| \leq d + 2$. Thus, $\mathrm{sd}(i, j) \leq 2d + 2$.

Now, we suppose that $k > d^2$. It follows there are two indices $i, j \leq k$ such that $i \in D_p$ and $j \in D_q$ such that $|p - q| \leq \frac{n}{d^2}$. Therefore, $|A_i \Delta A_j| \leq d + 2$ and $|B_i \Delta B_j| = |p - q| = \frac{n}{d^2}$ and $\mathrm{sd}(i, j) \leq d + 2 + \frac{n}{d^2}$. □

Now, we give a construction of an interval graph with symmetric difference $\Omega(\sqrt[4]{n})$. Let $d \in \mathbb{N}$ sufficiently large. We construct $\Theta(d^4)$ intervals on a line segment $[0, t]$ for $t = 20d^3$ such that the corresponding intersection graph G will have $\mathrm{sd}(G) \geq d$. There will be intervals of two types – short and long. See Fig. 2 for an illustration. Short intervals have length d and they start in each point $0, \ldots, t - d$, i.e.,

$$S = \{[i, i + d] \mid i \in \{0, \ldots, t - d\}\}.$$

Long intervals will have various lengths. For $i \geq 0$, let $\ell_i = 4d^2 \cdot (i + 1)$. For $0 \leq i \leq 2d - 1$, we define the i-th class of long intervals as

$$L_i = \{[a, b] \mid a, b \equiv i \pmod{2d}; \ell_i \leq b - a \leq \ell_i + 2d^2\},$$

i.e., the set L_i contains intervals such that they start and end in points congruent to i modulo $2d$ and their length is between ℓ_i and $\ell_i + 2d^2$. Let $I = S \cup \bigcup_{0 \leq i \leq 2d-1} L_i$ be the set of all constructed intervals. We start with two observations about I.

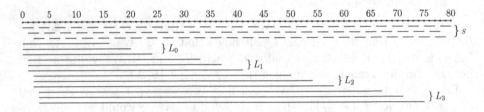

Fig. 2. An example of the interval graph lower bound construction for $d = 2$. For clarity, only the first three intervals are displayed from each set L_i.

Observation 8. *Any interval $[a, b]$ (for $a, b \in \mathbb{N}$) of the line segment $[0, t]$ of length $2d$ contains d short intervals.*

Observation 9. *There are at most $O(d^4)$ intervals in I.*

Proof. Clearly, there are $t - d + 1 = O(d^3)$ intervals in S. Note that for any $a \leq \frac{t}{2}, a \equiv i \pmod{2d}$, there are exactly $d + 1$ intervals of a form $[a, b]$ in L_i because of the length constraints of the long interval. Analogously, for any $b \geq \frac{t}{2}, b \equiv i \pmod{2d}$, there are exactly $d + 1$ intervals of a form $[a, b]$ in L_i. We remark this indeed holds even for L_{2d-1} as we set t to be large enough. There are no other intervals in L_i. Since there are $O(d^2)$ points $p \in [0, t]$ such that $p \equiv i \pmod{2d}$, it follows that $|L_i| \leq O(d^3)$. Therefore, there are at most $O(d^4)$ long intervals as there are $O(d)$ classes of long intervals. $\qquad\square$

The graph G is an intersection graph of I. Now, we are ready to prove Theorem 7, i.e., $\mathrm{sd}(G) \geq \Omega(\sqrt[4]{n})$.

Proof (Proof of Theorem 7). Let $r = [a, b]$ and $r' = [a', b']$ be two intervals in I. Without loss of generality let $a \leq a'$. First, suppose that the $r, r' \in S$, i.e., both of them are short. In this case it holds that $a < a'$ and $b < b'$. First, if

$a' > a + 2d$, then all d short interval of form $[i, i + d]$ for $i \in r$ do not intersect r'. Thus, $|N(r)\Delta N(r')| \geq d$.

Now, suppose that $a' \leq a + 2d$. Further, suppose that $b' \leq \frac{t}{2}$. Then as already observed, there are d intervals of the form $[b', c]$ in L_i for $b' \equiv i \pmod{2d}$. Since b' is not in r, we have that $|N(r)\Delta N(r')| \geq d$.

If $b' > \frac{t}{2}$, then $a > \frac{t}{2} - 3d$, as $b' - d = a' \leq a + 2d$. Analogously, it holds there are d intervals of the form $[c, a]$ in L_i for $a \equiv i \pmod{2d}$ as d and t is large enough.

Now, suppose that $r, r' \in L_i$ for some i. In this case $a, b, a', b' \equiv i \pmod{2d}$. Since $r \neq r'$, it follows that $|a - a'| \geq 2d$ or $|b - b'| \geq 2d$. Thus, at least one of the intervals r and r' has a private subinterval of length at least $2d$ and by Observation 8, we have that $|N(r)\Delta N(r')| \geq d$.

Let k be the difference of length of r and r'. For the remaining cases we will prove that $k \geq 4d$. Then, at least one of the intervals r and r' contains a private subinterval of length at least $2d$ and again by Observation 8, we have that $|N(r)\Delta N(r')| \geq d$. There are two remaining cases:

1. $r \in S, r' \in L_i$: Then, $|r| = d$ and $|r'| \geq 4d^2$.
2. $r \in L_i, r' \in L_j$ for $i < j$: Then, $|r| \leq 4d^2 \cdot (i+1) + 2d^2$ and $|r'| \geq 4d^2 \cdot (j+1)$. It follows that $k = 4d^2 \cdot (j - i) - 2d^2 \geq 2d^2$.

Thus, in both cases we have that $k \geq 4d$ for $d \geq 2$. We have showed that $|N(r)\Delta N(r')| \geq d$ for all cases. Thus by Observation 9, we conclude that $\mathrm{sd}(G) \geq \Omega(\sqrt[4]{n})$. $\qquad\square$

Acknowledgements. The research presented in this paper has been started during the KAMAK workshop in 2021. We are grateful to the organizers of this wonderful event.

References

1. Alecu, B., Atminas, A., Lozin, V.V.: Graph functionality. J. Comb. Theory Ser. B **147**, 139–158 (2021). https://doi.org/10.1016/j.jctb.2020.11.002
2. Alon, N., Haussler, D., Welzl, E.: Partitioning and geometric embedding of range spaces of finite Vapnik-Chervonenkis dimension. In: Soule, D. (ed.) Proceedings of the Third Annual Symposium on Computational Geometry, Waterloo, Ontario, Canada, 8–10 June 1987, pp. 331–340. ACM (1987)
3. Atminas, A., Collins, A., Lozin, V., Zamaraev, V.: Implicit representations and factorial properties of graphs. Discret. Math. **338**(2), 164–179 (2015). https://doi.org/10.1016/j.disc.2014.09.008
4. Balogh, J., Bollobás, B., Weinreich, D.: The speed of hereditary properties of graphs. J. Comb. Theory Ser. B **79**(2), 131–156 (2000). https://doi.org/10.1006/jctb.2000.1952, https://www.sciencedirect.com/science/article/pii/S009589560091952X
5. Dallard, C., Lozin, V., Milanič, M., Štorgel, K., Zamaraev, V.: Functionality of box intersection graphs (2023). https://doi.org/10.48550/ARXIV.2301.09493
6. Hatami, H., Hatami, P.: The implicit graph conjecture is false. In: 63rd IEEE Annual Symposium on Foundations of Computer Science, FOCS 2022, Denver, CO, USA, October 31–3 November 2022, pp. 1134–1137. IEEE (2022)

7. Kranakis, E., Krizanc, D., Ruf, B., Urrutia, J., Woeginger, G.: The VC-dimension of set systems defined by graphs. Discret. Appl. Math. **77**(3), 237–257 (1997). https://doi.org/10.1016/S0166-218X(96)00137-0
8. Matoušek, J., Nešetřil, J.: Invitation to Discrete Mathematics, 2 ed. Oxford University Press, Oxford (2009)

Cops and Robbers on Multi-Layer Graphs

Jessica Enright[1][iD], Kitty Meeks[1][iD], William Pettersson[1(✉)][iD],
and John Sylvester[1,2][iD]

[1] School of Computing Science, University of Glasgow, Glasgow, UK
{jessica.enright,kitty.meeks,william.pettersson}@glasgow.ac.uk,
john.sylvester@liverpool.ac.uk
[2] Department of Computer Science, University of Liverpool, Liverpool, UK

Abstract. We generalise the popular *cops and robbers* game to multi-layer graphs, where each cop and the robber are restricted to a single layer (or set of edges). We show that initial intuition about the best way to allocate cops to layers is not always correct, and prove that the multi-layer cop number is neither bounded from above nor below by any function of the cop numbers of the individual layers. We determine that it is NP-hard to decide if k cops are sufficient to catch the robber, even if each layer is a tree plus some isolated vertices. However, we give a polynomial time algorithm to determine if k cops can win when the robber layer is a tree. Additionally, we investigate a question of worst-case division of a simple graph into layers: given a simple graph G, what is the maximum number of cops required to catch a robber over all multi-layer graphs where each edge of G is in at least one layer and all layers are connected? For cliques, suitably dense random graphs, and graphs of bounded treewidth, we determine this parameter up to multiplicative constants. Lastly we consider a multi-layer variant of Meyniel's conjecture, and show the existence of an infinite family of graphs whose multi-layer cop number is bounded from below by a constant times $n/\log n$, where n is the number of vertices in the graph.

Keywords: Cops and robbers · multi-layer graphs · pursuit-evasion games · Meyniel's conjecture

1 Introduction

We investigate the game of cops and robbers played on multi-layer graphs. Cops and robbers is a 2-player adversarial game played on a graph introduced independently by Nowakowski and Winkler [22], and Quilliot [25]. At the start of the game, the cop player chooses a starting vertex position for each of a specified number of cops, and the robber player then chooses a starting vertex position for the robber. Then in subsequent rounds, the cop player first chooses none, some, or all cops and moves them along exactly one edge to a new vertex. The robber player then either moves the robber along an edge, or leaves the robber on its current vertex. The cop player wins if after some finite number of rounds a cop

© The Author(s), under exclusive license to Springer Nature Switzerland AG 2023
D. Paulusma and B. Ries (Eds.): WG 2023, LNCS 14093, pp. 319–333, 2023.
https://doi.org/10.1007/978-3-031-43380-1_23

occupies the same vertex as the robber, and the robber wins otherwise. Both players have perfect information about the graph and the locations of cops and robbers. Initially, research focussed on games with only one cop and one robber, and graphs on which the cop could win were classed as *copwin* graphs. Aigner and Fromme [1] introduced the idea of playing with multiple cops, and defined the *cop number* of a graph as the minimum number of cops required for the cop player to win on that graph. Many variants of the game have been studied, and for an in-depth background on cops and robbers, we direct the reader to [5].

In this paper, we play cops and robbers on multi-layer graphs where each cop and the robber will be associated with exactly one layer, and during their respective turns, will move only over the edges in their own layer. While we define multi-layer graphs formally in upcoming sections, roughly speaking, here a multi-layer graph is a single set of vertices with each layer being a different (though possibly overlapping) set of edges on those vertices. The variants we study could intuitively be based on the premise that the cops are assigned different modes of transport. For instance, a cop in a car may be able to move quickly down streets, while a cop on foot may be slower down a street, but be able to quickly cut between streets by moving through buildings or down narrow alleys.

Extending cops and robbers to multi-layer graphs creates some new variants, and generalises some existing variants. Fitzpatrick [14] introduced the *precinct* variant, which assigns to each cop a subset of the vertices (called their *beat*). In the precinct variant, a cop can never leave their beat. This can be modelled as multi-layer cops and robbers by restricting each layer to a given beat. Fitzpatrick [14] mainly considers the case were a beat is an isometric path, we allow more arbitrary (though usually spanning and connected) beats/layers. Clarke [11] studies the problem of covering a graph with a number of cop-win subgraphs to upper bound the cop number of a graph — again such constructions can be modelled as multi-layer graphs with the edges of each layer forming a cop-win graph. Another commonly studied variant of cops and robbers defines a speed s (which may be infinite) such that the robber can move along a path of up to s edges on their turn [6, Section 3.2]. These can also be modelled as multi-layer graphs by adding edges between any pair of vertices of distance at most s that only belong to the layer the robber is occupying.

1.1 Further Related Work

Temporal graphs, in which edges are active only at certain time steps, are sometimes modelled as multi-layered graphs. There has been some work on cops and robbers on temporal graphs, though generally yielding quite a different game to the ones we consider here as a cop is not restricted to one layer. In particular, [3] considers cops and robbers on temporal graphs and when the full temporal graph is known they give a $O(n^3T)$ algorithm to determine the outcome of the game where T is the number of timesteps.

Variants of cops and robbers are also studied for their relationships to other parameters of graphs. For instance, the cop number of a graph G is at most one plus half the treewidth of G [17]. And if one considers the "helicopter" variant of cops and robbers, the treewidth of a graph is strictly less than the helicopter cop number of the graph [29]. Toruńczyk [32] generalises many graph parameters, including treewidth, clique-width, degeneracy, rank-width, and twin-width, through the use of variants of cops and robbers. We introduce our multi-layer variants of cops and robbers partially in the hopes of spurring research towards multi-layer graph parameters using similar techniques.

Recently Lehner, resolving a conjecture by Schröeder [27], showed the cop number of a toroidal graph is at most three [19]. There is also an interesting connection between cop number and the genus of the host graph [1,8,26,27]. It remains open whether any such connection can be made in the multi-layer setting.

1.2 Outline and Contributions

In Sect. 2 we define multi-layer graphs and multi-layer cops and robbers.

In Sect. 3 we develop several examples which highlight several counter-intuitive facts and properties of the multi-layer cops and robbers game. In particular, we show the multi-layer cop number is not bounded from above or below by a non-trivial function of the cop numbers of the individual cop layers.

In Sect. 4 we study the computational complexity of some multi-layer cops and robbers problems. We show that deciding if a given number of cops can catch a robber is NP-hard even if each layer is a tree plus some isolated vertices, but that if only the robber layer is required to be a tree the problem is FPT in the number of cops and the number of layers of the graph.

In Sect. 5 we consider an extremal version of multi-layer cop number over all divisions into layers of a single-layer graph. In particular, for a given single-layer graph G what is the maximum multi-layer cop number of any multi-layer graph $\mathcal{G}$ when all edges of G are present in at least one layer of $\mathcal{G}$.

In Sect. 6 we consider Meyniel's conjecture, which states that the single-layer cop number is $\mathcal{O}(\sqrt{|V|})$ and is a central open question in cops and robbers. We investigate whether a multi-layer analogue of Meyniel's conjecture can hold and, determine the worst case multi-layer cop number up to a multiplicative $\mathcal{O}(\log n)$ factor. This contrasts with the situation on simple graphs, where the worst-case is only known up to a multiplicative $n^{1/2-o(1)}$ factor.

Finally in Sect. 7 we reflect and conclude with some open problems.

Due to space limitations, most proofs have been omitted. For a complete version with all proofs, please refer to [13].

2 Definitions and Notation

We write $[n]$ to mean the set of integers $\{1, \ldots, n\}$, and given a set V we write $\binom{V}{2}$ to mean all possible 2-element subsets (i.e., edges) of V. A simple graph is then defined as $G := (V, E)$ where $E \subseteq \binom{V}{2}$. For a vertex $v \in V$ we let $d_G(v) := |\{u : uv \in E\}|$ be the degree of vertex v in G, and $\delta(G) := \min_{v \in V(G)} d_G(v)$ denote the minimum degree in a graph G. If, for all $v \in V$, $d_G(v) = r$ for some integer r, we say that G is r-regular. If the exact value of r is not important, we may just say that G is regular. If, instead, for all $v \in V$, $d_G(v) \in \{r, r + 1\}$ for some integer r, we say that G is almost-regular. The *distance* between two vertices u and v in a graph is the length of a shortest path between u and v.

A multi-layer graph $(V, \{E_1, \ldots, E_\tau\})$ consists of a vertex set V and a collection $\{E_1, \ldots, E_\tau\}$, for some integer $\tau \geqslant 1$, of edge sets (or *layers*), where for each i, $E_i \subseteq \binom{V}{2}$. We often slightly abuse terminology and refer to a layer E_i as a graph; when we do this, we specifically refer to the graph (V, E_i) (i.e., we always include every vertex in the original multi-layer graph, even if such a vertex is isolated in (V, E_i)). For instance, we often restrict ourselves to multi-layer graphs where, for each $i \in [\tau]$, the simple graph (V, E_i) is connected. We will say that each layer is *connected* to represent this notion. Given a multi-layer graph $(V, \{E_1, \ldots E_\tau\})$ let the *flattened* version of a multi-layer graph, written as $\mathsf{fl}(\mathcal{G})$, be the simple graph $G = (V, E_1 \cup \cdots \cup E_\tau)$.

Cops and robbers is typically played on a simple graph, with one player controlling some number of cops and the other player controlling the robber. On each turn, the cop player can move none, some, or all of the cops, however each cop can only move along a single edge incident to their current vertex. The robber player can then choose to move the robber along one edge, or have the robber stay still. The goal for the cop player is to end their turn with the robber on the same vertex as at least one cop, while the aim for the robber is to avoid capture indefinitely. If a cop player has a winning strategy on a graph G with k cops but not with $k-1$ cops, we say that the graph G has cop number k, denoted $\mathsf{c}(G) = k$, and that G is k-copwin. Given a multi-layer graph $(V, \{E_1, \ldots, E_\tau\})$, we will say the cop number of layer E_i to mean the cop number of the graph (V, E_i).

As this paper deals with both simple and multi-layer graphs, as well as cops and robbers variants played on these graphs, we will use *single-layer* as an adjective to denote when we are referring to either specifically a simple graph, or to cops and robbers played on a single-layer (i.e., simple) graph. This extends to parameters such as the cop number as well.

In this paper we consider the cops and robbers game on multi-layer graphs and so it will be convenient to define multi-layer graphs with a distinguished layer for the robber. More formally, for an integer $\tau \geqslant 1$, we use the notation $\mathcal{G} = (V, \{C_1, \ldots, C_\tau\}, R)$ to denote a multi-layer graph with vertex set V and collection $\{C_1, \ldots, C_\tau, R\}$ of layers, where $\{C_1, \ldots, C_\tau\}$ are the cop layers and R is the robber layer. In the cops and robber game on $\mathcal{G}$ each cop is allocated to a single-layer from $\{C_1, \ldots, C_\tau\}$, and the robber to R, and each cop (and the robber) will then only move along edges in their respective layer. We do not allow any cop or the robber to move between layers. We note that this is a slight abuse of notation, and that both $(V, \{C_1, \ldots, C_\tau, R\})$ and $(V, \{C_1, \ldots, C_\tau\}, R)$ both denote a multi-layer graphs with the the same collection $\{C_1, \ldots, C_\tau\} \cup \{R\}$ of edge sets, the latter has designated layers for the robber/cops whereas the former does not. We will use E_i to denote edge sets in multi-layer graphs that do not have a cop or robber labels.

A setting that appears often is $R = C_1 \cup \cdots \cup C_\tau$, where the robber can use any edge that exists in a cop layer. This setting is given by the multi-layer graph $\mathcal{G} := (V, \{C_1, \ldots, C_\tau\}, C_1 \cup \cdots \cup C_\tau)$, but for readability we will instead use $\mathcal{G} := (V, \{C_1, \ldots, C_\tau\}, *)$ to denote this.

We define several variants of cops and robbers on multi-layer graphs, however in each of them we have an *allocation* $\mathbf{k} := (k_1, \ldots, k_\tau)$ of cops to layers, such that there are k_i cops on layer C_i. We will often use $k := \sum_i k_i$ to refer to the total number of cops in a game.

We now define multi-layer cops and robbers: a two player game played with an allocation $\mathbf{k}$ on a multi-layer graph $\mathcal{G} = (V, \{C_1, \ldots, C_\tau\}, R)$. The two players are the cop player and the robber player. Each cop is assigned a layer such that there are exactly k_i cops in layer C_i. The game begins with the cop player assigning each cop to some vertex, and then the robber player assigns the robber to some vertex. The game then continues with each player taking turns in sequence, beginning with the cop player. On the cop player's turn, the cop player may move each cop along one edge in that cop's layer. The cop player is allowed to move none, some, or all of the cops. The robber player then takes their turn, either moving the robber along one edge in the robber layer or letting the robber stay on its current vertex. This game ends as a victory for the cop player if, at any point during the game, the robber is on a vertex that is also occupied by one or more cops. The robber wins if they can evade capture indefinitely.

We can now begin defining our problems, starting with ALLOCATED MULTI-LAYER COPS AND ROBBER.

ALLOCATED MULTI-LAYER COPS AND ROBBER
Input: A tuple $(\mathcal{G}, \mathbf{k})$ where $\mathcal{G} = (V, \{C_1, \ldots, C_\tau\}, R)$ is a multi-layer graph and $\mathbf{k}$ is an allocation of cops to layers.
Question: Does the cop player have a winning strategy when playing multi-layer cops and robbers on $\mathcal{G}$ with allocation $\mathbf{k}$?

We also consider a variant in which the cop player has a given number k of cops, but gets to choose the layers to which the cops are allocated.

MULTI-LAYER COPS AND ROBBER
Input: A tuple $(\mathcal{G}, k)$ where $\mathcal{G} = (V, \{C_1, \ldots, C_\tau\}, R)$ is a multi-layer graph and $k \geqslant 1$ is an integer.
Question: Is there an allocation $\mathbf{k}$ with $\sum_i k_i = k$ such that $(\mathcal{G}, \mathbf{k})$ is yes-instance for ALLOCATED MULTI-LAYER COPS AND ROBBER?

Lastly we consider MULTI-LAYER COPS AND ROBBER WITH FREE LAYER CHOICE, a variant of MULTI-LAYER COPS AND ROBBER in which, before the game is played, the layers in the multi-layer graph are not assigned to being either cop layers or robber layers. Instead the layers are simply labelled E_1 through E_τ, and in this variant the cop player first allocates each cop to one layer, and then the robber player is free to allocate the robber to any layer.

MULTI-LAYER COPS AND ROBBER WITH FREE LAYER CHOICE

Input: A tuple $(\mathcal{G}, k)$ where $\mathcal{G} = (V, \{E_1, \ldots, E_\tau\})$ is a multi-layer graph and $k \geqslant 1$ is an integer.

Question: Is there an allocation $\mathbf{k}$ with $\sum_i k_i = k$ such that for every j, $((V, \{E_1, \ldots, E_\tau\}, E_j), \mathbf{k})$ is a yes-instance for MULTI-LAYER COPS AND ROBBER?

We say that the multi-layer cop number of a multi-layer graph $\mathcal{G}$ is k if $(\mathcal{G}, k)$ is a yes-instance for MULTI-LAYER COPS AND ROBBER but $(\mathcal{G}, k - 1)$ is a no-instance for MULTI-LAYER COPS AND ROBBER. We will denote this with $\mathsf{mc}(\mathcal{G})$. We round out this section with a number of basic observations.

Proposition 1. *Let* $\mathcal{G} = (V, \{C_1, \ldots, C_\tau\}, R)$ *and* $\mathcal{G}' = (V, \{C_1, \ldots, C_\tau\}, R')$ *be any two multi-layer graphs where* $R \subseteq R' \subseteq \binom{V}{2}$. *If* $(\mathcal{G}, k)$ *is a no-instance to* MULTI-LAYER COPS AND ROBBER, *then* $(\mathcal{G}', k)$ *is a no-instance to* MULTI-LAYER COPS AND ROBBER. *Consequently,* $\mathsf{mc}(\mathcal{G}) \leqslant \mathsf{mc}(\mathcal{G}')$.

Proof. To win, the robber on $\mathcal{G}'$ uses the strategy from $\mathcal{G}$. The robber can execute this strategy as any edge in R' is in R. Since the cop layers have no added edges, the strategy must be robber-win as else the cops would win on $\mathcal{G}$.   □

Proposition 2. *Let* $\mathcal{G} = (V, \{C_1, \ldots, C_\tau\}, R)$ *and* $\mathcal{G}' = (V, \{C'_1, \ldots, C'_\tau\}, R)$ *be any two multi-layer graphs that satisfy* $C_i \subseteq C'_i$ *for every* $i \in [\tau]$. *If* $(\mathcal{G}, k)$ *is a yes-instance to* MULTI-LAYER COPS AND ROBBER, *then* $(\mathcal{G}', k)$ *is also a yes-instance to* MULTI-LAYER COPS AND ROBBER.

Proof. To win, the cops on $\mathcal{G}'$ use the strategy from $\mathcal{G}$. As no edge has been removed from $\mathcal{G}$ to create $\mathcal{G}'$, this must still result in the cops winning.   □

Proposition 3. *Let* $\mathcal{G} = (V, \{C_1, \ldots, C_\tau\}, *)$ *be a multi-layer graph. If* $(\mathcal{G}, k)$ *is a yes-instance for* MULTI-LAYER COPS AND ROBBER, *then, letting* $E_i = C_i$ *for each* $i \in [\tau]$, $((V, \{E_1, \ldots, E_\tau\}), k)$ *is a yes-instance for* MULTI-LAYER COPS AND ROBBER WITH FREE LAYER CHOICE.

Proof. Immediate from the problem definitions and Proposition 1.   □

3   Counter Examples and Anti-Monotonicity Results

In this section we provide some concrete examples of cops and robbers on multi-layer graphs illustrating some peculiarities of the game that may seem counter-intuitive. We begin with the following that states that it is sometimes beneficial to put multiple cops on the same layer, and leave other layers empty.

Theorem 1. *For any* $n \geqslant 4$ *there exists a multi-layer graph* $(V, \{C_H, C_V\}, *)$ *on* n *vertices such that a cop player can win with two cops if both cops are on* C_H, *or if both cops are on* C_V, *but the robber player can win if one cop is on* C_V *and the other is on* C_H.

It is natural to ask if, given some multi-layer graph $\mathcal{G} = (V, \{C_1, \ldots, C_\tau\}, R)$, the multi-cop number of $\mathcal{G}$ is bounded from below by the minimum cop-number of a single cop layer; namely, does $\mathsf{mc}(\mathcal{G}) \geqslant \min_i \mathsf{c}((V, C_i))$ hold? Observe that, if $|V| = n$ and we let S_n denote the star graph on n vertices, any multi-layer graph $\mathcal{G} = (V, \{E(S_n), C_2, \ldots, C_\tau\}, R)$ has cop number 1, as the cop can start on the centre of the star and reach any other vertex in one move. This is not enough resolve the question directly, however in the next result we build on this idea to show a general bound of the form $\mathsf{mc}(\mathcal{G}) = \Omega(\min_i \mathsf{c}((V, C_i)))$ does not hold.

Proposition 4. *For any $c \geqslant 2$ there exist graphs $G_1 = (V, E_1)$ and $G_2 = (V, E_2)$ such that $\mathsf{c}(G_1), \mathsf{c}(G_2) \geqslant c$ and $\mathsf{mc}((V, \{E_1, E_2\}, *)) = 2$.*

The idea of the proof is to take two n-vertex graphs with cop number c and add a $n - 1$ pendent vertices from a single vertex in each graph (u_n and v_n respectively). The graphs are then identified as cop layers in such a way that a cop at u_n can police half the vertices, and a cop at v_n can cover the other half. See Fig. 1 for an illustration. In fact, in such a construction the two cops will catch the robber after at most one cop move. As a result, the edges present in the robber layer are irrelevant and we have the following corollary.

Corollary 1. *For any $c \geqslant 2$ there exist graphs $G_1 = (V, E_1)$ and $G_2 = (V, E_2)$ such that $\mathsf{c}(G_1), \mathsf{c}(G_2) \geqslant c$, and for any set of edges $R \subseteq \binom{V}{2}$,*

$$\mathsf{mc}((V, \{E_1, E_2\}, R)) \leqslant 2.$$

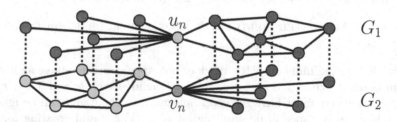

G_1

u_n

v_n

G_2

Fig. 1. Illustration of the construction in the proof of Proposition 4. The dotted edges signify an identification of the two end points of that edge.

We now consider the reverse inequality: is the multi-layer cop number bounded from above by a function of the cop numbers of the individual layers? If, in a multi-layer graph $\mathcal{G} = (V, \{C_1, \ldots, C_\tau\}, R)$, the robber layer is a subset of one of the cop layers, i.e. $R \subset C_i$ for some $i \in [\tau]$, then $\mathsf{mc}(\mathcal{G}) \leqslant \mathsf{c}((V, C_i))$ as the cop player can allocate $\mathsf{c}((V, C_i))$ cops to layer i, ignoring all other cop layers. The same reasoning gives an upper bound of $\sum_{i \in [\tau]} \mathsf{c}((V, C_i))$ on the cop number in the 'free choice layer' variant of the game. Thus, in this special case an upper bound that depends only on the cop numbers of individual layers does exist. However the next result shows that this is not the case in general.

Theorem 2. *For any positive integer k, there exists a multi-layer graph $\mathcal{G} = (V, \{C_1, C_2\}, R)$ on $O(k^3)$ vertices such that each of (V, R), (V, C_1), and (V, C_2) are connected, $\mathsf{c}((V, R)) \leqslant 3$, $\mathsf{c}((V, C_i)) \leqslant 2$ for $i \in \{1, 2\}$, and $\mathsf{mc}(\mathcal{G}) \geqslant k$.*

4   Complexity Results

In this section we will examine multi-layer cops and robbers from a computational complexity viewpoint. For a background on computational complexity, we point the reader to [30]. First note that as determining the cop-number of a simple graph is EXPTIME-complete [18], MULTI-LAYER COPS AND ROBBER WITH FREE LAYER CHOICE is also EXPTIME-complete by the obvious reduction that creates a multi-layer graph with one layer from a simple graph. The same reduction, and the fact that, unless the strong exponential time hypothesis fails[1], determining if a graph is k-copwin requires $\Omega(n^{k-o(1)})$ time [9], we also get that MULTI-LAYER COPS AND ROBBER WITH FREE LAYER CHOICE requires $\Omega(n^{k-o(1)})$ time.

An algorithm that determines whether a simple graph G is k-copwin in $O(kn^{k+2})$ time is given in [23]. Petr, Portier, and Versteegen show this by first constructing a *state graph* — a directed graph H wherein each vertex of H corresponds to a state of a game of cops and robbers played on the original graph G. They then give an $O(kn^{k+2})$ algorithm for finding all cop-win vertices of H, where a vertex is cop-win if the corresponding state either is a winning state for the cops, or can only lead to a winning state for the cops. We adapt their construction by only creating arcs of H where the move of a given cop or robber is allowed (i.e., the edge in the multi-layer graph exists in the same layer as the cop or robber that is moving). By doing this we obtain the following.

Theorem 3. ALLOCATED MULTI-LAYER COPS AND ROBBER *can be solved in* $O(k^2 n^{2k+2})$.

Note that τ, the number of layers, does not appear in the above as if $\tau \leqslant k$ then any dependence on τ is absorbed by the dependence on k, and if $k < \tau$ then at least $\tau - k$ layers must have zero cops allocated to them and can be ignored.

By taking an instance of dominating set $G = (V, E)$, and creating for each vertex $v \in V$ a layer E_v containing every edge incident to v, we create an instance of MULTI-LAYER COPS AND ROBBER WITH FREE LAYER CHOICE that has a winning strategy for k cops if and only if G has a dominating set of size k, leading to the following.

Theorem 4. MULTI-LAYER COPS AND ROBBER WITH FREE LAYER CHOICE *is NP-hard, even if each layer is a tree plus some isolated vertices.*

Note that the input size to MULTI-LAYER COPS AND ROBBER WITH FREE LAYER CHOICE is the number of bits required to represent both the underlying graph and each of the layers.

---

[1] See [12, Chapter 14] for background on the strong exponential time hypothesis.

If it is only the robber that is limited to a tree, however, then determining if k cops can win is FPT in the number of cops and number of layers in the graph. In particular, this result applies even if the layers are not connected. We obtain this result by characterising whether the robber can win based on the existence of edges in the robber layer that cops can easily patrol.

Theorem 5. *Given a multi-layer graph* $\mathcal{G} = (V, \{C_1, \ldots, C_\tau\}, R)$, *if R is a tree, then* MULTI-LAYER COPS AND ROBBER *on* $\mathcal{G}$ *can be solved in time* $O(f(k, \tau) \cdot poly(n))$, *where k is the number of cops, τ is the number of layers of $\mathcal{G}$, f is a computable function independent of n, and $poly(n)$ is a fixed polynomial in n.*

The next result follows immediately from the proof technique used to prove Theorem 5.

Corollary 2. *Given a multi-layer graph* $\mathcal{G} = (V, \{C_1, \ldots, C_\tau\}, R)$, *if R is a tree, and each cop layer is connected, then* $\mathsf{mc}(\mathcal{G}) \leqslant 2$.

5 Extremal Multi-Layer Cop-Number

In this section we study, for a given simple connected graph $G = (V, E)$, the extremal multi-layer cop number of G. This is the multi-layer cop number maximised over the set of all multi-layer graphs with connected cop-layers, which when flattened give G. More formally, for given connected graph $G = (V, E)$, if we define the set

$$\mathcal{L}(G) = \{(V, \{C_1, \ldots, C_\tau\}, *) : E = C_1 \cup \cdots \cup C_\tau$$
$$\text{and for each } i \in [\tau], (V, C_i) \text{ is connected}\},$$

then the extremal multi-layer cop-number of G is given by

$$\mathsf{emc}_\tau(G) = \max_{\mathcal{G} \in \mathcal{L}} \mathsf{mc}(\mathcal{G}).$$

We generalise two tools for bounding the cop number of graphs to the setting of multi-layer graphs; $(1, k)$-existentially closed graphs [7] and bounds by domination number. See the arXiv version of this paper [13] for more details on the former method; here we will now outline our use of dominating sets.

Let $\mathcal{G} = (V, \{E_1, \ldots, E_\tau\})$ be a multi-layer graph (without designated layers). A *multi-layer dominating set* in $\mathcal{G}$ is a set $D \subseteq V \times \{1, \ldots, \tau\}$ of vertex-layer pairs such that for every $v \in V$, either $(v, i) \in D$ for some i, or there is a $(w, i) \in D$ such that $w \in V$ and $vw \in E_i$. We define the *domination number* $\gamma(\mathcal{G})$ of $\mathcal{G}$ to be the size of a smallest multi-layer dominating set in $\mathcal{G}$. Note that if $\mathcal{G}$ has a single-layer this definition aligns with the traditional notion of dominating set, which justifies the overloaded notation. It is a folklore result that the cop number is at most the size of any dominating set in the graph, this also holds in the multi-layer setting.

Theorem 6. *Let* $\mathcal{G} := (V, \{E_1, \ldots, E_\tau\})$ *be any multi-layer graph and* $\mathcal{G}' := (V, \{E_1, \ldots, E_\tau\}, \binom{V}{2})$. *Then,* $\mathsf{mc}(\mathcal{G}') \leqslant \gamma(\mathcal{G})$.

Proof. Let D be any multi-layer dominating set of size $|D| = \gamma(\mathcal{G})$ and for each $(v, i) \in D$ place one cop in layer i at the vertex v. The result now follows as if the robber is at an any vertex then they are adjacent to a cop in some layer and so the robber will be caught after the robber first move.   □

We now introduce the parameter $\delta(\mathcal{G})$ which is an analogue of minimum degree for a multi-layer graph $\mathcal{G} = (V, \{E_1, \ldots, E_\tau\})$. This is given by

$$\delta(\mathcal{G}) := \min_{v \in V} \sum_{i \in [\tau]} d_{(V, E_i)}(v). \tag{1}$$

Using this notion we prove a bound on the domination number of a multi-layer graph, the proof is based on a classic application of the probabilistic method [2, Theorem 1.2.2].

Theorem 7. *Let $\mathcal{G} = (V, \{E_1, \ldots, E_\tau\})$ be any multi-layer graph. Then,*

$$\gamma(\mathcal{G}) \leqslant \frac{n\tau}{\tau + \delta(\mathcal{G})} \cdot \left(\ln \left(\frac{\tau + \delta(\mathcal{G})}{\tau} \right) + 1 \right).$$

Note that there are least two other sensible definitions of 'multi-layer minimum degree', namely the minimum degree of each layer $\min_{i \in [\tau]} \min_{v \in V} d_{(V, E_i)}(v)$, and minimum number of neighbours within any layer $\delta(\mathsf{fl}(\mathcal{G}))$. Our definition of $\delta(\mathcal{G})$ above in (1) can be thought of as the 'minimum number of edges incident in any layer', this is arguably a less natural notion than $\delta(\mathsf{fl}(\mathcal{G}))$ however it gives a better bound in our application (Theorem 7), in particular.

Proposition 5. *For any multi-layer graph $\mathcal{G}$ we have $\delta(\mathsf{fl}(\mathcal{G})) \leqslant \delta(\mathcal{G})$.*

Returning to extremal multi-layer cop numbers, for a complete graph we obtain Theorem 8. The upper bound we arrive at by placing all τ cops on a single vertex v; as each edge of K_n must be in some layer, there is no vertex that is not incident with v in some layer. The lower bound requires more work and relies on constructing cop layers with no overlap by combining colour classes of an edge colouring of the clique due to Sofier [31].

Theorem 8. *Let $n \geqslant 1$, $1 \leqslant \tau < \lfloor \frac{n}{2} \rfloor$ be integers. Then, $\lceil \frac{\tau}{10} \rceil \leqslant \mathsf{emc}_\tau(K_n) \leqslant \tau$.*

We now consider the extremal multi-layer cop number of the binomial random graph $G_{n,p}$. For any integer $n \geqslant 1$, this is the probability distribution over all n-vertex simple graphs generated by sampling each possible edge independently with probability $0 < p = p(n) < 1$, see [4] for more details. The following result shows that, for a suitably dense binomial random graph $G_{n,p}$, with probability tending to 1 as $n \to \infty$, $\mathsf{emc}_\tau(G_{n,p}) = \Theta(\tau \log(n)/p)$. The single-layer cop number of $G_{n,p}$ in the same range is known to be $\Theta(\log(n)/p)$ [7], so in some sense our result generalises this result.

Theorem 9. *For $\varepsilon > 0$, if $n^{1/2+\varepsilon} \leqslant np = o(n)$, and $1 \leqslant \tau \leqslant n^\varepsilon$ then,*

$$\mathbb{P}\left(\frac{\varepsilon}{10} \cdot \frac{\tau \cdot \ln n}{p} \leqslant \mathsf{emc}_\tau(G_{n,p}) \leqslant 10 \cdot \frac{\tau \cdot \ln n}{p}\right) \geqslant 1 - e^{-\Omega(\sqrt{n})}.$$

The upper bound in the proof of Theorem 9 follows from Theorems 6 and 7, whereas the lower bound follows by independently choosing cop layers which are each distributed as a random graph with edge probability $\Theta(p/\tau)$ and then applying a generalised form of the existential closure technique developed in [7]. See [24] for results on the cop number of $G_{n,p}$ for other ranges of p.

The extremal multi-layer cop number of a graph G is also bounded from above by the treewidth of G.

Theorem 10. *For any graph $G := (V,E)$, $\mathsf{emc}_\tau(G) \leqslant \mathsf{tw}(G)$. Furthermore, these cops can placed in any layers and still capture the robber.*

6   Multi-Layer Analogue of Meyniel's Conjecture

For the classical cop number, Meyniel's Conjecture [15] states that $\mathcal{O}(\sqrt{n})$ cops are sufficient to win cops and robbers on any graph G. After a sequence of results [10,15,16,20,28] the current best bound stands at $n \cdot 2^{-(1-o(1))\sqrt{\log_2 n}}$, see [5, Chapter 3] for a more detailed overview.

It is natural to explore analogues of Meyniel's Conjecture for the multi-layer cop number. Namely, what is the minimum number of cops needed to patrol any multi-layer graph with τ connected layers? Our results for the clique show that if τ is allowed to be arbitrary then no bound better than $\mathcal{O}(n)$ can hold. We conjecture that this is not the case when the number of layers is bounded.

Conjecture 1. For any fixed integer $\tau \geqslant 1$ and collection $(V,E_1),\ldots,(V,E_\tau)$ of connected graphs where $|V| = n$, we have

$$\mathsf{mc}((V,\{E_1,\ldots,E_\tau\},*)) = o(n).$$

Observe that the connected assumption is necessary in Conjecture 1 if we do not have divergent minimum degree, as shown by the example with two cop layers given by two edge disjoint matchings who's union forms an even cycle. This conjecture might seem very modest in comparison to Meyniel's conjecture, however the following result shows that it would be almost tight.

Theorem 11. *For any positive integer n there is a n-vertex multi-layer graph $\mathcal{G} = (V,\{C_1,C_2,C_3\},*)$ such that $|V| = \Theta(n)$, each cop layer is connected and has cop-number 2, and*

$$\mathsf{mc}(\mathcal{G}) = \Omega\left(\frac{n}{\log n}\right).$$

The construction in Theorem 11 starts with a 3-edge coloured cubic expander graph X on N vertices, where each color class is a cop layer. The vertices of X are then connected to the leaves of a star that has been subdivided $\Theta(\log N)$

many times — these can be used by all cops. The idea is that k cops can police at most $2k$ vertices of X within $\Theta(\log N)$ steps as it takes each cop this long to change location in X (via the arms of the star). If $k = \Theta(N)$ is chosen to be a suitably small but constant fraction of N, then even with the vertices policed by the cops removed there is still an expander subgraph of X not adjacent to any cops. The robber can then use this expander subgraph to change position before any cop can threaten them.

Many of the current approaches to Meyniel's Conjecture use some variation the fact that a single cop can guard any shortest path between any two vertices. For example the first step of the approach in [28] is to iteratively remove long geodesics until the graph has small diameter (following this a more sophisticated argument matching randomly placed cops to possible robber trajectories is applied). What makes Conjecture 1 difficult to approach is that, even for two layers, a shortest path in the flattened graph $\mathsf{fl}(\mathcal{G})$ may not live within a single cop layer. We note that [16] use a different approach based on expansion, their approach is more versatile however the authors were unable to apply it in the multi-layer setting.

This suggests a new or more refined approach is needed. However, by a direct application of Theorems 6 and 7, using a simple dominating set approach we can prove Conjecture 1 for multi-layer graphs with diverging minimum degree.

Proposition 6. *For any n-vertex multi-layer graph $\mathcal{G} = (V, \{E_1, \ldots, E_\tau\})$ satisfying $\delta(\mathcal{G})/\tau \to \infty$ as $n \to \infty$, we have $\mathsf{mc}\Big(\big(V, \{E_1, \ldots, E_\tau\}, \binom{V}{2}\big)\Big) = o(n)$.*

7   Conclusion and Open Problems

We studied the game of cops and robbers on multi-layer graphs, via several different approaches, including concrete strategies for certain graphs, the construction of counter-intuitive examples, algorithmic and hardness results, and the use of probabilistic methods and expanders for extremal constructions. We find that the multi-layer cop number cannot be bound from above or below by (non-constant) functions of the cop numbers of the individual layers. We bound an extremal variant for cliques and dense binomial random graphs (extending some tools from the single layer case along the way). We also find that a naive transfer of Meyniel's conjecture to the multi-layer setting is not true: there are multi-layer graphs which have multi-layer cop number in $\Omega(n/\log n)$. Algorithmically, we find that even if each layer is a tree plus some isolated vertices, the free layer choice variant of the problem remains NP-hard. Positively, we find that the problem can be resolved by an algorithm that is FPT in the number of cops and layers if the layer the robber resides in is a tree.

We are hopeful that our contribution will spark future work in multi-layer variants of cops and robbers, and suggest a number of possible open questions:

We were not able to generalise some frequently used tools from single-layer cops and robbers: for example, we have no useful notion of a corner, or a retract, nor dismantleability - we are hopeful that such tools may exist.

We have made some progress on the parameterised complexity of our problems, but have only considered a limited set of parameters and have not considered any parameter that constraints the nature of interaction between the layers: if we, for example, require that the layers are very similar alongside other restrictions does that impact the computational complexity of our problems?

Single-layer cops and robbers has been very successful as a tool for defining useful graph parameters of simple graphs, and we ask whether multi-layer cops and robbers could be used to define algorithmically useful graph parameters.

Some of our bounds and extremal results are unlikely to be tight in number of layers or with respect to other graph characteristics: can they be improved?

Is the extremal multi-layer cop number of $G_{n,p}$ always $\Theta(\tau \cdot \mathsf{c}(G_{n,p}))$ w.h.p. for any p? Of particular is whether this holds even in the 'zig-zag' regime [21]?

While we showed that a naive adaptation of Meyniel's conjecture to our multi-layer setting fails, it is still possible that $o(|V|)$ cops are sufficient for a bounded number of connected layers. We have shown this for a special case related to degree: is it true in general?

Finally, while we introduced a particular notion of multi-layer dominating set for our use in proving other results (inspired by similar ideas in single-layer cops and robbers), we suggest that this multi-layer graph characteristic may also be interesting in its own right, in particular for algorithms for other problems on multi-layer graphs.

Acknowledgements. This work was supported by the Engineering and Physical Sciences Research Council [EP/T004878/1].

References

1. Aigner, M., Fromme, M.: A game of cops and robbers. Discret. Appl. Math. **8**(1), 1–12 (1984)
2. Alon, N., Spencer, J.H.: The Probabilistic Method, Third Edition. Wiley-Interscience series in discrete mathematics and optimization. Wiley, Hoboken (2008)
3. Balev, S., Laredo Jiménez, J.L., Lamprou, I., Pigné, Y., Sanlaville, E.: Cops and robbers on dynamic graphs: offline and online case. In: Richa, A.W., Scheideler, C. (eds.) SIROCCO 2020. LNCS, vol. 12156, pp. 203–219. Springer, Cham (2020). https://doi.org/10.1007/978-3-030-54921-3_12
4. Bollobás, B.: Random graphs, volume 73 of Cambridge Studies in Advanced Mathematics, second edition. Cambridge University Press, Cambridge (2001)
5. Bonato, A., Nowakowski, R.J.: The Game of Cops and Robbers on Graphs. Student Mathematical Library. American Mathematical Society, New York (2011)
6. Bonato, A., Pralat, P.: Graph Searching Games and Probabilistic Methods. Discrete Mathematics and Its Applications. CRC Press, London, England, December 2017

7. Bonato, A., Pralat, P., Wang, C.: Pursuit-evasion in models of complex networks. Internet Math. **4**(4), 419–436 (2007)
8. Bowler, N.J., Erde, J., Lehner, F., Pitz, M.: Bounding the cop number of a graph by its genus. SIAM J. Discret. Math. **35**(4), 2459–2489 (2021)
9. Brandt, S., Pettie, S., Uitto, J.: Fine-grained lower bounds on cops and robbers. In: Azar, Y., Bast, H., Herman, G. (eds.), 26th Annual European Symposium on Algorithms (ESA 2018), volume 112 of Leibniz International Proceedings in Informatics (LIPIcs), pp. 9:1–9:12, Dagstuhl, Germany, 2018. Schloss Dagstuhl-Leibniz-Zentrum fuer Informatik (2018)
10. Chiniforooshan, E.: A better bound for the cop number of general graphs. J. Graph Theory **58**(1), 45–48 (2008)
11. Clarke, N.E.B.: Constrained cops and robber. ProQuest LLC, Ann Arbor, MI, 2002. Thesis (Ph.D.)-Dalhousie University (Canada) (2002)
12. Cygan, M., et al.: Parameterized Algorithms. Springer, Cham (2015). https://doi.org/10.1007/978-3-319-21275-3
13. Enright, J., Meeks, K., Pettersson, W., Sylvester, J.: Cops and robbers on multi-layer graphs. arXiv:2303.03962 (2023)
14. Fitzpatrick, S.L.: Aspects of domination and dynamic domination. ProQuest LLC, Ann Arbor, MI, 1997. Thesis (Ph.D.)-Dalhousie University (Canada) (1997)
15. Frankl, P.: Cops and robbers in graphs with large girth and Cayley graphs. Discret. Appl. Math. **17**(3), 301–305 (1987)
16. Frieze, A.M., Krivelevich, M., Loh, P.-S.: Variations on cops and robbers. J. Graph Theory **69**(4), 383–402 (2012)
17. Joret, G., Kaminski, M., Theis, D.O.: The cops and robber game on graphs with forbidden (induced) subgraphs. Contrib. Discret. Math. **5**(2) (2010)
18. Kinnersley, W.B.: Cops and robbers is exptime-complete. J. Comb. Theory Ser. B **111**, 201–220 (2015)
19. Lehner, F.: On the cop number of toroidal graphs. J. Comb. Theory, Ser. B **151**, 250–262 (2021)
20. Linyuan, L., Peng, X.: On Meyniel's conjecture of the cop number. J. Graph Theory **71**(2), 192–205 (2012)
21. Luczak, T., Pralat, P.: Chasing robbers on random graphs: zigzag theorem. Random Struct. Algorithms **37**(4), 516–524 (2010)
22. Nowakowski, R.J., Winkler, P.: Vertex-to-vertex pursuit in a graph. Discret. Math. **43**(2–3), 235–239 (1983)
23. Petr, J., Portier, J., Versteegen, L.: A faster algorithm for cops and robbers. Discret. Appl. Math. **320**, 11–14 (2022)
24. Pralat, P., Wormald, N.C.: Meyniel's conjecture holds for random graphs. Random Struct. Algorithms **48**(2), 396–421 (2016)
25. Quilliot, A.: Jeux et pointes fixes sur les graphes. PhD thesis, Ph. D. Dissertation, Université de Paris VI (1978)
26. Quilliot, A.: A short note about pursuit games played on a graph with a given genus. J. Comb. Theory Ser. B **38**(1), 89–92 (1985)
27. Schroeder, B.S.W.: The copnumber of a graph is bounded by $\lfloor \frac{3}{2} \text{ genus } (G) \rfloor + 3$. In: Categorical perspectives (Kent, OH, 1998), Trends Math., pp. 243–263. Birkhäuser Boston, Boston, MA (2001)
28. Scott, A., Sudakov, B.: A bound for the cops and robbers problem. SIAM J. Discret. Math. **25**(3), 1438–1442 (2011)
29. Seymour, P.D., Thomas, R.: Graph searching and a min-max theorem for tree-width. J. Comb. Theory Ser. B **58**(1), 22–33 (1993)

30. Sipser, M.: Introduction to the Theory of Computation. Cengage Learning, Boston (2012)
31. Soifer, A.: The Mathematical Coloring Book. Springer, New York (2009). https://doi.org/10.1007/978-0-387-74642-5
32. Toruńczyk, S.: Flip-width: cops and robber on dense graphs. In: 2023 IEEE 64th Annual Symposium on Foundations of Computer Science (FOCS 2023), page to appear. IEEE (2023)

Parameterized Complexity
of Broadcasting in Graphs

Fedor V. Fomin[1], Pierre Fraigniaud[2], and Petr A. Golovach[1(✉)]

[1] Department of Informatics, University of Bergen, Bergen, Norway
{Fedor.Fomin,Petr.Golovach}@uib.no
[2] Institut de Recherche en Informatique Fondamentale,
Université Paris Cité and CNRS, Paris, France
pierre.fraigniaud@irif.fr

Abstract. The task of the broadcast problem is, given a graph G and a source vertex s, to compute the minimum number of rounds required to disseminate a piece of information from s to all vertices in the graph. It is assumed that, at each round, an informed vertex can transmit the information to at most one of its neighbors. The broadcast problem is known to be NP-hard. We show that the problem is FPT when parametrized by the size k of a feedback edge set, or by the size k of a vertex cover, or by $k = n - t$, where t is the input deadline for the broadcast protocol to complete.

Keywords: broadcasting · telephone model · parameterized complexity

1 Introduction

The aim of *broadcasting* in a network is to transmit a message from a given source node of the network to all the other nodes. Let $G = (V, E)$ be a connected simple graph modeling the network, and let $s \in V$ be the source of a message M. The standard *telephone model* [21] assumes that the communication proceeds in synchronous rounds. At any given round, any node $u \in V$ aware of M can forward M to at most one neighbor v of u. The minimum number of rounds for broadcasting a message from s in G to all other vertices is denoted by $b(G, s)$, and we let $b(G) = \max_{s \in V} b(G, s)$ be the broadcast time of graph G. As the number of informed nodes (i.e., nodes aware of the message) can at most double at each round, $b(G, s) \geq \lceil \log_2 n \rceil$ in n-node networks. On the other hand, since G is connected, at least one uninformed node receives the message at any given round, and thus $b(G) \leq n - 1$. Both bounds are tight, as witnessed by the complete graph K_n and the path P_n, respectively. The problem of computing

The research leading to these results has received funding from the Research Council of Norway via the project BWCA (grant no. 314528) and from the ANR project DUCAT. The full version of the paper is available in [12].

the broadcast time $b(G, s)$ for a given graph G and a given source $s \in V$ is NP-hard [30]. Also, the results of [27] imply that it is NP-complete to decide whether $b(G, s) \leq t$ for graphs with $n = 2^t$ vertices.

Three lines of research have emerged since the early days of studying broadcasting in the telephone model. One line is devoted to determining the broadcast time of specific classes of graphs judged important for their desirable properties as interconnection networks (e.g., hypercubes, de Bruijn graphs, Cube Connected Cycles, etc.). We refer to the surveys [15, 22] for this matter. Another line of research takes inspiration from extremal graph theory. It aims at constructing n-node graphs G with optimal broadcast time $b(G) = \lceil \log_2 n \rceil$ and minimizing the number of edges sufficient to guarantee this property. Let $B(n)$ be the minimum number of edges of n-node graphs with broadcast time $\lceil \log_2 n \rceil$. It is known [18] that $B(n) = \Theta(n\, L(n))$ where $L(n)$ denotes the number of consecutive leading 1s in the binary representation of $n - 1$. On the other hand, it is still not known whether $B(\cdot)$ is non-decreasing for $2^t \leq n < 2^{t+1}$, for every $t \in \mathbb{N}$. We are interested in a third, more recent line of research, namely the design of algorithms computing efficient broadcast protocols. Note that a protocol for broadcasting from a source s in a graph G can merely be represented as a spanning tree T rooted at s, with an ordering of all the children of each node in the tree.

Polynomial-time algorithms are known for trees [30] and some classes of tree-like graphs [4, 17, 20]. Several (polynomial-time) approximation algorithms have been designed for the broadcast problem. In particular, the algorithm in [25] computes, for every graph G and every source s, a broadcast protocol from s performing in $2\, b(G, s) + O(\sqrt{n})$ rounds, hence this algorithm has approximation ratio $2 + o(1)$ for graphs with broadcast time $\gg \sqrt{n}$, but $\tilde{\Theta}(\sqrt{n})$ in general. Later, a series of papers tighten the approximation ratio, from $O(\log^2 n / \log \log n)$ [29], to $O(\log n)$ [1], and eventually $O(\log n / \log \log n)$ [10], which is, up to our knowledge the current best approximation ratio for the broadcast problem. Better approximation ratios can be obtained for specific classes of graphs [2, 3, 19].

Despite all the achievements obtained on the broadcast problem, it has not yet been approached from the parameterized complexity viewpoint [8]. There might be a good reason for that. Since at most 2^t vertices can have received the message after t communication rounds, an instance of the broadcast problem with time-bound t in an n-vertex graph is a no-instance whenever $n > 2^t$. It follows that the broadcast problem has a trivial kernel when parameterized by the broadcast time. This makes the natural parameterization by the broadcast time not very significant. Nevertheless, as we shall show in this paper, there is a parameterization below the natural upper bound for the number of rounds that leads to interesting conclusions.

Our Results. Let TELEPHONE BROADCAST be the following problem: given a connected graph $G = (V, E)$, a source vertex $s \in V$, and a nonnegative integer t, decide whether there is a broadcast protocol from s in G that ensures that all the vertices of G get the message in at most t rounds. We first show that TELEPHONE BROADCAST can be solved in a single-exponential time by an exact algorithm. Our algorithm is based on the dynamic programming over subsets [13].

Theorem 1 ($\star^1$). TELEPHONE BROADCAST *can be solved in* $3^n \cdot n^{\mathcal{O}(1)}$ *time for* n-*vertex graphs.*

Motivated by the fact that the complexity of TELEPHONE BROADCAST remains open in pretty simple tree-like graphs (e.g., cactus graphs, and therefore outerplanar graphs), we first consider the *cyclomatic* number as a parameter, i.e., the minimum size of a feedback edge set, that is, the size of the smallest set of edges whose deletion results to an acyclic graph. We show that TELEPHONE BROADCAST is FPT when parameterized by this parameter.

Theorem 2. TELEPHONE BROADCAST *can be solved in* $2^{\mathcal{O}(k \log k)} \cdot n^{\mathcal{O}(1)}$ *time for* n-*vertex graphs with cyclomatic number at most* k.

As far as we know, no NP-hardness result is known on graphs of treewidth at most $k \geq 2$. While we did not progress in that direction, we provide an interesting result for a stronger parameterization, namely the *vertex cover* number of a graph. (Note that, for all graphs, the treewidth never exceeds the vertex cover number.) As for the cyclomatic number, we do not only show that, for every fixed k, the broadcast time can be found in polynomial time on graphs with vertex cover at most k, but we prove a stronger result: the problem is FPT.

Theorem 3. TELEPHONE BROADCAST *can be solved in* $2^{\mathcal{O}(k2^k)} \cdot n^{\mathcal{O}(1)}$ *time for* n-*vertex graphs with a vertex cover of size at most* k.

Finally, we focus on graphs with very large broadcast time, for which the algorithm in [25] provides hope to derive very efficient broadcast protocols as this algorithm constructs a broadcast protocol performing in $2\,b(G) + O(\sqrt{n})$ rounds. While we were not able to address the problem over the whole range $\sqrt{n} \ll t \leq n - 1$, we were able to provide answers for the range $n - O(1) \leq t \leq n - 1$. More specifically, we consider the parameter $k = n - t$ and study the kernelization for the problem under such parametrization.

Theorem 4. TELEPHONE BROADCAST *admits a kernel with* $\mathcal{O}(k)$ *vertices in* n-*vertex graphs when parameterized by* $k = n - t$.

As a direct consequence of Theorem 4, TELEPHONE BROADCAST is FPT for the parameterization by $k = n - t$. Specifically the problem can be solved in $2^{O(k)} \cdot n^{O(1)}$ time.

Related Work. A classical generalization of the broadcast problem is the *multicast* problem, in which the message should only reach a given subset of target vertices in the input graph. Many of the previously mentioned approximation algorithms for the broadcast problem extend to the multicast problem, and, in particular, the algorithm in [10] is an $O(\log k / \log \log k)$-approximation algorithm for the multicast problem with k target nodes.

---

[1] The proofs of the statements labeled by ($\star$) are omitted and can be found in the full version of the paper [12].

Many variants of the telephone model have been considered in the literature, motivated by different network technologies. One typical example is the *line* model [11], in which a call between a vertex u and a vertex v is implemented by a path between u and v in the graph, with the constraint that all calls performed at the same round must be performed along edge-disjoint paths. (The intermediate nodes along the path do *not* receive the message, which "cut through" them.) Interestingly, the broadcast time of *every* n-node graph is exactly $\lceil \log_2 n \rceil$. The result extends to networks in which the paths are constructed by an underlying routing function [6]. The vertex-disjoint variant of the line model, i.e., the line model in which the calls performed at the same round must take place along vertex-disjoint paths, is significantly more complex. There is an $O(\log n / \log \log n)$-approximation algorithm for the vertex-disjoint line model [25], which naturally extend to an $O(\log n / \log OPT)$-approximation algorithm — see also [14] where an explicit $O(\log n / \log OPT)$-approximation algorithm is provided. It is also worth mentioning that the broadcast model has been also extensively studied in models aiming at capturing any type of node- or link-latencies, e.g., the message takes λ_e units of time to traverse edge e, and the algorithm in [1] also handles such constraints. Other variants take into account the size of the message, e.g., a message of L bits takes time $\alpha + \beta \cdot L$ to traverse an edge (see [23]). Under such a model, it might be efficient to split the original message into smaller packets and pipeline the broadcast of these packets through disjoint spanning trees [23,31].

2 Preliminaries

We refer to the book of Cygan et al. [8] for a detailed introduction to Parameterized Complexity. We consider only finite undirected graphs and refer to the textbook of Diestel [9] for basic notation. We always assume that the considered graphs are connected if it is not explicitly said to be otherwise. We use n and m to denote the number of vertices and edges if this does not create confusion. A set of edges S of a graph G is a *feedback edge set* if $G - S$ has no cycle. The *cyclomatic number* of a graph G is the minimum size of a feedback edge set. It is well-known, that for a connected graph G, the cyclomatic number is $m - n + 1$ and a feedback edge set can be found in linear time by constructing a spanning tree (see, e.g., [7,9]). A set of vertices S of a graph G is a *vertex cover* if each edge of G has at least one of its endpoints in G. The *vertex cover number* of G is the minimum size of a vertex cover. Note that for a vertex cover S, the set $I = V(G) \setminus S$ is an *independent set*, that is, any two distinct vertices of I are not adjacent.

Broadcasting. Let G be a graph and let $s \in V(G)$ be a source vertex from which a message is broadcasted. In general, a broadcasting protocol is a mapping that for each round $i \geq 1$, assigns to each vertex $v \in V(G)$ that is either a source or has received the message in rounds $1, \ldots, i - 1$, a neighbor u to which v sends the message in the i-th round. However, it is convenient to note that it can be assumed that each vertex v that got the message, in the next $d \leq d_G(v)$ rounds, transmits the message to some neighbors in a certain order in such a way that each vertex receives the message only once. This allows us to formally define a *broadcasting protocol* as a pair $(T, \{C(v) \mid v \in V(T)\})$, where T is a spanning tree of G rooted in s and for each $v \in V(T)$, $C(v)$ is an ordered set of children of v in T. As soon as v gets the message, v starts to send it to the children in T in the order defined by $C(v)$. For G and $s \in V(G)$, we use $b(G, s)$ to denote the minimum integer $t \geq 0$ such that there is a broadcasting protocol such that every vertex of G gets the message after t rounds. We say that a broadcasting protocol ensuring that every vertex gets a message in $b(s, G)$ rounds is *optimal*.

As it was proved by Proskurowski [28] and Slater, Cockayne, and Hedetniemi [30], $b(G, s)$ can be computed in linear time for trees by dynamic programming.

Lemma 1 ([28,30]). *For an n-vertex tree T and $s \in V(T)$, $b(T, s)$ can be computed in $\mathcal{O}(n)$ time.*

3   Telephone Broadcast Parameterized by the Cyclomatic Number

In this section, we sketch the proof of Theorem 2. We need some auxiliary results. Let T be a tree and let x and y be distinct leaves, that is, vertices of degree one in T. For an integer $h \geq 0$, we use $b_h(T, x, y)$ to denote the minimum number of rounds needed to broadcast the message from the source x to y in such a way that every vertex of T gets the message in at most h rounds. We assume that $b_h(T, x, y) = +\infty$ if $b(T, x) > h$. We prove that $b_h(T, x, y)$ can be computed in linear time similarly to $b(T, s)$ (see [28,30]). The difference is that it is more convenient to use recursion instead of dynamic programming.

Lemma 2 ($\star$). *For an n-vertex tree T with given distinct leaves x and y of T and an integer $h \geq 0$, $b_h(T, x, y)$ can be computed in $\mathcal{O}(n)$ time.*

We also need a subroutine computing the minimum number of rounds for broadcasting from two sources with the additional constraint that the second source starts sending the message with a delay. Let T be a tree and let x and y be distinct leaves of T. Let also $h \geq 0$ be an integer. We use $d_h(T, x, y)$ to denote the minimum rounds needed to broadcast the message from x and y to every vertex of $T - y$ in such a way that y can send the message starting from the $(h+1)$-th round (that is, we assume that y gets the message from outside in the h-th round).

Lemma 3 ($\star$). *For an n-vertex tree T with given distinct leaves x and y of T and an integer $h \geq 0$, $d_h(T, x, y)$ can be computed in $\mathcal{O}(n^2)$ time.*

Sketch of the Proof of Theorem 2. Let (G, s, t) be an instance of TELEPHONE BROADCAST. If G is a tree, then we can compute $b(G, s)$ in linear time using

Lemma 1. Assume that this is not the case, and let $k = m - n + 1 \geq 1$ be the cyclomatic number of G. We find in linear time a feedback edge set S of size k by finding an arbitrary spanning tree F of G and setting $S = E(G) \setminus E(F)$.

We iteratively construct the set U as follows. Initially, we set $U := W = \{s\} \cup \{v \in V(G) \mid v$ is an endpoint of an edge of $S\}$. Then while $G - U$ has a vertex v such that G has three internally vertex-disjoint paths joining v and U, we set $U := U \cup \{v\}$. The properties of U are summarized in the following claims.

Claim 1. *We have that $|U| \leq 4k$, and for each connected component F of $G - U$, F is a tree such that each vertex $x \in U$ has at most one neighbor in F and*

(i) either U has a unique vertex x that has a neighbor in F,
(ii) or U contain exactly two vertices x and y having neighbors in F.

If F is a connected component of $G - U$ satisfying (i) of Claim 1, then we say that F is a x-tree and x is its *anchor*. For a connected component F of $G - U$ satisfying (ii), we say that F is an (x, y)-tree and call x and y anchors of F. We also say that F is *anchored* in x (x and y, respectively). Because $G - S$ is a tree and $|U| \leq 4k - 1$, we immediately obtain the next property.

Claim 2. *For every distinct $x, y \in U$, $G - U$ has at most one (x, y)-tree. Furthermore, the graph H with $V(H) = U$ such that $xy \in E(H)$ if and only if $G - U$ has an (x, y)-tree is a forest. In particular, the total number of (x, y)-trees is at most $4k - 1$.*

To prove the theorem, we have to verify the existence of a broadcasting protocol $P = (T, \{C(v) \mid v \in V(T)\})$ that ensures that every vertex receives the message after at most t rounds. To do it, we guess the *scheme* of P restricted to U. Namely, we consider the graph G' obtained from G by the deletion of the vertices of x-trees for all $x \in U$ and for each vertex $x \in U$, we guess how the message is broadcasted to x and from x to the neighbors of x in G'. Notice that $T' = T[V(G')]$ is a tree by the definition of G'. Observe also that for each x-tree F for $x \in U$, the message is broadcasted to the vertices of F from x, because $s \in U$. In particular, this means that the parents of the vertices of U in T are in G'. For each $v \in U$ distinct from s, we guess its parent $p(v) \in V(G')$ in T' and assume that $p(s) = s$. Then for each $v \in V(G)$, we guess the ordered subset $R(v)$ of vertices of $N_{G'}(v) \setminus \{p(v)\}$ such that $R(v) = C(v) \cap N_{G'}(v)$. We guess $p(v)$ and $R(v)$ for $v \in U$ by considering all possible choices. To guess $R(v)$ for each $v \in U$, we first guess the (unordered) set $S(v)$ and then consider all possible orderings of the elements of $S(v)$. The selection of $p(v)$ and $S(v)$ is done by brute force. However, we are only interested in choices, where the selection of the neighbors $p(v)$ and $S(v)$ of v for $v \in U$ can be extended to a spanning tree T' of G'.

Let T' be an arbitrary spanning tree of G' rooted in s. Let T'' be the tree obtained from T' by the iterative deletion of leaves not included in U. Observe that T'' is a tree such that $U \subseteq V(T'')$ and each leaf of T'' is a vertex of U. By Claim 1, each edge of T'' is either an edge of $G[U]$ or is an edge of an (x, y)-path Q for distinct $x, y \in U$ such that the internal vertices of Q are the vertices of the (x, y)-tree F; in the second case, each edge of Q is in T''. Notice also that $s \in V(T'')$ and for each $v \in U$ distinct from s, the parent of v in T is the parent of v in T'' with respect to the source vertex s. Hence, our first step in constructing $p(v)$ and $S(v)$, is to consider all possible choices of T''. Observe that $G[U]$ has at most $\binom{4k}{2}$ edges and the total number of (x, y)-trees is at most $4k - 1$ by Claim 2. Because T'' is a tree, it contains $|U| - 1$ edges of $G[U]$ and (x, y)-paths Q in total. We obtain that we have $k^{\mathcal{O}(k)}$ possibilities to choose T''. From now, we assume that T'' is fixed.

The choice of T'' defines $p(v)$ for $v \in U \setminus \{s\}$. For each $v \in U$, we initiate the construction of $S(v)$ by including in the set the neighbors of v in T'' distinct from $p(v)$. We proceed with guessing of $S(v)$ by considering (x, y)-trees F for $x, y \in U$ such that the (x, y)-path with the internal vertices in F is not included in T''. Clearly, the vertices of every F of such a type should receive the message either via x, or via y, or via both x and y. Let F be an (x, y)-tree of this type. Denote by x' and y' the neighbors of x and y, respectively. We have that either $x' \in S(x)$ and $y' \notin S(y)$, or $x \notin S(x)$ and $y \in S(Y)$, or $x' \in S(x)$, $y' \in S(y)$ and $x' \neq y'$. Thus, we have three choices for F. By Claim 2, the total number of choices is $2^{\mathcal{O}(k)}$. We go over all the choices and include the vertices to the sets $S(v)$ for $v \in U$ with respect to them. By Claim 1, this concludes the construction of the sets $S(v)$. From now, we assume that $S(v)$ for $v \in U$ are fixed.

We construct the ordered sets $R(v)$ by considering all possible orderings of the elements of $S(v)$. The number of these orderings is $\Pi_{v \in U}(|S(v)|!)$. Recall that the sets $S(v)$ are sets of neighbors of v in a spanning tree of G'. This and Claims 1 and 2 imply that $\sum_{v \in U} |S(v)| \leq 2(|U|-1)+2(4k-1) \leq 16k$. Therefore, the total number of orderings is $\Pi_{v \in U}(|S(v)|!) = k^{\mathcal{O}(k)}$. This completes the construction of $R(v)$. Now we can assume that $p(v)$ for each $v \in U \setminus \{s\}$ and $R(v)$ for each $v \in U$ are given.

The final part of our algorithm is checking whether the guessed scheme for a broadcasting protocol can be extended to the protocol itself. This is done in two stages.

In the first stage, we compute for each $v \in U$, the minimum number $r(v)$ of a round in which v can receive the message and the ordered set $C(v)$. Initially, we set $r(s) = 0$ and set $X := \{s\}$. Then we iteratively either compute $C(v)$ for $v \in X$ or extend X by including a new vertex $v \in U \setminus X$ and computing $r(v)$. We proceed until we get $X = U$ and compute $C(v)$ for every $v \in U$. We also stop and discard the current choice of the scheme if we conclude that the choice cannot be extended to a broadcasting protocol terminating in at most t steps.

Suppose that there is $v \in X$ such that $C(v)$ is not constructed yet. Notice that $r(v)$ is already computed. To construct $C(v)$, we observe that for each v-tree F anchored in v, the vertices of F should receive the message via v. Hence, to construct $C(v)$, we extend $R(v)$ by inserting the neighbors of v in the v-trees. If there is no v-tree anchored in v, then we simply set $C(v) = R(v)$. Assume that this is not the case and let $F_1, \ldots, F_k$ be the v-trees anchored in v. Denote by $u_1, \ldots, u_k$ the neighbors of v in $T_1, \ldots, T_k$, respectively. Because the message is broadcasted from u to each u_i, we can assume that to broadcast the message from u_i to the other vertices of T_i, an optimal protocol requiring $b(T_i, u_i)$ rounds is used. We compute the values $b(T_i, u_i)$ for all $i \in \{1, \ldots, k\}$ and assume that $b(T_1, u_1) \geq \cdots \geq b(T_k, u_k)$.

If $r(v) + |R(v)| + k > t$, we discard the current choice of the scheme, because we cannot transmit the message to the neighbors of v in t rounds. Notice also that $r(v) + \max\{b(T_i, u_i) + i \mid i \in \{1, \ldots, k\}\}$ rounds are needed to transmit the message to the vertices of all v-trees. Hence, if $r(v) + \max\{b(T_i, u_i) + i \mid i \in \{1, \ldots, k\}\} > t$, we discard the considered scheme. From now on, we assume that $r(v) + |R(v)| + k \leq t$ and $r(v) + \max\{b(T_i, u_i) + i \mid i\{1 \in, \ldots, k\}\} \leq t$.

The main idea for constructing $C(v)$ is to ensure that the message is sent to the vertices of $R(v)$ as early as possible. To achieve this, we put $u_1, \ldots, u_k$ in $C(v)$ in such a way, that the message is sent to each u_i as late as possible. Since $|C(v)| = |R(v)| + k$, we represent $C(v)$ as an $|R(v)| + k$-element array whose elements are indexed $1, 2, \ldots, |R(v)| + k$. Because $b(T_1, u_1) \geq \cdots \geq b(T_k, u_k)$, we can assume that the ordering of the vertices $u_1, \ldots, u_k$ in $C(v)$ is $(u_1, \ldots, u_k)$. Therefore, we insert u_i in $C(v)$ consecutively for $i = k, k - 1, \ldots, 1$.

Suppose that $i \in \{1, \ldots, k\}$ and $u_{i+1}, \ldots, u_k$ are in $C(v)$. Denote by h_{i+1} the index of u_{i+1} assuming that $h_{k+1} = |R(v)| + k + 1$. We find maximum positive integer $h < h_{i+1}$ such that $r(v) + h + b(T_i, u_i) \leq t$ and set the index $h_i = h$ for u_i. In words, we find the maximum index that is prior to the index of u_{i+1} such that if we transmit the message from v to u_i in the h-th round after v got aware of the message, then the vertices of T_i still may get the message in t rounds. After placing $u_1, \ldots, u_k$ into the array, we place the vertices of $R(v)$ in the remaining $|R(v)|$ places following the order in $R(v)$. This completes the construction of $C(v)$.

Suppose that $U \setminus X \neq \emptyset$ and for each $v \in X$, $C(v)$ is given. We assume that for each $v \in X$, the elements of $C(v)$ are indexed $1, \ldots, |C(v)|$ according to the order. By the constriction of the schemes, there is $y \in U \setminus X$ such that y receives the message from some vertex $x \in X$ either directly or via some (x, y)-tree F anchored in x and y. We find such a vertex y, compute $r(y)$, and include y in X.

If there is $y \in U \setminus X$ such that $p(y) = x \in X$, then we set $r(y) = r(x) + h$, where h is the index of y in $C(v)$ and set $X := X \cup \{y\}$. Since $r(x)$ is the minimum number of a round when x gets the message, $r(y)$ is the minimum number of a round in which y gets the message. Suppose that such a vertex y does not exist. Then by construction of the considered scheme, there are $x \in X$ and $y \in U \setminus X$ such that the tree T'' which was used to construct the scheme contains an (x, y)-path whose internal vertices are in the (x, y)-tree F anchored in x and y. This means that the neighbor x' of x in F is included in $C(v)$. Let h be the index of x' in $C(v)$. We also have that $y' = p(y)$ is the unique neighbor of y in F. In other words, we have to transmit the message from x to y according to the scheme. To compute $r(y)$, we have to transmit the message as fast as possible. For this, we use Lemma 2. Notice that the vertices of F should receive the message in at most $t' = t - r(x) - h + 1$ rounds because x receives the message in the round $r(v)$ and $h - 1$ vertices of $C(v)$ get the message before x'. Let F' be the tree obtained from F by adding the vertices x, y and the edges xx', yy'. We compute $b_{t'}(F', x, y)$ using the algorithm from Lemma 2. If $b_{t'}(F', x, y) = +\infty$, we discard the considered scheme because y cannot receive the message in t rounds. Otherwise, we set $r(y) = r(x) + (h - 1) + b_{t'}(F', x, y)$ and set $X := X \cup \{y\}$.

This completes the first stage where we compute $r(v)$ and $C(v)$ for $v \in U$. Observe that if we completed this stage without discarding the considered choice of the scheme, we already have a partially constructed broadcasting protocol that ensures that (i) the vertices of v-trees for $v \in U$ get the message in at most t rounds, (ii) the vertices of (x, y)-trees that are assigned by the scheme to transmit the message from x to y receive the message in at most t rounds, and (iii) for each $v \in U$, $r(v)$ is the minimum number of a round when v can receive the message according to the scheme. By Claim 1, it remains to check whether the vertices of (x, y)-trees F that are not assigned by the scheme to transmit the message from x to y or vice versa can receive the message in at most t rounds. We do it using Lemmas 1 and 3.

Suppose that F is a (x, y)-tree anchored in $x, y \in U$ such that $p(x), p(y) \notin V(F)$, that is, F is not assigned by the scheme to transmit the message from x to y or vice versa. Let x' and y' be the neighbors in F of x and y, respectively. By the construction of the schemes, we have three cases: (i) $x' \in C(x)$ and $y' \notin C(y)$, (ii) $x' \notin C(x)$, $y' \in C(y)$, and (iii) $x' \in C(x)$, $y' \in C(y)$, and $x' \neq y'$.

In case (i), the vertices of F should receive the message via x. Clearly, we can use an optimal protocol for F with the source x' to transmit the message from x'. Let h be the index of x' in $C(x)$. We use Lemma 1, to verify whether $t - r(x) - h \geq b(F, x)$. If the inequality holds, we conclude that the message can be transmitted to the vertices of F in at most t rounds. Otherwise, we conclude that this is impossible and discard the scheme. Case (ii) is symmetric and the arguments are the same.

Suppose that $x' \in C(x)$, $y' \in C(y)$, and $x' \neq y'$. Then the vertices of F are receiving the message from both x and y. Denote by i and j the indexes of x' and y' in $C(x)$ and $C(y)$, respectively. By symmetry, we assume without loss of generality that $r(x) + i \leq r(y) + j$ and let $h = (r(y) + j) - (r(x) + i)$. Notice that the vertices of F start to get the message after the round $r(v) + i - 1$. Denote by F' the tree obtained from F by adding the vertices x, y and the edges xx', yy'. We use Lemma 3 and compute $d_h(F', x, y)$. If $d_h(F', x, y) \leq t - r(v) - i + 1$, then we obtain that the message can be transmitted to the vertices of F in at most t rounds. Otherwise, we cannot do it and discard the scheme.

This completes the description of the second stage of the verification of whether the considered scheme can be extended to a broadcasting protocol terminating in at most t rounds.

If we find a scheme that allows us to conclude that the message can be broadcasted in at most t rounds, we conclude that (G, s, t) is a yes-instance. Otherwise, if every scheme gets discarded, we return that (G, s, t) is a no-instance of TELEPHONE BROADCAST. This concludes the description of the algorithm. □

4 Telephone Broadcast Parameterized by the Vertex Cover Number

In this section, we briefly sketch the proof of Theorem 3. Recall that we aim to show that TELEPHONE BROADCAST is FPT on graphs with the vertex cover number at most k when the problem is parameterized by k. We start with some auxiliary claims about the broadcasting on a graph with a given vertex cover S.

Lemma 4 ($\star$). *Let G be a graph with at least one edge and $s \in V(G)$. Let also S be a vertex cover of G. Then there is an optimal broadcasting protocol for G with the source s such that the vertices of S receive the message in at most $2|S| - 1$ rounds.*

We also use the bound for the number of vertices of $I = V(G) \setminus S$ getting the message in the first p rounds.

Lemma 5 ($\star$). *Let G be a graph with at least one edge and $s \in V(G)$. Let also S be a vertex cover of G and $p \geq 1$ be an integer. Then for any broadcasting protocol for G with the source s, at most $p|S|$ vertices of $I = V(G) \setminus S$ receive the message in the first p rounds.*

Sketch of the Proof of Theorem 3. Let (G, s, t) be an instance of TELEPHONE BROADCAST and let $k > 0$ be an integer. We use the algorithm of Chen, Kanj, and Xia [5] to find in $1.2738^k \cdot n^{\mathcal{O}(1)}$ time a vertex cover S of G of size at most k. If the algorithm fails to find such a set, then we stop and return the answer that G has no vertex cover of size at most k. From now on, we assume that S is given and $I = V(G) \setminus S$. We use the well-known fact that I has a partition $\{L_1, \ldots, L_p\}$ into classes of false twins with $p \leq 2^k$. By Lemma 4, we can assume that the vertices of S receive the message in the first $2k - 1$ rounds. Then we apply Lemma 5 and guess the vertices of I which receive the message in the first $2k - 1$ rounds using the fact that the vertices of each L_i are indistinguishable. Our task boils down to deciding whether the remaining vertices of I can receive the message in the following $t - 2k + 1$ rounds. The crucial observation is that these vertices can get the message only from S. This allows us to encode a broadcasting protocol as a system of linear inequalities over $\mathbb{Z}$. For every $v \in S$ and every $i \in \{1, \ldots, p\}$, we introduce an integer-valued variable x_{vi} meaning that exactly x_{vi} neighbors of v in $L_i \setminus Y_i$ receive the message from v in the last $t - 2k + 1$ rounds. Notice that the number of variables is upper bounded by $k2^k$. Thus, we obtain the system of integer linear inequalities with at most $k2^k$ variables, which can be solved in $2^{\mathcal{O}(k2^k)} \cdot n^{\mathcal{O}(1)}$ time by the results of Lenstra [26] and Kannan [24] (see also [16]). $\square$

5 Kernelization for the Parameterization by $k = n - t$

In this section, we sketch the proof of Theorem 4. Recall that we parameterize TELEPHONE BROADCAST by $k = n - t$. Hence, it is convenient for us to denote the considered instances as triples (G, s, k) throughout the section instead of $(G, s, n - k)$. Let (G, s, k) be an instance of TELEPHONE BROADCAST. We exhaustively apply the following reduction rules in the order in which they are stated. The first rule is straightforward because $b(G, s) \leq n - 1$ and (G, s, k) is a yes-instance if $k \leq 1$. Also if $k > n$, then (G, s, k) is a no-instance.

Reduction Rule 1. *If $k \leq 1$, then return a trivial yes-instance, e.g., the instance with $G = (\{s\}, \emptyset)$ and $k = 0$, and stop. If $k > n$, then return a trivial no-instance, e.g., the instance with $G = (\{s, v\}, \{sv\})$ and $k = 1$, and stop.*

Observe that after applying Reduction Rule 1, $n \geq 2$, because if $n = 1$, then either $k \leq 1$ of $k > n$ and we would stop. Notice that if $d_G(s) = 1$, then the source s sends the message to its unique neighbor in the first round. This allows us to delete s and define a new source.

Reduction Rule 2. *If $d_G(s) = 1$, then let v be the neighbor of s, set $G := G - s$ and define $s := v$.*

Now we can assume that $d_G(s) \geq 2$. By the next rule, we delete certain pendent vertices.

Reduction Rule 3. *If there is a vertex $v \in V(G)$ such that for the set of vertices of degree one $W \subseteq N_G(v)$, it holds that $|W| \geq |V(G) \setminus W|$, then select an arbitrary $w \in W$ and set $G := G - w$.*

To state the following rule, we introduce an auxiliary notation. For a vertex v of a graph H, we define $\rho_H(v) = \max\{\text{dist}_H(v, u) \mid u \in V(H)\}$.

Reduction Rule 4. *If G has a bridge $e = uv$ such that $G - e$ has two connected components G_1 and G_2, where $s, u \in V(G_1)$, $v \in V(G_2)$, $d_G(u) = 2$, and $|V(G_1)| < \text{dist}_{G_1}(s, u) + \rho_{G_2}(v)$, then set $G := G/e$.*

From now, we can assume that Reduction Rules 1–4 are not applicable. We run the standard breadth-first search (BFS) algorithm on G from s (see, e.g., [7] for the description). The algorithm produces a spanning tree B of G of shortest paths and the partition of $V(G)$ into *BFS-levels* $L_0, \ldots, L_r$, where L_i is the set of vertices at distance i from s for every $i \in \{1, \ldots, r\}$.

Using the observation that $b(B, s) \geq b(G, s)$ and Lemma 1, we apply the following rule.

Reduction Rule 5. *Compute $b(B, s)$ and if $b(B, s) \leq n - k$, then return a trivial yes-instance and stop.*

Then we apply the final rule.

Reduction Rule 6. *If there is $v \in L_i$ for some $i \in \{0, \ldots, r - 1\}$ such that for $X = N_G(v) \cap L_{i+1}$ and for the (s, v)-path P in B, it holds that (i) $|X| \geq 2k + 1$ and (ii) the total number of vertices in nontrivial, i.e., having at least two vertices, connected components of $G - V(P)$ containing vertices of X is at least $4k - 2$, then return a trivial yes-instance and stop.*

The crucial property of the instance obtained by applying Reduction Rules 1–6 is given in the following lemma.

Lemma 6 (⋆). *Suppose that Reduction Rules 1–6 are not applicable to (G, s, k). Then $|V(G)| \leq 18k - 12$.*

By Lemma 6, if we do not stop during the exhaustive applications of Reduction Rules 1–6, then for the obtained instance (G, s, k), $|V(G)| \leq 18k - 12$. Hence, to complete the kernelization algorithm, we return (G, s, k).

It is straightforward to see that Reduction Rules 1–6 can be applied in polynomial time. In particular, BFS and finding bridges can be done in linear time by classical graph algorithms (see, e.g., the textbook [7]). Thus, the total running time of the kernelization algorithm is polynomial. This completes the the sketch of the proof of Theorem 4.

6 Conclusion

In our paper, we initiated the study of TELEPHONE BROADCAST from the parameterized complexity viewpoint. In this section, we discuss further directions of research.

We observed that TELEPHONE BROADCAST is trivially FPT when parameterized by t and Theorem 1 implies that the problem can be solved in $3^{2^t} \cdot n^{\mathcal{O}(1)}$ time. Is it possible to get a better running time for the parameterization by t?

In Theorem 4, we obtained a polynomial kernel for the parameterization by $k = n - t$, that is, for the parameterization below the trivial upper bound for $b(G, s)$. This naturally leads to the question about parameterization below some other bounds for this parameter. We note that the parameterization of TELEPHONE BROADCAST above the natural lower bound $b(G, s) \geq \log n$ leads to a para-NP-complete problem. To see this, observe that for graphs with $n = 2^t$ vertices, $b(G, s) \leq t$ if and only if G has a binomial spanning tree rooted in s, and it is NP-complete to decide whether G contains such a spanning tree [27].

In Theorems 2 and 3, we considered structural parameterizations of TELEPHONE BROADCAST by the cyclomatic and vertex cover numbers, respectively. It is interesting to consider other structural parameterizations. In particular, is TELEPHONE BROADCAST FPT when parameterized by the *feedback vertex number* and *treewidth* (we refer to [8] for the definitions)? For the parameterization by treewidth, the complexity status of TELEPHONE BROADCAST is open even for the case when the treewidth of the input graphs is at most two, that is, for series-parallel graphs.

References

1. Bar-Noy, A., Guha, S., Naor, J., Schieber, B.: Message multicasting in heterogeneous networks. SIAM J. Comput. **30**(2), 347–358 (2000)

2. Bhabak, P., Harutyunyan, H.A.: Constant approximation for broadcasting in k-cycle graph. In: Ganguly, S., Krishnamurti, R. (eds.) CALDAM 2015. LNCS, vol. 8959, pp. 21–32. Springer, Cham (2015). https://doi.org/10.1007/978-3-319-14974-5_3

3. Bhabak, P., Harutyunyan, H.A.: Approximation algorithm for the broadcast time in k-path graph. J. Interconnect. Netw. **19**(4), 1950006:1–1950006:22 (2019). https://doi.org/10.1142/S0219265919500063

4. Cevnik, M., Zerovnik, J.: Broadcasting on cactus graphs. J. Comb. Optim. **33**(1), 292–316 (2017)

5. Chen, J., Kanj, I.A., Xia, G.: Improved upper bounds for vertex cover. Theor. Comput. Sci. **411**(40–42), 3736–3756 (2010). https://doi.org/10.1016/j.tcs.2010.06.026

6. Cohen, J., Fraigniaud, P., König, J., Raspaud, A.: Optimized broadcasting and multicasting protocols in cut-through routed networks. IEEE Trans. Parallel Distrib. Syst. **9**(8), 788–802 (1998)

7. Cormen, T.H., Leiserson, C.E., Rivest, R.L., Stein, C.: Introduction to Algorithms. 3rd edn. MIT Press, Cambridge (2009). https://mitpress.mit.edu/books/introduction-algorithms

8. Cygan, M., et al.: Parameterized Algorithms. Springer, Cham (2015). https://doi.org/10.1007/978-3-319-21275-3

9. Diestel, R.: Graph Theory. GTM, vol. 173. Springer, Heidelberg (2017). https://doi.org/10.1007/978-3-662-53622-3

10. Elkin, M., Kortsarz, G.: Sublogarithmic approximation for telephone multicast. J. Comput. Syst. Sci. **72**(4), 648–659 (2006)

11. Farley, A.M.: Minimum-time line broadcast networks. Networks **10**(1), 59–70 (1980)

12. Fomin, F.V., Fraigniaud, P., Golovach, P.A.: Parameterized complexity of broadcasting in graphs. CoRR abs/2306.01536 (2023). https://doi.org/10.48550/arXiv.2306.01536

13. Fomin, F.V., Kratsch, D.: Exact Exponential Algorithms. TTCSAES, Springer, Heidelberg (2010). https://doi.org/10.1007/978-3-642-16533-7

14. Fraigniaud, P.: Approximation algorithms for minimum-time broadcast under the vertex-disjoint paths mode. In: auf der Heide, F.M. (ed.) ESA 2001. LNCS, vol. 2161, pp. 440–451. Springer, Heidelberg (2001). https://doi.org/10.1007/3-540-44676-1_37

15. Fraigniaud, P., Lazard, E.: Methods and problems of communication in usual networks. Discret. Appl. Math. **53**(1–3), 79–133 (1994)

16. Frank, A., Tardos, É.: An application of simultaneous diophantine approximation in combinatorial optimization. Comb. **7**(1), 49–65 (1987). https://doi.org/10.1007/BF02579200

17. Gholami, M.S., Harutyunyan, H.A., Maraachlian, E.: Optimal broadcasting in fully connected trees. J. Interconnect. Netw. **23**(1), 2150037:1–2150037:20 (2023). https://doi.org/10.1142/S0219265921500377

18. Grigni, M., Peleg, D.: Tight bounds on minimum broadcast networks. SIAM J. Discret. Math. **4**(2), 207–222 (1991)

19. Harutyunyan, H.A., Hovhannisyan, N.: Broadcasting in split graphs. In: Mavronicolas, M. (ed.) CIAC 2023. LNCS, vol. 13898, pp. 278–292. Springer, Cham (2023). https://doi.org/10.1007/978-3-031-30448-4_20

20. Harutyunyan, H.A., Maraachlian, E.: On broadcasting in unicyclic graphs. J. Comb. Optim. **16**(3), 307–322 (2008). https://doi.org/10.1007/s10878-008-9160-2

21. Hedetniemi, S.M., Hedetniemi, S.T., Liestman, A.L.: A survey of gossiping and broadcasting in communication networks. Networks **18**(4), 319–349 (1988). https://doi.org/10.1002/net.3230180406
22. Hromkovic, J., Klasing, R., Pelc, A., Ruzicka, P., Unger, W.: Dissemination of Information in Communication Networks - Broadcasting, Gossiping, Leader Election, and Fault-Tolerance. Texts in Theoretical Computer Science. An EATCS Series, Springer, Heidelberg (2005). https://doi.org/10.1007/b137871
23. Johnsson, S.L., Ho, C.: Optimum broadcasting and personalized communication in hypercubes. IEEE Trans. Comput. **38**(9), 1249–1268 (1989)
24. Kannan, R.: Minkowski's convex body theorem and integer programming. Math. Oper. Res. **12**(3), 415–440 (1987). https://doi.org/10.1287/moor.12.3.415
25. Kortsarz, G., Peleg, D.: Approximation algorithms for minimum-time broadcast. SIAM J. Discret. Math. **8**(3), 401–427 (1995)
26. Lenstra, H.W., Jr.: Integer programming with a fixed number of variables. Math. Oper. Res. **8**(4), 538–548 (1983). https://doi.org/10.1287/moor.8.4.538
27. Papadimitriou, C.H., Yannakakis, M.: The complexity of restricted spanning tree problems. J. ACM **29**(2), 285–309 (1982). https://doi.org/10.1145/322307.322309
28. Proskurowski, A.: Minimum broadcast trees. IEEE Trans. Comput. **30**(5), 363–366 (1981). https://doi.org/10.1109/TC.1981.1675796
29. Ravi, R.: Rapid rumor ramification: approximating the minimum broadcast time. In: 35th IEEE Symposium on Foundations of Computer Science (FOCS), pp. 202–213 (1994)
30. Slater, P.J., Cockayne, E.J., Hedetniemi, S.T.: Information dissemination in trees. SIAM J. Comput. **10**(4), 692–701 (1981). https://doi.org/10.1137/0210052
31. Stout, Q.F., Wagar, B.: Intensive hypercube communication prearranged communication in link-bound machines. J. Parallel Distrib. Comput. **10**(2), 167–181 (1990)

Turán's Theorem Through Algorithmic Lens

Fedor V. Fomin[1], Petr A. Golovach[1], Danil Sagunov[2], and Kirill Simonov[3(✉)]

[1] Department of Informatics, University of Bergen, Bergen, Norway
{fomin,petr.golovach}@ii.uib.no
[2] St. Petersburg Department of V.A. Steklov Institute of Mathematics,
St. Petersburg, Russia
[3] Hasso Plattner Institute, University of Potsdam, Potsdam, Germany
kirillsimonov@gmail.com

Abstract. The fundamental theorem of Turán from Extremal Graph Theory determines the exact bound on the number of edges $t_r(n)$ in an n-vertex graph that does not contain a clique of size $r + 1$. We establish an interesting link between Extremal Graph Theory and Algorithms by providing a simple compression algorithm that in linear time reduces the problem of finding a clique of size ℓ in an n-vertex graph G with $m \geq t_r(n) - k$ edges, where $\ell \leq r + 1$, to the problem of finding a maximum clique in a graph on at most $5k$ vertices. This also gives us an algorithm deciding in time $2.49^k \cdot (n + m)$ whether G has a clique of size ℓ. As a byproduct of the new compression algorithm, we give an algorithm that in time $2^{\mathcal{O}(td^2)} \cdot n^2$ decides whether a graph contains an independent set of size at least $n/(d+1)+t$. Here d is the average vertex degree of the graph G. The multivariate complexity analysis based on ETH indicates that the asymptotical dependence on several parameters in the running times of our algorithms is tight.

Keywords: Parameterized algorithms · Extremal graph theory · Turan's theorem · Above guarantee · Kernelization · Exponential time hypothesis

1 Introduction

In 1941, Pál Turán published a theorem that became one of the central results in extremal graph theory. The theorem bounds the number of edges in an undirected graph that does not contain a complete subgraph of a given size. For positive integers $r \leq n$, the *Turán's graph* $T_r(n)$ is the unique complete r-partite n-vertex graph where each part consists of $\lfloor \frac{n}{r} \rfloor$ or $\lceil \frac{n}{r} \rceil$ vertices. In other words, $T_r(n)$ is isomorphic to $K_{a_1,a_2,\dots,a_r}$, where $a_i = \lceil \frac{n}{r} \rceil$ if i is less than or equal to n modulo r and $a_i = \lfloor \frac{n}{r} \rfloor$ otherwise. We use $t_r(n)$ to denote the number of edges in $T_r(n)$.

The research leading to these results has received funding from the Research Council of Norway via the project BWCA (grant no. 314528) and DFG Research Group ADYN via grant DFG 411362735.

Theorem 1 (Turán's Theorem [29]). *Let* $r \leq n$. *Then any* K_{r+1}-*free* n-*vertex graph has at most* $t_r(n)$ *edges. The only* K_{r+1}-*free* n-*vertex graph with exactly* $t_r(n)$ *edges is* $T_r(n)$.

The theorem yields a polynomial time algorithm that for a given n-vertex graph G with at least $t_r(n)$ edges decides whether G contains a clique K_{r+1}. Indeed, if a graph G is isomorphic to $T_r(n)$, which is easily checkable in polynomial time, then it has no clique of size $r + 1$. Otherwise, by Turán's theorem, G contains K_{r+1}. There are constructive proofs of Turán's theorem that also allow to find a clique of size $r + 1$ in a graph with at least $t_r(n)$ edges.

The fascinating question is whether Turán's theorem could help to find efficiently larger cliques in sparser graphs. There are two natural approaches to defining a "sparser" graph and a "larger" clique. These approaches bring us to the following questions; addressing these questions is the primary motivation of our work.

First, what happens when the input graph has a bit less edges than the Turán's graph? More precisely,

> Is there an efficient algorithm that for some $k \geq 1$, decides whether an n-vertex graph with at least $t_r(n) - k$ edges contains a clique of size $r + 1$?

Second, could Turán's theorem be useful in finding a clique of size larger than $r + 1$ in an n-vertex graph with $t_r(n)$ edges? That is,

> Is there an efficient algorithm that for some $\ell > r$ decides whether an n-vertex graph with at least $t_r(n)$ edges contains a clique of size ℓ?

We provide answers to both questions, and more. We resolve the first question by showing a *simple* fixed-parameter tractable (FPT) algorithm where the parameter is k, the "distance" to the Turán's graph. Our algorithm builds on the cute ideas used by Erdős in his proof of Turán's theorem [11]. Viewing these ideas through algorithmic lens leads us to a simple preprocessing procedure, formally a linear-time polynomial compression. For the second question, unfortunately, the answer is negative.

Our Contribution. To explain our results, it is convenient to state the above questions in terms of the computational complexity of the following problem.

> TURÁN'S CLIQUE
> **Input:** An n-vertex graph G, positive integers $r, \ell \leq n$, and k such that $|E(G)| \geq t_r(n) - k$.
> **Question:** Is there a clique of size at least ℓ in G?

Our first result is the following theorem (Theorem 2). Let G be an n-vertex graph with $m \geq t_r(n) - k$ edges. Then there is an algorithm that for any $\ell \leq r + 1$, in time $2.49^k \cdot (n + m)$ either finds a clique of size at least ℓ in G or correctly reports that G does not have a clique of size ℓ. Thus for $\ell \leq r + 1$, TURÁN'S CLIQUE is FPT parameterized by k. More generally, we prove that the problem admits a compression of size linear in k. That is, we provide a linear-time procedure that reduces an instance (G, r, ℓ, k) of TURÁN'S CLIQUE to an equivalent instance (G', p) of the CLIQUE problem with at most $5k$ vertices. The difference between CLIQUE and TURÁN'S CLIQUE is that we do not impose any bound on the number of edges in the input graph of CLIQUE. This is why we use the term compression rather than kernelization,[1] and we argue that stating our reduction in terms of compression is far more natural and helpful. Indeed, after reducing the instance to the size linear in the parameter k, the difference between CLIQUE and TURÁN'S CLIQUE vanes, as even the total number of edges in the instance is automatically bounded by a function of the parameter. On the other hand, CLIQUE is a more general and well-studied problem than TURÁN'S CLIQUE.

Pipelined with the fastest known exact algorithm for MAXIMUM INDEPENDENT SET of running time $\mathcal{O}(1.1996^n)$ [30], our reduction provides the FPT algorithm for TURÁN'S CLIQUE parameterized by k. This algorithm is single-exponential in k and linear in $n + m$, and we also show that the existence of an algorithm subexponential in k would contradict Exponential Time Hypothesis (Corollary 5). Thus the running time of our algorithm is essentially tight, up to the constant in the base of the exponent.

The condition $\ell \leq r + 1$ required by our algorithm is, unfortunately, unavoidable. We prove (Theorem 4) that for any fixed $p \geq 2$, the problem of deciding whether an n-vertex graph with at least $t_r(n)$ edges contains a clique of size $\ell = r + p$ is NP-complete. Thus for any $p \geq 2$, TURÁN'S CLIQUE parameterized by k is para-NP-hard. (We refer to the book of Cygan et al. [6] for an introduction to parameterized complexity.)

While our hardness result rules out finding cliques of size $\ell > r + 1$ in graphs with $t_r(n)$ edges in FPT time, an interesting situation arises when the ratio $\xi := \lfloor \frac{n}{r} \rfloor$ is small. In the extreme case, when $n = r$, the n-vertex graph G with $t_r(n) = n(n-1)/2$ is a complete graph. In this case the problem becomes trivial.

To capture how far the desired clique is from the Turán's bound, we introduce the parameter

$$\tau = \begin{cases} 0, & \text{if } \ell \leq r, \\ \ell - r, & \text{otherwise.} \end{cases}$$

The above-mentioned compression algorithm into CLIQUE with at most $5k$ vertices yields almost "for free" a compression of TURÁN'S CLIQUE into CLIQUE with $\mathcal{O}(\tau \xi^2 + k)$ vertices. Hence for any ℓ, one can decide whether an n-vertex graph with $m \geq t_r(n) - k$ edges contains a clique of size ℓ in time $2^{\mathcal{O}(\tau \xi^2 + k)} \cdot (n + m)$. Thus the problem is FPT parameterized by $\tau + \xi + k$. This result has an interesting interpretation when we look for a large independent set in the complement of a graph. Turán's theorem, when applied to the complement $\overline{G}$ of a graph G, yields a bound

$$\alpha(G) \geq \frac{n}{d+1},$$

---

[1] A *kernel* is by definition a reduction to an instance of the same problem. See the book [14] for an introduction to kernelization.

where $\alpha(G)$ is the size of the largest independent set in G (the independence number of G), and d is the average vertex degree of G. This motivates us to define the following problem.

TURÁN'S INDEPENDENT SET

Input: An n-vertex graph G with average degree d, a positive integer t.

Question: Is there an independent set of size at least $\frac{n}{d+1} + t$ in G?

By Theorem 3, we have a simple algorithm (Corollary 3) that compresses an instance of TURÁN'S INDEPENDENT SET into an instance of INDEPENDENT SET with $\mathcal{O}(td^2)$ vertices. Pipelined with an exact algorithm computing a maximum independent set, the compression results in the algorithm solving TURÁN'S INDEPENDENT SET in time $2^{\mathcal{O}(td^2)} \cdot n^2$.

As we already mentioned, TURÁN'S CLIQUE is NP-complete for any fixed $\tau \geq 2$ and $k = 0$. We prove that the problem remains intractable being parameterized by any pair of the parameters from the triple $\{\tau, \xi, k\}$. More precisely, TURÁN'S CLIQUE is also NP-complete for any fixed $\xi \geq 1$ and $\tau = 0$, as well as for any fixed $\xi \geq 1$ and $k = 0$. These lower bounds are given in Theorem 4.

Given the algorithm of running time $2^{\mathcal{O}(\tau\xi^2+k)} \cdot (n+m)$ and the lower bounds for parameterization by any pair of the parameters from $\{\tau, \xi, k\}$, a natural question is, what is the optimal dependence of a TURÁN'S CLIQUE algorithm on $\{\tau, \xi, k\}$? We use the Exponential Time Hypothesis (ETH) of Impagliazzo, Paturi, and Zane [21] to address this question. Assuming ETH, we rule out the existence of algorithms solving TURÁN'S CLIQUE in time $f(\xi, \tau)^{o(k)} \cdot n^{f(\xi,\tau)}$, $f(\xi, k)^{o(\tau)} \cdot n^{f(\xi,k)}$, and $f(k, \tau)^{o(\sqrt{\xi})} \cdot n^{f(k,\tau)}$, for any function f of the respective parameters.

Related Work. CLIQUE is a notoriously difficult computational problem. It is one of Karp's 21 NP-complete problems [23] and by the work of Håstad, it is hard to approximate CLIQUE within a factor of $n^{1-\epsilon}$ [20]. CLIQUE parameterized by the solution size is W[1]-complete [8]. The problem plays the fundamental role in the W-hierarchy of Downey and Fellows, and serves as the starting point in the majority of parameterized hardness reductions. From the viewpoint of structural parameterized kernelization, CLIQUE does not admit a polynomial kernel when parameterized by the size of the vertex cover [3]. A notable portion of works in parameterized algorithms and kernelization is devoted to solving INDEPENDENT SET (equivalent to CLIQUE on the graph's complement) on specific graph classes like planar, H-minor-free graphs and nowhere-dense graphs [2,7,10,28].

Our algorithmic study of Turán's theorem fits into the paradigm of the "above guarantee" parameterization [26]. This approach was successfully applied to various problems, see e.g. [1,5,12,15–19,22,25,27].

Most relevant to our work is the work of Dvorak and Lidicky on independent set "above Brooks' theorem" [9]. By Brooks' theorem [4], every n-vertex graph of maximum degree at most $\Delta \geq 3$ and clique number at most Δ has an independent set of size at least n/Δ. Then the INDEPENDENT SET OVER BROOK'S BOUND problem is to decide whether an input graph G has an independent set of size at least $\frac{n}{\Delta} + p$. Dvorak and Lidicky [9, Corollary 3] proved that INDEPEN-

DENT SET OVER BROOK'S BOUND admits a kernel with at most $114p\Delta^3$ vertices. This kernel also implies an algorithm of running time $2^{\mathcal{O}(p\Delta^3)} \cdot n^{\mathcal{O}(1)}$. When average degree d is at most $\Delta - 1$, by Corollary 3, we have that INDEPENDENT SET OVER BROOK'S BOUND admits a compression into an instance of INDEPENDENT SET with $\mathcal{O}(p\Delta^2)$ vertices. Similarly, by Corollary 4, for $d \leq \Delta - 1$, INDEPENDENT SET OVER BROOK'S BOUND is solvable in time $2^{\mathcal{O}(p\Delta^2)} \cdot n^{\mathcal{O}(1)}$. When $d > \Delta - 1$, for example, on regular graphs, the result of Dvorak and Lidicky is non-comparable with our results.

2   Algorithms

While in the literature it is common to present Turán's theorem under the implicit assumption that n is divisible by r, here we make no such assumption. For that, it is useful to recall the precise value of $t_r(n)$ in the general setting, as observed by Turán [29].

Proposition 1 (Turán [29]). *For positive integers $r \leq n$,*

$$t_r(n) = \left(1 - \frac{1}{r}\right) \cdot \frac{n^2}{2} - \frac{s}{2} \cdot \left(1 - \frac{s}{r}\right)$$

where $s = n - r \cdot \lfloor \frac{n}{r} \rfloor$ is the remainder in the division of n by r.

Note that [29] uses the expression $t_r(n) = \frac{r-1}{2r} \cdot (n^2 - s^2) + \binom{s}{2}$, however it can be easily seen to be equivalent to the above.

We start with our main problem, where we look for a K_{r+1} in a graph that has slightly less than $t_r(n)$ edges. Later in this section, we show how to derive our other algorithmic results from the compression routine developed next.

2.1   Compression Algorithm for $\ell \leq r + 1$

First, we make a crucial observation on the structure of a TURÁN'S CLIQUE instance that will be the key part of our compression argument. Take a vertex v of maximum degree in G, partition $V(G)$ on $S = N_G(v)$ and $T = V(G) \setminus S$, and add all edges between S and T while removing all edges inside T. It can be argued that this operation does not decrease the number of edges in G while also preserving the property of being K_{r+1}-free. Performing this recursively yields that $T_r(n)$ has indeed the maximum number of edges for a K_{r+1}-free graph, and this is the gist of Erdős' proof of Turán's Theorem [11]. Now, we want to extend this argument to cover our above-guarantee case. Again, we start with the graph G and perform exactly the same recursive procedure to obtain the graph G'. While we cannot say that G' is equal to G, since the latter has slightly less than $t_r(n)$ edges, we can argue that every edge that gets changed from G to G' can be attributed to the "budget" k. Thus we arrive to the conclusion that G is different from G' at only $\mathcal{O}(k)$ places. The following lemma makes this intuition formal.

Lemma 1. *There is an $\mathcal{O}(m + k)$-time algorithm that for non-negative integers $k \geq 1$, $r \geq 2$ and an n-vertex graph G with $m \geq t_r(n) - k$ edges, finds a partition $V_1, V_2, \ldots, V_p$ of $V(G)$ with the following properties*

(i) *$p \geq r - k$;*

(ii) *For each $i \in \{1, \ldots, p\}$, there is a vertex $v_i \in V_i$ with $N_G(v_i) \supset V_{i+1} \cup V_{i+2} \cup \cdots \cup V_p$;*

(iii) *If $p \leq r$, then for the complete p-partite graph G' with parts $V_1, V_2, \ldots, V_p$, we have $|E(G')| \geq |E(G)|$ and $|E(G) \triangle E(G')| \leq 3k$. Moreover, all vertices covered by $E(G) \setminus E(G')$ are covered by $E(G') \setminus E(G)$ and $|E(G') \setminus E(G)| \leq 2k$.*

Let us clarify this technical definition. The lemma basically states that if a graph G has at least $t_r(n) - k$ edges, then it either has a clique of size $r + 1$, or it has at most $3k$ edit distance to a complete multipartite graph G' consisting of $p \in [r - k, r]$ parts. Moreover, G has a clique of size p untouched by the edit, i.e. this clique is present in the complete p-partite graph G' as well.

We should also note that Lemma 1 is close to the concept of stability of Turán's theorem. This concept received much attention in extremal graph theory (see e.g. recent work of Korándi et al. [24]), and appeals the structural properties of graphs having number of edges close to the Turán's number $t_r(n)$. Lemma 1 can also be seen as a stability version of Turán's theorem, but from the algorithmic point of view. We move on to the proof of the lemma.

Proof (of Lemma 1). First, we state the algorithm, which follows from the Erdős' proof of Turán's Theorem from [11]. We start with an empty graph G' defined on the same vertex set as G, and set $G_1 = G$. Then we select the vertex $v_1 \in V(G_1)$ as an arbitrary maximum-degree vertex in G_1, i.e. $\deg_{G_1}(v_1) = \max_{u \in V(G_1)} \deg_{G_1}(u)$. We put $V_1 = V(G_1) \setminus N_{G_1}(v_1)$ and add to G' all edges between V_1 and $V(G_1) \setminus V_1$.

We then put $G_2 := G_1 - V_1$ and, unless G_2 is empty, apply the same process to G_2. That is, we select $v_2 \in V(G_2)$ with $\deg_{G_2}(v_2) = \max_{u \in V(G_2)} \deg_{G_2}(u)$ and put $V_2 = V(G_2) \setminus N_{G_2}(v_2)$ and add all edges between V_2 and $V(G_2) \setminus V_2$ to G'. We repeat this process with $G_{i+1} := G_i - V_i$ until G_{i+1} is empty. The process has to stop eventually as each V_i is not empty. In this way three sequences are produced: $G = G_1, G_2, \ldots, G_p, G_{p+1}$, where G_1 is G and G_{p+1} is the empty graph; $v_1, v_2, \ldots, v_p$, and $V_1, V_2, \ldots, V_p$. Note that the sequences $\{v_i\}$ and $\{V_i\}$ satisfy property (ii) by construction. Observe that this procedure can be clearly performed in time $\mathcal{O}(n^2)$, and for any $r \geq 2$, $m + k = t_r(n) = \Theta(n^2)$, thus the algorithm takes time $\mathcal{O}(m + k)$.

Clearly, G' is a complete p-partite graph with parts $V_1, V_2, \ldots, V_p$ as in G' we added all edges between V_i and $V(G_i) \setminus V_i = (V_{i+1} \cup V_{i+2} \cup \ldots \cup V_p)$ for each $i \in \{1, \ldots, p\}$ and never added an edge between two vertices in the same V_i. Since a p-partite graph is always K_{p+1}-free, by Theorem 1 $|E(G')| \leq t_p(n)$.

Claim. $|E(G')| - |E(G)| \geq \sum_{i=1}^{p} |E(G[V_i])|$ and for each $u \in V(G)$, $\deg_G(u) \leq \deg_{G'}(u)$.

Proof (of Claim). For each $i \in \{1, \ldots, p\}$, denote by E_i the edges of G' added in the i-th step of the construction. Formally, $E_i = V_i \times (V_{i+1} \cup V_{i+2} \cup \ldots \cup V_p)$ for $i < p$ and $E_p = \emptyset$. We aim to show that $|E_i| - |E(G_i) \setminus E(G_{i+1})| \geq |E(G[V_i])|$. The first part of the claim will follow as $|E(G')| = \sum_{i=1}^{p} |E_i|$ and $|E(G)| = \sum_{i=1}^{p} |E(G_i) \setminus E(G_{i+1})|$.

Denote by d_i the degree of v_i in G_i. Since $N_{G_i}(v_i) = (V_{i+1} \cup V_{i+2} \cup \ldots \cup V_p)$, $|E_i| = d_i |V_i|$. As v_i is a maximum-degree vertex in G_i, $d_i \geq \deg_{G_i}(u)$ for every $u \in V_i$, so $|E_i| \geq \sum_{u \in V_i} \deg_{G_i}(u)$. Recall that $G_{i+1} = G_i - V_i$. Then

$$|E(G_i) \setminus E(G_{i+1})| = \sum_{u \in V_i} \deg_{G_i}(u) - |E(G_i[V_i])| = \sum_{u \in V_i} \deg_{G_i}(u) - |E(G[V_i])|$$

$$\leq |E_i| - |E(G[V_i])|,$$

and the first part of the claim follows.

To show the second part, note that for a vertex $u \in V_i$, $\deg_G(u) \leq \sum_{j=1}^{i-1} |V_j| + \deg_{G_i}(u)$. On the other hand, u is adjacent to every vertex in $V_1 \cup V_2 \cup \ldots \cup V_{i-1} \cup V_{i+1} \cup \ldots \cup V_p$ in G'. We have already seen that $|V_{i+1} \cup \ldots \cup V_p| \geq \deg_{G_i}(u)$. Thus, $\deg_G(u) \leq \deg_{G'}(u)$. Proof of the claim is complete.   □

The claim yields that $|E(G)| \leq t_p(n)$, so $t_p(n) \geq t_r(n) - k$. By Theorem 1, we have that $t_i(n) > t_{i-1}(n)$, as $T_{i-1}(n)$ is distinct from $T_i(n)$, so $t_i(n) \geq t_{i-1}(n) + 1$ for every $i \in [n]$. Hence if $r \geq p$ then $k \geq t_r(n) - t_p(n) \geq r - p$. It concludes the proof of (i).

It is left to prove (iii), i.e. that $|E(G) \triangle E(G')| \leq 3k$ under assumption $p \leq r$. First note that $E(G) \setminus E(G') = \bigcup E(G[V_i])$. Second, since $|E(G')| \leq t_p(n) \leq t_r(n)$ and $|E(G)| \geq t_r(n) - k$, $|E(G')| - |E(G)| \leq k$. By Claim, we have that $|E(G')| - |E(G)| \geq \sum |E(G[V_i])|$. Finally

$$|E(G) \triangle E(G')| = |E(G')| - |E(G)| + 2|E(G) \setminus E(G')|$$

$$= |E(G')| - |E(G)| + 2 \sum |E(G[V_i])| \leq 3k.$$

By Claim, each vertex covered by $E(G) \setminus E(G')$ is covered by $E(G') \setminus E(G)$. The total size of these edge sets is at most $3k$, while $|E(G') \setminus E(G)| - |E(G) \setminus E(G')| = |E(G')| - |E(G)| \leq k$. Hence, the size of $|E(G) \setminus E(G')|$ is at most $2k$. This concludes the proof of (iii) and of the lemma.   □

We are ready to prove our main algorithmic result. Let us recall that we seek a clique of size ℓ in an n-vertex graph with $t_r(n) - k$ edges, and that $\tau = \max\{\ell - r, 0\}$.

Theorem 2. TURÁN'S CLIQUE *with* $\tau \in \{0, 1\}$ *admits an* $\mathcal{O}(n + m)$-*time compression into* CLIQUE *on at most* $5k$ *vertices.*

Proof. Let (G, r, k, ℓ) be the input instance of TURÁN'S CLIQUE. If $r < 2$ or $n \leq 5k$, a trivial compression is returned. Apply the algorithm of Lemma 1 to (G, r, k, ℓ) and obtain the partition $V_1, V_2, \ldots, V_p$. Observe that this takes time $\mathcal{O}(m + k) = \mathcal{O}(n + m)$ since $n > 5k$. By the second property of Lemma 1,

$v_1, v_2, \ldots, v_p$ induce a clique in G, so if $p \geq \ell$ we conclude that (G, r, k, ℓ) is a yes-instance. Formally, the compression returns a trivial yes-instance of CLIQUE in this case.

We now have that $r - k \leq p \leq r$. Then the edit distance between G and the complete p-partite graph G' with parts $V_1, V_2, \ldots, V_p$ is at most $3k$. Denote by X the set of vertices covered by $E(G) \triangle E(G')$. Denote $R = E(G') \setminus E(G)$ and $A = E(G) \setminus E(G')$. We know that $|R| + |A| \leq 3k$, $|R| \leq 2k$ and $|R| \geq |A|$. By Lemma 1, R covers all vertices in X, so $|X| \leq 2|R|$.

Clearly, (G, r, k, ℓ) as an instance of TURÁN'S CLIQUE is equivalent to an instance (G, ℓ) of CLIQUE. We now apply the following two reduction rules exhaustively to (G, ℓ). Note that these rules are an adaption of the well-known two reduction rules for the general case of CLIQUE (see, e.g., [30]). Here the adapted rules employ the partition $V_1, V_2, \ldots, V_p$ explicitly.

Reduction rule 1. *If there is $i \in [p]$ such that $V_i \not\subseteq X$ and V_i is independent in G, remove V_i from G and reduce ℓ by one.*

Reduction rule 2. *For each $i \in [p]$ with $|V_i \setminus X| > 1$, remove all but one vertices in $V_i \setminus X$ from G.*

Since the reduction rules are applied independently to parts $V_1, V_2, \ldots, V_p$, and each rule is applied to each part at most once, clearly this can be performed in linear time. We now argue that these reduction rules always produce an equivalent instance of CLIQUE.

Claim. Reduction rule 1 and Reduction rule 2 are safe.

Proof. For Reduction rule 1, note that there is a vertex $v \in V_i \setminus X$ such that $N_G(v) = N_G(V_i) = V(G) \setminus V_i$. Since V_i is independent, for any vertex set C that induces a clique in G, we have $|C \cap V_i| \leq 1$. On the other hand, if $C \cap V_i = \emptyset$, $C \cup \{v\}$ also induces a clique in G as $C \subseteq N_G(v)$. Hence, any maximal clique in G contains exactly one vertex from V_i, so Reduction rule 1 is safe.

To see that Reduction rule 2 is safe, observe that $N_G(u) = N_G(v)$ for any two vertices $u, v \in V_i \setminus X$. Then no clique contains both u and v, and if $C \ni v$ induces a clique in G, $C \setminus \{v\} \cup \{u\}$ also induces a clique in G of the same size. Hence, v can be safely removed from G so Reduction rule 2 is safe. □

It is left to upperbound the size of G after the exhaustive application of reduction rules. In this process, some parts among $V_1, V_2, \ldots, V_p$ are removed from G. W.l.o.g. assume that the remaining parts are $V_1, V_2, \ldots, V_t$ for some $t \leq p$. Note that parts that have no common vertex with X are eliminated by Reduction rule 1, so $t \leq |X|$. On the other hand, by Reduction rule 2, we have $|V_i \setminus X| \leq 1$ for each $i \in [t]$.

Consider $i \in [t]$ with $|V_i \setminus X| = 1$. By Reduction rule 1, $G[V_i]$ contains at least one edge. Since V_i is independent in G', $E(G[V_i]) \subseteq A$. Hence, the number of $i \in [t]$ with $|V_i \setminus X| = 1$ is at most $|A|$. We obtain

$$|V(G)| = \sum_{i=1}^{t} |V_i| = \sum_{i=1}^{t} |V_i \cap X| + \sum_{i=1}^{t} |V_i \setminus X|$$
$$\leq |X| + |A| \leq 2|R| + |A| \leq |R| + (|R| + |A|) \leq 5k.$$

We obtained an instance of CLIQUE that is equivalent to (G, r, k, ℓ) and contains at most $5k$ vertices. The proof is complete. $\qquad\square$

Combining the polynomial compression of Theorem 2 with the algorithm of Xiao and Nagamochi [30] for INDEPENDENT SET running in $\mathcal{O}(1.1996^n)$, we obtain the following.

Corollary 1. TURÁN'S CLIQUE *with* $\tau \leq 1$ *is solvable in time* $2.49^k \cdot (n + m)$.

Proof. Take a given instance of TURÁN'S CLIQUE and compress it into an equivalent instance (G, ℓ) of CLIQUE with $|V(G)| \leq 5k$. Clearly, $(G, |V(G)| - \ell)$ is an instance of INDEPENDENT SET equivalent to (G, ℓ). Use the algorithm from [30] to solve this instance in $\mathcal{O}(1.1996^{|V(G)|})$ running time. Since $1.1996^5 < 2.49$, the running time of the whole algorithm is bounded by $2.49^k \cdot n^{\mathcal{O}(1)}$. $\qquad\square$

2.2   Looking for Larger Cliques

In this subsection we consider the situation when $\tau > 1$. As we will see in Theorem 4, an FPT algorithm is unlikely in this case, unless we take a stronger parameterization. Here we show that TURÁN'S CLIQUE is FPT parameterized by $\tau + \xi + k$. Recall that $\xi = \lfloor \frac{n}{r} \rfloor$. Theorem 4 argues that this particular choice of the parameter is necessary.

First, we show that the difference between $t_\ell(n)$ and $t_r(n)$ can be bounded in terms of τ and ξ. This will allow us to employ Theorem 2 for the new FPT algorithm by a simple change of the parameter. The proof of the next lemma is done via a careful counting argument.

Lemma 2. *Let* n, r, ℓ *be three positive integers with* $r < \ell \leq n$. *Let* $\xi = \lfloor \frac{n}{r} \rfloor$ *and* $\tau = \ell - r$. *Then for* $\tau = \mathcal{O}(r)$, $t_\ell(n) - t_r(n) = \Theta(\tau \xi^2)$.

Proof. Throughout the proof, we assume $\xi = \frac{n}{r}$ since this does not influence the desired Θ estimation. Let s_r be the remainder in the division of n by r and s_ℓ be the remainder in the division of n by ℓ. By Lemma 1,

$$t_\ell(n) - t_r(n) = \frac{\tau n^2}{2r\ell} + \left(\frac{s_r}{2} \cdot \left(1 - \frac{s_r}{r}\right) - \frac{s_\ell}{2} \cdot \left(1 - \frac{s_\ell}{\ell}\right) \right). \qquad (1)$$

The first summand in (1) is $\Theta(\xi^2 \tau)$. Indeed, since $\tau = \mathcal{O}(r)$ we have

$$\frac{\tau n^2}{2r\ell} = \frac{\tau}{2} \cdot \frac{n}{r} \cdot \frac{n}{r+\tau} = \frac{\xi^2 \tau}{2} \cdot \frac{r}{r+\tau} = \Theta(\xi^2 \tau). \qquad (2)$$

For the second summand,

$$\frac{s_r}{2} \cdot \left(1 - \frac{s_r}{r}\right) - \frac{s_\ell}{2} \cdot \left(1 - \frac{s_\ell}{\ell}\right) = \frac{\ell s_r(r - s_r) - r s_\ell(\ell - s_\ell)}{2r\ell} = \frac{(rs_\ell^2 - \ell s_r^2) + r\ell(s_r - s_\ell)}{2r\ell}$$

$$= \frac{(rs_\ell^2 - rs_r^2 - \tau s_r^2) + r\ell(s_r - s_\ell)}{2r\ell} \tag{3}$$

$$= \frac{r(s_\ell - s_r)(s_\ell + s_r) + r\ell(s_r - s_\ell)}{2r\ell} - \frac{\tau s_r^2}{2r\ell}$$

$$= \frac{(s_r - s_\ell)(\ell - (s_\ell + s_r))}{2\ell} - \frac{\tau s_r^2}{2r\ell}. \tag{4}$$

Since $n = \lfloor \frac{n}{\ell} \rfloor \cdot \ell + s_\ell$, we have that

$$s_r \equiv \left\lfloor \frac{n}{\ell} \right\rfloor \cdot \ell + s_\ell \pmod{r},$$

and

$$s_r \equiv \left\lfloor \frac{n}{\ell} \right\rfloor \cdot (r + \tau) + s_\ell \pmod{r}.$$

Hence,

$$s_r - s_\ell \equiv \left\lfloor \frac{n}{\ell} \right\rfloor \cdot \tau \pmod{r}.$$

By definition $s_r < r$, thus we get from the above that $s_r - s_\ell \leq \lfloor \frac{n}{\ell} \rfloor \cdot \tau \leq \xi\tau$.
Analogously,

$$s_\ell - s_r \equiv \left\lfloor \frac{n}{r} \right\rfloor \cdot (-\tau) \pmod{\ell}$$

Since $s_\ell - s_r > -r > -\ell$, we have that $s_\ell - s_r \geq \lfloor \frac{n}{r} \rfloor \cdot (-\tau) \geq -\xi\tau$. Therefore $|s_\ell - s_r| \leq \xi\tau$. It is easy to see that $|\ell - (s_\ell + s_r)| \leq \ell + (s_\ell + s_r) \leq 3\ell$. Finally, $\frac{\tau s_r^2}{2r\ell}$ is non-negative and is upper bounded by $\frac{\tau r^2}{2r\ell} \leq \frac{\tau}{2}$. Thus, the absolute value of (4), is at most

$$\frac{\xi\tau \cdot 3\ell}{2\ell} + \frac{\tau}{2} = \mathcal{O}(\xi\tau).$$

By putting together (2) and (4), we conclude that $t_\ell(n) - t_r(n) = \Theta(\xi^2\tau) + \mathcal{O}(\xi\tau) = \Theta(\xi^2\tau)$. $\square$

The following compression algorithm is a corollary of Lemma 2 and Theorem 2. It provides a compression of size linear in k and τ.

Theorem 3. Turán's Clique *admits a compression into* Clique *on* $\mathcal{O}(\tau\xi^2 + k)$ *vertices.*

Proof. Let (G, k, r, ℓ) be the given instance of Turán's Clique. If $\ell \leq r+1$, then the proof follows from Theorem 2. Otherwise, reduce (G, k, r, ℓ) to an equivalent instance $(G, k + t_\ell(n) - t_r(n), \ell, \ell)$ of Turán's Clique just by modifying the parameters. This is a valid instance since $|E(G)| \geq t_r(n) - k \geq t_\ell(n) - (t_\ell(n) + t_r(n) + k)$. Denote $k' = k + (t_\ell(n) - t_r(n))$. By Lemma 2, $k' = k + \mathcal{O}(\tau\xi^2)$. Apply polynomial compression of Theorem 2 to (G, k', ℓ, ℓ) into Clique with $\mathcal{O}(k')$, i.e. $\mathcal{O}(\tau\xi^2 + k)$, vertices. $\square$

Pipelined with a brute-force algorithm computing a maximum independent set in time $\mathcal{O}(2^n)$, Theorem 3 yields the following corollary.

Corollary 2. Turán's Clique *is solvable in time* $2^{\mathcal{O}(\tau\xi^2 + k)} \cdot (n + m)$.

2.3   Independent Set above Turán's Bound

Another interesting application of Theorem 3 concerns computing INDEPENDENT SET in graphs of small average degree. Recall that Turán's theorem, when applied to the complement $\overline{G}$ of a graph G, yields a bound

$$\alpha(G) \geq \frac{n}{d+1}.$$

Here $\alpha(G)$ is the size of the largest independent set in G (the independence number of G), and d is the average vertex degree of G. Then in TURÁN'S INDEPENDENT SET, the task is for an n-vertex graph G and positive integer t to decide whether there is an independent set of size at least $\frac{n}{d+1} + t$ in G.

Theorem 3 implies a compression of TURÁN'S INDEPENDENT SET into INDEPENDENT SET. In other words, we give a polynomial time algorithm that for an instance (G, t) of TURÁN'S INDEPENDENT SET constructs an equivalent instance (G', p) of INDEPENDENT SET with at most $\mathcal{O}(td^2)$ vertices. That is, the graph G has an independent set of size at least $\frac{n}{d+1} + t$ if and only if G' has an independent set of size p.

Corollary 3. TURÁN'S INDEPENDENT SET *admits a compression into* INDEPENDENT SET *on* $\mathcal{O}(td^2)$ *vertices.*

Proof. For simplicity, let us assume that n is divisible by $d + 1$. (For arguments here this assumption does not make an essential difference.) We select $r = \frac{n}{d+1}$, $\tau = t$, and $k = 0$. Then $d = \frac{n}{r} - 1 = \xi - 1$. The graph $\overline{G}$ has at most $nd/2$ edges, hence G has at least $\frac{n(n-1)}{2} - nd/2 = \frac{n(n-1)}{2} - n(\xi - 1)/2 \geq t_r(n)$ edges, see Lemma 1. An independent set of size $\frac{n}{d+1} + t$ in graph $\overline{G}$, corresponds in graph G to a clique of size $r + t$. Since Theorem 3 provides compression into a CLIQUE with $\mathcal{O}(\tau\xi^2 + k) = \mathcal{O}(\tau\xi^2)$ vertices, for independent set and graph $\overline{G}$ this corresponds to a compression into an instance of INDEPENDENT SET with $\mathcal{O}(td^2)$ vertices. □

By Corollary 3, we obtain the following corollary.

Corollary 4. TURÁN'S INDEPENDENT SET *is solvable in time* $2^{\mathcal{O}(td^2)} \cdot n^2$.

3   Lower Bounds

In this section, we investigate how the algorithms above are complemented by hardness results. First, observe that k has to be restricted, otherwise the TURÁN'S CLIQUE problem is not any different from CLIQUE. In fact, reducing from INDEPENDENT SET on sparse graphs, one can show that there is no $2^{o(k)}$-time algorithm for TURÁN'S CLIQUE even when $\tau \leq 1$. (The formal argument is presented in Theorem 5.) This implies that the $2^{\mathcal{O}(k)}$-time algorithm given by Corollary 1 is essentially tight.

Also, the difference between r and ℓ has to be restricted, as it can be easily seen that TURÁN'S CLIQUE admits no $n^{o(\ell)}$-time algorithm even when $k = 0$, assuming ETH. This is observed simply by considering the special case of TURÁN'S CLIQUE where $r = 1$, there the only restriction on G is that $|E(G)| \geq t_r(n) - k = 0$, meaning that the problem is as hard as CLIQUE. However, Theorem 4 shows that even for any fixed $\tau \geq 2$ and $k = 0$ TURÁN'S CLIQUE is NP-complete. This motivates Theorem 3, where the exponential part of the running time has shape $2^{\mathcal{O}(\tau\xi^2 k)}$. In the rest of this section, we further motivate the running time of Theorem 3. First, in Theorem 4 we show that not only setting τ and k to constants is not sufficient to overcome NP-hardness, but also that the same holds for any choice of two parameters out of $\{\tau, \xi, k\}$.

Theorem 4. TURÁN'S CLIQUE *is NP-complete. Moreover, it remains NP-complete in each of the following cases*

   (i) *for any fixed $\xi \geq 1$ and $\tau = 0$;*
  (ii) *for any fixed $\xi \geq 1$ and $k = 0$;*
 (iii) *for any fixed $\tau \geq 2$ and $k = 0$.*

Proof. Towards proving (i) and (ii), we provide a reduction from CLIQUE. Let $\xi \geq 1$ be a fixed constant. Let (G, ℓ) be a given instance of CLIQUE and let $n = |V(G)|$. We assume that $\ell \geq \xi$, otherwise we can solve (G, ℓ) in polynomial time. Construct a graph G' from G as follows. Start from $G' = G$ and $\ell' = \ell$. Then add $\max\{\xi\ell - n, 0\}$ isolated vertices to G'. Note that (G, k) and (G', k') are equivalent and $|V(G')| \geq \xi\ell'$. If we have $\xi\ell' \leq |V(G')| < (\xi + 1)\ell'$, we are done with the construction of G'. Otherwise, repeatedly add a universal vertex to G', increasing ℓ' by one, so $|V(G')| - (\xi + 1)\ell'$ decreases by ξ each time. We repeat this until $|V(G')|$ becomes less than $(\xi + 1)\ell'$. Since the gap between $\xi\ell'$ and $(\xi+1)\ell'$ is at least ξ at any moment, we derive that $\xi\ell' \leq |V(G')| < (\xi+1)\ell'$. The construction of G' is complete. Note that (G', ℓ') is an instance of CLIQUE equivalent to (G, ℓ). We added at most $\max\{n, \xi\ell\}$ vertices to G', hence this is a polynomial-time reduction.

By the above, $\lfloor |V(G')/\ell'| \rfloor = \xi$, so we can reduce (G', ℓ') to an equivalent instance $(G', \ell', \binom{|V(G')|}{2}, \ell')$ of TURÁN'S CLIQUE. Clearly, this instance has the required fixed value of ξ and $\tau = 0$. This proves (i). For (ii), we use the fact that $t_1(n) = 0$ for every $n > 0$ and reduce (G', ℓ') to $(G', 1, 0, \ell')$.

To show (iii), we need another reduction from CLIQUE. Let $\tau \geq 2$ be a fixed integer constant. Take an instance (G, ℓ) of CLIQUE with $\ell \geq 2\tau$. We denote $n = |V(G)|$. To construct G' from G, we start from a large complete $(\ell-1)$-partite graph with equal-sized parts. The size of each part equals x, so $|V(G')| = (\ell-1)x$. We denote $N = |V(G')|$ and choose the value of x later, for now we only need that $N \geq n$. Clearly, $|E(G')| = t_{\ell-1}(N)$ at this point. To embed G into G', we select arbitrary n vertices in G' and make them isolated. This removes at most $n(\ell - 2)x$ edges from G'. Then we identify these n isolated vertices with $V(G)$ and add edges of G between these vertices in G' correspondingly. This operation does not decrease $|E(G')|$. This completes the construction of G'. Since G' is isomorphic to a complete $(\ell-1)$-partite graph united disjointly with G, we have that (G, ℓ) and (G', ℓ) are equivalent instances of CLIQUE.

We now want to reduce (G', ℓ) to an instance $(G', \ell - \tau, 0, \ell)$ of TURÁN'S CLIQUE. To do so, we need $|E(G')| \geq t_{\ell-\tau}(N)$. By Lemma 2, $t_{\ell-1}(N) - t_{\ell-\tau}(N) \geq C \cdot (\tau - 1) \cdot \left(\frac{N}{\ell-\tau}\right)^2$ for some constant $C > 0$. Since $|E(G')| \geq t_{\ell-1}(N) - n(\ell-2)x$, we want to choose x such that

$$n(\ell-2)x \leq C \cdot (\tau - 1) \cdot \left(\frac{N}{\ell-\tau}\right)^2.$$

By substituting $N = (\ell - 1)x$, we derive that x should satisfy

$$\frac{n}{C} \cdot \frac{(\ell-2)(\ell-\tau)}{(\ell-1)^2} \cdot \frac{\ell-\tau}{\tau-1} \leq x.$$

Now simply pick as x the smallest integer that satisfies the above. Then $(G', \ell - \tau, 0, \ell)$ is an instance of TURÁN'S CLIQUE that is equivalent to the instance (G, k) of CLIQUE and is constructed in polynomial time.   □

Now, recall that Theorem 3 gives an FPT-algorithm for TURÁN'S CLIQUE that is single-exponential in $\tau\xi^2 + k$. The previous theorem argues that all three of τ, ξ, k have to be in the exponential part of the running time. However, that result does not say anything about what can be the best possible dependency on these parameters. The next Theorem 5 aims to give more precise lower bounds based on ETH, in particular it turns out that the dependency on τ and k cannot be subexponential unless ETH fails. First, we need to show the relation between the parameter ξ and the average degree of $\overline{G}$. The proof of the following proposition is available in the full version of the paper [13].

Proposition 2. *Let G be an n-vertex graph, $r \leq n$ be an integer, and denote $\xi = \lfloor\frac{n}{r}\rfloor$. Let $\overline{G}$ denote the complement of G and $\overline{d}$ denote the average degree of $\overline{G}$. Then $\overline{d} \leq \xi$ if $|E(G)| \geq t_r(n)$ and $|E(G)| \geq t_r(n)$ if $\overline{d} \leq \xi - 1$.*

We are ready to give lower bounds for algorithms solving TURÁN'S CLIQUE in terms of the parameters τ, ξ, and k.

Theorem 5. *Unless the Exponential Time Hypothesis fails, for any function f there is no $f(\xi,\tau)^{o(k)} \cdot n^{f(\xi,\tau)}$, $f(\xi,k)^{o(\tau)} \cdot n^{f(\xi,k)}$, or $f(k,\tau)^{o(\sqrt{\xi})} \cdot n^{f(k,\tau)}$ algorithm for TURÁN'S CLIQUE.*

The proof of this result is available in the full version of the paper [13]. It is based on the proof of Theorem 4, but is much more careful to details and contains some new ideas. Moreover, the proof of the first point of the theorem lets us observe that our $2.49^k \cdot (n + m)$-time algorithm for TURÁN'S CLIQUE with $\tau \leq 1$ is essentially tight.

Corollary 5. *Assuming ETH, there is no $2^{o(k)} \cdot n^{\mathcal{O}(1)}$ algorithm for TURÁN'S CLIQUE with $\ell \leq r + 1$.*

4   Conclusion

We conclude by summarizing natural questions left open by our work. Theorem 5 rules out (unless ETH fails) algorithms with running times subexponential in τ and k. However, when it comes to ξ, the dependency in the upper bound of Corollary 2 is $2^{\mathcal{O}(\tau\xi^2+k)} \cdot n^{\mathcal{O}(1)}$, while Theorem 5 only rules out the running time of $f(k,\tau)^{o(\sqrt{\xi})} \cdot n^{f(k,\tau)}$ under ETH. Thus, whether the correct dependence in ξ is single-exponential or subexponential, is left open. Similarly, the question whether TURÁN'S CLIQUE admits a compression into CLIQUE whose size is linear in ξ, τ, and k, is open. A weaker variant of this question (for the case $k = 0$) for TURÁN'S INDEPENDENT SET, whether it admits a compression or kernel linear in d and in t, is also open.

References

1. Alon, N., Gutin, G., Kim, E.J., Szeider, S., Yeo, A.: Solving MAX-r-SAT above a tight lower bound. Algorithmica **61**(3), 638–655 (2011)
2. Bodlaender, H.L., Fomin, F.V., Lokshtanov, D., Penninkx, E., Saurabh, S., Thilikos, D.M.: (Meta) kernelization. J. ACM **63**(5), 44:1–44:69 (2016). https://doi.org/10.1145/2973749
3. Bodlaender, H.L., Jansen, B.M.P., Kratsch, S.: Kernelization lower bounds by cross-composition. SIAM J. Discret. Math. **28**(1), 277–305 (2014). https://doi.org/10.1137/120880240
4. Brooks, L.R.: On colouring the nodes of a network. Proc. Camb. Philos. Soc. **37**, 194–197 (1941)
5. Crowston, R., Jones, M., Muciaccia, G., Philip, G., Rai, A., Saurabh, S.: Polynomial kernels for lambda-extendible properties parameterized above the Poljak-Turzik bound. In: IARCS Annual Conference on Foundations of Software Technology and Theoretical Computer Science (FSTTCS). Leibniz International Proceedings in Informatics (LIPIcs), vol. 24, pp. 43–54. Schloss Dagstuhl-Leibniz-Zentrum fuer Informatik, Dagstuhl, Germany (2013)
6. Cygan, M., et al.: Parameterized Algorithms. Springer, Cham (2015). https://doi.org/10.1007/978-3-319-21275-3
7. Demaine, E.D., Fomin, F.V., Hajiaghayi, M., Thilikos, D.M.: Subexponential parameterized algorithms on graphs of bounded genus and H-minor-free graphs. J. ACM **52**(6), 866–893 (2005)
8. Downey, R.G., Fellows, M.R.: Parameterized Complexity. Springer-Verlag, New York (1999)
9. Dvořák, Z., Lidický, B.: Independent sets near the lower bound in bounded degree graphs. In: Proceedings of the 34th International Symposium on Theoretical Aspects of Computer Science (STACS). Leibniz International Proceedings in Informatics (LIPIcs), vol. 66, pp. 28:1–28:13. Schloss Dagstuhl - Leibniz-Zentrum fuer Informatik (2017). https://doi.org/10.4230/LIPIcs.STACS.2017.28
10. Dvořák, Z., Mnich, M.: Large independent sets in triangle-free planar graphs. SIAM J. Discret. Math. **31**(2), 1355–1373 (2017). https://doi.org/10.1137/16M1061862
11. Erdős, P.: On the graph theorem of Turán. Mat. Lapok **21**, 249–251 (1970)
12. Fomin, F.V., Golovach, P.A., Lokshtanov, D., Panolan, F., Saurabh, S., Zehavi, M.: Going far from degeneracy. SIAM J. Discret. Math. **34**(3), 1587–1601 (2020). https://doi.org/10.1137/19M1290577

13. Fomin, F.V., Golovach, P.A., Sagunov, D., Simonov, K.: Turán's theorem through algorithmic lens (2023)
14. Fomin, F.V., Lokshtanov, D., Saurabh, S., Zehavi, M.: Kernelization: Theory of Parameterized Preprocessing. Cambridge University Press, Cambridge (2019)
15. Garg, S., Philip, G.: Raising the bar for vertex cover: fixed-parameter tractability above a higher guarantee. In: Proceedings of the Twenty-Seventh Annual ACM-SIAM Symposium on Discrete Algorithms (SODA), pp. 1152–1166. SIAM (2016). https://doi.org/10.1137/1.9781611974331.ch80
16. Gutin, G., van Iersel, L., Mnich, M., Yeo, A.: Every ternary permutation constraint satisfaction problem parameterized above average has a kernel with a quadratic number of variables. J. Comput. Syst. Sci. **78**(1), 151–163 (2012)
17. Gutin, G., Kim, E.J., Lampis, M., Mitsou, V.: Vertex cover problem parameterized above and below tight bounds. Theory Comput. Syst. **48**(2), 402–410 (2011)
18. Gutin, G.Z., Patel, V.: Parameterized traveling salesman problem: beating the average. SIAM J. Discret. Math. **30**(1), 220–238 (2016)
19. Gutin, G.Z., Rafiey, A., Szeider, S., Yeo, A.: The linear arrangement problem parameterized above guaranteed value. Theory Comput. Syst. **41**(3), 521–538 (2007). https://doi.org/10.1007/s00224-007-1330-6
20. Håstad, J.: Clique is hard to approximate within $n^{1-\epsilon}$. Acta Math. **182**(1), 105–142 (1999). https://doi.org/10.1007/BF02392825
21. Impagliazzo, R., Paturi, R., Zane, F.: Which problems have strongly exponential complexity. J. Comput. Syst. Sci. **63**(4), 512–530 (2001)
22. Jansen, B.M.P., Kozma, L., Nederlof, J.: Hamiltonicity below Dirac's condition. In: Sau, I., Thilikos, D.M. (eds.) WG 2019. LNCS, vol. 11789, pp. 27–39. Springer, Cham (2019). https://doi.org/10.1007/978-3-030-30786-8_3
23. Karp, R.M.: Reducibility among combinatorial problems. In: Complexity of Computer Computations, pp. 85–103. Plenum Press, New York (1972)
24. Korándi, D., Roberts, A., Scott, A.: Exact stability for turán's theorem. Adv. Comb. 31079 (2021)
25. Lokshtanov, D., Narayanaswamy, N.S., Raman, V., Ramanujan, M.S., Saurabh, S.: Faster parameterized algorithms using linear programming. ACM Trans. Algorithms **11**(2), 15:1–15:31 (2014). https://doi.org/10.1145/2566616
26. Mahajan, M., Raman, V.: Parameterizing above guaranteed values: MaxSat and MaxCut. J. Algorithms **31**(2), 335–354 (1999)
27. Mahajan, M., Raman, V., Sikdar, S.: Parameterizing above or below guaranteed values. J. Comput. Syst. Sci. **75**(2), 137–153 (2009)
28. Pilipczuk, M., Siebertz, S.: Kernelization and approximation of distance-r independent sets on nowhere dense graphs. Eur. J. Comb. **94**, 103309 (2021). https://doi.org/10.1016/j.ejc.2021.103309
29. Turán, P.: Eine Extremalaufgabe aus der Graphentheorie. Mat. Fiz. Lapok **48**, 436–452 (1941)
30. Xiao, M., Nagamochi, H.: Exact algorithms for maximum independent set. Inf. Comput. **255**, 126–146 (2017). https://doi.org/10.1016/j.ic.2017.06.001

On the Frank Number and Nowhere-Zero Flows on Graphs

Jan Goedgebeur[1,2] , Edita Máčajová[3] , and Jarne Renders[1(✉)]

[1] Department of Computer Science, KU Leuven Kulak, 8500 Kortrijk, Belgium
{jan.goedgebeur,jarne.renders}@kuleuven.be
[2] Department of Applied Mathematics, Computer Science and Statistics,
Ghent University, 9000 Ghent, Belgium
[3] Comenius University, Mlynská dolina, 842 48 Bratislava, Slovakia
macajova@dcs.fmph.uniba.sk

Abstract. An edge e of a graph G is called *deletable* for some orientation o if the restriction of o to $G - e$ is a strong orientation. In 2021, Hörsch and Szigeti proposed a new parameter for 3-edge-connected graphs, called the *Frank number*, which refines k-edge-connectivity. The Frank number is defined as the minimum number of orientations of G for which every edge of G is deletable in at least one of them. They showed that every 3-edge-connected graph has Frank number at most 7 and that in case these graphs are also 3-edge-colourable graphs the parameter is at most 3. Here we strengthen the latter result by showing that such graphs have Frank number 2, which also confirms a conjecture by Barát and Blázsik. Furthermore, we prove two sufficient conditions for cubic graphs to have Frank number 2 and use them in an algorithm to computationally show that the Petersen graph is the only cyclically 4-edge-connected cubic graph up to 36 vertices having Frank number greater than 2.

Keywords: Frank number · Connectivity · Orientation · Snark · Nowhere-zero flows

1 Introduction

An *orientation* o of a graph G is a directed graph with vertices $V(G)$ such that each edge $uv \in E(G)$ is replaced by exactly one of the arcs $u \to v$ or $v \to u$. An orientation is called *strong* if for every two distinct vertices u and v there exists an oriented uv-path, i.e. an oriented path with endpoints u and v. An edge e is *deletable* in a strong orientation o of G if the restriction of o to $E(G) - \{e\}$ is a strong orientation of $G - e$. The *cyclic edge connectivity* of a graph is the smallest number of edges k whose removal separates the graph into two components, each of which contains a cycle. Such a graph is called *cyclically k-edge-connected*.

In 2021, Hörsch and Szigeti [9] proposed a new parameter for 3-edge-connected graphs called the Frank number, which can be seen as a generalisation of a theorem by Nash-Williams [11] stating that a graph has a k-arc-connected orientation if it is $2k$-edge-connected. For a 3-edge-connected graph G, the *Frank number* – denoted by $fn(G)$ – is defined to be the minimum number k for which G admits k orientations such that every edge $e \in E(G)$ is deletable in at least one of them.

Hörsch and Szigeti proved in [9] that every 3-edge-connected graph G has $fn(G) \leq 7$ and that the Berge-Fulkerson conjecture [12] implies that $fn(G) \leq 5$. They conjectured that every 3-edge-connected graph G has $fn(G) \leq 3$ and showed that the Petersen graph has Frank number equal to 3. In this paper we will mainly investigate the following stronger problem: Is it true that the Petersen graph is the only cyclically 4-edge-connected cubic graph with Frank number greater than 2? Let G_1 and G_2 be cubic graphs. Create a new graph G by removing a vertex from each of G_1 and G_2 and adding three edges between the resulting 2-valent vertices in such a way that the edges have one vertex in G_1 and one vertex in G_2 and the result is cubic. We call G the *3-join* of G_1 and G_2. We can equivalently formulate the problem as follows.

Problem 1. Can every 3-edge-connected cubic graph G with $fn(G) > 2$ be created by a sequence of 3-joins from the Petersen graph?

Note that a cyclically 3-edge-connected graph uniquely decomposes into cyclically 4-edge connected graphs and that a 3-join of a graph with Frank number 3 with any other graph has Frank number at least 3.

Barát and Blázsik showed in [1] that for any 3-edge-connected graph G, there exists a 3-edge-connected cubic graph H with $fn(H) \geq fn(G)$. Hence, it is sufficient to study this problem in the cubic case.

Hörsch and Szigeti proved in [9] that every 3-edge-connected 3-edge-colourable graph has Frank number at most 3. In Sect. 2 we strengthen this result by showing that these graphs have Frank number equal to 2. Note that such graphs are always cubic.

Barát and Blázsik also verified that several well-known infinite families of 3-edge-connected graphs have Frank number 2. This includes wheel graphs, Möbius ladders, prisms, flower snarks and an infinite subset of the generalised Petersen graphs. Note that except for the wheel graphs and flower snarks, these families all consist of 3-edge-colourable graphs. They also conjectured that every 3-edge-connected hamiltonian cubic graph has Frank number 2. Since every hamiltonian cubic graph is 3-edge-colourable, our result that every 3-edge-connected 3-edge-colourable graph has Frank number 2 also proves this conjecture.

Our proof of this result uses nowhere-zero integer flows. We give a sufficient condition for an edge to be deletable in an orientation which is the underlying orientation of some all-positive nowhere-zero k-flow and construct two specific nowhere-zero 4-flows that show that the Frank number is 2.

Moreover, in Sect. 2 we give two sufficient conditions for cyclically 4-edge-connected cubic graphs to have Frank number 2. In Sect. 3 we propose a heuristic algorithm and an exact algorithm for determining whether the Frank number of a 3-edge-connected cubic graph is 2. The heuristic algorithm makes use of the sufficient conditions for cyclically 4-edge-connected graphs shown in the previous section. Using an implementation of these algorithms we show that the Petersen graph is the only cyclically 4-edge-connected *snark*, i.e. cubic graph which does not admit a 3-edge-colouring, up to and including 36 vertices with Frank number greater than 2.

Due to space constraints we had to omit several proofs. A full-length version of this article containing all proofs can be found on arXiv [6].

1.1 Preliminaries

For some integer k, a *k-flow* (o, f) on a graph consists of an orientation o of the edges of G and a valuation $f : E(G) \to \{0, \pm 1, \pm 2 \dots, \pm(k - 1)\}$ such that at every vertex the sum of the values on incoming edges equals the sum on the outgoing edges. A k-flow (o, f) is said to be *nowhere-zero* if the value of f is not 0 for any edge of $E(G)$. A nowhere-zero k-flow on G is said to be *all-positive* if the value $f(e)$ is positive for every edge e of G. Every nowhere-zero k-flow can be transformed to an all-positive nowhere-zero k-flow by changing the orientation of the edges with negative $f(e)$ and changing negative values of $f(e)$ to $-f(e)$.

Let (G, o) be a graph with orientation o. Let H be a subgraph of G. If the context is clear we write (H, o) to be the graph H with the orientation of o restricted to H. We define the set $D(G, o) \subset E(G)$ to be the set of all edges of G which are deletable in o. Let $u, v \in V(G)$, if the edge uv is oriented from u to v, we write $u \to v$.

2 Theoretical Results

Let (o, f) be an all-positive nowhere-zero k-flow on a cubic graph G. An edge e with $f(e) = 2$ is called a *strong 2-edge* if there is no 3-edge-cut containing the edge e (cycle-separating or not) such that the remaining edges of the cut have value 1 in f.

Lemma 1. *Let G be a 3-edge-connected graph and let (o, f) be an all-positive nowhere-zero k-flow on G. Then all edges of G which receive value 1 and all strong 2-edges in (o, f) are deletable in o.*

For the proof we make use of the fact that if (o, f) is a k-flow on G, then in every edge-cut the sum of the flow values on the edges oriented in one direction equals the sum of the flow values on the edges oriented in the other direction. Due to page limits we omit the full proof.

The following theorem can be shown using Lemma 1 and careful application of the fact that every nowhere-zero 4-flow can be expressed as a combination of two 2-flows.

Theorem 1. *If G is a graph admitting a nowhere-zero 4-flow, then $fn(G) = 2$.*

Due to page limits we omit the proof.

It is known that a cubic graph is 3-edge-colourable if and only if it admits a nowhere-zero 4-flow. Therefore we have the following corollary.

Corollary 1. *If G is a 3-edge-connected 3-edge-colourable graph, then $fn(G) = 2$.*

Since every hamiltonian cubic graph is 3-edge-colourable, we have also shown the following conjecture by Barát and Blázsik [1].

Corollary 2. *If G is a 3-edge-connected cubic graph admitting a hamiltonian cycle, then $fn(G) = 2$.*

The following lemmas and theorems give two sufficient conditions for a cyclically 4-edge-connected cubic graph to have Frank number 2. These will be used in the algorithm in Sect. 3.

Lemma 2. *Let o be a strong orientation of a cubic graph G. Let $e_1 = u_1v_1$ and $e_2 = u_2v_2$ be two nonadjacent edges in G such that o contains $u_1 \rightarrow v_1$ and $u_2 \rightarrow v_2$. Assume that both e_1 and e_2 are deletable in o. Create a cubic graph G' from G by subdividing the edges e_1 and e_2 with vertices x_1 and x_2, respectively, and adding a new edge between x_1 and x_2. Let o' be the orientation of G' containing $u_1 \rightarrow x_1$, $x_1 \rightarrow v_1$, $x_1 \rightarrow x_2$, $u_2 \rightarrow x_2$, $x_2 \rightarrow v_2$ and such that $o'(e) = o(e)$ for all the remaining edges of G'. Then*

$$D(G', o') \supseteq (D(G, o) - \{e_1, e_2\}) \cup \{x_1v_1, x_1x_2, u_2x_2\}.$$

Due to page limits we omit the proof.

Let C be a 2-factor of G with exactly two odd circuits, say N_1 and N_2, (and possibly some even circuits). Let x_1x_2 be an edge of G such that $x_i \in N_i$ for $i \in \{1, 2\}$. Let $F = G - C$. Let M be a maximum matching in $C - \{x_1, x_2\}$. For $i \in \{1, 2\}$ denote by u_i and v_i the vertices of N_i which are adjacent to x_i. Denote by z_i the edge of N_i incident with u_i and not incident with x_i, denote by y_i the edge of N_i incident with v_i and not incident with x_i. The vertices of the graph $F - \{x_1x_2\} \cup M$ have degree 2, so the components of this graph are circuits. An orientation of these circuits is *consistent on N_i* if the edges z_i and y_i are oriented in the same direction with regards to N_i, see Fig. 1.

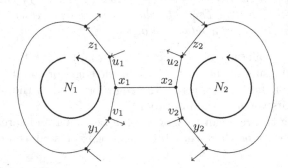

Fig. 1. Consistent orientation of the circuits from $F - \{x_1x_2\} \cup M$.

Theorem 2. *Let G be a cyclically 4-edge-connected cubic graph. Let C be a 2-factor of G with exactly two odd circuits, say N_1 and N_2, (and possibly some even circuits). Let $e = x_1x_2$ be an edge of G such that $x_1 \in N_1$ and $x_2 \in N_2$. For $i \in \{1, 2\}$ denote by u_i and v_i the vertices of N_i which are adjacent to x_i. Let $F = G - C$. Let M be a maximum matching in $C - \{x_1, x_2\}$. If there exists an orientation of the circuits in $F - \{x_1x_2\} \cup M$ such that the edges of $N_i \cap M$ incident with u_i and v_i are consistent on N_i for $i \in \{1, 2\}$, then $fn(G) = 2$.*

Due to space constains we omit the full proof.

The proof of the following Lemma is similar to that of Lemma 2, but is also omitted due to the page limit.

Lemma 3. *Let o be a strong orientation of a cubic graph G. Let $e_1 = u_1v_1$, $e_2 = u_2v_2$, and $f = w_2w_1$ be pairwise nonadjacent edges in G such that o contains $u_1 \to v_1$, $u_2 \to v_2$, and $w_2 \to w_1$. Let a cubic graph G' be created from G by performing the following steps:*

- *subdivide the edges e_1 and e_2 with the vertices x_1 and x_2, respectively,*
- *subdivide the edge w_1w_2 with the vertices y_1 and y_2 (in this order), and*
- *add the edges x_1y_1 and x_2y_2.*

Let o' be the orientation of G' containing $u_1 \to x_1$, $x_1 \to v_1$, $y_1 \to w_1$, $y_2 \to y_1$, $w_2 \to y_2$, $u_2 \to x_2$, $x_2 \to v_2$ and such that $o'(e) = o(e)$ for all the remaining edges of G' except for x_1y_1 and x_2y_2. Then

(a) *if o' contains $y_1 \to x_1$ and $x_2 \to y_2$, o' will be a strong orientation of G' and $D(G', o') \supseteq D(G, o) - \{e_1, e_2, f\} \cup \{u_1x_1, x_1y_1, y_1w_1, y_2w_2, x_2y_2, x_2v_2\}$ (Fig. 2(left));*
(b) *if o' contains $x_1 \to y_1$ and $y_2 \to x_2$, o' will be a strong orientation of G' and $D(G', o') \supseteq D(G, o) - \{e_1, e_2, f\} \cup \{x_1v_1, y_1y_2, u_2x_2\}$ (Fig. 2(right)).*

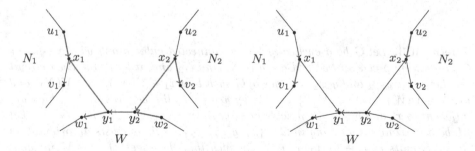

Fig. 2. A part of G' and orientation o' as defined in Lemma 3. The left-hand-side corresponds with the orientation of (a) and the right-hand-side corresponds with the orientation of (b). If the conditions of Lemma 3 are met the thick, blue edges will be deletable. (Color figure online)

Let C be a 2-factor of G with exactly two odd circuits, say N_1 and N_2 and at least one even circuit W. Let x_1y_1, y_1y_2 and y_2x_2 be edges of G such that

$x_i \in N_i$ and $y_i \in W$ for $i \in \{1,2\}$. Let $F = G - C$ and let M be a maximum matching in $C - \{x_1, x_2, y_1, y_2\}$. For $i \in \{1,2\}$ denote by u_i and v_i the vertices of N_i incident with x_i and by w_i the vertex of $W - \{y_1, y_2\}$ adjacent to y_i. Denote by z_i the edge of $N_i \cap M$ incident with u_i, by z_i' the edge of $N_i \cap M$ incident with v_i, by z the edge of $W \cap M$ incident with y_1 and by z' the edge of $W \cap M$ incident with y_2. The vertices of the graph $F - \{x_1 y_1, y_2 x_2\} \cup M$ have degree 2, so the components are circuits. An orientation of these circuits is *consistent on* N_i *and* W if the edges z_i and z_i' are oriented in the same direction with regards to N_i and the edges z and z' are oriented in the same direction with regards to W, see Fig. 3.

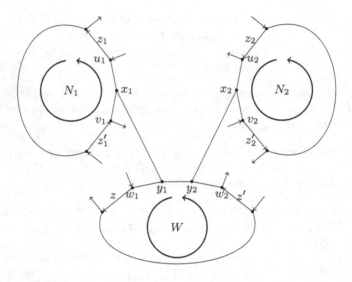

Fig. 3. Consistent orientation of the circuits from $F - \{x_1 y_1, y_2 x_2\} \cup M$.

Theorem 3. *Let G be a cyclically 4-edge-connected cubic graph with a 2-factor C containing precisely two odd circuits N_1 and N_2 and at least one even circuit W. Let $x_1 y_1$, $y_1 y_2$ and $y_2 x_2$ be edges of G such that $x_1 \in V(N_1)$, $x_2 \in V(N_2)$ and $y_1, y_2 \in V(W)$. For $i \in \{1,2\}$ denote by u_i and v_i the vertices of N_i which are adjacent to x_i and by w_i the neighbour of y_i in $W - \{y_1, y_2\}$. Let $F = G - C$. Let M be a maximum matching in $C - \{x_1, y_1, y_2, x_2\}$. If there exists an orientation of the circuits in $F - \{x_1 y_1, x_2 y_2\} \cup M$ such that the edges of $N_i \cap M$ incident with u_i and v_i are consistent on N_i for $i \in \{1,2\}$ and the edges of $W \cap M$ incident with y_1 and y_2 are consistent on W and $G \sim x_1 y_1 \sim x_2 y_2$ has no cycle-separating set of three edges $\{e_1, e_2, e_3\}$ with $e_1 \in \{u_1 v_1, u_2 v_2, w_1 w_2\}$ and $e_2, e_3 \in E(F - \{x_1 y_1, x_2 y_2\} \cup M)$, then $fn(G) \leq 2$.*

Due to page limits we omit the full proof.

3   Algorithm

We propose two algorithms for computationally verifying whether or not a given 3-edge-connected cubic graph has Frank number 2, i.e. a heuristic and an exact algorithm. Note that the Frank number for 3-edge-connected cubic graphs is always at least 2. Our algorithms are intended for graphs which are not 3-edge-colourable, since 3-edge-colourable graphs have Frank number 2 (cf. Corollary 1).

The first algorithm is a heuristic algorithm, which makes use of Theorem 2 and Theorem 3. Hence, it can only be used for cyclically 4-edge-connected cubic graphs. For every 2-factor in the input graph G, we verify if one of the configurations of these theorems is present. If that is the case, the graph has Frank number 2. The pseudocode of this algorithm can be found in Algorithm 1. In this algorithm we look at every 2-factor of G by generating every perfect matching and looking at its complement. We then count how many odd cycles there are in the 2-factor under investigation. If there are precisely two odd cycles, then we check for every edge connecting the two odd cycles whether or not the conditions of Theorem 2 hold. If they hold for one of these edges, we stop the algorithm and return that the graph has Frank number 2. If these conditions do not hold for any of these edges or if there are none, we check for all triples of edges x_1y_1, y_1y_2, y_2x_2, where x_1 and x_2 lie on the two odd cycles and y_1 and y_2 lie on some even cycle, whether the conditions of Theorem 3 hold. If they do, then G has Frank number 2 and we stop the algorithm.

The second algorithm is an exact algorithm for determining whether or not a 3-edge-connected cubic graph has Frank number 2. The pseudocode of this algorithm can be found in Algorithm 2. Due to space constraints, we omitted some technical subroutines from the pseudocode and focused on the main routines. These subroutines can be found in [6]. For a graph G, we start by looking at each of its strong orientations o and try to find a complementary orientation o' such that every edge is deletable in either o or o'. If there is a vertex in G for which none of its adjacent edges are deletable in o, then there exists no complementary orientation as no orientation of a cubic graph has three deletable edges incident to the same vertex. If o is suitable, we look for a complementary orientation using some tricks to reduce the search space. More precisely, we first we start with an empty *partial orientation*, i.e. a directed spanning subgraph of some orientation of G, and orient some edge. Note that we do not need to consider the opposite orientation of this edge, since an orientation of a graph in which all arcs are reversed has the same deletable edges as the original orientation. We then recursively orient edges of G that have not yet been oriented. After orienting an edge, the rules of Lemma 4 may enforce the orientation of edges which are not yet oriented. We orient them in this way before proceeding with the next edge. This heavily restricts the number edges which need to be added. As soon as a complementary orientation is found, we can stop the algorithm and return that the graph G has Frank number 2. If for all strong orientations of G no such complementary orientation is found, then the Frank number of G is higher than 2.

Since the heuristic algorithm is much faster than the exact algorithm, we will first apply the heuristic algorithm. After this we will apply the exact algorithm for those graphs for which the heuristic algorithm was unable to decide whether or not the Frank number is 2. In Sect. 3.1 we give more details on how many graphs pass this heuristic algorithm.

An implementation of these algorithms can be found on GitHub [5]. Our implementation uses bitvectors to store adjacency lists and lists of edges and bitoperations to efficiently manipulate these lists.

Theorem 4. *Let G be a cyclically 4-edge-connected cubic graph. If Algorithm 1 is applied to G and returns True, G has Frank number 2.*

Proof. Suppose the algorithm returns True for G. This happens in a specific iteration of the outer for loop corresponding to a perfect matching F. The complement of F is a 2-factor, say C, and since the algorithm returns True, C has precisely two odd cycles, say N_1 and N_2, and possibly some even cycles.

Suppose first that the algorithm returns True on Line 16. Then there is an edge $x_1 x_2$ in G with $x_1 \in V(N_1)$ and $x_2 \in V(N_2)$, a maximal matching M of $C - \{x_1, x_2\}$ and an orientation o of the cycles in $F - \{x_1 x_2\} \cup M$ such that o is consistent on N_1 and N_2. Now by Theorem 2 it follows that G has Frank number 2.

Now suppose that the algorithm returns True on Line 35. Then there are edges $x_1 y_1, y_1 y_2$ and $y_2 x_2$ such that $x_1 \in V(N_1)$, $x_2 \in V(N_2)$ and $y_1, y_2 \in V(W)$ where W is some even cycle in C. Since the algorithm returns True, there is a maximal matching M of $C - \{x_1, y_1, y_2, x_2\}$ and an orientation o of the cycles in $F - \{x_1 y_1, x_2 y_2\} \cup M$ such that o is consistent on N_1, N_2 and W. Denote the neighbours of x_1 and x_2 in C by u_1, v_1 and u_2, v_2, respectively and denote the neighbour of y_1 in $C - y_2$ by w_1 and the neighbour of y_2 in $C - y_1$ by w_2. Since no triple e, e_1, e_2, where $e \in \{u_1 x_1, w_1 y_1, u_2 x_2\}$, $e_1, e_2 \in E(F - \{x_1 y_1, x_2 y_2\} \cup M)$, is a cycle-separating edge-set of $G - \{x_1 y_1, x_2 y_2\}$, $G \sim x_1 y_1 \sim x_2 y_2$ has no cycle-separating edge-set $\{e, e_1, e_2\}$, where $e \in \{u_1 v_1, u_2 v_2, w_1 w_2\}$ and $e_1, e_2 \in E(F - \{x_1 y_1, x_2 y_2\} \cup M)$. Now by Theorem 3 it follows that G has Frank number 2. $\qquad \square$

We will use the following Lemma for the proof of the exact algorithm's correctness.

Lemma 4. *Let G be a cubic graph with $fn(G) = 2$ and let o and o' be two orientations of G such that every edge $e \in E(G)$ is deletable in either o or o'. Then the following hold for o':*

1. *every vertex has at least one incoming and one outgoing edge in o',*
2. *let $uv \notin D(G, o)$, then the remaining edges incident to u are one incoming and one outgoing in o',*
3. *let $uv, vw \notin D(G, o)$, then both are incoming to v or both are outgoing from v in o'.*

Due to page limits we omit the full proof.

Theorem 5. *Let G be a cubic graph. Algorithm 2 applied to G returns True if and only if G has Frank number 2.*

Algorithm 1. heuristicForFrankNumber2(Graph G)

1: **for** each perfect matching F **do**
2:     Store odd cycles of $C := G - F$ in $\mathcal{O} = \{N_1, \ldots, N_k\}$
3:     **if** $|\mathcal{O}|$ is not 2 **then**
4:         Continue with the next perfect matching
5:     **for all** vertices x_1 in N_1 **do**
6:         **if** x_1 has a neighbour x_2 in N_2 **then**
7:             // Test if Theorem 2 can be applied
8:             Store a maximal matching of $C - \{x_1, x_2\}$ in M
9:             Denote the neighbours of x_1 and x_2 in C by u_1, v_1 and u_2, v_2, respectively
10:             Create an empty partial orientation o of the cycles in $F - \{x_1, x_2\} \cup M$
11:             **for all** $x \in \{u_1, v_1, u_2, v_2\}$ **do**
12:                 **if** the cycle in $F - \{x_1, x_2\} \cup M$ containing x is not yet oriented **then**
13:                     Orient the cycle in $F - \{x_1 x_2\} \cup M$ containing x
14:                     Store this in o
15:             **if** o is consistent on N_1 and on N_2 **then**
16:                 **return** True // Theorem 2 applies
17:         **else if** x_1 has a neighbour y_1 on an even cycle W of C **then**
18:             **for** each neighbour y_2 of y_1 in C **do**
19:                 **if** y_2 has a neighbour x_2 in N_2 **then**
20:                     // Test if Theorem 3 can be applied
21:                     Store a maximal matching of $C - \{x_1, y_1, y_2, x_2\}$ in M
22:                     Denote the neighbours of x_1 and x_2 in C by u_1, v_1 and u_2, v_2
23:                     Denote the neighbour of y_1 in $C - y_2$ by w_1
24:                     Denote the neighbour of y_2 in $C - y_1$ by w_2
25:                     Create an empty partial orientation o of the cycles in
                        $F - \{x_1, y_1, y_2, x_2\} \cup M$
26:                     **for all** $x \in \{u_1, v_1, u_2, v_2, w_1, w_2\}$ **do**
27:                         **if** the cycle in $F - \{x_1, y_1, y_2, x_2\} \cup M$ containing x is not
                            oriented in o **then**
28:                             Orient the cycle in $F - \{x_1, y_1, y_2, x_2\} \cup M$ containing x
29:                             Store this in o
30:                     **if** o is consistent on N_1, N_2 and W **then**
31:                         // Check cycle-separating edge-set condition
32:                         **for all** pairs of edges e_1, e_2 in $F - \{x_1, y_1, y_2, x_2\} \cup M$ **do**
33:                             **for all** $e \in \{u_1 x_1, w_1 y_1, u_2 x_2\}$ **do**
34:                                 **if** $\{e, e_1, e_2\}$ is a cyclic edge-cut in $G - x_1 y_1 - x_2 y_2$
                                    then
35:                                     **return** True // Theorem 3 applies
36: **return** False

Algorithm 2. frankNumberIs2(Graph G)

1: **for all** orientations o of G **do**
2:     **if** o is not strong **then**
3:         Continue with next orientation
4:     Store deletable edges of o in a set D
5:     **for all** $v \in V(G)$ **do**
6:         **if** no edge incident to v is deletable **then**
7:             Continue with next orientation
8:     Create empty partial orientation o' of G
9:     Choose an edge xy in G and its orientation $x \to y$
10:    **if** not canAddArcsRecursively(G, D, o', $x \to y$) **then** // Algorithm 3
11:        Continue loop with next orientation
12:    **while** not all edges are oriented in o' **do**
13:        Store a copy of o' in o''
14:        Take an edge uv of G which is unoriented in o'
15:        **if** not canAddArcsRecursively(G, D, o', $u \to v$) **then**
16:            Reset o' using o''
17:            **if** not canAddArcsRecursively(G, D, o', $v \to u$) **then**
18:                Continue outer loop with next orientation
19:    **if** $D(G,o) \cup D(G,o') = E(G)$ **then**
20:        **return** True
21: **return** False

Proof. Suppose that frankNumberIs2(G) returns True. Then there exist two orientations o and o' for which $D(G,o) \cup D(G,o') = E(G)$. Hence, $fn(G) = 2$. Conversely, let $fn(G) = 2$. We will show that Algorithm 2 returns True. Let o_1 and o_2 be orientations of G such that every edge of G is deletable in either o_1 or o_2. If the algorithm returns True before we consider o_1 in the loop of Line 1, we are done. So, suppose we are in the iteration where o_1 is considered in the loop of Line 1. Without loss of generality assume that the orientation of xy we choose in Line 9 is in o_2. (If not, reverse all edges of o_2 to get an orientation with the same set of deletable edges.) Let o' be a partial orientation of G and assume that all oriented edges correspond to o_2. Let $u \to v$ be an arc of o_2. If $u \to v$ is present in o', then canAddArcsRecursively(G, $D(G,o)$, o', $u \to v$) (Algorithm 3) returns True and no extra edges become oriented in o'. If $u \to v$ is not present in o', it gets added on Line 8 of Algorithm 3, since the if-statement on Line 6 of Algorithm 3 will return True by Lemma 4. Note that this is the only place where an arc is added to o' in Algorithm 3. Hence, if we only call Algorithm 3 on arcs present in o_2, then all oriented edges of o' will always be oriented in the same way as in o_2. Now we will show that we only perform this call on arcs in o_2.

Again, suppose $u \to v$ is an arc in o_2, that it is not yet present in o' and that every oriented edge of o' has the same orientation as in o_2. Let u have two outgoing and no incoming arcs in o'. Let ux be the final unoriented edge incident to u. Then o_2 must have arc $x \to u$, otherwise it has three outgoing arcs from the same vertex. Let v have two incoming and no outgoing arcs in o'. Let vx be the final unoriented edge incident to v. Then o_2 must have arc $v \to x$, otherwise it has three incoming arcs to the same vertex.

Suppose uv is deletable in o_1. Let ux also be deletable in o_1. Denote the final edge incident to u by uy. Clearly, uy cannot be deletable in o_1, hence it is deletable in o_2. If o_2 contains $u \to x$, then uy is not deletable in o_2, hence, o_2 contains $x \to u$. Let vx be a deletable edge of o_1 and denote the final edge incident to v by vy. Since vy cannot be deletable in o_1, o_2 must contain arc $v \to x$. Suppose that the edges incident with u which are not uv are both not in $D(G, o_1)$. Then they must be oriented incoming to u in o_2. Similarly, if the edges incident with v which are not uv are both not in $D(G, o_1)$, they must both be outgoing from v in o_2.

Finally, suppose that uv is not a deletable edge in o_1. Suppose that o' still has one unoriented edge incident to u, say ux. If the other incident edges are one incoming and one outgoing from u, then o_2 contains the arc $u \to x$. Otherwise, uv cannot be deletable in o_2. Similarly, if o' still has one unoriented edge incident to v, say vx and the remaining incident edges are one incoming and one outgoing, then the arc $x \to v$ must be present in o_2. Otherwise, uv cannot be deletable in o_2. If ux is not deletable in o_1 $x \neq v$. Then o_2 contains the arc $u \to x$. Otherwise, not both of uv and ux can be deletable in o_2. Similarly, if vy is not deletable in o_1 and $y \neq u$, then o_2 must contain the arc $y \to v$. Otherwise, not both of uv and vy can be deletable in o_2.

It now inductively follows that during the execution of Algorithm 2 in the iteration of orientation o_1 on Line 19 that $o' = o_2$. Hence, the if-statement passes and the algorithm returns True.   □

Algorithm 3. canAddArcsRecursively(Graph G, Set D, Partial Orientation o', Arc $u \to v$)

1: // Check if $u \to v$ can be added and recursively orient edges for which the orientation is enforced by the rules of Lemma 4
2: **if** $u \to v$ is present in o' **then**
3:     **return** True
4: **if** $v \to u$ is present in o' **then**
5:     **return** False
6: **if** adding $u \to v$ violates rules of Lemma 4 **then** // Algorithm 4 in Appendix of [6]
7:     **return** False
8: Add $u \to v$ to o'
9: **if** the orientation of edges enforced by Lemma 4 yields a contradiction **then** // Algorithm 5 in Appendix of [6]
10:     **return** False
11: **return** True

3.1   Results

Since by Corollary 1 all 3-edge-connected 3-edge-colourable (cubic) graphs have Frank number 2, we will focus in this section on cubic graphs which are *not* 3-edge-colourable.

In [3] Brinkmann et al. determined all cyclically 4-edge-connected snarks up to order 34 and of girth at least 5 up to order 36. This was later extended with all cyclically 4-edge-connected snarks on 36 vertices as well [7]. These lists of snarks can be obtained from the House of Graphs [4] at: https://houseofgraphs. org/meta-directory/snarks. Using our implementation of Algorithms 1 and 2, we tested for all cyclically 4-edge-connected snarks up to 36 vertices if they have Frank number 2 or not. This led to the following result.

Proposition 1. *The Petersen graph is the only cyclically 4-edge-connected snark up to and including order 36 which has Frank number not equal to 2.*

This was done by first running our heuristic Algorithm 1 on these graphs. It turns out that there are few snarks in which neither the configuration of Theorem 2 nor the configuration of Theorem 3 are present. For example: for more than 99.97% of the cyclically 4-edge-connected snarks of order 36, Algorithm 1 is sufficient to determine that their Frank number is 2. Thus we only had to run our exact Algorithm 2 (which is significantly slower than the heuristic) on the graphs for which our heuristic algorithm failed. In total about 214 CPU days of computation time was required to prove Proposition 1 using Algorithm 1 and 2.

In [10] Jaeger defines a snark G to be a *strong snark* if for every edge $e \in E(G)$, $G \sim e$, i.e. the unique cubic graph such that $G - e$ is a subdivision of $G \sim e$, is not 3-edge-colourable. Hence, a strong snark containing a 2-factor which has precisely two odd cycles, has no edge e connecting those two odd cycles, i.e. the configuration of Theorem 2 cannot be present. Therefore, they might be good candidates for having Frank number greater than 2.

In [3] it was determined that there are 7 strong snarks on 34 vertices having girth at least 5, 25 strong snarks on 36 vertices having girth at least 5 and no strong snarks of girth at least 5 of smaller order. By Proposition 1, their Frank number is 2. In [2] it was determined that there are at least 298 strong snarks on 38 vertices having girth at least 5 and the authors of [2] speculate that this is the complete set. We found the following.

Observation 6. *The 298 strong snarks of order 38 determined in [2] have Frank number 2.*

These snarks can be obtained from the House of Graphs [4] by searching for the keywords "strong snark".

The configurations of Theorem 2 and Theorem 3 also cannot occur in snarks of *oddness* 4, i.e. the smallest number of odd cycles in a 2-factor of the graph is 4. Hence, these may also seem to be good candidates for having Frank number greater than 2. In [7,8] it was determined that the smallest snarks of girth at least 5 with oddness 4 and cyclic edge-connectivity 4 have order 44 and that there are precisely 31 such graphs of this order. We tested each of these and found the following.

Observation 7. *Let G be a snark of girth at least 5, oddness 4, cyclic edge-connectivity 4 and order 44. Then $fn(G) = 2$.*

These snarks of oddness 4 can be obtained from the House of Graphs [4] at https://houseofgraphs.org/meta-directory/snarks.

The correctness of our algorithm was shown in Theorem 4 and Theorem 5. We also performed several tests to verify that our implementations are correct. However, due to space constraints this had to be omitted. These tests can be found in [6].

References

1. Barát, J., Blázsik, Z.: Quest for graphs of Frank number 3 (2022). https://doi.org/10.48550/arXiv.2209.08804
2. Brinkmann, G., Goedgebeur, J.: Generation of cubic graphs and snarks with large girth. J. Graph Theory **86**(2), 255–272 (2017). https://doi.org/10.1002/jgt.22125
3. Brinkmann, G., Goedgebeur, J., Hägglund, J., Markström, K.: Generation and properties of snarks. J. Comb. Theory. Ser. B **103**(4), 468–488 (2013). https://doi.org/10.1016/j.jctb.2013.05.001
4. Coolsaet, K., D'hondt, S., Goedgebeur, J.: House of graphs 2.0: a database of interesting graphs and more. Discret. Appl. Math. **325**, 97–107 (2023). https://doi.org/10.1016/j.dam.2022.10.013
5. Goedgebeur, J., Máčajová, E., Renders, J.: Frank-Number (2023). https://github.com/JarneRenders/Frank-Number
6. Goedgebeur, J., Máčajová, E., Renders, J.: Frank number and nowhere-zero flows on graphs (2023). arXiv:2305.02133 [math.CO]
7. Goedgebeur, J., Máčajová, E., Škoviera, M.: Smallest snarks with oddness 4 and cyclic connectivity 4 have order 44. ARS Math. Contemp. **16**(2), 277–298 (2019). https://doi.org/10.26493/1855-3974.1601.e75
8. Goedgebeur, J., Máčajová, E., Škoviera, M.: The smallest nontrivial snarks of oddness 4. Discret. Appl. Math. **277**, 139–162 (2020). https://doi.org/10.1016/j.dam.2019.09.020
9. Hörsch, F., Szigeti, Z.: Connectivity of orientations of 3-edge-connected graphs. Eur. J. Comb. **94**, 103292 (2021). https://doi.org/10.1016/j.ejc.2020.103292
10. Jaeger, F.: A survey of the cycle double cover conjecture. In: Alspach, B.R., Godsil, C.D. (eds.) North-Holland mathematics studies, annals of discrete mathematics (27): cycles in graphs, vol. 115, pp. 1–12. North-Holland (1985). https://doi.org/10.1016/S0304-0208(08)72993-1
11. Nash-Williams, C.S.J.A.: On orientations, connectivity and odd-vertex-pairings in finite graphs. Canad. J. Math. **12**, 555–567 (1960). https://doi.org/10.4153/CJM-1960-049-6. publisher: Cambridge University Press
12. Seymour, P.D.: On multi-colourings of cubic graphs, and conjectures of Fulkerson and Tutte. Proc. London Math. Soc. **3**(3), 423–460 (1979). https://doi.org/10.1112/plms/s3-38.3.423

On the Minimum Number of Arcs
in 4-Dicritical Oriented Graphs

Frédéric Havet[1], Lucas Picasarri-Arrieta[1(✉)], and Clément Rambaud[1,2]

[1] Université Côte d'Azur, CNRS, Inria, I3S, Sophia Antipolis, France
{frederic.havet,lucas.picasarri-arrieta}@inria.fr
[2] DIENS, École Normale Supérieure, CNRS, PSL University, Paris, France
clement.rambaud@ens.psl.eu

Abstract. We prove that every 4-dicritical oriented graph on n vertices has at least $(\frac{10}{3} + \frac{1}{51})n - 1$ arcs.

Keywords: dichromatic number · oriented graphs · directed graphs · dicritical · density

1 Introduction

Let G be a graph. We denote by $V(G)$ its vertex set and by $E(G)$ its edge set; we set $n(G) = |V(G)|$ and $m(G) = |E(G)|$. A k-**colouring** of G is a function $\varphi : V(G) \to [k]$. It is **proper** if for every edge $uv \in E(G)$, $\varphi(u) \neq \varphi(v)$. The smallest integer k such that G has a proper k-colouring is the **chromatic number**, and is denoted by $\chi(G)$. Since χ is non decreasing with respect to the subgraph relation, it is natural to consider the minimal graphs (for this relation) which are not $(k-1)$-colourable. Following this idea, Dirac defined k-**critical** graphs as the graphs G with $\chi(G) = k$ and $\chi(H) < k$ for every proper subgraph H of G. A first property of k-critical graph is that their minimum degree is at least $k-1$. Indeed, if a vertex v has degree at most $k-2$, then a $(k-1)$-colouring of $G - v$ can be easily extended to G, contradicting the fact that $\chi(G) = k$. As a consequence, the number of edges in a k-critical graph is at least $\frac{k-1}{2}n$. This bound is tight for complete graphs and odd cycles, but Dirac [3] proved an inequality of the form $m \geq \frac{k-1+\varepsilon_k}{2}n - c_k$ for every n-vertex k-critical graph with m edges, for some c_k and $\varepsilon_k > 0$. This shows that, for n sufficiently large, the average degree of a k-critical graph is at least $k - 1 + \varepsilon_k$. This initiated the quest after the best lower bound on the number of edges in n-vertex k-critical graphs. This problem was almost completely solved by Kostochka and Yancey in 2014 [11].

Theorem 1 (Kostochka and Yancey [11]). *Every k-critical graph on n vertices has at least $\frac{1}{2}(k - \frac{2}{k-1})n - \frac{k(k-3)}{2(k-1)}$ edges. For every k, this bound is tight for infinitely many values of n.*

Kostochka and Yancey [12] also characterised k-critical graphs for which this inequality is an equality, and all of them contain a copy of K_{k-2}, the complete graph on $k - 2$ vertices. This motivated the following conjecture of Postle [13].

D. Paulusma and B. Ries (Eds.): WG 2023, LNCS 14093, pp. 376–387, 2023.
https://doi.org/10.1007/978-3-031-43380-1_27

Conjecture 2 (Postle [13]). For every integer $k \geq 4$, there exists $\varepsilon_k > 0$ such that every k-critical K_{k-2}-free graph G on n vertices has at least $\frac{1}{2}\left(k - \frac{2}{k-1} + \varepsilon_k\right)n - \frac{k(k-3)}{2(k-1)}$ edges.

For $k = 4$, the conjecture trivially holds as there is no K_2-free 4-critical graph. Moreover, this conjecture has been confirmed for $k = 5$ by Postle [13], for $k = 6$ by Gao and Postle [5], and for $k \geq 33$ by Gould, Larsen, and Postle [6].

Let D be a digraph. We denote by $V(D)$ its vertex set and by $A(D)$ its arc set; we set $n(D) = |V(D)|$ and $m(D) = |E(D)|$. A k-**colouring** of D is a function $\varphi : V(D) \rightarrow [k]$. It is a k-**dicolouring** if every directed cycle C in D is not monochromatic for φ (that is $|\varphi(V(C))| > 1$). Equivalently, it is a k-dicolouring if every colour class induces an acyclic subdigraph. The smallest integer k such that D has a k-dicolouring is the **dichromatic number** of D and is denoted by $\vec{\chi}(D)$.

A **digon** in D is a pair of opposite arcs between two vertices. Such a pair of arcs $\{uv, vu\}$ is denoted by $[u, v]$. We say that D is a **bidirected graph** if every pair of adjacent vertices forms a digon. In this case, D can be viewed as obtained from an undirected graph G by replacing each edge $\{u, v\}$ of G by the digon $[u, v]$. We say that D is a bidirected G, and we denote it by $\overleftrightarrow{G}$. Observe that $\chi(G) = \vec{\chi}(\overleftrightarrow{G})$. Thus every statement on proper colouring of undirected graphs can be seen as a statement on dicolouring of bidirected graphs.

Exactly as in the undirected case, one can define k-**dicritical** digraphs to be digraphs D with $\vec{\chi}(D) = k$ and $\vec{\chi}(H) < k$ for every proper subdigraph H of D. It is easy to check that if G is a k-critical graph, then $\overleftrightarrow{G}$ is k-dicritical. Kostochka and Stiebitz [10] conjectured that the k-dicritical digraphs with the minimum number of arcs are bidirected graphs. Thus they conjectured the following generalisation of Theorem 1 to digraphs.

Conjecture 3 (Kostochka and Stiebitz [10]). Let $k \geq 2$. Every k-dicritical digraph on n vertices has at least $(k - \frac{2}{k-1})n - \frac{k(k-3)}{k-1}$ arcs. Moreover, equality holds only if D is bidirected.

In the case $k = 2$, this conjecture is easy and weak as it states that a 2-dicritical digraph on n vertices has at least two arcs, while, for all $n \geq 2$, the unique 2-dicritical digraph of order n is the directed n-cycle which has n arcs. The case $k = 3$ of the conjecture has been confirmed by Kostochka and Stiebitz [10]. Using a Brooks-type result for digraphs due to Harutyunyan and Mohar [7], they proved the following: if D is a 3-dicritical digraph of order $n \geq 3$, then $m(D) \geq 2n$ and equality holds if and only if n is odd and D is a bidirected odd cycle. The conjecture has also been proved for $k = 4$ by Kostochka and Stiebitz [10]. However, the conjecture is open for every $k \geq 5$. Recently, this problem has been investigated by Aboulker and Vermande [2] who proved the weaker bound $(k - \frac{1}{2} - \frac{2}{k-1})n - \frac{k(k-3)}{k-1}$ for the number of arcs in an n-vertex k-dicritical digraph.

For integers k and n, let $d_k(n)$ denote the minimum number of arcs in a k-dicritical digraph of order n. By the above observations, $d_2(n) = n$ for all $n \geq 2$, and $d_3(n) \geq 2n$ for all possible n, and equality holds if and only if n is odd and $n \geq 3$. Moreover, if n is even then $d_3(n) = 2n + 1$ (see [1]).

Kostochka and Stiebitz [9] showed that if a k-critical graph G is triangle-free (that is has no cycle of length 3), then $m(G)/n(G) \geq k - o(k)$ as $k \to +\infty$. Informally, this means that the minimum average degree of a k-critical triangle-free graph is (asymptotically) twice the minimum average degree of a k-critical graph. Similarly to this undirected case, it is expected that the minimum number of arcs in a k-dicritical digraph of order n is larger than $d_k(n)$ if we impose this digraph to have no short directed cycles, and in particular if the digraph is an **oriented graph**, that is a digraph with no digon. Let $o_k(n)$ denote the minimum number of arcs in a k-dicritical oriented graph of order n (with the convention $o_k(n) = +\infty$ if there is no k-dicritical oriented graph of order n). Clearly $o_k(n) \geq d_k(n)$.

Conjecture 4 (Kostochka and Stiebitz [10]). For any $k \geq 3$, there is a constant $\alpha_k > 0$ such that $o_k(n) > (1 + \alpha_k)d_k(n)$ for n sufficiently large.

For $k = 3$, this conjecture has been recently confirmed by Aboulker, Bellitto, Havet, and Rambaud [1] who proved that $o_3(n) \geq (2 + \frac{1}{3})n + \frac{2}{3}$.

In view of Conjecture 2, Conjecture 4 can be generalized to $\overleftrightarrow{K_{k-2}}$-free digraphs.

Conjecture 5. For any $k \geq 4$, there is a constant $\beta_k > 0$ such that every k-dicritical $\overleftrightarrow{K_{k-2}}$-free digraph D on n vertices has at least $(1 + \beta_k)d_k(n)$ arcs.

Together with Conjecture 3, this conjecture would imply the following generalisation of Conjecture 2.

Conjecture 6. For every integer $k \geq 4$, there exists $\varepsilon_k > 0$ such that every k-dicritical $\overleftrightarrow{K_{k-2}}$-free digraph D on n vertices has at least $(k - \frac{2}{k-1} + \varepsilon_k)n - \frac{k(k-3)}{k-1}$ arcs.

A $\overleftrightarrow{K_2}$-free digraph is an oriented graph, and there are infinitely many 4-dicritical oriented graphs. Thus, while Conjecture 2 holds vacuously for $k = 4$, this is not the case for Conjecture 6. In this paper, we prove that Conjectures 4, 5, and 6 hold for $k = 4$.

Theorem 7. *If $\vec{G}$ is a 4-dicritical oriented graph, then*

$$m(\vec{G}) \geq \left(\frac{10}{3} + \frac{1}{51}\right)n(\vec{G}) - 1.$$

To prove Theorem 7, we use an approach similar to the proof of the case $k = 5$ of Conjecture 2 by Postle [13]. This proof is based on the potential method, which was first popularised by Kostochka and Yancey [11] when they proved Theorem 1. The idea is to prove a more general result on every 4-dicritical digraphs that takes into account the digons.

With a slight abuse, we call **digon** a subdigraph isomorphic to $\overleftrightarrow{K_2}$, the bidirected complete graph on two vertices. We also call **bidirected triangle** a subdigraph isomorphic to $\overleftrightarrow{K_3}$, the bidirected complete graph on three vertices. A **packing** of digons and bidirected triangles is a set of vertex-disjoint digons and bidirected triangles. To take into account the digons, we define a parameter $T(D)$ as follows.

$T(D) = \max\{d + 2t \mid \text{there exists a packing of } d \text{ digons and } t \text{ bidirected triangles}\}$

Clearly, $T(D) = 0$ if and only if D is an oriented graph.

Let ε, δ be fixed non-negative real numbers. We define the **potential** (with respect to ε and δ) of a digraph D to be

$$\rho(D) = \left(\frac{10}{3} + \varepsilon\right) n(D) - m(D) - \delta T(D).$$

Thus Theorem 7 can be rephrased as follows.

Theorem 7. *Set $\varepsilon = \frac{1}{51}$ and $\delta = 6\varepsilon = \frac{2}{17}$. If $\vec{G}$ is a 4-dicritical oriented graph, then $\rho(\vec{G}) \leq 1$.*

In fact, we prove a more general statement which holds for every 4-dicritical digraph (with or without digons), except for some exceptions called the 4-**Ore digraphs**. Those digraphs, which are formally defined in Sect. 2, are the bidirected graphs whose underlying graph is one of the 4-critical graphs reaching equality in Theorem 1. In particular, every 4-Ore digraph D has $\frac{10}{3}n(D) - \frac{4}{3}$ arcs. Moreover, the statement holds for all non-negative constants ε and δ satisfying the following inequalities:

- $\delta \geq 6\varepsilon$;
- $3\delta - \varepsilon \leq \frac{1}{3}$;

Theorem 8. *Let $\varepsilon, \delta \geq 0$ be constants satisfying the aforementioned inequalities. If D is a 4-dicritical digraph with n vertices, then*

(i) $\rho(D) \leq \frac{4}{3} + \varepsilon n - \delta\frac{2(n-1)}{3}$ if D is 4-Ore, and
(ii) $\rho(D) \leq 1$ otherwise.

In order to provide some intuition to the reader, let us briefly describe the main ideas of our proof. We will consider a minimum counterexample D to Theorem 8, and show that every subdigraph of D must have large potential. To do so, we need to construct some smaller 4-dicritical digraphs to leverage the minimality of D. These smaller 4-dicritical digraphs will be constructed by identifying some vertices of D. This is why, in the definition of the potential, we consider $T(D)$ instead of the number of digons: when identifying a set of vertices, the number of digons may be arbitrary larger in the resulting digraph, but $T(D)$ increases at most by 1. Using the fact that every subdigraph of D has large potential, we will prove that some subdigraphs are forbidden in D. Using this, we get the final contradiction by a discharging argument.

In addition to Theorem 7, Theorem 8 has also the following consequence when we take $\varepsilon = \delta = 0$.

Corollary 9. *If D is a 4-dicritical digraph, then $m(D) \geq \frac{10}{3}n(D) - \frac{4}{3}$. Moreover, equality holds if and only if D is 4-Ore, otherwise $m(D) \geq \frac{10}{3}n(D) - 1$.*

This is a slight improvement on a result of Kostochka and Stiebitz [10] who proved the inequality $m(D) \geq \frac{10}{3}n(D) - \frac{4}{3}$ without characterising the equality case.

Another interesting consequence of our result is the following bound on the number of vertices in a 4-dicritical oriented graph embedded on a fixed surface. Since a graph on n vertices embedded on a surface of Euler characteristic c has at most $3n - 3c$ edges, we immediately deduce the following from Theorem 7.

Corollary 10. *If $\vec{G}$ is a 4-dicritical oriented graph embedded on a surface of Euler characteristic c, then $n(\vec{G}) \leq \frac{17}{6}(1 - 3c)$.*

The previous best upper bound was $n(\vec{G}) \leq 4 - 9c$ [10].

In Sect. 2 we prove some first preliminary results on 4-Ore digraphs, before proving Theorem 8 in Sect. 3. For the sake of brevity, we skip the proofs of lemmas and claims. All the detailed proofs can be found in [8].

2   The 4-Ore Digraphs and Their Properties

We start with a few notations. We denote by $[\![x_1, \ldots, x_n]\!]$ the bidirected path with vertex set $\{x_1, \ldots, x_n\}$ in this order. If $x_1 = x_n$, $[\![x_1, \ldots, x_n]\!]$ denotes the bidirected cycle of order n with cyclic order $x_1, \ldots, x_n$. If D is a digraph, for any $X \subseteq V(D)$, $D - X$ is the subdigraph induced by $V(D) \setminus X$. We abbreviate $D - \{x\}$ into $D - x$. Moreover, for any $F \subseteq V(D) \times V(D)$, $D \setminus F$ is the subdigraph $(V(D), A(D) \setminus F)$ and $D \cup F$ is the digraph $(V(D), A(D) \cup F)$.

Let D_1, D_2 be two bidirected graphs, $[x, y] \subseteq A(D_1)$, and $z \in V(D_2)$. An **Ore-composition** D of D_1 and D_2 with **replaced digon** $[x, y]$ and **split vertex** z is a digraph obtained by removing $[x, y]$ of D_1 and z of D_2, and adding the set of arcs $\{xz_1 \mid zz_1 \in A(D_2) \text{ and } z_1 \in Z_1\}$, $\{z_1x \mid z_1z \in A(D_2) \text{ and } z_1 \in Z_1\}$, $\{yz_2 \mid zz_2 \in A(D_2) \text{ and } z_2 \in Z_2\}$, $\{z_2y \mid z_2z \in A(D_2) \text{ and } z_2 \in Z_2\}$, where (Z_1, Z_2) is a partition of $N_{D_2}(z)$ into non-empty sets. We call D_1 the **digon side** and D_2 the **split side** of the Ore-composition. The class of the **4-Ore digraphs** is the smallest class containing $\overleftrightarrow{K_4}$ which is stable under Ore-composition. See Fig. 1 for an example of a 4-Ore digraph. Observe that all the 4-Ore-digraphs are bidirected.

Proposition 11 (Dirac [4]). *4-Ore digraphs are 4-dicritical.*

Proof. One can easily show that a bidirected digraph is 4-dicritical if and only if its undirected underlying graph is 4-critical. Then the result follows from Theorem 1 in [4]. □

Lemma 12. *Let D be a 4-dicritical bidirected digraph and $v \in V(D)$. Let (N_1^+, N_2^+) and (N_1^-, N_2^-) be two partitions of $N(v)$. Consider D' the digraph with vertex set $V(D) \setminus \{v\} \cup \{v_1, v_2\}$ with $N^+(v_i) = N_i^+, N^-(v_i) = N_i^-$ for $i = 1, 2$ and $D'\langle V(D) \setminus \{v\}\rangle = D - v$. Then D' has a 3-dicolouring with v_1 and v_2 coloured the same except if $N_1^+ = N_1^-$ (that is D' is bidirected).*

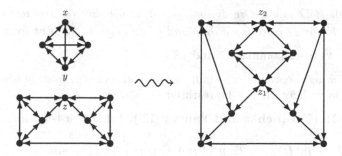

Fig. 1. An example of a 4-Ore digraph obtained by an Ore-composition of two smaller 4-Ore digraphs, with replaced digon $[x, y]$ and split vertex z.

Lemma 13. *Let D be a digraph. If v is a vertex of D, then $T(D-v) \geq T(D)-1$.*

Lemma 14. *If D_1, D_2 are two digraphs, and D is an Ore-composition of D_1 and D_2, then $T(D) \geq T(D_1) + T(D_2) - 2$. Moreover, if D_1 or D_2 is isomorphic to $\overleftrightarrow{K_4}$, then $T(D) \geq T(D_1) + T(D_2) - 1$.*

Lemma 15. *If D is 4-Ore, then $T(D) \geq \frac{2}{3}(n(D) - 1)$.*

Let D be a digraph. A **diamond** in D is a subdigraph isomorphic to $\overleftrightarrow{K_4}$ minus a digon $[u, v]$, with vertices different from u and v having degree 6 in D. An **emerald** in D is a subdigraph isomorphic to $\overleftrightarrow{K_3}$ whose vertices have degree 6 in D.

Let R be an induced subdigraph of D with $n(R) < n(D)$. The **boundary** of R in D, denoted by $\partial_D(R)$, or simply $\partial(R)$ when D is clear from the context, is the set of vertices of R having a neighbour in $V(D) \setminus R$. We say that R is **Ore-collapsible** if the boundary of R contains exactly two vertices u and v and $R \cup [u, v]$ is 4-Ore.

Lemma 16. *If D is 4-Ore and $v \in V(D)$, then there exists either an Ore-collapsible subdigraph of D disjoint from v or an emerald of D disjoint from v.*

Lemma 17. *If $D \neq \overleftrightarrow{K_4}$ is 4-Ore and T is a copy of $\overleftrightarrow{K_3}$ in D, then there exists either an Ore-collapsible subdigraph of D disjoint from T or an emerald of D disjoint from T.*

Lemma 18. *If R is an Ore-collapsible induced subdigraph of a 4-Ore digraph D, then there exists a diamond or an emerald of D whose vertices lie in $V(R)$.*

Lemma 19. *If D is a 4-Ore digraph and v is a vertex in D, then D contains a diamond or an emerald disjoint from v.*

Proof. Follows from Lemmas 16 and 18.

Lemma 20. *If D is a 4-Ore digraph and T is a bidirected triangle in D, then either $D = \overleftrightarrow{K_4}$ or D contains a diamond or an emerald disjoint from T.*

Proof. Follows from Lemmas 17 and 18.

The following theorem was formulated for undirected graphs, but by replacing every edge by a digon, it can be restated as follows:

Theorem 21 (Kostochka and Yancey [12]). *Let D be a 4-dicritical bidirected digraph.*
If $\frac{10}{3}n(D) - m(D) > 1$, then D is 4-Ore and $\frac{10}{3}n(D) - m(D) = \frac{4}{3}$.

Lemma 22. *If D is a 4-Ore digraph with n vertices, then $\rho(D) \leq \frac{4}{3} + \varepsilon n - \delta \frac{2(n-1)}{3}$.*

Proof. Follows from Theorem 21 and Lemma 15.

Lemma 23 (Kostochka and Yancey [12], Claim 16). *Let D be a 4-Ore digraph. If $R \subseteq D$ and $5 \leq n(R) < n(D)$, then $\frac{10}{3}n(R) - m(R) \geq \frac{10}{3}$.*

Lemma 24. *Let D be a 4-Ore digraph obtained from a copy J of $\overleftrightarrow{K_4}$ by successive Ore-compositions with 4-Ore digraphs, vertices and digons in J being always on the digon side. Let $[u, v]$ be a digon in $D\langle V(J)\rangle$. For every 3-dicolouring φ of $D \setminus [u, v]$, vertices in $V(J)$ receives distinct colours except u and v.*

Lemma 25. *Let D be a 4-Ore digraph obtained from a copy J of $\overleftrightarrow{K_4}$ by successive Ore-compositions with 4-Ore digraphs, vertices and digons in J being always on the digon side. Let v be a vertex in $V(J)$. For every 3-dicolouring φ of $D - v$, vertices in J receives distinct colours.*

3   Proof of Theorem 8

Let D be a 4-dicritical digraph, R be an induced subdigraph of D with $4 \leq n(R) < n(D)$ and φ a 3-dicolouring of R. The φ-**identification** of R in D, denoted by $D_\varphi(R)$ is the digraph obtained from D by identifying for each $i \in [3]$ the vertices coloured i in $V(R)$ to a vertex x_i, adding the digons $[x_i, x_j]$ for all $i, j \in [3]$ and then deleting loops and parallel arcs. Observe that $D_\varphi(R)$ is not 3-dicolourable. Indeed, assume for a contradiction that $D_\varphi(R)$ has a 3-dicolouring φ'. Since $V(R)$ induces a $\overleftrightarrow{K_3}$, we may assume without loss of generality that $\varphi'(x_i) = i$ for $i \in [3]$. Consider the 3-colouring φ'' of D defined by $\varphi''(v) = \varphi'(v)$ if $v \notin R$ and $\varphi''(v) = \varphi(v)$ if $v \in R$. One easily checks that φ'' is a 3-dicolouring of D, a contradiction to the fact that $\vec{\chi}(D) \geq 4$.

Now let W be a 4-dicritical subdigraph of $D_\varphi(R)$ and $X = \{x_1, x_2, x_3\}$. Then we say that $R' = D\langle(V(W) \setminus X) \cup R\rangle$ is the **dicritical extension** of R with **extender** W. We call $X_W = X \cap V(W)$ the **core** of the extension. Note that X_W is not empty, because W is not a subdigraph of D. Thus $1 \leq |X_W| \leq 3$. See Fig. 2 for an example of a φ-identification and a dicritical extension.

Let D be a counterexample to Theorem 8 with minimum number of vertices. By Lemma 22, D is not 4-Ore. Thus $\rho(D) > 1$.

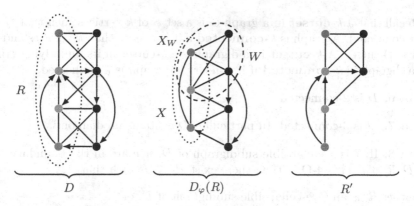

Fig. 2. A 4-dicritical digraph D together with an induced subdigraph R of D and φ a 3-dicolouring of R, the φ-identification $D_\varphi(R)$ of R in D and the dicritical extension R' of R with extender W and core X_W. For clarity, the digons are represented by undirected edges.

Claim 1. If $\tilde{D}$ is a 4-dicritical digraph with $n(\tilde{D}) < n(D)$, then $\rho(\tilde{D}) \leq \frac{4}{3} + 4\varepsilon - 2\delta$.

Claim 2. Let R be a subdigraph of D with $4 \leq n(R) < n(D)$. If R' is a dicritical extension of R with extender W and core X_W, then

$$\rho(R') \leq \rho(W) + \rho(R) - \left(\rho(\overleftrightarrow{K}_{|X_W|}) + \delta \cdot T(\overleftrightarrow{K}_{|X_W|})\right) + \delta \cdot (T(W) - T(W - X_W))$$

and in particular

$$\rho(R') \leq \rho(W) + \rho(R) - \frac{10}{3} - \varepsilon + \delta.$$

Claim 3. If R is a subdigraph of D with $4 \leq n(R) < n(D)$, then $\rho(R) \geq \rho(D) + 2 - 3\varepsilon + \delta > 3 - 3\varepsilon + \delta$.

As a consequence of Claim 3, any subdigraph (proper or not) of size at least 4 has potential at least $\rho(D)$.

We say that an induced subdigraph R of D is **collapsible** if, for every 3-dicolouring φ of R, its dicritical extension R' (with extender W and core X_W) is D, has core of size 1 (i.e. $|X_W| = 1$), and the border $\partial_D(R)$ of R is monochromatic in φ.

Claim 4. Let R be an induced subdigraph of D and φ a 3-dicolouring of R such that $\partial(R)$ is not monochromatic in φ. If D is a dicritical extension of R dicoloured by φ with extender W and core X_W with $|X_W| = 1$, then

$$\rho(R) \geq \rho(D) + 3 - 3\varepsilon + \delta.$$

Claim 5. If R is a subdigraph of D with $4 \leq n(R) < n(D)$ and R is not collapsible, then $\rho(R) \geq \rho(D) + \frac{8}{3} - \varepsilon - \delta > \frac{11}{3} - \varepsilon - \delta$.

Recall that a k-**cutset** in a graph G is a set S of k vertices such that $G - S$ is not connected. A graph is k-**connected** if it has more than k vertices and has no $(k-1)$-cutset. A k-**cutset** in a digraph is a k-cutset in its underlying graph, and a digraph is k-**connected** if its underlying graph is k-connected.

Claim 6. D is 2-connected.

Claim 7. D is 3-connected. In particular, D contains no diamond.

Claim 8. If R is a collapsible subdigraph of D, u, v are in the boundary of R and $D\langle R \rangle \cup [u, v]$ is 4-Ore, then there exists $R' \subseteq R$ such that

(i) either R' is an Ore-collapsible subdigraph of D, or
(ii) R' is an induced subdigraph of R, $n(R') < n(R)$, and there exist u', v' in $\partial_D(R')$ such that $R' \cup [u', v']$ is 4-Ore.

Claim 9. If R is a subdigraph of D with $n(R) < n(D)$ and $u, v \in V(R)$, then $R \cup [u, v]$ is 3-dicolourable. As a consequence, there is no collapsible subdigraph in D.

Claim 10. If R is a subdigraph of D with $n(R) < n(D)$ and $u, v, u', v' \in R$, then $R \cup \{uv, u'v'\}$ is 3-dicolourable. In particular, D contains no copy of $\overleftrightarrow{K_4}$ minus two arcs.

For any $v \in V(D)$, we denote by $n(v)$ its number of neighbours, that is $n(v) = |N^+(u) \cup N^-(v)|$, and by $d(v)$ its number of incident arcs, that is $d(v) = d^+(v) + d^-(v)$.

Claim 11. Vertices of degree 6 in D have either three or six neighbours.

Claim 12. There is no bidirected triangle containing two vertices of degree 6. In particular, D contains no emerald.

So now we know that D contains no emerald, and no diamond by Claim 7.

Claim 13. If R is an induced subdigraph of D with $4 \leq n(R) < n(D)$, then $\rho(R) \geq \rho(D) + 3 + 3\varepsilon - 3\delta$, except if $D - R$ contains a single vertex which has degree 6 in D.

In D, we say that a vertex v is a **simple in-neighbour** (resp. **simple out-neighbour**) if v is a in-neighbour (resp. out-neighbour) of u and $[u, v]$ is not a digon in D. If v is a simple in-neighbour or simple out-neighbour of u, we simply say that v is a **simple neighbour** of u.

Claim 14. Vertices of degree 7 have seven neighbours. In other words, every vertex of degree 7 has only simple neighbours.

The 8^+-**valency** of a vertex v, denoted by $\nu(v)$, is the number of arcs incident to v and a vertex of degree at least 8.

Let D_6 be the subdigraph of D induced by the vertices of degree 6 incident to digons. Let us describe the connected components of D_6 and their neighbourhoods. Remember that vertices of degree 7 are incident to no digon by Claim 14, and so they do not have neighbours in $V(D_6)$. If v is a vertex in D_6, we define its **neighbourhood valency** to be the sum of the 8^+-valency of its neighbours of degree at least 8. We denote the neighbourhood valency of v by $\nu_N(v)$.

Claim 15. If $[x, y]$ is a digon and both x and y have degree 6, then either

(i) the two neighbours of y distinct from x have degree at least 8, or
(ii) the two neighbours of x distinct from y have degree at least 8 and $\nu_N(x) \geq 4$.

Claim 16. Let C be a connected component of D_6. It is either

(i) a single vertex, or
(ii) a bidirected path on two vertices, or
(iii) a bidirected path on three vertices, whose extremities have neighbourhood valency at least 4, or
(iv) a star on four vertices, whose non-central vertices have neighbourhood valency at least 4.

An arc xy is said to be **out-chelou** if

(i) $yx \notin A(D)$, and
(ii) $d^+(x) = 3$, and
(iii) $d^-(y) = 3$, and
(iv) there exists $z \in N^-(y) \setminus N^+(y)$ distinct from x.

Symmetrically, we say that an arc xy is **in-chelou** if yx is out-chelou in the digraph obtained from D by reversing every arc. See Fig. 3 for an example of an out-chelou arc.

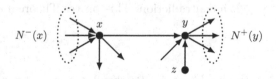

Fig. 3. An example of an out-chelou arc xy.

Claim 17. There is no out-chelou arc and no in-chelou arc in D.

We now use the discharging method. For every vertex v, let $\sigma(v) = \frac{\delta}{|C|}$ if v has degree 6 and is in a component C of D_6 of size at least 2, and $\sigma(v) = 0$ otherwise. Clearly $T(D)$ is at least the number of connected components of size at least 2 of D_6 so $\sum_{v \in V(D)} \sigma(v) \leq \delta T(D)$. We define the **initial charge** of v to be $w(v) = \frac{10}{3} + \varepsilon - \frac{d(v)}{2} - \sigma(v)$. We have

$$\rho(D) \leq \sum_{v \in V(D)} w(v).$$

We now redistribute this total charge according to the following rules:

(R1) A vertex of degree 6 incident to no digon sends $\frac{1}{12} - \frac{\varepsilon}{8}$ to each of its neighbours.

(R2) A vertex of degree 6 incident to digons sends $\frac{2}{d(v)-\nu(v)}(-\frac{10}{3} + \frac{d(v)}{2} - \varepsilon)$ to each neighbour v of degree at least 8 (so $\frac{1}{d(v)-\nu(v)}(-\frac{10}{3} + \frac{d(v)}{2} - \varepsilon)$ via each arc of the digon).

(R3) A vertex of degree 7 with $d^-(v) = 3$ (resp. $d^+(v) = 3$) sends $\frac{1}{12} - \frac{\varepsilon}{8}$ to each of its in-neighbours (resp. out-neighbours).

For every vertex v, let $w^*(v)$ be the final charge of v.

Claim 18. If v has degree at least 8, then $w^*(v) \leq 0$.

Claim 19. If v has degree 7, then $w^*(v) \leq 0$.

Claim 20. If v is a vertex of degree 6 incident to no digon, then $w^*(v) \leq 0$.

Claim 21. Let v be a vertex in D_6 having at least two neighbours of degree at least 8. Then $w^*(v) \leq 0$. Moreover, if v is not an isolated vertex in D_6 and $\nu_N(v) \geq 4$, then $w^*(v) \leq -\frac{1}{9} + \frac{5}{3}\varepsilon - \frac{\delta}{4}$.

Claim 22. If C is a connected component of D_6, then $\sum_{v \in V(C)} w^*(v) \leq 0$.

As a consequence of these last claims, we have $\rho(D) \leq \sum_{v \in V(D)} w(v) = \sum_{v \in V(D)} w^*(v) \leq 0 \leq 1$, a contradiction. This proves Theorem 8. $\qquad \square$

References

1. Aboulker, P., Bellitto, T., Havet, F., Rambaud, C.: On the minimum number of arcs in k-dicritical oriented graphs. arXiv preprint arXiv:2207.01051 (2022)
2. Aboulker, P., Vermande, Q.: Various bounds on the minimum number of arcs in a k-dicritical digraph. arXiv preprint arXiv:2208.02112 (2022)
3. Dirac, G.A.: A theorem of R. L. Brooks and a conjecture of H. Hadwiger. Proc. London Math. Soc. **3**(1), 161–195 (1957)
4. Dirac, G.A.: On the structure of 5-and 6-chromatic abstract graphs. J. für die reine und angew. Math. (Crelles J.) **1964**(214–215), 43–52 (1964)
5. Gao, W., Postle, L.: On the minimal edge density of K_4-free 6-critical graphs. arXiv:1811.02940 [math] (2018)

6. Gould, R.J., Larsen, V., Postle, L.: Structure in sparse k-critical graphs. J. Comb. Theory Ser. B **156**, 194–222 (2022)
7. Harutyunyan, A., Mohar, B.: Gallai's theorem for list coloring of digraphs. SIAM J. Discret. Math. **25**(1), 170–180 (2011)
8. Havet, F., Picasarri-Arrieta, L., Rambaud, C.: On the minimum number of arcs in 4-dicritical oriented graphs. arXiv preprint arXiv:2306.10784 (2023)
9. Kostochka, A., Stiebitz, M.: On the number of edges in colour-critical graphs and hypergraphs. Combinatorica **20**(4), 521–530 (2000)
10. Kostochka, A., Stiebitz, M.: The minimum number of edges in 4-critical digraphs of given order. Graphs Comb. **36**(3), 703–718 (2020)
11. Kostochka, A., Yancey, M.: Ore's conjecture on color-critical graphs is almost true. J. Comb. Theory, Ser. B **109**, 73–101 (2014)
12. Kostochka, A., Yancey, M.: A Brooks-type result for sparse critical graphs. Combinatorica **38**(4), 887–934 (2018)
13. Postle, L.: On the minimum number of edges in triangle-free 5-critical graphs. Eur. J. Comb. **66**, 264–280 (2017). selected papers of EuroComb15

Tight Algorithms for Connectivity Problems Parameterized by Modular-Treewidth

Falko Hegerfeld[ID] and Stefan Kratsch[✉][ID]

Institut für Informatik, Humboldt-Universität zu Berlin, Berlin, Germany
{hegerfeld,kratsch}@informatik.hu-berlin.de

Abstract. We study connectivity problems from a fine-grained parameterized perspective. Cygan et al. (TALG 2022) first obtained algorithms with single-exponential running time $\alpha^{\mathrm{tw}} n^{\mathcal{O}(1)}$ for connectivity problems parameterized by treewidth (tw) by introducing the cut-and-count-technique, which reduces the connectivity problems to locally checkable counting problems. In addition, the obtained bases α were proven to be optimal assuming the Strong Exponential-Time Hypothesis (SETH).

As only sparse graphs may admit small treewidth, these results are not applicable to graphs with dense structure. A well-known tool to capture dense structure is the *modular decomposition*, which recursively partitions the graph into *modules* whose members have the same neighborhood outside of the module. Contracting the modules, we obtain a *quotient graph* describing the adjacencies between modules. Measuring the treewidth of the quotient graph yields the parameter *modular-treewidth*, a natural intermediate step between treewidth and clique-width. While less general than clique-width, modular-treewidth has the advantage that it can be computed as easily as treewidth.

We obtain the first tight running times for connectivity problems parameterized by modular-treewidth. For some problems the obtained bounds are the same as relative to treewidth, showing that we can deal with a greater generality in input structure at no cost in complexity. We obtain the following randomized algorithms for graphs of modular-treewidth k, given an appropriate decomposition:

- STEINER TREE can be solved in time $3^k n^{\mathcal{O}(1)}$,
- CONNECTED DOMINATING SET can be solved in time $4^k n^{\mathcal{O}(1)}$,
- CONNECTED VERTEX COVER can be solved in time $5^k n^{\mathcal{O}(1)}$,
- FEEDBACK VERTEX SET can be solved in time $5^k n^{\mathcal{O}(1)}$.

The first two algorithms are tight due to known results and the last two algorithms are complemented by new tight lower bounds under SETH.

Keywords: connectivity · modular-treewidth · tight algorithms

The first author was partially supported by DFG Emmy Noether-grant (KR 4286/1).

D. Paulusma and B. Ries (Eds.): WG 2023, LNCS 14093, pp. 388–402, 2023.
https://doi.org/10.1007/978-3-031-43380-1_28

1   Introduction

Connectivity constraints are a very natural form of global constraints in the realm of graph problems. We study connectivity problems from a fine-grained parameterized perspective. The starting point is an influential paper of Cygan et al. [11] introducing the cut-and-count-technique which yields randomized algorithms with running time[1] $\mathcal{O}^*(\alpha^{tw})$, for some constant *base* $\alpha > 1$, for connectivity problems parameterized by *treewidth* (tw). The obtained bases α were proven to be optimal assuming the Strong Exponential-Time Hypothesis[2] (SETH) [10].

Since dense graphs cannot have small treewidth, the results for treewidth do not help for graphs with dense structure. A well-known tool to capture dense structure is the *modular decomposition* of a graph, which recursively partitions the graph into *modules* whose members have the same neighborhood outside of the module. Contracting these modules, we obtain a *quotient graph* describing the adjacencies between the modules. Having isolated the dense part to the modules, measuring the complexity of the quotient graph by standard graph parameters such as treewidth yields e.g. the parameter *modular-treewidth* (mod-tw), a natural intermediate step between treewidth and clique-width. While modular-treewidth is not as general as clique-width, the algorithms for computing treewidth transfer to modular-treewidth, yielding e.g. reasonable constant-factor approximations for modular-treewidth in single-exponential time, whereas for clique-width we are currently only able to obtain approximations with exponential error.

We obtain the first tight running times for connectivity problems parameterized by modular-treewidth. To do so, we lift the algorithms using the cut-and-count-technique from treewidth to modular-treewidth. A crucial observation is that all vertices inside a module will be connected by choosing a single vertex from a neighboring module. In some cases, this observation is strong enough to lift the treewidth-based algorithms to modular-treewidth for free, i.e., the base α of the running time does not increase, showing that we can deal with a greater generality in input structure at no cost in complexity for these problems.

Theorem 1 (informal). *There are one-sided error Monte-Carlo algorithms that, given a decomposition witnessing modular-treewidth k, can solve*

- STEINER TREE *in time* $\mathcal{O}^*(3^k)$,
- CONNECTED DOMINATING SET *in time* $\mathcal{O}^*(4^k)$.

These bases are optimal under SETH, by known results of Cygan et al. [10].

However, in other cases the interplay of the connectivity constraint and the remaining problem constraints does increase the complexity for modular-treewidth compared to treewidth. In these cases, we provide new algorithms adapting the cut-and-count-technique to this more intricate setting.

---

[1] The $\mathcal{O}^*$-notation hides polynomial factors in the input size.

[2] The hypothesis that for every $\delta < 1$, there is some q such that q-SATISFIABILITY cannot be solved in time $\mathcal{O}(2^{\delta n})$, where n is the number of variables.

Theorem 2 (informal). *There are one-sided error Monte-Carlo algorithms that, given a decomposition witnessing modular-treewidth* k, *can solve* CONNECTED VERTEX COVER *and* FEEDBACK VERTEX SET *in time* $\mathcal{O}^*(5^k)$.

Both problems can be solved in time $\mathcal{O}^*(3^k)$ parameterized by treewidth [11]. In contrast, VERTEX COVER (without the connectivity constraint) has complexity $\mathcal{O}^*(2^k)$ with respect to treewidth [22] and modular-treewidth simultaneously.

For these latter two problems, we provide new lower bounds to show that the bases are optimal under SETH. However, we do not need the full power of the modular decomposition to prove the lower bounds. The modular decomposition allows for *recursive* partitioning. When instead allowing for only a single level of partitioning and limited complexity inside the modules, we obtain parameters called *twinclass-pathwidth* (tc-pw) and *twinclass-treewidth*.

Theorem 3. *Unless SETH fails,* CONNECTED VERTEX COVER *and* FEEDBACK VERTEX SET *cannot be solved in time* $\mathcal{O}^*((5-\varepsilon)^{\text{tc-pw}})$ *for any* $\varepsilon > 0$.

As twinclass-pathwidth is a larger parameter than modular-treewidth, the lower bounds of Theorem 3 transfer to modular-treewidth.

The obtained results on connectivity problems parameterized by modular-treewidth are situated in the larger context of a research program aimed at determining the optimal running times for connectivity problems relative to width-parameters of differing generality, thus quantifying the price of generality in this setting. The known results are summarized in Table 1. Beyond the results for treewidth by Cygan et al. [10,11], Bojikian et al. [8] obtain tight results for the more restrictive *cutwidth* by either providing faster algorithms resulting from combining cut-and-count with the rank-based approach or by showing that the same lower bounds already hold for cutwidth. Hegerfeld and Kratsch [16] consider *clique-width* and obtain tight results for CONNECTED VERTEX COVER and CONNECTED DOMINATING SET. Their algorithms combine cut-and-count with several nontrivial techniques to speed up dynamic programming on clique-expressions, where the interaction between cut-and-count and clique-width can yield more involved states compared to modular-treewidth, as clique-width is more general. These algorithms are complemented by new lower bound constructions following similar high-level principles as for modular-treewidth, but allow for more flexibility in the gadget design due to the mentioned generality. However, the techniques of Hegerfeld and Kratsch [16] for clique-width yield tight results for fewer problems compared to the present work; in particular, the optimal bases for STEINER TREE and FEEDBACK VERTEX SET parameterized by clique-width are currently not known.

Related Work. We survey some more of the literature on parameterized algorithms for connectivity problems relative to dense width-parameters. Bergougnoux [2] has applied cut-and-count to several width-parameters based on structured neighborhoods such as clique-width, rank-width, or mim-width. Building upon the rank-based approach of Bodlaender et al. [6], Bergougnoux and Kanté [4] obtain single-exponential running times $\mathcal{O}^*(\alpha^{\text{cw}})$ for a large

Table 1. Optimal running times of connectivity problems with respect to various width-parameters listed in increasing generality. The results in the penultimate column are obtained in this paper. The "?" denotes cases, where an algorithm with single-exponential running time is known by Bergougnoux and Kanté [4], but a gap between the lower bound and algorithm remains.

Parameters	cutwidth	treewidth	modular-tw	clique-width
CONNECTED VERTEX COVER	$\mathcal{O}^*(2^k)$	$\mathcal{O}^*(3^k)$	$\mathcal{O}^*(5^k)$	$\mathcal{O}^*(6^k)$
CONNECTED DOMINATING SET	$\mathcal{O}^*(3^k)$	$\mathcal{O}^*(4^k)$	$\mathcal{O}^*(4^k)$	$\mathcal{O}^*(5^k)$
STEINER TREE	$\mathcal{O}^*(3^k)$	$\mathcal{O}^*(3^k)$	$\mathcal{O}^*(3^k)$	?
FEEDBACK VERTEX SET	$\mathcal{O}^*(2^k)$	$\mathcal{O}^*(3^k)$	$\mathcal{O}^*(5^k)$	?
References	[8]	[10, 11]	here	[16]

class of connectivity problems parameterized by clique-width (cw). The same authors [5] also generalize this approach to other dense width-parameters via structured neighborhoods. All these works deal with general CONNECTED (σ, ρ)-DOMINATING SET problems capturing a wide range of problems; this generality of problems (and parameters) comes at the cost of yielding running times that are far from optimal for specific problem-parameter combinations, e.g., the first article [2] is the most optimized for clique-width and obtains the running time $\mathcal{O}^*((2^{4+\omega})^{cw}) \geq \mathcal{O}^*(64^{cw})$, where ω is the matrix multiplication exponent [1], for CONNECTED DOMINATING SET. Bergougnoux et al. [3] obtain XP algorithms parameterized by mim-width for problems expressible in a logic that can also capture connectivity constraints. Beyond dense width-parameters, cut-and-count has also been applied to the parameters branchwidth [28] and treedepth [14, 26].

Our version of modular-treewidth was first used by Bodlaender and Jansen for MAXIMUM CUT [7]. Several papers [21, 24, 27] also use the name modular-treewidth, but use it to refer to what we call *twinclass-treewidth*. In particular, Lampis [21] obtains tight results under SETH for q-COLORING with respect to twinclass-treewidth and clique-width. Hegerfeld and Kratsch [15] obtain tight results for ODD CYCLE TRANSVERSAL parameterized by twinclass-pathwidth and clique-width and for DOMINATING SET parameterized by twinclass-cutwidth. Kratsch and Nelles [20] combine modular decompositions with tree-depth in various ways and obtain parameterized algorithms for various efficiently solvable problems.

Organization. In Sect. 2 we discuss the general preliminaries and in Sect. 3 the cut-and-count-technique. We sketch Theorem 1 in Sect. 4 and Theorem 2 in Sect. 5. Everything marked with $\star$ has a more detailed version in the full version [17]; in particular, the lower bounds of Theorem 3 are completely contained in the full version due to space constraints.

2 Preliminaries

For two integers a, b we write $a \equiv_c b$ to indicate equality modulo $c \in \mathbb{N}$. We use Iverson's bracket notation: for a boolean predicate p, we have that $[p]$ is 1 if p

is true and 0 otherwise. For a function f we denote by $f[v \mapsto \alpha]$ the function $(f \setminus \{(v, f(v))\}) \cup \{(v, \alpha)\}$, viewing f as a set. By $\mathbb{F}_2$ we denote the field of two elements. For $n_1, n_2 \in \mathbb{Z}$, we write $[n_1, n_2] = \{x \in \mathbb{Z} : n_1 \leq x \leq n_2\}$ and $[n_2] = [1, n_2]$. For a function $f \colon V \to \mathbb{Z}$ and a subset $W \subseteq V$, we write $f(W) = \sum_{v \in W} f(v)$. Note that for functions $g \colon A \to B$, where $B \not\subseteq \mathbb{Z}$, and a subset $A' \subseteq A$, we still denote the *image of A' under g* by $g(A') = \{g(v) : v \in A'\}$. If $f \colon A \to B$ is a function and $A' \subseteq A$, then $f|_{A'}$ denotes the *restriction* of f to A' and for a subset $B' \subseteq B$, we denote the *preimage of B' under f* by $f^{-1}(B') = \{a \in A : f(a) \in B'\}$. The *power set* of a set A is denoted by $\mathcal{P}(A)$.

We use common graph-theoretic notation and the essentials of parameterized complexity. For two disjoint vertex subsets $A, B \subseteq V$, we define $E_G(A, B) = \{\{a, b\} \in E(G) : a \in A, b \in B\}$ and adding a *join* between A and B means adding an edge between every vertex in A and every vertex in B. We denote the *number of connected components* of G by $\mathrm{cc}(G)$. A *cut* of G is a partition $V = V_L \cup V_R$, $V_L \cap V_R = \emptyset$, of its vertices into two parts.

Quotients and Twinclasses. Let Π be a partition of $V(G)$. The *quotient graph G/Π* is given by $V(G/\Pi) = \Pi$ and $E(G/\Pi) = \{\{B_1, B_2\} \subseteq \Pi : B_1 \neq B_2, \exists u \in B_1, v \in B_2 : \{u, v\} \in E(G)\}$. We say that two vertices u, v are *twins* if $N(u) \setminus \{v\} = N(v) \setminus \{u\}$. The equivalence classes of this relation are called *twinclasses* and we let $\Pi_{tc}(G)$ denote the partition of $V(G)$ into twinclasses. A twinclass of size at least 2 either induces an independent set (false twins) or a clique (true twins). We define the *twinclass-treewidth* and *twinclass-pathwidth* of G by $\mathrm{tc\text{-}tw}(G) = \mathrm{tw}(G/\Pi_{tc}(G))$ and $\mathrm{tc\text{-}pw}(G) = \mathrm{pw}(G/\Pi_{tc}(G))$, respectively. The parameters twinclass-treewidth and twinclass-pathwidth were considered before under the name modular treewidth and modular pathwidth [21,24,27]. We use the prefix *twinclass* instead of *modular* to distinguish from the quotient graph arising from a *modular partition* of G.

Modular Decomposition. A vertex set $M \subseteq V(G)$ is a *module* of G if $N(v) \setminus M = N(w) \setminus M$ for every pair $v, w \in M$ of vertices in M. The modules $\emptyset$, $V(G)$, and all singletons are called *trivial*; A graph is *prime* if it only admits trivial modules. A module M is *proper* if $M \neq V(G)$. For two disjoint modules $M_1, M_2 \in \mathcal{M}(G)$, either $\{\{v, w\} : v \in M_1, w \in M_2\} \subseteq E(G)$ or $\{\{v, w\} : v \in M_1, w \in M_2\} \cap E(G) = \emptyset$; in the first case, M_1 and M_2 are *adjacent* and in the second case, they are *nonadjacent*. A module M is *strong* if for every module M' of G we have $M \cap M' = \emptyset$, $M \subseteq M'$, or $M' \subseteq M$. The family of nonempty strong modules is denoted $\mathcal{M}_{\mathrm{tree}}(G)$, which can be arranged as the *modular decomposition tree* via the inclusion-relation. We freely switch between viewing $\mathcal{M}_{\mathrm{tree}}(G)$ as a set family or as the modular decomposition tree of G; in the latter case, we refer also to the modules as *nodes*. Every graph G containing at least two vertices can be uniquely partitioned into a set of inclusion-maximal nonempty strong modules $\Pi_{mod}(G) = \{M_1, \ldots, M_\ell\}$, with $\ell \geq 2$, called *canonical modular partition*; $\Pi_{mod}(G)$ is undefined for $|V(G)| \leq 1$. For $M \in \mathcal{M}_{\mathrm{tree}}(G)$ with $|M| \geq 2$, we write $\mathtt{children}(M) = \Pi_{mod}(G[M])$ as the sets in $\Pi_{mod}(G[M])$ are precisely the children of M in the modular decomposition tree; if $|M| = 1$, then $\texttt{children}(M) = \emptyset$. Forming the *quotient graph* $G^q_M = G[M]/\Pi_{mod}(G[M])$ *at* M, there are three possible cases:

Theorem 4 ([12]). *If $|M| \geq 2$, then exactly one of the following holds:*

- *Parallel node:* $G[M]$ *is not connected and* G^q_M *is an independent set,*
- *Series node:* the complement $\overline{G[M]}$ *is not connected and* G^q_M *is a clique,*
- *Prime node:* $\Pi_{mod}(G[M])$ *consists of the inclusion-maximal proper modules of* $G[M]$ *and* G^q_M *is prime.*

We define the family $\mathcal{H}_p(G) = \{G^q_M : M \in \mathcal{M}_{\text{tree}}(G), |M| \geq 2, G^q_M \text{ is prime}\}$ and the *modular-pathwidth* by $\text{mod-pw}(G) = \max(2, \max_{H \in \mathcal{H}_p(G)} \text{pw}(H))$ and the *modular-treewidth* by $\text{mod-tw}(G) = \max(2, \max_{H \in \mathcal{H}_p(G)} \text{tw}(H))$. The modular decomposition tree can be computed in time $\mathcal{O}(n + m)$, see e.g. Tedder et al. [29] or the survey by Habib and Paul [13]. Running a treewidth-algorithm, such as the approximation algorithm of Korhonen [19], on every graph in $\mathcal{H}_p(G)$ and observing that[3] $|\mathcal{M}_{\text{tree}}(G)| \leq 2n$, which also bounds the total number of vertices appearing in quotient graphs, we obtain the following.

Theorem 5. *There is an algorithm, that given an n-vertex graph G and an integer k, in time $2^{\mathcal{O}(k)}n$ either outputs a tree decomposition of width at most $2k + 1$ for every prime quotient graph $G^q_M \in \mathcal{H}_p(G)$ or determines that $\text{mod-tw}(G) > k$.*

Let $M \in \mathcal{M}_{\text{tree}}(G) \setminus \{V\}$ and $M^\uparrow \in \mathcal{M}_{\text{tree}}(G)$ be its *parent module*. We have $M \in \texttt{children}(M^\uparrow) = \Pi_{mod}(G[M^\uparrow])$, hence M appears as a vertex of the quotient graph $G^q_{M^\uparrow}$; we also denote this vertex by v^q_M. We define the *projection* at $M^\uparrow$ by $\pi_{M^\uparrow}: M^\uparrow \to V(G^q_{M^\uparrow})$ with $\pi_{M^\uparrow}(v) = v^q_M$ whenever $v \in M \in \Pi_{mod}(G[M^\uparrow])$.

Tree Decompositions ($\star$). The definition of (very nice) tree decompositions and treewidth is given in the full version of the paper. The very nice tree decompositions of Cygan et al. [11] augment the nice tree decompositions of Kloks [18] by empty root and leaf bags and every edge is introduced exactly once in an *introduce edge* bag.

Lemma 6 ([11]). *Any tree decomposition of G can be converted into a very nice tree decomposition of G with the same width in polynomial time.*

Given a very nice tree decomposition $(T^q_{M^\uparrow}, (\mathbb{B}^q_t)_{t \in V(T^q_{M^\uparrow})})$ of the quotient graph $G^q_{M^\uparrow}$, we associate to every node $t \in V(T^q_{M^\uparrow})$ a subgraph $G^q_t = (V^q_t, E^q_t)$ of $G^q_{M^\uparrow}$ in the standard way. Based on the vertex subsets of the quotient graph $G^q_{M^\uparrow}$, we define vertex subsets of the original graph $G[M^\uparrow]$ as follows: $\mathbb{B}_t = \pi^{-1}_{M^\uparrow}(\mathbb{B}^q_t) =$

---

[3] The modular decomposition tree has n leaves and every internal node has at least two children, hence $|\mathcal{M}_{\text{tree}}(G)| \leq 2n$.

$\bigcup_{v_M^q \in \mathbb{B}_t^q} M$ and $V_t = \pi_{M\uparrow}^{-1}(V_t^q) = \bigcup_{v_M^q \in V_t^q} M$. We also transfer the edge set as follows

$$E_t = \bigcup_{v_M^q \in V_t^q} E(G[M]) \cup \bigcup_{\{v_{M_1}^q, v_{M_2}^q\} \in E_t^q} \{\{u_1, u_2\} : u_1 \in M_1 \wedge u_2 \in M_2\},$$

allowing us to define the graph $G_t = (V_t, E_t)$ associated to any node $t \in V(T_{M\uparrow}^q)$.

Parameter Relationships $(\star)$. The standard definitions of clique-width, $\mathrm{cw}(G)$, and linear clique-width, $\mathrm{lin\text{-}cw}(G)$, can be found in the full version. We have the following relationships between the considered parameters.

Lemma 7 $(\star)$. *We have* $\mathrm{cw}(G) \leq \mathrm{mod\text{-}pw}(G) + 2 \leq \max(2, \mathrm{tc\text{-}pw}(G)) + 2$ *and* $\mathrm{mod\text{-}tw}(G) \leq \max(2, \mathrm{tc\text{-}tw}(G))$ *for every graph* G.

Note that Theorem 7 can only hold for modular-pathwidth and not modular-treewidth, as already for treewidth, Corneil and Rotics [9] show that for every k there exists a graph G_k with treewidth k and clique-width exponential in k.

Theorem 8 ([15]). *We have* $\mathrm{cw}(G) \leq \mathrm{lin\text{-}cw}(G) \leq \mathrm{tc\text{-}pw}(G) + 4 \leq \mathrm{pw}(G) + 4$.

Problem Definitions

Connected Vertex Cover

> **Input:** An undirected graph $G = (V, E)$, a cost function $\mathbf{c} \colon V \to \mathbb{N} \setminus \{0\}$ and an integer $\bar{b}$.
> **Question:** Is there a set $X \subseteq V$, $\mathbf{c}(X) \leq \bar{b}$, such that $G - X$ contains no edges and $G[X]$ is connected?

Connected Dominating Set

> **Input:** An undirected graph $G = (V, E)$, a cost function $\mathbf{c} \colon V \to \mathbb{N} \setminus \{0\}$ and an integer $\bar{b}$.
> **Question:** Is there a set $X \subseteq V$, $\mathbf{c}(X) \leq \bar{b}$, such that $N[X] = V$ and $G[X]$ is connected?

(Node) Steiner Tree

> **Input:** An undirected graph $G = (V, E)$, a set of terminals $K \subseteq V$, a cost function $\mathbf{c} \colon V \to \mathbb{N} \setminus \{0\}$ and an integer $\bar{b}$.
> **Question:** Is there a set $X \subseteq V$, $\mathbf{c}(X) \leq \bar{b}$, such that $K \subseteq X$ and $G[X]$ is connected?

Feedback Vertex Set

> **Input:** An undirected graph $G = (V, E)$, a cost function $\mathbf{c} \colon V \to \mathbb{N} \setminus \{0\}$ and an integer $\bar{b}$.
> **Question:** Is there a set $X \subseteq V$, $\mathbf{c}(X) \leq \bar{b}$, such that $G - X$ contains no cycles?

3   Cut and Count for Modular-Treewidth

Let $G = (V, E)$ denote a connected graph. For easy reference, we repeat the key definition and lemmas of the cut-and-count-technique [11] here. A cut (V_L, V_R) of an undirected graph $G = (V, E)$ is *consistent* if $u \in V_L$ and $v \in V_R$ implies $\{u, v\} \notin E$, i.e., $E_G(V_L, V_R) = \emptyset$. A *consistently cut subgraph* of G is a pair $(X, (X_L, X_R))$ such that $X \subseteq V$ and (X_L, X_R) is a consistent cut of $G[X]$. We denote the set of consistently cut subgraphs of G by $\mathcal{C}(G)$.

Lemma 9 ([11]). *Let $X \subseteq V$ be a subset of vertices. The number of consistently cut subgraphs $(X, (X_L, X_R))$ is equal to $2^{\mathrm{cc}(G[X])}$.*

Fix some $M^\uparrow \in \mathcal{M}_{\mathrm{tree}}(G)$ with $|M^\uparrow| \geq 2$ and $X \subseteq M^\uparrow$ with $|\pi_{M^\uparrow}(X)| \geq 2$, i.e., X intersects at least two child modules of $M^\uparrow$, for this section. A simple exchange argument shows that the connectivity of $G[X]$ is not affected by the precise intersection $X \cap M$, $M \in \mathtt{children}(M^\uparrow)$, but only whether $X \cap M$ is empty or not. This observation allows us to reduce checking the connectivity of $G[X]$ to the quotient graph at $M^\uparrow$, as $G_{M^\uparrow}^q$ is isomorphic to the induced subgraph of G obtained by picking one vertex from each child module of $M^\uparrow$.

Lemma 10 $(\star)$. *If $G[X]$ is connected, then for any $v_M^q \in \pi_{M^\uparrow}(X)$ and $\emptyset \neq Y \subseteq M$, the graph $G[(X \setminus M) \cup Y]$ is connected. Furthermore, $G[X]$ is connected if and only if $G_{M^\uparrow}^q[\pi_{M^\uparrow}(X)]$ is connected.*

Theorem 10 shows that we do not need to consider *heterogeneous* cuts, i.e., $(X, (X_L, X_R)) \in \mathcal{C}(G)$ with $X_L \cap M \neq \emptyset$ and $X_R \cap M \neq \emptyset$ for some module $M \in \Pi_{mod}(G)$, since we can assume that $|X \cap M| \leq 1$.

Definition 11. *Let $M^\uparrow \in \mathcal{M}_{\mathrm{tree}}(G)$. We say that a cut (X_L, X_R), with $X_L \cup X_R \subseteq M^\uparrow$, is $M^\uparrow$-homogeneous if $X_L \cap M = \emptyset$ or $X_R \cap M = \emptyset$ for every $M \in \mathtt{children}(M^\uparrow)$. We may just say that (X_L, X_R) is homogeneous when $M^\uparrow$ is clear from the context. We define for every subgraph G' of G the set $\mathcal{C}_{M^\uparrow}^{hom}(G') = \{(X, (X_L, X_R)) \in \mathcal{C}(G') : (X_L, X_R) \text{ is } M^\uparrow\text{-homogeneous}\}$.*

Combining Theorem 9 with Theorem 10, the connectivity of $G[X]$ can be determined by counting $M^\uparrow$-homogeneous consistent cuts of $G[X]$ modulo 4.

Lemma 12 $(\star)$. *We have*

$$|\{(X_L, X_R) : (X, (X_L, X_R)) \in \mathcal{C}_{M^\uparrow}^{hom}(G)\}| = 2^{\mathrm{cc}(G_{M^\uparrow}^q[\pi_{M^\uparrow}(X)])}.$$

Furthermore, $G[X]$ is connected if and only if $|\{(X_L, X_R) : (X, (X_L, X_R)) \in \mathcal{C}_{M^\uparrow}^{hom}(G)\}| \neq 0 \mod 4$.

With the isolation lemma we avoid unwanted cancellations in the cut-and-count-technique at the cost of introducing randomization.

Definition 13. *A function $\mathbf{w} : U \to \mathbb{Z}$ isolates a set family $\mathcal{F} \subseteq \mathcal{P}(U)$ if there is a unique $S' \in \mathcal{F}$ with $\mathbf{w}(S') = \min_{S \in \mathcal{F}} \mathbf{w}(S)$, where for subsets X of U we define $\mathbf{w}(X) = \sum_{u \in X} \mathbf{w}(u)$.*

Lemma 14 (Isolation Lemma, [25]). *Let* $\emptyset \neq \mathcal{F} \subseteq \mathcal{P}(U)$ *be a set family over a universe* U. *Let* $N \in \mathbb{N}$ *and for each* $u \in U$ *choose a weight* $\mathbf{w}(u) \in [N]$ *uniformly and independently at random. Then* $\mathbb{P}[\mathbf{w}$ *isolates* $\mathcal{F}] \geq 1 - |U|/N$.

4   Reduction to Treewidth (⋆)

We sketch the ideas behind Theorem 1. For both problems, STEINER TREE and CONNECTED DOMINATING SET, we use Theorem 10 to reduce the problems to a quotient graph and apply the treewidth-algorithms of Cygan et al. [11]. As Theorem 10 only applies to sets intersecting at least two modules, we separately search for solutions contained in a single module. We also handle the special cases of series and parallel nodes via special polynomial-time algorithms or recursing in the modular decomposition tree depending on the node type and problem.

Assuming that the topmost quotient graph $G_V^q = G/\Pi_{mod}(G)$ is prime and we are searching for solutions X intersecting at least two modules, i.e., $|\pi_V(X)| \geq 2$, we provide more details. First, consider such a STEINER TREE instance $(G, K, \mathbf{c}, \overline{b})$. Theorem 10 implies that the only sensible intersections are $X \cap M \in \{\emptyset, \{v_M\}, K \cap M\}$ for $M \in \Pi_{mod}(G)$, where v_M is a vertex of minimum cost inside M. In particular, we distinguish whether $K \cap M = \emptyset$ or $K \cap M \neq \emptyset$; in the former case, we can assume $X \cap M \in \{\emptyset, \{v_M\}\}$ and in the latter $X \cap M \in \{\emptyset, K \cap M\}$. This motivates the reduction to the quotient graph: we set $K^q = \pi_V(K)$ and $\mathbf{c}^q(v_M^q) = \mathbf{c}(K \cap M) = \sum_{v \in K \cap M} \mathbf{c}(v)$ if $K \cap M \neq \emptyset$ and $\mathbf{c}^q(v_M^q) = \mathbf{c}(v_M)$ otherwise, compressing the cost of $K \cap M$ into a single vertex or choosing a vertex of minimum cost respectively. Then, the instance $(G, K, \mathbf{c}, \overline{b})$ is equivalent to $(G_V^q, K^q, \mathbf{c}^q, \overline{b})$ and we can run a weighted variant of the STEINER TREE algorithm of Cygan et al. [11] on the latter instance.

The reduction for CONNECTED DOMINATING SET uses a very similar principle by considering the cheapest vertex inside each module, which works as a module is completely dominated as soon as we take at least one vertex in an adjacent module. However, for CONNECTED DOMINATING SET we might need to call the treewidth-algorithm due to more complicated recursions several times and not only once. This makes the algorithm more technical, as we have to be careful with the randomization to avoid increasing the error probability. By observing that an isolating weight function induces an isolating weight function for appropriate subinstances, we are able to maintain the error probability.

In the context of kernelization, Luo [23] uses similar reductions for STEINER TREE and CONNECTED DOMINATING SET parameterized by modular-width, however, these reductions do not consider the weighted setting and do not have to contend with randomization.

5   Dynamic Programming Algorithms

In this section, we prove Theorem 2, by presenting novel algorithms using the cut-and-count-technique for CONNECTED VERTEX COVER and FEEDBACK VERTEX SET.

5.1   Connected Vertex Cover

We assume that $G = (V, E)$ is connected and contains at least two vertices, hence V cannot be a parallel node. We only consider cost functions $\mathbf{c}$ that are polynomially bounded in $|V|$. To solve CONNECTED VERTEX COVER, we begin by computing some optimum (possibly nonconnected) vertex cover Y_M with respect to $\mathbf{c}|_M$ for every module $M \in \Pi_{mod}(G)$ such that $G[M]$ contains an edge. If $G[M]$ contains no edges, then we set $Y_M = \{v_M^*\}$, where $v_M^* \in M$ is a vertex minimizing the cost inside M, i.e., $v_M^* := \arg\min_{v \in M} \mathbf{c}(v)$. The vertex covers can be computed in time $\mathcal{O}^*(2^{\text{mod-tw}(G)})$ by using a straightforward algorithm presented in the full version.

Definition 15. Let $X \subseteq V$ be a vertex subset. We say that X is *nice* if for every module $M \in \Pi_{mod}(G)$ it holds that $X \cap M \in \{\emptyset, Y_M, M\}$.

Via exchange arguments, in particular Theorem 10, we show that it is sufficient to only consider nice vertex covers. This shows that only a constant number of states per module in the dynamic programming algorithm are necessary.

Lemma 16 ($\star$). *If there exists a connected vertex cover X of G with $|\pi_V(X)| \geq 2$, then there exists a connected vertex cover X' of G that is nice with $|\pi_V(X')| \geq 2$ and $\mathbf{c}(X') \leq \mathbf{c}(X)$.*

Some simple observations allow us to handle the edge case of connected vertex covers contained in a single module $M \in \Pi_{mod}(G)$ and series nodes. We proceed by looking for connected vertex covers X with $|\pi_V(X)| \geq 2$ when $G^q := G_V^q = G/\Pi_{mod}(G)$ is prime. We are given a very nice tree decomposition $(\mathcal{T}^q, (\mathbb{B}_t^q)_{t \in V(\mathcal{T}^q)})$ of $G^q := G_V^q = G/\Pi_{mod}(G)$ of width k. Making use of Theorem 16 and Theorem 12, we can employ the cut-and-count-technique and perform dynamic programming along the tree decomposition $\mathcal{T}^q$ and extend our partial solutions module by module. The cut-and-count-formulation of the problem is as follows. For any subgraph G' of G, we define the *relaxed solutions* $\mathcal{R}(G') = \{X \subseteq V(G') : X \text{ is a nice vertex cover of } G'\}$ and *the cut solutions* $\mathcal{Q}(G') = \{(X, (X_L, X_R)) \in \mathcal{C}_V^{hom}(G') : X \in \mathcal{R}(G')\}$. For the isolation lemma, cf. Theorem 14, we sample a weight function $\mathbf{w}: V \to [2n]$ uniformly at random. We track the cost $\mathbf{c}(X)$, the weight $\mathbf{w}(X)$, and the number of intersected modules $|\pi_V(X)|$ of each partial solution $(X, (X_L, X_R))$. Accordingly, we define $\mathcal{R}^{\overline{c}, \overline{w}, \overline{m}}(G') = \{X \in \mathcal{R}(G') : \mathbf{c}(X) = \overline{c}, \mathbf{w}(X) = \overline{w}, |\pi_V(X)| = \overline{m}\}$ and $\mathcal{Q}^{\overline{c}, \overline{w}, \overline{m}}(G') = \{(X, (X_L, X_R)) \in \mathcal{Q}(G') : X \in \mathcal{R}^{\overline{c}, \overline{w}, \overline{m}}(G')\}$ for all $\overline{c} \in [0, \mathbf{c}(V)], \overline{w} \in [0, \mathbf{w}(V)], \overline{m} \in [0, |\Pi_{mod}(G)|]$.

As discussed, to every node $t \in V(\mathcal{T}^q)$ we associate a subgraph $G_t^q = (V_t^q, E_t^q)$ of G^q in the standard way, which in turn gives rise to a subgraph $G_t = (V_t, E_t)$ of G. The subgraphs G_t grow module by module and are considered by the dynamic program, hence we define $\mathcal{R}_t^{\overline{c}, \overline{w}, \overline{m}} = \mathcal{R}^{\overline{c}, \overline{w}, \overline{m}}(G_t)$ and $\mathcal{Q}_t^{\overline{c}, \overline{w}, \overline{m}} = \mathcal{Q}^{\overline{c}, \overline{w}, \overline{m}}(G_t)$ for all $\overline{c}, \overline{w}$, and $\overline{m}$. We will compute the sizes of the sets $\mathcal{Q}_t^{\overline{c}, \overline{w}, \overline{m}}$ by dynamic programming over the tree decomposition $\mathcal{T}^q$, but to do so we need to parameterize the partial solutions by their state on the current bag.

Disregarding the side of the cut, Theorem 16 tells us that each module $M \in \Pi_{mod}(G)$ has one of three states for some $X \in \mathcal{R}_t^{\bar{c},\bar{w},\bar{m}}$, namely $X \cap M \in \{\emptyset, Y_M, M\}$. Since we are considering homogeneous cuts there are two possibilities if $X \cap M \neq \emptyset$; $X \cap M$ is contained in the left side of the cut or in the right side. Thus, there are five total choices. We define $\mathbf{states} = \{\mathbf{0}, \mathbf{1}_L, \mathbf{1}_R, \mathbf{A}_L, \mathbf{A}_R\}$ with $\mathbf{1}$ denoting that the partial solution contains at least one vertex, but not all, of the module and with $\mathbf{A}$ denoting that the partial solution contains all vertices of the module; the subscript denotes the side of the cut.

A function of the form $f \colon \mathbb{B}_t^q \to \mathbf{states}$ is called t-*signature*. For every node $t \in V(T^q)$, cost $\bar{c}$, weight $\bar{w}$, number of modules $\bar{m}$, and t-signature f, the family $\mathcal{A}_t^{\bar{c},\bar{w},\bar{m}}(f)$ consists of all $(X, (X_L, X_R)) \in \mathcal{Q}_t^{\bar{c},\bar{w},\bar{m}}$ that satisfy for all $v_M^q \in \mathbb{B}_t^q$:

$$f(v_M^q) = \mathbf{0} \leftrightarrow X \cap M = \emptyset,$$
$$f(v_M^q) = \mathbf{1}_L \leftrightarrow X_L \cap M = Y_M \neq M, \quad f(v_M^q) = \mathbf{1}_R \leftrightarrow X_R \cap M = Y_M \neq M,$$
$$f(v_M^q) = \mathbf{A}_L \leftrightarrow X_L \cap M = M, \qquad f(v_M^q) = \mathbf{A}_R \leftrightarrow X_R \cap M = M.$$

Recall that by considering homogeneous cuts, we have that $X_L \cap M = \emptyset$ or $X_R \cap M = \emptyset$ for every module $M \in \Pi_{mod}(G)$. We use the condition $Y_M \neq M$ for the states $\mathbf{1}_L$ and $\mathbf{1}_R$ to ensure a well-defined state for modules of size 1. Note that the sets $\mathcal{A}_t^{\bar{c},\bar{w},\bar{m}}(f)$, ranging over f, partition $\mathcal{Q}_t^{\bar{c},\bar{w},\bar{m}}$ due to the consideration of nice vertex covers and homogeneous cuts.

Our goal is to compute the size of $\mathcal{A}_{\hat{r}}^{\bar{c},\bar{w},\bar{m}}(\emptyset) = \mathcal{Q}_{\hat{r}}^{\bar{c},\bar{w},\bar{m}} = \mathcal{Q}^{\bar{c},\bar{w},\bar{m}}(G)$, where $\hat{r}$ is the root vertex of the tree decomposition T^q, modulo 4 for all $\bar{c}, \bar{w}, \bar{m}$. By Theorem 12, there is a connected vertex cover X of G with $\mathbf{c}(X) = \bar{c}$ and $\mathbf{w}(X) = \bar{w}$ if the result is nonzero. We present the recurrences for the various bag types to compute $A_t^{\bar{c},\bar{w},\bar{m}}(f) = |\mathcal{A}_t^{\bar{c},\bar{w},\bar{m}}(f)|$; if not stated otherwise, then $t \in V(T^q)$, $\bar{c} \in [0, \mathbf{c}(V)]$, $\bar{w} \in [0, \mathbf{w}(V)]$, $\bar{m} \in [0, |\Pi_{mod}(G)|]$, and f is a t-signature. We set $A_t^{\bar{c},\bar{w},\bar{m}}(f) = 0$ whenever at least one of $\bar{c}, \bar{w},$ or $\bar{m}$ is negative.

Leaf Bag: We have that $\mathbb{B}_t^q = \mathbb{B}_t = \emptyset$ and t has no children. The only possible t-signature is $\emptyset$ and the only possible partial solution is $(\emptyset, (\emptyset, \emptyset))$. Hence, we only need to check the tracker values: $A_t^{\bar{c},\bar{w},\bar{m}}(\emptyset) = 1$ if $\bar{c} = \bar{w} = \bar{m} = 0$ and 0 otherwise.

Introduce Vertex Bag: We have that $\mathbb{B}_t^q = \mathbb{B}_s^q \cup \{v_M^q\}$, where $s \in V(T^q)$ is the only child of t and $v_M^q \notin \mathbb{B}_s^q$. Hence, $\mathbb{B}_t = \mathbb{B}_s \cup M$. We have to consider all possible interactions of a partial solution with M, though since we are considering nice vertex covers these interactions are quite restricted. To formulate the recurrence, we let, as an exceptional case, f be an s-signature and not a t-signature. Since no edges of the quotient graph G^q incident to v_M^q are introduced yet, we only have to check some edge cases and update the trackers when introducing v_M^q:

$$A_t^{\bar{c},\bar{w},\bar{m}}(f[v_M^q \mapsto \mathbf{s}]) = \begin{cases} [G[M] \text{ is edgeless}] A_s^{\bar{c},\bar{w},\bar{m}}(f), & \mathbf{s} = \mathbf{0}, \\ [|M| > 1] A_s^{\bar{c}-\mathbf{c}(Y_M),\bar{w}-\mathbf{w}(Y_M),\bar{m}-1}(f), & \mathbf{s} \in \{\mathbf{1}_L, \mathbf{1}_R\}, \\ A_s^{\bar{c}-\mathbf{c}(M),\bar{w}-\mathbf{w}(M),\bar{m}-1}(f), & \mathbf{s} \in \{\mathbf{A}_L, \mathbf{A}_R\}. \end{cases}$$

Introduce Edge Bag: Let $\{v_{M_1}^q, v_{M_2}^q\}$ denote the introduced edge. We have that $\{v_{M_1}^q, v_{M_2}^q\} \subseteq \mathbb{B}_t^q = \mathbb{B}_s^q$. The edge $\{v_{M_1}^q, v_{M_2}^q\}$ corresponds to adding a join between the modules M_1 and M_2. We need to filter all solutions whose states at M_1 and M_2 are not consistent with M_1 and M_2 being adjacent. There are two possible reasons: either not all edges between M_1 and M_2 are covered, or the introduced edges go across the homogeneous cut. We implement this via the helper function cons: **states** $\times$ **states** $\rightarrow \{0,1\}$ defined by $\text{cons}(s_1, s_2) = [\{s_1, s_2\} \cap \{A_L, A_R\} \neq \emptyset][s_1 \in \{1_L, A_L\} \rightarrow s_2 \notin \{1_R, A_R\}][s_1 \in \{1_R, A_R\} \rightarrow s_2 \notin \{1_L, A_L\}]$. The recurrence is given by $A_t^{\overline{c}, \overline{w}, \overline{m}}(f) = \text{cons}(f(v_{M_1}^q), f(v_{M_2}^q)) A_s^{\overline{c}, \overline{w}, \overline{m}}(f)$.

Forget Vertex Bag: We have that $\mathbb{B}_t^q = \mathbb{B}_s^q \setminus \{v_M^q\}$, where $v_M^q \in \mathbb{B}_s^q$ and $s \in V(\mathcal{T}^q)$ is the only child of t. Here, we only need to forget the state at v_M^q and accumulate the contributions from the different states v_M^q could assume. As the states are disjoint no overcounting happens: $A_t^{\overline{c}, \overline{w}, \overline{m}}(f) = \sum_{s \in \textbf{states}} A_s^{\overline{c}, \overline{w}, \overline{m}}(f[v \mapsto s])$.

Join Bag: We have $\mathbb{B}_t^q = \mathbb{B}_{s_1}^q = \mathbb{B}_{s_2}^q$, where $s_1, s_2 \in V(\mathcal{T}^q)$ are the children of t. Two partial solutions, one at s_1, and the other at s_2, can be combined when the states agree on all $v_M^q \in \mathbb{B}_t^q$. Since we update the trackers already at introduce vertex bags, we need to take care that the values of the modules in the bag are not counted twice. For this sake, define $S^f = \bigcup_{v_M^q \in f^{-1}(\{1_L, 1_R\})} Y_M \cup \bigcup_{v_M^q \in f^{-1}(\{A_L, A_R\})} M$ for all t-signatures f. This definition satisfies $X \cap \mathbb{B}_t = S^f$ for all $(X, (X_L, X_R)) \in \mathcal{A}^{\overline{c}, \overline{w}, \overline{m}}(f)$. Then, the recurrence is given by

$$A_t^{\overline{c}, \overline{w}, \overline{m}}(f) = \sum_{\substack{\overline{c}_1 + \overline{c}_2 = \overline{c} + \mathbf{c}(S^f) \\ \overline{w}_1 + \overline{w}_2 = \overline{w} + \mathbf{w}(S^f)}} \sum_{\overline{m}_1 + \overline{m}_2 = \overline{m} + (|\mathbb{B}_t^q| - f^{-1}(\mathbf{0}))} A_{s_1}^{\overline{c}_1, \overline{w}_1, \overline{m}_1}(f) A_{s_2}^{\overline{c}_2, \overline{w}_2, \overline{m}_2}(f).$$

Theorem 17 ($\star$). *There exists a Monte-Carlo algorithm that given a tree decomposition of width at most k for every prime quotient graph $H \in \mathcal{H}_p(G)$, solves* CONNECTED VERTEX COVER *in time $\mathcal{O}^*(5^k)$. The algorithm cannot give false positives and may give false negatives with probability at most $1/2$.*

Proof (sketch). We compute the sets Y_M for all $M \in \Pi_{mod}(G)$ in time $\mathcal{O}^*(2^k)$ using a straightforward algorithm described in the full version [17]. For the remainder, we only need to consider the topmost quotient graph $G_V^q = G/\Pi_{mod}(G)$. In polynomial time, we can find an optimum connected vertex cover that is contained in a single module and also deal with the case where V is a parallel or series node. It remains to handle the case that G_V^q is prime. In that case, we run the presented dynamic programming algorithm along the very nice tree decomposition of G_V^q with the sampled weight function $\mathbf{w}: V \rightarrow [2n]$. The algorithm returns true if there are $\overline{c} \in [0, \overline{b}]$, $\overline{w} \in [0, \mathbf{w}(V)]$, $\overline{m} \in [2, |\Pi_{mod}(G)|]$ such that $A_{\hat{r}}^{\overline{c}, \overline{w}, \overline{m}}(\emptyset) \not\equiv_4 0$, where $\hat{r}$ is the root of the tree decomposition, otherwise the algorithm returns false.

We skip the correctness proofs of the recurrences here. Setting $\mathcal{S}^{\overline{c}, \overline{w}, \overline{m}} = \{X \in \mathcal{R}^{\overline{c}, \overline{w}, \overline{m}}(G) : G[X] \text{ is connected}\}$, we have that $A_{\hat{r}}^{\overline{c}, \overline{w}, \overline{m}}(\emptyset) = |\mathcal{Q}^{\overline{c}, \overline{w}, \overline{m}}(G)| = \sum_{X \in \mathcal{R}^{\overline{c}, \overline{w}, \overline{m}}(G)} 2^{cc(G[X])} \equiv_4 2|\mathcal{S}^{\overline{c}, \overline{w}, \overline{m}}|$ by Theorem 12, so the algorithm cannot return false negatives. Common isolation lemma arguments and Theorem 16 show that we return correctly with probability at least $1/2$.

By assumption the cost function $\mathbf{c}$ is polynomially bounded, hence there are $\mathcal{O}^*(5^k)$ table entries to compute. Furthermore, every recurrence can be computed in polynomial time, hence the running time of the algorithm follows. $\qquad\square$

5.2   Feedback Vertex Set ($\star$)

The algorithm for FEEDBACK VERTEX SET is the most technical part of this work; we give a summary of the main ideas. The first step is to solve the complementary problem INDUCED FOREST instead of FEEDBACK VERTEX SET as that matches the usage of the cut-and-count-technique better. By analyzing the structure of an induced forest X with respect to a module M, we see that there are four sensible possibilities for the intersection $X \cap M$: it is empty, a single vertex, an independent set, or an induced forest containing an edge.

In particular, the last possibility leads to many technical issues, as we must solve INDUCED FOREST for every module $M \in \mathcal{M}_{\mathrm{tree}}(G)$, whereas for CONNECTED VERTEX COVER only problems without connectivity constraints needed to be solved for all modules. Due to the randomization of cut-and-count, some subproblems might be solved incorrectly. We have to ensure that this does not cause an issue and that the error probability stays constant. In particular, we need to carefully define the subproblems as they rely on the output of the previous subproblems. We sample a global weight function once and, assuming that the weight function is isolating, we analyze where the restricted weight function remains isolating and hence which subproblems are solved correctly.

For the independent set state we distinguish whether a neighboring module is intersected (degree-1) or not (degree-0), as degree-2 or higher leads to a cycle. The degree-1 independent set state and the induced forest state behave the same with respect to any future neighboring module, as the intersection of X with this module has to be empty, otherwise X would contain a cycle. Hence, we would like to collapse these two states into a single one, this however causes issues in the join nodes. Instead we allow the induced forest state only for modules that are already forgotten; to be precise, when a degree-0 independent set is forgotten, we can safely exchange the independent set with an induced forest without introducing any cycles. Hence, the induced forest state will not affect the table sizes of the dynamic program.

Until now, we have not specified the cut sides of the modules. Since we can work with homogeneous cuts, any non-empty state naively turns into two states, one for the left side and one for the right side; this would yield 7 states in total. However, we can avoid specifying the cut side for the independent set states; in the degree-0 case, the cut side is independent of any other modules; in the degree-1 case, we can inherit the cut side from the unique non-empty neighboring module. Therefore, we obtain only the desired 5 states in total.

References

1. Alman, J., Williams, V.V.: A refined laser method and faster matrix multiplication. In: Marx, D. (ed.) Proceedings of the 2021 ACM-SIAM Symposium on Discrete Algorithms, SODA 2021, Virtual Conference, 10–13 January 2021, pp. 522–539. SIAM (2021). https://doi.org/10.1137/1.9781611976465.32
2. Bergougnoux, B.: Matrix decompositions and algorithmic applications to (hyper)graphs. Ph.D. thesis, University of Clermont Auvergne, Clermont-Ferrand, France (2019). https://tel.archives-ouvertes.fr/tel-02388683
3. Bergougnoux, B., Dreier, J., Jaffke, L.: A logic-based algorithmic meta-theorem for mim-width, pp. 3282–3304 (2023). https://doi.org/10.1137/1.9781611977554.ch125
4. Bergougnoux, B., Kanté, M.M.: Fast exact algorithms for some connectivity problems parameterized by clique-width. Theor. Comput. Sci. **782**, 30–53 (2019). https://doi.org/10.1016/j.tcs.2019.02.030
5. Bergougnoux, B., Kanté, M.M.: More applications of the d-neighbor equivalence: acyclicity and connectivity constraints. SIAM J. Discret. Math. **35**(3), 1881–1926 (2021). https://doi.org/10.1137/20M1350571
6. Bodlaender, H.L., Cygan, M., Kratsch, S., Nederlof, J.: Deterministic single exponential time algorithms for connectivity problems parameterized by treewidth. Inf. Comput. **243**, 86–111 (2015). https://doi.org/10.1016/j.ic.2014.12.008
7. Bodlaender, H.L., Jansen, K.: On the complexity of the maximum cut problem. Nord. J. Comput. **7**(1), 14–31 (2000)
8. Bojikian, N., Chekan, V., Hegerfeld, F., Kratsch, S.: Tight bounds for connectivity problems parameterized by cutwidth. In: Berenbrink, P., Bouyer, P., Dawar, A., Kanté, M.M. (eds.) 40th International Symposium on Theoretical Aspects of Computer Science, STACS 2023, 7–9 March 2023, Hamburg, Germany. LIPIcs, vol. 254, pp. 14:1–14:16. Schloss Dagstuhl - Leibniz-Zentrum für Informatik (2023). https://doi.org/10.4230/LIPIcs.STACS.2023.14
9. Corneil, D.G., Rotics, U.: On the relationship between clique-width and treewidth. SIAM J. Comput. **34**(4), 825–847 (2005). https://doi.org/10.1137/S0097539701385351
10. Cygan, M., Nederlof, J., Pilipczuk, M., Pilipczuk, M., van Rooij, J.M.M., Wojtaszczyk, J.O.: Solving connectivity problems parameterized by treewidth in single exponential time. CoRR abs/1103.0534 (2011)
11. Cygan, M., Nederlof, J., Pilipczuk, M., Pilipczuk, M., van Rooij, J.M.M., Wojtaszczyk, J.O.: Solving connectivity problems parameterized by treewidth in single exponential time. ACM Trans. Algorithms **18**(2), 17:1–17:31 (2022). https://doi.org/10.1145/3506707
12. Gallai, T.: Transitiv orientierbare graphen. Acta Math. Hungar. **18**(1–2), 25–66 (1967)
13. Habib, M., Paul, C.: A survey of the algorithmic aspects of modular decomposition. Comput. Sci. Rev. **4**(1), 41–59 (2010). https://doi.org/10.1016/j.cosrev.2010.01.001
14. Hegerfeld, F., Kratsch, S.: Solving connectivity problems parameterized by treedepth in single-exponential time and polynomial space. In: Paul, C., Bläser, M. (eds.) 37th International Symposium on Theoretical Aspects of Computer Science, STACS 2020, 10–13 March 2020, Montpellier, France. LIPIcs, vol. 154, pp. 29:1–29:16. Schloss Dagstuhl - Leibniz-Zentrum für Informatik (2020). https://doi.org/10.4230/LIPIcs.STACS.2020.29

15. Hegerfeld, F., Kratsch, S.: Towards exact structural thresholds for parameterized complexity. In: Dell, H., Nederlof, J. (eds.) 17th International Symposium on Parameterized and Exact Computation, IPEC 2022, 7–9 September 2022, Potsdam, Germany. LIPIcs, vol. 249, pp. 17:1–17:20. Schloss Dagstuhl - Leibniz-Zentrum für Informatik (2022). https://doi.org/10.4230/LIPIcs.IPEC.2022.17

16. Hegerfeld, F., Kratsch, S.: Tight algorithms for connectivity problems parameterized by clique-width. CoRR abs/2302.03627 (2023). https://doi.org/10.48550/arXiv.2302.03627, accepted at ESA 2023

17. Hegerfeld, F., Kratsch, S.: Tight algorithms for connectivity problems parameterized by modular-treewidth. CoRR abs/2302.14128 (2023). https://doi.org/10.48550/arXiv.2302.14128

18. Kloks, T. (ed.): Treewidth. LNCS, vol. 842. Springer, Heidelberg (1994). https://doi.org/10.1007/BFb0045375

19. Korhonen, T.: A single-exponential time 2-approximation algorithm for treewidth. In: 62nd IEEE Annual Symposium on Foundations of Computer Science, FOCS 2021, Denver, CO, USA, 7–10 February 2022, pp. 184–192. IEEE (2021). https://doi.org/10.1109/FOCS52979.2021.00026

20. Kratsch, S., Nelles, F.: Efficient parameterized algorithms on graphs with heterogeneous structure: combining tree-depth and modular-width. CoRR abs/2209.14429 (2022). https://doi.org/10.48550/arXiv.2209.14429

21. Lampis, M.: Finer tight bounds for coloring on clique-width. SIAM J. Discret. Math. **34**(3), 1538–1558 (2020). https://doi.org/10.1137/19M1280326

22. Lokshtanov, D., Marx, D., Saurabh, S.: Known algorithms on graphs of bounded treewidth are probably optimal. ACM Trans. Algorithms **14**(2), 13:1–13:30 (2018). https://doi.org/10.1145/3170442

23. Luo, W.: Polynomial turing compressions for some graph problems parameterized by modular-width. CoRR abs/2201.04678 (2022)

24. Mengel, S.: Parameterized compilation lower bounds for restricted CNF-formulas. In: Creignou, N., Le Berre, D. (eds.) SAT 2016. LNCS, vol. 9710, pp. 3–12. Springer, Cham (2016). https://doi.org/10.1007/978-3-319-40970-2_1

25. Mulmuley, K., Vazirani, U.V., Vazirani, V.V.: Matching is as easy as matrix inversion. Combinatorica **7**(1), 105–113 (1987). https://doi.org/10.1007/BF02579206

26. Nederlof, J., Pilipczuk, M., Swennenhuis, C.M.F., Węgrzycki, K.: Hamiltonian cycle parameterized by Treedepth in single exponential time and polynomial space. In: Adler, I., Müller, H. (eds.) WG 2020. LNCS, vol. 12301, pp. 27–39. Springer, Cham (2020). https://doi.org/10.1007/978-3-030-60440-0_3

27. Paulusma, D., Slivovsky, F., Szeider, S.: Model counting for CNF formulas of bounded modular treewidth. Algorithmica **76**(1), 168–194 (2016). https://doi.org/10.1007/s00453-015-0030-x

28. Pino, W.J.A., Bodlaender, H.L., van Rooij, J.M.M.: Cut and count and representative sets on branch decompositions. In: Guo, J., Hermelin, D. (eds.) 11th International Symposium on Parameterized and Exact Computation, IPEC 2016, 24–26 August 2016, Aarhus, Denmark. LIPIcs, vol. 63, pp. 27:1–27:12. Schloss Dagstuhl - Leibniz-Zentrum fuer Informatik (2016). https://doi.org/10.4230/LIPIcs.IPEC.2016.27

29. Tedder, M., Corneil, D., Habib, M., Paul, C.: Simpler linear-time modular decomposition via recursive factorizing permutations. In: Aceto, L., Damgård, I., Goldberg, L.A., Halldórsson, M.M., Ingólfsdóttir, A., Walukiewicz, I. (eds.) ICALP 2008. LNCS, vol. 5125, pp. 634–645. Springer, Heidelberg (2008). https://doi.org/10.1007/978-3-540-70575-8_52

Cops and Robber - When Capturing Is Not Surrounding

Paul Jungeblut[ID], Samuel Schneider, and Torsten Ueckerdt[✉]

Karlsruhe Institute of Technology, Karlsruhe, Germany
{paul.jungeblut,torsten.ueckerdt}@kit.edu,
samuel.schneider@student.kit.edu

Abstract. We consider "surrounding" versions of the classic Cops and Robber game. The game is played on a connected graph in which two players, one controlling a number of cops and the other controlling a robber, take alternating turns. In a turn, each player may move each of their pieces. The robber always moves between adjacent vertices. Regarding the moves of the cops we distinguish four versions that differ in whether the cops are on the vertices or the edges of the graph and whether the robber may move on/through them. The goal of the cops is to surround the robber, i.e., occupying all neighbors (vertex version) or incident edges (edge version) of the robber's current vertex. In contrast, the robber tries to avoid being surrounded indefinitely. Given a graph, the so-called *cop number* denotes the minimum number of cops required to eventually surround the robber.

We relate the different cop numbers of these versions and prove that none of them is bounded by a function of the classical cop number and the maximum degree of the graph, thereby refuting a conjecture by Crytser, Komarov and Mackey [Graphs and Combinatorics, 2020].

1 Introduction

Cops and Robber is a well-known combinatorial game played by two players on a graph $G = (V, E)$. The robber player controls a single robber, which we shall denote by r, whereas the cop player controls k cops, denoted $c_1, \ldots, c_k$, for some specified integer $k \geq 1$. The players take alternating turns, and in each turn may perform one move with each of their pieces (the single robber or the k cops). In the classical game (and also many of its variants) the vertices of G are the possible positions for the pieces, while the edges of G model the possible moves. Let us remark that no piece is forced to move, i.e., there is no *zugzwang*. On each vertex there can be any number of pieces.

The game begins with the cop player choosing vertices as the starting positions for the k cops $c_1, \ldots, c_k$. Then (seeing the cops' positions) the robber player places r on a vertex of G as well. The cop player wins if the cops *capture* the robber, which in the classical version means that at least one cop stands on the same vertex as the robber. On the other hand, the robber player wins if the robber can avoid being captured indefinitely.

D. Paulusma and B. Ries (Eds.): WG 2023, LNCS 14093, pp. 403–416, 2023.
https://doi.org/10.1007/978-3-031-43380-1_29

The *cop number* denoted by $c(G)$ of a given connected[1] graph $G = (V, E)$ is the smallest k for which k cops can capture the robber in a finite number of turns. Clearly, every graph satisfies $1 \leq c(G) \leq |V|$.

We consider several versions of the classical Cops and Robber game. In some of these the cops are placed on the edges of G and allowed to move to an *adjacent* edge (that is an edge sharing an endpoint) during their turn. In all our versions the robber acts as in the original game but loses the game if he is *surrounded*[2] by the cops, meaning that they have to occupy all adjacent vertices or incident edges. At all times, let us denote by v_r the vertex currently occupied by the robber. Specifically, we define the following versions of the game, each specifying the possible positions for the cops and the exact surrounding condition:

Vertex Version Cops are positioned on vertices of G (like the robber). They surround the robber if there is a cop on each neighbor of v_r. Let $c_V(G)$ denote the smallest number of cops needed to eventually surround the robber.

Edge Version Cops are positioned on edges of G. A cop on an edge e can move to any edge e' sharing an endpoint with e during its turn. The cops surround the robber if there is a cop on each edge incident to v_r. Let $c_E(G)$ denote the smallest number of cops needed to eventually surround the robber.

In both versions above, the robber sits on the vertices of G and moves along the edges of G. Due to the winning condition for the cops being a full surround, the robber may come very close to, say, a single cop without being threatened. As this can feel counterintuitive, let us additionally consider a restrictive version of each game where we constrain the possible moves for the robber when cops are close by. These *restrictive* versions are given by the following rules:

Restrictive Vertex Version After the robber's turn, there may not be any cop on v_r. In particular, the robber may not move onto a vertex occupied by a cop. Additionally, if a cop moves onto v_r, then in his next turn the robber must leave that vertex.

Restrictive Edge Version. The robber may not move along an edge that is currently occupied by a cop.

We denote the cop numbers of the restrictive versions by putting an additional "r" in the subscript, i.e., the smallest number of cops needed to eventually surround the robber in these versions is $c_{V,r}(G)$ and $c_{E,r}(G)$, respectively.

Clearly, the restrictive versions are favorable for cops as they only restrict the robber. Consequently, the corresponding cop numbers are always at most their non-restrictive counterparts. Thus, for every connected graph G we have

$$c_{V,r}(G) \leq c_V(G) \quad \text{and} \quad c_{E,r}(G) \leq c_E(G). \tag{1}$$

---

[1] Cops cannot move between different connected components, so the cop number of any graph is the sum over all components. We thus consider connected graphs only.

[2] To distinguish between the classical and our versions, we use the term *capture* to express that a cop occupies the same vertex as the robber. In contrast, a *surround* always means that all neighbors, respectively incident edges, are occupied.

A recent conjecture by Crytser, Komarov and Mackey [7] states that the cop number in the restrictive edge version can be bounded from above by the classical cop number and the maximum degree of the graph:

Conjecture 1 ([7]). *For every connected graph G we have $c_{E,\mathrm{r}}(G) \leq c(G) \cdot \Delta(G)$.*

Prałat [14] verified Conjecture 1 for the random graph $G(n, p)$, i.e., the graph on n vertices where each possible edge is chosen independently with probability p, for some ranges of p. Let us note that Conjecture 1, if true, would strengthen a theorem by Crytser, Komarov and Mackey [7] stating that $c_{E,\mathrm{r}}(G) \leq \gamma(G) \cdot \Delta(G)$, where $\gamma(G)$ denotes the size of a smallest dominating set in G.

1.1 Our Results

Our main contribution is to disprove Conjecture 1. In fact, we prove that there are graphs G for which none of the surrounding cop numbers can be bounded by any function of $c(G)$ and $\Delta(G)$. This proves that the classical game of Cops and Robber is sometimes fundamentally different from all its surrounding versions.

Theorem 2. *There is an infinite family of connected graphs G with classical cop number $c(G) = 2$ and $\Delta(G) = 3$ while neither $c_V(G)$, $c_{V,\mathrm{r}}(G)$, $c_E(G)$ nor $c_{E,\mathrm{r}}(G)$ can be bounded by any function of $c(G)$ and the maximum degree $\Delta(G)$.*

Additionally, we relate the different surrounding versions to each other. Equation (1) already gives an upper bound for the cop numbers in the restrictive versions in terms of their corresponding unrestrictive cop numbers. To complete the picture, our second contribution is to prove several lower and upper bounds for different combinations:

Theorem 3. *Each of the following holds (assuming G to be connected):*

1. $\forall G : c_V(G) \leq \Delta(G) \cdot c_{V,\mathrm{r}}(G)$ *and* $\exists G : c_V(G) \geq \Delta(G) \cdot c_{V,\mathrm{r}}(G)$
2. $\forall G : c_E(G) \leq \Delta(G) \cdot c_{E,\mathrm{r}}(G)$ *and* $\exists G : c_E(G) \geq \Delta(G)/4 \cdot c_{E,\mathrm{r}}(G)$
3. $\forall G : c_V(G) \leq 2 \cdot c_E(G)$ *and* $\exists G : c_V(G) \geq 2 \cdot (c_E(G) - 1)$
4. $\forall G : c_{V,\mathrm{r}}(G) \leq 2 \cdot c_{E,\mathrm{r}}(G)$ *and* $\exists G : c_{V,\mathrm{r}}(G) \geq c_{E,\mathrm{r}}(G)$
5. $\forall G : c_E(G) \leq \Delta(G) \cdot c_V(G)$ *and* $\exists G : c_E(G) \geq \Delta(G)/12 \cdot c_V(G)$
6. $\forall G : c_{E,\mathrm{r}}(G) \leq \Delta(G) \cdot c_{V,\mathrm{r}}(G)$ *and* $\exists G : c_{E,\mathrm{r}}(G) \geq \Delta(G)/48 \cdot c_{V,\mathrm{r}}(G)$

Note that all lower and upper bounds from Theorem 3 are tight up to a small additive or multiplicative constant. We prove the upper bounds in Sect. 2. The main idea is the same for all six inequalities: Given a winning strategy for a set of cops in one version we can simulate the strategy in any other version. Afterwards, in Sect. 3, we consider the lower bounds by constructing explicit families of graphs with the desired surrounding cop numbers. While some lower bounds already follow from standard graph families (like complete bipartite graphs), others need significantly more involved constructions. For example we construct a family of graphs from a set of $k - 1$ mutually orthogonal Latin squares (see Sect. 3.3 for a definition).

Trivial Bounds. Clearly, for the robber to be surrounded at a vertex v_r, at least $\deg(v_r)$ cops are required in all considered versions. This already gives that the minimum degree $\delta(G)$ of G is a lower bound on the cop numbers of G in these cases (stated in [6] for $c_{V,r}(G)$). In fact, the robber could restrict himself to a subgraph H of G of highest minimum degree, which gives the lower bound of $d(G) := \max\{\delta(H) \mid H \subseteq G\}$, also called the *degeneracy* of G. Moreover, in those versions in which the robber could simply start at a vertex of *highest* degree and *never move*, that is in all but the restrictive vertex version, we get the *maximum degree* $\Delta(G)$ of G as a lower bound (stated in [7] for $c_{E,r}(G)$):

Observation 4. *For every connected graph $G = (V, E)$ we have*

- $c_{V,r}(G) \geq d(G)$ *as well as*
- $c_V(G) \geq \Delta(G)$, $c_E(G) \geq \Delta(G)$ *and* $c_{E,r}(G) \geq \Delta(G)$.

1.2   Related Work

The game of Cops and Robber was introduced by Nowakowski and Winkler [13] as well as Quilliot [15] almost forty years ago. Both consider the case where a single cop chases the robber. The version with many cops and therefore also the notion of the cop number $c(G)$ was introduced shortly after by Aigner and Fromme [1], already proving that $c(G) \leq 3$ for all connected planar graphs G. Their version is nowadays considered the standard version of the game and we refer to it as the *classical version* throughout the paper. The most important open question regarding $c(G)$ is Meyniel's conjecture stating that a connected n-vertex graph G has $c(G) \in O(\sqrt{n})$ [2,8]. It is known to be EXPTIME-complete to decide whether $c(G) \leq k$ (for k being part of the input) [12].

By now, countless different versions of the game with their own cop numbers have been considered, see for example these books on the topic [3,4].

The restrictive vertex version was introduced by Burgess et al. [6]. They prove bounds for $c_{V,r}(G)$ in terms of the clique number $\omega(G)$, the independence number $\alpha(G)$ and the treewidth $\mathrm{tw}(G)$, as well as considering several interesting graph families. They also show that deciding whether $c_{V,r}(G) \leq k$ can be decided in polynomial time for every fixed value of k. The complexity is unknown for k being part of the input. Bradshaw and Hosseini [5] extend the study of $c_{V,r}(G)$ to graphs of bounded genus, proving (among other results) that $c_{V,r}(G) \leq 7$ for every connected planar graph G. See the bachelor's thesis of Schneider [16] for several further results on $c_{V,r}(G)$ (including a version with zugzwang).

The restrictive edge version was introduced even more recently by Crytser, Komarov and Mackey [7] (under the name *containment variant*). Besides stating Conjecture 1, which is verified for some graphs by Prałat [14], they give several bounds on $c_{E,r}(G)$ for different families of graphs.

To the best of our knowledge, $c_V(G)$ and $c_E(G)$ were not considered before.

In the light of the (restrictive) vertex and edge versions one might also define a *face version* for embedded planar graphs. Here the cops would occupy the faces and surround the robber if they occupy all faces incident to v_r. A restrictive face version could be that the robber must not move along an edge with either one or both incident faces being occupied by a cop. This version was introduced recently by Ha, Jungeblut and Ueckerdt [9]. Despite their similar motivation, the face versions seem to behave differently than the vertex or edge versions.

In each version, one might also add *zugzwang*, i.e., the obligation to actually move during one's turn. We are not aware of any results about this in the literature.

1.3 Outline of the Paper

Section 2 proves the upper bounds from Theorem 3. Then, in Sect. 3 we give constructions implying the corresponding lower bounds. Finally, in Sect. 4, we disprove Conjecture 1. Proofs of statements marked with ($\star$) are in the full version [10].

2 Relating the Different Versions

In this section we prove the upper bounds from Theorem 3. The main idea is always that a sufficiently large group of cops in one version can simulate a single cop in another version. We denote by $N_G(v)$ and $N_G[v]$ the open and closed neighborhood of vertex v in G, respectively.

Proof (only Item 1 of Theorem 3, others in the full version [10]). Let G be an arbitrary connected graph.

1. $c_V(G) \leq \Delta(G) \cdot c_{V,\mathrm{r}}(G)$: Let $\mathcal{S}_{V,\mathrm{r}}(G)$ be a winning strategy for $k \in \mathbb{N}$ restrictive vertex cops $c_1, \ldots, c_k$ in G. For $i \in \{1, \ldots, k\}$, replace c_i by a group of $\Delta(G)$ non-restrictive vertex cops $C_i := \{c_i^1, \ldots, c_i^{\Delta(G)}\}$. Initially all cops in C_i start at the same vertex as c_i and whenever c_i moves to an adjacent vertex, all cops in C_i copy its move.
   If the restrictive cops $c_1, \ldots, c_k$ arrive in a position where they surround the robber, then he is also surrounded by the groups of cops $C_1, \ldots, C_k$. It remains to consider the case that the robber ends their turn on vertex v currently occupied by some group C_i (a move that would be forbidden in the restrictive version). Then the cops in C_i can spread to the up to $\Delta(G)$ neighbors of v in G, thereby surrounding the robber.   □

Corollary 5. *For every graph G the surrounding cop numbers $c_V(G)$, $c_E(G)$, $c_{V,\mathrm{r}}(G)$ and $c_{E,\mathrm{r}}(G)$ are always within a factor of $2\Delta(G)$ of each other.*

Proof. In each of the six upper bounds stated in Theorem 3 the number of cops increases by at most a factor of $\Delta(G)$. In all cases this is obtained by simulating a winning strategy of one surrounding variant by (groups of) cops in another variant. The only cases where two such simulations need to be combined is when changing both, the cop type (vertex-cops/edge-cops) and the restrictiveness. It is easy to check that in all but one combination the number of cops increases by at most a factor of $2\Delta(G)$. The only exception is when a winning strategy for restricted vertex-cops is simulated by unrestricted edge-cops, where the number of cops would increase by a factor of $\Delta(G)^2$. However, looking at the proof of Theorem 3, we can see that both simulations replace a single cop by a group of $\Delta(G)$ cops. In this particular case it suffices to do this replacement just once.

We remark that all upper bounds proven above in Theorem 3 result from simulating a winning strategy of another surrounding version. In the next section we show that, surprisingly, these are indeed (asymptotically) tight. After all, it would seem natural that every version comes with its own unique winning strategy (more involved than just simulating one from a different version).

3   Explicit Graphs and Constructions

In this section we shall mention or construct several families of graphs with some extremal behavior for their corresponding classical and surrounding cop numbers. Together, these graphs prove all lower bounds stated in Theorem 3.

3.1   Complete Bipartite Graphs

We start by considering complete bipartite graphs. They already serve to directly prove two of the lower bounds from Theorem 3 and also appear again in proofs in the subsequent subsections.

Proposition 6 ($\star$). *For any complete bipartite graph G it holds that $c(G) = \min\{2, \delta(G)\}$, $c_{V,r}(G) = \delta(G)$ and $c_V(G) = c_{E,r}(G) = c_E(G) = \Delta(G)$.*

Let us consider two special cases of Proposition 6 for all $\Delta \in \mathbb{N}$: First, the star $K_{\Delta,1}$ has $c_{V,r}(K_{\Delta,1}) = 1$ while $c_V(K_{\Delta,1}) = \Delta$, thus proving the lower bound in Item 1 of Theorem 3. Second, the complete bipartite graph $K_{\Delta,\Delta}$ has $c_{V,r}(K_{\Delta,\Delta}) = c_{E,r}(K_{\Delta,\Delta}) = \Delta$, thus proving the lower bound in Item 4 of Theorem 3.

3.2   Regular Graphs with Leaves

Our first construction takes a connected k-regular graph H and attaches a set of ℓ new degree-1-vertices (*leaves*) to each vertex. Depending on H, k and ℓ we can give several bounds on the surrounding cop numbers of the resulting graph.

Lemma 7. *Let $H = (V_H, E_H)$ be a k-regular connected graph and let $G = (V_G, E_G)$ be the graph obtained from H by attaching to each vertex $v \in V_H$ a set of ℓ new leaves for some $\ell \geq 0$. Then each of the following holds:*

1. $c_V(G) \geq \begin{cases} k(k+\ell-1) & \text{if girth}(H) \geq 7 \\ (k+1)\ell & \text{always} \end{cases}$
2. $c_{V,r}(G) = \max\{c_{V,r}(H), k+1\}$

3. $c_E(G) \geq \begin{cases} k(k+\ell-1) & \text{if } \mathrm{girth}(H) \geq 6 \\ k\ell & \text{if } \mathrm{girth}(H) \geq 4 \\ \frac{1}{2}(k+1)\ell & \text{always} \end{cases}$

4. $c_{E,\mathrm{r}}(G) = \max\{c_{E,\mathrm{r}}(H), k+\ell\}$

Proof (only Item 1, others in the full version [10]). Note that most claimed inequalities hold trivially for the case that $\ell = 0$ (many lower bounds become 0 while others follow from $G = H$ in this case). Only the two cases requiring $\mathrm{girth}(H) \geq 6$, respectively $\mathrm{girth}(H) \geq 7$, are not directly clear. However, their proofs below hold for $\ell = 0$ as well. In all other cases, we implicitly assume $\ell \geq 1$ to avoid having to handle additional corner cases.

1. To prove the lower bounds on $c_V(G)$, consider any configuration of cops on the vertices of G. For a vertex $v \in V_H$ of G, let A_v be the set consisting of v and all leaves that are attached to it, i.e., $A_v = \{v\} \cup (N_G[v] \setminus V_H)$. We call a vertex $v \in V_H$ *safe* if there are fewer than ℓ cops on A_v in G. Note that if the robber ends his turn on a safe vertex, then the cops cannot surround him in their next turn. Let v_r be the current position of the robber. If the total number of cops is less than $(k+1)\ell$, then at least one of the $k+1$ vertices in the closed neighborhood $N_H[v_r]$ of v_r is safe, as no cop can be in A_v and A_w for $v \neq w$. Thus, the robber always has a safe vertex to move to (or to remain on), giving him a strategy to avoid being surrounded. It follows that $c_V(G) \geq (k+1)\ell$.

Now, if $\mathrm{girth}(H) \geq 7$ and the robber is on $v_r \in V_H$, then we consider for each neighbor v of v_r in V_H additionally the set $B_v = N_G[N_G(v) \setminus \{v_r\}]$, i.e., all vertices w with $\mathrm{dist}(w,v) \leq 2$ except from $N_G[v_r] \setminus \{v\}$. Since $\mathrm{girth}(H) \geq 7$, we have that $B_v \cap B_w = \emptyset$ for distinct $v, w \in N_H(v_r)$. Similar to above, we call $v \in N_H(v_r)$ *safe* if B_v contains fewer than $k+\ell-1$ cops. Again, if the robber ends his turn on a safe vertex, the cops cannot surround him in their next turn. If the total number of cops is less than $k(k+\ell-1)$, then at least one of the k neighbors of v_r in H is safe. This would give the robber a strategy to avoid being surrounded. It follows that $c_V(G) \geq k(k+\ell-1)$ in the case that $\mathrm{girth}(H) \geq 7$. $\square$

Applied to different host graphs H above Lemma 7 yields several interesting bounds. In particular, considering Theorem 3, Corollary 8 proves the lower bound in Item 2 for even Δ and Corollary 9 proves the lower bound in Item 3.

Corollary 8 ($\star$). *For every $\Delta \geq 2$ there is a connected graph G with $\Delta(G) = \Delta$ such that $c_{V,\mathrm{r}}(G) = \lfloor \frac{\Delta}{2} \rfloor + 1$, $c_V(G) = (\lfloor \frac{\Delta}{2} \rfloor + 1)\lceil \frac{\Delta}{2} \rceil$, $c_{E,\mathrm{r}}(G) = \Delta(G)$ and $c_E(G) = \lfloor \frac{\Delta}{2} \rfloor \lceil \frac{\Delta}{2} \rceil$.*

The host graph H in Corollary 8 is the complete bipartite graph $K_{k,k}$ with $k = \lfloor \frac{\Delta}{2} \rfloor$ with $h = \lceil \frac{\Delta}{2} \rceil$ leaves attached to each vertex.

Corollary 9 ($\star$). *For every $\Delta \geq 2$ there is a connected graph G with $\Delta(G) = \Delta$ such that $c_V(G) = 2(\Delta-1)$ and $c_E(G) = \Delta$.*

The host graph H in Corollary 9 is a single edge, i.e., the graph K_2 (therefore $k = 1$) with $h = \Delta - 1$ leaves attached to both vertices.

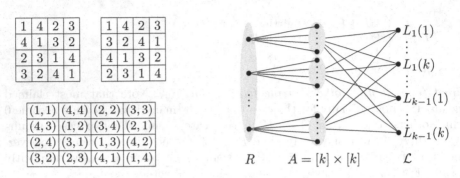

Fig. 1. Left: Two Latin squares and their juxtaposition, proving that they are orthogonal. Right: The graph G_k created from $k - 1$ MOLS of order k. The vertices in R correspond to the rows of A, the middle vertices correspond to the cells of A (ordered row by row in the drawing) and the vertices in $\mathcal{L}$ correspond to the parts of the MOLS.

3.3 Graphs from Mutually Orthogonal Latin Squares

A *Latin square* of order $k \geq 1$ is a $k \times k$ array filled with numbers from $[k] = \{1, \ldots, k\}$ such that each row and each column contains each number from $[k]$ exactly once. Formally, a Latin square L is a partition $L(1) \cup \cdots \cup L(k)$ of $A = [k] \times [k]$ such that for the i-th row ($i \in [k]$) $A[i, \cdot] = \{(i, j) \in A \mid j \in [k]\}$ and every number $n \in [k]$ we have $|A[i, \cdot] \cap L(n)| = 1$, and symmetrically for the columns. See the left of Fig. 1 for two different Latin squares.

Let L_1 and L_2 be two Latin squares of order k. Their *juxtaposition* $L_1 \otimes L_2$ is the Latin square of order k that contains in each cell the ordered pair of the entries of L_1 and L_2 in that cell. We say that L_1 and L_2 are *orthogonal* if each ordered pair appears exactly once in $L_1 \otimes L_2$, i.e., if for every two distinct $n_1, n_2 \in [k]$ we have $|L_1(n_1) \cap L_2(n_2)| = 1$. It is well-known that $k - 1$ *mutually orthogonal Latin squares (MOLS)* $L_1, \ldots, L_{k-1}$ (meaning that L_s and L_t are orthogonal whenever $s \neq t$) exist if and only if k is a prime power [11]. The two Latin squares in Fig. 1 (left) are indeed orthogonal, as can be seen by their juxtaposition below.

Construction of G_k. Let k be a prime power and $L_1, \ldots, L_{k-1}$ a set of $k - 1$ mutually orthogonal Latin squares of order k. Let $A = [k] \times [k]$ denote the set of all positions, $R = \{A[i, \cdot] \mid i \in [k]\}$ denote the set of all rows in A, and $\mathcal{L} = \{L_s(n) \mid s \in [k-1] \wedge n \in [k]\}$ denote the set of all parts of the Latin squares $L_1, \ldots, L_{k-1}$. Let $G_k = (V, E)$ be the graph with

$$V = A \cup R \cup \mathcal{L} \quad \text{and} \quad E = \{pS \mid p \in A, S \in R \cup \mathcal{L}, p \in S\}.$$

We observe that G_k is a k-regular bipartite graph with $|A| + |R \cup \mathcal{L}| = k^2 + (k + k(k - 1)) = 2k^2$ vertices with an edge between position $p \in A$ and a set $S \in R \cup \mathcal{L}$ if and only if p is in set S. See also the right of Fig. 1 for a schematic drawing.

Lemma 10 (⋆). *For a prime power k graph G_k has girth(G_k) ≥ 6.*

Lemma 11 (⋆). *For a prime power k graph G_k has $c(G_k) = k$, $c_{V,r}(G_k) \leq k+1$ and $c_E(G_k) \geq k(k-1)$.*

Remark 12. Burgess et al. [6] notice that for graphs G of many different families with a "large" value of $c_{V,r}(G)$ the classical cop number $c(G)$ was "low" (often even constant). In fact, they only provide a single family of graphs (constructed from finite projective planes) where $c(G) \approx c_{V,r}(G)$. They ask (Question 7 in [6]) whether graphs with large $c_{V,r}(G)$ inherently possess some property that implies that $c(G)$ is low. Our graph G_k from $k-1$ MOLS satisfies $c(G_k), c_{V,r}(G_k) \in \{k, k+1\}$. We interpret this as evidence that there is no such property.

3.4   Line Graphs of Complete Graphs

The *line graph* $L(G)$ of a given graph $G = (V, E)$ is the graph whose vertex set consists of the edges E of G and two vertices of $L(G)$ are connected if their corresponding edges in G share an endpoint. For $n \geq 3$, let K_n denote the complete graph on the set $[n] = \{1, \ldots, n\}$. For distinct $x, y \in [n]$, we denote by $\{x, y\}$ the vertex of $L(K_n)$ corresponding to the edge between x and y in K_n. Burgess et al. [6] showed that $c_{V,r}(L(K_n)) = 2(n-2) = \delta(L(K_n))$. This is obtained by placing the cops on all vertices $\{1, x\}$ for $x \in \{2, \ldots n\}$ and $\{2, y\}$ for $y \in \{3, \ldots, n-1\}$. The cops can surround the robber in their first move.

Lemma 13 (⋆). *For every $n \geq 3$ we have $c_V(L(K_n)) = 2(n-2)$, $c_E(L(K_n)) \geq n(n-2)/3$, and $c_{E,r}(L(K_n)) \geq n^2/12$.*

Stating the above bounds in terms of their maximum degree $\Delta := \Delta(L(K_n)) = 2(n-2)$, we obtain the claimed lower bounds in Items 5 and 6 of Theorem 3.

$$c_V(L(K_n)) = \Delta \qquad\qquad c_E(L(K_n)) \geq \frac{\Delta^2 + 4\Delta}{12}$$

$$c_{V,r}(L(K_n)) = \Delta \qquad\qquad c_{E,r}(L(K_n)) \geq \frac{\Delta^2 + 8\Delta + 16}{48}$$

4   When Capturing Is Not Surrounding

This section is devoted to the proof of Theorem 2, i.e., that none of the four surrounding cop numbers can be bounded by any function of the classical cop number and the maximum degree of the graph. In particular, we shall construct for infinitely many integers $k \geq 1$ a graph G_k with $c(G_k) = 2$ and $\Delta(G_k) = 3$, but $c_{V,r}(G_k) \geq k$. Theorem 3 then implies that also $c_V(G_k)$, $c_E(G_k)$ and $c_{E,r}(G_k)$ are unbounded for growing k.

The construction of G_k is quite intricate and we divide it into several steps.

The Graph H[s]. Let $s \geq 1$ be an integer and let $k = 2^{s-1}$. We start with a graph $H[s]$, which we call the *base graph*, with the following properties:

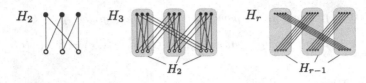

Fig. 2. Iterating $C_6 = H_2$ to obtain r-regular (bipartite) graphs H_r with girth$(H_r) \geq 5$.

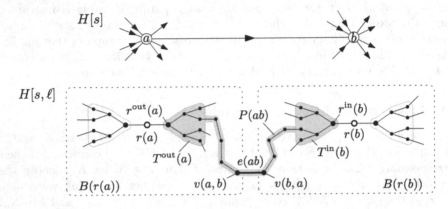

Fig. 3. Construction of $H[s, \ell]$ based on $H[s]$. A directed edge ab in $H[s]$ and the corresponding trees $T^{\text{out}}(a)$, $T^{\text{in}}(b)$, and path $P(ab)$ with middle edge $e(ab)$ in $H[s, \ell]$.

- $H[s]$ is $2k$-regular, i.e., every vertex of $H[s]$ has degree $2k$.
- $H[s]$ has girth at least 5.

There are many ways to construct such graphs $H[s]$, one being our graphs in Sect. 3.3 constructed from $2^s - 1$ mutually orthogonal Latin squares. An alternative construction (not relying on non-trivial tools) is an iteration of the 6-cycle as illustrated in Fig. 2. We additionally endow $H[s]$ with an orientation such that each vertex has exactly $k = 2^{s-1}$ outgoing and exactly $k = 2^{s-1}$ incoming edges. (For example, orient the edges according to an Eulerian tour in $H[s]$.)

The Graph $H[s, \ell]$. Let $\ell \geq 1$ be another integer[3]. We define a graph $H[s, \ell]$ on the basis of $H[s]$ and its orientation. See Fig. 3 for an illustration with $s = 3$ and $\ell = 4$. For each vertex a in $H[s]$ take a complete balanced binary tree $T(a)$ of height $s = \log_2(k) + 1$ with root $r(a)$ and $2^s = 2k$ leaves. Let $r^{\text{in}}(a)$ and $r^{\text{out}}(a)$ denote the two children of $r(a)$ in $T(a)$, and let $T^{\text{in}}(a)$ and $T^{\text{out}}(a)$ denote the subtrees rooted at $r^{\text{in}}(a)$ and $r^{\text{out}}(a)$, respectively. We associate each of the $k = 2^{s-1}$ leaves in $T^{\text{in}}(a)$ with one of the k incoming edges at a in $H[s]$, and each of the k leaves in $T^{\text{out}}(a)$ with one of the k outgoing edges at a in $H[s]$. Finally, for each edge ab in $H[s]$ oriented from a to b, connect the associated leaf in $T^{\text{out}}(a)$ with the associated leaf in $T^{\text{in}}(b)$ by a path $P(ab)$ of length $2\ell + 1$, i.e., on 2ℓ new inner vertices. This completes the construction of $H[s, \ell]$.

---

[3] We shall choose $\ell \gg s$ later. So you may think of s as "short" and of ℓ as "long".

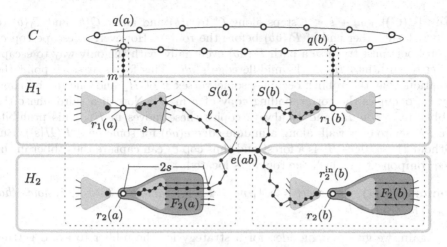

Fig. 4. Construction of $H[s, \ell, m]$ based on two copies of $H[s, \ell]$. A directed edge ab in $H[s]$ and the corresponding sets $S(a)$, $S(b)$, $F_2(a)$, etc. and edge $e(ab)$ in $H[s, \ell, m]$.

For each edge ab in $H[s]$, let $e(ab)$ denote the unique middle edge of $P(ab)$ in $H[s, \ell]$. I.e., $e(ab)$ connects a vertex in $B(r(a))$ which we denote by $v(a, b)$ with a vertex in $B(r(b))$ which we denote by $v(b, a)$; see Fig. 3.

The Graph $H[s, \ell, m]$. Let $m \geq 1$ be yet another integer[4]. We start with two vertex-disjoint copies H_1 and H_2 of $H[s, \ell]$ and transfer our notation (such as R, $r(a)$, $v(a, b)$, etc.) for $H[s, \ell]$ to H_i for $i \in \{1, 2\}$ by putting on the subscript i, e.g., R_1, $r_2(a)$, or $v_1(a, b)$. We connect H_1 and H_2 as follows: For each edge ab in $H[s]$ we identify the edge $e_1(ab)$ in H_1 with the edge $e_2(ab)$ in H_2 such that $v_1(a, b) = v_2(a, b)$ and $v_1(b, a) = v_2(b, a)$. Next, for each vertex a of $H[s]$ use the vertex $r_1(a)$ in H_1 as an endpoint for a new path $Q(a)$ of length m, and denote the other endpoint of $Q(a)$ by $q(a)$. Note that we do this only for the roots in H_1.

Finally, we connect the vertices $\{q(a) \mid a \in V(H[s])\}$ by a cycle C of length $|V(H[s])|$. This completes the construction of $H[s, \ell, m]$. See Fig. 4 for an illustration. Note that $H[s, \ell, m]$ has maximum degree 3 and degeneracy 2.

Lemma 14 ($\star$). *For every $s \geq 1$, $m \geq 1$, and $\ell > |V(H[s])| + m + s$, it holds that $c(H[s, \ell, m]) \leq 2$.*

Let us give a vague idea of the winning strategy for two cops. Two cops could easily capture the robber on cycle C (including the attached paths $Q(a)$ for $a \in V(H[s])$). Thus they force him to "flee" to H_1 at some point. In a second phase they can even force him to H_2. Loosely speaking, cop c_1 stays on C to prevent the robber from getting back on C, while cop c_2 always goes towards the robber in $H[s, \ell, m] - E(C)$. Whenever the robber traverses one of the long paths $P_1(ab)$ for some $ab \in E(H[s])$, say from $T_1(a)$ towards $T_1(b)$, then c_1 can

---

[4] We shall choose $\ell \gg m \gg s$ later. So you may think of m as "medium".

go in $|V(C)| + m + s < \ell$ steps along C to $q(b)$ and along $Q(b)$ and $T_1(b)$ to arrive at the other end of $P_1(ab)$ before the robber. However, to escape cop c_2 the robber must traverse a path $P_1(ab)$ eventually, with the only way to escape being to turn into H_2 at the middle edge $e(ab)$. This forces at some point the situation that the robber occupies some vertex v of H_2 and one of the cops, say c_1, occupies the corresponding copy of v in H_1. Now in a third phase the robber moves in H_2 and c_1 always copies these moves in H_1. This prohibits the robber to ever walk along a middle edge $e(ab)$ for some $ab \in E(H[s])$. But without these edges H_2 is a forest and thus cop c_2 can capture the robber in the tree component in which the robber is located.

Lemma 15 ($\star$). *For every $s \geq 1$, $m > 2s + 1$, and $\ell > 3s + 1$, it holds that $c_{V,\mathrm{r}}(H[s, \ell, m]) \geq k = 2^{s-1}$.*

Again, we give a vague idea for a strategy for the robber to avoid getting surrounded against $k - 1$ vertex cops. To this end, the robber always stays in H_2 and moves from a root $r_2(a)$ to the next $r_2(b)$ for which the edge ab in $H[s]$ is directed from a to b. In $H[s]$, vertex a has k outgoing neighbors and we show that at least one such neighbor b is always safe for the robber to escape to, meaning that the region labeled $F_2(b)$ in Fig. 4 is free of cops when the robber reaches $r_2(b)$. However, it is quite tricky to identify this safe neighbor. Indeed, the robber has to start moving in the "right" direction down the tree $T_2(a)$ always observing the cops' response, before he can be absolutely sure which outgoing neighbor b of a is safe. With suitable choices of s, ℓ and m, the robber is fast enough at $r_2(b)$ to then choose his next destination from there. The crucial point is that the cops can "join" the robber when he traverses the middle edge $e(ab)$, but can never ensure being on any vertex of $P_2(ab)$ one step *before* the robber; and thus never surround him.

Finally, Lemmas 14 and 15 and Theorem 3 immediately give the following corollary, which proves Theorem 2.

Corollary 16. *For any $s \geq 1$, $k = 2^{s-1}$, $m \geq 2s+1$, and $\ell \geq |V(H[s])|+m+s$, the graph $G_k = H[s, \ell, m]$ has $\Delta(G_k) = 3$ and*

$$c(G_k) \leq 2, \quad c_V(G_k) \geq k, \quad c_{V,\mathrm{r}}(G_k) \geq k, \quad c_{E,\mathrm{r}}(G_k) \geq \frac{k}{2} \quad and \quad c_E(G_k) \geq \frac{k}{2}.$$

5   Conclusion

We considered the cop numbers of four different surrounding versions of the well-known Cops and Robber game on a graph G, namely $c_V(G)$, $c_{V,\mathrm{r}}(G)$, $c_E(G)$ and $c_{E,\mathrm{r}}(G)$. Here index "V" denotes the vertex versions while index "E" denotes the edge versions, i.e., whether the cops occupy the vertices or the edges of the graph (recall that the robber always occupies a vertex). An additional index "r" stands for the corresponding restrictive versions, meaning that the robber must not end his turn on a cop or move along an edge occupied by a cop.

Only the two restrictive cop numbers have recently been considered in the literature, the vertex version $c_{V,r}(G)$ in [5,6] (denoted by $\sigma(G)$ and $s(G)$) and the edge version $c_{E,r}(G)$ in [7,14] (denoted by $\xi(G)$).

In this paper we related the four different versions to each other, showing that all of them lie (at most) within a factor of $2\Delta(G)$ of each other. For all combinations we presented explicit graph families showing that this is tight (up to small additive or multiplicative constants). It is an interesting open question to find out the exact constant factors for the lower and upper bounds in Theorem 3. We conjecture that all six presented upper bounds are tight (up to small *additive* constants). This would mean that optimal strategies for the cops in one surrounding version can indeed be obtained by simulating strategies from different surrounding versions.

As a second main result, we disproved a conjecture by Crytser, Komarov and Mackey [7] by constructing a family of graphs of maximum degree 3 and where the classical cop number is bounded whereas the cop number in all four surrounding versions is unbounded. It remains open to find other parameters that can be used to bound the surrounding cop numbers from above in combination with the classical cop number.

References

1. Aigner, M.S., Fromme, M.: A game of cops and robbers. Discret. Appl. Math. **8**(1), 1–12 (1984). https://doi.org/10.1016/0166-218X(84)90073-8
2. Baird, W., Bonato, A.: Meyniel's conjecture on the cop number: a survey (2013). https://arxiv.org/abs/1308.3385
3. Bonato, A.: An Invitation to Pursuit-Evasion Games and Graph Theory. American Mathematical Society, Providence (2022)
4. Bonato, A., Nowakowski, R.J.: The Game of Cops and Robbers on Graphs. American Mathematical Society, Providence (2011). https://doi.org/10.1090/stml/061
5. Bradshaw, P., Hosseini, S.A.: Surrounding cops and robbers on graphs of bounded genus (2019). https://arxiv.org/abs/1909.09916
6. Burgess, A.C., et al.: Cops that surround a robber. Discret. Appl. Math. **285**, 552–566 (2020). https://doi.org/10.1016/j.dam.2020.06.019
7. Crytser, D., Komarov, N., Mackey, J.: Containment: a variation of cops and robbers. Graphs Comb. **36**(3), 591–605 (2020). https://doi.org/10.1007/s00373-020-02140-5
8. Frankl, P.: Cops and robbers in graphs with large girth and Cayley graphs. Discret. Appl. Math. **17**(3), 301–305 (1987). https://doi.org/10.1016/0166-218X(87)90033-3
9. Ha, M.T., Jungeblut, P., Ueckerdt, T.: Primal-dual cops and robber. In: Seara, C., Huemer, C. (eds.) Proceedings of the 39th European Workshop on Computational Geometry (EuroCG) (2023). https://arxiv.org/abs/2301.05514
10. Jungeblut, P., Schneider, S., Ueckerdt, T.: Cops and Robber - When Capturing is not Surrounding (2023). https://doi.org/10.48550/arXiv.2302.10577
11. Keedwell, D.A., Dénes, J.: Latin Squares and their Applications. 2nd edn. Elsevier, Amsterdam (2015). https://doi.org/10.1016/C2014-0-03412-0
12. Kinnersley, W.B.: Cops and Robbers is EXPTIME-complete. J. Comb. Theory Ser. B **111**, 201–220 (2015). https://doi.org/10.1016/j.jctb.2014.11.002

13. Nowakowski, R.J., Winkler, P.: Vertex-to-vertex pursuit in a graph. Discret. Math. **43**(2–3), 235–239 (1983). https://doi.org/10.1016/0012-365X(83)90160-7

14. Prałat, P.: Containment game played on random graphs: another zig-zag theorem. Electron. J. Comb. **22**(2) (2015). https://doi.org/10.37236/4777

15. Quilliot, A.: Jeux et pointes fixes sur les graphes. Ph.D. thesis, Université de Paris VI (1978)

16. Schneider, S.: Surrounding cops and robbers. Bachelor's thesis, Karlsruhe Institute of Technology (2022). https://i11www.iti.kit.edu/_media/teaching/theses/ba_schneider22.pdf

Complexity Results for Matching Cut Problems in Graphs Without Long Induced Paths

Hoàng-Oanh Le[1] and Van Bang Le[2(⊠)]

[1] Independent Researcher, Berlin, Germany
HoangOanhLe@outlook.com
[2] Institut für Informatik, Universität Rostock, Rostock, Germany
van-bang.le@uni-rostock.de

Abstract. In a graph, a (perfect) matching cut is an edge cut that is a (perfect) matching. MATCHING CUT (MC), respectively, PERFECT MATCHING CUT (PMC), is the problem of deciding whether a given graph has a matching cut, respectively, a perfect matching cut. The DISCONNECTED PERFECT MATCHING problem (DPM) is to decide if a graph has a perfect matching that contains a matching cut. Solving an open problem recently posed in [Lucke, Paulusma, Ries (ISAAC 2022) & Feghali, Lucke, Paulusma, Ries (arXiv:2212.12317)], we show that PMC is NP-complete in graphs without induced 14-vertex path P_{14}. Our reduction also works simultaneously for MC and DPM, improving the previous hardness results of MC on P_{19}-free graphs and of DPM on P_{23}-free graphs to P_{14}-free graphs for both problems.

Actually, we prove a slightly stronger result: within P_{14}-free graphs, it is hard to distinguish between

(i) those without matching cuts and those in which every matching cut is a perfect matching cut;

(ii) those without perfect matching cuts and those in which every matching cut is a perfect matching cut;

(iii) those without disconnected perfect matchings and those in which every matching cut is a perfect matching cut.

Moreover, assuming the Exponential Time Hypothesis, none of these problems can be solved in time $2^{o(n)}$ for n-vertex P_{14}-free input graphs. As a corollary from (i), computing a matching cut with a maximum number of edges is hard, even when restricted to P_{14}-free graphs. This answers a question asked in [Lucke, Paulusma & Ries (arXiv:2207.07095)]. We also consider the problems in graphs without long induced cycles. It is known that MC is polynomially solvable in graphs without induced cycles of length at least 5 [Moshi (JGT 1989)]. We point out that the same holds for DPM.

Keywords: Matching cut · Maximum matching cut · Perfect matching cut · Disconnected perfect matching · H-free graph · Computational complexity

D. Paulusma and B. Ries (Eds.): WG 2023, LNCS 14093, pp. 417–431, 2023.
https://doi.org/10.1007/978-3-031-43380-1_30

1   Introduction and Results

In a graph $G = (V, E)$, a *cut* is a partition $V = X \cup Y$ of the vertex set into disjoint, non-empty sets X and Y. The set of all edges in G having an endvertex in X and the other endvertex in Y, written $E(X, Y)$, is called the *edge cut* of the cut (X, Y). A *matching cut* is an edge cut that is a (possibly empty) matching. Another way to define matching cuts is as follows; see [3,7]: a cut (X, Y) is a matching cut if and only if each vertex in X has at most one neighbor in Y and each vertex in Y has at most one neighbor in X. The classical NP-complete problem MATCHING CUT (MC) [3] asks if a given graph admits a matching cut.

An interesting special case, where the edge cut $E(X, Y)$ is a *perfect matching* of G, was considered in [8]. Such a matching cut is called a *perfect matching cut* and the PERFECT MATCHING CUT (PMC) problem asks whether a given graph admits a perfect matching cut. It was shown in [8] that this special case PMC of MC remains NP-complete.

A notion related to matching cut is *disconnected perfect matching* which has been considered recently in [1]: a disconnected perfect matching is a perfect matching that contains a matching cut. Observe that any perfect matching cut is a disconnected perfect matching but not the converse. Figure 1 provides some small examples for matching cuts, perfect matching cuts and disconnected perfect matchings.

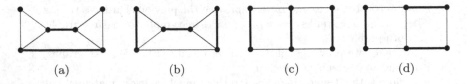

(a)   (b)   (c)   (d)

Fig. 1. Some example graphs; bold edges indicate a matching in question. (a): a matching cut. (b): a perfect matching that is neither a matching cut nor a disconnected perfect matching; this graph has no disconnected perfect matching, hence no perfect matching cut. (c): a perfect matching cut, hence a disconnected perfect matching. (d): a disconnected perfect matching that is not a perfect matching cut.

The related problem to MC and PMC, DISCONNECTED PERFECT MATCHING (DPM), asks if a given graph has a disconnected perfect matching; equivalently: if a given graph has a matching cut that is extendable to a perfect matching. It was shown in [1] that DPM is NP-complete. All these three problems have received much attention lately; see, e.g., [1,2,4,6,13–16] for recent results.

In this paper, we focus on the complexity of these three problems restricted to graphs without long induced paths and cycles. The current best known hardness results for MC and DPM in graphs without long induced paths are:

Theorem 1 ([14,15]). MC *remains* NP-*complete in* $\{4P_5, P_{19}\}$-*free graphs.*[1]

---

[1] Meanwhile the result has been improved to $\{3P_5, P_{15}\}$-free graphs [18].

Theorem 2 ([14,15]). DPM *remains* NP-*complete in* $\{4P_7, P_{23}\}$-*free graphs.*[2]

Prior to the present paper, no similar hardness result for PMC was known. Indeed, it was asked in [4,14,15], whether there is an integer t such that PMC is NP-complete in P_t-free graphs. Polynomial-time algorithms exist for MC and PMC in P_6-free graphs [14,15] and for DPM in P_5-free graphs [1].

For graphs without long induced cycles (including chordal graphs and chordal bipartite graphs), the only result we are aware of is that MC is polynomially solvable:

Theorem 3 ([21]). *There is a polynomial-time algorithm solving* MC *in graphs without induced cycles of length five and more.*

Previously, no similar polynomial-time results for DPM and PMC in long-hole-free graphs were known.

Our Contributions. We prove that PMC is NP-complete in graphs without induced path P_{14}, solving the open problem posed in [4,14,15]. For MC and DPM we improve the hardness results in Theorems 1 and 2 in graphs without induced path P_{19}, respectively, P_{23}, to graphs without induced path P_{14}. It is remarkable that all these hardness results for *three* problems will be obtained simultaneously by only *one* reduction, and can be stated in more details as follows.

Theorem 4. MC, PMC *and* DPM *are* NP-*complete in* $\{3P_6, 2P_7, P_{14}\}$-*free graphs. Moreover, under the ETH, no algorithm with runtime* $2^{o(n)}$ *can solve any of these problems for n-vertex* $\{3P_6, 2P_7, P_{14}\}$-*free input graphs.*

Actually, we prove the following slightly stronger result: within $\{3P_6, 2P_7, P_{14}\}$-free graphs, it is hard to distinguish between those without matching cuts (respectively perfect matching cuts, disconnected perfect matchings) and those in which every matching cut is a perfect matching cut. Moreover, under the ETH, this task cannot be solved in subexponential time in the vertex number of the input graph.

An interesting problem interposed between MC and PMC, called MAXIMUM MATCHING CUT (MAXMC), has recently been proposed in [15]. Here, given a graph G, we want to compute a matching cut of G (if any) with maximum number of edges. Formally, MAXMC in its decision version is as follows.

MAXIMUM MATCHING CUT (MAXMC)

Instance: A graph G and an integer k.
Question: Does G have a matching cut with k or more edges?

It has been asked in [15] what is the complexity of MAXMC on P_t-free graphs. Our next result answers this question.[3]

---

[2] Meanwhile the result has been improved to $\{3P_7, P_{19}\}$-free graphs [18].
[3] Meanwhile the complexity of MAXMC in H-free graphs has been completely determined [17].

Theorem 5. MAXMC *is* NP-*complete in* $\{3P_6, 2P_7, P_{14}\}$-*free graphs. Moreover, under the ETH, no algorithm with runtime* $2^{o(n)}$ *can solve* MAXMC *for n-vertex* $\{3P_6, 2P_7, P_{14}\}$-*free input graphs.*

On the positive side, we prove the following.

Theorem 6. *There is a polynomial-time algorithm solving* DPM *in graphs without induced cycle of length five and more.*

The paper is organized as follows. We recall some notion and notations in Sect. 2 which will be used. Then, we prove a slightly stronger result than Theorem 4 in Sect. 3 which then implies Theorem 5. The proof of Theorem 6 will be given in Sect. 4. Section 5 concludes the paper.

2 Preliminaries

For a set $\mathcal{H}$ of graphs, $\mathcal{H}$-free graphs are those in which no induced subgraph is isomorphic to a graph in $\mathcal{H}$. We denote by P_t the t-vertex path with $t - 1$ edges and by C_t the t-vertex cycle with t edges. C_3 is also called a triangle, and a *hole* is a C_t for some $t \geq 4$; C_t with $t \geq 5$ are *long holes*. The union $G + H$ of two vertex-disjoint graphs G and H is the graph with vertex set $V(G) \cup V(H)$ and edge set $E(G) \cup E(H)$; we write pG for the union of p copies of G. For a subset $S \subseteq V(G)$, let $G[S]$ denote the subgraph of G induced by S; $G - S$ stands for $G[V(G) \setminus S]$. By 'G contains an H' we mean G contains H as an induced subgraph.

Given a matching cut $M = (X, Y)$ of a graph G, a vertex set $S \subseteq V(G)$ is *monochromatic* if S belongs to the same *part* of M, i.e., $S \subseteq X$ or else $S \subseteq Y$. Notice that every clique different from the P_2 is monochromatic with respect to any matching cut.

Algorithmic lower bounds in this paper are conditional, based on the Exponential Time Hypothesis (ETH) [9]. The ETH asserts that no algorithm can solve 3SAT in subexponential time $2^{o(n)}$ for n-variable 3-CNF formulas. As shown by the Sparsification Lemma in [10], the hard cases of 3SAT already consist of sparse formulas with $m = O(n)$ clauses. Hence, the ETH implies that 3SAT cannot be solved in time $2^{o(n+m)}$.

Recall that an instance for 1-IN-3SAT is a 3-CNF formula $\phi = C_1 \wedge C_2 \wedge \cdots \wedge C_m$ over n variables, in which each clause C_j consists of three distinct literals. The problem asks whether there is a truth assignment of the variables such that every clause in ϕ has exactly one true literal. We call such an assignment an *1-in-3 assignment*. There is a polynomial reduction from 3SAT to 1-IN-3SAT ([20, Theorem 7.2]), which transforms an instance for 3SAT with n variables and m clauses to an equivalent instance for 1-IN-3SAT with $n + 4m$ variables and $3m$ clauses. Thus, assuming ETH, 1-IN-3SAT cannot be solved in time $2^{o(n+m)}$ on inputs with n variables and m clauses. We will need a restriction of 1-IN-3SAT, POSITIVE 1-IN-3SAT, in which each variable occurs positively. There is a well-known reduction from 1-IN-3SAT to POSITIVE 1-IN-3SAT, which transforms an instance for 1-IN-3SAT to an equivalent instance for POSITIVE 1-IN-3SAT, linear in the number of variables and clauses. Hence, we obtain: assuming ETH, POSITIVE 1-IN-3SAT cannot be solved in time $2^{o(n+m)}$ for inputs with n variables and m clauses.

3 Proof of Theorem 4 and Theorem 5

Recall that a perfect matching cut is in particular a matching cut, as well as a disconnected perfect matching. This observation leads to the following promise versions of MC, PMC and DPM. (We refer to [5] for background on promise problems.)

PROMISE-PMC MC

Instance: A graph G that either has no matching cut, or every matching cut is a perfect matching cut.

Question: Does G have a matching cut?

PROMISE-PMC PMC

Instance: A graph G that either has no perfect matching cut, or every matching cut is a perfect matching cut.

Question: Does G have a perfect matching cut?

PROMISE-PMC DPM

Instance: A graph G that either has no disconnected perfect matching, or every matching cut is a perfect matching cut.

Question: Does G have a disconnected perfect matching?

In all the promise versions above, we are allowed not to consider certain input graphs. In PROMISE-PMC MC, PROMISE-PMC PMC and PROMISE-PMC DPM, we are allowed to ignore graphs having a matching cut that is not a perfect matching cut, for which MC must answer 'yes', and PMC and DPM may answer 'yes' or 'no'.

We slightly improve Theorem 4 by showing the following result.

Theorem 7. PROMISE-PMC MC, PROMISE-PMC PMC *and* PROMISE-PMC DPM *are* NP-*complete, even when restricted to* $\{3P_6, 2P_7, P_{14}\}$-*free graphs. Moreover, under the ETH, no algorithm with runtime* $2^{o(n)}$ *can solve any of these problems for* n-*vertex* $\{3P_6, 2P_7, P_{14}\}$-*free input graphs.*

Clearly, Theorem 7 implies Theorem 4. Theorem 7 shows in particular that distinguishing between graphs without matching cuts and graphs in which every matching cut is a perfect matching cut is hard, and not only between those without matching cuts and those with matching cuts which is implied by the NP-completeness of MC. Similar implications of Theorem 7 can be derived for PMC and DPM.

Also, Theorem 7 implies Theorem 5. Indeed, if MC or PMC is NP-hard in a graph class then MAXMC is NP-hard in the same class as well.

3.1   The Reduction

We give a polynomial-time reduction from POSITIVE 1-IN-3SAT to PROMISE-PMC PMC (and to PROMISE-PMC MC, PROMISE-PMC DPM at the same time).

Let ϕ be a 3-CNF formula with m clauses C_j, $1 \leq j \leq m$, and n variables x_i, $1 \leq i \leq n$, in which each clause C_j consists of three distinct variables. We will construct a $\{3P_6, 2P_7, P_{14}\}$-free graph G such that G has a perfect matching cut if and only if ϕ admits an 1-in-3 assignment. Moreover, every matching cut of G, if any, is a perfect matching cut.

For each clause C_j consisting of three variables c_{j1}, c_{j2} and c_{j3}, let $G(C_j)$ be the graph depicted in Fig. 2. We call c_j and c'_j the *clause vertices*, and c_{j1}, c_{j2} and c_{j3} the *variable vertices*. Then, the graph G is obtained from all $G(C_j)$ by adding

- all possible edges between variable vertices c_{jk} and $c_{j'k'}$ of the same variable. Thus, for each variable x,

$$Q(x) = \{c_{jk} \mid 1 \leq j \leq m, 1 \leq k \leq 3, x \text{ occurs} \\ \text{in clause } C_j \text{ as } c_{jk}\}$$

is a clique in G,
- all possible edges between the $2m$ clause vertices c_j and c'_j. Thus,

$$F = \{c_j \mid 1 \leq j \leq m\} \cup \{c'_j \mid 1 \leq j \leq m\}$$

is a clique in G,
- all possible edges between the $3m$ vertices a_{jk}. Thus,

$$T = \{a_{jk} \mid 1 \leq j \leq m, 1 \leq k \leq 3\}$$

is a clique in G.

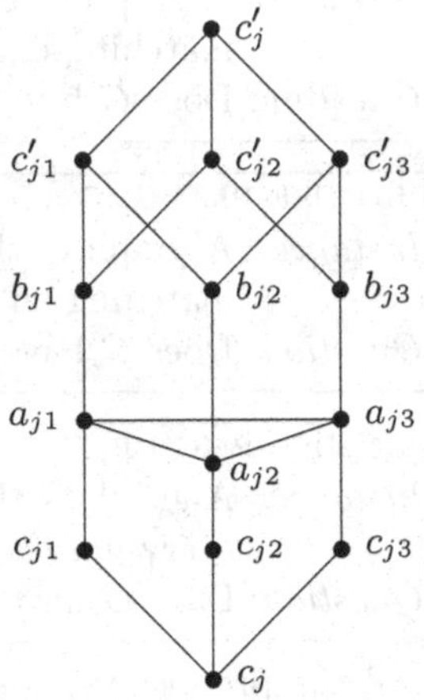

Fig. 2. The gadget $G(C_j)$.

The description of G is complete. As an example, the graph G from the formula ϕ with three clauses $C_1 = \{x, y, z\}, C_2 = \{u, z, y\}$ and $C_3 = \{z, v, w\}$ is depicted in Fig. 3.

Notice that no edge exists between the two cliques F and T. Notice also that $G - F - T$ has exactly $m + n$ components:

- For each $1 \leq j \leq m$, the 6-cycle $D_j : b_{j1}, c'_{j1}, b_{j2}, c'_{j3}, b_{j3}, c'_{j2}$ is a component of $G - F - T$, call it the clause 6-cycle (of clause C_j),
- For each variable x, the clique $Q(x)$ is a component of $G - F - T$, call it the variable clique (of variable x).

Lemma 1. *G is $\{3P_6, 2P_7, P_{14}\}$-free.*

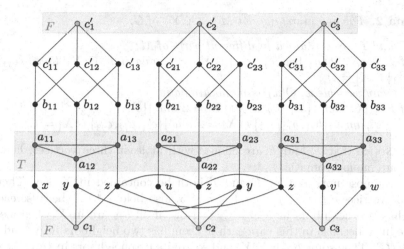

Fig. 3. The graph G from the formula ϕ with three clauses $C_1 = \{x, y, z\}, C_2 = \{u, z, y\}$ and $C_3 = \{z, v, w\}$. The 6 flax vertices $c_1, c_2, c_3, c'_1, c'_2, c'_3$ and the 9 teal vertices $a_{11}, a_{12}, a_{13}, a_{21}, a_{22}, a_{23}, a_{31}, a_{32}, a_{33}$ form the clique F and T, respectively.

Proof. First, observe that each component of $G - F - T$ is a clause 6-cycle D_j or a variable clique $Q(x)$. Hence,

$$G - F - T \text{ is } P_6\text{-free.} \tag{1}$$

Therefore, every induced P_6 in G must contain a vertex from the clique F or from the clique T. This shows that G is $3P_6$-free.

Observe next that, for each j, $c'_j \in F$ is the cut-vertex in $G - T$ separating the clause 6-cycle D_j and F, and $N(c'_j) \cap D_j = \{c'_{j1}, c'_{j2}, c'_{j3}\}$. Observe also that, for each x, $(G - T)[Q(x) \cup F]$ is a co-bipartite graph, the complement of a bipartite graph. Hence, it can be verified immediately that

$$G - T \text{ is } P_7\text{-free.} \tag{2}$$

Fact (2) implies that every induced P_7 in G must contain a vertex from the clique T. This shows that G is $2P_7$-free.

We now are ready to argue that G is P_{14}-free. Suppose not and let $P : v_1, v_2, \ldots, v_{14}$ be an induced P_{14} in G, with edges $v_i v_{i+1}$, $1 \leq i < 14$. For $i < j$, write $P[v_i, v_j]$ for the subpath of P between (and including) v_i and v_j. Then, by (2), each of $P[v_1, v_7]$ and $P[v_8, v_{14}]$ contains a vertex from the clique T. Since P has no chords, $P[v_1, v_7]$ has only the vertex v_7 in T and $P[v_8, v_{14}]$ has only the vertex v_8 in T. By (1), therefore, both $P[v_1, v_6]$ and $P[v_9, v_{14}]$ contain some vertex in the clique F, and thus P has a chord. This contradiction shows that G is P_{14}-free, as claimed. $\square$

We remark that there are many induced P_{13} in G; we briefly discuss the limit of our construction in the appendix.

Lemma 2. *For any matching cut $M = (X, Y)$ of G,*

(i) *F and T are contained in different parts of M;*
(ii) *if $F \subseteq X$, then $|\{c_{j1}, c_{j2}, c_{j3}\} \cap Y| = 1$, and if $F \subseteq Y$, then $|\{c_{j1}, c_{j2}, c_{j3}\} \cap X| = 1$;*
(iii) *for any variable x, $Q(x)$ is monochromatic;*
(iv) *if $F \subseteq X$, then $|\{b_{j1}, b_{j2}, b_{j3}\} \cap Y| = 2$ and $|\{c'_{j1}, c'_{j2}, c'_{j3}\} \cap Y| = 1$, and if $F \subseteq Y$, then $|\{b_{j1}, b_{j2}, b_{j3}\} \cap X| = 2$ and $|\{c'_{j1}, c'_{j2}, c'_{j3}\} \cap X| = 1$.*

Proof. Notice that F and T are cliques with at least three vertices, hence F and T are monochromatic.

(i): Suppose not, and let F and T both be contained in X, say. Then all variable vertices c_{jk}, $1 \le j \le m, 1 \le k \le 3$, also belong to X because each of them has two neighbors in $F \cup T \subseteq X$. Now, if all b_{jk} are in X, then also all c'_{jk} are in X because in this case each of them has two neighbors in X, and thus $X = V(G)$. Thus some b_{jk} is in Y, and so are its two neighbors in $\{c'_{j1}, c'_{j2}, c'_{j3}\}$. But then c'_j, which is in X, has two neighbors in Y. This contradiction shows that F and T must belong to different parts of M, hence (i).

(ii): By (i), let $F \subseteq X$ and $T \subseteq Y$, say. (The case $F \subseteq Y$ is symmetric.) Then, for any j, at most one of c_{j1}, c_{j2} and c_{j3} can be outside X. Assume that, for some j, all c_{j1}, c_{j2}, c_{j3} are in X. The assumption implies that all b_{j1}, b_{j2}, b_{j3} belong to Y, and then all $c'_{j1}, c'_{j2}, c'_{j3}$ belong to Y, too. But then c'_j, which is in X, has three neighbors in Y. This contradiction shows (ii).

(iii): Suppose that two variable vertices c_{jk} and $c_{j'k'}$ in some clique $Q(x)$ are in different parts of M. Then, as c_{jk} and $c_{j'k'}$ have neighbor c_j and $c_{j'}$, respectively, in the monochromatic clique F, c_{jk} has two neighbors in the part of $c_{j'k'}$ or $c_{j'k'}$ has two neighbors in the part of c_{jk}. This contradiction shows (iii).

(iv): This fact can be derived from (i) and (ii). □

Lemma 3. *Every matching cut of G, if any, is a perfect matching cut.*

Proof. Let $M = (X, Y)$ be a matching cut of G. By Lemma 2 (i), let $F \subseteq X$ and $T \subseteq Y$, say. We argue that every vertex in X has a neighbor (hence exactly one) in Y. Indeed, for each j,

- $c_j \in F \subseteq X$ has a neighbor $c_{jk} \in Y$ (by Lemma 2 (ii)),
- $c'_j \in F \subseteq X$ has a neighbor $c'_{jk} \in Y$ (by Lemma 2 (iv)),
- each $c_{jk} \in X$ has a neighbor $a_{jk} \in T \subseteq Y$ (by construction of G),
- each $b_{jk} \in X$ has a neighbor $a_{jk} \in T \subseteq Y$ (by construction of G),
- each $c'_{jk} \in X$ has a neighbor in $\{b_{j1}, b_{j2}, b_{j3}\} \cap Y$ (by Lemma 2 (iv)).

Similarly, it can be seen that every vertex in Y has a neighbor in X. □

Lemma 4. *If ϕ has an 1-in-3 assignment, then G has a perfect matching cut.*

Proof. Partition $V(G)$ into disjoint subsets X and Y as follows. (Figure 4 shows the partition for the example graph in Fig. 3 given the assignment $y = v = \text{True}$, $x = z = u = w = \text{False}$.) First,

- put F into X, and for all variables x which are assigned with False, put $Q(x)$ into X;
- for each $1 \leq j \leq m$, let c_{jk} with $k = k(j) \in \{1, 2, 3\}$ be the variable vertex, for which the variable x of c_{jk} is assigned with True. Then put b_{jk} and its two neighbors in $\{c'_{j1}, c'_{j2}, c'_{j3}\}$ into X.

Let $Y = V(G) \setminus X$. Then, it is not difficult to verify that $M = (X, Y)$ is a perfect matching cut of G. □

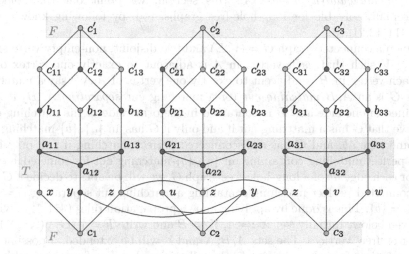

Fig. 4. The perfect matching cut (X, Y) of the example graph G in Fig. 3 given the assignment $y = v =$ True, $x = z = u = w =$ False. X and Y consist of the flax and teal vertices, respectively.

We now are ready to prove Theorem 7: First note that by Lemmas 1 and 3, G is $\{3P_6, 2P_7, P_{14}\}$-free and every matching cut of G (if any) is a perfect matching cut. In particular, every matching cut of G is extendable to a perfect matching.

Now, suppose ϕ has an 1-in-3 assignment. Then, by Lemma 4, G has a perfect matching cut. In particular, G has a disconnected perfect matching and, actually, a matching cut.

Conversely, let G have a matching cut $M = (X, Y)$, possibly a perfect matching cut or one that is contained in a perfect matching of G. Then, by Lemma 2 (i), we may assume that $F \subseteq X$, and set variable x to True if the corresponding variable clique $Q(x)$ is contained in Y and False if $Q(x)$ is contained in X. By Lemma 2 (iii), this assignment is well defined. Moreover, it is an 1-in-3 assignment for ϕ: consider a clause $C_j = \{x, y, z\}$ with $c_{j1} = x, c_{j2} = y$ and $c_{j3} = z$. By Lemma 2 (ii) and (iii), exactly one of $Q(x), Q(y)$ and $Q(z)$ is contained in Y, hence exactly one of x, y and z is assigned True.

Finally, note that G has $N = 14m$ vertices and recall that, assuming ETH, POSITIVE 1-IN-3SAT cannot be solved in $2^{o(m)}$ time. Thus, the ETH implies that no algorithm with runtime $2^{o(N)}$ exists for PROMISE-PMC MC, PROMISE-PMC PMC and PROMISE-PMC DPM, even when restricted to N-vertex $\{3P_6, 2P_7, P_{14}\}$-free graphs.

The proof of Theorem 7 is complete.

4   Proof of Theorem 6

Recall Theorem 3, MC is polynomially solvable for long-hole-free graphs (also called *quadrangulated graphs*). In this section, we point out that DPM is polynomially solvable for long-hole-free graphs, too, by following known approach [11, 12, 21].

Given a connected graph $G = (V, E)$ and two disjoint, non-empty vertex sets $A, B \subset V$ such that each vertex in A is adjacent to exactly one vertex of B and each vertex in B is adjacent to exactly one vertex of A. We say a matching cut of G is an A, B-*matching cut* (or a matching cut *separating A, B*) if A is contained in one side and B is contained in the other side of the matching cut. Observe that G has a matching cut if and only if G has an $\{a\}, \{b\}$-matching cut for some edge ab, and G has a disconnected perfect matching if and only if G has a perfect matching containing an $\{a\}, \{b\}$-matching cut for some edge ab.

For each edge ab of a long-hole-free graph G, we will be able to decide if G has a disconnected perfect matching containing a matching cut separating $A = \{a\}$ and $B = \{b\}$. This is done by applying known propagation rules ([11, 12]), which are given below. Initially, set $X := A$, $Y := B$ and write $F = V(G) \setminus (X \cup Y)$ for the set of 'free' vertices. The sets A, B, X and Y will be extended, if possible, by adding vertices from F according to the following rules. The first three rules will detect certain vertices that ensure that G cannot have an A, B-matching cut.

(R1) Let $v \in F$ be adjacent to a vertex in A. If v is
 – adjacent to a vertex in B, or
 – adjacent to (at least) two vertices in $Y \setminus B$,
then G has no A, B-matching cut.

(R2) Let $v \in F$ be adjacent to a vertex in B. If v is
 – adjacent to a vertex in A, or
 – adjacent to (at least) two vertices in $X \setminus A$,
then G has no A, B-matching cut.

(R3) If $v \in F$ is adjacent to (at least) two vertices in $X \setminus A$ and to (at least) two vertices in $Y \setminus B$, then G has no A, B-matching cut.

The correctness of (R1), (R2) and (R3) is quite obvious. We assume that, before each application of the rules (R4) and (R5) below, none of (R1), (R2) and (R3) is applicable.

(R4) Let $v \in F$ be adjacent to a vertex in A or to (at least) two vertices in $X \setminus A$. Then $X := X \cup \{v\}$, $F := F \setminus \{v\}$. If, moreover, v has a unique neighbor $w \in Y \setminus B$ then $A := A \cup \{v\}$, $B := B \cup \{w\}$.

(R5) Let $v \in F$ be adjacent to a vertex in B or to (at least) two vertices in $Y \setminus B$. Then $Y := Y \cup \{v\}$, $F := F \setminus \{v\}$. If, moreover, v has a unique neighbor $w \in X \setminus A$ then $B := B \cup \{v\}$, $A := A \cup \{w\}$.

We refer to [12] for the correctness of rules (R4) and (R5), and for the following facts.

Fact 1. *The total runtime for applying (R1) – (R5) until none of the rules is applicable is bounded by $O(nm)$.*

Fact 2. *Suppose none of (R1) – (R5) is applicable. Then*

- (X, Y) *is an A, B-matching cut of $G[X \cup Y]$, and any A, B-matching cut of G must contain X in one side and Y in the other side;*
- *for any vertex $v \in F$,*

$$N(v) \cap A = \emptyset, \ N(v) \cap B = \emptyset \ and \ |N(v) \cap X| \leq 1, \ |N(v) \cap Y| \leq 1.$$

We now are ready to prove Theorem 6: Let G be a connected, long-hole-free graph, and let ab be an edge of G. Set $A = \{a\}$ and $B = \{b\}$, and assume that none of (R1) – (R5) is applicable. Then, denoting $N(S)$ the set of vertices outside S adjacent to some vertex in S,

for any connected component S of $G[F]$, $|N(S) \cap X| = 0$ or $|N(S) \cap Y| = 0$.

For, otherwise choose two vertices $s, s' \in S$ with a neighbor $x \in N(s) \cap X$ and a neighbor $y \in N(s') \cap Y$ such that the distance between s and s' in S is as small as possible. Then s, s', x and y and a shortest s, s'-path in S, a chordless x, y-path in $G[X \cup Y]$ together would induce a long hole in G. (Observe that, by the definition of X and Y, $G[X \cup Y]$ is connected.)

Partition F into disjoint subsets F_X and F_Y as follows:

$$F_X = \bigcup \{S : S \text{ is a connected component of } G[F] \text{ with } N(S) \cap X \neq \emptyset\},$$

$$F_Y = \bigcup \{T : T \text{ is a connected component of } G[F] \text{ with } N(T) \cap Y \neq \emptyset\}.$$

Then, by the facts above and recall that G is connected,

$$F = F_X \cup F_Y \text{ and } F_X \cap F_Y = \emptyset.$$

Thus,

$$(X \cup F_X, Y \cup F_Y) \text{ is an } A, B\text{- matching cut of } G,$$

and it follows, that

G has a disconnected perfect matching containing an A, B-matching cut if and only if $G - A - B$ has a perfect matching.

Therefore, with Fact 1, in time $O(nm)$ we can decide whether G has a matching cut containing a given edge. Moreover, as a maximum matching can be computed in $O(\sqrt{n}m)$ time [19], we can decide in time $O(n\sqrt{n}m^2)$ whether G has a disconnected perfect matching containing an $\{a\}, \{b\}$-matching cut for a given edge ab. Since there are at most m edges to check, Theorem 6 follows.

5   Conclusion

We have shown that all three problems MC, PMC and DPM are NP-complete in P_{14}-free graphs. The hardness result for PMC solves an open problem posed in [4,14,15]. For MC and DPM, the hardness result improves the previously known one in P_{19}-free graphs, respectively, in P_{23}-free graphs, to P_{14}-free graphs. An obvious question is whether one of these problems remains NP-complete in P_t-free graphs for some $t < 14$.

We also pointed out that, like MC [21], DPM can be solved in polynomial time when restricted to long-hole-free graphs. We leave open the complexity of PMC restricted to long-hole-free graphs. More general, the chordality of a graph G is the length of a longest induced cycle in G. Chordal graphs and long-hole-free graphs (including weakly chordal and chordal bipartite graphs) have chordality 3 and 4, respectively. Notice that P_t-free graphs have chordality bounded by t, hence Theorem 4 implies that MC, PMC and DPM are NP-complete when restricted to graphs of chordality ≤ 14. We remark, however, that the graph constructed in the proof of Theorem 4 has chordality 8, and thus MC, PMC and DPM are NP-complete when restricted to graphs of chordality ≤ 8. Does there exist any class of graphs of chordality < 8 in which MC, PMC or DPM is NP-complete?

Acknowledgment. We thank the anonymous reviewers of WG 2023 for their very carefull reading. In particular, we thank all three reviewers for pointing out a small mistake in the earlier proof of Theorem 6.

A   Limits of Our Reduction in the Proof of Theorem 4

As remarked, the graph G constructed from an instance of POSITIVE 1-IN-3SAT contains many induced paths P_{13}. For example, refer to Fig. 3; see also Fig. 5–7:

- $b_{11}, c'_{11}, b_{12}, c'_{13}, b_{13}, a_{13}, a_{22}, c_{22} = z, c_{31} = z, c_3, c_1, c_{12} = y, c_{23} = y$;
- $b_{11}, c'_{11}, b_{12}, c'_{13}, b_{13}, a_{13}, a_{21}, b_{21}, c'_{21}, c'_2, c'_3, c'_{33}, b_{33}$;
- $b_{11}, c'_{11}, b_{12}, c'_{13}, b_{13}, a_{13}, a_{21}, b_{21}, c'_{21}, c'_2, c_2, c_{23} = y, c_{12} = y$.

It can be seen that all P_{13} in G contain a P_5 from a 6-cycle D_j : $b_{j1}, c'_{j1}, b_{j2}, c'_{j3}, b_{j3}, c'_{j2}$. We now are going to describe how the gadget $G(C_j)$ used in the construction of G depicted in Fig. 2 was found. This could be useful when one is trying to improve the construction with shorter induced paths.

A general idea in constructing a graph without long induced paths from a given CNF-formula is to ensure that long induced paths must go through some, say at most three, cliques. Assuming we want to reduce POSITIVE 1-IN-3SAT (or NAE 3SAT) to PMC, the following observation gives a hint how to get such a clique: Let G be a graph, in which the seven vertices $c, c_k, a_k, 1 \leq k \leq 3$, induce a tree with leaves a_1, a_2, a_3 and degree-2 vertices c_1, c_2, c_3 and the degree-3 vertex c. If G has a perfect matching cut, then a_1, a_2, a_3 must belong to the same part of the cut. Therefore, we can make $\{a_1, a_2, a_3\}$ adjacent to a clique and the resulting graph still has a perfect matching cut.

Now, a gadget $G(H; v)$ may be obtained from a suitable graph H with $v \in V(H)$ as follows. Let H be a graph having a vertex v of degree 3. Let b_1, b_2, b_3 be the neighbors of v in H. Let $G(H; v)$ be the graph obtained from $H - v$ by adding 7 new vertices $a_1, a_2, a_3, c_1, c_2, c_3$ and c, and edges cc_k, $c_k a_k$, $a_k b_k$, $1 \le k \le 3$, and $a_1 a_2, a_1 a_3$ and $a_2 a_3$. (Thus, contracting the triangle $a_1 a_2 a_3$ from $G(H; v) \setminus \{c, c_1, c_2, c_3\}$ we obtain the graph H.)

Observation 1. *Assuming, for any neighbor w of v in H, H has a perfect matching cut (X, Y) such that $v \in X$ and $w \in Y$. Then, for any neighbor d of c in $G(H; v)$, the graph $G(H; v)$ has a perfect matching cut (X', Y') such that $c \in X'$ and $d \in Y'$.*

Examples of graphs H in Observation 1 include the cube, the Petersen graph and the 10-vertex Heggernes-Telle graph in [8, Fig. 3.1]. Our gadget $G(C_j)$ depicted in Fig. 2 is obtained by taking the cube. Take the Petersen graph or the Heggernes-Telle graph will produce induced P_t for some $t \ge 15$. If there exists another graph H 'better' than the cube, then our construction will yield a P_t-free graph for some $10 \le t \le 13$.

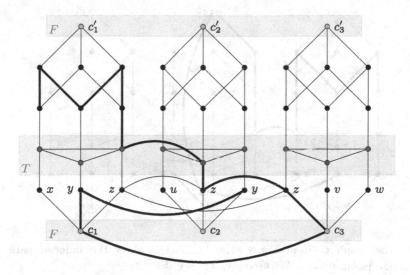

Fig. 5. The graph G from Fig. 3. The bold edges show the induced path P_{13}: $b_{11}, c'_{11}, b_{12}, c'_{13}, b_{13}, a_{13}, a_{22}, c_{22} = z, c_{31} = z, c_3, c_1, c_{12} = y, c_{23} = y$.

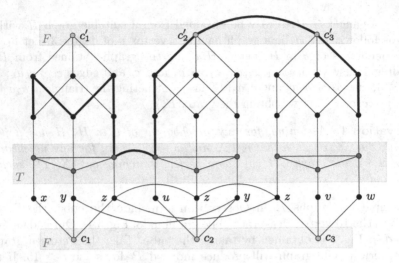

Fig. 6. The graph G from Fig. 3. The bold edges show the induced path P_{13}: $b_{11}, c'_{11}, b_{12}, c'_{13}, b_{13}, a_{13}, a_{21}, b_{21}, c'_{21}, c'_2, c'_3, c'_{33}, b_{33}$.

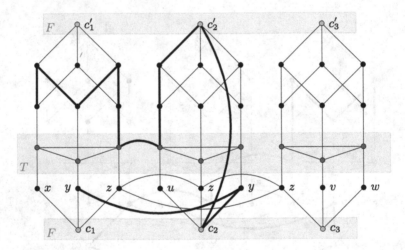

Fig. 7. The graph G from Fig. 3. The bold edges show the induced path P_{13}: $b_{11}, c'_{11}, b_{12}, c'_{13}, b_{13}, a_{13}, a_{21}, b_{21}, c'_{21}, c'_2, c_2, c_{23} = y, c_{12} = y$.

References

1. Bouquet, V., Picouleau, C.: The complexity of the perfect matching-cut problem. CoRR, abs/2011.03318v2 (2021). https://doi.org/10.48550/arXiv.2011.03318
2. Chen, C.-Y., Hsieh, S.-Y., Le, H.-O., Le, V.B., Peng, S.-L.: Matching cut in graphs with large minimum degree. Algorithmica **83**(5), 1238–1255 (2020). https://doi.org/10.1007/s00453-020-00782-8
3. Chvátal, V.: Recognizing decomposable graphs. J. Graph Theory **8**(1), 51–53 (1984). https://doi.org/10.1002/jgt.3190080106

4. Feghali, C., Lucke, F., Paulusma, D., Ries, B.: New hardness results for matching cut and disconnected perfect matching. CoRR, abs/2212.12317 (2022). https://doi.org/10.48550/arXiv.2212.12317

5. Goldreich, O.: On promise problems: a survey. In: Goldreich, O., Rosenberg, A.L., Selman, A.L. (eds.) Theoretical Computer Science. LNCS, vol. 3895, pp. 254–290. Springer, Heidelberg (2006). https://doi.org/10.1007/11685654_12

6. Golovach, P.A., Komusiewicz, C., Kratsch, D., Le, V.B.: Refined notions of parameterized enumeration kernels with applications to matching cut enumeration. STACS 2021, LIPIcs 187, 37:1–37:18 (2021). also. J. Comput. Syst. Sci., **123**, 76–102 (2022). https://doi.org/10.1016/j.jcss.2021.07.005

7. Graham, R.L.: On primitive graphs and optimal vertex assignments. Ann. N. Y. Acad. Sci. **175**(1), 170–186 (1970)

8. Pinar Heggernes and Jan Arne Telle: Partitioning graphs into generalized dominating sets. Nord. J. Comput. **5**(2), 128–142 (1998)

9. Impagliazzo, R., Paturi, R.: On the complexity of k-sat. J. Comput. Syst. Sci. **62**(2), 367–375 (2001). https://doi.org/10.1006/jcss.2000.1727

10. Impagliazzo, R., Paturi, R., Zane, F.: Which problems have strongly exponential complexity? J. Comput. Syst. Sci. **63**(4), 512–530 (2001). https://doi.org/10.1006/jcss.2001.1774

11. Dieter Kratsch and Van Bang Le: Algorithms solving the matching cut problem. Theor. Comput. Sci. **609**, 328–335 (2016). https://doi.org/10.1016/j.tcs.2015.10.016

12. Le, H.O., Le, V.B.: A complexity dichotomy for matching cut in (bipartite) graphs of fixed diameter. Theor. Comput. Sci. **770**, 69–78 (2019). https://doi.org/10.1016/j.tcs.2018.10.029

13. Le , V.B., Telle, J.A.: The perfect matching cut problem revisited. In: Proceedings of the WG 2021, LNCS, vol. 12911, pp. 182–194 (2021). Also, Theor. Comput. Sci. **931**, 117–130 (2022). https://doi.org/10.1016/j.tcs.2022.07.035

14. Lucke, F., Paulusma, D., Ries, B.: Finding matching cuts in H-free graphs. In: Bae, S.W., Park, H. (eds.) 33rd International Symposium on Algorithms and Computation, ISAAC (2022), volume 248 of LIPIcs, pp. 22:1–22:16 (2022). https://doi.org/10.4230/LIPIcs.ISAAC.2022.22

15. Lucke, F., Paulusma, D., Ries, B.: Finding matching cuts in H-free graphs. CoRR, abs/2207.07095 (2022). https://doi.org/10.48550/arXiv.2207.07095

16. Lucke, F., Paulusma, D., Ries, B.: On the complexity of matching cut for graphs of bounded radius and H-free graphs. Theor. Comput. Sci. **936**, 33–42 (2022). https://doi.org/10.1016/j.tcs.2022.09.014

17. Lucke, F., Paulusma, D., Ries, B.: Dichotomies for maximum matching cut: H-freeness, bounded diameter, bounded radius. In: MFCS 2023. CoRR, abs/2304.01099 (2023). https://doi.org/10.48550/arXiv.2304.01099

18. Lucke, F., Paulusma, D., Ries, B.: Finding matching cuts in H-free graphs. Algorithmica (2023). https://doi.org/10.1007/s00453-023-01137-9

19. Micali, S., Vazirani, V.V.: An $O(\sqrt{|V|} \cdot |E|)$ algorithm for finding maximum matching in general graphs. In: 21st Annual Symposium on Foundations of Computer Science, Syracuse, New York, USA, 13–15 October 1980, pp. 17–27. IEEE Computer Society (1980). https://doi.org/10.1109/SFCS.1980.12

20. Bernard, M., Moret, E.: Theory of Computation. Addison-Wesley-Longman, Boston (1998)

21. Moshi, A.M.: Matching cutsets in graphs. J. Graph Theory **13**(5), 527–536 (1989). https://doi.org/10.1002/jgt.3190130502

Upper Clique Transversals in Graphs

Martin Milanič[1][(✉)] and Yushi Uno[2]

[1] FAMNIT and IAM, University of Primorska, Koper, Slovenia
martin.milanic@upr.si
[2] Graduate School of Informatics, Osaka Metropolitan University, Sakai, Japan
yushi.uno@omu.ac.jp

Abstract. A *clique transversal* in a graph is a set of vertices intersecting all maximal cliques. The problem of determining the minimum size of a clique transversal has received considerable attention in the literature. In this paper, we initiate the study of the "upper" variant of this parameter, the *upper clique transversal number*, defined as the maximum size of a minimal clique transversal. We investigate this parameter from the algorithmic and complexity points of view, with a focus on various graph classes. We show that the corresponding decision problem is NP-complete in the classes of chordal graphs, chordal bipartite graphs, and line graphs of bipartite graphs, but solvable in linear time in the classes of split graphs and proper interval graphs.

Keywords: Clique transversal · Upper clique transversal number · Vertex cover

1 Introduction

A set of vertices of a graph G that meets all maximal cliques of G is called a *clique transversal* in G. Clique transversals in graphs have been studied by Payan in 1979 [36], by Andreae, Schughart, and Tuza in 1991 [4], by Erdős, Gallai, and Tuza in 1992 [20], and also extensively researched in the more recent literature (see, e.g., [3,5,10,13,15,19,23,28–31,37]). What most of these papers have in common is that they are interested in questions regarding the *clique transversal number* of a graph, that is, the minimum size of a clique transversal of the graph. For example, Chang, Farber, and Tuza showed in [13] that computing the clique transversal number for split graphs is NP-hard, and Guruswami and Pandu Rangan showed in [23] that the problem is NP-hard for cocomparability, planar, line, and total graphs, and solvable in polynomial time for Helly circular-arc graphs, strongly chordal graphs, chordal graphs of bounded clique size, and cographs.

In this paper, we initiate the study of the "upper" version of this graph invariant, the *upper clique transversal number*, denoted by $\tau_c^+(G)$ and defined as the maximum size of a minimal clique transversal, where a clique transversal in a graph G is said to be *minimal* if it does not contain any other clique transversal. The corresponding decision problem is defined as follows.

UPPER CLIQUE TRANSVERSAL (UCT)

Input: A graph G and an integer k.
Question: Does G contain a minimal clique transversal S such that $|S| \geq k$?

Our study contributes to the literature on upper variants of graph minimization problems, which already includes the upper vertex cover (also known as maximum minimal vertex cover; see [11,16,41]), upper feedback vertex set (also known as maximum minimal feedback vertex set; see [18,27]), upper edge cover (see [26]), upper domination (see [2,6,25]), and upper edge domination (see [33]).

Our Results. We provide a first set of results on the algorithmic complexity of UPPER CLIQUE TRANSVERSAL. Since clique transversals have been mostly studied in the class of chordal graphs and related classes, we also find it natural to first focus on this interesting graph class and its subclasses. In this respect, we provide an NP-completeness result as well as two very different linear-time algorithms. We show that UCT is NP-complete in the class of chordal graphs, but solvable in linear time in the classes of split graphs and proper interval graphs. Note that the result for split graphs is in contrast with the aforementioned NP-hardness result for computing the clique transversal number in the same class of graphs [13]. In addition, we provide NP-completeness proofs for two more subclasses of the class of perfect graphs, namely for chordal bipartite graphs, and for line graphs of bipartite graphs.

The diagram in Fig. 1 summarizes the relationships between various graph classes studied in this paper and indicates some boundaries of tractability of the UCT problem. We define those graph classes in the corresponding later sections in the paper. For further background and references on graph classes, we refer to [12].

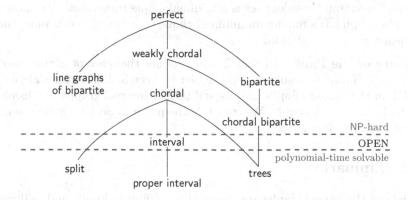

Fig. 1. The complexity of UCT in various graph classes studied in this paper.

Our Approach. Our approach is based on connections with a number of graph parameters. For example, the NP-completeness proofs for the classes of chordal bipartite graphs and of line graphs of bipartite graphs are based on the fact that for triangle-free graphs without isolated vertices, minimal clique transversals are exactly the minimal vertex covers, and they are closely related with minimal edge covers via the line graph operator. In particular, if G is a triangle-free graph without isolated vertices, then the upper clique transversal number of G equals the upper vertex cover number of G, that is, the maximum size of a minimal vertex cover. Since the upper vertex cover number of a graph G plus the independent domination number of G equals the order of G, there is also a connection with the independent dominating set problem. Let us note that, along with a linear-time algorithm for computing a minimum independent set in a tree [9], the above observations suffice to justify the polynomial-time solvability of the upper clique transversal problem on trees, as indicated in Fig. 1. The NP-completeness proof for the class of chordal graphs is based on a reduction from SPANNING STAR FOREST, the problem of computing a spanning subgraph with as many edges as possible that consists of disjoint stars; this problem, in turn, is known to be closely related to the dominating set problem.

The linear-time algorithm for computing the upper clique transversal number of proper interval graphs relies on a linear-time algorithm for the maximum induced matching problem in bipartite permutation graphs due to Chang [14]. More precisely, we prove that the upper clique transversal number of a given graph cannot exceed the maximum size of an induced matching of a derived bipartite graph, the *vertex-clique incidence graph*, and show, using new insights on the properties of the matching computed by Chang's algorithm, that for proper interval graphs, the two quantities are the same.

The linear-time algorithm for computing the upper clique transversal number of a split graph is based on a characterization of minimal clique transversals of split graphs. A clique transversal that is an independent set is also called a *strong independent set* (or *strong stable set*; see [32] for a survey). It is not difficult to see that every strong independent set is a minimal clique transversal. We show that every split graph has a maximum minimal clique transversal that is independent (and hence, a strong independent set).

Structure of the Paper. In Sect. 2 we introduce the relevant graph theoretic background. Hardness results are presented in Sect. 3. Linear-time algorithms for UCT in the classes of split graphs and proper interval graphs are developed in Sects. 4 and 5, respectively. We conclude the paper in Sect. 6. Some proofs are omitted due to lack of space.

2   Preliminaries

Throughout the paper, graphs are assumed to be finite, simple, and undirected. We use standard graph theory terminology, following West [39]. A graph G with vertex set V and edge set E is often denoted by $G = (V, E)$; we write $V(G)$ and $E(G)$ for V and E, respectively. The set of vertices adjacent to a vertex $v \in V$ is the *neighborhood* of v, denoted $N(v)$; its cardinality is the *degree* of v. The *closed neighborhood* is the set $N[v]$, defined as $N(v) \cup \{v\}$. An *independent set* in a graph is a set of pairwise non-adjacent vertices; a *clique* is a set of pairwise adjacent vertices. An independent set (resp., clique) in a graph G is *maximal* if it is not contained in any other independent set (resp., clique). A *clique transversal* in a graph is a subset of vertices that intersects all the maximal cliques of the graph. A *dominating set* in a graph $G = (V, E)$ is a set S of vertices such that every vertex not in S has a neighbor in S. An *independent dominating set* is a dominating set that is also an independent set. The *(independent) domination number* of a graph G is the minimum size of an (independent) dominating set in G. Note that a set S of vertices in a graph G is an independent dominating set if and only if S is a maximal independent set. In particular, the independent domination number of a graph is a well-defined invariant leading to a decision problem called INDEPENDENT DOMINATING SET.

The *clique number* of G is denoted by $\omega(G)$ and defined as the maximum size of a clique in G. An *upper clique transversal* of a graph G is a minimal clique transversal of maximum size. The *upper clique transversal number* of a graph G is denoted by $\tau_c^+(G)$ and defined as the maximum size of a minimal clique transversal in G. A *vertex cover* in G is a set $S \subseteq V(G)$ such that every edge $e \in E(G)$ has at least one endpoint in S. A vertex cover in G is *minimal* if it does not contain any other vertex cover. These notions are illustrated in Fig. 2. Note that if G is a triangle-free graph without isolated vertices, then the maximal cliques of G are exactly its edges, and hence the clique transversals of G coincide with its vertex covers.

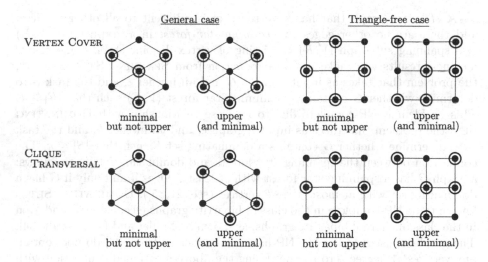

Fig. 2. Upper clique transversal and related notions.

3   Intractability of UCT for Some Graph Classes

In this section we prove that UPPER CLIQUE TRANSVERSAL is NP-complete in the classes of chordal graphs, chordal bipartite graphs, and line graphs of bipartite graphs. First, let us note that for the class of all graphs, we do not know whether the problem is in NP. If S is a minimal clique transversal in G such that $|S| \geq k$, then a natural way to verify this fact would be to certify separately that S is a clique transversal and that it is a minimal one. Assuming that S is a clique transversal, one can certify minimality simply by exhibiting for each vertex $u \in S$ a maximal clique C in G such that $C \cap S = \{u\}$. However, unless P = NP, we cannot verify the fact that S is a clique transversal in polynomial time. This follows from a result of Zang [40], showing that it is co-NP-complete to check, given a weakly chordal graph G and an independent set S, whether S is a clique transversal in G. A graph G is *weakly chordal* if neither G nor its complement contain an induced cycle of length at least five.

We do not know whether UPPER CLIQUE TRANSVERSAL is in NP when restricted to the class of weakly chordal graphs. However, for their subclasses chordal graphs and chordal bipartite graphs, membership of UCT in NP is a consequence of the following proposition.

Proposition 1. *Let $\mathcal{G}$ be a graph class such that every graph $G \in \mathcal{G}$ has at most polynomially many maximal cliques. Then,* UPPER CLIQUE TRANSVERSAL *is in* NP *for graphs in $\mathcal{G}$.*

A *star* is a graph that has a vertex that is adjacent to all other vertices, and there are no other edges. A *spanning star forest* in a graph $G = (V, E)$ is a spanning subgraph (V, F) consisting of vertex-disjoint stars. Some of our hardness results will make use of a reduction from SPANNING STAR FOREST, the problem that takes as input a graph G and an integer ℓ, and the task is to determine whether G contains a spanning star forest (V, F) such that $|F| \geq \ell$. This problem is NP-complete due to its close relationship with DOMINATING SET, the problem that takes as input a graph G and an integer k, and the task is to determine whether G contains a dominating set S such that $|S| \leq k$. The connection between the spanning star forests and dominating sets is as follows: a graph G has a spanning star forest with at least ℓ edges if and only if G has a dominating set with at most $|V| - \ell$ vertices (see [21,35]). DOMINATING SET is known to be NP-complete in the class of bipartite graphs (see, e.g., [8]) and even in the class of chordal bipartite graphs, as shown by Müller and Brandstädt [34]. The graphs constructed in the NP-hardness reduction from [34] do not contain any vertices of degree zero or one. Using the above-mentioned connection with SPANNING STAR FOREST, we obtain the following.

Theorem 1. SPANNING STAR FOREST *is* NP-*complete in the class of bipartite graphs with minimum degree at least* 2.

We present the hardness results in increasing order of difficulty of the proofs, starting with the class of chordal bipartite graphs. A *chordal bipartite* graph is a bipartite graph in which all induced cycles are of length four. Clearly, any chordal bipartite graph is triangle-free. Recall also that in any triangle-free graph G without isolated vertices, a set $S \subseteq V(G)$ is a minimal vertex cover if and only if it is a minimal clique transversal. Furthermore, in any graph G a set $S \subseteq V(G)$ is a minimal vertex cover if and only its complement $V(G) \setminus S$ is an independent dominating set. Using a reduction from the INDEPENDENT DOMINATING SET in chordal bipartite graphs (which is NP-complete [17]), we thus obtain the following.

Theorem 2. UPPER CLIQUE TRANSVERSAL *is* NP-*complete in the class of chordal bipartite graphs.*

We next consider the class of line graphs of bipartite graphs. The *line graph* of a graph G is the graph H with $V(H) = E(G)$ in which two distinct vertices are adjacent if and only if they share an endpoint as edges in G.

Lemma 1. *Let G be a triangle-free graph with minimum degree at least 2 and let H be the line graph of G. Then, the maximal cliques in H are exactly the sets E_v for $v \in V(G)$, where E_v is the set of edges in G that are incident with v.*

An *edge cover* of a graph G is a set F of edges such that every vertex of G is incident with some edge of F. Immediately from the definitions and Lemma 1 we obtain the following.

Lemma 2. *Let G be a triangle-free graph with minimum degree at least 2 and let H be the line graph of G. Then, a set $F \subseteq E(G)$ is a clique transversal in H if and only if F is an edge cover in G. Consequently, a set $F \subseteq E(G)$ is a minimal clique transversal in H if and only if F is a minimal edge cover in G.*

As shown by Hedetniemi [24], the maximum size of a minimal edge cover equals to the maximum number of edges in a spanning star forest, which is the number of vertices minus the domination number. Thus, using Proposition 1, Lemma 2, and a reduction from SPANNING STAR FOREST in the class of bipartite graphs with minimum degree at least 2 we obtain the following.

Theorem 3. UPPER CLIQUE TRANSVERSAL *is* NP-*complete in the class of line graphs of bipartite graphs.*

We now prove intractability of UCT in the class of chordal graphs. A graph is *chordal* if it does not contain any induced cycles on at least four vertices.

Theorem 4. UPPER CLIQUE TRANSVERSAL *is* NP-*complete in the class of chordal graphs.*

Proof (sketch). We reduce from SPANNING STAR FOREST. Let $G = (V, E)$ and ℓ be an input instance of SPANNING STAR FOREST. We may assume without loss of generality that G has an edge and that $\ell \geq 2$, since if any of these assumptions is violated, then it is trivial to verify if G has a spanning star forest with at least ℓ edges. We construct a chordal graph G' as follows. We start with a complete graph with vertex set V. For each edge $e = \{u, v\} \in E$, we introduce two new vertices x^e and y^e, and make x^e adjacent to u, to v, and to y^e. The obtained graph is G'. We thus have $V(G') = V \cup X \cup Y$, where $X = \{x^e : e \in E\}$ and $Y = \{y^e : e \in E\}$. See Fig. 3 for an example. Clearly, G' is chordal. Furthermore, let $k = \ell + |E|$.

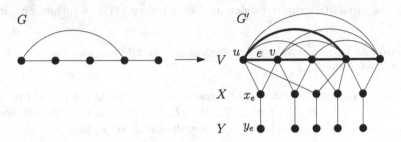

Fig. 3. Transforming G to G'.

To complete the proof, we show that G has a spanning star forest of size at least ℓ if and only if G' has a minimal clique transversal of size at least k.

First, assume that G has a spanning star forest (V, F) such that $|F| \geq \ell$. Since (V, F) is a spanning forest in which each component is a star, each edge of F is incident with a vertex of degree one in (V, F). Let S be a set obtained by selecting from each edge in F one vertex of degree one in (V, F). Then, every edge of F has one endpoint in S and the other one in $V \setminus S$. In particular, $|S| = |F| \geq \ell$. Let $S' = S \cup \{x^e : e \in E \setminus F\} \cup \{y^f : f \in F\}$ (see Fig. 4 for an example). The size of S' is at least $\ell + |E| = k$ and it can be shown that S' is a minimal clique transversal of G'.

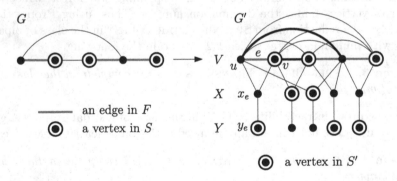

— an edge in F
◉ a vertex in S

◉ a vertex in S'

Fig. 4. Transforming a spanning star forest (V, F) in G into a minimal clique transversal S' in G'.

For the converse direction, let S' be a minimal clique transversal of G' such that $|S'| \geq k$. Let $S = S' \cap V$. It can be shown that we can associate to each vertex

$u \in S$ a vertex $v(u) \in V$ such that $e = \{u, v(u)\} \in E$ and $S' \cap \{u, v(u), x^e\} = \{u\}$. For each $u \in S$, let us denote by $e(u)$ the corresponding edge $\{u, v(u)\}$, and let $F = \{e(u) : u \in S\}$ (see Fig. 5). It can be shown that the set F satisfies $|F| = |S|$ and every vertex in S has degree one in (V, F). Therefore, the graph (V, F) is a spanning star forest of G.

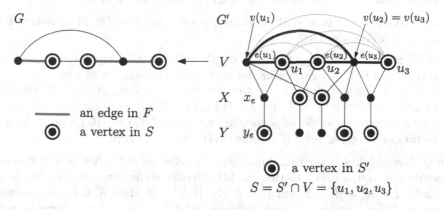

Fig. 5. Transforming a minimal clique transversal S' in G' into a spanning star forest (V, F) in G.

Since S' is a minimal clique transversal of G', for each edge $e \in E$ exactly one of x^e and y^e belongs to S'. Therefore, $|F| = |S| = |S'| - |E| \geq k - |E| = \ell$, and G has a spanning star forest of size at least ℓ. □

4 A Linear-Time Algorithm for UCT in Split Graphs

A *split graph* is a graph that has a *split partition*, that is, a partition of its vertex set into a clique and an independent set. We denote a split partition of a split graph G as (K, I) where K is a clique, I is an independent set, $K \cap I = \emptyset$, and $K \cup I = V(G)$. We may assume without loss of generality that I is a maximal independent set. In what follows, we repeatedly use the structure of maximal cliques of split graphs. If G is a split graph with a split partition (K, I), then the maximal cliques of G are as follows: the closed neighborhoods $N[v]$, for all $v \in I$, and the clique K, provided that it is a maximal clique, that is, every vertex in I has a non-neighbor in K.

Given a graph G and a set of vertices $S \subseteq V(G)$, we denote by $N(S)$ the set of all vertices in $V(G) \setminus S$ that have a neighbor in S. Moreover, given a vertex $v \in S$, an *S-private neighbor* of v is any vertex $w \in N(S)$ such that $N(w) \cap S = \{v\}$. The following proposition characterizes minimal clique transversals of split graphs.

Proposition 2. *Let G be a split graph with a split partition (K, I) such that I is a maximal independent set and let $S \subseteq V(G)$. Let $K' = K \cap S$ and $I' = I \cap S$. Then, S is a minimal clique transversal of G if and only if the following conditions hold:*

(i) $K' \neq \emptyset$ *if K is a maximal clique.*
(ii) $I' = I \setminus N(K')$.
(iii) *Every vertex in K' has a K'-private neighbor in I.*

Proposition 2 leads to the following result about maximum minimal clique transversals in split graphs. We denote by $\alpha(G)$ the *independence number* of a graph G, that is, the maximum size of an independent set in G.

Theorem 5. *Let G be a split graph with a split partition (K, I) such that I is a maximal independent set. Then:*

1. *If K is not a maximal clique in G, then I is a maximum minimal clique transversal in G; in particular, we have $\tau_c^+(G) = \alpha(G)$ in this case.*
2. *If K is a maximal clique in G, then for every vertex $v \in K$ with the smallest number of neighbors in I, the set $\{v\} \cup (I \setminus N(v))$ is a maximum minimal clique transversal in G; in particular, we have $\tau_c^+(G) = \alpha(G) - \delta_G(I, K) + 1$ in this case, where $\delta_G(I, K) = \min\{|N(v) \cap I| : v \in K\}$.*

Proof (sketch). Let S be a minimal clique transversal of G that is of maximum possible size and, subject to this condition, contains as few vertices from K as possible. Let $K' = K \cap S$ and $I' = I \cap S$. If $K' = \emptyset$, then K is not a maximal clique in G, and we have $S = I$, implying $\tau_c^+(G) = |S| = \alpha(G)$. Suppose now that $K' \neq \emptyset$. We first show that $|K'| = 1$. Suppose for a contradiction that $|K'| \geq 2$ and let $v \in K'$. Let I_v denote the set of K'-private neighbors of v in I and let $S' = (S \setminus \{v\}) \cup I_v$. Using Proposition 2, it can be verified that S' is a minimal clique transversal in G. Furthermore, since $v \in K'$, the set I_v is nonempty. This implies that $|S'| \geq |S|$; in particular, S' is a maximum minimal clique transversal in G. However, S' contains strictly fewer vertices from K than S, contradicting the choice of S. This shows that $|K'| = 1$, as claimed.

Let w be the unique vertex in K'. Since Condition (ii) from Proposition 2 holds for S, we have $I' = I \setminus N(w)$. Hence $S = \{w\} \cup (I \setminus N(w))$ and $|S| = 1 + |I| - |N(w) \cap I|$. Since $w \in K$, we have $|N(w) \cap I| \geq \delta_G(I, K)$ and hence $\tau_c^+(G) = |S| \leq \alpha(G) - \delta_G(I, K) + 1$. It can be verified that for every vertex $z \in K$ the set $X_z := \{z\} \cup (I \setminus N(z))$ satisfies Conditions (i)–(iii) from Proposition 2, and hence is a minimal clique transversal in G. Choosing z to be a vertex in K with the smallest number of neighbors in I, we obtain a set X_z of size $\alpha(G) - \delta_G(I, K) + 1$. Thus $\tau_c^+(G) \geq |X_z| = \alpha(G) - \delta_G(I, K) + 1$ and since we already proved that $\tau_c^+(G) \leq \alpha(G) - \delta_G(I, K) + 1$, any such X_z is optimal.

Since I is a maximal independent set and K is nonempty, we have $\delta_G(I, K) \geq 1$. Thus, $\tau_c^+(G) \leq \alpha(G)$. Suppose that K is not a maximal clique in G. Then I is a minimal clique transversal in G and therefore $\tau_c^+(G) \geq |I| = \alpha(G) \geq \tau_c^+(G)$. Hence equalities must hold throughout and I is a maximum minimal clique transversal. Finally, suppose that K is a maximal clique in G. Then every minimal clique transversal S in G satisfies $S \cap K \neq \emptyset$. In this case, the above analysis shows that for every vertex $v \in K$ with the smallest number of neighbors in I, the set $\{v\} \cup (I \setminus N(v))$ is a maximum minimal clique transversal in G. $\square$

Corollary 1. Upper Clique Transversal *can be solved in linear time in the class of split graphs.*

5 A Linear-Time Algorithm for UCT in Proper Interval Graphs

A graph $G = (V, E)$ is an *interval graph* if it has an *interval representation*, that is, if its vertices can be put in a one-to-one correspondence with a family $(I_v : v \in V)$ of closed intervals on the real line such that two distinct vertices u and v are adjacent if and only if the corresponding intervals I_u and I_v intersect. If G has a *proper interval representation*, that is, an interval representation in which no interval contains another, then G is said to be a *proper interval graph*.

Our approach towards a linear-time algorithm for UPPER CLIQUE TRANSVERSAL in the class of proper interval graphs is based on a relation between clique transversals in G and induced matchings in the so-called vertex-clique incidence graph of G. This relation is valid for arbitrary graphs.

UCT via Induced Matchings in the Vertex-Clique Incidence Graph

Given a graph $G = (V, E)$, we denote by B_G the *vertex-clique incidence graph* of G, a bipartite graph defined as follows. The vertex set of B_G consists of two disjoint sets X and Y such that $X = V$ and $Y = \mathcal{C}_G$, where $\mathcal{C}_G$ is the set of maximal cliques in G. The edge set of B_G consists of all pairs $x \in X$ and $C \in \mathcal{C}_G$ that satisfy $x \in C$. An *induced matching* in a graph G is a set M of pairwise disjoint edges such that the set of endpoints of edges in M induces no edges other than those in M. Given two disjoint sets of vertices A and B in a graph G, we say that A *dominates* B *in* G if every vertex in B has a neighbor in A. Given a matching M in a graph G and a vertex $v \in V(G)$, we say that v is M-*saturated* if it is an endpoint of an edge in M.

Clique transversals and minimal clique transversals of a graph G can be expressed in terms of the vertex-clique incidence graph as follows.

Lemma 3. *Let G be a graph, let $B_G = (X, Y; E)$ be its vertex-clique incidence graph, and let $S \subseteq V(G)$. Then:*

1. *S is a clique transversal in G if and only if S dominates Y in B_G.*
2. *S is a minimal clique transversal in G if and only if S dominates Y in B_G and there exists an induced matching M in B_G such that S is exactly the set of M-saturated vertices in X.*

The *induced matching number* of a graph G is the maximum size of an induced matching in G.

Corollary 2. *For every graph G, the upper clique transversal number of G is at most the induced matching number of B_G.*

As another consequence of Lemma 3, we obtain a sufficient condition for a set of vertices in a graph to be a minimal clique transversal of maximum size.

Corollary 3. *Let G be a graph, let $B_G = (X, Y; E)$ be its vertex-clique incidence graph, and let $S \subseteq V(G)$. Suppose that S dominates Y in B_G and there exists a maximum induced matching M in B_G such that S is exactly the set of M-saturated vertices in X. Then, S is a minimal clique transversal in G of maximum size.*

To apply Corollary 3 to proper interval graphs, we first state several characterizations of proper interval graphs in terms of their vertex-clique incidence graphs, establishing in particular a connection with bipartite permutation graphs.

Characterizing Proper Interval Graphs via Their Vertex-Clique Incidence Graphs

A bipartite graph $G = (X, Y; E)$ is said to be *biconvex* if there exists a *biconvex ordering* of (the vertex set of) G, that is, a pair $(<_X, <_Y)$ where $<_X$ is a linear ordering of X and $<_Y$ is a linear ordering of Y such that for every $x \in X$, the vertices in Y adjacent to x appear consecutively with respect to the ordering $<_Y$, and, similarly, for every $y \in Y$, the vertices in X adjacent to y appear consecutively with respect to the ordering $<_X$. Let $(<_X, <_Y)$ be a biconvex ordering of a biconvex graph $G = (X, Y; E)$. Two edges e and f of G are said to *cross* (each other) if there exist vertices $x_1, x_2 \in X$ and $y_1, y_2 \in Y$ such that $\{e, f\} = \{\{x_1, y_2\}, \{x_2, y_1\}\}$, $x_1 <_X x_2$, and $y_1 <_Y y_2$. A biconvex ordering $(<_X, <_Y)$ of a biconvex graph $G = (X, Y; E)$ is said to be *induced-crossing-free* if for any two crossing edges $e = \{x_1, y_2\}$ and $f = \{x_2, y_1\}$, either x_1 is adjacent to y_1 or x_2 is adjacent to y_2.

A *strongly induced-crossing-free ordering* (or simply a *strong ordering*) of G is a pair $(<_X, <_Y)$ of linear orderings of X and Y such that for any two crossing edges $e = \{x_1, y_2\}$ and $f = \{x_2, y_1\}$, vertex x_1 is adjacent to y_1 and vertex x_2 is adjacent to y_2. A *permutation graph* is a graph $G = (V, E)$ that admits a permutation model, that is, vertices of G can be ordered $v_1, \ldots, v_n$ such that there exists a permutation $(a_1, \ldots, a_n)$ of the set $\{1, \ldots, n\}$ such that for all $1 \le i < j \le n$, vertices v_i and v_j are adjacent in G if and only if $a_i > a_j$. A *bipartite permutation graph* is a graph that is both a bipartite graph and a permutation graph.

Theorem 6. *Let G be a graph. Then, the following statements are equivalent:*

1. *G is a proper interval graph.*
2. *B_G is a biconvex graph.*
3. *B_G is a bipartite permutation graph.*
4. *B_G has a strong ordering.*
5. *B_G has a strong biconvex ordering.*
6. *B_G has an induced-crossing-free biconvex ordering.*

The proof is based on showing that the vertex-clique incidence graph of every proper interval graph has a strong ordering, on characterizations of proper interval graphs and bipartite permutation graphs from [22] and [38], respectively, and on properties of biconvex graphs [1].

Maximum Induced Matchings in Bipartite Permutation Graphs, Revisited

Our goal is to show that the sufficient condition given by Corollary 3 is satisfied if G is a proper interval graph, namely, that there exists a maximum induced matching M in B_G such that the set S of M-saturated vertices in X dominates Y in B_G. We show the claimed property of B_G as follows. First, by applying Theorem 6, we infer that the graph B_G is a bipartite permutation graph. Second, by construction, no two distinct vertices in Y have comparable neighborhoods in X. It turns out that these two properties are already enough to guarantee the desired conclusion. We show this by a careful analysis of the linear-time algorithm due to Chang from [14] for computing a maximum induced matching in bipartite permutation graphs.

Theorem 7. *Given a bipartite permutation graph $G = (X, Y; E)$, there is a linear-time algorithm that computes a maximum induced matching M in G such that, if no two vertices in Y have comparable neighborhoods in G, then the set of M-saturated vertices in X dominates Y.*

Solving UCT in Proper Interval Graphs in Linear Time

We now have everything ready to prove the announced result.

Theorem 8. UPPER CLIQUE TRANSVERSAL *can be solved in linear time in the class of proper interval graphs.*

Proof. The algorithm proceeds in three steps. In the first step, we compute from the input graph $G = (V, E)$ its vertex-clique incidence graph B_G, with parts $X = V$ and $Y = \mathcal{C}_G$. By Theorem 6, the graph B_G is a bipartite permutation graph. In the second step of the algorithm, we compute a maximum induced matching M of B_G using Theorem 7. Finally, the algorithm returns the set of M-saturated vertices in X.

By construction, the set M_X returned by the algorithm is a subset of X, and thus a set of vertices of G. Since the vertices of Y are precisely the maximal cliques of G, no two vertices in Y have comparable neighborhoods in B_G. Therefore, by Theorem 7, the set M_X dominates Y. By Corollary 3, M_X is a maximum minimal clique transversal in G. Computing the vertex-clique incidence graph B_G can be done in linear time (see [7]). Since B_G is a bipartite permutation graph, a maximum induced matching of B_G can be computed in linear time (see Theorem 7). The set of M-saturated vertices in X can also be computed in linear time. Thus, the overall time complexity of the algorithm is linear.   □

The above proof also shows the following.

Theorem 9. *For every proper interval graph G, the upper clique transversal number of G is equal to the induced matching number of B_G.*

It can be shown that the result of Theorem 9 does not generalize to the class of interval graphs. In fact, there exist interval graphs with arbitrarily large difference between the induced matching number of their vertex-clique incidence graph and the upper clique transversal number of the graph (for example, the double stars).

6   Conclusion

We performed a systematic study of the complexity of UPPER CLIQUE TRANSVERSAL in various graph classes, showing, on the one hand, NP-completeness of the problem in the classes of chordal graphs, chordal bipartite graphs, and line graphs of bipartite graphs, and, on the other hand, linear-time solvability in the classes of split graphs and proper interval graphs. Our work leaves open several questions:

- What is the complexity of computing a minimal clique transversal in a given graph?
- What is the complexity of UPPER CLIQUE TRANSVERSAL in the class of interval graphs?
- For what graphs G does the upper clique transversal number equal to the induced matching number of the vertex-clique incidence graph? While not all interval graphs have the stated property, Theorem 9 shows that the property is satisfied by every proper interval graph. But there is more; for example, all cycles have the property.
- The upper clique transversal number is a trivial upper bound for the clique transversal number; however, the ratio between these two parameters can be arbitrarily large in general. For instance, in the complete bipartite graph $K_{1,q}$ the former one has value q while the latter one has value 1. For which graph classes is the ratio (or even the difference) between the clique transversal number and the upper clique transversal number bounded?

Acknowledgements. We are grateful to Nikolaos Melissinos, Haiko Müller, and the anonymous reviewers for their helpful comments. The work of the first named author is supported in part by the Slovenian Research Agency (I0-0035, research program P1-0285 and research projects N1-0102, N1-0160, J1-3001, J1-3002, J1-3003, J1-4008, and J1-4084). Part of the work was done while the author was visiting Osaka Prefecture University in Japan, under the operation Mobility of Slovene higher education teachers 2018–2021, co-financed by the Republic of Slovenia and the European Union under the European Social Fund. The second named author is partially supported by JSPS KAKENHI Grant Number JP17K00017, 20H05964, and 21K11757, Japan.

References

1. Abbas, N., Stewart, L.K.: Biconvex graphs: ordering and algorithms. Discrete Appl. Math. **103**(1–3), 1–19 (2000)

2. AbouEisha, H., Hussain, S., Lozin, V., Monnot, J., Ries, B., Zamaraev, V.: Upper domination: towards a dichotomy through boundary properties. Algorithmica **80**(10), 2799–2817 (2018)
3. Andreae, T., Flotow, C.: On covering all cliques of a chordal graph. Discrete Math. **149**(1–3), 299–302 (1996)
4. Andreae, T., Schughart, M., Tuza, Z.: Clique-transversal sets of line graphs and complements of line graphs. Discrete Math. **88**(1), 11–20 (1991)
5. Balachandran, V., Nagavamsi, P., Rangan, C.P.: Clique transversal and clique independence on comparability graphs. Inform. Process. Lett. **58**(4), 181–184 (1996)
6. Bazgan, C., et al.: The many facets of upper domination. Theoret. Comput. Sci. **717**, 2–25 (2018)
7. Berry, A., Pogorelcnik, R.: A simple algorithm to generate the minimal separators and the maximal cliques of a chordal graph. Inform. Process. Lett. **111**(11), 508–511 (2011)
8. Bertossi, A.A.: Dominating sets for split and bipartite graphs. Inform. Process. Lett. **19**(1), 37–40 (1984)
9. Beyer, T., Proskurowski, A., Hedetniemi, S., Mitchell, S.: Independent domination in trees. In: Proceedings of the Eighth Southeastern Conference on Combinatorics, Graph Theory and Computing (Louisiana State Univ., Baton Rouge, La., 1977), pp. 321–328. Congressus Numerantium, No. XIX (1977)
10. Bonomo, F., Durán, G., Safe, M.D., Wagler, A.K.: Clique-perfectness of complements of line graphs. Discrete Appl. Math. **186**, 19–44 (2015)
11. Boria, N., Della Croce, F., Paschos, V.T.: On the MAX MIN VERTEX COVER problem. Discrete Appl. Math. **196**, 62–71 (2015)
12. Brandstädt, A., Le, V.B., Spinrad, J.P.: Graph classes: a survey. In: Society for Industrial and Applied Mathematics (SIAM), Philadelphia, PA, SIAM Monographs on Discrete Mathematics and Applications (1999)
13. Chang, G.J., Farber, M., Tuza, Z.: Algorithmic aspects of neighborhood numbers. SIAM J. Discrete Math. **6**(1), 24–29 (1993)
14. Chang, J.M.: Induced matchings in asteroidal triple-free graphs. Discrete Appl. Math. **132**(1–3), 67–78 (2003)
15. Cooper, J.W., Grzesik, A., Král, D.: Optimal-size clique transversals in chordal graphs. J. Graph Theory **89**(4), 479–493 (2018)
16. Damaschke, P.: Parameterized algorithms for double hypergraph dualization with rank limitation and maximum minimal vertex cover. Discrete Optim. **8**(1), 18–24 (2011)
17. Damaschke, P., Müller, H., Kratsch, D.: Domination in convex and chordal bipartite graphs. Inform. Process. Lett. **36**(5), 231–236 (1990)
18. Dublois, L., Hanaka, T., Khosravian Ghadikolaei, M., Lampis, M., Melissinos, N.: (In) approximability of maximum minimal FVS. J. Comput. System Sci. **124**, 26–40 (2022)
19. Eades, P., Keil, M., Manuel, P.D., Miller, M.: Two minimum dominating sets with minimum intersection in chordal graphs. Nordic J. Comput. **3**(3), 220–237 (1996)
20. Erdős, P., Gallai, T., Tuza, Z.: Covering the cliques of a graph with vertices. Discrete Math. **108**, 279–289 (1992)
21. Ferneyhough, S., Haas, R., Hanson, D., MacGillivray, G.: Star forests, dominating sets and Ramsey-type problems. Discrete Math. **245**(1–3), 255–262 (2002)
22. Gardi, F.: The Roberts characterization of proper and unit interval graphs. Discrete Math. **307**(22), 2906–2908 (2007)
23. Guruswami, V., Pandu Rangan, C.: Algorithmic aspects of clique-transversal and clique-independent sets. Discrete Appl. Math. **100**(3), 183–202 (2000)

24. Hedetniemi, S.T.: A max-min relationship between matchings and domination in graphs. Congr. Numer. **40**, 23–34 (1983)
25. Jacobson, M.S., Peters, K.: Chordal graphs and upper irredundance, upper domination and independence. Discrete Math. **86**(1–3), 59–69 (1990)
26. Khoshkhah, K., Ghadikolaei, M.K., Monnot, J., Sikora, F.: Weighted upper edge cover: complexity and approximability. J. Graph Algorithms Appl. **24**(2), 65–88 (2020)
27. Lampis, M., Melissinos, N., Vasilakis, M.: Parameterized max min feedback vertex set. In: Proceedings of 48th International Symposium on Mathematical Foundations of Computer Science (MFCS 2023) (2023). arXiv:2302.09604
28. Lee, C.M.: Algorithmic aspects of some variations of clique transversal and clique independent sets on graphs. Algorithms (Basel) **14**(1), 22 (2021)
29. Lee, C.M., Chang, M.S.: Distance-hereditary graphs are clique-perfect. Discrete Appl. Math. **154**(3), 525–536 (2006)
30. Lin, M.C., Vasiliev, S.: Approximation algorithms for clique transversals on some graph classes. Inform. Process. Lett. **115**(9), 667–670 (2015)
31. Liu, K., Lu, M.: Complete-subgraph-transversal-sets problem on bounded treewidth graphs. J. Comb. Optim. **41**(4), 923–933 (2021)
32. Milanič, M.: Strong cliques and stable sets. In: Topics in algorithmic graph theory, Encyclopedia of Mathematics and its Applications, vol. 178, pp. 207–227. Cambridge University Press, Cambridge (2021)
33. Monnot, J., Fernau, H., Manlove, D.: Algorithmic aspects of upper edge domination. Theoret. Comput. Sci. **877**, 46–57 (2021)
34. Müller, H., Brandstädt, A.: The NP-completeness of Steiner tree and dominating set for chordal bipartite graphs. Theoret. Comput. Sci. **53**(2–3), 257–265 (1987)
35. Nguyen, C.T., Shen, J., Hou, M., Sheng, L., Miller, W., Zhang, L.: Approximating the spanning star forest problem and its application to genomic sequence alignment. SIAM J. Comput. **38**(3), 946–962 (2008)
36. Payan, C.: Remarks on cliques and dominating sets in graphs. Ars Combin. **7**, 181–189 (1979)
37. Shan, E., Liang, Z., Kang, L.: Clique-transversal sets and clique-coloring in planar graphs. Eur. J. Combin. **36**, 367–376 (2014)
38. Spinrad, J., Brandstädt, A., Stewart, L.: Bipartite permutation graphs. Discrete Appl. Math. **18**(3), 279–292 (1987)
39. West, D.B.: Introduction to graph theory. Prentice Hall Inc, Upper Saddle River, NJ (1996)
40. Zang, W.: Generalizations of Grillet's theorem on maximal stable sets and maximal cliques in graphs. Discrete Math. **143**(1–3), 259–268 (1995)
41. Zehavi, M.: Maximum minimal vertex cover parameterized by vertex cover. SIAM J. Discrete Math. **31**(4), 2440–2456 (2017)

Critical Relaxed Stable Matchings
with Two-Sided Ties

Meghana Nasre[1], Prajakta Nimbhorkar[2], and Keshav Ranjan[1(✉)]

[1] IIT Madras, Chennai, India
meghana@cse.iitm.ac.in, ranjankeshav08@gmail.com
[2] Chennai Mathematical Institute and UMI ReLaX, Chennai, India
prajakta@cmi.ac.in

Abstract. We consider the stable marriage problem in the presence of ties in preferences and critical vertices. The input to our problem is a bipartite graph $G = (\mathcal{A} \cup \mathcal{B}, E)$ where $\mathcal{A}$ and $\mathcal{B}$ denote sets of vertices which need to be matched. Each vertex has a preference ordering over its neighbours possibly containing ties. In addition, a subset of vertices in $\mathcal{A} \cup \mathcal{B}$ are marked as critical and the goal is to output a matching that matches as many critical vertices as possible. Such matchings are called critical matchings in the literature and in our setting, we seek to compute a matching that is critical as well as optimal with respect to the preferences of the vertices.

Stability, which is a well-accepted notion of optimality in the presence of two-sided preferences, is generalized to weak-stability in the presence of ties. It is well known that in the presence of critical vertices, a matching that is critical as well as weakly stable may not exist. Popularity is another well-investigated notion of optimality for the two-sided preference list setting, however, in the presence of ties (even with no critical vertices), a popular matching need not exist. We, therefore, consider the notion of relaxed stability which was introduced and studied by Krishnaa et. al. (SAGT 2020). We show that in our setting a critical matching which is relaxed stable always exists although computing a maximum-sized relaxed stable matching turns out to be NP-hard. Our main contribution is a $\frac{3}{2}$-approximation to the maximum-sized critical relaxed stable matching for the stable marriage problem where ties as well as critical vertices are present on both the sides of the bipartition.

Keywords: Stable Matching · Ties in Preferences · Critical · Relaxed Stable · Approximation Algorithm

1   Introduction

We study the stable marriage problem in the presence of *ties* in preferences and *critical vertices*. Formally, the input to our problem is a bipartite graph $G = (\mathcal{A} \cup \mathcal{B}, E)$, where $\mathcal{A}$ and $\mathcal{B}$ are two sets of vertices and E denotes the set of all the acceptable vertex-pairs. Each vertex $u \in \mathcal{A} \cup \mathcal{B}$ ranks a subset of vertices in the other partition (its neighbours in G) in the order of its preference possibly

© The Author(s), under exclusive license to Springer Nature Switzerland AG 2023
D. Paulusma and B. Ries (Eds.): WG 2023, LNCS 14093, pp. 447–461, 2023.
https://doi.org/10.1007/978-3-031-43380-1_32

involving ties – this ordering is denoted as $\mathsf{Pref}(u)$. For a vertex u let v_1 and v_2 be two of its neighbours in G. The vertex u strictly prefers v_1 over v_2 (denoted as $v_1 \succ_u v_2$) if the rank of the edge (u, v_1) is smaller than the rank of the edge (u, v_2). The vertex u is tied between v_1 and v_2 (denoted as $v_1 =_u v_2$) if the ranks on the edges (u, v_1) and (u, v_2) are the same. We use $v_1 \succeq_u v_2$ to denote that the rank of v_1 is at least as good as the rank of v_2 in $\mathsf{Pref}(u)$. In addition, the input consists of a set $\mathcal{C} \subseteq (\mathcal{A} \cup \mathcal{B})$ of *critical vertices*. Our goal is to compute an assignment which minimizes the number of unassigned critical vertices.

Formally, an assignment or a *matching* $M \subseteq E$ in G is a set of edges that do not share an end-point. For each vertex $u \in \mathcal{A} \cup \mathcal{B}$, we denote by $M(u)$, the neighbour of u that is assigned to u in M. In the presence of critical vertices, we consider that the most important attribute of a matching is to match as many critical vertices as possible. A matching M is *critical* [11] if there is no matching that matches more critical vertices than M. In this work, we are interested in computing a critical matching that is *optimal* with respect to the preferences of the vertices in an instance of our setting.

Critical vertices or lower-quota positions naturally arise in applications like the Hospitals/Residents problem [7], where rural hospitals must be prioritized to ensure sufficient staffing. Another example is the problem of assigning sailors to billets [28] in the US Navy, where some critical billets cannot be left vacant [25,29]. Ties in preferences is yet another important practical consideration in matching problems and has been extensively investigated in the literature [2,8,9,13,18,19, 24]. However, there is a limited investigation (see for example [5]) of matching problems with ties as well critical vertices and ours is the first work that allows ties as well as critical vertices on both sides of the bipartition.

Stability, which is the de-facto notion of optimality for two-sided preferences, is defined by the absence of a blocking pair. Informally, an assignment is stable if no unassigned pair wishes to deviate from it.

Definition 1 (Stable Matchings). *Given a matching M, a pair $(a, b) \in E \setminus M$ is called a blocking pair w.r.t. M if (i) either a is unmatched or $b \succ_a M(a)$ and (ii) either b is unmatched or $a \succ_b M(b)$. A matching M is stable if there is no blocking pair w.r.t. M.*

When all preferences are strict, that is, there are no ties, every instance of the stable marriage problem admits a stable matching, and it can be computed using the well-known Gale and Shapley algorithm [3]. In addition, it is also known [26,27] that all stable matchings have the same size.

Stable Matchings in the Presence of Ties: When preferences are allowed to have ties, the notion of stability defined above is called as *weak stability* (referred to as stability in the rest of the paper). We remark that, for a pair (a, b) to block a matching M, both a and b prefer each other *strictly* over their current partners in M. Every instance of the stable marriage problem with ties admits a stable matching, and it can be efficiently computed. However, unlike in the case of strict lists, all the stable matchings need not have the same size, and the problem of computing a maximum or minimum size stable matching is NP-hard [18] under severe restrictions – e.g. when ties occur at the end of preference lists and only on one side of the bipartition, there is at most one tie per list, and each tie is of length two.

Stable/Popular Matching in the Presence of Critical Vertices: When we have critical vertices as a part of the input, a stable matching which is also critical, may not exist – for example, consider an instance of the stable matching problem with strict lists obtained by arbitrarily breaking ties in the preference lists of all agents in the example shown in Fig. 1. Any critical matching in the instance must match b_2 with a_1, resulting in the blocking edge (a_1, b_1). Since stability and criticality are not simultaneously guaranteed, an alternate notion of optimality, namely *popularity* [4], is extensively investigated in the literature [11,20,22] for the case of strict lists. The goal is to compute a matching which is *popular* amongst the set of critical matchings. Informally, a matching M is *popular* in a set of matchings if no majority of vertices wish to deviate from M to any other matching in that set. It is known [11,22] that an instance with strict preference lists *always* admits a matching which is popular amongst critical matchings, and such a matching can be computed efficiently. Hence, it is natural to consider popularity in the presence of critical vertices and ties.

However, popular matchings are not guaranteed to exist even when ties are present in the preferences only on one side of the bipartition, without any critical vertices. Moreover, in the presence of ties, deciding whether a popular matching exists is NP-hard [1]. In light of this, we explore the notion of *relaxed stability*.

Relaxed Stability in the Presence of Ties and Critical Vertices: The notion of relaxed stability was introduced and studied by Krishnaa *et al.* [14] for the Hospitals/Residents problem with lower quotas (HRLQ). In their setting, preferences are assumed to be strict. The HRLQ setting is a many-to-one matching problem where a hospital h can accept at most $q^+(h)$ many residents and has $q^-(h) \leq q^+(h)$ many critical positions. To satisfy the critical positions at a hospital, certain residents may be *forced* to be matched to the hospital. The notion of relaxed stability allows only such residents to participate in blocking pairs. In addition, if a resident matched to h participates in a blocking pair then the hospital h should not be *surplus*, that is $|M(h)| \leq q^-(h)$.

In the HRLQ setting, preferences are strict, hospitals have capacities, and critical positions are allowed only for hospitals. In contrast, we allow ties in preferences as well as critical vertices to appear on *both sides* of the bipartition. However, our setting is one-to-one.

We now define the notion of relaxed stability (RSM) for our setting. Intuitively, a matching M is an RSM if every blocking pair (a, b) w.r.t. M is *justified* by either a or b or both. A vertex a justifies the blocking pair if $M(a)$ is a critical vertex. That is, $M(a)$ forces a to be matched to a lower-preferred vertex than b. Similarly, the vertex b can justify the blocking pair (a, b).

Definition 2 (Relaxed stability in our setting). *A matching M is RSM if for every blocking pair (a, b) w.r.t. M at least one of the following holds:*

1. *a is matched and $b' = M(a)$ is critical, or*
2. *b is matched and $a' = M(b)$ is critical.*

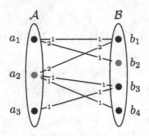

Fig. 1. Red vertices are critical, black vertices are non-critical. The numbers on the edges denote the ranks of the respective end-points. The instance does not admit any critical stable matching because b_2 remains unmatched in every stable matching. $M_1 = \{(a_1, b_2), (a_2, b_1), (a_3, b_3)\}$ is critical but not RSM because the blocking edge (a_2, b_4) is not justified. $M_2 = \{(a_1, b_2), (a_2, b_4), (a_3, b_3)\}$ is CRITICAL-RSM because the only blocking edge (a_1, b_1) is justified. (Color figure online)

A matching M is called *a critical relaxed stable matching* (CRITICAL-RSM) if it is *critical* as well as *relaxed stable*. In the instance shown in Fig. 1, the matching M_1 is critical but not RSM whereas M_2 is CRITICAL-RSM.

Our first contribution is to show that a CRITICAL-RSM always exists in our setting. We remark that when $\mathcal{C} = \emptyset$, an instance of our setting is the same as the stable marriage setting with ties but without critical vertices, and hence the set of CRITICAL-RSM is the same as the set of stable matchings. This immediately implies that computing a maximum size critical RSM is NP-hard [18] and hard to approximate within any factor smaller than $\frac{21}{19}$ [6]. For the problem of computing a maximum-sized stable matching when ties appear on both sides of the bipartition, the current best approximation factor [13,19,24] is $\frac{3}{2}$. The main result (Theorem 1) provides the same approximation size guarantee for a maximum sized CRITICAL-RSM in our setting.

Theorem 1. *Let $G = (\mathcal{A} \cup \mathcal{B}, E)$ be an instance of the stable marriage problem where ties and critical vertices can appear in both the bipartitions of G. Then G always admits a CRITICAL-RSM M such that $|M| \geq \frac{2}{3}|M'|$, where M' is a maximum size CRITICAL-RSM in G. Moreover, M can be computed in polynomial-time.*

Related Work: As mentioned earlier, the generalizations of the stable marriage problem to allow either ties in preferences or critical vertices/lower-quota positions has been extensively investigated. Very recently, Goko *et al.* [5] and Makino *et al.* [17] have considered the instances with both ties and critical vertices. They study the Hospitals/Residents problem with lower-quotas where ties appear on both sides. In their setting, only one side of the bipartition can have critical vertices. They define a matching with maximum satisfaction ratio, which for our one-to-one setting, coincides with critical matchings. However, their goal is to compute a matching that matches the maximum possible critical vertices amongst all stable matchings.

For strict preferences and lower-quotas/critical vertices, various notions like envyfreeness [15,30], popularity [11,20,22,23], and relaxed stability [14,15] have been studied. Relaxed stability and popularity do not define the same set of matchings even in the one-to-one strict-list setting and critical vertices restricted to one side only (see full version [21]) for the details. Hamada et al. [7] consider the problem of computing a matching with minimum number of blocking pairs or blocking residents, and give approximation algorithms for the same.

For the stable marriage problem with ties (without critical vertices) there is a long line of investigation [2,9,10,12,13,19,24] in order to improve the approximation ratio of the maximum size stable matching under various restricted settings. The best-known approximation algorithm for the case when ties are allowed only in one bipartition of the graph is by Lam and Plaxton [16] whereas the best-known for the case where ties are allowed on both sides is by [13,19,24]. We use Király's algorithm [13] in our work.

2 Preliminaries

Our algorithm described in the next section combines the ideas in (i) Király's algorithm [13] for computing a stable matching in instances where ties appear on both sides and (ii) Multi-level algorithm for computing popular critical matching [23] for strict preferences. We give an overview of the algorithms and also define terminology useful for our algorithm.

Overview of Király's Algorithm [13]. Király's algorithm [13] is a proposal-based algorithm where vertices in $\mathcal{A}$ propose and vertices in $\mathcal{B}$ accept or reject. We need the term *uncertain proposal* from [13] which is defined below.

Definition 3 (Uncertain Proposal). *Let b be some k^{th} rank neighbour of a in $\mathsf{Pref}(a)$. During the course of the algorithm, the proposal from a to b is uncertain if there exists another k^{th} rank neighbour b' of a which is unproposed by a and unmatched in the matching. Once a proposal (a, b) is uncertain, it remains uncertain until b rejects a.*

Each time an $a \in \mathcal{A}$ proposes to its *favourite* neighbour b (we define favourite neighbour formally in Definition 4), the vertex b accepts/rejects as follows:

1. If b is unmatched then b immediately accepts the proposal.
2. If b is matched, say to a', and (a', b) is an uncertain proposal, then b rejects a' and accepts the proposal from a, irrespective of the ranks of a and a' in $\mathsf{Pref}(b)$. In this case, b is *marked* by a'.
3. If b is matched, say to a', and (a', b) is not an uncertain proposal, then
   (i) if $a \succ_b a'$ then b rejects a' and accepts the proposal from a, or
   (ii) if $a' \succeq_b a$ then b rejects a.

The reason for a' marking the vertex b in (2) is as follows: In this case, b rejects the uncertain proposal from a' and accepts a *irrespective* of b's preference between a and a'. Later, when a' gets its chance to propose, and if none of the neighbours of a' at the rank of b accept the proposal from a', then a' will propose to the marked vertex b before proposing to the next lower-ranked neighbours. In contrast in (3)(i) above, when the proposal (a', b) is not uncertain and $a \succ_b a'$ then a' does not mark b. Note that a vertex $b \in \mathcal{B}$ can be part of an uncertain proposal at most once. Once a vertex receives its first proposal, it will remain matched and thereafter cannot be part of any uncertain proposal. Thus, any $b \in \mathcal{B}$ can be marked at most once during the course of the algorithm.

Now, we define the favourite neighbour of a vertex a, which is an adaptation of the definition in [13].

Definition 4 (Favourite neighbour of a). *Assume that k is the best rank at which some unproposed or marked neighbours of a exist in $\mathsf{Pref}(a)$. Then b is the favourite neighbour of a if one of the following conditions holds:*

(i) there exists at least one unmatched neighbour of a at the kth rank and b has the lowest index among all such unmatched neighbours, or
(ii) all the kth ranked neighbours of a are matched and b is the lowest index among all such neighbours which are unproposed by a, or
(iii) all the kth ranked neighbours are already proposed by a and b has the lowest index among all the vertices which are marked by a.

Király's algorithm begins with every vertex $a \in \mathcal{A}$ being active. As long as there exists an active vertex which is unmatched and has not exhausted its preference list, the vertex proposes to its favourite neighbour. If $a \in \mathcal{A}$ remains unmatched after exhausting its preference list, it achieves a '$*$' status and starts proposing to vertices in $\mathsf{Pref}(a)$ with $*$ status. The $*$ status of a vertex a can be interpreted as improving the rank of a in $\mathsf{Pref}(b)$ by 0.5 for any neighbour b of a. Thus, the $*$ status vertex is used to decide between vertices in a tie, but does not affect strict preferences. It is shown in [13] that the resulting matching is a $\frac{3}{2}$-approximation of a maximum size stable matching.

Overview of the Popular Critical Matching Algorithm [23]. Now, we briefly describe the algorithm in [23] for computing the maximum size popular critical matching in the one-to-one strict list setting. Let s and t denote the number of critical vertices in $\mathcal{A}$ and $\mathcal{B}$, respectively. The algorithm in [23] is a multi-level algorithm which first matches as many critical vertices from $\mathcal{B}$ as possible. This is achieved by restricting unmatched vertices in $\mathcal{A}$ at levels $0, \ldots, t-1$ to propose only to critical vertices on the $\mathcal{B}$-side. At the level t, each vertex $a \in \mathcal{A}$ is allowed to propose *all* its neighbours. If a vertex $a \in \mathcal{A}$ remains unmatched even after exhausting its preference list at level t, a raises its level to $t + 1$ and proposes to its neighbours until it is matched or it exhausts its preference list at the level $t + 1$. If a critical vertex a remains unmatched then a raises its level above $t + 1$ and continues proposing to all its neighbours until it is matched, or it exhausts its preference list at the highest level $s + t + 1$. A vertex b which receives the proposal always prefers a higher level vertex a over any lower level vertex a' irrespective of the ranks of a and a' in $\mathsf{Pref}(b)$. It is shown in [23] that the resulting matching is a maximum size popular matching among all the critical matchings.

3   Algorithm for Computing CRITICAL-RSM

Our algorithm (see Algorithm 1) is a combination of Király's algorithm and the popular critical matching algorithm discussed in the previous section. In each level, vertices in $\mathcal{A}$ propose and vertices in $\mathcal{B}$ accept or reject. The set of vertices from $\mathcal{B}$ that $a \in \mathcal{A}$ proposes to depends on the level of a. Furthermore, depending on the level of a, the preference list at that level may be strict or may contain ties. Throughout Algorithm 1, b uses its original preference list $\mathsf{Pref}(b)$ which possibly contains ties. For a vertex $a \in \mathcal{A}$, let $\mathsf{PrefS}(a)$ denote a *strict* preference list obtained by breaking ties in $\mathsf{Pref}(a)$ in such a way that the vertices in ties are ordered by increasing order of their indices. Furthermore, let $\mathsf{PrefSC}(a)$ be the strict list obtained from $\mathsf{PrefS}(a)$ by omitting all the non-critical vertices from $\mathsf{PrefS}(a)$. For example, assume $\mathsf{Pref}(a) = (b_2, b_1), b_5, (b_3, b_4)$ where b_4 and b_5 are critical vertices. Here, a ranks b_1 and b_2 as rank-1, b_5 as rank-2 and b_3 and b_4 as rank-3. We have $\mathsf{PrefS}(a) = b_1, b_2, b_5, b_3, b_4$ and $\mathsf{PrefSC}(a) = b_5, b_4$ where comma separated vertices denote a strict ordering.

Initially, all the vertices in $\mathcal{A}$ have their levels set to 0. A vertex a at level ℓ is denoted as a^ℓ. At a level less than t, each $a \in \mathcal{A}$ proposes to vertices in $\mathsf{PrefSC}(a)$ (see Lines 4–8 of Algorithm 1). Each time it remains unmatched, it proposes to its *most preferred* neighbour b. The most preferred neighbour in $\mathsf{PrefSC}(a)$ or $\mathsf{PrefS}(a)$ is the best-ranked neighbour b to whom a has not yet proposed at the current level. If a remains unmatched after proposing to all its neighbours in $\mathsf{PrefSC}(a)$ at a level $\ell < t - 1$, then a raises its level to $\ell + 1$ and again proposes to vertices in $\mathsf{PrefSC}(a)$. In this part of the algorithm, we invoke `CriticalPropose()` which encodes the level-based accept/reject by b. A vertex $b \in \mathcal{B}$ prefers a_i^ℓ over $a_j^{\ell'}$ if :

(i) either $\ell > \ell'$ (ranks of a_i and a_j in $\mathsf{Pref}(b)$ do not matter) or
(ii) $\ell = \ell'$ and $a_i \succ_b a_j$.

If vertex a remains unmatched after exhausting $\mathsf{PrefSC}(a)$ at level $t - 1$, a attains level t where it uses its original preference list $\mathsf{Pref}(a)$ which may contain ties. At level t our algorithm executes Király's algorithm [13]. This corresponds to Lines 10–13 of Algorithm 1. Király's algorithm is encoded in the procedure `TiesPropose()`. Since we have ties on both sides of the graph, at this level, we need the notion of a favourite neighbour and uncertain proposal defined in Sect. 2. If the vertex a remains unmatched after exhausting $\mathsf{Pref}(a)$ at level t, it attains the $*$ status, and for this, we have the sub-level t^*. The interpretation of the $*$ status is the same as discussed in Sect. 2.

If a *critical* vertex a remains unmatched after exhausting its preference list $\mathsf{Pref}(a)$ at level t^*, a raises its level to $t + 1$, and starts proposing to vertices in $\mathsf{PrefS}(a)$ (see Lines 16–20 of Algorithm 1). It continues to do so until either it is matched or it has exhausted $\mathsf{PrefS}(a)$ at level $s + t$. In contrast, if a non-critical vertex a remains unmatched after exhausting its preference list $\mathsf{Pref}(a)$ at level t^*, a does not propose any further. Recall that $\mathsf{PrefS}(a)$ is a strict preference list containing all the neighbours (not restricted to critical vertices). Here, Algorithm 1, again invokes `CriticalPropose()` for the level-based accept/reject by b. The algorithm terminates when either (i) all the vertices in $\mathcal{A}$ are matched or

---

Algorithm 1: Critical relaxed stable matching in $G = (\mathcal{A} \cup \mathcal{B}, E)$

---

1 Set $M = \emptyset$, Initialize a queue $Q = \{a^0 \; : \; a \in \mathcal{A}\}$
2 **while** Q *is not empty* **do**
3   Let $a^\ell = dequeue(Q)$ // a is unmatched
4   **if** $\ell < t$ **then**
5   **if** a^ℓ *has not exhausted* $\mathsf{PrefSC}(a)$ **then**
6              $\mathtt{CriticalPropose}(a^\ell, \mathsf{PrefSC}(a), M, Q)$
7   **else**
8              $\ell = \ell + 1$ and add a^ℓ to Q
9   **else if** $\ell == t$ *or* $\ell == t^*$ **then**
10   **if** $\exists \, b' \in \mathsf{Pref}(a)$ *which is marked or unproposed by* a^ℓ **then**
11             $\mathtt{TiesPropose}(a^\ell, \mathsf{Pref}(a), M, Q)$
12   **else**
13   **if** $\ell == t$ **then** $\ell = t^*$ and add a^ℓ to Q
14   **if** $\ell == t^*$ *and* a *is critical* **then** $\ell = t + 1$ and add a^ℓ to Q
15   **else**
         // a is critical
16   **if** a^ℓ *has not exhausted* $\mathsf{PrefS}(a)$ **then**
17              $\mathtt{CriticalPropose}(a^\ell, \mathsf{PrefS}(a), M, Q)$
18   **else**
19   **if** $\ell < s + t$ *and* a *is critical* **then**
20                 $\ell = \ell + 1$ and add a^ℓ to Q
21 **return** M

(ii) all unmatched critical $a \in \mathcal{A}$ have exhausted $\mathsf{PrefS}(a)$ at level $s + t$ and all unmatched non-critical $a \in \mathcal{A}$ have exhausted $\mathsf{Pref}(a)$ at level t^*. We note that $s + t = |\mathcal{C}| = O(n)$, where $n = |\mathcal{A} \cup \mathcal{B}|$ and each edge of G is explored at most $s + t + 3$ times (at most three times at level t, the Király's step, and at most once at every other level). Thus, the running time of our algorithm is $O(n \cdot |E|)$.

It is worth noting that in our algorithm, not all vertices in $\mathcal{A}$ propose at all levels. Similarly, not all vertices in $\mathcal{B}$ receive proposals from vertices at all levels. In other words, only critical vertices in $\mathcal{B}$ are allowed to receive proposals from vertices in $\mathcal{A}$ at levels at most $t - 1$, and only critical vertices in $\mathcal{A}$ are allowed to propose at levels above t. Also, note that when a vertex in $\mathcal{A}$ transitions to a higher level, it proposes to possibly a superset of vertices that it proposes to in the lower level (recall that $\mathsf{Pref}(a)$ and its strict counterpart $\mathsf{PrefS}(a)$ are both a superset of $\mathsf{PrefSC}(a)$). Therefore, we have the following useful observation.

Observation 1. *If a vertex* $b \in \mathcal{B}$ *receives a proposal from some* $a' \in \mathcal{A}$ *at a level* z *then* b *receives proposals from all its neighbours who exhausted their preference list at level* z.

4   Correctness of Our Algorithm

We prove that the matching M output by Algorithm 1 is

Procedure CriticalPropose(a^ℓ, List(a), M, Q)

---

1 Let b be the most preferred unproposed vertex by a^ℓ in List(a)
2 **if** b *is unmatched in M* **then**
3 | $M = M \cup \{(a^\ell, b)\}$
4 **else**
5 | Let $a_j^y = M(b)$
6 | **if** *($\ell > y$) or ($\ell == y$ and $a \succ_b a_j$)* **then**
7 | | $M = M \setminus \{(a_j^y, b)\} \cup \{(a^\ell, b)\}$ and add a_j^y to Q
8 | **else** add a^ℓ to Q

Procedure TiesPropose(a^ℓ, List(a), M, Q)

---

1 Let b be the favourite neighbour of a^ℓ in List(a) at rank k
2 **if** b *was marked by a^ℓ* **then** a^ℓ unmarks b
3 **if** b *is unmatched* **then**
4 | $M = M \cup \{(a^\ell, b)\}$
5 | **if** *there exists an unmatched b'' at rank k in* Pref(a) **then**
6 | | Set (a^ℓ, b) as uncertain proposal   // $\ell = t$ as b'' is unmatched
7 **else if** b *is part of an uncertain proposal (a_j^y, b)* **then**
8 | $M = (M \setminus \{(a_j^y, b)\}) \cup \{(a^\ell, b)\}$ // Here, $y = t$
9 | a_j^y marks b and add a_j^y to Q
10 **else if** b *is not part of an uncertain proposal* **then**
11 | Let $a_j^y = M(b)$
12 | **if** $\ell == t$ **then**
13 | | **if** *($y < t$) or (($y == t$ or $y == t^*$) and $a \succ_b a_j$)* **then**
14 | | | $M = M \setminus \{(a_j^y, b)\} \cup \{(a^\ell, b)\}$ and add a_j^y to Q
15 | | **else** add a^ℓ to Q
16 | **if** $\ell == t^*$ **then**
17 | | **if** *($y < t$) or ($y == t$ and $a \succeq_b a_j$) or ($y == t^*$ and $a \succ_b a_j$)* **then**
18 | | | $M = M \setminus \{(a_j^y, b)\} \cup \{(a^\ell, b)\}$ and add a_j^y to Q
19 | | **else** add a^ℓ to Q

(I) Critical as well as relaxed stable (RSM) and
(II) A $\frac{3}{2}$ approximation to the maximum size CRITICAL-RSM in G.

We define a partition of the vertices in $\mathcal{A} \cup \mathcal{B}$ based on the levels of vertices in $\mathcal{A}$ and the matching M. This partition is useful to establish the correctness of our algorithm.

Partition of Vertices: The vertex set $\mathcal{A}$ is partitioned into $\mathcal{A}_0 \cup \mathcal{A}_1 \cup \ldots \cup \mathcal{A}_t \cup \ldots \cup \mathcal{A}_{s+t}$, and the vertex set $\mathcal{B}$ is partitioned into $\mathcal{B}_0 \cup \mathcal{B}_1 \cup \ldots \cup \mathcal{B}_t \cup \ldots \cup \mathcal{B}_{s+t}$. For every matched vertex $a \in \mathcal{A}$ there exists $x \in \{0, \ldots, s+t\}$ such that $(a^x, b) \in M$. We use x to partition the vertex set. Note that if $(a^{t^*}, b) \in M$ then for the purpose of partitioning we consider $t^* = t$ as t^* is a sub-level of the level t.

– **Matched vertices in $\mathcal{A} \cup \mathcal{B}$**: Let $a \in \mathcal{A}, b \in \mathcal{B}$ and $(a^x, b) \in M$ for some $x \in \{0, \ldots, s+t\}$. We add a to $\mathcal{A}_x$ and b to $\mathcal{B}_x$.
– **Unmatched vertices in $\mathcal{A} \cup \mathcal{B}$**:
  - If a non-critical vertex $a \in \mathcal{A}$ is unmatched in M then we add a to $\mathcal{A}_t$.
  - If a critical vertex $a \in \mathcal{A}$ is unmatched in M then we add a to $\mathcal{A}_{s+t}$.
  - If a non-critical vertex $b \in \mathcal{B}$ is unmatched in M then we add b to $\mathcal{B}_t$.
  - If a critical vertex $b \in \mathcal{B}$ is unmatched in M then we add b to $\mathcal{B}_0$.

It is convenient to visualize the partitions as shown in Fig. 2. This particular drawing of the graph G is denoted by G_M throughout the rest of the section. It is useful to assume that the edges in G_M are implicitly directed from $\mathcal{A}$ to $\mathcal{B}$. By construction, the edges of M (shown in blue colour) are horizontal whereas the unmatched edges (shown as solid black edges) can be horizontal, upwards or downwards. We state the properties of the vertices and edges in G_M with respect to this partition in Property 1 (see the full version [21] for justification).

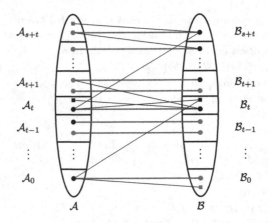

Fig. 2. The graph G_M. Red vertices are critical and black vertices are non-critical. Matched vertices are represented by circles, and unmatched vertices are represented by squares. The blue horizontal lines represent matched edges in M. Solid black lines represent edges which are not matched in M. Note that no edge in G_M is of the form $\mathcal{A}_x \times \mathcal{A}_y$ for $y \leq x - 2$. (Color figure online)

Property 1. Let $a \in \mathcal{A}$ and $b \in \mathcal{B}$. Then the following hold in graph G_M.

1. If $a \in \bigcup_{x=t+1}^{s+t} \mathcal{A}_x$ then a is critical. Thus, $|\bigcup_{x=t+1}^{s+t} \mathcal{A}_x| \leq s$.
2. If $b \in \bigcup_{x=0}^{t-1} \mathcal{B}_x$ then b is critical. Thus, $|\bigcup_{x=0}^{t-1} \mathcal{B}_x| \leq t$.
3. If a is critical and is unmatched in M then $a \in \mathcal{A}_{s+t}$ and all the neighbours of a are matched and present in $\mathcal{B}_{s+t}$ only.
4. If a is not critical and is unmatched in M then $a \in \mathcal{A}_t$ and all the neighbours of a are matched and present in $\mathcal{B}_x$ for $x \geq t$.

5. If b is critical and is unmatched in M then $b \in \mathcal{B}_0$ and all the neighbours of b are present in $\mathcal{A}_0$ only.
6. If b is not critical and is unmatched in M then $b \in \mathcal{B}_t$ and all the neighbours of b are present in $\mathcal{A}_x$ for $x \leq t$.

Let $(a, b) \in E$ be an edge such that $a \in \mathcal{A}_x$ and $b \in \mathcal{B}_y$. We say that such an edge is of the form $\mathcal{A}_x \times \mathcal{B}_y$. Lemma 1 below gives an important property about the edges which cannot be present in G_M. An edge of the form $\mathcal{A}_x \times \mathcal{B}_y$ with $x > y + 1$ is referred to as a *steep downward* edge.

Lemma 1. *The graph G_M does not contain steep downward edges. That is, there is no edge in G_M of the form $\mathcal{A}_x \times \mathcal{B}_y$ such that $x > y + 1$.*

Proof. Let (a, b) be any edge in G_M such that $a \in \mathcal{A}_x$ and $b \in \mathcal{B}_y$. If b is unmatched, then irrespective of whether b is critical or not by Property 1(5) and Property 1(6), we have $x \leq y$. Now suppose that b is matched and $(a', b) \in M$. If $a = a'$ then by construction of G_M, $(a, b) \in \mathcal{A}_x \times \mathcal{B}_x$. If $a \neq a'$, then we use Claim 1, which is given below. It is immediate from this claim that b is in $\mathcal{B}_y$ for $y \geq x - 1$. □

Claim 1. *Let $(a, b) \in E \setminus M$ and b be matched in M to $\tilde{a}$ at level y, that is, $M(b) = \tilde{a}^y$. If the level x of a is at least 2 then $y \geq x - 1$.*

Proof. Suppose for contradiction that there exists $\tilde{a} \in \mathcal{A}$ such that $(\tilde{a}^y, b) \in M$ for $y < x - 1$. The fact that $(a, b) \in E$ and a achieves the level x implies that a remains unmatched after a^{x-1} exhausted its preference list $\mathsf{Pref}(a)$, $\mathsf{PrefS}(a)$ or $\mathsf{PrefSC}(a)$ as appropriate. Since b is matched to a vertex at level $y < x - 1$, and a^{x-1} exhausted its preference list, by Observation 1, b received a proposal from a^{x-1}. At this time, b must accept this proposal by rejecting $\tilde{a}^y$ because $y < x - 1$. This implies $(\tilde{a}^y, b) \notin M$ which contradicts our assumption that $(\tilde{a}^y, b) \in M$ for $y < x - 1$. □

Lemma 2. *Let (a, b) be a blocking pair w.r.t. M. Then the corresponding edge in G_M is an upward edge.*

Proof. For the blocking pair (a, b) let a and b be at levels x and y, respectively. First, suppose that b is a critical vertex. Since (a, b) is a blocking pair, irrespective of whether a is matched or unmatched, a^x must have proposed to the critical vertex b. Thus, b cannot remain unmatched. This implies $M(b)$ exists. We consider the following two cases:

1. The proposal by a to b results in (a, b) to be uncertain: Note that a^x is rejected by b because b receives another proposal, and hence a^x marks b. Since (a, b) is a blocking pair, $M(a)$ is ranked lower than b. However, before proposing to any vertex ranked strictly lower than b, a^x must propose to the marked vertex b. At this point, either b is matched to a better preferred partner than a which contradicts that (a, b) blocks M, otherwise, b accepts the proposal from a^x. Thus, a^x is matched to either b or to some other vertex on the same rank as b. This implies (a, b) is not a blocking edge.

2. The proposal by a to b does not result in (a, b) to be uncertain: The fact that $a \succ_b M(b)$ implies $M(b)$ must be at a level y such that $y > x$. Thus, (a, b) edge is an upward edge in G_M.

   Now, suppose that b is a non-critical vertex. Then by Property 1(2), $b \in \mathcal{B}_y$ for $y \geq t$. If $x < t$, then (a, b) is an upward edge, and we are done. Hence, assume that $x \geq t$. Since $x \geq t$, a^x is proposes to *all* of its neighbours. Again, since (a, b) is a blocking pair, irrespective of whether a is matched or unmatched, a^x must have proposed to b. Thus, b cannot be unmatched. Now, either b is matched to a better-preferred partner than a, which contradicts that (a, b) is a blocking pair or $M(b)$ is at a higher level than a and hence (a, b) is an upward edge.   □

Lemma 3 below shows that the matching M output by Algorithm 1 is critical.

Lemma 3. *The output matching M is critical for G.*

Proof sketch: We prove the criticality of M by using the level structure of the graph G_M. The idea is to show that there is no alternating path ρ in G_M with respect to M such that the number of critical vertices matched in $M \oplus \rho$ is more than the number of critical vertices matched in M. We prove the criticality for the individual parts, that is, for $\mathcal{A}$-part and for $\mathcal{B}$-part. In other words, we show that M matches maximum possible critical nodes from $\mathcal{A}$-side, and maximum possible critical nodes from the $\mathcal{B}$-side. This immediately implies that M matches the maximum possible critical nodes that can be matched in *any* matching. Hence, M is critical. For the $\mathcal{A}$-part we show that the path $\rho = \langle u_0, v_1, u_1, \ldots \rangle$ begins at the highest level $s + t$ with an unmatched critical vertex $u_0 \in \mathcal{A}$. Using Property 1(5), we also show that at least the first two vertices on the $\mathcal{A}$-side (u_0 and u_1) on ρ are at the same level $s + t$. Then we argue that the other end of ρ must be at a level below $t + 1$. Since there are no steep downward edges (Lemma 1), the path contains at least one vertex from each level $t+1, \ldots, s+t-1$. Thus, we have at least $s + 1$ many vertices in $\mathcal{A}_{t+1} \cup \ldots \cup \mathcal{A}_{s+t}$. This contradicts Property 1(1). Proof for the $\mathcal{B}$-part is analogous. See full version [21] for the complete proof.   □

Lemma 4. *The output matching M of Algorithm 1 is RSM for G.*

Proof. If there is no blocking pair w.r.t. M then we are done. Hence, assume that (a, b) is a blocking pair w.r.t. M. By Lemma 2, (a, b) is an upward edge. We consider two cases based on the level of b.

Case 1: $b \in \mathcal{B}_y$ **for** $y \leq t$. Clearly, $a \in \mathcal{A}_x$ for $x \leq t-1$. Thus, by the construction of G_M, a is matched, and hence $M(a)$ exists. Clearly, $M(a)$ is at level at most $t-1$. By Property 1(2), $M(a)$ is critical. Hence, the blocking pair (a, b) is justified by Condition 1 of Definition 2.

Case 2: $b \in \mathcal{B}_y$ **for** $y > t$. By construction of G_M, b is matched. Thus, $M(b)$ exists and $M(b) \in \mathcal{A}_x$ for $x \geq t + 1$. By Property 1(1), $M(b)$ is critical. Hence, the blocking pair (a, b) is justified by Condition 2 of Definition 2.   □

Lemma 5. *Let M' be any maximum size CRITICAL-RSM and M be the output of Algorithm 1 for an instance of our problem. Then $|M| \geq \frac{2}{3} \cdot |M'|$.*

Proof. We prove that $M \oplus M'$ does not admit any 1-length or 3-length augmenting path w.r.t. M. This immediately implies that $|M| \geq \frac{2}{3} \cdot |M'|$. If a is unmatched (critical or otherwise), we know from Property 1(3) and Property 1(4) that no neighbour b of a is unmatched in M. Thus, M is a maximal matching.

For contradiction assume that $M \oplus M'$ contains a 3-length augmenting path $\rho = \langle a_1, b, a, b_1 \rangle$ w.r.t. M. Here $(a, b) \in M$ and the other two edges are in M'. We show that (a, b) blocks M' and the blocking pair is not justified. This will contradict relaxed stability of M'. We first establish the levels of the vertices.

Levels of Vertices: The fact that a_1 remains unmatched in M implies that $a_1^{t^*}$ exhausted $\mathsf{Pref}(a_1)$. Thus, a_1 is at level at least t^*. Since b_1 remains unmatched in M, a did not exhaust $\mathsf{Pref}(a)$ at the level t. Thus, a is at level at most t. We claim that a_1 is not at level $t + 1$ or higher. If a_1 is at level $x \geq t + 1$ then a_1^x must have proposed to b as a_1 is unmatched in M. Since a is at level at most t, b must reject a and accept a_1 – a contradiction to $(a, b) \in M$. Thus, we conclude that a_1 is at level t^*. Now, if a is at level $y < t$ then b must reject a and accept a_1 as a_1 at level t^* proposed to it. Recall that t^* is a sub-level of t used in the algorithm, and t^* does not appear as a separate level in G_M. Thus, the vertices $a, a_1 \in \mathcal{A}_t$.

The Pair (a, b) Blocks M': If $a_1 \succ_b a$, then b would have accepted the proposal of a_1^t by rejecting a^t. Thus, $a \succeq_b a_1$. Since $a_1^{t^*}$ was rejected by b, it implies $M(b) = a$ and a_1 cannot be in tie for b, otherwise b would not have rejected a $*$ status vertex over a non $*$ status vertex. Thus, $a \succ_b a_1$. Now, we show that $b \succ_a b_1$. Suppose not. Then, if $b_1 \succ_a b$, then a^t must have proposed to b_1 before b and got matched to it – a contradiction that b_1 is unmatched. Hence, assume that $b =_a b_1$. In this case, when a^t proposes to b, the vertex b must also be unmatched; otherwise, b cannot be a favourite neighbour of a^t. This implies that a_1 proposes to b only *after* a proposes to b. Since b_1 was unmatched when a proposed to b, the proposal from a to b was uncertain. We claim that b must reject the proposal by a after the proposal (a, b) becomes uncertain due to a proposal by some vertex, possibly a_1^t. Such a vertex must exist because a_1^t proposed to b after (a, b) becomes uncertain. Since a has an unmatched neighbour b_1 at the same rank, a must have proposed b_1 before proposing to b again. This implies b_1 is matched, a contradiction. Thus, $b \succ_a b_1$; hence (a, b) blocks M'.

The Blocking Pair (a, b) is not Justified: In order to prove this, we show $b_1 = M'(a)$ and $a_1 = M'(b)$ are both non-critical. Note that b_1 is unmatched in M, hence if it is critical then $b_1 \in \mathcal{B}_0$ and the number of critical vertices on $\mathcal{B}$-side is at least 1 (that is $t \geq 1$). This implies that a cannot be at a level ≥ 1 since it has not yet proposed to at least one critical neighbour, namely b_1. Thus, b_1 is not critical. We finish the proof by showing that a_1 is also not critical. Note that a_1 is unmatched in M, hence, if it is critical then $a_1 \in \mathcal{A}_{s+t}$ and $s > 0$. This is a contradiction that $a_1 \in \mathcal{A}_t$. Thus, a_1 is not critical.

This finishes the proof that the claimed 3-length augmenting path w.r.t. M does not exist establishing the size guarantee. $\square$

Using Lemma 3, Lemma 4 and Lemma 5, we establish Theorem 1.

5   Conclusion

In this work, we consider the problem of computing a matching in the stable marriage problem where ties and critical vertices can appear on both sides of the bipartition. We investigate a recently introduced notion of optimality called relaxed stability for our setting. We show that every instance of our problem admits a Relaxed Stable Matching (RSM) which is also critical. It follows from the known results [6,18] that computing a maximum size critical RSM is NP-hard and hard to approximate within any factor smaller than $\frac{21}{19}$. We present a polynomial-time algorithm to compute a $\frac{3}{2}$-approximation of the maximum size critical RSM.

References

1. Biró, P., Irving, R.W., Manlove, D.F.: Popular matchings in the marriage and roommates problems. In: Calamoneri, T., Diaz, J. (eds.) CIAC 2010. LNCS, vol. 6078, pp. 97–108. Springer, Heidelberg (2010). https://doi.org/10.1007/978-3-642-13073-1_10
2. Dean, B., Jalasutram, R.: Factor revealing LPs and stable matching with ties and incomplete lists. In: Proceedings of the 3rd International Workshop on Matching Under Preferences, pp. 42–53 (2015)
3. Gale, D., Shapley, L.S.: College admissions and the stability of marriage. Am. Math. Mon. **69**(1), 9–15 (1962)
4. Gärdenfors, P.: Match making: assignments based on bilateral preferences. Behav. Sci. **20**(3), 166–173 (1975)
5. Goko, H., Makino, K., Miyazaki, S., Yokoi, Y.: Maximally satisfying lower quotas in the hospitals/residents problem with ties. In: 39th International Symposium on Theoretical Aspects of Computer Science (2022)
6. Halldórsson, M.M., Iwama, K., Miyazaki, S., Yanagisawa, H.: Improved approximation results for the stable marriage problem. ACM Trans. Algorithm (TALG) **3**(3), 30-es (2007)
7. Hamada, K., Iwama, K., Miyazaki, S.: The hospitals/residents problem with lower quotas. Algorithmica **74**(1), 440–465 (2016)
8. Hamada, K., Miyazaki, S., Yanagisawa, H.: Strategy-proof approximation algorithms for the stable marriage problem with ties and incomplete lists. In: International Symposium on Algorithms and Computation (2019)
9. Huang, C.C., Kavitha, T.: Improved approximation algorithms for two variants of the stable marriage problem with ties. Math. Program. **154**, 353–380 (2015)
10. Iwama, K., Miyazaki, S., Yanagisawa, H.: A 25/17-approximation algorithm for the stable marriage problem with one-sided ties. Algorithmica **68**(3), 758–775 (2014)
11. Kavitha, T.: Matchings, critical nodes, and popular solutions. In: 41st IARCS Annual Conference on Foundations of Software Technology and Theoretical Computer Science (FSTTCS 2021) (2021)
12. Király, Z.: Better and simpler approximation algorithms for the stable marriage problem. Algorithmica **60**(1), 3–20 (2011)
13. Király, Z.: Linear time local approximation algorithm for maximum stable marriage. Algorithms **6**(3), 471–484 (2013)

14. Krishnaa, P., Limaye, G., Nasre, M., Nimbhorkar, P.: Envy-freeness and relaxed stability: hardness and approximation algorithms. In: Harks, T., Klimm, M. (eds.) SAGT 2020. LNCS, vol. 12283, pp. 193–208. Springer, Cham (2020). https://doi. org/10.1007/978-3-030-57980-7_13
15. Krishnaa, P., Limaye, G., Nasre, M., Nimbhorkar, P.: Envy-freeness and relaxed stability: hardness and approximation algorithms. J. Comb. Optim. 45(1), 1–30 (2023)
16. Lam, C.K., Plaxton, C.G.: A (1+ 1/e)-approximation algorithm for maximum stable matching with one-sided ties and incomplete lists. In: Proceedings of the Thirtieth Annual ACM-SIAM Symposium on Discrete Algorithms, pp. 2823–2840. SIAM (2019)
17. Makino, K., Miyazaki, S., Yokoi, Y.: Incomplete list setting of the hospitals/residents problem with maximally satisfying lower quotas. In: Kanellopoulos, P., Kyropoulou, M., Voudouris, A. (eds.) SAGT 2022. Lecture Notes in Computer Science, vol. 13584, pp. 544–561. Springer, Cham (2022). https://doi.org/10.1007/ 978-3-031-15714-1_31
18. Manlove, D.F., Irving, R.W., Iwama, K., Miyazaki, S., Morita, Y.: Hard variants of stable marriage. Theoret. Comput. Sci. 276(1–2), 261–279 (2002)
19. McDermid, E.: A 3/2-approximation algorithm for general stable marriage. In: Albers, S., Marchetti-Spaccamela, A., Matias, Y., Nikoletseas, S., Thomas, W. (eds.) ICALP 2009. LNCS, vol. 5555, pp. 689–700. Springer, Heidelberg (2009). https://doi.org/10.1007/978-3-642-02927-1_57
20. Nasre, M., Nimbhorkar, P.: Popular matchings with lower quotas. In: 37th IARCS Annual Conference on Foundations of Software Technology and Theoretical Computer Science (FSTTCS 2017), pp. 44:1–44:15 (2017)
21. Nasre, M., Nimbhorkar, P., Ranjan, K.: Critical relaxed stable matchings with two-sided ties. arXiv preprint arXiv:2303.12325 (2023)
22. Nasre, M., Nimbhorkar, P., Ranjan, K., Sarkar, A.: Popular matchings in the hospital-residents problem with two-sided lower quotas. In: 41st IARCS Annual Conference on Foundations of Software Technology and Theoretical Computer Science (FSTTCS 2021), vol. 213, pp. 30:1–30:21 (2021)
23. Nasre, M., Nimbhorkar, P., Ranjan, K., Sarkar, A.: Popular critical matchings in the many-to-many setting. arXiv:2206.12394 (2023)
24. Paluch, K.: Faster and simpler approximation of stable matchings. Algorithms 7(2), 189–202 (2014)
25. Robards, P.A.: Applying two-sided matching processes to the united states navy enlisted assignment process. Technical report, NAVAL POSTGRADUATE SCHOOL MONTEREY CA (2001)
26. Roth, A.E.: Stability and polarization of interests in job matching. Econometrica: J. Econometric Soc. 52, 47–57 (1984)
27. Roth, A.E.: On the allocation of residents to rural hospitals: a general property of two-sided matching markets. Econometrica: J. Econometric Soc. 54, 425–427 (1986)
28. Tan, S.J., Yeong, C.M.: Designing economics experiments to demonstrate the advantages of an electronic employment market in a large military organization. Technical report, NAVAL POSTGRADUATE SCHOOL MONTEREY CA (2001)
29. Yang, W., Sycara, K.: Two-sided matching for the us navy detailing process with market complication. Technical report, Technical Report CMU-RI-TR-03-49, Robotics Institute, Carnegie-Mellon University (2003)
30. Yokoi, Y.: Envy-free matchings with lower quotas. Algorithmica 82(2), 188–211 (2020)

Graph Search Trees and Their Leaves

Robert Scheffler$^{(\boxtimes)}$ (iD)

Institute of Mathematics, Brandenburg University of Technology, Cottbus, Germany
robert.scheffler@b-tu.de

Abstract. Graph searches and their respective search trees are widely used in algorithmic graph theory. The problem whether a given spanning tree can be a graph search tree has been considered for different searches, graph classes and search tree paradigms. Similarly, the question whether a particular vertex can be visited last by some search has been studied extensively in recent years. We combine these two problems by considering the question whether a vertex can be a leaf of a graph search tree. We show that for particular search trees, including DFS trees, this problem is easy if we allow the leaf to be the first vertex of the search ordering. We contrast this result by showing that the problem becomes hard for many searches, including DFS and BFS, if we forbid the leaf to be the first vertex. Additionally, we present several structural and algorithmic results for search tree leaves of chordal graphs.

Keywords: Graph search · Graph search trees · Leaves

1 Introduction

Graph searches are an extensively used concept in algorithmic graph theory. The searches BFS and DFS belong to the most basic algorithms and are used in a wide range of applications as subroutines. The same holds for more sophisticated searches as LBFS, LDFS, and MCS (see, e.g., [4,8,14]).

An important structure closely related to a graph search is the corresponding search tree. Such a tree contains all the vertices of the graph and for every vertex different from the start vertex exactly one edge to a vertex preceding it in the search ordering. Those trees can be of particular interest as for instance the tree obtained by a BFS contains the shortest paths from the root to all other vertices in the graph and DFS trees are used for fast planarity testing [19]. Furthermore, trees generated by LBFS were used to design a linear-time implementation of the search LDFS for chordal graphs [3].

The problem of deciding whether a given spanning tree of a graph can be obtained by a particular search was introduced by Hagerup [17] in 1985, who presented a linear-time algorithm that recognizes DFS trees. In the same year, Hagerup and Nowak [18] gave a similar result for the BFS tree recognition. In 2021, Beisegel et al. [2] presented a more general framework for the search tree recognition problem. They introduced the term $\mathcal{F}$-tree for search trees where a vertex is connected to its first visited neighbor, i.e., BFS-like trees, and $\mathcal{L}$-trees for search trees where a vertex is connected to its most recently visited neighbor, i.e., DFS-like trees. They showed, among other things, that $\mathcal{F}$-tree recognition is NP-hard for LBFS, LDFS, and MCS on weakly chordal graphs, while the problem can be solved in polynomial time for all three searches on chordal graphs. These results are complemented in [29], where it is shown that the recognition of $\mathcal{F}$-trees of DFS and $\mathcal{L}$-trees of BFS is NP-hard, a strong contrast to the polynomial results for $\mathcal{F}$-trees of BFS and $\mathcal{L}$-trees of DFS.

Another feature of a graph search that was used several times within algorithms are its end-vertices, i.e., the vertices that can be visited last by the search. Some of these end-vertices have nice properties. One example are the end-vertices of LBFS on chordal graphs. These vertices are simplicial, a fact that was used by Rose et al. [27] to design a linear-time recognition algorithm for chordal graphs. Furthermore, the end-vertices of LBFS are strongly related to dominating pairs of AT-free graphs [13] and transitive orientations of comparability graphs [16]. Thus, it is well motivated to consider the end-vertex problem, i.e., the question whether a given vertex of a graph is an end-vertex of a particular search. Introduced in 2010 by Corneil et al. [11], the problem has gained much attention by several researchers, leading to a wide range of hardness results and algorithms for different searches on different graph classes (see, e.g., [1,6,15,22,25,33]).

If we compare the known complexity results for the end-vertex problem and the recognition problem of $\mathcal{F}$-trees, we notice strong similarities between these two problems. Motivated by that fact, a generalization of both problems, called *Partial Search Order Problem*, was introduced in [28]. This problem asks whether a given partial order on a graph's vertex set can be linearly extended by a search ordering. Another way to combine the end-vertex problem with the search tree recognition problems is motivated by the following observation: If a vertex is the end-vertex of some search ordering, then it is a leaf in the respective search tree, no matter whether we consider the $\mathcal{F}$-tree or the $\mathcal{L}$-tree. Therefore, we ask whether a given vertex can be a leaf of a search tree constructed by a particular search. Note that this problem was first suggested in 2020 by Michel Habib. Here, we study its complexity for $\mathcal{F}$-trees and $\mathcal{L}$-trees of several searches, including BFS, DFS, LBFS, LDFS, and MCS.

Our Contribution. We consider two different types of leaves of search trees. A leaf is a *root leaf* of a search tree if it is the start vertex of the respective search ordering. All other leaves of a search tree are called *branch leaves*. We show that it is easy for all the searches considered here to identify the possible root leaves both for $\mathcal{F}$-trees and for $\mathcal{L}$-trees. For some searches, including DFS, these results imply directly that the general problem of recognizing leaves of $\mathcal{L}$-trees is easy. This is contrasted by the result that, at least for DFS, the recognition of branch leaves of $\mathcal{L}$-trees is NP-hard. We show that the same holds for $\mathcal{F}$-tree branch leaves of several searches, including DFS and BFS. In contrast, the leaves of $\mathcal{L}$-trees of BFS can be recognized in polynomial time for bipartite graphs. This is quite surprising since the $\mathcal{L}$-tree recognition problem of BFS is NP-hard on bipartite graphs [29] while $\mathcal{F}$-trees of BFS can be recognized efficiently on general graphs [18]. In the final section we consider chordal graphs and show that on this graph class the branch leaves of almost all considered searches can be recognized in linear time.

Due to lack of space, the proofs of some results are omitted here. They can be found in the full version [30].

2   Preliminaries

General Notation. The graphs considered in this paper are finite, undirected, simple and connected. Given a graph G, we denote by $V(G)$ the *set of vertices* and by $E(G)$ the *set of edges*. The terms $n(G)$ and $m(G)$ describe the number of vertices and edges of G, respectively, i.e., $n(G) = |V(G)|$ and $m(G) = |E(G)|$. For a vertex $v \in V(G)$, we denote by $N_G(v)$ the *(open) neighborhood* of v in G, i.e., the set $N_G(v) = \{u \in V \mid uv \in E\}$ where uv denotes an edge between u and v. The *closed neighborhood* of a vertex v is the union of the open neighborhood of v with the set $\{v\}$ and is denoted by $N_G[v]$. Given a set $S \subseteq V(G)$, the term $G[S]$ describes the subgraph of G that is induced by S.

The *distance* $dist_G(v, w)$ of two vertices v and w in G is the length (i.e., the number of edges) of the shortest v-w-path in G. The *eccentricity* $ecc_G(v)$ of a vertex v in G is the largest distance of v to any other vertex in G. The *diameter* $diam(G)$ of G is the largest eccentricity of a vertex in G and the *radius* $rad(G)$ of G is the smallest eccentricity of a vertex in G. A vertex v with $ecc_G(v) = rad(G)$ is called *central vertex* of G. The set $N_G^\ell(v)$ contains all vertices whose distance to the vertex v in G is equal to ℓ.

A *vertex ordering* of G is a bijection $\sigma : \{1, 2, \ldots, |V(G)|\} \to V(G)$. We denote by $\sigma^{-1}(v)$ the position of vertex $v \in V(G)$. Given two vertices u and v in G we say that u is *to the left* (resp. *to the right*) of v if $\sigma^{-1}(u) < \sigma^{-1}(v)$ (resp. $\sigma^{-1}(u) > \sigma^{-1}(v)$) and we denote this by $u \prec_\sigma v$ (resp. $u \succ_\sigma v$).

A *clique* in a graph G is a set of pairwise adjacent vertices and an *independent set* in G is a set of pairwise nonadjacent vertices. A clique C is *dominating* if any vertex of G is either in C or has a neighbor in C. A vertex v is *simplicial* if its neighborhood induces a clique. A vertex v of a connected graph G is a *cut vertex* if $G - v$ is not connected. Two vertices u and w form a *two-pair* if any induced path between u and w has length two.

A graph is *bipartite* if its vertex set can be partitioned into two independent sets X and Y. A graph is *weakly chordal* if G contains neither an induced cycle of the length ≥ 5 nor the complement of such an induced cycle. A graph is *chordal* if it does not contain an induced cycle of length ≥ 4. A vertex ordering σ of a graph G is a *perfect elimination ordering* if any vertex v is simplicial in the graph $G[S(v)]$ with $S(v) := \{w \mid w \prec_\sigma v\}$. A graph G has a PEO if and only if G is chordal [26]. A *split graph* G is a graph whose vertex set can be partitioned into sets C and I, such that C is a clique in G and I is an independent set in G. It is easy to see that any split graph is chordal.

Algorithm 1: Label Search($\prec_{\mathcal{A}}$)

Input: A graph G
Output: A search ordering σ of G
1 **begin**
2   **foreach** $v \in V(G)$ **do** $label(v) \leftarrow \emptyset$ **for** $i \leftarrow 1$ **to** $n(G)$ **do**
3      $Eligible \leftarrow \{x \in V(G) \mid x$ unnumbered and $\nexists$ unnumbered $y \in V(G)$
4   such that $label(x) \prec_{\mathcal{A}} label(y)\}$;
5   let v be an arbitrary vertex in $Eligible$;
6      $\sigma(i) \leftarrow v$;   /* assigns to v the number i */
7   **foreach** *unnumbered vertex* $w \in N(v)$ **do** $label(w) \leftarrow label(w) \cup \{i\}$

A *tree* is an acyclic connected graph. A *spanning tree* of a graph G is an acyclic connected subgraph of G which contains all vertices of G. A tree together with a distinguished *root vertex* r is said to be *rooted*. In such a rooted tree T, a vertex v is the *parent* of vertex w if v is an element of the unique path from w to the root r and the edge vw is contained in T. A vertex w is called the *child* of v if v is the parent of w.

Searches, Search Trees and Their Leaves. In the most general sense, a *graph search* $\mathcal{A}$ is a function that maps every graph G to a set $\mathcal{A}(G)$ of vertex orderings of G. The elements of the set $\mathcal{A}(G)$ are the $\mathcal{A}$-*orderings of* G. The graph searches considered in this paper can be formalized adapting a framework introduced by Corneil et al. [10] (a similar framework is given in [23]). This framework uses subsets of $\mathbb{N}^+$ as vertex labels. Whenever a vertex is numbered, its index in the search ordering is added to the labels of its unnumbered neighbors. The search $\mathcal{A}$ is defined via a strict partial order $\prec_{\mathcal{A}}$ on the elements of $\mathcal{P}(\mathbb{N}^+)$ (see Algorithm 1). The respective $\mathcal{A}$-orderings are exactly those vertex orderings that can be found by this framework using the partial label order $\prec_{\mathcal{A}}$.

In the following, we define the searches considered in this paper by presenting suitable partial orders $\prec_{\mathcal{A}}$ (see [10]). The *Generic Search* (GS) is equal to the Label Search($\prec_{GS}$) where $A \prec_{GS} B$ if and only if $A = \emptyset$ and $B \neq \emptyset$. Thus, any vertex with a numbered neighbor can be numbered next.

The partial label order $\prec_{BFS}$ for *Breadth First Search* (BFS) is defined as follows: $A \prec_{BFS} B$ if and only if $A = \emptyset$ and $B \neq \emptyset$ or $\min(A) > \min(B)$. For the *Lexicographic Breadth First Search* (LBFS) [27] we consider the partial order $\prec_{LBFS}$ with $A \prec_{LBFS} B$ if and only if $A \subsetneq B$ or $\min(A \setminus B) > \min(B \setminus A)$. Both BFS and LBFS are *layered*, i.e., the sets $N_G^{\ell}(r)$ are consecutive within orderings starting in r. We sometimes use the term *layer* if we refer to a set $N_G^{\ell}(r)$.

The partial label order $\prec_{DFS}$ for *Depth First Search* (DFS) is defined as follows: $A \prec_{DFS} B$ if and only if $A = \emptyset$ and $B \neq \emptyset$ or $\max(A) < \max(B)$. For the *Lexicographic Depth First Search* [12] we use the strict partial order $\prec_{LDFS}$ where $A \prec_{LDFS} B$ if and only if $A \subsetneq B$ or $\max(A \setminus B) < \max(B \setminus A)$.

The *Maximum Cardinality Search* (MCS) [32] uses the partial order $\prec_{MCS}$ with $A \prec_{MCS} B$ if and only if $|A| < |B|$. The *Maximal Neighborhood Search* (MNS) [12] is defined using $\prec_{MNS}$ with $A \prec_{MNS} B$ if and only if $A \subsetneq B$. It follows directly from these partial label orders, that any LBFS, LDFS, and MCS ordering is also an MNS ordering. Furthermore, the orderings of all presented searches are GS orderings.

Searches as BFS and DFS are often used to compute corresponding graph search trees. Beisegel et al. [2] formalized the different concepts of search trees as follows.

Definition 2.1 (Beisegel et al. [2]). *Let σ be a GS ordering of a connected graph G. The $\mathcal{F}$-tree of σ is the spanning tree of G containing the edge from each vertex v with $\sigma^{-1}(v) > 1$ to its leftmost neighbor in σ.*

The $\mathcal{L}$-tree of σ is the spanning tree containing the edge from each vertex v with $\sigma^{-1}(v) > 1$ to its rightmost neighbor w in σ with $w \prec_\sigma v$.

In this paper, we consider the leaves of these search trees. For both $\mathcal{F}$-trees and $\mathcal{L}$-trees, we distinguish two different types of leaves.

Definition 2.2. *Let σ be a GS ordering of a connected graph G. A vertex $v \in V(G)$ is an $\mathcal{F}$-leaf ($\mathcal{L}$-leaf) of σ if v is a leaf in the $\mathcal{F}$-tree ($\mathcal{L}$-tree) of σ. If v is the first vertex of σ, then it is the $\mathcal{F}$-root leaf ($\mathcal{L}$-root leaf) of σ, otherwise it is an $\mathcal{F}$-branch leaf ($\mathcal{L}$-branch leaf) of σ.*

As the graph with exactly one vertex has no leaf in its spanning tree, we will consider only graphs with at least two vertices.

3   Root Leaves

We start this section with the simple observation that $\mathcal{F}$-root leaves of GS orderings of a graph G are quite boring as they are exactly the leaves of G.

Observation 3.1. *Let G be a connected graph with $n(G) \geq 2$. The following conditions are equivalent for a vertex $v \in V(G)$.*

(i)   Vertex v is the $\mathcal{F}$-root leaf of some GS ordering of G.
(ii) Vertex v is the $\mathcal{F}$-root leaf of every GS ordering of G starting in v.
(iii) Vertex v is a leaf of G.

Next we consider the $\mathcal{L}$-root leaves of GS, DFS, and MCS. They are exactly those vertices of the graph that are not cut vertices. The same even holds for $\mathcal{F}$-branch leaves and $\mathcal{L}$-branch leaves of GS.

Theorem 3.2. *Let G be a connected graph with $n(G) \geq 2$. The following conditions are equivalent for a vertex $v \in V(G)$.*

(i) Vertex v is the $\mathcal{L}$-root leaf of some DFS ordering of G starting in v.
(ii) Vertex v is the $\mathcal{L}$-root leaf of every DFS ordering of G starting in v.

  (iii) *Vertex v is the $\mathcal{L}$-root leaf of some MCS ordering of G.*
  (iv) *Vertex v is the $\mathcal{L}$-root leaf of some GS ordering of G.*
   (v) *Vertex v is an $\mathcal{L}$-branch leaf of some GS ordering of G.*
  (vi) *Vertex v is an $\mathcal{F}$-branch leaf of some GS ordering of G.*
 (vii) *Vertex v is the end-vertex of some GS ordering of G.*
(viii) *Vertex v is not a cut vertex of G.*

   Note that DFS differs from GS and MCS in this result. While for the latter three searches it is possible that a vertex is not the $\mathcal{L}$-root leaf of a search ordering starting with that vertex, this is not possible for DFS.

   Since DFS, LDFS, MCS, and MNS orderings are also GS orderings, Theorem 3.2 directly implies that we can characterize the $\mathcal{L}$-leaves of these orderings.

Theorem 3.3. *For any search $\mathcal{A} \in \{GS, DFS, LDFS, MCS, MNS\}$ and any vertex v of a connected graph G with $n(G) \geq 2$, the following statements are equivalent.*

  (i) *Vertex v is the $\mathcal{L}$-root leaf of some $\mathcal{A}$-ordering of G.*
 (ii) *Vertex v is an $\mathcal{L}$-leaf of some $\mathcal{A}$-ordering of G.*
(iii) *Vertex v is not a cut vertex of G.*

   As we can check in linear time whether a vertex is a cut vertex, we can also recognize $\mathcal{L}$-leaves of GS, DFS, LDFS, MCS, and MNS within this time bound. However, we will see in Corollary 4.2 that at least for DFS the recognition of $\mathcal{L}$-branch leaves is NP-complete.

   The characterization of $\mathcal{L}$-root leaves given in Theorem 3.2 does not work for BFS as the following theorem shows.

Theorem 3.4. *Let G be a connected graph with $n(G) \geq 2$. A vertex $v \in V(G)$ is the $\mathcal{L}$-root leaf of some BFS ordering of G if and only if $G[N_G(v)]$ is connected.*

4   NP-Hardness of Branch Leaf Recognition

Branch Leaves of DFS. DFS $\mathcal{L}$-trees can be recognized in linear time [17,20]. As we have seen in Theorem 3.3, this also holds for DFS $\mathcal{L}$-leaves. In contrast, recognizing DFS $\mathcal{L}$-branch leaves of a graph is as hard as the recognition of DFS end-vertices since the two concepts are equivalent.

Theorem 4.1. *A vertex $v \in V(G)$ of a graph G is an $\mathcal{L}$-branch leaf of some DFS ordering of G if and only if v is the end-vertex of some DFS ordering of G.*

   Charbit et al. [6] gave sufficient conditions on a graph class $\mathcal{G}$ such that the end-vertex problem of DFS is NP-complete on $\mathcal{G}$. Due to Theorem 4.1, we can replace the term end-vertex in their result by the term $\mathcal{L}$-branch leaf.

Corollary 4.2. *Let $\mathcal{G}$ be a graph class that is closed under the insertion of universal vertices. If the Hamiltonian path problem is NP-complete on $\mathcal{G}$, then the problem of deciding whether a vertex of a graph $G \in \mathcal{G}$ is an $\mathcal{L}$-branch leaf of some DFS ordering of G is NP-complete. In particular, the problem is NP-complete on split graphs.*

A similar result can be given for $\mathcal{F}$-branch leaves of DFS. By adapting the proof given in [29] that $\mathcal{F}$-trees of DFS are hard to recognize, we can show that the same holds for $\mathcal{F}$-branch leaves of DFS.

Theorem 4.3. *Let $\mathcal{G}$ be a graph class that is closed under the insertion of universal vertices and leaves. If the Hamiltonian path problem is NP-complete on $\mathcal{G}$, then the problem of deciding whether a vertex of a graph $G \in \mathcal{G}$ is an $\mathcal{F}$-branch leaf of some DFS ordering of G is NP-complete. In particular, the problem is NP-complete on chordal graphs.*

If we compare Corollary 4.2 and Theorem 4.3, then we see that for $\mathcal{L}$-branch leaves it is sufficient that the graph class $\mathcal{G}$ is closed under the addition of universal vertices while for $\mathcal{F}$-branch leaves we have the additional condition that $\mathcal{G}$ is closed under the addition of leaves. We cannot omit this constraint (unless P = NP) as the $\mathcal{F}$-branch leaf recognition problem of DFS can be solved in polynomial time on split graphs (see Corollary 5.13).

Branch Leaves of BFS. The end-vertex problem of BFS is NP-complete, even if the graph is bipartite and the start vertex of the BFS ordering is fixed [6]. This fact can be used to show that recognizing BFS $\mathcal{F}$-branch leaves is also NP-complete.

Theorem 4.4. *It is NP-complete to decide whether a vertex of a bipartite graph G is an $\mathcal{F}$-branch leaf of some BFS ordering of G.*

In contrast to this result, there is a simple characterization of BFS $\mathcal{L}$-branch leaves of bipartite graphs.

Theorem 4.5. *Let G be a connected bipartite graph with $n(G) \geq 2$. A vertex $v \in V(G)$ is an $\mathcal{L}$-branch leaf of some BFS ordering of G if and only if there is an $r \in V(G) \setminus \{v\}$ such that $dist_G(r, w) = dist_{G-v}(r, w)$ for all $w \in V(G) \setminus \{v\}$.*

Proof. Assume that there is a vertex $r \in V(G) \setminus \{v\}$ such that $dist_G(r, w) = dist_{G-v}(r, w)$ for all $w \in V(G) \setminus \{v\}$. Let $(r = w_0, \ldots, w_k = v)$ be a shortest path from r to v, i.e., v has distance k to r. It is easy to see that there is a BFS ordering σ of G in which every vertex w_i is the first vertex of the i-th layer. Let T be the $\mathcal{L}$-tree of σ and let x be a vertex in the $(k+1)$-th layer. Due to the condition on r, there is a shortest path from r to x in G that does not use vertex v. Therefore, x has a neighbor y in the k-th layer that is not v. Since $v \prec_\sigma y \prec_\sigma x$, vertex v is not the parent of x in T. Since G is bipartite, the layers of σ are independent sets and, thus, v is neither the parent of any vertex in the k-th layer. Hence, v is a leaf of T.

Now assume that v is a branch leaf of the $\mathcal{L}$-tree T of the BFS ordering σ. Let r be the start vertex of σ. Let w be a vertex different from v and r. Consider the r-w-path P in T. Since G is bipartite, the edges of G and, thus, the edges of T only connect vertices of consecutive layers. Furthermore, every vertex has a neighbor in its preceding layer. Thus, P has $dist_G(r,w)$ edges. Since v is a leaf of T, P does not contain v. Therefore, P is also contained in $G - v$ and $dist_{G-v}(r,w) = dist_G(r,w)$. $\qquad\square$

To check whether the condition of Theorem 4.5 is fulfilled, we simply make two all-pair-shortest paths computations and compare the results. This can be done in $\mathcal{O}(n(G) \cdot m(G))$ by using $\mathcal{O}(n(G))$ many BFS computations.

Corollary 4.6. *Given a connected bipartite graph G and a vertex $v \in V(G)$, we can decide in time $\mathcal{O}(n(G) \cdot m(G))$ whether v is the $\mathcal{L}$-branch leaf of some BFS ordering of G.*

The results of Theorem 4.4 and Corollary 4.6 are quite surprising since the $\mathcal{L}$-tree recognition problem of BFS is NP-hard on bipartite graphs [29] while the $\mathcal{F}$-tree recognition problem of BFS can be solved in linear time [18,24].

Branch Leaves of MNS-like Searches. For several subsearches of MNS, the recognition problem of $\mathcal{F}$-branch leaves is NP-complete on weakly chordal graphs.

Theorem 4.7. *Let $\mathcal{A}$ be one of the following searches: LBFS, LDFS, MCS, MNS. It is NP-complete to decide whether a vertex of a weakly chordal graph G is an $\mathcal{F}$-branch leaf of some $\mathcal{A}$-ordering.*

Proof. The proof of the theorem is inspired by the NP-completeness proof of the $\mathcal{F}$-tree recognition problem of MNS given by Beisegel et al. [2]. We construct a polynomial-time reduction from 3-SAT. Let $\mathcal{I}$ be an instance of 3-SAT. W.l.o.g. we may assume that $\mathcal{I}$ contains at least two clauses. We construct the corresponding graph $G(\mathcal{I})$ as follows. Let $X = \{x_1, \ldots, x_k, \overline{x}_1, \ldots, \overline{x}_k\}$ be the set of vertices representing the literals of $\mathcal{I}$. The graph $G(\mathcal{I})[X]$ forms the complement of the matching in which x_i is matched to $\overline{x}_i$ for every $i \in \{1, \ldots, k\}$. Let $C = \{c_1, \ldots, c_\ell\}$ be the set of vertices representing the clauses of $\mathcal{I}$. The set C forms an independent set in $G(\mathcal{I})$ and every clause vertex c_i is adjacent to each vertex of X whose corresponding literal is contained in the clause associated with c_i. Additionally, we add a universal vertex t.

Assume $G(\mathcal{I})$ has a fulfilling assignment $\mathcal{B}$. Then we create the following $\mathcal{A}$-ordering σ. We first number all literal vertices of literals that are set to true in $\mathcal{B}$ and then we number t. Since these vertices form a clique, this ordering is a prefix of an $\mathcal{A}$-ordering. We number the remaining vertices following an arbitrary $\mathcal{A}$-ordering. As $\mathcal{B}$ is fulfilling, all clause vertices and all literal vertices have a neighbor that is to the left of t in σ. Thus, t is an $\mathcal{F}$-branch leaf of σ.

Now assume that t is an $\mathcal{F}$-branch leaf of the $\mathcal{A}$-ordering σ of $G(\mathcal{I})$. Let S be the set of literal vertices that are to the left of t in σ. Since t is universal and the edges $x_i\overline{x}_i$ are missing, the set S contains at most one literal vertex for every variable. Thus, we can define an assignment $\mathcal{B}$ by giving all literals whose vertices are contained in S the value true. If some variable value is not fixed, then we choose an arbitrary value for the variable. If a clause vertex has a parent in the $\mathcal{F}$-tree T of σ, then this parent is an element of S since t is a leaf in T. If the clause vertex c_i does not have a parent in T, then c_i is the first vertex of σ. Since there are at least two clause vertices, the second vertex of σ is not t but a literal vertex adjacent to c_i. Therefore, every clause vertex has a neighbor in S and, thus, $\mathcal{B}$ is a fulfilling assignment.

To see that $G(\mathcal{I})$ is weakly chordal, we first observe that every pair $(x_i, \overline{x}_i)$ forms a two-pair in $G(\mathcal{I})$. Spinrad and Sritharan [31] showed that the graph that results from the addition of an edge between a two-pair is weakly chordal if and only if the initial graph is weakly chordal. If we add all the edges $x_i \overline{x}_i$, then the resulting graph is a split graphs and, thus, $G(\mathcal{I})$ is weakly chordal. $\square$

5   Branch Leaves and Chordal Graphs

Branch Leaves of MNS-Like Searches. MNS and all of its subsearches compute PEOs of chordal graphs [12,32]. Thus, any $\mathcal{F}$-tree or $\mathcal{L}$-tree of an MNS ordering is also an $\mathcal{F}$-tree or $\mathcal{L}$-tree of some PEO. Beisegel et al. [2] showed that this also holds the other way around for a large family of graph searches including LBFS, LDFS, MCS, and MNS, i.e., the rooted $\mathcal{F}$-trees and rooted $\mathcal{L}$-trees of these searches on chordal graphs are exactly the rooted $\mathcal{F}$-trees and rooted $\mathcal{L}$-trees of PEOs, respectively. Therefore, we will only characterize $\mathcal{F}$-branch leaves and $\mathcal{L}$-branch leaves of PEOs.

We start by showing that the $\mathcal{L}$-branch leaves of PEOs of a chordal graph are exactly the graph's simplicial vertices.

Theorem 5.1. *Let G be a connected chordal graph with $n(G) \geq 2$. A vertex $v \in V(G)$ is an $\mathcal{L}$-branch leaf of some PEO of G if and only if v is simplicial.*

Proof. If v is a simplicial vertex, then there is a PEO σ that ends with v. Vertex v is an $\mathcal{L}$-branch leaf of σ.

For the other direction, let σ be a PEO and let v be a non-simplicial vertex of G. Hence, not all neighbors of v are to the left of v in σ. Let w be the leftmost neighbor of v in σ that is to the right of v in σ. Let x be the parent of w in the $\mathcal{L}$-tree T of σ. If x is not equal v, then it holds $v \prec_\sigma x \prec_\sigma w$. As σ is a PEO and $vw, xw \in E(G)$, the edge vx is also in $E(G)$; a contradiction to the choice of w. Hence, v is the parent of w in T and v is not an $\mathcal{L}$-branch leaf of σ. $\square$

Since we can decide in linear time whether a vertex is simplicial [1], we can recognize $\mathcal{L}$-branch leaves of PEOs in linear time.

Corollary 5.2. *Given a connected chordal graph G and a vertex $v \in V(G)$, we can decide in time $\mathcal{O}(n(G) + m(G))$ whether v is the $\mathcal{L}$-branch leaf of some PEO of G. Therefore, we can also decide in time $\mathcal{O}(n(G) + m(G))$ whether v is the $\mathcal{L}$-branch leaf of some LBFS, LDFS, MCS, or MNS ordering.*

Obviously, simplicial vertices are also $\mathcal{F}$-branch leaves of PEOs. However, there are further $\mathcal{F}$-branch leaves.

Theorem 5.3. *Let G be a connected chordal graph with $n(G) \geq 2$. A vertex $v \in V(G)$ is an $\mathcal{F}$-branch leaf of some PEO of G if and only if the graph $G[N_G(v)]$ has a dominating clique.*

Proof. First assume that $G[N_G(v)]$ has a dominating clique C. It is obvious that there is an LBFS ordering σ of G that starts with the vertices of C. The ordering σ is a PEO. Since all neighbors of v have a neighbor in C or are elements of C, v is an $\mathcal{F}$-branch leaf of σ.

Now let v be an $\mathcal{F}$-branch leaf of the PEO σ. Let S be the set of neighbors of v that are to the left of v in σ. The set S induces a clique of G. Thus, if $S = N_G(v)$, then $G[N_G(v)]$ is a clique and we are done. Hence, we may assume that there is a vertex $w \in N_G(v) \setminus S$. As w is not a child of v in the $\mathcal{F}$-tree of σ, there is a vertex $x \in N_G(w)$ with $x \prec_\sigma v$. Since σ is a PEO, vertex x is a neighbor of v and, thus, $x \in S$. Thus, any neighbor of v that is not in S has a neighbor in S and, hence, S induces a dominating clique of $G[N_G(v)]$. $\square$

To decide the complexity of the $\mathcal{F}$-branch leaf recognition problem of PEOs, we examine the complexity of deciding the existence of a dominating clique in a chordal graph. Kratsch et al. [21] showed that such a clique exists if and only if the diameter of the graph is at most three.

Theorem 5.4 (Kratsch et al. [21]). *A chordal graph G has a dominating clique if and only if the diameter of G is at most three.*

As the diameter of a graph can be determined by computing $n(G)$ many BFS orderings, we can decide the existence of a dominating clique in a chordal graph in polynomial time. Although it is unlikely that the diameter of a chordal graph can be computed in linear time,[1] we can improve our algorithm to decide the existence of a dominating clique in linear time. To this end, we can use the following result of Corneil et al. [9].

Theorem 5.5 (Corneil et al. [9]). *Let G be a chordal graph and let $v \in V(G)$ be the end-vertex of some LBFS ordering of G. If $ecc(v) < diam(G)$, then $ecc(v)$ is even and $ecc(v) = diam(G) - 1$.*

Combining Theorems 5.3 to 5.5, we can give a linear-time recognition algorithm for $\mathcal{F}$-branch leaves of PEOs.

Corollary 5.6. *Given a chordal graph G and a vertex $v \in V(G)$, we can decide in time $\mathcal{O}(n(G) + m(G))$ whether v is an $\mathcal{F}$-branch leaf of some PEO of G. Therefore, we can also decide in time $\mathcal{O}(n(G) + m(G))$, whether v is the $\mathcal{F}$-branch leaf of some LBFS, LDFS, MCS, or MNS ordering.*

---

[1] Even on split graphs, the diameter cannot be computed in subquadratic time unless the Strong Exponential Time Hypothesis fails [5].

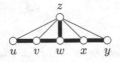

Fig. 1. The given graph G is chordal. There is no dominating clique in the graph $G[N_G(z)]$. However, the given spanning tree is the $\mathcal{F}$-tree of the BFS ordering (w, v, x, z, u, y) and, thus, z is a $\mathcal{F}$-branch leaf of BFS.

Proof. Due to Theorems 5.3 and 5.4, it is sufficient to check whether $G' = G[N_G(v)]$ has diameter 3. We compute an LBFS ordering σ of G' in linear time. Let v be the end-vertex of σ. We compute the eccentricity of v in G' in linear time by starting a BFS in v. If $ecc_{G'}(v) > 3$, then the diameter of G' is larger than 3. If $ecc_{G'}(v) = 3$, then, by Theorem 5.5, $diam(G') = 3$. If $ecc_{G'}(v) < 3$, then $diam(G') \leq ecc_{G'}(v) + 1 \leq 3$, due to Theorem 5.5. $\qquad\square$

Branch Leaves of BFS. The condition given in Theorem 5.3 is also sufficient for a vertex to be a BFS $\mathcal{F}$-branch leaf since every LBFS ordering is also a BFS ordering. However, it is not necessary as can be seen in Fig. 1. To characterize BFS $\mathcal{F}$-branch leaves of chordal graphs, we start with the following two lemmas.

Lemma 5.7. *Let G be a chordal graph and let r be a vertex in $V(G)$. Let x and y be two vertices in $N_G^i(r)$. If there is a vertex $z \in N_G^{i+1}(r)$ which is adjacent to both x and y, then $xy \in E(G)$.*

Lemma 5.8. *Let G be a chordal graph and let r be a vertex in $V(G)$. Let x and y be two vertices in $N_G^i(r)$. If $xy \in E(G)$, then $N_G(x) \cap N_G^{i-1}(r) \subseteq N_G(y)$ or $N_G(y) \cap N_G^{i-1}(r) \subseteq N_G(x)$.*

The next lemma makes a statement about the distances of neighbors of a vertex in a chordal graph.

Lemma 5.9. *Let G be a connected chordal graph and let $v \in V(G)$. For any $x, y \in N_G(v)$, the distance between x and y in $G - v$ is equal to the distance between x and y in $G[N_G(v)]$.*

Using Lemmas 5.7 to 5.9, we characterize BFS $\mathcal{F}$-branch leaves of chordal graphs.

Theorem 5.10. *Let G be a connected chordal graph with $n(G) \geq 2$. A vertex $v \in V(G)$ is an $\mathcal{F}$-branch leaf of some BFS ordering of G if and only if the radius of $G[N_G(v)]$ is at most two.*

Proof. First assume that $G[N_G(v)]$ has radius two and let w be a central vertex of $G[N_G(v)]$. There is a BFS ordering σ that starts with w followed by all neighbors of w that are not v. Vertex v is an $\mathcal{F}$-branch leaf of σ since all neighbors of v have some neighbor in $N_G[w] \setminus \{v\}$ or are equal to w.

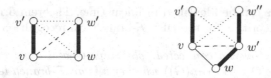

Fig. 2. Two cases of the proof of Theorem 5.10. The vertical arrangement of the vertices represent their layers. Thick edges are edges of the $\mathcal{F}$-tree. Dotted edges are not present. Dashed edges are implied by either Lemma 5.7 or Lemma 5.8.

Now assume that $v \in N_G^i(r)$ is an $\mathcal{F}$-branch leaf of some BFS ordering σ of G starting with r. Let T be the $\mathcal{F}$-tree of σ rooted in r and let v' be the parent of v in T. Since T is an BFS $\mathcal{F}$-tree rooted in r, it holds that $v' \in N_G^{i-1}(r)$.

We claim that in $G-v$ vertex v' has a distance of at most two to every element of $N_G(v)$. Let $w \in N_G(v) \setminus \{v'\}$. If $v'w \in E(G)$, then v' and w have distance one in $G-v$. Therefore, we may assume in the following that $v'w \notin E(G)$. Then Lemma 5.7 implies that $w \notin N_G^{i-1}(r)$. Furthermore, the parent of w in T, say w', is different from v and v'. If $v'w' \in E(G)$, then v' and w have distance two in $G-v$ via the path (v',w',w). Thus, we may also assume that $v'w' \notin E(G)$.

First assume that $w \in N_G^i(r)$ (see left part of Fig. 2). Then $w' \in N_G^{i-1}(r)$. Lemma 5.8 implies that w' is adjacent to v because vw, vv', and $ww' \in E(G)$ but $v'w \notin E(G)$. Now the non-existence of $v'w'$ contradicts Lemma 5.7.

Now assume that $w \in N_G^{i+1}(r)$ (see right part of Fig. 2). Then, due to Lemma 5.7, $vw' \in E(G)$. Since $v'w' \notin E(G)$, the parent of w', say w'', is different from v'. Since w' is the parent of w, it holds that $w' \prec_\sigma v$. This implies that $w'' \prec_\sigma v'$. Therefore, w'' is not adjacent to v since, otherwise, v' would not be the parent of v. The non-existence of both $v'w'$ and vw'' contradicts Lemma 5.8.

Summarizing, v' has distance at most two in $G-v$ to any neighbor of v. By Lemma 5.9, v' has distance at most two in $G[N_G(v)]$ to any neighbor of v in G and, thus, $G[N_G(v)]$ has radius at most two.   □

Chepoi and Dragan [7] presented a linear-time algorithm that computes a central vertex of a chordal graph. As the eccentricity of such a vertex can be computed in linear time using BFS, we can compute the radius of a chordal graph and, in particular, of $G[N_G(v)]$ in linear time. Thus, Theorem 5.10 implies a linear-time algorithm for the BFS $\mathcal{F}$-branch leaf recognition on chordal graphs.

Corollary 5.11. *Given a connected chordal graph G and a vertex $v \in V(G)$, we can decide in time $\mathcal{O}(n(G) + m(G))$ whether v is an $\mathcal{F}$-branch leaf of some BFS ordering of G.*

Branch Leaves of DFS. As we have seen in Theorem 4.3, the $\mathcal{F}$-branch leaf recognition problem of DFS is NP-complete on chordal graphs. However, there is a simple characterization of DFS $\mathcal{F}$-branch leaves of split graphs.

Theorem 5.12. *Let G be a connected split graph with $n(G) \geq 2$. A vertex $v \in V(G)$ is an $\mathcal{F}$-branch leaf of some DFS ordering if and only if v is not a cut vertex of G.*

As cut vertices can be identified in linear time, Theorem 5.12 leads directly to a linear-time algorithm for the DFS $\mathcal{F}$-branch leaf recognition on split graphs.

Corollary 5.13. *Given a connected split graph G and a vertex $v \in V(G)$, we can decide in time $\mathcal{O}(n(G) + m(G))$ whether v is an $\mathcal{F}$-branch leaf of some DFS ordering of G.*

In contrast to this result, it is NP-hard to decide whether a vertex of a split graph is a DFS $\mathcal{L}$-branch leaf (see Corollary 4.2). Thus, the $\mathcal{L}$-branch leaf recognition of DFS seems to be harder than the $\mathcal{F}$-branch leaf recognition of DFS, a surprising contrast to the hardness of the DFS $\mathcal{F}$-tree recognition problem [29] and the easiness of the DFS $\mathcal{L}$-tree recognition problem [17,20]. Recall that we have made a similar observation for the complexity of the branch leaf and tree recognition of BFS (see Theorem 4.4 and Corollary 4.6).

References

1. Beisegel, J., et al.: On the end-vertex problem of graph searches. Discret. Math. Theor. Comput. Sci. **21**(1) (2019). https://doi.org/10.23638/DMTCS-21-1-13
2. Beisegel, J., et al.: The recognition problem of graph search trees. SIAM J. Discret. Math. **35**(2), 1418–1446 (2021). https://doi.org/10.1137/20M1313301
3. Beisegel, J., Köhler, E., Scheffler, R., Strehler, M.: Linear time LexDFS on chordal graphs. In: Grandoni, F., Herman, G., Sanders, P. (eds.) 28th Annual European Symposium on Algorithms (ESA 2020). LIPIcs, vol. 173, pp. 13:1–13:13. Schloss Dagstuhl-Leibniz-Zentrum für Informatik, Dagstuhl (2020). https://doi.org/10.4230/LIPIcs.ESA.2020.13
4. Berry, A., Blair, J.R., Heggernes, P., Peyton, B.W.: Maximum cardinality search for computing minimal triangulations of graphs. Algorithmica **39**(4), 287–298 (2004). https://doi.org/10.1007/s00453-004-1084-3
5. Borassi, M., Crescenzi, P., Habib, M.: Into the square: On the complexity of some quadratic-time solvable problems. In: Crescenzi, P., Loreti, M. (eds.) Proceedings of ICTCS 2015, the 16th Italian Conference on Theoretical Computer Science, ENTCS, vol. 322, pp. 51–67 (2016). https://doi.org/10.1016/j.entcs.2016.03.005
6. Charbit, P., Habib, M., Mamcarz, A.: Influence of the tie-break rule on the end-vertex problem. Discrete Math. Theor. Comput. Sci. **16**(2), 57 (2014). https://doi.org/10.46298/dmtcs.2081
7. Chepoi, V., Dragan, F.: A linear-time algorithm for finding a central vertex of a chordal graph. In: van Leeuwen, J. (ed.) ESA 1994. LNCS, vol. 855, pp. 159–170. Springer, Heidelberg (1994). https://doi.org/10.1007/BFb0049406
8. Corneil, D.G., Dalton, B., Habib, M.: LDFS-based certifying algorithm for the minimum path cover problem on cocomparability graphs. SIAM J. Comput. **42**(3), 792–807 (2013). https://doi.org/10.1137/11083856X
9. Corneil, D.G., Dragan, F.F., Habib, M., Paul, C.: Diameter determination on restricted graph families. Discret. Appl. Math. **113**(2), 143–166 (2001). https://doi.org/10.1016/S0166-218X(00)00281-X
10. Corneil, D.G., Dusart, J., Habib, M., Mamcarz, A., De Montgolfier, F.: A tie-break model for graph search. Discret. Appl. Math. **199**, 89–100 (2016). https://doi.org/10.1016/j.dam.2015.06.011

11. Corneil, D.G., Köhler, E., Lanlignel, J.M.: On end-vertices of lexicographic breadth first searches. Discret. Appl. Math. **158**(5), 434–443 (2010). https://doi.org/10.1016/j.dam.2009.10.001

12. Corneil, D.G., Krueger, R.M.: A unified view of graph searching. SIAM J. Discret. Math. **22**(4), 1259–1276 (2008). https://doi.org/10.1137/050623498

13. Corneil, D.G., Olariu, S., Stewart, L.: Linear time algorithms for dominating pairs in asteroidal triple-free graphs. SIAM J. Comput. **28**(4), 1284–1297 (1999). https://doi.org/10.1137/S0097539795282377

14. Corneil, D.G., Olariu, S., Stewart, L.: The LBFS structure and recognition of interval graphs. SIAM J. Discret. Math. **23**(4), 1905–1953 (2009). https://doi.org/10.1137/S0895480100373455

15. Gorzny, J., Huang, J.: End-vertices of LBFS of (AT-free) bigraphs. Discret. Appl. Math. **225**, 87–94 (2017). https://doi.org/10.1016/j.dam.2017.02.027

16. Habib, M., McConnell, R., Paul, C., Viennot, L.: Lex-BFS and partition refinement, with applications to transitive orientation, interval graph recognition and consecutive ones testing. Theoret. Comput. Sci. **234**(1–2), 59–84 (2000). https://doi.org/10.1016/S0304-3975(97)00241-7

17. Hagerup, T.: Biconnected graph assembly and recognition of DFS trees. Technical Report, A 85/03, Universität des Saarlandes (1985). https://doi.org/10.22028/D291-26437

18. Hagerup, T., Nowak, M.: Recognition of spanning trees defined by graph searches. Technical Report, A 85/08, Universität des Saarlandes (1985)

19. Hopcroft, J., Tarjan, R.E.: Efficient planarity testing. J. ACM **21**(4), 549–568 (1974). https://doi.org/10.1145/321850.321852

20. Korach, E., Ostfeld, Z.: DFS tree construction: algorithms and characterizations. In: van Leeuwen, J. (ed.) WG 1988. LNCS, vol. 344, pp. 87–106. Springer, Heidelberg (1989). https://doi.org/10.1007/3-540-50728-0_37

21. Kratsch, D., Damaschke, P., Lubiw, A.: Dominating cliques in chordal graphs. Discret. Math. **128**(1), 269–275 (1994). https://doi.org/10.1016/0012-365X(94)90118-X

22. Kratsch, D., Liedloff, M., Meister, D.: End-vertices of graph search algorithms. In: Paschos, V.T., Widmayer, P. (eds.) CIAC 2015. LNCS, vol. 9079, pp. 300–312. Springer, Cham (2015). https://doi.org/10.1007/978-3-319-18173-8_22

23. Krueger, R., Simonet, G., Berry, A.: A general label search to investigate classical graph search algorithms. Discret. Appl. Math. **159**(2–3), 128–142 (2011). https://doi.org/10.1016/j.dam.2010.02.011

24. Manber, U.: Recognizing breadth-first search trees in linear time. Inf. Process. Lett. **34**(4), 167–171 (1990). https://doi.org/10.1016/0020-0190(90)90155-Q

25. Rong, G., Cao, Y., Wang, J., Wang, Z.: Graph searches and their end vertices. Algorithmica **84**, 2642–2666 (2022). https://doi.org/10.1007/s00453-022-00981-5

26. Rose, D.J.: Triangulated graphs and the elimination process. J. Math. Anal. Appl. **32**(3), 597–609 (1970). https://doi.org/10.1016/0022-247X(70)90282-9

27. Rose, D.J., Tarjan, R.E., Lueker, G.S.: Algorithmic aspects of vertex elimination on graphs. SIAM J. Comput. **5**(2), 266–283 (1976). https://doi.org/10.1137/0205021

28. Scheffler, R.: Linearizing partial search orders. In: Bekos, M.A., Kaufmann, M. (eds.) WG 2022. LNCS, vol. 13453, pp. 425–438. Springer, Cham (2022). https://doi.org/10.1007/978-3-031-15914-5_31

29. Scheffler, R.: On the recognition of search trees generated by BFS and DFS. Theoret. Comput. Sci. **936**, 116–128 (2022). https://doi.org/10.1016/j.tcs.2022.09.018

30. Scheffler, R.: Graph search trees and their leaves. Preprint on arXiv (2023). https://doi.org/10.48550/arXiv.2307.07279

31. Spinrad, J., Sritharan, R.: Algorithms for weakly triangulated graphs. Discret. Appl. Math. **59**(2), 181–191 (1995). https://doi.org/10.1016/0166-218X(93)E0161-Q
32. Tarjan, R.E., Yannakakis, M.: Simple linear-time algorithms to test chordality of graphs, test acyclicity of hypergraphs, and selectively reduce acyclic hypergraphs. SIAM J. Comput. **13**(3), 566–579 (1984). https://doi.org/10.1137/0213035
33. Zou, M., Wang, Z., Wang, J., Cao, Y.: End vertices of graph searches on bipartite graphs. Inf. Proc. Lett. **173**, 106176 (2022). https://doi.org/10.1016/j.ipl.2021.106176

Author Index

D. Paulusma and B. Ries (Eds.): WG 2023, LNCS 14093, pp. 477–478, 2023.
https://doi.org/10.1007/978-3-031-43380-1

Printed in the United States
by Baker & Taylor Publisher Services

Lecture Notes in Computer Science

Lecture Notes in Artificial Intelligence   **14646**

Founding Editor

Jörg Siekmann

Series Editors

Randy Goebel, *University of Alberta, Edmonton, Canada*
Wolfgang Wahlster, *DFKI, Berlin, Germany*
Zhi-Hua Zhou, *Nanjing University, Nanjing, China*

The series Lecture Notes in Artificial Intelligence (LNAI) was established in 1988 as a topical subseries of LNCS devoted to artificial intelligence.

The series publishes state-of-the-art research results at a high level. As with the LNCS mother series, the mission of the series is to serve the international R & D community by providing an invaluable service, mainly focused on the publication of conference and workshop proceedings and postproceedings.

De-Nian Yang · Xing Xie · Vincent S. Tseng ·
Jian Pei · Jen-Wei Huang · Jerry Chun-Wei Lin
Editors

Advances in Knowledge Discovery and Data Mining

28th Pacific-Asia Conference
on Knowledge Discovery and Data Mining, PAKDD 2024
Taipei, Taiwan, May 7–10, 2024
Proceedings, Part II

Editors
De-Nian Yang ⓘD
Academia Sinica
Taipei, Taiwan

Vincent S. Tseng ⓘD
National Yang Ming Chiao Tung University
Hsinchu, Taiwan

Jen-Wei Huang ⓘD
National Cheng Kung University
Tainan, Taiwan

Xing Xie ⓘD
Microsoft Research Asia
Beijing, China

Jian Pei ⓘD
Duke University
Durham, NC, USA

Jerry Chun-Wei Lin ⓘD
Silesian University of Technology
Gliwice, Poland

ISSN 0302-9743 ISSN 1611-3349 (electronic)
Lecture Notes in Artificial Intelligence
ISBN 978-981-97-2252-5 ISBN 978-981-97-2253-2 (eBook)
https://doi.org/10.1007/978-981-97-2253-2

LNCS Sublibrary: SL7 – Artificial Intelligence

This Springer imprint is published by the registered company Springer Nature Singapore Pte Ltd.
The registered company address is: 152 Beach Road, #21-01/04 Gateway East, Singapore 189721, Singapore

Paper in this product is recyclable.

General Chairs' Preface

On behalf of the Organizing Committee, we were delighted to welcome attendees to the 28th Pacific-Asia Conference on Knowledge Discovery and Data Mining (PAKDD 2024). Since its inception in 1997, PAKDD has long established itself as one of the leading international conferences on data mining and knowledge discovery. PAKDD provides an international forum for researchers and industry practitioners to share their new ideas, original research results, and practical development experiences across all areas of Knowledge Discovery and Data Mining (KDD). This year, after its two previous editions in Taipei (2002) and Tainan (2014), PAKDD was held in Taiwan for the third time in the fascinating city of Taipei, during May 7–10, 2024. Moreover, PAKDD 2024 was held as a fully physical conference since the COVID-19 pandemic was contained.

We extend our sincere gratitude to the researchers who submitted their work to the PAKDD 2024 main conference, high-quality tutorials, and workshops on cutting-edge topics. The conference program was further enriched with seven high-quality tutorials and five workshops on cutting-edge topics. We would like to deliver our sincere thanks for their efforts in research, as well as in preparing high-quality presentations. We also express our appreciation to all the collaborators and sponsors for their trust and cooperation. We were honored to have three distinguished keynote speakers joining the conference: Ed H. Chi (Google DeepMind), Vipin Kumar (University of Minnesota), and Huan Liu (Arizona State University), each with high reputations in their respective areas. We enjoyed their participation and talks, which made the conference one of the best academic platforms for knowledge discovery and data mining. We would like to express our sincere gratitude for the contributions of the Steering Committee members, Organizing Committee members, Program Committee members, and anonymous reviewers, led by Program Committee Chairs De-Nian Yang and Xing Xie. It is through their untiring efforts that the conference had an excellent technical program. We are also thankful to the other Organizing Committee members: Workshop Chairs, Chuan-Kang Ting and Xiaoli Li; Tutorial Chairs, Jiun-Long Huang and Philippe Fournier-Viger; Publicity Chairs, Mi-Yen Yeh and Rage Uday Kiran; Industrial Chairs, Kun-Ta Chuang, Wei-Chao Chen and Richie Tsai; Proceedings Chairs, Jen-Wei Huang and Jerry Chun-Wei Lin; Registration Chairs, Chih-Ya Shen and Hong-Han Shuai; Web and Content Chairs, Cheng-Te Li and Shan-Hung Wu; Local Arrangement Chairs, Yi-Ling Chen, Kuan-Ting Lai, Yi-Ting Chen, and Ya-Wen Teng. We feel indebted to the PAKDD Steering Committee for their constant guidance and sponsorship of manuscripts. We are also grateful to the hosting organizations, National Yang Ming Chiao Tung University and Academia Sinica, and all our sponsors for continuously providing institutional and financial support to PAKDD 2024.

May 2024   Vincent S. Tseng
                                                                                        Jian Pei

PC Chairs' Preface

It is our great pleasure to present the 28th Pacific-Asia Conference on Knowledge Discovery and Data Mining (PAKDD 2024) as Program Committee Chairs. PAKDD is one of the longest-established and leading international conferences in the areas of data mining and knowledge discovery. It provides an international forum for researchers and industry practitioners to share their new ideas, original research results, and practical development experiences in all KDD-related areas, including data mining, data warehousing, machine learning, artificial intelligence, databases, statistics, knowledge engineering, big data technologies, and foundations.

This year, PAKDD received a record number of 720 submissions, among which 86 submissions were rejected at a preliminary stage due to policy violations. There were 595 Program Committee members and 101 Senior Program Committee members involved in the double-blind reviewing process. For submissions entering the double-blind review process, each one received at least three quality reviews from PC members. Furthermore, each valid submission received one meta-review from the assigned SPC member, who also led the discussion with the PC members. The PC Co-chairs then considered the recommendations and meta-reviews from SPC members and looked into each submission as well as its reviews and PC discussions to make the final decision.

As a result of the highly competitive selection process, 175 submissions were accepted and recommended to be published, with 133 oral-presentation papers and 42 poster-presentation papers. We would like to thank all SPC and PC members whose diligence produced a high-quality program for PAKDD 2024. The conference program also featured three keynote speeches from distinguished data mining researchers, eight invited industrial talks, five cutting-edge workshops, and seven comprehensive tutorials.

We wish to sincerely thank all SPC members, PC members, and external reviewers for their invaluable efforts in ensuring a timely, fair, and highly effective paper review and selection procedure. We hope that readers of the proceedings will find the PAKDD 2024 technical program both interesting and rewarding.

May 2024

De-Nian Yang
Xing Xie

Organization

Organizing Committee

Honorary Chairs

Philip S. Yu   University of Illinois at Chicago, USA
Ming-Syan Chen   National Taiwan University, Taiwan

General Chairs

Vincent S. Tseng   National Yang Ming Chiao Tung University,
                                      Taiwan
Jian Pei   Duke University, USA

Program Committee Chairs

De-Nian Yang   Academia Sinica, Taiwan
Xing Xie   Microsoft Research Asia, China

Workshop Chairs

Chuan-Kang Ting   National Tsing Hua University, Taiwan
Xiaoli Li   A*STAR, Singapore

Tutorial Chairs

Jiun-Long Huang   National Yang Ming Chiao Tung University,
                                      Taiwan
Philippe Fournier-Viger   Shenzhen University, China

Publicity Chairs

Mi-Yen Yeh   Academia Sinica, Taiwan
Rage Uday Kiran   University of Aizu, Japan

Industrial Chairs

Kun-Ta Chuang   National Cheng Kung University, Taiwan
Wei-Chao Chen   Inventec Corp./Skywatch Innovation, Taiwan
Richie Tsai   Taiwan AI Academy, Taiwan

Proceedings Chairs

Jen-Wei Huang   National Cheng Kung University, Taiwan
Jerry Chun-Wei Lin   Silesian University of Technology, Poland

Registration Chairs

Chih-Ya Shen   National Tsing Hua University, Taiwan
Hong-Han Shuai   National Yang Ming Chiao Tung University,
                                   Taiwan

Web and Content Chairs

Shan-Hung Wu   National Tsing Hua University, Taiwan
Cheng-Te Li   National Cheng Kung University, Taiwan

Local Arrangement Chairs

Yi-Ling Chen   National Taiwan University of Science and
                                   Technology, Taiwan
Kuan-Ting Lai   National Taipei University of Technology, Taiwan
Yi-Ting Chen   National Yang Ming Chiao Tung University,
                                   Taiwan
Ya-Wen Teng   Academia Sinica, Taiwan

Steering Committee

Chair

Longbing Cao   Macquarie University, Australia

Vice Chair

Gill Dobbie   University of Auckland, New Zealand

Treasurer

Longbing Cao   Macquarie University, Australia

Members

Ramesh Agrawal   Jawaharlal Nehru University, India
Gill Dobbie   University of Auckland, New Zealand
João Gama   University of Porto, Portugal
Zhiguo Gong   University of Macau, Macau SAR
Hisashi Kashima   Kyoto University, Japan
Hady W. Lauw   Singapore Management University, Singapore
Jae-Gil Lee   KAIST, Korea
Dinh Phung   Monash University, Australia
Kyuseok Shim   Seoul National University, Korea
Geoff Webb   Monash University, Australia
Raymond Chi-Wing Wong   Hong Kong University of Science and
                                Technology, Hong Kong SAR
Min-Ling Zhang   Southeast University, China

Life Members

Longbing Cao   Macquarie University, Australia
Ming-Syan Chen   National Taiwan University, Taiwan
David Cheung   University of Hong Kong, China
Joshua Z. Huang   Chinese Academy of Sciences, China
Masaru Kitsuregawa   Tokyo University, Japan
Rao Kotagiri   University of Melbourne, Australia
Ee-Peng Lim   Singapore Management University, Singapore
Huan Liu   Arizona State University, USA
Hiroshi Motoda   AFOSR/AOARD and Osaka University, Japan
Jian Pei   Duke University, USA
P. Krishna Reddy   IIIT Hyderabad, India
Jaideep Srivastava   University of Minnesota, USA
Thanaruk Theeramunkong   Thammasat University, Thailand
Tu-Bao Ho   JAIST, Japan
Vincent S. Tseng   National Yang Ming Chiao Tung University,
                                Taiwan
Takashi Washio   Osaka University, Japan
Kyu-Young Whang   KAIST, Korea
Graham Williams   Australian National University, Australia
Chengqi Zhang   University of Technology Sydney, Australia

| Ning Zhong | Maebashi Institute of Technology, Japan |
| Zhi-Hua Zhou | Nanjing University, China |

Past Members

Arbee L. P. Chen	Asia University, Taiwan
Hongjun Lu	Hong Kong University of Science and Technology, Hong Kong SAR
Takao Terano	Tokyo Institute of Technology, Japan

Senior Program Committee

Aijun An	York University, Canada
Aris Anagnostopoulos	Sapienza Università di Roma, Italy
Ting Bai	Beijing University of Posts and Telecommunications, China
Elisa Bertino	Purdue University, USA
Arnab Bhattacharya	IIT Kanpur, India
Albert Bifet	Université Paris-Saclay, France
Ludovico Boratto	Università degli Studi di Cagliari, Italy
Ricardo Campello	University of Southern Denmark, Denmark
Longbing Cao	University of Technology Sydney, Australia
Tru Cao	UTHealth, USA
Tanmoy Chakraborty	IIT Delhi, India
Jeffrey Chan	RMIT University, Australia
Pin-Yu Chen	IBM T. J. Watson Research Center, USA
Bin Cui	Peking University, China
Anirban Dasgupta	IIT Gandhinagar, India
Wei Ding	University of Massachusetts Boston, USA
Eibe Frank	University of Waikato, New Zealand
Chen Gong	Nanjing University of Science and Technology, China
Jingrui He	UIUC, USA
Tzung-Pei Hong	National University of Kaohsiung, Taiwan
Qinghua Hu	Tianjin University, China
Hong Huang	Huazhong University of Science and Technology, China
Jen-Wei Huang	National Cheng Kung University, Taiwan
Tsuyoshi Ide	IBM T. J. Watson Research Center, USA
Xiaowei Jia	University of Pittsburgh, USA
Zhe Jiang	University of Florida, USA

Toshihiro Kamishima	National Institute of Advanced Industrial Science and Technology, Japan
Murat Kantarcioglu	University of Texas at Dallas, USA
Hung-Yu Kao	National Cheng Kung University, Taiwan
Kamalakar Karlapalem	IIIT Hyderabad, India
Anuj Karpatne	Virginia Tech, USA
Hisashi Kashima	Kyoto University, Japan
Sang-Wook Kim	Hanyang University, Korea
Yun Sing Koh	University of Auckland, New Zealand
Hady Lauw	Singapore Management University, Singapore
Byung Suk Lee	University of Vermont, USA
Jae-Gil Lee	KAIST, Korea
Wang-Chien Lee	Pennsylvania State University, USA
Chaozhuo Li	Microsoft Research Asia, China
Gang Li	Deakin University, Australia
Jiuyong Li	University of South Australia, Australia
Jundong Li	University of Virginia, USA
Ming Li	Nanjing University, China
Sheng Li	University of Virginia, USA
Ying Li	AwanTunai, Singapore
Yu-Feng Li	Nanjing University, China
Hao Liao	Shenzhen University, China
Ee-peng Lim	Singapore Management University, Singapore
Jerry Chun-Wei Lin	Silesian University of Technology, Poland
Shou-De Lin	National Taiwan University, Taiwan
Hongyan Liu	Tsinghua University, China
Wei Liu	University of Technology Sydney, Australia
Chang-Tien Lu	Virginia Tech, USA
Yuan Luo	Northwestern University, USA
Wagner Meira Jr.	UFMG, Brazil
Alexandros Ntoulas	University of Athens, Greece
Satoshi Oyama	Nagoya City University, Japan
Guansong Pang	Singapore Management University, Singapore
Panagiotis Papapetrou	Stockholm University, Sweden
Wen-Chih Peng	National Yang Ming Chiao Tung University, Taiwan
Dzung Phan	IBM T. J. Watson Research Center, USA
Uday Rage	University of Aizu, Japan
Rajeev Raman	University of Leicester, UK
P. Krishna Reddy	IIIT Hyderabad, India
Thomas Seidl	LMU München, Germany
Neil Shah	Snap Inc., USA

Yingxia Shao	Beijing University of Posts and Telecommunications, China
Victor S. Sheng	Texas Tech University, USA
Kyuseok Shim	Seoul National University, Korea
Arlei Silva	Rice University, USA
Jaideep Srivastava	University of Minnesota, USA
Masashi Sugiyama	RIKEN/University of Tokyo, Japan
Ju Sun	University of Minnesota, USA
Jiliang Tang	Michigan State University, USA
Hanghang Tong	UIUC, USA
Ranga Raju Vatsavai	North Carolina State University, USA
Hao Wang	Nanyang Technological University, Singapore
Hao Wang	Xidian University, China
Jianyong Wang	Tsinghua University, China
Tim Weninger	University of Notre Dame, USA
Raymond Chi-Wing Wong	Hong Kong University of Science and Technology, Hong Kong SAR
Jia Wu	Macquarie University, Australia
Xindong Wu	Hefei University of Technology, China
Xintao Wu	University of Arkansas, USA
Yiqun Xie	University of Maryland, USA
Yue Xu	Queensland University of Technology, Australia
Lina Yao	University of New South Wales, Australia
Han-Jia Ye	Nanjing University, China
Mi-Yen Yeh	Academia Sinica, Taiwan
Hongzhi Yin	University of Queensland, Australia
Min-Ling Zhang	Southeast University, China
Ping Zhang	Ohio State University, USA
Zhao Zhang	Hefei University of Technology, China
Zhongfei Zhang	Binghamton University, USA
Xiangyu Zhao	City University of Hong Kong, Hong Kong SAR
Yanchang Zhao	CSIRO, Australia
Jiayu Zhou	Michigan State University, USA
Xiao Zhou	Renmin University of China, China
Xiaofang Zhou	Hong Kong University of Science and Technology, Hong Kong SAR
Feida Zhu	Singapore Management University, Singapore
Fuzhen Zhuang	Beihang University, China

Program Committee

Zubin Abraham	Robert Bosch, USA
Pedro Henriques Abreu	CISUC, Portugal
Muhammad Abulaish	South Asian University, India
Bijaya Adhikari	University of Iowa, USA
Karan Aggarwal	Amazon, USA
Chowdhury Farhan Ahmed	University of Dhaka, Bangladesh
Ulrich Aïvodji	ÉTS Montréal, Canada
Esra Akbas	Georgia State University, USA
Shafiq Alam	Massey University Auckland, New Zealand
Giuseppe Albi	Università degli Studi di Pavia, Italy
David Anastasiu	Santa Clara University, USA
Xiang Ao	Chinese Academy of Sciences, China
Elena-Simona Apostol	Uppsala University, Sweden
Sunil Aryal	Deakin University, Australia
Jees Augustine	Microsoft, USA
Konstantin Avrachenkov	Inria, France
Goonmeet Bajaj	Ohio State University, USA
Jean Paul Barddal	PUCPR, Brazil
Srikanta Bedathur	IIT Delhi, India
Sadok Ben Yahia	University of Southern Denmark, Denmark
Alessandro Berti	Università di Pisa, Italy
Siddhartha Bhattacharyya	University of Illinois at Chicago, USA
Ranran Bian	University of Sydney, Australia
Song Bian	Chinese University of Hong Kong, Hong Kong SAR
Giovanni Maria Biancofiore	Politecnico di Bari, Italy
Fernando Bobillo	University of Zaragoza, Spain
Adrian M. P. Brasoveanu	Modul Technology GmbH, Austria
Krisztian Buza	Budapest University of Technology and Economics, Hungary
Luca Cagliero	Politecnico di Torino, Italy
Jean-Paul Calbimonte	University of Applied Sciences and Arts Western Switzerland, Switzerland
K. Selçuk Candan	Arizona State University, USA
Fuyuan Cao	Shanxi University, China
Huiping Cao	New Mexico State University, USA
Jian Cao	Shanghai Jiao Tong University, China
Yan Cao	University of Texas at Dallas, USA
Yang Cao	Hokkaido University, Japan
Yuanjiang Cao	Macquarie University, Australia

Sharma Chakravarthy	University of Texas at Arlington, USA
Harry Kai-Ho Chan	University of Sheffield, UK
Zhangming Chan	Alibaba Group, China
Snigdhansu Chatterjee	University of Minnesota, USA
Mandar Chaudhary	eBay, USA
Chen Chen	University of Virginia, USA
Chun-Hao Chen	National Kaohsiung University of Science and Technology, Taiwan
Enhong Chen	University of Science and Technology of China, China
Fanglan Chen	Virginia Tech, USA
Feng Chen	University of Texas at Dallas, USA
Hongyang Chen	Zhejiang Lab, China
Jia Chen	University of California Riverside, USA
Jinjun Chen	Swinburne University of Technology, Australia
Lingwei Chen	Wright State University, USA
Ping Chen	University of Massachusetts Boston, USA
Shang-Tse Chen	National Taiwan University, Taiwan
Shengyu Chen	University of Pittsburgh, USA
Songcan Chen	Nanjing University of Aeronautics and Astronautics, China
Tao Chen	China University of Geosciences, China
Tianwen Chen	Hong Kong University of Science and Technology, Hong Kong SAR
Tong Chen	University of Queensland, Australia
Weitong Chen	University of Adelaide, Australia
Yi-Hui Chen	Chang Gung University, Taiwan
Yile Chen	Nanyang Technological University, Singapore
Yi-Ling Chen	National Taiwan University of Science and Technology, Taiwan
Yi-Shin Chen	National Tsing Hua University, Taiwan
Yi-Ting Chen	National Yang Ming Chiao Tung University, Taiwan
Zheng Chen	Osaka University, Japan
Zhengzhang Chen	NEC Laboratories America, USA
Zhiyuan Chen	UMBC, USA
Zhong Chen	Southern Illinois University, USA
Peng Cheng	East China Normal University, China
Abdelghani Chibani	Université Paris-Est Créteil, France
Jingyuan Chou	University of Virginia, USA
Lingyang Chu	McMaster University, Canada
Kun-Ta Chuang	National Cheng Kung University, Taiwan

Robert Churchill	Georgetown University, USA
Chaoran Cui	Shandong University of Finance and Economics, China
Alfredo Cuzzocrea	Università della Calabria, Italy
Bi-Ru Dai	National Taiwan University of Science and Technology, Taiwan
Honghua Dai	Zhengzhou University, China
Claudia d'Amato	University of Bari, Italy
Chuangyin Dang	City University of Hong Kong, China
Mrinal Das	IIT Palakkad, India
Debanjan Datta	Virginia Tech, USA
Cyril de Runz	Université de Tours, France
Jeremiah Deng	University of Otago, New Zealand
Ke Deng	RMIT University, Australia
Zhaohong Deng	Jiangnan University, China
Anne Denton	North Dakota State University, USA
Shridhar Devamane	KLE Institute of Technology, India
Djellel Difallah	New York University, USA
Ling Ding	Tianjin University, China
Shifei Ding	China University of Mining and Technology, China
Yao-Xiang Ding	Zhejiang University, China
Yifan Ding	University of Notre Dame, USA
Ying Ding	University of Texas at Austin, USA
Lamine Diop	EPITA, France
Nemanja Djuric	Aurora Innovation, USA
Gillian Dobbie	University of Auckland, New Zealand
Josep Domingo-Ferrer	Universitat Rovira i Virgili, Spain
Bo Dong	Amazon, USA
Yushun Dong	University of Virginia, USA
Bo Du	Wuhan University, China
Silin Du	Tsinghua University, China
Jiuding Duan	Allianz Global Investors, Japan
Lei Duan	Sichuan University, China
Walid Durani	LMU München, Germany
Sourav Dutta	Huawei Research Centre, Ireland
Mohamad El-Hajj	MacEwan University, Canada
Ya Ju Fan	Lawrence Livermore National Laboratory, USA
Zipei Fan	Jilin University, China
Majid Farhadloo	University of Minnesota, USA
Fabio Fassetti	Università della Calabria, Italy
Zhiquan Feng	National Cheng Kung University, Taiwan

Len Feremans	Universiteit Antwerpen, Belgium
Edouard Fouché	Karlsruher Institut für Technologie, Germany
Dongqi Fu	UIUC, USA
Yanjie Fu	University of Central Florida, USA
Ken-ichi Fukui	Osaka University, Japan
Matjaž Gams	Jožef Stefan Institute, Slovenia
Amir Gandomi	University of Technology Sydney, Australia
Aryya Gangopadhyay	UMBC, USA
Dashan Gao	Hong Kong University of Science and Technology, China
Wei Gao	Nanjing University, China
Yifeng Gao	University of Texas Rio Grande Valley, USA
Yunjun Gao	Zhejiang University, China
Paolo Garza	Politecnico di Torino, Italy
Chang Ge	University of Minnesota, USA
Xin Geng	Southeast University, China
Flavio Giobergia	Politecnico di Torino, Italy
Rosalba Giugno	Università degli Studi di Verona, Italy
Aris Gkoulalas-Divanis	Merative, USA
Djordje Gligorijevic	Temple University, USA
Daniela Godoy	UNICEN, Argentina
Heitor Gomes	Victoria University of Wellington, New Zealand
Maciej Grzenda	Warsaw University of Technology, Poland
Lei Gu	Nanjing University of Posts and Telecommunications, China
Yong Guan	Iowa State University, USA
Riccardo Guidotti	Università di Pisa, Italy
Ekta Gujral	University of California Riverside, USA
Guimu Guo	Rowan University, USA
Ting Guo	University of Technology Sydney, Australia
Xingzhi Guo	Stony Brook University, USA
Ch. Md. Rakin Haider	Purdue University, USA
Benjamin Halstead	University of Auckland, New Zealand
Jinkun Han	Georgia State University, USA
Lu Han	Nanjing University, China
Yufei Han	Inria, France
Daisuke Hatano	RIKEN, Japan
Kohei Hatano	Kyushu University/RIKEN AIP, Japan
Shogo Hayashi	BizReach, Japan
Erhu He	University of Pittsburgh, USA
Guoliang He	Wuhan University, China
Pengfei He	Michigan State University, USA

Yi He	Old Dominion University, USA
Shen-Shyang Ho	Rowan University, USA
William Hsu	Kansas State University, USA
Haoji Hu	University of Minnesota, USA
Hongsheng Hu	CSIRO, Australia
Liang Hu	Tongji University, China
Shizhe Hu	Zhengzhou University, China
Wei Hu	Nanjing University, China
Mengdi Huai	Iowa State University, USA
Chao Huang	University of Hong Kong, Hong Kong SAR
Congrui Huang	Microsoft, China
Guangyan Huang	Deakin University, Australia
Jimmy Huang	York University, Canada
Jinbin Huang	Hong Kong Baptist University, Hong Kong SAR
Kai Huang	Hong Kong University of Science and Technology, China
Ling Huang	South China Agricultural University, China
Ting-Ji Huang	Nanjing University, China
Xin Huang	Hong Kong Baptist University, Hong Kong SAR
Zhenya Huang	University of Science and Technology of China, China
Chih-Chieh Hung	National Chung Hsing University, Taiwan
Hui-Ju Hung	Pennsylvania State University, USA
Nam Huynh	JAIST, Japan
Akihiro Inokuchi	Kwansei Gakuin University, Japan
Atsushi Inoue	Eastern Washington University, USA
Nevo Itzhak	Ben-Gurion University, Israel
Tomoya Iwakura	Fujitsu Laboratories Ltd., Japan
Divyesh Jadav	IBM T. J. Watson Research Center, USA
Shubham Jain	Visa Research, USA
Bijay Prasad Jaysawal	National Cheng Kung University, Taiwan
Kishlay Jha	University of Iowa, USA
Taoran Ji	Texas A&M University - Corpus Christi, USA
Songlei Jian	NUDT, China
Gaoxia Jiang	Shanxi University, China
Hansi Jiang	SAS Institute Inc., USA
Jiaxin Jiang	National University of Singapore, Singapore
Min Jiang	Xiamen University, China
Renhe Jiang	University of Tokyo, Japan
Yuli Jiang	Chinese University of Hong Kong, Hong Kong SAR
Bo Jin	Dalian University of Technology, China

Ming Jin	Monash University, Australia
Ruoming Jin	Kent State University, USA
Wei Jin	University of North Texas, USA
Mingxuan Ju	University of Notre Dame, USA
Wei Ju	Peking University, China
Vana Kalogeraki	Athens University of Economics and Business, Greece
Bo Kang	Ghent University, Belgium
Jian Kang	University of Rochester, USA
Ashwin Viswanathan Kannan	Amazon, USA
Tomi Kauppinen	Aalto University School of Science, Finland
Jungeun Kim	Kongju National University, Korea
Kyoung-Sook Kim	National Institute of Advanced Industrial Science and Technology, Japan
Primož Kocbek	University of Maribor, Slovenia
Aritra Konar	Katholieke Universiteit Leuven, Belgium
Youyong Kong	Southeast University, China
Olivera Kotevska	Oak Ridge National Laboratory, USA
P. Radha Krishna	NIT Warangal, India
Adit Krishnan	UIUC, USA
Gokul Krishnan	IIT Madras, India
Peer Kröger	CAU, Germany
Marzena Kryszkiewicz	Warsaw University of Technology, Poland
Chuan-Wei Kuo	National Yang Ming Chiao Tung University, Taiwan
Kuan-Ting Lai	National Taipei University of Technology, Taiwan
Long Lan	NUDT, China
Duc-Trong Le	Vietnam National University, Vietnam
Tuan Le	New Mexico State University, USA
Chul-Ho Lee	Texas State University, USA
Ickjai Lee	James Cook University, Australia
Ki Yong Lee	Sookmyung Women's University, Korea
Ki-Hoon Lee	Kwangwoon University, Korea
Roy Ka-Wei Lee	Singapore University of Technology and Design, Singapore
Yue-Shi Lee	Ming Chuan University, Taiwan
Dino Lenco	INRAE, France
Carson Leung	University of Manitoba, Canada
Boyu Li	University of Technology Sydney, Australia
Chaojie Li	University of New South Wales, Australia
Cheng-Te Li	National Cheng Kung University, Taiwan
Chongshou Li	Southwest Jiaotong University, China

Fengxin Li	Renmin University of China, China
Guozhong Li	King Abdullah University of Science and Technology, Saudi Arabia
Huaxiong Li	Nanjing University, China
Jianxin Li	Beihang University, China
Lei Li	Hong Kong University of Science and Technology (Guangzhou), China
Peipei Li	Hefei University of Technology, China
Qian Li	Curtin University, Australia
Rong-Hua Li	Beijing Institute of Technology, China
Shao-Yuan Li	Nanjing University of Aeronautics and Astronautics, China
Shuai Li	Cambridge University, UK
Shuang Li	Beijing Institute of Technology, China
Tianrui Li	Southwest Jiaotong University, China
Wengen Li	Tongji University, China
Wentao Li	Hong Kong University of Science and Technology (Guangzhou), China
Xin-Ye Li	Bytedance, China
Xiucheng Li	Harbin Institute of Technology, China
Xuelong Li	Northwestern Polytechnical University, China
Yidong Li	Beijing Jiaotong University, China
Yinxiao Li	Meta Platforms, USA
Yuefeng Li	Queensland University of Technology, Australia
Yun Li	Nanjing University of Posts and Telecommunications, China
Panagiotis Liakos	University of Athens, Greece
Xiang Lian	Kent State University, USA
Shen Liang	Université Paris Cité, France
Qing Liao	Harbin Institute of Technology (Shenzhen), China
Sungsu Lim	Chungnam National University, Korea
Dandan Lin	Shenzhen Institute of Computing Sciences, China
Yijun Lin	University of Minnesota, USA
Ying-Jia Lin	National Cheng Kung University, Taiwan
Baodi Liu	China University of Petroleum (East China), China
Chien-Liang Liu	National Yang Ming Chiao Tung University, Taiwan
Guiquan Liu	University of Science and Technology of China, China
Jin Liu	Shanghai Maritime University, China
Jinfei Liu	Emory University, USA
Kunpeng Liu	Portland State University, USA

Ning Liu	Shandong University, China
Qi Liu	University of Science and Technology of China, China
Qing Liu	Zhejiang University, China
Qun Liu	Louisiana State University, USA
Shenghua Liu	Chinese Academy of Sciences, China
Weifeng Liu	China University of Petroleum (East China), China
Yang Liu	Wilfrid Laurier University, Canada
Yao Liu	University of New South Wales, Australia
Yixin Liu	Monash University, Australia
Zheng Liu	Nanjing University of Posts and Telecommunications, China
Cheng Long	Nanyang Technological University, Singapore
Haibing Lu	Santa Clara University, USA
Wenpeng Lu	Qilu University of Technology, China
Simone Ludwig	North Dakota State University, USA
Dongsheng Luo	Florida International University, USA
Ping Luo	Chinese Academy of Sciences, China
Wei Luo	Deakin University, Australia
Xiao Luo	UCLA, USA
Xin Luo	Shandong University, China
Yong Luo	Wuhan University, China
Fenglong Ma	Pennsylvania State University, USA
Huifang Ma	Northwest Normal University, China
Jing Ma	Hong Kong Baptist University, Hong Kong SAR
Qianli Ma	South China University of Technology, China
Yi-Fan Ma	Nanjing University, China
Rich Maclin	University of Minnesota, USA
Son Mai	Queen's University Belfast, UK
Arun Maiya	Institute for Defense Analyses, USA
Bradley Malin	Vanderbilt University Medical Center, USA
Giuseppe Manco	Consiglio Nazionale delle Ricerche, Italy
Naresh Manwani	IIIT Hyderabad, India
Francesco Marcelloni	Università di Pisa, Italy
Leandro Marinho	UFCG, Brazil
Koji Maruhashi	Fujitsu Laboratories Ltd., Japan
Florent Masseglia	Inria, France
Mohammad Masud	United Arab Emirates University, United Arab Emirates
Sarah Masud	IIIT Delhi, India
Costas Mavromatis	University of Minnesota, USA

Mario Prado-Romero	Gran Sasso Science Institute, Italy
Bardh Prenkaj	Sapienza Università di Roma, Italy
Jianzhong Qi	University of Melbourne, Australia
Buyue Qian	Xi'an Jiaotong University, China
Huajie Qian	Columbia University, USA
Hezhe Qiao	Singapore Management University, Singapore
Biao Qin	Renmin University of China, China
Zengchang Qin	Beihang University, China
Tho Quan	Ho Chi Minh City University of Technology, Vietnam
Miloš Radovanović	University of Novi Sad, Serbia
Thilina Ranbaduge	Australian National University, Australia
Chotirat Ratanamahatana	Chulalongkorn University, Thailand
Chandra Reddy	IBM T. J. Watson Research Center, USA
Ryan Rossi	Adobe Research, USA
Morteza Saberi	University of Technology Sydney, Australia
Akira Sakai	Fujitsu Laboratories Ltd., Japan
David Sánchez	Universitat Rovira i Virgili, Spain
Maria Luisa Sapino	Università degli Studi di Torino, Italy
Hernan Sarmiento	UChile & IMFD, Chile
Badrul Sarwar	CloudAEye, USA
Nader Shakibay Senobari	University of California Riverside, USA
Nasrin Shabani	Macquarie University, Australia
Ankit Sharma	University of Minnesota, USA
Chandra N. Shekar	RGUKT RK Valley, India
Chih-Ya Shen	National Tsing Hua University, Taiwan
Wei Shen	Nankai University, China
Yu Shen	Peking University, China
Zhi-Yu Shen	Nanjing University, China
Chuan Shi	Beijing University of Posts and Telecommunications, China
Yue Shi	Meta Platforms, USA
Zhenwei Shi	Beihang University, China
Motoki Shiga	Tohoku University, Japan
Kijung Shin	KAIST, Korea
Kai Shu	Illinois Institute of Technology, USA
Hong-Han Shuai	National Yang Ming Chiao Tung University, Taiwan
Zeren Shui	University of Minnesota, USA
Satyaki Sikdar	Indiana University, USA
Dan Simovici	University of Massachusetts Boston, USA
Apoorva Singh	IIT Patna, India

Bikash Chandra Singh	Islamic University, Bangladesh
Stavros Sintos	University of Illinois at Chicago, USA
Krishnamoorthy Sivakumar	Washington State University, USA
Andrzej Skowron	University of Warsaw, Poland
Andy Song	RMIT University, Australia
Dongjin Song	University of Connecticut, USA
Arnaud Soulet	Université de Tours, France
Ja-Hwung Su	National University of Kaohsiung, Taiwan
Victor Suciu	University of Wisconsin, USA
Liang Sun	Alibaba Group, USA
Xin Sun	Technische Universität München, Germany
Yuqing Sun	Shandong University, China
Hirofumi Suzuki	Fujitsu Laboratories Ltd., Japan
Anika Tabassum	Oak Ridge National Laboratory, USA
Yasuo Tabei	RIKEN, Japan
Chih-Hua Tai	National Taipei University, Taiwan
Hiroshi Takahashi	NTT, Japan
Atsuhiro Takasu	National Institute of Informatics, Japan
Yanchao Tan	Fuzhou University, China
Chang Tang	China University of Geosciences, China
Lu-An Tang	NEC Laboratories America, USA
Qiang Tang	Luxembourg Institute of Science and Technology, Luxembourg
Yiming Tang	Hefei University of Technology, China
Ying-Peng Tang	Nanjing University of Aeronautics and Astronautics, China
Xiaohui (Daniel) Tao	University of Southern Queensland, Australia
Vahid Taslimitehrani	PhysioSigns Inc., USA
Maguelonne Teisseire	INRAE, France
Ya-Wen Teng	Academia Sinica, Taiwan
Masahiro Terabe	Chugai Pharmaceutical Co. Ltd., Japan
Kia Teymourian	University of Texas at Austin, USA
Qing Tian	Nanjing University of Information Science and Technology, China
Yijun Tian	University of Notre Dame, USA
Maksim Tkachenko	Singapore Management University, Singapore
Yongxin Tong	Beihang University, China
Vicenç Torra	University of Umeå, Sweden
Nhu-Thuat Tran	Singapore Management University, Singapore
Yash Travadi	University of Minnesota, USA
Quoc-Tuan Truong	Amazon, USA

Yi-Ju Tseng   National Yang Ming Chiao Tung University,
                                  Taiwan
Turki Turki   King Abdulaziz University, Saudi Arabia
Ruo-Chun Tzeng   KTH Royal Institute of Technology, Sweden
Leong Hou U   University of Macau, Macau SAR
Jeffrey Ullman   Stanford University, USA
Rohini Uppuluri   Glassdoor, USA
Satya Valluri   Databricks, USA
Dinusha Vatsalan   Macquarie University, Australia
Bruno Veloso   FEP - University of Porto and INESC TEC,
                                  Portugal
Anushka Vidanage   Australian National University, Australia
Herna Viktor   University of Ottawa, Canada
Michalis Vlachos   University of Lausanne, Switzerland
Sheng Wan   Nanjing University of Science and Technology,
                                  China
Beilun Wang   Southeast University, China
Changdong Wang   Sun Yat-sen University, China
Chih-Hang Wang   Academia Sinica, Taiwan
Chuan-Ju Wang   Academia Sinica, Taiwan
Guoyin Wang   Chongqing University of Posts and
                                  Telecommunications, China
Hongjun Wang   Southwest Jiaotong University, China
Hongtao Wang   North China Electric Power University, China
Jianwu Wang   UMBC, USA
Jie Wang   Southwest Jiaotong University, China
Jin Wang   Megagon Labs, USA
Jingyuan Wang   Beihang University, China
Jun Wang   Shandong University, China
Lizhen Wang   Yunnan University, China
Peng Wang   Southeast University, China
Pengyang Wang   University of Macau, Macau SAR
Sen Wang   University of Queensland, Australia
Senzhang Wang   Central South University, China
Shoujin Wang   Macquarie University, Australia
Sibo Wang   Chinese University of Hong Kong, Hong Kong
                                  SAR
Suhang Wang   Pennsylvania State University, USA
Wei Wang   Fudan University, China
Wei Wang   Hong Kong University of Science and
                                  Technology (Guangzhou), China
Weicheng Wang   Hong Kong University of Science and
                                  Technology, Hong Kong SAR

Wei-Yao Wang	National Yang Ming Chiao Tung University, Taiwan
Wendy Hui Wang	Stevens Institute of Technology, USA
Xiao Wang	Beihang University, China
Xiaoyang Wang	University of New South Wales, Australia
Xin Wang	University of Calgary, Canada
Xinyuan Wang	George Mason University, USA
Yanhao Wang	East China Normal University, China
Yuanlong Wang	Ohio State University, USA
Yuping Wang	Xidian University, China
Yuxiang Wang	Hangzhou Dianzi University, China
Hua Wei	Arizona State University, USA
Zhewei Wei	Renmin University of China, China
Yimin Wen	Guilin University of Electronic Technology, China
Brendon Woodford	University of Otago, New Zealand
Cheng-Wei Wu	National Ilan University, Taiwan
Fan Wu	Central South University, China
Fangzhao Wu	Microsoft Research Asia, China
Jiansheng Wu	Nanjing University of Posts and Telecommunications, China
Jin-Hui Wu	Nanjing University, China
Jun Wu	UIUC, USA
Ou Wu	Tianjin University, China
Shan-Hung Wu	National Tsing Hua University, Taiwan
Shu Wu	Chinese Academy of Sciences, China
Wensheng Wu	University of Southern California, USA
Yun-Ang Wu	National Taiwan University, Taiwan
Wenjie Xi	George Mason University, USA
Lingyun Xiang	Changsha University of Science and Technology, China
Ruliang Xiao	Fujian Normal University, China
Yanghua Xiao	Fudan University, China
Sihong Xie	Lehigh University, USA
Zheng Xie	Nanjing University, China
Bo Xiong	Universität Stuttgart, Germany
Haoyi Xiong	Baidu, Inc., China
Bo Xu	Donghua University, China
Bo Xu	Dalian University of Technology, China
Guandong Xu	University of Technology Sydney, Australia
Hongzuo Xu	NUDT, China
Ji Xu	Guizhou University, China

Tong Xu	University of Science and Technology of China, China
Yuanbo Xu	Jilin University, China
Hui Xue	Southeast University, China
Qiao Xue	Nanjing University of Aeronautics and Astronautics, China
Akihiro Yamaguchi	Toshiba Corporation, Japan
Bo Yang	Jilin University, China
Liangwei Yang	University of Illinois at Chicago, USA
Liu Yang	Tianjin University, China
Shaofu Yang	Southeast University, China
Shiyu Yang	Guangzhou University, China
Wanqi Yang	Nanjing Normal University, China
Xiaoling Yang	Southwest Jiaotong University, China
Xiaowei Yang	South China University of Technology, China
Yan Yang	Southwest Jiaotong University, China
Yiyang Yang	Guangdong University of Technology, China
Yu Yang	City University of Hong Kong, Hong Kong SAR
Yu-Bin Yang	Nanjing University, China
Junjie Yao	East China Normal University, China
Wei Ye	Tongji University, China
Yanfang Ye	University of Notre Dame, USA
Kalidas Yeturu	IIT Tirupati, India
Ilkay Yildiz Potter	BioSensics LLC, USA
Minghao Yin	Northeast Normal University, China
Ziqi Yin	Nanyang Technological University, Singapore
Jia-Ching Ying	National Chung Hsing University, Taiwan
Tetsuya Yoshida	Nara Women's University, Japan
Hang Yu	Shanghai University, China
Jifan Yu	Tsinghua University, China
Yanwei Yu	Ocean University of China, China
Yongsheng Yu	Macquarie University, Australia
Long Yuan	Nanjing University of Science and Technology, China
Lin Yue	University of Newcastle, Australia
Xiaodong Yue	Shanghai University, China
Nayyar Zaidi	Monash University, Australia
Chengxi Zang	Cornell University, USA
Alexey Zaytsev	Skoltech, Russia
Yifeng Zeng	Northumbria University, UK
Petros Zerfos	IBM T. J. Watson Research Center, USA
De-Chuan Zhan	Nanjing University, China

Huixin Zhan	Texas Tech University, USA
Daokun Zhang	Monash University, Australia
Dongxiang Zhang	Zhejiang University, China
Guoxi Zhang	Beijing Institute of General Artificial Intelligence, China
Hao Zhang	Chinese University of Hong Kong, Hong Kong SAR
Huaxiang Zhang	Shandong Normal University, China
Ji Zhang	University of Southern Queensland, Australia
Jianfei Zhang	Université de Sherbrooke, Canada
Lei Zhang	Anhui University, China
Li Zhang	University of Texas Rio Grande Valley, USA
Lin Zhang	IDEA Education, China
Mengjie Zhang	Victoria University of Wellington, New Zealand
Nan Zhang	Wenzhou University, China
Quangui Zhang	Liaoning Technical University, China
Shichao Zhang	Central South University, China
Tianlin Zhang	University of Manchester, UK
Wei Emma Zhang	University of Adelaide, Australia
Wenbin Zhang	Florida International University, USA
Wentao Zhang	Mila, Canada
Xiaobo Zhang	Southwest Jiaotong University, China
Xuyun Zhang	Macquarie University, Australia
Yaqian Zhang	University of Waikato, New Zealand
Yikai Zhang	Guangzhou University, China
Yiqun Zhang	Guangdong University of Technology, China
Yudong Zhang	Nanjing Normal University, China
Zhiwei Zhang	Beijing Institute of Technology, China
Zike Zhang	Hangzhou Normal University, China
Zili Zhang	Southwest University, China
Chen Zhao	Baylor University, USA
Jiaqi Zhao	China University of Mining and Technology, China
Kaiqi Zhao	University of Auckland, New Zealand
Pengfei Zhao	BNU-HKBU United International College, China
Pengpeng Zhao	Soochow University, China
Ying Zhao	Tsinghua University, China
Zhongying Zhao	Shandong University of Science and Technology, China
Guanjie Zheng	Shanghai Jiao Tong University, China
Lecheng Zheng	UIUC, USA
Weiguo Zheng	Fudan University, China

Aoying Zhou	East China Normal University, China
Bing Zhou	Sam Houston State University, USA
Nianjun Zhou	IBM T. J. Watson Research Center, USA
Qinghai Zhou	UIUC, USA
Xiangmin Zhou	RMIT University, Australia
Xiaoping Zhou	Beijing University of Civil Engineering and Architecture, China
Xun Zhou	University of Iowa, USA
Jonathan Zhu	Wheaton College, USA
Ronghang Zhu	University of Georgia, China
Xingquan Zhu	Florida Atlantic University, USA
Ye Zhu	Deakin University, Australia
Yihang Zhu	University of Leicester, UK
Yuanyuan Zhu	Wuhan University, China
Ziwei Zhu	George Mason University, USA

External Reviewers

Zihan Li	University of Massachusetts Boston, USA
Ting Yu	Zhejiang Lab, China

Sponsoring Organizations

Accton

ACSI

Appier

Chunghwa Telecom Co., Ltd

DOIT, Taipei

ISCOM

Metaage

NSTC

PEGATRON
Pegatron

Quanta Computer

TWS

WAVENET
潮 網 科 技
Wavenet Co., Ltd

Contents – Part II

Deep Learning

AdaPQ: Adaptive Exploration Product Quantization with Adversary-Aware Block Size Selection Toward Compression Efficiency

Yan-Ting Ye[✉], Ting-An Chen, and Ming-Syan Chen

National Taiwan University, Taipei, Taiwan
{ytye,tachen}@arbor.ee.ntu.edu.tw, mschen@ntu.edu.tw

Abstract. *Product Quantization (PQ)* has received an increasing research attention due to the effectiveness on bit-width compression for memory efficiency. PQ is developed to divide weight values into blocks and adopt clustering algorithms dividing them into groups assigned with quantized values accordingly. Existing research mainly focused on the clustering strategy design with a minimal error between the original weights and the quantized values for the performance maintenance. However, the block division, i.e., the selection of block size, determines the choices of number of clusters and compression rate which has not been fully studied. Therefore, this paper proposes a novel scheme *AdaPQ* with the process, *Adaptive Exploration Product Quantization*, to first flexibly construct varying block sizes by padding the filter weights, which enlarges the search space of quantization result of PQ and avoids being suffered from a sub-optimal solution. Afterward, we further design a strategy, *Adversary-aware Block Size Selection*, to select an appropriate block size for each layer by evaluating the sensitivity on performance under perturbation for obtaining a minor performance loss under a high compression rate. Experimental results show that AdaPQ achieves a higher accuracy under a similar compression rate compared with the state-of-the-art.

Keywords: Model compression · Product quantization

1 Introduction

In recent research, neural network models, such as ResNet [1] and MobileNet [37], have been demonstrated with remarkable performances across various applications [2,3]. However, due to a large number of model parameters and numerous multiply-accumulate (MAC) operations, it poses significant challenges on deployment of these models onto resource-limited edge devices [4]. Accordingly, several techniques have been developed to address the issues of high memory and computational costs, including quantization [5,11,12], knowledge distillation [23–25], and pruning [6–8]. In particular, quantization has obtained an increasing attention since its efficacy on memory reduction and inference acceleration by decreasing the bit-widths of model weights [5].

© The Author(s), under exclusive license to Springer Nature Singapore Pte Ltd. 2024
D.-N. Yang et al. (Eds.): PAKDD 2024, LNAI 14646, pp. 3–14, 2024.
https://doi.org/10.1007/978-981-97-2253-2_1

Scalar quantization (SQ) is a process which divides the range of numerical values into a set of discrete levels and maps the original values to the nearest level individually [5,10]. However, the *one-to-one* mapping causes inefficiency due to a large amount of network parameters [30]. Therefore, another branch of research *product quantization (PQ)* has been developed to address this issue. Instead of the projection of single values, product quantization process divides similar values into groups and then represents each group with a value. [16] is the very beginning PQ work proposed to split weights into block units and then operate the k-means algorithm [31] to cluster the blocks. The centroids of clusters are adopted as the representative values. Therefore, it only needs the storage for the footprints of the cluster indices and centers rather than individual quantized values, which derives a higher compression rate and more memory reduction than the prior works in scalar quantization [17,19].

According to the demonstrated breakthrough on the model compression rate in [16], more research focuses on studying PQ subsequently [17,19,20]. Q-CNN [20] focuses on storing the partial results of clustering to save computation costs which are derived from the linear combination of output feature maps. iPQ [17] iteratively quantize and finetune the model weights by Expectation-Maximization (EM) algorithm [18] to find the cluster centroids to minimize the output feature map error during quantization. Recently, QuantNoise [19] addresses the biased gradient issue during quantization by randomly selecting weights to maintain the precision of the gradients during backpropagation.

In general, the existing PQ research is developed based on the process of weight block division and clustering. However, there are two main issues in PQ not fully studied. The first is *the constraint of the block size*. The block size is selected so that the size of filter weights is divisible by the block size, which limits the choices of number of clusters, i.e., the choices of the compression rate and may lead to compression inefficiency. Moreover, another issue studied in this paper is *the fixed block size*. The block size is fixed for each network layer, which results in uniform compression rate for different layers. Nevertheless, the sensitivities to the compression for separate layers are varying. That is, the performance degradation is different in each layer. Accordingly, current quantization algorithms based on the fixed block size may suffer from a large accuracy loss resulting from the sensitive layers, which may derive a sub-optimal result.

In this paper, to mitigate the problems of compression inefficiency and a notable accuracy loss from the choices of the block size, we propose a novel PQ algorithm, **Adaptive Exploration Product Quantization with Adversary-aware Block Size Selection Toward Compression Efficiency (AdaPQ)**. To address the first issue, the constraint of the block size, we propose **Adaptive Exploration Quantization**. We adaptively expanding the filter sizes by padding to flexibly explore the clustering (quantization) results under varying block sizes for each layer based on the new filters. Without the constraint of the block size, there are more choices of number of clusters, which enables higher compression efficiency. Moreover, for the issue of fixed block size, we design a strategy of block size selection, **Adversary-aware Block Size Selection**, to construct a larger search space with variant choices of block sizes for each network layer to explore the quantization configuration closer to the optimum to achieve a minimal accuracy loss at a high compression rate. We compute the perturbation during PQ process to evaluate the sensitivity of each layer and further determine the block size. The layers with smaller perturbation are more robust and can be represented with fewer values, i.e., with a larger block size.

Our contributions are summarized as follows.

1. We make the first attempt to address the issue of fixed network block sizes to break the limitation of compression rate of existing product quantization (PQ) research.
2. We propose mixed block size PQ and a simple PQ block size selection scheme to achieve a better tradeoff between compression rate and performance, which is orthogonal to existing PQ approaches.
3. Experiments demonstrate that AdaPQ can achieve an outstanding compression rate without an accuracy loss. For example, a 32-bit ResNet18 on tiny-ImageNet is compressed with 50.90x rate and 57.94% accuracy, 1.56% accuracy increment compared with the state-of-the-art.

2 Related Works

Scalar Quantization. There are numerous of studies on Scalar Quantization (SQ) [5,10,27]. Jacob *et al.* [5] propose a complete 8-bit quantization pipeline, including quantization-aware training and efficient integer-only inference. DoReFaNet [10] explores lower-bit quantization and uses quantized gradient during training to further reduce the computation cost. HAWQ [11] and its following work [28] achieve a better trade-off between compression rate and accuracy by using a mixed precision quantization scheme. HAQ [12] employs reinforcement learning (RL) to combine information of both hardware and software implementation, enabling a hardware-friendly computation process.

Product Quantization. Product Quantization (PQ) is developed to improve the compression efficiency of SQ. Gong *et al.* [16] is regarded as the first PQ utilized for model compression. They propose to adopt the K-means algorithm [38] to cluster model weights as the quantized values. The clustering enables a more efficient compression process than SQ which quantizes each individual weight value. iPQ [17] quantizes and fine-tunes the model weights iteratively with the Expectation-Maximization (EM) algorithm [18] to find the cluster centroids and indices with a minimal error of output feature maps during quantization. QuantNoise [19] points out that the biased gradients from using STE during quantization-aware training cause performance loss, especially under a high compression rate. Hence, they randomly select weights to keep full-precision to mitigate the bias. However, existing PQ research employs a consistent block size setting for each layer, which leads to a limited compression rate. Moreover, the choice of block size depends on the size of filter weights, which induces a small search space of compression rates.

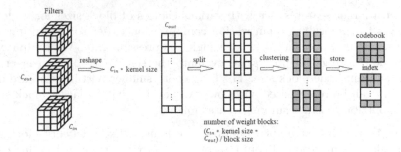

Fig. 1. The basic PQ procedure.

3   Preliminary

We first introduce the PQ procedure and notations. The primary concept behind PQ in model compression is to represent weights with few indices. This is achieved by partitioning weights and clustering them. Since PQ only needs to store the indices and the codebook, the compression rate can achieve remarkably high (over 100x), making PQ receive an increasing research attention.

Clustering. The first step of PQ is splitting the weight matrix into bocks. Consider applying PQ on a single layer in a neural network. Suppose the weight matrix W has a size (C_{in}, C_{out}) and the block size b is given, W is first split into C_{out} vectors along the output channel dimension and then each vector is further split into blocks with size b. Now we have $S = C_{out} * \frac{C_{in}}{b}$ blocks, denote as w_j, $j = 1...S$.

Then a k-means algorithm [31] is applied to these blocks to determine cluster centers and the cluster index of each blocks. Given the bit-width of the cluster indices n, i.e. there are 2^n clusters, the quantized weight now contains cluster centers $q_1...q_{2^n}$ and cluster indices $i_1...i_S$. The objective is to minimize the weight distortion:

$$\min \sum_{j=1}^{S} ||w_j - q_{i_j}||_2^2 \tag{1}$$

Inference. After clustering, we can recover the weight matrix $\hat{W}$ from the stored codebook and index. We retrieve weight blocks $\hat{w}_j = q_j$, $j = i_1...i_S$ then concatenate and reshape them to size (C_{in}, C_{out}). Then we can inference with the recovered weight matrix $\hat{W}$.

Block Size Constraint. Each weight block should have the same size in order to perform k-means clustering. Given a weight matrix's size, it must be divisible by the block size. The available block sizes for a given weight matrix is $\{b|C_{in}\%b = 0\}$. This is called the block size constraint.

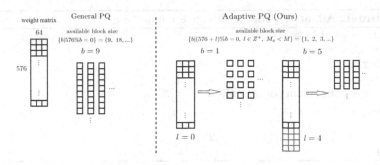

Fig. 2. The difference of the block size search spaces between previous PQ methods and our AdaPQ.

4   Methodology

Section 4.1 will introduce that our proposed scheme AdaPQ with an adaptive padding method that allows PQ with varying block sizes to select from. In Sect. 4.2, we will illustrate the evaluation of the sensitivity of each layer under separate block sizes and the optimization of the size configurations for more accuracy recovery within a desired compression rate.

4.1   Adaptive Exploration Quantization

To relax the block size constraint and enable PQ with more choices of block sizes to explore the optimal solution of compression result, we adaptively add padding values to filter weights for the PQ process illustrated in Fig. 2, where l represents the padding length. The block size is varying with different numbers of padding values. The available block sizes for any given layer are now $\{b|C_{in}\%(b+l) = 0, \, l \in Z^+\} = Z^+$. Compared to the ones mentioned in Sect. 3, more block sizes are able to be searched now. Accordingly, more compression rates can be explored. In other words, it is more likely to search a compression result closer to the optimum with the minimal performance degradation. Note that the padded values are employed merely in obtaining the cluster centroids, i.e., the quantized weight values, during the learning process of PQ. During inference, the padding values are removed so that no additional memory storage is required.

A potential problem behind the PQ algorithm with padding is that the padding values can affect the searching result of cluster centroids (demonstrated and discussed in ablation study, the last subsection in Sect. 5). To avoid the impact of padding value, we modify the clustering procedure in our Adaptive Exploration Quantization. We ignore padding values when computing the distance and updating centroids. For weight blocks without padding, the distance computation remains unchanged (see Lines 3-4 in Algorithm 1). Otherwise, we ignore the padded values and their corresponding elements in the calculation of cluster centroids (see Lines 5-6 in Algorithm 1). During centroid updates, we average the elements of filters without padding values to determine the new centroids (see Lines 6–7 in Algorithm 2).

Algorithm 1. Adaptive padding distance computing

    Input: weight blocks $\tilde{w}_1...\tilde{w}_S$
1: current cluster centers $q_1...q_{2^n}$
2: padding length l
    Output:
3: **if** $\tilde{w}_i$ does not contain padding **then**
4:      $d_{ij} = ||\tilde{w}_i - q_j||_2^2, \ j = 1...2^n$
5: **else**
6:      $d_{ij} = ||\tilde{w}_i[: l] - q_j[: l]||_2^2, \ j = 1...2^n$
7: **end if**

Algorithm 2. Adaptive padding centroid update

    Input: weight blocks $\tilde{w}_1...\tilde{w}_S$
1: current cluster indices $i_1...i_S$
2: padding length l
    Output:
3: **for** cluster index $c = 1...2^n$ **do**
4:      Suppose in cluster c, there are a weight blocks without padding (denoted $\tilde{w}_{c_i}, \ i = 1...a$) and b weight blocks with padding (denoted $\tilde{w}_{c_i}, \ i = (a+1)...(a+b)$).
5:      Compute new cluster center q_c.
6:      $q_c[: l] = \sum_{i=1}^{a+b} \tilde{w}_{c_i}[: l]/(a + b)$
7:      $q_c[l :] = \sum_{i=a+1}^{a+b} \tilde{w}_{c_i}[l :]/(b)$
8: **end for**

4.2   Adversary-Aware Block Size Selection

In previous PQ works [16,17,19], the block size is a predefined hyperparameter, and each layer uses the same block size. Uniform block sizes, however, may not be the best quantization configuration. In neural network models, it is discovered that different layers can have varying influences on the performance [11,12] in previous scalar quantization research. The sensitivities toward quantization error are also different across layers. Assigning each layer a suitable compression rate, i.e., the block size, is an effective approach achieving a higher compression rate while a minimal performance loss. Therefore, we propose Adversary-aware Block Size Selection in PQ process for block size selection within each network layer. We evaluate the weight distortion of the quantized model as the sensitivity toward quantization layerwisely given a memory budget.

The procedure of our Adversary-aware Block Size Selection is outlined in Algorithm 3. We first apply PQ with a uniform block size to compute the weight distortion. We set initial minimum and maximum block sizes. The layer with the highest distortion, i.e. highest sensitivity, will have the minimum block size; And the layer with the lowest distortion will have the maximum block size. Block sizes of remaining layers are computed by linear interpolation with their distortion and the current maximum and minimum block sizes. With a given bit-width of cluster index n and block size b, the size of a quantized layer is $(\lceil \frac{C_{in}}{b} \rceil * C_{out} * n + 2^n * b * 32)$ bits. We can compute the size of quantized models accordingly. We

---

Algorithm 3. Adversary-aware Block Size Selection

    Input: weight matrix W_l, $l = 1...L$
1: model size constraint M
    Output:
2: Compute weight perturbation $e_l = ||W_l - \hat{W}_l||_2^2$, $l = 1...L$ under a uniform block size scheme, and let $e_{min} = \min_l e_l$, $e_{max} = \max_l e_l$
3: Set the minimum block size $b_{min} = 1$ and the maximum block size $b_{max} = 1$, quantized model size $m = inf$
4: **while** $m > M$ **do**
5:     $b_{max} = b_{max} + 1$
6:     Determine block size $b_l = b_{min} + \frac{e_{max} - e_l}{e_{max} - e_{min}} * (b_{max} - b_{min})$, $l = 1...L$
7:     Compute the quantized model size m_{new} with $b_1...b_L$
8:     **if** $m_{new} > m$ **then**
9:         $b_{min} = b_{min} + 1$
10:        $b_{max} = b_{min} + 1$
11:        $m = inf$
12:     **else**
13:        $m = m_{new}$
14:     **end if**
15: **end while**

---

keep increasing the maximum block size until the memory budget requirement is met. If the maximum block size exceeds the valid block size range, which means the model size become dominated by the cluster center and start increasing, we increase the minimum block size instead.

5   Experiments

Benchmark Datasets and Architectures. We conduct several experiments on three benchmark datasets, CIFAR-10 [33], CIFAR-100 [33], and tiny-ImageNet dataset [35], to validate the effectiveness of AdaPQ. CIFAR-10 dataset contains 32×32 colored images in 10 classes. There are 50,000 training images and 10,000 for testing. CIFAR-100 dataset also contains the same amount of 32×32 colored images as CIFAR-10 but in 100 classes. Tiny-ImageNet is a subset of ImageNet [36] with 10,000 64×64 images in 200 classes in bith training and testing sets. For model architectures, we evaluate AdaPQ and baselines on the benchmarks ResNet18 [1] and MobileNet-v2 [37].

Experimental Settings. Before quantization, models are trained from scratch on three benchmark datasets separately with SGD optimization in 100 epochs. The learning rate is 0.01, and the batch size is 128. A multi-step learning rate decay is applied during training for better convergence. All experiments are conducted on an NVIDIA RTX3060 GPU.

Table 1. Comparisons with baseline PQ methods.

(a) Resnet18

Quantization scheme	CIFAR-10			CIFAR-100			tiny-ImageNet		
	Compression	Top-1(%)	Top-1 drop(%)	Compression	Top-1(%)	Top-1 drop(%)	Compression	Top-1(%)	Top-1 drop(%)
Full precision model	1.00x	94.62	–	1.00x	75.44	–	1.00x	62.92	–
iPQ(small block) [17]	31.97x	93.49	1.13	31.97x	72.30	3.14	31.81x	58.67	4.25
QN(small block) [19]	32.10x	93.58	1.04	32.10x	73.91	1.53	31.95x	58.36	4.56
Ours	33.08x	**94.05**	0.57	34.61x	73.35	2.09	30.21x	58.70	4.22
Ours+QN	33.08x	**94.16**	0.46	34.61x	**73.99**	1.45	30.21x	**60.22**	**2.70**
iPQ(large block) [17]	51.74x	91.78	2.84	51.74x	71.03	4.41	51.19x	54.28	8.64
QN(large block) [19]	53.32x	93.44	1.18	53.32x	72.16	3.28	53.13x	56.38	6.54
Ours	53.31x	**93.65**	0.97	51.87x	72.18	3.26	50.90x	55.78	7.14
Ours+QN	53.31x	**93.98**	0.64	51.87x	**72.76**	2.68	50.90x	**57.94**	**4.98**

(b) Mobilenet-v2

Quantization scheme	CIFAR-10			CIFAR-100		
	Compression	Top-1(%)	Top-1 drop(%)	Compression	Top-1(%)	Top-1 drop(%)
Full precision model	1.00x	90.43	–	1.00x	67.49	–
iPQ(small block) [17]	20.08x	*	*	20.08x	*	*
QN(small block) [19]	20.08x	*	*	20.08x	*	*
Ours	22.67x	**87.81**	2.62	20.43x	**61.60**	5.89

Comparisons with the State-of-the-Art. Table 1 illustrates the quantization results of our proposed AdaPQ compared with the previous works, Quant-Noise [17] and iPQ [19], where the compression rates are obtained from dividing the original model sizes by the quantized model sizes. For comparisons, we present the accuracy improvements under similar compression rates. Table 1 (a) shows that our proposed AdaPQ based on the basic PQ procedure obtains 0.5%-2% accuracy increment compared with previous works on benchmark datasets under Resnet18 architecture. It is worth mentioning that when the previous methods, iPQ [17] and QN [17], are applied to the lightweight architecture Mobilenet-v2, their performances deteriorate during the quantization procedure mainly due to the small weight block size. Therefore, these prior works fail to inference data as shown with the notation * in Table 1 (b). By contrast, it reveals that our AdaPQ with the block size exploration and optimization processes can search a quantization configuration with higher compression rate and no significant accuracy loss accordingly.

Effectiveness of Search Space Expansion Under Adaptive Exploration Quantization. Adaptive Exploration Quantization is proposed to explore more compression results by padding values to filter weights to find a solution close to the optimum (introduced in Sect. 4.1). Figure 3 presents the quantization results of PQ with and without considering the padding. The horizontal axis shows the compression rate, and the vertical axis presents accuracy. In particular, the black dots in the figure manifest the quantization results that are explored without padding. The compression results depend on the block sizes which are limited to the factors of the filter size. However, the PQ process under padding on the other hand, can have much more choices of block sizes and search more results with higher compression efficiency. The additional search results are marked in

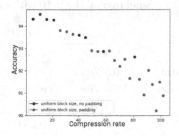

Fig. 3. Quantization results of Resnet18 on CIFAR-10 with generalized PQ. Black dots represent the performances without padding, and green dots manifest the results using adaptive padding. (Color figure online)

Table 2. Accuracy improvement of mixed block size PQ. We train Resnet18 (a) and Mobilenet-v2 (b) on CIFAR-10, CIFAR-100, and tinyImageNet. Models quantized with uniform block size and mixed block size have a close compression rate. So the accuracy difference can show an improvement.

(a) Uniform block size and mixed block size PQ for Resnet18.

Quantization scheme	CIFAR-10			CIFAR-100			tiny-ImageNet		
	Compression	Top-1(%)	Top-1 drop(%)	Compression	Top-1(%)	Top-1 drop(%)	Compression	Top-1(%)	Top-1 drop(%)
full precision model	1.00x	94.62	–	1.00x	75.44	–	1.00x	62.92	–
Ours, uniform block size	53.87x	92.87	1.75	41.96x	72.82	2.62	41.81x	56.32	6.60
Ours, mixed block size	53.31x	**93.65**	**0.97**	40.56x	**73.03**	**2.41**	42.92x	**57.27**	**5.65**
Ours, uniform block size	101.76x	90.88	3.74	80.75x	67.88	7.56	72.83x	51.66	11.26
Ours, mixed block size	100.90x	**92.37**	**2.25**	81.85x	**70.58**	**4.86**	72.50x	**53.45**	**9.47**

(b) Uniform block size and mixed block size PQ for Mobilenet-v2.

Quantization scheme	CIFAR-10			CIFAR-100			tiny-ImageNet		
	Compression	Top-1(%)	Top-1 drop(%)	Compression	Top-1(%)	Top-1 drop(%)	Compression	Top-1(%)	Top-1 drop(%)
full precision model	1.00x	90.43	–	1.00x	67.49	–	1.00x	49.48	–
Ours, uniform block size	11.77x	88.27	2.16	10.30x	64.37	3.12	10.42x	42.09	7.39
Ours, mixed block size	11.77x	**89.05**	**1.38**	10.06x	**64.56**	**2.93**	10.58x	**43.37**	**6.11**
Ours, uniform block size	22.44x	85.39	5.04	18.99x	60.86	6.63	18.90x	38.96	10.52
Ours, mixed block size	22.57x	**88.33**	**2.10**	19.05x	**63.26**	**4.23**	19.26x	**40.74**	**8.74**

green in Fig. 3. From the results, we can observe that Adaptive Exploration Quantization can effectively explore more compression rates without the block size constraint, especially the higher compression rates from 60x to 100x.

Effectiveness of Performance Improvements by Adversary-Aware Block Size Selection. Table 2 shows the accuracy improvements by using mixed block size configurations which are found under Adversary-aware Block Size Selection (introduced in Sect. 4.2). At the similar compression rates, models quantized with mixed block sizes increase accuracy by 1–3% compared with the performances under the setting of uniform block size for each network layers. In other words, the enhancement demonstrates the effectiveness of the mixed block size by leveraging the sensitivity of each layer.

Table 3. Quantiation result of Resnet18 on CIFAR-10 with different padding methods.

Padding method	Compression	Top-1(%)	Top-1 drop(%)
Non-quantized	1.00x	94.62	–
OOD padding (uniform)	101.76x	89.73	4.89
OOD padding (random)	101.76x	85.43	9.19
Zero padding	101.76x	90.11	4.51
Mean padding	101.76x	90.34	4.28
Adaptive padding (Ours)	101.76x	**90.88**	**3.74**

Ablation Study on Padding Values. We employ different padding values in AdaPQ to evaluate the effects. The experiment is conducted on CIFAR-10 using Resnet18. The block size is uniformly 25. We use zero padding, mean padding, uniformly out-of-distribution padding, and randomly out-of-distribution padding. From Table 3, we can observe that if the padding value is too far away from the weight distribution, e.g., OOD padding, the performance will deteriorate significantly. Zero padding and mean padding can achieve better performances than OOD padding. However, there still exists performance degradation due to the bias of the clustering result. In contrast, our proposed adaptive padding (detailed in Algorithm 1 in Sect. 4.1), can achieve the highest accuracy at the same compression rate, since we skip the padding values when calculating and updating the clustering distances and centroids. In other words, the converged clustering result can be guaranteed as same as the PQ process without padding to mitigate the problem of bias clustering result derived from padding.

6   Conclusion

In this paper, we propose an algorithm, AdaPQ, to improve the compression efficiency of existing PQ research. To mitigate the problem of the block size constraint, we employ padding technique to adaptively expanding the filter weights to explore a larger search space of block sizes. Furthermore, instead of adoption of fixed block size for the whole network, we dynamically select the appropriate size for each layer by evaluating the sensitiveness, i.e., the distortion of quantized weights under perturbation. Experimental results demonstrate that the removal of the block size constraint and the adversary-aware block size selection can effectively enhance the performance under the similar compression rate.

Acknowledgement. This work was supported in part by the National Science and Technology Council, Taiwan, under grant NSTC-112-2223-E-002-015, and by the Ministry of Education, Taiwan, under grant MOE 112L9009.

References

1. He, K., Zhang, X., Ren, S., Sun, J.: Deep residual learning for image recognition. In: 2016 IEEE Conference on Computer Vision and Pattern Recognition (CVPR), Las Vegas, NV, USA, pp. 770–778 (2016). https://doi.org/10.1109/CVPR.2016.90
2. Minaee, S., Boykov, Y., Porikli, F., Plaza, A., Kehtarnavaz, N., Terzopoulos, D.: Image segmentation using deep learning: a survey. IEEE Trans. Pattern Anal. Mach. Intell. **44**(7), 3523–3542 (2022)
3. Anwar, S., Barnes, N.: Real image denoising with feature attention. In: Proceedings of the IEEE/CVF International Conference on Computer Vision (ICCV), pp. 3155–3164 (2019)
4. Li, E., Zeng, L., Zhou, Z., Chen, X.: Edge AI: on-demand accelerating deep neural network inference via edge computing. IEEE Trans. Wireless Commun. **19**(1), 447–457 (2019)
5. Jacob, B., et al.: Quantization and training of neural networks for efficient integer-arithmetic-only inference. In: Proceedings of the IEEE Conference on Computer Vision and Pattern Recognition (CVPR), pp. 2704–2713 (2018)
6. Anwar, S., Hwang, K., Sung, W.: Structured pruning of deep convolutional neural networks (2015). https://doi.org/10.48550/arXiv.1512.08571
7. Luo, J., Wu, J., Lin, W.: ThiNet: a filter level pruning method for deep neural network compression. In: Proceedings of the IEEE International Conference on Computer Vision (ICCV) (2017)
8. Blalock, D., Gonzalez, O., Jose, J., Frankle, J., Guttag, J.: What is the state of neural network pruning? Proc. Mach. Learn. Syst. **2**, 129–146 (2020)
9. Uhlich, S., et al.: Mixed Precision DNNs: all you need is a good parametrization (2020). https://doi.org/10.48550/arXiv.1905.11452
10. Zhou, S., Wu, Y., Ni, Z., Zhou, X., Wen, H., Zou, Y.: DoReFa-Net: training low bitwidth convolutional neural networks with low bitwidth gradients (2018). https://doi.org/10.48550/arXiv.1606.06160
11. Dong, Z., Yao, Z., Gholami, A., Mahoney, M., Keutzer, K.: HAWQ: Hessian Aware Quantization of neural networks with mixed-precision. In: Proceedings of The IEEE/CVF International Conference on Computer Vision (ICCV) 2019, pp. 293–302 (2019)
12. Wang, K., Liu, Z., Lin, Y., Lin, J., Han, S.: HAQ: Hardware-aware automated quantization with mixed precision. In: Proceedings of the IEEE/CVF Conference on Computer Vision and Pattern Recognition (CVPR) 2019, pp. 8604–8612 (2019)
13. Lin, X., Zhao, C., Pan, W.: Towards accurate binary convolutional neural network. In: Advances in Neural Information Processing Systems 2017, vol. 30, pp. 345–353 (2017)
14. Lin, M., et al.: Rotated binary neural network. In: Advances in Neural Information Processing Systems 2020, vol. 33, pp. 7474–7485 (2020)
15. Jégou, H., Douze, M., Schmid, C.: Product quantization for nearest neighbor search. IEEE Trans. Pattern Anal. Mach. Intell. **33**, 117–128 (2011)
16. Gong, Y., Liu, L., Yang, M., Bourdev, L.: Compressing Deep Convolutional Networks using Vector Quantization (2014). https://doi.org/10.48550/arXiv.1412.6115
17. Stock, P., Joulin, A., Gribonval, R., Graham, B., Jégou, H.: And the bit goes down: revisiting the quantization of neural networks. In: International Conference on Learning Representations. (2020)

18. Moon, T.K.: The expectation-maximization algorithm. IEEE Sig. Process. Mag. **13**, 47–60 (1996)
19. Stock, P., et al.: Training with quantization noise for extreme model compression. In: International Conference on Learning Representations 2021 (2021)
20. Wu, J., Leng, C., Wang, Y., Hu, Q., Cheng, J.: Quantized convolutional neural networks for mobile devices. In: Proceedings of the IEEE Conference on Computer Vision and Pattern Recognition (CVPR) 2016, pp. 4820–4828 (2016)
21. Yoshua, B., Léonard, N., Courville, A.: Estimating or propagating gradients through stochastic neurons for conditional computation (2013). https://doi.org/10.48550/arXiv.1308.3432
22. Wu, Y., Lee, H., Lin, Y., Chien, S.: Accelerator design for vector quantized convolutional neural network. In: 2019 IEEE International Conference on Artificial Intelligence Circuits and Systems (AICAS), pp. 46–50 (2019)
23. Hinton, G., Vinyals, O., Dean, J.,: Distilling the knowledge in a neural network. In: NIPS 2014 Deep Learning Workshop (2014)
24. Lopes, R., Fenu, S., Starner, T.: Data-free knowledge distillation for deep neural networks (2017). https://doi.org/10.48550/arXiv.1710.07535
25. Gou, J., et al.: Knowledge distillation: a survey. Int. J. Comput. Vision **129**(6), pp. 1789–1819 (2021). https://doi.org/10.1007/s11263-021-01453-z
26. Gholami, A., Kim, S., Dong, Z., Yao, Z., Mahoney, M., Keutzer, K.: A survey of quantization methods for efficient neural network inference (2021). https://doi.org/10.48550/arXiv.2103.13630
27. Sung, W., Shin, S., Hwang, K.: Resiliency of deep neural networks under quantization (2016). https://doi.org/10.48550/arXiv.1511.06488
28. Yao, Z., et al.: HAWQ-V3: dyadic neural network quantization. In: Proceedings of the 38th International Conference on Machine Learning, vol. 139, pp. 11875–11886 (2021)
29. Johnson, J., Douze, M., Jégou, H.: Billion-scale similarity search with GPUs (2017) https://doi.org/10.48550/arXiv.1702.08734
30. Han, S., Mao, H., Dally, W.: Deep compression: compressing deep neural network with pruning, trained quantization and huffman coding. In: 4th International Conference on Learning Representations (ICLR) (2016)
31. MacQueen, J., et al.: Some methods for classification and analysis of multivariate observations. In: Proceedings of the Fifth Berkeley Symposium on Mathematical Statistics and Probability, vol. 1, pp. 281–297 (1967)
32. Ren, P., et al.: A comprehensive survey of neural architecture search: challenges and solutions. ACM Comput. Surv. **54**(4) (2021). https://doi.org/10.1145/3447582
33. Krizhevsky, A., Nair, V., Hinton, G.: CIFAR-10 (Canadian Institute for Advanced Research). http://www.cs.toronto.edu/~kriz/cifar.html
34. Krizhevsky, A., Nair, V., Hinton, G.: CIFAR-100 (Canadian Institute for Advanced Research). http://www.cs.toronto.edu/~kriz/cifar.html
35. Le, Y., Yang, X.: Tiny ImageNet Visual Recognition Challenge (2015)
36. Deng, J., Dong, W., Socher, R., Li, L., Li, K., Fei-Fei, L.: ImageNet: a large-scale hierarchical image database. In: 2009 IEEE Conference on Computer Vision and Pattern Recognition, pp. 248–255 (2009)
37. Howard, A., et al.: MobileNets: efficient convolutional neural networks for mobile vision applications (2017). https://doi.org/10.48550/arXiv.1704.04861
38. Pelleg, D., Moore, A.: Accelerating exact k-means algorithms with geometric reasoning. In: Proceedings of the fifth ACM Sigkdd International Conference on Knowledge Discovery And Data Mining, pp. 277–281 (1999)

Ranking Enhanced Supervised Contrastive Learning for Regression

Ziheng Zhou, Ying Zhao[✉], Haojia Zuo, and Wenguang Chen

Tsinghua University, Beijing, China
{zhouzihe18,zuohj19}@mails.tsinghua.edu.cn, {yingz,cwg}@tsinghua.edu.cn

Abstract. Supervised contrastive learning has shown promising results in image classification tasks where the representations are pulled together if they share same labels or otherwise pushed apart. Such dispersion process in the representation space benefits the downstream classification tasks. However, when applied to regression tasks directly, such dispersion lacks guidance of the relationship among target labels (i.e. the label distances), which leads to the disalignment between representation distances and label distances. Achieving such alignment without compromising the dispersion of learned representations is challenging. In this paper, we propose a Ranking Enhanced Supervised Contrastive Loss (RESupCon) to empower the representation dispersion process with ranking alignment between representation distances and label distances in a controlled fashion. We demonstrate the effectiveness of our method in image regression tasks on four real-world datasets with various interests, including meteorological, medical and human facial data. Experimental results of our method show that representations with better ranking are learned and improvements are made over other baselines in terms of RMSE on all four datasets.

Keywords: contrastive learning · representation learning · regression · representation order

1 Introduction

Recent years have seen rapid progress in the field of representation learning, where self-supervised learning methods such as SimCLR [4] and the MoCo family [5,10] have gained popularity and shown remarkable abilities to extract features automatically from unlabeled data. Generally speaking, representation learning maps data samples to the feature space and learns their positions based on a certain policy, e.g. samples augmented from the same image should be neighbors in the feature space. Specifically, contrastive learning serves to compare similar and dissimilar samples to which different policies are applied [12]. In

This work was supported by the National Key Research and Development Program of China (2022YFC3004102) and Qinghai Kunlun Talents Program.

the context of supervised learning, modifications can be made upon those self-supervised methods in order to ultilize the extra knowledge provided by some kind of supervision. For example, Supervised Contrastive Learning (SupCon) [11] extends SimCLR and takes not only augmentation but also the class labels into account. Contrastive Language-Image Pre-training (CLIP) [17] encourages alignment between two modalities by conducting contrastive training on image-text-paired datasets. Both methods report their promising abilities on the task of image classification. In theory, Graf et al. [8] prove that SupCon forces the representations to collapse according to classes and to be pushed away from those of different classes. Such dispersion in high dimensional space could benefit downstream classifiers as features of different classes become more divisible.

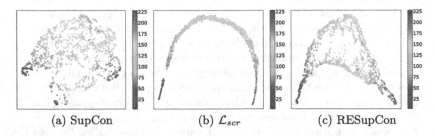

(a) SupCon   (b) $\mathcal{L}_{scr}$ (c) RESupCon

Fig. 1. UMAP [13] visualization of the representations learned by SupCon, $\mathcal{L}_{scr}$ and RESupCon on BoneAge validation dataset. Each dot represents the representation of an input image and its color represents the corresponding target label.

However, the advantages that Supervised Contrastive Learning can offer are mainly confined to classification tasks. Its effectiveness in regression tasks could be doubted, given the continuous nature of real-valued labels. A naive way to apply SupCon to regression tasks is to simply redefine the strategy of selecting positive and negative pairs, e.g. binning on the label values to construct artificial classes. In this setting, the class representations learned by SupCon lack a discernible order, whereas the relative relationships between real-valued labels are intrinsically meaningful in regression tasks. As shown in Fig. 1(a), representations trained with SupCon are dispersed, however ones with similar target labels are often separated, suggesting that the relationships between label values are not well preserved in the feature space. On the other hand, enforcing the ranking of representations in the feature space, either by using a ranking loss alone as shown in Fig. 1 (b) or by contrasting samples against each other based on their rankings in the target space as in [20], compromises the feature space dispersion that is initially achieved by contrastive learning.

In this paper, we propose the Ranking Enhanced Supervised Contrastive Loss (RESupCon), a simple yet effective method to improve the performance of regression by regulating the feature space learned by supervised contrastive learning. As shown in Fig. 1 (c), representations learned by our RESupCon method are well organized in accordance with the labels and still retain the dispersion in contrast with using the ranking loss ($\mathcal{L}_{scr}$) alone.

The contributions of our work are:

- We define the Spearman correlation ranking loss for measuring and optimizing the alignment between the ranking of representation distances and that of label distances, which helps to generate an orderly representation space.
- We propose RESupCon, a supervised contrastive learning method for regression under the guidance of the Spearman correlation ranking loss, which is capable of learning representations with ranking while retaining dispersion in the representation space.
- We demonstrate the superiority of our method over state-of-the-art baseline methods on four real world datasets through extensive experiments. Further experiments also show that the effectiveness of our method comes from the controlled combination of the dispersion process and the ranking alignment, whereas using either of them alone does not lead to optimal representations for downstream regression tasks.

The rest of paper is organized as follows. We first give a brief overview of related work in Sect. 2. Then, we introduce the preliminary knowledge on both unsupervised and supervised contrastive learning in Sect. 3. Next, we present our proposed method in Sect. 4. Experimental results of our proposed method on four benchmark datasets are shown in Sect. 5. Finally, we make concluding remarks in Sect. 6.

2   Related Work

Supervised Contrastive Learning for Classification. Self-supervised contrastive learning methods, e.g. SimCLR [4] and the MoCo family [5,10], extract informative representations by contrasting similar samples ("positive pair"), which in common practice are produced by augmenting the same input, with dissimilar ones ("negative pairs"). SupCon [11], further makes use of labels while keeping the contrastive basis. Samples with the same label are defined as positive pairs, and vice versa. SupCon outperforms end-to-end classification with raw Cross-Entropy loss on datasets. [8] show that SupCon encourages features to collapse by classes and then be dispersed in the feature space, increasing the feature diversity of distinct labels. [3] explore the coarse-to-fine transfer learning ability of SupCon on the basis of the aforementioned geometric properties.

Supervised Contrastive Learning for Regression. Despite the success of SupCon on classification, it is non-trivial to transplant supervised contrastive learning onto the field of regression. Given the continuous nature of real-valued labels, related work focus on strengthening the relationship between feature similarities and label distances, which is absent in the concept of SupCon. AdaCon [7] modifies the SupCon loss by adding a term to force the feature similarities of negative samples to reduce according to their label distances. AdaCon reports a minor improvement over end-to-end regression with raw L1 and L2 losses.

[20] argue that the learned representations should be ordered in accordance with label values to benefit regression, and propose RNC as a solution. However, RNC selects any pair of samples as positive pairs, which somewhat compromises the dispersion.

Ranking as a Loss. Learning to Rank (LTR), which targets at building ranking models from training data, has long been a research concern in the field of information retrieval [16]. Ranking (and sorting) operations are typically non-differentiable which cannot be used as a loss to directly optimize a model via back-propagation. Various differentiable approaches have been proposed as proxies to ranking, e.g. [6,19], often at the cost of computational complexity. [1] propose a fast differentiable approximate algorithm for ranking and sorting in $O(n \log n)$ time, enabling the use of ranking as a practical loss component in deep learning.

3   Preliminaries

We give a brief introduction to the self-supervised method SimCLR [4], as well as the supervised method SupCon [11]. This line of work forms the basis of our proposed method.

Contrastive learning as hinted by its name, contrasts "positive" samples against "negativ" ones. In absence of labeled data, SimCLR constructs positive and negative samples by means of data augmentation. Given input $\mathbf{x} \in X$, the same data augmentation operator is applied on $\mathbf{x}$ twice, producing two "views" of the input, $\tilde{\mathbf{x}}_i, \tilde{\mathbf{x}}_j$, which are then passed through a network to obtain projected vectors[1] $\mathbf{z}_i, \mathbf{z}_j$. Following the procedure, a batch of N inputs is processed into a two-viewed batch $\{\tilde{\mathbf{x}}_k\}_{k=1...2N}$ and then $\{\mathbf{z}_k\}_{k=1...2N}$. SimCLR designates $\tilde{\mathbf{x}}_i$ as the "anchor", relative to which positive and negative samples are defined: $\tilde{\mathbf{x}}_j$ is defined positive while every other sample in the batch is defined negative. The SimCLR loss is defined as the following,

$$\mathcal{L}_{SimCLR} = -\sum_{i \in I} \log \frac{\exp\left(s(\mathbf{z}_i, \mathbf{z}_{j(i)})/\tau\right)}{\sum_{k \in A(i)} \exp\left(s(\mathbf{z}_i, \mathbf{z}_k)/\tau\right)}, \tag{1}$$

where $I \equiv \{1...2N\}$, $A(i) \equiv I \setminus \{i\}$, $s(\cdot, \cdot)$ is the similarity function (e.g. the cosine similarity), $j(i)$ is the index of the positive counterpart of sample i, and τ is the temperature hyper-parameter.

On the basis of SimCLR, altering the definition of positive and negative pairs yields variants of contrastive losses, among which the SupCon loss is the first to utilize labels, formulated as the following,

---

[1] Note that they are outputs of the projection head which is omitted after the contrastive training. Network before the projection head is called the encoder, whose outputs we denote by "representations". We generally refer to both as "feature".

$$\mathcal{L}_{SupCon} = \sum_{i \in I} \frac{-1}{|P(i)|} \sum_{j \in P(i)} \log \frac{\exp\left(s(\mathbf{z}_i, \mathbf{z}_j)/\tau\right)}{\sum_{k \in A(i)} \exp\left(s(\mathbf{z}_i, \mathbf{z}_k)/\tau\right)}, \tag{2}$$

where $P(i)$ is the index set of batch samples that share the same label as sample i. It differs from $\mathcal{L}_{SimCLR}$ in that $|P(i)|$ could be greater than 1, and thereby the cross-entropy loss is revised accordingly with a non-one-hot ground truth.

4   Methodology

4.1   Motivation

Graf et al. [8] inspect into the geometry of the SupCon loss and prove that minimizing the SupCon loss is equivalent to forming a regular simplex with the projected vectors $\{\mathbf{z}_i\}$. In other words, SupCon forces the features to collapse according to their corresponding labels, and meanwhile be dispersed on the unit sphere. In general, contrastive learning leads to *the dispersion of learned features*, along with uniformity in SimCLR or class collapse in SupCon. Such dispersion enhances the divisibility of features in the high dimensional space, reducing the need for a complicated downstream classifier.

As aforementioned, applying SupCon on regression tasks directly using pre-defined bins on target labels would produce a representation space that are not well organized by their label distances. Zha et al. [20] prove that the alignment between the ranking of label distances and that of representation distances matters for regression tasks. However in [20], with any pair of samples can be selected as positive pairs, the dispersion is somewhat compromised.

Inspired by the pros and cons of these works, we believe if the dispersion procedure of SupCon could be guided by label distances, the undesirable distributions of representations as shown in Fig. 1 (a) could be mitigated. Towards this end, we propose to enhance SupCon by adding a ranking correlation term to its loss function, which is formly defined as Ranking Enhanced Supervised Contrastive Loss (RESupCon), to ensure that the distances between representations learned by SupCon are aligned with their label distances and still maintain the efficacy in learning distinguishable and scattered representations.

4.2   Ranking Enhanced Supervised Contrastive Learning (RESupCon)

Overall Structure. For a regression task, as shown in Fig. 2, we aim to train a neural network composed of two components: a representation encoder network $f : X \to \mathcal{R}^{de}$ and a regression prediction network $g : \mathcal{R}^{de} \to \mathcal{R}^{dt}$ to predict the target $y \in \mathcal{R}^{dt}$ with input $\mathbf{x} \in X$. The encoder network can use any backbone structure to extract representations corresponding to $\mathbf{x}$. The regression prediction network can simply be a linear regressor. Same as in other works, we deploy the two stage training strategy to firstly train the upstream encoder network and then fix the encoder to train the downstream regressor.

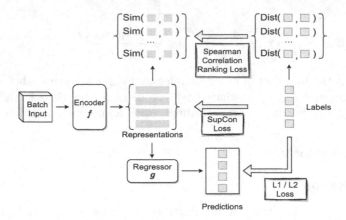

Fig. 2. Overall structure

For the upstream encoder training, RESupCon loss is imposed, which is formally defined as the following,

$$\mathcal{L}_{RESupCon} = \mathcal{L}_{SupConM} + \lambda \mathcal{L}_{scr}, \tag{3}$$

where $\mathcal{L}_{scr}$ is the Spearman correlation ranking loss, λ is the weight for $\mathcal{L}_{scr}$ and $\mathcal{L}_{SupConM}$ is the SupCon loss with a modified $P(i)$. Here, $P(i)$ is the index set of batch samples satisfying $|y_i - y_j| < \theta$, where y_i and y_j are the labels corresponding to sample i and sample j, θ is the predefined bin size. The Spearman correlation ranking loss $\mathcal{L}_{scr}$ is defined in the next subsection.

The representations learned with the Spearman correlation ranking loss shall exhibit a good correlation between their representation distances and label distances while still maintaining the efficacy of supervised contrastive learning that pulls together representations within the same bin and pushes apart other representations. These representations with better ranking correlation and dispersion shall then be used directly to train the downstream regressor. Hence, instead of imposing the loss on the projected vectors generated by a projection head, we impose both $\mathcal{L}_{SupConM}$ and $\mathcal{L}_{scr}$ directly on the features generated by the encoder and these features are used as the representations to train the downstream regressor subsequently.

Spearman Correlation Ranking Loss. How well the ranking of the representation distances are in accordance with the ranking of the label distances can be represented using the Spearman's rank correlation coefficient. The Spearman's rank correlation coefficient is defined as the following,

$$r_s(A, B) = \frac{Cov(R(A), R(B))}{\sigma_{R(A)} \sigma_{R(B)}}, \tag{4}$$

where A, B are two sequence variables and $R(A)$ and $R(B)$ are the rank values of A and B, respectively. Note that Spearman's rank correlation coefficient is actually the Pearson correlation coefficient between the ranks of two variables. Intuitively, a high Spearman's rank correlation coefficient (maximum at 1) between two variables indicates that they have similar or identical ranks. Conversely, a low correlation coefficient suggests dissimilar or completely opposite ranks of the two variables, with a correlation coefficient of -1 indicating perfect opposition. Formally, the Spearman correlation ranking loss is defined as the following,

$$\mathcal{L}_{scr} = -\frac{1}{2N} \sum_{i \in I} r_s(d_z(i), d_y(i)), \tag{5}$$

where $i \in I \equiv \{1...2N\}$ is the index set of augmented samples in the two-viewed batch. Taking sample i as an anchor, $d_z(i)$ is the distance function that returns the sequence of distances between anchor representation $\mathbf{z}_i$ and all other representations $\mathbf{z}_j$ $(j \in I)$. Distances between representations are computed using negative cosine similarity. $d_y(i)$ is the distance function that returns the sequence of distances between anchor label y_i and all other labels $y_j(j \in I)$. Distance between labels is computed using absolute difference. $d_z(i)$ and $d_y(i)$ are defined as the following,

$$\begin{cases} d_z(i) = [-s(\mathbf{z}_i, \mathbf{z}_1), -s(\mathbf{z}_i, \mathbf{z}_2), ..., -s(\mathbf{z}_i, \mathbf{z}_{2N})] \\ d_y(i) = [|y_i - y_1|, |y_i - y_2|, ..., |y_i - y_{2N}|] \end{cases}. \tag{6}$$

Since the Spearman's rank correlation coefficient between two variables is mathematically non-differentiable, we use the implementation of the differentiable ranking algorithm proposed by [1] to approximate it.

Time Complexity. We analyze the time complexity of RESupCon in the serialized scenario. As formulated in Eq. 2, SupCon involves $P(i)$ operations for positive samples and $O(2N - 1)$ operations for negative samples per anchor choice i, where $2N$ is the two-viewed batch size. A summation is done over the $2N$ anchor choices, thus giving us a total of $O(\sum_{i=1}^{2N}(P(i) + 2N - 1)) = O(N^2 + NL)$ operations for the SupCon loss term, where L denotes the expected number of samples in a batch that is annotated with the same label. As stated by [1], the Spearman correlation ranking loss has a time complexity of $O(n \log n)$, where $n = 2N - 1$ in our case of in-batch ranking, per anchor. Therefore the Spearman correlation ranking loss term yields a time complexity of $O(N^2 \log N)$ for a batch. To summarize, the time complexity of the RESupCon loss is $O(N^2 + NL + N^2 \log N) = O(N^2 \log N)$ for a batch of size $2N$.

5   Experiments

5.1   Datasets

We evaluate our method on four regression datasets in various area of interests, e.g. meteorological data, medical data and human facial data.

AgeDB-32. AgeDB [15] dataset contains a total of 16,488 in-the-wild human facial images and their corresponding ages as labels ranging from 1 to 101. We split it into a 12,208-image training set, a 2,140-image validation set and a 2140-image test set. We filtered out images with channels less than 3 and resize all the images to 32×32 before feeding into the network. We name this dataset AgeDB-32.

TCIR. TCIR [2] is a publicly available dataset for tropical cyclone (TC) Intensity Regression (TCIR). We use 36,566 frames of TCs from 2003-2013 as training data, 3,245 frames from 2014 as validation data, and 7,570 frames from 2015-2016 as testing data. Infrared and passive microwave rain-rate channels are used as input and the corresponding intensity from best track intensities (in kts) are used as labels. Each image contains 201×201 pixels and the central area of 65×65 pixels is cropped as the input.

WIKI. IMDB-WIKI [18] is another dataset that contains 523,051 human facial images and corresponding ages as labels. In our experiment, we only use its WIKI dataset, which contains 62,328 images and is split into a 49,692-image training set, a 6,335-image validation set and a 6,301-image test set.

BoneAge. The 2017 RSNA Pediatric Bone Age Challenge Dataset [9] contains X-ray images of hands and their skeletal age annotations. The dataset is split into a 12,611-image training set, a 1,425-imgage validation set and a 200-image test set. Skeletal ages range from 0 to 216 months.

Data Preprocessing and Augmentation: For TCIR, normalization is applied on each channel of a single sample to have zero mean and unit standard deviation. Random rotation augmentation is used during training. For all other datasets, we normalize the dataset with means and standard deviations calculated on the training set. Random resized cropping, random horizontal flipping, random color jittering and random gray-scaling are applied during training. The input size is resized to 32×32 for TCIR, AgeDB-32 and WIKI, and 64×64 for BoneAge.

5.2   Baselines and Settings

We compare our method with various baselines.

We use ResNet-18 as backbone and use the cosine similarity function for all contrastive losses including ours. Baselines are configured as follows.

E2E-L1 and E2E-L2. Features generated by the encoder are L2-normalized and connected directly to a downstream linear regressor and are trained in an end-to-end fashion using L1 and L2 losses.

SupCon. As recommended in [11], we append a two-layer projection head (Linear, ReLU, Linear) to the encoder and apply the SupCon loss on the L2-normalized output. Every distinct value is considered as a class when calculating the SupCon loss. After training the upstream network, a linear layer is appended to the encoder and is trained to produce a single value regression output.

AdaCon. We implement the model and loss as described in [7]. AdaCon adopts the same projection head mechanism as SupCon. We also set every distinct integer label to be a class, as AdaCon relies on a certain binning policy to convert regression to classification.

RNC. We implement the model and loss as described in [20]. According to the authors, RNC does not include a projection head in the model, same as our method. The L2-normalized representations trained with the RNC loss are used for downstream regressors directly. We train the model following the two-stage scheme of upstream contrastive training and downstream linear probing.

We implement all methods in PyTorch and conduct experiments on RTX 2080Ti GPUs. For E2E-L1 and E2E-L2, models are trained for 400 epochs. For the contrastive methods, upstream training runs for 400 epochs followed by a downstream training (linear probing) stage for 100 epochs. Best model is selected according to the loss on the validation set. All models are optimized with SGD. Hyper-parameters required for the optimizer, namely learning rate, momentum and weight decay, are chosen by conducting grid searches using NNI [14] on all methods to find their best settings respectively (on validation set). The temperature τ is set to 0.1. All experiments are repeated five times and the average results on test sets are reported.

5.3   Overall Performance

As outlined above, we train the models using one of two approaches: an end-to-end manner employing L1 (E2E-L1) and L2 (E2E-L2) losses, or a two-stage scheme that involves upstream contrastive training followed by downstream linear probing (with either L1 or L2 losses). The overall regression performance as measured by RMSE and R^2 is reported in Table 1. Our results show that SupCon underperforms when compared to end-to-end regression with either L1 or L2 losses. This suggests that failing to capture the relationship between representation similarities and label distances is suboptimal for regression tasks. AdaCon performs better than SupCon, but it does not constantly surpass the performance of E2E-L1 or E2E-L2. RNC outperforms the end-to-end methods on three datasets, while slightly worse on the WIKI dataset. It is worth mentioning that training with the Spearman correlation ranking loss ($\mathcal{L}_{scr}$) alone does not outperform end-to-end methods. In contrast, our proposed method RESupCon exhibits superior performance when compared to all baselines in terms of both RMSE and R^2, regardless of the loss used in the downstream phase. Specifically, RESupCon (+L2) improves over SupCon (+L2) in terms of RMSE by 3.9%,

Table 1. RMSE and R^2 results on various test datasets

Method	AgeDB-32		TCIR		WIKI		BoneAge	
	RMSE↓	R^2↑	RMSE↓	R^2↑	RMSE↓	R^2↑	RMSE↓	R^2↑
E2E-L1	9.93	0.628	11.57	0.852	10.76	0.590	14.70	0.883
E2E-L2	10.04	0.620	11.24	0.861	10.73	0.593	15.07	0.877
SupCon(+L1)	10.04	0.620	11.64	0.850	11.24	0.552	17.60	0.832
SupCon(+L2)	10.06	0.618	11.46	0.855	11.05	0.568	17.32	0.837
AdaCon(+L1)	9.91	0.630	11.41	0.856	11.02	0.570	15.53	0.870
AdaCon(+L2)	9.97	0.626	11.57	0.852	10.91	0.579	16.34	0.856
RNC(+L1)	9.84	0.634	11.21	0.861	10.88	0.580	14.14	0.891
RNC(+L2)	9.78	0.640	11.23	0.861	10.87	0.582	14.21	0.891
$\mathcal{L}_{scr}$(+L1)	10.19	0.609	12.15	0.837	11.08	0.566	15.25	0.874
$\mathcal{L}_{scr}$(+L2)	10.25	0.604	11.38	0.857	10.95	0.576	14.98	0.878
RESupCon(+L1)	**9.64**	**0.650**	11.07	0.865	10.66	0.597	13.86	0.896
RESupCon(+L2)	9.67	0.647	**10.98**	**0.867**	**10.60**	**0.603**	**13.40**	**0.903**

4.2%, 4.1% and 22.6% on four datasets respectively, and improves over E2E-L2 on RMSE by 3.7%, 2.3%, 1.2% and 11% respectively.

5.4 Comparison on Spearman's Rank Correlation Coefficients

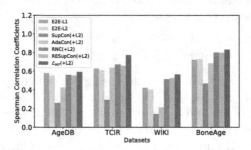

Fig. 3. Spearman's rank correlation coefficients between representation and label distances on validation sets.

Here we further analyze the correlation between representation and label distances. We feed the validation set $\{\mathbf{x}_i, y_i\}$ into the trained encoder network (or up to the penultimate layer for end-to-end models), and collect the output representations $\{\mathbf{z}_i\}$. The Spearman's rank correlation coefficient between representation and label distances, $r_s(d_z(i), d_y(i))$, is calculated upon each sample i and averaged over the whole validation set. Figure 3 presents the results on all datasets and methods. As shown in the figure, E2E-L1 and E2E-L2 losses already result in good ranking at the penultimate layer. It is worth noting that SupCon method harms ranking greatly, confirming the statement that SupCon only disperses representations with little concern about the continuous nature of labels. AdaCon improves the ranking over SupCon but still fails to catch up with E2E-L1 or E2E-L2. RNC and RESupCon both concentrate on the ranking and succeed to retain the coefficient, even higher than that of E2E-L1 and E2E-L2 in most cases. It is worth noting that, training with $\mathcal{L}_{scr}$ alone achieves the highest Spearman's rank correlation coefficient but does not achieve the best downstream performance. This indicates that a good ranking itself does not lead to optimal representations, without the help of dispersion gained from contrastive learning.

5.5 Parameter Study and Loss Curve

A parameter study is conducted on all four datasts on the weight of $\mathcal{L}_{scr}$. For illustration, we show the prediction performance in RMSE on AgeDB validation set with respect to different weights in Fig. 4 (a). The RMSE begins to drop initially as λ increases until a certain value, in this case $\lambda = 4$, and becomes worse afterwards. It complies with our assumption that emphasis on ranking alone or using SupCon alone does not achieve the optimal representation space.

The loss curves of both $\mathcal{L}_{SupConM}$ and $\mathcal{L}_{scr}$ of the RESupCon method on AgeDB training set and validation set and the loss curve of the SupCon method on AgeDB validation set are shown in Fig. 4 (b). It shows that $\mathcal{L}_{SupConM}$ trained alone and $\mathcal{L}_{SupConM}$ within RESupCon converge to a similar value on the validation set, while $\mathcal{L}_{scr}$ also decreases along training in a similar pace. The above observation implies that $\mathcal{L}_{SupConM}$ and $\mathcal{L}_{scr}$ can indeed be optimized simultaneously.

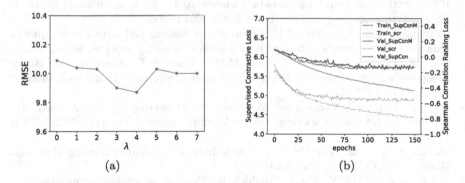

(a)   (b)

Fig. 4. (a) Parameter Study on λ using AgeDB-32 validation dataset. (b)Training and validation loss curves on AgeDB-32.

6   Conclusion

We propose a new method RESupCon to improve the alignment between representations and labels and to leverage contrastive learning in a controlled fashion, designed to not only learn a representation space with better ranking, but also preserve the dispersion of the learned representations. We verify that the properties of *dispersion* and *ranking* of the representation space do complement each other when conducting contrastive learning on datasets with ordinal labels. RESupCon improves performance on regression tasks by combining and balancing both properties rather than focusing on either one in isolation.

References

1. Blondel, M., Teboul, O., Berthet, Q., Djolonga, J.: Fast differentiable sorting and ranking. In: International Conference on Machine Learning, pp. 950–959. PMLR (2020)
2. Chen, B., Chen, B.F., Lin, H.T.: Rotation-blended CNNs on a new open dataset for tropical cyclone image-to-intensity regression. In: Proceedings of the 24th ACM SIGKDD International Conference on Knowledge Discovery & Data Mining, pp. 90–99 (2018)
3. Chen, M., Fu, D.Y., Narayan, A., Zhang, M., Song, Z., Fatahalian, K., Ré, C.: Perfectly balanced: improving transfer and robustness of supervised contrastive learning. In: International Conference on Machine Learning, pp. 3090–3122. PMLR (2022)
4. Chen, T., Kornblith, S., Norouzi, M., Hinton, G.: A simple framework for contrastive learning of visual representations. In: International Conference on Machine Learning, pp. 1597–1607. PMLR (2020)
5. Chen, X., Xie, S., He, K.: An empirical study of training self-supervised vision transformers. In: 2021 IEEE/CVF International Conference on Computer Vision, ICCV 2021, Montreal, QC, Canada, October 10-17, 2021, pp. 9620–9629. IEEE (2021). https://doi.org/10.1109/ICCV48922.2021.00950
6. Cuturi, M., Teboul, O., Vert, J.P.: Differentiable ranking and sorting using optimal transport. In: Advances in Neural Information Processing Systems. vol. 32 (2019)
7. Dai, W., Li, X., Chiu, W.H.K., Kuo, M.D., Cheng, K.T.: Adaptive contrast for image regression in computer-aided disease assessment. IEEE Trans. Med. Imaging **41**(5), 1255–1268 (2021)
8. Graf, F., Hofer, C., Niethammer, M., Kwitt, R.: Dissecting supervised contrastive learning. In: International Conference on Machine Learning, pp. 3821–3830. PMLR (2021)
9. Halabi, S.S., et al.: The RSNA pediatric bone age machine learning challenge. Radiology **290**(2), 498–503 (2019)
10. He, K., Fan, H., Wu, Y., Xie, S., Girshick, R.: Momentum contrast for unsupervised visual representation learning. In: Proceedings of the IEEE/CVF conference on computer vision and pattern recognition, pp. 9729–9738 (2020)
11. Khosla, P., et al.: Supervised contrastive learning. Adv. Neural. Inf. Process. Syst. **33**, 18661–18673 (2020)
12. Le-Khac, P.H., Healy, G., Smeaton, A.F.: Contrastive representation learning: a framework and review. IEEE Access **8**, 193907–193934 (2020)

13. McInnes, L., Healy, J., Melville, J.: UMAP: uniform manifold approximation and projection for dimension reduction. arXiv:1802.03426 (2020)
14. Microsoft: Neural Network Intelligence (2021). https://github.com/microsoft/nni. Accessed 02 Aug 2023
15. Moschoglou, S., Papaioannou, A., Sagonas, C., Deng, J., Kotsia, I., Zafeiriou, S.: AgeDB: the first manually collected, in-the-wild age database. In: Proceedings of the IEEE Conference on Computer Vision and Pattern Recognition Workshop, vol. 2, p. 5 (2017)
16. Qin, T., Liu, T.Y., Xu, J., Li, H.: LETOR: a benchmark collection for research on learning to rank for information retrieval. Inf. Retrieval **13**, 346–374 (2010)
17. Radford, A., et al.: Learning transferable visual models from natural language supervision. In: International Conference on Machine Learning, pp. 8748–8763. PMLR (2021)
18. Rothe, R., Timofte, R., Van Gool, L.: DEX: deep expectation of apparent age from a single image. In: Proceedings of the IEEE international conference on computer vision workshops, pp. 10–15 (2015)
19. Taylor, M., Guiver, J., Robertson, S., Minka, T.: SoftRank: optimizing non-smooth rank metrics. In: Proceedings of the 2008 International Conference on Web Search and Data Mining, pp. 77–86 (2008)
20. Zha, K., Cao, P., Son, J., Yang, Y., Katabi, D.: Rank-n-contrast: learning continuous representations for regression. In: Thirty-seventh Conference on Neural Information Processing Systems (2023)

Treatment Effect Estimation Under Unknown Interference

Xiaofeng Lin[✉], Guoxi Zhang, Xiaotian Lu, and Hisashi Kashima

Graduate School of Informatics, Kyoto University, Kyoto, Japan
{lxf,guoxi,lu}@ml.ist.i.kyoto-u.ac.jp, kashima@i.kyoto-u.ac.jp

Abstract. Causal inference is a powerful tool for effective decision-making in various areas, such as medicine and commerce. For example, it allows businesses to determine whether an advertisement has a role in influencing a customer to buy the advertised product. The influence of an advertisement on a particular customer is considered the advertisement's individual treatment effect (ITE). This study estimates ITE from data in which units are potentially connected. In this case, the outcome for a unit can be influenced by treatments to other units, resulting in inaccurate ITE estimation, a phenomenon known as interference. Existing methods for ITE estimation that address interference rely on knowledge of connections between units. However, these methods are not applicable when this connection information is missing due to privacy concerns, a scenario known as unknown interference. To overcome this limitation, this study proposes a method that designs a graph structure learner, which infers the structure of interference by imposing an L_0-norm regularization on the number of potential connections. The inferred structure is then fed into a graph convolution network to model interference received by units. We carry out extensive experiments on several datasets to verify the effectiveness of the proposed method in addressing unknown interference.

Keywords: Causal inference Individual · Treatment Effect Estimation · Interference

1 Introduction

Estimating treatment effects is crucial in various fields, including medicine [22], education [20], and e-commerce [18]. In commerce, it helps assess whether advertisements influence customer purchases. Treatment effects can be estimated at different levels: average treatment effect (ATE) for the average overall population effect of a treatment and individual treatment effect (ITE) for specific individual effects.

This study focuses on estimating treatment effects using observational data, which includes records of treatment assignments, outcomes, and covariates of units. In particular, we consider scenarios where units are potentially connected. In this case, the outcome of a unit can be influenced by the treatments assigned

D.-N. Yang et al. (Eds.): PAKDD 2024, LNAI 14646, pp. 28–42, 2024.
https://doi.org/10.1007/978-981-97-2253-2_3

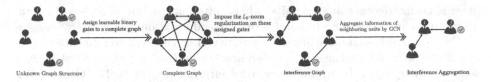

Fig. 1. Overview of the proposed method that models unknown interference. A unit with a checkmark (✓) represents a treated unit, e.g., a unit that receives an advertisement, and a unit without a checkmark represents a control unit.

to other units. Consider an advertising example, users who do not see an advertisement can receive influence from their friends who did, so their responses are indirectly affected by the advertisement, a concept known as *interference* [25]. If such interference is not properly modeled, the estimation of the treatment effects will be inaccurate. However, the potential connections among units are usually hidden, owing to issues such as privacy preservation, a scenario known as *unknown interference* [25]. Existing studies [2,25] estimated ATE under unknown interference but lacked a solution for ITE, which is crucial in practical applications.

This study proposes an end-to-end method to model unknown interference, aiming to automatically identify significant connections that cause interference (Fig. 1). We utilize a graph structure learner (GSL) to construct an *interference graph* by initially creating a complete graph, as any pair of units could potentially be interconnected. The GSL assigns learnable binary gates to the edges in this complete graph to discern which edges cause interference. During training, an L_0-norm regularization is applied to these binary gates and minimized using an approximate solution [15]. Additionally, the GSL assigns varying weights to the remaining edges by employing a method similar to graph attention mechanisms [27,29]. To accurately model interference, our approach utilizes the inferred interference graph, edge weights, and a graph convolution network (GCN) [28] to aggregate interference received by units. The model architecture is shown in Fig. 2, with further details provided in Sect. 4. Upon the training of the model is complete, the proposed model is used for ITE and ATE estimation, as elaborated in Sect. 4.4.

The contributions of this study can be summarized as follows:

- This study overcomes a novel challenge: ITE estimation under unknown interference.
- This study proposes a new method to address unknown interference.
- Extensive experiments verify the efficacy of the proposed methods for ITE and ATE estimation in the presence of unknown interference.

2   Related Work

Many existing methods estimate ITE and ATE without accounting for interference [6,7,10,12,23]. They estimate treatment effects under the assumption that there is no interference among units. This is often unrealistic in real-world data where units are interconnected and propagate information.

Previous studies on modeling interference can be categorized into three types based on their assumptions of interference: *group-level interference* [9,14,26], *pairwise interference* [1,4], and *networked interference* [13,16,17]. Group-level interference assumes that interference exists within subgroups of units but not across them, which is a partial interference and does not align well with real-world scenarios. Pairwise interference presumes that units can be influenced by their immediate neighbors, limiting their ability to capture the broader propagation of interference. Networked interference is the most general assumption, allowing interference to propagate through a graph of units. Ma et al. [17] applied an aggregation mechanism of GCNs [28] to model the propagation of interference on graphs, enabling their model to capture networked interference. Ma et al. [16] proposed a hypergraph-GCN-based method for modeling high-order interference on hypergraphs. Lin et al. [13] modeled interference across different views of multi-view graphs using a heterogeneous GCN-based method. However, these GCN-based methods need knowledge about graph structure, which is not always available due to privacy concerns.

Several existing studies have considered the issue of unknown interference, but they still have certain limitations. Rakesh et al. [19] constructed a graph based on the similarity of unit covariates to fill in the missing graph information. Their approach cannot discover the interference among dissimilar units and only considers pairwise interference. Bhattacharya et al. [2] assumed partial interference rather than networked interference and proposed a greedy network structure search method to find a partial interference structure. Furthermore, this study only estimated ATE. In a different approach, Sävje et al. [25] did not construct interference structures but proposed an expected average treatment effect (EATE) estimation method under unknown interference. However, this method still cannot estimate ITE, which is often crucial for many applications, such as advertising pushing. To overcome their limitations, this study proposes an end-to-end method that models the networked interference and estimates ITE without relying on information about graph structures. In our experiments, we compare the proposed method with applying the existing GCN-based methods [17] on the similarity graph [19] of units.

3 Preliminaries

Let $x_i \in \mathbb{R}^d$ be the covariates (e.g., age) of a unit i, $t_i \in \{0,1\}$ be the treatment assigned to the unit i (e.g., an advertisement), X be the covariates of all units, and T be the treatments of all units. We use non-bold, italicized, and capitalized letters to denote random variables, such as X_i, and use subscript $-i$ to denote all the other units except i, such as T_{-i}. This study considers that units are potentially connected, and we use s_i to represent the summary of neighbor information (i.e., information of units connected to the unit i) for the unit i. Potential outcomes can be indexed only by the individual treatment and the neighbor information [4], denoted as $y_i^{T_i}(S_i)$. Let $y_i^t(s_i)$ be the simplified expression for $y_i^{T_i=t}(S_i = s_i)$. Let $y_i^{\text{obs}} \in \mathbb{R}$ be the observed outcome of the unit i under the actual value of t_i and s_i. The covariates, treatments, and observed outcomes of units constitute observational data $\{(\boldsymbol{x}_i, t_i, y_i^{\text{obs}})\}_{i=1}^N$.

Conventionally, the ITE is defined in a scenario where there is no interference among units [21], under the so-called stable unit treatment value assumption (SUTVA) [21]. Such a definition of the ITE characterizes the intrinsic treatment effect of every unit. We can formulate such an ITE by masking out neighbor information S in the potential outcomes using $\tau(X_i = \boldsymbol{x}_i, S_i = \boldsymbol{0}) = \mathbb{E}\left[Y_i^1(S_i = \boldsymbol{0})|X_i = \boldsymbol{x}_i\right] - \mathbb{E}\left[Y_i^0(S_i = \boldsymbol{0})|X_i = \boldsymbol{x}_i\right]$.

We now introduce graphs. A graph contains information about both units and their relationships, represented by connections or edges among units. Let $\boldsymbol{A} \in \{0, 1\}^{N \times N}$ be the adjacency matrix of the graph. If there is a direct edge from a unit j to a unit i, i.e., the unit j is a neighbor of the unit i, $A_{ij} = 1$; otherwise, $A_{ij} = 0$. In this study, we assume that the units are potentially connected through an interference graph $\boldsymbol{A}^{\text{itf}}$. However, due to observational limitations and privacy protection issues, $\boldsymbol{A}^{\text{itf}}$ is unknown.

In the situation where units are associated with unknown graphs, the SUTVA cannot hold, as connected units often propagate information. Therefore, this study assumes that there exists unknown interference, where the outcome of a unit is not only influenced by its own treatment and covariates but also influenced by those of its neighbors on the unknown interference graph $\boldsymbol{A}^{\text{itf}}$, i.e., $S \neq \boldsymbol{0}$. In this case, the ITE under $S_i = \boldsymbol{s}_i$ can be formulated using the following equation, similar to the definition of ITE under interference in the existing study [17]: $\tau(X_i = \boldsymbol{x}_i, S_i = \boldsymbol{s}_i) = \mathbb{E}\left[Y_i^1(S_i = \boldsymbol{s}_i)|X_i = \boldsymbol{x}_i\right] - \mathbb{E}\left[Y_i^0(S_i = \boldsymbol{s}_i)|X_i = \boldsymbol{x}_i\right]$.

Apart from interference, the issue of confounders is a well-known challenge in observational data [17]. Confounders are parts of covariates and affect both the treatment assignment and outcome. Confounders result in an imbalance in the population distributions with different treatment assignments [10]. Without properly addressing confounders, ITE estimation will be biased.

4 Proposed Method: Treatment Effect Estimation Under Unknown Interference

This study aims to estimate both the ITE without interference $\tau(X_i = \boldsymbol{x}_i, S_i = \boldsymbol{0})$ and ITE with unknown interference $\tau(X_i = \boldsymbol{x}_i, S_i = \boldsymbol{s}_i)$ from observational data $\{(\boldsymbol{x}_i, t_i, y_i^{\text{obs}})\}_{i=1}^N$. To this end, we propose a method called **T**reatment **E**ffect Estimator under **U**nknown **I**nterference (UNITE). We prove that ITE is identifiable in Appendix A. A diagram for the proposed UNITE is provided in Fig. 2. We released codes of UNITE at https://github.com/LINXF208/UNITE. UNITE contains four components: a covariate representation learner ϕ, a GSL, an aggregation function ψ, and two outcome predictors. The first component ϕ learns representations of covariates to address confounders, and it also makes the representations interference-free in order to estimate $\tau(X_i = \boldsymbol{x}_i, S_i = \boldsymbol{0})$. The second component GSL infers an interference graph from observed data.

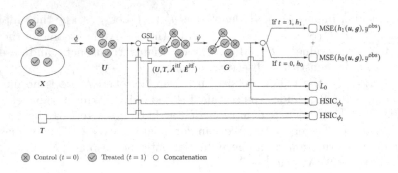

Fig. 2. Model architecture of UNITE.

The third component ψ utilizes a multi-layered GCN to aggregate interference received by units and transform aggregated results into unit representations, referred to as interference representations. Finally, the last component predicts potential outcomes using the covariate and interference representations.

4.1 Covariate Representation Learner

The covariate representation learner ϕ serves two primary purposes. Firstly, it aims to mitigate the imbalance in the distributions of different treatment groups caused by confounders. To this end, ϕ maps covariates to a representation space where treatment assignments and covariate representations become approximately independent [17]. Secondly, ϕ also aim to generate interference-free covariate representations for estimating $\tau(X_i = \boldsymbol{x}_i, S_i = \boldsymbol{0})$. This is similar to finding a representation space where interference representations and covariate representations are approximately independent.

To be specific, let ϕ be a covariate representation learner, $\boldsymbol{u}_i = \phi(\boldsymbol{x}_i; \boldsymbol{W}_\phi)$ be the covariate representation of a unit i, $\boldsymbol{W}_\phi$ be the paramaters of the ϕ, and $\boldsymbol{U}$ be the covariate representations of all units.

We now introduce a Hibelt-Schmidt independence criterion (HSIC) regularization [5], a criterion for testing the independence of two random variables, with values ranging from 0 to 1. If two random variables are independent, the HSIC equals 0. Thus, minimizing the HSIC between treatment assignments and covariate representations, and minimizing the HSIC between interference representations and covariate representations can achieve the goals of ϕ. We use an efficient approach for calculating HSIC [5], as described in Appendix B. Let HSIC_{Φ_1} denote the HSIC between covariate representations and treatment assignments, HSIC_{Φ_2} denote the HSIC between covariate and interference representations, and $\mathcal{L}_\phi = \beta_1 \text{HSIC}_{\phi_1} + \beta_2 \text{HSIC}_{\phi_2}$ denote the entire HSIC regularization, where β_1 and β_2 are hyperparameters.

4.2   Graph Structure Learner

The challenge of GSL is that, although a unit might be potentially connected to all other units in the data, the outcome of a unit usually receives influence from only parts of other units. Furthermore, the level of influence can be different for different influencers. For example, suppose we want to estimate the effect of a promotion for a specific computer on all the students in a university. A student can receive more influence from peers who take the same classes than students he or she does not meet on campus. The level of influence received will depend on the proximity between the influencer and the influenced units. Thus, the goal of our GSL is to infer the structure of the interference graph A^{itf} and the level of interference E^{itf}, which are scalar weights for edges of A^{itf}.

Inspired by Ye et al. [29], we infer the interference graph from the complete graph of units by associating binary gates to edges. Let $A^{\text{com}} \in \{1\}^{N \times N}$ be the adjacency matrix of the complete graph, $z_{ij} \in \{0, 1\}$ be a binary gate between unit i and j, and Z be all the binary gates. $z_{ij} = 1$ if the unit i receives interference from the unit j, and $z_{ij} = 0$ otherwise.[1] Then, the A^{itf} is given by $A^{\text{itf}} = A^{\text{com}} \odot Z$, where the $\odot$ represents the Hadamard product. To infer the edges that actually cause interference, the UNITE imposes the L_0-norm regularization on the gates, which is defined as $\mathcal{L}_{\text{GSL}} = \frac{1}{N^2} \sum_{(i,j)} 1_{[z_{ij} \neq 0]}$, where $1_{[\cdot]}$ is the indicator function. The gates are computed using the information of units available to the GSL.

However, direct optimization of the $\mathcal{L}_{\text{GSL}}$ is intractable. As the interference graph A^{itf} is determined by the Z and used by other modules of the UNITE, direct optimization of the $\mathcal{L}_{\text{GSL}}$ requires enumerating all possible binary matrices in $\{0, 1\}^{N \times N}$. Thus, we apply a hard concrete estimator technique proposed by Louizos et al. [15] for approximating the L_0-norm regularization. This technique is based on an observation that, for any distribution over the binary gates $\Pr(Z)$, we have $\min_Z \mathcal{L}_{\text{GSL}} \leq \mathbb{E}_{Z \sim \Pr(Z)}[\mathcal{L}_{\text{GSL}}]$, which means that the minimum of the $\mathcal{L}_{\text{GSL}}$ is upper bounded by its expectation over a sampling distribution $\Pr(Z)$. We approximate z_{ij} by sampling $\hat{z}_{ij}$ from a hard concrete distribution and use $\hat{z}_{ij}$ to compute a sample for the interference graph $\hat{A}^{\text{itf}}$ during learning.

Specifically, the hard concrete distribution $\Pr(Z; \log \alpha)$ can be considered as a continuous approximation to the Bernoulli distribution, which admits using a reparameterization trick for efficient optimization. Suppose v is a random variable that follows the uniform distribution over $[0, 1]$, i.e. $v \sim \text{Unif}(0, 1)$. Then, we can express a sample of $\hat{z}_{ij}$ using a sample of v as follows:

$$\hat{z}_{ij} = \min\left\{1, \max\left\{0, \text{sigmoid}((\log v - \log(1-v) + \log \alpha_{ij})/\eta_1)(\eta_3 - \eta_2) + \eta_2\right\}\right\}, \tag{1}$$

where $\eta_1 = 2/3, \eta_2 = -0.1, \eta_3 = 1.1$ are typical parameters for the hard concrete distribution [15]. Here, $\log \alpha_{ij}$ can be approximated by a function of the information for unit i and unit j available to the GSL. Let $p_i, p_j \in \mathbb{R}^{d'}$ be the information of the unit i and j, which is the concatenation of corresponding unit covariate

---

[1] Note that we assume the edges of the interference graph to be directed.

representation $\boldsymbol{u}$ and treatment t. Here, $\log \alpha_{ij} = \left[\boldsymbol{p}_i^{\mathrm{T}} \boldsymbol{W}_{\mathrm{GSL}} \| \boldsymbol{p}_j^{\mathrm{T}} \boldsymbol{W}_{\mathrm{GSL}}\right] \boldsymbol{o}_{\mathrm{GSL}}$, where $\boldsymbol{o}_{\mathrm{GSL}} \in \mathbb{R}^{2d''}$ and $\boldsymbol{W}_{\mathrm{GSL}} \in \mathbb{R}^{d' \times d''}$ are parameters to be learned. $\|$ means the concatenation operation for two vectors. In this case, during training $\mathcal{L}_{\mathrm{GSL}}$ can be approximated by $\hat{\mathcal{L}}_{\mathrm{GSL}} = \frac{\lambda}{N^2} \sum_{(i,j)} \mathrm{sigmoid}\left(\log \alpha_{ij} - \eta_1 \log \frac{-\eta_2}{\eta_3}\right)$, where λ is a hyperparameter. Ideally, most of the elements in the $\hat{\boldsymbol{Z}}$ will be zeros under the L_0-norm regularization, i.e., L_{GSL}; thus, the model will preserve only a few crucial edges. Units j with remaining edges, i.e., $\hat{A}_{ij}^{\mathrm{itf}} > 0$ are considered to cause interference to a unit i, where $\hat{A}_{ij}^{\mathrm{itf}} = 1_{[\hat{z}_{ij}>0]}$.

A sample $\hat{z}_{ij}$ is a continuous approximation of the binary gate to the edge from the unit j to i. The larger $\hat{z}_{ij}$ is, the more confident the proposed UNITE is about the unit j causing interference to unit i. In other words, a higher value of $\hat{z}_{ij}$ implies a higher importance for the interference from the j to i. However, there usually exist discrepancies in the number of remaining edges and neighbors for different units. In this case, if we use $\hat{z}_{ij}$ to aggregate interference and compute interference representations, there will be significant numerical discrepancies in the representations of different units. This can impact the performance and stability of the proposed method. To overcome this issue, we approximate the level of interference $\hat{e}_{ij}$ between unit i and j by normalizing $\hat{z}_{ij}$ by $\hat{e}_{ij} = \frac{\hat{z}_{ij}}{\sum_{\hat{z}_{ik}>0} \hat{z}_{ik}}$.[2] After normalizing the $\hat{z}_{ij}$, the value of $\hat{e}_{ij}$ is the level of interference relative to other remaining neighbors of i. This is similar to graph attention mechanisms [27,29].

Once the training of the GSL under the L_{GSL} is complete, most of the elements in the $\hat{\boldsymbol{Z}}$ are zeros; therefore, only a few units j with remaining edges (i.e., $\hat{A}_{ij}^{\mathrm{itf}} > 0$) are considered to cause interference to a unit i. In the test phase, we stop sampling v from $\mathrm{Unif}(0,1)$. Then, $\hat{z}_{ij}$ is generated by $\hat{z}_{ij} = \min\{1, \max\{0, \mathrm{sigmoid}((\log \alpha_{ij})/\eta_1)(\eta_3 - \eta_2) + \eta_2\}\}$.

4.3   Aggregation Function

Given a sampled interference graph $\hat{\boldsymbol{A}}^{\mathrm{itf}}$, the level of interference $\hat{\boldsymbol{E}}^{\mathrm{itf}}$, covariate representations $\boldsymbol{U}$, and treatments $\boldsymbol{T}$, the aggregation function ψ computes unit representations $\boldsymbol{G}$ as the summary of the neighbor information: $\boldsymbol{G} = \psi(\boldsymbol{U}, \boldsymbol{T}, \hat{\boldsymbol{A}}^{\mathrm{itf}} \odot \hat{\boldsymbol{E}}^{\mathrm{itf}})$. Here, we use $\boldsymbol{W}_{\psi}$ to denote all parameters of the ψ. For a unit i, its representation $\boldsymbol{g}_i$ is supposed to capture the interference of its potential neighbors, called the interference representation of the unit i.

The ψ of the UNITE employs a multi-layered GCN [28] to aggregate and model interference. A GCN layer can aggregate information from unit neighbors into a representation for each unit via a message-propagation mechanism. Let $\boldsymbol{g}_i^{(l)}$ be the representation of unit i computed by the l-th layer, and let $\boldsymbol{G}^{(l)}$ be the representation of all units computed by this layer. We use $\boldsymbol{g}_i^{(0)}$ and $\boldsymbol{G}^{(0)}$ as

---

[2] As the information of unit i itself is always important for computing the level of interference, we set $\hat{z}_{ii}$ to 1.

input to the first layer of GCN, which is the concatenation of the covariate representations U and treatments T. The message-passing result, i.e., the calculation of the l-th GCN layer for a unit i can be expressed as follows:

$$g_i^{(l)} = \sigma\left((\hat{A}^{\text{itf}} \odot \hat{E}^{\text{itf}})G^{(l-1)}W_\psi^{(l)}\right),$$

where $W_\psi^{(l)}$ is a parameter matrix for the l-th layer, and $\sigma(\cdot)$ is a non-linear activation function. This equation shows that $g_i^{(l)}$ is a weighted sum of the representations of neighbors of the unit i, weighted by their level of interference. A single GCN layer captures one-hop neighbor information, and to capture networked interference, we stack multiple GCN layers.

4.4 Outcome Predictors and ITE Estimators

Inspired by prior works [10,17,23], given U and G, we use feedforward neural networks to parameterize predictors h_0 and h_1 for predicting potential outcomes of $t = 0$ and $t = 1$. Denote by W_h for their parameters. The outcome predictors are trained by minimizing the mean square error (MSE) between their outputs and the observed outcomes, denoted by $\mathcal{L}_h = \frac{1}{n}\sum_{i=1}^n \left(h_{t_i}(u_i, g_i) - y_i^{\text{obs}}\right)^2$.

To learn the model parameters $(W_\phi, W_{\text{GSL}}, o_{\text{GSL}}, W_\psi, \text{ and } W_h)$ of UNITE, we now introduce the model optimization. As it is intractable to directly optimize the L_0-norm regularization of the GSL directly, we sample $\hat{Z}$ using Eq. (1) to construct samples of interference graphs during training. We optimizes the above parameters by minimizing the loss function $\hat{\mathcal{L}}$, defined as

$$\hat{\mathcal{L}}(W_\phi, W_{\text{GSL}}, o_{GSL}, W_\psi, W_h) = \mathbb{E}_{\hat{Z}}[\mathcal{L}_\phi + \mathcal{L}_\psi + \mathcal{L}_h] + \hat{\mathcal{L}}_{\text{GSL}}.$$

With the outcome predictors, we can estimate the ITE under interference using $\hat{\tau}(X = x_i, S_i = s_i) = h_1(u_i, g_i) - h_0(u_i, g_i)$, and estimate the interference-free ITE using $\hat{\tau}(X_i = x_i, S_i = 0) = h_1(u_i, 0) - h_0(u_i, 0)$. In addition, ATE estimation with or without interference is performed by averaging the outputs of $\hat{\tau}(X_i = x_i, S_i = s_i)$ or $\hat{\tau}(X_i = x_i, S_i = 0)$.

5 Experiments

We carried out experiments on three public datasets and compared the performance of the proposed methods with that of various existing methods to verify the effectiveness of the proposed method.

5.1 Experiment Settings

Datasets. We evaluated the proposed method using the following datasets.

Jobs Dataset [11]: The Jobs dataset is a widely used benchmark in the causal inference for observational studies [19,23]. It comprises a random control trial (RCT) subset, known as the LaLonde experimental sample, which has 297 treated and 425 control units, and the PSID comparison group, which includes 2,490 control units [24]. The dataset includes information about workers (covariates), the status of job training participation (treatment), and employment status (outcomes), gathered from real-world observations. The objective is to estimate the effect of job training on unemployment. Following Rakesh [19] who considered interference in the Jobs dataset, we also consider interference for this dataset, although it lacks information about connections. We averaged results over ten different training/validation/testing splits with ratios of 64/16/20.

Amazon Datasets [8]: Rakesh [19] used the co-purchase graph from the Amazon datasets [8] to explore the effect of positive and negative reviews on product sales and interference issues. To study these effects separately, Rakesh [19] divided the dataset into Amazon$^-$ (comprising only negative reviews) and Amazon$^+$ (comprising only positive reviews). Units in these datasets are connected through directed graphs. In Amazon$^-$, there are 14,538 units with 15,011 connections. For the Amazon$^+$ dataset, we extract the first 10,000 units with 3,000 connections. The treatment t (with values 0 or 1) depends on the number of negative (or positive) reviews a unit has: $t = 1$ if a unit has more than three negative (or positive) reviews, and $t = 0$ if it has fewer than three [19]. Each unit's covariate vector x consists of 300 features generated from their reviews using the doc2vec method. The ITE under interference (i.e., $\tau(X_i = x_i, S_i = s_i)$) for each unit is estimated using matching methods [19]. We follow Ma [17], conducting three repeated experiments and averaging the results for the Amazon$^-$ and Amazon$^+$ datasets, with a train/validation/test ratio of 80/5/15.

Baselines. We compare UNITE with various baseline methods, and they can be divided into the following categories:

No Graph: We compare UNITE with traditional methods, which do not consider graphs and ignore interference. BNN [10] addresses confounders by minimizing distribution discrepancies of covariate representation between different treatment groups but does not account for interference. We consider two BNN structures: BNN-4-0 with four representation layers and a linear output layer, and BNN-2-2 with two representation layers, two hidden prediction layers, and a linear output layer. CFR [23] employs a strategy similar to BNN, using the Maximum Mean Distance (MMD) and Wasserstein distance to penalize distribution discrepancies in the representation space. We consider two CFR schemes: CFR-MMD, using MMD for distribution discrepancy, and CFR-Wass, using the Wasserstein distance. TARNet [23] has the same model architecture as the CFR but removes the distribution discrepancy penalty term.

GCN-Based Methods [17]: GCN-based methods use GCNs [28] to model interference. To verify their performance under unknown interference, we construct k-NN graphs (with $k=10$) based on cosine similarity between units, which is inspired by Chen et al. [3] and Rakesh et al. [19]. The same graph structures are used throughout the training, validation, and testing phases. Two regularization schemes are considered: GCNϕ, which uses HSIC to balance outputs of representation layers, and GCNψ, which uses HSIC to balance outputs of GCN layers. We use GCNϕ/ψ + k-NN to denote GCN-based methods with k-NN graphs.

Metrics. To verify the quality of $\hat{\tau}(X = x, S_i = 0)$ estimation, we calculate $\hat{R}^{\text{pol}}(\pi_f)$ in the balanced and interference-free RCT subset, following prior work [23]: $\hat{R}^{\text{pol}}(\pi_f) = 1 - (\mathbb{E}[y_1^f | \pi_f(x) = 1, t = 1] \cdot p(\pi_f = 1) + \mathbb{E}[y_0^f | \pi_f(x) = 0, t = 0] \cdot p(\pi_f = 0))$. Here, $\pi_f(x) = 1$ if $\hat{\tau}(X = x, S_i = 0) > 0$, and $\pi_f(x) = 0$, otherwise. We also calculate ϵ^{ATT}, which represents the error in ATE estimation for the treated group (ATT) in the balanced and interference-free RCT subset: $\epsilon^{\text{ATT}} = \left| \frac{1}{|\mathbf{R}^t|} \sum_{i \in \mathbf{R}^t} \tau(X_i = x_i, S_i = 0) - \frac{1}{|\mathbf{R}^t|} \sum_{i \in \mathbf{R}^t} \hat{\tau}(X_i = x_i, S_i = 0) \right|$, where $\mathbf{R}^t$ represents the set of treated units of the RCT subset. For Amazon$^-$, and Amazon$^+$ datasets, we calculate ϵ^{PEHE}, which represents the error in estimation of $\tau(X_i = x_i, S_i = s_i)$ and is defined as $\epsilon^{\text{PEHE}} = \frac{1}{n} \sum_{i=1}^{n} (\hat{\tau}(X_i = x_i, S_i = s_i) - \hat{\tau}(X_i = x_i, S_i = s_i))^2$. Further, we compute ϵ^{ATE}, which represents the error in ATE estimation under interference, and is defined as $\epsilon^{\text{ATE}} = |\frac{1}{n} \sum_{i=1}^{n} \tau(X_i = x_i, S_i = s_i) - \frac{1}{n} \sum_{i=1}^{n} \hat{\tau}(X_i = x_i, S_i = s_i)|$.

The implementation details are elaborated in Appendix C.

5.2   Results

To verify the claim that existing unknown interference and the proposed methods can address unknown interference, we conducted experiments on the Jobs, Amazon$^-$, and Amazon$^+$ datasets. The Amazon$^-$ and Amazon$^+$ datasets contain interference and provide graph structures [17,19]. We hid the graph structure to simulate the unknown interference. Table 1 shows that the proposed UNITE yields better results than the baseline methods on the Jobs. Comparing the results of the UNITE on the Jobs dataset with those of the baseline methods (such as CFR) that ignore interference, we can observe that considering addressing interference can help improve the performance of the ITE and ATE estimation. It implies that there might be interference between units even if there is no observed information on the connection between units. The results also reveal that UNITE achieves lower errors in ITE estimation on the Amazon$^-$ and Amazon$^+$ datasets than the other baseline methods. Meanwhile, the error in

Table 1. Results on the test sets of the Jobs, Amazon$^-$, and Amazon$^+$ datasets under unknown interference. The proposed UNITE outperforms the baseline methods in the ITE estimation under unknown interference.

Method	Jobs		Amazon$^-$		Amazon$^+$	
	$\hat{R}^{\text{pol}}(\pi_f)$	ϵ^{ATT}	ϵ^{PEHE}	ϵ^{ATE}	ϵ^{PEHE}	ϵ^{ATE}
TARNet	0.22 ± 0.04	0.05 ± 0.02	1.79 ± 0.01	0.24 ± 0.01	1.80 ± 0.02	0.06 ± 0.00
BNN-2-2	0.26 ± 0.07	0.07 ± 0.01	2.01 ± 0.02	0.17 ± 0.01	1.93 ± 0.01	0.06 ± 0.01
BNN-4-0	0.23 ± 0.05	0.07 ± 0.00	1.96 ± 0.00	$\mathbf{0.06 \pm 0.04}$	1.92 ± 0.00	$\mathbf{0.04 \pm 0.01}$
CFR-MMD	0.21 ± 0.05	0.10 ± 0.05	1.79 ± 0.01	0.25 ± 0.02	1.84 ± 0.00	0.07 ± 0.01
CFR-Wass	0.21 ± 0.04	0.09 ± 0.05	1.79 ± 0.01	0.24 ± 0.01	1.85 ± 0.02	0.05 ± 0.00
GCN$_\phi$ + k-NN	0.22 ± 0.04	0.06 ± 0.01	1.87 ± 0.04	$\mathbf{0.06 \pm 0.01}$	1.85 ± 0.03	$\mathbf{0.04 \pm 0.03}$
GCN$_\psi$ + k-NN	0.22 ± 0.04	0.06 ± 0.01	1.90 ± 0.03	0.19 ± 0.03	1.85 ± 0.01	0.17 ± 0.02
UNITE (ours)	$\mathbf{0.19 \pm 0.03}$	$\mathbf{0.04 \pm 0.02}$	$\mathbf{1.57 \pm 0.06}$	0.07 ± 0.03	$\mathbf{1.72 \pm 0.01}$	0.07 ± 0.02

ATE estimation is comparable to the other baseline methods. These results verify that the proposed methods can address unknown interference. In addition, these results indicate that the GCN-based methods with k-NN graphs cannot correctly address the unknown interference, as they cannot achieve better performance in the ITE estimation than other baseline methods, such as CFR-MMD.

We also conducted ablation experiments, which are presented in Appendix D.

6   Conclusion

In this study, we proposed UNITE, which models unknown interference and estimates the ITE in the presence of unknown interference. We performed and evaluated the proposed method on three public datasets to verify the effectiveness of the proposed method. The results indicate that UNITE is powerful in addressing unknown interference.

Currently, the proposed method is limited to the binary treatment setting; extending the setting to multi-valued treatments is a possible future work.

Acknowledgements. This study was supported by JSPS KAKENHI Grant Number 20H04244.

A   Identifiability of the Expectation of Potential Outcomes

Here, we show that the expectation of potential outcomes $Y_i^t(s_i)$ can be identified from observed data. To this end, we need some assumptions.

Inspired by Chen and Ji [3] who learn graph structures using the features of units, we have the following assumption.

Assumption 1 *(A1). The unknown structure $\mathbf{A}^{\mathrm{itf}}$ can be discovered from the $\mathbf{X}$ and $\mathbf{T}$ using a graph structure learner $GSL(\cdot)$, i.e., $\mathbf{A}^{\mathrm{itf}} = GSL(\mathbf{X}, \mathbf{T})$.*

Similar to the existing studies on addressing interference [16,17], we assume the interference can be aggregated, as the following assumption.

Assumption 2 *(A2). There exists an aggregation function $\psi(\cdot)$, which can aggregate information of other units on the graph $\mathbf{A}^{\mathrm{itf}}$ while outputting the $\mathbf{s}$, i.e., $s_i = \psi(\mathbf{T}_{-i}, \mathbf{X}_{-i}, \mathbf{A}^{\mathrm{itf}})$.*

We extend the neighbor interference assumption [4] to the networked interference [17].

Assumption 3 *(A3). For $\forall i$, $\forall \mathbf{T}_{-i}, \mathbf{T}'_{-i}, \forall \mathbf{X}_{-i}, \mathbf{X}'_{-i}$, and $\forall \mathbf{A}^{\mathrm{itf}}, \mathbf{A}^{\mathrm{itf}'}$: when $s_i = s'_i$, i.e., $\psi(\mathbf{T}_{-i}, \mathbf{X}_{-i}, \mathbf{A}^{\mathrm{itf}}) = \psi(\mathbf{T}'_{-i}, \mathbf{X}'_{-i}, \mathbf{A}^{\mathrm{itf}'})$, $Y_i^t(S_i = s_i) = Y_i^t(S_i = s'_i)$ holds.*

We use the consistency assumption which is similar to the consistency assumption in the existing study on interference [4].

Assumption 4 *(A4)*. $Y_i^{obs} = Y_i^{t_i}(S_i = s_i)$ *on the graph* $\boldsymbol{A}$ *for a unit* i *under* t_i *and* s_i.

We take a similar unconfoundedness assumption of the existing study on addressing interference [4].

Assumption 5 *(A5)*. *There is no hidden confounder. For any unit* i, *given the covariates, the treatment assignment and output of the aggregation function are independent of potential outcomes, i.e.,* $T_i, S_i \perp\!\!\!\perp Y_i^1(s_i), Y_i^0(s_i)|X_i$.

We now prove the identification of the expectation of potential outcomes $Y_i^t(s_i)$ ($t = 1$ or $t = 0$) as follows:

$$\mathbb{E}[Y_i^{obs}|X_i = \boldsymbol{x}_i, T_i = t, X_{-i} = \boldsymbol{X}_{-i}, T_{-i} = \boldsymbol{T}_{-i}, X = \boldsymbol{X}, T = \boldsymbol{T}]$$

$$= \mathbb{E}[Y_i^{obs}|X_i = \boldsymbol{x}_i, T_i = t, X_{-i} = \boldsymbol{X}_{-i}, T_{-i} = \boldsymbol{T}_{-i}, A^{\text{itf}} = \boldsymbol{A}^{\text{itf}}] \quad (A1)$$

$$= \mathbb{E}[Y_i^{obs}|X_i = \boldsymbol{x}_i, T_i = t, S_i = s_i] \quad\quad (A2)$$

$$= \mathbb{E}[Y_i^t(s_i)|X_i = \boldsymbol{x}_i, T_i = t, S_i = s_i] \quad\quad (A3, A4)$$

$$= \mathbb{E}[Y_i^t(s_i)|X_i = \boldsymbol{x}_i]. \quad\quad (A5)$$

Inspired by the above proof, if unknown interference is properly modeled, the potential outcomes can be modeled.

B HSIC

The calculation of HSIC is as follows:

$$\text{HSIC}(\boldsymbol{B}, \boldsymbol{C}) = \frac{1}{(n-1)^2} \text{tr}(\boldsymbol{KHLH}), \quad \boldsymbol{H} = \boldsymbol{I}_n - \frac{1}{n}\boldsymbol{1}_n \boldsymbol{1}_n^{\text{T}}, \quad (2)$$

where $\boldsymbol{B} \in \mathbb{R}^{n \times d_1}$ and $\boldsymbol{C} \in \mathbb{R}^{n \times d_2}$ denote two different matrices or vectors, $\boldsymbol{I}_n$ is the identity matrix, $\boldsymbol{1}_n$ is a vector of all ones, and $\cdot^{\text{T}}$ is the transposition operation. $\boldsymbol{K}$ and $\boldsymbol{L}$ are Gaussian kernels applied to $\boldsymbol{B}$ and $\boldsymbol{C}$, respectively:

$$K_{ij} = \exp\left(-\frac{\|\boldsymbol{b}_i - \boldsymbol{b}_j\|_2^2}{2}\right), \quad L_{ij} = \exp\left(-\frac{\|\boldsymbol{c}_i - \boldsymbol{c}_j\|_2^2}{2}\right), \quad (3)$$

where vectors $\boldsymbol{b}_i$ (or $\boldsymbol{b}_j$) and $\boldsymbol{c}_i$ (or $\boldsymbol{c}_j$) represent the elements of the i-th (or j-th) row of $\boldsymbol{B}$ and $\boldsymbol{C}$, respectively.

C Implementation Details

Following Ma et al. [17], the entire $\boldsymbol{X}$ and $\boldsymbol{T}$ are given during the training, validation, and testing phases. However, only the observed outcomes of the units in the training set are provided during the training phase.

The three hidden layers of the representation layers of UNITE have $(128, 64, 64)$-dimensions for the Jobs, Amazon$^-$, and Amazon$^+$ datasets. The hidden layer of the GSL of the UNITE has a 128 dimension for the Jobs dataset and 64 for the Amazon$^-$ and Amazon$^+$ datasets. The hidden layers of the GCN layers of the UNITE have $(64, 32)$-dimensions for the Jobs dataset and $(64, 64, 32)$-dimensions for the Amazon$^-$ and Amazon$^+$ datasets. The hidden layers of the prediction networks of the UNITE have $(128, 64, 64)$-dimensions for the Jobs dataset and $(128, 64, 32)$-dimensions for the Amazon$^-$ and Amazon$^+$ datasets.

In addition, we train our models with the GPU of the NVIDIA RTX A5000. The UNITE utilizes the Adam optimizer with 500 training iterations for the Jobs dataset and 2,000 training iterations for the Amazon$^-$, and Amazon$^+$ datasets. The learning rate is set to $\lambda_{lr} = 0.0005$ for the Jobs dataset and $\lambda_{lr} = 0.001$ for the Amazon$^-$, and Amazon$^+$ datasets, and the weight decay γ is set to $\gamma = 0.01$ for the Jobs dataset and $\gamma = 0.001$ for the Amazon$^-$, and Amazon$^+$ datasets.

We use the grid search method to search for hyperparameters by checking the results on the validation set. The train batch size of the UNITE is the full batch of training units for the Jobs and is searched from $\{128, 256, 512, 1024\}$ for the Amazon$^-$, and Amazon$^+$ datasets. The β_1 and β_2 of the UNITE are searched from $\{0.01, 0.05, 0.1, 0.15, 0.2\}$ for the Jobs and $\{0.1, 0.5, 1.0, 1.5, 2.0\}$ for the Amazon$^-$ and Amazon$^+$ datasets. The λ of the UNITE is 0.0005 for the Jobs datasets, 0.02 for the Amazon$^-$ datasets, and 0.01 for the Amazon$^+$ datasets. Dropout is applied for the UNITE (the dropout rate is searched from $\{0.0, 0.1, 0.5\}$). Moreover, after 400 iterations, early stopping is applied for the Amazon$^-$, and Amazon$^+$ datasets to avoid overfitting. The UNITE-KG uses the same hyperparameters as the UNITE, but it sets the $\lambda = 0$.

Here, all the baseline methods use the default hyperparameter or search for hyperparameters from the ranges suggested in the literature. To avoid overfitting, all the baseline methods apply early stopping on the Amazon$^-$, and Amazon$^+$ datasets.

For the Amazon$^-$, and Amazon$^+$ datasets, as the y values span a wide range, z-score normalization is applied during training and testing.

D   Ablation Experiments

To investigate the importance of different regularization terms, we conduct ablation experiments on the Jobs and Amazon$^-$ datasets. Table 2 shows the results of the ablation experiments. We can observe that removing the HSIC$_{\phi_1}$, or HSIC$_{\phi_2}$, or L_0 results in performance degradation for the ITE estimation. This verifies that these regularization terms are important for the ITE estimation of UNITE. In addition, the results on the Jobs dataset show the performance degradation in the interference-free ITE and ATE estimation when UNITE removes the HSIC$_{\phi_2}$. This implies that the HSIC$_{\phi_2}$ is important to estimate the interference-free ITE and ATE estimation.

Table 2. Results of the ablation experiments on the Jobs and Amazon$^-$ datasets.

Method	Jobs		Amazon$^-$	
	$\hat{R}^{pol}(\pi_f)$	ϵ^{ATT}	ϵ^{PEHE}	ϵ^{ATE}
UNITE	$\mathbf{0.19 \pm 0.03}$	$\mathbf{0.04 \pm 0.02}$	$\mathbf{1.57 \pm 0.06}$	0.07 ± 0.03
w/o HSIC$_{\phi_1}$	0.22 ± 0.03	0.08 ± 0.04	1.61 ± 0.08	0.38 ± 0.06
w/o HSIC$_{\phi_2}$	0.21 ± 0.03	0.08 ± 0.04	1.58 ± 0.07	0.07 ± 0.04
w/o L_0	0.21 ± 0.03	0.08 ± 0.05	1.60 ± 0.04	$\mathbf{0.05 \pm 0.03}$

References

1. Aronow, P.M., Samii, C.: Estimating average causal effects under general interference, with application to a social network experiment. Ann. Appl. Stat. **11**, 1912–1947 (2017)
2. Bhattacharya, R., Malinsky, D., Shpitser, I.: Causal inference under interference and network uncertainty. In: Proceedings of the 35th Uncertainty in Artificial Intelligence Conference, vol. 2019 (2019)
3. Chen, Y., Wu, L., Zaki, M.: Iterative deep graph learning for graph neural networks: Better and robust node embeddings. In: Advances in Neural Information Processing Systems, vol. 33, pp. 19314–19326 (2020)
4. Forastiere, L., Airoldi, E.M., Mealli, F.: Identification and estimation of treatment and interference effects in observational studies on networks. J. Am. Stat. Assoc. **116**(534), 901–918 (2021)
5. Gretton, A., Bousquet, O., Smola, A., Schölkopf, B.: Measuring statistical dependence with Hilbert-Schmidt norms. In: Proceedings of the 16th International Conference on Algorithmic Learning Theory, pp. 63–77 (2005)
6. Guo, R., Li, J., Li, Y., Candan, K.S., Raglin, A., Liu, H.: Ignite: a minimax game toward learning individual treatment effects from networked observational data. In: Proceedings of the Twenty-Ninth International Conference on International Joint Conferences on Artificial Intelligence, pp. 4534–4540 (2021)
7. Guo, R., Li, J., Liu, H.: Learning individual causal effects from networked observational data. In: Proceedings of the 13th International Conference on Web Search and Data Mining, pp. 232–240 (2020)
8. He, R., McAuley, J.: Ups and downs: modeling the visual evolution of fashion trends with one-class collaborative filtering. In: Proceedings of the 25th International Conference on World Wide Web, pp. 507–517 (2016)
9. Hudgens, M.G., Halloran, M.E.: Toward causal inference with interference. J. Am. Stat. Assoc. **103**(482), 832–842 (2008)
10. Johansson, F., Shalit, U., Sontag, D.: Learning representations for counterfactual inference. In: Proceedings of the 33rd International Conference on Machine Learning, vol. 48, pp. 3020–3029 (2016)
11. LaLonde, R.J.: Evaluating the econometric evaluations of training programs with experimental data. Am. Econ. Rev. 604–620 (1986)
12. Li, Q., Wang, Z., Liu, S., Li, G., Xu, G.: Deep treatment-adaptive network for causal inference. Int. J. Very Large Data Bases, 1–16 (2022)

13. Lin, X., Zhang, G., Lu, X., Bao, H., Takeuchi, K., Kashima, H.: Estimating treatment effects under heterogeneous interference. In: Koutra, D., Plant, C., Gomez Rodriguez, M., Baralis, E., Bonchi, F. (eds.) ECML PKDD 2023. LNCS, vol. 14169, pp. 576–592. Springer, Cham (2023). https://doi.org/10.1007/978-3-031-43412-9_34
14. Liu, L., Hudgens, M.G.: Large sample randomization inference of causal effects in the presence of interference. J. Am. Stat. Assoc. **109**(505), 288–301 (2014)
15. Louizos, C., Welling, M., Kingma, D.P.: Learning sparse neural networks through L_0 regularization. In: Proceedings of the 6th International Conference on Learning Representations (2018)
16. Ma, J., Wan, M., Yang, L., Li, J., Hecht, B., Teevan, J.: Learning causal effects on hypergraphs. In: Proceedings of the 28th ACM SIGKDD Conference on Knowledge Discovery and Data Mining, pp. 1202–1212 (2022)
17. Ma, Y., Tresp, V.: Causal inference under networked interference and intervention policy enhancement. In: Proceedings of the 24th International Conference on Artificial Intelligence and Statistics, vol. 130, pp. 3700–3708 (2021)
18. Nabi, R., Pfeiffer, J., Charles, D., Kıcıman, E.: Causal inference in the presence of interference in sponsored search advertising. Front. Big Data **5** (2022)
19. Rakesh, V., Guo, R., Moraffah, R., Agarwal, N., Liu, H.: Linked causal variational autoencoder for inferring paired spillover effects. In: Proceedings of the 27th ACM International Conference on Information and Knowledge Management, pp. 1679–1682 (2018)
20. Raudenbush, S.W., Schwartz, D.: Randomized experiments in education, with implications for multilevel causal inference. Annu. Rev. Stat. Appl. **7**(1) (2020)
21. Rubin, D.B.: Randomization analysis of experimental data: the fisher randomization test comment. J. Am. Stat. Assoc. **75**(371), 591–593 (1980)
22. Schnitzer, M.E.: Estimands and estimation of COVID-19 vaccine effectiveness under the test-negative design: connections to causal inference. Epidemiology **33**(3), 325 (2022)
23. Shalit, U., Johansson, F.D., Sontag, D.: Estimating individual treatment effect: generalization bounds and algorithms. In: Proceedings of the 34th International Conference on Machine Learning, vol. 70, pp. 3076–3085 (2017)
24. Smith, J.A., Todd, P.E.: Does matching overcome LaLonde's critique of nonexperimental estimators? J. Econometrics **125**(1–2), 305–353 (2005)
25. Sävje, F., Aronow, P.M., Hudgens, M.G.: Average treatment effects in the presence of unknown interference. Ann. Stat. **49**(2), 673–701 (2021)
26. Tchetgen, E.J.T., VanderWeele, T.J.: On causal inference in the presence of interference. Stat. Methods Med. Res. **21**(1), 55–75 (2012)
27. Veličković, P., Cucurull, G., Casanova, A., Romero, A., Liò, P., Bengio, Y.: Graph attention networks. In: Proceedings of the 6th International Conference on Learning Representations (2018)
28. Welling, M., Kipf, T.N.: Semi-supervised classification with graph convolutional networks. In: Proceedings of the 4th International Conference on Learning Representations (2016)
29. Ye, Y., Ji, S.: Sparse graph attention networks. IEEE Trans. Knowl. Data Eng. (2021)

A New Loss for Image Retrieval: Class Anchor Margin

Alexandru Ghiţă and Radu Tudor Ionescu[✉]

Department of Computer Science, University of Bucharest, Bucharest, Romania
`raducu.ionescu@gmail.com`

Abstract. The performance of neural networks in content-based image retrieval (CBIR) is highly influenced by the chosen loss (objective) function. The majority of objective functions for neural models can be divided into metric learning and statistical learning. Metric learning approaches require a pair mining strategy that often lacks efficiency, while statistical learning approaches are not generating highly compact features due to their indirect feature optimization. To this end, we propose a novel repeller-attractor loss that falls in the metric learning paradigm, yet directly optimizes for the L_2 metric without the need of generating pairs. Our loss is formed of three components. One leading objective ensures that the learned features are attracted to each designated learnable class anchor. The second loss component regulates the anchors and forces them to be separable by a margin, while the third objective ensures that the anchors do not collapse to zero. Furthermore, we develop a more efficient two-stage retrieval system by harnessing the learned class anchors during the first stage of the retrieval process, eliminating the need of comparing the query with every image in the database. We establish a set of three datasets (CIFAR-100, Food-101, and ImageNet-200) and evaluate the proposed objective on the CBIR task, by using both convolutional and transformer architectures. Compared to existing objective functions, our empirical evidence shows that the proposed objective is generating superior and more consistent results.

Keywords: Loss Function · Image Retrieval · Deep Neural Networks

1 Introduction

Content-based image retrieval (CBIR) is a challenging task that aims to retrieve images from a database that are most similar to a query image. The state-of-the-art image retrieval systems addressing the aforementioned task are currently based on deep neural networks [3,7,18,29,30,41]. Although neural models obtained impressive performance levels compared to handcrafted CBIR models [27], one of the main challenges that remain to be solved in training neural networks for retrieval problems is the choice of the objective function. Indeed, a well-chosen objective function should enhance the discriminative power of the

D.-N. Yang et al. (Eds.): PAKDD 2024, LNAI 14646, pp. 43–54, 2024.
https://doi.org/10.1007/978-981-97-2253-2_4

learned embeddings, i.e. the resulting features should exhibit small differences for images representing the same object and large differences for images representing different objects. This would implicitly make the embeddings more suitable for image retrieval.

The majority of loss functions used to optimize neural models can be separated into two major categories: statistical learning and metric learning. The most popular category is represented by statistical learning, which includes objective functions such as the cross-entropy loss [24] or the hinge loss [5]. The optimization problem is usually based on minimizing a particular probability distribution. Consequently, objective functions in this category achieve the desired properties as an indirect effect during optimization. Hence, objective functions based on statistical learning are more suitable for modeling multi-class classification tasks rather than retrieval tasks. The second category corresponds to metric learning, which comprises objective functions such as contrastive loss [12], triplet loss [32], and quadruplet loss [4]. Objective functions in this category operate directly in the embedding space and optimize for the desired distance metric. However, they usually require forming tuples of examples to compute the loss, which can be a costly step in terms of training time. To reduce the extra time required to build tuples, researchers resorted to hard example mining schemes [10,11,13,32,34,40]. Still, statistical learning remains more time efficient.

In this context, we propose a novel repeller-attractor objective function that falls in the metric learning paradigm. Our loss directly optimizes for the L_2 metric without needing to generate pairs, thus alleviating the necessity of performing the costly hard example mining. The proposed objective function achieves this through three interacting components, which are expressed with respect to a set of learnable embeddings, each representing a class anchor. The leading loss function ensures that the data embeddings are attracted to the designated class anchor. The second loss function regulates the anchors and forces them to be separable by a margin, while the third objective ensures that the anchors do not collapse to the origin. In addition, we propose a two-stage retrieval method that compares the query embedding with the class anchors in the first stage, then continues by comparing the query embedding with image embeddings assigned to the nearest class anchors.

We carry out experiments on three datasets (CIFAR-100 [17], Food-101 [2], and ImageNet-200 [31]) to compare the proposed loss function against representative statistical and deep metric learning objectives. We evaluate the objectives on the CBIR task, by using both convolutional and transformer architectures, such as residual networks (ResNets) [14] and shifted windows (Swin) transformers [21]. Compared to existing loss functions, our empirical results show that the proposed objective is generating higher and more consistent performance levels across the considered evaluation scenarios. Furthermore, we conduct an ablation study to demonstrate the influence of each loss component on the overall performance.

In summary, our contribution is twofold:

- We introduce a novel repeller-attractor objective function that directly optimizes for the L_2 metric, alleviating the need to generate pairs via hard example mining or alternative mining strategies.
- We conduct comprehensive experiments to compare the proposed loss function with popular loss choices on multiple datasets.

2 Related Work

As related work, we refer to studies introducing new loss functions, which are related to our first contribution, and new content-based image retrieval methods, which are connected to our second contribution.

Loss Functions. For retrieval systems based on neural networks, the choice of the objective function is the most important factor determining the geometry of the resulting feature space [35]. We hereby discuss related work proposing various loss functions aimed at generating effective embedding spaces.

Metric learning objective functions directly optimize a desired metric and are usually based on pairs or tuples of known data samples [4,33,36,43]. One of the earliest works on metric learning proposed the contrastive loss [12], which was introduced as a method to reduce the dimensionality of the input space, while preserving the separation of feature clusters. The idea behind contrastive loss is to generate an attractor-repeller system that is trained on positive and negative pairs generated from the available data. The repelling can happen only if the distance between a negative pair is smaller than a margin m. In the context of face identification, another successful metric learning approach is triplet loss [32]. It obtains the desired properties of the embedding space by generating triplets of anchor, positive and negative examples. For each triplet, the proposed objective enforces the distance between the anchor and the positive example to be larger than the distance between the anchor and the negative example, by a margin m. Other approaches introduced objectives that directly optimize the AUC [9], recall [26] or AP [30]. The main issues when optimizing with loss functions based on metric learning are the usually slow convergence [33] and the difficulty of generating useful example pairs or tuples [34]. In contrast, our method does not require mining strategies and, as shown in the experiments, it converges much faster than competing losses. The usual example mining strategies are hard, semi-hard, and online negative mining [32,40]. In hard negative mining, for each anchor image, we need to construct pairs with the farthest positive example and the closest negative example. This adds an extra computational step at the beginning of every training epoch, extending the training time. Similar problems arise in the context of semi-hard negative mining, while the difference consists in the mode in which the negatives are sampled.

Statistical learning objective functions indirectly optimize the learned features of the neural network. Popular objectives are based on some variation of the cross-entropy loss [6,8,19,20,37,38] or the cosine loss [1]. By optimizing such functions, the model is forced to generate features that are close to the direction of the class center. For example, ArcFace [6] reduces the optimization space to an n-dimensional hypersphere by normalizing both the embedding generated by the encoder, and the corresponding class weight from the classification head, using an Euclidean norm.

Hybrid objective functions promise to obtain better embeddings by minimizing a statistical objective function in conjunction with a metric learning objective [15,22]. For example, Center Loss [39] minimizes the intra-class distances of the learned features by using cross-entropy in conjunction with an attractor for each sample to its corresponding class center. Another approach [44] similar to Center Loss [39] is based on predefined evenly distributed class centers. A more complex approach [3] is to combine the standard cross-entropy with a cosine classifier and a mean squared error regression term to jointly enhance global and local features.

Both contrastive and triplet loss objectives suffer from the need of employing pair mining strategies, but in our case, mining strategies are not required. The positive pairs are built online for each batch, between each input feature vector and the dedicated class anchor, the number of positive pairs thus being equal to the number of examples. The negative pairs are constructed only between the class centers, thus alleviating the need of searching for a good negative mining strategy, while also significantly reducing the number of negative pairs. To the best of our knowledge, there are no alternative objective functions for CBIR that use dedicated self-repelling learnable class anchors acting as attraction poles for feature vectors belonging to the respective classes.

Content-Based Image Retrieval Methods. CBIR systems are aimed at finding similar images with a given query image, matching the images based on the similarity of their scenes or the contained objects. Images are encoded using a descriptor (or encoder), and a system is used to sort a database of encoded images based on a similarity measure between queries and images. In the context of content-based image retrieval, there are two types of image descriptors. On the one hand, there are general descriptors [29], where a whole image is represented as a feature vector. On the other hand, there are local descriptors [41] where portions of an image are represented as feature vectors. Hybrid descriptors [3] are also used to combine both global and local features. To improve the quality of the results retrieved by learned global descriptors, an additional verification step is often employed. This step is meant to re-rank the retrieved images by a precise evaluation of each candidate image [28]. The re-ranking step is usually performed with the help of an additional system, and in some cases, it can be directly integrated with the general descriptor [18]. In the CBIR task, one can search for visually similar images as a whole, or search for images that contain similar regions [27] of a query image. In this work, we focus on building global descriptors that match whole images. Further approaches based on metric learning, statistical learning, or hand-engineered features are discussed in the recent survey of Dubey et al. [7]. Different from other CBIR methods, we propose a novel two-stage retrieval system that leverages the use of the class anchors learned through the proposed loss function to make the retrieval process more efficient and effective.

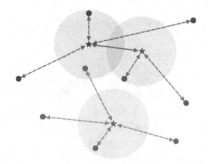

Fig. 1. An example showing the behavior of the attractor-repeller loss components for three classes. The stars represent the class anchors C. Faded circles around class anchors represent the sphere of radius m around each anchor. Solid circles represent feature vectors generated by the encoder f_θ. Dashed arrows between feature vectors and class anchors represent the attraction of the force generated by the attractor $\mathcal{L}_A$. The solid red arrow between class anchors represents the repelling force generated by the repeller $\mathcal{L}_R$. Best viewed in color. (Color figure online)

3   Method

Overview. Our objective function consists of three components, each playing a different role in obtaining a discriminative embedding space. All three loss components are formulated with respect to a set of learnable class anchors (centroids). The first loss component acts as an attraction force between each input embedding and its corresponding class anchor. Its main role is to draw the embeddings representing the same object to the corresponding centroid, thus creating embedding clusters of similar images. Each center can be seen as a magnet with a positive charge and its associated embeddings as magnets with negative charges, thus creating attraction forces between anchors and data samples of the same class. The second loss component acts as a repelling force between class anchors. In this case, the class centers can be seen as magnets with similar charges. If brought together, they will repel each other, and if they lie at a certain distance, the repelling force stops. The last component acts similarly to the second one, with the difference that an additional magnet is introduced and fixed at the origin of the embedding space. Its main effect is to push the class centers away from the origin.

Notations. Let $\mathbf{x}_i \in \mathbb{R}^{h \times w \times c}$ be an input image and $y_i \in \mathbb{N}$ its associated class label, $\forall i \in \{1, 2, ..., l\}$. We aim to optimize a neural encoder f_θ which is parameterized by the learnable weights θ to produce a discriminative embedding space. Let $\mathbf{e}_i \in \mathbb{R}^n$ be the n-dimensional embedding vector of the input image $\mathbf{x}_i$ generated by f_θ, i.e. $\mathbf{e}_i = f_\theta(\mathbf{x}_i)$. In order to employ our novel loss function, we need to introduce a set of learnable class anchors $C = \{\mathbf{c}_1, \mathbf{c}_2, ..., \mathbf{c}_t\}$, where $\mathbf{c}_j \in \mathbb{R}^n$ resides in the embedding space of f_θ, and t is the total number of classes.

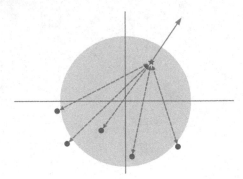

Fig. 2. Contribution of the minimum norm loss $\mathcal{L}_N$ imposed on the class anchors. The blue star represents a class anchor. Solid circles represent embedding vectors generated by the encoder f_θ. Dashed arrows represent the attraction force generated by the attractor $\mathcal{L}_A$. The solid gray line represents the direction in which the anchor is pushed away from the origin due to the component $\mathcal{L}_N$. Best viewed in color.

Loss Components. With the above notations, we can now formally define our first component, the attractor loss $\mathcal{L}_A$, as follows:

$$\mathcal{L}_A(\mathbf{x}_i, C) = \frac{1}{2}\|\mathbf{e}_i - \mathbf{c}_{y_i}\|_2^2. \tag{1}$$

The main goal of this component of the objective is to cluster feature vectors as close as possible to their designated class anchor by minimizing the distance between $\mathbf{e}_i$ and the corresponding class anchor $\mathbf{c}_{y_i}$. Its effect is to enforce low intra-class variance. However, obtaining low intra-class variance is only a part of what we aim to achieve. The first objective has little influence over the inter-class similarity, reducing it only indirectly. Therefore, another objective is required to repel samples from different classes. As such, we introduce the repeller loss $\mathcal{L}_R$, which is defined as follows:

$$\mathcal{L}_R(C) = \frac{1}{2} \sum_{y,y' \in Y, y \neq y'} \{\max(0, 2 \cdot m - \|\mathbf{c}_y - \mathbf{c}_{y'}\|)\}^2, \tag{2}$$

where y and y' are two distinct labels from the set of ground-truth labels Y, and $m > 0$ is a margin representing the radius of an n-dimensional sphere around each anchor, in which no other anchor should lie. The goal of this component is to push anchors away from each other during training to ensure high inter-class distances. The margin m is used to limit the repelling force to an acceptable margin value. If we do not set a maximum margin, then the repelling force can push the anchors too far apart, and the encoder could struggle to learn features that satisfy the attractor loss defined in Eq. (1).

A toy example of the attractor-repeller mechanism is depicted in Fig. 1. Notice how the optimization based on the attractor-repeller objective tends to pull data samples from the same class together (due to the attractor), and push samples from different classes away (due to the repeller). However, when the training begins, all data samples start from a location close to the origin of the embedding space, essentially having a strong tendency to pull the class anchors to the origin. To ensure that the anchors do not collapse to the origin (as observed in some of our preliminary experiments), we introduce an additional objective that imposes a minimum norm on the class anchors. The minimum norm loss $\mathcal{L}_N$ is defined as:

$$\mathcal{L}_N(C) = \frac{1}{2} \sum_{y \in Y} \left\{ \max\left(0, p - \|\mathbf{c}_y\|\right) \right\}^2, \tag{3}$$

where p is the minimum norm that each anchor must have. This objective contributes to our full loss function as long as at least one class anchor is within a distance of p from the origin. Figure 2 provides a visual interpretation of the effect induced by the minimum norm loss. Notice how the depicted class anchor is pushed away the origin (due to the minimum norm loss), while the data samples belonging to the respective class move along with their anchor (due to the attractor loss).

Assembling the three loss components presented above into a single objective leads to the proposed class anchor margin (CAM) loss $\mathcal{L}_{CAM}$, which is formally defined as follows:

$$\mathcal{L}_{CAM}(\mathbf{x}, C) = \mathcal{L}_A(\mathbf{x}, C) + \mathcal{L}_R(C) + \mathcal{L}_N(C). \tag{4}$$

Notice that only $\mathcal{L}_A$ is directly influenced by the training examples, while $\mathcal{L}_R$ and $\mathcal{L}_N$ operate only on the class anchors. Hence, negative mining strategies are not required at all.

4   Experiments

In the experiments, we compare our class anchor margin loss with the cross-entropy and the contrastive learning losses on three datasets, considering both convolutional and transformer models.

4.1   Datasets

We perform experiments on three datasets: CIFAR-100 [17], Food-101 [2], and ImageNet-200 [31]. CIFAR-100 contains 50,000 training images and 10,000 test images belonging to 100 classes. Food-101 is composed of 101,000 images from 101 food categories. The official split has 750 training images and 250 test images per category. ImageNet-200 is a subset of ImageNet-1K, which contains 100,000 training images, 25,000 validation images and 25,000 test images from 200 classes.

Table 1. Retrieval performance levels of ResNet-18 (RN-18), ResNet-50 (RN-50), ResNet-101 (RN-101) and Swin-T models on the CIFAR-100 [17], Food-101 [2] and ImageNet-200 [31] datasets, while comparing the cross-entropy (CE) loss, the contrastive learning (CL) loss, and the proposed class anchor margin (CAM) loss. The best score for each architecture and each metric is highlighted in bold.

	Loss	CIFAR-100			Food-101			ImageNet-200		
		mAP	P@20	P@100	mAP	P@20	P@100	mAP	P@20	P@100
RN-18	CE	.249±.003	.473±.005	.396±.005	.234±.002	.547±.002	.459±.001	.130±.001	.340±.001	.262±.001
	CL	.220±.013	.341±.010	.303±.011	.025±.001	.042±.004	.038±.003	.070±.012	.139±.017	.117±.016
	CAM	**.622**±.005	**.560**±.007	**.553**±.007	**.751**±.001	**.676**±.003	**.669**±.003	**.508**±.004	**.425**±.004	**.418**±.005
RN-50	CE	.211±.006	.454±.007	.366±.006	.158±.004	.471±.005	.370±.005	.088±.003	.292±.006	.209±.005
	CL	.164±.016	.271±.017	.240±.016	.019±.000	.030±.000	.028±.001	.005±.000	.008±.001	.007±.000
	CAM	**.640**±.008	**.581**±.009	**.578**±.009	**.765**±.005	**.697**±.008	**.697**±.009	**.543**±.003	**.472**±.002	**.468**±.002
RN-101	CE	.236±.009	.482±.008	.397±.009	.160±.003	.479±.008	.376±.007	.093±.002	.299±.002	.216±.002
	CL	.034±.028	.069±.051	.056±.044	.014±.002	.018±.004	.017±.003	.006±.001	.007±.002	.007±.002
	CAM	**.629**±.006	**.575**±.008	**.573**±.008	**.758**±.007	**.690**±.007	**.693**±.006	**.529**±.005	**.458**±.006	**.455**±.006
Swin-T	CE	.661±.004	**.808**±.001	.770±.001	.617±.006	.817±.002	.777±.003	.560±.006	.743±.001	.707±.003
	CL	.490±.010	.629±.007	.584±.008	.495±.004	.667±.001	.633±.002	.223±.002	.397±.001	.338±.001
	CAM	**.808**±.004	.794±.003	**.795**±.004	**.873**±.001	**.858**±.001	**.861**±.001	**.761**±.002	**.745**±.003	**.749**±.004

4.2 Experimental Setup

As underlying neural architectures, we employ three ResNet [14] model variations (ResNet-18, ResNet-50, ResNet-101) and a Swin transformer [21] model (Swin-T). We rely on the PyTorch [25] library together with Hydra [42] to implement and test the models.

We apply random weight initialization for all models, except for the Swin transformer, which starts from the weights pre-trained on the ImageNet Large Scale Visual Recognition Challenge (ILSVRC) dataset [31]. We employ the Adam [16] optimizer to train all models, regardless of their architecture. For the residual neural models, we set the learning rate to 10^{-3}, while for the Swin transformer, we use a learning rate of 10^{-4}. The residual nets are trained from scratch for 100 epochs, while the Swin transformer is fine-tuned for 30 epochs. For the lighter models (ResNet-18 and ResNet-50), we use a mini-batch size of 512. Due to memory constraints, the mini-batch size is set to 64 for the Swin transformer, and 128 for ResNet-101. Residual models are trained with a linear learning rate decay with a factor of 0.5. We use a patience of 6 epochs in all experiments. Fine-tuning the Swin transformer does not employ a learning rate decay. Input images are normalized to have all pixel values in the range of $[0, 1]$ by dividing the values by 255. The inputs of the Swin transformer are standardized with the image statistics from ILSVRC [31].

We use several data augmentation techniques such as random crop with a padding of 4 pixels, random horizontal flip, color jitter, and random affine transformations (rotations, translations). Moreover, for the Swin transformer, we add the augmentations described in [23].

For the models optimized either with the cross-entropy loss or our class anchor margin loss, the target metric is the validation accuracy. When using

Table 2. Classification and retrieval results while ablating the proposed class anchor initialization and various components of our CAM loss. Results are shown for a ResNet-18 model on CIFAR-100.

$\mathcal{L}_R$	$\mathcal{L}_N$	Anchors init	Accuracy	mAP
×	×	random	17.46%	0.110
		base vectors	39.24%	0.378
×	✓	random	17.88%	0.112
		base vectors	40.90%	0.386
✓	×	random	58.34%	0.558
		base vectors	59.55%	0.570
✓	✓	random	58.75%	0.539
		base vectors	61.94%	0.622

our loss, we set the parameter m for the component $\mathcal{L}_R$ to 2, and the parameter p for the component $\mathcal{L}_N$ to 1, across all datasets and models. Since the models based on the contrastive loss optimize in the feature space, we have used the 1-nearest neighbors (1-NN) accuracy, which is computed for the closest feature vector retrieved from the gallery.

As evaluation measures for the retrieval experiments, we report the mean Average Precision (mAP) and the precision@k on the test data, where $k \in \{20, 100\}$ is the retrieval rank. For the classification experiments, we use the classification accuracy on the test set. We run each experiment in 5 trials and report the average score and the standard deviation.

4.3   Results

Main Results. We first evaluate the performance of the ResNet-18, ResNet-50, ResNet-101 and Swin-T models on the CBIR task, while using three alternative losses, including our own, to optimize the respective models. The results obtained on the CIFAR-100 [17], Food-101 [2], and ImageNet-200 [31] datasets are reported in Table 1. First, we observe that our loss function produces better results on the majority of datasets and models. Furthermore, as the rank k increases from 20 to 100, we notice that our loss produces more consistent results, essentially maintaining the performance level as k increases.

Ablation Study. In Table 2, we demonstrate the influence of each additional loss component on the overall performance of ResNet-18 on the CIFAR-100 dataset, by ablating the respective components from the proposed objective. We emphasize that the component $\mathcal{L}_A$ is mandatory to make our objective work properly. Hence, we only ablate the other loss components, namely $\mathcal{L}_R$ and $\mathcal{L}_N$.

In addition, we investigate different class center initialization heuristics. As such, we conduct experiments to compare the random initialization of class centers and the base vector initialization. The latter strategy is based on initializing class anchors as scaled base vectors, such that each class center has no intersection with any other class center in the n-dimensional sphere of radius m, where n is the size of the embedding space.

As observed in Table 2, the class center initialization has a major impact on the overall performance. For each conducted experiment, we notice a significant performance gain for the base vector initialization strategy. Regarding the loss components, we observe that removing both $\mathcal{L}_R$ and $\mathcal{L}_N$ from the objective leads to very low performance. Adding only the component $\mathcal{L}_N$ influences only the overall accuracy, but the mAP is still low, since $\mathcal{L}_N$ can only impact each anchor's position with respect to the origin. Adding only the component $\mathcal{L}_R$ greatly improves the performance, proving that $\mathcal{L}_R$ is crucial for learning the desired task. Using both $\mathcal{L}_R$ and $\mathcal{L}_N$ further improves the results, justifying the proposed design.

5 Conclusion

In this paper, we proposed a novel loss function based on class anchors to optimize convolutional networks and transformers for object retrieval in images. We conducted comprehensive experiments using four neural models on four image datasets, demonstrating the benefits of our loss function against conventional losses based on statistical learning and contrastive learning. We also performed an ablation study to showcase the influence of the proposed components, empirically justifying our design choices.

In future work, we aim to extend the applicability of our approach to other data types, beyond images. We also aim to explore new tasks and find out when our loss is likely to outperform the commonly used cross-entropy.

References

1. Barz, B., Denzler, J.: Deep learning on small datasets without pre-training using cosine loss. In: Proceedings of WACV, pp. 1360–1369. IEEE (2020)
2. Bossard, L., Guillaumin, M., Van Gool, L.: Food-101 – mining discriminative components with random forests. In: Fleet, D., Pajdla, T., Schiele, B., Tuytelaars, T. (eds.) ECCV 2014. LNCS, vol. 8694, pp. 446–461. Springer, Cham (2014). https://doi.org/10.1007/978-3-319-10599-4_29
3. Cao, B., Araujo, A., Sim, J.: Unifying deep local and global features for image search. In: Vedaldi, A., Bischof, H., Brox, T., Frahm, J.-M. (eds.) ECCV 2020. LNCS, vol. 12365, pp. 726–743. Springer, Cham (2020). https://doi.org/10.1007/978-3-030-58565-5_43
4. Chen, W., Chen, X., Zhang, J., Huang, K.: Beyond triplet loss: a deep quadruplet network for person re-identification. In: Proceedings of CVPR, pp. 1320–1329. IEEE (2017)
5. Cortes, C., Vapnik, V.: Support-vector networks. Mach. Learn. **20**(3), 273–297 (1995)

6. Deng, J., Guo, J., Xue, N., Zafeiriou, S.: ArcFace: additive angular margin loss for deep face recognition. In: Proceedings of CVPR, pp. 4685–4694. IEEE (2019)

7. Dubey, S.R.: A decade survey of content based image retrieval using deep learning. IEEE Trans. Circuits Syst. Video Technol. **32**(5), 2687–2704 (2021)

8. Elezi, I., et al.: the group loss++: a deeper look into group loss for deep metric learning. IEEE Trans. Pattern Anal. Mach. Intell. **45**(2), 2505–2518 (2022)

9. Gajić, B., Amato, A., Baldrich, R., van de Weijer, J., Gatta, C.: Area under the ROC curve maximization for metric learning. In: Proceedings of CVPR, pp. 2807–2816. IEEE (2022)

10. Georgescu, M.I., Duţă, G.E., Ionescu, R.T.: Teacher-student training and triplet loss to reduce the effect of drastic face occlusion: application to emotion recognition, gender identification and age estimation. Mach. Vis. Appl. **33**(1), 12 (2022)

11. Georgescu, M.I., Ionescu, R.T.: Teacher-student training and triplet loss for facial expression recognition under occlusion. In: Proceedings of ICPR, pp. 2288–2295. IEEE (2021)

12. Hadsell, R., Chopra, S., LeCun, Y.: Dimensionality reduction by learning an invariant mapping. In: Proceedings of CVPR, vol. 2, pp. 1735–1742. IEEE (2006)

13. Harwood, B., Kumar, V.B., Carneiro, G., Reid, I., Drummond, T.: Smart mining for deep metric learning. In: Proceedings of ICCV, pp. 2821–2829. IEEE (2017)

14. He, K., Zhang, X., Ren, S., Sun, J.: Deep residual learning for image recognition. In: Proceedings of CVPR, pp. 770–778. IEEE (2016)

15. Khosla, P., et al.: Supervised contrastive learning. In: Proceedings of NeurIPS, vol. 33, pp. 18661–18673. Curran Associates, Inc. (2020)

16. Kingma, D.P., Ba, J.: Adam: a method for stochastic optimization. In: Proceedings of ICLR (2015)

17. Krizhevsky, A.: Learning multiple layers of features from tiny images. Technical report, University of Toronto (2009)

18. Lee, S., Seong, H., Lee, S., Kim, E.: Correlation verification for image retrieval. In: Proceedings of CVPR, pp. 5374–5384. IEEE (2022)

19. Liu, W., Wen, Y., Yu, Z., Li, M., Raj, B., Song, L.: SphereFace: deep hypersphere embedding for face recognition. In: Proceedings of CVPR, pp. 6738–6746. IEEE, Los Alamitos, CA, USA (2017)

20. Liu, W., Wen, Y., Yu, Z., Yang, M.: Large-margin softmax loss for convolutional neural networks. In: Proceedings of ICML, pp. 507–516. JMLR.org (2016)

21. Liu, Z., et al.: Swin Transformer: Hierarchical vision transformer using shifted windows. In: Proceedings of ICCV, pp. 10012–10022. IEEE (2021)

22. Min, W., Mei, S., Li, Z., Jiang, S.: A two-stage triplet network training framework for image retrieval. IEEE Trans. Multimedia **22**(12), 3128–3138 (2020)

23. Muller, S.G., Hutter, F.: TrivialAugment: tuning-free yet state-of-the-art data augmentation. In: Proceedings of ICCV, pp. 754–762. IEEE, Los Alamitos, CA, USA (2021)

24. Murphy, K.P.: Machine Learning: A Probabilistic Perspective. The MIT Press, Cambridge (2012)

25. Paszke, A., et al.: PyTorch: an imperative style, high-performance deep learning library. In: Proceedings of NeurIPS, pp. 8024–8035. Curran Associates, Inc. (2019)

26. Patel, Y., Tolias, G., Matas, J.: Recall@k surrogate loss with large batches and similarity mixup. In: Proceedings of CVPR, pp. 7502–7511. IEEE (2022)

27. Philbin, J., Chum, O., Isard, M., Sivic, J., Zisserman, A.: Object retrieval with large vocabularies and fast spatial matching. In: Proceedings of CVPR, pp. 1–8. IEEE (2007)

28. Polley, S., Mondal, S., Mannam, V.S., Kumar, K., Patra, S., Nürnberger, A.: X-vision: explainable image retrieval by re-ranking in semantic space. In: Proceedings of CIKM, pp. 4955–4959. Association for Computing Machinery, New York, NY, USA (2022)

29. Radenović, F., Tolias, G., Chum, O.: Fine-tuning CNN image retrieval with no human annotation. IEEE Trans. Pattern Anal. Mach. Intell. **41**(7), 1655–1668 (2019)

30. Revaud, J., Almazán, J., Rezende, R.S., Souza, C.R.d.: Learning with average precision: training image retrieval with a listwise loss. In: Proceedings of ICCV, pp. 5107–5116. IEEE (2019)

31. Russakovsky, O., et al.: ImageNet large scale visual recognition challenge. Int. J. Comput. Vision **115**, 211–252 (2015)

32. Schroff, F., Kalenichenko, D., Philbin, J.: FaceNet: a unified embedding for face recognition and clustering. In: Proceedings of CVPR, pp. 815–823. IEEE (2015)

33. Sohn, K.: Improved deep metric learning with multi-class N-pair loss objective. In: Proceedings of NIPS, vol. 29. Curran Associates, Inc. (2016)

34. Suh, Y., Han, B., Kim, W., Lee, K.M.: Stochastic class-based hard example mining for deep metric learning. In: Proceedings of CVPR, pp. 7244–7252. IEEE (2019)

35. Tang, Y., Bai, W., Li, G., Liu, X., Zhang, Y.: CROLoss: towards a customizable loss for retrieval models in recommender systems. In: Proceedings of CIKM, pp. 1916–1924. Association for Computing Machinery, New York, NY, USA (2022)

36. Balntas, V., Riba, E., Ponsa, D., Mikolajczyk, K.: Learning local feature descriptors with triplets and shallow convolutional neural networks. In: Proceedings of BMVC, pp. 119.1–119.11. BMVA Press (2016)

37. Wang, F., Cheng, J., Liu, W., Liu, H.: Additive margin softmax for face verification. IEEE Sig. Process. Lett. **25**(7), 926–930 (2018)

38. Wang, H., et al.: CosFace: large margin cosine loss for deep face recognition. In: Proceedings of CVPR, pp. 5265–5274. IEEE (2018)

39. Wen, Y., Zhang, K., Li, Z., Qiao, Yu.: A discriminative feature learning approach for deep face recognition. In: Leibe, B., Matas, J., Sebe, N., Welling, M. (eds.) ECCV 2016. LNCS, vol. 9911, pp. 499–515. Springer, Cham (2016). https://doi.org/10.1007/978-3-319-46478-7_31

40. Wu, C.Y., Manmatha, R., Smola, A.J., Krähenbühl, P.: Sampling matters in deep embedding learning. In: Proceedings of ICCV, pp. 2859–2867. IEEE (2017)

41. Wu, H., Wang, M., Zhou, W., Li, H.: Learning deep local features with multiple dynamic attentions for large-scale image retrieval. In: Proceedings of ICCV, pp. 11416–11425. IEEE (2021)

42. Yadan, O.: Hydra - a framework for elegantly configuring complex applications. Github (2019). https://github.com/facebookresearch/hydra

43. Yu, B., Tao, D.: Deep metric learning with tuplet margin loss. In: Proceedings of ICCV, pp. 6489–6498. IEEE (2019)

44. Zhu, Q., Zhang, P., Wang, Z., Ye, X.: A new loss function for CNN classifier based on predefined evenly-distributed class centroids. IEEE Access **8**, 10888–10895 (2019)

Personalized EDM Subject Generation via Co-factored User-Subject Embedding

Yu-Hsiu Chen, Zhi Rui Tam⬥, and Hong-Han Shuai(✉)⬥

National Yang Ming Chiao Tung University, Hsinchu, Taiwan
{yhchen.cm06g,ray.eed08g}@nctu.edu.tw, hhshuai@nycu.edu.tw

Abstract. This paper introduces the Co-Factored User-Subject Embedding based Personalized EDM Subject Generation Framework (COUPES), a model for creating personalized Electronic Direct Mail (EDM) subjects. COUPES adapts to individual content and style preferences using a dual-encoder structure to process product descriptions and template features. It employs a soft template-based selective encoder and matrix co-factorization for nuanced user embeddings. Experiments show that COUPES excels in generating engaging, personalized subjects and reconstructing recommendation ratings, proving its effectiveness in personalized marketing and recommendation systems.

Keywords: Summarization · Recommendation · Tensor co-factorization

1 Introduction

In the competitive landscape of e-commerce advertising, Email Direct Marketing (EDM) has proven its mettle, contributing significantly to the conversion of prospects into paying customers and enhancing open/conversion rates. With an estimated revenue exceeding $10 billion[1], personalizing EDM subject lines is a critical lever for capturing user attention, with personalized subjects boosting open rates by an impressive 37% over non-personalized ones[2].

Despite the effectiveness of personalized subjects, the prevailing practice involves manual creation by promotion specialists, a method that does not scale well with the voluminous and dynamic nature of e-commerce. With the advance of deep learning in natural language processing, studies have spurred research into automating the generation of engaging headlines. For instance, some researchers propose to inject specific styles, such as humor and romance, into news headlines or generate interrogative ones to attract readers' attention [5,10,22,29,30], disregarding the diverse style preferences of users.

Addressing this, Personalized Headline Generation (PHG) has surfaced as a novel research direction, aiming to tailor content to user-specific information inferred from click history [1,6,21]. Yet, this approach presents a formidable

---

[1] https://www.statista.com/statistics/812060/email-marketing-revenue-worldwide/.
[2] http://bit.ly/2kc7OD7.

© The Author(s), under exclusive license to Springer Nature Singapore Pte Ltd. 2024
D.-N. Yang et al. (Eds.): PAKDD 2024, LNAI 14646, pp. 55–67, 2024.
https://doi.org/10.1007/978-981-97-2253-2_5

challenge: the generation of personalized headlines without explicit annotations for personal text styles. Current methods often default to predefined rules or human-curated word lists, which fail to accommodate the breadth of individual user styles, particularly for new users with sparse interaction records [13,19].

Our work seeks to transcend these limitations by introducing a sophisticated approach that does not rely on predefined style classifications or representative word lists. We propose the Co-Factored Embeddings based Personalized EDM Subject Generation Framework, a model designed to navigate the complexities of personalized subject generation. This framework embodies two encoders to process description and template features, a bi-directional selective encoder for optimal information extraction, and a matrix factorization network for crafting user embeddings that reflect individual style preferences. These elements converge in an RNN decoder that generates EDM subjects aligned with user preferences, without the constraints of known styles or fixed word lists.

Furthermore, our approach harnesses user embeddings from browsing histories to enable collaborative filtering, recommending items and generating titles for products. We substantiate our model's efficacy with a newly compiled dataset containing 17,617 products with descriptions and 278,876 ratings from 21,379 users, demonstrating state-of-the-art performance in personalized subject generation. This represents not only a leap in the automation of EDM but also a new frontier in personalized digital marketing strategies.

The main contributions of our work are summarized as follows:

- We introduce the Co-Factored Embeddings based Personalized EDM Subject Generation Framework, a Seq2Seq-based model tailored for generating personalized EDM subjects. This represents the first endeavor, to our knowledge, to apply a data-driven approach specifically for this purpose, marking a significant departure from manual curation methods.
- Our model innovatively applies matrix factorization to infer users' preferences from click-through data, effectively utilizing this information not only for subject generation but also to enable a collaborative filtering mechanism within a recommendation system.
- Extensive experiments demonstrate that our framework significantly outperforms benchmarks in generating relevant and personalized subjects as measured by both qualitative and quantitative metrics. Furthermore, our approach shows adeptness in recommending items that align closely with users' preferences, showcasing personalization and recommendation.

2   Related Work

Text summarization techniques, fundamental in natural language processing, are widely-used in a variety of applications [2,8,28]. The methods are broadly classified into extractive and abstractive methods. Extractive methods focus on selecting and combining sentences from source articles into cohesive paragraphs [9,18]. In contrast, abstractive methods, such as those developed by [4], rewrite summaries, often producing content not explicitly present in the source text. The advent of sequence-to-sequence (Seq2Seq) frameworks [7,17] has marked a significant advancement in abstractive summarization. Techniques like the pointer network [24] and the coverage mechanism [20] enhance detail accuracy and minimize word repetition. Meanwhile, template-based summarization utilizes predefined templates for abstractive summarization. For instance, soft template-based models [3,26] offer a more flexible approach by selecting soft templates from specific training articles. However, the integration of user preferences in subject generation remains an underexplored area. Studies like [13,14] address this by generating reviews based on user preferences, but often require predefined word lists and fail to capture the diversity of user styles.

Recent research in Personalized Headline Generation (PHG) attempts to inject specific styles into headlines, but often overlooks the diverse and individualized style preferences of users [5,10,22,29,30]. To mitigate the gap, several works advocate for tailoring content to user-specific information inferred from click history [1,6,21]. For instance, [1] constructs a new dataset from Microsoft News user logs, which contains user-specific titles based on individual reading interests and news content. They further address personalized news headline generation by proposing a framework that learns user preferences from behavioral data to personalize text generation. Despite advancements, creating personalized headlines without explicit annotations for personal text styles remains challenging.

3   Proposed Model

Problem Formulation. Suppose we have a corpus D with m article-template-subject triples a-t-s, and each triple contains an article a, a template t and a subject s. Article a consists of l words as $\{a_1, a_2, ..., a_l\}$, where $a_i \in D$. Template t and subject s consist of $p \leq l, q \leq l$ words as $\{t_1, l_2, ..., t_p\}$ and $\{s_1, s_2, ..., s_q\}$ individually, where $t_i, s_i \in D$. In addition, the user preferences are presented by an n-by-m binary matrix $\mathbf{R}$. $\mathbf{R}_{ij} = 1$ if user i has clicked into the item j. The personalized subject generation task intends to: 1) retrieve template t given article a, and then further summarize a subject $\hat{s}$ given a and t by attending to user u's preference feature vector, 2) predict the rest of ratings in $\hat{\mathbf{R}}$ given a part of the ratings in $\mathbf{R}$.

To generate personalized EDM subjects, there are three challenges required to be addressed. First, different users prefer different styles of headlines, while it is difficult to define all styles. One possible solution is to select a subject line among users' clicking history and exploit template-based summarization method to generate a personalized subject. However, this approach is limited since only one clicked subject line is used. Second, to summarize the article, articles are encoded into a latent vector. Therefore, it is necessary to design a loss for disentangling the style and content so that the summarization only changes the style and preserves the content. Third, the clicking history of users may be sparse for new users, which leads to the "cold start" problem. Therefore, to overcome these issues, it requires combining the generation and recommendation elegantly.

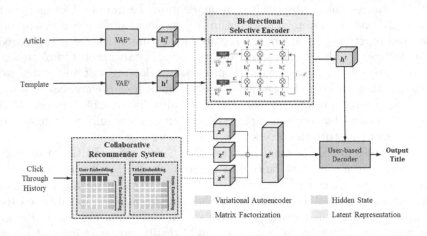

Fig. 1. The user-based encoder-decoder architecture.

Keep the goals in mind, in this paper, we propose a new framework including 5 key modules: **Retrieve and re-rank**, **Variational feature autoencoder**, **Bi-directional selective encoder**, **Collaborative recommender system**, and an **User-based decoder**. The overview model architecture is illustrated as shown in Fig. 1. Specifically, **Retrieve and re-rank** aims to return a few candidate templates from the training corpus and re-rank by semantic relationship to identify the best one. To jointly address the first and second issues, **Variational feature autoencoder** encodes the description to a Gaussian distribution. On the other hand, it encodes template's feature latent to a Gumbel distribution to preserve the categorical nature of templates. Afterward, **Bi-directional selective encoder** mutually selects important information from the source article and template hidden state to generate an article representation of summarization. With the user-item click through history as implicit feedback, **Collaborative recommender system** learns the users' preference representation to address the third issue. Finally, with the generated article, template, user feature latent, and the hidden states fused by **Bi-directional selective encoder**, the **User-based decoder** use an ordinary RNN to generate a personalized subject.

3.1 Retrieve and Re-rank

The module selects the closest candidate templates from the training corpus based on word frequency and semantic distance, assuming similar sentences have similar summary patterns. Utilizing Lucene[3], as per the default settings in [3], it

---

[3] https://lucene.apache.org.

retrieves similar paragraphs and their summaries as candidate templates, choosing the top 30 search results for each paragraph.

To identify the best template, we calculate the embedding space cosine similarity between two titles, using a Word2Vec embedding [16] trained on Wikipedia. The similarity score s_i between a template $\mathbf{z}_i^t$ and a gold subject z^s is given by:

$$s_i = (\mathbf{z}_i^t)^{\mathrm{T}} \mathbf{z}^s.$$

We select the subject with the highest score as the template, i.e., $\arg\max_i s_i$.

Notably, unlike the typical ROUGE score evaluation, which only captures word-level similarity, we use Word2Vec similarity [25] to better capture semantic meanings and select our desired template.

3.2   Variational Encoder and Bi-directional Selective Encoder

To better capture the latent representation of the article and template, we use Seq2seq Variational Autoencoder (VAE) [11] to derive the encoded representations of articles and templates. For a dataset $\mathbf{X} = \{\mathbf{x_i}\}_{i=1}^{N}$ containing N independent and identically distributed (i.i.d.) samples of a variable $\mathbf{x}$ alongside an unobserved continuous random variable $\mathbf{z}$, our objective is to generate new data $\mathbf{x}$ from the latent variable $\mathbf{z}$. These latent variables are assumed to arise from a prior distribution $p\theta(\mathbf{z})$, while the data $\mathbf{x}$ are assumed to be generated from a likelihood distribution $p_\theta(\mathbf{x}|\mathbf{z})$. Given that our inputs are sequences of words, we adopt a sequential variant of the Variational Autoencoder (VAE), specifically a variational Seq2Seq model. For the jth timestep of the nth data point, with input $x^{(n)}$ and target $y^{(n)}$, we formulate the loss function to maximize the variational lower bound, incorporating both the reconstruction loss and a regularization term derived from KL divergence, i.e.,

$$\mathcal{L}_j^{(n)}(\boldsymbol{\theta}, \boldsymbol{\phi}) = \underbrace{\mathbb{E}_{z \sim q_\phi(z|x^{(n)})} \left[\log p_\theta \left(y^{(n)} \mid z \right) \right]}_{\text{Reconstruction loss}} - \underbrace{D_{\mathrm{KL}} \left(q_\phi(z|x^{(n)}) \| p(z) \right)}_{\text{Regularization term}},$$

where q_ϕ and p_θ correspond to the encoder and decoder networks with weights ϕ and θ respectively. $q_\phi(\mathbf{z}|\mathbf{x}_i)$ is the variational distribution to approximate the true posterior $p_\theta(\mathbf{z}|\mathbf{x}_i)$. Here, BiLSTM cells are used within both encoder and decoder of the VAEs, allowing us to obtain the latent encodings $\mathbf{z}^\mathbf{a}$ and $\mathbf{z}^\mathbf{t}$ for articles and templates, respectively.

Moreover, our work utilizes a Bi-directional selective encoder for personalized summarization, merging template patterns with article sentences. We introduce a Template-to-Article gate and an Article-to-Template gate to filter and weigh the article's BiLSTM hidden states $\mathbf{h}_i^a$ using the template representation $\mathbf{h}^t$, producing a gated vector $\mathbf{g}_i^t$:

$$\mathbf{g}_i^t = \sigma(\mathbf{W}_{ah}\mathbf{h}_i^a + \mathbf{W}_{th}\mathbf{h}^t + \mathbf{b}_a),$$

$$\mathbf{h}_i^{gt} = \mathbf{h}_i^a \odot \mathbf{g}_i^t,$$

and a confidence score d^t to compute the final representation $\mathbf{h}_i^d$:

$$\mathbf{h}_i^d = d^t \mathbf{h}_i^{gt} + (1 - d^t) \mathbf{h}_i^a.$$

3.3   User-Subject Co-factor System

To capture user preferences for personalization, we employ a collaborative recommender system for generating user embeddings. While a direct approach might select subject lines from users' click histories or use template-based summarization for personalization as suggested in [6], it may not effectively capture the preferences of users with sparse click histories. We incorporate two matrix factorization (MF) modules within a co-factor model. The first MF module recommends items based on a user's click-through history, treating it as implicit feedback. The second MF module recommends titles to products, ensuring a product's relevance to its subject line.

For training with implicit feedback, we utilize negative sampling, randomly selecting negative instances to enhance efficiency. Given M users and N items, the user-item interaction matrix $\mathbf{R} \in \mathbb{R}^{M \times N}$, derived from implicit feedback, is defined as:

$$R_{ui} = \begin{cases} 1, & \text{if interaction (user } u, \text{ item } i) \text{ is observed;} \\ 0, & \text{otherwise.} \end{cases}$$

Here, $R_{ui} = 1$ indicates an interaction between user u and item i, while $R_{ui} = 0$ does not imply a negative preference but possibly unseen items. Learning from such data poses challenges due to the partial insight it provides into user preferences, with unobserved entries potentially representing missing data and observed ones indicating interest.

Matrix Factorization (MF) maps both users and items into a joint latent feature space of d dimension such that interactions are modeled as inner products in that space. Let $\mathbf{p}_u$ and $\mathbf{q}_i$ be the factor vector for user u and item i, respectively, matrix factorization method estimates an interaction r_{ui} as the inner product of $\mathbf{p}_u$ and $\mathbf{q}_i$:

$$\hat{R}_{ui} = \mathbf{p}_u^T \mathbf{q}_i = \sum_{k=1}^{K} p_{uk} q_{ik},$$

where K denotes the dimension of the latent space. The loss function of the first matrix factorization model can be expressed as follows:

$$\mathcal{L}_{mf} = \sum_{(u,i) \in \mathcal{R} \cup \mathcal{R}^-} \left(r_{ui} - \mathbf{p}_u^T \mathbf{q}_i \right)^2 + \lambda_p \sum_u \|\mathbf{p}_u\|^2 + \lambda_q \sum_p \|\mathbf{q}_i\|^2,$$

where $\mathcal{R}$ is the set of observed click-through interactions, and $\mathcal{R}^-$ is the set of negative instances, which is sampled from unobserved interactions. The last two terms are l_2 regularization terms.

Similarly, we use the same method to construct the loss function of the second MF recommending preferable titles to a certain item. Let $\mathbf{p}_t$ and $\mathbf{q}_i$ denote the latent vector for title t and item i, respectively; MF estimates an interaction r_{ui} as the inner product of $\mathbf{p}_t$ and $\mathbf{q}_i$, i.e., $\hat{y}_{ti} = \mathbf{p}_t^T \mathbf{q}_i = \sum_{k=1}^{K} p_{tk} q_{ik}$. The loss function of title matrix factorization then becomes:

$$\mathcal{L}_{tmf} = \sum_{(u,i)\in\mathcal{R}\cup\mathcal{R}-} \left(r_{ti} - \mathbf{p}_t^T \mathbf{q}_i\right)^2.$$

Subsequently, we obtain the user preference latent representation $\mathbf{p_u}$ and concatenate it with the template latent $\mathbf{z^t}$ and article latent $\mathbf{z^a}$ to form a comprehensive personalized representation $\mathbf{z^u}$, encompassing article, template, and user features.

3.4 User-Based Decoder

Following the encoder's selection of key information, a decoder RNN generates the subject line, initiated with the final personalized representation $\mathbf{z^u}$ as its starting state. At each time step t, the decoder updates its state based on the previous word and state:

$$\mathbf{h}_t^c = \text{RNN}(\mathbf{w}t-1, \mathbf{h}^c t-1).$$

Using concatenate attention, the context vector $\mathbf{c}t$ is computed, and combined with the decoder's state into $\mathbf{h}_t^o$, which is then used to predict the next word in the sequence. The model is trained by minimizing the negative log-likelihood of the generated sequence compared to the gold standard summary:

$$\mathcal{L}_s = -\frac{1}{M}\sum_{i=1}^{M}\sum_{j=1}^{L}\log p(\mathbf{w}_j^{*(i)}|\mathbf{w}_{j-1}^{(i)}, \mathbf{x}^{a(i)}, \mathbf{x}^{t(i)}),$$

where M is the number of sequences, and $\mathbf{x}^a$ and $\mathbf{x}^t$ represent the article and template.

The total loss of the proposed model is shown as follows:

$$\mathcal{L} = \mathcal{L}_{cycle}^a + \mathcal{L}_{cycle}^t + \mathcal{L}_{mf} + \mathcal{L}_{tmf} + \mathcal{L}_s.$$

The first two terms $\mathcal{L}_{cycle}^a$ and $\mathcal{L}_{cycle}^t$ denote the loss of Seq2Seq VAE, $\mathcal{L}_{mf}$ and $\mathcal{L}_{tmf}$ is the matrix factorization loss, and $\mathcal{L}_s$ is the summarization loss.

4 Experimental Results

Dataset. Due to the lack of public dataset for personalized EDM, we collect a new one from the well-known travel experience e-commerce platform in Asia that sales tour packages. The raw data contains the paired tuple (article, subject) of tour package DMs, together with the ID list of users who click the subject. Statistics are summarized in Table 1. Specifically, there are 17617 products associated with the EDM subject and a short paragraph of the product description.

Table 1. Dataset statistics.

# products	17617	# title	17617
# users	21379	# word/title	12.8
# item	5365	#word/article	155.2
# item/user	13.04	# vocabulary	71625

Each description contains 885 Traditional Chinese characters on average and is tokenized by CKIP [23].

For the user clicking histories, we collect the data from December 2018 to July 2019, which contains 26,662,557 user-item interaction with 1,271,297 users and 9,702 items. We observe that users interact with items sequentially between a small period. This is usually the situation that they are first attracted by the subject of a tour package and continue to browse related packages, which may have the same tour destination with the first one. To approximately collect items whose subjects are most attractive to the user, we simply take the first item viewed by the user for each viewing sequence as the ground truth. After filtering out users with less than 10 items viewed and items without descriptions, there are 278,876 records with 21,379 users and 5,365 items. For summarization task, we randomly split the dataset into 14095 products for training, 1761 for testing and 1761 for validation. The input vocabularies are collected from the training data, which have 71625 words in total. For recommendation task, we randomly select 80% of each user's records for training, and the remaining data are for testing.

Evaluation Metrics. For relevance between generated and reference sentences, we adopt ROUGE score, which is a commonly-used metric for text summarization. We report the recall, precision and F-measure, where F-measure is the geometric mean of the precision and recall of ROUGE-1, ROUGE-2, and ROUGE-L. Moreover, the performance of the recommender system is evaluated by Recall@K and NDCG@K. Recall@K indicates the percentage of the recommended items among items relevant to the user. NDCG@K is short for Normalized Discounted Cumulative Gain at K, which takes the position of the recommended items' order into account.

Baselines. Here, we compare our model with several methods which are popular in text summarization as follows:

- *Lead*-1 is an extractive approach which selects the first sentence in review as a summary.
- *S2S+Att* is a sequence-to-sequence model with attention implemented by the OpenNMT [12].
- *BiSET* [26] adopts a selective network to bidirectionally select important information from template and article.
- *TemPEST* [6] uses selective network to bidirectionally select important information from template and article, and also adopts users' click-through titles

Table 2. The recommendation performance comparison of all methods in terms of *Precision@k*, *Recall@k*, and *NDCG@k*. The best performance is marked as **bold**.

Models	Precision@5	Precision@10	Recall@5	Recall@10	NDCG@5	NDCG@10
CDAE	0.021	0.018	0.061	0.036	0.022	0.020
GATE	0.058	0.047	0.110	0.161	0.068	0.053
TemPEST	0.069	0.054	0.119	**0.189**	0.075	0.063
Ours	**0.253**	**0.173**	**0.125**	0.166	**0.306**	**0.231**

to perform personalization, which is the state-of-the-art personalize summarization model.

- *COUPES* is the proposed model. We use two-layer BiLSTM for both TSE and USE network, and the hidden state size of 500. The learning rate and dropout rate are set to 0.001 and 0.3 respectively.

For the recommendation baselines, since the goal of CREPES is to improve the clicking rate of users, we evaluate the recommender system with CREPES and the following baselines.

- *CDAE* [27] is short for collaborative denoising autoencoder, which uses denoising autoencoder to learn latent representation from user-item feedback.
- *GATE* is the gated attentive autoencoder [15] with source articles input.
- *TemPEST* [6] uses selective network to bidirectionally select important information from template and article, and also adopts users' click-through titles to perform personalization, which is the state-of-the-art personalize summarization model.
- *COUPES* is the proposed model.

4.1 Quantitative Results

Recommendation Results. Since the goal of our model is to improve the clicking rate of users, we evaluate the recommender system with our model and the above baselines. Table 2 compares our model with baselines in terms of *Precision@k*, *Recall@k*, and *NDCG@k*. our model outperforms TemPEST, which indicates that the user-specific article representation generated by our model better captures the feature of the item description than the fused embeddings in TemPEST. Moreover, in terms of *Precision@k* and *NDCG@k*, our model outperforms other baseline models a lot, which demonstrates that the recommender mechanism truly improves the recommendation results.

Summarization Results. Table 3 presents the results of ROUGE scores, which manifests that our model performs comparable results in terms of ROUGE-2 score and outperforms in ROUGE-L.

For ROUGE score, ROUGE-L scores outperform other baselines in this task, while the ROUGE-2 scores performs comparable results. This indicates that the generated sentence contains important words in the reference subject without

Table 3. ROUGE scores of all methods on the test set. R in the table denotes ROUGE, R is the Recall, P is the Precision. The best performance is marked as **bold**.

Models	R-1 F1(R/P)	R-2 F1(R/P)	R-L F1(R/P)
Lead-1	48.05 (47.91/49.13)	4.13 (3.93/4.68)	6.80 (6.67/7.89)
S2S+attn	56.91 (56.79/57.47)	3.92 (3.96/4.00)	7.91 (7.80/8.47)
BiSET	55.54 (55.61/56.01)	4.70 (4.81/4.72)	8.70 (8.78/9.15)
TemPEST	**65.47 (65.34/66.32)**	3.44 (3.43/3.86)	8.33 (8.21/9.17)
Ours	46.38 (46.26/47.12)	**4.70 (4.72/4.85)**	**9.24 (9.13/9.96)**

directly copy that since ROUGE-1 and ROUGE-2 is effected by the copying of the real answer. Moreover, since the ROUGE-1 score can directly reflects the performance of copying the original ground truth. The ROUGE-1 score of the purposed model is comparatively low in this task. Note that all ROUGE-1 scores are pretty high since subjects contain a lot of proper nouns like place names to describe the products. Therefore, the extractive method *Lead*-1 performs relatively well comparing to other summarization datasets.

Since the evaluation metric has only a gold summary with a certain style. Imagine that we simply rewrite subjects from the ground truth answer and make them transfer to another style preserving the informativeness. The ROUGE metric cannot capture this kind of information well and hence the scores will decrease. In short, our model attempts to strike a balance between the personalized styling and the faithfulness.

4.2 Effect of Template

Our research also explores the impact of template latent factors on final results. Experiments indicate that while templates play a minor role in enhancing quantitative measures like ROUGE scores, they are significant in controlling the output's stylistic aspects. A notable impact is the presence of parentheses in the results. The dataset includes samples with and without parentheses in sentences. Selecting a template with parentheses tends to strongly influence the outcome, leading to results that adhere to this format. As illustrated in Table 4, we present two cases using different templates. In each case, the first template, chosen by our model described earlier, contrasts with a second, randomly selected template from our dataset. The comparison in 4 demonstrates varying writing styles and tones. The first template, marked by an exclamation point and parentheses, tends to produce more exaggerated results. Consequently, the subjects generated mirror this style, featuring exclamatory sentences with parentheses. Conversely, the second template is more informative and lacks parentheses, resulting in smoother, shorter, and more straightforward subjects.

Table 4. Example of sample subject lines generated by our model conditioned on different related templates.The Chinese descriptions on top are the real inputs of this dataset and translated to the English descriptions.

Article	這個武士培訓計劃將幫助您實現童年的夢想。穿上真正的武士服裝，練習如何正確使用武士刀。有各式各樣的服裝選擇。培訓課程結束後將頒發紀念證書。磨F你的精神，准備好掌握一些武士技巧! 行程特色‧接受正式武士戰鬥訓練‧穿上正統武士服裝與武器‧受訓後可獲得武士培訓證書‧使用對象: 孩童: 06-15 歲成人: 16-99 F‧使用期限: 依照旅客訂購日期 This warrior training program can help you realize your dream in the childhood. Put on real warrior clothes, and learn to use the samurai knife properly. There are various choices of clothes. You will receive a certificate after the training program. Sharpen your spiritual senses and ready to grasp some warrior skills! Highlights: take real warrior combating training course, put on real warrior clothes and gears, get an certificate after training. Children from 6 to 15, adults from 16 to 99. Expiration Date: by purchase.
Template 1	【東京時代道場】侍! 成F武士‧學習武士道體驗 [Tokyo Era Dojo] Slash! Way To Become A Warrior. The Bushido Training Experience.
Subject 1	【東京武士體驗】舉起武士刀‧接受訓練體驗成F真正的武士吧 [Tokyo Bushido Experience] Raise Your Weapons! Get Training And Be A Real Warrior!
Template 2	一日F‧埃斯坦西亞克里斯蒂娜牧場 One Day Tour: Estancia Cristina
Subject 2	東京武士體驗‧舉起武士刀的武士體驗 Tokyo Bushido Experience An experience to raise your samurai knife

5 Conclusions and Future Work

In this study, we developed a model for generating user-specific EDM subjects and providing personalized recommendations. Our Co-Factored Embeddings based Personalized EDM Subject Generation Framework enhances personalization for users with limited interaction histories. Our results show it effectively creates relevant and preferred subject lines and recommends items aligning with user preferences. In the future, we plan to focus on refining the model to better interpret complex user behaviors and integrating real-time data for dynamic personalization.

References

1. Ao, X., Wang, X., Luo, L., Qiao, Y., He, Q., Xie, X.: PENS: a dataset and generic framework for personalized news headline generation. In: ACL/IJCNLP (2021)
2. Bražinskas, A., Lapata, M., Titov, I.: Unsupervised opinion summarization as copycat-review generation. In: ACL (2020)
3. Cao, Z., Li, W., Li, S., Wei, F.: Retrieve, rerank and rewrite: soft template based neural summarization. In: ACL (2018)

4. Cao, Z., Wei, F., Li, W., Li, S.: Faithful to the original: fact aware neural abstractive summarization. In: AAAI (2018)
5. Chen, C.Y., Wu, D., Ku, L.W.: HonestBait: forward references for attractive but faithful headline generation. In: Rogers, A., Boyd-Graber, J., Okazaki, N. (eds.) Findings of ACL (2023)
6. Chen, Y.H., Chen, P.Y., Shuai, H.H., Peng, W.C.: Tempest: soft template-based personalized EDM subject generation through collaborative summarization. In: AAAI (2020)
7. Chopra, S., Auli, M., Rush, A.M.: Abstractive sentence summarization with attentive recurrent neural networks. In: NAACL (2016)
8. Gao, S., et al.: Dialogue summarization with static-dynamic structure fusion graph. In: ACL (2023)
9. Jadhav, A., Rajan, V.: Extractive summarization with swap-net: sentences and words from alternating pointer networks. In: ACL (2018)
10. Jin, D., Jin, Z., Zhou, J.T., Orii, L., Szolovits, P.: Hooks in the headline: learning to generate headlines with controlled styles. In: Proceedings of the 58th Annual Meeting of the Association for Computational Linguistics, pp. 5082–5093. Association for Computational Linguistics, Online (Jul 2020). https://doi.org/10.18653/v1/2020.acl-main.456, https://aclanthology.org/2020.acl-main.456
11. Kingma, D.P., Welling, M.: Auto-encoding variational bayes. Statistics (2014)
12. Klein, G., Kim, Y., Deng, Y., Senellart, J., Rush, A.M.: OpenNMT: open-source toolkit for neural machine translation. In: ACL (2017)
13. Li, J., Li, H., Zong, C.: Towards personalized review summarization via user-aware sequence network. In: AAAI (2019)
14. Liu, T., Li, H., Zhu, J., Zhang, J., Zong, C.: Review headline generation with user embedding. In: China National Conference on Chinese Computational Linguistics (2018)
15. Ma, C., Kang, P., Wu, B., Wang, Q., Liu, X.: Gated attentive-autoencoder for content-aware recommendation. In: WSDM (2019)
16. Mikolov, T., Chen, K., Corrado, G., Dean, J.: Efficient estimation of word representations in vector space. In: ICLR (2013)
17. Nallapati, R., Zhou, B., dos Santos, C., Gulcehre, C., Xiang, B.: Abstractive text summarization using sequence-to-sequence RNNs and beyond. In: The special interest group on natural language learning (2016)
18. Narayan, S., Cohen, S.B., Lapata, M.: Ranking sentences for extractive summarization with reinforcement learning. In: NAACL (2018)
19. Ni, J., McAuley, J.: Personalized review generation by expanding phrases and attending on aspect-aware representations. In: ACL (2018)
20. See, A., Liu, P.J., Manning, C.D.: Get to the point: summarization with pointer-generator networks. In: ACL (2017)
21. Song, Y.Z., Chen, Y.S., Wang, L., Shuai, H.H.: General then Personal: decoupling and Pre-training for Personalized Headline Generation. Transactions of the Association for Computational Linguistics (2023)
22. Song, Y.Z., Shuai, H.H., Yeh, S.L., Wu, Y.L., Ku, L.W., Peng, W.C.: Attractive or faithful? Popularity-reinforced learning for inspired headline generation. AAAI 34(05), 8910–8917 (2020)
23. Sproat, R., Emerson, T.: The first international Chinese word segmentation bakeoff. In: The Special Interest Group of the Association for Computational Linguistics (2003)
24. Vinyals, O., Fortunato, M., Jaitly, N.: Pointer networks. In: NIPS (2015)

25. Vulić, I., Moens, M.F.: Monolingual and cross-lingual information retrieval models based on (bilingual) word embeddings. In: The ACM Special Interest Group on Information Retrieval (2015)
26. Wang, K., Quan, X., Wang, R.: BiSET: bi-directional selective encoding with template for abstractive summarization. In: ACL (2019)
27. Wu, Y., DuBois, C., Zheng, A.X., Ester, M.: Collaborative denoising auto-encoders for top-n recommender systems. In: WSDM (2016)
28. Yi-Ting, C., Song, Y.Z., Chen, Y.S., Shuai, H.H.: Beyond detection: a defend-and-summarize strategy for robust and interpretable rumor analysis on social media. In: EMNLP (2023)
29. Zhan, J., Gao, Y., Bai, Y., Liu, Q.: Stage-wise stylistic headline generation: style generation and summarized content insertion. In: IJCAI (2022)
30. Zhang, R., et al.: Question headline generation for news articles. In: CIKM (2018)

Spatial-Temporal Bipartite Graph Attention Network for Traffic Forecasting

Dimuthu Lakmal[✉], Kushani Perera, Renata Borovica-Gajic, and Shanika Karunasekera

University of Melbourne, Melbourne, VIC 3010, Australia
dkariyawasan@student.unimelb.edu.au,
{kushani.perera,renata.borovica,karus}@unimelb.edu.au

Abstract. Accurate traffic forecasting is pivotal for an efficient data-driven transportation system. The intricate nature of spatial-temporal dependencies and non-linearity present in traffic data has posed a significant challenge to the modeling of accurate traffic forecasting systems. Lately, there has been a significant effort to develop complex Spatial-Temporal Graph Neural Networks (STGNN) that predominantly utilize various Graph Neural Networks (GNN) and attention-based encoder-decoder architectures due to their ability to capture non-linear dependencies in spatial and temporal domains effectively. However, conventional GNNs limit explicit propagation of past information among nodes, while attention-based models such as transformers do not support finer-grained attention score distribution. In this study, we address the aforementioned issues and introduce a novel STGNN namely, Spatio-Temporal Bipartite Graph Attention Network (STBGAT) that allows explicit modeling of past information propagation among nodes. Further, we present a heterogeneous cross-attention mechanism in a transformer to compute finer-grained feature-wise attention distribution enabling the model to capture richer and more expressive temporal dependencies. Our experiments reveal that the proposed architecture outperforms the state-of-the-art approaches proposed in recent literature.

Keywords: Graph Attention Network · Traffic Forecasting · Transformers · Spatial Graph Attention Networks

1 Introduction

Transportation systems have become complex with the rapid growth of infrastructure and people's needs. Thus, relevant stakeholders continuously invest in implementing intelligent transportation systems (ITS) aiming for more efficient, accurate, and data-driven traffic management solutions. Accurate and real-time traffic condition forecasting is one of the core components of ITS. Traffic condition forecasting systems are designed to predict future traffic conditions given the historical traffic condition observations. Particularly, we focus on forecasting

D.-N. Yang et al. (Eds.): PAKDD 2024, LNAI 14646, pp. 68–80, 2024.
https://doi.org/10.1007/978-981-97-2253-2_6

traffic flow, one of the main traffic condition measurements. Typically, traffic flow in a specific road section is influenced by not only its own historical traffic conditions, but also the traffic conditions in adjacent connected road sections. *Hence, it is crucial to consider the propagation of traffic information through the spatial structure of the road network when forecasting traffic flow. Moreover, intricate temporal dynamics in road networks have made long-term (30~60 min) traffic forecasting even more challenging* [28].

To address the aforementioned challenges, researchers have formulated traffic flow forecasting as a spatial-temporal graph modeling problem and proposed various types of Spatial-Temporal Graph Neural Networks (STGNN) [12,29]. Even though recent efforts have attained substantial improvements [9] in accuracy compared to early versions of STGNNs [27], they are not sophisticated enough to effectively discover and leverage intricate spatial and temporal dependencies. This study primarily focuses on two major deficiencies of existing traffic forecasting STGNNs. First, the effects of traffic conditions on roads take time to gradually propagate to their adjacent roads through the network. However, existing approaches fail to ascertain how the traffic flow of a specific road at a given time is impacted by previous traffic conditions on adjacent roads. Second, the majority of existing approaches relied on raw historical observations as input features and have not included and assessed alternative feature sequences, such as averaged traffic flow sequences which could reveal more temporal and spatial dependencies [29].

To address these shortcomings, we present a novel spatial-temporal graph neural network (STBGAT), that consists of a bipartite graph attention network and a transformer with a heterogeneous cross-attention mechanism (Source code is available here: https://github.com/DimuthuLakmal/STBGAT). We conducted a comprehensive set of experiments using five different traffic datasets to evaluate the performance of the proposed model. Those experiments revealed that STBGAT significantly outperforms the current state-of-the-art models. The main contributions of this study can be summarized as follows:

- We propose a novel Bipartite Graph Attention Network for past neighborhood information propagation towards center nodes. This mechanism ensures the impact of the previous traffic conditions on adjacent roads is explicitly considered.
- We introduce a heterogeneous cross-attention mechanism in the transformer model which enables the decoder to assign separate feature-wise attention scores to the encoder outputs. This architecture allows for the integration of multiple encoders, each handling different input sequences. It will alleviate the impact of noise and missing values in each feature sequence while revealing more temporal and spatial dependencies.

2   Related Work

Prior to recent advancements in Graph Neural Networks (GNN), researchers have widely adopted classic statistical time series algorithms and machine learning models to make predictions [21,25]. However, these models are only capable of analyzing temporal dynamics in traffic data leaving spatial correlations unused. In contrast, Spatial-Temporal Graph Neural Networks (STGNN) have significantly improved accuracy by efficiently capturing and modeling both temporal and spatial dynamics in road networks [24,27]. Since the introduction of STGCN architecture by Yu et al. [27], various highly complex STGNN architectures have been proposed in the literature attaining substantially improved accuracy [9,29].

Various graph neural network architectures have been adopted as the spatial module in STGNNs, ranging from recurrent GNN to Graph Attention Neural Networks (GAT) [7,12,17]. GCN [11] is one of the prominent works in the Graph Neural Network domain that led to rapid success in STGNNs [26]. The superiority in efficiency, flexibility, and accuracy of GCN and its variants over prior GNN architectures have resulted in wide adoption of GCNs in STGNN models. On the other hand, Graph Attention Network (GAT) [23] outperforms GCNs as it uses an attention mechanism in the data propagation process within the graph. In this study, we develop the proposed bipartite graph by extending the default GAT implementation.

Further, various architectures have been proposed for the temporal module in STGNNs. There are three frequently used neural network architectures in the temporal module: 1) RNN-based, 2) CNN-based, and 3) Attention-based [1,14,16]. Compared to the other two, the attention mechanism has emerged as a highly compelling approach in temporal sequence modeling. We develop the proposed temporal module based on a transformer which is an encoder-decoder architecture relying on an attention mechanism [22]. None of the recent STGNN approaches have proposed explicit modeling of past information propagation from neighbors due to the additional complexity it imposes on these models. Further, only a few attempted to incorporate features beyond raw traffic flow values as inputs [5], and these attempts were insufficient to distinctly discern the significance of each feature in making predictions.

3 Definitions and Problem Statement

3.1 Definitions

Traffic Road Network. We represent a traffic road network with a directed graph $G = (V, E, A)$ where $V = v_1, ..., v_N$ is a set of N nodes representing traffic sensors; E is a set of edges among nodes; $A \in R_{N \times N}$ is a weighted adjacency matrix representing connectivity among nodes and edge weights between any of two connected nodes.

Traffic Flow Matrix. $X_t \in N \times C$ denotes the traffic condition feature matrix at time step t. N represents the number of nodes in the network and C represents the number of traffic condition-related features including traffic flow value associated with each individual node. $X_t' \in N \times 1$ denotes the traffic flow matrix at time step t.

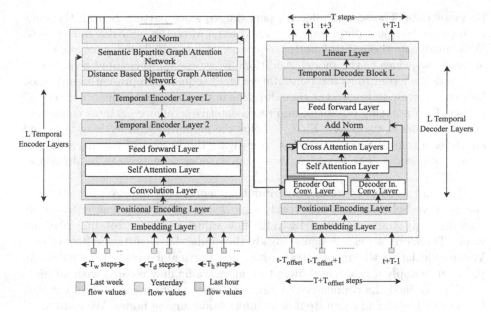

Fig. 1. Overall architecture of STBGAT

3.2 Problem Statement

Given the historical observation spanning a specific time frame, the problem is to establish a mapping function that can output the sequence of future traffic flow values having a predefined length. Let us assume the number of historical observations ending at the current step t is T_1 and the length of the prediction sequence is T_2. Then the mapping function f can be formally expressed as,

$$[X_{(t-T_1+1)}, X_{(t-T_1+2)}, ..., X_t; G] \xrightarrow{f} [X'_{(t+1)}, X'_{(t+2)}, ..., X'_{(t+T_2)}; G]$$

4 Methodology

4.1 Data Inputs and Data Preprocessing

This section focuses on briefing data inputs and the non-trivial data preprocessing steps we followed in this study.

If the target sequence is the traffic flow in the next hour from the current time, the most recent available sequence will be the traffic flow from the last hour, and it is incorporated as a part of the input sequence to the encoder. Several past studies have utilized additional sequences of historical observations as inputs, where these sequences may closely resemble either the pattern of the target sequence or the latest historical observation sequence [19]. We identified two effective periodic sequence patterns that tend to yield better results across all the datasets that we experimented with. The first is the sequence of traffic flow from the same hour as the hour just preceding the target hour, but on the previous date. The second effective periodic sequence pattern is from the same hour as the hour just preceding the target hour, but on the same day in last week. We concatenate these three shorter sequences into one single sequence and use as an input sequence to the model. Additionally, we use another input sequence to the model, referred to as a representative input sequence, representing averaged behavior corresponding to the time duration of the raw input sequence described earlier. Incorporating repetitive and representative sequence patterns could benefit the model in two ways. First, it helps the model to identify long-term and short-term trends. Second, it helps to mitigate the impact of missing values in shorter sequences. The total length of each encoder sequence can be defined as $T_e = T_{last\_week} + T_{last\_day} + T_{last\_hour}$.

To construct the representative input sequence, it is required to determine the average behavior at each weekly time index. All traffic flow datasets we tested in this study consist of 12 traffic flow values per hour, totaling 2016 per week. Therefore, we can assign a weekly time index for each traffic flow record. When calculating the average traffic flow at a particular weekly time index, we judiciously apply a rule-based filter to remove traffic flow records with noises.

We redefined the connectivity within the sensor network in certain datasets for more efficient and accurate flow of information among nodes. We introduced two types of connectivity producing two different connectivity graphs namely: distance-based bipartite graph and semantic bipartite graph. The distance-based graph is defined based on the assumption that two sensors in close geographic proximity to each other exhibit significant correlations between their recorded traffic flow values. To accommodate this assumption, we calculated edge attributes based on the shortest distance between nodes using Dijkstra's algorithm [10]. Then we dropped edges that exceeded a predefined distance threshold.

We defined a second graph based on the time series semantics among nodes. This helps to identify nodes that have similar behavior, but are not connected in the geographically connected graph defined above. For instance, in a scenario where two sensor nodes are located near two different schools, but are physically distant from each other, it may be important to propagate information between those two sensor nodes to identify common short-term temporal behaviors. For each weekly index described above, we picked a certain number of most similar nodes for every node in the graph. Then, a single global semantic graph is constructed assigning the set of nodes as neighbors of each node in the graph, which have the highest number of short-term similar behaviors with each center node. This calculation is based on the semantic distances among representative time series of nodes, each of which consists of 12 time steps. Dynamic Time Warping (DTW) algorithm is used to measure the similarity between two sequences [3].

4.2   Encoder Decoder Architecture

In this section, we brief the overall architecture of STBGAT model depicted in Fig. 1. The model follows transformer encoder-decoder architecture with some optimization done focusing on the traffic forecasting problem. The model can accommodate multiple encoders at once to facilitate feature extraction from

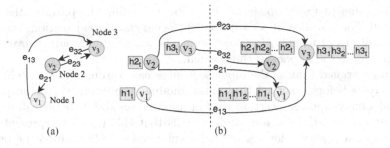

(a)   (b)

Fig. 2. Formation of bipartite graph. Figure 2a shows the example base graph with three nodes for the bipartite graph depicted in Fig. 2b.

multiple types of input sequences. In this study, we employ two encoders to process the two input sequences described in Sect. 4.1. The embedding layer maps the inputs to a higher dimensional vector space. The positional embedding layer fuses time step information and makes sure that the self-attention module is aware of the positional information.

The output of the positional encoding layer is then processed through a temporal encoder layer stack comprised of L number of layers. Each encoder layer includes a convolution layer similar to the one used in the temporal trend-aware multi-head self-attention layer described in [4]. The processed output of the convolution layer is then fed into a self-attention layer followed by a feed-forward layer. The output of the last temporal encoder layer is subsequently passed into the spatial module that consists of two bipartite graph attention networks enabling information propagation among nodes. Bipartite graphs are constructed as explained in Sect. 4.1. Finally, the output from bipartite GATs and the output from the final temporal encoder layer are summed together to produce the final encoder output. The decoder uses the output from the encoder as the historical context in the process of cross-attention calculation.

The embedding layer, positional encoding layer, self-attention layer, and feed-forward layer of the decoder are similar to the ones used in the encoder. Further, the decoder input convolution layer is also similar to the convolution layer used in the encoder layer. We introduce a heterogeneous feature-wise cross-attention mechanism as described in Sect. 4.4. Latent vector output given by the last decoder layer L is projected to an output with a single dimension by the linear layer of the decoder.

4.3 Bipartite Graph Attention Layer

In this section, we explain one of the main contributions of this paper that solves past information propagation problem described in Sect. 1. Traffic flow recorded at a sensor at a specific time step is influenced not only by the traffic flow on neighboring roads at the exact time step but also by previous traffic conditions on neighboring roads. The dependency on traffic conditions during previous time steps arises due to the propagation delay between different parts of the road network [9]. For instance, if a road accident occurs on a road section, then the traffic conditions of neighboring road segments will gradually adjust over time to accommodate the impact of the accident.

It can be argued that as the input sequences pass through a set of temporal encoder layers before reaching the spatial module, the latent vector output at each time step comprises a certain level of information about its preceding temporal context. Nonetheless, our study reveals that this implicit representation of the temporal context alone is not sufficient to effectively address the propagation delay in road networks. Instead, we propose to use a bipartite graph G_t that consists of two sets of nodes, (u_t, v_{T_e}) where u_t represents nodes consisting of outputs of temporal encoder module at time step t, and each node in v_{T_e} consists of concatenated output of temporal encoder from time step $t = 1$ to $t = T_e$. The edge attributes of the bipartite graph are equal to the corresponding edge attributes in the regular graph. The formulation of the bipartite graph from a regular graph is shown in Fig. 2. The implementation of Bipartite GAT can be outlined in three equations from Eq. 1 to Eq. 3.

$$e_{ij} = LeakyReLU(a^T(Wh_i\|Wh_j\|W_E E_{ij}))$$ (1)

$$\alpha_{ij} = \frac{exp(e_{ij})}{\sum_{k \in N_i} exp(e_{ik})}$$ (2)

$$h_i^{imm} = \|_{k=1}^K \sigma(\sum_{j \in N_i} \alpha_{ij}^k W^k h_j)$$ (3)

In Eq. 1, W, W_E and a denote learnable weights while E_{ij} denotes edge attribute associated with the edge connecting nodes i and j. Node i belongs to u_t and node j belongs to v_{T_e}. h_i represents the value of the central node i while h_j represents the value of neighbor node j. α_{ij} in Eq. 2 is the attention score calculated for neighbor node j. The final output for central node i is derived using Eq. 3. A single GAT layer consists of K heads. Thus, the final output h_i^{imm} is formed by either a concatenating or averaging of the outputs generated by K heads. Equation 3 shows only the concatenation operation over K heads.

4.4   Heterogeneous Cross Attention Layers

In this section, we present the second contribution of this paper that enables integrating multiple feature sequences. The transformer decoder consists of a cross-attention component that calculates attention values for elements in the encoder output sequence with respect to the decoder input sequence. This mechanism allows the decoder to focus on relevant information in the encoder output when generating the decoder output. The naive implementation of cross-attention only accepts the encoder output sequence as a single sequence and is not granular enough to calculate feature-wise attention. In contrast, we propose a heterogeneous cross-attention mechanism that is capable of calculating

Table 1. Prediction accuracy results (PEMS04-08)

		VAR	SVR	LSTM	DC-RNN	ST-GCN	GMAN	AST-GNN	PDF-ormer	PDF-ormer(L)	ST-BGAT
PEMS04	MAE	23.75	28.66	26.81	23.65	22.27	19.14	18.60	18.39	18.40	**18.17**
	RMSE	36.66	44.59	40.74	37.12	35.02	31.60	31.03	30.01	30.25	**28.23**
	MAPE	18.10	19.15	22.33	14.75	13.87	13.19	12.63	12.13	12.23	**12.02**
PEMS07	MAE	101.20	32.97	29.71	23.63	22.90	20.97	20.62	19.83	N/A	**18.34**
	RMSE	155.14	50.15	45.32	36.51	35.44	34.10	34.02	32.87	N/A	**30.86**
	MAPE	39.69	15.43	14.14	12.28	11.98	9.05	8.86	8.53	N/A	**7.63**
PEMS08	MAE	22.32	23.25	22.19	18.19	17.84	15.31	13.29	13.58	12.51	**12.39**
	RMSE	33.83	36.15	33.59	28.18	27.12	24.92	23.33	23.51	22.10	**21.02**
	MAPE	14.47	14.71	18.74	11.24	11.21	10.13	9.03	9.05	8.55	**8.43**

distinct attention distributions for different feature sequences. This mechanism allows for more precise modeling of temporal dynamics and relationships present in different feature sequences. STBGAT produces two separate encoder output sequences in parallel using the two input sequences described in Sect. 4.1. Following this, two cross-attention distributions are generated based on encoder output sequences. The final output of the cross-attention layer will be calculated according to the Eq. 4. In Eq. 4, LayerNorm refers to layer normalization. $X_{self-attn}$ refers to self-attention output calculated over decoder input. x_f is the encoder output calculated for the sequence of feature type f.

$$h_{cross} = LayerNorm(X_{self-attn} + \sum_{f \in F} CrossAttn(x_f)) \tag{4}$$

5 Experiments

5.1 Experiment Setup

We evaluate our model on two groups of datasets that are widely used in the literature. The first group consists of three datasets; PEMS04, PEMS07, and PEMS08 [2] while the second group consists of two datasets; PEMS-BAY and METR-LA [8].

We evaluate STBGAT against a variety of baselines proposed in the literature on the aforementioned two dataset groups. Tested baseline models are listed below.

- **PEMS04-08**: *VAR* [18], *SVR, LSTM* [6], *DCRNN* [15], *STGCN* [27], *GMAN* [29], *ASTGNN* [4], *PDFormer* [9]
- **PEMS-BAY, METR-LA**: *VAR* [18], *SVR, FC-LSTM* [6], *DCRNN* [15], *STGCN* [27], *GMAN* [29], *STGM* [13], *STEP* [20]

The default PDFormer only supports input sequences with a maximum length of 12. However, to ensure a fair comparison, we adapted the default PDFormer to

Table 2. Prediction accuracy results (PEMS-BAY, METR-LA)

		VAR	SVR	LSTM	DC-RNN	ST-GCN	GMAN	STGM	STEP	ST-BGAT
PEMS-BAY	MAE	2.93	3.28	2.37	2.07	2.49	1.86	1.86	1.79	**1.75**
	RMSE	5.44	7.08	4.96	4.74	5.69	4.32	4.37	4.20	**3.60**
	MAPE	6.50	8.00	5.70	4.90	5.79	4.37	4.34	4.18	**4.01**
METR-LA	MAE	6.52	6.72	4.37	3.60	4.59	3.44	**3.23**	3.37	3.94
	RMSE	10.11	13.76	8.69	7.60	9.40	7.35	7.10	**6.99**	7.25
	MAPE	15.80	16.70	14.00	10.50	12.70	10.07	**9.39**	9.61	9.79

accept a 36-length input sequence that includes repetitive patterns. It is referred to as PDFormer(L) in Table 1. In contrast, other recent architectures STEP and ASTGNN support longer sequences by default.

Three evaluation matrices are used namely, Mean Absolute Error (MAE), Root Mean Square Error (RMSE), and Mean Absolute Percentage Error (MAPE). Masked versions of MAE, RMSE, and MAPE matrices are used to alleviate the effect of missing values in the dataset. Experiments on each model are repeated 3 times and we report mean values for said matrices.

The test results of the STBGAT model presented in Table 1 and Table 2 are generated by passing input sequences consisting of repetitive patterns. The reported error values are the averaged error values computed across the entire prediction sequence length.

5.2   Comparison of Performance

The overall performance of baseline models and STBGAT is summarized in Table 1 and Table 2. It is important to highlight that the PDFormer(L) model fails to run on PEMS07 dataset due to memory overflow which suggests that it is not suitable for large road networks (tested on 128GB of RAM). The best results for each metric reported in each dataset are highlighted in bold. Our model outperforms all baselines in every performance metric across four datasets; PEMS07, PEMS08, PEMS04, and PEMS-BAY. However, the STBGAT model exhibits lower performance on the METR-LA dataset, particularly in MAE metric. We discover that this is attributed to the high discrepancy between train and test data distributions. Moreover, STBGAT model significantly outperforms all baselines in terms of RMSE metric. This metric is a useful indicator of performance when large errors between ground truth and predicted values are undesirable. Hence, it suggests that STBGAT can better approximate sudden fluctuations in traffic flow.

The most notable observation across the experiments is the substantial performance enhancement achieved by spatial-temporal models in comparison to temporal prediction models alone. This accentuates the importance of discover-

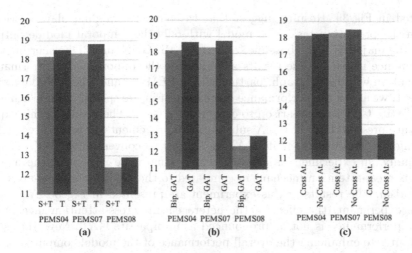

Fig. 3. Ablation study results (MAE Values)

ing and exploiting spatial dependencies in road network graphs. LSTM exhibits the best performance among temporal prediction models leveraging its ability to identify long temporal dependencies compared to other temporal models. Based on our experiments, the spatial module can be considered as an enhancement to improve the accuracy of the temporal module of spatial-temporal models. Hence, having a comprehensive temporal module is also important to have better performance. DCRNN and STGCN models consist of RNN and convolution components in the temporal module that could hinder performance in modeling temporal dependency effectively. In contrast, GMAN, ASTGNN, STEP, STGM, and PDFormer use various attention mechanisms that help achieve superior performance over traditional RNN-based models. In addition to the performance enhancements achieved through the two novel concepts presented here, two other existing concepts contributed to STBGAT's performance. First, STBGAT use CNN layers in temporal module for local context recognition which is also used in ASTGNN, PDFormer and STGM. Second, STBGAT effectively harnesses both short-term and long-term spatial dependencies as it uses two types of bipartite graphs in the encoder. A comparable approach is also utilized in PDFormer and STGM models. The contribution of the two novel concepts presented in this paper is discussed in the next section.

5.3   Ablation Study

Additional experiments are carried out to investigate the effectiveness of different components of STBGAT using PEMS04, PEMS07 and PEMS08 datasets. We first study the prediction capability of the temporal module by training the model without the spatial component of the model. We then test the impact and contribution of the spatial component to predictions. The evaluation results are depicted in Fig. 3a. Results suggest that the temporal module plays a critical role in the prediction task. The model with only the temporal module outperforms the majority of the baselines except for ASTGNN and PDFormer in every performance metric. These results also indicate the importance of information propagation within the graph, particularly in PEMS07 and PEMS08 datasets.

Next, we assess the performance enhancement in the spatial module achieved by bipartite GAT in comparison to conventional GAT. The results of this experiment are presented in Fig. 3b. A substantial improvement can be observed when utilizing the proposed bipartite GAT in contrast to conventional GAT.

Finally, an experiment is conducted to evaluate the impact of the heterogeneous cross-attention mechanism compared to the traditional cross-attention mechanism. The results of this experiment are presented in Fig. 3c. According to the experiment, the effect of the heterogeneous cross-attention mechanism on the performance is not as pronounced as in bipartite GAT. Nevertheless, it contributes to enhancing the overall performance of the model compared to the conventional cross-attention mechanism.

6   Conclusion and Future Works

In this paper, we introduce a novel spatial-temporal graph neural network architecture for traffic forecasting that outperforms the latest state-of-the-art baselines across four real-world datasets. We proposed two novel concepts in this paper; bipartite graph attention network and heterogeneous cross-attention mechanism. The first concept enhances the spatial information propagation while the second concept improves the temporal dependency analysis of the model. The ablation study demonstrates the effectiveness of these two novel concepts in modeling spatial and temporal dynamics. As future work, the model can be extended and utilized in various downstream tasks in domains such as traffic analysis and social media.

Acknowledgements. This research was supported by The University of Melbourne's Research Computing Services and the Petascale Campus Initiative.

References

1. Bai, L., Yao, L., Kanhere, S.S., Wang, X., Liu, W., Yang, Z.: Spatio-temporal graph convolutional and recurrent networks for citywide passenger demand prediction. In: Proceedings of the 28th ACM CIKM, pp. 2293–2296 (2019)
2. Chen, C., Petty, K., Skabardonis, A., Varaiya, P., Jia, Z.: Freeway performance measurement system: mining loop detector data. TRR **1748**(1), 96–102 (2001)
3. Giorgino, T.: Computing and visualizing dynamic time warping alignments in r: the dtw package. J. Stat. Softw. **31**, 1–24 (2009)
4. Guo, S., Lin, Y., Wan, H., Li, X., Cong, G.: Learning dynamics and heterogeneity of spatial-temporal graph data for traffic forecasting. IEEE TKDE **34**(11), 5415–5428 (2021)
5. He, H., Ye, K., Xu, C.Z.: Multi-feature urban traffic prediction based on unconstrained graph attention network. In: 2021 IEEE BigData, pp. 1409–1417 (2021)

6. Hochreiter, S., Schmidhuber, J.: Long short-term memory. Neural Comput. **9**(8), 1735–1780 (1997)
7. Huang, R., Huang, C., Liu, Y., Dai, G., Kong, W.: Lsgcn: Long short-term traffic prediction with graph convolutional networks. In: IJCAI, vol. 7, pp. 2355–2361 (2020)
8. Jagadish, H.V., Gehrke, J., Labrinidis, A., Papakonstantinou, Y., Patel, J.M., Ramakrishnan, R., Shahabi, C.: Big data and its technical challenges. Commun. ACM **57**(7), 86–94 (2014)
9. Jiang, J., Han, C., Zhao, W.X., Wang, J.: Pdformer: propagation delay-aware dynamic long-range transformer for traffic flow prediction. arXiv preprint arXiv:2301.07945 (2023)
10. Johnson, D.B.: A note on dijkstra's shortest path algorithm. JACM **20**(3), 385–388 (1973)
11. Kipf, T.N., Welling, M.: Semi-supervised classification with graph convolutional networks. arXiv preprint arXiv:1609.02907 (2016)
12. Kong, X., Xing, W., Wei, X., Bao, P., Zhang, J., Lu, W.: Stgat: spatial-temporal graph attention networks for traffic flow forecasting. IEEE Access **8**, 134363–134372 (2020)
13. Lablack, M., Shen, Y.: Spatio-temporal graph mixformer for traffic forecasting. Expert Syst. Appl. **228**, 120281 (2023)
14. Li, W., Wang, X., Zhang, Y., Wu, Q.: Traffic flow prediction over muti-sensor data correlation with graph convolution network. Neurocomputing **427**, 50–63 (2021)
15. Li, Y., Yu, R., Shahabi, C., Liu, Y.: Diffusion convolutional recurrent neural network: Data-driven traffic forecasting. arXiv preprint arXiv:1707.01926 (2017)
16. Li, Y., Moura, J.M.: Forecaster: a graph transformer for forecasting spatial and time-dependent data. In: ECAI 2020, pp. 1293–1300. IOS Press (2020)
17. Lu, Z., Lv, W., Cao, Y., Xie, Z., Peng, H., Du, B.: Lstm variants meet graph neural networks for road speed prediction. Neurocomputing **400**, 34–45 (2020)
18. Lütkepohl, H.: Vector autoregressive models. Handbook of research methods and applications in empirical macroeconomics 30 (2013)
19. Roy, A., Roy, K.K., Ali, A.A., Amin, M.A., Rahman, A.M.: Unified spatio-temporal modeling for traffic forecasting using graph neural network. In: 2021 IJCNN, pp. 1–8. IEEE (2021)
20. Shao, Z., Zhang, Z., Wang, F., Xu, Y.: Pre-training enhanced spatial-temporal graph neural network for multivariate time series forecasting. In: Proc. 28th ACM SIGKDD Conf. Know. Disc. Data Min., pp. 1567–1577 (2022)
21. Tian, Y., Zhang, K., Li, J., Lin, X., Yang, B.: Lstm-based traffic flow prediction with missing data. Neurocomputing **318**, 297–305 (2018)
22. Vaswani, A., et al.: Attention is all you need. Adv. NIPS **30** (2017)
23. Velickovic, P., Cucurull, G., Casanova, A., Romero, A., Lio, P., Bengio, Y., et al.: Graph attention networks. Stat **1050**(20), 10–48550 (2017)
24. Wang, X., et al.: Traffic flow prediction via spatial temporal graph neural network. In: Proc. web conf. 2020, pp. 1082–1092 (2020)
25. Williams, B.M., Hoel, L.A.: Modeling and forecasting vehicular traffic flow as a seasonal arima process: theoretical basis and empirical results. J. Trans. Eng. **129**(6), 664–672 (2003)
26. Wu, Z., Pan, S., Chen, F., Long, G., Zhang, C., Philip, S.Y.: A comprehensive survey on graph neural networks. IEEE TNNLS **32**(1), 4–24 (2020)
27. Yu, B., Yin, H., Zhu, Z.: Spatio-temporal graph convolutional networks: A deep learning framework for traffic forecasting

28. Yu, H., Wu, Z., Wang, S., Wang, Y., Ma, X.: Spatiotemporal recurrent convolutional networks for traffic prediction in transportation networks. Sensors **17**(7), 1501 (2017)
29. Zheng, C., Fan, X., Wang, C., Qi, J.: Gman: a graph multi-attention network for traffic prediction. In: Proc. of the AAAI Conf. on Art. Intell., vol. 34, pp. 1234–1241 (2020)

CMed-GPT: Prompt Tuning for Entity-Aware Chinese Medical Dialogue Generation

Zhijie Qu$^{(\boxtimes)}$, Juan Li, Zerui Ma, and Jianqiang Li

Beijing University of Technology, Beijing 100124, China
quzhijie@emails.bjut.edu.cn

Abstract. Medical dialogue generation relies on natural language generation techniques to enable online medical consultations. Recently, the widespread adoption of large-scale models in the field of natural language processing has facilitated rapid advancements in this technology. Existing medical dialogue models are mostly based on BERT and pre-trained on English corpora, but there is a lack of high-performing models on the task of Chinese medical dialogue generation. To solve the above problem, this paper proposes CMed-GPT, which is the GPT pre-training language model based on Chinese medical domain text. The model is available in two versions, namely, base and large, with corresponding perplexity values of 8.64 and 8.01. Additionally, we incorporate lexical and entity embeddings into the dialogue text in a uniform manner to meet the requirements of downstream dialogue generation tasks. By applying both fine-tuning and p-tuning to CMed-GPT, we lowered the PPL from 8.44 to 7.35. This study not only confirms the exceptional performance of the CMed-GPT model in generating Chinese biomedical text but also highlights the advantages of p-tuning over traditional fine-tuning with prefix prompts. Furthermore, we validate the significance of incorporating external information in medical dialogue generation, which enhances the quality of dialogue generation.

Keywords: Chinese medical dialogue · Pre-trained language model · P-tuning

1 Introduction

In the context of the current global health crisis, telemedicine's role as a supplement to traditional healthcare has grown in significance. Telemedicine can not only help address the imbalance in the distribution of medical resources but also alleviate the problem of resource scarcity. Additionally, it can enhance medical efficiency and convenience, facilitate real-time communication between patients and doctors, reduce treatment duration, and improve overall medical efficiency. Due to the rapid advancement of Artificial Intelligence (AI) systems, an increasing number of medical researchers are eager to advance intelligent AI dialogue systems into virtual medical practitioners. These virtual medical practitioners can engage in patient consultations, understand their medical conditions and complete medical histories, and offer well-informed clinical recommendations. Consequently, in recent years, He [1], Liu [2], Li [3], Wei [4], Xu [5] and

D.-N. Yang et al. (Eds.): PAKDD 2024, LNAI 14646, pp. 81–92, 2024.
https://doi.org/10.1007/978-981-97-2253-2_7

others have proposed task-oriented medical dialogue models to play the role of virtual medical practitioners and engage in one-on-one consultations with patients. However, the presence of specialized phrases and formal medical expressions in Chinese medical conversations makes medical dialogue systems more difficult to implement than task-oriented dialogue systems (TDS).

Pre-training followed by fine-tuning has emerged as the prevailing approach for transferring the capabilities of large models to downstream tasks within the NLP domain [6], including medical dialogue generation. In the biomedical field, researchers have begun investigating the application of pretrained language models. Unfortunately, general pretrained models perform poorly in the medical domain [7, 8], likely due to the presence of a significant number of domain-specific terms that are difficult for these models to comprehend. To enhance pretrained model performance in the medical domain, a series of expert models based on medical datasets have emerged. Even without fine-tuning, PubMedBERT [8], BioBERT [9], SciBERT [10], and DilBERT [11] models outperform BERT in downstream tasks. This represents an outstanding development, as researchers appear to have identified the future research direction for telemedicine: medical pre-training models and fine-tuning.

However, most prior research has focused on BERT-pretrained models and English medical dialogue datasets, leading to underwhelming performance in the generation of Chinese medical dialogues. After analyzing the causes, the main problems with the existing methods are as follows: Regarding medical dialogue datasets, most of the dialogue datasets are in English not Chinese. Regarding models, BERT, a representative bidirectional language model with its bidirectional attention mechanism, may not be as suitable as unidirectional models like GPT for natural language generation (NLG) tasks. Regarding medical domain knowledge, it is more specialized compared to other domains, characterized by complex, less frequent vocabulary, and abundant technical jargon. Undoubtly, these three factors pose significant challenges for the models.

This paper makes the following major contributions to solving the above problems:

1. We collect medical dialogue datasets in Chinese and proceed with pre-training of GPT on the Chinese medical dialogue datasets to develop a Chinese generalized pre-trained language model, namely CMed-GPT, for the biomedical domain.
2. To enhance the model's understanding of medical dialogue, we executed fine-tuning and p-tuning on the pre-trained CMed-GPT model employing downstream medical dialogue data. This enhanced the model's comprehension while simultaneously reducing the number of parameters during the fine-tuning stage.
3. To enable the model to better understand the entities in the medical domain, we combine lexical and entity embeddings with dialog history to obtain the entity-aware medical dialog generation models KB-FT-CMed-GPT and KB-PT-CMed-GPT. The results demonstrate the effectiveness of our method.

The paper is organized as follows: the related works on dialogue generation is shown in Sect. 2. Datasets are shown in Sect. 3. Details of the proposed Chinese medical dialogue generation model is introduced in Sect. 4. Experimental and results are presented in Sect. 5. Finally, conclusion is given in Sect. 6.

2 Related Work

For recent research on dialogue generation, pre-training on large-scale datasets and fine-tuning on downstream tasks have proved to be a successful paradigm and have become the standard models. Two of the typical models are BERT and GPT.

BERT is a bidirectional encoder representation language model based on the transformer architecture. In subsequent studies, the improvement points of BERT are mainly focused on two parts: the construction of large-scale corpus data [1] and model pre-training [12]. In the field of biomedicine, scholars have proposed several approaches to enhance the performance of BERT models by pre-training them with English medical domain datasets. These approaches include the following three ideas: 1) Continued pre-training using medical domain data, such as BioBERT [9] and DilBERT [11]; 2) Pre-training from scratch using medical domain data, such as SciBERT [10] and PubMedBERT [8]; 3) Pre-training based on self-supervised tasks specifically designed for the medical domain, such as MC-BERT [13], SMedBERT [14] and BERT-MK [15]. While various biomedical pre-trained language models based on BERT have achieved great success in natural language understanding and classification tasks, few scholars have worked on the task of generating BERT models due to its Transformer encoder model architecture that limits NLG.

GPT is an autoregressive language model based on Transformer's decoder. Compared to the BERT, GPT has significantly larger training corpora and model parameters such as GPT-2 [16] and GPT-3 [17]. The modeling structure of GPT makes it outstanding for generative tasks. With the emergence of ChatGPT, a new fine-tuning paradigm based on pre-trained language models, p-tuning, has become a major mainstream approach to improve the performance of GPT on downstream tasks. For domain-specific generative tasks, the inclusion of prompt allows the model to generate text that is more consistent with the user's intent. In the field of biomedicine, while GPT is well-suited for medical dialogue generation, there are very few pretrained language models based on GPT for medical text generation. Previous work on pretraining GPT in the biomedical literature is DARE [18]. However, they pretrained GPT on an extremely limited dataset consisting of only 0.5 million PubMed abstracts and used it solely for data augmentation in relation extraction tasks. Generative pre-trained language modeling for Chinese is a research gap. With the increase of Chinese medical dialogue datasets in recent years, it is a matter of concern how the GPT model performs in the field of Chinese medical dialogue generation.

Therefore, we propose the CMed-GPT specifically designed for Chinese medical dialogue scenarios. During the pre-training phase of the model, we continue to train the GPT model using Chinese medical datasets. In the fine-tuning phase, we perform p-tuning of CMed-GPT by integrating lexical and entity information into the dialogue text in a uniform manner. This approach not only further demonstrates the effectiveness of domain-specific pretraining on improving model performance but also significantly reduces the parameter count compared to fine-tuning. Furthermore, the uniform incorporation of lexical and entity information in this paper, as opposed to the discrete approach, enhances the model's understanding ability and generates higher-quality dialogue responses that better align with user intent.

3 Datasets

Datasets are crucial for pre-trained language models. This section aims to provide a comprehensive description of the datasets utilized in this study. The datasets used for model pre-training include Chinese medical dialogue datasets and medical books.

Chinese Medical Dialogue Datasets. The dataset consists of seven single and multi-turn medical dialogue datasets from different sources. It includes five multi-turn medical dialogue datasets labeled with entities such as symptoms, diseases, drugs, examinations, etc., as well as two medical dialogue datasets without entity labeling. The above datasets were obtained from online consulting medical websites, smart conversation clinic competitions and previous studies. The data is shown in Table 1:

Table 1. Chinese medical dialogue datasets.

Dataset	Diseases	Dialogues	Utterances	Tokens	Entities
CHIP-MDCFNPC	–	8000	247,520	4259,819	5494,944
CMDD	4	2064	86,874	868,738	651,553
cMQA-master	–	260,000	520,000	67,080,000	–
IMCS-IR	6	3052	122,080	1,599,248	1,903,227
MedDG	12	17,864	385,862	6829,764	4692,086
ChunYu	15	12,842	317,197	3,362,292	4091,846
COVID-DDC	1	1088	9465	405,128	–

Chinese Medical Book Text Dataset. To enhance the comprehension capabilities of pre-trained models, this paper additionally incorporates open-source medical book text data. The datasets were used to further pre-train existing models, resulting in a pre-trained Chinese language model for the medical field. The book text encompasses pharmacology, diagnostics, pathology, etc., totaling approximately 655.8MB of Chinese biomedical text and roughly 11.2 million tokens.

4 Method

4.1 Pre-training Model

Model Structure. We primarily continue pretraining on GPT2-Chinese[1], which follows the structure of a Transformer decoder [19] (GPT structure) and employs a multi-head unidirectional attention mechanism. Given the dialogue history $X = X_1, X_2, X_3 \ldots X_i \ldots X_k$, where X_i is either a doctor's or a patient's utterance. The utterance $X_i = [u_1^i, u_2^i \ldots u_j^i \ldots u_s^i]$ has s tokens, when predicting the token u_j^i, the model will mask the tokens after u_j^i. The model structure (see Fig. 1).

---

[1] https://github.com/Morizeyao/GPT2-Chinese.

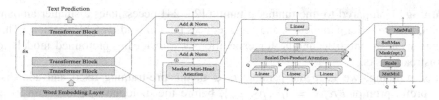

Fig. 1. Pre-trained language model structure.

As shown in Fig. 1, the model structure is divided into three layers, including the word embedding layer, transformer block layer, and text prediction layer. The GPT model architecture consists of six transformer block layers, with its core component being the multi-head attention structure. The input matrix h_0 is transformed using linear transformation matrices $WQ/WK/WV$ to obtain Q, K, and V, which are used to compute the output of the self-attention mechanism.

$$head_i(Q, K, V) = soft \max\left(\frac{Q_i K_i^T}{\sqrt{d_k}}\right) V \tag{1}$$

Multi-head Attention consists of multiple self-attention layers:

$$Multihead(Q, K, V) = Concat(head_1, head_2..., head_h) W \tag{2}$$

Model Training Objectives. Given a dialogue history $X = X_1, X_2, X_3...X_i...X_k$, the objective of training the model is to minimize the following likelihood function:

$$\min\left(-\frac{1}{K} \sum_{i=1}^{K} \sum_{j=1}^{S} \log P(u_j^i | u_1^i...u_{j-1}^i)\right) \tag{3}$$

The final loss is:

$$Loss = CrossEntorpyLoss(H) \tag{4}$$

where H represents the vector after passing through the text prediction layer.

In the following, fine-tuning, p-tuning, and the model with lexical and entity embeddings, the training objectives for these models all involve the next token prediction task, and the formulas are as shown in Eq. 4.

4.2 Medical Dialogue Generation Model

Fine-Tuning. Fine-tuning is an effective method for transferring the capabilities of pretrained models to downstream tasks. Therefore, this paper incorporates fine-tuning as one of the techniques to evaluate the performance of CMed-GPT. Prior studies have extensively employed fine-tuning on models such as BERT or GPT [20–22], and this paper refrains from delving further into this topic.

P-Tuning. Compared to fine-tuning, p-tuning [23, 24] necessitates a smaller amount of data, fewer model parameters, and thus, reduces the demand on GPU memory. This renders it an essential technique for expeditiously transferring pretrained models to downstream tasks.

For p-tuning, we keep the parameters φ of the pre-training fixed and add discrete prefix prompt $Prefix = \{p_1, p_2...p_m\}$, before the dialogue text to obtain $Z = [Prefix; X; Y]$. The goal of the model fine-tuning is to maximize the likelihood of Y, $P_{\varphi\theta_p} = P(Y|[Prefix; X])$, where the parameter $\theta_p \ll \varphi$. Assuming that the pre-trained model has a vocabulary size of V, V_p prompt tokens, and a hidden layer dimension of H, the model's word embedding matrix would be of size $E \in V \times H$, and the prompt's word embedding matrix would also be of size $E_p \in V_p \times H$. Considering the input $X = X_1, X_2, X_3...X_i...X_k$ comprising utterances $X_i = [u_1^i, u_2^i...u_j^i...u_s^i]$, where $u_j^i \in V$, and the prompt prefix $Prefix = \{p_1, p_2...p_m\}$, where $p_m \in V_p$. During the training process, the prompt tokens are concatenated with the input text before forwarding it. This concatenation transforms the input X into $Prefix, X_1, X_2, X_3...X_i...X_k$. The loss is then computed based on this modified input text. It is important to note that the model's parameter gradients are set to false, while only the prompt parameter gradients are set to true. As a result, during the backward pass, the gradients for the model parameters are set to 0, while only the prompt parameters are updated.

Lexical and Entity Embeddings. In the medical domain, there are many specialized and less common entities (long-tail entities) that are relatively rare in general corpora. This rarity can lead to inaccurate word vector representations for these entities. By enhancing the representation of entities, it helps the model better understand the context of the conversation and improves the accuracy of the model's responses [25]. In this paper, we represent the entity positions using one-hot vectors. Consequently, the entity information E_i is encoded as an entity vector.

$$E_i = e_1^i, ..., e_s^i \tag{5}$$

where s represents the length of a utterance, $e_j^i = 0|1$.

In addition to one-hot encoding entity positions, we also enhance the text representation using lexical embedding. In general, entities in medical text are predominantly nouns (n), while the severity of symptoms is primarily adjectives (adj). These lexical have significant importance in medical text. Therefore, we use "jieba" for lexical tagging in the text, identifying nouns, adjectives, and verbs and assigning them numerical labels, such as nouns: 0, adjectives: 1, verbs: 2. These labels are encoded into vectors and introduced into the text vectors. The modified model structure is shown in Fig. 2. Compared to the original structure, we use a positional encoding-like approach to embed lexical and entity encodings within the input embedding. This modification is intended to enhance the model's ability to comprehend medical text.

Assuming the original word vector representation is E_w and the position vector is E_p the input vector for the block is $E = E_w + E_p$, After adding the lexical vector E_t and the entity vector E_e :

$$E = E_w + E_p + E_t + E_e \tag{6}$$

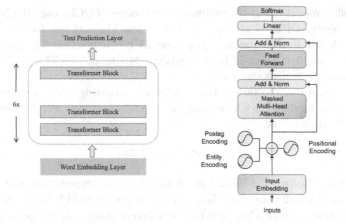

Fig. 2. Model structure for fusion entity and lexical embeddings.

5 Experiments

In this section, we present the experimental setup and related experimental results for the pre-training model CMed-GPT and fine-tuning, p-tuning.

5.1 Experimental Setting

The model utilizes all the data from Sect. 3 for pre-training, excluding CMDD and IMCS-IR, which is divided into training and test sets at a ratio of 100:1. The AdamW [26] training optimizer was employed with β1 and β2 values set to 0.9 and 0.95, respectively. Weight decay was set at 0.1, and gradient norm clipping was performed at 0.5. A warmup cosine scheduler was used with an initial learning rate of 0.0001, a warmup step of 2000, followed by a cosine decay to a minimum learning rate of 5e-6. The minimum learning rate was maintained after 100,000 steps and continued training until completion, totaling 3 epochs. The maximum text length was set to 512, and if the text exceeded this length, it was truncated from the end to 512. The batch size was set at 32. The entire model was trained using float16 mixed precision, which accelerates training speed while maintaining accuracy compared to float32. Lastly, the base model was trained for 20 days on a 4-card machine with NVIDIA TITAN RTX 24GB, while the large model was trained for 30 days.

To evaluate the model's performance in downstream tasks, we conducted p-tuning/fine-tuning using CMDD and IMCS-IR2 data. The dataset was partitioned into training and test sets, with a ratio of 8:2. The hyperparameters for fine-tuning remained consistent with the pretraining phase, except for a modification in the initial learning rate, set to 5e-5. The model underwent training for 6 epochs. For p-tuning, different quantities of prompt tokens {1, 25, 50, 75, 100} were employed in experimental trials.

5.2 Experimental Results

Pre-training Model. The pretrained language model for the Chinese medical domain proposed in this paper hasn't found an equivalent pretrained model. Here, we compared it with four other models: Bert-base-chinese (see footnote 2), Chinese-GPT2-base (see footnote 3), Chinese-GPT2-large (see footnote 3), and Medbert (see footnote 4). Bert-base-chinese: A BERT model pretrained on general Chinese text data; Chinese-GPT2: A GPT model pretrained on general Chinese text data; Medbert: A BERT model pretrained on Chinese medical domain text data.

Given the nature of the task being a language modeling task, perplexity (PPL) was employed as the metric for evaluating the model's performance. The formula for calculating PPL is as follows:

$$PPL = e^{loss}, e \approx 2.78 \tag{7}$$

The results are shown in the Table 2, and it was observed that general models, whether BERT or GPT2, perform poorly on Chinese medical text, with PPL values greater than 20. However, Medbert, which has been fine-tuned on medical text, outperforms general BERT models, further emphasizing the importance of continued training on medical domain text.

Table 2. PPL of CMed-GPT compared to other models in the medical domain test set

Model	Parameters	Test PPL	Size
Bert-base-chinese[2]	6 layers, 12 head, 768 hidden size	22.63	110M
Chinese-GPT2-base[3]	6 layers, 12 head, 768 hidden size	20.82	110M
Chinese-GPT2-large (see footnote 3)	12 layers, 12 head, 768 hidden size	18.64	200M
Medbert[4]	6 layers, 12 head, 768 hidden size	9.37	110M
CMed-GPT-base(Ours)	6 layers, 12 head, 768 hidden size	8.64	110M
CMed-GPT-Large(Ours)	12 layers, 12 head, 768 hidden size	8.01	200M

In addition, CMed-GPT-base also outperforms an equivalently sized Medbert. This can be attributed to the inherent suitability of unidirectional language models for generation tasks. Furthermore, increasing the model size (from 6 to 12 layers) leads to a noticeable improvement in performance (Fig. 3).

Fine-Tuning/P-Tuning. In the fine-tuning and p-tuning phases, we utilized CMDD dataset and MCS-IR dataset. The results are shown in Table 3, where CMed-GPT-base is the pre-trained model, KB-FT-CMed-GPT is the CMed-GPT-base post-fine-tuning model and KB-PT-CMed-GPT is the CMed-GPT-base post-p-tuning model. It was observed that after fine-tuning/p-tuning on the downstream dialog task datasets, the PPL decreased from above 8 to below 8, indicating an improvement of approximately 10%. Additionally, p-tuning achieves comparable or slightly superior results compared to fine-tuning. This suggests that the prefix prompts used in p-tuning had a significant

---

[2] https://huggingface.co/bert-base-chinese.
[3] https://github.com/Morizeyao/GPT2-Chinese.
[4] https://github.com/trueto/medbert.

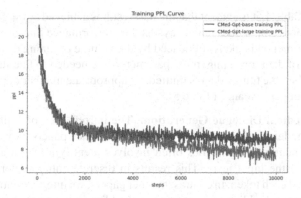

Fig. 3. Train PPL curves for CMed-Gpt-base/large

impact. Compared to fine-tuning, p-tuning required fewer parameters and less time for training.

Table 3. CMed-GPT-base fine-tuning/p-tuning results on CMDD and IMCS-IR.

Dataset	CMed-GPT-base	KB-FT-CMed-GPT	KB-PT-CMed-GPT
CMDD	8.44	7.68	7.58
IMCS-IR	8.79	7.75	7.62

In the p-tuning phase, we experimented with different numbers of prompt tokens to assess their impact on the results. For this experiment, we only used the CMDD dataset, and the results are as follows:

Table 4. PPL of the model with different prompt token.

Dataset	1	25	50	75	100
CMDD	8.32	7.85	7.68	7.69	7.93

Table 5. PPL of the model after lexical/entity embedding for different datasets.

Dataset	P-tuning	Lexical	Entity	Lexical and entity	Entity splicing method[5]
CMDD	7.58	7.5	7.46	7.35	7.89
IMCS-IR	7.62	7.58	7.53	7.43	7.93

---

[5] https://arxiv.org/abs/2212.06049.

Based on the Table 4, it is evident that the best result is achieved at v = 50, suggesting that the addition of prompts does not always lead to performance improvements for the model. This phenomenon is likely influenced by the volume of tuning data. When there is a large amount of data for tuning, more parameters are needed to fine-tune and fit the data. However, when the tuning data is limited, an appropriate number of prompt tokens can improve model performance (Table 5).

Entity-Aware Medical Dialogue Generation. The introduction of entity and lexical enhanced GPT models has been widely discussed in other papers as well. However, the approach of incorporating entities often involves identifying them and discretely appending them to the input text[5]. This results in discrete entity vectors that cannot uniformly represent each token. In contrast to other papers, we integrate entity and lexical vectors with the text embedding and positional embedding in an encoded manner, similar to the transformer input format. The uniformly spliced matrix is fed into our model after p-tuning. Experimental results indicate that incorporating entity and lexical information further improves the model's performance.

The inclusion of lexical and entity vectors in medical text has strengthened the representation of certain important words in the text. This makes it easier for the model to generate accurate responses during subsequent conversations.

6 Conclusion

In this paper, we first conducted a comprehensive compilation and analysis of existing Chinese medical dialogue datasets and medical book data. Subsequently, using this valuable collection of Chinese medical text data, we successfully pretrained a powerful Chinese medical pre-trained language model, named CMed-GPT. This model not only possesses extensive medical knowledge but also has the ability to understand and generate medical dialogue text.

To address the common long-tail entity problem in medical text, we propose an innovative approach that effectively integrates lexical and entity information during model training. Additionally, we devised discrete pseudo-token prefixes, which were appended to the text data. Through a series of rigorous experiments, we have concluded that our proposed CMed-GPT, including both the base and large versions, exhibits outstanding performance in the medical domain.

Furthermore, our CMed-GPT model can be easily applied to various downstream tasks in the medical field. By performing simple p-tuning, which involves adjusting a small number of parameters, the model's performance on specific tasks can be significantly improved. This provides a powerful and efficient tool for medical information processing.

To summarize, this study not only successfully constructs a pre-trained language model CMed-GPT for the Chinese medical domain, but also proposes an effective method for entity information incorporation and demonstrates its excellent performance in medical text processing tasks. In the future, we will continue to improve the model and explore more application areas to better meet the needs of medical information processing and contribute to advancements in the medical field.

References

1. He, X., et al.: MedDialog: Two large-scale medical dialogue datasets (2020). arXiv preprint arXiv:2004.03329
2. Liu, W., Tang, J., Qin, J., Xu, L., Liang, X.: MedDG: A large-scale medical consultation dataset for building medical dialogue system (2020). arXiv preprint arXiv:2010.07497
3. Li, D., et al.: Semi-supervised variational reasoning for medical dialogue generation. In: Proceedings of the 44th International ACM SIGIR Conference on Research and Development in Information Retrieval, pp. 544–554. Association for Computing Machinery, New York (2021)
4. Wei, Z., et al.: Task-oriented dialogue system for automatic Diagnosis. In: Proceedings of the 56th Annual Meeting of the Association for Computational Linguistics, pp. 201–207. Association for Computational Linguistics, Melbourne (2018)
5. Xu, L., Zhou Q., Gong, K., Liang, X., Tang, J., Lin, L.: End-to-end knowledge-routed relational dialogue system for automatic diagnosis. In: Proceedings of the AAAI Conference on Artificial Intelligence, pp. 7346–7353. Association for the Advancement of Artificial Intelligence (2019)
6. Wang, A., Singh, A., Michael, J., Hill, F., Levy, O., Bowman, S.R.: GLUE: a multi-task benchmark and analysis platform for natural language understanding. In: Proceedings of the 2018 EMNLP Workshop BlackboxNLP: Analyzing and Interpreting Neural Networks for NLP, pp. 353–355. Association for Computational Linguistics, Brussels (2018)
7. Peng, Y., Yan, S., Lu, Z.: Transfer learning in biomedical natural language processing: an evaluation of BERT and ELMo on ten benchmarking datasets. In: Proceedings of the 18th BioNLP Workshop and Shared Task, pp.58–65. Association for Computational Linguistics, Florence (2019)
8. Gu, Y., et al.: Domain-specific language model pretraining for biomedical natural language processing. ACM Trans. Comput. Healthc. 3(1), 1–23 (2022)
9. Lee, J., et al.: BioBERT: a pre-trained biomedical language representation model for biomedical text mining. J. Leukoc. Biol. 36(4), 1234–1240 (2020)
10. Beltagy, I., Lo, K., Cohan, A.: SciBERT: a pretrained language model for scientific text. In: Proceedings of the 2019 Conference on Empirical Methods in Natural Language Processing and the 9th International Joint Conference on Natural Language Processing, pp.3615–3620. Association for Computational Linguistics, Hong Kong (2019)
11. Roitero, K., et al.: DiLBERT: cheap embeddings for disease related medical NLP. IEEE Access 9(9), 2169–3536 (2021)
12. Liu, Y., et al.: RoBERTa: A robustly optimized BERT pretraining approach (2019). arXiv preprint arXiv:1907.11692
13. Zhang, N., Jia, Q., Yin, K., Dong, L., Gao, F., Hua, N.: Conceptualized Representation Learning for Chinese Biomedical Text Mining (2020). arXiv preprint arXiv:2008.10813
14. Zhang, T., Cai, Z., Wang, C., Qiu, M., Yang, B., He, X.: SMedBERT: a knowledge-enhanced pre-trained language model with structured semantics for medical text mining. In: Proceedings of the 59th Annual Meeting of the Association for Computational Linguistics and the 11th International Joint Conference on Natural Language Processing, pp.5882–5893. Association for Computational Linguistics (2021)
15. He, B., et al.: BERT-MK: integrating graph contextualized knowledge into pre-trained language models. In: Findings of the Association for Computational Linguistics: EMNLP 2020, pp.2281–2290. Association for Computational Linguistics (2020)
16. Radford, A., et al.: Language models are unsupervised multitask learners. GPT-2 OpenAI blog (2019)

17. Brown, T.B., et al.: Language models are few-shot learners. In: Proceedings of the 34th International Conference on Neural Information Processing Systems, pp.1877–1901. Curran Associates Inc, Red Hook (2020)
18. Papanikolaou, Y., Pierleoni, A.: DARE: Data augmented relation extraction with GPT-2 (2020). arXiv preprint arXiv:2004.13845
19. Vaswani, A., et al.: Attention is all you need. In: Proceedings of the 31st International Conference on Neural Information Processing Systems, pp.6000–6010. Curran Associates Inc., Red Hook (2017)
20. Loshchilov, I., Hutter, H.: Decoupled weight decay regularization. In: International Conference on Learning Representations (2017)
21. Peng, X., et al.: Fine-Tuning a transformer-based language model to avoid generating non-normative text (2020). arXiv preprint arXiv:2001.08764v1
22. Davier, M.V., Training optimus prime, M.D.: Generating medical certification items by Fine-Tuning OpenAI's gpt2 transformer model (2019). arXiv preprint arXiv:1908.08594
23. Tsai, D.C.L., et al.: Short answer questions generation by Fine-Tuning BERT and GPT-2. In: 29th International Conference on Computers in Education Conference, pp. 509–515. Asia-Pacific Society for Computers in Education (2021)
24. Li, X., Liang, P.: Prefix-Tuning: optimizing continuous prompts for generation. In: Proceedings of the 59th Annual Meeting of the Association for Computational Linguistics and the 11th International Joint Conference on Natural Language Processing, pp.4582–4597. Association for Computational Linguistics (2021)
25. Lester, B., et al.: The power of scale for parameter-efficient prompt tuning. In: Proceedings of the 2021 Conference on Empirical Methods in Natural Language Processing, pp.3045–3059. Association for Computational Linguistics (2021)
26. Cui, L., et al.: Knowledge enhanced fine-tuning for better handling unseen entities in dialogue generation. In: Proceedings of the 2021 Conference on Empirical Methods in Natural Language Processing, pp.2328–2337. Association for Computational Linguistics (2021)

MvRNA: A New Multi-view Deep Neural Network for Predicting Parkinson's Disease

Lin Chen, Yuxin Zhou, Xiaobo Zhang$^{(\boxtimes)}$, Zhehao Zhang, and Hailong Zheng

School of Computing and Artificial Intelligence, Southwest Jiaotong University, Chengdu, China
{chenlynn,yuzs,zhzhang,zhl}@my.swjtu.edu.cn, zhangxb@swjtu.edu.cn

Abstract. Magnetic Resonance Imaging (MRI) is a critical medical diagnostic tool that assists experts in precisely identifying lesions. However, due to its high-dimensional nature, it requires substantial storage resources. If only one MRI slice were to be used, a significant amount of information might be lost. To address these issues, we propose segmenting 3D MRI data and training these slices separately. We propose a new Multi-view Learning neural network based on ResNet and an Attention mechanism, called MvRNA. ResNet18 is selected as the backbone network, and the Squeeze-and-Excitation network is applied between blocks to extract features from slices. Additionally, we propose a new BWH (Basic Block with Hybrid Dilated Convolution) module to capture a broader range of receptive fields, thus acquiring additional spatial features. We obtained data from Parkinson's Progression Markers Initiative (PPMI) and applied our method to distinguish between Healthy Control, Prodromal, and Parkinson's disease patients. The experimental results demonstrate that our method achieved an accuracy of 81.84%.

Keywords: Parkinson's disease (PD) · Multi-view Learning · 3D MRI · ResNet · Squeeze-and-Excitation Attention · Hybrid Dilated Convolution

1 Introduction

Parkinson's disease (PD) is a prevalent neurodegenerative condition. Its primary pathological feature involves the degeneration and demise of dopaminergic neurons in the substantia nigra of the midbrain, resulting in a notable reduction in dopamine content within the striatum. The intricate nature of PD leads to over 90% of patients enduring a period exceeding a year before receiving an official diagnosis. This diagnostic delay brings significant harm to patients, emphasizing the critical importance of timely detection and treatment. In recent years, with the advancement of deep learning, computer-aided diagnoses [1] have also

---

L. Chen and Y. Zhou contribute equally to this work.

© The Author(s), under exclusive license to Springer Nature Singapore Pte Ltd. 2024
D.-N. Yang et al. (Eds.): PAKDD 2024, LNAI 14646, pp. 93–104, 2024.
https://doi.org/10.1007/978-981-97-2253-2_8

made significant progress. The combination of machine learning techniques and medical knowledge offers a promising direction for neurological disease research.

Numerous novel neural network algorithms [2,3] have been proposed for classifying medical subjects in various medical scenarios and have achieved commendable results. However, as research has evolved, a critical issue has garnered attention: a single two-dimensional medical image slice loses a significant amount of contextual information, which is detrimental to volume-level medical research. To model three-dimensional spaces and acquire more complex feature representations, 3D convolutional neural networks (CNNs) have been introduced [4]. Nevertheless, it comes with new limitations, such as increased requirements for computing resources and time due to more intricate structures.

As a distinct learning paradigm, multi-view learning [5] involves acquiring data characteristics from various perspectives. These unique viewpoints offer complementary and consistent information. Specifically, when data is missing from one perspective, other views can act as supplements to alleviate the problem of feature loss [6]. To simultaneously reduce the heavy usage of computational resources and capture as much multi-dimensional information from the images as possible, inspired by multi-view learning, we position 3D MRI images within a three-dimensional coordinate system. We slice the original images along the x, y, and z axes separately and achieve our goal through a split-and-merge strategy. Slices from different axes are sent separately to the network for training, and we concatenate the obtained features to enable the prediction of PD. To enhance the model's sensitivity to the data, we integrate Squeeze-and-Excitation (SE) [7] blocks to improve the identification of channel features. In more detail, this process captures important features between slices. Furthermore, we have made improvements to the basic blocks and enhanced the spatial feature extraction capabilities by introducing hybrid dilated convolutions [8].

In summary, the main contributions of this work are summarized as follows:

- We propose a split-merge strategy for 3D MRI data. This strategy preserves all data information while minimizing computational resource usage.
- MvRNA, a new Multi-view learning neural network based on ResNet and Attention mechanism is introduced. This new framework can learn meaningful feature representations that are less noise-sensitive.
- We improved the basic block and introduced a basic block with hybrid dilated convolutions (BWH), which enhances the perception of image spatial features.
- We conducted a series of processing steps on the PPMI dataset [9], and numerous experiments have demonstrated that our method significantly enhances the accuracy of PD prediction.

2   Related Work

In this section, we will introduce Convolutional Neural Networks, the concept of Multi-view Learning, Attention Mechanisms with a particular focus on the Squeeze-and-Excitation network, and Hybrid Dilated Convolutions.

Convolutional Neural Networks (CNNs). With the development of machine learning, more and more CNNs have been proposed. AlexNet [10], VGG [11], and GoogleNet [12] have achieved good results on image prediction tasks. Utilizing CNNs, [13,14] have effectively employed 2D slices from 3D images to classify Healthy Control (HC) and PD. In order to train the deep network more easily, ResNet [15] was proposed, which introduced residual connections to solve the problem of gradient disappearance. Building upon this concept, an efficient ResNet-50-based CNN model for pneumonia prediction using medical images was proposed [16]. Res2Net [17] was developed by constructing hierarchical residual connections within a single residual block, thereby achieving multi-scale feature representation at a fine-grained level and expanding the receptive field of each network layer.

Multi-view Learning. Multi-view learning [5] is a machine learning paradigm that leverages information from multiple perspectives or data sources to enhance the generalization ability of models. Given that different views often contain complementary information, merging them results in a more accurate feature representation. In the context of medical data, Masked Multi-view with Swin Transformers (SwinMM) [18] has been developed to extract hidden multi-view information from 3D medical datasets. Additionally, the introduction of Multi-scale attention (MSA) [19] has expanded the horizons of multi-view imaging integration, enabling robust medical image segmentation by collecting global correspondences of multi-scale feature representations. A feature representation method aimed at enhancing multimodal MRI data through the integration of multi-view information was proposed [20], a multiple kernel learning algorithm was employed to combine multiple features for enhancing prediction performance in tasks related to mild cognitive impairment prediction.

Attention Mechanism. Attention mechanisms have gained widespread application in the domain of deep neural networks, empowering networks to dynamically focus on crucial elements while disregarding less relevant details. This innovative approach has shown success in PD prediction tasks, as evidenced by [5,21]. A novel network architecture [22] combined BAM [23] and SA [24] blocks with a 3D ResNet18 backbone network to achieve PD prediction, highlighting the potential of integrating attention mechanisms for enhanced model performance. One noteworthy attention mechanism that prioritizes channel importance is the Squeeze-and-Excitation Network (SENet) [7], consisting of two key components: Squeeze and Excitation. SENet successfully enhances model performance with minimal computational cost. Additionally, [25] employed SENet as a bridge between the convolutional layer and the transition layer to design a novel brain tumor prediction network model based on DenseNet-201. [26] introduced the 3D SE-DenseNet for the prediction of hepatocellular carcinoma grading, utilizing enhanced clinical MRI data from two different clinical centers. The combination of SENet with CNN, as demonstrated by [27,28], has significantly improved prediction accuracy.

Hybrid Dilated Convolution (HDC). As we all know, with an increase in the number of network layers, deeper features are captured, but concurrently, the resolution decreases. Dilated convolution [29] preserves the resolution by using zero-padding in the convolution kernel, but it can result in a gridding effect. To tackle this issue, hybrid dilated convolution (HDC) [8] was introduced. By overlapping dilated convolutions with different rates, it becomes possible to capture more contextual information while retaining image resolution. [30] proposed a neural network with the HDC for pixel-level crack detection on concrete bridges, which effectively improved the detection accuracy of blurred cracks. [31] improved the decoding process using HDC, obtaining more receptive fields and obtaining higher Dice scores.

3   Methods

In this section, we introduce the MvRNA framework. First of all, we position 3D MRI in a three-dimensional coordinate system and slice the images along the x, y, and z axes to obtain a multi-view dataset. Subsequently, we employ ResNet18 as the basic network. Combined with HDC, we introduce a novel BWH feature extraction module. This module not only preserves medical image resolution but also expands the receptive field. Additionally, we incorporate the SE block into our framework, enhancing feature representation capability. Finally, we concatenate features of varying dimensions to achieve precise PD prediction. The structure of MvRNA is shown in Fig. 1.

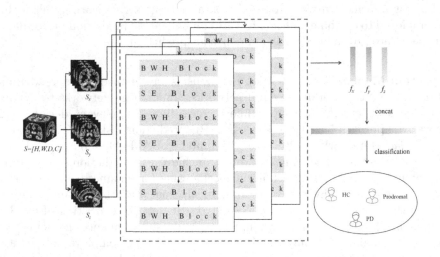

Fig. 1. The overall structure of MvRNA framework.

3.1 Data Representation Based on Multiple Views

While 3D CNN can perform feature extraction while preserving the original data structure, it involves a substantial computational burden. Hence, we propose a split-merge strategy to address this issue. Splitting involves dividing the original 3D MRI into 2D slice sets from different angles, and these sets form the multi-view dataset. Each preprocessed MRI data is represented in the [x, y, z] format. Therefore, we position the medical image within a three-dimensional coordinate system and perform slicing operations along the x-axis, y-axis, and z-axis, respectively.

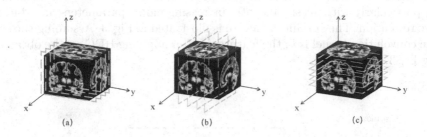

Fig. 2. Schematic diagram of dimensional splitting of 3D data along the x-axis(a), y-axis(b) and z-axis(c) respectively.

As shown in Fig. 2, the dotted box illustrates the partition plane. We define a single original image as S = (H, W, D, C), where H, W, D, and C respectively represent the height, width, depth, and channels (C = 1). The split-merge strategy is depicted in Eq. 1.

$$S_i = \{(H', W', C') \mid C' = seg(i), i \in \{x, y, z\}\} \tag{1}$$

$$S = S_x \cup S_y \cup S_z \tag{2}$$

where S_i represents the new data obtained after slicing along a single axis i, H', W', and C' are the updated height, width, and number of channels respectively, and $seg(\cdot)$ represents segmentation.

Taking the x-axis division as an example (Fig. 2(a)), the original 3D data has dimensions of [113, 137, 113, 1]. Following the x-axis segmentation, we obtain S_1, a 2D dataset with 113 channels and dimensions of 113 in height and 137 in width, which can be expressed as $S_x = [113,137,113]$. The segmentation for the y-axis (Fig. 2(b)) and z-axis (Fig. 2(c)) follows the same process as for the x-axis. To preserve information integrity and avoid potential issues arising from dimensionality splitting, datasets from all three perspectives are concatenated to create a multi-view representation dataset, referred to as 'merge', refer to Eq. 2.

3.2   ResNet18 with BWH

ResNet18 preserves shallow features in deeper network layers through the use of skip connections. As convolutional layers are added, deeper features are captured, but the resolution of the image decreases. To address this issue, dilated convolution was introduced. However, dilated convolution can introduce a gridding effect. To mitigate this, [8] proposed hybrid dilated convolution. This method involves the continuous use of multiple dilation rates that meet specific conditions for dilated convolution, thereby achieving full pixel utilization. Inspired by this, we introduced a novel feature extraction module: the Basic block with Hybrid dilated convolution (BWH). We embed the HDC block between two basic blocks to extract additional spatial features before each downsampling step, particularly without significantly increasing model parameters or reducing image resolution. The specific structure is illustrated in Fig. 3. Assuming the size of the convolution kernel is k, the feature f_h after applying HDC can be obtained using Eq. 3.

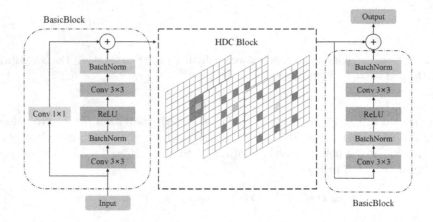

Fig. 3. Structure diagram of BasicBlock with HDC block (BWH).

$$f_h(x; k, r) = \frac{1}{k} \sum_{i=1}^{k} h(x, r) \tag{3}$$

where x represents the feature vector of the input, $r=[r_1,...,r_k]$ is the expansion rate list and h(,) represents the operation of HDC.

The features obtained through the basic block can be calculated using Eq. 4.

$$f_b(x; W_i) = g(W_i \cdot x) + x \tag{4}$$

where W_i represents the weight parameters of the convolution, and $g(\cdot)$ represents the convolution operation and activation function. Finally, we can obtain the feature representation f_{bwh} after the BWH block by using Eq. 5.

$$f_{bwh} = f_b(f_h(f_b(x)))$$ (5)

3.3 Channel Attention Implemented Using SENet

Since we sliced a 3D subject into a 2D slide set, the channel dimension of the image is converted from the original 1D to the number of slices generated by the partition plane.

As described in Sect. 3.1, the dimension of the image is transformed from 1 dimension to 113 dimensions. Therefore, it is crucial to extract features across channels. SENet is an attention mechanism specifically designed for channels, allowing it to adaptively focus on features in channels beneficial for classification while ignoring unimportant channels. As a result, we embed the SE block in the downsampling process, as shown in Fig. 1, and combine it with the proposed BWH to achieve efficient feature extraction from both spatial and channel perspectives.

The feature extraction process described above is shown in Algorithm 1.

Algorithm 1. MvRNA Feature Extraction Algorithm.

Require: 3D MRI X with dimension [H, W, D, C]
    where H:height,W:width, D:depth, C:channel
Ensure: Extracted features f.
 1: **for** dimension $\in X$ **do:**
 2:      Transform X, convert dimension to channel.
 3:      **for** single-view $\in X$ **do:**
 4:          Convolution, Normalization;
 5:          ReLU, Max Pooling;
 6:          **while** n < num(block) **do:**
 7:              **while** n < num(BasicBlock) **do:**
 8:                  Extract features with BWH block;
 9:              **end while**
10:             Extract features with SE block;
11:          **end while**
12:      **end for**
13: **end for**
14: **return** f_i;
15: Concatenate feature vectors $f_i \rightarrow f$;
16: **return** f

---

4 Experiments and Results

4.1 Dataset

The T1-weighted MRI dataset is acquired from the PPMI database (www.ppmi-info.org/data) for this study. We collect a total of 348 instances from the website to create the dataset. The dataset comprises 124 instances from healthy control, 77 from individuals with prodromal Parkinson's disease, and 147 from diagnosed Parkinson's disease patients.

Each 3D MRI data includes dimensions of height, width, and depth and is formatted as NiFTI files. Since the data come from different research centers, they need to be normalized into a unified space, image registration operation is performed on the raw data using MATLAB. All the data are normalized to the MNIPD25-T1MPRAGE-1 mm template space [32]. After image registration, each image can be represented as [113, 137, 113, 1], where each parameter corresponds to height, width, depth, and channel, respectively. We divide all MRI images downloaded from PPMI into training set, validation set and test set and divide them according to the ratio of 8:1:1. Each MRI data instance is transformed along different axes to generate various view sets. Specifically, segmenting 3D MRI data along the x-axis results in sagittal (SAG) channels with 113 channels, along the y-axis results in coronal (COR) channels with 137 channels, and along the z-axis results in an axial (AXI) channel set with 113 channels. Comprehensive statistical results are provided in Table 1 to describe this data. It's important to note that the channel data listed in Table 1 only pertains to a single MRI data instance.

Table 1. Segmentation of a single 3D MRI.

Class	SAG Channel	COR Channel	AXI Channel
HC	113	137	113
Prodromal	113	137	113
PD	113	137	113

4.2 Experimental Settings

The experiments were conducted using Python on a GPU server (equipped with an Intel(R) Xeon CPU E5-2620 V4 @2.10 GHz, 128 GB RAM, and an NVIDIA TITAN Xp GPU with 12 GB memory). In our experiments, Experiments are conducted in 20 iterations, and due to the complexity of the data, the batch size is set to 8. Furthermore, the most effective number of epochs is 100, and the best learning rate is 1e-6. Additionally, we utilize the Adam optimizer to expedite the model's convergence. We set the weight decay to 5e-4, a value determined through experimentation. Additionally, we selected Precision, Recall, F1-score and Accuracy as model evaluation metrics.

4.3 Experimental Results and Analysis

In this section, we introduce the comparison models and the comparison results between each model and our proposed method under the same experimental conditions.

Our model is compared with 3D VGG16, 3D DenseNet121, and 3D ResNet18 equipped with BAM and SA. The models share the same parameter settings; however, due to the complexity of the data and limitations in computational resources, the batch size for 3D VGG16 is set to 4. The changes in training loss and validation loss for all models throughout the entire training process are shown in Fig. 4 and Fig. 5, respectively.

Table 2 displays the comparative performance of our proposed model and the baseline, utilizing the same dataset. The figures presented in the Table 2 represent the average values of the performance metrics and the best performance indicators have been bolded. Our experimental results yielded an accuracy of 81.84%, showcasing the best prediction performance when compared to the baseline model. Moreover, our method attains a precision of 0.85, recall of 0.79, and F1-score of 0.81, all of which represent the best performance.

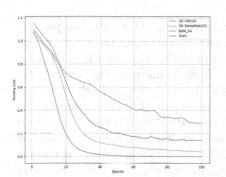
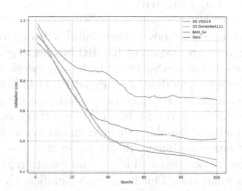

Fig. 4. The training loss of all models.   **Fig. 5.** The validation loss of all models.

Table 2. Comparison of results between different models.

Model	Precision	Recall	F-score	Accuracy
3D VGG16	0.83	0.74	0.74	78.42%
3D DenseNet121	0.83	0.76	0.76	77.88%
3D ResNet18 + BAM + SA	0.82	0.79	0.79	81.06%
MvRNA(ours)	**0.85**	**0.79**	**0.81**	**81.84%**

4.4   Ablation Experiment

To verify the effectiveness of adding the HDC (Hierarchical Dilated Convolution) module for the task of predicting Parkinson's disease, an ablation experiment was conducted. The results, as shown in Table 3, indicate that after incorporating the HDC module, the model achieved better results across various metrics due to the increased range of the receptive field. This demonstrates the efficacy of the HDC module in enhancing the performance of Parkinson's disease prediction algorithms.

Table 3. Comparison of results between different models.

Model	Precision	Recall	F-score	Accuracy
MvRNA(without HDC)	0.79	0.75	0.75	76.76%
MvRNA	**0.85**	**0.79**	**0.81**	**81.84%**

5   Conclusion

We present a novel deep neural network for predicting PD. Firstly, we perform axial segmentation of 3D MRI images to generate a multi-view dataset. Secondly, we employ ResNet18 as the base network and introduce the novel BWH feature extraction module in combination with HDC. This module not only preserves the resolution of medical images but also extends the receptive field. Furthermore, we integrate the SE block into our framework to enhance feature representation. Finally, we combine these features to achieve accurate PD prediction. The results of our experiments demonstrate that the novel MvRNA framework maximizes the utilization of all feature information in 3D MRI data while reducing computational load. This framework significantly enhances the accuracy of Parkinson's disease recognition. It's important to note that, unlike other studies, our approach addresses the multi-categorization problem associated with Parkinson's disease.

Acknowledgements. This work was supported by the National Natural Science Foundation of China (No. 61976247), the Key Research and Development Program in Sichuan Province of China (No. 2023YFS0404), the Central guiding Local Science and Technology Development Fund (No. 23ZYZYTS0189), the Fundamental Research Funds for the Central Universities (Nos. 2682022KJ045 and 2682023ZTPY081), and the Open Research Fund Program of Data Recovery Key Laboratory of Sichuan Province (No. DRN2203).

References

1. Dmitriev, K., Marino, J., Baker, K., Kaufman, A.E.: Visual analytics of a computer-aided diagnosis system for pancreatic lesions. IEEE Trans. Visual Comput. Graphics **27**(3), 2174–2185 (2021). https://doi.org/10.1109/TVCG.2019. 2947037

2. Han, L., Kamdar, M.R.: MRI to MGMT: predicting methylation status in glioblastoma patients using convolutional recurrent neural networks. In: Pacific Symposium on Biocomputing Pacific Symposium on Biocomputing, vol. 23, p. 331 (2018)

3. Tang, Z., Xu, Y., Jin, L., Aibaidula, A., Shen, D.: Deep learning of imaging phenotype and genotype for predicting overall survival time of glioblastoma patients. IEEE Trans. Med. Imaging **PP**(99), 1 (2020)

4. Thuseethan, S., Rajasegarar, S., Yearwood, J.: Detecting micro-expression intensity changes from videos based on hybrid deep CNN. In: Yang, Q., Zhou, Z.-H., Gong, Z., Zhang, M.-L., Huang, S.-J. (eds.) PAKDD 2019. LNCS (LNAI), vol. 11441, pp. 387–399. Springer, Cham (2019). https://doi.org/10.1007/978-3-030-16142-2_30

5. Xu, C., Tao, D., Xu, C.: A survey on multi-view learning. Computer Science (2013)

6. Zhou, W., Wang, H., Yang, Y.: Consensus Graph Learning for Incomplete Multi-view Clustering. In: Yang, Q., Zhou, Z.-H., Gong, Z., Zhang, M.-L., Huang, S.-J. (eds.) PAKDD 2019. LNCS (LNAI), vol. 11439, pp. 529–540. Springer, Cham (2019). https://doi.org/10.1007/978-3-030-16148-4_41

7. Hu, J., Shen, L., Sun, G.: Squeeze-and-excitation networks. In: 2018 IEEE/CVF Conference on Computer Vision and Pattern Recognition, pp. 7132–7141 (2018). https://doi.org/10.1109/CVPR.2018.00745

8. Wang, P., et al.: Understanding convolution for semantic segmentation. In: 2018 IEEE Winter Conference on Applications of Computer Vision (WACV), pp. 1451–1460 (2018). https://doi.org/10.1109/WACV.2018.00163

9. Marek, K., Jennings, D., Lasch, S., Siderowf, A., Taylor, P.: The Parkinson progression marker initiative (PPMI). Prog. Neurobiol. **95**(4), 629–635 (2011)

10. Krizhevsky, A., Sutskever, I., Hinton, G.: Imagenet classification with deep convolutional neural networks. In: Advances in Neural Information Processing Systems, vol. 25, no. 2 (2012)

11. Simonyan, K., Zisserman, A.: Very deep convolutional networks for large-scale image recognition. arXiv e-prints (2014)

12. Szegedy, C., et al.: Going deeper with convolutions. In: 2015 IEEE Conference on Computer Vision and Pattern Recognition (CVPR), pp. 1–9 (2015). https://doi.org/10.1109/CVPR.2015.7298594

13. Zhu, S.: Early diagnosis of Parkinsons disease by analyzing magnetic resonance imaging brain scans and patient characteristics (2022)

14. Erdaş, Ç.B., Sümer, E.: A deep learning method to detect Parkinson's disease from MRI slices. SN Comput. Sci. **3**(2), 1–7 (2022). https://doi.org/10.1007/s42979-022-01018-y

15. He, K., Zhang, X., Ren, S., Sun, J.: Deep residual learning for image recognition. In: 2016 IEEE Conference on Computer Vision and Pattern Recognition (CVPR), pp. 770–778 (2016). https://doi.org/10.1109/CVPR.2016.90

16. Chhabra, M., Kumar, R.: An efficient ResNet-50 based intelligent deep learning model to predict pneumonia from medical images. In: 2022 International Conference on Sustainable Computing and Data Communication Systems (ICSCDS), pp. 1714–1721 (2022). https://doi.org/10.1109/ICSCDS53736.2022.9760995

17. Gao, S.H., Cheng, M.M., Zhao, K., Zhang, X.Y., Yang, M.H., Torr, P.: Res2Net: a new multi-scale backbone architecture. IEEE Trans. Pattern Anal. Mach. Intell. **43**(2), 652–662 (2021). https://doi.org/10.1109/TPAMI.2019.2938758
18. Wang, Y., et al.: SwinMM: masked Multi-view with Swin transformers for 3D medical image segmentation. In: Greenspan, H., et al. (ed.) MICCAI 2023. MICCAI 2023. LNCS, vol. 14222. Springer, Cham. (2023). https://doi.org/10.1007/978-3-031-43898-1_47
19. Liu, D., Gao, Y., Zhangli, Q., Yan, Z., Zhou, M., Metaxas, D.: Transfusion: multi-view divergent fusion for medical image segmentation with transformers. In: Wang, L., Dou, Q., Fletcher, P.T., Speidel, S., Li, S. (eds.) MICCAI 2022. LNCS, vol. 13435. Springer, Cham (2022). https://doi.org/10.1007/978-3-031-16443-9_47
20. Liu, J., Pan, Y., Wu, F.X., Wang, J.: Enhancing the feature representation of multi-modal MRI data by combining multi-view information for MCI classification. Neurocomputing **400**, 322–332 (2020)
21. Xue, Z., Zhang, T., Lin, L.: Progress prediction of Parkinson's disease based on graph wavelet transform and attention weighted random forest. Expert Syst. Appl. **203**, 117483 (2022)
22. Zhang, Y., Lei, H., Huang, Z., Li, Z., Liu, C.M., Lei, B.: Parkinson's disease classification with self-supervised learning and attention mechanism. In: 2022 26th International Conference on Pattern Recognition (ICPR), pp. 4601–4607 (2022). https://doi.org/10.1109/ICPR56361.2022.9956213
23. Park, J., Woo, S., Lee, J.Y., Kweon, I.S.: BAM: Bottleneck attention module (2018)
24. Zhang, Q.L., Yang, Y.B.: SA-Net: shuffle attention for deep convolutional neural networks. In: ICASSP 2021 - 2021 IEEE International Conference on Acoustics, Speech and Signal Processing (ICASSP), pp. 2235–2239 (2021). https://doi.org/10.1109/ICASSP39728.2021.9414568
25. Liu, M., Yang, J.: Image classification of brain tumor based on channel attention mechanism. J. Phys: Conf. Ser. **2035**(1), 012029 (2021). https://doi.org/10.1088/1742-6596/2035/1/012029
26. Zhou, Q., et al.: Grading of hepatocellular carcinoma using 3d SE-DenseNet in dynamic enhanced MR images. Comput. Biol. Med. **107**, 47–57 (2019). https://doi.org/10.1016/j.compbiomed.2019.01.026
27. Linqi, J., Chunyu, N., Jingyang, L.: Glioma classification framework based on SE-ResNeXt network and its optimization. IET Image Processing **2**(16), 596–605 (2022)
28. Luo, M., et al.: A multi-granularity convolutional neural network model with temporal information and attention mechanism for efficient diabetes medical cost prediction. Comput. Biol. Med. **151**, 106246 (2022). https://doi.org/10.1016/j.compbiomed.2022.106246
29. Yu, F., Koltun, V.: Multi-scale context aggregation by dilated convolutions. In: ICLR (2016)
30. Jiang, W., Liu, M., Peng, Y., Wu, L., Wang, Y.: HDCB-Net: a neural network with the hybrid dilated convolution for pixel-level crack detection on concrete bridges. IEEE Trans. Industr. Inf. **17**(8), 5485–5494 (2021). https://doi.org/10.1109/TII.2020.3033170
31. Zhao, X., et al.: D2A U-NET: automatic segmentation of COVID-19 CT slices based on dual attention and hybrid dilated convolution. Comput. Biol. Med. **135**, 104526 (2021)
32. Xiao, Y., Fonov, V., Chakravarty, M.M., Beriault, S., Collins, D.L.: A dataset of multi-contrast population-averaged brain MRI atlases of a Parkinsons disease cohort. Data Brief **12**(C), 370–379 (2017)

Path-Aware Cross-Attention Network for Question Answering

Ziye Luo[1], Ying Xiong[1], and Buzhou Tang[1,2(✉)]

[1] Department of Computer Science, Harbin Institute of Technology, Shenzhen, China
tangbuzhou@gmail.com
[2] Peng Cheng Laboratory, Shenzhen, China

Abstract. Reasoning is an essential ability in QA systems, and the integration of this ability into QA systems has been the subject of considerable research. A prevalent strategy involves incorporating domain knowledge graphs using Graph Neural Networks (GNNs) to augment the performance of pre-trained language models. However, this approach primarily focuses on individual nodes and fails to leverage the extensive relational information present within the graph fully. In this paper, we present a novel model called Path-Aware Cross-Attention Network (PCN), which incorporates meta-paths containing relational information into the model. The PCN features a multi-layered, bidirectional cross-attention mechanism that facilitates information exchange between the textual representation and the path representation at each layer. By integrating rich inference information into the language model and contextual semantic information into the path representation, this mechanism enhances the overall effectiveness of the model. Furthermore, we incorporate a self-learning mechanism for path scoring, enabling weighted evaluation. The performance of our model is assessed across three benchmark datasets, covering the domains of commonsense question answering (CommonsenseQA, OpenbookQA) and medical question answering (MedQA-USMLE). The experimental results validate the efficacy of our proposed model.

Keywords: Meta Path · Cross-Attention · Self-Learning

1 Introduction

Question answering requires confronting questions that may be ambiguous, vague, or require reasoning to obtain an answer. It is an important challenge in the field of natural language processing. In recent years, pre-trained language models [11,12] have been remarkably successful for most of the tasks in natural language processing. Language models seem to encode implicit factual knowledge, but they still struggle to accurately capture the domain knowledge used to accomplish complex reasoning. One possible reason is that this domain knowledge is implicit in the natural language text and is not explicitly informed.

In contrast, language models tend to remember explicit co-occurrence relations between words.

To enhance the reasoning ability of models, a structured knowledge graph with inference capabilities has been introduced in previous research [10,22,26,27] and investigated how it can be used to enhance the inference capabilities of Pre-trained language models (PTM). A part of the current research [22,26,27] aims at enhancing Q&A model inference by constructing graph encoders to integrate subgraph knowledge. This approach often relies on fusing shallow knowledge between nodes in the GNN and language model and requires the construction of complex network structures. In addition, graph encoders constructed in this way tend to focus more on the information of the nodes than the relation information of the edges. In order to focus on the large amount of relational information in the graph, [7,19] processed the relational information in the graph by extracting meta-paths from the knowledge graph and constructing different meta-path encoders. However, the interactions between their language model and meta-path encoders are too crude and neither reflects the model's selection process for inference paths.

When combining knowledge graphs with pre-trained language models, the following three issues need to be considered: (1) How to combine structured information from knowledge graphs with pre-trained language models that deal with unstructured text (2) The feature space of a textual representation is completely different from that of a graphical representation. How to do feature space alignment (3) There are many noisy nodes in the knowledge graph. How do we remove the noise and select the valid information.

For these three problems, we have carried out a study. To address problem one, we employ a form of meta-path to encode relational paths in the knowledge graph. To solve problem two, We introduce a cross-attention mechanism between the text encoder and the path encoder, allowing structured knowledge and contextual semantic information to be passed between the two modalities. How to select the useful information is important. We propose a self-learning mechanism that weights and scores the meta-path representation. The results of the experiments prove the superiority of our proposed model.

Our contributions are as follows:

1. We introduce a cross-attention mechanism that facilitates the interaction between contextual semantic information and path inference information.
2. We put forth a weighted scoring method as part of our self-learning mechanism, offering valuable assistance in selecting pertinent and valid path information from a vast array of available paths.
3. Our approach has yielded remarkable results on prominent commonsense QA datasets, along with the medical QA dataset, demonstrating the effectiveness of our approach.

2   Related Work

Knowledge graphs have been widely used to enhance QA systems due to their ability to provide structured reasoning information. In previous studies, some works have attempted to encode knowledge graph information as unstructured text to directly augment the input information of pre-trained language models, (e.g., Knowledgeable Reader [16], KagNet [10]). Others use the textual representation provided by the language model as instruction information to do inference with the graph encoder represented by GNN (e.g., MHGRN [4]).

PTM+GNN: Some previous works [4,10,20] try to use PTM+GNN approach for bi-directional inference. QA-GNN [26] proposes to use the output vectors of the language model as graph neural network nodes so that they can pass information to the graph encoder and guide the convolution. However, their method uses pooled representation vectors, which limits the information exchange and propagation, and the information propagation is only from LM to GNN. GreaseLM [27] proposes a bidirectional information propagation mechanism to fuse the information from GNN and LM through an MLP, but such an approach is too crude to screen the information. JointLK [22] proposes a bidirectional attention and dynamic cropping mechanism, which achieves good results. However, the space complexity of their approach is too high and the computation is too time-consuming.

PTM+Meta-Path: GNN-based graph encoders tend to be node-centric during inference, downplaying the relational information in the graph. To overcome this drawback, relationship-centric meta-path encoders have emerged, and good results have been achieved in SAFE [7] by encoding meta-paths with unique one-hot embedding and scoring them by MLP. Their method has good interpretability but does not exclude noise paths and fails to exchange information between the two modalities. QAT [19] improves the lack of interaction by feeding the meta-path representation and the text representation into the Transformer directly, but their interaction is brute and dependent on the performance of the Transformer.

3   Task Definition

The commonsense QA or medical QA task described in this paper is a multiple-choice problem. First, a natural language question Q and n options $c_1, c_2, ..., c_n$ are given, and we are asked to choose the most sensible answer from the given n options as the answer. To obtain background knowledge, we need to introduce domain knowledge graphs. A general knowledge graph can be represented as a multi-relational graph $G = (V, R, E)$, where V is the set of all concept nodes, R is the set of relation types, and $E \subseteq \{V \times R \times V\}$ is the set of edges that connect two concept nodes in V.

4   Method

In this section, We present the overall structure of our model. The model structure is shown in Fig. 1. The model consists of four components: text encoder, path encoder, PACA module, and self-learning module.

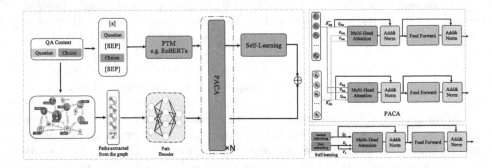

Fig. 1. Overall structure of the PCN (left), detailed structure of PACA (top right), detailed structure of Self-Learning (lower right). Initially, we extract the graph path and employ an MLP for encoding (§4.1). Simultaneously, the domain Pre-trained LM (PTM) encodes the QA content (§4.1). Subsequently, the PACA module plays a pivotal role by integrating these two information representations, facilitating the exchange of modal information and enabling reasoning capabilities (§4.2). Lastly, the self-learning module incorporates the path representation, enabling the selection and scoring of valid paths (§4.3).

4.1   Text Encoder and Path Encoder

Text Encoder. We obtain the representation of the question-choice content based on the pre-trained language model. Given the sequence of tokens $[q; c_i]$ of the input question-choice pair, we obtain the textual representation of the pair via the domain pre-trained language model, denoted as:

$$R_{lm} = f_{enc}(text[q; c_i]) \tag{1}$$

where f_{enc} denotes the output of the language model and $[;]$ represents the concatenation of the question text and the answer text.

Path Encoder. We follow the data processing method of QAT [19] to extract the meta-paths from the knowledge graph. Specifically, we first link the entities in the question texts and answer texts to nodes on the knowledge graph. Based on the linked concepts in the question and each answer candidate, we further extract their neighboring nodes from G and the relational edges connecting nodes to compose a subgraph G^{q,c_i}. We extract all the paths from the entity nodes in the question to the ones in the answer and encode the edge features with one-hot vectors. Specifically, one-hop edges and two-hop paths are encoded as:

$$P_1 = g_{\theta_1}([\phi(h), r, \phi(t), \delta_{h,t}]); \quad P_2 = g_{\theta_2}([\phi(h), r_1, \phi(v_1), r_2, \phi(t), \delta_{h,t}]) \quad (2)$$

Where $g_{\theta_1}(\cdot)$ and g_{θ_2} are MLP. $\delta_{h,t} = f_t - f_h$, where f_t, f_h are the tail and head node features, respectively. $\phi(\cdot) : V \to T_v$ is a one-hot encoder that maps a node $v \in V$ to a node type in T_v.

In our experiments, we use meta-paths with two hops and less, since paths with three hops and more tend to be irrelevant and rare. These paths are denoted as:

$$R_{KG} = \bigcup_{k=1}^{2} \{p_i | p_i \in P_k\} \quad (3)$$

4.2 Path-Aware Cross-Attention

Theoretically, integrating the knowledge of meta-paths and QA texts based on the contextual semantic information provided by the pre-trained language model and the inference knowledge provided by the meta-paths can provide more accurate answers. However, the data formats and data quality of these two data sources are different and need to be transformed and standardized before they can be fused. To distinguish between the two source datasets, we add learnable embedding representations to the output representations of the two encoders:

$$\hat{R}_{lm} = R_{lm} + e_{lm}; \quad \hat{R}_{kg} = R_{kg} + e_{kg}; \quad (4)$$

where both e_{lm} and e_{kg} are learnable embedding representations. In order to make inference information and contextual information fully integrated, we propose a path-aware cross-attention mechanism. The detail of the path-aware cross-attention (PACA for short) structure is shown in Fig. 1:

We perform a linear transformation on the text representation, utilizing the multiplication of $W_{Q_{lm}}$, $W_{K_{lm}}$, and $W_{V_{lm}}$, resulting in the derivation of three distinct vector representations, namely Q_{lm}, K_{lm}, and V_{lm}.

$$Q_{lm} = \widehat{R_{lm}}W_{Q_{lm}}; \quad K_{lm} = \widehat{R_{lm}}W_{K_{lm}}; \quad V_{lm} = \widehat{R_{lm}}W_{V_{lm}}; \quad (5)$$

Likewise, the meta-path representation $\widehat{R_{kg}}$ undergoes multiplication with $W_{Q_{kg}}$, $W_{K_{kg}}$, and $W_{V_{kg}}$ to yield three distinct vectors, namely Q_{kg}, K_{kg}, and V_{kg}, representing the meta-path information.

We employ the query vector Q_{lm} to perform multi-headed attention in conjunction with the meta-path key vector K_{kg}. Subsequently, the attention weights and the value vector V_{kg} are combined through weighting and summation processes to derive the conclusive contextual representation Z_{lm}, which is followed by an LN layer and a Skip connection to get $\widehat{Z_{lm}}$. This representation not only encompasses the inherent contextual semantic information but also captures the inference logic of the paths, thereby enriching the overall comprehension.

$$Z_{lm} = MultiHead(Q_{lm}, K_{kg}, V_{kg}); \quad \widehat{Z_{lm}} = FFN(LN(Z_{lm})) + Z_{lm} \quad (6)$$

In the same manner, we obtain the representation of the path:

$$Z_{kg} = MultiHead(Q_{kg}, K_{lm}, V_{lm}); \quad \widehat{Z_{kg}} = FFN(LN(Z_{kg})) + Z_{kg} \quad (7)$$

We use multiple layers of PACA to enhance the deep interaction between the two modalities, and the experiments in Table 3 show that a certain number of layers of PACA enhances the information propagation between the two modalities.

4.3 Self-learning Based Path Scoring Method

The path-aware cross-attention-based mechanism provides better information interaction capability. However, not all of the extracted meta-paths inherently possess valid inference information. To address this, we introduce a self-learning-based scoring approach, enabling the evaluation and scoring of the extracted paths based on their reliability.

As depicted in Fig. 1, we establish a one-dimensional learnable vector, which serves as the query vector. Additionally, we employ linear transformations to the path vector to derive the key and value vectors. Through the utilization of the attention mechanism, the query vector and key vector collectively calculate the attention weights. Subsequently, the value vector is weighted and consolidated based on the attention weights, resulting in an aggregated representation.

$$Q_L = E_L W_Q; \quad K_L = Z_{kg} W_K; \quad V_L = Z_{kg} W_V \quad (8)$$

$$Z'_{kg} = MHA(Q_L, K_L, V_L) \quad (9)$$

$$S^{ph}_{q,c_i} = MLP_1(FFN(LN(Z'_{kg})) + Z'_{kg}) \quad (10)$$

where S^{ph}_{q,c_i} is the path scoring of the question-answer pair (q, c_i) and E_L is the learnable vector. W_Q, W_k, W_v are three linear transformation matrices.

4.4 Learning and Inference

The scoring of a question-option pair encompasses two components: the language model score and the meta-path score. In Sect. 4.3, we elaborate on the derivation of the meta-path scores. As for the language model score, it is computed using the following approach:

$$S^{lm}_{q,c_i} = MLP_1(MLP_2(h_{<cls>})) \quad (11)$$

where $h_{<cls>}$ is the first token representation in Z_{lm}, that is, the representation of $<cls>$. Ultimately, the scoring of a question-option pair is represented as:

$$S_{q,c_i} = S_{q,c_i}^{lm} + S_{q,c_i}^{ph} \tag{12}$$

For a question-option pair, if it is labeled as positive, the final score is assigned a value of 1; otherwise, it is set to 0. To obtain the final probability, we apply softmax normalization across all question-choice pairs. The model is then trained end-to-end using the cross-entropy loss, ensuring comprehensive optimization.

During the inference stage, the option c_i with the highest score among all available options is selected as the predicted true option.

5   Experiment

5.1   Dataset

We evaluate our model on three multiple-choice question-answering datasets across two domains: CommonsenseQA [23] and OpenBookQA [15] as commonsense QA benchmarks, and MedQA-USMLE [8] as a clinical QA benchmark.

CommonsenseQA is a 5-way multiple choice QA task that requires reasoning with commonsense knowledge, containing 12,102 questions. The test set of CommonsenseQA is not publicly available, and model predictions can only be evaluated once every two weeks via the official leaderboard. Hence, we perform main experiments on the in-house (IH) data split used in [23].

OpenBookQA is a 4-way multiple choice QA task that requires reasoning with elementary science knowledge, containing 5,957 questions. We use the official data split from [15].

MedQA-USMLE is a 4-way multiple-choice question-answering dataset, which requires biomedical and clinical knowledge. The questions are originally from practice tests for the United States Medical License Exams (USMLE). The dataset contains 12,723 questions. We use the original data splits from [8].

5.2   Baseline Models

We compare our model with the following baseline methods, including a fine-tuned PTM, 4 PTM+GNN Encoder, and 2 latest PTM+Meta-Path Encoder.

Fine-Tuned PTM. Without adding any external knowledge, we directly fine-tune the PTM. For the commonsense QA task, we use language models such as RoBERTa-L [12]. For medical Q&A, we use Sap-BERT [11] to compare various methods.

PTM+GNN. Joint scoring using GNN and PTM. That is, LM is used as a text encoder, and GNN or RN [20] is used as a KG encoder. We compare several representative methods: (1) Relationship network (RN) (2) QA-GNN [26] (3) GreaseLM [27] (4) JointLK [22].

PTM+Meta-Path. Joint scoring using Path Encoder and PTM. That is, encoding meta-paths and interacting with pre-trained language models using different inference methods. We compare two representative methods: (1) SAFE [7] (2) QAT [19].

5.3   Main Result

Table 1. Performance comparison on CommonsenseQA in-house split and test accuracy on OpenBookQA. For CommonsenseQA, we follow the data division method of Lin et al. (2019) and report the in-house Dev (IHdev) and Test (IHtest) accuracy (mean and standard deviation of four runs).

Method	CommonsenseQA		OpenBookQA
	IHdev-Acc. (%)	IHtest-Acc. (%)	Test-Acc. (%)
RoBERTa-Large(w/o KG)	73.1($\pm$0.5)	68.7($\pm$0.6)	64.8($\pm$2.4)
RN [20]	74.6($\pm$0.9)	69.1($\pm$0.2)	65.2($\pm$1.2)
QAGNN [26]	76.5($\pm$0.2)	73.4($\pm$0.9)	67.8($\pm$2.8)
GreaseLM [27]	78.5($\pm$0.5)	74.2($\pm$0.4)	-
JointLK [22]	77.9($\pm$0.3)	74.4($\pm$0.8)	70.3($\pm$0.8)
SAFE [7]	-	74.5($\pm$NA)	-
QAT [19]	79.5($\pm$0.4)	75.4($\pm$0.3)	71.2($\pm$0.8)
PCN(Ours)	79.1($\pm$0.6)	75.6($\pm$0.7)	71.5($\pm$1.7)

CommonsenseQA and OpenBookQA. The experimental results in Table 1 show that the proposed model approach performs best on CommonsenseQA (IHtest) and OpenBookQA datasets, verifying the effectiveness of the model under Commonsense question answering. Our proposed PCN, compared to the vanilla pre-trained language model RoBERTa-Large, has improved by 6.9% in CommonsenseQA and 6.7% in OpenBookQA, demonstrating that meta-paths can integrate effective inference information to enhance the performance of PTM on commonsense question answering task. Additionally, among all PTM+GNN approaches, JointLK performs best thanks to the excellent interaction and graph pruning techniques. However, our PCN outperforms them by 1.2% on CSQA and OBQA, respectively, illustrating the superiority of meta-paths as auxiliary inference information. Our approach also performs best among all PTM+Path Encoder models. QAT has achieved the best result among PTM+Path Encoder approaches in the past with its excellent path encoder. Our approach adds path-aware cross-attention and self-learning mechanisms to the path representation of QAT and achieves the best results. This fully demonstrates the superiority of our approach.

MedQA-USMLE. To evaluate the domain generalizability of our method, we carry out experiments on MedQA-USMLE with results shown in Table 2. The table demonstrates that our method surpasses previous methods in achieving the best results on this dataset. This result illustrates the excellent domain generalization ability of our method.

Table 2. Test accuracy comparison on MedQA-USMLE.

Method	Accuracy (%)
SapBERT-Base [11]	37.2
QA-GNN [26]	38.0
GreaseLM [27]	38.5
QAT [19]	39.3
PCN (Ours)	39.7

6 Analysis

6.1 Ablation Studies

To thoroughly analyze the role of each component in the model, we develop ablation experiments on the test set on OpenBookQA:

We find that the final performance of the model is reduced by at least 2.4% when either path or text is removed from the attention of the module. When we completely break the interaction between these two modules, the model's score decreases significantly, to 69.4%. When the self-learning module is removed, the prediction accuracy drops by 1.6%. In order to verify that the meta-path works, we remove the meta-path encoder and the subsequent interaction module, and the model's performance is sharply reduced by 8%. We also conduct experiments on the number of layers of interaction to find the most appropriate interaction depth and find that the best performance is achieved when the number of layers is 2, which is roughly in line with previous experiments [5,26].

6.2 Model Interpretability

To explore the specific inference process of our proposed PACA module and Self-Learning module, we analyse the Attention weights in the modules. A concrete example is illustrated in Fig. 2.

Table 3. Ablation study of our model components, using the OpenBookQA test set. We choose one of the four random seeds for the ablation experiment.

Model	Test Acc.(%)
PCN (ours)	73.0
w/o meta-path	64.8
w/o PACA	69.4
w/o PACA (Path)	70.2
w/o PACA (Text)	70.6
w/o self-learning	71.4

PACA Layers	Test acc
$L = 1$	71.6
$L = 2$	73.0
$L = 3$	70.6
$L = 4$	70.6

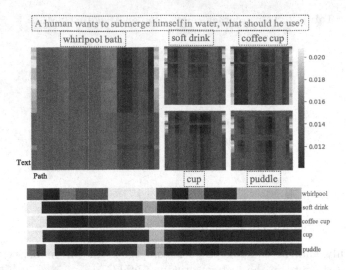

Fig. 2. Attention visualisation in PACA and Self-Learning. (above) The Attention matrix for text and paths in PACA. (below) The Attention matrix of the self-learning vector and paths

As the diagram shows, for the question "A human wants to submerge himself in water, what should he use?", our model has a larger bright range for the correct option "whirlpool bath". This means for the correct option the model finds more correlations and supporting details between the two representations, text and path, which enriches the inference process and enhances the robustness of the model.

6.3   Quantitative Analisis

To further analyse the capabilities of the model, we explore (1) what types of problems the model's performance improvement is reflected in (2) whether the model can solve more complex problems (e.g. problems containing negation and complex long sentences). We compare the best results of the previous meta-paths QAT with our model, which are shown in Table 4.

Negation. To investigate the model's reasoning ability in negative sentences, we extract negative sentences from the test set in commonsenseQA. We treat sentences containing negative terms as negative sentences. Based on the results in Table 4, our PCN outperforms the previous PTM+GNN method in reasoning about negation. Although we were slightly less effective than QAT on dev, we perform on par with QAT on negation, indicating that our model has a stable judgment ability for negative samples. We conjecture that it is the PACA mechanism that allows the model to understand negation in sentences well.

Question with Fewer/More Entities. The complexity of a problem can usually be measured according to the number of entities extracted from the problem.

Table 4. Performance on questions with negative words and fewer/more entities. The questions are retrieved from the CommonsenseQA IDev set. We choose one of the four random seeds for the experiment.

Method	Number	PTM+GNN	QAT	Ours
ITese-Acc (Overall (%))	1221	77.9	79.8	79.7
Questions w/negation	133	75.9	79.0	79.0
Questions w/<7 entities	723	76.2	79.9	79.6
Questions w/>7 entities	498	77.9	79.7	79.8

Following the NER method in QAGNN [26], we obtain the number of entities in the problem, use a threshold of 7 entities, and divide them into two categories: containing fewer entities (<7) and more entities (>7). As Table 4 shows, our method showcases notable enhancements in scenarios involving both fewer and greater than 7 entities. Furthermore, when compared to QAT, our approach exhibits a slight improvement in cases where the entity count exceeds 7. This shows that our method is robust and better at removing noisy information when faced with complex problems.

7 Conclusion

We introduce a novel QA model that leverages a path-aware cross-attention mechanism to facilitate inference. Our model incorporates meta-paths, which possess rich inference information, as auxiliary information. Through multiple layers of cross-attention, the textual representation obtains inference information about the relationship, while the contextual semantics is encoded into the meta-path. Additionally, we introduce a self-learning scoring mechanism that dynamically assigns weights to the meta-paths. We perform a series of exploratory experiments on three distinct datasets encompassing two domains. The experimental findings strongly validate the success and effectiveness of our proposed approach.

Acknowledgements. We thank the reviewers for their insightful comments and valuable suggestions. This study is partially supported by National Key R&D Program of China (2021ZD0113402), National Natural Science Foundation of China (62276082), Major Key Project of PCL (PCL2021A06), Shenzhen Soft Science Research Program Project (RKX20220705152815035), Shenzhen Science and Technology Research and Development Fund for Sustainable Development Project (GXWD20231128103819001, No. KCXFZ20201221173613036) and the Fundamental Research Fund for the Central Universities (HIT.DZJJ.2023117).

References

1. Alsentzer, E., et al.: Publicly available clinical BERT embeddings. In: Proceedings of the 2nd Clinical Natural Language Processing Workshop (2019)
2. Chen, D., Li, Y., Yang, M., Zheng, H.T., Shen, Y.: Knowledge-aware textual entailment with graph attention network. In: Proceedings of the 28th ACM International Conference on Information and Knowledge Management (2019)
3. Devlin, J., Chang, M.W., Lee, K., Toutanova, K.: BERT: pre-training of deep bidirectional transformers for language understanding (2019)
4. Feng, Y., Chen, X., Lin, B.Y., Wang, P., Yan, J., Ren, X.: Scalable multi-hop relational reasoning for knowledge-aware question answering. In: EMNLP (2020)
5. Guan, X., Cao, B., Gao, Q., Yin, Z., Liu, B., Cao, J.: CORN: co-reasoning network for commonsense question answering. In: Proceedings of the 29th International Conference on Computational Linguistics, Gyeongju, Republic of Korea, October 2022. International Committee on Computational Linguistics (2022)
6. Gururangan, S., et al.: Don't stop pretraining: adapt language models to domains and tasks. In: Proceedings of ACL (2020)
7. Jiang, J., Zhou, K., Wen, J.R., Zhao, X.: *great truths are always simple*: a rather simple knowledge encoder for enhancing the commonsense reasoning capacity of pre-trained models. In: Findings of the Association for Computational Linguistics (2022)
8. Jin, D., Pan, E., Oufattole, N., Weng, W.H., Fang, H., Szolovits, P.: What disease does this patient have? A large-scale open domain question answering dataset from medical exams. Appl. Sci. **11**(14), 6421 (2021)
9. Lee, J., et al.: BioBERT: a pre-trained biomedical language representation model for biomedical text mining. Bioinformatics **36**(4), 1234–1240 (2020)
10. Lin, B.Y., Chen, X., Chen, J., Ren, X.: KagNet: knowledge-aware graph networks for commonsense reasoning. In: EMNLP-IJCNLP 2019 (2019)
11. Liu, F., Shareghi, E., Meng, Z., Basaldella, M., Collier, N.: Self-alignment pretraining for biomedical entity representations. In: Proceedings of the 2021 Conference of the North American Chapter of the Association for Computational Linguistics: Human Language Technologies (2021)
12. Liu, Y., et al.: RoBERTa: a robustly optimized BERT pretraining approach. arXiv preprint arXiv:1907.11692 (2019)
13. Lv, S., et al.: Graph-based reasoning over heterogeneous external knowledge for commonsense question answering. In: AAAI 2020 (2020)
14. Lv, S., et al.: Graph-based reasoning over heterogeneous external knowledge for commonsense question answering. Proc. AAAI Conf. Artif. Intell. **34**(05), 8449–8456 (2020)
15. Mihaylov, T., Clark, P., Khot, T., Sabharwal, A.: Can a suit of armor conduct electricity? A new dataset for open book question answering. In: Proceedings of the 2018 Conference on Empirical Methods in Natural Language Processing (2018)
16. Mihaylov, T., Frank, A.: Knowledgeable reader: enhancing cloze-style reading comprehension with external commonsense knowledge. In: Proceedings of the 56th Annual Meeting of the Association for Computational Linguistics (2018)
17. Mihaylov, T., Frank, A.: Knowledgeable reader: enhancing cloze-style reading comprehension with external commonsense knowledge. In: ACL (2018)
18. Pan, X., et al.: Improving question answering with external knowledge. In: Proceedings of the 2nd Workshop on Machine Reading for Question Answering, November 2019

19. Park, J., et al.: Relation-aware language-graph transformer for question answering (2022)
20. Santoro, A., et al.: A simple neural network module for relational reasoning. In: Guyon, I., et al. (eds.) Advances in Neural Information Processing Systems, vol. 30 (2017)
21. Schlichtkrull, M., Kipf, T.N., Bloem, P., Van Den Berg, R., Titov, I., Welling, M.: Modeling relational data with graph convolutional networks. In: The Semantic Web: 15th International Conference, ESWC 2018 (2018)
22. Sun, Y., Shi, Q., Qi, L., Zhang, Y.: JointLK: joint reasoning with language models and knowledge graphs for commonsense question answering. In: Proceedings of the 2022 Conference of the North American Chapter of the Association for Computational Linguistics: Human Language Technologies (2022)
23. Talmor, A., Herzig, J., Lourie, N., Berant, J.: CommonsenseQA: a question answering challenge targeting commonsense knowledge. In: Proceedings of the 2019 Conference of the North American Chapter of the Association for Computational Linguistics: Human Language Technologies (2019)
24. Wang, P., Peng, N., Ilievski, F., Szekely, P., Ren, X.: Connecting the dots: a knowledgeable path generator for commonsense question answering. In: Findings of the Association for Computational Linguistics, EMNLP 2020, November 2020 (2020)
25. Yang, A., et al.: Enhancing pre-trained language representations with rich knowledge for machine reading comprehension. In: Proceedings of the 57th Annual Meeting of the Association for Computational Linguistics (2019)
26. Yasunaga, M., Ren, H., Bosselut, A., Liang, P., Leskovec, J.: QA-GNN: reasoning with language models and knowledge graphs for question answering. In: North American Chapter of the Association for Computational Linguistics (NAACL) (2021)
27. Zhang, X., et al.: GreaseLM: graph reasoning enhanced language models. In: International Conference on Learning Representations (2021)

StyleAutoEncoder for Manipulating Image Attributes Using Pre-trained StyleGAN

Andrzej Bedychaj[✉][iD], Jacek Tabor[iD], and Marek Śmieja[iD]

Jagiellonian University, Kraków, Poland
andrzej.bedychaj@student.uj.edu.pl, marek.smieja@uj.edu.pl

Abstract. Deep conditional generative models are excellent tools for creating high-quality images and editing their attributes. However, training modern generative models from scratch is very expensive and requires large computational resources. In this paper, we introduce StyleAutoEncoder (StyleAE), a lightweight AutoEncoder module, which works as a plugin for pre-trained generative models and allows for manipulating the requested attributes of images. The proposed method offers a cost-effective solution for training deep generative models with limited computational resources, making it a promising technique for a wide range of applications. We evaluate StyleAE by combining it with StyleGAN, which is currently one of the top generative models. Our experiments demonstrate that StyleAE is at least as effective in manipulating image attributes as the state-of-the-art algorithms based on invertible normalizing flows. However, it is simpler, faster, and gives more freedom in designing neural architecture.

1 Introduction

Generative models, such as generative adversarial networks (GAN) [10], variational AutoEncoders (VAE) [20], diffusion models [23], and flow-based generative models [8], have gained popularity due to their ability to create high-quality images, videos, and texts. These models are trained using supervised or unsupervised learning techniques on large datasets. They have transformed the field of artificial intelligence and are used to create innovative applications across various research fields [9,13,22,24,26,28].

StyleGAN [15–17] is one of the most popular generative models used for creating high-quality images, known for its ability to control various aspects of the generated image such as pose, expression, and style. However, the latent space of StyleGAN is highly entangled [16], meaning that the different attributes of the generated image are not easily separable. This makes it challenging to manipulate individual attributes without affecting others, limiting the controllability of the generated images. Furthermore, manipulating the latent space of StyleGAN is a challenging task due to the complex and high-dimensional nature of the model, which limits its practical applications.

© The Author(s), under exclusive license to Springer Nature Singapore Pte Ltd. 2024
D.-N. Yang et al. (Eds.): PAKDD 2024, LNAI 14646, pp. 118–130, 2024.
https://doi.org/10.1007/978-981-97-2253-2_10

Fig. 1. Single-attribute manipulation with StyleAE in StyleGAN's latent space.

There are various methods for simplifying the latent space of generative models. Unsupervised methods include algorithms such as Interface GAN [24], GANSpace [11] and InfoGAN [5], which aim to learn a disentangled representation of the data without requiring any labeled data. Supervised methods like PluGeN [25,29] or StyleFlow [3], on the other hand, require labeled data and involve training an auxiliary model to predict a particular attribute, such as the pose or shape of the generated image. These approaches are essential for improving the practical applications of generative models and making them more useful for a wider range of applications.

While flow-based models such as StyleFlow and PluGeN have shown promising results in disentangling the latent space of StyleGAN, they also have some limitations. One significant limitation is the difficulty in scaling these models to high-resolution images due to their computationally intensive nature [18]. Moreover, invertible models require a large amount of training data to learn the complex data distributions [14], which may be challenging to obtain in some cases. Finally, flow-based models can be sensitive to hyperparameters [27], making them difficult to develop optimal performance.

This paper presents a novel approach, called StyleAE, for modifying image attributes using a combination of AutoEncoders with StyleGAN. StyleAE simplifies the latent space of StyleGAN so that the values of target attributes can be effectively changed. While StyleAE achieves at least as good results as the state-of-the-art flow-based methods, it is computationally more efficient and does not require so large amount of training data, which makes it a practical approach for various applications (see Fig. 1 for sample results).

We conducted an assessment of StyleAE through extensive experiments on datasets containing images of human and animal faces, benchmarking it against state-of-the-art flow-based models. Our findings demonstrate that StyleAE's effectiveness in manipulating the latent space of StyleGAN is on par with that of flow-based models. Our research provides crucial insights into the unique strengths and limitations of StyleAE model, highlighting the potential of our algorithm to improve the effectiveness of latent space manipulation for StyleGAN and other generative models. Furthermore, our approach exhibits superior time complexity, making it a more feasible solution for a wide range of applications.

2 Related Work

Conditional VAE (cVAE) introduced label information integration into generative models but lacks assured latent code and label independence, impacting generation quality; on the other hand, Conditional GAN (cGAN) produces higher-quality outputs but involves more complex training and falls short in manipulating existing examples [13,19].

Fader Networks [21] address this limitation by combining cVAE and cGAN components, utilizing an encoder-decoder architecture and a discriminator predicting attributes. However, Fader Networks struggle with attribute disentanglement, and their training is more challenging than standard GANs. CAGlow [22] takes a different approach, using Glow for conditional image generation based on joint probabilistic density modelling. However, it does not reduce data dimensionality, limiting its applicability to more complex data. Competitive approaches like HifaFace [9], Pie [26], and GANSpace [11] have also been explored.

InterFaceGAN [24] and Hijack-GAN [28] manipulate facial semantics through linear models and a proxy model for latent space traversal. Recent approaches focus on manipulating latent codes of pre-trained networks, where data complexity is less restrictive, making flow models applicable. StyleFlow [3] and PluGeN [29] use normalizing flow modules on GAN latent spaces, employing conditional CNF and NICE, respectively. While StyleFlow is tailored for StyleGAN, PluGeN achieves important results across various models and domains. Further extensions of PluGeN on face images (dubbed PluGeN4Faces) [25] allowed for a significant improvement in the disentanglement between the attributes of the face and the identity of the person.

Our work takes a distinct approach, achieving results comparable to these methods while providing superior attribute decomposition and structural consistency for image datasets, coupled with significantly improved computational efficiency.

3   Methodology

In this section, we give a description of the proposed StyleAE approach to disentangling the latent space of generative models. Before that, we recall the StyleGAN and AutoEncoder architectures, which are the main building blocks of StyleAE .

3.1   Preliminaries

StyleGAN [15–17]*:* is a cutting-edge generative model lauded for its capacity to produce high-quality, realistic images. Its architecture comprises two key elements. Initially, the latent vector z, generated from a standard Gaussian distribution $\mathcal{N}(0, I)$, undergoes mapping to the style space vector w through a series of fully connected layers. Subsequently, this style vector is injected into the following convolutional blocks of the synthesis network, progressively crafting the image in the desired resolution.

While the latent vector z serves as the foundation for image creation, manipulating the image via the style vector w is notably more convenient. As the replicated style vector influences various blocks of the synthesis network, it enables the control of the image's style at different abstraction levels, offering users versatile means to manipulate generated images. However, achieving control over specific attributes without impacting others requires disentangling the style space. In the subsequent sections, we detail how we employ an AutoEncoder to modify the latent space of StyleGAN, enhancing its separability and controllability.

AutoEncoder: architecture comprises an encoder function, mapping input data to a compressed representation in the latent space, and a decoder function, mapping this representation back to the original input space. This is represented as:

$$z = \mathcal{E}(x) \text{ and } x' = \mathcal{D}(z), \quad (1)$$

where $\mathcal{E}$ is the encoder, $\mathcal{D}$ is the decoder, x is the input data, z is the latent space representation, and x' is the reconstructed output.

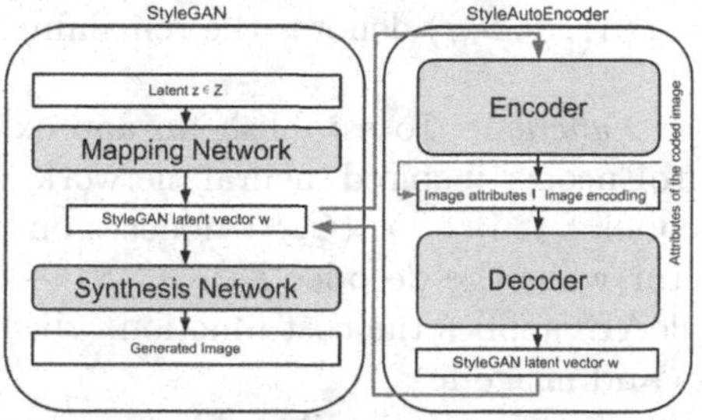

Fig. 2. Architecture design of StylebreakAE. StyleAE maps the style code w of the pre-trained StyleGAN into a target space, where labelled attributes are modelled by individual coordinates.

Training involves minimizing the reconstruction error, typically mean squared error (MSE) or binary cross-entropy (BCE), between x and x'. If the latent space's dimensionality is much lower than the input resolution, the AutoEncoder learns essential features in the compressed representation z. In our research, as we do not emphasize image compression, we do not reduce the dimensionality.

3.2 StyleAutoEncoder

Our goal is to simplify and enhance the manipulation of image attributes by modifying the latent space of StyleGAN using an AutoEncoder. To this end, we develop StyleAE, an AutoEncoder plugin to StyleGAN, which allows for convenient manipulation of the requested attributes and preserving the quality of StyleGAN, see Fig. 2.

Structure of the Target Space: We assume that every instance $x \in X$ is associated with a K dimensional vector of labels $y = (y_1, \ldots, y_K)$. The labels can represent a combination of discrete and continuous features. Our objective is to find a representation of images, where each label is encoded as a separate coordinate. More precisely, let the k-th latent variable c_k correspond to the attribute y_k. By modifying the value of c_k, we would like to change the value of k-th attribute in the image. Since labels do not fully describe the image, additional M variables $(s_1, \ldots, s_M)$ are included to encode the remaining information. Therefore, the latent vector of our new target space is defined as $(c_1, \ldots, c_K, s_1, \ldots, s_M)$.

To construct such a target space, we operate on the representation given by a pre-trained generative model. Although our approach theoretically applies to arbitrary generative models, we consistently use StyleGAN as a base model and fix our attention to synthesis network $G : W \to \mathbb{R}^N$, which maps style vectors to the images. We focus on finding a mapping between the style space W and the target space (C, S), where $C = (C_1, \ldots, C_k)$ describes labelled attributes and $S = (S_1, \ldots, S_M)$ denotes the remaining variables.

Loss Function: To establish an approximately invertible mapping, we use an AutoEncoder-inspired neural network dubbed StyleAE. More precisely, the encoder $\mathcal{E} : W \to (C, S)$ focuses on retrieving the attributes from the style vector while the decoder $\mathcal{D} : (C, S) \to W$ is used to recover the input data. StyleAE applies the cost function, which consists of two components: attribute loss and image loss.

To explain the details behind our loss, let w be the style vector representing the image x with attributes y, $(c, s) = \mathcal{E}(w)$ be the target representation, and $\hat{w} = \mathcal{D}(c, s)$ be the recovered style code. The image loss

$$d_I(x, G(\hat{w})) = \|x - G(\hat{w})\|^2 \tag{2}$$

aims at reconstructing the image from AutoEncoder representation while the attribute loss

$$\sum_{k=1}^{K} d_A(c_k, y_k), \tag{3}$$

aligns target coordinates c_k with image attributes. While the mean-square error (MSE) is the obvious choice for implementing attribute loss d_A, we found that for binary attributes $y_k \in \{0, 1\}$ an alternative loss can be beneficial. Namely, for positive label $y_k = 1$, we calculate the distance between the value c_k returned by AE and the interval $[y_k, \infty) = [1, \infty)$ as follows:

$$d_A^S(c_k, 1) = \max(0, 1 - c_k) \tag{4}$$

This allows us to encode a different style for a binary value, e.g. different type of facial hair. For negative label $y_k = 0$, we use typical MSE since this corresponds to the absence of an attribute.

3.3   Discussion

Image Editing: Attribute manipulation in an image x involves obtaining the style vector w. While a classification mechanism can assist by tagging generated images with desired attributes, this method was applied to our human facial features dataset, categorized externally using the Microsoft Face API.

However, for datasets lacking pre-tagged samples, a mechanism to retrieve w is essential. Literature suggests various methods [2], often using an iterative approach to approximate the StyleGAN latent vector w for a given image x.

Fig. 3. Examples of attributes modification for all of the tested models on FFHQ dataset. One can observe that StyleAE correctly modifies the requested attributes and is less invasive to the remaining characteristics of the image than the competitive flow-based methods.

This process, though, can be time-consuming and does not guarantee optimal results.

In our experiments, we used the method proposed in [1] to retrieve w for the AFHQv2 dataset, which did not initially come with the required latent vectors for the images.

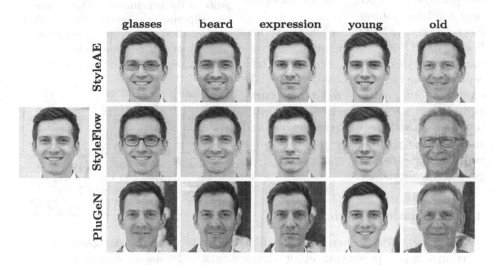

Fig. 4. Attribute modification on a sample image generated from StyleGAN. The generated images by all models exhibit successful changes in the manipulated attributes while maintaining the overall coherence of the image. Our findings indicate that the performance of StyleAE method is comparable to state-of-the-art flow-based models in producing effective attribute manipulation.

Related Models: The proposed simplification of the StyleGAN latent space utilizes an AutoEncoder architecture, offering advantages over flow-based models like PluGeN [29] and StyleFlow [3].

Unlike PluGeN and StyleFlow, which rely on complex architectures for reversible transformations due to challenges posed by flow models, the AutoEncoder is simpler and computationally more efficient. With only two components - Encoder and Decoder, AutoEncoders facilitate easier training and demand fewer computational resources than flow-based models. Additionally, AutoEncoders autonomously learn a disentangled data representation without explicit supervision [4], enhancing adaptability. In contrast, flow-based models often require intricate objective functions, posing challenges in optimization. AutoEncoders also excel in scenarios with limited labelled data, being effectively trainable with smaller datasets.

It's crucial to emphasize that StyleAE presents a unique and innovative approach, benefiting from its inherent simplicity and ease. Its architecture, based on AutoEncoders, allows for task-specific cost functions independent of distribution assumptions, contributing to a straightforward and efficient design.

4 Experiments

In this section, we present the results for manipulating image attributes through the proposed StyleAE . We use two publicly available datasets:
Flickr Faces (FFHQ) [16] and Animal Faces (AFHQv2) [6]. Our method is compared to StyleFlow and PluGeN, which represent the current state-of-the-art. We compare our method regarding structural coherence, effectiveness in generating images with requested changes, and time efficiency. In all experiments, we combine the considered methods with the StyleGAN backbone.

Table 1. Perceptual MSE describing the distance between embeddings of input and modified images. We utilized an ArcFace model to extract the embedding of each image. One may observe that StyleAE obtains significantly lower MSE than state-of-the-art models. Our qualitative experiments also demonstrate that our method preserves a substantial number of other visual facial features when manipulating just one of them.

Attribute	StyleAE	PluGeN	StyleFlow
man	73.241	*56.955*	53.999
woman	36.650	28.300	*23.507*
no glasses	*20.532*	49.325	32.092
glasses	*48.298*	52.424	57.537
no beard	*26.638*	57.297	52.511
beard	*4.099*	5.106	5.422
no smile	47.496	*36.251*	36.471
smile	*13.697*	22.987	17.308

4.1 Evaluation Metrics

The goal of attribute manipulation is to accurately modify designated image attributes while preserving other characteristics. To assess accuracy, we use classification accuracy from an independent multi-label face attribute classifier, trained on datasets not used in training the evaluated models.

To evaluate potential impacts on other image features, we employ three metrics: mean square error (MSE), peak signal-to-noise ratio (PSNR), and structural similarity index (SSIM). PSNR, a logarithmic-scale-modified MSE, and SSIM, assessing visible structures, offer insights, with higher values indicating better performance. Additionally, we calculate perceptual MSE (p-MSE) on image embeddings from a pre-trained ArcFace model [7] sourced from the Python library arcface.

We intentionally avoided using the Frechet Inception Distance (FID) measure due to its unsuitability for attribute manipulation settings. FID primarily compares the distribution of generated images to real ones, which may not precisely reflect changes in specific attributes. As image attribute manipulation alters the distribution of generated data, FID scores can increase, even if image quality and diversity remain constant.

4.2   Models Implementation

StyleAE is a neural network comprising three fully connected layers in the encoder and decoder, each with 512 neurons and PReLU activation. Inputting a 512-dimensional style vector w to the encoder yields a decoded projection $\hat{w}$ from the decoder. The omission of further latent vector compression aligns with flow-based models for a fair comparison.

Training StyleAE with the Adam optimizer at a learning rate of 0.0001 spans 100 epochs. To foster effective learning in both proper reconstruction and desirable attribute organization, we gradually increase the attribute loss weight factor from 0 to 0.3 over the initial 30 epochs.

Baseline comparisons include two popular attribute manipulation plugins: StyleFlow [3] and PluGeN [29]. Both rely on flow-based models: StyleFlow using CNF and PluGeN using NICE. We use publicly available checkpoints for evaluation, avoiding retraining PluGeN or StyleFlow ourselves.

Table 2. Accuracy in modifying consecutive image attributes. We assessed the classifier's predictive accuracy for a specific attribute, incorporating it in the final phase of vector modification for fair comparison across methods. Results indicate our plugin's efficiency, comparable to flow-based models in achieving the goal of attribute modification.

Attribute	StyleAE	PluGeN	StyleFlow
man	0.92	0.91	*0.95*
woman	*0.96*	0.89	0.91
no glasses	0.74	*0.78*	0.67
glasses	0.90	*0.94*	0.88
no beard	*0.96*	0.95	*0.96*
beard	*0.78*	0.55	0.67
no smile	0.99	*1.0*	0.96
smile	*1.0*	*1.0*	0.99

4.3   Manipulation of Facial Features

Setup: In the first experiment, we use the FFHQ dataset, which contains 70 000 high-quality images of resolution 1024 × 1024. All considered methods were trained on 10 000 images generated by StyleGAN. Eight attributes of these images were labelled using the Microsoft Face API (gender, pitch, yaw, eyeglasses, age, facial hair, expression, and baldness).

While the previous studies employing flow-based models utilized the Microsoft Face API for evaluating the accuracy of attribute manipulation, we decided to develop our own classification network due to alterations in the licensing of the Microsoft model. Our classifier is based on the ResNet18 architecture [12] and consisted of 8 target classes aligned with Microsoft's classification system.

Table 3. Average training and inference time. Time required for training and inference on a single NVIDIA GeForce RTX 3080 GPU for each method. Results highlight the substantial speed advantage of our plugin over state-of-the-art flow-based models.

Time	StyleAE	PluGeN	StyleFlow
Training^a	~15 min	~30 min	~1.5 h
Inference^b	~20 s	~5 min	~1 h

^{a} Average time of 1 training epoch.
^{b} Average time taken by generation of 500 images.

Since every method can use different scales to represent the intensity of attributes being modelled, we employed an attribute classifier to apply a minimal modification to obtain the requested value of the attribute. In other words, we gradually modify the attribute until the classifier recognizes the attribute of the generated image with sufficient confidence. If we cannot obtain the requested modification, the classifier returns failure. All the metrics comparing original images with the modified ones, including MSE, PSNR and SSIM, are calculated on minimally modified images.

Results: Sample results of attribute manipulation, presented in Fig. 3 and 4, suggest that StyleAE correctly modifies the requested attributes while preserving the remaining characteristics of the image to a high extent. To support this conclusion with quantitative assessment, we analyze the classification accuracy shown in Table 2 and the remaining metrics describing the difference between the original and modified images, see Tables 1 and 4.

As can be seen from Table 2, in most cases, StyleAE obtains comparable accuracy to PluGeN and better performance than StyleFlow. Closer inspection reveals that it is more accurate at modifying beard attributes, presents very good performance on gender and smile attributes, and slightly lower scores on the glasses feature. Taking into account image differences, reported in Tables 1 and 4, we can observe that StyleAE bet-

Table 4. Structural reconstruction qua-breaklity measures. MSE and PSNR estimate absolute errors, while SSIM considers perceived changes in structural information. Our results indicate that StyleAE maintains greater structural similarity between modifications and base images compared to state-of-the-art flow-based models.

Attribute	Measure	StyleAE	PluGeN	StyleFlow
man	PSNR ↑	17.947	*19.445*	19.607
	SSIM ↑	0.684	*0.733*	0.709
	MSE ↓	0.138	*0.130*	0.153
woman	PSNR ↑	*20.485*	19.026	19.576
	SSIM ↑	0.761	0.733	*0.822*
	MSE ↓	0.105	0.121	*0.083*
no glasses	PSNR ↑	*22.810*	18.764	17.552
	SSIM ↑	*0.830*	0.733	0.800
	MSE ↓	0.075	0.121	*0.055*
glasses	PSNR ↑	*18.790*	18.210	16.840
	SSIM ↑	*0.731*	0.706	0.698
	MSE ↓	*0.121*	0.125	0.156
no beard	PSNR ↑	*19.433*	20.389	28.463
	SSIM ↑	*0.763*	0.728	0.706
	MSE ↓	0.108	0.110	*0.051*
beard	PSNR ↑	*21.425*	20.259	20.769
	SSIM ↑	*0.798*	0.769	0.713
	MSE ↓	*0.088*	0.109	0.092
no smile	PSNR ↑	*19.869*	19.108	19.463
	SSIM ↑	*0.750*	0.739	0.706
	MSE ↓	*0.104*	0.118	0.105
smile	PSNR ↑	*22.817*	22.474	22.701
	SSIM ↑	0.831	0.801	*0.833*
	MSE ↓	*0.073*	0.089	0.076

ter preserves most of the remaining image characteristics. While the competitive approaches excel at modifying attributes with a significant impact on the image structure (such as gender), StyleAE showcases superior performance in manipulating more subtle attributes (such as facial expressions or an addition of the eyeglasses). One can see the comparison of such modifications in Fig. 3. In the case smile attribute the results are comparable. The outcomes demonstrate that our approach can generate images with the desired changes without considerably altering other aspects of the image, as evident from Fig. 4. This highlights the ability of our method to simplify the latent space and produce more meaningful and controllable images.

Moreover, our method benefits significantly from the simplicity of the AutoEncoder approach. It requires fewer parameters and less complex architecture compared to other models, making it less time-consuming and easier to train. In fact, our model achieved comparable results with state-of-the-art flow-based models with only 100 epochs of training. On the other hand, these models require significantly more complicated setups and longer training times, as reported in respective papers. Furthermore, our method was more efficient in generating the same number of images compared to both models, as shown in Table 3.

Table 5. Examples of attributes modification for AFHQv2 dataset. Pose, shape and fur colour seem to be inherited from the input image. The style transfer is not ideal i.e., the quality of the input image features is not reliably copied. Finner traversing over the requested latent attribute could solve that particular issue.

| Input | Cat | Dog | Wild | Input | Cat | Dog | Wild |

4.4 Evaluation on Animal Faces

For the assessment of StyleAE additional potential, a qualitative evaluation was performed utilizing the AFHQv2 dataset. This dataset comprises high-quality images featuring animal faces, categorized into specific classes, namely cats, dogs, and wild animals.

Setup: Given the unavailability of suitable classification tools, the training of StyleAE on generated images, similar to the FFHQ dataset, was unfeasible. In order to overcome this obstacle, we applied a projection technique, as presented in [1], to convert real images of animal faces from the AFHQv2 dataset into latent vectors of StyleGAN. These transformed vectors, marked with labels denoting the original animal class, were employed to train our model. In this specific experiment, the training of StyleAE was conducted for 100 epochs, utilizing the $d_A^S(c_k, 1)$ method described in Eq. (4).

It is essential to highlight that a direct comparison between our results and those of other methods was not feasible in this setting, because previous methods were not trained nor evaluated on the AFHQv2 dataset. Retraining the other models on the reconstructed StyleGAN latent vectors projections dataset would entail potential risks associated with the need to optimize their parameters for our specific task.

Results: In this experiment, we aimed to explore the feasibility of achieving style transfer, specifically in terms of animal type, through the modification of racial attributes.

Our empirical outcomes illustrate the effectiveness of StyleAE plugin. This approach adeptly facilitates the transfer of specific animal classes onto generated animal faces, all while preserving integral structural characteristics like fur color and animal posture, as shown in Table 5. This outcome attests to the robustness and effectiveness of our method in producing images that conform to the desired attributes while retaining essential features.

The generated images display a high level of diversity and realism, highlighting the versatility of our approach. Notably, these results stand out considering the challenges inherent in the animal faces dataset, which encompasses a wide array of shapes and textures.

The results of this experiment suggest that it has the potential to be applied to a variety of image-generation tasks, including those involving complex and diverse datasets.

5   Conclusion

This paper presents StyleAE, a novel method utilizing AutoEncoders to modify StyleGAN latent space efficiently. StyleAE is computationally efficient and capable of generating high-quality images with controllable features across diverse datasets.

Our experiments show that StyleAE achieves comparable attribute modification accuracy to state-of-the-art flow-based models while being less intrusive to other image characteristics. The model's simplicity and time efficiency are key advantages.

Future research could focus on enhancing StyleAE's latent space disentanglement for more precise image control. Exploring advanced optimization methods for model fine-tuning and assessing StyleAE's efficacy in various generative model settings are promising directions.

Acknowledgements. This research has been supported by the flagship project entitled "Artificial Intelligence Computing Center Core Facility" from the Priority Research Area Digi World under the Strategic Programme Excellence Initiative at Jagiellonian University. The work of M. Śmieja was supported by the National Science Centre (Poland), grant no. 2022/45/B/ST6/01117.

References

1. Abdal, R., Qin, Y., Wonka, P.: Image2StyleGAN: how to embed images into the StyleGAN latent space? CoRR abs/1904.03189 (2019)
2. Abdal, R., Zhu, P., Femiani, J., Mitra, N.J., Wonka, P.: CLIP2StyleGAN: unsupervised extraction of StyleGAN edit directions. CoRR abs/2112.05219 (2021)
3. Abdal, R., Zhu, P., Mitra, N.J., Wonka, P.: StyleFlow: attribute-conditioned exploration of StyleGAN-generated images using conditional continuous normalizing flows. CoRR abs/2008.02401 (2020)
4. Cha, J., Thiyagalingam, J.: Disentangling autoencoders (DAE) (2022)
5. Chen, X., Duan, Y., Houthooft, R., Schulman, J., Sutskever, I., Abbeel, P.: InfoGAN: interpretable representation learning by information maximizing generative adversarial nets (2016)
6. Choi, Y., Uh, Y., Yoo, J., Ha, J.: StarGAN v2: diverse image synthesis for multiple domains. CoRR abs/1912.01865 (2019)
7. Deng, J., Guo, J., Zafeiriou, S.: ArcFace: additive angular margin loss for deep face recognition. CoRR abs/1801.07698 (2018)
8. Dinh, L., Sohl-Dickstein, J., Bengio, S.: Density estimation using real NVP (2016)
9. Gao, Y., et al.: High-fidelity and arbitrary face editing. CoRR abs/2103.15814 (2021)
10. Goodfellow, I.J., et al.: Generative adversarial networks (2014)
11. Härkönen, E., Hertzmann, A., Lehtinen, J., Paris, S.: GANSpace: discovering interpretable GAN controls. CoRR abs/2004.02546 (2020)
12. He, K., Zhang, X., Ren, S., Sun, J.: Deep residual learning for image recognition. CoRR abs/1512.03385 (2015)
13. He, Z., Zuo, W., Kan, M., Shan, S., Chen, X.: AttGAN: facial attribute editing by only changing what you want. IEEE Trans. Image Process. **28**(11), 5464–5478 (2019)
14. Ho, J., Chen, X., Srinivas, A., Duan, Y., Abbeel, P.: Flow++: improving flow-based generative models with variational dequantization and architecture design. CoRR abs/1902.00275 (2019)
15. Karras, T., et al.: Alias-free generative adversarial networks. In: Proceedings of the NeurIPS (2021)
16. Karras, T., Laine, S., Aila, T.: A style-based generator architecture for generative adversarial networks. CoRR abs/1812.04948 (2018)
17. Karras, T., Laine, S., Aittala, M., Hellsten, J., Lehtinen, J., Aila, T.: Analyzing and improving the image quality of StyleGAN. CoRR abs/1912.04958 (2019)
18. Kingma, D.P., Dhariwal, P.: Glow: generative flow with invertible 1×1 convolutions (2018)
19. Kingma, D.P., Rezende, D.J., Mohamed, S., Welling, M.: Semi-supervised learning with deep generative models. CoRR abs/1406.5298 (2014)
20. Kingma, D.P., Welling, M.: Auto-encoding variational bayes. arXiv preprint arXiv:1312.6114 (2013)

21. Lample, G., Zeghidour, N., Usunier, N., Bordes, A., Denoyer, L., Ranzato, M.: Fader networks: manipulating images by sliding attributes. CoRR abs/1706.00409 (2017)
22. Liu, R., Liu, Y., Gong, X., Wang, X., Li, H.: Conditional adversarial generative flow for controllable image synthesis. CoRR abs/1904.01782 (2019)
23. Preechakul, K., Chatthee, N., Wizadwongsa, S., Suwajanakorn, S.: Diffusion autoencoders: toward a meaningful and decodable representation. In: CVPR (2022)
24. Shen, Y., Yang, C., Tang, X., Zhou, B.: InterFaceGAN: interpreting the disentangled face representation learned by GANs. CoRR abs/2005.09635 (2020)
25. Suwała, A., Wójcik, B., Proszewska, M., Tabor, J., Spurek, P., Śmieja, M.: Face identity-aware disentanglement in StyleGAN. In: Proceedings of the IEEE/CVF Winter Conference on Applications of Computer Vision, pp. 5222–5231 (2024)
26. Tewari, A., et al.: PIE: portrait image embedding for semantic control. ACM Trans. Graph. **39**(6), 1–14 (2020)
27. Vidal, A., Wu Fung, S., Tenorio, L., Osher, S., Nurbekyan, L.: Taming hyperparameter tuning in continuous normalizing flows using the JKO scheme. Sci. Rep. **13**, 4501 (2023)
28. Wang, H., Yu, N., Fritz, M.: Hijack-GAN: unintended-use of pretrained, black-box GANs. CoRR abs/2011.14107 (2020)
29. Wołczyk, M., et al.: PluGeN: multi-label conditional generation from pre-trained models. In: AAAI 2022 (2022)

SEE: Spherical Embedding Expansion for Improving Deep Metric Learning

Binh Minh Le[iD] and Simon S. Woo[(⊠)][iD]

Dept. of Computer Science & Engineering, Sungkyunkwan University, Suwon, South Korea
{bmle,swoo}@g.skku.edu

Abstract. The primary goal of deep metric learning is to construct a comprehensive embedding space that can effectively represent samples originating from both intra- and inter-classes. Although extensive prior work has explored diverse metric functions and innovative training strategies, much of this work relies on default training data. Consequently, the potential variations inherent within this data remain largely unexplored, constraining the model's robustness to unseen images. In this context, we introduce the Spherical Embedding Expansion (SEE) method. SEE aims to uncover the latent semantic variations in training data. Especially, our method augments the embedding space with synthetic representations based on Max-Mahalanobis distribution (MMD) centers, which maximize the dispersion of these synthetic features without increasing computational costs. We evaluated the efficacy of SEE on four renowned standard benchmarks for the image retrieval task. The results demonstrate that SEE consistently enhances the performance of conventional methods when integrated with them, setting a new benchmark for deep metric learning performance across all settings. Particularly, the proposed method reveals its potency, especially when training with a low-dimensional embedding space and a large number of classes.

Keywords: Deep Metric Learning · Max-Mahalanobis Distribution

1 Introduction

Learning to create a semantic embedding space that possesses both discriminative and generalized properties has been extensively studied across a variety of machine learning tasks. Such tasks encompass image retrieval [18], face verification [6], person re-identification [4], few-shot learning [27], and representation learning [12]. Consequently, deep metric learning, facilitated by neural networks, has garnered significant attention. Its objective is to learn an efficient embedding space in which semantically similar sample are pulled close together, while dissimilar ones are pushed far apart. To this end, various training loss functions, which are broadly categorized into pair-based and proxy-based methods, have been proposed.

In addition to refining the loss function, the development of sampling strategies is also pivotal in enhancing performance. Prevailing methods [38] emphasize the mining of hard samples. However, this often results in a biased model, as it overlooks the majority of easy samples [38,42]. To address this critical issue, recent research [8,41,42] has suggested the use of generative adversarial networks or autoencoders to synthesize

D.-N. Yang et al. (Eds.): PAKDD 2024, LNAI 14646, pp. 131–143, 2024.
https://doi.org/10.1007/978-981-97-2253-2_11

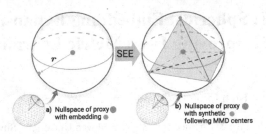

Fig. 1. Motivation of SEE. SEE aims to discover the latent space of training data by synthesis new embedding vectors ($n_{\text{aug}} = 3$) derived from the nullspace $\mathcal{S}^{d-2}_{||r||}$ of a proxy. The synthesize samples follows the MMD properties that enrich for the representation space, benefiting for optimization procedure.

challenging samples using easy ones. Although promising, these approaches have drawbacks, such as model size and optimization issues. Other studies [13, 19] have attempted to synthesize these challenging samples directly from the original embedding, yet they are predominantly constrained to paired-based techniques.

In this paper, we introduce a novel proxy-based synthesis technique in the embedding space of deep metric learning, termed as Spherical Embedding Expansion (SEE). As depicted in Fig. 1, given an embedding and its corresponding proxy anchor, our approach initially explores the proxy's null space, which is represented as a sphere with a radius of r. This ensures consistent distances of the synthetic samples to the anchor. Subsequently, the synthetic embeddings are generated according to the Max-Mahalanobis distribution (MMD) mean vectors [25] (hereinafter referred to as *MMD centers*), allowing for enabling a comprehensive exploration of the embedding space. Our method is straightforward and seamlessly integrates with existing proxy-based metric learning losses. Notably, implementing our approach neither alters the embedding network architecture nor impacts its training speed. Nonetheless, it enhances overall performance, especially in scenarios with low-dimensional spaces, having a large number of classes. Our contributions in this paper are summarized as follows[1]:

- We propose a novel method that augments the embedding space during training by constructing synthetic feature points aligned with MMD centers.
- Through seamless integration, SEE improves proxy-based metric learning losses across numerous backbones and benchmarks without adding parameters.
- SEE excels at densely navigating embedding space, significantly boosting performance, particularly in low-dimensional spaces with datasets that have a large number of training classes.

---

[1] Our code is available at https://github.com/leminhbinh0209/Spherical-Expansion.

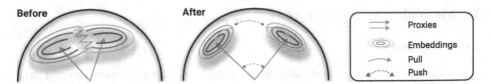

Fig. 2. A schematic representation of our learning objective. Left: Training with a constrained dataset can result in under-represented regions that fuse the representations of two distinct classes. **Middle:** SEE enhances intra-class samples, leading to denser clustering within each class while ensuring distinct separations between different classes.

2 Related Works

Deep Metric Learning. Metric learning losses have been developed to enhance the similarity of feature representations within a designated metric space. Initial methodologies employed pair-based loss functions, striving to refine the embedding space by diminishing the distance between samples of the same class (positive pairs) and amplifying the distance for samples from distinct classes (negative pairs) [5]. More recent advancements in pair-based loss functions have considered relationships between pairs [18,35,36]. However, these methods encounter extreme computational complexity during training, and the difficulty in drawing sufficient informative tuples, as well as the risk of overfitting [14,38]. To overcome these limitations, proxy-based approaches have been introduced in recent studies [1,6,17]. A proxy is a representative of a class or a subset of the training dataset, and it is optimized simultaneously with the networks. Thus, samples from the same classes are pushed near to their proxies but distant from those of other classes. However, while the objective of metric learning is to produce a well-structured embedding space with fast convergence, it heavily relies on the limited training set and may not encompass the entire spectrum of variations present in the validation set. Our proposed method generates samples to train with augmented information by fully discovering the subspace of proxies in the embedding space.

Sample Generation. Recently, the generation of challenging samples that can potentially augment pair-based metric learning losses has garnered considerable interest of researchers [8,41,42]. The central objective of these studies revolves around the creation of hard samples enriched with supplemental semantic information derived from an abundant pool of easy negative samples. Among the pioneering works in this domain are DAML [8] and HTG [41] that both leverage GAN networks. Subsequently, Zheng *et al.* [42] introduced an autoencoder-based hardness-aware deep metric learning framework the generate label-preserving synthetic samples with adjustable degrees of difficulty. Later on, researchers [13,19] introduced linear synthesis and symmetrical synthesis methods. Notably, while these techniques do not rely on a secondary network, their utility is primarily confined to pair-based metric learning losses. On the other hand, our proposed method supports various proxy-based approaches, which have consistently showcased their superiority over their pair-based counterparts.

3   Method

3.1   Preliminary

Consider a deep neural network, denoted as $f : \mathcal{D} \rightarrow \mathcal{Z}$, which maps an input data space $\mathcal{D}$ to an embedding space $\mathcal{Z}$ belonging to a unit d-dimensional hypersphere $\mathcal{S}^{d-1}$. Let $y \in \mathcal{Y} = \{1, ..., C\}$ be the label of an embedding feature z. We define a set of normalized proxies as $\boldsymbol{w} = \{w_1, w_2, ...w_C\}$ and formulate a general proxies-based loss function for metric learning as follows:

$$\mathcal{L}_{\mathrm{ML}} = \mathop{\mathbb{E}}_{(z,y)\sim(\mathcal{Z},\mathcal{Y})} \ell(z|y, \boldsymbol{w}). \tag{1}$$

In Eq. 1, the normalized softmax loss [33] and its variations [6,29,34] are widely used as classification loss ℓ due to their interpretability and performance.

3.2   Spherical Embedding Expansion

Motivation. Our primary purpose of metric learning is to construct a robust and efficient embedding space for *unseen* samples. A common approach is to apply data augmentation techniques such as Mixup [40]. However, these techniques require forwarding augmented inputs to obtain augmented representations. In contrast, we introduce a plug-and-play module, Spherical Embedding Expansion (SEE), which operates in the embedding space $\mathcal{Z}$. This method facilitates a more efficient augmentation process by allowing for the forwarding of un-augmented inputs and performing augmentations directly on the output representations. The conceptual illustration of SEE is provided in Fig. 2. In fact, the main motivation of our work is to address the following requirements: *Given an embedding vector z and its corresponding proxy w_y, how can we efficiently synthesize n_{aug} additional embedding vectors z_i^* that satisfy the following conditions: (1) The distances between the synthetic vectors and w_y, denoted as $d_{(y,i)}$, remain unchanged. (2) The distances between any two synthetic vectors, denoted as $d_{(i,j)}$, are maximized, resulting in optimal dispersion of synthetic vectors in the space.*

The first requirement ensures that the synthetic vectors maintain similar quality to the original input and do not become outliers, or too close to their proxies. The second condition aims to diversify the distribution of the synthetic vectors in the embedding space, enabling the proxies of other classes more challenging and pushing those classes further away from their proxies.

Method. To ensure the first requirement, we define a $||r||$-radius $(d-1)$-dimensional hypersphere as: $\mathcal{S}_{||r||}^{d-2} = \{\mu|\mu \perp w_y \wedge ||\mu|| = ||r||\}$, where $r = z - \langle w_y, z \rangle \cdot w_y$. This space $\mathcal{S}_{||r||}^{d-2}$ represents the null space of w_y, and r is the projection of z onto this defined null space. As a result, for any $\mu \in \mathcal{S}_{||r||}^{d-2}$, a synthetic vector formed by $z^* = \langle w_y, z \rangle \cdot w_y + \mu$ will satisfy $d(y, i) = d_y$. In practice, basis of this space can be constructed using Gram-Schmidt process.

To generate a set of synthetic vectors z^*, one approach is to randomly sample n_{aug} vectors μ from the hyper-spherical space $\mathcal{S}^{d-2}_{||r||}$ and translate them to z^*. However, randomly sampling n_{aug} vectors when $n_{\text{aug}} \ll d$ may not efficiently utilize the space. Conversely, if we choose a large value of n_{aug}, it will scale up the mini-batch size and affect computational efficiency. Hence, to fully utilize the space $\mathcal{S}^{d-2}_{||r||}$ while maintaining efficiency, we need to satisfy the second requirement. This requirement aims to maximize the distance between any two synthetic vectors and achieve optimal dispersion in the embedding space. Inspired by the above analysis, we propose the Max-Mahalanobis center sampling method to induce high-density regions in the hyper-spherical space $\mathcal{S}^{d-2}_{||r||}$, where the MMD [25] is a mixture of Gaussian distributions with an identity covariance matrix and K preset centers denoted as $\boldsymbol{\mu}^* = \{\mu_j^*\}_{[K]}$. The MMD centers are created based on the criterion $\boldsymbol{\mu}^* = \arg\min_\mu \max_{i \neq j} \langle \mu_i, \mu_j \rangle$. This criterion aims to maximize the smallest angle between any two centers, resulting in the most dispersion of the centers across the entire hyper-spherical space [25]. Previous work [25] introduced a fixed set of $\boldsymbol{\mu}^*$. However, in our case, the centers vary depending on $r = \mu_1^*$, which is the image of z in the defined null space as illustrated in Fig. 1. Additionally, the set of centers must satisfy the constraint $\mu_i^* \perp w_y$. To overcome this challenge, we propose a novel algorithm outlined in Algorithm 1 to generate optimal expanded vectors following the centers of MMD within the constrained space $\mathcal{S}^{d-2}_{||r||}$. The main difference between Algorithm 1 and the *GenerateOptMeans* algorithm in [25] are the initialization of $\mu_1^* = r/||r||$ vs. $\mu_1^* = e_1$ (one-hot vector), and subsequent MMD centers formalized in lines $2^{nd} - 5^{th}$. Next, in the 7th line, centers are re-scaled so that $||\mu_i^*|| = ||r||$. By using Algorithm 1, one can easily prove that the set $\{\mu_i^*\}_{n_{\text{aug}}+1}$ are MMD centers, *i.e.*, with $C = ||r||^2$,

$$\mu_i^{*T} \mu_j^* = \begin{cases} C, & i = j \\ -C/n_{\text{aug}}, & i \neq j \end{cases}, \tag{2}$$

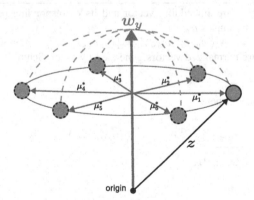

Fig. 3. Illustration of feature point generation. The solid boundary represents the original feature point, while the dotted boundary indicates its synthetic counterparts. For illustration, we depict $\mathcal{S}^{d-2}_{||r||}$ (which represents a hypersphere) as a 2D blue circle.

but they are more flexible than [25] in terms of initialization of μ_1^*. Hence, the *GenerateOptMeans* algorithm in [25] is solely used as a regularization and inapplicable to synthesize new embedding vectors in our case. Consequently, the optimal sampled vectors, as shown in Fig. 3, are produced as (see line 8th of Algorithm 1): $z_i^* = \langle w_y, z \rangle \cdot w_y + \mu_i^*$.

Although the synthetic embedding vectors can diversity its metric space, early applying the expansion can hinder model's optimization. Inspired by curriculum training scheme [2, 16], we selectively apply our method on top-k embedding vector z's such that its d_y in top-k smallest in one mini-batch, denoted as $\mathcal{M}_k$, where k is monotonously increasing after epochs. Therefore, at epoch t^{th}, we have the loss function for synthetic vectors as follows:

$$\mathcal{L}_{\text{SEE}} = \underset{(z,y) \sim (\boldsymbol{\mathcal{Z}}, \boldsymbol{\mathcal{Y}}), d_y \in \mathcal{M}_k}{\mathbb{E}} \left[\sum_{n_{\text{aug}}} \ell(z_i^* | y, \boldsymbol{w}) \right]. \tag{3}$$

Overall Objective. Although our approach can help learning model to be more robust by diversely and optimally exploring the embedding space, it is important to note that a metric learning loss still play crucial roles as it utilizes the ground-truth labels for supervised training. The overall training loss for our proposed approach is formulated as follows:

$$\mathcal{L} = \mathcal{L}_{\text{ML}} + \lambda \mathcal{L}_{\text{SEE}}, \tag{4}$$

where λ is a hyper-parameter that balances the contribution of the original embedding and the synthetic vectors. It is important to note that our proposed approach does not require any modification to the loss function. It can be used as a plug-and-play module in the training process, introducing negligible computational cost.

Algorithm 1. Generate optimal synthetic samples following MMD centers.

Require: $z, w_y \in \mathcal{S}^{d-1}$ are embedding vector and its corresponding proxy vector, and the number of expansion samples n_{aug}.

1: Initialization: Let $r = z - \langle w_y, z \rangle \cdot w_y$, $V = \{v_0, v_1, ..., v_{n_{\text{aug}}+1}\}$, in which $v_0 = w_y$, $v_1 = r/||r||$, and $v_{i>1}$ are normalized vectors generated by Gram-Schmidt process sequentially. Let $\mu_1^* = v_1$.

2: **for** $k = 2$ to $n_{\text{aug}} + 1$ **do**

3:     $\mu_k^* = \sum_{i=1}^{k} \alpha_{ki} v_i$, where

4:     $\begin{cases} \alpha_{k1} = -1/n_{\text{aug}} \\ \alpha_{kj} = -\left(1 + n_{\text{aug}} \cdot \sum_{i=1}^{j-1} \alpha_{ki} \alpha_{ji}\right) / (n_{\text{aug}} \cdot \alpha_{jj}) \\ \alpha_{kk} = \sqrt{1 - \sum_{i=1}^{k-1} \alpha_{ki}^2} \end{cases}$

5: **end for**

6: **for** $k = 1$ to $n_{\text{aug}} + 1$ **do**

7:     $\mu_k^* = ||r|| \cdot \mu_k^*$

8:     $z_k^* = \langle w_y, z \rangle \cdot w_y + \mu_k^*$

9: **end for**

10: **return** The optimal expanded vectors $z_i^*, i \in \overline{2, ..., n_{\text{aug}} + 1}$.

Discussion. Incorporating SEE into a deep metric model yields two pronounced effects. Firstly, the synthetic embeddings foster a more generalized model; trivial samples, augmented with additional information, now facilitate a more comprehensive exploration of under-represented regions. Secondly, the generation of hard negative samples effectively distances the representations from the anchors of other classes. This results in an amplified inter-class differentiation and a more compact intra-class representation.

Consider the normalized softmax loss in Eq. 1 as a simple example. This loss can be expressed in a more generalized form as:

$$\ell(z|y, \boldsymbol{w}) = -\tau \log \frac{e^{-d_y/\tau}}{\epsilon \cdot e^{-d_y/\tau} + \sum_{j \neq y}^{C} e^{-d_j/\tau}} = \tau \log[\epsilon + \sum_{j \neq y} e^{(d_y - d_j)/\tau}]$$

$$= \tau \text{Softplus} \left[\text{LSE}_{j \neq y} (d_y - d_j)/\tau \right], \tag{5}$$

where $\epsilon \geq 0$, and d_j represents the distance between z and the proxy w_j, such as $d_j = -\langle w_j, z \rangle = -\cos\theta_j$, and Softplus$(x) = \log(\epsilon + e^x)$. As illustrated in Fig. 3, the synthetic feature points z_i^* maintain consistent distances to their respective proxy anchors; that is, all d_ys are identical. Furthermore, the Log-Sum-Exp (LSE) function serves as a smooth approximation to the maximum function [23]. Consequently, by probing various directions using MMD centers with optimal dispersion, our SEE is adept at effectively pushing the most challenging negative anchors (represented by the smallest d_j) with every synthetic feature point.

4 Experiments

4.1 Experiment Setting

Datasets. We use the following four popular benchmark datasets for evaluating our method: 1) CUB-200-2011 (CUB) [32], 2) Cars-196 (Cars) [20], 3) Stanford Online Product (SOP) [24] , 4) In-shop Clothes Retrieval (In-Shop) [21].

Metrics. We utilize the Recall@k metric which is used to quantify the fraction of query images that have, within their k-nearest neighbors in the embedding space, at least one instance belonging to the same class.

Networks. To ensure fair comparison with prior works, we adopt ResNet50 [15] (R) with an embedding size of $d = 512$ and three versions of vision transformer architecture: DeiT-S [30] (D), DINO [3] (DN), and ViT-S [7] (V), each with embedding sizes $d = 128$ and $d = 384$. All the backbones are pre-trained on the ImageNet dataset. The linear projection layer used for patch embedding in the variants of the vision transformer is kept frozen during training.

Training. The embedding models are optimized using the AdamW optimizer [22], with a learning rate of 10^{-5} for ViT-S and DeiT-S, and 5×10^{-6} for DINO models. For the ResNet50 backbone, we adopt the training settings from [17]. Training images are randomly resized and cropped to 224×224 using bicubic interpolation and are subjected to random horizontal flips. For test images, we resize them to 256×256 and then apply a center crop of size 224×224. Across our experiments, we use a consistent set of hyperparameters: $\lambda = 0.2$, with $n_{aug} = 16$ for CNN backbones ($d = 512$), and $n_{aug} = 8$ and $n_{aug} = 16$ for ViT backbones ($d = 128$ and $d = 384$, respectively). Our training settings is consistent with prior work without additional training epoch.

Table 1. Performance of metric learning methods on the four datasets. The employed network architectures are represented using abbreviations: R signifies ResNet50 [15], B denotes Inception with BatchNorm, De corresponds to DeiT [30], DN is representative of DINO [3], and V stands for ViT [7]. It should be highlighted that ViT [7] is pre-trained on ImageNet-21k. † indicates models using larger input images.

Methods	Arch.	CUB R@1	R@2	R@4	Cars R@1	R@2	R@4	SOP R@1	R@10	R@100	In-Shop R@1	R@10	R@20
Backbone architecture: CNN													
NSoftmax [39]	R^{128}	56.5	69.6	79.9	81.6	88.7	93.4	75.2	88.7	95.2	86.6	96.8	97.8
MIC [28]	R^{128}	66.1	76.8	85.6	82.6	89.1	93.2	77.2	89.4	94.6	88.2	97.0	-
XBM [37]	R^{128}	-	-	-	-	-	-	80.6	91.6	96.2	91.3	97.8	98.4
XBM [37]	B^{512}	65.8	75.9	84.0	82.0	88.7	93.1	79.5	90.8	96.1	89.9	97.6	98.4
HTL [11]	B^{512}	57.1	68.8	78.7	81.4	88.0	92.7	74.8	88.3	94.8	80.9	94.3	95.8
MS [36]	B^{512}	65.7	77.0	86.3	84.1	90.4	94.0	78.2	90.5	96.0	89.7	97.9	98.5
SoftTriple [26]	B^{512}	65.4	76.4	84.5	84.5	90.7	94.5	78.6	86.6	91.8	-	-	-
PA [17]	B^{512}	68.4	79.2	86.8	86.1	91.7	95.0	79.1	90.8	96.2	91.5	98.1	98.8
NSoftmax [39]	R^{512}	61.3	73.9	83.5	84.2	90.4	94.4	78.2	90.6	96.2	86.6	97.5	98.4
†ProxyNCA++ [29]	R^{512}	69.0	79.8	87.3	86.5	92.5	95.7	80.7	92.0	96.7	90.4	98.1	98.8
Hyp [10]	R^{512}	65.5	76.2	84.9	81.9	88.8	93.1	79.9	91.5	96.5	90.1	98.0	98.7
SEE (ours)	R^{512}	69.3	79.0	87.3	88.5	93.4	95.9	80.3	91.5	96.5	92.8	98.3	98.8
Backbone architecture: ViT													
IRT$_R$ [9]	De^{128}	72.6	81.9	88.7	-	-	-	83.4	93.0	97.0	91.1	98.1	98.6
Hyp [10]	De^{128}	74.7	84.5	90.1	82.1	89.1	93.4	83.0	93.4	97.5	90.9	97.9	98.6
SEE (ours)	De^{128}	75.1	84.1	90.1	85.2	91.5	94.8	83.0	93.1	97.2	91.2	98.0	98.6
Hyp [10]	DN^{128}	78.3	86.0	91.2	86.0	91.9	95.2	84.6	94.1	97.7	92.6	98.4	99.0
SEE (ours)	DN^{128}	78.8	86.5	91.6	89.0	93.6	96.3	84.8	94.1	97.5	92.6	98.6	99.0
Hyp [10]	V^{128}	84.0	90.2	94.2	82.7	89.7	93.9	85.5	94.9	98.1	92.7	98.4	98.9
SEE (ours)	V^{128}	84.1	90.2	93.5	86.8	91.7	95.1	85.9	94.7	97.9	92.8	98.6	99.1
IRT$_R$ [9]	De^{384}	76.6	85.0	91.1	-	-	-	84.2	93.7	97.3	91.9	98.1	98.9
DeiT-S [30]	De^{384}	70.6	81.3	88.7	52.8	65.1	76.2	58.3	73.9	85.9	37.9	64.7	72.1
Hyp [10]	De^{384}	77.8	86.6	91.9	86.4	92.2	95.5	83.3	93.5	97.4	90.5	97.8	98.5
SEE (ours)	De^{384}	78.3	86.5	91.9	88.8	93.7	96.3	83.6	93.4	97.4	91.7	98.1	98.7
DNO [3]	DN^{384}	70.8	81.1	88.8	42.9	53.9	64.2	63.4	78.1	88.3	46.1	71.1	77.5
Hyp [10]	DN^{384}	80.9	87.6	92.4	89.2	94.1	96.7	85.1	94.4	97.8	92.4	98.4	98.9
SEE (ours)	DN^{384}	81.9	88.8	92.9	91.5	95.2	97.3	85.5	94.6	97.9	93.0	98.5	99.1
ViT-S [20]	V^{384}	83.1	90.4	94.4	47.8	60.2	72.2	62.1	77.7	89.0	43.2	70.2	76.7
Hyp [10]	V^{384}	85.6	91.4	94.8	86.5	92.1	95.3	85.9	94.9	98.1	92.5	98.3	98.8
SEE (ours)	V^{384}	85.8	91.4	94.6	88.8	93.8	96.4	86.3	95.0	98.2	93.2	98.6	99.1

4.2 Quantitative Results

In our evaluation, we benchmark the efficacy of our proposed approach against existing state-of-the-art methods across four canonical datasets. For these experiments, we employ the proxy anchor loss as our metric learning objective $\mathcal{L}_{ML}$. To ensure an equitable comparison, the results are systematically tabulated in Table 1, segmented based on the backbone architecture, specifically, CNN to ViT, and the embedding dimensions, namely 128, 384, and 512.

Our experimental findings underscore the effectiveness of our proposed methodology. When compared with other CNN-based techniques, our approach, utilizing ResNet50 as the backbone, consistently outperforms competitors across multiple datasets, with the exception of SOP. It is crucial to mention that ProxyNCA++ [29] employs a more expansive input size, and XBM [37] leverages an extensive memory bank to augment their training process. Nevertheless, our method still surpasses them in most datasets. Specifically, we observe improvements of 2% and 2.4% on the Cars and In-shop datasets, respectively, when compared to ProxyNCA++.

In the context of ViT-based experiments, our proposed methodology exhibits a discernible advantage over competing baselines, particularly with respect to R@1 scores, spanning various embedding dimensions. For instance, on the Cars dataset in comparison with the Hyp method [10], our approach registers performance gains of 3.5%, 3%, and 4.1% when employing the De^{128}, DN^{128}, and V^{128} backbones, respectively. It is noteworthy that, when utilizing the DN^{128} backbone, our method surpasses all other approaches, even those configured with 512-dimensional CNN settings. Furthermore, even on challenging datasets such as SOP or In-shop, our technique continues to demonstrate marked improvements, even at reduced embedding dimensions like 128 and 384. This enhancement can be attributed to the optimal dispersion of synthetic feature points within the embedding space.

Table 2. Recall@k accuracy for proxy-based losses integrated with our synthesis method on the Cars dataset, using CNN-based networks with 512-dimensional embeddings.

Method	Arch.	R@1	R@2	R@4	R@8	R@16
NSoftmax [39]	R^{512}	84.2	90.4	94.4	96.9	-
NSoftmax+SEE	R^{512}	86.5	92.0	95.4	97.4	98.7
CosFace [34]	R^{512}	86.9	92.3	95.3	97.4	98.6
CosFace+SEE	R^{512}	87.1	92.5	95.4	97.5	98.7
ArcFace [6]	R^{512}	86.8	92.1	95.3	97.3	98.7
ArcFace+SEE	R^{512}	87.6	92.8	95.9	97.6	98.7
†ProxyNCA++ [29]	R^{512}	86.5	92.5	95.7	97.7	-
†ProxyNCA++ +SEE	R^{512}	88.3	93.4	96.4	98.0	99.0
PA [17]	B^{512}	86.1	91.7	95.0	97.0	98.3
PA+SEE	B^{512}	86.2	91.9	95.2	97.2	98.4
PA [17]	R^{512}	87.7	92.7	95.5	97.3	98.4
PA+SEE	R^{512}	88.5	93.4	95.9	97.5	98.8

Table 3. Recall@1 accuracy of Proxy-Anchor (PA) [17] across various augmentation types, including our method, using the DeiT [30] network architecture with 128-dimensional embeddings.

Methods	CUB	Cars	SOP	In-Shop
PA[17]	74.7	84.3	82.3	90.4
PA+Mixup	75.0	86.2	82.2	91.0
PA+ManifoldMixup	74.4	85.4	81.9	90.5
PA+SEE (random)	74.6	85.1	82.4	91.2
PA+SEE	75.1	85.2	83.0	91.2

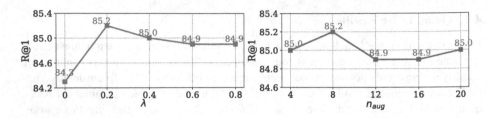

Fig. 4. Ablation studies: augmentation loss weight λ and number of synthesis points n_{aug}. Note that $\lambda = 0$ denotes not using our SEE.

4.3 Ablation Studies

Proxy-based Losses. In this section, we demonstrate the enhancements achieved by our proposed synthesis method when applied to various proxy-based metric learning losses, specifically when integrated with CNN-based architectures. The detailed results are presented in Table 2. As evident from the table, our synthesis approach consistently enhances performance across different proxy-based losses, achieving up to a 1.8% improvement in Recall@1 accuracy. Furthermore, this enhancement persists even as we increase the number of neighbors, k, in Recall@k.

Augmentation Types. In Table 3, we juxtapose our SEE with alternative augmentation strategies, notably Mixup [40] and ManifoldMixup [31], and our variation employing a randomized μ, when integrated with the Proxy-anchor [17] loss using De^{128}. Evidently, each methodology exhibits enhancements over the baseline Proxy-anchor. However, our proposal, characterized by the optimal dispersion of μ, consistently yields performance amplifications across the quartet of datasets. It is worth acknowledging that Mixup manifests a pronounced improvement on the Cars dataset. Yet, for other fine-grained datasets, namely CUB, SOP, and In-shop, its efficacy pales in comparison to our approach. This can be attributed to our method's adeptness at augmenting based on the feature relationship within an image relative to its corresponding anchor.

Impact of Hyperparameters. We explore the influence of two critical hyperparameters on the efficacy of our method: the loss weight of SEE, denoted by λ, and the number of synthetic feature points, n_{aug}. Note that, as mentioned, the radius r depends on the distance between the embedding vector and its proxy, so we do not include it in our studies. For this analysis, we employ the De^{128} backbone and evaluate the performance on the Cars dataset. As depicted in Fig. 4, our method exhibits significant robustness to variations in these hyperparameters, implying consistent performance irrespective of their specific settings.

5 Conclusion

In this work, we have introduced a spherical embedding expansion technique for augmentation within the embedding space, designed to complement existing proxy-based metric learning losses. Within this space, we augment a sample around its anchor by adhering to the MMD centers situated within the anchor's nullspace, thereby ensuring a thorough exploration. Our proposed method is streamlined and straightforward, obviating the need to modify model architecture or incur computational overhead. Empirical results reveal that our approach considerably enhances the efficacy of established proxy-based losses across a range of model architectures and benchmark datasets. Notably, it yields substantial improvements in challenging learning scenarios characterized by low-dimensional spaces and a vast number of classes.

Acknowledgements. This work was partly supported by Institute for Information & communication Technology Planning & evaluation (IITP) grants funded by the Korean government MSIT: (No. 2022-0-01199, Graduate School of Convergence Security at Sungkyunkwan University), (No. 2022-0-01045, Self-directed Multi-Modal Intelligence for solving unknown, open domain problems), (No. 2022-0-00688, AI Platform to Fully Adapt and Reflect Privacy-Policy Changes), (No. 2021-0-02068, Artificial Intelligence Innovation Hub), (No. 2019-0-00421, AI Graduate School Support Program at Sungkyunkwan University), and (No. RS-2023-00230337, Advanced and Proactive AI Platform Research and Development Against Malicious Deepfakes). Lastly, this work was supported by Korea Internet & Security Agency (KISA) grant funded by the Korea government (PIPC) (No.RS-2023-00231200, Development of personal video information privacy protection technology capable of AI learning in an autonomous driving environment).

References

1. Aziere, N., Todorovic, S.: Ensemble deep manifold similarity learning using hard proxies. In: Proceedings of the IEEE/CVF Conference on Computer Vision and Pattern Recognition, pp. 7299–7307 (2019)
2. Bengio, Y., Louradour, J., Collobert, R., Weston, J.: Curriculum learning. In: Proceedings of the 26th Annual International Conference on Machine Learning, pp. 41–48 (2009)
3. Caron, M., et al.: Emerging properties in self-supervised vision transformers. In: Proceedings of the IEEE/CVF International Conference on Computer Vision, pp. 9650–9660 (2021)
4. Chen, W., Chen, X., Zhang, J., Huang, K.: Beyond triplet loss: a deep quadruplet network for person re-identification. In: Proceedings of the IEEE Conference on Computer Vision and Pattern Recognition, pp. 403–412 (2017)
5. Chopra, S., Hadsell, R., LeCun, Y.: Learning a similarity metric discriminatively, with application to face verification. In: 2005 IEEE Computer Society Conference on Computer Vision and Pattern Recognition (CVPR 2005), vol. 1, pp. 539–546. IEEE (2005)
6. Deng, J., Guo, J., Xue, N., Zafeiriou, S.: ArcFace: additive angular margin loss for deep face recognition. In: Proceedings of the IEEE/CVF Conference on Computer Vision and Pattern Recognition, pp. 4690–4699 (2019)
7. Dosovitskiy, A., et al.: An image is worth 16×16 words: transformers for image recognition at scale. arXiv preprint arXiv:2010.11929 (2020)
8. Duan, Y., Zheng, W., Lin, X., Lu, J., Zhou, J.: Deep adversarial metric learning. In: Proceedings of the IEEE Conference on Computer Vision and Pattern Recognition, pp. 2780–2789 (2018)
9. El-Nouby, A., Neverova, N., Laptev, I., Jégou, H.: Training vision transformers for image retrieval. arXiv preprint arXiv:2102.05644 (2021)
10. Ermolov, A., Mirvakhabova, L., Khrulkov, V., Sebe, N., Oseledets, I.: Hyperbolic vision transformers: combining improvements in metric learning. In: Proceedings of the IEEE/CVF Conference on Computer Vision and Pattern Recognition, pp. 7409–7419 (2022)

11. Ge, W., Huang, W., Dong, D., Scott, M.R.: Deep metric learning with hierarchical triplet loss. In: Ferrari, V., Hebert, M., Sminchisescu, C., Weiss, Y. (eds.) ECCV 2018. LNCS, vol. 11210, pp. 272–288. Springer, Cham (2018). https://doi.org/10.1007/978-3-030-01231-1_17
12. Grill, J.B., et al.: Bootstrap your own latent-a new approach to self-supervised learning. Adv. Neural. Inf. Process. Syst. **33**, 21271–21284 (2020)
13. Gu, G., Ko, B.: Symmetrical synthesis for deep metric learning. In: Proceedings of the AAAI Conference on Artificial Intelligence, vol. 34, pp. 10853–10860 (2020)
14. Harwood, B., Kumar BG, V., Carneiro, G., Reid, I., Drummond, T.: Smart mining for deep metric learning. In: Proceedings of the IEEE International Conference on Computer Vision, pp. 2821–2829 (2017)
15. He, K., Zhang, X., Ren, S., Sun, J.: Deep residual learning for image recognition. In: Proceedings of the IEEE Conference on Computer Vision and Pattern Recognition, pp. 770–778 (2016)
16. Huang, Y., et al.: CurricularFace: adaptive curriculum learning loss for deep face recognition. In: Proceedings of the IEEE/CVF Conference on Computer Vision and Pattern Recognition, pp. 5901–5910 (2020)
17. Kim, S., Kim, D., Cho, M., Kwak, S.: Proxy anchor loss for deep metric learning. In: Proceedings of the IEEE/CVF Conference on Computer Vision and Pattern Recognition, pp. 3238–3247 (2020)
18. Kim, S., Seo, M., Laptev, I., Cho, M., Kwak, S.: Deep metric learning beyond binary supervision. In: Proceedings of the IEEE/CVF Conference on Computer Vision and Pattern Recognition, pp. 2288–2297 (2019)
19. Ko, B., Gu, G.: Embedding expansion: augmentation in embedding space for deep metric learning. In: Proceedings of the IEEE/CVF Conference on Computer Vision and Pattern Recognition, pp. 7255–7264 (2020)
20. Krause, J., Stark, M., Deng, J., Fei-Fei, L.: 3D object representations for fine-grained categorization. In: Proceedings of the IEEE International Conference on Computer Vision Workshops, pp. 554–561 (2013)
21. Liu, Z., Luo, P., Qiu, S., Wang, X., Tang, X.: DeepFashion: powering robust clothes recognition and retrieval with rich annotations. In: Proceedings of the IEEE Conference on Computer Vision and Pattern Recognition, pp. 1096–1104 (2016)
22. Loshchilov, I., Hutter, F.: Decoupled weight decay regularization. arXiv preprint arXiv:1711.05101 (2017)
23. Nielsen, F., Sun, K.: Guaranteed bounds on the Kullback-Leibler divergence of univariate mixtures. IEEE Signal Process. Lett. **23**(11), 1543–1546 (2016)
24. Oh Song, H., Xiang, Y., Jegelka, S., Savarese, S.: Deep metric learning via lifted structured feature embedding. In: Proceedings of the IEEE Conference on Computer Vision and Pattern Recognition, pp. 4004–4012 (2016)
25. Pang, T., Du, C., Zhu, J.: Max-mahalanobis linear discriminant analysis networks. In: International Conference on Machine Learning, pp. 4016–4025. PMLR (2018)
26. Qian, Q., Shang, L., Sun, B., Hu, J., Li, H., Jin, R.: SoftTriple Loss: deep metric learning without triplet sampling. In: Proceedings of the IEEE/CVF International Conference on Computer Vision, pp. 6450–6458 (2019)
27. Qiao, L., Shi, Y., Li, J., Wang, Y., Huang, T., Tian, Y.: Transductive episodic-wise adaptive metric for few-shot learning. In: Proceedings of the IEEE/CVF International Conference on Computer Vision, pp. 3603–3612 (2019)
28. Roth, K., Brattoli, B., Ommer, B.: MIC: mining interclass characteristics for improved metric learning. In: Proceedings of the IEEE/CVF International Conference on Computer Vision, pp. 8000–8009 (2019)

29. Teh, E.W., DeVries, T., Taylor, G.W.: ProxyNCA++: revisiting and revitalizing proxy neighborhood component analysis. In: Vedaldi, A., Bischof, H., Brox, T., Frahm, J.-M. (eds.) ECCV 2020. LNCS, vol. 12369, pp. 448–464. Springer, Cham (2020). https://doi.org/10.1007/978-3-030-58586-0_27
30. Touvron, H., Cord, M., Douze, M., Massa, F., Sablayrolles, A., Jégou, H.: Training data-efficient image transformers & distillation through attention. In: International Conference on Machine Learning, pp. 10347–10357. PMLR (2021)
31. Verma, V., Lamb, A., Beckham, C., Najafi, A., Mitliagkas, I., Lopez-Paz, D., Bengio, Y.: Manifold mixup: Better representations by interpolating hidden states. In: International Conference on Machine Learning, pp. 6438–6447. PMLR (2019)
32. Wah, C., Branson, S., Welinder, P., Perona, P., Belongie, S.: The caltech-UCSD birds-200-2011 dataset (2011)
33. Wang, F., Xiang, X., Cheng, J., Yuille, A.L.: NormFace: L2 hypersphere embedding for face verification. In: Proceedings of the 25th ACM International Conference on Multimedia, pp. 1041–1049 (2017)
34. Wang, H., et al.: CosFace: large margin cosine loss for deep face recognition. In: Proceedings of the IEEE Conference on Computer Vision and Pattern Recognition, pp. 5265–5274 (2018)
35. Wang, J., et al.: Learning fine-grained image similarity with deep ranking. In: Proceedings of the IEEE Conference on Computer Vision and Pattern Recognition, pp. 1386–1393 (2014)
36. Wang, X., Han, X., Huang, W., Dong, D., Scott, M.R.: Multi-similarity loss with general pair weighting for deep metric learning. In: Proceedings of the IEEE/CVF Conference on Computer Vision and Pattern Recognition, pp. 5022–5030 (2019)
37. Wang, X., Zhang, H., Huang, W., Scott, M.R.: Cross-batch memory for embedding learning. In: Proceedings of the IEEE/CVF Conference on Computer Vision and Pattern Recognition, pp. 6388–6397 (2020)
38. Wu, C.Y., Manmatha, R., Smola, A.J., Krahenbuhl, P.: Sampling matters in deep embedding learning. In: Proceedings of the IEEE International Conference on Computer Vision, pp. 2840–2848 (2017)
39. Zhai, A., Wu, H.Y.: Classification is a strong baseline for deep metric learning. arXiv preprint arXiv:1811.12649 (2018)
40. Zhang, H., Cisse, M., Dauphin, Y.N., Lopez-Paz, D.: mixup: beyond empirical risk minimization. arXiv preprint arXiv:1710.09412 (2017)
41. Zhao, Yiru, Jin, Zhongming, Qi, Guo-jun, Lu, Hongtao, Hua, Xian-sheng: An adversarial approach to hard triplet generation. In: Ferrari, Vittorio, Hebert, Martial, Sminchisescu, Cristian, Weiss, Yair (eds.) ECCV 2018. LNCS, vol. 11213, pp. 508–524. Springer, Cham (2018). https://doi.org/10.1007/978-3-030-01240-3_31
42. Zheng, W., Chen, Z., Lu, J., Zhou, J.: Hardness-aware deep metric learning. In: Proceedings of the IEEE/CVF Conference on Computer Vision and Pattern Recognition, pp. 72–81 (2019)

Multi-modal Recurrent Graph Neural Networks for Spatiotemporal Forecasting

Nicholas Majeske and Ariful Azad[✉]

Department of Intelligent Systems Engineering, Indiana University Bloomington, Bloomington, IN, USA
{nmajeske,azad}@iu.edu

Abstract. The spatial and temporal dynamics of many real-world systems present a significant challenge to multi-variate forecasting where features of both forms, as well as their inter-dependencies, must be modeled correctly. State-of-the-art approaches utilize a limited set of exogenous features (outside the forecast variable) to model temporal dynamics and Graph Neural Networks, with pre-defined or learned networks, to model spatial dynamics. While much work has been done to model dependencies, existing approaches do not adequately capture the explicit and implicit modalities present in real-world systems. To address these limitations we propose MMR-GNN, a spatiotemporal model capable of (a) augmenting pre-defined (or absent) networks into optimal dependency structures (b) fusing multiple explicit modalities and (c) learning multiple implicit modalities. We show improvement over existing methods using several hydrology and traffic datasets. Our code is publicly available at https://github.com/HipGraph/MMR-GNN.

Keywords: forecasting · spatiotemporal · graph · clustering · traffic · hydrology

1 Introduction

Spatiotemporal systems such as streamflow [13], traffic flow [1, 11], and energy systems [14, 21] possess complex dynamics that span both space and time. Components of these systems often influence one another giving rise to an explicit or implicit dependency structure captured by a graph or network (e.g., road networks, river networks, etc.). In more complex cases, these networks generate *explicit multi-modal signals* such as static and dynamic node features, dynamic dependency structure, and dynamic graph features. Even further, *implicit multi-modal signals* can be observed within these features and must be captured for accurate forecasting. This paper presents MMR-GNN, a machine learning (ML) model capable of forecasting highly complex spatiotemporal systems containing explicit and implicit multi-modal signals.

As an example of multi-modal spatiotemporal systems, we consider the Wabash River basin which spans Indiana and parts of Ohio and Illinois in the USA (see Fig. 1(a)). In this dataset, various hydrological and meteorological features (streamflow, soil water content, temperature, etc.) were recorded from 1929 to 2013 at 1276 unique regions (sub-basins). This basin can be modeled as a graph where sub-basins

© The Author(s), under exclusive license to Springer Nature Singapore Pte Ltd. 2024
D.-N. Yang et al. (Eds.): PAKDD 2024, LNAI 14646, pp. 144–157, 2024.
https://doi.org/10.1007/978-981-97-2253-2_12

denote nodes and streamflow denotes the connectivity pattern. Each node in this graph has *explicit multi-modal signals* including spatiotemporal features such as streamflow, precipitation, snowfall, etc., static features, such as location in latitude-longitude, and shared dynamic features such as seasonality in time-of-year as shown in Fig. 1(b). Furthermore, sub-basin streamflow is the result of broad and localized properties including season, geology, local weather, and tributary streams from neighboring sub-basins. These properties give rise to *implicit multi-modal signals* where particular sub-basins show high feature correlation and form natural clusters (see Fig. 1(c)). Based on these observations in the Wabash River basin, we define a comprehensive system with three challenging aspects:

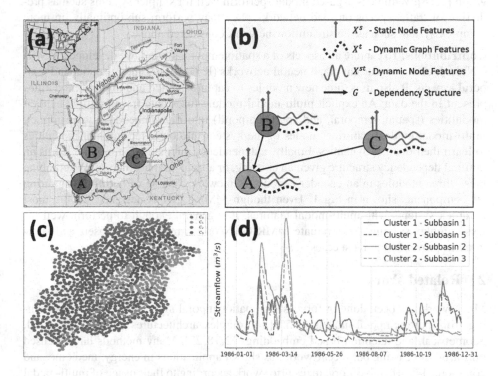

Fig. 1. (a) The Wabash River basin. Three of the 1276 sub-basins are marked with A, B, and C. (b) Three sub-basins (nodes) impact each other through a streamflow network G, generate dynamic node features X^{st} (e.g., streamflow and precipitation), posses static node features X^s (e.g., location), and exhibit dynamic graph features X^t (e.g., seasonality). These features represent *explicit multi-modal signals* of the system. (c) 1276 sub-basins are grouped into four clusters where each cluster includes highly correlated sub-basins (e.g., red and orange clusters capture several major rivers). The strong intra-cluster feature correlation represents *implicit multi-modal signals* of the system. (d) Streamflow of two sub-basins in the orange cluster and two sub-basins in the blue cluster demonstrate this correlation which MMR-GNN can capture. (Color figure online)

- *Explicit multi-modality.* The system has explicit multi-modal spatial, temporal, and spatiotemporal signals as shown in Fig. 1(b).
- *Implicit multi-modality.* The system exhibits implicit modality where node groups exhibit strong feature correlation and form natural clusters as shown in Fig. 1(c).
- *Implicit and explicit dependencies.* Dependency among nodes can be both explicit (e.g., a streamflow network) and implicit (similar streamflow in distant sub-basins).

Despite tremendous progress in spatiotemporal ML, no state-of-the-art (SOTA) method can efficiently capture all explicit and implicit modalities and dependencies in a single model. Most recent methods are based on graph neural networks (GNNs) trained to predict future values of time series at different nodes of a given or learned graph [2,18]. While GNN-based models perform well for simpler systems such as prediction of traffic speeds on road networks, they may perform sub-optimally on more complex system such as the streamflow network considered above.

Contributions. To capture all aspects of a spatiotemporal system, we developed MMR-GNN which uses recurrent graph neural networks (R-GNN) to capture spatial and temporal aspects. It also uses three new modules to capture explicit and implicit modalities present in the data. An explicit multi-modal module fuses data from multiple explicit modalities (spatial, temporal, and spatiotemporal) into rich embeddings. An implicit multi-modal module partitions nodes of the system into observed implicit modalities to learn their distinct dynamics. Finally, a dependency augmentation module learns an optimal dependency structure given an existing, or absent, graph structure. We combine these three modules in an encoder-decoder framework capable of faithfully capturing all components shown in Fig. 1. Even though MMR-GNN is designed for the most complex systems with multi-modal features, it is general enough to perform well for less-complex systems. We evaluate MMR-GNN on six real-world datasets and show good improvement in most cases.

2   Related Work

Much work has been done in forecasting spatiotemporal systems with methods ranging from pure temporal models [3,5,8] to complex architectures featuring temporal, geometrical, and attention-based embeddings [2,18,23]. Many methods have focused on traffic forecasting but recent work has shown applications in energy, medicine, and commerce [4,24]. Table 1 organizes prior work according to their usage of multi-modal variables, graph structures, and graph convolutions.

Multi-modal Variable Correlation. Methods that can capture correlations among multiple time-series include GRU, TCN [3], DCRNN [11], T-GCN [22], A3T-GCN [1], ST-GCN [20], AST-GCN [7], AGCRN [2], and Graph WaveNet [19]. Most methods can model multi-variate relationships but several recent methods are still limited to the uni-variate case [4,12,23]. Less common among prior methods is the incorporation of features from spatial and temporal modalities which may also be highly correlated. The methods EA-LSTM [10] and MTGNN [18] integrate spatial features via the LSTM input gate and as priors for a graph learning module, respectively. GMAN [23] combines spatial and temporal features via a spatiotemporal embedding module but temporal features are limited to just two categorical variables. The literature appears to lack any comprehensive module for learning correlations across multiple explicit modalities.

Table 1. Related work grouped into three categories based on their usage of (a) variables from multiple explicit modalities, (b) graph dependency structures, and (c) graph convolution for geometrical embedding. Model components captured by our method are indicated in the last column.

Model components	Examples	MMR-GNN
Use multi-variate features	GRU, EA-LSTM [10], TCN [3], DCRNN [11], T-GCN [22], A3T-GCN [1], ST-GCN [20], AST-GCN [7], AGCRN [2], Graph WaveNet [19], MTGNN [18]	✓
Use spatial features	EA-LSTM, GMAN [23], MTGNN	✓
Use temporal features	GMAN	✓
Graphs not used	RNN, GRU, LSTM, EA-LSTM, TCN	✗
Pre-defined graphs	T/A3T-GCN, ST/AST-GCN, DCRNN, MTGNN	✓
Learned graphs	StemGNN [4], SCINet [12], GMAN, MTGNN, AGCRN	✓
GNN: one encoder	DCRNN, StemGNN, SCINet, GMAN, MTGNN	✗
GNN: shared encoders	AGCRN	✓
GNN: partitioned encoders	MMR-GNN (this paper)	✓

Dependency Structure Modeling. Proper modeling of the inter-node dependency structure is pivotal for accurate forecasting but these structures are often undefined or non-optimal. These scenarios have inspired researchers to construct graphs heuristically [11] or from the minimization of an objective function [18]. We distinguish prior work according to their inter-node dependency modeling including (a) no dependency (b) pre-defined dependency and (c) dependency learned through minimization of loss. Many methods utilize a pre-defined graph [1,7,20,22] but they are ultimately limited by the optimality of that graph in forecasting. Given that there is no guarantee of optimality, it is unsurprising that recent work has moved towards learning the dependency structure entirely [2,4,12,18,19,23]. SCINet [12] proposed the interaction module, StemmGNN [4] proposed self-attention on GRU embeddings, while Graph WaveNet, MTGNN, and AGCRN proposed approximating the dependency via non-linear activations on learned node embeddings. Our method aims to capture the benefits of both aspects by augmenting pre-defined graphs (if they exist) to improve forecasting.

Graph Convolution Methods. GNN-based methods utilize various forms of graph convolution including classical GCN [9], Chebyshev polynomial expansion [6], and novel forms such as SCINet's interaction module. However, these approaches mostly use a single parameter set when embedding across all nodes of the system. This presents a limitation for complex spatiotemporal systems where node subsets can exhibit significantly different dynamics in the form of multiple implicit modalities (see Fig. 1). AGCRN [2] addresses this issue by assigning a unique set of parameters (weights and biases) to each node using a full-rank parameter tensor. To avoid over-fitting, the full-rank parameter tensor is approximated via factorization with factors placed in d-dimensional space representing d modalities. We argue that a multi-modal learning module, constrained to the binary mapping of node subsets to individual modalities instead of unconstrained mixture of modalities, improves performance and interpretability.

3   Methods

3.1   Problem Formulation

As shown in Fig. 1(b), we model a spatiotemporal system as N inter-dependent entities (e.g., traffic sensors, stream gauges, etc.) and their spatial and temporal features. The dependency structure is represented by a graph $G = (V, E)$, where V is the set of nodes (entities) with $|V| = N$, and E is the set of edges (dependencies). Every node has F^s static spatial features (e.g., latitude, longitude, etc.) and F^{st} spatiotemporal features that vary with time (e.g., streamflow, traffic speed, etc.). Additionally, the overall system has F^t dynamic features shared by all nodes (e.g., day of year, season, etc.). We represent all features of a system as $\mathcal{X} = \{X^s, X^t, X^{st}\}$, where $X^s \in \mathbb{R}^{N \times F^s}$ denotes spatial features, $X^t = \{X_0^t, X_1^t, ..., X_\tau^t, ...\}$ denotes temporal features with $X_\tau^t \in \mathbb{R}^{F^t}$, and $X^{st} = \{X_0^{st}, X_1^{st}, ..., X_\tau^{st}, ...\}$ denotes spatiotemporal features with $X_\tau^{st} \in \mathbb{R}^{N \times F^{st}}$. We select one spatiotemporal feature $Y \in X^{st}$ as the target variable and solve Eq. 1. That is, given graph G and S historical time-steps of all variables $\mathcal{X}$, we aim to learn a function $\mathcal{F}_\Theta$ that can forecast T future time-steps of Y:

$$\arg \min_{\Theta} = \mathcal{L}(Y_{(\tau+1):(\tau+T)}, \mathcal{F}_\Theta(\mathcal{X}_{\tau-S+1:\tau}; G)) \tag{1}$$

where Θ denotes learnable model parameters and $\mathcal{L}$ denotes forecasting loss. This model is flexible enough to capture most spatiotemporal methods proposed in the literature. For example, if we consider one spatiotemporal feature and do not include spatial or temporal features (i.e., $F^{st} = 1, F^s = 0, F^t = 0$), Eq. 1 is equivalent to the multi-step traffic forecasting problem in AGCRN [2].

3.2   Model Design

In this section, we cover the design of MMR-GNN starting with (a) its high-level encoder-decoder architecture (b) its module for fusing explicit modalities (X^s, X^t, and X^{st}) into rich embeddings (c) its module for partitioning node-space and learning implicit modalities and (d) its graph augmentation module for learning an optimal dependency structure G^*. Note that symbol C defines the mapping from nodes to implicit modalities and is discussed later in section *Implicit Multi-Modality Learning Module*.

$$stGRU(\mathcal{X}_{\tau-S+1:\tau}, G, C) \rightarrow \mathcal{E}^{enc} \in \mathbb{R}^{N \times H}$$
$$stGRU(\mathcal{E}^{enc}, G, C) \rightarrow \mathcal{E}^{dec} \in \mathbb{R}^{T \times N \times H} \tag{2}$$
$$mLinear(\mathcal{E}^{dec}, C) \rightarrow \hat{Y} \in \mathbb{R}^{T \times N}$$

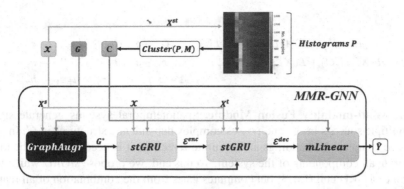

Fig. 2. Overview of MMR-GNN's architecture. Prior to the forward-pass, nodes are clustered from histograms of the spatiotemporal features. The first step of forward-pass is a graph augmentation module called *GraphAugr* that constructs a graph from node embeddings, prunes it, and then joins it to an existing graph (if present). The augmented graph, all features (spatial, temporal, and spatiotemporal), and clusters are then fed into the multi-modal encoder cell called *stGRU*. Summary embeddings, future temporal features, augmented graph, and clusters are fed into the auto-regressive decoder, another *stGRU* cell. Finally, decoded embeddings and clusters are fed into the final projection layer called *mLinear* to produce forecastings.

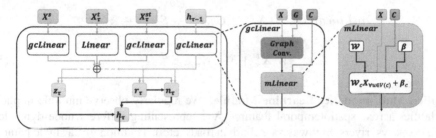

Fig. 3. (left) Overview of the stGRU cell. The forward pass projects and then fuses spatial, temporal, and spatiotemporal features. Fused features and previous hidden state are used to compute reset and update gates. Fused features, previous hidden state, and reset gate are used to compute the new gate. Finally, previous hidden state, update gate, and new gate are used to compute the new hidden state. (middle) Overview of the *gcLinear* layer. The forward-pass first performs graph convolution on features according to the given graph. Convolved features and clusters are then fed into the multi-modal linear projection layer *mLinear* to compute embeddings. (right) The *mLinear* layer projects nodes of each cluster using its assigned weights and biases.

Encoder-Decoder Framework. We utilize an encoder-decoder framework (Fig. 2) to encode S historical time-steps into summary embeddings and causally decode them into T future time-steps via auto-regression. Equation 2 details the forward pass which involves (a) encoding $\mathcal{X}_{\tau-S+1:\tau}$ into embedding $\mathcal{E}^{enc}$ (b) auto-regressively decoding $\mathcal{E}^{enc}$ into embedding $\mathcal{E}^{dec}$ then (c) projecting $\mathcal{E}^{dec}$ from embedding to output dimension $\hat{Y}$.

$$\sigma(gcL(\boldsymbol{X}^s, G, C) + L(\boldsymbol{X}^t_\tau) + gcL(\boldsymbol{X}^{st}_\tau, G, C) + gcL(\boldsymbol{h}_{\tau-1}, G, C)) \rightarrow \boldsymbol{r}_\tau, \boldsymbol{z}_\tau$$

$$tanh(gcL(\boldsymbol{X}^s, G, C) + L(\boldsymbol{X}^t_\tau) + gcL(\boldsymbol{X}^{st}_\tau, G, C) + \boldsymbol{r}_\tau \odot gcL(\boldsymbol{h}_{\tau-1}, G, C)) \rightarrow \boldsymbol{n}_\tau \qquad (3)$$

$$(1 - \boldsymbol{z}_\tau) \odot \boldsymbol{n}_t + \boldsymbol{z}_\tau \odot \boldsymbol{h}_{\tau-1} \rightarrow \boldsymbol{h}_\tau$$

Explicit Multi-modality Fusion Module. Spatiotemporal systems generate signals from multiple sources in $\mathcal{X}_\tau$ and posses complex dependency structures in G. In order to achieve complete embeddings and accurate forecasts, it is necessary that the model incorporate all components of the system. To this end, we propose *stGRU* which takes the form of a GRU cell (Eq. 3) but computes gates from the combination of all features (see Fig. 3). For spatial and spatiotemporal features ($\boldsymbol{X}^s$, $\boldsymbol{X}^{st}$), we use our custom graph convolution layer *gcLinear* which combines graph convolution and linear projection to embed dependencies. The standard linear projection layer is used for temporal features $\boldsymbol{X}^t_\tau$ since they do not exist at node-level. Equation 3 uses $gcL(\cdot)$ and $L(\cdot)$ to denote *gcLinear* and *Linear* layers. The forward pass of layers *gcLinear* and *Linear* are defined in Eq. 4 and 5 respectively where $\boldsymbol{X}^{s-st}_\tau$ indicates spatial or spatiotemporal features. The *gcLinear* layer applies Chebyshev graph convolution [6] and a new linear projection layer *mLinear* discussed in the next section.

$$mLinear(GCN(\boldsymbol{X}^{s-st}_\tau, G), C) \rightarrow \boldsymbol{\mathcal{E}}^{s-st}_\tau \in \mathbb{R}^{N \times H} \qquad (4)$$

$$W\boldsymbol{X}^t_\tau + b \rightarrow \boldsymbol{\mathcal{E}}^t_\tau \in \mathbb{R}^H \qquad (5)$$

Implicit Multi-modality Learning Module. We regularly observe multiple implicit modalities across spatiotemporal features $\boldsymbol{X}^{st}$ representing different node dynamics (e.g., creeks vs. rivers, highway vs. suburban roads, etc.). Traditionally, a single parameter set $\theta = (W \in \mathbb{R}^{H \times F}, b \in \mathbb{R}^H)$ would be used to capture all node dynamics but this is an unnecessary constraint. Instead, we propose to separately learn the dynamics of M implicit modalities by partitioning the nodes into M clusters. We derive clustering C according to similarity of node feature distribution by applying agglomerative clustering on the histograms of $\boldsymbol{X}^{st}$. Multiple parameter sets $\Theta = (\mathcal{W} \in \mathbb{R}^{M \times H \times F}, \beta \in \mathbb{R}^{M \times H})$ are then used to learn the unique dynamics of M implicit modalities. We implement implicit modality learning as a new linear projection layer *mLinear* (Eq. 6) and use it for linear projection in *gcLinear* and final projection in the greater encoder-decoder.

$$\mathcal{W}_c\boldsymbol{X}_{\forall u \in V(c)} + \beta_c \rightarrow \boldsymbol{\mathcal{E}}_c \in \mathbb{R}^{|V(c)| \times H} \qquad (6)$$

Dependency Augmentation Module. An integral part of any spatiotemporal system is its dependency structure (e.g., streamflow network, road network, etc.) represented by graph G. However, this structure is often undefined and some methods [11] apply heuristics to infer G as a pre-processing step. Moreover, neither existing nor inferred graphs are guaranteed to be optimal in forecasting. To address these challenges, we propose *GraphAugr* which aims to learn an optimal dependency structure G^*.

Table 2. Details of datasets including sample count, graph size, and feature dimensions.

Dataset	Time-steps	Nodes	F^s	F^t	F^{st}	G	Resolution	Default prediction horizon
Little River	13,515	8	2	1	4	✓	1 day	1 time-step (1 day)
Wabash River	31,046	1,276	11	1	5	✓	1 day	1 time-step (1 day)
METR-LA	34,272	207	2	1	1	✓	5 min	12 time-steps (1 h)
E-METR-LA	34,272	207	12	1	5	✓	5 min	12 time-steps (1 h)
PEMS-BAY	52,116	325	2	1	1	✓	5 min	12 time-steps (1 h)
E-PEMS-BAY	52,116	325	12	1	5	✓	5 min	12 time-steps (1 h)

$$Softmax(\boldsymbol{E} \cdot \boldsymbol{E}^T) \rightarrow \boldsymbol{S} \in \mathbb{R}^{N \times N}$$
$$\Phi(\boldsymbol{W} \odot \boldsymbol{S} + \boldsymbol{B}) \rightarrow \hat{G} \qquad G \cup \hat{G} \rightarrow G^* \tag{7}$$

We define the forward-pass of *GraphAugr* in Eq. 7 where $\boldsymbol{E} \in \mathbb{R}^{N \times d}$ are learnable node embeddings, $\boldsymbol{S} \in \mathbb{R}^{N \times N}$ is the computed similarity matrix, matrices $\boldsymbol{W}, \boldsymbol{B} \in \mathbb{R}^{N \times N}$ are weights and biases on similarity, and $\Phi(\cdot)$ is a pruning function. The *GraphAugr* layer is generalized to cover a broad span of use-cases including (a) using an existing graph alone (b) augmenting an existing graph with learned edges or (c) constructing a graph entirely. Weights and biases $\boldsymbol{W}, \boldsymbol{B}$ allow for fine control over edge construction. For example, $\boldsymbol{W}$ can be used as a mask to restrict certain edges (e.g., intra-highway edges) and $\boldsymbol{B}$ may be used to favor certain edges (e.g., existing edges). Finally, by feeding G^* to subsequent layers (*gcLinear* in encoder and decoder) we constrain graph construction to minimize forecasting loss, and hence, be optimal in forecasting.

4 Experiments

For evaluation, we consider six public datasets spanning hydrology, and traffic. We summarize these datasets in Table 2.

Hydrological Data. In hydrology, we consider the datasets Little River and Wabash River [13]. These datasets record multiple hydrological and meteorological features across unique regions of their watersheds. They also include multiple spatial features and pre-defined graphs in the form of streamflow networks. For these datasets, we forecast next-day streamflow across all sub-basins of the watershed given the previous 7 d of meteorological and hydrological records.

Traffic Data. In traffic, we consider METR-LA and PEMS-BAY [11], which are well-studied in the literature. A limitation of METR-LA and PEMS-BAY is that they contain only traffic speed and are restricted to uni-variate modeling. We extend these datasets by adding four spatiotemporal features (Samples, Percent Observed, Total Flow, and Average Occupancy) and ten spatial features acquired from [17]. We also create new pre-defined graphs (detailed in the appendix) using these new spatial features. These extended datasets are called E-METR-LA and E-PEMS-BAY. The task

Table 3. MAE, RMSE, and MAPE scores of MMR-GNN and other baseline models for two datasets. Emboldened numbers indicate the best performance while underlined numbers indicate second-best. An 'N/A' means the model was incompatible with a dataset or exceeded GPU runtime/memory.

Model	Wabash River			E-PEMS-BAY		
	MAE	RMSE	MAPE	MAE	RMSE	MAPE
GRU	4.01 ± 0.08	10.23 ± 0.03	27.60 ± 2.16	2.04 ± 0.07	4.03 ± 0.11	4.45 ± 0.15
TCN	4.04 ± 0.24	10.37 ± 0.04	27.21 ± 1.30	2.20 ± 0.02	4.52 ± 0.01	4.84 ± 0.03
FEDformer	4.26 ± 0.01	9.75 ± 0.01	36.15 ± 0.04	2.85 ± 0.03	4.96 ± 0.03	5.81 ± 0.05
LTSFDLinear	4.05 ± 0.00	10.77 ± 0.00	30.13 ± 0.00	2.37 ± 0.00	4.75 ± 0.00	5.08 ± 0.00
T-GCN	7.04 ± 0.13	14.268 ± 0.11	35.27 ± 0.70	2.67 ± 0.01	4.58 ± 0.015	5.67 ± 0.03
A3T-GCN	7.14 ± 0.12	14.43 ± 0.11	35.47 ± 1.32	N/A	N/A	N/A
ST-GCN	4.28 ± 0.15	10.24 ± 0.13	33.15 ± 2.48	N/A	N/A	N/A
AST-GCN	N/A	N/A	N/A	N/A	N/A	N/A
GMAN	N/A	N/A	N/A	N/A	N/A	N/A
StemGNN	9.60 ± 10.21	17.46 ± 12.55	38.30 ± 19.07	2.03 ± 0.06	4.03 ± 0.07	4.40 ± 0.14
MTGNN	N/A	N/A	N/A	2.48 ± 0.08	5.24 ± 0.19	5.73 ± 0.20
AGCRN	<u>2.98 ± 0.01</u>	<u>9.03 ± 0.10</u>	<u>17.59 ± 0.68</u>	<u>2.05 ± 0.04</u>	4.16 ± 0.09	<u>4.38 ± 0.14</u>
SCINet	6.31 ± 6.29	13.61 ± 8.92	32.96 ± 13.38	2.11 ± 0.02	<u>3.96 ± 0.02</u>	4.47 ± 0.04
MMR-GNN	**2.96 ± 0.04**	**8.85 ± 0.10**	**17.14 ± 0.25**	**2.01 ± 0.04**	**3.85 ± 0.06**	**4.32 ± 0.09**

for these datasets is to forecast the next hour (12 time-steps) of traffic speed given the previous hour.

Data Processing. We apply the same method of imputation and normalization to each dataset. In the event of missing values (found in LittleRiver, E-METR-LA, and E-PEMS-BAY), we impute with the periodic mean computed at each node from available samples. We standardize each node feature according to their mean and standard deviation computed from the training set. Each model is trained on standardized inputs/outputs with final forecasts computed from inverse standardization on model output.

Performance Metrics. To quantify forecast accuracy, we use Mean Absolute Error (MAE), Root Mean Square Error (RMSE), and Mean Absolute Percentage Error (MAPE). We ran all experiments 10 times using 10 random initialization seeds and report their mean and standard deviation.

4.1 Model Baselines

For each dataset, we evaluate MMR-GNN against thirteen models found throughout the literature. The components of these baselines are listed in Table 1 separated according to pure temporal methods, GNN-based methods with pre-defined graphs, and GNN-based methods with learned graphs. Implementation of GRU was acquired from [13], while T-GCN, A3T-GCN, and GMAN were acquired from PyTorch Geometric Temporal [16].

Table 4. (a) RMSE scores of MMR-GNN and other baseline models for remaining datasets. Emboldened numbers indicate the best performance while underlined numbers indicate second-best. An 'N/A' means the model was incompatible with a dataset or exceeded GPU run-time/memory. (b) MAE scores for three models at multiple output horizons.

(a) RMSE scores for other datatsets

Model	Little River	METR-LA	E-METR-LA
GRU	0.67 ± 0.01	<u>11.08</u> ± 0.081	5.11 ± 0.02
TCN	0.66 ± 0.01	11.49 ± 0.05	5.74 ± 0.02
FEDformer	0.55 ± 0.00	12.53 ± 0.04	6.51 ± 0.05
LTSFDLinear	0.64 ± 0.00	11.59 ± 0.00	5.73 ± 0.00
T-GCN	0.79 ± 0.02	13.6 ± 0.01	5.66 ± 0.01
A3T-GCN	0.87 ± 0.01	N/A	5.69 ± 0.01
ST-GCN	0.55 ± 0.01	N/A	N/A
AST-GCN	0.54 ± 0.02	11.60 ± 0.14	5.12 ± 0.05
GMAN	0.65 ± 0.03	12.91 ± 0.16	N/A
StemGNN	0.68 ± 0.03	11.31 ± 0.08	5.16 ± 0.07
MTGNN	0.5 ± 0.01	11.34 ± 0.05	<u>4.57</u> ± 0.03
AGCRN	**0.46** ± 0.01	11.78 ± 0.08	4.76 ± 0.02
SCINet	0.6 ± 0.01	12.345 ± 0.72	4.96 ± 0.03
MMR-GNN	**0.46** ± 0.01	**10.91** ± 0.11	**4.56** ± 0.03

(b) Impact of output horizon

		Wabash River
Model	Horizon	MAE
FEDformer	1	4.26
LTSF-DLinear		<u>4.04</u>
MMR-GNN		**2.96**
FEDformer	14	16.90
LTSF-DLinear		<u>14.82</u>
MMR-GNN		**13.05**
FEDformer	28	20.19
LTSF-DLinear		<u>18.84</u>
MMR-GNN		**16.53**

All other models were acquired from their published GitHub repositories. We train and evaluate all models on an Nvidia A100 GPU with 40 GB memory.

4.2 Primary Results

We break down the primary results into two tables. Table 3 shows MAE, RMSE, and MAPE of all models on our two largest multi-modal datasets. Due to space limitations, Table 4a shows only RMSE for three other datasets. In all tables, we group models by (a) pure temporal and transformers, (b) GNN-based models with existing graphs, and (c) GNN-based models with learned graphs. A3T-GCN, GMAN, and MTGNN often show prohibitive scaling with either excessive runtime and/or memory requirements.

We observe that GNN models using pre-defined graphs (middle group of Table 3) struggle to out-perform their pure temporal counter-parts. In comparison, most GNN models using learned graphs perform on-par or better than pure temporal models. This suggests pre-defined graphs are not necessarily optimal in forecasting and can even degrade performance. Looking at MMR-GNN in Tables 3 and 4a, we see consistent improvement with lower errors than the best baseline model across all datasets.

Effect of Forecast Horizon. Thus far, we mostly considered short-term predictions as mentioned in Table 2. It is well documented that transformers [25] and some decou-

Table 5. Alternative RNN cells

Cell	MAE	RMSE	MAPE
gcRNN	3.019 ± 0.056	9.106 ± 0.218	18.901 ± 0.471
gcGRU	**2.962** ± 0.044	**8.812** ± 0.184	**17.231** ± 0.287
gcLSTM	2.977 ± 0.060	8.988 ± 0.245	17.269 ± 0.524

Table 6. Alternative fusion method

Fusion	MAE	RMSE	MAPE
Addition	3.00 ± 0.066	9.091 ± 0.268	17.127 ± 0.441
Attention	**2.979** ± 0.0545	**9.003** ± 0.174	**16.977** ± 0.309

Table 7. Alternative clustering algorithms

Method	MAE	RMSE	MAPE
Random	2.984 ± 0.051	8.901 ± 0.189	17.648 ± 0.417
KMeans	2.977 ± 0.058	8.866 ± 0.154	17.608 ± 0.295
Agglomerative	**2.911** ± 0.022	**8.601** ± 0.046	**17.269** ± 0.007

pled linear models such as LTSF-DLinear [21] perform better in long-term predictions. Table 4b shows that MMR-GNN outperforms FEDformer and LTSF-DLinear in mid to long-term forecasting of streamflow.

4.3 Ablation Study

We conduct a variety of experiments to test different components of MMR-GNN. These tests include: (1) substitution of our proposed *stGRU* cell with vanilla GRU, (2) sampling the number of implicit modalities, and (3) sampling the space of graph augmentation. We perform all tests on the Wabash River dataset since it includes a pre-defined graph and has the greatest complexity in terms of available features and scale.

Explicit Multi-modality Fusion. We proposed the *stGRU* cell to learn from the multiple explicit modalities present in spatiotemporal systems and test its contribution here. Table 8a shows performance when using the vanilla GRU cell or our *stGRU* cell. We see improvement across all metrics demonstrating that features from additional modalities are informative to forecasting.

Implicit Modality Learning. We test the effect of increasing the number of implicit modalities learned by our multi-modal projection layer *mLinear*. Table 8b shows MMR-GNN's performance when increasing implicit modality count M starting from the traditional single modality (single parameter set) case. By increasing M, we partition the system into smaller problems, and thus, expect performance to improve to a point of saturation where over-fitting occurs. This effect is seen in Table 8b where the optimal modality is found at $M = 8$ according to two metrics.

Table 8. Ablation studies

(l) Graph augmentation rate

G	Density	MAE	RMSE	MAPE
✗	0%	3.392	10.538	2.184
✓	25%	3.111	9.315	**2.177**
✓	50%	2.953	8.872	2.218
✓	75%	2.969	8.862	2.186
✓	99%	**2.897**	**8.614**	2.189

(b) Implicit modality count

M	MAE	RMSE	MAPE
1	3.066	9.076	**2.179**
4	3.103	9.336	2.192
8	**2.873**	**8.647**	2.217
12	3.038	9.118	**2.179**
16	3.038	9.123	2.212

(a) stGRU vs. vanilla GRU

stGRU	MAE	RMSE	MAPE
✗	3.035	9.038	2.247
✓	**2.937**	**8.763**	**2.180**

Graph Augmentation. Here we aim to understand the contribution of learned graph augmentation by testing various scenarios including (a) no pre-defined graph or augmentation (b) pre-defined graph without augmentation and (c) increasing rates of augmentation on the pre-defined graph. Table 8c shows monotonic improvement for RMSE with increased graph augmentation.

Other Ablation Studies. We also test the impact of other RNN cells (Table 5), attention for fusion of explicit modalities (Table 6), and various clustering algorithms when finding implicit modality (Table 7).

5 Conclusions

This paper presents MMR-GNN, a network capable of learning from the multiple explicit and implicit modalities present in many spatiotemporal systems. For multi-modal data, MMR-GNN shows consistent improvement over SOTA methods but we see one limitation that, if addressed, could further improve accuracy. The projection layer mLinear depends on the quality of computed clusters which may be erroneous and inhibit implicit modality learning. A possible solution is to cluster on the learned node embeddings (from graph augmentation) allowing implicit modality assignment to be learned indirectly. While this does not remove potentially erroneous clustering algorithms, it allows the model to adapt towards an optimal clustering. Our future work will aim to address this limitation and expand MMR-GNN into more application domains.

Acknowledgements. This research is supported by the Applied Mathematics Program of the DOE Office of Advanced Scientific Computing Research under contracts numbered DE-SC0022098 and DE-SC0023349 and by the NSF OAC-2339607 grant.

7 Appendix

Road Network Inference. E-METR-LA and E-PEMS-BAY are extensions of the quintessential METR-LA and PEMS-BAY datasets adding new spatial and spatiotemporal features. Following [11], we wanted to offer pre-defined graphs for each dataset and chose to infer them using new spatial features. We apply *GraphAugr* to construct a graph heuristically using exponentiated Minkowski ($p = 2$) distance to define node similarity and W, B to define rules for edge construction. Each traffic sensor of E-METR-LA and E-PEMS-BAY includes latitude, longitude, freeway name, and freeway bearing. We use latitude and longitude to define node similarity S and freeway/bearing to define a mask W that restricts to intra-highway edges of the same bearing.

Using OpenStreetMap [15], we found many sensor pairs placed closely together but recording traffic in opposite bearings (e.g., U.S. Route 101 in CA with North/South bearing). Using distance-based similarity alone leads to connections between these and many other unrelated sensors. With the previous similarity and restrictions, the modified similarity becomes $S^* = W \odot S$ which we prune using *K-NN* with $k = 2$. The original graph of METR-LA and our new graph for E-METR-LA are shown in Fig. 4.

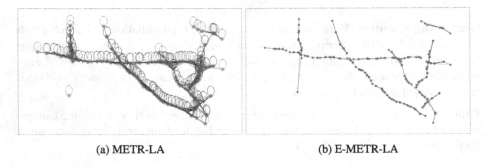

(a) METR-LA   (b) E-METR-LA

Fig. 4. The pre-defined road network from METR-LA (a) and our new road network for E-METR-LA (b)

References

1. Bai, J., et al.: A3t-GCN: attention temporal graph convolutional network for traffic forecasting. ISPRS Int. J. Geo Inf. **10**(7), 485 (2021)
2. Bai, L., Yao, L., Li, C., Wang, X., Wang, C.: Adaptive graph convolutional recurrent network for traffic forecasting. In: Advances in Neural Information Processing Systems, vol. 33, pp. 17804–17815 (2020)
3. Bai, S., Kolter, J.Z., Koltun, V.: An empirical evaluation of generic convolutional and recurrent networks for sequence modeling. arXiv preprint arXiv:1803.01271 (2018)
4. Cao, D., et al.: Spectral temporal graph neural network for multivariate time-series forecasting. In: Advances in Neural Information Processing Systems, vol. 33, pp. 17766–17778 (2020)
5. Contreras, J., Espinola, R., Nogales, F.J., Conejo, A.J.: ARIMA models to predict next-day electricity prices. IEEE Trans. Power Syst. **18**(3), 1014–1020 (2003)
6. Defferrard, M., Bresson, X., Vandergheynst, P.: Convolutional neural networks on graphs with fast localized spectral filtering. In: Advances in Neural Information Processing Systems, vol. 29 (2016)

7. Guo, S., Lin, Y., Feng, N., Song, C., Wan, H.: Attention based spatial-temporal graph convolutional networks for traffic flow forecasting. In: Proceedings of the AAAI Conference on Artificial Intelligence, vol. 33, pp. 922–929 (2019)
8. Hochreiter, S., Schmidhuber, J.: Long short-term memory. Neural Comput. 9(8), 1735–1780 (1997)
9. Kipf, T.N., Welling, M.: Semi-supervised classification with graph convolutional networks. arXiv preprint arXiv:1609.02907 (2016)
10. Kratzert, F., Klotz, D., Shalev, G., Klambauer, G., Hochreiter, S., Nearing, G.: Towards learning universal, regional, and local hydrological behaviors via machine learning applied to large-sample datasets. Hydrol. Earth Syst. Sci. 23(12), 5089–5110 (2019)
11. Li, Y., Yu, R., Shahabi, C., Liu, Y.: Diffusion convolutional recurrent neural network: data-driven traffic forecasting. arXiv preprint arXiv:1707.01926 (2017)
12. Liu, M., et al.: Scinet: time series modeling and forecasting with sample convolution and interaction. In: Thirty-Sixth Conference on Neural Information Processing Systems (NeurIPS) (2022)
13. Majeske, N., Zhang, X., Sabaj, M., Gong, L., Zhu, C., Azad, A.: Inductive predictions of hydrologic events using a long short-term memory network and the soil and water assessment tool. Environ. Model. Softw. 152, 105400 (2022)
14. NREL: Solar power data for integration studies (2006). https://www.nrel.gov/grid/solar-power-data.html
15. OpenStreetMap contributors: Planet dump retrieved from https://planet.osm.org (2017). https://www.openstreetmap.org
16. Rozemberczki, B., et al.: PyTorch geometric temporal: spatiotemporal signal processing with neural machine learning models. In: Proceedings of the 30th ACM International Conference on Information and Knowledge Management, pp. 4564–4573 (2021)
17. Varaiya, P.P.: Freeway performance measurement system (pems), pems 7.0. Technical report (2007)
18. Wu, Z., Pan, S., Long, G., Jiang, J., Chang, X., Zhang, C.: Connecting the dots: multivariate time series forecasting with graph neural networks. In: Proceedings of the 26th ACM SIGKDD International Conference on Knowledge Discovery and Data Mining, pp. 753–763 (2020)
19. Wu, Z., Pan, S., Long, G., Jiang, J., Zhang, C.: Graph wavenet for deep spatial-temporal graph modeling. arXiv preprint arXiv:1906.00121 (2019)
20. Yu, B., Yin, H., Zhu, Z.: Spatio-temporal graph convolutional networks: a deep learning framework for traffic forecasting. arXiv preprint arXiv:1709.04875 (2017)
21. Zeng, A., Chen, M., Zhang, L., Xu, Q.: Are transformers effective for time series forecasting? (2023)
22. Zhao, L., et al.: t-GCN: a temporal graph convolutional network for traffic prediction. IEEE Trans. Intell. Transp. Syst. 21(9), 3848–3858 (2019)
23. Zheng, C., Fan, X., Wang, C., Qi, J.: Gman: a graph multi-attention network for traffic prediction. In: Proceedings of the AAAI Conference on Artificial Intelligence, vol. 34, pp. 1234–1241 (2020)
24. Zhou, H., et al.: Informer: Beyond efficient transformer for long sequence time-series forecasting. In: Proceedings of the AAAI Conference on Artificial Intelligence, vol. 35, pp. 11106–11115 (2021)
25. Zhou, T., Ma, Z., Wen, Q., Wang, X., Sun, L., Jin, R.: FEDformer: frequency enhanced decomposed transformer for long-term series forecasting. In: Proceedings of 39th International Conference on Machine Learning (ICML 2022) (2022)

Layer-Wise Sparse Training of Transformer via Convolutional Flood Filling

Bokyeong Yoon[ID], Yoonsang Han[ID], and Gordon Euhyun Moon[✉][ID]

Sogang University, Seoul, Republic of Korea
{bkyoon,han14931,ehmoon}@sogang.ac.kr

Abstract. Sparsifying the Transformer has garnered considerable interest, as training the Transformer is very computationally demanding. Prior efforts to sparsify the Transformer have either used a fixed pattern or data-driven approach to reduce the number of operations involving the computation of multi-head attention, which is the main bottleneck of the Transformer. However, existing methods suffer from inevitable problems, including potential loss of essential sequence features and an increase in the model size. In this paper, we propose a novel sparsification scheme for the Transformer that integrates convolution filters and the flood filling method to efficiently capture the layer-wise sparse pattern in attention operations. Our sparsification approach significantly reduces the computational complexity and memory footprint of the Transformer during training. Efficient implementations of the layer-wise sparsified attention algorithm on GPUs are developed, demonstrating our SPION that achieves up to 2.78× speedup over existing state-of-the-art sparse Transformer models and maintain high evaluation quality.

Keywords: Deep Learning · Sparse Transformer · Convolutional Flood Filling

1 Introduction

The Transformer is a state-of-the-art deep neural network developed for addressing sequence tasks, originally proposed by Vaswani et al. [18]. One of the main advantages of the Transformer is that, given a sequence of input data points (e.g., a sentence of word tokens), it is able to compute the multi-head attention (MHA) operation in parallel, thereby quickly and accurately capturing long-term dependencies of data points. However, as the sequence length increases, the overall computational cost and memory space required for training the Transformer also increase quadratically [1]. Especially, a MHA sub-layer in the Transformer occupies a substantial portion of the total execution time and becomes the main bottleneck as the sequence length increases. The MHA operation requires a large number of dot-product operations to compute the similarity between all data points in the sequence. However, the dot-product operation inherently has limitations in memory bandwidth since it performs only two floating-point operations

© The Author(s), under exclusive license to Springer Nature Singapore Pte Ltd. 2024
D.-N. Yang et al. (Eds.): PAKDD 2024, LNAI 14646, pp. 158–170, 2024.
https://doi.org/10.1007/978-981-97-2253-2_13

for each pair of data elements read from memory. Hence, in order to mitigate computational complexity and improve data locality of the Transformer, several approaches have addressed the sparsification of computations associated with the MHA operation [2,16,19–21]. However, previous approaches suffer from two primary limitations. First, when the Transformer adopts identical fixed patterns of non-zero entries in the attention matrices across all layers during training, it becomes difficult to effectively capture key features within the sequence. Second, when the Transformer employs additional parameters to learn the sparsity pattern in the attention matrices, both the model size and the computational overhead increase. To address the limitations of previous approaches, we focus on developing a specialized sparse attention scheme that significantly reduces memory consumption and efficiently handles variations of sparse patterns across different types of Transformers and datasets.

In this paper, we present a new sparsity-aware layer-wise Transformer (called **SPION**) that dynamically captures the variations in sparse pattern within the MHA operation for each layer. SPION judiciously explores the sparsity pattern in the MHA operation based on a novel convolutional flood filling method. To precisely detect the characteristics of the sparse pattern in the attention matrices, SPION identifies the shape of sparse pattern by utilizing a convolution filter and the degree of sparsity through a flood filling-based scheme. During the generation of the sparse pattern for each layer, we construct a sparsity pattern matrix with a blocked structure to enhance data locality for the MHA operation that involves sparse matrix multiplication. Furthermore, by capturing layer-specific sparse pattern for each layer, SPION performs layer-wise MHA computations iteratively until convergence is achieved. As the sparse MHA operations contribute significantly to the overall training workload, we develop an efficient GPU implementation for sparse MHA to achieve high performance with quality of results.

We conduct an extensive comparative evaluation of SPION across various classification tasks, including image and text classification, as well as document retrieval involving lengthy sequences. Experimental results demonstrate that our SPION model achieves up to a $5.91\times$ reduction in operations for MHA computation and up to a $2.78\times$ training speedup compared to existing state-of-the-art sparse Transformers while maintaining a better quality of solution.

2   Background and Related Work

2.1   Transformer

The encoder-only Transformer, which is one of the variants of the Transformer model, is widely used for various classification tasks using text and image datasets [5,6,17]. In the encoder-only Transformer, each encoder layer consists of a MHA sub-layer and a feed-forward sub-layer. For each encoder layer, the query ($Q \in \mathbb{R}^{L \times D}$), key ($K \in \mathbb{R}^{L \times D}$), and value ($V \in \mathbb{R}^{L \times D}$) are obtained by performing linear transformations on the input embedding. Hereafter, we denote L, D and H as the length of input sequence, the size of the embedding for each data point in the sequence, and the number of heads, respectively. To efficiently perform MHA computation, each Q, K, and V matrix is divided into H sub-matrices (multi-heads) along the D dimension.

$$A = softmax \left(\frac{Q \times K^T}{\sqrt{(D/H)}} \right) \times V \tag{1}$$

The MHA computation is defined by Eq. 1. After computing the attention for each head, all matrices A_0 through A_{H-1} are concatenated to form the final attention matrix A. Then, the attention matrix A is passed through the feed-forward sub-layer to produce new embedding vectors, which is then fed into the next encoder layer. Therefore, in terms of computational complexity, the main bottleneck of the encoder layer is associated with processing the MHA sub-layer, which involves a large number of matrix-matrix multiplications across multiple heads. More specifically, the number of operations required for computing the attention matrix A is $2L^2(2D + 1) - L(D + 1)$, indicating a quadratic increase in operations as the input sequence length (L) increases.

Sparse MHA. Given the long input sequences, several sparse attention techniques have been proposed to reduce the computational complexity involved in multiplying Q and K^T in the MHA operation. The basic intuition behind sparse attention techniques is that a subset of data points in the long sequence can effectively represent the entire input sequence. In other words, only the highly correlated necessary elements (i.e., data points) in Q and K can be utilized to reduce the computational workload. Therefore, when employing sparse attention, the number of operations required to compute $Q \times K^T$ is $C(2D - 1)$, where $C \ll L^2$ represents the number of critical elements in the resulting matrix. In contrast, without sparsification, the original computation of $Q \times K^T$ requires $L^2(2D - 1)$ operations.

2.2 Related Work on Sparse Attention

Many previous efforts to achieve efficient sparse Transformers have sought to sparsify the MHA operation both before and during model training/fine-tuning.

Fixed Sparse Pattern. One of the strategies for performing sparse MHA is to use a predetermined sparsity patterns, where only specific data points in the input sequence are selected to perform the MHA operation before training the model. Several variants of the Transformer model adapt the sliding windows approach in which the attention operations are performed using only the neighboring data points in the matrices Q and K. The Sparse Transformer [3] originally employs the sliding windows attention to sparsify the MHA operation. The Longformer [2] is an extension to the Sparse Transformer and introduces dilated sliding windows, which extend the receptive field for computing similarity by skipping one data point at a time while performing the sliding windows attention. Furthermore, ETC [1] and BigBird [20] incorporate global attention that performs similarity calculations between a given data point and all other data points in the input sequence.

The main advantage of utilizing the fixed sparsity pattern is the reduction in computational overhead and memory footprint. However, the primary problem with a fixed pattern is that it may lose the fine-grained important features and dependencies in the input sequence. For example, the attention mechanisms utilized in Longformer and BigBird have limitations when applied to image classification tasks. Since image data is typically shaped in two dimensions (except for RGB channels), neighboring image pixels (or patches) arranged vertically do not appear side by side in sequential data. Consequently, the sliding window attention-based Longformer is unable to recognize connections between vertically adjacent data points, as it attends only to neighbors within the sequence. Moreover, the global attention used in BigBird only focuses on specific tokens. However, it is possible that there can be other data points besides these specific tokens that are important to all the other data points in a sequence. In such cases, models based on fixed sparse patterns may lose critical information from the data points in the sequence.

Data-Driven Sparse Pattern. Sparse patterns in the MHA operation can also be generated by leveraging a data-driven approach, which clusters and sorts the data points of the input sequence during training. Most recently, the LSG Attention [4] extends pretrained RoBERTa [11] for fine-tuning various language related tasks and outperforms Longformer and BigBird models. LSG Attention dynamically captures data patterns by generating sparse masks by utilizing local, sparse, and global attention mechanisms. However, since LSG Attention generates sparse masks at every training step, it results in additional computational overhead during training. Reformer [9] utilizes locality-sensitive hashing to calculate a hash-based similarity and cluster multiple data points into chunks. Similarly, Routing Transformer [16] performs k-means clustering given data points. In the process of training the model, these clustering-based attention approaches also require learning additional functions to identify the relevant dependencies of data points in the input sequence. However, even though utilizing data-driven sparsification techniques during training produces a high quality model, this approach requires additional parameters and operations to learn the sparse patterns, resulting in larger memory space and higher computational cost.

3   Motivation: Analysis of Sparse Patterns in MHA

In order to identify common sparsity patterns in the MHA operation, we conducted experiments on the encoder-only Transformer [6] pretrained with the ImageNet-21k and ImageNet-1k datasets. Hereafter, we denote A^s as the attention score matrix obtained after computing $softmax(Q \times K^T / \sqrt{(D/H)})$ in MHA operation. Figure 1 shows A^s from different encoder layers during the inference. Since the sparsity patterns of multiple A^s within the same encoder layer typically

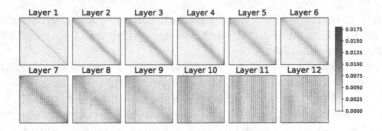

Fig. 1. Sparsity patterns in the attention score matrices across different encoder layers during inference of the encoder-only Transformer for image classification.

show similar patterns, we averaged the A^s across multiple heads in each encoder layer. The results clearly show that most of the elements in A^s are close to zero, indicating that only a few data points in the input sequence are correlated to each other. In practice, a number of studies have shown that considering only the critical data points does not adversely affect the convergence of the model [2,14,20]. The characteristics of attention score matrices A^s described below motivate us to develop a new layer-wise sparse attention algorithm.

Shape of Sparse Pattern. As shown in Fig. 1, the attention score matrices produced by the different encoder layers exhibit distinct sparsity patterns. For example, in the first to ninth encoder layers, the diagonal elements have relatively large values, similar to a band matrix that stores nonzeros within the diagonal band of the matrix. It is obvious that the MHA operation relies on the self-attention scheme and therefore, the resulting values of the dot-product between linearly transformed vectors for the same data points tend to be larger compared to the resulting outputs produced with different data points. In addition to the diagonal sparsity pattern, encoder layers 10, 11 and 12 show a vertical sparsity pattern, with nonzeros mostly stored in specific columns. This vertical sparsity pattern emerges when the attention operation focuses on the similarity between all data points in Q and particular data points in K. In light of these observations, applying the same fixed sparse pattern to all layers may lead to the exclusion of unique essential features that need to be captured individually at different layers. Hence, it is crucial to consider layer-wise sparsification of the MHA based on the sparse pattern observed across different layers. Furthermore, it is necessary to generate domain-specific flexible sparse patterns for various tasks and datasets.

Degree of Sparsity. Across different encoder layers, there exists variation not only in the shape of the sparse pattern but also in the number of nonzero elements in attention score matrices. For example, layer 12 has a higher number of nonzero elements compared to layer 1, indicating that layer 12 extensively computes the MHA operation using a larger number of data points in the sequence. Hence, it is crucial to consider varying degrees of sparsity for every encoder layer.

Considering the irregular distribution of non-zero entries in different attention matrices is essential for effectively reducing computational operations while preserving key features across distinct encoder layers.

4   SPION: Layer-Wise Sparse Attention in Transformer

In this section, we provide a high-level overview and details of our new SPION that dynamically sparsifies the MHA operation, incorporating the major considerations described in Sect. 3.

4.1   Overview of SPION

In SPION, the overall training process is decoupled into three phases: dense-attention training, sparsity pattern generation, and sparse-attention training. Our SPION is capable of sparsifying the MHA operation after training the model for a few steps with dense-attention training. The dense-attention training follows the same training procedure as the original Transformer, without sparsifying the MHA operation. The original dense MHA continues until the attention score matrix A^s of each encoder layer exhibits a specific sparsity pattern. In order to determine the end of the dense-attention training or the start of the sparse-attention training, for each step, we first measure the Frobenius distance between the A^s produced in the previous step $i-1$ and the current step i as defined in Eq. 2.

$$distance_i = \left| \sqrt{\sum (A_{i-1}^s)^2} - \sqrt{\sum (A_i^s)^2} \right| \tag{2}$$

Then, we compare the previous $distance_{i-1}$ with the current $distance_i$ to ensure whether a common pattern of nonzeros has emerged. Intuitively, when the difference between $distance_{i-1}$ and $distance_i$ is very small, it is possible to assume that A^s ends up with a specialized sparsity pattern. Hence, if the difference in Frobenius distance between the previous step and the current step is less than a threshold value, we cease the dense-attention training phase. Thereafter, we dynamically generates the sparsity pattern matrix P for each encoder layer based on our novel convolutional flood-filling scheme, as described in Sect. 4.2.

After identifying the sparsity pattern, SPION proceeds with the sparse-attention training phase until convergence by adapting the sparsity pattern. Given the matrices Q and K, along with the sparsity pattern matrix P, the SDDMM (Sampled Dense-Dense Matrix Multiplication) operation is utilized to accelerate producing the sparsified attention score matrix. Next, the sparse attention score matrix is used to apply the sparse softmax function. After computing the sparse softmax operation, since the sparsified attention score matrix S^s remains sparse, we utilize SpMM (Sparse-Dense Matrix Multiplication) operation to obtain final attention matrix S by multiplying the sparse matrix S^s with the dense matrix V. To accelerate both SDDMM and SpMM operations on GPUs, we utilize the cusparseSDDMM() and cusparseSpMM() functions provided by the NVIDIA cuSPARSE library [13]. Moreover, we implement a custom CUDA kernel for the sparse softmax function by leveraging warp-level reduction to accelerate the sparse-attention training phase.

4.2 Sparsity Pattern Generation with Convolutional Flood Fill Algorithm

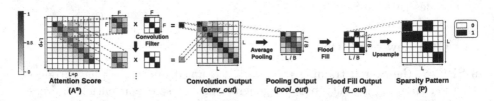

Fig. 2. Overview of our convolutional flood filling method for generating the sparsity pattern in the attention score matrix.

To precisely identify the sparsity patterns in the attention score matrix A^s, we develop a new convolutional flood fill algorithm that extensively explores the shape, degree and locality of sparsity patterns in A^s for each encoder layer. Figure 2 shows the overview for generating the sparsity pattern in A^s. An initial step is to apply a diagonal convolution filter to A^s in order to identify the shape of sparsity pattern. If the diagonal elements in A^s have larger values compared to the others, applying a diagonal convolution filter increases the values of the diagonal elements in the convolution output ($conv\_out$). This leads to the emergence of a diagonal sparsity pattern. Otherwise, if the off-diagonal elements, especially the vertical ones, in A^s have larger values compared to the others, applying a diagonal convolution filter results in a vertical sparsity pattern in $conv\_out$ matrix. In order to ensure that A^s and $conv\_out$ have the same size ($L \times L$), we adopt zero-padding to A^s during computing the convolution operation defined in Eq. 3.

$$conv\_out(i,j) = \sum_{f=1}^{F} A^s(i+f, j+f) \times filter(f,f) \qquad (3)$$

After generating the $conv\_out$ through a diagonal convolution operation, our algorithm performs average pooling on the $conv\_out$ using a kernel/block of size ($B \times B$) as defined in Eq. 4.

$$pool\_out\left(\frac{i}{B}, \frac{j}{B}\right) = \frac{1}{B^2} \sum_{p=1}^{B} \sum_{q=1}^{B} conv\_out(i+p, j+q) \qquad (4)$$

Instead of analyzing the sparsity pattern of A^s element by element, applying average pooling enables capturing block sparsity pattern, which considers both the critical data points and their surrounding data points. Hence, since the output of average pooling (*pool_out*) has a smaller size ($L/B \times L/B$) compared to the attention score matrix A^s, *pool_out* can be considered as a block-wise abstract sparsity representation of A^s.

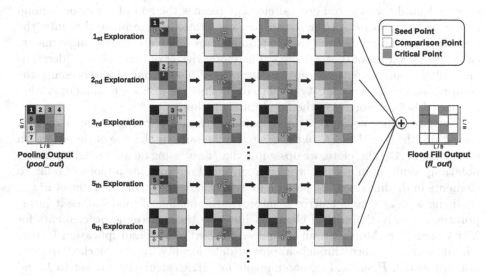

Fig. 3. Walk-through example of the identification of critical elements using the flood fill algorithm. (Colour figure online)

In order to dynamically explore the critical elements in the *pool_out*, we develop a novel method inspired by the flood fill algorithm. The flood fill algorithm is originally developed to determine the area connected to a given cell/pixel in a multi-dimensional array and a bitmap image [7]. Unlike the traditional flood fill algorithm, which compares all neighbors of a current element to find the element with the largest value, our scheme only compares to the neighboring elements on the right, below, and diagonally below, as shown in Fig. 3. Essentially, this is because the *pool_out* matrix follows the sequential order from left to right and top to bottom. In the process of capturing the pattern, it is necessary to sequentially follow the important features starting from the top-left corner to the bottom of the matrix. All the elements in the first row and all the elements in the first column of the *pool_out* are used as the seed starting points. From the particular seed point, our algorithm compares the values of elements to its right, below, and diagonally below to extract the element with the largest value in order to check whether the neighbors of the current element are relevant. If the neighboring elements are not relevant to the current element, our algorithm moves to the element diagonally below to continue comparing the neighbors. When the largest value is greater than a predefined threshold, we determine the corresponding element as the critical element (green colored elements in Fig. 3). Here, the threshold is determined by calculating the $\alpha\%$ quantile of *pool_out*. By utilizing the threshold, our flood fill algorithm ensures that the values of selected critical elements are sufficiently large. After determining the critical elements, our algorithm recursively compares the values of elements to the right, below, and diagonally below the critical element while avoiding duplicate comparisons with elements that have already been selected as critical points. This process is repeated until the selected critical element reaches the end of a row or column in the matrix $pool\_out$. After conducting explorations from all seed points, the resulting critical points from each exploration are combined into a single matrix called the flood fill output (fl_out). Therefore, the fl_out can be considered as an explicit sparsity pattern captured from $pool\_out$. fl_out also represents the compressed block-level sparsity pattern of A^s since the average pooling operation is applied before processing the flood fill algorithm.

Next, to utilize newly generated sparse pattern of fl_out in the sparse MHA operation, the size of the fl_out needs to be the same as the size of the attention score matrix A^s. Therefore, we upsample the fl_out using nearest-neighbor interpolation, resulting in each nonzero element in fl_out forming a block of nonzero elements in the final sparsity pattern matrix (P), as shown on the right of Fig. 2. Utilizing a block sparse matrix P improves the quality of model since it incorporates not only the critical elements but also their surrounding elements for MHA operation. Moreover, an optimized blocked matrix multiplication further enhances performance through improved data locality. In the blocked sparsity pattern matrix P, critical elements involving MHA calculation are set to 1, and the rest of the non-critical elements are set to 0. Finally, during the sparse MHA, only the elements of A^s that have the same indices with a value of 1 in P will be computed.

5   Experimental Evaluation

This section provides both performance and quality assessments of our SPION implementation. Our SPION is compared with various state-of-the-art sparse Transformer models. All the experiments were run on four NVIDIA RTX A5000 GPUs. Our GPU implementation integrates optimized CUDA kernels and CUDA libraries for sparse-attention training with PyTorch, utilizing the ctypes library (CDLL) provided in Python.

Datasets and Tasks. We evaluated our SPION on four tasks: image classification, text classification, ListOps, and document retrieval. For image classification, we utilized the CIFAR-10 [10] and iNaturalist 2018 [8] datasets, with the former having images of 32×32 pixels and the latter resized to this resolution, resulting in a sequence length of 1,024. In text classification, we used the AG News [22] and Yelp Reviews [22] datasets, both having a maximum sequence length of 1,024. The ListOps task, based on the dataset from Nangia et al. [12], involves classifying answers from 0 to 9 from sequences of numbers and operators, with a sequence length of 2,048. For the document retrieval task, we used the AAN dataset [15], classifying whether two given documents are related or not, with a sequence length of 4,096. All the text data is evaluated at the character level.

Models Compared. We compared our SPION with four state-of-the-art sparse Transformers: LSG Attention [4], Longformer [2], BigBird [20] and Reformer [9]. In addition, we evaluated variations of our SPION model: SPION-C, SPION-F and SPION-CF. Specifically, the SPION-C model omits the flood filling scheme and selects the top few percent of block elements after convolution filter and average pooling (*pool_out*), facilitating an adjustable sparsity ratio during generating the P matrix. In contrast, the SPION-F model bypasses the convolution filter, applying the flood fill-based algorithm immediately after the average pooling. The SPION-CF integrates both the convolution filter and the flood fill-based approach while generating a sparsity pattern. For all models, the sizes of configuration (window, bucket, block) were determined by the maximum sequence length of dataset. For example, Longformer used 64 or 128, while all other compared models, including our SPION variants, used 32 or 64.

We set the embedding dimension (D) to a size of 64. For image classification and document retrieval tasks, we used two layers, whereas for text classification and ListOps tasks, we employed four layers. The batch size was determined by 512 for image classification and text classification, 256 for ListOps, and 32 for document retrieval. Across all experiments, the size of the convolution filter in SPION was fixed at (31×31). As for the threshold in the flood fill algorithm, we configured α to 96 for image classification and text classification tasks, 98 for ListOps, and 99.5 for the document retrieval task. Note that all experimental results presented in this section are averaged over three distinct runs.

5.1 Performance Evaluation

Convergence. Table 1 shows the accuracy of six models on four different tasks. Our SPION-CF consistently achieved the highest accuracy in all tasks, surpassing the highest accuracy obtained from other compared models by +0.812%, +0.115%, +0.697%, +2.735%, +0.707%, and +0.205% for the evaluation tasks. It is interesting to see that among the SPION variants, incorporating both the convolution filter and flood-filling scheme led to higher accuracy for all tasks. This indicates that the convolution filter and flood-filling method synergize with each other. Additionally, we observed that the flood filling method shows more significant effect on accuracy compared to the convolution filter. This result demonstrates that considering the connectivity between elements is an important factor when capturing sparse patterns.

Speedup and Memory Reduction. Table 2 shows the time and memory required to train each model across four tasks, using the CIFAR-10 dataset for image classification and Yelp-review dataset for text classification. Our SPION-CF, compared to BigBird, achieved speedups of up to 2.78× per training step. Particularly in longer sequence tasks like ListOps and document retrieval,

Table 1. Classification accuracy (%) of various tasks

Model	Image Classification		Text Classification		ListOps	Document Retrieval
	CIFAR-10	iNaturalist	Yelp-Review	AG-News		
LSG Attention	43.435	53.758	80.449	79.371	39.137	80.538
Longformer	42.845	54.269	80.858	75.465	39.620	80.595
BigBird	41.978	52.964	76.813	74.177	39.658	80.396
Reformer	40.824	51.298	74.625	67.164	38.058	78.891
SPION-C	41.355	53.759	76.971	72.796	40.290	80.266
SPION-F	43.846	53.553	81.339	81.645	40.160	80.423
SPION-CF	**44.247**	**54.384**	**81.555**	**82.106**	**40.365**	**80.800**

SPION showed greater efficiency. Against LSG-Attention, SPION excelled in tasks with shorter sequences, and also required less training time for longer sequences. Table 2 further shows SPION-CF reducing memory footprint by over 7.24× in comparison to BigBird on ListOps task. Our SPION variants consistently had the smallest memory footprints. Remarkably, despite memory savings, SPION-CF maintained the highest accuracy across all tasks.

Table 2. Comparison of elapsed time (ms) and memory usage (GB) per step for training on various classification tasks

Model	Image Classification		Text Classification		ListOps		Document Retrieval	
	Time	Memory	Time	Memory	Time	Memory	Time	Memory
LSG Attention	97.064	12.390	233.947	31.532	197.602	31.835	104.699	11.652
Longformer	154.511	26.418	385.044	63.022	480.873	91.038	208.290	29.645
BigBird	217.889	35.458	503.207	82.698	539.533	104.134	276.082	34.300
Reformer	138.936	18.545	354.041	49.067	309.772	50.614	144.285	19.590
SPION-C	87.471	6.536	217.360	13.770	**187.382**	**14.259**	106.608	5.854
SPION-F	**85.298**	6.550	212.343	**13.745**	197.928	15.267	106.111	5.877
SPION-CF	86.257	**6.530**	**206.387**	13.769	194.256	14.368	**103.395**	**5.799**

5.2 Computational Complexity Analysis

Figure 4 shows the FLOPs (Floating Point Operations) for computing sparse attention matrix S for various sparse Transformers. Note that all sparse Transformers maintain a total of C non-zero elements in the sparse attention matrix S. Therefore, the total operations required for computing S for each head in all compared sparse Transformers is the same as $2C(2D + 1) - L(D + 1)$. As shown in Fig. 4, our SPION-CF achieves up to a 5.91× reduction in FLOPs compared to BigBird on the document retrieval task. This result demonstrates the

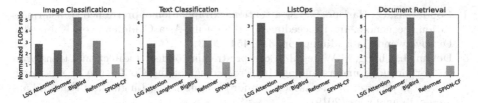

Fig. 4. Comparison of FLOPs required for computing the attention score matrix on various tasks. All the results are normalized to SPION-CF.

effectiveness of our convolutional flood-fill attention scheme, which dynamically captures relevant elements specific to the dataset and task. Furthermore, unlike the Longformer and BigBird models that apply the same fixed sparse pattern for every layer, our layer-wise sparse MHA enables the avoidance of unnecessary computations with non-relevant elements at each layer.

6 Conclusion

Due to the compute-intensive nature of the Transformer, that must perform a large number of MHA operations, we develop a novel sparsification scheme. This scheme leverages convolution filters and the flood fill algorithm to reduce the overall complexity of MHA operations. Our sparsification technique dynamically identifies the critical elements in the attention score matrix on a layer-wise basis. This method is applicable to many other Transformer models, not limited to the encoder-only Transformer. Experimental results on various datasets demonstrate that our sparse MHA approach significantly reduces training time and memory usage compared to state-of-the-art sparse Transformers, while achieving better accuracy on various classification tasks.

Acknowledgments. This research was supported by the MSIT(Ministry of Science and ICT), Korea, under the ITRC(Information Technology Research Center) support program(RS-2024-00259099) supervised by the IITP(Institute for Information & Communications Technology Planning & Evaluation), and in part by the National Research Foundation of Korea (NRF) grant funded by the Korea government (MSIT) (No. NRF-2021R1G1A1092597).

References

1. Ainslie, J., Ontanon, S., et al.: Etc: encoding long and structured inputs in transformers. In: Proceedings of the 2020 Conference on Empirical Methods in Natural Language Processing (EMNLP), pp. 268–284 (2020)
2. Beltagy, I., Peters, M.E., Cohan, A.: Longformer: the long-document transformer. arXiv preprint arXiv:2004.05150 (2020)
3. Child, R., Gray, S., Radford, A., Sutskever, I.: Generating long sequences with sparse transformers. arXiv preprint arXiv:1904.10509 (2019)

4. Condevaux, C., Harispe, S.: Lsg attention: extrapolation of pretrained transformers to long sequences. In: Proceedings of the Pacific-Asia Conference on Knowledge Discovery and Data Mining, pp. 443–454 (2023)
5. Devlin, J., Chang, M.W., et al.: Bert: pre-training of deep bidirectional transformers for language understanding. In: Proceedings of the Conference of the North American Chapter of the Association for Computational Linguistics: Human Language Technologies (2019)
6. Dosovitskiy, A., Beyer, L., et al.: An image is worth 16×16 words: transformers for image recognition at scale. In: International Conference on Learning Representations (2021)
7. Goldman, R.: Graphics gems, p. 304 (1990)
8. iNaturalist 2018 competition dataset. (2018)
9. Kitaev, N., Kaiser, Ł., Levskaya, A.: Reformer: the efficient transformer. In: Proceedings of the International Conference on Learning Representations (2020)
10. Krizhevsky, A., Hinton, G., et al.: Learning multiple layers of features from tiny images (2009)
11. Liu, Y., Ott, M., et al.: Roberta: a robustly optimized bert pretraining approach. arXiv preprint arXiv:1907.11692 (2019)
12. Nangia, N., Bowman, S.R.: Listops: a diagnostic dataset for latent tree learning. arXiv preprint arXiv:1804.06028 (2018)
13. Nvidia: the api reference guide for cusparse, the cuda sparse matrix library. Technical report (2023). https://docs.nvidia.com/cuda/cusparse/index.html
14. Qiu, J., Ma, H., et al.: Blockwise self-attention for long document understanding. In: Findings of the Association for Computational Linguistics: EMNLP 2020, pp. 2555–2565 (2020)
15. Radev, D.R., Muthukrishnan, P., et al.: The ACL anthology network corpus. Lang. Res. Eval. **47**, 919–944 (2013)
16. Roy, A., Saffar, M., et al.: Efficient content-based sparse attention with routing transformers. Trans. Assoc. Comput. Linguist. **9**, 53–68 (2021)
17. Tay, Y., Dehghani, M., et al.: Efficient transformers: a survey. ACM Comput. Surv. **55**(6), 1–28 (2022)
18. Vaswani, A., Shazeer, N., et al.: Attention is all you need. Adv. Neural Inf. Process. Syst. **30** (2017)
19. Wang, S., Li, B.Z., et al.: Linformer: self-attention with linear complexity. arXiv preprint arXiv:2006.04768 (2020)
20. Zaheer, M., Guruganesh, G., et al.: Big bird: transformers for longer sequences. Adv. Neural. Inf. Process. Syst. **33**, 17283–17297 (2020)
21. Zhang, H., Gong, Y., et al.: Poolingformer: long document modeling with pooling attention. In: International Conference on Machine Learning, pp. 12437–12446. PMLR (2021)
22. Zhang, X., Zhao, J., LeCun, Y.: Character-level convolutional networks for text classification. Adv. Neural Inf. Process. Syst. **28** (2015)

Towards Cost-Efficient Federated Multi-agent RL with Learnable Aggregation

Yi Zhang[1,2], Sen Wang[1], Zhi Chen[1], Xuwei Xu[1,2], Stano Funiak[2], and Jiajun Liu[1,2(✉)]

[1] School of Electrical Engineering and Computer Science, The University of Queensland, St Lucia, Brisbane, QLD 4066, Australia
{uqyzha91,sen.wang,zhi.chen,xuwei.xu}@uq.edu.au
[2] DATA61, Commonwealth Scientific and Industrial Research Organisation (CSIRO), Pullenvale, QLD 4069, Australia
{stano.funiak,ryan.liu}@data61.csiro.au

Abstract. Multi-agent reinforcement learning (MARL) often adopts centralized training with a decentralized execution (CTDE) framework to facilitate cooperation among agents. When it comes to deploying MARL algorithms in real-world scenarios, CTDE requires gradient transmission and parameter synchronization for each training step, which can incur disastrous communication overhead. To enhance communication efficiency, federated MARL is proposed to average the gradients periodically during communication. However, such straightforward averaging leads to poor coordination and slow convergence arising from the non-*i.i.d.* problem which is evidenced by our theoretical analysis. To address the two challenges, we propose a federated MARL framework, termed cost-efficient federated multi-agent reinforcement learning with learnable aggregation (FMRL-LA). Specifically, we use asynchronous critics to optimize communication efficiency by filtering out redundant local updates based on the estimation of agent utilities. A centralized aggregator rectifies these estimations conditioned on global information to improve cooperation and reduce non-*i.i.d.* impact by maximizing the composite system objectives. For a comprehensive evaluation, we extend a challenging multi-agent autonomous driving environment to the federated learning paradigm, comparing our method to competitive MARL baselines. Our findings indicate that FMRL-LA can adeptly balance performance and efficiency. Code and appendix can be found in https://github.com/ArronDZhang/FMRL_LA.

Keywords: Multi-agent reinforcement learning · Federated learning

1 Introduction

Federated reinforcement learning (RL) [4,9,11] has exhibited immense potential in integrating deep reinforcement learning models into a client-server paradigm. It has been proven effective in balancing communication efficiency and privacy

D.-N. Yang et al. (Eds.): PAKDD 2024, LNAI 14646, pp. 171–183, 2024.
https://doi.org/10.1007/978-981-97-2253-2_14

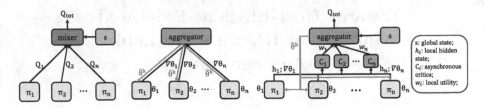

Fig. 1. Framework Comparison. π_i: local policy whose parameters are θ_i; Q_i: local value function; Q_{tot}: joint value function; $\nabla\theta_i$: local gradients; $\bar{\theta}^k$: global model's parameters at round k; h_i: local hidden state; w_i: local utility. The left subplot indicates the CTDE framework where the agents share parameters. The middle one represents the current vanilla FMARL where the agents maintain a global model through periodical averaging and the right figure refers to our FMRL-LA framework which selectively chooses and weights the involved agents to maximize the system utility.

preservation. With the burgeoning rise of the Internet of Things [20] that requires agent cooperation, and the prevalent use of multi-agent systems (MAS) [15,17], it is desirable to develop federated multi-agent reinforcement learning (MARL).

In MARL, centralized training with decentralized execution (CTDE) [12,21, 24] is a conventional learning regime lying between independent learning [28] and fully centralized learning [27]. This middle-ground strategy can not only mitigate the non-stationarity caused by agents' simultaneous decision-making but also prevent state and action spaces from expanding exponentially with agent number. Nevertheless, the training phase of CTDE requires continual communication between agents and servers. Thus, simply incorporating CTDE into federated learning (FL) [16] will lead to intractable communication overhead and bandwidth burdens. On the other hand, agent interactions with their local environments make their experiences non-independent and identically distributed (non-$i.i.d.$).

While recent efforts like FMARL [29] and Fed-MADRL [23] have marked advances in federated MARL [4], they typically assume an implicit $i.i.d.$ in agent interactions and lack server-side coordination. Furthermore, these methods tend to optimize singular, task-oriented objectives, $e.g.$, the average speed in multi-vehicle autonomous systems [29] and the system throughput in wireless communications [23]. Such settings may be impractical for complex real-world settings with composite objectives. For instance, in autonomous driving [6,13,19], apart from communication efficiency, we also consider factors like success rate in reaching destinations, overall safety, and average vehicle speed.

In response to these challenges, we introduce Cost-Efficient Federated MARL with Learnable Aggregation (FMRL-LA). It decouples the CTDE by separating the training steps of the server and the client. On the server side, we propose two components for learnable aggregation: 1. Asynchronous critics evaluate the utility of learning agents, guiding selection for optimal system communication. 2. A centralized aggregator integrates global information with agent utilities to periodically update the global model, thus maximizing composite system targets. This design facilitates FMRL-LA to improve coordination under the federated paradigm. Delving deeper into the non-$i.i.d.$ challenge posed by federated MARL, we theoretically delineate its adverse effects, providing a convergence upper bound. We further prove that the proposed learnable aggregation can mitigate the challenge. The comparison of different frameworks is exhibited in Fig. 1.

To conduct experiments with FMRL-LA, we resort to real-world multi-agent environment simulations based on MetaDrive [13], an intricate autonomous driving benchmark out of its flexibility across diverse scenarios. We extend it to support a client-server learning paradigm, incorporating communication efficiency. To further enhance the practicality, in addition to the existing navigation tasks, we design a multi-vehicle cooperative exploration task. Notably, we have integrated baselines from the representative methods of cooperative MARL [28] and communication-inclusive MARL [7], as well as the state-of-the-art method [19] using MetaDrive. Our experimental evaluations in navigation and exploration tasks underscore that FMRL-LA can optimize system performance and efficiency simultaneously, delivering a balanced performance across the metrics corresponding to composite objectives.

2 Preliminary

Cooperative MARL. Cooperative MARL can be formulated as Decentralized Partially Observable Markov Decision Processes (Dec-POMDPs) [12,31], described by a tuple $G = \langle n, S, O, A, P, r, Z, \gamma \rangle$, where n is the number of agents, and S, O denote the state and observation spaces. A, the joint action space, is the product of all agents' action spaces, $i.e.$, $A = \prod_{i=1}^{n} A_i$, where i is the agent index. We use lowercase s, o, a to represent an element in the corresponding space. The environments' dynamics are characterized by the transition function $P(s'|s,a) : S \times A \times S \to [0,1]$. The system has a shared team reward function $r(s,a) : S \times A \to \mathbb{R}$. In the aspect of each agent, due to the partially observable setting, at time step t, its observation o_t is drawn by applying the function Z to the current state s_t. Thus, $o_i^t = Z_i(s^t) : S \to O$. γ is the discount factor. The solution of a Dec-POMDP is a joint policy $\bar{\pi} = (\pi_1, \pi_2, ..., \pi_n)$, where π_i stands for the policy of agent i and we use θ_i, $\bar{\theta}$ to represent the parameters of agent i and the joint policy, respectively. Each agent policy is trained with the agent's experience comprised of a collection of agent observation-action history denoted as $\xi_i = \{(o^t, a^t, o^{t+1})\}_{t=0}^{T}$, where T denotes the time horizon. In addition, we use $\xi = \{(s^t, a^t, s^{t+1}, r^t)\}_{t=0}^{T}$ to represent one global team episode. The goal of MARL is to learn a joint policy that can maximize the expected cumulative reward, $i.e.$, $\pi^* = \arg\max_{\bar{\pi}} \mathbb{E}_{\tau \sim \bar{\pi}}[R_T(\tau)]$, where $R_T(\tau) = \sum_{t=0}^{T} \gamma^t r^{(t)}$.

Federated MARL. We use τ to represent the number of local updates. K is the termination condition of the training process, which is usually set as maximum communication rounds [4]. ψ denotes the system communication efficiency. We use the parameter θ to represent policy π. $F(\cdot)$ represents the global objective function, while $F_i(\cdot)$ stands for the local objective function for each agent i.

Their relationship between the global objective and the locals in [4,23,29] are the same: $F(x) = \frac{1}{n} \sum_{i=1}^{n} F_i(x)$. In round k, all agent policies are synchronized as $\bar{\theta}^k$, which is drawn from the server. Then, each agent interacts with the environment concurrently to accumulate local experience for updating the local policy indicated by $\{\theta_i^{k,\tau_i}\}_{i=1}^{n}$. Next, the parameters $\{\theta_i^{k,\tau_i}\}_{i=1}^{n}$ or stochastic gradients $\{g(\theta_i^{k,j}; \xi_i^{k,j})\}_{j=1}^{\tau_i}$ for $i \in 1, 2, \cdots, n$ will be uploaded to the server. To sum up, the update rules on the server and client side are:

$$\bar{\theta}^{k+1} = \bar{\theta}^k - \eta \frac{1}{n} \sum_{i=1}^{n} \sum_{j=1}^{\tau_i^k} g(\theta_i^{k,j}), \quad \theta_i^{k+1,j} = \begin{cases} \bar{\theta}^{k+1}, & j \bmod \tau_i = 0, \\ \theta_i^{k,j} - \eta g(\theta_i^{k,j}), & \text{otherwise.} \end{cases} \quad (1)$$

To indicate the convergence of the algorithm, we use the expected averaged gradient norm to guarantee convergence to a stationary point [2,25,26]:

$$\mathbb{E}[\frac{1}{K} \sum_{k=0}^{K-1} ||\nabla F(\bar{\theta}^k)||^2] \leq \epsilon, \quad (2)$$

where $||\cdot||$ is the ℓ_2-norm and ϵ is used to describe the sub-optimality. When the above condition holds, we say the algorithm achieves an ϵ-suboptimal solution.

3 Federated MARL with Learnable Aggregation

Server Side. When federated MARL [23,29] adopts Eq. (1) as the update rule for the server, it implicitly assumes that the agents are homogeneous. However, in real-world environments, the agents are diverse in various aspects such as computing capability, network connection, and local observation distributions, which results in heterogeneous agents with non-*i.i.d.* experience distribution.

To deal with these issues, we introduce **Asynchronous Critics** to dynamically evaluate the agent utilities in each communication round. Each critic corresponds to one learning agent. Its goal is to maximize the return of the current agent. The inputs are hidden information h_i^k, accumulated rewards in recent communication round $r^k := \sum_{j=\tau_i^{k-1}}^{\tau_i^k} r_i^j$ and in agent history $\sum_{j=0}^{\tau_i^k} r_i^j$. The output is a prediction of the agent's local utility:

$$w_i^k = C_i \left(h_i^k, r^k, \sum_k r^k \right), \quad (3)$$

where C_i is the asynchronous critic network of agent i. The output w_i^k can be zero. The corresponding agent does not need to upload its training parameters to the server in the current communication round, achieving client selection.

Next, the agent utilities are passed through a **Centralized Aggregator** to facilitate coordination. It works similarly to the mixing network in value decomposition methods such as [21,24], which takes the local utility function as input and facilitates agent coordination by maximizing the system utility

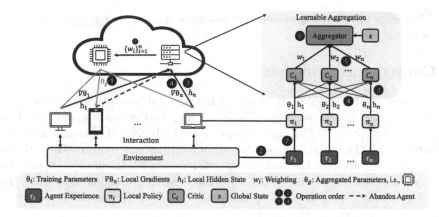

Fig. 2. The workflow of our proposed framework.

condition on the global state. The RL loss is back-propagated to the critics for improving the local utility estimation:

$$Q_{tot} = Mix\left(w_1^k, w_2^k, \cdots, w_n^k | s\right),\tag{4}$$

where Q_{tot} denotes the system utility, which is reflected by the composite objectives. The server aggregates the gradients based on the local utilities to update the global policy. Thus, the update rule of the server is:

$$\bar{\theta}^{k+1} = \bar{\theta}^k - \eta \sum_{i=1}^{n} w_i^k \sum_{j=1}^{\tau_i^k} g(\theta_i^{k,j}),\tag{5}$$

while the update rule for the clients remains the same as the client side of Eq. 1.

Client Side. Considering the generalization of our method, we choose an independent reinforcement learning algorithm and take the hidden states as additional outputs. During the communication, the upload process of the clients can be divided into two stages. In the first stage, the agents upload their rewards and hidden information to the asynchronous critics for local utility estimation and agent selection, optimizing communication efficiency. In the second stage, the selected agents upload their gradients to the server for aggregation.

Framework Design. Compared to the update rule for FMARL [23,29] represented by Eq. (1), we adopt a weighted aggregation for global policy update implemented by the learnable aggregation module. The workflow of FMRL-LA is illustrated by Fig. 2. Specifically, 1. the server broadcasts the global model $\bar{\theta}^k$ to each agent; 2. The agents learn local behavior policies $\{\pi_i\}_{i=1}^n$ by interacting with the environment and maintaining hidden states; 3 and 4. During client-server communication, agents conduct the two-stage upload described in the above subsection; 5. The centralized aggregator maximizes the composite system objectives condition on the global states to facilitate coordination; 6. The global model is updated based on the local utilities. The corresponding pseudo code is in Appendix A.1.

4   Convergence Analysis

In FL theory, a substantial body of research is devoted to exploring convergence properties under diverse settings. These settings predominantly fall into two categories, $i.i.d.$ [4,16,23,29], and non-$i.i.d.$ [10,14,18,26]. While $i.i.d.$ settings facilitate robust theoretical results, non-$i.i.d.$ settings are more realistic about data distribution. Despite the theoretical progresses in FL schemes in supervised learning, the influence of non-$i.i.d.$ in federated MARL remains uncharted. To address this issue, we conduct our theoretical analysis in the following paragraphs.

We begin with showing the convergence under the ideal $i.i.d.$ setting. To do that, we first list out the following assumptions:

Assumption 1 *(Lipschitz continuity). The local loss functions at the client side are Lipschitz continuous, which means* $||\nabla F_i(\theta_1) - \nabla F_i(\theta_2)|| \le L||\theta_1 - \theta_2||, \forall i \in \{1, 2, ..., n\}.$

Assumption 2 *(Unbiased gradients and bounded variance under i.i.d.). The stochastic gradients at the client side are unbiased estimators of the global gradient, i.e.,* $\mathbb{E}_\xi[g_i(\theta)] = \nabla F(\theta)$ *and* $\mathbb{E}_\xi[||g_i(\theta) - \nabla F(\theta)||^2] \le \mu||\nabla F(\theta)||^2 + \sigma^2, \forall i \in \{1, 2, ..., n\}, \mu$ *and* σ^2 *are non-negative.*

Assumption 3 *(Unbiased local gradients and bounded variance under non-i.i.d.). The stochastic gradient at each client is an unbiased estimator of the local gradient, i.e.,* $\mathbb{E}_\xi[g_i(\theta)] = \nabla F_i(\theta)$ *and* $\mathbb{E}_\xi[||g_i(\theta|\xi) - \nabla F_i(\theta)||^2] \le \mu||\nabla F_i(\theta)||^2 + \sigma^2, \forall i \in \{1, 2, ..., n\}, \mu$ *and* σ^2 *are non-negative.*

Assumption 4 *(Bounded Dissimilarity). For any sets of weights* $\{w_i^k \ge 0\}_{i=1}^n, \sum_{i=1}^n w_i^k \le M^k, M^k$ *is finite,* $\forall k \in [0, K]$, *there exist constants* $\beta^2 \ge 1, \kappa^2 \ge 0$ *such that* $\sum_{i=1}^n w_i^k||\nabla F_i(\theta)||^2 \le \beta^2\{||\sum_{i=1}^n w_i^k \nabla F_i(\theta)||^2,$ $||\sum_{i=1}^n \frac{1}{n}\nabla F_i(\theta)||^2\}_{min} + \kappa^2, \forall k \in [0, K].$ *If local loss functions are identical to each other, then we have* $\beta^2 = 1, \kappa^2 = 0.$

Assumption 1 is Lipschitz continuity, a common assumption in the convergence analysis in FL theory. Assumption 2 states that the local stochastic gradient is an unbiased estimation of the local gradient and the variance of the deviation is bounded to support our exploration under a $i.i.d.$ setting. Assumption 3, on the other hand, is the gradient bias and variance assumption under non-$i.i.d.$ setting. Assumption 4 is inspired by FedNova [26], which bounds the dissimilarities on the weighted norm of local gradients.

We provide the convergence bound under the $i.i.d.$ and non-$i.i.d.$ settings as Theorem 1 and Theorem 2, respectively. We further provide a special case to show that there is room for tightening the bound with the learnable aggregation mechanism as Theorem 3 in the Appendix A.4. In addition, the proof of these theorems as well as more theoretical details are provided in the Appendix A.4.

Theorem 1. *Suppose $\{\theta_i^{k,j}\}$ and $\{\bar{\theta}^k\}$ are parameters' sequences generated by Eq.(1). The federated MARL is conducted under Assumptions 1 and 2. If the total communication rounds K is large enough, which can be divided by τ, and the learning rate η satisfies:*

$$\left\{ L\eta < 1, 2L^2\eta^2\tau(2\mu + 1 + \tau) < 1 \right\}, \tag{6}$$

then the expected gradient norm after K iterations is bounded by:

$$\mathbb{E}[\frac{1}{K}\sum_{k=1}^{K} ||\nabla F(\bar{\theta}^k)||^2] \leq \frac{2[F(\bar{\theta}^1) - F(\bar{\theta}^K)]}{\eta K} + \frac{\eta L\sigma^2}{n} + \eta^2 L^2\sigma^2(\tau + 1), \tag{7}$$

where $\bar{\theta}^1$ stands for one lower bound of the objective function.

Theorem 2. *Suppose $\{\bar{\theta}^k\}$ are parameters' sequences generated by the weighted gradients Eq.(5), while the $\{\theta_i^{k,j}\}$ remains the same. The federated MARL is conducted under Assumptions 1, 3 and 4. If the total communication rounds K is large enough, and the learning rate η satisfies (6), then the expected gradient norm after K iterations is bounded by:*

$$\mathbb{E}[\frac{1}{K}\sum_{k=1}^{K} ||\nabla F(\bar{\theta}^k)||^2] \leq \frac{4\left(E\left[F\left(\bar{\theta}^1\right)\right] - E\left[F\left(\bar{\theta}^K\right)\right]\right)}{K\eta}$$

$$+ 4\left(C + D + E + F + \mu\eta C\sum_{k=0}^{K}\frac{1}{K}\sum_{i=1}^{n} w_i^2\tau_i^k\right), \tag{8}$$

where $\bar{A} = \frac{1}{K}\sum_{i=1}^{K} A$, $B = 2L^2\eta^2\tau(2\mu + 1 + \tau)$, $C = \frac{\eta^2\sigma^2L^2}{\mu L\eta\tau\beta^2 + 2B\beta^2}$, $D = \frac{(1-2\mu L\eta\tau\beta^2)\kappa^2}{(2\mu L\eta\tau\beta^2 + 4B\beta^2)(1+4\beta^2)}$, $E = \frac{\mu L\eta\tau\kappa^2}{2\mu L\eta\tau\beta^2 + 4B\beta^2}$, and $F = \frac{L\eta\sigma^2}{2K}\sum_{k=1}^{K}\left(M^k\right)^2$.

Discussion. The result of Theorem 1 is an ideal upper bound where the distribution of each client is *i.i.d.* More generally, in Theorem 2, we provide another upper bound to illustrate the impact introduced by the non-*i.i.d.* issue. Comparing the bounds in these two theorems, it is obvious that the convergence bound in the non-*i.i.d.* setting is greater than that in the *i.i.d.* setting.

Special Cases. When $w_i^k \equiv \frac{1}{n}$, the convergence upper bound degenerates to the same as FMARL [29], which derives the same upper bound as in its Theorem

Fig. 3. The six extended scenarios used in our evaluation.

2. When $w_i^k \equiv \frac{1}{n}$ and $\tau_i^k \equiv 1$, the convergence upper bound further degenerates to the same as PASGD, which coincides with the conclusions drawn from [25].

Discussion with Federated Learning in Supervised Learning.   We compare our method with FedNova [26] – a general federated method targeting supervised learning. It induces several federated learning methods in a general form and targets the problem of an unbalanced number of local updates by regularizing the weights for one-period local gradients with the number of local updates. However, in MARL, the different number of policy iterations may not be a more significant reason than the diversity of local environments to the non-$i.i.d.$ issue. In other words, this issue cannot be rectified by simply regularizing the weights of local gradients by the number of local policy iterations, which necessitates the importance of our learnable aggregation mechanism.

5   Experiments

Baseline Methods.   We present a comparative analysis of our proposed method alongside strong baselines covering a wide range of related fields, namely conventional MARL (**IPPO** [28]), communication-based MARL (**RIAL and DIAL** [7]), the state-of-the-art method **CoPO** [19] and **FMARL** [29] in a multi-agent autonomous driving simulation environment, MetaDrive [13]. We provide a detailed introduction and adaptation of these methods in the Appendix A.6.

Implementation by Extending MetaDrive.   The MetaDrive benchmark represents a flexible and lightweight simulation benchmark for autonomous driving, encompassing a variety of tasks that serve as a reasonable abstraction of real-world environments. In this paper, we focus on its multi-agent tasks. The agents adopt the conventional MARL suite, including parameter sharing and disregarding communication overhead. We add six challenging scenarios whose maps are depicted in Fig. 3. Originally, MetaDrive used parameter sharing for all the methods, so we first expand this benchmark into a client-server learning setting by adopting a non-parameter sharing scheme and simulating a virtual server. This server only periodically collects local gradients and hidden states from the clients, aggregates the gradients for the update of the global model, and then sends it back to the agents. When an existing vehicle terminates and a new vehicle spawns, it accepts the latest global model from the server to prevent the "cold start" problem. To enrich the testing bed and serve as a real-world simulation, in addition to the existing navigation task, we extend a cooperative exploration task where vehicles cooperatively explore the specified destinations.

Evaluation Metrics.   MetaDrive provides three evaluation metrics: success rate, efficiency, and safety, which respectively reflect navigation capability (the success ratio of vehicles relative to the total number of vehicles), navigation efficiency (the differences between the successes and failures within a unit of time), and safety driving (the number of crushes within an episode). In our cooperative exploration task, we adapt the navigation success rate to the exploration success rate. For realistic concerns, we also record the communication overhead which

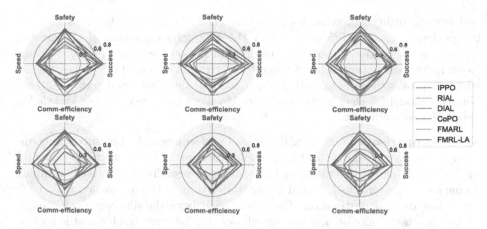

Fig. 4. The system performance and efficiency comparison with baselines in six scenarios of the cooperative navigation tasks.

is reflected by the number of parameters exchanged between the agents and the server. The **system utility** is derived from the weighted average of these metrics, where the weights reflect the specific requirements of particular scenarios. In our experiments, we employ a simple average to gauge overall performance.

In summary, in the **cooperative navigation** task, our evaluation metrics including the navigation success rate (*Success*), safety (*Safety*), overall agents' speed (*Speed*), and communication efficiency (*Comm-efficiency*). As for the **cooperative exploration**, our evaluation metrics including the exploration success rate (*Explore*), safety (*Safety*), overall agents' speed (*Speed*), communication efficiency (*Comm-efficiency*).

Main Results Analysis. The experiment results on cooperative navigation and exploration across six scenarios can be found in Fig. 4 and Fig. 5, respectively. More detailed results related to the performance on two tasks can be found in Table 2 to Table 5 in the Appendix A.7.

Our Performance. In both tasks, FMRL-LA achieves or is comparable to the best success, speed, safety, and system utility. From the perspective of communication efficiency, since RIAL [7] uses a simple actor-critic algorithm with few parameters, its communication efficiency serves as a reference of the upper bound for the methods with a relatively complex algorithm on the client side. Actually, more than half of the baselines adapt PPO [22] as the clients' algorithm. Among them, our method exhibits the capacity to dynamically harmonize the system performance and communication efficiency.

In detail, we focus on the performance of IPPO, FMARL, and our FMRL-LA in both tasks. The three methods have nearly the same client-side algorithms but differ from each other on the server side. IPPO only conducts direct training parameter averaging, while the FMARL adds a weight decay mechanism during the averaging. And FMRL-LA dynamically learns the aggregation weights. From IPPO to FMARL, then to FMRL-LA, the performance of success, safety, speed, and system utility follow an ascending manner. We believe that it is because the performance of IPPO is bound by the averaging capability of all learning agents while FMARL can enlarge the bound to some extent by weight decay. Nonetheless, the potential performance of FMRL-LA is bound by the best agent, which improves the generalization of our method since we can deploy a suitable behavior model for the clients in advance if we can make use of prior knowledge about the environments.

Task Comparison. Comparing the overall performance of all methods on cooperative navigation and cooperative exploration on six scenarios, we can find navigation is more difficult than exploration, especially in relatively complex scenarios, *e.g.*, scenario 4, 5, and 6. We notice that in the navigation task, each agent has its own destination. Therefore, we believe the different performance on the two tasks may be because the relationship between the local utilities and the system utility is easier to capture in exploration than in navigation.

Scenario Difficulty. In both tasks, if we compare the performance pair by pair such as scenario 1 and scenario 5, we observe that generally, the more building blocks involved in a scenario, the more challenging it is. Then, looking into the performance of scenario 1 and 2, both of them consists of four building blocks while scenario 2 contains one more roundabout than scenario 1. If we compare the safety of CoPO, FMARL, and our FMRL-LA, three robust methods in the two tasks on these six scenarios, we can find that roundabout tends to result in more crushes. Further, if we compare the performance of scenarios 1 and 4 on two tasks, we observe that though the two scenarios both contain one roundabout and the same number of building blocks, the performance of our method and other baselines is generally better in scenario 1. Considering the difference in these two scenarios, we hypothesize that it is due to the influence of wide turn. For intuitive methods like IPPO and RIAL, it is difficult for them to avoid crushing or driving out of the roads during the wide turns. On the other hand, the safety of CoPO in scenarios 3 and 4 is relatively high, it may benefit from its explicit modeling of the surrounding agents.

Ablation Study.   To investigate the effectiveness of our design and components in FMRL-LA, we conduct an ablation study about the usage of asynchronous critics and a centralized aggregator as well as an alternative design for the centralized aggregator. We intuitively use the average of multiple metrics as the system utility. From Table 1 we can observe that if we directly use critics without the centralized aggregator, the performance is unstable, resulting in large standard errors. In scenario 4, the performance w.r.t. system utility is worse than federated IPPO. We believe that without the coordination of the centralized aggregator, the server cannot filter out less valuable agents, so their parameters can depreciate the update of the global model in the current round. Meanwhile, the asynchronous critics are useful in our method since the variant that only uses an aggregator performs worse than the full model. We believe that accepting information from all involved agents can stagnate the learning of an aggregator due to redundant information. When we change the centralized

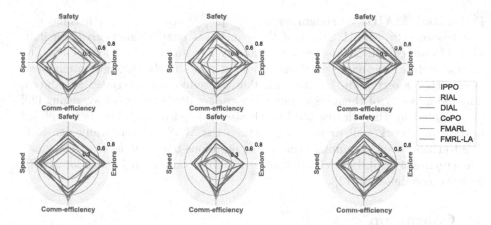

Fig. 5. The system performance and efficiency comparison with baselines on six scenarios of the cooperative exploration tasks.

Table 1. Ablation study on the effectiveness of our critical components on navigation task. The **system utility** is provided.

Experiment	Scenario1	Scenario2	Scenario3	Scenario4	Scenario5	Scenario6
IPPO	49.96±3.06	45.13±3.68	51.49±2.73	33.12±5.29	34.41±6.17	34.33±5.97
FMRL-LA w/o aggregator	54.75±6.00	49.28±5.23	52.06±4.71	40.43±7.02	32.85±8.92	40.22±7.74
FMRL-LA w/o critics	52.69±3.55	48.46±4.65	55.49±3.18	45.35±5.26	38.49±6.19	48.07±5.57
FMRL-LA w/ vdn-aggregator	57.98±2.97	55.84±4.39	57.26±3.69	52.50±5.80	46.98±6.81	50.86±5.09
FMRL-LA	**59.84±2.82**	**62.30±4.29**	**63.16±3.33**	**57.27±4.70**	**50.28±6.67**	**56.42±5.63**

aggregator to a VDN-based [24] one, it yields an inferior performance compared to our QMIX-based [21] design, which suggests the non-linear modeling of the relationship between the agents and the server is more suitable for complex realistic environments than a simple sum as in VDN.

Client Selection Analysis.   To verify that FMRL-LA can effectively select involved agents to save communication costs, we also conduct experiments with varying numbers of agents to show the client selection effect. The experiment results and elaboration can be found in the Appendix A.7.

6   Related Work

Cooperative MARL.   Cooperative MARL has widespread applications in real-world scenarios [13,30]. Current methods are mainly developed in game scenarios [3,15,21] where the methods can focus on technical design rather than practical details. These environments support parameter sharing (PS) [5] and CTDE regime [8,15] to enable multiple agents to be trained on one device and facilitate cooperation, respectively. However, when it is the stage to consider practical MARL in realistic environments [1,19], either PS or CTDE cannot be simply applied due to privacy concerns and communication overhead.

Federated MARL.   Federated MARL [23] appears to be a feasible way towards realistic MARL. Most of these methods enable agents to learn individual behavior policies and set a virtual server to maintain a global policy. The agents' policies are aggregated and updated periodically through communication with the server [4]. In this way, the communication overhead is reduced, and the majority of them aggregate the local gradients by direct averaging [29] or weighted by the relative mini-batch size [23]. Although this oversimplified update may work well under $i.i.d.$ setting, the MASs are naturally non-$i.i.d.$ due to the interaction among agents. The notorious non-$i.i.d.$ issue can stagnate convergence [14, 26]. Besides, without centralized training, it is hard for MARL to learn coordination [12].

7   Conclusion

We aim to adapt MARL for real-world applications by introducing a hybrid distributed, client-server learning framework that takes into account communication and computation overhead. Our framework offers theoretical guarantees even under the influence of non-$i.i.d.$ distribution of agents in local environments. To empirically validate the efficacy of our proposed Cost-Efficient Federated Multi-Agent Reinforcement Learning with Learnable Aggregation (FMRL-LA) method, we modify an existing multi-agent autonomous driving simulation environment to conform to a client-server scheme. Experimental results emphasize the superior performance against baseline methods.

Acknowledgements. This work is supported by project DE200101610 funded by Australian Research Council and CSIRO's Science Leader project R-91559.

References

1. Abegaz, M., Erbad, A., Nahom, H., Albaseer, A., Abdallah, M., Guizani, M.: Multi-agent federated reinforcement learning for resource allocation in UAV-enabled internet of medical things networks. IoT-J (2023)
2. Bottou, L., Curtis, F.E., Nocedal, J.: Optimization methods for large-scale machine learning. SIAM (2018)
3. Chaudhuri, R., Mukherjee, K., Narayanam, R., Vallam, R.D.: Collaborative reinforcement learning framework to model evolution of cooperation in sequential social dilemmas. In: PAKDD (2021)
4. Chen, T., Zhang, K., Giannakis, G.B., Başar, T.: Communication-efficient policy gradient methods for distributed reinforcement learning. TCNS (2021)
5. Christianos, F., Papoudakis, G., Rahman, A., Albrecht, S.V.: Scaling multi-agent reinforcement learning with selective parameter sharing. In: ICML (2021)
6. Du, X., Wang, J., Chen, S.: Multi-agent meta-reinforcement learning with coordination and reward shaping for traffic signal control. In: PAKDD (2023)
7. Foerster, J., Assael, I.A., De Freitas, N., Whiteson, S.: Learning to communicate with deep multi-agent reinforcement learning. In: NeurIPS (2016)
8. Hu, S., Zhu, F., Chang, X., Liang, X.: UPDeT: universal multi-agent reinforcement learning via policy decoupling with transformers. In: ICLR (2021)

9. Jin, H., Peng, Y., Yang, W., Wang, S., Zhang, Z.: Federated reinforcement learning with environment heterogeneity. In: AISTATS (2022)
10. Karimireddy, S.P., Kale, S., Mohri, M., Reddi, S., Stich, S., Suresh, A.T.: Scaffold: stochastic controlled averaging for federated learning. In: ICML (2020)
11. Khodadadian, S., Sharma, P., Joshi, G., Maguluri, S.T.: Federated reinforcement learning: linear speedup under Markovian sampling. In: ICML (2022)
12. Kuba, J.G., Chen, R., Wen, M., Wen, Y., Sun, F., Wang, J., Yang, Y.: Trust region policy optimisation in multi-agent reinforcement learning. In: ICLR (2022)
13. Li, Q., Peng, Z., Feng, L., Zhang, Q., Xue, Z., Zhou, B.: MetaDrive: composing diverse driving scenarios for generalizable reinforcement learning. TPAMI (2022)
14. Li, T., Sahu, A.K., Zaheer, M., Sanjabi, M., Talwalkar, A., Smith, V.: Federated optimization in heterogeneous networks. In: MLSys (2020)
15. Lowe, R., Wu, Y., Tamar, A., Harb, J., Abbeel, P., Mordatch, I.: Multi-agent actor-critic for mixed cooperative-competitive environments. In: NeurIPS (2017)
16. McMahan, B., Moore, E., Ramage, D., Hampson, S., y Arcas, B.A.: Communication-efficient learning of deep networks from decentralized data. In: AISTATS (2017)
17. Mo, J., Xie, H.: A multi-player MAB approach for distributed selection problems. In: PAKDD (2023)
18. Pang, Y., Zhang, H., Deng, J.D., Peng, L., Teng, F.: Rule-based collaborative learning with heterogeneous local learning models. In: PAKDD (2022)
19. Peng, Z., Hui, K.M., Liu, C., Zhou, B.: Learning to simulate self-driven particles system with coordinated policy optimization. In: NeurIPS (2021)
20. Pinto Neto, E.C., Sadeghi, S., Zhang, X., Dadkhah, S.: Federated reinforcement learning in IoT: applications, opportunities and open challenges. Appl. Sci. (2023)
21. Rashid, T., Samvelyan, M., De Witt, C.S., Farquhar, G., Foerster, J., Whiteson, S.: Monotonic value function factorisation for deep multi-agent reinforcement learning. JMLR (2020)
22. Schulman, J., Wolski, F., Dhariwal, P., Radford, A., Klimov, O.: Proximal policy optimization algorithms. arXiv:1707.06347 (2017)
23. Song, Y., Chang, H.H., Liu, L.: Federated dynamic spectrum access through multi-agent deep reinforcement learning. In: GLOBECOM (2022)
24. Sunehag, P., et al.: Value-decomposition networks for cooperative multi-agent learning. arXiv:1706.05296 (2017)
25. Wang, J., Joshi, G.: Cooperative SGD: a unified framework for the design and analysis of local-update SGD algorithms. JMLR (2021)
26. Wang, J., Liu, Q., Liang, H., Joshi, G., Poor, H.V.: Tackling the objective inconsistency problem in heterogeneous federated optimization. In: NeurIPS (2020)
27. Wen, M., et al.: Multi-agent reinforcement learning is a sequence modeling problem. Front. Comput. Sci. (2022)
28. de Witt, C.S., et al.: Is independent learning all you need in the starcraft multi-agent challenge? arXiv:2011.09533 (2020)
29. Xu, X., Li, R., Zhao, Z., Zhang, H.: The gradient convergence bound of federated multi-agent reinforcement learning with efficient communication. TWC (2023)
30. Yu, C., Velu, A., Vinitsky, E., Gao, J., Wang, Y., Bayen, A., Wu, Y.: The surprising effectiveness of PPO in cooperative multi-agent games. In: NeurIPS (2022)
31. Zhou, X., Matsubara, S., Liu, Y., Liu, Q.: Bribery in rating systems: a game-theoretic perspective. In: PAKDD (2022)

LongStory: Coherent, Complete and Length Controlled Long Story Generation

Kyeongman Park, Nakyeong Yang, and Kyomin Jung[(✉)]

Seoul National University, Seoul, Republic of Korea
{zzangmane,kjung,yny0506}@snu.ac.kr

Abstract. A human author can write any length of story without losing coherence. Also, they always bring the story to a proper ending, an ability that current language models lack. In this work, we present the LongStory for coherent, complete, and length-controlled long story generation. LongStory introduces two novel methodologies: (1) the long and short-term contexts weight calibrator (CWC) and (2) long story structural positions (LSP). The CWC adjusts weights for long-term context Memory and short-term context Cheating, acknowledging their distinct roles. The LSP employs discourse tokens to convey the structural positions of a long story. Trained on three datasets with varied average story lengths, LongStory outperforms other baselines, including the strong story generator Plotmachine, in coherence, completeness, relevance, and repetitiveness. We also perform zero-shot tests on each dataset to assess the model's ability to predict outcomes beyond its training data and validate our methodology by comparing its performance with variants of our model.

Keywords: Long Story generation · Completeness · CWC · LSP

1 Introduction

The story generation task is one of the most challenging problems in natural language processing since it requires writing long lengths with a consistent context. Existing studies have tried to solve this problem but have failed to generate variable lengths of stories since they have only considered fixed lengths when generating stories.

Longformer [1] has tried to solve this problem of handling longer sequences by combining global and local attention in sliding windows. However, they only focused on increasing the input context window size. Their maximum generation length is only 1,024 tokens. Short-length problems in text generation also occur in recent large language models such as GPT-4 [4]. Plug-and-blend [2] has also covered a similar problem, but the length limitation remains since it only aimed to control a few sentences.

To address this challenge, a recursive paragraph generation approach is necessary to compose long stories, given the length limitations imposed by existing

D.-N. Yang et al. (Eds.): PAKDD 2024, LNAI 14646, pp. 184–196, 2024.
https://doi.org/10.1007/978-981-97-2253-2_15

language models. However, this recursive generation of stories may cause undesirable forgetting of the previous context since the information leak may occur in the recursive process of information transfer (*coherence*) [13]. Also, existing studies have mentioned that language models tend to repeat the same story in a recursive generation setting. (*variety* and *repetitiveness*) [6,33].

Thus, previous works [6,12,13] have tried to define coherence and repetitiveness metrics and used them to measure the ability of story generators.

However, their attempts are also limited only to short sentence generation problems and have never considered **completeness**. Completeness is the ability to conclude a story of any length properly. It is a significant metric not only for evaluating story generation but also for a wide range of open-domain generation and dialogue system tasks. Therefore, in our model evaluation, we consider not only coherence and repetitiveness but also prioritize completeness as a critical measure of performance.

In this paper, we tackle this challenging problem by introducing a novel long story generator, *Coherent, Complete and Length Controlled Long story Generator* (LongStory), which covers from a few hundred tokens to 10k tokens length, not limited to only a few sentences. The LongStory utilizes two novel methods: (1) long and short-term contexts weight calibrator (CWC) and (2) long story structural positions (LSP) to tackle coherence and completeness.

Specifically, we implement the CWC using BERT-tiny to calibrate the weight to which long-term and short-term contexts are utilized since both contexts contribute differently in every paragraph when writing a story. For the LSP, we use discourse tokens, which means the order of paragraphs, to give information about the structural positions of paragraphs to the story generator (e.g., $< front >$, $< middle >$, $< ending >$, $< next\_is\_ending >$). Our model uses more abundant discourse tokens than the previous study [6] since a detailed understanding of story structure is essential to generate much longer posts.

We use three diverse story generation datasets with varying average lengths to train our model on representations of different story lengths. We introduce quantitative metrics for coherence, completeness, and repetitiveness, evaluating the model's performance against other baselines, including the established story generator Plotmachine [6]. The experimental results for three story generation datasets demonstrate that our model outperforms the other baselines in coherence, completeness, and relevance. Surprisingly, our model also shows better results in repetitiveness, suggesting that our methods are effective for the variety. Furthermore, we performed zero-shot tests on each dataset to assess how well our model predicts outcomes in settings beyond its training data. Additionally, our analysis of the augmented CWC version suggests that elevating relevance does not always translate to improvements in coherence and completeness.

The main contributions of this paper are (1). a new open-domain metric called completeness. (2) a new challenge, *Coherent, Complete and Length Controlled Long story Generation* (LongStory), incorporating CWC and LSP methodologies, and (3) the presentation of datasets with varying average document lengths along with zero-shot tests for them.

2   Related Works

Many contemporary automatic story generation models have employed prompt engineering strategies, manipulating input prompts for large pretrained language models like GPT-3 [2,7,13]. Despite the effectiveness of these black-box models, they could not directly enhance the internal structure for optimized performance. In contrast, our approach focuses on directly improving the structural aspects of existing language models.

2.1   Neural Story Generation

Story generators using neural networks have been developed in various ways [3,9,18,25,30]. The main approaches have included outline-based generation [3,6,7,13,25,27], event graph-based [8,15], goal-oriented methodologies [16,17], and common sense reasoning [18,19,32]. Outline-based generation methods often involve interpolating plot keywords [6,25]. Our model also uses keywords to plan plots and create relevant stories. Event graph-based approaches have proven highly effective, but only if given a detailed plot of the entire article. Our model uses keywords for the entire story but not the plot of the story, thus performing a more challenging task. Goal-oriented methodologies have aimed to make the story's characters achieve given goals, akin to controlling the ending of a story [5,12]. Common sense inference methods have focused on generating realistic texts by training on common sense datasets.

Length-Controllable Story Generation. Many models have generated text within a fixed range of length. The length has been mostly determined when selecting the training datasets. For example, Plotmachine [6] has fixed the number of paragraphs it generates, while EtriCA [8] and Fusion [3] also have set their generation lengths by their training dataset. Re3 [7] has generated narratives by applying pre-trained GPT-3 [9], and the length has been adjusted by manipulating the prompts to GPT-3. However, Re3 has not solely focused on controlling length. As such, it has not been specifically designed to ensure that the models generate without losing coherence and completeness with various lengths.

2.2   Recursive Models

Due to finite parameters, there has always been a limit to the length a language model can generate at once. While Longformer [1] and GPT-4 [4] have introduced 5K or even 100K context windows, their output length remains only a few thousand tokens at most. In contrast, human writers can produce thousands of pages without encountering such limitations. This reality implies the necessity of employing a language model recursively [6,7,25].

2.3   Autometic Metrics

In NLP generation tasks, traditional evaluation methods commonly involve metrics such as ROUGE [21], BLEU [22]. These metrics predominantly rely on n-gram matching between labels and predictions. However, applying these evaluation methods to the story generation task may not be sufficient. The reason lies in the fact that a well-crafted story may not necessarily exhibit similarity to the label; in fact, a story resembling the label might score poorly in terms of diversity and creativity [23,26]. In our case, we additionally employ *coherence*, *completeness*, and *repetitiveness* as the primary evaluation metrics along with the ROUGE scores. While *coherence* and *repetitiveness* is a well-established metric in many natural language generation tasks [5–7,13,24,25], we are the first to use *completeness* as a metric (Fig. 1).

3   Methodology

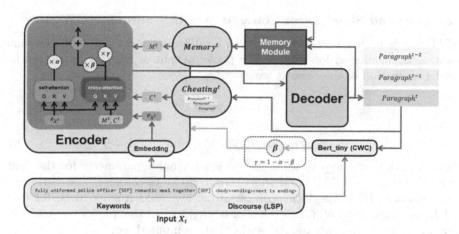

Fig. 1. Model architecture. LongStory takes the keywords of the entire story and the discourse tokens(LSP) representing the order of the target paragraphs as input. The BERT-tiny serves as the long and short-term context weight calibrator (CWC), determining the degree to which long-term and short-term contexts are employed. The CWC takes the discourse tokens and the last generated paragraph as inputs and outputs the optimal β and γ(defined as 1-α-β) for every paragraph. While the α is a hyperparameter applied to input embedding, β is a learnable parameter for the long-term context *Memory*(M^t) and short-term context *Cheating*(C^t)

3.1   Task Description

Our challenge can be described as follows: Given keywords $K = (k_1, k_2, k_3...)$, discourse tokens $D_t = (d_1^t, d_2^t, d_3^t...)$ (each k_i, d_i^t is a token), and a separation token [SEP], an input $X^t = (k_1, [SEP], k_2, [SEP], k_3, ..., [SEP], d_1^t, d_2^t, d_3^t...)$. The model should generate each t-th paragraph recursively: $Y^t = (y_1^t, y_2^t, y_3^t, ...)$ (each y_i^t is a sentence, a set of tokens) from given contexts of previous paragraphs and the input X^t, which is coherent and completive. The model should minimize negative log-likelihood loss L_{LM} of each paragraph t:

$$L_{LM} = -\sum_{i=1}^{n} log P(y_j^t | y_{<j}^t, X^t) \tag{1}$$

$$logP(y_j^t|y_{<j}^t, X^t) = softmax(H_j^t W + b) \tag{2}$$

$$H_j^t = LM(X^t, M^t, C^t) \tag{3}$$

where LM is an attention-based language model, H_j^t is the j-th position of the language model's last hidden state, M^t is the *Memory* which saves the long-term context by the t-th paragraph, and C^t is *Cheating* which keeps the short term context from the very last paragraphs it generated. The M^t ensures a sustained understanding of the broader context, while C^t contributes to short-term continuity.

3.2 Long and Short Term Contexts Weight Calibrator(CWC)

BERT-tiny is inserted into our model as a calibrator and trained on the same negative log likelihood loss, L_{LM}. The pooler output of BERT-tiny is used for the determination of the context weight, β, γ.

$$B_k^t = Bert(D^t, Y^{t-1}) \tag{4}$$

$$\beta = (1 - \alpha) \cdot \sigma(B_1^t W + b), \quad \gamma = 1 - \alpha - \beta \tag{5}$$

where $\alpha \in [0, 1]$ is a hyperparameter[1], β is a learnable parameter for the long-term context *Memory*(M^t) and short-term context *Cheating*(C^t), σ is the sigmoid function, B_k^t is the k-th position of the last hidden state of BERT-tiny, and B_1^t is the position of $[CLS]$ token of it, often called the 'pooler output.' The long and short-term contexts, M^t and C^t, are computed as:

$$g^t = \sigma(W_1 M^{t-1} + W_2 h^{t-1}), \quad \hat{M}^t = tanh(W_3 M^{t-1} + W_4 h^{t-1}) \tag{6}$$

$$M^t = (1 - g^t) \odot M^{t-1} + g^t \odot \hat{M}^t \tag{7}$$

$$C^t = tanh(Emb(y_{n-c}^{t-1}, ..., y_{n-1}^{t-1}, y_n^{t-1})) \tag{8}$$

where Emb denotes a positional embedding layer that involves the element-wise sum of word embeddings and positional encoding vectors, h^{t-1} is the average embedding of Y^{t-1}, and c is a hyperparameter determining the size of the cheating window. We utilized the same computation process of initializing and updating M^t as introduced in Plotmachine. Finally, we define the computation of an attention block within a LM as follows:

$$LM(X^t, M^t, C^t) = \alpha Attention(Q, K, V) + \beta Attention(Q, M^t, M^t) \\ + \gamma Attention(Q, C^t, C^t) \tag{9}$$

---

[1] We covered a test where alpha is also a learnable parameter rather than a constant hyperparameter in Sect. 4.3.2. In this test, the BERT-tiny determines α, β, and γ independently, and finally divides by their sum, so that each variable adds up to 1.

where $Q = Emb(X^t)W^Q$, $K = Emb(X^t)W^K$, $V = Emb(X^t)W^V$, and *Attention* is a typical transformer attention block [20]. By weighting the two contexts, the model can generate cohesive and contextually connected text.

3.3 Long Story Structural Positions (LSP)

We must provide structural positional information for each paragraph [31]. All discourse tokens are of the following types $< intro >$, $< body >$, $< tail >$, $< front >$, $< middle >$, $< ending >$, $< next\_is\_ending >$. $< intro >$ corresponds to the first paragraph, $< tail >$ to the last paragraph, and $< body >$ to everything in between as introduced in the previous work [6]. However, since our task is to generate much more paragraphs, $< body >$ token is not sufficient to give representation to our model. Therefore, four tokens ($< front >$, $< middle >$, $< ending >$, $< next - is - ending >$) are added: $< front >$ for the first 1/3, $< middle >$ for the next 1/3, and $< ending >$ for the last 1/3. $< next - is - ending >$ is for the paragraph immediately before the end, as its name implies.

3.4 Base Pretrained Model

We finetune a Pretrained Language Model(PLM) based on Transformer [20] for this task. Although our algorithm is model-agnostic so that it can apply any of PLM, we choose BART [10] because (1) it had learned denoising sequences to efficiently extract information from a given input, appropriate for our task as the model should extract representations of the inputs and contexts. (2) it is free to download from online[2]. However, our model holds the potential for even more excellent performance when applied to larger language models in the future.

4 Experiments

4.1 Experiments Set-Up

4.1.1 Datasets

We train our models on Writing Prompts [3], Booksum [14], and Reedsy Prompts[3] datasets. For Booksum, we use the 'Chapter' column as a dataset. Since our task is to generate various lengths of stories with coherence and completeness, we must prepare datasets of different lengths. We use the same train/valid/test split from the original papers [3,14], but since Reedsy Prompts is not used in any other works, we split it into train/valid/test sets of 60k/4k/4k. The NLTK library's sentence tokenizer is utilized for document segmentation into sentences during truncation, with each sentence sequentially compiled into a paragraph, ensuring a maximum length of 200 tokens per paragraph. For each

---

[2] https://huggingface.co/facebook/bart-large.
[3] We crawl this dataset from https://blog.reedsy.com/short-stories/..

dataset, we extract keywords using the RAKE algorithm [11] to make keyword-story pairs. These keywords are used as input for the model to write each paragraph of a story (Table 1).

Table 1. The average number of tokens, paragraphs per document, and the number of documents used in experiments.

	Avg # of Tokens	Avg # of Paragraphs	# of documents
WritingPrompts	768	3.4	300k
Booksum	6065	27.7	12k
ReedsyPrompts	2426	12.3	68k

4.1.2   Baselines

We mainly compare our model to Plotmachine, the only model that performs paragraph-level recursive story generation. We also compare the performance of our model against GPT-2 [28] and BART with the recursive generation setting, no applied memory, and cheating contexts; thus, it does not utilize the CWC. Plotmachine receives only the discourse tokens from the original paper, while other baselines are given the same as ours. We also compare our model with the ablated version. The *LongStory with no memory* (LongStory$^{\neg M}$), *LongStory with no cheating* (LongStory$^{\neg C}$) and *LongStory with no new discourse* (LongStory$^{\neg D}$) show the effectiveness of context vectors and newly added discourse tokens.

4.1.3   Implementation Details

Experiments are run on an NVIDIA RTX A5000 GPU server. For training steps, we use the teacher-forcing method so each label is used for not only Y_t but also contexts. We set the hyperparameter $\alpha = \frac{1}{2}$ because providing the model with sufficient inductive bias for input embeddings was beneficial for coherence and completeness. We also configured the size of the cheating window c up to the last three paragraphs it generated. For generation, we use beam sampling with $top\_k = 50$, $top\_p = 0.95$, $no\_repeat\_ngram\_size = 3$ and $repetition\_penalty = 3.5$ [18] for every length of paragraphs. To reproduce Plotmachine, we refer to the author's github repository.[4]

4.2   Experimental Results

---

[4] https://github.com/hrashkin/plotmachines.

Table 2. ROUGE-1,2,L scores, in-self-BLEU n-gram scores, Coherence, Completeness scores, and the average length of a paragraph per 5,10,20,30,50 number of paragraphs. These are results from the combined test datasets of Writing Prompts, Reedsy Prompts, and Booksum. We bold the highest model scores in each column and underline the second highest model scores.

	R-1	R-2	R-L	B2	B-3	B-4	B-5	Coherence	Completeness	AvgL
5 Paragraphs										
Golden Label				0.28	0.15	0.07	0.03	82.62	62.5	161
Plotmachine	29.85	7.15	28.16	**0.47**	**0.33**	**0.23**	**0.16**	60.19	2.47	192
BART	38.77	13.64	37.35	0.61	0.49	0.41	0.33	54.93	2.39	167
GPT-2	33.41	9.25	31.82	0.55	0.42	0.32	0.24	_63.08_	1.13	193
LongStory	39.41	13.50	37.84	_0.49_	_0.36_	_0.27_	_0.20_	**65.13**	**21.15**	158
LongStory$^{\neg M}$	_39.69_	13.66	**38.20**	0.51	0.38	0.29	0.22	59.38	-3.04	167
LongStory$^{\neg C}$	39.35	**14.02**	37.90	0.62	0.51	0.42	0.35	62.59	8.53	170
LongStory$^{\neg D}$	**39.69**	_13.67_	_38.08_	0.50	0.37	0.28	0.21	62.27	_9.21_	166
10 Paragraphs										
Golden Label				0.37	0.19	0.08	0.03	80.44	55.5	164
Plotmachine	30.34	7.74	28.87	0.62	0.47	_0.34_	_0.25_	64.17	3	195
BART	36.28	11.71	34.83	0.67	0.53	0.43	0.34	60.46	2.65	172
GPT-2	32.32	8.84	30.74	0.64	0.49	0.36	0.27	**66.06**	3.85	196
LongStory	_37.46_	_12.11_	35.82	**0.59**	**0.43**	**0.31**	**0.23**	_64.8_	**21.15**	166
LongStory$^{\neg M}$	37.32	12.05	35.72	_0.61_	_0.45_	0.34	0.25	58.4	3.59	171
LongStory$^{\neg C}$	37.06	12.11	35.49	0.67	0.53	0.43	0.34	63.24	6.56	170
LongStory$^{\neg D}$	**37.54**	**12.30**	**35.86**	0.60	0.44	0.33	0.25	63.4	_11.12_	171
20 Paragraghs										
Golden Label				0.45	0.24	0.12	0.05	80.13	57.66	166
Plotmachine	31.45	9.49	29.96	0.71	0.57	0.44	0.34	64.26	4.75	194
BART	35.43	12.29	34.02	0.74	0.62	0.51	0.42	61.37	-1.53	169
GPT-2	33.28	10.52	31.60	0.72	0.58	0.46	0.35	**65.8**	0.7	198
LongStory	_36.85_	12.69	_35.22_	**0.68**	**0.53**	**0.40**	**0.31**	_64.6_	**6.73**	169
LongStory$^{\neg M}$	36.72	_12.78_	35.17	0.70	0.55	0.43	0.33	58.3	-1.96	170
LongStory$^{\neg C}$	35.44	12.35	33.89	0.74	0.62	0.51	0.46	64.51	0.23	170
LongStory$^{\neg D}$	**37.03**	**12.81**	**35.35**	_0.69_	_0.54_	_0.41_	_0.32_	61.14	_6.38_	170
30 Paragraphs										
Golden Label				0.48	0.25	0.11	0.04	80.19	58.64	164
Plotmachine	28.85	8.89	27.58	0.75	0.62	0.49	0.38	63.24	6.7	193
BART	30.62	10.53	29.25	0.78	0.67	0.55	0.46	61.65	-0.61	164
GPT-2	29.43	9.78	28.06	0.76	0.62	0.49	0.37	**65.13**	-0.26	199
LongStory	_32.24_	**10.97**	_30.90_	0.73	0.58	0.45	0.35	_65_	**23.9**	163
LongStory$^{\neg M}$	31.81	_10.96_	30.23	0.74	0.60	0.47	_0.37_	57.76	_19.43_	167
LongStory$^{\neg C}$	29.96	10.16	28.76	0.79	0.67	0.56	0.46	64.32	-0.1	165
LongStory$^{\neg D}$	**32.42**	10.95	**31.02**	_0.73_	_0.59_	_0.47_	0.48	60.4	11.58	168
50 Paragraph										
Golden Label				0.53	0.30	0.15	0.06	78.06	52.4	166
Plotmachine	29.01	9.92	28.14	_0.81_	0.70	0.59	0.49	_67.95_	-9.08	167
BART	27.83	10.07	27.05	0.84	0.75	0.65	0.56	66.73	-5.66	167
GPT-2	28.03	9.94	27.17	0.84	0.73	0.62	0.52	67.73	3.9	198
LongStory	**31.81**	**11.65**	**30.87**	0.79	0.66	0.54	0.43	63.67	**31.76**	161
LongStory$^{\neg M}$	30.97	_11.51_	29.84	0.81	_0.69_	0.58	_0.48_	63.06	-10.47	168
LongStory$^{\neg C}$	27.87	9.94	27.12	0.84	0.74	0.64	0.54	**68.99**	20	170
LongStory$^{\neg D}$	_30.99_	10.80	_30.09_	0.81	0.69	_0.58_	0.48	63.25	21.68	167

4.2.1   Coherence and Completeness

For the Coherence scorer, we divide two consecutive paragraphs into true pairs and two randomly selected paragraphs into fake pairs. We then train GPT-2 to use these constructed datasets on a classification task to distinguish whether the two paragraphs are true or fake folds [12,18], and convert the result of it into a score $\in [0,1]$ using sigmoid function, as a Coherence score. We do the same for Completeness scorer, except for defining the true fold with the last paragraphs and the fake fold with the non-last paragraphs. It's important to highlight that the reported Completeness scores are calculated as the *last paragraph's Completeness score* minus the *average Completeness scores of non-last paragraphs*. This approach is chosen because an ideal story should generate a last paragraph appropriately and treat the body distinctly. Regarding Coherence, as shown in Table 2, our model is the best or second best in the 5,10,19,30 paragraph length sample. Plotmachine had a higher Coherence score than BART but lower than our model or GPT-2. In the ablated analysis, LongStory$^{\neg C}$ almost always outperforms LongStory$^{\neg M}$ and LongStory$^{\neg D}$ on Coherence. This suggests that *Memory* is central to capturing the natural flow between paragraphs. In the Completeness, our model is the best for all lengths. LongStory$^{\neg M}$ and LongStory$^{\neg C}$ have no absolute advantage in this column, so it is difficult to know exactly how *Memory* and *Cheating* affect Completeness. Still, since our model performs the best, we can say that a balanced mix of the two contexts improves performance. The results for LongStory$^{\neg D}$ also suggest that the newly added discourse tokens help improve Completeness.

4.2.2   Relevance

We also show the average ROUGE scores between the generated documents and the golden labels to evaluate how well the model reproduces the golden label from the keywords. This serves as a metric to assess the relevance of the keywords to the predictions. As shown in Table 2, our model is best for results with 50 paragraph lengths and second best in many cases. Our model outperforms Plotmachine, GPT-2, and BART for all lengths of results. In the ablated analysis, LongStory$^{\neg D}$ scores the best, for the most part. This shows that the newly added order tokens have little or even a negative effect on relevance. However, in the repeatability test in the next section, LongStory$^{\neg D}$ performed worse than our model, which suggests that it reproduced the overall repetitive n-gram. For the most part, LongStory$^{\neg M}$ performed better than LongStory$^{\neg C}$, which means that *Cheating* was more effective than *Memory* in relevance.

4.2.3   Repetitiveness

To see how much repetition occurs within a single output, we calculate the n-gram BLEU score by taking one paragraph as a hypothesis and the rest as a reference and averaging the results. We call the averaged results the in-SELF

BLEU score[5]. The higher in-SELF BLEU score means that there are greater repetitions within the text and less diversity. As shown in Table 2, Our model performs best on the in-self-BLEU score for all but five paragraphs length. For five paragraph lengths, Plotmachine performs best, and ours is second highest. However, for longer results, our model is always better. In the ablated analysis, LongStory$^{\neg M}$ outperforms LongStory$^{\neg C}$ and BART. This suggests that *Cheating* prevented the model from over-attending to past context or repeating n-grams irrelevant to the overall context. The better performance of our model over LongStory$^{\neg D}$ suggests that the order tokens we added positively affected reducing repetition.

4.3   Further Analysis

4.3.1   Zero-Shot Test

To see how well LongStory can have representations of the others, we do zero-shot tests on three models: WP, BK, and RP. As shown in Table 3, Writing Prompt is large enough to have some representation on other types of test datasets, while Booksum and ReedsyPrompts are not. Total outperforms all others, underscoring the effectiveness of our methodology in integrating these diverse datasets.

Table 3. Zero-shot test results averaged across test datasets of 5, 10, 20, 30, and 50 paragraphs. WP, RP, and BK are our models trained on Writing Prompts, Reedsy Prompts, and Booksum only, respectively. Total is the model trained on all three datasets combined. We bold the best results for each dataset.

	Writing Prompts			Booksum			Reedsy Prompts		
	RP	BK	Total	WP	RP	Total	WP	BK	Total
Coherence	38.55	43.75	**65.78**	56.84	48.62	**58.70**	57.56	47.79	**62.11**
Completeness	−0.1	1.51	**27.06**	1.67	−4.03	**12.37**	7.75	0.93	**21.29**

4.3.2   Augmented CWC

We experiment with calibrating not only memory and cheating but also the proportion of input embeddings in the attention block. Table 4 shows that the augmented version is better than ours on ROUGE-1,2, L scores but worse on Coherence and Completeness. This indicates that models excelling in relevance may not necessarily outperform in natural flow.

---

[5] Note that in-self-BLEU score is not the same as self-BLEU score [6]. The self-BLEU score has taken one whole generated document as a hypothesis and the others as references, which cannot represent inner repetitiveness.

Table 4. A comparison of the Augmented version of our model with the averaged scores of ROUGE-1, ROUGE-2, ROUGE-L, Coherence, and Completeness across test datasets of 5, 10, 20, 30, and 50 paragraphs. We bold the best results for each dataset.

	R-1	R-2	R-L	Coherence	Completeness
Aug-LongStory	**38.25**	**13.06**	**36.65**	61.01	5.25
LongStory	38.04	12.90	36.46	**64.92**	**18.36**

5   Conclusion

We introduce a novel task, Length Controllable Story Generation, aimed at recursively producing paragraph-by-paragraph stories of considerable length while ensuring coherence and completeness. To achieve this, we employ the long- and short-term contexts weight calibrator (CWC) and long story structural positions (LSP). The model is trained on three distinct datasets with varying average lengths, enabling it to learn representations of different lengths. Quantitative analysis demonstrates that our model excels in generating longer stories that exhibit coherence, completeness, relevance, and reduced repetitiveness compared to other baseline models.

Acknowledgements. We thank anonymous reviewers for their constructive and insightful comments. K. Jung is with ASRI, Seoul National University, Korea. This work was supported by Institute of Information & communications Technology Planning & Evaluation (IITP) grant funded by the Korea government (MSIT) [NO.2022-0-00184, Development and Study of AI Technologies to Inexpensively Conform to Evolving Policy on Ethics & NO.2021-0-01343, Artificial Intelligence Graduate School Program (Seoul National University)].

References

1. Beltagy, I., Peters, M.E., Cohan, A.: Longformer: the long-document transformer. arXiv preprint arXiv:2004.05150 (2020)
2. Lin, Z., Riedl, M.O.: Plug-and-blend: a framework for plug-and-play controllable story generation with sketches. In: Proceedings of the AAAI Conference on Artificial Intelligence and Interactive Digital Entertainment, vol. 17, No. 1, pp. 58–65 (2021)
3. Fan, A., Lewis, M., Dauphin, Y.: Hierarchical neural story generation. arXiv preprint arXiv:1805.04833 (2018)
4. OpenAI. GPT-4 Technical Report (2023)
5. Peng, N., Ghazvininejad, M., May, J., Knight, K.: Towards controllable story generation. In: Proceedings of the First Workshop on Storytelling, pp. 43–49 (2018)
6. Rashkin, H., Celikyilmaz, A., Choi, Y., Gao, J.: PlotmaChines: outline-conditioned generation with dynamic plot state tracking. arXiv preprint arXiv:2004.14967 (2020)
7. Yang, K., Peng, N., Tian, Y., Klein, D.: Re3: generating longer stories with recursive reprompting and revision. arXiv preprint arXiv:2210.06774 (2022)

8. Tang, C., Lin, C., Huang, H., Guerin, F., Zhang, Z.: EtriCA: event-triggered context-aware story generation augmented by cross attention. arXiv preprint arXiv:2210.12463 (2022)
9. Brown, T., et al.: Language models are few-shot learners. In: Advances in Neural Information Processing Systems, vol. 33, pp. 1877–1901 (2020)
10. Lewis, M., et al.: Bart: denoising sequence-to-sequence pre-training for natural language generation, translation, and comprehension. arXiv preprint arXiv:1910.13461 (2019)
11. Rose, S., Engel, D., Cramer, N., Cowley, W.: Automatic keyword extraction from individual documents. Text Mining: Applications and Theory, 1–20 (2010)
12. Wang, S., Durrett, G., Erk, K.: Narrative interpolation for generating and understanding stories.arXiv preprint arXiv:2008.07466 (2020)
13. Yang, K., Klein, D., Peng, N., Tian, Y.: DOC: improving long story coherence with detailed outline control. arXiv preprint arXiv:2212.10077 (2022)
14. Kryściński, W., Rajani, N., Agarwal, D., Xiong, C., Radev, D.: Booksum: a collection of datasets for long-form narrative summarization. arXiv preprint arXiv:2105.08209 (2021)
15. Yao, L., Peng, N., Weischedel, R., Knight, K., Zhao, D., Yan, R.: Plan-and-write: towards better automatic storytelling. InP: roceedings of the AAAI Conference on Artificial Intelligence, vol. 33, no. 01, pp. 7378–7385 (2019)
16. Alabdulkarim, A., Li, W., Martin, L.J., Riedl, M.O.: Goal-directed story generation: augmenting generative language models with reinforcement learning. arXiv preprint arXiv:2112.08593 (2021)
17. Pradyumna, T., Murtaza, D., Lara, J. M., Mehta, A., Harrison, B.: Controllable neural story plot generation via reward shaping. In: Proceedings of the International Joint Conference Artificial Intelligence, pp. 5982–5988 (2019)
18. Guan, J., Huang, F., Zhao, Z., Zhu, X., Huang, M.: A knowledge-enhanced pre-training model for commonsense story generation. In: Transactions of the Association for Computational Linguistics, vol. 8, pp. 93–108 (2020)
19. Peng, X., Li, S., Wiegreffe, S., Riedl, M.: Inferring the reader: guiding automated story generation with commonsense reasoning. arXiv preprint arXiv:2105.01311 (2021)
20. Vaswani, A., et al.: Attention is all you need. In: Advances in Neural Information Processing Systems, vol. 30 (2017)
21. Lin, C.Y.: Rouge: a package for automatic evaluation of summaries. In: Text summarization branches out, pp. 74–81 (2004)
22. Papineni, K., Roukos, S., Ward, T., Zhu, W.J.: Bleu: a method for automatic evaluation of machine translation. In: Proceedings of the 40th annual meeting of the Association for Computational Linguistics, pp. 311–318 (2002)
23. Safovich, Y., Azaria, A.: Fiction sentence expansion and enhancement via focused objective and novelty curve sampling. In: 2020 IEEE 32nd International Conference on Tools with Artificial Intelligence (ICTAI), pp. 835–843. IEEE (2020)
24. Li, J., Bing, L., Qiu, L., Chen, D., Zhao, D., Yan, R.: Learning to write stories with thematic consistency and wording novelty. In: Proceedings of the AAAI Conference on Artificial Intelligence, vol. 33, vol. 01, pp. 1715–1722 (2019)
25. Hu, Z., Chan, H.P., Liu, J., Xiao, X., Wu, H., Huang, L.: Planet: dynamic content planning in autoregressive transformers for long-form text generation. arXiv preprint arXiv:2203.09100 (2022)
26. Yang, K., Klein, D.: FUDGE: controlled text generation with future discriminators. arXiv preprint arXiv:2104.05218 (2021)

27. Sakaguchi, K., Bhagavatula, C., Bras, R. L., Tandon, N., Clark, P., Choi, Y.: proscript: partially ordered scripts generation via pre-trained language models. arXiv preprint arXiv:2104.08251 (2021)
28. Budzianowski, P., Vulić, I.: Hello, it's GPT-2–how can I help you? towards the use of pretrained language models for task-oriented dialogue systems. arXiv preprint arXiv:1907.05774 (2019)
29. Welleck, S., Kulikov, I., Kim, J., Pang, R.Y., Cho, K.: Consistency of a recurrent language model with respect to incomplete decoding. arXiv preprint arXiv:2002.02492 (2020)
30. Zellers, R., et al.: Defending against neural fake news. In: Advances in Neural Information Processing Systems, vol. 32 (2019)
31. Guan, J., Mao, X., Fan, C., Liu, Z., Ding, W., Huang, M.: Long text generation by modeling sentence-level and discourse-level coherence. arXiv preprint arXiv:2105.08963 (2021)
32. Lewis, P., et al.: Retrieval-augmented generation for knowledge-intensive NLP tasks. In: Advances in Neural Information Processing Systems, vol. 33, pp. 9459–9474 (2020)
33. McCoy, R.T., Smolensky, P., Linzen, T., Gao, J., Celikyilmaz, A.: How much do language models copy from their training data? evaluating linguistic novelty in text generation using raven. Trans. Assoc. Comput. Linguist. **11**, 652–670 (2023)

Relation-Aware Label Smoothing
for Self-KD

Jeongho Kim[1] and Simon S. Woo[2]([envelope])

[1] Korea Advanced Institute of Science and Technology, Daejeon, South Korea
rlawjdghek@kaist.ac.kr
[2] Sungkyunkwan University, Seoul, South Korea
swoo@g.skku.edu

Abstract. Knowledge distillation (KD) is widely used to improve models' performances by transferring a larger teacher's knowledge to a smaller student model. However, KD has a disadvantage where a pre-trained teacher model is required, which can lead to training inefficiency. Therefore, self-knowledge distillation, enhancing the student by itself, has been proposed. Although self-knowledge distillation shows remarkable performance improvement with fewer resources than conventional teacher-student based KD approaches, existing self-KD methods still require additional time and memory for training. We propose Relation-Aware Label Smoothing for Self-Knowledge Distillation (RAS-KD) that regularizes the student model itself by utilizing the inter-class relationships between class representative vectors with a light-weight auxiliary classifier. Compared to existing self-KD methods that only consider the instance-level knowledge, we show that proposed global-level knowledge is sufficient to achieve competitive performance while being extremely efficient training cost. Also, we achieve extra performance improvement through instance-level supervision. We demonstrate RAS-KD outperforms existing self-KD approaches in various tasks with negligible additional cost.

1 Introduction

Recently, deep neural networks (DNNs) based approaches have achieved remarkable performance across various research areas. However, high-performing DNN models tend to become huge with a large number of parameters and complex internal model architectures. Typically, training such models requires prohibitive computing resources. However, such resources are not always available in practice. To address such issues, several approaches such as weight pruning have been proposed to compress or reduce the model architecture while maintaining its high performance. Also, knowledge distillation (KD) has been proposed by Hinton *et al.* [4]. Although conventional KD [4] achieved remarkable results in various research fields, it also has the unavoidable disadvantage of requiring a

---

Supplementary Information The online version contains supplementary material available at https://doi.org/10.1007/978-981-97-2253-2_16.

high-performing pre-trained teacher network. Further, it can cause the following issues: 1) training the student model can lead to inefficiency because training the teacher model should be performed first. 2) Also, finding the best teacher networks is not always trivial, considering the relative size or model structure of the student networks. Such inefficiency and cost issues from the teacher model have opened up a new research area, self-knowledge distillation (self-KD), a self-improving training strategy without the pre-trained teacher model. Yun et al. [14] proposed the class-wise self-KD, which utilizes the output distribution of intra-class data for supervision, and Kim et al. [6] used the inference results of the previous epoch as a regularizer of the current epoch to improve performance. Also, Zhang et al. [15] introduced the softmax representation from intermediate branch networks, and Ji et al. utilized a larger auxiliary self-teacher network based on BIFPN [10] structure. However, those self-KD approaches cause training inefficiency because of their additional forward pass [14] and require large memory to save all the previous results [6]. Also, they use additional networks such as branch networks [15] or an auxiliary self-teacher network [5], which burden the training even more.

In this paper, we propose a novel Relation-aware Label Smoothing for Self-KD (RAS-KD), which shows efficient training cost via global-level supervision, while showing competitive performance and achieve additional performance improvement with instance-level soft targets from a lightweight auxiliary classifier. Compared to other self-KD methods, our soft targets use a negligible number of parameters, meaning that the temporal and spatial cost are nearly the same as the vanilla training method that uses the cross entropy loss only using hard-labels. We demonstrate the effectiveness of RAS-KD through extensive experiments with other baselines, achieving SOTA performance, not only on a large-scale image classification but also on object detection and transfer learning tasks.

2   Related Work

Many approaches have been proposed to prevent overfitting and enhance the generalization ability of DNNs. In particular, label smoothing (LS) [9] is one of the popular methods to mitigate an overconfidence issue by smoothing the one-hot encoded hard-label. Also, based on the powerful teacher network in knowledge distillation, a variety of approaches [8,12] has been proposed. While the aforementioned distillation methods effectively conjugate and distill the representations of the teacher models, their remarkable results were achieved by exploiting a pre-trained larger network, leading to training inefficiency. Recently, to alleviate the burden of conventional teacher-student distillation frameworks, self-KD has been proposed, aiming to improve performance without a larger teacher model. Zhang et al. [15] introduced more powerful soft targets from the ensemble of students and several branch networks. Moreover, Ji et al. [5] also utilized a BIFPN-based auxiliary self-teacher network to transfer the refined feature-map and soft target to the student network. However, during training, the entire network to train is larger than the original student network due to the

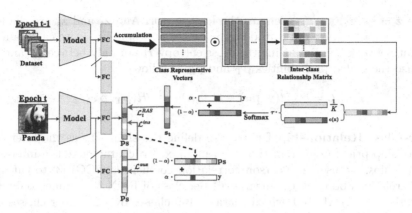

Fig. 1. Overview of RAS-KD. First, class representative vectors are obtained by averaging the entire predictions for each class. Next, to acquire an inter-class relationship prior matrix used for transferring information of class relationship, RAS-KD utilizes the Pearson correlation to map inter-class relationships. Then, we provide more weights to similar classes by multiplying τ to relationship vectors to emphasize the differing relationships between relevant classes and dissimilar classes. Moreover, the auxiliary classifier provides diversity via instance-level logit supervision.

intermediate branch networks or the auxiliary self-teacher network, consuming more VRAM and training time as in conventional KD. Another approach is to match the output of different input images. Class-wise self-KD (CS-KD) was proposed by Yun *et al.* [14], to minimize the KL divergence between two images in the same class. However, this approach requires a forward process to acquire additional supervision, requiring an additional computational cost. Also, Kim *et al.* [6] proposed progressive self-knowledge distillation (PS-KD), which utilizes a predictive probability of the student network at the previous epoch to compose soft targets. While existing self-KD approaches solely rely on instance-level soft targets, leading to expensive overhead, we propose to use the global-level knowledge to obtain better performance with more efficient computational memory and training time.

3 Our Approach

Soft Targets. In KD, instead of directly using the hard-label (i.e., one-hot vector) as target prediction, the student can learn from the soft targets consisting of the hard-label $\mathbf{y}$ and the teacher's predictive probabilities $\mathbf{p_T}$. Hinton *et al.* [4] proposed a temperature τ, which is used to scale the softmax output $\mathbf{p_T}$ of the teacher and $\mathbf{p_S}$ of the student. In particular, τ has an effect of smoothing the distribution to handle the classes with low probabilities better as follows:

$$\hat{p}_i = \phi(\mathbf{z}; \tau)_i = \frac{exp(z_i/\tau)}{\sum_j exp(z_j/\tau)}, \tag{1}$$

where $\mathbf{z} = [z_1, z_2, \ldots, z_K]$ denotes the logit vector. And $\mathbf{z_t}$ and $\mathbf{z_s}$ are the logit vector from a teacher and a student, respectively. Using Eq. 1, we can train the student network by minimizing the cross entropy $\mathcal{H}$ not only between $\mathbf{y}$ and $\mathbf{p_S}$, but also the scaled probabilities $\hat{\mathbf{p}}_\mathbf{T}$ and $\hat{\mathbf{p}}_\mathbf{S}$ as follows:

$$\mathcal{L} = \alpha \cdot \mathcal{H}(\mathbf{y}, \mathbf{p_S}) + (1 - \alpha) \cdot \tau^2 \cdot \mathcal{H}(\hat{\mathbf{p}}_\mathbf{T}, \hat{\mathbf{p}}_\mathbf{S}). \tag{2}$$

Inter-Class Relationship Prior. We define the similarity as the inter-class relationship prior, $\mathbf{c}(\mathbf{x})$ from the given input $\mathbf{x}$. To characterize the inter-class relationships, we use the Pearson correlation coefficient (PCC) [3] to calculate the correlations between the mean logit per class of ResNet18 trained only with hard-label on CIFAR-100, which contains 100 classes over 20 super-classes consisting of 5 sub-classes. To provide better visualization, we group the 5 sub-classes within the same super-class, and plot the heatmap of PCC in Fig. 2. As shown, we can observe that the classes within the same super-classes clearly demonstrate strongly correlated relationships with high intensity forming squares along the diagonal. Therefore, this result confirms that the logits from the trained vector can represent relationships between classes similar to each other. Moreover, this can be regarded as utilizing the attention scores of the self-attention mechanism [11] directly in the learning process. Thus, we use the inter-class relationship prior to better transfer knowledge in our RAS-KD.

Smoothed Distribution. The next important component in our work is generating a smoothed distribution over all classes to construct soft targets. As Yuan *et al.* [13] show that LS regularization is an ad-hoc version of KD, this has a similar effect on the teacher to provide a uniform distribution for distilling knowledge. Surprisingly, they show that even a poorly-trained teacher network with much lower accuracy can enhance the student's performance with the noisy logits over all classes produced by a teacher, where the noisy logits have a similar effect of having the uniform distribution on distilling knowledge. From K classes, we can create, $\frac{1}{K}$, the uniform distribution as a smoothing method. Therefore, we consider the LS as an

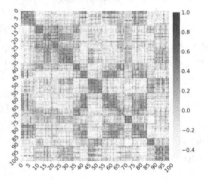

Fig. 2. Heatmap of the Pearson correlation matrix on CIFAR-100, which contains 100 classes over 20 super-classes and 5 sub-classes for each super-class.

effective method to improve performance, while focusing more on classes with high probabilities, which is different from other KDs.

Definition of Overall Knowledge. In this work, we define knowledge from the trained network in a more constructive way incorporating 1) inter-class relationship prior and 2) smoothed distribution over all the classes. From the ground-truth label information $\mathbf{y}$, we can have the final soft target vector $\mathbf{s}$ for the student network as follows:

$$\mathbf{s} = a_1 \cdot \mathbf{y} + a_2 \cdot \mathbf{c}(\mathbf{x}) + a_3 \cdot \frac{1}{K}, \tag{3}$$

where $\mathbf{c}(\mathbf{x})$ is the inter-class relationship prior from the given input image $\mathbf{x}$, and $\frac{1}{K}$ is a smoothing vector over K classes, and a_1, a_2, and a_3 are the weight coefficients.

3.1   RAS-KD

Inter-Class Relationship Prior Matrix. Let $\mathbf{Z_t} = [\bar{\mathbf{z}}_t^1, \dots, \bar{\mathbf{z}}_t^K]^T \in \mathbb{R}^{K \times K}$ be a matrix of the mean logits from the student network for each class during t-th epoch. From $\mathbf{Z_t}$, we can obtain the representative inter-class relationships, similarity score from the Pearson correlation, σ_t^{ij} between i-th and j-th class $(1 \leq i, j \leq K)$. This is the same as the matrix $\sum_{\mathbf{Z_t Z_t}} = [\sigma_t^1, \sigma_t^2, \dots, \sigma_t^K]^T \in \mathbb{R}^{K \times K}$, where a vector $\sigma_t^i = [\sigma_t^{i1}, \sigma_t^{i2}, \dots, \sigma_t^{iK}]$ $(-1 \leq \sigma_t^{ij} \leq 1)$. For example, classes that are significantly different (i.e., weak or negative correlations) from the i-th class would have correlation values close to -1. Next, we set the lower bound to 0 from -1 so that we can treat such weak or irrelevant classes as noise. Hence, we obtain all the elements in $\sum_{\mathbf{Z_t Z_t}}$ to be in $[0, 1]$, which is denoted as $\sum'_{\mathbf{Z_t Z_t}} = [\sigma_t'^1, \sigma_t'^2, \dots, \sigma_t'^K]^T$. We define this matrix $\sum'_{\mathbf{Z_t Z_t}}$ as an inter-class relationship prior matrix in RAS-KD.

Relation-aware Label Smoothing. The conventional KD utilizes τ to control the probability distribution to smooth the logits. However, we apply τ to emphasize the inter-class relationships more, boosting the probabilities of highly correlated classes. Therefore, unlike dividing τ in Hinton *et al.*'s KD [4], we divide $1/\tau$ to $\sum'_{\mathbf{Z_t Z_t}}$, to make the range of values to be changed from $[0, 1]$ to $[0, \tau]$ so that the final similarity distribution would have a higher peak, providing more weights to similar classes. Next, we apply the softmax function to obtain the probability distribution. Therefore, our final soft target vector $\mathbf{s_t}$ for i-th class and the relation-aware label smoothing objective function $(\mathcal{L}^{RAS})$ at t-th epoch are defined as follows:

$$\mathbf{s_t} = \alpha \cdot \mathbf{y} + (1 - \alpha) \cdot \phi(\sigma_{t-1}'^i; 1/\tau), \tag{4}$$

$$\mathcal{L}_t^{RAS} = \mathcal{H}(\mathbf{s_t}, \mathbf{ps}) = \alpha \cdot \mathcal{H}(\mathbf{y}, \mathbf{ps}) + (1 - \alpha) \cdot \mathcal{H}(\phi(\sigma_{t-1}'^i; 1/\tau), \mathbf{ps}), \tag{5}$$

where α is a weighting hyper-parameter. For training, we initialize $\mathbf{Z_0}$ to the zero-matrix at the first epoch. Also, by taking the softmax to $\sigma_{t-1}'^i$, the soft target vector $\mathbf{s_t}$ in Eq. 4 provides a label smoothing effect. Indeed, we demonstrate Eq. 4 has an equivalent form to Eq. 3.

Table 1. Avg. top-1 and top-5 accuracies (%) on fine-grained image classification datasets. The best top-1 and top-5 performances are highlighted in bold, and the second-best performances are underlined for each model, respectively.

ResNet18	CUB-200	Stanford Cars	FGVC-Aircraft
CE	53.17 (77.16)	72.20 (90.14)	66.46 (89.58)
LS	57.14 (78.06)	76.08 (91.98)	74.91 (91.90)
BYOT	62.03 (83.57)	82.67 (95.85)	76.48 (92.83)
CS-KD	64.24 (85.34)	**86.97 (97.34)**	73.21 (92.56)
FRSKD	65.4 (85.00)	84.22 (96.38)	**79.51 (94.93)**
PS-KD	50.34 (75.04)	73.81 (91.46)	68.10 (89.89)
RAS-KD w/o $\mathcal{L}^{ins,aux}$	65.67 (85.55)	85.13 (96.40)	77.34 (94.37)
RAS-KD	**66.91 (86.31)**	86.03 (97.00)	78.88 (93.85)
ResNext50	**CUB-200**	**Stanford Cars**	**FGVC-Aircraft**
CE	52.78 (78.84)	79.34 (94.83)	71.26 (91.15)
LS	59.89 (80.83)	84.35 (95.86)	72.82 (90.70)
BYOT	63.05 (84.12)	86.27 (96.41)	78.25 (94.08)
CS-KD	64.55 (86.14)	85.59 (93.89)	76.63 (93.87)
FRSKD	63.36 (85.05)	85.06 (96.21)	80.17 (94.12)
PS-KD	52.62 (78.71)	79.85 (94.12)	76.87 (93.22)
RAS-KD w/o $\mathcal{L}^{ins,aux}$	66.36 (86.43)	86.46 (72.22)	79.65 (94.57)
RAS-KD	**67.07 (87.16)**	**86.95 (97.00)**	**80.71 (94.33)**
ResNext101	**CUB-200**	**Stanford Cars**	**FGVC-Aircraft**
CE	59.01 (82.57)	83.12 (95.89)	78.34 (93.67)
LS	62.85 (82.60)	86.60 (96.49)	71.95 (90.94)
BYOT	66.10 (**88.75**)	87.26 (97.11)	80.32 (94.92)
CS-KD	67.86 (87.36)	87.46 (97.86)	80.94 (**95.11**)
FRSKD	–	–	–
PS-KD	56.01 (79.67)	84.12 (95.91)	77.22 (93.64)
RAS-KD w/o $\mathcal{L}^{ins,aux}$	67.38 (86.74)	87.39 (97.30)	80.01 (94.42)
RAS-KD	**69.31 (**88.04**)**	**89.62 (98.23)**	**82.00 (94.81)**
ViT-Base	**CUB-200**	**Stanford Cars**	**FGVC-Aircraft**
CE	80.03 (95.58)	75.09 (93.35)	70.56 (92.13)
LS	79.56 (94.96)	80.07 (95.56)	77.07 (94.33)
CS-KD	83.21 (96.57)	86.42 (97.51)	81.49 (95.77)
PS-KD	80.34 (95.63)	74.44 (93.01)	69.72 (91.62)
RAS-KD w/o $\mathcal{L}^{ins,aux}$	83.75 (96.63)	87.58 (97.89)	81.90 (95.61)
RAS-KD	**84.77 (96.78)**	**87.77 (97.93)**	**82.98 (95.49)**

Instance-level Soft Target. Proposed soft target in Eq. 4 successfully regularizes the network itself, while it considers the representative similarity between the classes. In other words, data samples in the same class are supervised with the same global target during training. Meanwhile, existing works such as PS-KD [6] or CS-KD [14] show the promised performance via instance-level logit supervision. To this end, we introduce an auxiliary classifier with negligible cost, to provide the instance-level soft targets. As shown in Fig. 1, we attach a fully connected layer as an auxiliary classifier trained with the auxiliary loss ($\mathcal{L}^{aux}$) through distilling the knowledge of the student network:

$$\mathcal{L}^{aux} = \alpha \cdot \mathcal{H}(\mathbf{y}, \mathbf{p_S'}) + (1 - \alpha) \cdot \mathcal{H}(\mathbf{p_S}, \mathbf{p_S'}), \tag{6}$$

where α is a weighting hyper-parameter with the same value as in Eq. 4 and $\mathbf{p}'_S$ is a predictive probability of the auxiliary classifier. Moreover, we constrain the auxiliary classifier from directly imitating the original forward path by detaching the gradient from the auxiliary classifier to the feature extractor. Finally, $\mathbf{p}'_S$ provides the instance-level loss ($\mathcal{L}^{ins}$) to the student network via KL divergence:

$$\mathcal{L}^{ins} = \mathcal{H}(\mathbf{p}'_S, \mathbf{p}_S). \tag{7}$$

Therefore, our framework is trained with the following objective function at t-th epoch:

$$\mathcal{L}^{RAS-KD} = \beta \cdot \mathcal{L}_t^{RAS} + (1 - \beta) \cdot \mathcal{L}^{ins} + \mathcal{L}^{aux}, \tag{8}$$

where β is a weighting hyper-parameter. While the proposed $\mathcal{L}^{ins}$ appears to be similar to Eq. 6, $\mathcal{L}^{ins}$ functions differently from Eq. 6. Inspired by PS-KD, which exploits the logits from the previous epoch, we introduce a lightweight auxiliary classifier. In particular, our proposed auxiliary classifier shares the feature extractor but applies a gradient stop to the feature extractor.

4   Experimental Results

We compare our method with four popular self-KD approaches: BYOT, CS-KD, TF-KD, FRSKD, PS-KD, and LS.

Fine-grained Classification. As shown in Table 1, RAS-KD achieves the best top-1 accuracy in all datasets, except for the ResNet18 on Stanford Cars dataset. Note that in the case of ViT-Base [1], due to architectural differences, we only measured the accuracy of the response-based KD methods, excluding cases that involve the utilization of intermediate feature maps such as BYOT and FRSKD. Specifically, in the FGVC-Aircraft dataset, we achieve performance improvement across all the existing self-KD approaches up to 2.46%. Note that RAS-KD requires only a few parameters and computational cost, while other methods

Table 2. Avg. top-1 accuracies (%) of ours, label smoothing (LS) and existing self-KD methods on CIFAR-100. We rerun all the results three times, and report the average of the results. We highlight the best scores in bold, and the second-best performance in underlining for each model, respectively.

Model	ResNet18	ResNet50	DenseNet121	ResNeXt29
CE	75.55	78.07	79.58	80.59
LS	78.21	79.31	79.86	81.44
BYOT	78.51	**80.56**	80.27	81.51
CS-KD	78.19	78.69	79.76	82.16
TF-KD	76.59	78.45	80.24	82.23
FRSKD	77.71	74.72	–	–
PS-KD	78.89	79.33	**81.29**	82.27
RAS-KD w/o $\mathcal{L}^{ins,aux}$	<u>79.17</u>	79.56	80.31	<u>82.58</u>
RAS-KD	**79.28**	<u>79.86</u>	<u>80.89</u>	**82.8**

need additional memory to save all predictions or store the network to re-forward images to the model. Furthermore, we visualize the t-SNE results based on the outputs of ResNet18 trained with cross entropy (CE), LS, and RAS-KD in Fig. 3.

Clearly, CE shows the scattered and overlapped distributions compared to other approaches for both datasets, where such distributions make it difficult for the model to find decision boundaries that separate classes. While LS has better boundaries than CE, t-SNE results from RAS-KD show more compact clusters. This clearly shows that our RAS-KD effectively reduces the intra-class variation, while maximizing the inter-class boundaries, demonstrating that RAS-KD successfully learns more useful knowledge from soft targets from our approach.

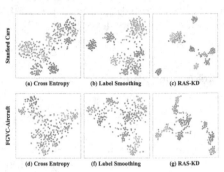

Fig. 3. t-SNE visualization results of two baselines and our RAS-KD. For clear visualization, we use the first 10 classes of Stanford Cars and FGVC-Aircraft and use the outputs from ResNet18.

CIFAR-100 Classification. We compare our method with three popular self-KD approaches: CS-KD, TF-KD, and PS-KD, and additional regularization method, LS. For self-KD methods, we use the same hyperparameter settings reported in their respective research, and we use the smoothing parameter ϵ of 0.1 for LS. As shown in Table 2, our RAS-

Fig. 4. Heatmaps of the Pearson correlation matrix on CIFAR-100. RAS-KD shows much lower correlation between classes in different super-classes.

KD outperforms other approaches in most of the experimental settings. Compared to LS, RAS-KD achieves higher performance, demonstrating the effectiveness of using inter-class relationships. Even though BYOT shows the highest performance in ResNet50, it requires more resources to build a powerful ensemble model, which works as a teacher network.

Data Augmentation. Also, we exploit data augmentation methods for evaluation. As shown in Table 3, overall, RAS-KD achieves the highest accuracies in all of the results, except for the DenseNet121 with CutMix.

Effectiveness of Inter-Class Relationship Prior. To illustrate that our RAS-KD effectively learns the inter-class relationship during training, we present PCC heatmaps in Sect. 3 to visualize the correlations of ResNet18 trained with LS and RAS-KD in Fig. 4. Compared to CE, LS (Fig. 4 (b)) smooths most

Table 3. Top-1 accuracies (%). of four baseline models with CutOut and CutMix augmentation-based regularization on CIFAR-100 dataset. Bold values illustrate the best result and the highest increase.

Method	ResNet18	ResNet50	DenseNet121	ResNeXt29
CE	75.55	78.07	79.58	80.59
LS	78.21	79.31	79.86	81.44
RAS-KD	**79.28**	**79.86**	**80.89**	**82.8**
CE + CutOut	76.92 (↑**1.37**)	79.15 (↑**1.08**)	80.06 (↑0.48)	81.76 (↑**1.17**)
LS + CutOut	78.72 (↑0.51)	79.90 (↑0.59)	81.18 (↑0.55)	81.62 (↑0.18)
RAS-KD + CutOut	**79.51** (↑0.23)	**80.49** (↑0.63)	**81.83** (↑**0.94**)	**83.22** (↑0.42)
CE + CutMix	78.68 (↑**3.13**)	80.54 (↑**2.47**)	81.68 (↑2.1)	82.94 (↑**2.35**)
LS + CutMix	79.70 (↑1.49)	81.04 (↑1.73)	**82.49** (↑**2.63**)	83.67 (↑2.23)
RAS-KD + CutMix	**80.51** (↑1.23)	**81.34** (↑1.48)	82.32 (↑1.43)	**84.19** (↑1.39)

Table 4. Top-1 accuracies (%) of label smoothing (LS) and self-KD methods on ImageNet. We highlight the best results in bold and underline the second-best results.

ImageNet	CE	LS	BYOT	TF-KD	CS-KD	FRSKD	PS-KD	RAS-KD
ResNet50	75.51	<u>75.88</u>	75.24	75.12	75.62	–	75.73	**76.16**
ResNet101	77.66	77.7	–	<u>77.98</u>	77.35	–	77.78	**78.29**
ResNeXt101	79.17	79.78	–	**80.26**	79.36	–	79.76	<u>80.02</u>

Table 5. **Performance of mAP (%) scores** with 0.5 of IoU threshold. The best result is highlighted in bold.

Backbone	CE	LS	CS-KD	PS-KD	RAS-KD
ResNet50	78.5	78.5	78.3	78.5	**79.4**
ResNet101	79.8	80.0	80.2	80.2	**81.1**

of the correlation values. On the other hand, we can observe that our RAS-KD (Fig. 4 (c)) not only significantly decreases the similarity score between classes in different-super classes (non-diagonal region), but also increases the correlations within the same super class from the distinct squares along the diagonal.

ImageNet Classification. As shown in Table 4, RAS-KD achieves the SOTA performance on ImageNet [7] dataset, compared to other self-KD methods. In ResNet101, RAS-KD further improves accuracy up to 0.63% compared to CS-KD and PS-KD. In addition, memory-based methods such as TF-KD and PS-KD, which have to store predictive outputs of the teacher network or previous epoch, need more GPU memory to load the fixed models and require re-forwarding data to the model. On the other hand, the training algorithm with RAS-KD requires the same cost as CE, as shown in the ablation section.

Object Detection (Transfer Learning Task). Transfer learning provides a practical and efficient approach to improving performance by leveraging a pre-trained model. Our proposed RAS-KD can serve as a great backbone network for various computer vision tasks. To evaluate the performance on transfer learning, we replace the feature extractor pre-trained with ImageNet using vanilla CE to other training methods and then fine-tune the entire model with PascalVOC [2] dataset. In Table 5, we present the mean Average Precision (mAP) for five pre-trained backbone networks trained by existing KD methods. Compared to existing self-KD methods, our RAS-KD improves the mAP score by 0.9%.

Table 6. Ablation results on τ and α. We gradually increase τ from Config. A to D, while adjusting α from E to G with no instance-level supervision (i.e., $\beta = 1.0$). We color the best results from the two settings in red and blue, respectively.

Configuration	ResNet18	ResNet50	ResNeXt29	DenseNet121
A. $\alpha = 0.8, \tau = 1$	78.82	78.3	81.87	80.30
B. $\alpha = 0.8, \tau = 1.5$	79.17	79.56	82.58	80.31
C. $\alpha = 0.8, \tau = 3$	78.75	79.50	82.03	80.21
D. $\alpha = 0.8, \tau = 5$	78.38	78.31	82.31	79.76
E. $\alpha = 0.7, \tau = 1$	78.48	78.39	81.82	79.86
F. $\alpha = 0.8, \tau = 1$	**78.82**	78.30	**81.87**	**80.30**
G. $\alpha = 0.9, \tau = 1$	78.53	**78.78**	81.84	80.00

5　Ablation Study

Table 7. Ablation results on β. We gradually increase β values from 0.6 to 0.9 with fixed α and τ to 0.8 and 1.5, respectively. We can observe that beyond a certain threshold value of β (i.e., 0.7), no significant difference in performance is discernible.

$\alpha = 0.8, \tau = 1.5$	Model	$\beta = 0.6$	$\beta = 0.7$	$\beta = 0.8$	$\beta = 0.9$
CIFAR-100	ResNet18	78.90	79.17	**79.28**	79.27
	ResNet50	79.41	79.45	**79.86**	78.63
	ResNext29	82.47	82.75	**82.80**	82.70
	DenseNet121	80.51	80.79	80.89	**81.00**
CUB-200	ResNet18	65.53	66.67	**66.91**	66.15
	ResNext50	63.34	67.04	67.07	**67.12**
Stanford Cars	ResNet18	85.73	85.33	**86.03**	85.67
	ResNext50	85.44	**86.98**	86.95	86.96
FGVC-Aircraft	ResNet18	78.12	**79.20**	78.88	78.61
	ResNext50	78.73	79.98	**80.71**	78.49

Our RAS-KD employs three main components in producing the soft targets: 1) inter-class relationships prior, 2) label smoothing, as shown in Eq. 3, and 3) logits from the auxiliary classifier. Since the auxiliary classifier is supervised by the student network from the first two terms, we only conduct ablation experiments to explore how the first and second component affects performance improvement, while setting $\beta = 1$ instead of 0.8. We first adjust τ to $1, 1.5, 3$, and 5. If τ increases, the soft targets provide more weight to the classes with closer similarity. We report the results according to different τ from Config. A to D in Table 6. All baseline models consistently show the best performance with $\tau = 1.5$, bit we observe an accuracy drop as τ increases from 1.5. On the other hand, when $\alpha = 0.7$, we can observe the performance degradation compared to $\alpha = 0.8$ or 0.9. This indicates that too less weighting on hard-labels (more weighting on the soft targets) can hinder the optimization. With the best configuration for $\mathcal{L}^{RAS}$ (Config. B), we extend $\mathcal{L}^{RAS}$ to $\mathcal{L}^{RAS-KD}$ ($\beta = 0.8$) by including the additional instance-level supervision. As shown in Table 2, the proposed auxiliary classifier achieves performance improvement, which demonstrates that global and instance-level supervision regularizes the student independently and effectively. Moreover, we have observed a significant improvement up to 2.0% in top-1 accuracy in fine-grained datasets through the instance-level supervision, as shown in Table 1. Lastly, we conduct additional ablation experiments for the sensitivity of the β. As reported in Table 7, we can observe that across the four datasets, there is a noticeable performance degradation at a common value of $\beta = 0.6$, while $\beta > 0.7$, does not result in significant changes. If β is too small, the weight of hard-label decreases, leading to unstable training of the main classifier.

Training Efficiency of RAS-KD.
We compare our RAS-KD with the existing methods to time and space efficiency. For both experiments, we used ResNet18 architecture trained on the CUB-200 dataset. As shown in Fig. 5(a), existing self-KD approaches spend more time on training the model. In particular, CS-KD requires more time than other approaches due to the additional forward process. PS-KD and BYOT also require extra

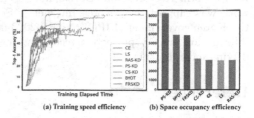

(a) Training speed efficiency   (b) Space occupancy efficiency

Fig. 5. Time and space efficiency of RAS-KD vs. other methods. RAS-KD costs much less training time and VRAM compared to other methods.

time, because they utilize additional computation or use additional branch networks. On the other hand, our RAS-KD has similar training time performance to CE and LS because it only requires little computation for calculating the class representative vectors and gradient of one fully connected layer for the auxiliary classifier. We also show the space efficiency of RAS-KD by utilizing the VRAM occupancy, as shown in Fig. 5(b). Especially, PS-KD consumes significant time and space due to the previous model being stored separately. However, RAS-KD has an advantage in terms of memory footprints, because of requiring minimal memory to compute the class representative vectors and one linear layer for the auxiliary classifier.

6 Conclusion

We proposed the novel Relation-aware Label Smoothing for Self-Knowledge distillation (RAS-KD). RAS-KD leverages the inter-class relationship matrix obtained by class representative vectors from the previous epoch of the model. For global-level supervision, our soft targets can be categorized into: 1) one-hot encoded hard-label, 2) inter-class relationship prior, and 3) smoothed distribution over all classes. With the global-level approach, we can significantly reduce the temporal and spatial training cost. Moreover, we provide the diverse logits via instance-level supervision from an auxiliary classifier with negligible parameters. Through extensive experiments, we show that RAS-KD outperforms existing self-KD methods on several vision tasks.

Acknowledgements. This work was partly supported by Institute for Information & communication Technology Planning & evaluation (IITP) grants funded by the Korean government MSIT: (No. 2022-0-01199, Graduate School of Convergence Security at Sungkyunkwan University), (No. 2022-0-01045, Self-directed Multi-Modal Intelligence for solving unknown, open domain problems), (No. 2022-0-00688, AI Platform to Fully Adapt and Reflect Privacy-Policy Changes), (No. 2021-0-02068, Artificial Intelligence Innovation Hub), (No. 2019-0-00421, AI Graduate School Support Program at Sungkyunkwan University), and (No. RS-2023-00230337, Advanced and Proactive AI Platform Research and Development Against Malicious Deepfakes). Lastly, this work was supported by Korea Internet & Security Agency (KISA) grant funded by the Korea government (PIPC) (No.RS-2023-00231200, Development of personal video information privacy protection technology capable of AI learning in an autonomous driving environment).

References

1. Dosovitskiy, A., et al.: An image is worth 16×16 words: transformers for image recognition at scale. arXiv preprint arXiv:2010.11929 (2020)
2. Everingham, M., Van Gool, L., Williams, C.K., Winn, J., Zisserman, A.: The pascal visual object classes (voc) challenge. Int. J. Comput. Vision **88**(2), 303–338 (2010)
3. Freedman, D., Pisani, R., Purves, R.: Statistics (international student edition). Pisani, R. Purves, 4th edn. WW Norton & Company, New York (2007)
4. Hinton, G., Vinyals, O., Dean, J.: Distilling the knowledge in a neural network. arXiv preprint arXiv:1503.02531 (2015)
5. Ji, M., Shin, S., Hwang, S., Park, G., Moon, I.C.: Refine myself by teaching myself: feature refinement via self-knowledge distillation. In: Proceedings of the IEEE/CVF Conference on Computer Vision and Pattern Recognition, pp. 10664–10673 (2021)
6. Kim, K., Ji, B., Yoon, D., Hwang, S.: Self-knowledge distillation with progressive refinement of targets. In: Proceedings of the IEEE/CVF International Conference on Computer Vision, pp. 6567–6576 (2021)
7. Krizhevsky, A., Sutskever, I., Hinton, G.E.: Imagenet classification with deep convolutional neural networks. Adv. Neural. Inf. Process. Syst. **25**, 1097–1105 (2012)
8. Park, W., Kim, D., Lu, Y., Cho, M.: Relational knowledge distillation. In: Proceedings of the IEEE/CVF Conference on Computer Vision and Pattern Recognition, pp. 3967–3976 (2019)

9. Szegedy, C., Vanhoucke, V., Ioffe, S., Shlens, J., Wojna, Z.: Rethinking the inception architecture for computer vision. In: Proceedings of the IEEE Conference on Computer Vision and Pattern Recognition, pp. 2818–2826 (2016)
10. Tan, M., Pang, R., Le, Q.V.: Efficientdet: scalable and efficient object detection. In: Proceedings of the IEEE/CVF Conference on Computer Vision and Pattern Recognition, pp. 10781–10790 (2020)
11. Vaswani, A., et al.: Attention is all you need. Adv. Neural Inf. Process. Syst. **30** (2017)
12. Yang, J., Martinez, B., Bulat, A., Tzimiropoulos, G., et al.: Knowledge distillation via softmax regression representation learning. In: International Conference on Learning Representations (ICLR) (2021)
13. Yuan, L., Tay, F.E., Li, G., Wang, T., Feng, J.: Revisiting knowledge distillation via label smoothing regularization. In: Proceedings of the IEEE/CVF Conference on Computer Vision and Pattern Recognition, pp. 3903–3911 (2020)
14. Yun, S., Park, J., Lee, K., Shin, J.: Regularizing class-wise predictions via self-knowledge distillation. In: Proceedings of the IEEE/CVF Conference on Computer Vision and Pattern Recognition, pp. 13876–13885 (2020)
15. Zhang, L., Song, J., Gao, A., Chen, J., Bao, C., Ma, K.: Be your own teacher: improve the performance of convolutional neural networks via self distillation. In: Proceedings of the IEEE/CVF International Conference on Computer Vision, pp. 3713–3722 (2019)

Bi-CryptoNets: Leveraging Different-Level Privacy for Encrypted Inference

Man-Jie Yuan[1,2], Zheng Zou[1,2], and Wei Gao[1,2(✉)]

[1] National Key Laboratory for Novel Software Technology, Nanjing University, Nanjing, China
{yuanmj,zouz}@lamda.nju.edu.cn
[2] School of Artificial Intelligence, Nanjing University, Nanjing, China
gaow@lamda.nju.edu.cn

Abstract. Privacy-preserving neural networks have attracted increasing attention in recent years, and various algorithms have been developed to keep the balance between accuracy, computational complexity and information security from the cryptographic view. This work takes a different view from the input data and structure of neural networks. We decompose the input data (e.g., some images) into sensitive and insensitive segments according to importance and privacy. The sensitive segment includes some important and private information such as human faces and we take strong homomorphic encryption to keep security, whereas the insensitive one contains some background and we add perturbations. We propose the bi-CryptoNets, i.e., plaintext and ciphertext branches, to deal with two segments, respectively, and ciphertext branch could utilize the information from plaintext branch by unidirectional connections. We adopt knowledge distillation for our bi-CryptoNets by transferring representations from a well-trained teacher neural network. Empirical studies show the effectiveness and decrease of inference latency for our bi-CryptoNets.

Keywords: Neural network · Encryption · Privacy-preserving inference

1 Introduction

Recent years have witnessed increasing attention on privacy-preserving neural networks [4,14,25], which can be viewed as a promising security solution to the emerging Machine Learning as a Service (MLaaS) [13,15,30]. Specifically, some clients can upload their encrypted data to the powerful cloud infrastructures, and then obtain machine learning inference services; the cloud server performs inference without seeing clients' sensitive raw data by cryptographic primitives, and preserve data privacy.

Privacy-preserving neural networks are accompanied with heavy computational costs because of homomorphic encryption [2,12], and various algorithms and techniques have been developed to keep the balance between accuracy, information security and computation complexity. For example, Brutzkus et al. proposed the LoLa network for fast inference over a single image based on well-designed packing methods [4], and Lou and Jiang introduced the circuit-based network SHE with better accuracies [27], implemented with the TFHE scheme [7]. Dathathri et al. presented the compiler CHET to optimize data-flow graph of HE computations for privacy-preserving neural networks [10]. Yin et al. presented a comprehensive survey on privacy-preserving networks [38].

© The Author(s), under exclusive license to Springer Nature Singapore Pte Ltd. 2024
D.-N. Yang et al. (Eds.): PAKDD 2024, LNAI 14646, pp. 210–222, 2024.
https://doi.org/10.1007/978-981-97-2253-2_17

Previous studies mostly encrypted the entire input data and treated indiscriminately. In some applications, however, input data may consist of sensitive and insensitive segments according to different importance and privacy. As shown in Fig. 1, a fighter is more sensitive than background such as sky and mountains in a military picture, and a human face is more private than landscape in a photo.

Fig. 1. An illustration for input data (e.g., some images (Download from ILSVRC dataset [37].)), which consists of two segments with different importance and privacy.

This work presents new privacy-preserving neural network from the view of input data and network structure, and the main contributions can be summarized as follows:

- We decompose input data (e.g., images) into sensitive and insensitive segments according to importance and privacy. We adopt strong homomorphic encryption to keep the security of sensitive segment, yet mingle some perturbations [11,37] to insensitive segment. This could reduce computational overhead and perform private inference of low latency, without unnecessary encryption on insensitive segment.
- We propose the bi-CryptoNets, i.e., ciphertext and plaintext branches, to deal with sensitive and insensitive segments, respectively. The ciphertext branch could use information from plaintext branch by unidirectional connections, but the converse direction does not hold because of the spread of ciphertexts. We integrate features for the final predictions, from the outputs of both branches.
- We present the feature-based knowledge distillation to improve the performance of our bi-CryptoNets, from a teacher of convolutional neural network trained on the entire data without decomposition.
- We present empirical studies to validate the effectiveness and decrease of inference latency for our bi-CryptoNets. We could improve inference accuracy by $0.2\%-2.1\%$, and reduce inference latency by $1.15\times-3.43\times$ on a single image, and decrease the amortized latency by $4.59\times-13.7\times$ on a batch of images.

The rest of this work is organized as follows: Sect. 2 introduces relevant work. Section 3 presents our bi-CryptoNets. Section 4 proposes the feature-based knowledge distillation. Section 5 conducts experiments, and Sect. 6 concludes with future work.

2   Relevant Work

Homomorphic Encryption. Homomorphic Encryption (HE) is a cryptosystem that allows operations on encrypted data without requiring access to a secret key [12]. In addition to encryption function E and decryption function D, HE scheme provides two operators $\oplus$ and $\otimes$ such that, for every pair of plaintexts x_1 and x_2,

$$D(E(x_1) \oplus E(x_2)) = x_1 + x_2 , \quad D(E(x_1) \otimes E(x_2)) = x_1 \times x_2 ,$$

where $+$ and $\times$ are the standard addition and multiplication, respectively. Hence, we could directly perform private addition, multiplication and polynomial functions over encrypted data, by using $\oplus$ and $\otimes$ operators without knowing true values in plaintexts.

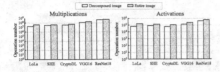

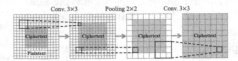

Fig. 2. Counts of HE multiplications and activations for decomposed and entire image.

Fig. 3. The fast spread of ciphertext via CNN layers.

The commonly-used HE cryptosystems include CKKS [6], BGV [3] and TFHE [7] for privacy-preserving machine learning. The BGV and TFHE schemes support integer and bits computations, while the CKKS scheme supports fixed-point computations. For CKKS, a ciphertext is an element in the ring $\mathcal{R}_p^2$, where $\mathcal{R}_p = \mathbb{Z}_p[x]/(x^N + 1)$ is the residue ring of polynomial ring, with polynomial degree N. In addition to HE, a recent study introduced a new data encryption scheme by incorporating crucial ingredients of learning algorithm, specifically the Gini impurity in random forests [36].

Privacy-Preserving Neural Networks. Much attention has been paid on the privacy-preserving neural networks in recent years. For example, Gilad-Bachrach et al. proposed the first CryptoNets to show the feasibility of private CNN inference by HE scheme [14]. Brutzkus et al. proposed the LoLa to optimize the implementation of matrix-vector multiplication by different packing methods, and achieve fast inference for single image [4]. In those studies, the ReLU activation has been replaced by squared activation because the used HE schemes merely support polynomial functions. Lou and Jiang presented a circuit-based quantized neural network SHE implemented by TFHE scheme, which could perform ReLU and max pooling by comparison circuits to obtain good accuracies [27]. Recent studies utilized the CKKS scheme to avoid quantization of network parameters for better accuracies [9,25].

Knowledge Distillation. Knowledge distillation focuses on transferring knowledge from a pretrained teacher model to a student model [22,26]. Traditional methods aimed to match the output distributions of two models by minimizing the Kullback-Leibler divergence loss [22]. Subsequently, a variety of innovative distillation techniques have been developed. Romero et al. proposed Fitnets [31], which matches the feature activations, while Zagoruyko and Komodakis suggested matching attention maps between two models [40]. Gou et al. gave an extensive review on knowledge distillation [17].

Threat Model. Our threat model follows previous studies on private inference [1,8,28]. A honest but curious cloud-based machine learning service hosts a network, which is trained on plaintext data (such as public datasets) and hence the network weights are not encrypted [14]. To ensure the confidentiality of client's data, the client could encrypt the sensitive segment of data by using HE scheme and send data to the cloud server for performing private inference service without decrypting data or accessing the client's private key. Only the client could decrypt the inference result by using the private key.

3   Our Bi-CryptoNets

In this section, we will present new privacy-preserving neural network according to different-level privacy of input data. Our motivation is to decompose input data into *sensitive* and *insensitive* segments according to their privacy. The sensitive segment includes some important and private information such as human faces in an image, whereas the insensitive one contains some background information, which is not so private yet beneficial to learning algorithm.

Based on such decomposition, we could take the strongest homomorphic encryption to keep data security for sensitive segment, while mingle with some perturbations [11, 37] to the data of insensitive segment. This is quite different from previous private inference [14, 25], where homomorphic encryption is applied to the entire input data.

Notice that we can not directly use some previous neural networks to tackle such decomposition with much smaller computational cost, as shown in Fig. 2. We compare with three private networks LoLa [4], SHE [27], CryptoDL [20], and two conventional networks VGG16 and ResNet18. Prior networks can not reduce HE operations because ciphertexts could quickly spread over the entire image via several convolution and pooling operations, as shown in Fig. 3.

Our idea is quite simple and intuitive for the decomposition of input data. We construct a bi-branch neural network to deal with the sensitive and insensitive segments, which are called *ciphertext* and *plaintext* branch, respectively. The ciphertext branch can make use of features from plaintext branch by unidirectional connections, while the converse direction does not hold because of the quick spread of ciphertexts. We take the feature integration for final predictions from the outputs of two branches of network.

Figure 4 presents an overview for our new privacy-preserving neural network, and it is short for *bi-CryptoNets*. We will go to the details of bi-CryptoNets in the following.

3.1   The Bi-branch of Neural Network

We construct ciphertext and plaintext branches to deal with the sensitive and insensitive segments of an input instance, respectively. The plaintext branch deals with the entire input instance, where the insensitive segment is mingled with some perturbations [11, 37], while the sensitive segment is simply filled with zero. The ciphertext branch tackles the sensitive segment with the homomorphic encryption due to its privacy.

The HE operations are restricted in the ciphertext branch to avoid the ciphertext spread on plaintext branch. This could keep the proportion of HE operations in a stable level, rather than previous close approximation to 1 as depth increases [14]. Hence, our bi-branch structure could decease HE operations. Moreover, plaintext branch can be computed with full precision for performing inference, rather than quantized homomorphic ciphertext in prior privacy-preserving neural networks [20].

3.2   The Unidirectional Connections

It is well-known that one great success of deep learning lies in the strong features or representations by plentiful co-adaptations of neural network [16, 21, 34]. Our bi-branch neural network reduces some co-adaptations of features obviously, to restrict

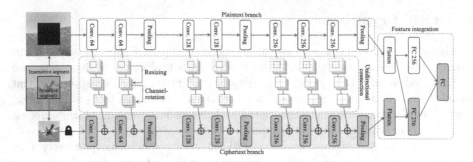

Fig. 4. The overview of our proposed bi-CryptoNets, where grey layers are computed with encrypted inputs, and other layers are computed with plaintext inputs.

the spread of ciphertexts. Hence, it is necessary to strengthen features' representations by exploiting some correlations in bi-branches.

Notice that the ciphertext branch can make use of features from plaintext branch, while the converse direction does not hold because of the spread of ciphertexts. In this work, we consider the simple addition to utilize features from plaintext branch as convolutional neural network [19,29,35]. Concatenation is another effective way to utilize features [23,24], while it yields more HE operations and computational costs.

One reason for addition is of small computational costs, since it yields relatively few homomorphic additions between plaintext and ciphertext, as shown in Table 1. We compare with addition (Add_{PC}) and multiplication (Mul_{PC}) between a plaintext and a ciphertext, as well as addition (Add_{CC}) and multiplication (Mul_{CC}) between two ciphertexts. We compare with four operations under three popular HE schemes: BGV [3], CKKS [6] and TFHE [7], and it is evident that the Add_{PC} has the smallest latency.

Table 1. The comparisons of latency (ms) for three schemes and four operations.

HE Scheme	Add_{PC}	Add_{CC}	Mul_{PC}	Mul_{CC}
BGV	**0.049**	0.077	2.055	3.379
CKKS	**0.039**	0.077	0.173	0.390
TFHE	**56.03**	256.8	1018	1585

For unidirectional connections, we first resize the feature map of plaintext branch to fit the addition with ciphertext branch, and then introduce a channel-rotation for ciphertext channel to extract some information from multiple plaintext channels. Specifically, suppose there are L layers in ciphertext and plaintext branches, and denote by $\mathcal{F}_{c,l}$ and $\mathcal{F}_{p,l}$ their respective l-th layer. For every $l \in [L]$, we consider the following two steps:

i) Resizing feature map of plaintext branch

We begin with the resizing function

$$\boldsymbol{y}_{\text{p},l} = \text{Resize}_l(\boldsymbol{x}_{\text{p},l}) \colon \mathbb{R}^{h_{\text{p},l} \times w_{\text{p},l} \times \text{ch}_{\text{p},l}} \longrightarrow \mathbb{R}^{h_{\text{c},l} \times w_{\text{c},l} \times \text{ch}_{\text{p},l}}$$

where $\boldsymbol{x}_{\text{p},l} \in \mathbb{R}^{h_{\text{p},l} \times w_{\text{p},l} \times \text{ch}_{\text{p},l}}$ denotes the output feature maps from the l-th plaintext layer $\mathcal{F}_{\text{p},l}$ with $h_{\text{p},l}$ rows, $w_{\text{p},l}$ columns and $\text{ch}_{\text{p},l}$ channels, and $h_{\text{c},l}$ and $w_{\text{c},l}$ denote the number of row and column in ciphertext branch.

For resizing function, a simple yet effective choice is the cropping [32], which maintains the center features of size $h_{c,l} \times w_{c,l} \times ch_{p,l}$ because those features can capture more important information around the sensitive segment. We can also select pooling or convolution operations for resizing function [19].

ii) Channel-rotation for ciphertext channel

We now introduce the channel-rotation function as follows:

$$z_{p,l} = \mathrm{CRot}_l(y_{p,l}) = \left[\mathrm{CRot}_l^1(y_{p,l}), \cdots, \mathrm{CRot}_l^i(y_{p,l}), \cdots, \mathrm{CRot}_l^{ch_{c,l}}(y_{p,l}) \right]$$

where $ch_{c,l}$ is number of channels in ciphertext branch, and $\mathrm{CRot}_l^i(y_{p,l})$ denotes the i-th channel of $\mathrm{CRot}_l(x)$, that is,

$$\mathrm{CRot}_l^i(y_{p,l}) = \sum_{j=1}^{ch_{p,l}} W_{\mathrm{CRot},l}^{i,j} \cdot y_{p,l}^j \quad \text{for} \quad i \in [ch_{c,l}] \, .$$

Here, $y_{p,l}^j \in \mathbb{R}^{h_{c,l} \times w_{c,l}}$ denotes the j-th channel of $y_{p,l}$ for $j \in [ch_{p,l}]$, and $W_{\mathrm{CRot},l} = (W_{\mathrm{CRot},l}^{i,j})_{ch_{c,l} \times ch_{p,l}}$ is the weight matrix for channel-rotation in the l-th layer.

The channel-rotation is helpful for ciphertext channels to extract useful information automatically from plaintext channels. This is because channel-rotation could compress, rotate and scale different channels of plaintext feature maps, without the requirements of channel-wise alignments and identical numbers and magnitudes of channels. Moreover, each channel, in connection with ciphertext branch, can be viewed as a linear combination of multiple plaintext channels, rather than one plaintext channel. Therefore, the ciphertext branch can get better information from plaintext branch and achieve better performance with the help of channel-rotation.

After resizing and channel-rotation, the ciphertext branch can make use of features from plaintext branch by addition. The unidirectional connections can be written as

$$x_{c,l+1} = \mathcal{F}_{c,l+1}(x_{c,l} + z_{p,l}) = \mathcal{F}_{c,l+1}(x_{c,l} + \mathrm{CRot}_l(\mathrm{Resize}_l(x_{p,l}))) \, ,$$

where $x_{c,l} \in \mathbb{R}^{h_{c,l} \times w_{c,l} \times ch_{c,l}}$ denotes the output feature maps of $\mathcal{F}_{c,l}$ with $h_{c,l}$ rows, $w_{c,l}$ columns and $ch_{c,l}$ channels.

Finally, it is feasible to improve computational efficiency by implementing plaintext and ciphertext branches in parallel, because computing the plaintext branch and unidirectional connections is much faster than that of ciphertext branch.

3.3 The Feature Integration

We now design a two-layer neural network to integrate the outputs from ciphertext and plaintext branches. In the first layer, we split all neurons into two halves, one half for ciphertext yet the other for plaintext. The plaintext neurons are only connected with the outputs from plaintext branch, whereas the ciphertext neurons have full connections. We take full connections in the second layer.

Specifically, there are n_1 (even) and n_2 neurons in the first and second layer, respectively. Let $x_c \in \mathbb{R}^{n_c}$ and $x_p \in \mathbb{R}^{n_p}$ denote the flattened outputs from ciphertext and plaintext branches with n_c and n_p features, respectively. Then, the outputs of ciphertext and plaintext neurons in the first layer can be given by, respectively,

$$x_p^{\langle 1 \rangle} = \sigma(W_{p,1}' x_p + b_1') , \quad x_c^{\langle 1 \rangle} = \sigma(W_{c,1} x_c + W_{p,1} x_p + b_1) ,$$

where $\sigma(\cdot)$ is an activation function, $b_1, b_1' \in \mathbb{R}^{n_1/2}$ are bias vectors, $W_{c,1} \in \mathbb{R}^{n_c \times n_1/2}$ and $W_{p,1} \in \mathbb{R}^{n_p \times n_1/2}$ are the weight for ciphertext neurons, yet $W_{p,1}' \in \mathbb{R}^{n_p \times n_1/2}$ is the weight for plaintext neurons. The final output in the second layer can be given by

$$x_{out} = \sigma(W_{c,2} x_c^{\langle 1 \rangle} + W_{p,2} x_p^{\langle 1 \rangle} + b_2) ,$$

where $b_2 \in \mathbb{R}^{n_2}$ is a bias vector, and $W_{c,2} \in \mathbb{R}^{n_1/2 \times n_2}$ and $W_{p,2} \in \mathbb{R}^{n_1/2 \times n_2}$ are the weight matrices for ciphertext and plaintext neurons, respectively. Here, we consider two-layer neural network for simplicity, and similar constructions could be made for deeper neural networks based on the splitting of plaintext and ciphertext neurons.

4   Knowledge Distillation for Bi-CryptoNets

Knowledge distillation has been an effective way to improve learning performance by distilling knowledge from a teacher network [17,22]. This section develops a feature-based knowledge distillation for bi-CryptoNets, as shown in Fig. 5. The basic idea is to supplement the representations of two branches of bi-CryptoNets by imitating a teacher network.

For the teacher network, we learn a classical convolutional neural network from training data over the entire images without decomposition. Hence, the teacher network has plentiful connections between different neurons, and strengthens the intrinsic correlations for better performance.

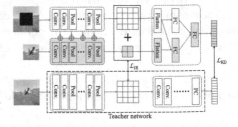

Fig. 5. The overview of our feature-based knowledge distillation.

We learn the representations of our two branches from the corresponding intermediate representations of teacher network, and also learn the final outputs of our bi-CryptoNets from that of teacher network. This is partially motivated from previous knowledge distillations on internal representations [31].

Specifically, we first pad the output of ciphertext branch with 0 to match the size of plaintext output and add them together, as in convolutional neural network literature [16,18]. We then train two branches of bi-CryptoNets to learn teacher's intermediate representations by minimizing the following loss function:

$$\mathcal{L}_{IR}(W_T^h, W_c, W_p) = \left\| \mathcal{F}_T^h(x, W_T^h) - (\text{Pad}(\mathcal{F}_c(x, W_c)) + \mathcal{F}_p(x, W_p)) \right\|_2^2 , \quad (1)$$

where $\mathcal{F}_T^h(\cdot; W_T^h)$ is the first h layers of teacher network of parameter W_T^h, and $\mathcal{F}_c(\cdot; W_c)$ and $\mathcal{F}_p(\cdot; W_p)$ denote the ciphertext and plaintext branches of parameter W_c and W_p, respectively, and $\text{Pad}(\cdot)$ is the zero-padding function.

Here, we denote by h the corresponding h-th layer in the teacher network that has the same output size as plaintext branch's output. Such loss can help the sum of outputs from two branches to approximate the teacher's intermediate output. In this manner, two branches could obtain stronger representations, and this makes it much easier for further learning in the following layers of our bi-CryptoNets.

After training two branches, we then perform knowledge distillation to the whole network. The student network is trained such that its output is similar to that of the teacher and to the true labels. Given bi-CryptoNets' output a_S and teacher's output a_T, we first get softened output as

$$p_T^\tau = \text{softmax}(a_T/\tau) \quad \text{and} \quad p_S^\tau = \text{softmax}(a_S/\tau) ,$$

where $\tau > 1$ is a relaxation parameter. We try to optimize the following loss function:

$$\mathcal{L}_{\text{KD}}(W_\text{T}, W_{\text{BCN}}) = \mathcal{H}(y_\text{True}, p_s) + \lambda \mathcal{H}(p_T^\tau, p_s^\tau) , \tag{2}$$

where y_True is the true label, $p_s = \text{softmax}(a_S)$, $\mathcal{H}$ refers to the cross-entropy, W_T and W_{BCN} are the parameters of teacher network and our bi-CryptoNets, respectively, and $\lambda \in [0, 1]$ is a hyperparameter balancing two cross-entropies. With our proposed method, two branches in our bi-CryptoNets can learn better representations, enhancing overall performance. More details can be found in [39].

5 Experiments

This section presents empirical studies on datasets[1] MNIST, CIFAR-10 and CIFAR-100, which have been well-studied in previous private inference studies [4,25,27]. We develop different backbones for our bi-CryptoNets according to different datasets. We adopt the 3-layer CNN [4,14] as backbone for MNIST, and we take 11-layer CNN [5], VGG-16 [33] and ResNet-18 [19] for CIFAR-10 and CIFAR-100. Figure 4 presents our bi-CryptoNets with VGG-16 backbone, and more details are shown in [39].

We take the CKKS scheme for sensitive segment with polynomial degree $N = 2^{14}$, and mingle with Gaussian noise for insensitive segment. We further improve packing method for our bi-CryptoNets to perform inference for multiple images simultaneously.

For simplicity, we focus on the regular images, where the area of sensitive segment is restricted to a quarter in the center of every image, and we could take some techniques of zooming, stretching and rotating to adjust those irregular images.

Ablation Studies
We conduct ablation studies to verify the effectiveness of feature-based knowledge distillation and the structures of our bi-CryptoNets. Table 2 presents the details of experimental results of inference accuracies on three datasets.

---

[1] Downloaded from yann.lecun.com/exdb/mnist and www.cs.toronto.edu/~kriz/cifar.

Table 2. The accuracies (%) on MNIST, CIFAR-10 and CIFAR-100 datasets (KD refers to the conventional knowledge distillation; FKD refers to our feature-based knowledge distillation).

Models	Knowledge Distillation	MNIST	CIFAR-10			CIFAR-100	
		CNN-3	CNN-11	VGG-16	ResNet-18	VGG-16	ResNet-18
Backbone network (w/o decomposition)	w/o KD	99.21	90.99	93.42	94.30	72.23	74.00
bi-CryptoNets	w/o KD	98.60	81.91	90.36	92.04	64.85	69.46
bi-CryptoNets	KD	98.61	80.03	88.91	92.43	64.96	65.67
bi-CryptoNets (w/o uni. connections)	FKD	98.89	84.69	91.78	91.73	70.61	70.29
bi-CryptoNets (w/o channel-rotations)	FKD	98.90	90.09	92.14	91.86	71.43	71.40
bi-CryptoNets	**FKD**	**99.15**	**90.30**	**93.27**	**93.91**	**72.35**	**73.33**

We first exploit the influence of feature-based knowledge distillation in our bi-CryptoNets. We consider two different methods: bi-CryptoNets without knowledge distillation and bi-CryptoNets with conventional knowledge distillation. As can be seen from Table 2, it is observable that our feature-based knowledge distillation could effectively improve the inference accuracy by $0.54\%-8.39\%$, in comparison to bi-CryptoNets without knowledge distillation; it also achieves better inference accuracy by $1.45\%-10.27\%$ than that of conventional knowledge distillation, which fails to stably improve accuracy with our bi-branch structure.

We then study the influence of unidirectional connections in bi-CryptoNets, and consider two variants: our bi-CryptoNets without unidirectional connections and bi-CryptoNets with resizing yet without channel-rotations. As can be seen from Table 2, the unidirectional connections are helpful for ciphertext branch to extract useful information from plaintext branch. This is because the unidirectional connections with only resizing can enhance inference accuracy by $0.13\%-5.40\%$, and it could further improve accuracy by $0.26\%-5.61\%$ with both resizing and channel-rotations.

We finally compare the inference accuracies of our bi-CryptoNets with the backbone network, which is trained and tested on the entire input images without decomposition. From Table 2, it is observable that our bi-CryptoNets with feature-based knowledge distillation achieves comparable or even better inference accuracies than the corresponding backbone networks without data decomposition; therefore, our proposed bi-CryptoNets could effectively compensate for the information loss of plaintext and ciphertext branches under the help of feature-based knowledge distillation.

Experimental Comparisons

We compare our bi-CryptoNets with the state-of-the-art schemes on privacy-preserving neural networks [4,8,9,20,25,27]. We also implement the structure of backbone network with square activation for fair comparisons. We take the batch size 20 and 32 for MNIST and CIFAR-10, respectively. We employ commonly-used criteria in experiments, i.e., inference accuracy, inference latency and the number of homomorphic

Table 3. The experimental comparisons on MNIST.

Scheme	HEOPs	Add$_{CC}$	Mul$_{PC}$	Act$_C$	Latency (s)	Amortized Latency (s)	Acc (%)
FCryptoNets [8]	63K	38K	24K	945	39.1	2.0	98.71
CryptoDL [20]	4.7M	2.3M	2.3M	1.6K	320	16.0	99.52
LoLa [4]	573	393	178	2	2.2	2.2	98.95
SHE [27]	23K	19K	945	3K	9.3	9.3	99.54
EVA [9]	8K	4K	4K	3	121.5	121.5	99.05
VDSCNN [25]	4K	2K	2K	48	105	105	99.19
Backbone CNN-3	1962	973	984	5	7.2	1.4	98.95
bi-CryptoNets (CNN-3)	830	406	418	6	2.1	0.1	99.15

operations as in [8,28]. We also consider the important amortized inference latency in a batch of images [25], and concern the number of activations for ciphertexts [4,9,27]. Tables 3 and 4 summarize experimental comparisons on MNIST and CIFAR-10, respectively. Similar results could be made for CIFAR-100, presented in our full work [39].

From Table 3, our bi-CryptoNets can reduce HE operations by 2.35× and inference latency by 3.43× in contrast to backbone network. Our bi-CryptoNets could decrease the amortized latency by 13.7×, and achieve better accuracy (about 0.2%) than backbone because of our knowledge distillation and precise inference in the plaintext branch.

From Table 4, our bi-CryptoNets reduces HE operations by 1.43×, and inference latency by 1.15× in contrast to backbone network. Our bi-CryptoNets could decrease the amortized latency by 4.59× and improve accuracy by 2.1%. We also implement our bi-CryptoNets with VGG-16 and ResNet-18 as backbones, and deeper networks yield better inference accuracies, i.e., 93.27% for VGG-16 and 93.91% for ResNet-18.

From Tables 3 and 4, our bi-CryptoNets takes a good balance between inference accuracies and computation cost in comparison with other neural networks. Our bi-CryptoNets achieves lower inference latency and amortized latency than other methods except for LoLa, where light-weight neural network is implemented with complicated packing method and smaller HE operations. Our bi-CryptoNets takes relatively-good inference accuracies except for SHE, CryptoDL and VDSCNN, where deeper neural networks are adopted with larger computational overhead. Based on deeper backbones such as VGG-16 and ResNet-18, our bi-CryptoNets could achieve better inference accuracy, comparable inference latency and smaller amortized latency.

Table 4. The experimental comparisons on CIFAR-10.

Scheme	HEOPs	Add$_{CC}$	Mul$_{PC}$	Act$_C$	Latency (s)	Amortized Latency (s)	Acc (%)
FCryptoNets [8]	701M	350M	350M	64K	22372	1398	76.72
CryptoDL [20]	2.4G	1.2G	1.2G	212K	11686	731	91.50
LoLa [4]	70K	61K	9K	2	730	730	76.50
SHE [27]	4.4M	4.4M	13K	16K	2258	2258	92.54
EVA [9]	135K	67K	67K	9	3062	3062	81.50
VDSCNN [25]	18K	8K	9K	752	2271	2271	91.31
Backbone CNN-11	1.0M	493K	521K	246	1823	228	88.21
bi-CryptoNets (CNN-11)	709K	341K	368K	246	1587	49	90.30
bi-CryptoNets (VGG-16)	3.4M	1.5M	1.9M	1.1K	2962	92	93.27
bi-CryptoNets (ResNet-18)	15.5M	6.9M	8.6M	2.8K	6760	211	93.91

6 Conclusion

Numerous privacy-preserving neural networks have been developed to keep the balance between accuracy, efficiency and security. We take a different view from the input data and network structure. We decompose the input data into sensitive and insensitive segments, and propose the bi-CryptoNets, i.e., plaintext and ciphertext branches, to deal with two segments, respectively. We also introduce feature-based knowledge distillation to strengthen the representations of our network. Empirical studies verify the effectiveness of our bi-CryptoNets. An interesting future work is to exploit multiple levels of privacy of input data and multiple branches of neural network, and it is also interesting to generalize our idea to other settings such as multi-party computation.

Acknowledgements. The authors want to thank the reviewers for their helpful comments and suggestions. This research was supported by National Key R&D Program of China (2021ZD0112802), NSFC (62376119) and CAAI-Huawei MindSpore Open Fund.

References

1. Boemer, F., Costache, A., Cammarota, R., Wierzynski, C.: nGraph-HE2: a high-throughput framework for neural network inference on encrypted data. In: WAHC@CCS, pp. 45–56 (2019)
2. Boulemtafes, A., Derhab, A., Challal, Y.: A review of privacy-preserving techniques for deep learning. Neurocomputing **384**, 21–45 (2020)
3. Brakerski, Z., Gentry, C., Vaikuntanathan, V.: (leveled) fully homomorphic encryption without bootstrapping. ACM Trans. Comput. Theory **6**(3), 1–36 (2014)
4. Brutzkus, A., Gilad-Bachrach, R., Elisha, O.: Low latency privacy preserving inference. In: ICML, pp. 812–821 (2019)
5. Chabanne, H., Wargny, A., Milgram, J., Morel, C., Prouff, E.: Privacy-preserving classification on deep neural network. Cryptol. ePrint Arch. (2017)

6. Cheon, J.H., Kim, A., Kim, M., Song, Y.: Homomorphic encryption for arithmetic of approximate numbers. In: Takagi, T., Peyrin, T. (eds.) ASIACRYPT 2017. LNCS, vol. 10624, pp. 409–437. Springer, Cham (2017). https://doi.org/10.1007/978-3-319-70694-8_15

7. Chillotti, I., Gama, N., Georgieva, M., Izabachène, M.: TFHE: fast fully homomorphic encryption over the torus. J. Cryptol. **33**(1), 34–91 (2020)

8. Chou, E., Beal, J., Levy, D., Yeung, S., Haque, A., Fei-Fei, L.: Faster CryptoNets: leveraging sparsity for real-world encrypted inference. CoRR/abstract **1811.09953** (2018)

9. Dathathri, R., Kostova, B., Saarikivi, O., Dai, W., Laine, K., Musuvathi, M.: EVA: an encrypted vector arithmetic language and compiler for efficient homomorphic computation. In: PLDI, pp. 546–561 (2020)

10. Dathathri, R., et al.: CHET: an optimizing compiler for fully-homomorphic neural-network inferencing. In: PLDI, pp. 142–156 (2019)

11. Dwork, C., McSherry, F., Nissim, K., Smith, A.: Calibrating noise to sensitivity in private data analysis. J. Priv. Confidentiality **7**(3), 17–51 (2016)

12. Gentry, C.: Fully homomorphic encryption using ideal lattices. In: STOC, pp. 169–178 (2009)

13. Ghodsi, Z., Jha, N., Reagen, B., Garg, S.: Circa: stochastic ReLUs for private deep learning. In: NeurIPS, pp. 2241–2252 (2021)

14. Gilad-Bachrach, R., Dowlin, N., Laine, K., Lauter, K., Naehrig, M., Wernsing, J.: CryptoNets: applying neural networks to encrypted data with high throughput and accuracy. In: ICML, pp. 201–210 (2016)

15. Gong, X., Chen, Y., Yang, W., Mei, G., Wang, Q.: InverseNet: augmenting model extraction attacks with training data inversion. In: IJCAI, pp. 2439–2447 (2021)

16. Goodfellow, I., Bengio, Y., Courville, A.: Deep Learning. MIT Press, Cambridge (2016)

17. Gou, J., Yu, B., Maybank, S., Tao, D.: Knowledge distillation: a survey. Int. J. Comput. Vis. **129**(6), 1789–1819 (2021)

18. Hashemi, M.: Enlarging smaller images before inputting into convolutional neural network: zero-padding vs. interpolation. J. Big Data **6**(1), 1–13 (2019)

19. He, K., Zhang, X., Ren, S., Sun, J.: Deep residual learning for image recognition. In: CVPR, pp. 770–778 (2016)

20. Hesamifard, E., Takabi, H., Ghasemi, M.: Deep neural networks classification over encrypted data. In: CODASPY, pp. 97–108 (2019)

21. Hinton, G., Srivastava, N., Krizhevsky, A., Sutskever, I., Salakhutdinov, R.: Improving neural networks by preventing co-adaptation of feature detectors. CoRR/abstract **1207.0580** (2012)

22. Hinton, G., Vinyals, O., Dean, J.: Distilling the knowledge in a neural network. CoRR/abstract **1503.02531** (2015)

23. Huang, G., Liu, Z., Maaten, L., Weinberger, K.: Densely connected convolutional networks. In: CVPR, pp. 2261–2269 (2017)

24. Iandola, F., Moskewicz, M., Ashraf, K., Han, S., Dally, W., Keutzer, K.: SqueezeNet: AlexNet-level accuracy with 50x fewer parameters and <1MB model size. CoRR/abstract **1602.07360** (2016)

25. Lee, E., et al.: Low-complexity deep convolutional neural networks on fully homomorphic encryption using multiplexed parallel convolutions. In: ICML, pp. 12403–12422 (2022)

26. Li, Z., et al.: Curriculum temperature for knowledge distillation. In: AAAI, pp. 1504–1512 (2023)

27. Lou, Q., Jiang, L.: SHE: a fast and accurate deep neural network for encrypted data. In: NeurIPS, pp. 10035–10043 (2019)

28. Lou, Q., Lu, W., Hong, C., Jiang, L.: Falcon: fast spectral inference on encrypted data. In: NeurIPS, pp. 2364–2374 (2020)

29. Radosavovic, I., Kosaraju, R., Girshick, R., He, K., Dollár, P.: Designing network design spaces. In: CVPR, pp. 10425–10433 (2020)

30. Ribeiro, M., Grolinger, K., Capretz, M.: MLaaS: machine learning as a service. In: ICMLA, pp. 896–902 (2015)
31. Romero, A., Ballas, N., Kahou, S., Chassang, A., Gatta, C., Bengio, Y.: FitNets: hints for thin deep nets. In: ICLR (2015)
32. Ronneberger, O., Fischer, P., Brox, T.: U-Net: convolutional networks for biomedical image segmentation. In: Navab, N., Hornegger, J., Wells, W.M., Frangi, A.F. (eds.) MICCAI 2015. LNCS, vol. 9351, pp. 234–241. Springer, Cham (2015). https://doi.org/10.1007/978-3-319-24574-4_28
33. Simonyan, K., Zisserman, A.: Very deep convolutional networks for large-scale image recognition. CoRR/abstract **1409.1556** (2014)
34. Srivastava, N., Hinton, G., Krizhevsky, A., Sutskever, I., Salakhutdinov, R.: Dropout: a simple way to prevent neural networks from overfitting. JMLR **15**(1), 1929–1958 (2014)
35. Tan, M., Le, Q.: EfficientNet: rethinking model scaling for convolutional neural networks. In: ICML, pp. 6105–6114 (2019)
36. Xie, X.R., Yuan, M.J., Bai, X.T., Gao, W., Zhou, Z.H.: On the Gini-impurity preservation for privacy random forests. In: NeurIPS (2023)
37. Yang, K., Yau, J., Fei-Fei, L., Deng, J., Russakovsky, O.: A study of face obfuscation in imagenet. In: ICML, pp. 25313–25330 (2022)
38. Yin, X., Zhu, Y., Hu, J.: A comprehensive survey of privacy-preserving federated learning: a taxonomy, review, and future directions. ACM Comput. Surv. **54**(6), 131:1–131:36 (2021)
39. Yuan, M.J., Zou, Z., Gao, W.: Bi-cryptonets: leveraging different-level privacy for encrypted inference. CoRR/abstract **2402.01296** (2024)
40. Zagoruyko, S., Komodakis, N.: Paying more attention to attention: improving the performance of convolutional neural networks via attention transfer. In: ICLR (2017)

Enhancing YOLOv7 for Plant Organs Detection Using Attention-Gate Mechanism

Hanane Ariouat[1]([✉]), Youcef Sklab[1], Marc Pignal[2], Florian Jabbour[2], Régine Vignes Lebbe[2], Edi Prifti[1,3], Jean-Daniel Zucker[1,3], and Eric Chenin[1]

[1] IRD, Sorbonne Université, UMMISCO, 93143 Bondy, France
hanane.ariouat@ird.fr
[2] Institut de Systématique, Evolution, Biodiversité (ISYEB), Muséum national d'Histoire naturelle, CNRS, Sorbonne Université, Université des Antilles, EPHE, 57, rue Cuvier, CP39, 75005 Paris, France
[3] Sorbonne Université, INSERM, Nutrition et Obesities; Systemic Approaches, NutriOmique, AP-HP, Paris, France

Abstract. Herbarium scans are valuable raw data for studying how plants adapt to climate change and respond to various factors. Characterization of plant traits from these images is important for investigating such questions, thereby supporting plant taxonomy and biodiversity description. However, processing these images for meaningful data extraction is challenging due to scale variance, complex backgrounds that contain annotations, and the variability in specimen color, shape, and orientation of specimens. In addition, the plant organs often appear compressed, deformed, or damaged, with overlapping occurrences that are common in scans. Traditional organ recognition techniques, while adept at segmenting discernible plant characteristics, are limited in herbarium scanning applications. Two automated methods for plant organ identification have been previously reported. However, they show limited effectiveness, especially for small organs. In this study we introduce YOLOv7-ag model, which is a novel model based on the YOLOv7 that incorporates an attention-gate mechanism, which enhances the detection of plant organs, including stems, leaves, flowers, fruits, and seeds. YOLOv7-ag significantly outperforms previous state of the art as well as the original YOLOv7 and YOLOv8 models with a precision and recall rate of 99.2% and 98.0%, respectively, particularly in identifying small plant organs.

Keywords: Object Detection · Attention-gate · YOLOv7 · Herbarium specimen

1 Introduction

Plants are a crucial component of our plant's biodiversity, by shaping landscapes and being the foundation of trophic networks, and by playing a major role in

---

Our code is available at https://github.com/IA-E-Col/YOLOv7-ag
H. Ariouat and Y. Sklab—Both authors contributed equally to this work.

the balance between carbon dioxyde and oxygen [4]. Natural history collections are house to several hundreds years of data on the evolution of the environment and biodiversity [3]. Herbarium collections are gaining interest in the scientific community since they can lead to our improved understanding and mitigate serious threats to biodiversity thanks to the information they carry [1,2,5–8]. The collections of several herbaria have recently been extensively digitized [9], creating several virtual collections (online biodiversity data portals) and making possible the use of computer-based processing techniques, such as Deep Learning (DL) to describe plant morphology, such as leaf size, perimeter and width, and counting organs [10], for assessing crop quality after harvest [11] and/or for disease classification [12]. More specifically, with object detection techniques, organs or prominent areas can be automatically detected and measured [15].

The objective of this work is to improve existing computer vision techniques in the task of predicting bounding boxes and classify them in a set of categories, such as plant organs. Here we focus on a subset of digitized herbarium specimens from the Paris Museum collection. More specifically, we focus on the images of the scanned 2D herbarium sheets and aim at identifying plants' organs (leaf, flower, fruit, stem, root, seed) for further analyses. Recently, there has been important progress in the field of object detection, and most of the existing AI-based models produce good results for medium and big objects [19]. However, since there are possibly a large number and overlapping plant organs of different sizes in the images, using existing models to detect objects in herbarium scans may not be appropriate, particularly, for small organs detection.

In this article, we propose an extension of the YOLOv7 model [20], which improves plant organ identification, particularly the smaller organs. This new model called YOLOv7-ag integrates an attention mechanism: the attention-gate [21]. YOLOv7-ag better exploits spatial information like the arrangement of organs and the presence of other organs in the spatial perimeter of a given organ. Using an attention mechanism essentially improves identifying regions of interest in plants, which is critical in the detection of the targeted organs.

2   Related Work

A herbarium is a collection of dried plant specimens that are arranged on herbarium sheets [8]. A typical herbarium sheet scan might contain non-vegetal elements such as stamps, scale bars, color palettes, annotation labels, barcodes, and envelopes (cf. Fig. 1). A first analysis of the herbarium scans allowed us to notice that they include several elements that bring significant difficulties for object detection techniques based on conventional networks. For instance, they have important scale variance and complex backgrounds with several non-vegetable elements. Moreover, we noticed that the color, shape, and orientation of specimen images vary greatly and in several images the plant's organs are condensed and deformed/damaged. Most often, organs such as leaves, flowers and fruits are overlapped on a reduced space of the herbarium scan.

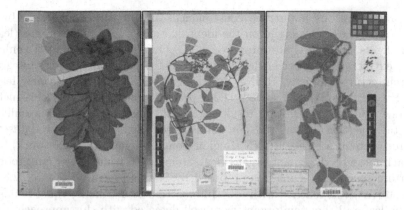

Fig. 1. Typical herbarium sheet scans, including plant material and non-biological material, with sometimes overlapping, incomplete, compressed and deformed (or even missing) objects.

YOLO-based algorithms have been used in numerous works on object detection. Zhang et al. [10] focused on identifying and detecting fruits with two different algorithms: OrangeYolo to detect the orange fruits and OrangeSort to track them. Both algorithms were based on YOLOv3 algorithm. Zhao et al. [23] used YOLOv7 and enhanced it for object detection for maritime UAV images. The authors added one head at the detection phase to allow the network to identify small people or objects and implemented a SimAM attention module to allow the network to find the scene's attention region. Almost 59% of precision was reached. James et al. [25] provided a whole pipeline for head density estimation for sorghum from RGB images, collected via unmanned aerial vehicle (UAV). The authors used YOLO-v5 and Faster RCNN algorithms. Based on YOLOv4, Zhang et al. [24] proposed the ViT-YOLO (Vision Transformer) object detection approach for drone captured images (UAV images, VisDrone2019-Det dataset). The authors enhanced YOLOv4 by adding a multi-head attention block using the MHSA-Darknet, which is an upgraded Darknet backbone. Xu et al. [26] proposed TrichomeYolo to identify the density and length of maize leaf trichomes from scanning electron microscopy images. They used YOLOv5 as the base model and integrated a Transfomer into it, which reached 92.5 % accuracy. In the same line of YOLO-based approaches, Zhou et al. [19] enhanced YOLOv5 to better detect small objects in infrared imaging. The authors proposed an image Super-Resolution reconstruction based on Generative Adversarial Network (SRGAN), Self-Adaption Squeeze-and-Excitation (SASE) module and a YOLO based network, called a multi-receptive field adaptive channel attention network.

Despite the existence of a large number of object detection approaches in the literature, there is a lack of work that addresses plant organ's detection on herbarium scans. Identifying these organs can considerably help provide valuable information to investigate several research questions such as long-term phenological investigations about the consequences of climatic change [22]. In this regard, Triki et al. [8] focused on the identification of four plant organs (leaves, flowers, fruits, and buds) and proposed a YOLOv3 based algorithm. To enhance feature extraction and better detect small plant organs, the authors integrated two architectures, namely, ResNet and DenseNet, and included a new feature scale. By doing so, they reached 94.2% precision rate. Although this method appears to provide interesting results, we could not assess its applicability on our dataset as the authors did not publish their models, nor their dataset. In the same context, a Faster RCNN (Region Convolutional Neural Network) based approach for detecting plant organs (leaf, stem, seed, fruit, flower, and root) on herbarium specimens scans was proposed by Souhaib et al. [5]. Only 22.8% of precision (AP) was reached with IoU (Intersection over Unions) equal to 0.5 and only the leaf and the stem organs were well detected with this approach.

2.1   Attention-Gate Mechanism

Attention mechanisms are used to focus on specific regions of interest (spatial information) in images that are deemed more relevant or important for the prediction task. We believe that using these mechanisms in object detection would improve the ability of the neural network to recognize plant organs, particularly in the regions of the images where there is a high concentration of organs. In this paper, we focus on the attention-gate mechanism (*cf.* Fig. 2), which is frequently used in natural image analysis, knowledge graph, and classification tasks [27]. Attention-gate directs the model's focus to crucial locations while suppressing the activation of features in irrelevant regions in an input image. It is able to focus on targets of varied sizes and shapes through an automated learning process.

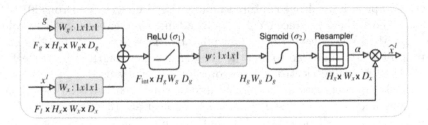

Fig. 2. Attention-gate block structure diagram as proposed by [21]. $\oplus$ and $\otimes$ represent element wise sum and matrix multiplication respectively, while $1 \propto 1$ represents a pointwise convolution.

In the Fig. 2, attention coefficients (α) are computed in order to scale the input features (x^l). Spatial regions are selected by analysing both the activation and contextual information provided by the gating signal (g), which is collected from a coarser scale. Grid re-sampling of attention coefficients is done using trilinear interpolation. The activations and contextual information provided by the gating signal (g), which is acquired from a coarser scale, are analyzed in order to select the spatial regions. Then, a trilinear interpolation is utilized in the process of resampling the attention coefficients. The features that are propagated through the skip connections (x^l) are filtered by attention gates and in order to achieve feature selectivity, contextual information (gating) gathered at coarser scales is used (g). More precisely, models trained with Attention-gate intuitively learn to highlight prominent features that are helpful for a particular task while suppressing irrelevant regions in an input image.

3 YOLOv7 with Attention-Gate Mechanism

We added two instances of the attention-gate module into two parts of the YOLOv7 model, as illustrated in Fig. 3. To the best of our knowledge, the attention-gate mechanism had not previously been incorporated into an object detection algorithm. YOLOv7-ag is structured in three main parts i) Backbone ii) Head and iii) Detection. The first one consists of the backbone and is the same as the one proposed in YOLOv7 [20]. It aims to extract the feature maps and consists of a succession of several blocks, namely, CBS, ELAN and MPConv. The Fig. 4 illustrates the layers that constitute these blocks. The second part of the model architecture as described in [20] is called the Head. It mainly consists of a succession of several blocks (CBS, ELAN-H, SPPCSP, MPConv), concatenation (Concat) and up-sampling (UPSample) blocks. With stride 1 and kernel sizes 3 or 1, ELAN and ELAN-H blocks in YOLOv7 [20] (see Fig. 4) have 8 stages (or block groups) of CBS (Convolution, Batch normalization, and activation function SILU). The MPConv block contains five layers, one of which is max-pooling and three are CBS. The SPP block, which is a mixture of three max-pooling, is one of the nine blocks that forms the SPPCSP blocks.

YOLOv7 uses a combination of low-level and high-level characteristics through a summation or concatenation operation. However, doing so without making any distinctions may sometimes result in feature mismatches and performance deterioration. Our aim is to integrate learnable weights to determine the significance/importance of various input features. In our approach, we modified the HEAD block by adding the attention-gate mechanism, followed by a CBS block. More specifically, we integrated two attention-gate blocks into the YOLOv7 architecture[1], which allow exploiting the spatial information coming from the upper layers (which is collected from a coarser scale) to compute a coefficient matrix, called attention coefficient. It is a condensed representation of the contextual information derived from the spatial information coming from the upper layers. Then, applying the computed attention coefficient on the deep layers, (which provide feature maps) allows to determine which features coming from the upper layers are determining in the feature maps. This aims to consider more the relevant features of the target contained in the shallow network and

---

[1] After the backbone network and before the detection layer.

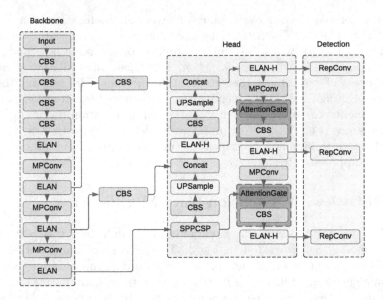

Fig. 3. YOLOv7-ag: YOLOv7 with attention-gate.

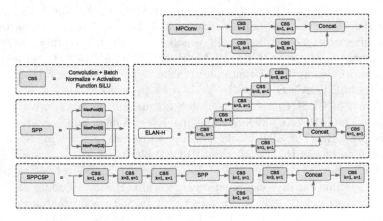

Fig. 4. YOLOv7 component blocks

give less importance to the irrelevant information, improving the detection performance of the algorithm. Because self-attention across n entities needs $O(n^2d)$ memory and computation [13], we believe that the simplest configuration that complies with the aforementioned considerations would be to include the attention mechanism in the two head's lowest resolution feature maps. For the third part of the model architecture, which aims at predicting layers at 3 different scales[2], we kept the detection part as it was originally introduced in [20].

---

[2] The three scales are designed to detect small, medium, and large objects, respectively. They are represented by feature maps that are extracted at different depths of the

4   Experiments

We conducted experiments using a dataset of several herbarium specimen scans that contains plants with organs of varying sizes and diverse shapes. We compared the results of YOLOv7-ag with four models: Faster R-CNN [5], YOLOv7 [20], YOLOv8 [28] and YOLOv8-ag. YOLOv8-ag is the version of YOLOv8 with the attention-gate mechanism, that we integrated similarly to our approach in YOLOv7-ag. We trained the YOLO models on two NVIDIA A100 GPUs with 80 GB memory for 800 epochs. The optimizer used was Adam, with a default weight decay of 0.0005 and a momentum of 0.937. The batch size was set to 16, and the initial learning rate was 0.001. The input image size was 640×640. We also compared the obtained results from YOLOv7-ag with the plant organ detection model of Younis et al. [5].

4.1   Experiment Materials

For the training data, we used the publicly available dataset, published by Younis et al. [5], which contains 635 images. The height of the images approximately ranges between 3500 and 1.600 pixels and the width ranges between 300 and 500 pixels. Taking into account their diversity observed in our dataset of herbarium scanned images (over 10 million images), the Younis et al. dataset is relatively small, which needs to be increased to cover a large spectrum of possible cases and improve the performance of our model. We therefore used a rotational data augmentation technique, with various angles. For each image we performed 5 rotations at 5 angles [45, 180, 225, 270, 315] and adjusted the bonding boxes accordingly. Note that we deactivated YOLOv7's (and YOLOv8's) default augmentation pipeline as it performs such transformations that produce images that aren't present in our herbarium scans (e.g., mosaic augmentation, cutout augmentation, hsv color augmentation, etc.). At the end we increased the initial dataset by a factor of 5, resulting in 3810 images, from which 3000 were allocated to the training set, and the remaining 784 to the validation set. For the test dataset, we used 26 images, representing various species with distinct sizes and shapes.

4.2   Evaluation Metrics

The prediction task in object detection is a bounding box along with the object at hand and its corresponding label (class). To evaluate the performance of YOLOv7-ag, we used a set of metrics widely used in object detection tasks: i) Precision, ii) Recall and iii) mean Average Precision (mAP). The Intersection over Union (IoU) metric is used to quantify how the predicted box is aligned compared to the actual ground truth of the bounding box[3].

---

neural network, thus allowing for precise detection across a varied range of object sizes.

[3] Ground truth represents the desired bounding box as output of the object detection algorithm.

$$IoU = \frac{Area\ of\ overlap}{Area\ of\ union} \tag{1}$$

We consider the values bigger than 0.5 for the prediction frame and the target frame's intersection over union (IoU), so $IoU > 0.5$, as the criterion for evaluating target detection.

4.3 Experimental Results

The results (cf. Table 1) demonstrate that our model YOLOv7-ag, has superior performance compared to YOLOv7, as well as YOLOv8, YOLOv8-ag, and FasterR-CNN. This superiority is visible across all the used metrics, including precision, recall, and mean average precision (mAP).

Table 1. Precision and Recall Metrics for All Classes on the Validation Dataset.

Method	Precision	Recall	mAP0.5
FasterR-CNN [5]	9.7%	NM	22.5%
YOLOv7 [28]	97.3%	96.2%	97%
YOLOv7-ag	**99.19%**	**98%**	**99.1%**
YOLOv8 [20]	94.5%	92.8%	96.1%
YOLOv8-ag	93%	93.6%	94.7%

Table 2. Precision Metrics by Class on the Validation Dataset

Method	leaf	seed	root	flower	fruit	stem
FasterR-CNN [5]	26.5%	0%	9.4%	4.7%	7.8%	9.9%
YOLOv7 [20]	99.2%	97.4%	97%	95%	97.5%	97%
YOLOv7-ag	**99.8%**	**98.1%**	**98.9%**	**98.5%**	**99.3%**	**99.5%**
YOLOv8 [28]	95.3%	97.4%	95%	90.1%	92.6%	97%
YOLOv8-ag	95.9%	97.4%	96.1%	92.7%	94.8%	81.5%

Table 2 further details the precision values for different organ classes, underscoring the particularly enhanced performance of YOLOv8-ag. It is also important to note that the integration of the attention-gate mechanism into YOLOv8 contributes to an improvement in its performance metrics as well (cf. Table 2). However, it still inferior to that of YOLOv7-ag. A key strength of YOLOv7-ag is its ability to detect smaller objects, even with varying degrees of occlusion, demonstrating its advanced capabilities in handling complex visual scenarios. When analyzing the precision for different plant organ classes individually,

YOLOv7-ag showed higher values, surpassing other models for every class as detailed in Table 2. The recall rates are also higher for leaves, seeds, roots, flowers, fruits, and stems, recorded at 99.6%, 100%, 100%, 93.9%, 95.9%, and 98.8% respectively.

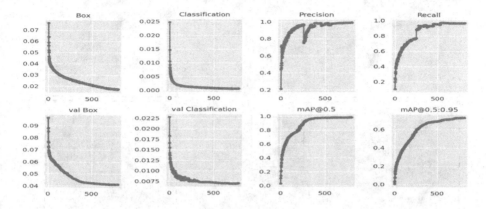

Fig. 5. Plots of box loss, classification loss, precision, recall and mean average precision (mAP) over the training epochs for the training and validation set for YOLOv7-ag.

The Fig. 5 shows various performance measures for both the training and validation set for YOLOv7-ag. The box loss represents how well the model can locate the center of an object and how well the predicted bounding box covers an object. Classification loss gives an idea of how well the model can predict the correct class of a given object. As illustrated in Fig. 5, before plateauing after roughly 500 epochs, the model quickly increased in terms of precision, recall, and mean average precision. The validation data's box, objectness, and classification losses all displayed a sharp decline.

The Fig. 6 shows a comparison of the prediction of the two models YOLOv7 and YOLOv7-ag, as they are the two models with the best performance, on the same test image. YOLOv7-ag can identify more organs, such as leaves, flowers, stems, etc., with greater accuracy than YOLOv7. It is particularly better at detecting small organs. However, both models struggle to identify leaves when there are many overlapping leaves. Additionally, it proves challenging for models to accurately identify stems obscured by leaves in an image. We observed that YOLOv7 faced difficulties in accurately distinguishing between fruits and flowers. The use of attention-gate mechanisms enhances the detection of plant organs in herbarium images by focusing the model's attention on relevant features. It helps in distinguishing between overlapping or closely situated organs, by emphasizing unique characteristics and reducing background noise. It also aid in handling the inherent variability in plant images and provide contextual understanding, crucial for accurately identifying and classifying the plant organs based on their surrounding features. Interestingly, both models have very

Fig. 6. Prediction example (YOLOv7 on the top and YOLOv7-ag on the bottom). 25 leaves (YOLOv7-ag) vs. 18 leaves (YOLOv7) in box A, 26 leaves leaves (YOLOv7-ag) vs. 17 leaves (YOLOv7) in box B, and 5 flowers (YOLOv7-ag) vs. 4 flowers (YOLOv7) in box A, 2 flowers (YOLOv7-ag) vs. 0 flowers (YOLOv7) in box B and 3 stems (YOLOv7-ag) vs. 1 stem (YOLOv7) in box A.

close inference times, with YOLOv7 at approximately 85 ms and YOLOv7-ag at 90 ms, while YOLOv8 and YOLOv8-ag have approximately 22 ms and 26 ms respectively.

5   Conclusion

Plant organ identification poses significant problems for general object detectors due to the tiny target size and the variety of forms and shapes. In response to these issues, we proposed the YOLOv7-ag model which outperformed existing plant organs detectors and achieved 99% precision in identifying plant organs. The accuracy and recall rate of YOLOv7-ag exceeded those of YOLOV7 by 2% and 3%, respectively, in the experiment, and significantly improved the stability of the results. Nevertheless, the detection of plant organs still requires further improvements. The most promising avenues for boosting performance involve refining the data annotation process and modifying the attention block within the original architecture to operate across multiple levels.

Recognizing the bounding boxes assigned to each organ type, in the herbarium images, could allow for targeted datasets for subsequent recognition of organ properties. It is thus a crucial step towards the automatic morphological annotation of millions of herbarium specimens, and future extended research interfaces on collection databases.

References

1. Ariouat, H., et al.: Extracting masks from herbarium specimen images based on object detection and image segmentation techniques. Biodiv. Inf. Sci. Stand. **7**, e112161 (2023)
2. Sahraoui, M., Sklab, Y., Pignal, M., Lebbe, R.V., Guigue, V.: Leveraging Multi-modality for Biodiversity Data: exploring joint representations of species descriptions and specimen images using CLIP. Biodivers. Inf. Sci. Stand. **7**(2023), e112666 (2023)
3. Meredith, L.: Roles of natural history collections. Ann. Mo. Bot. Gard. **4**(83), 536–545 (1996)
4. Raven, P.H.: Saving plants, saving ourselves. Plants People Planet **1**, 8–13 (2019)
5. Younis, S., Schmidt, M., Weiland, C., Dressler, S., Seeger, B., Hickler, T.: Detection and annotation of plant organs from digitised herbarium scans using deep learning. Biodiv. Data J. **8**, e57090 (2020)
6. Besnard, G., et al.: Herbarium-based science in the twenty-first century. Botany Lett. **165**, 323–327 (2018)
7. Soltis, P.: Digitization of herbaria enables novel research. Am. J. Bot. **104**, 1281–1284 (2017)
8. Abdelaziz, T., Bassem, B., Walid, M.: A deep learning-based approach for detecting plant organs from digitized herbarium specimen images. Eco. Inform. **69**, 101590 (2022)
9. Patrick, W., et al.: Large-scale digitization of herbarium specimens: development and usage of an automated, high-throughput conveyor system. Taxon **67**, 165–178 (2018)
10. Wenli, Z., et al.: Deep-learning-based in-field citrus fruit detection and tracking. Horticult. Res. **9**, uhac003 (2022)
11. Jiang, Y., Li, C.: Convolutional neural networks for image-based high-throughput plant phenotyping: a review. Plant Phenom. **9** (2020)
12. Borhani, Y., Khoramdel, J., Najafi, E.: A deep learning based approach for automated plant disease classification using vision transformer. Sci. Rep. **12**, 11554 (2022)
13. Ashish, V., et al.: Attention is all you need. CoRR abs/1706.03762 (2017)
14. Sue Han, L., Chee Seng, C., Simon, J., Paolo, R.: How deep learning extracts and learns leaf features for plant classification. Pattern Recogn. **71**, 1–13 (2017)
15. Mochida, K., et al.: Computer vision-based phenotyping for improvement of plant productivity: a machine learning perspective. GigaScience **8**, giy153 (2018)
16. Shaoqing, R., Kaiming, H., Ross, B.G., Jian, S.: Faster R-CNN: towards real-time object detection with region proposal networks. Computer Vision and Pattern Recognition abs/1506.01497 (2015)

17. Kaiming, H., Georgia, G., Piotr, D., Ross, B.G.: Mask R-CNN. Computer Vision and Pattern Recognition, vol. abs/1703.06870 (2017)
18. Redmon, J., Divvala, S., Girshick, R., Farhadi, A.: You only look once: unified, real-time object detection. In: Proceedings of the IEEE Conference on Computer Vision and Pattern Recognition (2016)
19. Xiao, Z., Lang, J., Shuai, L., Tingting, Z., Xingang, M.: YOLO-SASE: an improved YOLO algorithm for the small targets detection in complex backgrounds. Computer Vision and Pattern Recognition, vol. abs/2207.02696 (2022)
20. Chien-Yao, W., Alexey, B., Hong-Yuan Mark, L.: YOLOv7: trainable bag-of-freebies sets new state-of-the-art for real-time object detectors. Computer Vision and Pattern Recognition, vol. abs/2207.02696 (2022)
21. Ozan, O., et al.: Attention U-Net: learning where to look for the pancreas. Computer Vision and Pattern Recognition, vol. abs/1804.03999 (2018)
22. Lang, P.M., Willems, F., Scheepens, J.F., Burbano, H., Bossdorf, O.: Using herbaria to study global environmental change. New Phytol. **2021**, 110–122 (2019)
23. Zhao, H., Zhang, H., Zhao, Y.: YOLOv7-Sea: object detection of maritime UAV images based on improved YOLOv7. In: Proceedings of the IEEE/CVF Winter Conference on Applications of Computer Vision (WACV) Workshops (2023)
24. Zixiao, Z., et al.: ViT-YOLO: transformer-based YOLO for object detection. In: IEEE/CVF International Conference on Computer Vision Workshops, ICCVW 2021, Montreal, BC, Canada, 11–17 October 2021
25. James, C., et al.: From prototype to inference: a pipeline to apply deep learning in sorghum panicle detection. Plant Phenomics **5**, 0017 (2023)
26. Jie, X., et al.: TrichomeYOLO: a neural network for automatic maize trichome counting. Plant Phenom. **5**, 0024 (2023)
27. Zhaoyang, N., Guoqiang, Z., Hui, Y.: A review on the attention mechanism of deep learning. Plant Neurocomput. **452**, 48–62 (2021)
28. Dillon, R., Jordan, K., Jacqueline, H., Ahmad, D.: Real-time flying object detection with YOLOv8. Computer Vision and Pattern Recognition (2023)

On Dark Knowledge for Distilling Generators

Chi Hong[1], Robert Birke[3], Pin-Yu Chen[4], and Lydia Y. Chen[1,2(✉)]

[1] Delft University of Technology, Delft, Netherlands
c.hong@tudelft.nl, lydiaychen@ieee.org
[2] University of Neuchatel, Neuchatel, Switzerland
[3] University of Torino, Turin, Italy
robert.birke@unito.it
[4] IBM Research, New York, USA
pin-yu.chen@ibm.com

Abstract. Knowledge distillation has been applied on generative models, such as Variational Autoencoder (VAE) and Generative Adversarial Networks (GANs). To distill the knowledge, the synthetic outputs of a teacher generator are used to train a student model. While the dark knowledge, i.e., the probabilistic output, is well explored in distilling classifiers, little is known about the existence of an equivalent dark knowledge for generative models and its extractability. In this paper, we derive the first kind of empirical risk bound for distilling generative models from a Bayesian perspective. Through our analysis, we show the existence of the dark knowledge for generative models, i.e., Bayes probability distribution of a synthetic output from a given input, which achieves lower empirical risk bound than merely using the synthetic output of the generators. Furthermore, we propose a **D**ark **K**nowledge based **D**istillation, DKtill, which trains the student generator based on the (approximate) dark knowledge. Our extensive evaluation on distilling VAE, conditional GANs, and translation GANs on Facades and CelebA datasets show that the FID of student generators trained by DKtill combining dark knowledge are lower than student generators trained only by the synthetic outputs by up to 42.66%, and 78.99%, respectively.

Keywords: Knowledge distillation · Generators · Risk bounds

1 Introduction

Generative models, such as Variational autoencoders (VAE) and generative adversarial networks (GANs), are increasingly applied to synthesize images. To practically deploy those models on edge devices, namely the generators, it is critical to distill/compress the models first due to the memory and computation constraints. Knowledge distillation via teacher and student models can effectively compress the machine learning models, especially for classification models [2, 4, 6]. The student model tries to imitate the (non-compressed) teacher model through the input and output pairs from the teacher model such that the same learning

D.-N. Yang et al. (Eds.): PAKDD 2024, LNAI 14646, pp. 235–247, 2024.
https://doi.org/10.1007/978-981-97-2253-2_19

efficacy can be achieved via a smaller model. Dark knowledge [21] is shown particularly critical for the distillation quality of classifiers. Recently, some related studies [1,3,19,19] focus on distilling the generator of GANs. This prior art empirically distills the teacher generator by directly leveraging the synthetic outputs. While the studies in [13,16,20] provide theoretical analysis for distilling classifiers, e.g., distillation bound, little is on the theoretical understanding of distilling generative models, for instance, the empirical risk bound and the existence of dark knowledge of the generators.

In this paper, we rethink the generator distillation from a theoretical Bayesian perspective. We hypothesize the existence the "dark knowledge", which is embedded in the synthetic output, e.g., image, but not directly observable. To such an end, we model the synthetic outputs of a generator as a conditional Bayes probability distribution, $\mathbb{P}(y|x)$, where y is the synthetic output and x is the input of the generator. We derive two types of empirical risks for distillation: regular empirical risk using only synthetic inputs, and Bayes empirical risk using the proposed conditional probability distribution. We show that the variance of the Bayes empirical risk is lower than the variance of regular empirical risk, demonstrating better generalization capability. We thus refer to the knowledge of $\mathbb{P}(y|x)$ as dark knowledge. Our analysis can be applied on both probabilistic generative models, e.g., VAE, and non-probabilistic models, e.g., GANs. To derive such Bayes empirical risk bound for GANs, we approximate the conditional probability distribution and quantify its impact in the distillation bound.

Motivated by the effectiveness of such dark knowledge, we further propose the **Dark Knowledge Distillation**, DKtill, algorithm to train a student generator model through the (approximate) dark knowledge. Specifically, we try to minimizing the difference between teacher and student outputs by controlling the tensor of the second last layer of their networks, which is an approximated conditional Bayes probability. We extensively evaluate DKtill on distilling VAE, conditional GANs and translational GANs for Facades, and CelebA datasets. Our results show that the student generators from DKtill achieve lower FID than the ones trained on only synthetic images by up to 78.99% and a slightly higher FID than the ones from the teacher model by 17.77%.

Our key contributions for this work can be summarized as follows. *i)* Having Theorem 1 and Proposition 1 as the distillation empirical risk analysis to show the effectiveness of dark knowledge in distilling generative models. *ii)* Deriving Proposition 2 as the approximate distillation empirical risk analysis to capture the impact of approximating dark knowledge in distilling the GANs. *i)* Proposing DKtill which is a dark knowledge based distillation algorithm for training student generative models. *iv)* Achieving higher distillation quality of DKtill on three different generative models and 3 different datasets, compared to distillation algorithms which do not use the proposed (approximated) dark knowledge.

2   Preliminary

We consider general generative models for synthesizing images, including non-probabilistic generators and probabilistic generators. In existing generator distillation works [1,3,19] the basic idea is as follows. First, they provide the same inputs x into a trained teacher model $\mathcal{T}$ and a student model f where x is the generator input. x has different formats in different generative models. For vanilla GANs or VAE, x is a noise vector. For conditional GANs, x is a noise and a conditional label. x can also be a picture in image translation GANs. Second, the distance between the two model outputs $\mathcal{T}(x)$ and $f(x)$ is minimized. In this way, the student f can mimic the input-output mapping of $\mathcal{T}$ and extract the knowledge from the teacher's synthetic image outputs $\{\mathcal{T}(x)\}$.

In classifier distillation [4], the final predicted labels of a teacher are not sufficient to train a student. The logits or the softmax outputs of the teacher model are needed to train the student so as to capture the underlying "dark knowledge", which is the probability the teacher assigns to "wrong" labels. Some basic settings and terms of generator distillation used in this paper are introduced in the following.

Synthetic Images. In generator distillation, we input x into a trained teacher[1] $\mathcal{T}$ and obtain the corresponding synthetic output $y = \mathcal{T}(x)$. After N inferences, a training dataset $D = \{(x_n, y_n)\}_{n=1}^{N} \sim \mathbb{P}^N$ is obtained to train the student model f, where x_n is an input sample and $y_n = \mathcal{T}(x_n)$ is the corresponding synthetic image. To simplify the notation, without loss of generality, y_n is a single channel image and $y_n^{ij} \in \{0, ..., C\}$ is a pixel of the image, where C indicates the number of possible colors[2]. Let $i \in [1, ..., I]$ and $j \in [1, ..., J]$ be the pixel indexes of an image of width I and height J.

Distillation Optimization. To distill the knowledge from the teacher model $\mathcal{T}$, a student generator which can be defined as $f : \mathcal{X} \to \mathbb{R}^{I \times J \times C}$ is trained to minimize the following risk:

$$R(f) = \mathop{\mathbb{E}}_{(x,y) \sim \mathbb{P}} [\ell(y, f(x))]. \tag{1}$$

According to the definition of f, the mapping result $f(x)$ is an $I \times J \times C$ tensor, and $f^{ij}(x) \in \mathbb{R}^C$ corresponds to a pixel. The vector $f^{ij}(x)$ represents the weights of taking different colors in the pixel. For each training pair (x, y) collected from $\mathcal{T}$, the loss of $f(x)$ takes the form $\ell(y, f(x)) = \sum_i \sum_j \ell^{ij}(y^{ij}, f^{ij}(x))$. The term $\ell^{ij}(y^{ij}, f^{ij}(x))$ is the loss value of predicting $f^{ij}(x)$ when the true pixel color is y^{ij}. To do distillation, the student needs to mimic the teacher's outputs. Therefore $\ell^{ij}(y^{ij}, f^{ij}(x))$ should be a loss that minimizes the distance between $y = \mathcal{T}(x)$ and $f(x)$, e.g., the softmax cross-entropy loss.

Distillation Using Only Synthetic Images. Having different definitions of $\ell(y, f(x))$ in $R(f)$ leads to different distillation methods. A common way in most existing works is to define $\ell(y, f(x))$ directly using the synthetic output images $\{y\}$ from the teacher. Given $y = \mathcal{T}(x)$, the basic idea for training the student f is to maximize the element $f^{ij,y^{ij}}(x)$ in the output probabilities $f^{ij}(x)$ of the student for each pixel i, j.

---

[1] We interchangeably use teacher or target model/generator.

[2] The analysis of this paper can be straightforwardly extended to three channel images.

3   Theoretical Analysis of Dark Knowledge in Distilling the Generator

This section introduces the novel concept of dark knowledge in distilling generators and demonstrates that harvesting this dark knowledge can train a student generator f which generalizes better. In generator distillation, the goal is to transfer the knowledge of the teacher $\mathcal{T}$ as much as possible to a student f. We use the risk (generalization error) of the student f shown in Eq. 1 to evaluate the quality of the distilled student model. The risk describes the generalization ability of f on unseen new data sampled from $\mathbb{P}$ other than the observed training dataset $D = \{(x_n, y_n)\}_{n=1}^{N} \sim \mathbb{P}^N$. A lower risk value means having a better distilled f. In the following, we first introduce the concept of dark knowledge in generative models and then derive the distillation empirical risks.

3.1   Dark Knowledge of Generators

From a Bayesian viewpoint, each synthetic image y from the teacher is generated by a conditional distribution $\mathbb{P}(y|x)$ where the distribution $\mathbb{P}$ is potentially determined by the trained teacher generator $\mathcal{T}$. Obviously, the knowledge of $\mathbb{P}(y|x)$ cannot be fully extracted from the teacher's synthetic image output. A synthetic image y is only a sample drawn from the distribution $\mathbb{P}(y|x)$. Thus, in generator distillation we call the underlying distribution $\mathbb{P}(y|x)$ the "dark knowledge" of the teacher $\mathcal{T}$. The next section demonstrates that using the dark knowledge $\mathbb{P}(y|x)$ to define $\ell(y, f(x))$ so as to have a precise empirical estimation of $R(f)$ can facilitate the training of a student in generator distillation.

We consider two ways of distilling teacher generator: i) by solely the synthetic image outputs of $\mathcal{T}$, and ii) by using the underlying dark knowledge $\mathbb{P}(y|x)$ of $\mathcal{T}$. We analyze the generalization error of the student f under both distillation ways by comparing their (empirical) risks. According to the introduction in Sect. 2, in generator distillation, $R(f)$ should be minimized by the algorithm where Eq. 1 shows the general definition of $R(f)$. However, calculating the exact value of $R(f)$ is intractable because $\mathbb{P}$ is unknown or it has no explicit expression. Hence empirical risk definitions are required to approximate $R(f)$. The aforementioned two distillation ways leverage two different corresponding empirical risk definitions. These will be illustrated in detail in the following.

3.2   Distillation Empirical Risk

Distillation with Sole Synthetic Images. First, we introduce the distillation empirical risk of using $\mathcal{T}$'s synthetic image outputs. In this way of distillation, the dataset $D = \{(x_n, y_n)\}_{n=1}^{N}$ collected from the teacher $\mathcal{T}$ is used to train the student f. We have the following distillation empirical risk definition to approximate $R(f)$:

$$\hat{R}(f; D) = \frac{1}{N} \sum_{n \in [N]} \sum_{i} \sum_{j} \ell^{ij}(y_n^{ij}, f^{ij}(x_n)) = \frac{1}{N} \sum_{n \in [N]} \sum_{i} \sum_{j} e_{y_n^{ij}}^{\top} \ell^{ij}(f^{ij}(x_n)), \quad (2)$$

where $e_{y_n^{ij}}$ is the one-hot vector encoding of the color value of y_n^{ij}, and $\ell^{ij}(f^{ij}(x)) = [\ell^{ij}(1, f^{ij}(x)), ..., \ell^{ij}(C, f^{ij}(x))] \in \mathbb{R}^C$ is a vector of loss values for each possible color of the pixel with index i, j. As shown in Eq. 2, this empirical risk definition only needs the synthetic image y_n from $\mathcal{T}$ to decide the value of $e_{y_n^{ij}}$. Then we can calculate $\hat{R}(f; D)$ with no additional information from $\mathcal{T}$.

Distillation with Dark Knowledge. Here we assume that besides the final synthetic image output y we also get the conditional probability $\mathbb{P}(y|x)$ (the dark knowledge) of a given input x from $\mathcal{T}$. Consequently, we can define another distillation empirical risk $\hat{R}_\alpha(f, D)$ to approximate $R(f)$:

$$\hat{R}_\alpha(f, D) = \frac{1}{N} \sum_{n \in [N]} \sum_i \sum_j p^{ij}(x_n)^\top \ell^{ij}((f^{ij}(x_n)) \tag{3}$$

where the conditional probability of pixel y^{ij} color is denoted by

$$p^{ij}(x) = [\mathbb{P}(y^{ij}|x)]_{y_{ij} \in [C]} \tag{4}$$

Connection to the Empirical Risk. $\hat{R}(f; D)$ (Eq. 2) can be seen as an approximation of $\hat{R}_\alpha(f, D)$ (Eq. 3). Considering that: $R(f) = \underset{(x,y) \sim \mathbb{P}}{\mathbb{E}}[\ell(y, f(x))] = \underset{x}{\mathbb{E}}[\underset{y|x}{\mathbb{E}}[\ell(y, f(x))]]$, we know that $\hat{R}_\alpha(f, D)$ is an empirical estimate of $\underset{x}{\mathbb{E}}[\underset{y|x}{\mathbb{E}}[\ell(y, f(x))]]$ over the random variable x. If we further use the one-hot encoding $e_{y_n^{ij}}^\top$ to approximate $p^{ij}(x_n)$ in Eq. 3, we have Eq. 2. To distinguish these two empirical risks, we call $\hat{R}(f; D)$ (Eq. 2) as the *distillation empirical risk with synthetic images* and $\hat{R}_\alpha(f, D)$ (Eq. 3) as the *distillation empirical risk with dark knowledge*.

3.3   Generalization of the Student Generator

Given a teacher $\mathcal{T}$, a student model f can be trained by optimizing the distillation empirical risk with synthetic images (Eq. 2) or the distillation empirical risk with dark knowledge (Eq. 3). In this section, we analyze the generalization ability of the student f under the two different distillation empirical risk bounds. In the following, we show that a student generator trained with Eq. 3 is expected to generalise better than trained under Eq. 2, which means using the dark knowledge can facilitate generator distillation. Although both Eq. 2 and Eq. 3 are unbiased estimates of $R(f)$, the variances of Eq. 2 and Eq. 3 over the observed training dataset D are different. This is formally demonstrated by Theorem 1 (proof in the supplementary).

Theorem 1. *Let D be a training dataset sampled from $\mathbb{P}^N$. $\mathbb{V}$ represents the variance of a random variable. For any fixed hypothesis $f : \mathcal{X} \to \mathbb{R}^{I \times J \times C}$,*

$$\mathbb{V}_{D \sim \mathbb{P}^N}\left[\hat{R}_\alpha(f; D)\right] \leq \mathbb{V}_{D \sim \mathbb{P}^N}[\hat{R}(f; D)]$$

The two variances in Theorem 1 equal to each other when $p^{ij}(x)$ is concentrated on one single color and the probability on all other colors is zero. However, this case rarely happens. The benefit of having lower variance is that the student f generalises better as shown in the following. Applying Theorem 6 of [14], we have the following bound for the distillation empirical risk with dark knowledge.

Proposition 1. Let $D \sim \mathbb{P}^N$ and $\mathcal{F}$ be a class of hypotheses $f : \mathcal{X} \to \mathbb{R}^{I \times J \times C}$ with induced class $\mathcal{H} \subset [0,1]^{\mathcal{X}}$ of $h(x) = \sum_i \sum_j p^{ij}(x)^\top \ell^{ij}((f^{ij}(x))$. Suppose $\mathcal{H}$ has uniform covering number $\mathcal{N}_\infty$. For any $\delta \in (0,1)$ and set $\mathcal{M}_N = \mathcal{N}_\infty(1/N, \mathcal{H}, 2N)$. Then with probability at least $1 - \delta$ over D we have:

$$R(f) \le \hat{R}_\alpha(f; D) + \mathcal{O}\left(\sqrt{\mathbb{V}_N(f)/N} \cdot \sqrt{\log{(\mathcal{M}_N/\delta)}} + \log{(\mathcal{M}_N/\delta)}/N\right)$$

where $\mathbb{V}_N(f)$ is the empirical variance of $\{h(x_n)\}_{n=1}^N$.

Note that using the same procedure we can derive a similar bound for the distillation empirical risk with synthetic images (Eq. 2). Considering Theorem 1 and the two bounds, we know that the bound for Eq. 3 has lower variance penalty. Increasing the training dataset size N (number of queries on $\mathcal{T}$), Eq. 3 has a risk bound with a better rate of convergence. Thus, a student model trained by minimizing the distillation empirical risk with dark knowledge generalises better compared to one using the distillation empirical risk with synthetic images. We conclude that a student generator f trained by the dark knowledge $\mathbb{P}(y|x)$ is better than using sole synthetic image outputs $\{y\}$ from $\mathcal{T}$.

3.4   Impact of Probability Approximation

In the aforementioned analysis, we showed that having the dark knowledge $\mathbb{P}(y|x)$ from the teacher $\mathcal{T}$ allows for a more precise empirical approximation of $R(f)$. Thus it benefits the training of the student f in generator distillation. However, as mentioned before, doing distillation with dark knowledge requires the conditional probability $\mathbb{P}(y|x)$ for any given input x. In ideal cases, the intermediate layer output of $\mathcal{T}$, e.g., the second last layer output in VAE, can provide $\mathbb{P}(y|x)$. Then, it can be used to train the student model. Unfortunately, for some teacher models $\mathcal{T}$, e.g., GANs, the intermediate layer output cannot directly show $\mathbb{P}(y|x)$. We refer to the generators that can provide $\mathbb{P}(y|x)$ in the intermediate layer output as *probabilistic generators* and the generators that cannot show the probability as *non-probabilistic generators*. To distill the dark knowledge from non-probabilistic generators, it is required to approximate $\mathbb{P}(y|x)$. Let $\tilde{\mathbb{P}}(y|x)$ be an approximated probability of $\mathbb{P}(y|x)$. In this section, we study the impact on the student generalization error using such an approximated distribution to do distillation. Using the approximated probability to train the student f, referring to Eq. 3 we have the following distillation empirical risk:

$$\tilde{R}_\alpha(f, D) = \frac{1}{N} \sum_{n \in [N]} \sum_i \sum_j \tilde{p}^{ij}(x_n)^\top \ell^{ij}((f^{ij}(x_n)), \tag{5}$$

where $\tilde{p}^{ij}(x_n)$ is the approximation of $p^{ij}(x_n)$. According to Eq. 1 and Eq. 4 we can rewrite the population risk $R(f)$ as:

$$R(f) = \mathop{\mathbb{E}}_{x}[\mathop{\mathbb{E}}_{y|x}[\ell(y, f(x))]] = \mathop{\mathbb{E}}_{x}[\sum_i \sum_j p^{ij}(x)^\top \ell^{ij}(f^{ij}(x))]. \qquad (6)$$

The following Proposition 2 (proof in the supplementary) reveals the connection between the approximation of $\mathbb{P}(y|x)$ and the generalization error of f.

Proposition 2. *If the loss ℓ is bounded, we train the student model by minimizing the empirical risk shown in Eq. 5. For any hypothesis $f : \mathcal{X} \to \mathbb{R}^{I \times J \times C}$,*

$$\mathbb{E}\left[(\tilde{R}_\alpha(f; D) - R(f))^2\right] \leq \frac{1}{N} \cdot \mathbb{V}[\sum_i \sum_j \tilde{p}^{ij}(x)^\top \ell^{ij}((f^{ij}(x))] +$$
$$\mathcal{O}(\mathbb{E}[\sum_i \sum_j \left\| \tilde{p}^{ij}(x) - p^{ij}(x) \right\|_2])^2. \qquad (7)$$

On the right side of Eq. (7), when N is big, the second term dominates the upper bound of the gap $\tilde{R}(f; D) - R(f)$. That means minimizing the distance between the approximated probability $\tilde{\mathbb{P}}(y|x)$ and the ground truth probability $\mathbb{P}(y|x)$ yields a tighter upper bound of the risk gap $\tilde{R}(f; D) - R(f)$. Hence, a tighter approximation leads to a better student model.

4   DKtill: **Extracting Dark Knowledge for Training Student Generator**

In the previous section, we theoretically demonstrate that using the underlying dark knowledge of a teacher $\mathcal{T}$ can improve the generator distillation and train a student f that generalizes better. However, to make use of the underlying dark knowledge in distillation, we first need to know how to extract and use the dark knowledge from a teacher generator $\mathcal{T}$. In this section, we propose two methods to extract the dark knowledge based on the class of the generator: one for probabilistic generators and one for non-probabilistic generators. For verifying the effectiveness of the extracted dark knowledge, we propose a generator distillation algorithm DKtill. When distilling $\mathcal{T}$, DKtill makes the student f to mimic the extracted (approximated) $\mathbb{P}(y|x)$ besides the synthetic images $y = \mathcal{T}(x)$ (see the supplementary for more implementation details).

4.1   Extracting from Probabilistic Generators

Probabilistic generators, e.g., variational auto-encoder (VAE) [9], commonly assume the existence of latent variables. More in detail, they assume that a synthetic image y is generated by some random process involving some latent random variables. In such generators, to do distillation with dark knowledge, the conditional probability $\mathbb{P}(y|x)$ can be calculated by some middle layer outputs of the teacher generator. In the following, we take VAE as an example for dark knowledge extraction in probabilistic generators. The training method of VAE is a kind of variational inference that uses $q(x|y)$ to approximate the intractable true posterior $p(x|y)$ and maximize the evidence lower bound. For VAE distillation, we focus on its trained decoder. In the decoder, the synthetic image is produced by the conditional distribution $\mathbb{P}(y|x) = \mathcal{N}(y; \mu, \sigma^2 I)$, where the distribution parameters μ and σ are the middle layer outputs of the decoder neural network. Thus, given any input x, we can obtain $\mathbb{P}(y|x)$ by getting the output value of μ and σ from the middle layers.

4.2 Extracting from Non-probabilistic Generators

Non-probabilistic generators, e.g., GANs, do not assume any probability relationship between output image y and latent random variables. y is directly produced by a neural network mapping $\mathcal{T}(x)$, where $\mathcal{T}$ is the teacher generator. Different from probabilistic generators, in non-probabilistic generators we cannot derive the knowledge of $\mathbb{P}(y|x)$ by some middle layer outputs of $\mathcal{T}$. In the following, we take GANs as an example to show extracting dark knowledge $\mathbb{P}(y|x)$ from non-probabilistic generators. Given a GANs teacher generator $\mathcal{T}$, we let g_α be the second last layer and g_β be the last layer of $\mathcal{T}$. Given an input x into the teacher generator, the intermediate output of g_α is represented by α. Then the corresponding synthetic image can be represented as $y = g_\beta(\alpha)$. To get the dark knowledge, we can apply a differentiable probabilistic network (e.g., MLP) g_γ to replace the original last layer g_β. The input of g_γ is g_α's output, α and the output of g_γ is a tensor that takes the shape corresponding to the underlying Bayes probabilities. The shape of $g_\gamma(\alpha)$ is decided by I, J and C where I and J are decided by the synthetic image size and C depends on the color range. The final synthetic image is now sampled from the tensor $g_\gamma(\alpha)$. If $\mathcal{T}$ is untrained, we can train it from scratch using this new architecture. Thus, after the training, we can easily get $\mathbb{P}(y|x)$ from $g_\gamma(\alpha)$. If the given $\mathcal{T}$ is pre-trained, we can just fine-tune the parameters of g_γ using the optimization objective, $\text{argmin}_{g_\gamma} ||g_\beta(\alpha) - \hat{y}||_1$, where $\hat{y} \sim Discrete(g_\gamma(\alpha))$ is the synthetic image sampled from $g_\gamma(\alpha)$.

5   Empirical Illustration

5.1   Setting

Datasets: We consider the datasets, Facades [5] and CelebA [12]. In the case of conditional generation, we focus on the gender class, i.e., male and female, in the CelebA dataset. We downsampled the CelebA images from originally 178×218 to 64×64 pixels. The input images size of Facades is 256×256.

Networks: The network structures of all generators are based on convolutional neural networks [9,17]. For VAE, we compress a 27.2 MB teacher into a 2.4 MB student. As for conditional GANs, a 2.8 MB student is distilled from a 14.4 MB teacher. In the image translation experiment, Pix2pix is compressed from 217.8 MB into 14.0 MB.

Distillation Process and Baseline: For both DKtill and the baseline (abbreviated as "image" [3]), we train the student generators, by minimizing the distance of the generated images to the teacher from the same random inputs (details in Appendix). However, instead of only using generated images (baseline), DKtill also uses in parallel the information of dark knowledge (underlying distribution) for loss minimization. Note that although the baseline distillation is originally designed for Image Translation, we adopt it for VAE and conditional GANs.

Evaluation Metrics: we use Fréchet Inception Distance (FID) to evaluate the quality of the images produced by the distilled student model. Lower values of FID indicate higher quality of generated data.

5.2 Distilling Probabilistic Generators

Here, we distill VAE (the teacher generator) using the baseline and DKtill. Figure 1 shows the comparison results based on FID for VAE and conditional GANs (Fig. 1a). We also illustrate the FID of images generated by the teacher VAE, i.e., the horizontal line in Fig. 1a. The FID value decreases during the training, from 30.61 to 28.76, and 22.61 to 16.49, for

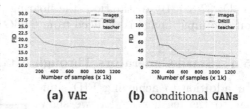

(a) VAE **(b) conditional GANs**

Fig. 1. FID of distilling VAE and conditional GANs on CelebA.

baseline and DKtill respectively, demonstrating lower distance to ground truth images. When looking at the final FID (after 128K examples), DKtill is 42.66% lower than the baseline image and 62.14% higher than the teacher generator. Here FID for the teacher VAE is 10.17, which is much better than both students. This is within our expectation as the teacher model is 10X the size of the students. In general, we can see that DKtill, using the dark knowledge provided by the intermediate output of teacher VAE, can significantly improve the quality of distillation. As the number of examples increases, the student learns the teacher's mapping and distribution better. Using dark knowledge can achieve lower FID and thus it can distill better.

5.3 Distilling Non-probabilistic Generators

Here we implement two tasks for non-probabilistic generators: the conditional GANs and image translation as the baseline [3]. Figure 1b evaluates the distillation on conditional GANs.We can see that FID shares the same trend as distilling the probabilistic generator VAE. Thus, our proposed method

Fig. 2. Pix2pix image translation GANs on Facades.

DKtill is able to effectively extract knowledge from non-probabilistic teacher generators too, given the approximate nature of dark knowledge. The reason why FID of DKtill gets much closer to the teacher than Fig. 1a is that the teacher model is only 5X bigger in parameter size than the students. When looking at the final FID (after 128K examples), DKtill is 78.99% lower than the baseline image and only 17.77% higher than the teacher generator.

Let us zoom into the distillation results of the image translation task in Fig. 2. The goal is to train the translation GANs so that given input (as the "input" in visualization), the Pix2pix network approximately maps it into the ground truth image. From the results, we observe that even with 6% size of the teacher model (FID: 35), DKtill is able to distill the image translation generator at high quality (FID: 35) with dark knowledge, especially compared with the baseline (FID: 37.88). By distillation, the floating point operations per second (FLOPs) on one input image is reduced from $6.06G$ (Teacher) to $0.83G$ (Student).

5.4   Small Generators Through DKtill

In addition to achieving FID similar to the teacher model, we show another advantage of exploring dark knowledge, i.e., smaller student generator. We first solicit the synthetic images generated by the following model in Fig. 3: the ground truth, teacher VAE, 27.2 MB student VAE trained by baseline, 27.2 MB student VAE from DKtill, and 2.4 MB student VAE from DKtill. Without any surprise, DKtill achieves better distillation quality when using bigger networks, comparing Fig. 3d and Fig. 3e. Another observation is that, smaller student (2.4 MB) VAE trained by DKtill, achieve better image quality than baseline with 27.2 MB, comparing Fig. 3c and Fig. 3e.

The conditional GANs result is also presented in Fig. 4 (on gender class, i.e., male and female for the left 3 and right 3 pictures). Note that here we do not show the corresponding real images of the synthetic ones since conditional GANs do not have a latent code encoder as VAE. Thus, having the latent code of a real image for reproducing some corresponding synthetic ones is difficult. Given the same small network with 2.8 MB for DKtill

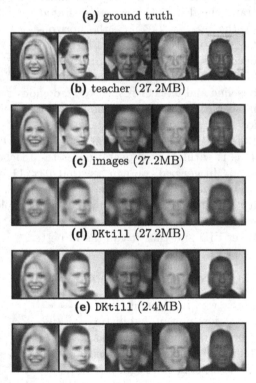

(a) ground truth

(b) teacher (27.2MB)

(c) images (27.2MB)

(d) DKtill (27.2MB)

(e) DKtill (2.4MB)

Fig. 3. VAE on CelebA.

and the baseline, `DKtill` shows better generated image quality than baseline. These results again prove the existence of dark knowledge for generators and its benefit for distilling generators.

6 Related Work

Knowledge distillation [2,4,6,8,11], which transfers the knowledge from a big network to a small one, enables the light-weight deep learning. Generally, knowledge distillation is studied on classifiers for model compression, e.g., [4] compresses the knowledge of an ensemble into a single model which is much easier to deploy. Such techniques are used to train surrogate models [10,18,22], even without the necessity of knowing the target model parameters.

Besides, knowledge distillation has also been applied on `GANs` [1, 3,19,19]. KDGAN [19] simultaneously optimizes the distillation and adversarial losses between a classifier, a teacher, and a discriminator, to learn the true data distribution at the equilibrium. To further improve the performance, [3] include a student discriminator to measure the distances between real data, and the synthetic data generated by student and teacher generators. Recently, there are theoretical analyses on knowledge distillation for classifiers [7,13,15,16,20,21].

(a) teacher (14.4MB)

(b) images (2.8MB)

(c) `DKtill` (2.8MB)

Fig. 4. Conditional `GANs` on `CelebA`.

On the one hand, [15] first provides the theoretical analysis of self-distillation, fitting a nonlinear function on Hilbert space and L2 regularization. This analysis sheds light on the relation between self-distillation rounds and (under-)over- fitting. On the other hand, based on neural tangent kernel, [7] provides a transfer risk bound for the linearized model of the wide neural networks, revealing the impact of soft (hard) labels for (im)perfect teachers according to the designed data inefficiency metric. Furthermore, [21] explores the bias-variance trade-off brought by distillation with soft labels. According to their analysis, novel weighted soft labels are inspired to help the network adaptively handle the trade-off. However, none of the existing work provides a generalization analysis on generators.

7 Conclusion

In this paper, we model the knowledge distillation for generative models from a Bayesian perspective, identifying dark knowledge and its influence on the generalization ability of student models, i.e., lower empirical risk. Furthermore, we propose a dark knowledge based distillation optimization, DKtill, to train student generators on both non-probability-based and probability-based generative models. Evaluation results on three datasets across different scenarios show that synthetic images from DKtill achieve lower FID by up to 78.99%, in contrast to using images only. DKtill also generates images of similar quality as the teacher model, using smaller and more compact generator networks.

Acknowledgements. This work has been supported by the Spoke "FutureHPC & BigData" of the ICSC - Centro Nazionale di Ricerca in "High Performance Computing, Big Data and Quantum Computing", funded by EU - NextGenerationEU and the EPI project funded by EuroHPC JU under G.A. 101036168.

References

1. Aguinaldo, A., Chiang, P., Gain, A., Patil, A., Pearson, K., Feizi, S.: Compressing GANs using knowledge distillation. CoRR **abs/1902.00159** (2019)
2. Chandrasekaran, V., Chaudhuri, K., Giacomelli, I., Jha, S., Yan, S.: Exploring connections between active learning and model extraction. In: USENIX Security (2020)
3. Chen, H., et al.: Distilling portable generative adversarial networks for image translation. In: AAAI (2020)
4. Hinton, G.E., Vinyals, O., Dean, J.: Distilling the knowledge in a neural network. CoRR abs/1503.02531 (2015)
5. Isola, P., Zhu, J.Y., Zhou, T., Efros, A.A.: Image-to-image translation with conditional adversarial networks. In: CVPR, pp. 1125–1134 (2017)
6. Jagielski, M., Carlini, N., Berthelot, D., Kurakin, A., Papernot, N.: High accuracy and high fidelity extraction of neural networks. In: USENIX Security (2020)
7. Ji, G., Zhu, Z.: Knowledge distillation in wide neural networks: risk bound, data efficiency and imperfect teacher. In: NeurIPS 2020 (2020)
8. Kanwal, N., Eftestøl, T., Khoraminia, F., Zuiverloon, T.C., Engan, K.: Vision transformers for small histological datasets learned through knowledge distillation. In: PAKDD (2023)
9. Kingma, D.P., Welling, M.: Auto-encoding variational Bayes. In: Bengio, Y., LeCun, Y. (eds.) ICLR (2014)
10. Krishna, K., Tomar, G.S., Parikh, A.P., Papernot, N., Iyyer, M.: Thieves on sesame street! Model extraction of BERT-based APIs. In: ICLR (2020)
11. Liu, Z., Zhu, Y., Gao, Z., Sheng, X., Xu, L.: Itrievalkd: an iterative retrieval framework assisted with knowledge distillation for noisy text-to-image retrieval. In: PAKDD (2023)
12. Liu, Z., Luo, P., Wang, X., Tang, X.: Deep learning face attributes in the wild. In: (ICCV) (2015)
13. Lopez-Paz, D., Bottou, L., Schölkopf, B., Vapnik, V.: Unifying distillation and privileged information. In: Bengio, Y., LeCun, Y. (eds.) ICLR (2016)
14. Maurer, A., Pontil, M.: Empirical Bernstein bounds and sample variance penalization. In: COLT 2009 - The 22nd Conference on Learning Theory (2009)
15. Mobahi, H., Farajtabar, M., Bartlett, P.L.: Self-distillation amplifies regularization in Hilbert space. In: NeurIPS (2020)
16. Phuong, M., Lampert, C.H.: Towards understanding knowledge distillation. CoRR abs/2105.13093 (2021)

17. Pu, Y., Gan, Z., Henao, R., Yuan, X., Li, C., Stevens, A., Carin, L.: Variational autoencoder for deep learning of images, labels and captions. In: NIPS **29** (2016)
18. Truong, J., Maini, P., Walls, R.J., Papernot, N.: Data-free model extraction. In: CVPR (2021)
19. Wang, X., Zhang, R., Sun, Y., Qi, J.: KDGAN: knowledge distillation with generative adversarial networks. In: NeurIPS, pp. 783–794 (2018)
20. Zhang, Z., Sabuncu, M.R.: Self-distillation as instance-specific label smoothing. In: NeurIPS (2020)
21. Zhou, H., et al.: Rethinking soft labels for knowledge distillation: a bias-variance tradeoff perspective. In: 9th International Conference on Learning Representations, ICLR 2021 (2021)
22. Zhou, M., Wu, J., Liu, Y., Liu, S., Zhu, C.: Dast: data-free substitute training for adversarial attacks. In: CVPR, pp. 231–240. IEEE (2020)

RPH-PGD: Randomly Projected Hessian for Perturbed Gradient Descent

Chi-Chang Li, Jay Huang, Wing-Kai Hon📧, and Che-Rung Lee(📧)📧

National Tsing Hua University, Hsinchu, Taiwan
{wkhon,cherung}@cs.nthu.edu.tw

Abstract. The perturbed gradient descent (PGD) method, which adds random noises in the search directions, has been widely used in solving large-scale optimization problems, owing to its capability to escape from saddle points. However, it is inefficient sometimes for two reasons. First, the random noises may not point to a descent direction, so PGD may still stagnate around saddle points. Second, the size of random noises, which is controlled by the radius of the perturbation ball, may not be properly configured, so the convergence is slow. In this paper, we proposed a method, called RPH-PGD (Randomly Projected Hessian for Perturbed Gradient Descent), to improve the performance of PGD. The randomly projected Hessian (RPH) is created by projecting the Hessian matrix into a relatively small subspace which contains rich information about the eigenvectors of the original Hessian matrix. RPH-PGD utilizes the eigenvalues and eigenvectors of the randomly projected Hessian to identify the negative curvatures and uses the matrix itself to estimate the changes of Hessian matrices, which is necessary information for dynamically adjusting the radius during the computation. In addition, RPH-PGD employs the finite difference method to approximate the product of the Hessian and vectors, instead of constructing the Hessian explicitly. The amortized analysis shows the time complexity of RPH-PGD is only slightly higher than that of PGD. The experimental results show RPH-PGD does not only converge faster than PGD, but also converges in cases that PGD cannot.

Keywords: Optimization · Gradient Descent · Projected Hessian

1 Introduction

The gradient descent is an iterative method that models the objective function as a hyperplane and searches along the steepest descent direction from one point to another. Although it is a kind of greedy strategy, and can only converge to a local minimum for non-convex problems, empirically it performs well and has been widely used in solving large-scale optimization problems. Because of its simplicity and effectiveness, many commonly used methods for machine learning or deep learning, such as stochastic gradient descent (SGD) [14], AdaGrad [6], and Adam [11], are based on its principle and extend with different variations.

D.-N. Yang et al. (Eds.): PAKDD 2024, LNAI 14646, pp. 248–259, 2024.
https://doi.org/10.1007/978-981-97-2253-2_20

However, gradient descent based methods can mis-converge to some non-optimal points, called saddle points, which appear often in high-dimensional and non-convex optimization problems. A saddle point is a first-order stationary point, whose gradient is equal to zero, but the eigenvalues of its Hessian are mixed of different signs, which means it is not an extremum. Theoretically, if the gradient descent method converges to a saddle point, it will get stuck and cannot escape from the saddle point.

Several methods [1,3,4] have been proposed to solve the saddle point problem for gradient descent. The major challenge is that the proposed method needs to keep the simplicity and effectiveness of gradient descent. Perturbed Gradient Descent (PGD) method [7] that adds noises in the search directions is one of the promising solutions to escape from the saddle points. Theoretically, Chi Jin et al. [10] have shown that the perturbed gradient descent can converge to local minima in a number of iterations that is almost dimension-free.

Nevertheless, PGD is inefficient sometimes for two reasons. First, the added random noises may not point to a descent direction, so it cannot escape from the saddle points efficiently. Second, both the analysis and algorithms of PGD rely on the understanding of the local geometry of underlying functions, which can be characterized by the eigenvalues of the Hessian matrix. One of the critical hyper-parameters is the radius r of perturbation ball [10], from which the perturbations are sampled. If r is too small, it cannot escape from saddle points; if r is too large, it may not converge. Although the analysis suggests some proper values of the radius r [10], the computation of r is too expensive to perform in practice. Moreover, the value of radius r is changing during the computation as the gradient descent moves to different regions of the function. Finding one constant value of r for the entire optimization process is mostly impossible to satisfy different situations.

The major challenge is how to obtain a good estimation of the Hessian without paying a high computation cost. The exact computation of Hessian matrix takes $O(Tn^2)$ time and $O(n^2)$ space, where n is the number of variables and T is the time for function evaluation. The cost of eigenvalue decomposition is $O(n^3)$. For large-scale problems, such time/space complexity is too high to use for real-world applications. A linear time algorithm is usually required for solving large-scale problems.

In this paper, we present an algorithm, called RPH-PGD (Randomly Projected Hessian for PGD), to improve the convergence of PGD. The projected Hessian is a much smaller matrix compared to the original Hessian. It is created by projecting the Hessian into a small subspace, which may contain rich information about the eigenvectors of the matrix. From the RPH, one can compute the approximate eigenvalues/eigenvectors which can be used to explore the negative curvatures, the descent directions at the saddle points. The RPH can also be used to adjust the radius r of the perturbation ball, which is a hyper-parameter related to the changes of Hessian matrices. In addition, RPH-PGD employs the finite difference method to approximate the product of the Hessian and vectors, instead of constructing the Hessian explicitly. The experimental results show

RPH-PGD does not only converge faster than PGD, but also converges in cases that PGD cannot. The amortized analysis shows the time complexity of RPH-PGD is only slightly higher than that of PGD.

2 Preliminary

2.1 Notation

The notation, definitions, and assumptions of the paper are summarized here. Most of them are the same as those in [8,10]. We use uppercase letters for matrices and lowercase letters for vectors and scalars. We use $\| \cdot \|$ to denote the Euclidean norm for vectors and matrices, and $\| \cdot \|_F$ for Frobenius norm for matrices. For a matrix A, we use A^* to represent A's conjugate transpose.

Let the objective function be f. The gradient is ∇f and the Hessian matrix is $\nabla^2 f$. We will use g as an approximation to ∇f and use H for $\nabla^2 f$. We use I for identity matrix and o for zero vector. For example, $\mathcal{N}(o, I)$ means the normal distribution with zero mean and identity covariance.

In this paper, we assume the optimization problem to solve is a minimization problem. We assume the Hessian matrix in problems is symmetric/Hermitian, which is true for most of the cases. Moreover, we assume the function $f : \mathbb{R}^n \rightarrow \mathbb{R}$ is twice-differentiable. The function f is called ℓ -gradient Lipschitz smooth :

$$\forall x_1, x_2 \in \mathbb{R}^n, \|\nabla f(x_1) - \nabla f(x_2)\| \leq \ell \|x_1 - x_2\| \tag{1}$$

Moreover, f is ρ-Hessian Lipschitz smooth if

$$\forall x_1, x_2 \in \mathbb{R}^n, \|\nabla^2 f(x_1) - \nabla^2 f(x_2)\| \leq \rho \|x_1 - x_2\| \tag{2}$$

For the domain, we also define some commonly used terms. First, $x \in \mathbb{R}^n$ is called an ϵ-first-order stationary point if $\|\nabla f(x)\| \leq \epsilon$. Second, for a stationary point x, we called x is a δ -approximate local minimum if

$$\exists \delta > 0 : f(x) \leq f(y), \forall y : \|y - x\| \leq \delta; \tag{3}$$

x is a δ -approximate local maximum:

$$\exists \delta > 0 : f(x) \geq f(y), \forall y : \|y - x\| \leq \delta; \tag{4}$$

Otherwise, x is a saddle point.

Last are the definitions of the search directions. At a function point x, a vector d points to a *descent direction* if

$$d^T \nabla f(x) < 0. \tag{5}$$

Furthermore, d is in the direction of *negative curvature* if

$$d^T \nabla^2 f(x)d < 0. \tag{6}$$

For a saddle point, because $\nabla f(x) = 0$, there is no descent direction. However, because $\nabla^2 f(x)$ is indefinite, which means its eigenvalues are mixed of different signs, the eigenvectors corresponding to the negative eigenvalues, as well as their linear combinations, are in the directions of negative curvature.

2.2 Methods to Escape from Saddle Points

There are many studies proposing methods to solve the saddle point problems. In [7], authors showed that stochastic gradient descent (SGD) can converge to a local minimum even with exponentially many local minima and saddle points. In [12], authors indicate the normalized gradient descent with noises can even converge faster. Chi Jin [10] et al. gave a comprehensive study on the convergence of PGD.

Many algorithms also leverage the second-order derivative to escape the saddle points. In [1,4], the authors used the method of approximating the Hessian-vector product. The approximation is based on the first-order Taylor expanding for multi-variable functions,

$$\nabla f(x + \delta d) \approx \nabla f(x) + \delta \nabla^2 f(x)d,$$

where ∇f is the gradient, $\nabla^2 f$ is the Hessian, d is a vector, and δ is a small non-zero scalar. Therefore, the Hessian vector product can be approximated by

$$\nabla^2 f(x)d \approx \frac{\nabla f(x + \delta d) - \nabla f(x)}{\delta}. \tag{7}$$

The same result can also be obtained from the finite difference method. Equation (7) shows that the computational cost of estimating the Hessian vector product is merely two gradient evaluations, which is as cheap as computing stochastic gradients. [2]

2.3 Perturbed Gradient Descent

In [10], Chi Jin et al. provide an analysis of the perturbed gradient descent method. The key idea is to regard all possible positions of x after perturbation as a ball. If the radius r is chosen correctly, it can be proven that the area where x gets stuck is very small compared to the whole sphere. Thus, it is of high probability to escape from saddle points.

Algorithm 1 sketches the perturbed gradient method (PGD), in which d is the dimension of the objective function. The perturbation ξ is sampled from a zero-mean Gaussian with covariance matrix $(r^2/n)I$ so that the expectation of the perturbation equals to r, $\mathbb{E}\|\xi\|^2 = r^2$. The hyper-parameter r is called the radius of the perturbation ball. In [10], authors show with a properly configured radius, the convergence of PGD can be as fast as gradient descent. The suggested formula is

$$r \leftarrow \epsilon\sqrt{R} \tag{8}$$

Input: initial guess x_0, step size η, dimension n, perturbation radius r.
Output: a local minimum x^*;
for $t = 0, 1, ...\tau - 1$ **do**
     if x_t *is trapped into a saddle point* **then**
          sample ξ_t from $\mathcal{N}(o, (r^2/n)I)$
          $x_{t+1} \leftarrow x_t - \eta\xi_t$
     end
     $x_{t+1} \leftarrow x_t - \eta\nabla f(x_t)$.
end
return x_τ;

Algorithm 1: Perturbed Gradient Descent (PGD)

where

$$R = 1 + \min\left(1/\epsilon^2 + \ell/\sqrt{\rho\epsilon}, n/\epsilon^2\right). \tag{9}$$

The value ϵ is the tolerance; The variable ℓ is used to define the ℓ-gradient Lipschitz smooth, as show in (1); ρ is for the ρ-Hessian Lipschitz smooth, defined in (2); and n is the dimension.

The perturbation only needs to be added to the search direction when x is trapped to a saddle point. In [10], authors evaluate this condition using a heuristic, which is if $\|\nabla f\| \leq \epsilon$ for some number of iterations t_{inter}, then x might get stuck into a saddle point. However, such strategy may fail sometimes because the added random noise may not point to a descent direction. PGD may wander around the saddle point and cause stagnation.

The advantage of this method is that it is simple and cheap, so it can be combined with any other algorithm. The disadvantage is that the radius r is an additional hyper-parameter, whose theoretical value is defined in (8) and (9). In practice, it is not easy to choose a proper r. Moreover, the radius is changing as the method moves to different regions of the function. So giving a constant number of r cannot fit all the needs during the computation.

3 Algorithms

In this section, we first introduce the algorithm for computing the randomly projected Hessian (RPH), followed by the main algorithm of RPH-PGD.

3.1 Randomly Projected Hessian

Let Q be an orthogonal basis for subspace, and A be a Hessian matrix. The projected Hessian is in the form Q^*AQ, which is also called the Rayleigh quotient. The projected Hessian has been widely used in the constrained optimization problem [13] to avoid the steps moving into the tangent cone of constraints. The idea is to project the Hessian matrix into the null space spanning by the gradients of active constraints. If the subspace is of full column rank, the projected Hessian can still keep its definiteness. Moreover, the dimension of the projected Hessian will be reduced to the dimension of the subspace.

Input: A function g to compute ∇f;
   the dimension of subspace k.
Output: H: projected Hessian;
   Q: orthogonal basis for the generated subspace.
Generate an $n \times k$ Gaussian random matrix Ω;
Approximate $W \leftarrow \nabla^2 f \Omega$ using g and (7);
Compute the QR decomposition of W, $QR \leftarrow W$;
Compute $H \leftarrow R(Q^*\Omega)^{-1}$;
return H, Q
   Algorithm 2: Randomly Projected Hessian (RPH)

Here we want to generate a projected Hessian that has similar spectrum as the original Hessian. Therefore, the subspace must contain good approximations to the eigenvectors of the original Hessian. Therefore, we will leverage the randomized eigen-decomposition (RED) [8] to construct the subspace and the projected Hessian.

Algorithm 2 shows the procedure of computing the randomly projected Hessian. As can be seen, it is similar to the RED, except in two places. First, for matrix-matrix multiplication, RPH uses the finite difference method, as shown in (7), to approximate the matrix-vector multiplication. Hessian matrix is the second-order derivative of the objective function. Therefore, without forming the Hessian matrix explicitly, one can use the directional derivative of the gradient to approximate the Hessian-vector multiplication.

Second, the procedure does not compute the eigenvalue and eigenvector approximations. It stops at the generation of the projected Hessian. In fact, the small matrix H is not exactly a projected Hessian. Let $A = \nabla^2 f$ be the Hessian matrix. From Step 4 in Algorithm 2, if R and A are non-singular, and $W = A\Omega$, one has

$$H = R(Q^*\Omega)^{-1} = R(Q^*A^{-1}W)^{-1} = R(Q^*A^{-1}QR)^{-1} = (Q^*A^{-1}Q)^{-1}. \quad (10)$$

Thus, H is the inverse of the Rayleigh quotient of A^{-1}, not A. Theoretically, such Rayleigh quotient produces better approximations to the eigenvalues of small magnitude [9], because the small eigenvalues in A are usually clustered and they will be well separated in A^{-1}. The well separation of eigenvalues implies better conditions for computing eigenvectors. In addition, the matrix Q, whose column vectors form an orthogonal basis for the product $A\Omega$, should contain good approximations to the eigenvectors of A. Combining those two factors, H is a good surrogate of the Hessian in terms of eigenvalues/eigenvectors.

From (7), one can see the computation cost equals one more gradient evaluation, $\nabla f(x + \delta d)$, because $\nabla f(x)$ is already available. Since there are k Hessian vector multiplications, the total computational cost of Step 2 is $O(kT)$, where T is the time for evaluating gradients. The time complexity of Algorithm 2 is

$$O(kT + nk^2 + k^3). \quad (11)$$

If $n \gg k$ and $T = O(n)$, the time complexity of Algorithm 2 is $O(n)$.

Input: A function g to compute ∇f;
        the dimension of subspace k.
        maximum eigenvalue in the previous iteration λ'.
Output: H: projected Hessian of $(A - \lambda' I)$;
          Q: orthogonal basis for the generated subspace.
Generate an $n \times k$ Gaussian random matrix Ω;
Approximate $W \leftarrow (\nabla^2 f - \lambda' I)\Omega$ using g and (7);
Compute the QR decomposition of W, $QR \leftarrow W$;
Compute $H \leftarrow R(Q^* \Omega)^{-1}$;
return H, Q
     Algorithm 3: Shifted Randomly Projected Hessian (SRPH)

3.2   Shifted Randomly Projected Hessian

A small subspace will make computation much faster, but the descent direction may not be sampled every time the projected Hessian is calculated. Algorithm 3 shows a solution that shifts the diagonal elements of ∇f by a constant λ'. It can be seen that Algorithm 2 is a special case of Algorithm 3, in which $\lambda' = 0$. By using Algorithm 3, we can ensure that the vectors in W contains better approximations to the eigenvectors of A's smallest eigenvalues, since now their magnitude becomes larger. Note H in Algorithm 3 is the inverse of the projected Hessian to $(A - \lambda' I)^{-1}$, so when computing the eigenvalues of A, we need to add λ' back.

3.3   RPH-PGD

Algorithm 4 gives a high-level description of the RPH-PGD, which is based on the PGD method (Algorithm 1). The major difference between them is when the original PGD sticks in the saddle points for a long time, a new procedure (Step 6 to Step 20) will be triggered. It first computes the shifted randomly projected Hessian (Algorithm 3), and uses it to escape the saddle points more efficiently.

There are two major mechanisms to assist PGD to escape the saddle points. First, RPH-PGD computes its eigenvalue decomposition $H = V \Lambda V^*$, and find the largest eigenvalue $\lambda_{\max}$, the smallest eigenvalue $\lambda_{\min}$, and the eigenvector $v_{\min}$ corresponding $\lambda_{\min}$.

$$H v_{\min} = \lambda_{min} v_{min}.$$

Since we only need the smallest and the largest eigenvalue approximation, the dimension of the subspace k can be small, such as 5 or 10. As a result, the exact eigenvalue decomposition of H is cheap to compute.

To obtain the direction of negative curvature, we need to project $v_{\min}$ back to the space of n. One common way is using the Ritz vector,

$$u_{\min} = Q v_{\min},$$

where Q is the orthogonal basis of generated by RPH. Based on [9], if Q is close to the eigen-space of $\nabla^2 f$, the Ritz vector will give a good eigenvector

Input: x_0: initial guess; τ: number of steps;
   r: initial radius; η: step size; ϵ: tolerance;
   t_{inter}: time interval;
   t_{rp}: random project interval.
Output: minimum point x^*
for $t = 0, 1, \ldots, \tau - 1$ **do**
 if $\|\nabla f(x_t)\| \leq \epsilon$ *more than* t_{inter} *iterations* **then**
  sample ξ_t from $\mathcal{N}(o, (r^2/n)I)$
  $x_{t+1} \leftarrow x_t - \eta \xi_t$
 end
 $\lambda' = 0$
 if $\|\nabla f(x_t)\| \leq \epsilon$ *more than* t_{rp} *iterations* **then**
  Compute the shifted randomly projected Hessian H
  Compute $\lambda_{\max}$, $\lambda_{\min}$, and $u_{\min}$ of H
  Update $\lambda_{\max} \leftarrow \lambda_{\max} + \lambda'$, $\lambda_{\min} \leftarrow \lambda_{\min} + \lambda'$, $\lambda' \leftarrow \lambda_{\max}$
  if $\lambda_{\min} > \beta$ **then**
   return x_t
  end
  if $\lambda_{\min} < 0$ **then**
   $x_{t+1} \leftarrow x_t + \lambda_{\max}^{-1} u_{\min}$
  end
  Update the radius r using H and ∇f
 end
 $x_{t+1} \leftarrow x_t - \eta \nabla f(x_t)$.
end
return x_τ

Algorithm 4: Randomly Projected Hessian for PGD (RPH-PGD)

approximation to the Hessian. Then, the largest and the smallest eigenvalues need to add back the value of the last largest eigenvalue since H is shifted (Algorithm 3).

Step 13 and Step 14 show that if $\lambda_{\min}$ is large enough ($\lambda_{\min} > \beta > 0$), which means the Hessian is positive definite, the algorithm reaches an optimal solution, so RPH-PGD stops earlier.

If $\lambda_{\min}$ is negative (Step 16 and Step 17), it means H is indefinite, or the Hessian is indefinite. We can move along the direction of negative curvature, along which the function value can be decreased. Suppose x_0 is a saddle point. The Taylor expansion of f around x_0 is

$$f(x_0 + d) \approx f(x_0) + d^* \nabla f(x_0) + \frac{1}{2} d^* \nabla^2 f(x_0) d.$$

Since x_0 is a saddle point, $\nabla f(x_0) = 0$. Therefore,

$$f(x_0 + d) \approx f(x_0) + \frac{1}{2} d^* \nabla^2 f(x_0) d. \tag{12}$$

If d is a direction in negative curvature, which means $d^* \nabla^2 f(x_0) d < 0$, the function value at $x_0 + d$ is less than that at x_0.

The computed $u_{\min}$ is a unit vector by default, so we need to decide a step length. The rule of thumb is that if the function changes rapidly, we should not move too far. The norm of Hessian matrix, which is $\lambda_{\max}$ here, can help to determine the sensitivity of the functions. If $\lambda_{\max}$ is large, the step length should be small. Therefore, we use $\lambda_{\max}^{-1}$ as the step length.

The second method to help PGD is the dynamic update of the radius of the perturbation ball r. The analysis of PGD [10] shows when the choice of r satisfies (8), the efficiency of PGD can be guaranteed. However, the values of ℓ and ρ are intrinsic for the given function f, and are hardly to know in advance. Here we use the changes of gradient and the projected Hessian to estimate the values of ℓ and ρ. Suppose that $x_i \neq x_j$.

$$\ell \leftarrow \frac{\|\nabla f(x_i) - \nabla f(x_j)\|}{\|x_i - x_j\|}, \text{ and } \rho \leftarrow \frac{\|H_i - H_j\|}{\|x_i - x_j\|} \tag{13}$$

where i, j are indices of iterations. Since H is small, the computation of (2) is simple and fast. They are local measurements of the parameter ℓ and ρ. Based on the same proof, if r can be set based on which, the efficiency of PGD can be guaranteed locally.

The remaining problem is how good the estimation is. Let A_i and A_j be the Hessian matrices in the ith iteration and the jth iteration respectively, and H_i and H_j are their randomly projected Hessian matrices. Their relation can be characterized as

$$A_i - Q_i H_i Q_i^* = E_i, \quad A_j - Q_j H_j Q_j^* = E_j$$

where Q_i and Q_j are the orthogonal bases for the randomly generated subspaces, and E_i and E_j represent their differences. Here we only show a special case that $Q_i = Q_j$.

$$\|H_i - H_j\|_F = \|Q_i H_i Q_i^* - Q_j H_j Q_j^*\|_F = \le \|A_i - A_j\|_F + \|E_i\|_F + \|E_j\|_F$$

The last line is by triangle inequality for norm. Therefore,

$$\|H_i - H_j\|_2 \le \|H_i - H_j\|_F \le \|A_i - A_j\|_F + 2\|E\|_F,$$

where $\|E\|_F = \max\{\|E_i\|_F, \|E_j\|_F\}$. The expectation of $\|E\|_F$ is

$$\mathbb{E}\|E\|_F \le \mathbb{E}(\sqrt{n}\|E\|_2) < \left[\sqrt{n} + \frac{4n\sqrt{k}}{k-2}\right]\sigma_2.$$

The difference between $\|H_i - H_j\|$ and $\|A_i - A_j\|$ may not be too small, because the additional term $\|E\|$ can be large. However, empirically, this method works well.

The time complexity of the additional steps is bounded by the computation of randomly projected Hessian,

$$O(kT + nk^2 + k^3),$$

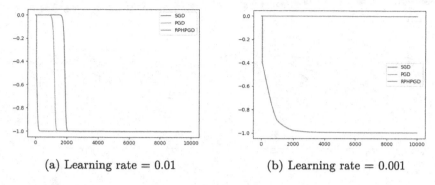

(a) Learning rate = 0.01   (b) Learning rate = 0.001

Fig. 1. Convergence of SGD, PGD, and RPH-PGD for function f defined in (14). The x-axis is the number of iterations, and the y-axis is the function value. (a) shows the convergence of three methods for learning rate = 0.01; (b) shows the convergence of three methods for learning rate = 0.001.

seeing the analysis of (11). However, they do not perform often, because t_{rp} is larger than t_{inter}. Suppose that it is performed every k iteration. The amortized time complexity is

$$O(T + nk + k^2).$$

Since the time of computing gradient T dominates the others, the time complexity of RPH-PGD is the same as that of PGD.

4   Experiments

In this section, we evaluate the effectiveness of RPH-PGD, and compare it with other PGD and SGD. We run our experiments on a machine with i5-6300HQ CPU, 8GB 2133MHz RAM, and Windows 10. We used Python 3.9.13 for coding.

The test problem in experiments is

$$f(x_1, x_2) = \frac{1}{16}x_1^4 - \frac{1}{2}x_1^2 + \frac{9}{8}x_2^2 \tag{14}$$

which has a saddle point at $(0, 0)$. The minimum value of this problem is at $(2, 0)$ and $(-2, 0)$, whose value is -1. This is a good benchmark problem for evaluating the ability of optimization methods to escape saddle points [15]. The initial point is at $(0, 0)$.

The convergence of three methods is shown in Fig. 1. We used two learning rates 0.01 and 0.001 to evaluate their convergence, as shown in subfigure (a) and subfigure (b) respectively. As can be seen, when the learning is large, such as 0.01, all three methods can still converge. Although the convergence rates (their slope) of three methods are similar, the convergence of RPH-PGD is much faster than that of SDG and PGD. The major reason is RPH-PGD starts to converge in the beginning, and the convergence of PGD starts much slower than that of

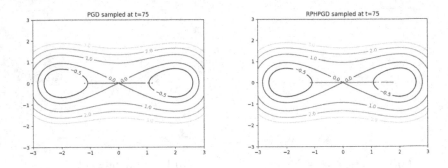

Fig. 2. The traces of PGD (left) and RPH-PGD (right) for function f defined in (14).

RPH-PGD, and the convergence of SDG starts even slower than that of PGD. For SGD, it does not have the ability to detect and avoid the saddle points, and the gradient near $(0,0)$ is close to 0, so SGD can only move slowly from the saddle point. For PGD, although it can detect the saddle points, and uses perturbation to escape them, its strategy is not strong enough owing to the lack of accurate information about the Hessian matrix.

We can further examine the convergence of PGD and RPH-PGD by comparing their traces in the function contour, as shown in Fig. 2. They show the traces at iteration 75. As can be seen, the traces of PGD show PGD moves slower than RPH-PGD toward to the minimization points.

When the learning rate is small, such as 0.001, SGD and PGD stagnate at the original point. Although the convergence of RPH-PGD also becomes slower, it reaches the minimum points eventually.

5   Conclusion and Future Work

In this paper, we present a new algorithm, called RPH-PGD to improve the convergence of PGD. It uses randomly projected Hessian to provide information about the second-order derivative, which is critical to reveal the properties of local geometry. When PGD fails, RPH-PGD employs the eigenvalues/eigenvectors to explore the negative curvature and utilizes the projected Hessian to adjust the radius of perturbation ball. Experiments show that RPH-PGD can significantly improve the convergence of PGD.

There are several future directions to explore. First, the current implementation of RPH-PGD can work only on some toy problems, as the ones shown in Sect. 4. To run it for some large-scale problems, we need to have a better implementation. Also, we will try it on more real- world applications to demonstrate its power. Second, the randomized projected Hessian algorithm is easier to be parallelized than subspace methods, because the subspace generation and orthogonalization can be done in parallel, such as using TSQR [5]. We want to explore those directions to further sharpen the implementation. Last, although

RPH-PGD performs well empirically, we want to investigate it convergence and stability analytically, which would build a more solid foundation for future directions.

References

1. Agarwal, N., Allen-Zhu, Z., Bullins, B., Hazan, E., Ma, T.: Finding approximate local minima faster than gradient descent. In: Proceedings of the 49th Annual ACM SIGACT Symposium on Theory of Computing, pp. 1195–1199. STOC 2017, Association for Computing Machinery, New York, NY, USA (2017)
2. Allen-Zhu, Z.: Natasha 2: faster non-convex optimization than SGD. In: Bengio, S., Wallach, H., Larochelle, H., Grauman, K., Cesa-Bianchi, N., Garnett, R. (eds.) Advances in Neural Information Processing Systems, vol. 31. Curran Associates, Inc. (2018)
3. Allen-Zhu, Z., Li, Y.: NEON2: finding local minima via first-order Oracles. In: Bengio, S., Wallach, H., Larochelle, H., Grauman, K., Cesa-Bianchi, N., Garnett, R. (eds.) Advances in Neural Information Processing Systems, vol. 31. Curran Associates, Inc. (2018)
4. Carmon, Y., Duchi, J.C., Hinder, O., Sidford, A.: Accelerated methods for non-convex optimization. SIAM J. Optim. 28(2), 1751–1772 (2018)
5. Demmel, J., Grigori, L., Hoemmen, M., Langou, J.: Communication-optimal parallel and sequential QR and LU factorizations. SIAM J. Sci. Comput. 34(1), A206–A239 (2012)
6. Duchi, J., Hazan, E., Singer, Y.: Adaptive subgradient methods for online learning and stochastic optimization. J. Mach. Learn. Res. 12(61), 2121–2159 (2011)
7. Ge, R., Huang, F., Jin, C., Yuan, Y.: Escaping from saddle points: online stochastic gradient for tensor decomposition. J. Mach. Learn. Res. 40(2015)
8. Halko, N., Martinsson, P.G., Tropp, J.A.: Finding structure with randomness: probabilistic algorithms for constructing approximate matrix decompositions. SIAM Rev. 53(2), 217–288 (2011)
9. Jia, Z., Stewart, G.W.: An analysis of the Rayleigh-Ritz method for approximating Eigenspaces. Math. Comput. 70, 637–647 (2001)
10. Jin, C., Netrapalli, P., Ge, R., Kakade, S.M., Jordan, M.I.: On nonconvex optimization for machine learning: gradients, stochasticity, and saddle points. J. ACM 68(2), 1–29 (2021). https://doi.org/10.1145/3418526
11. Kingma, D.P., Ba, J.: Adam: a method for stochastic optimization (2014). https://doi.org/10.48550/ARXIV.1412.6980, https://arxiv.org/abs/1412.6980
12. Levy, K.Y.: The power of normalization: faster evasion of saddle points. CoRR abs/1611.04831 (2016). http://arxiv.org/abs/1611.04831
13. Nocedal, J., Wright, S.J.: Numerical Optimization, 2e edn. Springer, New York (2006). https://doi.org/10.1007/978-0-387-40065-5
14. Robbins, H., Monro, S.: A stochastic approximation method. Ann. Math. Stat. 22(3), 400–407 (1951). https://doi.org/10.1214/aoms/1177729586
15. Zhang, C., Li, T.: Escape saddle points by a simple gradient-descent based algorithm. In: Ranzato, M., Beygelzimer, A., Dauphin, Y., Liang, P., Vaughan, J.W. (eds.) Advances in Neural Information Processing Systems, vol. 34, pp. 8545–8556. Curran Associates, Inc. (2021)

Transformer based Multitask Learning
for Image Captioning and Object Detection

Debolena Basak[✉], P. K. Srijith, and Maunendra Sankar Desarkar

Indian Institute of Technology Hyderabad, Sangareddy, India
ai20resch11003@iith.ac.in, {srijith,maunendra}@cse.iith.ac.in

Abstract. In several real-world scenarios like autonomous navigation and mobility, to obtain a better visual understanding of the surroundings, image captioning and object detection play a crucial role. This work introduces a novel multitask learning framework that combines image captioning and object detection into a joint model. We propose **TICOD**, **T**ransformer-based **I**mage **C**aptioning and **O**bject **D**etection model for jointly training both tasks by combining the losses obtained from image captioning and object detection networks. By leveraging joint training, the model benefits from the complementary information shared between the two tasks, leading to improved performance for image captioning. Our approach utilizes a transformer-based architecture that enables end-to-end network integration for image captioning and object detection and performs both tasks jointly. We evaluate the effectiveness of our approach through comprehensive experiments on the MS-COCO dataset. Our model outperforms the baselines from image captioning literature by achieving a 3.65% improvement in BERTScore.

Keywords: Transformer · Multitask Learning · Image Captioning · Object Detection

1 Introduction

In autonomous navigation and human mobility assistance systems, image captioning and object detection play a vital role in obtaining a better visual understanding of the surroundings. Applications such as human mobility assistance systems would require the detection of objects including their positions and description of the environment in natural language to help in the mobility of visually impaired people.

Most image captioning models follow the encoder-decoder framework along with the visual attention mechanism. The encoder processes input images and encodes them into fixed-length feature vectors, and the decoder utilizes these encoded image features to generate word-by-word descriptions [1, 23, 29, 34]. Most mainstream image captioning models [1, 5] use a two-step training method, which firstly extracts image regional features using a pre-trained object detector like Faster R-CNN [26], then feeds these feature vectors into an encoder-decoder framework for image captioning. However, this approach has a few inherent shortcomings: 1) the object detection model in the first step is trained on specific visual datasets like the Visual Genome, and the visual

D.-N. Yang et al. (Eds.): PAKDD 2024, LNAI 14646, pp. 260–272, 2024.
https://doi.org/10.1007/978-981-97-2253-2_21

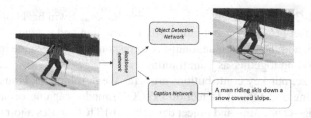

Fig. 1. A high-level framework of our proposed method. Our model TICOD has three major components, which we call as—(a) the *backbone network*, (b) the *object detection network*, and (c) the *caption network*.

representation is not optimized towards the image captioning dataset used (commonly MS-COCO). This may lead to an error propagation problem if the object detector fails to recognize certain important visual information [21], 2) the time-intensive nature of extracting region features causes state-of-the-art models to rely on cached visual features (usually pre-computed) for training and evaluation, imposing constraints on model designs and resulting in run-time inference inefficiencies during prediction [13,31,33]. This introduces a two-stage process where a pre-trained object detection model and an image captioning model are used in sequence to extract features and predict the captions. The learning typically updates the image captioning model. This impedes the end-to-end training from image pixels to descriptions in image captioning models, limiting their applicability in real-world scenarios [13,31].

Inspired by the NLP domain, the transformer architecture has shown its potential in computer vision (CV) tasks [7,20] and multimodal tasks [25]. Considering the drawbacks of pre-trained CNN and object detector in the encoder and advantages of vision transformers, we integrate the task of image captioning as a single-stage approach, which can also perform object detection parallelly, as shown in Fig. 1. The key idea is *multitask learning across object detection and image captioning* that enables the model to develop a better representation learning capability. This shared representation learning enables the model to leverage the knowledge gained from each task to effectively align the backbone representations, enhancing the overall learning capacity. The model parameters are learned by optimizing a joint loss that combines the losses from both tasks. A key advantage of this approach is that if we want to generate a caption, we can simply enable the captioning network, turning off the object detection network, which helps us to get output without introducing any additional latency. On the other hand, if we need more detailed information, we can enable the detection network to provide us with objects' details along with the captions simultaneously. Our model's generated captions and detected objects can be utilized to generate synthetic data using LLMs like GPT-4. For instance, a recent work [18] has used COCO captions and detection annotations to feed into *text-only* GPT-4 to generate *multimodal instruction-following data*, which includes *conversation*, *detailed description* and *complex reasoning* data. This synthetic data has been used for training the multimodal chatbot LLaVA [18].

We use Swin Transformer [20] as the backbone network for extracting image features. We use GPT2 [24] to decode the image features extracted from the Swin transformer and generate the captions of the corresponding image. For the object detection part, Cascade R-CNN [2] framework is used with the same Swin backbone. We use the MS-COCO dataset [16,32] for bounding box information and category labels to train object detection, and the caption annotations corresponding to the same image are used for image captioning together as joint training.

To summarize, our key contributions are – (a) We developed a Transformer-based Image Captioning and Object Detection (**TICOD**) model capable of simultaneously performing image captioning and object detection, (b) TICOD uses a joint loss function that combines the losses from object detection and caption networks while maintaining a trade-off between them, (c) We demonstrate that our multitask method outperforms the task-specific models on image captioning by 3.65% in BERTScore [35] and produces comparable performance in object detection.

2   Related Work

Image Captioning: Most of the existing image captioning works can be broadly categorized into CNN-RNN based models [1,29,34] and CNN-Transformer based models [5,21,23]. M^2 Transformer [5] used a mesh-like connection between each of the encoder and the decoder blocks to extract features from all levels of the encoder. It used Faster R-CNN [26] pre-trained on the Visual Genome dataset. The disadvantages of such an approach have been already discussed. After Vision Transformer (ViT) [7] and its variants [20,25,27] became popular in CV tasks, people began to explore it for image captioning as well. ClipCap [22] performs image captioning using a CLIP [25] encoder and GPT2 [24] decoder. Oscar [14] and VinVL [36] use BERT [6] but provide additional object tags for supervision, which limits their practical applicability in real-world scenarios. These approaches [14,36] are constrained to datasets that provide access to object detectors or annotations. *PureT* [31] used Swin Transformer [20] as a backbone network to extract image features for image captioning. They keep the Swin backbone pre-trained weights frozen in their experiments. However, we train end-to-end to leverage the information obtained from the captions to influence the Swin backbone weights.

Object Detection and Transformer-Based Vision Backbones: Object detection research has seen a breakthrough with the introduction of CNN-based models like R-CNN [10], Fast R-CNN [9], and Faster R-CNN [26]. Later, the remarkable success of the Transformer architecture [28] in the NLP domain has motivated researchers to explore its application in computer vision. Transformers were first introduced for vision problems in Vision Transformers (ViT) [7]. Several works on ViT and its variants followed up [11,27,30]. However, ViT required large-scale training datasets like JFT-300M to perform optimally. DeiT [27] addresses this limitation by introducing training strategies that enable ViT to work effectively with smaller datasets like ImageNet-1K. While ViT shows promising results in image classification, it is not suitable as a general-purpose backbone network for dense vision tasks or high-resolution input images [20]. In parallel to Swin Transformer [20], other researchers have also modified the ViT architecture to improve image classification [11]. But Swin Transformer [20] demonstrates that they achieve the best speed-accuracy trade-off among the above-mentioned methods on image classification, though it focuses on a general-purpose backbone. Also,

Swin Transformer has linear complexity to image size unlike [30] with quadratic complexity.

Multitask Learning: Early works on multitask image captioning such as [8] incorporate a simple multi-label classification as an auxiliary task for image captioning. [37] expands on this by introducing two auxiliary tasks: multi-label classification and syntax generation. Both papers highlight the benefits of auxiliary tasks in enhancing the performance of image captioning and use CNN-LSTM as an encoder-decoder. In the proposed approach, we use a more complex object detection as the auxiliary task and use Swin Transformer in our architecture. Recently, there has been a growing interest in developing models which can solve multiple tasks together. For instance, Pix2seq-v2 [4] proposes a prompt-based approach in an encoder-decoder framework based on Transformers to solve four tasks, namely, object detection, instance segmentation, keypoint detection, and image captioning.

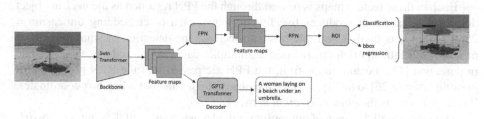

Fig. 2. Architectural overview of the proposed **T**ransformer-based **I**mage **C**aptioning and **O**bject Detection (**TICOD**) model.

3 Proposed Method

This work aims to leverage multitask learning to train objection detection and image captioning tasks jointly. The overall architecture of our model is shown in Fig. 2. The complete architecture has three major components: (a) an initial image feature extractor, (b) an object identification network, and (c) a caption generation network. As the name suggests, the image feature extractor extracts features from the input image. These features are then passed to the two networks for object detection and caption generation. Swin Transformer [20] is used for the image feature extractor owing to its superior performance in various image understanding tasks [15, 20, 31].

The object detection network involves passing the extracted image features through a sequence of Feature Pyramid Network (FPN) [17], Region Proposal Network (RPN) [26], and Region of Interest (RoI) pooling layer [9] to get the objects and their bounding boxes detected. It can be seen that this flow of encoding the image representation to finally detecting the objects follows the overall framework of Faster R-CNN [26]. This specific instantiation of Faster R-CNN uses Swin Transformer as the backbone to extract the image feature maps. On the other hand, the caption generation network passes the extracted image features through a GPT-2 [24] architecture to get the final captions. These two networks operate in parallel to solve the tasks of object detection and caption generation and the losses for the tasks are combined into a multitask loss. The use of a common feature extractor for both the tasks, and a combined loss function to guide the training process enables the two tasks to influence the learning of each other and improve their individual performance.

As shown in Fig. 2, our model consists of a common encoder (i.e., image feature extractor) and two parallel decoders (i.e., object identification network and caption generation network), trained as a single model with a joint loss function. Given an input image, it is first divided into image patches with patch size $= 4 \times 4$. It is then passed through a Swin backbone, producing four feature maps of the following dimensions:

$$\left\{ \left(bs \times C \times \frac{H}{4} \times \frac{W}{4}\right), \left(bs \times 2C \times \frac{H}{8} \times \frac{W}{8}\right), \left(bs \times 4C \times \frac{H}{16} \times \frac{W}{16}\right), \left(bs \times 8C \times \frac{H}{32} \times \frac{W}{32}\right) \right\}$$

where, bs denotes the batch size, C denotes the channel dimensions, i.e., the patch embedding dimension in each Swin block, and H, W denote the height and width of the image, respectively. We call this as the *backbone network*.

Each of these feature maps is passed through the FPN regarded as the *neck* in object detection literature. It produces five feature maps, each of embedding dimension = 256 [17]. FPN is used at multiple levels to increase the detection of small objects. It provides semantically rich information at multiple scales, which reduces the error rate in detection [17]. Feature maps from the FPN are passed into the RPN [26] and RoI pooling layer [9,26] to finally produce the object classes and bounding box coordinates. We refer to this as the *object detection network*.

The other parallel branch of our multitask model, which we call the *caption network*, consists of a GPT2 [24] Transformer as the decoder. It takes the last feature map of the Swin backbone as input along with a <start> token and generates captions word by word in an auto-regressive manner.

Attention: We use Multi-head Self Attention (MSA) [28] in the decoder to calculate the relationship between tokens in a sequence and cross-attention for the relationship between the tokens and image grid features. We also adopt Window MSA/Shifted Window MSA (W-MSA/SW-MSA) proposed in the Swin Transformer work [20]. They are used in the encoder to model the relationship between the image patches.

3.1 Objective Function

For training the caption network, we use the standard language modeling loss, which is the Cross-Entropy loss ($L^C(\theta, \phi)$) computed over all the samples. For a given sample, the loss function aims to predict the next token given the context and can be mathematically formulated as follows:

$$L^C(\theta, \phi) = -\sum_{t=1}^{T} log(p(y_t^* \mid y_{1:t-1}^*, x)) \tag{1}$$

where θ and ϕ represent the parameters of Swin Transformer backbone and GPT2 respectively, $y_{1:T}^*$ is the target ground-truth sequence, and x is the input image.

For objection detection, we consider a loss ($L^O(\theta, \psi)$) which consists of the Cross-Entropy Loss for object classification and smooth $L1$ Loss [9] for bounding box regression computed over all the samples. Here. ψ represents the parameters of the Faster

R-CNN/Cascade R-CNN network used for object detection. Since the *object detection network* consists of the Region Proposal Network (RPN) and Region of Interest (RoI) pooling layer, L^O can be further subdivided as -

$$L^O(\theta, \psi) = L^{RPN}_{cls} + L^{RPN}_{reg} + L^{RoI}_{cls} + L^{RoI}_{reg} \tag{2}$$

where L^{RPN} and L^{RoI} denote the losses from RPN and RoI layers respectively. L^{RPN}_{cls} and L^{RPN}_{reg} denote classification and bounding box (bbox) regression losses from the RPN network. L^{RoI}_{cls} and L^{RoI}_{reg} refer to the classification loss and bbox regression loss from the RoI network. L^{RPN}_{cls} is a classification log loss over two classes—object or non-object, i.e., *background* class [26]. L^{RPN}_{reg} is a Smooth L_1 Loss defined in [9] arising from the difference between the ground-truth bbox coordinates and the predicted bbox coordinates [26]. L^{RoI}_{cls} is the log loss over $(C + 1)$ categories—C object classes and a *background* class, and L^{RoI}_{reg} is again a similar Smooth L_1 Loss over the bbox coordinates [9].

For training our model. We use the joint multitask learning objective function, which combine the image captioning and object detection losses as

$$L^T(\theta, \phi, \psi) = L^O(\theta, \psi) + \lambda \cdot L^C(\theta, \phi) \tag{3}$$

where λ is the weightage to be given to the captioning loss. The λ value is chosen to maximize the evaluation scores of both image captioning and object identification. It is determined empirically using a validation data as demonstrated in Table 4. We found the most suitable values of λ to be 0.1 and 0.2 for TICOD-small and TICOD-large, respectively.

4   Experimental Setup

Dataset: We use the MS-COCO 2017 dataset [16,32] containing 118K training and 5K validation images. COCO has five captions per image and separate annotations for object categories and bounding boxes. For comparison with image captioning works, we follow the standard "Karpathy" split.

Evaluation Metrics: *Image Captioning*–We evaluate our captioning performance using the standard metrics used in Natural Language Processing, viz, BLEU, CIDEr, METEOR, ROUGE-L, and SPICE. The metrics above only evaluate the generated captions by matching them with the reference captions at the lexical level. Since these metrics try to find exact matches, the scores might not be a true measure of the quality of the captions, as the presence of synonymous words may lower the scores. So, we also measure the scores based on BERTScore [35], which uses contextual embeddings to find a semantic similarity measure between the candidate sentence and the reference sentence. BERTScore has been shown to exhibit a superior correlation with human judgments [38], and provide strong model selection performance [35].

Object Detection–We use the standard evaluation metrics—mAP as the mean of APs@[.5 : .05 : .95], AP@IoU = 0.50, and AP@IoU = 0.75. We also report the APs across scales, i.e., APs@small, medium, and large objects.

Implementation Details: We keep the same settings as in Swin Transformer [20] work. AdamW optimizer is used with an initial learning rate of 10^{-4}, weight decay of 0.05, and batch size of 2 per GPU. For training, we use 4 Nvidia V100 GPUs and for inference, we use a single V100 GPU. During inference, captions are generated using beam search with beam size $= 5$. For the image captioning part, we take the encoding from the last feature map of the Swin backbone. The last feature map of Swin-T and Swin-B models have embedding dimensions of $8 \times 96 = 768$ and $8 \times 128 = 1024$, respectively [20], which matches the embedding dimensions of GPT2-small ($dim = 768$) and GPT2-medium ($dim = 1024$). The combined loss from the object detection and the caption networks is backpropagated to update the model weights of the backbone network.

Architecture Details: The authors of Swin Transformer [20] have proposed several variants of the Swin backbone: Swin - tiny, small, base, and large. For our experiments, we have used –

- Swin-tiny [20] with GPT2-small [24] for image captioning and Swin-tiny backbone in Faster R-CNN framework [26] for object detection. We call this *TICOD-small*.
- Swin-base [20] with GPT2-medium [24] for image captioning and Swin-base backbone in Cascade R-CNN [2] for object detection. We call this as *TICOD-large*.

Our model TICOD-large performs better than the other variant. The Swin Transformer acts as a common backbone for image captioning and object detection in our multitask model.

5 Results

5.1 Comparison and Analysis

We compare our multitask model with task-specific baselines from the literature on image captioning and object detection. For image captioning, we compare our work with some recent models like – ClipCap [22], Meshed-memory ($\mathcal{M}^2$) Transformer [5], and PureT [31] on the MS-COCO [32] offline test split. We also compare our work with Pix2seq-V2 [4], a recently proposed multitask system that jointly learns object detection and image captioning along with two other vision tasks. Table 1 reports the performances of these models and our proposed model.

Table 1. Comparison on MS-COCO [16,32] dataset.

Methods	B1	B2	B3	B4	Meteor	RougeL	CIDEr	Spice	BERTScore	mAP	AP$_{50}$	AP$_{75}$	AP$_S$	AP$_M$	AP$_L$
ClipCap (MLP+GPT2 fine-tuning) [22]	70.9	54.4	41.2	31.5	27.7	54.8	106.7	20.5	68.001	–	–	–	–	–	–
ClipCap (Transformer) [22]	74.6	58.5	44.5	33.5	27.6	55.9	112.8	21.0	68.326	–	–	–	–	–	–
$\mathcal{M}^2$- Transformer [5]	80.8	–	–	39.1	29.2	58.6	131.2	22.6	64.556	–	–	–	–	–	–
PureT [31]	**82.1**	**67.3**	**53.0**	**40.9**	**30.2**	**60.1**	**138.1**	**24.2**	68.303	–	–	–	–	–	–
BUTD [1]	77.2	–	–	36.2	27.0	56.4	113.5	20.3	–	–	–	–	–	–	–
Pix2Seq-V2 [4]	–	–	–	34.9	–	–	–	–	–	46.5	–	–	–	–	–
Swin-B (Cascade R-CNN) [20]	–	–	–	–	–	–	–	–	–	51.9	**70.9**	56.5	35.4	55.2	**67.3**
DETR-R101 [3]	–	–	–	–	–	–	–	–	–	43.5	63.8	46.4	21.9	48.0	61.8
Faster R-CNN R101-FPN [26]	–	–	–	–	–	–	–	–	–	42.0	62.5	45.9	25.2	45.6	54.6
TICOD-large (Ours)	75.6	59.0	45.5	35.3	28.3	56.7	115.3	21.1	**70.794**	**52.1**	70.6	**56.7**	**34.8**	**55.3**	67.2

From Table 1, it can be seen that the PureT [31] model outperforms all other models in terms of the BLEU (B1 to B4), Meteor, RougeL, CIDEr, and Spice metrics. However, these are all lexical similarity based metrics, and try to match the exact words present in the generated and ground-truth captions. Although a good score in terms of these metrics indicate good match between the generated and the reference captions, a lower score does not necessarily indicate a poor match. This is because any concept or thought can be expressed in multiple ways with very less overlap in the words used in these *parallel expressions*. This drawback is addressed in the BERTScore metric [35], where the semantic embeddings of the texts are compared to decide the performance score. Our model achieves comparable performances with PureT and $\mathcal{M}^2$ transformer in terms of the lexical overlap based scores. At the same time, it achieves **3.65%** and **9.66%** improvements in BERTScore over these two models respectively, indicating that the proposed model can generate better quality captions at a semantic level. The proposed TICOD-large model also outperforms both model variants of ClipCap [22] in terms of all the metrics as well as BERTScore. The mentioned BERTScores in this table are calculated using Deberta-xlarge-mnli model [12] as it best correlates with human evaluation [35]. Our model also outperforms the popular BUTD model [1] in terms of CIDEr, Meteor, RougeL, Spice.

In addition, we also calculate BERTScore with other models like – Roberta-large [19] and Deberta-xlarge-mnli with idf (Inverse Document Frequency). BERTScore calculated using these three models for the baselines and the proposed multitask model are illustrated in Table 2. Clearly, our proposed multitask model out-performs all the baselines in terms of BERTScore calculated using all three mentioned methods. Comparable scores based on lexical overlap-based metrics and superior scores based on embedding-based methods indicate that the proposed model can generate good-quality captions for the input images.

Table 2. Comparison of captioning performance of our proposed multitask model with some image captioning baselines from literature, in terms of BERTScore [35].

Methods	Deberta-xlarge-mnli [12]	Roberta-large [19]	Deberta-xlarge-mnli with idf
$\mathcal{M}^2$- Transformer [5]	64.56	63.11	59.40
ClipCap (MLP+GPT2 fine-tuning) [22]	68.00	65.27	59.13
ClipCap (Transformer) [22]	68.33	66.25	59.37
PureT [31]	68.30	67.69	63.22
TICOD-large ($\lambda = 0.2$) (proposed model)	70.79	68.06	63.23
TICOD-large ($\lambda = 0.5$)	71.69	68.98	63.40

We report the object detection evaluation scores also in Table 1 for convenience of comparison. Our objective of this work is to perform image captioning and object detection simultaneously by improving the performance of image captioning due to joint training with a carefully constructed joint loss function. We demonstrate that we achieve superior image captioning performance in terms of BERTScore while maintaining a comparable performance in object detection. Since our model is developed upon

Swin Transformer architecture [20], we compare our model's performance on object detection with Swin Transformer [20] in the Cascade R-CNN [2] framework. The comparisons presented in Table 1, show that TICOD has better performance in terms of mAP, AP@0.75, AP@small, and AP@medium objects. This, in turn, shows that the caption generation task has positively influenced the object detection task through the joint training, and has resulted in improved performance for object detection. We also compare our model with other popular baselines like DETR [3] and Faster R-CNN [26]. We compare with Pix2seq-v2 [4] on their reported object detection and image captioning scores, and we can see a clear improvement in performance using our approach for both tasks. This is because Pix2seq-v2 doesn't use detection-specific architecture but instead uses language modeling to solve "core" vision tasks. Due to the significant departure from conventional architectures, the model needs further improvement to challenge the current SOTA of task-specific models [4]. It also has a slower inference speed (particularly for longer sequences) than the specialized systems, as the approach is based on autoregressive modeling [4].

Qualitative Comparison: While discussing the caption generation performance of the different models, based on the values of the evaluation metrics, we argued that TICOD generates good-quality captions. However, it may use words that are semantically similar but lexically different from the tokens in the corresponding ground-truth captions. Figure 3 presents some example cases to elaborate on this point. It shows a few images with detected objects by our proposed model and their corresponding captions, along with the captions generated by PureT model [31], $\mathcal{M}^2$ Transformer [5] and the ground-truths. In Fig. 3a, our model generates the correct caption whereas $\mathcal{M}^2$ incorrectly generates *green* tennis racket. The generation also doesn't end properly as it produces *on a* at the end. PureT generated *on a table*, which is incorrect. In Fig. 3b, our model generates *red stop sign* which is more detailed than $\mathcal{M}^2$ and PureT model's captions. In Fig. 3c, the captions are all similar. However, in Fig. 3d, we notice that our model has slightly under-performed as it generates *A couple of small birds* instead of *Two birds* produced by PureT. $\mathcal{M}^2$ also under-performed by producing *A small bird*. By qualitatively analyzing the captions produced by our model, we have observed that, while in most cases our model produces better or equivalent captions, there are some cases where our model has slightly regressed performance.

5.2   Ablation Studies

To evaluate the effectiveness of our proposed approach, we perform ablation studies on the COCO dataset [16,32] using two object detection frameworks – Faster R-CNN [26] and Cascade R-CNN [2]. We also check the model's performance by keeping different parts of the network frozen.

Effect of Backbone Size and Object Detection Framework: Table 3 shows the performance of our model with different backbone sizes and object detection frameworks. As observed, our model performs better when Cascade R-CNN is used with Swin-base and GPT2-medium. There is a performance gain of $\sim$ **3%** Bleu-1 score, **21.67%** Bleu-4 and **21.09%** CIDEr (**+18.5** CIDEr) over Faster-RCNN with Swin-tiny and GPT2-small with frozen backbone and object detection network (fine-tuning only the decoder network). Also, there is an improvement of **6.63%** Bleu-1, **27%** Bleu-4, and **28%** CIDEr

(a)

(b)

(c)

(d)

GT1: A tennis ball sitting on top of a tennis racket.
GT2: A tennis ball is sitting on a tennis racket.
$\mathcal{M}^2$:A tennis ball on a green tennis racket on a table.
PureT: A tennis racket and a tennis ball on a table.
Ours: A tennis ball sitting on top of a racquet.

GT1: A red stop sign sitting on the side of a road.
GT2: Stop sign on a street of a cemetary.
$\mathcal{M}^2$: A stop sign on the side of a street.
PureT: A stop sign on the side of a street.
Ours: A *red* stop sign sitting on the side of a road.

GT1: Two people are sitting on a bench together in front of water.
GT2: A couple is sitting on a bench in front of the water.
$\mathcal{M}^2$: Two people sitting on a bench near the water.
PureT: Two people sitting on a bench looking at the ocean.
Ours: Two people sitting on a bench facing the ocean.

GT1: A couple of small birds standing on top of a table.
GT2: Two little sparrows standing on a table by a knife.
$\mathcal{M}^2$: A small bird sitting on a table next to a knife.
PureT: Two birds sitting on top of a plate of food.
Ours: A couple of small birds standing on top of a table.

Fig. 3. Examples of captions generated by $\mathcal{M}^2$-Transformer [5], PureT [31], our model, and their corresponding ground-truths(GT) [32]. The images also display the detected object categories and their scores as predicted by our proposed model.

(**+ 25.2** CIDEr) when the backbone and decoder networks are finetuned. Clearly, with larger backbone and decoder sizes, there is performance improvement.

Effect of Frozen Layers: We perform experiments by keeping different components of the network frozen. The trainable components of the network are mentioned in the fourth column of Table 3. The remaining components of the model were frozen. We tried three combinations of ⟨frozen-finetuned⟩ components: (i) Swin backbone, FPN, RPN, ROI layers frozen while finetuning only the GPT2 decoder, (ii) finetuning Swin backbone and GPT2 while keeping the remaining layers required for object detection as frozen, and (iii) finetuning all components by keeping the caption network turned off. It is observed from Table 3 that case (ii) has a performance improvement of **2.74%** CIDEr for Faster R-CNN method and **8.57%** CIDEr for Cascade R-CNN. BERTScore has also improved by **3.65%** for Cascade R-CNN. We also observe that for object detection, the mAP improves from 51.9 to **52.1** when the backbone network is trainable, which demonstrates the positive impact of joint training.

For both the upper and lower halves of Table 3, the last rows represent the setup where the GPT-2 network is frozen, and the object detection network is finetuned. In the proposed TICOD model, the caption network is passed an image embedding. However, the GPT-2 in the caption network is initialized with pre-trained GPT-2 parameters that do not recognize that image embedding as input. Hence, without finetuning the

Table 3. Ablation study with different methods, backbones and decoder sizes on COCO [16, 32] dataset.

Methods	Backbone	Decoder	Finetuned networks	B1	B2	B3	B4	Meteor	RougeL	CIDEr	Spice	BERTScore	mAP	AP$_{50}$	AP$_{75}$	AP$_S$	AP$_M$	AP$_L$
Faster R-CNN [26]	Swin-T	GPT2 (small)	GPT2	70.4	51.5	36.8	26.3	23.7	50.9	87.7	16.7	66.473	**46.0**	**68.1**	**50.3**	**31.2**	**49.2**	**60.1**
		GPT2 (small)	Swin,GPT2	**70.9**	**52.5**	**38.2**	**27.8**	**24.1**	**51.3**	**90.1**	**17.1**	**66.934**	45.3	67.7	49.9	29.5	48.7	59.0
		✗	Swin,FPN, RPN, ROI	–	–	–	–	–	–	–	–	–	**46.0**	**68.1**	**50.3**	**31.2**	**49.2**	**60.1**
Cascade R-CNN [2]	Swin-B	GPT2 (med)	GPT2	72.5	55.3	41.8	32.0	27.2	54.6	106.2	19.7	68.301	51.9	**70.9**	56.5	**35.4**	55.2	**67.3**
		GPT2 (med)	Swin,GPT2	**75.6**	**59.0**	**45.5**	**35.3**	**28.3**	**56.7**	**115.3**	**21.1**	**70.794**	**52.1**	70.6	**56.7**	34.8	**55.3**	67.2
		✗	Swin,FPN, RPN, ROI	–	–	–	–	–	–	–	–	–	51.9	**70.9**	56.5	**35.4**	55.2	**67.3**

caption-generation network, the generated captions for this setup would be meaningless. Accordingly, the values of the evaluation metrics for image captioning are filled by "–"s for these rows.

The object detection scores in the first and third rows are identical in both halves of the table. The first row corresponds to a setup where the object detection network is frozen, while the third row involves finetuning the parameters of the object detection network. The similarity in scores arises from the fact that the object detection network's parameters are initialized with a pre-trained checkpoint optimized for this dataset. Further finetuning leads to a drop in performance on the validation set. As a result, the initial parameters (epoch 0) are retained, resulting in similar metric values for these rows for the object detection task.

Task-specific Performance: We examine the impact of multitasking on performance compared to task-specific models through experiments with two baselines: (i) object detection alone without the caption network, and (ii) captioning alone without the object detection network. Results in Table 3 indicate improved image captioning performance without compromising object detection.

Choosing Lambda: We conduct experiments to finetune the hyperparameter λ in Eq. 3. Table 4 demonstrates that 0.1 and 0.2 are the most suitable values for TICOD-small and TICOD-large, respectively. For TICOD-small, the difference in image captioning scores between $\lambda = 0.1$ and 10 is negligible, and as object detection performance degrades with increasing lambda, scores for $\lambda = 0.2$ and 0.5 are not reported.

Table 4. Hyperparameter tuning: Illustration of model performance with different λ values.

Methods	lambda (λ)	B1	B2	B3	B4	Meteor	RougeL	CIDEr	Spice	BERTScore	mAP	AP$_{50}$	AP$_{75}$	AP$_S$	AP$_M$	AP$_L$
TICOD-small	0.01	69.7	51.3	36.6	25.8	22.8	50.1	82.1	16.2	65.728	**45.4**	67.3	**50.2**	**31.1**	**48.8**	**59.1**
	0.1	**70.9**	**52.5**	**38.2**	**27.8**	**24.1**	**51.3**	**90.1**	**17.1**	**66.934**	45.3	**67.7**	49.9	29.5	48.7	59.0
	10	70.8	52.8	38.1	27.4	23.8	51.3	88.6	17.2	66.698	37.7	62.3	40.6	25.8	42.0	46.1
TICOD-large	0.01	72.8	56.9	41.8	32.1	26.7	54.2	100.7	18.6	67.801	51.9	70.5	56.6	**36.1**	55.2	**67.4**
	0.1	74.3	57.3	43.6	33.5	27.0	55.1	107.3	19.4	69.821	51.9	**70.6**	56.5	**36.1**	**55.3**	67.2
	0.2	75.6	59.0	45.5	35.3	28.3	56.7	115.3	21.1	70.794	**52.1**	**70.6**	**56.7**	34.8	**55.3**	67.2
	0.5	**76.5**	60.4	**46.9**	**36.6**	**28.9**	**57.6**	119.6	**21.6**	**71.686**	51.6	70.3	56.3	33.6	55.0	**67.4**
	10	76.2	**60.5**	45.7	36.1	28.5	57.1	119.4	21.5	71.221	47.2	67.8	50.4	30.2	51.5	61.1

6   Conclusion

In this work, we presented TICOD, a multitask framework for object detection and image captioning. Empirically, we show that joint learning helps improve image captioning by improving the image representations in the backbone. Swin Transformer is not pre-trained on a vision-language objective, yet we demonstrate that we can use it directly with GPT2 and show superior image captioning performance in BERTScore while maintaining a comparable performance in object detection. Our proposed framework is customizable as Swin Transformer and GPT2 can be replaced with newer specialized SOTA detection and large language models, which will further improve performance over general-purpose multitask models like Pix2seq-V2 [4].

Acknowledgements. This work was supported by DST National Mission on Interdisciplinary Cyber-Physical Systems (NM-ICPS), Technology Innovation Hub on Autonomous Navigation and Data Acquisition Systems: TiHAN Foundations at Indian Institute of Technology (IIT) Hyderabad, India. We also acknowledge the support from Japan International Cooperation Agency (JICA). We express gratitude to Suvodip Dey for his valuable insights and reviews on this work.

References

1. Anderson, P., et al.: Bottom-up and top-down attention for image captioning and visual question answering. In: Proceedings of the IEEE Conference on CVPR (2018)
2. Cai, Z., Vasconcelos, N.: Cascade r-cnn: delving into high quality object detection. In: Proceedings of the IEEE Conference on CVPR (2018)
3. Carion, N., Massa, F., Synnaeve, G., Usunier, N., Kirillov, A., Zagoruyko, S.: End-to-end object detection with transformers. In: Vedaldi, A., Bischof, H., Brox, T., Frahm, J.-M. (eds.) Computer Vision. ECCV 2020. LNCS, vol. 12346, pp. 213–229. Springer, Cham (2020). https://doi.org/10.1007/978-3-030-58452-8_13
4. Chen, T., et al.: A unified sequence interface for vision tasks. In: NeurIPS (2022)
5. Cornia, M., et al.: Meshed-memory transformer for image captioning. In: CVPR (2020)
6. Devlin, J., Chang, M.W., Lee, K., Toutanova, K.: Bert: pre-training of deep bidirectional transformers for language understanding. arXiv preprint arXiv:1810.04805 (2018)
7. Dosovitskiy, A., et al.: An image is worth 16x16 words: transformers for image recognition at scale. arXiv preprint arXiv:2010.11929 (2020)
8. Fariha, A.: Automatic image captioning using multitask learning. In: NeurIPS, vol. 20 (2016)
9. Girshick, R.: Fast r-cnn. In: International Conference on Computer Vision (ICCV) (2015)
10. Girshick, R., et al.: Rich feature hierarchies for accurate object detection and semantic segmentation. In: Proceedings of the IEEE Conference on CVPR (2014)
11. Han, K., et al.: Transformer in transformer. In: Advances in NeurIPS, pp. 15908–15919 (2021)
12. He, P., et al.: Deberta: decoding-enhanced bert with disentangled attention. In: ICLR (2021)
13. Jiang, H., et al.: In defense of grid features for visual question answering. In: CVPR (2020)
14. Li, X., et al.: Oscar: Object-semantics aligned pre-training for vision-language tasks. In: Computer Vision. ECCV 2020. Springer (2020)
15. Liang, J., et al.: Swinir: image restoration using Swin transformer. In: ICCV (2021)
16. Lin, T.-Y., et al.: Microsoft COCO: common objects in context. In: Fleet, D., Pajdla, T., Schiele, B., Tuytelaars, T. (eds.) Computer Vision. ECCV 2014. LNCS, vol. 8693, pp. 740–755. Springer, Cham (2014). https://doi.org/10.1007/978-3-319-10602-1_48

17. Lin, T.Y., et al.: Feature pyramid networks for object detection. In: CVPR (2017)
18. Liu, H., et al.: Visual instruction tuning. arXiv preprint arXiv:2304.08485 (2023)
19. Liu, Y., et al.: Roberta: a robustly optimized bert pretraining approach. arXiv preprint arXiv:1907.11692 (2019)
20. Liu, Z., et al.: Swin transformer: hierarchical vision transformer using shifted windows. In: Proceedings of the IEEE/CVF ICCV, pp. 10012–10022 (2021)
21. Luo, Y., et al.: Dual-level collaborative transformer for image captioning. Proc. AAAI Conf. Artif. Intell. **35**(3), 2286–2293 (2021). https://doi.org/10.1609/aaai.v35i3.16328
22. Mokady, R., Hertz, A., Bermano, A.H.: Clipcap: clip prefix for image captioning. arXiv preprint arXiv:2111.09734 (2021)
23. Pan, Y., et al.: X-linear attention networks for image captioning. In: CVPR (2020)
24. Radford, A., et al.: Language models are unsupervised multitask learners. OpenAI blog (2019)
25. Radford, A., et al.: Learning transferable visual models from natural language supervision. In: International Conference on Machine Learning, pp. 8748–8763. PMLR (2021)
26. Ren, S., et al.: Faster r-cnn: towards real-time object detection with region proposal networks. In: NeurIPS, vol. 28 (2015)
27. Touvron, H., et al.: Training data-efficient image transformers and distillation through attention. In: International Conference on Machine Learning, pp. 10347–10357. PMLR (2021)
28. Vaswani, A., et al.: Attention is all you need. In: NeurIPS, vol. 30 (2017)
29. Vinyals, O., et al: Show and tell: a neural image caption generator. In: CVPR (2015)
30. Wang, W., et al.: Pyramid vision transformer: a versatile backbone for dense prediction without convolutions. In: Proceedings of the IEEE/CVF ICCV, pp. 568–578 (2021)
31. Wang, Y., Xu, J., Sun, Y.: End-to-end transformer based model for image captioning. Proc. AAAI Conf. Artif. Intell. **36**(3), 2585–2594 (2022). https://doi.org/10.1609/aaai.v36i3.20160
32. Xinlei Chen, H.F., et al.: Microsoft coco captions: data collection and evaluation server. arXiv preprint arXiv:1504.00325 (2015)
33. Xu, H., et al.: E2e-vlp: end-to-end vision-language pre-training enhanced by visual learning. In: Proceedings of the 59th Annual Meeting of the ACL and the 11th IJCNLP (2021)
34. Xu, K., et al.: Show, attend and tell: neural image caption generation with visual attention. In: Proceedings of the 32nd ICML, vol. 37, pp. 2048–2057. PMLR (2015)
35. Zhang, K., et al.: Bertscore: evaluating text generation with bert. In: ICLR (2020)
36. Zhang, P., et al.: Vinvl: revisiting visual representations in vision-language models. In: Proceedings of the IEEE/CVF Conference on CVPR (2021)
37. Zhao, W., et al.: A multi-task learning approach for image captioning. In: IJCAI 2018
38. Zhuo, T.Y., et al.: Rethinking round-trip translation for machine translation evaluation. In: Findings of the Association for Computational Linguistics: ACL (2023)

Communicative and Cooperative Learning for Multi-agent Indoor Navigation

Fengda Zhu, Vincent CS Lee[✉], and Rui Liu

Monash University, Clayton, Australia
`vincent.cs.lee@monash.edu`

Abstract. The ability to cooperate and work as a team is one of the "holy grail" goals of intelligent robots. To address the importance of communication in multi-agent reinforcement learning (MARL), we propose a Cooperative Indoor Navigation (CIN) task, where agents cooperatively navigate to reach a goal in a 3D indoor room with realistic observation inputs. This navigation task is more challenging and closer to real-world robotic applications than previous multi-agent tasks since each agent can observe only part of the environment from its first-person view. Therefore, this task requires the communication and cooperation of agents to accomplish. To research the CIN task, we collect a large-scale dataset with challenging demonstration trajectories. The code and data of the CIN task have been released. The prior methods of MARL primarily emphasized the learning of policies for multiple agents but paid little attention to the communication model, resulting in their inability to perform optimally in the CIN task. In this paper, we propose a MARL model with a communication mechanism to address the CIN task. In our experiments, we discover that our proposed model outperforms previous MARL methods and communication is the key to addressing the CIN task. Our quantitative results shows that our proposed MARL method outperforms the baseline by 6% on SPL. And our qualitative results demonstrates that the agent with the communication mechanism is able to explore the whole environment sufficiently so that navigate efficiently.

Keywords: Multi-agent Reinforcement Learning · Cooperative Navigation

1 Introduction

Cooperative multi-agent problems are ubiquitous in real-world applications, such as multiplayer games [14,24], multi-robot control [39], language communication [16]. These applications focus on solving the decision-making problem of multiple autonomous agents within a common environment, which are modeled as the multi-agent reinforcement learning (MARL) [17]. Many game-based environments have been proposed to research the cooperative multi-agent problems, such as grounded communication environment [21], StarCraft II [25], multi-agent emergence environments [4], soccer shooting [18], etc.

Many game-based Multi-Agent Reinforcement Learning environments diverge from real-world robotic scenarios. Agents in these environments often receive noise-free, pre-processed data instead of experiencing authentic, first-person observations.

© The Author(s), under exclusive license to Springer Nature Singapore Pte Ltd. 2024
D.-N. Yang et al. (Eds.): PAKDD 2024, LNAI 14646, pp. 273–285, 2024.
https://doi.org/10.1007/978-981-97-2253-2_22

Additionally, the transition function relies on straightforward rules without simulating real-world physics or accounting for actual-world interferences. This significant domain gap hinders the application of current MARL models [24, 37] in real-world scenarios.

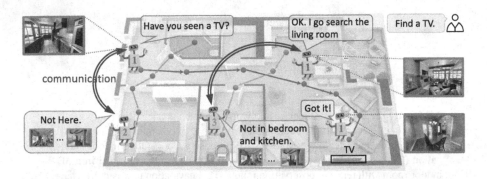

Fig. 1. A demonstration of our Cooperative Indoor Navigation (CIN) task. During the process of navigation, agents exchange the observed semantic information and explore the room toward different directions to find the target efficiently.

To overcome the limitations of previous multi-agent environments, we propose a novel task, Cooperative Indoor Navigation (CIN), where multiple agents are required to navigate to reach targets in a 3D indoor room. To the best of our knowledge, our CIN is the first multi-agent real-world navigation environment. Each agent in CIN observes an RGB-D image from its own first-person perspective and make actions to navigate. Thus, our CIN environment is closer to real-world robotic scenarios, whose realistic observation facilitates the learned agents to be deployed to real-world robotic applications. An overview of the CIN task with the communication mechanism is shown in Fig. 1. Three agents start from different positions and are asked to find a TV in this room. Each agent receives a first-person photo-realistic observation from their perspective respectively. To reach the target, they explore the entire room and communicate with each other in exchanging their discoveries. Through active exploration and cooperative communication, the No. 1 agent finally finds the target TV. To fully investigate the CIN task, we construct a large-scale dataset consisting of 24 million episodes for training, validation, and testing splits. Compared with other embodied navigation datasets [27, 32], our dataset is more challenging because we provide lengthier trajectory data.

The environments of the CIN task bring new challenges. For example, perceiving embodied vision information and Previous MARL methods [1, 12, 24, 31] adopt the Centralized Training Decentralized Execution (CTDE) framework, which ignores the communication among agents. To tackle this communication difficulty, we propose a new cooperative multi-agent communication model to enable the agents to exchange the latent embedding step-by-step during the whole navigation process. Our model first embeds the visual image by a Convolutional Neural Network as a visual embedding. The visual embedding and the word embedding of the global target are concatenated as latent embedding. Then our model adopts a Long short-term memory layer to embed the latent embeddings in multiple steps into a historical latent embedding. For each step, agents exchange their historical embedding with each other and make the action prediction via a policy network.

Our environment validates that our model outperforms previous MARL methods quantitatively and qualitatively. We discover that without sufficient communication, the navigation performance will not increase by increasing the number of agents. And the communication module we propose could improve the overall success rate in reaching the target. We compare the navigation trajectories between the agents with inter-agent communication and without inter-agent communication and find that our method enables the agents to sufficiently explore the semantic knowledge of the entire room and achieve high performance. We experiment on four kinds variants of our model and find that communicating through historical latent embeddings and using them as input to generate the historical latent embeddings of the next step achieves the best performance (Table 1).

Table 1. Compared with existing MARL environments (N is the number of agents). The ego-centric view means that the agent has a local observation of the environment from its perspective rather than receiving global information. The embodied view represents whether an agent observes the 3D environment from the first-person perspective (we only compare the navigable environments).

Input	Environment	Multi-agent	Embodied	Observation Overlap	Physics Engine
Array	Hanabi [5]	✓	–	$\frac{N-1}{N}$	–
	Diplomacy [22]	✓	–	$\frac{N-1}{N}$	–
	EGCL [21]	✓	✗	100%	–
	DOTA2 [7]	✓	✗	100%	Rubikon
	SMAC [25]	✓	✗	81.0%–89.8%	Havok
	Hide-and-seek [4]	✓	✗	$\frac{1}{N}$-100%	MuJoCo
	Soccer [18]	✓	✗	100%	MuJoCo
Syn image	MARLÖ [23]	✓	✓	$\frac{1}{N}$	hybrid voxel
	AI2-THOR [15]	✗	✓	$\frac{1}{N}$	Unity
Real image	Habitat [27]	✗	✓	$\frac{1}{N}$	Bullet
	CIN (Ours)	✓	✓	$\frac{1}{N}$	Bullet

2 Related Work

Multi-agent Environments have been proposed to research multi-agent problems. In Table 1, we compare the differences between our CIN environment with previous multi-agent environments. To the best of our knowledge, we claim that our CIN environment is the first multi-agent environment that offers realistic image input. Previous works [4, 7, 25] get clean and formatted array data via programming interfaces.

A particular challenge in the MARL problem is partial observability, where each agent could only observe a part of the global state. However, our investigation reveals that some of the previous MARL environments [4, 13, 25], even though claimed to be partially observable, have large observation overlaps. Little observations overlap makes our benchmark more challenging than the previous environments because little information sharing makes cooperation difficult. Our environment adopts the Bullet engine to simulate physics activities such as acceleration and collision.

Multi-agent Reinforcement Learning extends the problem of reinforcement learning [20,30] into multi-agent scenario and brings new challenges. Most MARL problems [19] fall into the centralized training with decentralized execution (CTDE) architecture [36]. Some works are aiming at improving the network structure and developing an individual function with better representation and transfer capability [11,12]. Besides, the utilization of state information varies. IPPO [33] incorporates the global state information barely by sharing network parameters among critics of individual agents. While MAPPO [38] constructs a centralized value function upon agents which takes the aggregated global state information as inputs. We discovered in this paper that exchanging historical information between agent facilitate the accuracy and efficiency of multi-agent navigation. And we have done extensive ablation experiments to validate this conclusion.

Embodied Navigation. Simulations such as Matterport3D simulator [3], Gibson simulator [35] and Habitat [27] propose high-resolution photo-realistic panoramic view to simulate the more realistic environment. Rendering frame rate is also important to embodied simulators since it is critical to training efficiency. MINOS [26] runs more than 100 frames per second (FPS), which is 10 times faster than its previous works. Habitat [27] runs more than 1000 FPS on 512×512 RGB+depth image, making it become the fastest simulator among existing simulators. Some complex tasks may require a robot to interact with objects, such as picking up a cup, moving a chair or opening a door. AI2-THOR [15], iGibson [34] and RoboTHOR [10] provide interactive environments to train such a skill. Multi-agent reinforcement learning [17] is a rising problem of cooperation and competition among agents. Based on the Habitat simulator, we construct a multi-agent environment to research realistic MARL problems.

3 Cooperative Indoor Navigation Task

3.1 Task Definition

Our Cooperative Indoor Navigation (CIN) task requires multiple agents $E = \{e_1, ..., e_n\}$ to navigate to the target g_i within a common embodied environment starting from different locations. For each step, the agent observes an observation and makes an action decision. An action could be "turn left", "turn right", "step forward" to navigate, or "found" to declare the agent has found the target object and stops. The episode of this agent is considered successful if the "found" action is selected while the agent is located with the threshold toward the target object.

We propose two sub-tasks, shared-target navigation, and individual-target navigation. *Shared-target navigation* is a task where all agents are asked to find a shared target g. The task is considered successful if any agent reaches g and considered a failure if any agent fails within its episode. *Individual-target navigation* is a task where each agent i has its own target g_i, and the task is considered successful only when all agents successfully find their own target.

3.2 Multi-agent Indoor Navigation Environment

The CIN environment is built based on the Habitat [27] simulator. Our environment renders the 3D assets of a house and provides a common photo-realistic embodied environment for multiple navigation agents. Our implementation of the CIN environment is shown in Fig. 2. The CIN environment creates B sub-environments for decentralized execution to sample data for training, where B is the size of the minibatch. Each sub-environment creates N processes, where the N is the number of agents. Each process has a copy of a Habitat simulator, and each simulator individually simulates the state of an agent and renders the RGB-D image observation for an agent. The CIN sub-environment synchronizes the processes and interacts with a copy of a multi-agent navigation model. In the decentralized execution, the parameters of the multi-agent navigation model are shared across all sub-environments. The multi-agent navigation model predicts actions for each agent for each step. The predicted actions are sent to the CIN environment and then distributed to each process to execute. The CIN sub-environment calculates the global reward based on the global state and sends the global reward and observations to the model. For each step, the global reward, observations for all agents, and actions that agents predict are stored in the episode memory. We sample the episodes from the episode memory to optimize a centralized model by stochastic gradient descent (SGD). The model after a step of SGD optimization is copied to each CIN sub-environment to update the execution model.

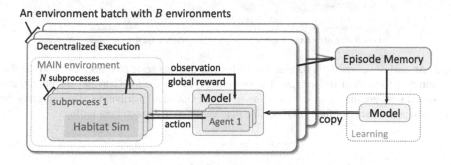

Fig. 2. Training overview of our CIN environment.

3.3 Data Collection

We use the room textures and other 3D assets provided by Matterport3D [8] to build the CIN environment. Matterport3D consists of 10,800 panoramic views constructed from 194,400 RGB-D images of 90 building-scale scenes, where 61 scenes for training, 11 for validation, and 18 for testing following the standard split [8]. Providing episode data for learning and testing, where an episode is defined by a randomly sampled starting position and target position within a navigable environment. And we constrain the length of an episode to be between 2m and 20m, which ensures that each episode is neither too trivial nor too hard.

We compare the distribution of the average trajectory length within a room with MultiON [32] and the object navigation data in the Habitat Challenge [6] in Fig. 3(a). Due to the different structures of the house, the episode data from each house have different average length. We find that with the same room setting, our average trajectory length is longer than both MultiON and Habitat, proving that our data is more challenging. Our dataset provide 24M episodes, 10 times more than the data scale of the Habitat dataset. The Fig. 3(b) shows the average distance that the agents need to navigate to successfully accomplish task. It reveals the gap of the difficulty among the two sub-tasks and the single-agent navigation task accompany with the agent amount using our dataset. With the increase of the agent amounts, the difficulty of individual-target task is significantly increasing while the shared-target task is reducing.

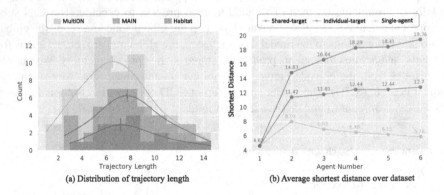

Fig. 3. An analysis of our CIN dataset. (a) analysis of the data scale of trajectories and the distribution of the length of trajectories. The curve in (a) stands for the Gaussian smoothing, and the vertical lines are the mean values of the smoothed Gaussian distributions. (b) compares the navigation performance of the IPPO baseline in three-different environment settings.

4 Cooperative Indoor Navigation Models

4.1 Preliminaries

We model our CIN problem as a multi-agent reinforcement learning problem in a partially-observed Markov decision process (POMDP). $P(s'|s, a)$ is the transition probability that transforms the current state space S to the next state space S' conditions on the a global action $a \in A$. We follow the centralized-training decentralized-execution framework that parameterize the shared policy of each agent as π_θ. For each step t, the agent i receive its partial observation $o_{t,i}$ and choose its action by $a_{t,i} = \pi_{\theta_i}(o_{t,i})$. The global action $a_t = a_{1,t}, ..., a_{n,t}$ All agents share the same global reward function $r(s,a) : S \times A \to \mathbb{R}$. And the $\gamma \in [0,1)$ is a discount factor that defines the length of the horizon. We optimize the parameter θ by minimizing the optimization objective $J(\theta) = \mathbb{E}\left[\sum_t \gamma^t r(s_t, a_t)\right]$ by PPO algorithm [29].

4.2   Framework

In this section, we are going to introduce our cooperative communicative navigation model, as shown in Fig. 4. The framework first embeds the target as an embedding feature and extracts visual features using a CNN network:

$$f_{i,t} = \text{CNN}(o_{i,t}), \tag{1}$$

where the f_i denotes visual embedding feature of the ith agent and the t denotes the tth timestep. The parameters of the embedding layer and the CNN layer are shared between agents to ensure the ability of generalization. Then the target feature and visual feature are concatenated to feed the Recurrent Neural Network (RNN) [9] module:

$$g_{i,t} = \text{RNN}_i(f_{i,t}). \tag{2}$$

The historical feature from the RNN is sent to two fully connected layers.

$$a_{i,t} = W_i^\pi(g_{i,t}), \\ v_{i,t} = W_i^v(g_{i,t}). \tag{3}$$

The W_π outputs the decision and the W_i^v predicts a value to estimate the effectiveness of the current situation. Note that the parameters of the RNN layer and the fully connected layers are not shared between agents. The action prediction is supervised by the clipped policy gradient loss [29], and the value prediction is supervised by the temporal difference of the bellman equation.

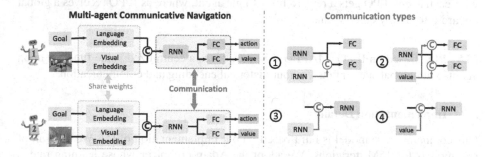

Fig. 4. A demonstration of our multi-agent communicative navigation framework using a two-agent model. The dual-directed orange arrow indicates weight sharing between orange boxes, while the orange arrow represents communication. On the right, we show four communication variants. (Color figure online)

Cooperative Communication. As shown in Fig. 4, the agents cooperatively communicate by exchanging the observed semantic information between the blue block and the green block. The information exchange is implemented by exchanging a embedding feature vector. The feature vector that an agent sent is named a "message". The agent that receives the message is named the "receiver" and the agent that sends the message is named the "sender". The gradient is not back-propagated from the 'receiver' to the 'sender' since it causes severe instability in training, which makes the performance of the learned navigation model to be almost zero. Different exchanging structure led to different experimental results. We design four types of communicative variants and compare their performance in our experiments, as demonstrated in Fig. 4 (right): 1) the sequential communication model (Comm-S), 2) value communication model (Comm-V), 3) recurrent message communication model (Comm-RM), and 4) recurrent value communication model (Comm-RV).

5   Experiment

5.1   Benchmarking CIN with MARL Models

We implement several multi-agent models to investigate the performance of multi-agent models on the CIN task.

Random Navigator with Oracle Founder. We implement a random baseline that randomly samples the action of "turn left", "turn right" and "go forward". This baseline model is used to validate if our dataset is too easy or has severe bias.

Multi-single Agent. This PPO-based [29] model, initially trained in a single-agent paradigm, is tested in a multi-agent setup. Our study investigates how altering the number of agents impacts navigation performance in this context.

IPPO. The IPPO [1] model learns the global reward and share network parameters with each agent. PPO gets a reward for a single agent, where as IPPO receives a global reward affected by other agents.

MAPPO. Based on IPPO, MAPPO [38] introduces a centralized value function upon agents with global state inputs without historical encoding and communication.

5.2   Implementation Details

Our communicative model is built based on our implementation of [1]. We train all of our models for 15M iterations. We adopt the Adam optimizer whose learning rate is 2.5×10^{-4}. The discount factor $\gamma = 0.99$ and the TD(λ) factor in GAE [28] is 0.95. The loss weight of the value function is 0.5. The hyperparameter for PPO clip is set to 0.2. Our model is trained on 8 GPUs (7 GPUs for rendering image inputs and 1 GPU for optimization) for 36 h.

5.3   Evaluation Metrics

Distance indicates the average distance forward the target when the agent stops.

Success Rate measures if the agent successfully finds the target when it yields 'found'. The agent is regarded as 'success' only if it is located within 1m towards the target.

SPL, short for Success weighted by Path Length [2], evaluates the accuracy and efficiency simultaneously. The SPL is calculated by $\frac{1}{N} \sum_{i=1}^{N} S_i \frac{p_i}{l_i}$, where the N is all testing samples, S_i is the success indicator, p_i is the shortest path length, and the l_i is the actual path length in testing.

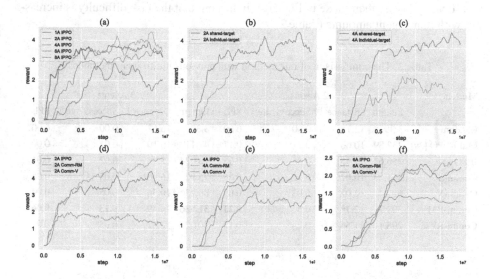

Fig. 5. The reward curves during the training process in the CIN environment: (a) ablates the agent number based on the IPPO baseline. (b) and (c) compare the shared-target task with the individual-target task. (d), (e) and (f) plot the reward curves of IPPO, Comm-RM, and Comm-V during the training process.

5.4   Quantitative and Qualitative Results

Ablation Study of Agent Amount. Here, we investigate if the amount of agents in cooperative navigation affects the navigation performance. Figure 5(a) shows the ablation results of the number of agents. The figure shows that the navigation setting with 2 agents achieves the best performance. The 2-agent model outperforms the 1-agent model indicating that multiple agents are able to find the target more accurately and efficiently than using a single agent. If the number of agents continues to increase, the navigation performance will progressively deteriorate, and the 8-agent model yields the poorest results. Increasing the number of agents can narrow down the search area for finding a target; however, optimizing 8 agents with a global reward is challenging. The actions of other agents can easily influence the gradients of each agent, making it difficult for the global reward to provide clear guidance to individual agents.

The Difficulty of Two Sub-tasks. Here, we compare the shared-target task and the individual-target task and investigate the differences between these two tasks. Figure 5(b), (c) ablate the difficulty of two sub-tasks with different settings. We train the IPPO baselines on individual-target tasks and shared-target tasks respectively. The blue curve represents the average reward from the model training for the shared-target navigation task, while the yellow curve represents the average reward for the individual-target navigation task. The model in the shared-target task significantly outperforms the model in the individual-target task by reward, which reveals that the individual-target task is harder than the shared-target task. Moreover, the reward curves in Fig. 5(b) are significantly higher than those in Fig. 5(c), indicating that the task difficulty is increasing with more agent amounts (Table 2).

Table 2. The comparison of our method with previous multi-agent methods.

Models	2 Agents				3 Agents			
	Length	Distance ↓	Success rate ↑	SPL ↑	Length	Distance ↓	Success rate ↑	SPL ↑
Random	3.39	12.75	0.00	0.00	3.30	16.67	0.00	0.00
Multi-PPO [29]	232.89	10.66	0.21	0.17	47.11	16.10	0.01	0.00
IPPO [1]	256.67	15.55	0.08	0.06	137.24	15.94	0.02	0.01
Comm-S	351.03	**10.02**	0.12	0.06	75.92	16.16	0.05	0.05
Comm-V	68.14	12.02	0.03	0.03	80.23	**10.36**	0.05	0.04
Comm-RM	309.7	12.78	**0.3**	**0.23**	312.86	14.59	**0.13**	**0.06**
Comm-RV	298.1	11.56	0.24	0.17	301.2	12.32	0.08	0.05

Effectiveness of Communication. We train the model with a historical communication mechanism (Comm-RM), the model with a value communication mechanism (Comm-V), and the IPPO baseline (IPPO) in different settings, including 2 agents, 4 agents, and 6 agents. The results are shown in Fig. 5(d), (e), (f). We discover that the reward curves of Comm-V and Comm-RM outperform the reward curves of other methods in all three figures. Moreover, there is a clear performance gap between the Comm-RM and Comm-V in all three figures, indicating that the historical communication mechanism is able to improve the performance of multi-agent navigation.

We detailed analysis of three baseline models (Random, Multi-PPO, and IPPO) and four variants of our proposed methods (Comm-S, Comm-V, Comm-RM, Comm-RV) in the shared-target task, as shown in Table 2. First of all, the Multi-PPO setting outperforms the IPPO method in the 2-agent setting while failing in the 3-agent setting. Secondly, we discover that our proposed method with communication mechanisms significantly outperforms the cooperative and non-cooperative baselines, which reveals that communication is a key to solving the multi-agent navigation problem. We find that the model with adopts historical information exchange (Comm-RM and Comm-RV), as shown in Fig. 4, outperforms the communication methods without considering the historical information such as Comm-S and Comm-V by the success rate and SPL.

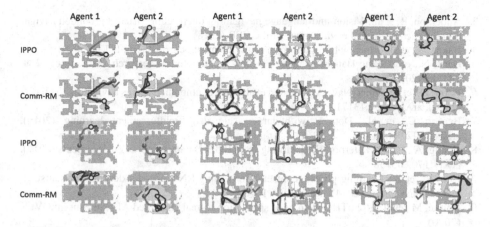

Fig. 6. The trajectory visualization results of the IPPO agent and the communicative agent in the testing environment. The red circle with a red flag shows where the target is located. The yellow circle is the starting position of an agent. The green line indicates the shortest path, and the blue line is the actual navigation path. (Color figure online)

Visualization for Navigation Process. We visualize the navigation trajectories of the IPPO model and our Comm-RM model in the shared-target task. The visualization results are shown in Fig. 6. Green lines represent the shortest path from start to end, while the blue line traces the actual navigation path of the agents. Notably, the blue lines in the second and fourth rows are notably longer than those in the first and third rows. This suggests that cooperative communication enhances the ability of the agents to explore larger areas and cover greater distances. Additionally, in the Comm-RM model, the navigation trajectories of different agents show minimal overlap, indicating effective communication that helps them avoid traversing through already-explored areas.

6   Conclusion

In this paper, we introduce the Cooperative Indoor Navigation (CIN) task for studying the multi-agent problem in realistic environments. We create a large-scale dataset for CIN research and highlight its advantages. Our analysis reveals that traditional MARL methods struggle with CIN challenges like limited observation overlap and high variance in embodied image views. We find that incorporating historical communication messages notably improves multi-agent navigation in CIN. Future work will focus on cooperative navigation using CIN and ongoing dataset updates.

References

1. Aiello, M., et al.: IPPO: A privacy-aware architecture for decentralized data-sharing (2020). arXiv:2001.06420
2. Anderson, P., et al.: On evaluation of embodied navigation agents (2018). arXiv:1807.06757

3. Anderson, P., et al.: Vision-and-language navigation: interpreting visually-grounded navigation instructions in real environments. In: CVPR (2018)
4. Baker, B., et al.: Emergent tool use from multi-agent autocurricula. In: ICLR (2020)
5. Bard, N., et al.: The Hanabi challenge: a new frontier for AI research. Artif. Intell. **280**, 103216 (2020)
6. Batra, D., et al.: ObjectNav revisited: On evaluation of embodied agents navigating to objects (2020). arXiv:2006.13171
7. Berner, C., et al.: Dota 2 with large scale deep reinforcement learning (2019). arXiv:1912.06680
8. Chang, A., et al.: Matterport3D: Learning from RGB-D data in indoor environments (2017). arXiv:1709.06158
9. Cho, K., et al.: Learning phrase representations using RNN encoder-decoder for statistical machine translation (2014). arXiv:1406.1078
10. Deitke, M., et al.: RoboTHOR: an open simulation-to-real embodied AI platform. In: CVPR (2020)
11. Foerster, J., Farquhar, G., Afouras, T., Nardelli, N., Whiteson, S.: Counterfactual multi-agent policy gradients. In: AAAI (2018)
12. Hu, S., Zhu, F., Chang, X., Liang, X.: UPDeT: universal multi-agent RL via policy decoupling with transformers. In: ICLR (2021)
13. Ikram, K., Mondragón, E., Alonso, E., Garcia-Ortiz, M.: HexaJungle: a marl simulator to study the emergence of language (2021)
14. Khan, M.J., Ahmed, S.H., Sukthankar, G.: Transformer-based value function decomposition for cooperative multi-agent reinforcement learning in starCraft. In: Proceedings of the AAAI Conference on Artificial Intelligence and Interactive Digital Entertainment. vol. 18, pp. 113–119 (2022)
15. Kolve, E., Mottaghi, R., Gordon, D., Zhu, Y., Gupta, A., Farhadi, A.: AI2-THOR: An interactive 3D environment for visual AI (2017). arXiv:1712.05474
16. Lin, T., Huh, J., Stauffer, C., Lim, S.N., Isola, P.: Learning to ground multi-agent communication with autoencoders. NeurIPS **34**, 15230–15242 (2021)
17. Littman, M.L.: Markov games as a framework for multi-agent reinforcement learning. In: ICML (1994)
18. Liu, S., Lever, G., Merel, J., Tunyasuvunakool, S., Heess, N., Graepel, T.: Emergent coordination through competition. In: ICLR (2019)
19. Mahajan, A., Rashid, T., Samvelyan, M., Whiteson, S.: MAVEN: Multi-agent variational exploration (2019). arXiv:1910.07483
20. Mnih, V., et al.: Playing Atari with deep reinforcement learning (2013). arXiv:1312.5602
21. Mordatch, I., Abbeel, P.: Emergence of grounded compositional language in multi-agent populations. In: AAAI (2017)
22. Paquette, P., et al.: No press diplomacy: Modeling multi-agent gameplay (2019). arXiv:1909.02128
23. Pérez-Liébana, D., et al.: The multi-agent reinforcement learning in malmÖ (marlÖ) competition (2019). arXiv:1901.08129
24. Rashid, T., Samvelyan, M., Schroeder, C., Farquhar, G., Foerster, J., Whiteson, S.: QMIX: Monotonic value function factorisation for deep multi-agent reinforcement learning. In: ICML (2018)
25. Samvelyan, M., et al.: The StarCraft multi-agent challenge (2019). arXiv:1902.04043
26. Savva, M., Chang, A.X., Dosovitskiy, A., Funkhouser, T.A., Koltun, V.: MINOS: Multimodal indoor simulator for navigation in complex environments (2017). arXiv:1712.03931
27. Savva, M., et al.: Habitat: a platform for embodied AI research. In: ICCV (2019)
28. Schulman, J., Moritz, P., Levine, S., Jordan, M., Abbeel, P.: High-dimensional continuous control using generalized advantage estimation. In: ICLR 2016 (2016)

29. Schulman, J., Wolski, F., Dhariwal, P., Radford, A., Klimov, O.: Proximal policy optimization algorithms (2017). arXiv:1707.06347
30. Sutton, R.S., McAllester, D.A., Singh, S.P., Mansour, Y., et al.: Policy gradient methods for reinforcement learning with function approximation. In: NeurIPS (1999)
31. Wang, T., Gupta, T., Mahajan, A., Peng, B., Whiteson, S., Zhang, C.: RODE: Learning roles to decompose multi-agent tasks (2020). arXiv:2010.01523
32. Wani, S., Patel, S., Jain, U., Chang, A.X., Savva, M.: MultiON: Benchmarking semantic map memory using multi-object navigation (2020). arXiv:2012.03912
33. de Witt, C.S., et al.: Is independent learning all you need in the StarCraft multi-agent challenge? CoRR (2020)
34. Xia, F., et al.: Interactive Gibson benchmark: a benchmark for interactive navigation in cluttered environments. IEEE Robot. Autom. Lett. 5(2), 713–720 (2020)
35. Xia, F., Zamir, A.R., He, Z., Sax, A., Malik, J., Savarese, S.: Gibson Env: real-world perception for embodied agents. In: CVPR (2018)
36. Yang, Y., et al.: Multi-agent determinantal Q-learning. In: ICML (2020)
37. Yu, C., Velu, A., Vinitsky, E., Wang, Y., Bayen, A.M., Wu, Y.: The surprising effectiveness of MAPPO in cooperative, multi-agent games (2021). arXiv:2103.01955
38. Yu, C., Velu, A., Vinitsky, E., Wang, Y., Bayen, A.M., Wu, Y.: The surprising effectiveness of MAPPO in cooperative, multi-agent games. CoRR (2021)
39. Zabounidis, R., Campbell, J., Stepputtis, S., Hughes, D., Sycara, K.P.: Concept learning for interpretable multi-agent reinforcement learning. In: Conference on Robot Learning, pp. 1828–1837. PMLR (2023)

Enhancing Continuous Domain Adaptation with Multi-path Transfer Curriculum

Hanbing Liu⑩, Jingge Wang⑩, Xuan Zhang⑩, Ye Guo⑩, and Yang Li$^{(\boxtimes)}$⑩

Tsinghua Shenzhen International Graudate School (SIGS), Tsinghua University,
Beijing, China
{liuhb21,wjg22}@mails.tsinghua.edu.cn,
{xuanzhang,guo-ye,yangli}@sz.tsinghua.edu.cn

Abstract. Addressing the large distribution gap between training and testing data has long been a challenge in machine learning, giving rise to fields such as transfer learning and domain adaptation. Recently, Continuous Domain Adaptation (CDA) has emerged as an effective technique, closing this gap by utilizing a series of intermediate domains. This paper contributes a novel CDA method, W-MPOT, which rigorously addresses the domain ordering and error accumulation problems overlooked by previous studies. Specifically, we construct a transfer curriculum over the source and intermediate domains based on Wasserstein distance, motivated by theoretical analysis of CDA. Then we transfer the source model to the target domain through multiple valid paths in the curriculum using a modified version of continuous optimal transport. A bidirectional path consistency constraint is introduced to mitigate the impact of accumulated mapping errors during continuous transfer. We extensively evaluate W-MPOT on multiple datasets, achieving up to 54.1% accuracy improvement on multi-session Alzheimer MR image classification and 94.7% MSE reduction on battery capacity estimation.

Keywords: Continuous Domain Adaptation · Wasserstein distance · Transfer curriculum · Optimal Transport · Path Consistency regularization

1 Introduction

Domain shift is a common challenge in many real life applications [15]. For example, in medical imaging, models trained on the data from one institution may not generalize well to another institution with different imaging hardware. Similarly, in battery capacity monitoring, models trained on lab-collected data may perform poorly under diverse operation environments. Obtaining the annotation for

---

Hanbing Liu, Jingge Wang, Xuan Zhang, and Yang Li are from the Shenzhen Key Laboratory of Ubiquitous Data Enabling, SIGS, Tsinghua University.

Supplementary Information The online version contains supplementary material available at https://doi.org/10.1007/978-981-97-2253-2_23.

the new domains, however, is often very costly or infeasible. To address this challenge, Unsupervised Domain Adaptation (UDA) has been proposed, leveraging the labeled data from the source domain to improve the performance of learning models on the unlabeled target domain [5]. In particular, UDA aims to align the distributions of the source and target domains using labeled source data and unlabeled target data, typically by learning domain-invariant representations [13] or adversarial learning schemes. Nevertheless, one challenge of UDA is its limited effectiveness when confronted with significant domain shift. Studies conducted by Zhao et al. [17] have shed light on the relationship between the domain shift and generalization error in UDA. They have shown that the effectiveness of UDA is bounded by the distributional divergence between the source and target domain, so the performance of the adapted model on the target domain may not be satisfactory with a substantial domain shift. In addressing this challenge, many works studied the problem of Continuous Domain Adaptation (CDA) [16].

Instead of directly adapting the model from the source to the target domain, CDA captures the underlying domain continuity leveraging a stream of observed intermediate domains, and gradually bridges the substantial domain gap by adapting the model progressively. There are various applications of CDA requiring continuous domains with indexed metadata [6,9]. For example, in medical data analysis, age acts as a continuous metadata for disease diagnosis across patients of different age groups. In the online battery capacity estimation problem, the state of health (SoH) of the battery acts as a continuous metadata that differs across batteries. CDA has attracted a great deal of attention and gained a rapid performance boost by self-training [7,18], pseudo-labeling [8], adversarial algorithms, optimal transport (OT) [10] and so on. In particular, Oritiz et al. [10] designed an efficient forward-backward splitting optimization algorithm for continuous optimal transport (COT) and demonstrated the efficacy of OT in reducing the domain shift and improving the performance of adaptation models.

While there have been significant advances in CDA, it still faces two critical challenges, determining the transfer order of intermediate domains in the absence of continuous metadata and mitigating cumulative errors throughout the continuous adaptation process. For the first issue, metadata could be missing or incorrect, and sometimes metadata alone can not fully explain the difference between data distributions. The proper ordering of intermediate domains is significant for CDA in transferring knowledge all the way to the target domain. Indeed, it is necessary to order the intermediate domains to facilitate continuous transfer without relying on explicit metadata. As a divergence measurement that takes into account the geometry of the data distributions, Wasserstein distance (w-distance) [14] plays an important role in deriving the generalization bound in domain adaptation, which implies the effectiveness of reducing domain divergence in the Wasserstein space. In this work, we propose a transfer curriculum in the Wasserstein space, aimed at determining the optimal sequence of intermediate domains for better knowledge transfer. The incorporation of Wasserstein-based transfer curriculum provides a principled and effective way to order the intermediate domains, enabling more precise and controlled knowledge transfer. A more comprehensive discussion regarding the process of selecting the appropriate transfer sequence, which ultimately leads to a tighter generalization bound, will be provided in the method section.

For the second issue, as the model progressively adapts to new domains, errors can accumulate and degrade the overall performance. The accumulation of errors can occur during CDA due to the successive estimation of pseudo-labels or intermediate adaptation results, e.g., the error accumulates during each projection of source domain data based on the estimated optimal transport map in [10]. Fourier domain filtering possesses the capability to mitigate cumulative errors [4], but its efficacy is limited to fixed frequencies of errors, thereby exhibiting inadequacies in terms of flexibility and robustness. To tackle this challenge, we introduce a path consistency regularization scheme for OT-based CDA inspired by [10]. Our multi-path regularization scheme enforces consistency among multiple transfer paths, effectively reducing the impact of accumulated errors and improving the robustness and stability of the transferred model.

In summary, the main contribution of this work lies in four aspects:

1) **W-MPOT**: The paper proposes a novel CDA framework, named W-MPOT, which incorporates a Wasserstein-based transfer curriculum and multi-path consistency regularization, providing a principled and effective solution for CDA in scenarios where explicit metadata is unavailable.
2) **Wasserstein-based Transfer Curriculum**: The method employs w-distance to devise the transfer curriculum, providing theoretical proofs and generalization upper bounds on the error incurred by improper sorting based on w-distance.
3) **Multi-Path Optimal Transport**: The paper introduces a multi-path domain adaptation method based on Optimal Transport, namely MPOT, to enforce consistency among multiple adaptation paths. By mitigating the impact of accumulated errors during continuous transfer, MPOT significantly enhances the overall performance and stability of the adaptation process.
4) **Comprehensive Empirical Validation**: We conduct a thorough set of experiments to validate the motivation and effectiveness of our proposed methods. These experiments cover various domains, including *ADNI, Battery Charging-discharging Capacity and Rotated MNIST* datasets, and demonstrate the superiority of our approach compared to alternative methods.

2 Methodology

2.1 Preliminary

We employ optimal transport (OT) to map the source domain into the target domain. OT provides a measurement of the divergence between two probability distributions by finding an optimal transportation plan that minimizes the total cost of mapping one distribution to the other [2]. Detailed explanations regarding OT can be found in Supplementary A.

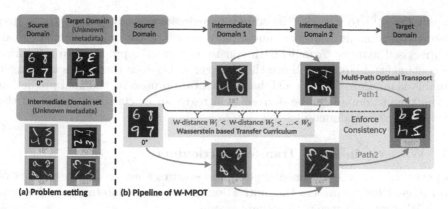

Fig. 1. Illustration of our proposed W-MPOT. (a) In the *Rotated MNIST* example, we are given a source domain of $0\,°C$ and a target/intermediate domain set with unknown degrees. (b) In our proposed W-MPOT, we address the challenge of unknown domain metadata (angles) and perform CDA. We utilize Wasserstein based transfer curriculum to sort intermediate domains and employ MPOT to enforce consistency, thereby enhancing the transfer effectiveness.

Consider the scenario where a labeled source domain and a collection of unlabeled auxiliary domains are available. Let $\mathcal{X} \subset \mathbb{R}^d$ be the feature space and $\mathcal{Y} \subset \mathbb{R}$ be the label space. Denote the source domain and target domain as $D_S = \{(\boldsymbol{x}_j, y_j)\}_{j=1}^{N_S}$, $D_T = \{\boldsymbol{x}_j\}_{j=1}^{N_T}$, where N_S and N_T are the number of samples in the source domain and target domain, respectively. The candidate set of intermediate domains is detonated as $\mathcal{D}_I = \{D_{I_1}, D_{I_2}, \ldots, D_{I_K}\}$, where each domain is denoted as $D_{I_k} = \{\boldsymbol{x}_j\}_{j=1}^{N_{I_k}}$, $k = 1, \ldots, K$ and N_{I_k} denotes the number of unlabeled data. Let μ_S, μ_{I_k}, $\mu_T \in \mathcal{P}(\mathcal{X})$ be the probability measures on $\mathbb{R}^d$ for the source, intermediate, and target domains, respectively. The objective is to make predictions $\{\widehat{y}_j\}_{j=1}^{N_T}$ for samples in the target domain.

While the order of intermediate domains in CDA can be determined using available metadata, the challenge arises when the metadata is absent, requiring further efforts to determine the proper order for CDA. Given the K candidate intermediate domains, a transfer curriculum defines an ordered sequence of N domains from $\mathcal{D}_I$, denoted as $\widehat{\mathcal{D}}_I$, to be used as intermediate domains for the CDA.

2.2   Method Framework

The framework of our proposed W-MPOT, comprising the Wasserstein-based transfer curriculum module and the Multi-Path Optimal Transport (MPOT) module, is depicted in Fig. 1. The Wasserstein-based transfer curriculum module selects an optimal intermediate domain sequence $\widehat{\mathcal{D}}_I$ for a given target domain D_T. Specifically, after calculating the w-distance [14] between each intermediate domain and the source domain, we sort intermediate domains closer to the source than the target domain by their w-distances. In the MPOT module, we adopt COT [10] to transfer the source domain knowledge through the sorted intermediate domain sequence, and finally adapt to the target domain in an unsupervised manner. A multi-path consistency term is proposed to regularize the continuous adaptation. Since the divergence between the source and target domains is minimized by the OT-based adaptation process, the final prediction for the target domain can be derived by a regressor or classifier trained on the transported source domain.

2.3   Wasserstein-Based Transfer Curriculum

Properly ordering the intermediate domains ensures a more effective transfer of knowledge. Here we present the motivation for arranging intermediate domains in the Wasserstein space with a simple example.

Given the labeled source domain D_S , the target domain D_T and two candidate intermediate domains D_{I_1} and D_{I_2}. We assume that the optimal transfer order is $D_S \rightarrow D_{I_1} \rightarrow D_{I_2} \rightarrow D_T$, i.e., the intermediate domain D_{I_1} is closer to D_S than to D_{T_2}. Apparently, another possible transfer order would be $D_S \rightarrow D_{I_2} \rightarrow D_{I_1} \rightarrow D_T$. We first present the generalization bound proposed by [12] which relates the source and target errors using w-distance. Let μ_{I_1}, $\mu_{I_2} \in \mathcal{P}(\mathcal{X})$ be the probability measures for domains D_{I_1}, D_{I_2}, respectively. h and f denote the predicted hypothesis and the true labeling function, respectively. $\epsilon_\mu(h, f) = \mathbb{E}_{x \sim \mu}[|h(x) - f(x)|]$ and ϵ is the combined error of the ideal hypothesis h^*. Assume that all hypotheses in the hypothesis set H satisfies the A-Lipschitz continuous condition for some A.

By simply applying the lemma proposed in [12] to each source-target pairs (D_S, D_{I_1}), (D_{I_1}, D_{I_2}) and (D_{I_2}, D_T), the generalization bound of the transfer path $D_S \rightarrow D_{I_1} \rightarrow D_{I_2} \rightarrow D_T$ could be derived as the following equation. For a detailed explanation of the lemma, please refer to Supplementary B.

$$\epsilon_{\mu_T}(h, f) \leq \epsilon_{\mu_S}(h, f) + 2\,A \cdot \mathcal{W}_1\left(\mu_S, \mu_{I_1}\right)$$
$$+ 2\,A \cdot \mathcal{W}_1\left(\mu_{I_1}, \mu_{I_2}\right) + 2\,A \cdot \mathcal{W}_1\left(\mu_{I_2}, \mu_T\right) + \epsilon, \tag{1}$$

Similarly, for another transfer path $D_S \rightarrow D_{I_2} \rightarrow D_{I_1} \rightarrow D_T$, the generalization bound is

$$\epsilon_{\mu_T}(h, f) \leq \epsilon_{\mu_S}(h, f) + 2\,A \cdot \mathcal{W}_1\left(\mu_S, \mu_{I_2}\right)$$
$$+ 2\,A \cdot \mathcal{W}_1\left(\mu_{I_2}, \mu_{I_1}\right) + 2\,A \cdot \mathcal{W}_1\left(\mu_{I_1}, \mu_T\right) + \epsilon, \tag{2}$$

where ϵ is the same. According to the optimal transfer path assumption, it is straightforward to get $\mathcal{W}_1\left(\mu_S, \mu_{I_2}\right) > \mathcal{W}_1\left(\mu_S, \mu_{I_1}\right)$, and $\mathcal{W}_1\left(\mu_{I_1}, \mu_T\right) > \mathcal{W}_1\left(\mu_{I_2}, \mu_T\right)$. Therefore, better domain transfer order $D_S \rightarrow D_{I_1} \rightarrow D_{I_2} \rightarrow D_T$ will lead to tighter generalization bound. The optimal transfer path order will achieve better performance on the target domain than the other, which justifies the use of the w-distance in our transfer curriculum.

This result leads to our Wasserstein-based transfer curriculum to select and sort for intermediate domains. We use the following form of w-distance to measure the closeness between each intermediate domain D_{I_k} and the source domain D_S,

$$W_k = \min_{\gamma} \langle \gamma_k, \mathbf{M}_k \rangle_F + \lambda \cdot \Omega(\gamma_k)$$

$$\text{s.t. } \gamma_k \mathbf{1} = \mu_S \quad \gamma_k^T \mathbf{1} = \mu_{I_k}, k = 1, \ldots, K,$$

(3)

where $\gamma_k \geq 0$ is the transport matrix and $\mathbf{M}_k$ is the cost matrix between the distribution μ_S and μ_{I_k}. W_k measures the similarity between each intermediate domain and the source domain, i.e., the greater the W_k is, the farther the intermediate domain is from the source domain. Note that any intermediate domain that is further from the source domain than the target domain is discarded. The remaining N domains in the intermediate domain set are then sorted in order of W_k and we could obtain a domain sequence $\widehat{\mathcal{D}}_I = \widehat{D}_{I_1} \rightarrow \widehat{D}_{I_2} \rightarrow \ldots \rightarrow \widehat{D}_{I_N}$, where $W_1 \leq W_2 \ldots \leq W_N$. As a result, the Wasserstein-based transfer curriculum generates a sorted transfer sequence $\widehat{\mathcal{D}}_I$ that represents the desired order of the domains, arranged from those closer to the source domain to those farther away. By utilizing the w-distance to sort multiple intermediate domains, we eliminate the need for meta-information.

2.4 Multi-path Optimal Transport

We apply OT-based domain adaptation iteratively across each intermediate domain indexed by the proposed curriculum. In order to make predictions for the target domain D_T, MPOT constructs sequential transport plans, denoted as γ_m, through a series of successive steps from the source to the target domain. The steps are as follows. Given the sorted sequence of intermediate domain $\widehat{\mathcal{D}}_I$, the source domain is initially mapped to the first intermediate domain $\widehat{D}_{I_1}$ using direct OT [2] and we can derive the first transport plan γ_0. Following this, the barycentric mapping of the source domain to the first intermediate domain $\widehat{D}_{I_1}$ can be defined by a weighted barycenter of its neighbors, $\mathcal{B}_{I_1}(D_S) = N_S \cdot \gamma_0 \cdot \widehat{D}_{I_1}$. The distribution of the mapped source domain on $\widehat{D}_{I_1}$ is denoted as μ_{SI_1}, which is consistent with the distribution of the target intermediate domain μ_{I_1}.

Then for the following intermediate domains, $\widehat{D}_{I_n}, n \in 2, \ldots, N$, the probabilistic coupling γ_{n-1} between the domain $\widehat{D}_{I_{n-1}}$ and the subsequent domain $\widehat{D}_{I_n}$ is calculated using COT [10],

$$\gamma_{n-1} = \operatorname*{argmin}_{\gamma \in \mathbb{R}^{N_S \times N_T}} \left\langle \gamma, \mathbf{M}^{[n-1,n]} \right\rangle + \lambda \Omega(\gamma) + \eta_t R_t(\gamma)$$

$$\text{s.t. } \gamma \mathbf{1} = \mu_{SI_{n-1}} \quad \gamma^T \mathbf{1} = \mu_{I_n} \quad \gamma \geq 0,$$

(4)

where $\mathbf{M}^{[n-1,n]}$ is the cost matrix defining the cost to move mass from the distribution of $\mu_{SI_{n-1}}$ to μ_{I_n}. $\Omega(\cdot)$ denotes the entropic regularization term $\Omega(\gamma) = \sum_{i,j} \gamma_{i,j} \log(\gamma_{i,j})$, and $\lambda > 0$ is the weight of entropic regularization. $R_t(\cdot)$ is a time regularizer with coefficients $\eta_t > 0$, which aims to enforce smoothness and coherence across consecutive time steps [10].

Upon the completion of the sequential transfer from the source domain to all intermediate domains, the transport plan γ_N from $\widehat{D}_{I_N}$ to the target domain D_T will be computed using MPOT with the equation as follows,

$$\gamma_N = \operatorname*{argmin}_{\gamma \in \mathbb{R}^{N_S \times N_T}} \left\langle \gamma, \mathbf{M}^{[N,N+1]} \right\rangle + \lambda\Omega(\gamma) + \eta_t R_t(\gamma) + \eta_p R_p(\gamma, \gamma_{p_2})$$

$$\text{s.t. } \gamma\mathbf{1} = \mu_{SI_N} \quad \gamma^T \mathbf{1} = \mu_T \quad \gamma \geq 0, \tag{5}$$

where $\mathbf{M}^{[N,N+1]}$ is the cost matrix between the distribution μ_{SI_N} and μ_T. $\eta_t > 0$ and $\eta_p > 0$ are the coefficients to adjust the weights. We further introduce a path consistency regularizer $R_p(\cdot)$ by comparing it with another transfer path,

$$R_p(\gamma, \gamma_{p_2}) = \left\| N_S \cdot \gamma \cdot \widehat{D}_{I_n} - N_S \cdot \gamma_{p_2} \cdot \widehat{D}_{I_n} \right\|_F^2, \tag{6}$$

where γ_{p_2} is the transport plan of the second possible path which is utilized to refine the γ of Path 1. During each adaptation step across intermediate domains in the curriculum, the incremental steps introduce minor inaccuracies, potentially impacting overall transfer performance over time. By employing $R_p(\cdot)$, we exploit the complementary information present in diverse paths to alleviate the accumulation of errors. The continuous adaptation results of the original COT and our proposed MPOT on the simulated half-moon dataset are shown in Fig. 2.

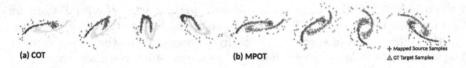

(a) COT   (b) MPOT   + Mapped Source Samples   △ GT Target Samples

Fig. 2. The Effect of Multi-Path regularization on the Optimal Transport of source domain sample. Visual experiments are conducted on simulated half-moon data to compare the migration effects of (a) COT with our proposed (b) MPOT. The source domain has an angle of $0\,^\circ$C, and the angles on the graph increase by $36\,^\circ$C from left to right. (a) and (b) depict the mapping results obtained by applying the COT and MPOT methods respectively to sequentially map the source domain to the target domains. The triangles represent the ground truth target domain data, while the plus signs represent the mapped source domain data after the adaptation.

During continuous adaptation, the COT method experiences a substantial increase in cumulative error, leading to deteriorating adaptation results as the rotation angle increases. In contrast, our proposed method, MPOT, effectively addresses the issue of cumulative errors by leveraging multiple transfer paths.

Notablely, the paths interact in a bidirectional manner within $R_p(\cdot)$: not only does Path 2 impose constraints on Path 1, but Path 1 likewise restricts Path 2. This creates a reciprocal dynamic and both paths can be jointly optimized simultaneously. Moreover, the above OT problem (5) fits into the forward-backward splitting algorithm [1] as in [10], whose solution could be efficiently computed with the Sinkhorn algorithm [3]. More details about the solution for the Sinkhorn algorithm are given in Supplementary C. Let $\mathcal{J}(\cdot)$ be the combination of

---

Algorithm 1: Bidirectional Optimization algorithm in MPOT

Input: Transport matrix of Path 1 γ_{p_1}, Transport matrix of Path 2 γ_{p_2}, step size α, cost matrix of Path 1 $\mathbf{M}^{[1]}$, cost matrix of Path 2 $\mathbf{M}^{[2]}$, weight of Path 1 λ_1 and Path 2 λ_2, iteration times c

Output: Refined transport matrix from N-th intermediate domain to target domain γ'_N

1 **Initialize:** $\gamma_0 \in (0, +\infty)^{N_S \times N_T}$
2 **for** $c \leftarrow 0, 1, \ldots$ **do**
3 $\quad \mathbf{M}_c^{[1]} = \alpha \mathbf{M}^{[N,N+1]} + \alpha \nabla \mathcal{J}(\gamma_c, \gamma_{p_2})$
4 $\quad \mathbf{M}_c^{[2]} = \alpha \mathbf{M}^{[N,N+1]} + \alpha \nabla \mathcal{J}(\gamma_c, \gamma_{p_1})$
5 $\quad \mathbf{M}_c = \lambda_1 \mathbf{M}_c^{[1]} + \lambda_2 \mathbf{M}_c^{[2]}$
6 $\quad \gamma_{c+1} = \text{Sinkhorn}\left(\mathbf{M}_c, 1 + \alpha\lambda, \mu_{SI_N}, \mu_T\right)$
7 $\gamma'_N = \gamma_\infty$

---

regularization terms. The details of the bidirectional optimization procedure are depicted in the subsequent Algorithm 1.

Once the refined transport matrix, denoted as γ'_N, has been computed, the barycentric mapping from the mapped source domain to the target domain can be obtained by using $\mathcal{B}_T(D_S^{I_N}) = N_S \cdot \gamma'_N \cdot D_T$, where $D_S^{I_j}$ denotes the domain mapped from the source domain D_S to $\widehat{D}_{I_j}$. Consequently, a classifier or regressor can be trained on $\mathcal{B}_T(D_S^{I_N})$ using the available source labels and can be directly deployed on the target domain for various applications.

3   Experimental Results

3.1   Datasets and Experimental Configurations

Experiments are conducted on three datasets, each offering unique characteristics and challenges. Details regarding the implementation are presented in Supplementary D.

ADNI. The Alzheimer's Disease Neuroimaging Initiative (ADNI) dataset is a crucial resource for Alzheimer's disease research, housing a repository of MRI images that furnish in-depth structural insights into the brain [11]. The 3D MRI data is sliced along the direction of the vertical spinal column, resulting in 2D MRI images. The 2D MRI images are categorized into a source domain (ages 50–70, 190 samples) and intermediate domains (ages 70–72, 72–74, 74–76, 76–78, 78–82, 82–92, 50 samples each). The primary task is classifying MRI images into five disease categories. In our ADNI MRI image experiments (128×128 dimensions), we performed image normalization to a 0–1 range. Subsequently, we utilized a pre-trained VGG16 model, originally trained on the ImageNet dataset, to extract image features. The VGG16 model's output consists of a

$512 \times 4 \times 4$ feature map, which is condensed into a 16-dimensional feature vector by averaging across the channel dimension.

Battery Charging-Discharging Capacity. The capacity of lithium-ion batteries holds paramount significance in the context of power systems and electric vehicles. Laboratory experiments are conducted via a pulse test [19] to collect voltage-capacity data pairs on batteries during charging and discharging processes at different state of charge (SoC) levels. The dataset includes a source domain (5% SoC) and nine intermediate domains (10%–50% SoC), each with 67 samples. It is a regression problem evaluated with Mean Squared Error (MSE).

Rotated MNIST. The Rotated MNIST dataset is a variation of the MNIST dataset, featuring rotated digit images. It is divided into five domains with different rotation angles. The source domain (0 °C) and intermediate domains (18, 36, 54, 72, and 90 °C) each contain 1000 samples.

3.2 Analysis of Wasserstein-Based Transfer Curriculum

This section assesses the effectiveness of our proposed Wasserstein-based transfer curriculum, comparing it with two alternative domain adaptation methods. We examine Directly Optimal Transport (DOT), which transfers data directly from the source to the target domain, and Continuously Optimal Transport (COT), a progressive adaptation approach. COT is explored in two configurations: COT+Metadata, utilizing genuine domain metadata for sorting intermediate domains, and COT+W-dis, employing our Wasserstein-based transfer curriculum in the absence of metadata. In our experimental design, we maintain consistency between the source and target domains across all methods.

As shown in Fig. 3, the superior performance of COT over DOT, coupled with the comparable performance between COT utilizing w-distance and COT with true metadata, shows the effect of intermediate domain sorting. The impact of varying the number of intermediate domains differs across datasets, which is determined by the inherent characteristics of the data itself. These findings highlight the benefits of incorporating COT and the Wasserstein distributional geometric relationships.

3.3 Adaptation Comparison Results

A comprehensive comparison of our proposed method, W-MPOT, is conducted with several classic approaches in the realm of CDA. We present the comparison results in Table 1.

In our experimental setup, we maintain a fixed number of two intermediate domains, and each experiment is repeated 100 times to ensure robustness, with the average results reported. In the bottom section of Table 1, we showcase the outcomes of our proposed algorithm, W-MPOT, applied in three distinct scenarios: p2→p1, p1→p2, and p1 + p2. These scenarios involve using one path

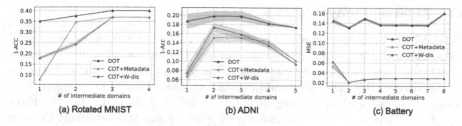

(a) Rotated MNIST   (b) ADNI   (c) Battery

Fig. 3. Domain ordering Results. The results for the (a)Rotated MNIST, (b)ADNI, and (c)Battery Charging-discharging Capacity datasets are shown in this figure. The evaluation metric for Rotated MNIST and ADNI is accuracy, while for the Battery dataset is MSE. The methods compared are DOT (Direct Optimal Transport) and COT (Continuous Optimal Transport). Two sorting approaches are utilized: metadata-based sorting and Wasserstein based transfer curriculum. For ease of comparison, the y-coordinate of the first two datasets represents $1 - ACC$, where lower values indicate superior performance. Each configuration is evaluated 100 times, and the shaded areas represent the variance.

to refine the other and utilizing both paths for a new refined path. Bilateral experiments are conducted by applying a mutual constraint mechanism using regularizers from both paths.

The results indicate the superior performance of W-MPOT over other methods across all three datasets, underscoring its efficacy in mitigating continuous domain shifts. Notably, W-MPOT using both path 1 and path2 shows the optimal performance, providing strong evidence for the effectiveness of the added regularization term $R_p(\cdot)$. The results of W-MPOT in other two scenarios: p2→p1 and p1→p2, showing similar performance, suggest that the regularization approach is robust and does not heavily rely on the specific choice of the second path. By leveraging both paths with mutual constraints, W-MPOT successfully improves the overall robustness of the model.

Table 1. MSE or Accuracy for three datasets of different algorithms

Method	ADNI (↑)	Battery (↓)	ROT MNIST (↑)
Source Model	41.2	0.3731	48.1
CMA [6]	55.4	0.3842	65.3
EAML [9]	68.3	0.2045	70.4
AGST [18]	57.3	0.3534	76.2
Gradual ST [7]	64.5	0.1068	87.9
CDOT [10]	82.6	0.0209	75.6
W-MPOT(p2→p1)	86.7	0.0199	88.3
W-MPOT(p1→p2)	86.5	0.0197	87.2
W-MPOT(p1 + p2)	**88.3**	**0.0185**	**89.1**

3.4   Ablation Study

We conducted ablation studies on the *Battery Charging-discharging Capacity* dataset to investigate the effect of domain partitioning, Wasserstein-based sorting strategy, and path consistency regularization separately. Delimited domains consistently exhibit lower mean and variance of MSE compared to random batch sampling in Fig. 4(a), indicating increased stability. This highlights the necessity of considering distinct domains for improved predictive accuracy. Comparing Unordered COT to Ordered COT reveals consistently lower MSE values in Fig. 4(b), emphasizing the value of the Wasserstein-based transfer curriculum for superior performance in battery capacity prediction tasks. The vital role of the path consistency regularization term in achieving accurate and robust domain mappings is demonstrated by the comparison of MPOT (with $R_p(\cdot)$) and COT (without $R_p(\cdot)$) in Fig. 4(c).

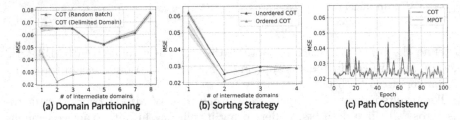

Fig. 4. Ablation study Results. The experiments from a to b involved domain partitioning, sorting strategy, and path consistency. For each number of intermediate domains in a and b, the experiment was randomly repeated 100 times. The solid line represents the mean MSE, while the shaded area represents the variance. The experiment in c is conducted by fixing the number of intermediate domains to 2 and randomly sampling different intermediate domains 100 times.

4   Conclusion

This research introduces the W-MPOT framework for CDA, effectively addressing substantial domain shifts and missing metadata. It comprises the Wasserstein-based Transfer Curriculum for domain ordering and MPOT for cumulative errors during adaptation. Experimental results across diverse datasets demonstrate its superior performance, highlighting the practicality and potential of these methods in handling substantial domain shift challenges. This work advances the field of CDA and offers insights into addressing domain shift effectively, especially in domains like healthcare and energy storage. Future research could focus on establishing generalization bounds and employing reinforcement learning to optimize the selection of intermediate domains, making domain adaptation a sequential decision-making process.

Acknowledgement. This work was supported in part by the Natural Science Foundation of China (Grant 62371270), Shenzhen Key Laboratory of Ubiquitous Data Enabling (No.ZDSYS20220527171406015), and Tsinghua Shenzhen International Graduate School Interdisciplinary Innovative Fund (JC2021006).

References

1. Bùi, M.N., Combettes, P.L.: Bregman forward-backward operator splitting. Set-Valued Variat. Anal. **29**, 583–603 (2021)
2. Courty, N., Flamary, R., Tuia, D., Rakotomamonjy, A.: Optimal transport for domain adaptation (2016)
3. Cuturi, M.: Sinkhorn distances: lightspeed computation of optimal transport. Adv. Neural Inf. Process. Syst. **26** (2013)
4. Durak, L., Aldirmaz, S.: Adaptive fractional fourier domain filtering. Signal Process. **90**(4), 1188–1196 (2010)
5. Ganin, Y., Lempitsky, V.: Unsupervised domain adaptation by backpropagation. In: International Conference on Machine Learning, pp. 1180–1189. PMLR (2015)
6. Hoffman, J., Darrell, T., Saenko, K.: Continuous manifold based adaptation for evolving visual domains. In: Proceedings of the IEEE Conference on Computer Vision and Pattern Recognition, pp. 867–874 (2014)
7. Kumar, A., Ma, T., Liang, P.: Understanding self-training for gradual domain adaptation. In: International Conference on Machine Learning, pp. 5468–5479. PMLR (2020)
8. Liang, J., Hu, D., Feng, J.: Do we really need to access the source data? source hypothesis transfer for unsupervised domain adaptation. In: International Conference on Machine Learning, pp. 6028–6039. PMLR (2020)
9. Liu, H., Long, M., Wang, J., Wang, Y.: Learning to adapt to evolving domains. Adv. Neural. Inf. Process. Syst. **33**, 22338–22348 (2020)
10. Ortiz-Jiménez, G., El Gheche, M., Simou, E., Maretić, H.P., Frossard, P.: Forward-backward splitting for optimal transport based problems. In: ICASSP 2020-2020 IEEE International Conference on Acoustics, Speech and Signal Processing (ICASSP), pp. 5405–5409. IEEE (2020)
11. Petersen, R.C., et al.: Alzheimer's disease neuroimaging initiative (adni): clinical characterization. Neurology **74**(3), 201–209 (2010)
12. Shen, J., Qu, Y., Zhang, W., Yu, Y.: Wasserstein distance guided representation learning for domain adaptation. In: Proceedings of the AAAI Conference on Artificial Intelligence, vol. 32 (2018)
13. Sun, B., Feng, J., Saenko, K.: Correlation alignment for unsupervised domain adaptation. In: Domain Adaptation in Computer Vision Applications, pp. 153–171 (2017)
14. Villani, C., Villani, C.: The wasserstein distances. In: Optimal Transport: Old and New, pp. 93–111 (2009)
15. Wang, Y., et al.: An empirical study of selection bias in pinterest ads retrieval. In: Proceedings of the 29th ACM SIGKDD Conference on Knowledge Discovery and Data Mining, pp. 5174–5183 (2023)
16. Xu, Y., Jiang, Z., Men, A., Liu, Y., Chen, Q.: Delving into the continuous domain adaptation. In: Proceedings of the 30th ACM International Conference on Multimedia, pp. 6039–6049 (2022)

17. Zhao, H., Des Combes, R.T., Zhang, K., Gordon, G.: On learning invariant representations for domain adaptation. In: International Conference on Machine Learning, pp. 7523–7532. PMLR (2019)
18. Zhou, S., Wang, L., Zhang, S., Wang, Z., Zhu, W.: Active gradual domain adaptation: dataset and approach. IEEE Trans. Multimedia **24**, 1210–1220 (2022)
19. Zhou, Z., et al.: A fast screening framework for second-life batteries based on an improved bisecting k-means algorithm combined with fast pulse test. J. Energy Storage **31**, 101739 (2020)

Graphs and Networks

Enhancing Network Role Modeling: Introducing Attributed Multiplex Structural Role Embedding for Complex Networks

Lili Wang[1]($\boxtimes$), Chenghan Huang[2], Ruiye Yao[3], Chongyang Gao[4],
Weicheng Ma[1], and Soroush Vosoughi[1]

[1] Dartmouth College, Hanover, NH 03755, USA
{lili.wang.gr,soroush}@dartmouth.edu
[2] Millennium Management, LLC, New York, NY 10022, USA
[3] University of California, Davis, Davis, CA 95616, USA
[4] Northwestern University, Evanston, IL 60208, USA

Abstract. Numerous studies have focused on defining node roles within networks, producing network embeddings that maintain the structural role proximity of nodes. Yet, these approaches often fall short when applied to complex real-world networks, such as Twitter, where nodes have varying types of relationships (e.g., following, retweeting, replying) and possess relevant attributes impacting their network role (e.g., user profiles). To address these limitations, this study presents a novel method for attributed (for dealing with attributed nodes) multiplex (for dealing with networks with different types of edges) structural role embedding. This approach uses an autoencoder mechanism to concurrently encode node structure, relationships, and attributes, thus successfully modeling nodes' roles within networks. Our method's effectiveness is shown through quantitative and qualitative analyses conducted on synthetic networks, outperforming established benchmarks in identifying node roles within multiplex and attributed networks. Additionally, we have assembled a robust real-world multiplex network composed of almost all verified Twitter users comprised of retweet, reply, and followership interactions between these users, representing three different layers in our multiplex network. This network serves as a practical environment to evaluate our method's capability to map the structural roles of users within real-world attributed multiplex networks. Using a verified dataset of influential users as a reference, we show our method excels over the existing benchmarks in learning structural roles on large-scale, real-world attributed multiplex networks, exemplified by our Twitter network.

1 Introduction

In recent years, social media has increasingly become a staple component of the public sphere, playing a vital role in a multitude of political, social, and cultural movements. This significant influence mandates the evolution of more robust tools to enhance our understanding of their operations. A particular aspect of

D.-N. Yang et al. (Eds.): PAKDD 2024, LNAI 14646, pp. 301–313, 2024.
https://doi.org/10.1007/978-981-97-2253-2_24

social media that necessitates further exploration is users' various roles within these networks. Effective information retrieval and recommendation in online social networks often depend on strategies that compare users on these platforms. Accurate user similarity computations can notably enhance cold-start recommendation systems. Beyond this, user representations are also crucial in pinpointing influential entities within expansive social networks, which isn't simply about tallying followers. Yet, going beyond rudimentary techniques, such as relying merely on user metadata, poses a challenge. Defining and formalizing user roles on social media is complex, given that these roles organically evolve as users participate in actions like sharing, replying, and following. These roles, in their essence, are multifaceted and determined by a user's position within diverse networks, such as followership, retweeting, and reply networks.

Existing methods, referred to as structural role network embedding, facilitate the identification of nodes' roles in networks. Here, "structural role" refers to a node's functional position, for example, as a hub or a bridge between communities. Distant or even disconnected nodes might exhibit identical or closely related structural roles. Methods such as struc2vec [16] and GraphWave [3] represent nodes as vectors encapsulating role-specific information. Such representations are pivotal for network analysis; for instance, the cosine similarity between two vectors can indicate the structural role similarity of the nodes they represent.

Nevertheless, existing methods primarily focus on singular-relationship networks and overlook node attributes. Contrarily, social media roles rely on a fusion of users' personal information and interactions within networks comprising various types of relationships. In other words, the roles are defined on multiplex and attributed networks. The domain of attributed multiplex structural role embedding remains relatively unexplored. Addressing this void, we introduce a formal definition of attributed multiplex structural roles and advance a novel method for learning their representation. Our paper puts forth the following contributions: (1) We introduce the first method for embedding attributed multiplex structural roles in networks. (2) To gauge the effectiveness of our embedding method both qualitatively and quantitatively, we have created a series of synthetic datasets. Our model consistently outperforms several established benchmarks in all evaluations. (3) We have curated a large real-world multiplex dataset containing nearly every verified Twitter account, summing up to approximately 360K accounts. This dataset encapsulates their posts, followers, retweets, and replies interactions. With this vast dataset, we validate the efficacy of our method in identifying structural roles in expansive, real-world attributed multiplex networks.

2   Related Work

Network embeddings convert nodes in a network into vector representations, with different embeddings capturing unique aspects of the nodes (e.g., see [6,7,12,14, 19,22–26]). Most of these embeddings rely on local homogeneous methods such as DeepWalk [14], node2vec [6], LINE [19], HOPE [12], GraRep [1], and SDNE [21]. These methods chiefly determine node similarity through their shared neighbors.

In contrast, structural role embedding methods like struc2vec [16], Graph-Wave [3], SEGK [11], RESD [31], and DRNE [20] aim to capture the specific

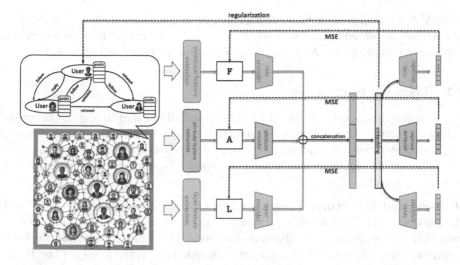

Fig. 1. The framework of our proposed method.

"roles" nodes play within a network. For instance, nodes, even if they are distant or disconnected, might have similar roles, such as serving as the hub of a community or acting as a bridge between two distinct groups. However, a limitation of these methods is their primary design for homogeneous networks, which have a singular edge type and lack node attributes.

Multiplex network embedding methods, such as MNE [29] and MVE [15], were developed to cater to the need for embeddings that can capture diverse relationships. Concurrently, there was the rise of heterogeneous network embedding techniques like metapath2vec [2], HIN2Vec [5], and JUST [8]. The next evolution incorporated node attributes into multiplex embeddings, illustrated by methods like DMGI [13] and URAMN [30]. Notably, these multiplex embedding methods integrated attributes but did not address structural roles.

There is a gap in the existing literature. While there are methods addressing attributed multiplex embedding, they focus on a granular level of embeddings. On the other hand, embeddings that preserve structural roles are still tied to homogeneous networks. This leaves a significant void when analyzing user roles on social media platforms with the nuanced lens of attributed multiplex structural role embedding. Our proposed approach endeavors to fill this crucial gap.

3   Methodology

An attributed multiplex network $\mathbf{G}$ is defined as: $\mathbf{G} = (\mathbf{V}, \mathbf{E}, \mathcal{T}_E, \mathcal{W}_E, \mathcal{A}_V)$, in which edges E are associated with two mapping functions $\phi(\mathbf{E}) : \mathbf{E} \to \mathcal{T}_E, \mathcal{T}_E \in \mathbb{N}$, $\psi(\mathbf{E}) : \mathbf{E} \to \mathcal{W}_E, \mathcal{W}_E \in \mathbb{R}$, and nodes $\mathbf{V}$ are associated with a mapping function $\eta(\mathbf{V}) : \mathbf{V} \to \mathcal{A}_V, \mathcal{A}_V \in \mathbb{R}^m$. The $\mathcal{T}_E$ denotes the possible set of relation types of the edges, the $\mathcal{W}_E$ denotes the edge weights and the $\mathcal{A}_V$ denotes the possible set of attributes of the nodes. Furthermore, for a node $v \in \mathbf{V}, \mathcal{N}_k(v) =$

$\{u \in \mathbf{V} \mid \exists$ a path from v to u, with length $k\}$ is the set of its k-hop neighbors. For a node v, we consider the task of learning a structural role preserving d-dimensional embedding, $\mathbf{X}_v \in \mathbb{R}^d, d \ll |\mathbf{V}|$, in an unsupervised fashion.

3.1   Learning Framework

Our learning framework, inspired by RESD and illustrated in Fig. 1, comprises four key components: role-related structural feature extraction, layer feature extraction, attribute feature extraction, and an auto-encoder mechanism. The auto-encoder mechanism jointly encodes these three types of features into a singular embedding.

Role-related Structural Feature Extraction. This module aims to extract the pure structural role identity of nodes. To achieve this, we adopt Forman curvature [4], an essential tool in discrete differential geometry, quantifying the local geometric properties inherent to a network. As a discrete analog to the traditional concept of curvature, it measures the interconnectedness and relationships between vertices, edges, and faces, encapsulating local and global geometric features.

Considering that the mathematical representation of Forman-Ricci curvature is highly technical and less applicable to the current context of networks, we adopt a simplified version [17] suitable for graph-based applications. In this version, the Forman curvature of an edge $e(u_1, u_2)$ is defined as:

$$
\begin{aligned}
\mathbf{F}(e) = 2 - \psi(e) &\sum_{v_1 \in \mathcal{N}_1(u_1), v_1 \neq u_2} \frac{1}{\sqrt{\psi(e)\psi((u_1, v_1))}} \\
-\psi(e) &\sum_{v_2 \in \mathcal{N}_1(u_2), v_2 \neq u_1} \frac{1}{\sqrt{\psi(e)\psi((u_2, v_2))}}
\end{aligned}
\tag{1}
$$

To extract the role-related structural feature F_v of node v, we generate a histogram that compiles the Forman curvature values of all $\mathbf{F}((v, u))$ $((v, u) \in E)$.

Role-related Attribute Feature Extraction. This module is designed to extract attribute features for each node, taking their roles into account. The fundamental concept here is to amalgamate the attributes of adjacent nodes towards the central node. However, considering roles, nodes positioned further from the center should have diminished importance. To account for this, we introduce a decay parameter $p < 1$ in the subsequent equation, which serves to exponentially decrease the significance of nodes situated farther from the center:

$$
\mathbf{A}_v = \sum_{k=1}^{K} p^k \mathcal{A}_u, u \in \mathcal{N}_k(v)
\tag{2}
$$

In this equation, $\mathcal{A}_v$ symbolizes the extracted attribute features of node v. The attributes of its k-hop neighbors u are aggregated in proportion to their distance from v. K denotes the longest distance we consider.

Role-related Layer Feature Extraction. Similar to attribute feature extraction, this module aims to extract layer features for each node. To achieve this, we enumerate all the k-hop edges of each layer surrounding the central node:

$$L_{v,i,k} = \sum_{u_1 \in \mathcal{N}_{k-1}(v), u_2 \in \mathcal{N}_k(v), (u_1,u_2) \in \mathbf{E}} \mathbb{1}\left(\phi\left(u_1, u_2\right) = i\right) \tag{3}$$

In this context, $L_{v,i,k}$ represents the extracted layer features of node v for layer i, considering k-hop neighbors.

Autoencoder. After extracting these three types of features, we utilize an Autoencoder mechanism to learn the multiplex and attributed role embedding for each node. Given the extracted features $\mathbf{F}$, the structural encoder is defined as:

$$\mathbf{h}_F^i = \tanh\left(\mathbf{W}_F^i \mathbf{h}_F^{i-1} + \mathbf{b}_F^i\right), i = 1, 2, \ldots, |Layer| \tag{4}$$

where $\mathbf{W}_F^i$ and $\mathbf{b}_F^i$ represent the weights and bias of i-th layer, respectively, and the $\mathbf{h}_F^0$ is initialized with $\mathbf{F}$ as the input of the encoder. The attribute and layer encoders are defined similarly. The outputs of the three encoders are then concatenated into a single vector and fed into a fully connected layer to generate the embeddings:

$$\mathbf{h}_C = \mathbf{h}_F^{|Layer|} \oplus \mathbf{h}_A^{|Layer|} \oplus \mathbf{h}_L^{|Layer|} \tag{5}$$

$$\mathbf{X} = \tanh\left(\mathbf{W}_E \mathbf{h}_C + \mathbf{b}_E\right) \tag{6}$$

The structural, attribute, and layer feature decoders are defined similarly to their corresponding encoders. For illustration, we provide the structure of the structural decoder as an example:

$$\hat{\mathbf{h}}_F^i = \tanh\left(\hat{\mathbf{W}}_F^i \hat{\mathbf{h}}_F^{i-1} + \hat{\mathbf{b}}_F^i\right), i = 1, 2, \ldots, |Layer| \tag{7}$$

where $\hat{\mathbf{W}}_F^i$ and $\hat{\mathbf{b}}_F^i$ represent the weights and bias of i-th layer, respectively, and the $\hat{\mathbf{h}}_F^0$ is initialized with $\mathbf{X}$ as the input of the decoder.

To optimize our autoencoder, we employ four loss functions: structural loss ($\mathcal{L}_{str}$), attribute loss ($\mathcal{L}_{att}$), layer loss ($\mathcal{L}_{layer}$), and regularization loss ($\mathcal{L}_r$). The first three are employed to ensure that the output of the decoders can accurately reconstruct the three input features $\mathbf{F}$, $\mathbf{A}$, and $\mathbf{L}$:

$$\mathcal{L}_{str} = \sum_{v \in \mathbf{V}} \left\| \mathbf{F}_v - \hat{\mathbf{h}}_{F,v}^{|Layer|} \right\|_2^2 t \quad [\mathcal{L}_{att} = \sum_{v \in \mathbf{V}} \left\| \mathbf{A}_v - \hat{\mathbf{h}}_{A,v}^{|Layer|} \right\|_2^2] \tag{8}$$

$$\mathcal{L}_{layer} = \sum_{v \in \mathbf{V}} \left\| \mathbf{L}_v - \hat{\mathbf{h}}_{L,v}^{|Layer|} \right\|_2^2$$

The regularization loss is designed to mitigate overfitting and to ensure that the learned embeddings $\mathbf{X}$ are able to recover at least some local structural properties of each node:

$$\mathcal{L}_r = \sum_{v \in V} \left\| \mathbf{q}(v) - \mathbf{g}(\mathbf{X}_v) \right\|_2^2 \tag{9}$$

where g is a non-linear function learned by a Multi-Layer perceptron model applied to the embedding of each node, and q is a function that reflects local structural information of each node (e.g., degree, centrality, etc.). This paper uses a simple concatenation of degrees in each layer as q. The overall loss $\mathcal{L}_{total}$ is defined as a weighted combination of all four losses:

$$\mathcal{L}_{total} = \mathcal{L}_r + \lambda_1 \mathcal{L}_{str} + \lambda_2 \mathcal{L}_{att} + \lambda_3 \mathcal{L}_{layer} \tag{10}$$

3.2   Model Training

In optimizing our model, we aim to minimize the global loss $\mathcal{L}_{total}$ to obtain the corresponding embeddings X. The Adam optimizer [9] is utilized for parameter optimization. The hyperparameters are set as follows: an embedding size of 128, learning rate of 0.001, batch size of 32, 500 training epochs, and loss weights λ_1, λ_2, and λ_3 all set to 3. For consistency across experiments, the random seed is initialized to 0. Experiments are conducted on an RTX A6000.

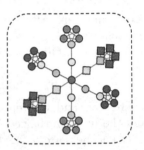

Fig. 2. An attributed multiplex barbell graph with two types of edges (blue and red) and two types of node attributes (square and circle). The colors of nodes represent roles which are not known beforehand. (Color figure online)

4   Experiments

Here, we evaluate our method on node visualization, clustering, and classification tasks.

4.1   Visualization of Attributed Multiplex Role Embeddings

We evaluate generated embeddings through visualization on a synthetic attributed multiplex barbell graph. This graph consists of three barbell graphs in two layers, each consisting of two complete subgraphs connected by a "bridge". Figure 2 shows the barbell graph used in our experiments. We use different edge colors to denote different layers and node shapes to denote different node attributes. The structurally equivalent nodes have the same color. For baselines, we use node2vec, struc2vec, and DMGI as the representative methods for homogeneous local proximity, structural role, and attributed multiplex local proximity methods, respectively. Other methods in the same category show similar embedding distributions with these three.

Figure 3(a) displays the embeddings generated by node2vec. As observed, node2vec fails to capture structural role identities as well as relational or attribute information. Instead, node2vec arranges the nodes of the six complete subgraphs somewhat haphazardly, resulting in two groups of reds, two groups of blue, and two groups of purple. Figure 3(b) presents the outcomes from struc2vec. Predictably, this method overlooks both relational and attribute data, embedding nodes solely based on their roles in the homogeneous version of the graph. This yields four distinct groups: (1) Subgraph components, depicted

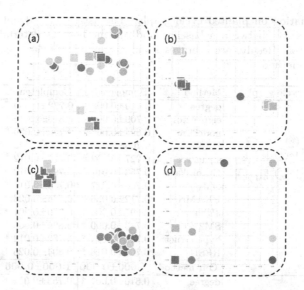

Fig. 3. Embeddings of nodes in the multiplex barbell by (a) node2vec (b) struc2vec (c) DMGI (d) ours. (Color figure online)

by red and blue circles encompassed by purple squares; (2) Subgraph connectors, shown as yellow and cyan circles encased by light sky blue squares; (3) Bridges, represented by green and orange circles surrounded by bisque squares; (4) The center, illustrated by a single indigo circle. Figure 3(c) demonstrates the embeddings of DMGI. While this method adeptly captures attribute information, it neglects other information important for identifying roles. It clusters nodes with similar attributes together. Lastly, Fig. 3(d) showcases the embeddings from our proposed method, which adeptly discerns the attributed multiplex structural roles between the nodes.

4.2 Node Clustering

We next evaluate the embeddings generated for the task of node clustering within intricate synthetic networks. Our analysis builds upon the synthetic datasets and settings employed by GraphWave [3], modifying them to a multiplex and attributed version. These datasets are characterized by a series of cycles adorned with basic shapes (e.g., "house", "fan", and "star") and accompanied by their ground-truth labels. In our adaptation, each graph comprises three layers of edges (colored black, blue, and red) and two distinct node attributes (represented as square and circle) as illustrated in Fig. 4. We also introduce a "perturbed" variant to gauge the robustness of models against noise. This variant augments the original graph with an additional 5% of edges inserted randomly. For the "house" and "house perturbed" configurations, we implement an 18-node cycle, interspersed uniformly with 3 red&circle, 3 blue&circle, and 3 blue&square "houses". In the scenarios labeled "varied" and "varied perturbed", an 18-node

Table 1. Evaluation on planted cycle graphs. The asterisk (*) denotes statistically significant differences between a baseline and our method, as determined by the t-test with $p < 0.05$. Results are averages from 10 runs. The bold text highlights the top-performing method.

Shapes along a cycle graph		Method	Homogeneity	Completeness	Rand index
Attributed House		degree	0.620±0.000 *	0.772±0.000 *	0.852±0.000 *
		eignvector	0.700±0.000 *	0.810±0.000 *	0.898±0.000 *
		node2vec	0.694±0.000 *	0.750±0.000 *	0.922±0.000 *
		DRNE	0.766±0.020 *	0.821±0.032 *	0.938±0.006 *
		struc2vec	0.727±0.012 *	0.771±0.012 *	0.928±0.007 *
		GraphWave	0.753±0.011 *	0.792±0.010 *	0.937±0.004 *
		DMGI	0.700±0.017 *	0.733±0.017 *	0.928±0.004 *
		URAMN	0.712±0.026 *	0.734±0.023 *	0.930±0.005 *
		JUST	0.695±0.004 *	0.750±0.001 *	0.923±0.001 *
		SM2Vec	0.713±0.000 *	0.891±0.000 *	0.885±0.000 *
		hyperstruct	0.747±0.011 *	0.782±0.014 *	0.937±0.003 *
		RESD	0.764±0.018 *	0.804±0.025 *	0.939±0.005 *
		Our method	**1.000±0.000**	**1.000±0.000**	**1.000±0.000**
House Perturbed		degree	0.610±0.007 *	0.753±0.011 *	0.856±0.004 *
		eignvector	0.661±0.025 *	0.748±0.027 *	0.892±0.015 *
		node2vec	0.698±0.004 *	0.744±0.004 *	0.925±0.002 *
		DRNE	0.755±0.030 *	0.803±0.037 *	0.937±0.007 *
		struc2vec	0.769±0.034 *	0.809±0.038 *	0.939±0.008*
		GraphWave	0.764±0.007 *	0.824±0.001 *	0.935±0.004 *
		DMGI	0.706±0.013 *	0.743±0.011 *	0.928±0.004 *
		URAMN	0.709±0.017 *	0.751±0.022 *	0.927±0.003 *
		JUST	0.701±0.004 *	0.745±0.005 *	0.925±0.002 *
		SM2Vec	0.739±0.020 *	0.828±0.018 *	0.920±0.010 *
		hyperstruct	0.733±0.020 *	0.777±0.019 *	0.932±0.006 *
		RESD	0.764±0.024 *	0.825±0.017 *	0.931±0.011 *
		Our method	**0.842±0.034**	**0.921±0.035**	**0.953±0.006**
Attributed Varied		degree	0.574±0.000 *	0.767±0.000 *	0.853±0.000 *
		eignvector	0.750±0.000 *	0.826±0.000 *	0.939±0.000 *
		node2vec	0.770±0.004 *	0.783±0.004 *	0.958±0.001 *
		DRNE	0.744±0.023 *	0.780±0.020 *	0.939±0.010 *
		struc2vec	0.751±0.008 *	0.805±0.014 *	0.942±0.002 *
		GraphWave	0.781±0.019 *	0.798±0.011 *	0.953±0.006 *
		DMGI	0.629±0.019 *	0.634±0.013 *	0.935±0.004 *
		URAMN	0.688±0.014 *	0.703±0.015 *	0.938±0.005 *
		JUST	0.768±0.003 *	0.781±0.004 *	0.956±0.001 *
		SM2Vec	0.688±0.000 *	0.867±0.000 *	0.893±0.000 *
		hyperstruct	0.746±0.007 *	0.758±0.013 *	0.946±0.002 *
		RESD	0.792±0.005 *	0.838±0.016 *	0.951±0.002 *
		Our method	**1.000±0.000**	**1.000±0.000**	**1.000±0.000**
Varied Perturbed		degree	0.557±0.011 *	0.720±0.012 *	0.866±0.007 *
		eignvector	0.666±0.032 *	0.731±0.017 *	0.923±0.019 *
		node2vec	0.757±0.004 *	0.758±0.007 *	0.956±0.001 *
		DRNE	0.706±0.018 *	0.737±0.016 *	0.937±0.005 *
		struc2vec	0.739±0.013 *	0.791±0.014 *	0.934±0.009 *
		GraphWave	0.725±0.007 *	0.788±0.001 *	0.933±0.002 *
		DMGI	0.628±0.013 *	0.634±0.007 *	0.935±0.003 *
		URAMN	0.687±0.015 *	0.700±0.014 *	0.937±0.004 *
		JUST	0.755±0.005 *	0.757±0.007 *	0.955±0.001 *
		SM2Vec	0.696±0.018 *	0.802±0.014 *	0.916±0.007 *
		hyperstruct	0.758±0.025 *	0.758±0.022 *	0.953±0.005 *
		RESD	0.763±0.013 *	0.842±0.009 *	0.944±0.008 *
		Our method	**0.873±0.008**	**0.911±0.008**	**0.973±0.002**

cycle is populated with a blend of "house", "fan", and "star" shapes (with 3 red&circle, 3 blue&circle, and 3 blue&square nodes for each form).

For the baselines, we select twelve representative unsupervised methods representing four distinct categories: (1) degree centrality and eigenvector centrality for centrality metrics; (2) node2vec as the representative microscopic homogeneous network embedding; (3) DRNE, struc2vec, GraphWave, SM2Vec [27], hyperstruct [28], and RESD for structural role embedding; (4) JUST, DMGI, and URAMN for multiplex embedding. It is noteworthy that DMGI and URAMN also accommodate attributed networks. Note that many well-known recent works on Graph Neural Networks are supervised or semisupervised methods, and they are not typically designed for structural roles, so we do not include them as baselines.

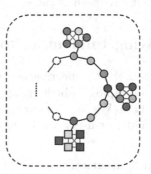

Fig. 4. A cycle planted with multiplex "houses" with three types of edges (black, blue, and red) and two types of node attributes (square and circle). Node colors represent roles (not known beforehand).

We use agglomerative clustering with a "complete" linkage strategy to classify the embeddings and subsequently report on the average homogeneity, completeness, and rand index. Each experiment is repeated 10 times and averaged.

Table 1 shows the results, including the standard deviations and statistical significance. Our method not only bests all 12 baselines in all four tasks across all three metrics but does so with statistical significance ($p < 0.05$). Especially in the "house" and "varied" scenarios, our model attains flawless scores, underscoring its adeptness at accurately distinguishing attributed multiplex roles while also showcasing its innate robustness to disruptive noise.

4.3 Scalability

For a given set of hyper-parameters, the time complexity of our model is linear to the number of nodes. To better illustrate the scalability of our model, we learn node representations using our method with the default parameters on Erdos-Renyi graphs with increasing sizes from 100 to 100,000 nodes with average degrees of 10. For each Erdos-Renyi graph, we uniformly split the edges into three different multiplex layers. We run the tests on an RTX 2080 GPU for 500 epochs. Figure 5 shows the log_{10} running time vs the log_{10} nodes. For a network with

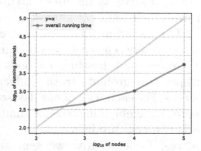

Fig. 5. Scalability of our method on Erdos-Renyi graphs with an average degree of 10. The grey line indicates $y = x$.

1,000,000 nodes and 10,000,000 edges, our method can still finish training in ten hours, which is faster than most of the existing embedding methods, and indicates that our method can easily be scaled to large networks.

5   Studying Influential Roles on Social Media

To test our method's effectiveness on extensive real-world multiplex networks, we use our multiplex embedding approach to derive representations of a subset of Twitter users. We then analyze the capability of our embeddings in predicting influential nodes or users within our network.

5.1   Data Collection

We used a multiplex network of almost all verified users on Twitter (in 2021). Our network is comprised of 362,055 nodes (users) and has 106,396,154 *follow* edges, 15,454,983 *retweet* edges, and 7,031,786 *reply* edges. It is worth noting that algorithms such as GraphWave, SM2Vec, and RESD are not designed for handling large-scale networks. Due to their reliance on matrix operations like SVD, these algorithms grapple with high spatial and temporal complexities. In the context of our network's size, using these methods would necessitate over 1 TB of memory. Consequently, we excluded them from our baselines.

5.2   Real-World Multiplex Role Similarity Ranking

We utilized a benchmark dataset of influential Twitter accounts provided by the Statista Research Department [18]. This dataset assigns an influence index score to each account. To construct attributes, we used Doc2Vec [10] on the latest 100 tweets from each user and averaged these embeddings to serve as the user's attributes (capturing their topics of interest).

Our experiments aimed to assess the correlation between account embeddings derived from our and baseline methods and their respective ground truth influence index scores. For each user in our dataset, we compute its embedding cosine similarity with other users, producing a set of similarity scores. Concurrently, we generate another set of similarity scores based on the ground truth, wherein similarity is gauged by the absolute difference of influence index scores.

Table 2. Average Spearman's correlations between user similarity lists generated through embeddings and the ground-truth influence scores.

Method	Correlation
degree	0.13
eigenvector	0.16
node2vec	0.09
DRNE	0.12
struc2vec	0.14
DMGI	0.09
URAMN	0.08
JUST	0.08
hyperstruct	0.13
Ours (w/o att)	**0.24**
Ours (w/ att)	**0.26**

The relationship between these two sets of scores is then quantified using Spearman's rank correlation. This process is repeated for all accounts in the ground truth dataset.

In Table 2, the average Spearman's correlation for each method is presented. Notably, methods that utilize structural information, including eigenvector centrality, degree centrality, struc2vec, hyperstruct, and our proposed technique, exhibit better performance than local proximity-based approaches like node2vec, URAMN, and JUST. Our approach, both with and without attributes, consistently outperforms other methods, showing an improvement of over 50% relative to the next best performer, eigenvector centrality. This highlights the proficiency of our method in capturing the structural role representations of accounts on platforms such as Twitter, distinguishing it from existing techniques.

5.3   Real-World Multiplex Role Visualization

We demonstrate how our model handles multiplex information by visualizing various networks: the retweet network (Fig. 6 (a)), reply network (Fig. 6 (b)), followership network (Fig. 6 (c)), and an integrated multiplex network (Fig. 6 (d)) that includes all three relationships. The red points represent the most influential accounts as identified by Statista. While our model in single-layer networks (retweet, reply, or followership) struggles to cluster these influential users effectively, the results from the integrated multiplex network successfully group them together (in the bottom-right corner in Fig. 6 (d)). This suggests that individual networks such as reply, retweet, and followership may not be enough to identify the influence of users on such complex networks. Whereas our model manages to effectively capture the influence information, emphasizing the significance of multiplex roles in such networks.

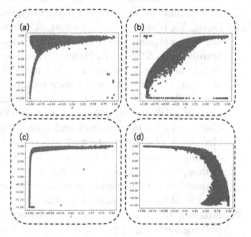

Fig. 6. The embeddings from our verified user Twitter dataset for (a) the retweet, (b) reply, (c) followership, and (d) the integrated networks of our model. The red points represent the most influential accounts. (Color figure online)

6   Conclusion

We present an unsupervised attributed multiplex structural role network embedding method tailored for large-scale networks. This method, utilizing an autoencoder framework, integrates node structures, relationships, and attributes, optimizing embeddings in a unified manner. Our evaluations across various synthetic networks demonstrate our method's superiority over leading baselines in discerning complex node roles in attributed multiplex networks. It also showcases remarkable scalability, efficiently handling large networks. Our analysis of a tri-layered Twitter multiplex network, encompassing verified accounts and their interactions, highlights our method's applicability to extensive social media platforms. A case study using a dataset of influential Twitter users, with similarity ranking and visual assessments, shows our embeddings are more effective at identifying key users than alternatives. These results suggest our approach provides deeper, more nuanced insights into multiplex network structures.

References

1. Cao, S., Lu, W., Xu, Q.: Grarep: learning graph representations with global structural information. In: Proceedings of the 24th ACM International on Conference on Information and Knowledge Management, pp. 891–900. ACM (2015)
2. Dong, Y., Chawla, N.V., Swami, A.: metapath2vec: Scalable representation learning for heterogeneous networks. In: Proceedings of the 23rd ACM SIGKDD (2017)
3. Donnat, C., Zitnik, M., Hallac, D., Leskovec, J.: Learning structural node embeddings via diffusion wavelets. In: Proceedings of the 24th KDD (2018)
4. Forman, R.: Discrete and computational geometry (2003)
5. Fu, T., Lee, W.C., Lei, Z.: Hin2vec: explore meta-paths in heterogeneous information networks for representation learning. In: Proceedings of the 2017 ACM on Conference on Information and Knowledge Management, pp. 1797–1806 (2017)
6. Grover, A., Leskovec, J.: node2vec: scalable feature learning for networks. In: Proceedings of the 22nd KDD, pp. 855–864. ACM (2016)
7. Huang, C., Wang, L., Cao, X., Ma, W., Vosoughi, S.: Learning dynamic graph embeddings using random walk with temporal backtracking. In: NeurIPS 2022 Temporal Graph Learning Workshop (2022)
8. Hussein, R., Yang, D., Cudré-Mauroux, P.: Are meta-paths necessary? Revisiting heterogeneous graph embeddings. In: Proceedings of the 27th ACM International Conference on Information and Knowledge Management, pp. 437–446 (2018)
9. Kingma, D.P., Ba, J.: Adam: a method for stochastic optimization. arXiv preprint arXiv:1412.6980 (2014)
10. Le, Q., Mikolov, T.: Distributed representations of sentences and documents. In: International Conference on Machine Learning, pp. 1188–1196. PMLR (2014)
11. Nikolentzos, G., Vazirgiannis, M.: Learning structural node representations using graph kernels. IEEE TKDE **33**(5), 2045–2056 (2019)
12. Ou, M., Cui, P., Pei, J., Zhang, Z., Zhu, W.: Asymmetric transitivity preserving graph embedding. In: Proceedings of the 22nd ACM SIGKDD International Conference on Knowledge Discovery and Data Mining, pp. 1105–1114 (2016)
13. Park, C., Kim, D., Han, J., Yu, H.: Unsupervised attributed multiplex network embedding. In: Proceedings of the AAAI Conference on Artificial Intelligence, vol. 34, pp. 5371–5378 (2020)
14. Perozzi, B., Al-Rfou, R., Skiena, S.: DeepWalk: online learning of social representations. In: Proceedings of the 20th KDD, pp. 701–710. ACM (2014)
15. Qu, M., Tang, J., Shang, J., Ren, X., Zhang, M., Han, J.: An attention-based collaboration framework for multi-view network representation learning. In: Proceedings of the 2017 CIKM (2017)

16. Ribeiro, L.F., Saverese, P.H., Figueiredo, D.R.: struc2vec: learning node representations from structural identity. In: the 23rd KDD, pp. 385–394 (2017)
17. Sreejith, R., Mohanraj, K., Jost, J., Saucan, E., Samal, A.: Forman curvature for complex networks. J. Stat. Mech. Theory Exp. **2016**(6), 063206 (2016)
18. Statista: Most popular influential twitter users in 2020. https://www.statista.com/statistics/1100266/top-influential-twitter-users/ (2022)
19. Tang, J., Qu, M., Wang, M., Zhang, M., Yan, J., Mei, Q.: Line: large-scale information network embedding. In: Proceedings of the 24th International Conference on World Wide Web, pp. 1067–1077. International World Wide Web Conferences Steering Committee (2015)
20. Tu, K., Cui, P., Wang, X., Yu, P.S., Zhu, W.: Deep recursive network embedding with regular equivalence. In: Proceedings of the 24th ACM SIGKDD International Conference on Knowledge Discovery & Data Mining, pp. 2357–2366 (2018)
21. Wang, D., Cui, P., Zhu, W.: Structural deep network embedding. In: Proceedings of the 22nd ACM SIGKDD International Conference on Knowledge Discovery and Data Mining, pp. 1225–1234. ACM (2016)
22. Wang, L., Gao, C., Huang, C., Liu, R., Ma, W., Vosoughi, S.: Embedding heterogeneous networks into hyperbolic space without meta-path. In: Proceedings of the AAAI Conference on Artificial Intelligence, vol. 35, pp. 10147–10155 (2021)
23. Wang, L., Huang, C., Cao, X., Ma, W., Vosoughi, S.: Graph-level embedding for time-evolving graphs. In: Companion Proceedings of the ACM Web Conference 2023, pp. 5–8 (2023)
24. Wang, L., Huang, C., Lu, Y., Ma, W., Liu, R., Vosoughi, S.: Dynamic structural role node embedding for user modeling in evolving networks. ACM Trans. Inf. Syst. **40**, 1–21 (2021)
25. Wang, L., Huang, C., Ma, W., Cao, X., Vosoughi, S.: Graph embedding via diffusion-wavelets-based node feature distribution characterization. In: Proceedings of the 30th ACM International Conference on Information & Knowledge Management, pp. 3478–3482 (2021)
26. Wang, L., Huang, C., Ma, W., Liu, R., Vosoughi, S.: Hyperbolic node embedding for temporal networks. Data Mining Knowl. Disc. **35**, 1–35 (2021)
27. Wang, L., Huang, C., Ma, W., Lu, Y., Vosoughi, S.: Embedding node structural role identity using stress majorization. In: Proceedings of the 30th ACM International Conference on Information & Knowledge Management, pp. 3473–3477 (2021)
28. Wang, L., Lu, Y., Huang, C., Vosoughi, S.: Embedding node structural role identity into hyperbolic space. In: Proceedings of the 29th ACM International Conference on Information & Knowledge Management, pp. 2253–2256 (2020)
29. Zhang, H., Qiu, L., Yi, L., Song, Y.: Scalable multiplex network embedding. IJCAI **18**, 3082–3088 (2018)
30. Zhang, R., Zimek, A., Schneider-Kamp, P.: Unsupervised representation learning on attributed multiplex network. In: Proceedings of the 31st ACM International Conference on Information & Knowledge Management, pp. 2610–2619 (2022)
31. Zhang, W., Guo, X., Wang, W., Tian, Q., Pan, L., Jiao, P.: Role-based network embedding via structural features reconstruction with degree-regularized constraint. Knowl. Based Syst. **218**, 106872 (2021)

Query-Decision Regression Between Shortest Path and Minimum Steiner Tree

Guangmo Tong$^{(\boxtimes)}$, Peng Zhao , and Mina Samizadeh

University of Delaware, Newark, USA
{amotong,pzhao,minasmz}@udel.edu

Abstract. Considering a graph with unknown weights, can we find the shortest path for a pair of nodes if we know the minimal Steiner trees associated with some subset of nodes? That is, with respect to a fixed latent decision-making system (e.g., a weighted graph), we seek to solve one optimization problem (e.g., the shortest path problem) by leveraging information associated with another optimization problem (e.g., the minimal Steiner tree problem). In this paper, we study such a prototype problem called *query-decision regression with task shifts*, focusing on the shortest path problem and the minimum Steiner tree problem. We provide theoretical insights regarding the design of realizable hypothesis spaces for building scoring models, and present two principled learning frameworks. Our experimental studies show that such problems can be solved to a decent extent with statistical significance.

Keywords: Statistical Learning · Data-driven Optimization · Combinatorial Optimization

1 Introduction

In its most general sense, a decision-making problem seeks to find the best decision for an input query in terms of an objective function that quantifies the decision qualities [6]. Traditionally, the objective function is given a prior, and we thus focus primarily on its optimization hardness. However, real-world systems are often subject to uncertainties, making the latent objective function not completely known to us [11]; this creates room for data-driven approaches to play a key role in building decision-making pipelines [17].

Query-decision Regression with Task Shifts (QRTS). When facing an unknown objective function, one can adopt the learn-and-optimize framework where we first learn the unknown objective function from data and then solve the target optimization problem based on the learned function [3], which can be dated back to Bengio's work twenty years ago [2]. Nevertheless, the learn-and-optimize framework suffers from the fact that the learning process is often driven by the *average* accuracy while good optimization effects demand *worst-case* guarantees [7]. Query-decision regression (QR), as an alternative decision-making diagram, seeks to infer good decisions by learning directly from successful

D.-N. Yang et al. (Eds.): PAKDD 2024, LNAI 14646, pp. 314–326, 2024.
https://doi.org/10.1007/978-981-97-2253-2_25

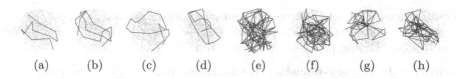

(a)　　　(b)　　　(c)　　　(d)　　　(e)　　　(f)　　　(g)　　　(h)

Fig. 1. Associated with a fixed weighted graph, (a)–(d) show the shortest paths of four pairs of nodes, and (e)–(h) show the minimal Steiner trees for four node subsets. Can we leverage the information in (a)–(d) to compute the solutions in (e)–(h), or vice versa?

optimization results. The feasibility of such a diagram has been proved by a few existing works [4]. Such a success points out an interesting way of generalizing QR called query-decision regression with task shifts (QRTS): assuming that there are two query-decision tasks associated with a latent system, can we solve one task (i.e., the target task) by using optimization results associated with the other task (i.e., the source task) – Fig. 1? Proving the feasibility of such problems is theoretically appealing, as it suggests that one can translate the optimal solutions between different optimization problems.

Contribution. This paper presents the first study on QRTS over stochastic graphs for two specific problems, the shortest path problem and the minimum Steiner tree problem. Taking QRTS as a statistical learning problem, we provide theoretical analysis regarding the creation of realizable hypothesis spaces for designing score functions, seeking to integrate the latent decision objective into the learning pipeline for better optimization effects. Based on the proposed hypothesis space, we design two principled methods QRTS-P and QRTS-D, where P stands for point estimation and D stands for distribution learning. In particular, QRTS-P is designed based on the principle of point estimation that implicitly searches for the best mean graph, while QRTS-D leverages distribution learning to compute the pattern of the edge weights that can best fit the samples. We present empirical studies using graphs of classic families. As one of the main contributions of this paper, our results confirm that QRTS can be solved to a satisfactory extent, which may be the first piece of evidence showing that one can successfully translate knowledge between different optimization problems sharing the same underlying system. The appendix and supplementary materials[1] include technical proofs, more discussions on experiments, source code, and data.

2   Preliminaries

We consider a countable family $\mathcal{G}$ of weighted directed graphs sharing the same graph structure $G = (V, E)$. For each weighted graph $g \in \mathcal{G}$, let $g() : E \to \mathbb{R}^+$ be its weight function. Without loss of generality, we assume that G has no multiple edge when the edge directions are omitted; therefore, each graph $g \in \mathcal{G}$ can also

---

[1] https://github.com/cdslabamotong/QRTS.

be taken as an undirected graph, without causing confusion regarding the edge weights. Associated with $\mathcal{G}$, there is an *unknown* distribution $\mathcal{D}_{\mathcal{G}}$ over $\mathcal{G}$. We consider optimization problems in the following form.

Definition 1. (Query-decision Optimization). *Let $\mathcal{X} \subseteq 2^V$ be a query space and $\mathcal{Y} \subseteq 2^E$ be a decision space. In addition, let $f(x, y, g) \in \mathbb{R}^+$ be the decision value associated with a query $x \in \mathcal{X}$, a decision $y \in \mathcal{Y}$, and a graph $g \in \mathcal{G}$ (either directed or undirected). For a given query $x \in \mathcal{X}$, we seek to find the decision that can minimize the expected decision value:*

$$\arg\min_{y \in \mathcal{Y}} F_{f, \mathcal{D}_{\mathcal{G}}}(x, y) \text{ where } F_{f, \mathcal{D}_{\mathcal{G}}}(x, y) := \mathbb{E}_{g \sim \mathcal{D}_{\mathcal{G}}}[f(x, y, g)]. \tag{1}$$

Such a task is specified by a three-tuple $(f, \mathcal{X}, \mathcal{Y})$. It reduces to the deterministic case with $|\mathcal{G}| = 1$ (e.g., Fig. 1).

When the distribution $\mathcal{D}_{\mathcal{G}}$ is known to us, the above problems fall into stochastic combinatorial optimization [18]. In addressing the case when $\mathcal{D}_{\mathcal{G}}$ is unknown, query-decision regression emphasizes the scenario when there is no proper data to learn $\mathcal{D}_{\mathcal{G}}$, and it is motivated by the aspiration to *learn directly from successful optimization results*. Since the optimization problem in question (i.e., Eq. (1)) can be computationally hard under common complexity assumptions (e.g., NP$\neq$P), we assume that an approximate solution is observed. Accordingly, we will utilize samples in the form of

$$D_{f, \alpha} = \left\{ (x_j, y_j) \mid F_{f, \mathcal{D}_{\mathcal{G}}}(x_j, y_j) \le \alpha \cdot \min_{y \in \mathcal{Y}} F_{f, \mathcal{D}_{\mathcal{G}}}(x_j, y) \right\}, \tag{2}$$

where the quality of the observed decision is controlled by a nominal ratio $\alpha \ge 1$. With such, we formulate query-decision regression with/without task shifts as statistical learning problems.

Definition 2. (Query-decision Regression (QR)). *Associated with a query-decision optimization problem $(f, \mathcal{X}, \mathcal{Y})$ and a distribution $\mathcal{D}_{\mathcal{X}}$ over $\mathcal{X}$, given a collection $D_{f, \alpha} = \{(x_j, y_j)\}$ of query-decision pairs with x_j being iid from $\mathcal{D}_{\mathcal{X}}$, we aim to learn a decision-making model $M : \mathcal{X} \to \mathcal{Y}$ that can predict high-quality decisions for future queries.*

Definition 3. (Query-decision Regression with Task Shifts (QRTS)). *Consider a source task $(f_S, \mathcal{X}_S, \mathcal{Y}_S)$ and a target task $(f_T, \mathcal{X}_T, \mathcal{Y}_T)$ sharing the same $\mathcal{G}$ and $\mathcal{D}_{\mathcal{G}}$. Suppose that we are provided with query-decision samples $D_{f_S, \alpha}$ associated with the source task. We aim to learn a decision-making model $M : \mathcal{X}_T \to \mathcal{Y}_T$ for the target task, with the goal of maximizing the optimization effect:*

$$\mathcal{L}(M, \mathcal{D}_{\mathcal{X}_T}) := \mathbb{E}_{x \sim \mathcal{D}_{\mathcal{X}_T}} \left[\frac{\min_{y \in \mathcal{Y}_T} F_{f_T, \mathcal{D}_{\mathcal{G}}}(x, y)}{F_{f_T, \mathcal{D}_{\mathcal{G}}}(x, M(x))} \right], \tag{3}$$

where $\mathcal{D}_{\mathcal{X}_T}$ is the query distribution of the target task.

*Remark 1. (**Technical Challenge**).* In principle, QR falls into the setting of supervised learning, in the sense that it attempts to learn a mapping using labeled data. Therefore, standard supervised learning methods can solve such problems with statistical significance more or less, although they may not be the optimal methods for specific tasks [21]. QRTS reduces to QR when the source task is identical to the target one, but it is arguably more challenging: one can no longer apply standard supervised learning methods because the query-decision mapping we seek to infer is associated with the target task but the samples are of the source task. To the best of our knowledge, no existing method can be directly applied to solve problems like the one in Fig. 1.

3   A Warm-Up Method: QRTS-P

In this section, we present a simple and intuitive method called QRTS-P for solving QRTS. To illustrate the idea, we notice that the latent objective function $F_{f,\mathcal{D}_G}$ can be expressed as a function of the mean weights of the edges, which is due to the linearity of expectation.

Example 1. Suppose that the considered query-decision optimization problem is the stochastic shortest path problem. For each node pair $x = (u, v)$, let $\mathcal{Y}_x$ be the set of all paths from u to v. The latent objective function can thus be expressed as

$$F_{f,\mathcal{D}_G}(x, y) = \begin{cases} \sum_{e \in y} \mathbb{E}_{g \sim \mathcal{D}_G}[g(e)] & \text{if } y \in \mathcal{Y}_x \\ +\infty & \text{otherwise.} \end{cases} \tag{4}$$

Similarly, for the stochastic minimal Steiner tree problem [22], which finds the min-weight subgraph that connects a given set of nodes, we may define $\mathcal{Y}_x$ as the set of valid Steiner trees of a node set x, and with such, the latent objective has the identical form as Eq. (4).

In abstract, let $\mathcal{Y}_{f,x} \subseteq \mathcal{Y}_f$ be the set of the feasible solutions associated with a query x, and $\mathbb{1}_S \in \{0, 1\}$ be the set indicator function, i.e., $\mathbb{1}_S(x) = 1 \iff x \in S$. The query-decision optimization now has the generic form of

$$\arg\min_{y \in \mathcal{Y}_{f,x}} \sum_{e \in y} \mathbb{E}_{g \sim \mathcal{D}_G}[g(e)] = \sum_{e \in E} \mathbb{E}_{g \sim \mathcal{D}_G}[g(e)] \mathbb{1}_y(e) \tag{5}$$

Such an abstraction suggests that it would be sufficient for solving the target task if one can find the mean graph induced by $\mathcal{D}_G$, which essentially asks for good estimations of $\{\mathbb{E}_{g \sim \mathcal{D}_G}[g(e)] | e \in E\}$ – leading to a point estimation problem [13]. In what follows, we will see how samples $D_{fs,\alpha}$ associated with the source task can be helpful for such a purpose.

For each $e \in E$, let $w_e \in \mathbb{R}^+$ be the sought-after estimation of $\mathbb{E}_{g \sim \mathcal{D}_G}[g(e)]$. Since each sample (x_j, y_j) in $D_{fs,\alpha}$ is an α-approximation, in light of Eq. (5), a desired set $\{w_e\} := \{w_e | e \in E\}$ should satisfy the linear constraint

$$\alpha \cdot \min_{y \in \mathcal{Y}_{fs,x_j}} \sum_{e \in E} w_e \mathbb{1}_y \geq \sum_{e \in E} w_e \mathbb{1}_{y_j}, \tag{6}$$

which means that the sample decision y_j is also an α-approximation in the mean graph induced by $\{w_e\}$. Applying the standard large-margin training to Eq.

(6) [19], a robust estimation can be inferred by solving the following quadratic program

$$\min_{w_e, \eta_j} \sum_{e \in E} w_e^2 + C \cdot \sum_j \eta_j \text{ s.t. } \alpha \cdot \underbrace{\min_{y \in \mathcal{Y}_{f_S, x_j}} \sum_{e \in E} w_e \mathbb{1}_y(e)}_{\text{source inference}} - \sum_{e \in E} w_e \mathbb{1}_{y_j}(e) \geq -\eta_j, \forall j. \quad (7)$$

where C is a hyperparamter. Optimization problems in the above form have been widely discussed for training structured prediction models, and they can be solved efficiently as long as the source inference problem in Eq. (7) can be effectively solved for a given $\{w_e\}$ [15]. We adopt the cutting plane algorithm in our experiments and defer the details to the Appendix. With the learned weights $\{w_e\}$, the inference for a query $x^* \in \mathcal{X}_T$ of the target problem can be computed through

$$\textbf{target inference:} \quad \min_{y \in \mathcal{Y}_{f_T, x^*}} \sum_{e \in E} w_e \mathbb{1}_y(e) \quad (8)$$

We denote such an approach as QRTS-P. The source and target inferences will be discussed later in Remark 2, as they are special cases of later problems.

4   A Probabilistic Perspective: QRTS-D

In this section, we present a more involved method called QRTS-D for solving QRTS. It turns out that QRTS-D subtly subsumes QRTS-P as a special case.

4.1   Overall Framework

QRTS-D follows the standard scoring model, where we assign each decision a score and make a prediction by selecting the decision with *the lowest score*:

$$\textbf{score function:} \quad h : \mathcal{X}_T \times \mathcal{Y}_T \to \mathbb{R} \quad (9)$$
$$\textbf{inference:} \quad \arg\min_{y \in \mathcal{Y}_T} h(x, y).$$

Such a framework is expected to solve QRTS well, provided that for each pair $(x, y) \in \mathcal{X}_T \times \mathcal{Y}_T$, a low score $h(x, y)$ can imply a small objective value $F_{f_T, \mathcal{D}_\mathcal{G}}(x, y)$. Implementing such an idea hinges on three integral parts: **a)** a hypothesis space H of h; **b)** training methods to search for the best score function within H based on the empirical evidence $D_{f_S, \alpha}$; **c)** algorithms for solving the inference problem. With such a framework, we will first discuss insights for designing a desired hypothesis space and then present training methods.

4.2   Hypothesis Design

In designing a desired score function h, the key observation is that the true objective function $F_{f_T, \mathcal{D}_\mathcal{G}}$ of the target task is a perfect score function, in that the inference over $F_{f_T, \mathcal{D}_\mathcal{G}}(x, y)$ recovers the exact optimal solution. While $\mathcal{D}_\mathcal{G}$ is unknown to us, the technique of importance sampling offers a means of deriving a parameterized approximation [20]. In particular, for any empirical distribution $\mathcal{D}_{\mathcal{G}}^{em}$ over $\mathcal{G}$, we have

$$F_{f_T,\mathcal{D}_{\mathcal{G}}}(x,y) = \int_{g\in\mathcal{G}} \frac{\mathcal{D}_{\mathcal{G}}[g]}{\mathcal{D}_{\mathcal{G}}^{em}[g]} f_T(x,y,g)d\mathcal{D}_{\mathcal{G}}^{em}, \tag{10}$$

which immediately implies the function approximation guarantee between $F_{f_T,\mathcal{D}_{\mathcal{G}}}$ and an affine combination of f_T.

Theorem 1. *Let* $\|\cdot\|$ *denote the function distance with respect to the Lebesgue measure associated with any distribution* $\mathcal{D}$ *over* $\mathcal{X}_T \times \mathcal{Y}_T$. *For each* $\epsilon \geq 0$, $\lim_{K\to\infty} \Pr_{g_i\sim\mathcal{D}_{\mathcal{G}}^{em}} \left[\inf_{w_i\in\mathbb{R}} \left\| F_{f_T,\mathcal{D}_{\mathcal{G}}}(x,y) - \sum_{i=1}^{K} w_i f_T(x,y,g_i) \right\| \leq \epsilon \right] = 1$.

Theorem 1 justifies the following hypothesis space for the score function of which the complexity is controlled by its dimension $K \in \mathbb{Z}$.

$$H_{K,\mathcal{D}_{\mathcal{G}}^{em}} := \left\{ h_{w,\{g_i\}}(x,y) := \sum_{i=1}^{K} w_i f_T(x,y,g_i) | g_i \sim \mathcal{D}_{\mathcal{G}}^{em}, w = (w_1,...,w_K) \in \mathbb{R}^K \right\}.$$

These score functions are very reminiscent of the principled kernel machines [8], with the distinction that our kernel function, namely f_T, is inherited from the latent optimization problem rather than standard kernels [16]. For such a score function $h_{w,\{g_i\}}(x,y)$, the inference process is further specialized as

$$\textbf{target inference}: M_{w,\{g_i\}}(x) := \min_{y\in\mathcal{Y}_T} \sum_i w_i f_T(x,y,g_i). \tag{11}$$

With the construction of $H_{K,\mathcal{D}_{\mathcal{G}}^{em}}$, a realizable space can be achieved provided that the dimension K is sufficiently large, which allows us to characterize the generalization loss Eq. (3).

Theorem 2. *Suppose that a* β-*approximation is adopted to solve the target inference problem Eq. (11). Let* $D_\infty \in \mathbb{R}$ *be the* ∞-*order Rényi divergence between* $\mathcal{D}_{\mathcal{G}}$ *and* $\mathcal{D}_{\mathcal{G}}^{em}$. *For each* $\epsilon \geq 0$ *and* $\delta > 0$, *there exists*

$$C = O\left(\frac{\ln|\mathcal{X}_T| + \ln|\mathcal{Y}_T|}{\epsilon^2} \cdot \ln\frac{1}{\delta} \cdot \exp(D_\infty) \right)$$

such that when $K \geq C$, *with probability at least* $1 - \delta$ *over the selection of* $\{g_i\}$, *we have* $\sup_{w\in\mathbb{R}^K} \mathcal{L}(M_{w,\{g_i\}}, \mathcal{D}_{\mathcal{X}_T}) \geq \beta \cdot \frac{1-\epsilon}{1+\epsilon}$.

Theorem 2 suggests that a high dimension (i.e., K) may be needed when a) the spaces are large and/or b) the deviation between $\mathcal{D}_{\mathcal{G}}$ and $\mathcal{D}_{\mathcal{G}}^{em}$ is high, which is intuitive. The proof of Theorem 2 leverages point-wise concentration to acquire the desired guarantees, while Theorem 1 is proved through concentrations in function spaces.

4.3   QRTS-D

With the design of $H_{K,\mathcal{D}_{\mathcal{G}}^{em}}$, we now present methods for computing a concrete score function $h_{w,\{g_i\}}$, which is to decide a collection $\{g_i\}$ of subgraphs as well as the associated weights $\boldsymbol{w}$. In light of Eq. (10), $\boldsymbol{w}_i$ can be viewed as the importance of graph g_i. We will not restrict ourselves to a specific choice of $\mathcal{D}_{\mathcal{G}}^{em}$ and thus assume that a nominal parametric family $\mathcal{D}_{\mathcal{G},\theta}^{em}$ is adopted, with an extra subscription θ added to denote the parameter set. Assuming that the hyperparameter K is given, the framework of QRTS-D loops over three phases: **a) graph sampling**, to sample $\{g_1, ..., g_K\}$ iid from $\mathcal{D}_{\mathcal{G},\theta}^{em}$; **b) importance learning**, to compute $\boldsymbol{w}$ for $\{g_i\}$; **c) distribution tuning**, to update $\mathcal{D}_{\mathcal{G},\theta}^{em}$. The first phase is trivial, and we will therefore focus on the other two phases.

Importance Learning. In computing the weights $\boldsymbol{w}$ for a given $\{g_i\}$, we have reached a key point to attack the challenges mentioned in Remark 1: the function approximation guarantee in Theorem 1 holds not only for the target task but also for the source task. That is, $\sum_{i=1}^{K} w_i f_T(x, y, g_i)$ is a desired score function (for solving the target task) if and only if $\sum_{i=1}^{K} w_i f_S(x, y, g_i)$ can well approximate the true objective function $F_{f_S, \mathcal{D}_{\mathcal{G}}}(x, y)$ of the source task. Therefore, since the samples in $D_{f_S, \alpha} = \{(x_j, y_j)\}$ are α-approximations to the source task, the ideal weights $\boldsymbol{w}$ should satisfy

$$\alpha \min_{y \in \mathcal{Y}_S} \sum_{i=1}^{K} w_i f_S(x_j, y, g_i) \geq \sum_{i=1}^{K} w_i f_S(x_j, y_j, g_i).$$

In this way, we have been able to leverage the samples from the source task to decide the best $\boldsymbol{w}$ associated with $\{g_i\}$. This owes to the fact that our design $H_{K, \mathcal{D}_{\mathcal{G}}^{em}}$ allows us to separate $\mathcal{D}_{\mathcal{G}}$ from the task-dependent kernels (i.e., f_S and f_T), which is otherwise not possible if we parametrized the score function (i.e., Eq. 9) using naive methods (e.g., neural networks). Following the same logic behind the translation from Eq. (6) to Eq. (7), the above constraints lead to a similar optimization program:

$$\min_{\boldsymbol{w}, \eta_i} \|\boldsymbol{w}\|^2 + C \cdot \sum_i \eta_i \text{ s.t. } \underbrace{\min_{y \in \mathcal{Y}_S} \alpha \sum_{i=1}^{K} w_i f_S(x_j, y, g_i)}_{\text{source inference}} - \sum_{i=1}^{K} w_i f_S(x_j, y_j, g_i) \geq -\eta_j, \forall j.$$

$$(12)$$

The above program shares the same type with Eq. (7), and we again defer the optimization details to the appendix.

Distribution Tuning. With the importance vector $\boldsymbol{w}$ learned based on the subgraphs $\{g_i\}$ sampled from the current θ, we seek to fine-tune $\mathcal{D}_{\mathcal{G},\theta}^{em}$ to make it aligned more with the latent distribution $\mathcal{D}_{\mathcal{G}}$, which is desired as suggested by the proof of Theorem 1. Inspired by Eq. (10), the true likelihood associated with g_i is approximated by $w_i^* := w_i \mathcal{D}_{\mathcal{G},\theta}^{em}[g_i]$. Consequently, one possible way to reshape $\mathcal{D}_{\mathcal{G},\theta}^{em}$ is to find the θ^* that can minimize the discrepancy between $\mathcal{D}_{\mathcal{G},\theta^*}^{em}$ and $\mathcal{D}_{\boldsymbol{w}^*}$, i.e., $\theta^* = \arg\min_\theta D(\mathcal{D}_{\mathcal{G},\theta}^{em} \| \mathcal{D}_{\boldsymbol{w}^*})_{|\{g_i\}}$, where $\mathcal{D}_{\boldsymbol{w}^*}$ is the discrete distribution over $\{g_i\}$ defined by normalizing $(w_1^*, ..., w_K^*)$, and the distance measure $D(\|)$ can be selected at the convenience of the choice of the $\mathcal{D}_{\mathcal{G},\theta}^{em}$ – for example, cosine similarity or cross-entropy. For such problems, standard methods can be directly applied when $\mathcal{D}_{\mathcal{G},\theta}^{em}$ is parameterized by common distribution families; first- and second-order methods can be readily used if $\mathcal{D}_{\mathcal{G},\theta}^{em}$ has a complex form such as neural networks.

Algorithm 1. QRTS-D

1: **Input:** $D_{f_S,\alpha} = \{x_y, y_i\}, C, K, T, \alpha, \mathcal{D}_{\mathcal{G},\theta}^{em}$;
2: **Output:** $w = (w_1, ..., w_K)$ and $\{g_1, ..., g_K\}$
3: Initialize θ, $t = 0$;
4: **repeat**
5:     $\{g_1, ..., g_K\}$ iid from $\mathcal{D}_{\mathcal{G},\theta}^{em}$;
6:     Compute w via Eq. (12) based on $D_{f_S,\alpha}$ and C;
7:     Update θ via $\theta^* = \arg\min_\theta D(\mathcal{D}_{\mathcal{G},\theta}^{em} \| \mathcal{D}_{w^*})_{|\{g_i\}}$;
8:     $t = t + 1$
9: **until** $t = T$
10: **Return** $\{g_1, ..., g_K\}$ and w

The QRTS-D method is conceptually simple, as summarized in Alg. 1. Similar to QRTS-P, using such a method requires algorithms for solving the source and target inferences in Eqs. (11) and (12). In what follows, we discuss such issues as well as the possibility of enhancing QRTS-D using QRTS-P.

*Remark 2 (**Source and Target Inferences**).* For QRTS-P, the source (resp., target) inference problem is nothing but to solve the source (resp., target) query-decision optimization task in its deterministic case. For QRTS-D, the inference problems are to solve the source and target tasks over a weighted combination of deterministic graphs. For the shortest path problem, such inference problems can be solved in polynomial time; for the minimum Steiner tree problem, such inference problems admit 2-approximation [22].

*Remark 3 (**QRTS-P vs QRTS-D**).* As one may have noticed, QRTS-P is a natural special case of the importance learning phase of QRTS-D, in the sense that each w_e in QRTS-P corresponds to the importance of the subgraph with one edge (i.e., e). In other words, QRTS-P can be viewed as the QRTS-D where the support of $\mathcal{D}_{\mathcal{G},\theta}^{em}$ is the span of single-edge subgraphs with unit weights. Notably, the dimension of QRTS-P is fixed and thus limited by the number of edges, while the dimension K of QRTS-D can be made arbitrarily large. For this reason, QRTS-P may be preferred if the sample size is small, while QRTS-D can better handle large sample sets, which is evidenced by our experimental studies.

*Remark 4 (**QRTS-PD**).* In QRTS-D, the initialization of $\mathcal{D}_{\mathcal{G},\theta}^{em}$ is an open issue, and this creates the possibility of integrating QRTS-P into QRTS-D by initializing $\mathcal{D}_{\mathcal{G},\theta}^{em}$ using the weights $\{w_e\}$ learned from QRTS-P. This leads to another approach called QRTS-PD consisting of three steps: **a)** run QRTS-P to acquire the estimations $\{w_e\}$; **b)** stabilize θ based on $\{w_e\}$ through maximum likelihood estimation, i.e., $\min_\theta - \sum_{e \in E} \log \sum_{g \in \mathcal{G}} \mathcal{D}_{\mathcal{G},\theta}^{em}[g|g(e) = w_e]$; **c)** run QRTS-D. From such a perspective, QRTS-PD can be taken as a continuation of QRTS-P to further improve the generalization performance by building models that are more expressive.

5 Empirical Studies

In this section, we present empirical studies demonstrating that QRTS can be solved with statistical significance using the presented methods.

5.1 Experimental Settings

Source and Target Tasks. We specifically focus on two query-decision optimization tasks: shortest path and minimum Steiner tree [9]. Depending on the selection of the source and target tasks, we have two possible settings: *Path-to-Tree* and *Tree-to-Path*. The source and target inferences can be approximated effectively, as discussed in Remark 2. These algorithms are also used to generate samples of query-decision pairs (i.e., Eq. (2)).

Table 1. Results for Path-to-Tree on Kro, Col and BA. Each cell shows the mean ratio together with the standard deviation (std). The top three results in each column are highlighted.

Train Size		Kro			Col			BA		
		60	240	2400	60	240	2400	60	240	2400
QRTS-P		**4.0**$_{(0.3)}$	3.4$_{(0.5)}$	2.4$_{(0.3)}$	291$_{(78)}$	**150**$_{(29)}$	128$_{(7.4)}$	181$_{(28)}$	144$_{(41)}$	63$_{(35)}$
QRTS -PD⁻	60	4.4$_{(0.3)}$	3.4$_{(0.8)}$	2.3$_{(0.4)}$	**250**$_{(82)}$	367$_{(56)}$	123$_{(2.3)}$	209$_{(22)}$	195$_{(22)}$	40$_{(13)}$
	240	4.6$_{(0.7)}$	3.7$_{(0.1)}$	2.7$_{(0.2)}$	330$_{(65)}$	227$_{(14)}$	117$_{(6.1)}$	**129**$_{(22)}$	149$_{(19)}$	35$_{(12)}$
	2400	4.3$_{(0.9)}$	**3.2**$_{(0.5)}$	2.4$_{(0.1)}$	214$_{(68)}$	183$_{(69)}$	**88**$_{(12)}$	139$_{(38)}$	**131**$_{(44)}$	**27**$_{(7.2)}$
QRTS -PD-1	60	**3.9**$_{(0.7)}$	3.4$_{(0.8)}$	**2.3**$_{(0.4)}$	361$_{(55)}$	225$_{(11)}$	115$_{(5.1)}$	177$_{(31)}$	130$_{(34)}$	43$_{(14)}$
	240	4.2$_{(0.4)}$	3.2$_{(0.5)}$	2.4$_{(0.2)}$	350$_{(88)}$	245$_{(31)}$	129$_{(16)}$	166$_{(34)}$	132$_{(49)}$	34$_{(13)}$
	2400	4.3$_{(0.5)}$	**3.1**$_{(0.3)}$	**2.3**$_{(0.4)}$	**261**$_{(93)}$	186$_{(24)}$	106$_{(6.2)}$	**160**$_{(29)}$	**119**$_{(3)}$	34$_{(7.3)}$
QRTS -PD-3	60	4.3$_{(1.1)}$	3.2$_{(0.6)}$	2.6$_{(0.5)}$	431$_{(27)}$	**167**$_{(99)}$	110$_{(15)}$	391$_{(46)}$	144$_{(13)}$	60$_{(13)}$
	240	4.3$_{(0.9)}$	3.3$_{(0.6)}$	2.3$_{(0.4)}$	317$_{(87)}$	183$_{(35)}$	126$_{(11)}$	202$_{(38)}$	138$_{(17)}$	**30**$_{(17)}$
	2400	**4.0**$_{(0.4)}$	**3.2**$_{(0.4)}$	**2.2**$_{(0.2)}$	324$_{(38)}$	**107**$_{(12)}$	113$_{(6.1)}$	184$_{(49)}$	**120**$_{(2.4)}$	**23**$_{(2.7)}$
Unit & Rand		5.2$_{(0.2)}$ & 10.4$_{(0.3)}$			990$_{(68)}$ & 2231$_{(45)}$			349$_{(4.7)}$ & 749$_{(13)}$		

Graph, True Distribution, and Samples. We adopt a collection of graphs of classic types: a Kronecker graph (**Kro**) [14], a road network of Colorado (**Col**) [5], a Barabasi-Albert graph (**BA**) [1], and two Watts-Strogatz graphs with different densities (**WS-dense** and **WS-sparse**) [23]. The statistics of these graphs can be found in the appendix. To have a diverse graph pattern, we generate the ground truth distribution $\mathcal{D}_\mathcal{G}$ by assigning each edge a Weibull distribution [12] with parameters randomly selected from $\{1, ..., 20\}$. For each graph and each problem instance, we generate a pool of 10,000 query-decision pairs.

QRTS Methods. We use QRTS-PD-1 (resp., QRTS-PD-3) to denote the QRTS-PD method when one (resp., three) iterations over the three phases are used. Based on QRTS-PD-1, we implement QRTS-PD⁻ that foregoes the distribution tuning phase. For these methods, the model dimensions K are selected from $\{60, 240, 2400\}$. We utilize the one-slack cutting plane algorithm [10] for the large-margin training (i.e., Eqs. (7) and (12)). The empirical distribution $\mathcal{D}_{\mathcal{G}}^{em}$ is parameterized by assigning each edge an exponential distribution.

Baselines. We set up two baselines **Unit** and **Rand**. Unit computes the predictions based on the graph with unit-weight edges. Rand computes the decision based on the graph with random weights. Unit essentially leverages only the graph structure to compute predictions. We note that none of the methods in existing papers can be directly applied to QRTS (Remark 1).

Training and Testing. In each run, the training size is selected from $\{60, 240, 2400\}$, and the testing size is 1000, where samples are randomly selected from the sample pool. Given a testing set $\{x_i, y_i\}$ of the target task, the performance is measured by the ratio $\sum_i F_{f_T, \mathcal{D}_{\mathcal{G}}}(x_i, y_i^*) / \sum_i F_{f_T, \mathcal{D}_{\mathcal{G}}}(x_i, y_i)$, where y_i^* is the predicted decision associated with x_i; a lower ratio implies better performance. We report the average ratios and the standard deviations over five runs for each method.

Table 2. Results on Tree-to-Path. Each cell shows the mean ratio together with the standard deviation (std). Small stds (< 0.1) are denoted as 0.0. The top three results in each column are highlighted.

Train Size		Kro			Col			BA		
		60	240	2400	60	240	2400	60	240	2400
QRTS-P		**1.46** (0.0)	**1.37** (0.0)	1.60 (0.0)	9.8 (0.2)	6.6 (0.4)	**6.1** (0.1)	1.6 (0.1)	1.4 (0.2)	1.4 (0.1)
QRTS -PD⁻	60	**1.44** (0.1)	1.41 (0.1)	1.39 (0.0)	11 (5.8)	8.6 (1.5)	7.1 (0.6)	1.9 (0.4)	1.4 (0.2)	1.5 (0.1)
	240	**1.42** (0.1)	1.46 (0.0)	1.39 (0.0)	9.9 (4.2)	6.8 (3.6)	6.5 (0.4)	1.7 (0.3)	1.5 (0.2)	1.5 (0.1)
	2400	1.48 (0.1)	1.45 (0.0)	1.38 (0.0)	**8.9** (3.1)	**6.0** (1.3)	6.2 (0.9)	**1.5** (0.2)	**1.3** (0.1)	**1.3** (0.1)
QRTS -PD-1	60	1.55 (0.0)	1.42 (0.1)	1.37 (0.0)	6.6 (0.6)	6.3 (2.4)	6.8 (0.6)	1.6 (0.3)	1.5 (0.0)	1.4 (0.1)
	240	1.56 (0.1)	1.44 (0.1)	1.36 (0.0)	11 (3.2)	7.6 (1.4)	6.6 (0.0)	1.6 (0.3)	1.5 (0.3)	1.5 (0.1)
	2400	1.52 (0.1)	1.39 (0.1)	1.37 (0.0)	14 (2.4)	7.7 (3.9)	7.2 (0.3)	1.7 (0.1)	1.5 (0.3)	1.7 (0.1)
QRTS -PD-3	60	1.53 (0.1)	**1.41** (0.1)	**1.34** (0.1)	13 (4.2)	**5.9** (1.5)	**5.5** (0.9)	1.7 (0.2)	1.5 (0.1)	1.4 (0.1)
	240	1.50 (0.1)	1.42 (0.0)	**1.36** (0.0)	**9.4** (1.4)	**6.6** (0.2)	**5.9** (0.1)	1.6 (0.1)	**1.4** (0.2)	**1.2** (0.1)
	2400	1.47 (0.1)	**1.41** (0.1)	**1.32** (0.0)	**7.8** (0.6)	8.5 (1.4)	7.7 (2.0)	**1.3** (0.1)	**1.3** (0.1)	**1.3** (0.2)
Unit & Rand		1.57 (0.1) & 1.78 (0.1)			9.2 (0.1) & 19 (0.1)			1.57(0.0) & 2.2(0.3)		

5.2 Analysis

The results on Kro, Col, and BA are given in Tables 1 and 2. The results on WS-sparse and WS-dense can be found in the Appendix. The main observations are listed below, and the minor observations are given in the appendix.

O1: The Proposed Methods Behave Reasonably with Promising Performance. First, we observe that the proposed methods perform significantly better when more samples are given, which suggests that they are able to

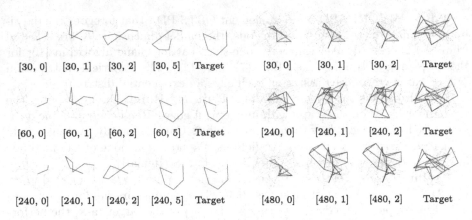

Fig. 2. The left (resp., right) shows the visualizations of the solution to one testing query for the Tree-to-Path (resp., Path-to-Tree) problem under QRTS-P on Kro. The figure labeled by $[a, b]$ shows the result with training size a after b iterations in the cutting plane algorithm. Each row shows the results under one training size, where the last figure shows the optimal solution.

infer meaningful information from the samples toward solving the target task. On the other hand, all the proposed methods are clearly better than Rand, implying that the model efficacy is non-trivial. In addition, they easily outperform Unit by an evident margin in most cases. For example, for Path-to-Tree on BA in Table 1, the best ratio achieved by QRTS-PD-3 is 23, while Unit and Random cannot produce a ratio smaller than 300.

O2: QRTS-P Offers an Effective Initialization for QRTS-D. With very few exceptions, QRTS-PD performs much better than QRTS-P under the same sample size, which confirms that QRTS-P can indeed be improved by using importance learning through re-sampling, which echos Remark 4. This is especially true when the sample size is large; for example, for Path-to-Tree on BA with 2400 samples, methods based on QRTS-PD with a dimension of 2400 are at least twice better than QRTS-P in terms of the performance ratio, demonstrating that QRTS-PD of a high dimension can better consume large datasets.

O3: Distribution Tuning is Helpful After Multiple Iterations. Since QRTS-D can be used without the distribution tuning phase, we are wondering if the distribution turning phase is necessary. By comparing QRTS-PD-1 with QRTS-PD$^-$, we see that the distribution tuning phase can be useful in many cases, but its efficacy is not very significant. However, combined with the results of QRTS-PD-3, we observe that the distribution turning phase can better reinforce the optimization effect when multiple iterations are used. Finally, by comparing QRTS-PD-1 and QRTS-PD-3, we find that training more iterations is useful mostly when the model dimension is large, which is especially the case for Tree-to-Path (Table 2).

O4: The Learning Process is Smooth. Fig. 2 visualizes the learning process of QRTS-P for two example testing queries. One can see that QRTS-P tends to select solutions with fewer edges under the initial random weights w, and it gradually finds better solutions (possibly with more edges) when better weights are learned. We have such observations for most of the samples, which suggest that the proposed method works the way it is supposed to. More visualizations can be found in the appendix.

6 Future Directions

Although we observed that the approximation ratio becomes better with the increase in training size and training iteration, it is not always the case that the solution converges to the optimal one – for example, Fig. 2-Right and the visualizations in the Appendix. This is reasonable because two solutions may have similar costs but with very different edge sets. Depending on the needs of the agents, other metrics can be adopted, which may require new designs of the hypothesis space and training methods. Another important future direction is to enhance the proposed method through (deep) representation learning.

Acknowledgement. This project is supported in part by National Science Foundation under Award Career IIS-2144285.

References

1. Barabási, A.L., Albert, R.: Emergence of scaling in random networks. Science **286**(5439), 509–512 (1999)
2. Bengio, Y.: Using a financial training criterion rather than a prediction criterion. Int. J. Neural Syst. **8**(04), 433–443 (1997)
3. Bertsimas, D., Kallus, N.: From predictive to prescriptive analytics. Manage. Sci. **66**(3), 1025–1044 (2020)
4. Chen, C., Seff, A., Kornhauser, A., Xiao, J.: Deepdriving: learning affordance for direct perception in autonomous driving. In: Proceedings of the IEEE International Conference on Computer Vision, pp. 2722–2730 (2015)
5. Demetrescu, C., Goldberg, A.V., Johnson, D.S.: Implementation challenge for shortest paths. Encyclopedia Algorithms **15**, 54 (2008)
6. Edwards, W.: The theory of decision making. Psychol. Bull. **51**(4), 380 (1954)
7. Ford, B., Nguyen, T., Tambe, M., Sintov, N., Fave, F.D.: Beware the soothsayer: from attack prediction accuracy to predictive reliability in security games. In: Khouzani, M.H.R., Panaousis, E., Theodorakopoulos, G. (eds.) GameSec 2015. LNCS, vol. 9406, pp. 35–56. Springer, Cham (2015). https://doi.org/10.1007/978-3-319-25594-1_3
8. Hofmann, T., Schölkopf, B., Smola, A.J.: Kernel methods in machine learning (2008)
9. Hwang, F.K., Richards, D.S.: Steiner tree problems. Networks **22**(1), 55–89 (1992)
10. Joachims, T., Finley, T., Yu, C.N.J.: Cutting-plane training of structural svms. Mach. Learn. **77**(1), 27–59 (2009)

11. Kochenderfer, M.J.: Decision making under uncertainty: theory and application. MIT press (2015)
12. Lai, C., Murthy, D., Xie, M.: Weibull distributions. Wiley Interdisciplinary Reviews: Computational Statistics **3**(3), 282–287 (2011)
13. Lehmann, E.L., Casella, G.: Theory of point estimation. Springer Science & Business Media (2006)
14. Leskovec, J., Chakrabarti, D., Kleinberg, J., Faloutsos, C., Ghahramani, Z.: Kronecker graphs: an approach to modeling networks. J. Mach. Learn. Res. **11**(2) (2010)
15. Lucchi, A., Li, Y., Fua, P.: Learning for structured prediction using approximate subgradient descent with working sets. In: Proceedings of the CVPR (2013)
16. Murphy, K.P.: Machine learning: a probabilistic perspective. MIT press (2012)
17. Provost, F., Fawcett, T.: Data science and its relationship to big data and data-driven decision making. Big Data **1**(1), 51–59 (2013)
18. Rajkumar, A., Agarwal, S.: Online decision-making in general combinatorial spaces. Advances in Neural Information Processing Systems 27 (2014)
19. Suthaharan, S., Suthaharan, S.: Support vector machine. Machine learning models and algorithms for big data classification: thinking with examples for effective learning, pp. 207–235 (2016)
20. Tokdar, S.T., Kass, R.E.: Importance sampling: a review. Wiley Interdisciplinary Reviews: Computational Statistics **2**(1), 54–60 (2010)
21. Tong, G.: Usco-solver: solving undetermined stochastic combinatorial optimization problems. NeurIPS **34**, 1646–1659 (2021)
22. Vazirani, V.V.: Approximation algorithms, vol. 1. Springer (2001)
23. Watts, D.J., Strogatz, S.H.: Collective dynamics of 'small-world' networks. Nature **393**(6684), 440–442 (1998)

Enhancing Policy Gradient for Traveling Salesman Problem with Data Augmented Behavior Cloning

Yunchao Zhang[1], Kewen Liao[2] (iD), Zhibin Liao[3], and Longkun Guo[1,4](✉)(iD)

[1] School of Computer Science, Qilu University of Technology
(Shandong Academy of Sciences), Jinan 250316, China
[2] HilstLab, Peter Faber Business School, Australian Catholic University,
North Sydney 2060, Australia
Kewen.Liao@acu.edu.au
[3] Australian Institute for Machine Learning, University of Adelaide,
Adelaide 5005, Australia
zhibin.liao@adelaide.edu.au
[4] School of Mathematics and Statistics, Fuzhou University, Fuzhou 350116, China
lkguo@fzu.edu.cn

Abstract. The use of deep reinforcement learning (DRL) techniques to solve classical combinatorial optimization problems like the Traveling Salesman Problem (TSP) has garnered considerable attention due to its advantage of flexible and fast model-based inference. However, DRL training often suffers low efficiency and scalability, which hinders model generalization. This paper proposes a simple yet effective pre-training method that utilizes behavior cloning to initialize neural network parameters for policy gradient DRL. To alleviate the need for large amounts of demonstrations in behavior cloning, we exploit the symmetry of TSP solutions for augmentation. Our method is demonstrated by enhancing the state-of-the-art policy gradient models Attention and POMO for the TSP. Experimental results show that the optimality gap of the solution is significantly reduced while the DRL training time is greatly shortened. This also enables effective and efficient solving of larger TSP instances.

Keywords: Traveling salesman problem · behavior cloning · policy gradient · deep reinforcement learning

1 Introduction

The traveling salesman problem (TSP) is a well-known classical combinatorial optimization problem in which a salesman aims to visit a list of cities once, starting from an origin city and returning to it, intending to minimize the tour length. TSP problems and variants are commonly computationally NP-hard even in the Euclidean space [18] which requires exponential time to find an exact solution. TSP and its variants have seen numerous real-world applications [16] in logistics,

D.-N. Yang et al. (Eds.): PAKDD 2024, LNAI 14646, pp. 327–338, 2024.
https://doi.org/10.1007/978-981-97-2253-2_26

transportation, and circuit design. Many traditional algorithms have been developed for solving TSPs, including exact algorithm [14], heuristic algorithm [7], and approximate algorithm [30]. Similar to the TSP, traditional algorithms for solving combinatorial optimization problems usually require expert knowledge of mathematical modeling and optimization associated with the structure and constraints of the problem. In addition, these algorithms are limited to solving specific types of problems and are not generalizable to a variety of combinatorial optimization problems. In practical applications, it is often not necessary to obtain an exact solution to a combinatorial optimization problem, but instead, practitioners are seeking approximate feasible solutions that are efficient.

In recent years, deep reinforcement learning (DRL) has emerged as a promising approach [4] to tackle combinatorial optimization. It leverages the capability of neural networks to automatically learn intricate patterns and mappings, bypassing the need for manually designed processes. DRL-based methods demonstrated effectiveness in problem-solving along with generalizability to related combinatorial optimization problems. Meanwhile, there are several challenges when using DRL to solve combinatorial optimization problems. DRL training poses significant overhead such as high computational cost and low sample efficiency, resulting in a considerably prolonged training time that prohibits scalability. Policy gradient [29] has been a popular reinforcement learning method for combinatorial optimization due to its simplicity of integration and property of faster convergence. Its policy network is designed to directly optimize the policy of an agent by computing gradients of expected rewards. Imitation learning [9] such as behavior cloning is a type of machine learning algorithm that learns behavior policies from demonstrations. It has been adopted for robot control, autonomous driving, and game playing [27]. AlphaGo [24] utilized imitation learning to learn behaviors from human players and then it can play against itself to make improved sequential decisions through reinforcement learning. The algorithm proposed in [21] also leveraged imitation learning and reinforcement learning and demonstrated the effectiveness in its application of controlling robotic arms. These successful examples represent innovative combinations of two learning approaches to overcome the limitation of using each approach alone. Inspired by their works, we improve the state-of-the-art policy gradient DRL methods for the TSP by proposing an effective pre-training method with behavior cloning to initialize the policy network. We also exploit the symmetry of TSP solutions in pre-training. Our main contributions are summarized as follows:

- We propose to employ behavior cloning with generated demonstration data to pre-train a policy network before it gets optimized by the policy gradient algorithm. Our pre-training method significantly enhances the state-of-the-art policy gradient DRL model's performance and convergence.
- We utilize the symmetry of TSP solutions to conduct augmentation on the demonstration data. This addresses the issue of lacking sufficient demonstration data for achieving effective behavior cloning.

- Our method enables the use of a pre-trained model developed from a smaller TSP instance to initialize the learning model for a larger TSP instance with quality solution and faster convergence.

2   Related Work

Machine learning methods for solving the TSP are concentrated on policy gradient reinforcement learning. In the following, we briefly cover the traditional combinatorial optimization methods and tools, early-developed ML methods, and finally the policy gradient.

Traditional Approaches. The solvers Concorde [6] and LKH3 [8] use linear programming and heuristics to exactly solve the TSP. OR Tools [19] is a software library that provides optimization tools to solve a range of optimization problems including TSP variants. Unlike Concorde and LKH3, solutions from OR Tools are not guaranteed to be optimal. Farthest Insertion is a naive greedy algorithm that is suitable for small to medium-sized problem instances but also does not guarantee solution optimality. 2-opt [1] is a popular local search-based approximation algorithm that is both simple and efficient.

Early Development. Vinyals et al. [28] proposed a Pointer Network (Ptr-Net) model based on a sequence-to-sequence architecture [25] that utilizes attention pointers to select input to form output. The solution is constructed in an autoregressive approach, and the model is trained using supervised learning. Dai et al. [10] utilized a combination of Deep Q-Network (DQN) [22,26] and graph embedding Structure2Vec [5]. The graph embedding network provides an output of the Q value based on the current state and then selects the action greedily according to the Q value, gradually constructing a solution.

Policy Gradient. Bello et al. [3] proposed to use reinforcement learning to further train the Ptr-Net. They suggested using the negative tour length as the reward and the policy gradient algorithm REINFORCE [29] to optimize the parameters of the Ptr-Net. Kool et al. [12] proposed a Transformer-based model – Attention Model (AM) for the TSP. The model maps the graph embedding and selected node embeddings to the context embedding. They also utilize the REINFORCE algorithm but in addition with a simple baseline and a deterministic greedy rollout strategy to train the model. The AM is currently one of the most effective deep learning models in terms of both solution quality and efficiency. Different from the full attention mechanism adopted by the above-mentioned Transformer-based model, the sparse attention mechanism takes up significantly less memory and trains notably faster [35]. Ma et al. [15] proposed a novel model called the Graph Pointer Network (GPN) by combining the Graph Neural Network (GNN) [33] and the Ptr-Net. They introduced a graph embedding layer on the input to build the Pointer Network, which captures the relationship between nodes. Kwon et al. [13] introduced Policy Optimization with Multiple Optima (POMO), which is another state-of-the-art method for the TSP. The POMO improves the solution representation of the TSP, adapts the REINFORCE algorithm, and mandates distinct rollouts for each optimal solution. The POMO's low-variance baseline also enhances the speed and stability of the DRL training, while increasing its resilience to local minima. Paulo et al. [17] proposed using the policy gradient algorithm to learn a local search heuristic algorithm based on 2-opt operators. Xin et al. [31] proposed a Multi-Decoder Attention Model (MDAM) based on AM to train various policies, effectively increasing the chances of finding optimal solutions. They also separately discovered in [32] that the existing methods did not fully comply with the Bellman optimality principle [2]. Therefore, a solution was proposed to remove visited nodes and recalculate node embeddings at each selection step. Yang et al. [34] proposed a new Random baseline in RL, which stands out for its contribution of reducing the average training time by 16%.

3   Behavior Cloning Enhanced Policy Gradient

The input of the TSP in two-dimensional Euclidean space consists of the coordinates of n nodes, which are denoted as $s = \{(x_1, y_1), (x_2, y_2), ..., (x_n, y_n)\}$. The aim is to determine a permutation π of n nodes that results in the shortest tour by visiting each node once except the starting node. Let $\pi(i)$ represent the i-th node of a TSP tour, and $\pi(< i)$ represent the partial solution before reaching the i-th node. In addition, tour length is denoted as $L(\pi|s)$, which is the distance traveled from the starting node to all other nodes and then back to the starting node. The reinforcement learning algorithm [3] proposed by Bello et al. utilizes the parameter θ of a neural network for defining the policy network $p_\theta(\pi|s)$. The policy network represents the probability of π for a given TSP instance s. Using the chain rule, the neural network factorizes the probability into a product of conditional probabilities:

$$p_\theta(\pi \mid s) = \prod_{i=1}^{n} p_\theta(\pi(i) \mid \pi(< i), s) \tag{1}$$

In the following, we will provide a detailed explanation of how to perform data augmentation on the demonstration data required by behavior cloning and how to pre-train the policy network $p_\theta(\pi|s)$ using behavior cloning. Then, we will elaborate on how to further train the policy network with reinforcement learning.

3.1   Behavior Cloning Pre-training

Behavior cloning [20] is a supervised imitation learning method that utilizes expert demonstrations to train policy networks. This method does not require random exploration or reward design, and it can produce relatively good results quickly. Demonstrations are composed of corresponding tuples of (states, actions), that is $D = \{(s_1, \pi_1), ..., (s_m, \pi_m)\}$ where m denotes the number of instances. The downside of behavior cloning is that it requires a substantial amount of demonstrations to be effective. However, for the TSP, even the most efficient solver takes up substantial time, especially for solving large-scale problem instances (e.g., n in the magnitude above thousands). As a consequence, sufficient demonstrations would be costly or impossible to obtain.

Data Augmentation. In fact, TSP solutions themselves possess some interesting properties. Specifically, every TSP solution/tour takes the form of a closed loop on a two-dimensional plane, whereas, in deep learning, this solution is encoded in a vector structure. Consequently, when constructing the demonstration data, we only designate a single node as the starting point for the solution, while each node in the loop can be the starting point. For instance, in behavior cloning, the initial node (x_1, y_1) is consistently designated as the starting point $\pi(1)$ for the partial solution π in solving the TSP. The autoregressive model will then be influenced by this choice, with the trained policy showing a tendency to select the initial node (x_1, y_1) as the starting point $\pi(1)$ for π in subsequent solution iterations. To address the above issue while augmenting the demonstration data, we utilize the symmetry of any TSP solution (i.e., let each node be a starting node $\pi(1)$) as follows:

$$\pi = \{j, ..., n, 1, ..., j - 1\} \Leftrightarrow \pi' = \{j + 1, ..., n, 1, ..., j\}, \tag{2}$$

where for $j \in (1, 2, ..., n)$. As a result, we can generate up to n different representations of a solution, so the size of the demonstration data can be significantly augmented.

Pre-training. After obtaining sufficient demonstration data, the policy network can then be effectively initialized in a supervised way. We employ an expert policy p_E from behavior cloning and train a learning policy $p_\theta(\pi|s)$, with the objective to approximate the expert policy p_E by minimizing the difference between the predictions generated by the learning policy $p_\theta(\pi|s)$ and the expert actions from p_E. As displayed in the left part of Fig. 1, we use the coordinates s of n nodes and a partial solution $\pi(< i)$ as a demonstration input to the policy network $p_\theta(\pi|s)$. The policy network then outputs probability values of n nodes. We select a node as $\pi(i)$ based on the demonstrations and record the corresponding probability value p_i. After an episode ends, we obtain the probabilities $(p_1, ..., p_n)$ of n nodes. We use maximum likelihood as the loss function to maximize the probability of selecting the nodes based on the demonstrations. The loss function of the policy network is represented as follows:

$$\mathcal{L}(\theta) = -\sum_{i=1}^{n} \log p_\theta(\pi(i)|\pi(< i), s). \tag{3}$$

Using Adam optimization [11] on the above loss function can direct the policy $p_\theta(\pi|s)$ to be closer to the expert policy p_E.

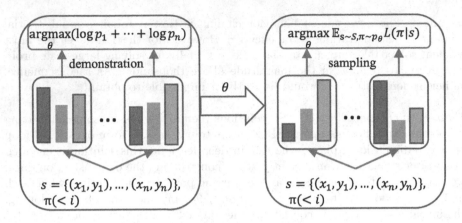

Fig. 1. Behavior cloning enhanced policy gradient.

3.2   Policy Gradient Training

Due to the limited collection of state-action pairs through expert demonstration and the non-explorative nature of behavior cloning, the pre-trained model is unable to explore new solutions or policies on its own. Therefore, the policy network trained through behavior cloning often cannot generalize well to new instances, i.e., distribution drift [23] exists. Although behavior cloning converges faster than the policy gradient algorithm, its prediction accuracy is often much lower due to limited solution exploration. To address the aforementioned problems, we utilize the advantage of behavior cloning to initialize better policy parameters, and further train the policy network model through policy gradient, which can continuously improve the policy through interaction with the environment. As shown in the right part of Fig. 1, after the policy network is initialized, we use it to generate the probability values of unselected nodes each time and then sample the nodes. At the end of the episode, we obtain solution π and its corresponding probability values. Our objective is to obtain the optimal policy by maximizing the expected return, which is equivalent to minimizing the tour length $L(\pi|s)$. The algorithm relies on refining the policy network $p_\theta(\pi|s)$ to achieve the optimal policy. Regarding the TSP, instances (i.e., node coordinates) are randomly sampled from S, thus the objective function can be represented as $J(\theta) = \mathbb{E}_{s \sim S, \pi \sim p_\theta(\pi|s)} L(\pi \mid s)$, and the policy gradient is defined as:

$$\nabla_\theta J(\theta) = \mathbb{E}_{s \sim S, \pi \sim p_\theta(\pi|s)} \left[(L(\pi \mid s) \nabla_\theta \log p_\theta(\pi \mid s) \right]. \tag{4}$$

We use the REINFORCE algorithm [29] with a baseline to optimize the objective function. The REINFORCE algorithm uses the Monte Carlo method to approximate the policy gradient (4). Here, $b(s)$ represents the baseline that is independent of the solution π and is used to estimate the tour length to reduce the variance of the gradient and speed up the training process. The algorithm generates B TSP instances from distribution S, and samples a solution π_j in each instance:

Algorithm 1. Behavior Cloning Enhanced Policy Gradient

1: **procedure** TRAINING(demonstrations D, training set S, steps per epoch T, batch size B, number of nodes of the instance N)
2:　　Perform data augmentation on D with formula (2)
3:　　Initialize pre-training network parameter θ:
4:　　**for** step $= 1, ..., T$ **do**
5:　　　　Set $s_j, \pi_j \leftarrow$ SAMPLEINPUT(D) for $j \in \{1, ..., B\}$
6:　　　　$\mathcal{L}(\theta) = -\sum_{j=1}^{B} \sum_{i=1}^{N} \log p_\theta(\pi_j(i)|\pi_j(<i), s_j)$
7:　　　　$\theta = $ ADAM $(\theta, \mathcal{L}(\theta))$
8:　　**end for**
9:　　Policy gradient with pre-trained network parameter θ:
10:　　**for** step $= 1, ..., T$ **do**
11:　　　　$s_j \leftarrow$ SAMPLEINPUT(S) for $j \in \{1, ..., B\}$
12:　　　　$\pi_j \leftarrow$ SAMPLESOLUTION($p_\theta(.|s_j)$) for $j \in \{1, ..., B\}$
13:　　　　Use (5) as the loss function $\nabla_\theta J(\theta)$
14:　　　　$\theta = $ ADAM $(\theta, \nabla_\theta J(\theta))$
15:　　**end for**
16:　　**return** θ
17: **end procedure**

$$\nabla_\theta J(\theta) \approx \frac{1}{B} \sum_{j=1}^{B} (L(\pi_j \mid s_j) - b(s_j)) \nabla_\theta \log p_\theta (\pi_j \mid s_j). \tag{5}$$

The details of the entire algorithm with behavior cloning pre-training and policy gradient are described in Algorithm 1. Finally, as the number of nodes n increases, generating sufficient demonstrations according to the problem size becomes virtually impossible, hence making the solution of large-scale TSP highly challenging. Fortunately, we can use a TSP model pre-trained on smaller TSP instances to be further trained on larger TSPs via policy gradient training. This is possible because the pre-training phase has already initialized the training model parameters with expert policies, thereby the time-consuming random exploration steps in reinforcement learning can be greatly reduced.

4　Experiments

Experiments were conducted on a server with an Intel Xeon Processor (Icelake) CPU, an Nvidia A100 GPU (40 G), and CentOS 7.9 64bit. By convention, TSP node coordinates are uniformly sampled within a square of unit side length. For each TSP of size n, algorithms (covered in the related work) were tested on 10,000 randomly generated TSP instances (as the test set w.r.t. n), with their average output tour length serving as the evaluation metric for model effectiveness. The pre-training method presented in this paper was applied to two state-of-the-art policy gradient algorithms AM [12] and POMO [13], which are the most effective deep learning models in terms of both solution quality and

Table 1. A comprehensive testing results on the TSP

Method	n=20		n=50		n=100		n=250		n=500	
	Len.	Gap	Len.	Gap	Len.	Gap	Len.	Gap	Len.	Gap
Concorde	3.83	0.00%	5.69	0.00%	7.77	0.00%	11.89	0.00%	16.55	0.00%
LKH3	3.83	0.00%	5.69	0.00%	7.77	0.00%	11.89	0.00%	16.55	0.00%
OR Tools	3.86	0.94%	5.85	2.87%	8.06	3.86%	12.86	8.15%	17.45	5.43%
Farthest Insertion	3.92	2.28%	6.01	5.55%	8.35	7.59%	13.03	9.55%	18.29	10.50%
2-opt	3.96	3.37%	6.13	7.66%	8.52	9.69%	13.25	11.46%	18.60	12.39%
Ptr-Net [28]	3.88	1.28%	7.66	34.53%	10.81	39.18%	-		-	
Rl Ptr-Net [3]	3.89	1.54%	5.95	4.50%	8.30	6.86%	-		-	
S2V-DQN [10]	3.89	1.42%	5.99	5.20%	8.31	6.99%	13.08	10.00%	18.43	11.35%
GPN [15]	3.87	0.94%	5.95	4.55%	8.34	7.42%	13.54	13.88%	19.53	17.99%
2-OPT DL [17]	3.84	0.23%	5.70	0.21%	7.83	0.87%	-		-	
MDAM [31]	3.84	0.29%	5.73	0.63%	7.93	2.10%	-		-	
ASW-TAM [32]	3.84	0.23%	5.76	1.16%	8.01	3.13%	-		-	
Rand Baseline [34]	3.84	0.23%	5.78	1.58%	7.87	1.29%	-		-	
AM [12]	3.85	0.50%	5.80	1.86%	8.12	4.54%	13.10	10.18%	19.85	19.94%
*AM-Ours	3.84	0.23%	5.77	1.33%	7.99	2.87%	12.66	6.48%	18.30	10.57%
POMO [13]	3.83	0.04%	5.70	0.21%	7.80	0.46%	12.42	4.46%	20.51	23.92%
*POMO-Ours	**3.83**	**0.00%**	**5.69**	**0.04%**	**7.78**	**0.17%**	**12.07**	**1.51%**	**17.63**	**6.53%**

efficiency. We also conducted ablation studies for both pre-training and reinforcement learning phases to evaluate the contributions of different components in the model.

4.1 Experimental Results

For testing on the TSP, we conducted a comprehensive study involving multiple traditional algorithms and state-of-the-art deep learning models. Due to the randomness of TSP instances, we report the average tour length (Len) to two decimal places, but for calculating the optimality gap (Gap), we use the average tour length rounded to three decimal places. Note that traditional methods such as Concorde and LKH3 produced optimal solutions from the solver. However, when $n = 500$, the solver takes approximately 38.6 h to solve 10,000 instances, whereas the trained POMO takes only 6 min for its inference. This demonstrates that the traditional methods often cannot achieve both solution speed and accuracy simultaneously. For TSP sizes up to $n = 100$, most models were able to attain a satisfactory solution. Due to the training complexity, Ptr-Net, Rl Ptr-Net, MDAM, and ASW-TAM cannot be generalized to larger TSP instances (i.e., $n = 250$ or 500). In addition, S2V-DQN, GPN, AM, and POMO can only generalize using the TSP100-trained RL models due to the excessive time in training larger models. However, with our pre-training that results in faster RL, we were able to obtain TSP250-trained RL models AM-Ours and POMO-Ours within the time limit and we can then generalize them to $n = 500$. As shown

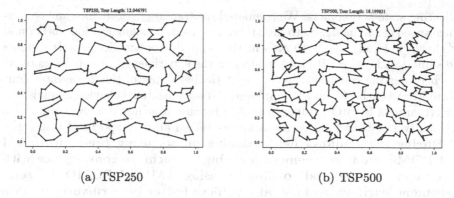

Fig. 2. The two graphics respectively display the solutions when $n = 250/500$. The tour lengths are marked above each graphic.

in Table 1, we obtained significantly improved results on both AM and POMO, especially on the larger TSPs. In particular, the POMO with our pre-training method (i.e., POMO-Ours) reached a solution that is close to optimal. Figure 2 displays the solutions to our large-scale TSP instances.

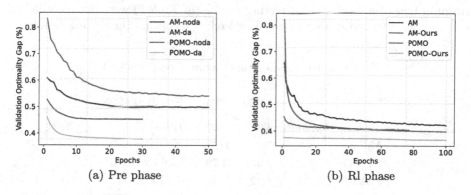

Fig. 3. Training performances of AM and POMO in both pre-training (pre) and reinforcement learning (rl) phases.

4.2   Ablation Study

During the pre-training phase, models that use data augmentation demonstrate faster convergence speed and better performance compared to those that do not use data augmentation. We first performed behavior cloning pre-training alone (as in Sect. 3.1) on AM and POMO. Specifically, we used the Concorde solver [6] to generate exact demonstrations by collecting state-action pairs. We compared the training performance and convergence speed of two models with pre-training using data augmentation (da) or not. As shown in Figure 3(a), we present the validation optimality gap (It exhibits the same trend across all problem sizes in the experiment.) as epochs increase, which proves the conclusion we made at the beginning of this paragraph. Furthermore, when both models employed data-augmented behavior cloning, the performance of POMO is superior to AM because it achieved a lower validation optimality gap while they both quickly converged after 30 epochs of training.

During the reinforcement learning phase, using pre-trained AM and POMO for reinforcement learning can achieve convergence with fewer epochs compared to directly using AM and POMO for reinforcement learning, and can also achieve better performance. We then applied data-augmented pre-training to AM and POMO and contrasted the performances of policy gradient training with and without pre-training. As shown in Fig. 3(b), we further showcase the comparative RL training, wherein pre-trained models have shown a lower validation optimality gap compared to their respective models. Notably, the POMO with pre-training already reached a good performance from the start and then it converged slowly. In addition, as shown in Table 2, we present the RL training time of models trained with and without our pretraining method, as well as the Random Baseline [34], to further demonstrate its advantage. Trainings taking longer than a week were terminated and their training times (in Table 2) and testing results (in Table 1) were not recorded. If not specified otherwise, we use greedy decoding to obtain the testing results.

Table 2. RL training times comparison (in hours)

Method	TSP20	TSP50	TSP100	TSP250
Random Baseline	4.37	9.50	17.50	-
AM	5.20	11.25	20.83	-
*AM-Ours	0.39	**2.79**	**15.64**	**67.52**
POMO	1.42	11.70	170.23	-
*POMO-Ours	**0.35**	3.60	56.10	127.50

5   Conclusion

In this paper, we first introduced pre-training with data-augmented behavior cloning for solving combinatorial optimization problems based on policy gradient reinforcement learning. We conducted a specific study on the classical TSP and demonstrated that our method has significantly progressed in narrowing the optimality gap, reducing the training time, and solving larger instances. For future work, we will apply our method for solving other important TSP variants to demonstrate model generalizability. We will also investigate more scalable solutions and conduct comprehensive experiments to validate the scalability.

Acknowledgments. This work is supported by the Taishan Scholars Young Expert Project of Shandong Province (No. tsqn202211215) and the National Science Foundation of China (Nos. 12271098 and 61772005). The corresponding author is Longkun Guo (lkguo@fzu.edu.cn).

References

1. Aarts, E.H., Lenstra, J.K.: Local search in combinatorial optimization. Princeton University Press (2003)
2. Bellman, R.: Dynamic programming treatment of the travelling salesman problem. J. ACM (JACM) **9**(1), 61–63 (1962)
3. Bello, I., Pham, H., Le, Q.V., Norouzi, M., Bengio, S.: Neural combinatorial optimization with reinforcement learning. In: Proceedings of International Conference on Learning Representations (ICLR) (2017)
4. Bengio, Y., Lodi, A., Prouvost, A.: Machine learning for combinatorial optimization: a methodological tour d'horizon. Eur. J. Oper. Res. **290**(2), 405–421 (2021)
5. Dai, H., Dai, B., Song, L.: Discriminative embeddings of latent variable models for structured data. In: International Conference on Machine Learning, pp. 2702–2711. PMLR (2016)
6. Applegate, D., Robert Bixby, V.C., Cook, W.: Concorde TSP Solver (2006). https://www.math.uwaterloo.ca/tsp/concorde/index.html
7. Halim, A.H., Ismail, I.: Combinatorial optimization: comparison of heuristic algorithms in travelling salesman problem. Arch. Comput. Methods Eng. **26**, 367–380 (2019)
8. Helsgaun, K.: An extension of the lin-kernighan-helsgaun TSP solver for constrained traveling salesman and vehicle routing problems: Technical report (2017)
9. Hussein, A., Gaber, M.M., Elyan, E., Jayne, C.: Imitation learning: a survey of learning methods. ACM Comput. Surv. (CSUR) **50**(2), 1–35 (2017)
10. Khalil, E., Dai, H., Zhang, Y., Dilkina, B., Song, L.: Learning combinatorial optimization algorithms over graphs. Adv. Neural. Inf. Process. Syst. **30** (2017)
11. Kingma, D.P., Ba, J.: Adam: a method for stochastic optimization. In: Proceedings of International Conference on Learning Representations (ICLR) (2015)
12. Kool, W., van Hoof, H., Welling, M.: Attention, learn to solve routing problems! In: International Conference on Learning Representations (2019)
13. Kwon, Y.D., Choo, J., Kim, B., Yoon, I., Gwon, Y., Min, S.: Pomo: Policy optimization with multiple optima for reinforcement learning. Adv. Neural. Inf. Process. Syst. **33**, 21188–21198 (2020)
14. Lawler, E.L., Wood, D.E.: Branch-and-bound methods: a survey. Oper. Res. **14**(4), 699–719 (1966)
15. Ma, Q., Ge, S., He, D., Thaker, D., Drori, I.: Combinatorial optimization by graph pointer networks and hierarchical reinforcement learning. In: AAAI Workshop on Deep Learning on Graphs: Methodologies and Applications (2020)
16. Matai, R., Singh, S.P., Mittal, M.L.: Traveling salesman problem: an overview of applications, formulations, and solution approaches. Traveling Salesman Problem, Theory and Applications **1** (2010)

17. d O Costa, P.R., Rhuggenaath, J., Zhang, Y., Akcay, A.: Learning 2-opt heuristics for the traveling salesman problem via deep reinforcement learning. In: Asian Conference on Machine Learning, pp. 465–480. PMLR (2020)

18. Papadimitriou, C.H.: The euclidean travelling salesman problem is np-complete. Theoret. Comput. Sci. 4(3), 237–244 (1977)

19. Perron, L., Furnon, V.: Or-tools (2022). https://developers.google.com/optimization/

20. Pomerleau, D.A.: Alvinn: an autonomous land vehicle in a neural network. Adv. Neural. Inf. Process. Syst. **1** (1988)

21. Rajeswaran, A., et al.: Learning complex dexterous manipulation with deep reinforcement learning and demonstrations. In: Proceedings of Robotics: Science and Systems. Pittsburgh, Pennsylvania (June 2018)

22. Riedmiller, M.: Neural fitted Q iteration – first experiences with a data efficient neural reinforcement learning method. In: Gama, J., Camacho, R., Brazdil, P.B., Jorge, A.M., Torgo, L. (eds.) ECML 2005. LNCS (LNAI), vol. 3720, pp. 317–328. Springer, Heidelberg (2005). https://doi.org/10.1007/11564096_32

23. Ross, S., Gordon, G., Bagnell, D.: A reduction of imitation learning and structured prediction to no-regret online learning. In: Proceedings of the Fourteenth International Conference on Artificial Intelligence and Statistics, pp. 627–635. JMLR Workshop and Conference Proceedings (2011)

24. Silver, D., et al.: Mastering the game of go with deep neural networks and tree search. Nature **529**(7587), 484–489 (2016)

25. Sutskever, I., Vinyals, O., Le, Q.V.: Sequence to sequence learning with neural networks. Adv. Neural. Inf. Process. Syst. **27** (2014)

26. Thrun, S., Littman, M.L.: Reinforcement learning: an introduction. AI Mag. **21**(1), 103–103 (2000)

27. Torabi, F., Warnell, G., Stone, P.: Recent advances in imitation learning from observation. In: Proceedings of the Twenty-Eighth International Joint Conference on Artificial Intelligence, pp. 6324–6331 (2019)

28. Vinyals, O., Fortunato, M., Jaitly, N.: Pointer networks. Adv. Neural. Inf. Process. Syst. **28** (2015)

29. Williams, R.J.: Simple statistical gradient-following algorithms for connectionist reinforcement learning. Reinforc. Learn., 5–32 (1992)

30. Williamson, D.P., Shmoys, D.B.: The design of approximation algorithms. Cambridge University Press (2011)

31. Xin, L., Song, W., Cao, Z., Zhang, J.: Multi-decoder attention model with embedding glimpse for solving vehicle routing problems. In: Proceedings of the AAAI Conference on Artificial Intelligence, vol. 35, pp. 12042–12049 (2021)

32. Xin, L., Song, W., Cao, Z., Zhang, J.: Step-wise deep learning models for solving routing problems. IEEE Trans. Industr. Inf. **17**(7), 4861–4871 (2021)

33. Xu, K., Hu, W., Leskovec, J., Jegelka, S.: How powerful are graph neural networks? In: International Conference on Learning Representations (2019)

34. Yang, H., Gu, M.: A new baseline of policy gradient for traveling salesman problem. In: 2022 IEEE 9th International Conference on Data Science and Advanced Analytics (DSAA), pp. 1–7. IEEE (2022)

35. Zaheer, M., et al.: Big bird: transformers for longer sequences. Adv. Neural. Inf. Process. Syst. **33** (2020)

Leveraging Transfer Learning for Enhancing Graph Optimization Problem Solving

Hui-Ju Hung[1]([✉]), Wang-Chien Lee[1], Chih-Ya Shen[2], Fang He[1], and Zhen Lei[1]

[1] The Pennsylvania State University, University Park, PA 16802, USA
{hzh131,wul2,fxh35,zlei}@psu.edu
[2] National Tsing Hua University, Hsinchu 300044, Taiwan
chihya@cs.nthu.edu.tw

Abstract. Reinforcement learning to solve graph optimization problems has attracted increasing attention recently. Typically, these models require extensive training over numerous graph instances to develop generalizable strategies across diverse graph types, demanding significant computational resources and time. Instead of tackling these problems one by one, we propose to employ transfer learning to utilize knowledge gained from solving one graph optimization problem to aid in solving another. Our proposed framework, dubbed the *State Extraction with Transfer-learning (SET)*, focuses on quickly adapting a model trained for a specific graph optimization task to a new but related problem by considering the distributional differences among the objective values between the graph optimization problems. We conduct a series of experimental evaluations on graphs that are both synthetically generated and sourced from real-world data. The results demonstrate that SET outperforms other algorithmic and learning-based baselines. Additionally, our analysis of knowledge transferability provides insights into the effectiveness of applying models trained on one graph optimization task to another. Our study is one of the first studies exploring transfer learning in the context of graph optimization problems.

Keywords: graph optimization problem · reinforcement learning · transfer learning · node representation

1 Introduction

Proven to be NP-hard, many graph optimization problems do not have polynomial-time algorithms to solve them. Yet, the quest to solve these complex problems is of great importance, as they find practical applications in diverse areas like transportation, scheduling, and social networking [14,30]. Traditionally, research efforts in solving graph optimization problems are mostly garnered in the theory and algorithm design communities. In recent years, however, research interests in solving graph optimization problems with various *learning* approaches such as supervised, unsupervised, or reinforcement learning have been growing [6,13,19,28,32]. In the following, we informally describe the learning task.

D.-N. Yang et al. (Eds.): PAKDD 2024, LNAI 14646, pp. 339–351, 2024.
https://doi.org/10.1007/978-981-97-2253-2_27

Table 1. Comparison between *No-Transfer* and *Transfer* in solving MC

Approach	No-Transfer	Transfer
Average objective	4.2578	4.2734
Number of training iterations to converge	4673	1443

Learning to Solve Graph Optimization Problems. To solve a graph optimization problem, a typical algorithmic approach is to develop a greedy algorithm that begins with an empty solution and sequentially incorporates the node with the highest score from the candidates until a predefined termination criterion is met. Instead of adopting a manually designed scoring function, a machine learning model may be trained to serve as the scoring function in the greedy algorithm.

Many models have been introduced to tackle various graph optimization problems and achieve commendable results [3,6,9,13,36]. These models typically undergo intensive training on numerous graph instances to automatically derive effective strategies for a variety of graph types, making the extensive training duration a challenge. An important research question arises in exploring ideas to address the challenge: *Do models designed for distinct graph optimization problems share common knowledge that can be used in other problem settings? In other words, is it possible to transfer the knowledge, e.g., inherent structural information of the graph instances, acquired from solving one graph optimization problem to another?*

To investigate whether *knowledge* acquired from one optimization problem can be beneficial for another, we introduce a motivating study, under the context of reinforcement learning, to examine the transferability of solutions between distinct problems. Specifically, we explore if insights gained from solving the Dense Subgraph (DSG) problem can aid in addressing the Maximum Clique (MC) problem. For clarity, we differentiate between two methodologies: *Transfer* and *No-transfer*, which share the same model structure constructed with GCN [34]. *Transfer* trains a model with the data collected for solving DSG first and adjusts the model with a few data collected for solving MC. In contrast, *No-transfer* trains the MC model from scratch with only the data collected for solving MC. Our comparative analysis assesses the *objective value* for the MC problem across a test set of 128 graphs, each with 20 nodes, achieved by the converged *Transfer* and *No-transfer* models.

The results averaged over three trials is shown in Table 1. *Transfer* achieves objective value of 4.2734, slightly better than the objective value of 4.2578 under *No-transfer*. More importantly, *Transfer* converges much faster, with 1443 training iterations, than *No-transfer* (4673 iterations). These findings suggest that parameters optimized for DSG can indeed enhance the training efficiency for MC, demonstrating the potential of transferring knowledge learned from one optimization challenge to benefit another.[1]

---

[1] Notably, the efficacy of *knowledge transfer* extends beyond this specific pair of graph optimization problems and has been observed in several other problem combinations.

The aforementioned example compellingly illustrates that transfer learning can facilitate solving one optimization problem by applying knowledge from another within the context of reinforcement learning frameworks. This success prompts a deeper inquiry: *What specific information is learned and subsequently utilized to address a new optimization challenge?* Most models claim that they are able to accurately encapsulate structural information, which contributes to their superior performance. However, the precise nature of the information, i.e., what facets of structure or patterns are captured and how they are leveraged to navigate graph optimization problems, remains largely unexplored.

In this work, we introduce the State Extraction with Transfer-learning (SET) framework, designed to generalize a model across different graph optimization problems. The core idea of SET is to train a model predominantly on one optimization problem, which can then swiftly adapt to solve another, thereby indicating the generalizability of the structural information captured. SET achieves rapid adaptation by acknowledging and resolving the distributional differences among the objective values across multiple graph optimization problems. We validate the effectiveness of SET on various graph optimization problems. We also conduct an analysis to show the knowledge transferability of applying models trained on one graph optimization task to another.

The contributions of this paper are summarized as follows.

- Existing work exploiting reinforcement learning requires a large amount of data (or a large number of training iterations) to train a model or solve a graph optimization problem. To alleviate this issue, we propose to exploit transfer learning to reduce the number of training iterations while maintaining a comparable performance.
- We introduce the State Extraction with Transfer-learning (SET) framework, which transfers the useful structural features learned from one graph optimization problem to another via transferring model parameters. Moreover, SET is able to quickly adapt the model to address another graph optimization problem by considering the distributional differences of objective values across multiple graph optimization problems.
- Our evaluation demonstrates the efficacy of SET in finding near-optimal solutions efficiently in significantly less number of training iterations. SET outperforms the algorithmic and learning-based baselines in various graph optimization problems, where learning-based ones are fully trained by the data collected from the target problem.

The rest of this paper is organized as follows. Section 2 reviews the related work. Section 3 proposes the SET framework. Section 4 reports the evaluation. Finally, Sect. 5 concludes the paper.

2   Related Work

2.1   Transfer Learning

Transfer learning, aiming to boost model performance by drawing on knowledge from related domains or datasets, has been widely adopted in various fields such as healthcare, natural language processing, computer vision, and robotics [12, 16, 20, 37]. To facilitate a task *transfer* from a source domain to a target domain, the most commonly employed method involves the transferal of model parameters, i.e., the parameters fine-tuned on the source task are utilized as the starting point for training the model on the target task.

Generally speaking, a model is composed of two parts [11, 29, 33]: the feature extractor, which identifies crucial features for the tasks, and the predictor, which bases its predictions on these extracted features. There are two primary forms of parameter adoption [29, 33]: 1) Linear Probing: where only the parameters in the predictor are modified using the target task's training data and the parameters in the feature extractor are not. 2) Fine-tuning: where adjustments are made to the parameters in both the feature extractor and the predictor with the target task's training data. Dai et al. have employed adversarial domain adaptation to bridge the distributional gap between source and target domains, thereby aiding the transfer of knowledge [7]. Similarly, Evci et al. have taken advantage of intermediate representations from all layers of the source model to construct a classification head for the target domain [10]. It should be noted, however, that the integration of these techniques into graph optimization problems has been relatively limited.

2.2   Reinforcement Learning for Graph Optimization Problems

Reinforcement learning (RL) to solve optimization problems has seen a notable surge in interest. These graph optimization problems are often delineated by the nature of their solutions, which may be either sequences or subsets of nodes. Among the variety of problems, those requiring a sequence of nodes, such as the Traveling Salesman Problem (TSP) and its variants, have garnered significant attention in the literature [3, 9].[2] For instance, Bello et al. leverage a *pointer network*, an architecture incorporating recurrent neural networks with an attention mechanism, to tackle TSP [3]. Dorigo et al. introduce a multi-agent system based on reinforcement learning to approach the TSP [9]. Expanding beyond the TSP, Nazari et al. address the Capacitated Vehicle Routing Problem, another routing problem that optimizes the paths of several vehicles in tandem, using RL techniques [24].

RL has also been adopted to address graph optimization problems where the goal is to identify a subset of nodes that meet certain criteria, such as the Minimum Vertex Cover (MVC) problem [6, 13, 35]. Dai et al. introduced a Q-learning-based method that leverages instances without optimal solutions [6]. Expanding on Dai et al.'s work, Yang and Shen enhanced the RL paradigm with graph augmentation techniques for tackling both the Maximum Independent Set (MIS) and MVC problems [35].[3] Furthermore, Groshev et al. utilized an alternative approach, namely *imitation learning*, to train a scoring function that assesses the combination of a state and an action [13].

---

[2] TSP finds the shortest route that visits each node once and returns to the origin.

[3] MIS involves selecting the largest set of vertices where no two are adjacent.

In summary, there has been considerable work leveraging RL to tackle graph optimization problems, yet these efforts typically train models for one specific problem at a time. In contrast, our study explores a new direction that employs transfer learning in the context of graph optimization problems.

3   The State Extraction with Transfer-Learning Framework

3.1   Framework Structure

Several studies have applied RL to solve various graph optimization problems, training independent models for each specific problem [3,6,8,13]. Achieving good solution qualities, these models capture structural and state information helpful to different graph optimization problems. Building on this, we hypothesize that *models trained for different graph optimization problems may capture some common structural and state information.* By testing this hypothesis through Centered Kernel Alignment (CKA) analysis [22], we find that models independently trained from scratch for the DSG, MVC, and MC problems indeed capture shared structural and state information (detailed in Sect. 4.2), which motivates us to explore the feasibility of transfer learning in solving graph optimization problems.

To effectively implement transfer learning in graph optimization problems, understanding the model's objective is crucial. As detailed later in Sect. 3.2, in an RL model developed from scratch, the model's predictions need to be aligned with the accumulated rewards, which is a reflection of the objective values of the tackled graph optimization problem. Given that accumulated rewards are distinct across different graph optimization problems, transferring knowledge from a source graph optimization problem (*source problem* for short) to another target graph optimization problem (*target problem* for short) necessitates the model's proficiency in obtaining not only the graph's structural and state information but also the distributional differences between the source and target problems' accumulated reward functions. Capturing the distributional differences is critical for a model to re-calibrate its predictions from being attuned to the source problem's reward structure to aligning with that of the target problem. To expedite this reward adjustment, we introduce the State Extraction with Transfer-learning (SET) framework. SET exploits the power of transfer learning to extract pivotal features from the resolution of a different graph optimization problem. The essence of this approach is to incorporate the above distributional differences into a loss function, thus facilitating a more efficient transition in learning applicable information to the target problem.

The SET framework is illustrated in Fig. 1. The model structure has two components, *feature extractor* and *predictor*. The feature extractor, responsible for capturing graph and state information as would be done in a training-from-scratch scenario, is constructed using GCN [34].[4] The predictors, comprising

---

[4] Alternative layer-wise GNN methodologies such as GAT [31], GraphSage [15], or GBP [5] could also be used to build the feature extractors.

stacked fully connected layers, are tasked to make predictions based on the extracted features. For both the source and target problems, we utilize the same feature extractor and predictor architecture. Despite this consistency, the notable differences between the source and target problems can lead to a mismatch in the accumulated reward distributions, which leads to differences in the model's parameters, especially those in the predictor. Consequently, a model fine-tuned on data from the source problem may not accurately estimate the accumulated rewards for the target problem. The SET framework is designed to rectify this issue and facilitate problem alignment.

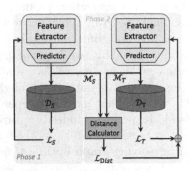
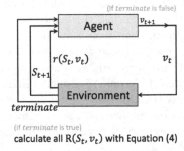

Fig. 1. SET framework illustration   **Fig. 2.** RL process illustration

SET has two distinct phases. During the first phase, the feature extractor and the predictor for the source problem (i.e., the left green box in Fig. 1) are trained to address the source problem. To align the model to approximate the accumulated rewards of the source problem, the training of the first phase leverages a loss function which measures the *mean square error* (MSE) between the accumulated reward and the model's predictions. The loss function is mathematically represented as follows.

$$\mathcal{L}_S = \Sigma_{(S_t, v_t) \in \mathcal{D}_S} MSE(R(S_t, v_t), s^S(S_t, v_t)), \tag{1}$$

where $\mathcal{D}_S$ denotes the dataset of (state, action) pairs compiled through the RL process, MSE is the mean square error function, $R(S_t, v_t)$ is the accumulated reward associated with state S_t and action v_t, and $s^S(S_t, v_t)$ is the score predicted by the model.

In the second phase, the parameters of the feature extractor for the source problem are transferred to initialize the feature extractor for the target problem (i.e., the right green box in Fig. 1). Then, we fine-tune both the feature extractor and the predictor for the target problem with the data collected in solving the target problem. The loss function for this phase (shown below) is similar to that used for the source problem, focusing on the *mean square error* (MSE) between the accumulated reward and the prediction made by the model.

$$\mathcal{L}_T = \Sigma_{(S_t, v_t) \in \mathcal{D}_T} MSE(R(S_t, v_t), s^T(S_t, v_t)), \tag{2}$$

where $\mathcal{D}_T$ represents the training data generated from the target problem's RL process, and $s^T(S_t, v_t)$ denotes the prediction of the target model.

Additionally, the learning process needs to account for $\mathcal{L}_{Dist}$, which is for the distributional differences of model parameters between the source and target problems, as defined below.

$$\mathcal{L}_{Dist} = WS(dist(\mathcal{M}_S), dist(\mathcal{M}_T)), \tag{3}$$

where WS denotes the Wasserstein distance [21] that quantifies the divergence in distributions, and $\mathcal{M}_S$ and $\mathcal{M}_T$ denote the model parameters trained for the source and target problems, respectively. In SET, we train the source and target problems' models separately for training efficiency. Specifically, for addressing a new graph optimization problem, only the second phase of training is required.

3.2   Reinforcement Learning Process

Reinforcement learning (RL) is a paradigm of learning where an agent is trained to make sequential decisions within an environment, targeting on maximize obtained rewards [17]. When applying RL to the domain of graph optimization problems [3, 6, 13, 35], three foundational components are imperative: 1) The *environment setting* corresponds to the definitions of an action, like the *next state* and *obtained reward* after performing an action, and whether an agent reaches the *termination condition*. 2) The *framework* refers to how the learned algorithm operates. For example, a commonly used greedy framework performs one action at a time and never reverses any previous decisions. Each time a new action is performed, a new *intermediate solution* is created based on the previous intermediate solution. Here, an intermediate solution is a subset of nodes $S \subseteq V$, and the action is to add a node to the current intermediate solution. 3) The *model* is responsible for encoding the information of the graph structure and intermediate solution in the embeddings, which serves as the basis for the learned algorithm to make decisions. In the interaction with the environment, the agent's sequence of actions progresses until a termination criterion is met. The obtained feedback, in the form of rewards, informs and propels the agent towards optimal decision strategies.

The training data is collected during the interaction process with the environment. The training data comprises pairs of trials and their associated rewards, where a trial is a sequence of intermediate solutions derived from the actions taken, and the rewards serve as indicators of the efficacy of each solution. Formally, we represent a trial as $<S_1, v_1, S_2, v_2, ..., S_n, v_n>$, where S_t denotes the state or intermediate solution at step t, and v_t represents the action taken or the node added at step t. Consequently, a trial of length n encompasses n state-action pairs.

Utilizing the training data, our objective is to calibrate the model such that it effectively functions an evaluative scoring function. This function takes a state S and an action v as inputs and produces a corresponding score $Q(S, v)$, with higher scores indicating more advantageous actions. To facilitate the training of SET, we distinguish between two types of rewards: the immediate rewards and the accumulated rewards. The immediate rewards, expressed as $r(S_t, v_t)$, represent the instantaneous feedback received from the environment subsequent to executing an action. This immediate feedback serves as an indicator of the action's utility in the given state.

Training the model solely to approximate immediate rewards can result in a myopic perspective that neglects the potential benefits of subsequent actions. To overcome this limitation, it is imperative that the model is trained to estimate the accumulated rewards rather than just the immediate ones. For a given state S and action v, the accumulated reward $R(S_t, v_t)$ is a cumulative function of the immediate rewards $r(S_i, v_i)$ from the current step $i = t$ to the end of the trial $i = n$. The accumulated reward $R(S_t, v_t)$ is formulated as follows.

$$R(S_t, v_t) = r(S_t, v_t) + \sum_{i=1}^{n-t} \gamma^i \cdot r(S_{t+i}, v_{t+i}), \tag{4}$$

where S_t represents the state at step t, v_t denotes the action executed at step v, and n is the trial's length. In Eq. (4), the term $r(S_t, v_t)$ is the immediate reward at step t, while the summation $\sum_{i=1}^{n-t} \gamma^i \cdot r(S_{t+i}, v_{t+i})$ accounts for the weighted rewards of future steps, with the discount factor γ ranging between 0 and 1. A higher value of γ reflects a stronger emphasis on the rewards of later actions. The RL process is illustrated in Fig. 2.

Table 2. Formulations and RL settings for the explored graph optimization problems

Problem	Formulation	State S	Action v	Immediate Reward	Termination
Minimum Vertex Cover (MVC)	Given: a graph $G = (V, E)$ Objective: find a minimal $S \subseteq V$ such that every $(u, v) \in E$ has at least u or v in S	The subset of selected nodes	Add v into S	-1	All edges are adjacent to S
Maximal Clique (MC)	Given: a graph $G = (V, E)$ Objective: find a maximal subset $S \subseteq V$, such that the subgraph induced by S is a clique	The subset of selected nodes	Add v into S	$+1$ if S is a clique, -1 otherwise	S is not a clique
Dense Sub-Graph (DSG)	Given: a graph $G = (V, E)$, an integer k Objective: find a k-node $S \subseteq V$, such that the subgraph induced by S has the most edges.	The subset of selected nodes	Add v into S	Number of newly-added edges among S	S involve k nodes

The training of our model hinges on minimizing the *mean square error* (MSE) between the true accumulated rewards $R(S_{t+i}, v_{t+i})$ in Eq. (4) and the predicted scores by the model $Q(S_{t+i}, v_{t+i})$. We utilize the Adam optimizer for training. The specific reward configurations and formulations for the graph optimization problems we address are detailed in Table 2. In the cases of the Dense Subgraph

(DSG) and Minimum Vertex Cover (MVC) problems, nodes are incrementally incorporated into the intermediate solution S until S qualifies as a feasible solution of the graph optimization problem. For DSG, feasibility is attained when the solution includes k nodes, whereas for MVC, it is when all edges in the graph are covered. Regarding the Maximum Clique (MC) problem, nodes are added to S until adding any further node makes S infeasible, i.e., S no longer constitutes a clique. Consequently, the training dataset comprises $n-1$ node additions with a reward of $+1$, indicating the maintenance of a clique, and one final node addition with a reward of -1, reflecting the formation of a non-clique.

4   Experiments

Graph Optimization Problems. To evaluate the efficacy of our SET framework, we conduct experiments targeting two graph optimization problems: the *Minimum Vertex Cover* (MVC) problem and the *Maximal Clique* (MC) problem. In the evalation, we use the *Dense Sub-Graph* (DSG) problem as the source graph optimization problem. Their problem formulations are detailed in Table 2.

Compared Baselines. To benchmark the SET framework's performance, we compare it against a suite of established algorithmic and learning-based approaches: 1) The optimal (OPT) approach: The optimal solution is obtained by solving a Mixed Integer Linear Programming with CPLEX. 2) The approximation (APPR) algorithms: For MVC, APPR follows the 2-approximation algorithm in guarantee [18]. For MC, APPR follows the $O(|V|/(\log|V|)^2)$-approximation algorithm in [4]. 3) DQN [6]: DQN is an RL method with a message-passing mechanism for state representation. 4) LIGD [35]: LIGD is an RL method that enhances training by diversifying the graph instances encountered during training. 5) LwD [1]: LwD reduces the search space by assessing the likelihood of each node belonging to the optimal solution, applicable to only locally decomposable problems. Note that LwD is only applicable to MC, as DSG and MVC do not meet the locality criteria. 6) ERD [19]: ERD is an unsupervised learning method inspired by Erdos' probabilistic method. 7) HAM [27]: HAM is an unsupervised learning method that exploits the Hamiltonian cost function. 8) REL [32]: REL is an unsupervised learning approach built upon relaxing the objective and constraint functions. 9) NT: NT shares the same structure as SET but trains the target model from scratch.

Metrics. The solution quality is measured by *objective value*. For DSG, the objective value is the number of edges (u, v) with u and v both in S. For MVC and MC, the objective value is the number of nodes within S, i.e., $|S|$. Since our experiments involve both maximization and minimization problems, we report the *approximation ratio* rather than the objective value. For both problem types, an approximation ratio is a value greater than 1, with a lower ratio indicative of superior solution quality. The approximation ratio is 1 if the optimal solution is obtained. To assess training efficiency, we report the *number of training iterations* required to obtain a converged model. In each iteration, solutions for 128 training graphs are generated and utilized in the training process.

Default Settings. Throughout our experiments, we use the following early-stop condition: training is halted if there is no improvement in the objective value or validation loss for w validation iterations. Unless otherwise stated, the default settings are as follows: the dimensionality of input embeddings is set to 64; the number of GCN layers in the feature extractor is 5; the number of fully connected layers in the predictor is 4; the batch size is 16, and the early-stop window size w is 50.[5] We also apply L2 regularization with the weight λ as 10^{-5}. The reported results are the mean of 3 trials. All experiments are conducted on an HP DL580 G9 server equipped with dual six-core Intel i7-8700K CPUs, a pair of GeForce RTX 2080 Graphics Cards, and 64GB of RAM.

4.1   Effectiveness and Efficiency

Testing on Real Networks. We compare the performance of SET and other baselines on several real networks: 1) Facebook, which comprises 4039 nodes and 88234 edges [23]; 2) GitHub, which includes 37700 nodes and 289003 edges, though we down-sample it to 5000 nodes [25]; and 3) LastFM, which consists of 7624 nodes and 27806 edges [26]. For learning-based methods, we utilize training data collected from 100-node graphs generated by the Barabási-Albert (BA) model [2]. The results are presented in Table 3. Due to the computational limitations on larger graphs, the optimal (OPT) solution is not always obtainable. Hence, we calculate the approximation ratio using the best-known objective values. LIGD and REL do not return results within a day and thus are marked with '*' in the table. LwD and HAM, being specific to the MC problem, have their fields for MVC indicated as '–'. As Table 3 demonstrates, SET and NT reach similar performance and consistently outperform all other baselines across these networks, showcasing SET's ability to capture the structural and state information that could be generalized to large-scale graphs. Also, SET requires fewer training iterations than NT, as shown in the next experiment.

Table 3. Testing on large and real networks ('*': out-of-time, '-': not applicable)

		SET	NT	DQN	LIGD	ERD	REL	HAM	APPR	LwD
Facebook	MVC	1.0000	1.0011	1.2135	*	1.2118	*	–	1.1356	–
	MC	1.0009	1.0000	1.2649	*	1.2647	*	∞	1.6152	1.6583
GitHub	MVC	1.0000	1.0032	1.2226	*	1.2235	*	–	1.4165	–
	MC	1.0000	1.0041	1.5192	*	1.5166	*	∞	1.7278	1.7268
LastFM	MVC	1.0021	1.0000	1.2552	*	1.2589	*	–	1.4531	–
	MC	1.0015	1.0000	1.8633	*	1.8654	*	∞	1.7316	1.7265

---

[5] We perform sensitivity tests on those parameters, and select the best-performing settings as the defaults. The sensitivity tests are eliminated for the sake of space.

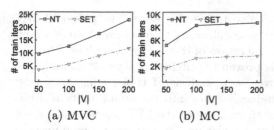

Fig. 3. Number of training iterations needed for models trained with various sizes of graphs

(a) MVC   (b) MC

Table 4. Measuring similarities of node embeddings of models trained for various problems

	DSG	MVC	MC	RAN
DSG	–	0.31	0.42	0.0001
MVC	–	–	0.29	0.0003
MC	–	–	–	0.0004

Testing on Synthetic Graphs. In addition, we evaluate the performance of SET and NT on graphs of varying sizes generated using the BA model, where both training and testing graphs are of the same size. For each graph size, the testing dataset has 128 graphs generated by BA. While the objective values of SET and NT are closely aligned, mirroring the results observed with real networks, we opt not to detail these values here for the sake of space. Figure 3 shows the number of training iterations required for both SET and NT when the target graph optimization problem is MVC and MC, respectively. Notably, SET consistently requires fewer training iterations than NT, underscoring its efficiency in reducing both training time and iterations.

4.2 Transferability of Features

We utilize Centered Kernel Alignment (CKA) [22] to analyze models that have been independently trained for various graph optimization problems. CKA serves as an effective tool for measuring the similarity between sets of high-dimensional data representations, where higher CKA values signify greater similarity or a higher degree of shared features. Specifically, we use CKA to evaluate node representations learned from scratch by NT on 32 sampled graphs, each comprising 20 nodes, for each examined graph optimization problem. The corresponding CKA values for pairs of these problems are detailed in Table 4. Furthermore, we compare the node embeddings from each model against a *random* baseline RAN, specifically, a set of node embeddings filled with randomly generated values. This comparison is to establish a benchmark against two unrelated sets of node embeddings. As shown in Table 4, the CKA scores among the graph optimization problems (DSG, MVC, and MC) are markedly higher than those between the problems and RAN, suggesting that the models for DSG, MVC, and MC share common structural and state information. Particularly, the highest CKA score between DSG and MC can be attributed to their similar objectives, both focusing on the retrieval of k-node dense subsets.

5 Conclusion

The utilization of RL in graph optimization problems has seen rising interest, but the requirement for extensive training remains a challenge. In this work, we harness the power of transfer learning to tackle this issue. The proposed SET framework is adept at rapidly adapting to different graph optimization problems by accounting for the distributional differences of their objective values. Through extensive experimentation, SET has proven to outclass various algorithmic and learning-based competitors. Moreover, our analysis of transferring knowledge across different tasks provides valuable insights into the adaptability of SET. This evidence suggests that models trained under the SET framework are not only effective for specific problems but can also be efficiently repurposed for new, related problems.

References

1. Ahn, S., Seo, Y., Shin, J.: Learning what to defer for maximum independent sets. In: ICML (2020)
2. Albert, R., Barabási, A.L.: Statistical mechanics of complex networks. Rev. Mod. Phys. (2002)
3. Bello, I., Pham, H., Le, Q.V., Norouzi, M., Bengio, S.: Neural combinatorial optimization with reinforcement learning. arXiv preprint arXiv:1611.09940 (2016)
4. Boppana, R., Halldórsson, M.M.: Approximating maximum independent sets by excluding subgraphs. BIT Numerical Mathematics (1992)
5. Chen, M., Wei, Z., Ding, B., Li, Y., Yuan, Y., Du, X., Wen, J.R.: Scalable graph neural networks via bidirectional propagation. In: NeurIPS (2020)
6. Dai, H., Khalil, E., Zhang, Y., Dilkina, B., Song, L.: Learning combinatorial optimization algorithms over graphs. In: NeurIPS (2017)
7. Dai, Q., Wu, X.M., Xiao, J., Shen, X., Wang, D.: Graph transfer learning via adversarial domain adaptation with graph convolution. IEEE Trans. Knowl. Data Eng. (2022)
8. Deudon, M., Cournut, P., Lacoste, A., Adulyasak, Y., Rousseau, L.M.: Learning heuristics for the TSP by policy gradient. In: CPAIOR (2018)
9. Dorigo, M., Gambardella, L.M.: Ant colony system: a cooperative learning approach to the traveling salesman problem. IEEE Trans. Evolutionary Comput. (1997)
10. Evci, U., Dumoulin, V., Larochelle, H., Mozer, M.C.: Head2toe: utilizing intermediate representations for better transfer learning. In: ICML (2022)
11. Ganin, Y., et al.: Domain-adversarial training of neural networks. J. Mach. Learn. Res. (2016)
12. Gopalakrishnan, K., Khaitan, S., Choudhary, A., Agrawal, A.: Deep convolutional neural networks with transfer learning for computer vision-based data-driven pavement distress detection. Construct. Building Mater. (2017)
13. Groshev, E., Tamar, A., Goldstein, M., Srivastava, S., Abbeel, P.: Learning generalized reactive policies using deep neural networks. In: ICAPS (2018)
14. Guze, S.: Graph theory approach to the vulnerability of transportation networks. Algorithms (2019)
15. Hamilton, W., Ying, Z., Leskovec, J.: Inductive representation learning on large graphs. In: NeurIPS (2017)
16. Hua, J., Zeng, L., Li, G., Ju, Z.: Learning for a robot: deep reinforcement learning, imitation learning, transfer learning. Sensors (2021)
17. Kaelbling, L.P., Littman, M.L., Moore, A.W.: Reinforcement learning: a survey. J. Artifi. Intell. Res. (1996)

18. Karakostas, G.: A better approximation ratio for the vertex cover problem. In: ICALP (2005)
19. Karalias, N., Loukas, A.: Erdos goes neural: an unsupervised learning framework for combinatorial optimization on graphs. In: NeurIPS (2020)
20. Kim, H., Cosa-Linan, A., Santhanam, N., Jannesari, M., Maros, M.E., Ganslandt, T.: Transfer learning for medical image classification: a literature review. BMC Med. Imaging (2022)
21. Kolouri, S., Nadjahi, K., Simsekli, U., Badeau, R., Rohde, G.: Generalized sliced wasserstein distances. In: NeurIPS (2019)
22. Kornblith, S., Norouzi, M., Lee, H., Hinton, G.: Similarity of neural network representations revisited. In: ICML (2019)
23. Leskovec, J., Mcauley, J.: Learning to discover social circles in ego networks. In: NeurIPS (2012)
24. Nazari, M., Oroojlooy, A., Snyder, L., Takác, M.: Reinforcement learning for solving the vehicle routing problem. In: NeurIPS (2018)
25. Rozemberczki, B., Allen, C., Sarkar, R.: Multi-scale attributed node embedding. J. Complex Netw. (2021)
26. Rozemberczki, B., Sarkar, R.: Characteristic functions on graphs: Birds of a feather, from statistical descriptors to parametric models. In: CIKM (2020)
27. Schuetz, M.J., Brubaker, J.K., Katzgraber, H.: Combinatorial optimization with physics-inspired graph neural networks. Nat. Mach. Intell. (2022)
28. Selsam, D., Lamm, M., Bünz, B., Liang, P., de Moura, L., Dill, D.L.: Learning a sat solver from single-bit supervision. arXiv preprint arXiv:1802.03685 (2018)
29. Shen, Z., Liu, Z., Qin, J., Savvides, M., Cheng, K.T.: Partial is better than all: revisiting fine-tuning strategy for few-shot learning. In: AAAI (2021)
30. Stastny, J., Skorpil, V., Balogh, Z., Klein, R.: Job shop scheduling problem optimization by means of graph-based algorithm. Appli. Sci. (2021)
31. Veličković, P., Cucurull, G., Casanova, A., Romero, A., Lio, P., Bengio, Y.: Graph attention networks. In: ICLR (2018)
32. Wang, H.P., Wu, N., Yang, H., Hao, C., Li, P.: Unsupervised learning for combinatorial optimization with principled objective relaxation. In: NeurIPS (2022)
33. Wang, H., Yue, T., Ye, X., He, Z., Li, B., Li, Y.: Revisit finetuning strategy for few-shot learning to transfer the emddings. In: ICLR (2023)
34. Welling, M., Kipf, T.N.: Semi-supervised classification with graph convolutional networks. In: ICLR (2017)
35. Yang, C.H., Shen, C.Y.: Enhancing machine learning approaches for graph optimization problems with diversifying graph augmentation. In: SIGKDD (2022)
36. Zhang, W., Dietterich, T.G.: Solving combinatorial optimization tasks by reinforcement learning: a general methodology applied to resource-constrained scheduling. J. Artifi. Intell. Res. (2000)
37. Zhu, Z., Lin, K., Jain, A.K., Zhou, J.: Transfer learning in deep reinforcement learning: A survey. IEEE Trans. Pattern Anal. Mach. Intell. (2023)

SD-Attack: Targeted Spectral Attacks on Graphs

Xianren Zhang[1], Jing Ma[2], Yushun Dong[3], Chen Chen[3], Min Gao[4],
and Jundong Li[3(✉)]

[1] The Pennsylvania State University, State College, PA, USA
`xzz5508@psu.edu`
[2] Case Western Reserve University, Cleveland, OH, USA
`jing.ma5@case.edu`
[3] University of Virginia, Charlottesville, VA, USA
`{yd6eb,zrh6du,jundong}@virginia.edu`
[4] Chongqing University, Chongqing, China
`gaomin@cqu.edu.cn`

Abstract. Graph learning (GL) models have been applied in various predictive tasks on graph data. But, similarly to other machine learning models, GL models are also vulnerable to adversarial attacks. As a powerful attack method on graphs, spectral attack jeopardizes the eigenvalues or eigenvectors of the graph topology-related matrices (e.g., graph adjacency matrix and graph Laplacian matrix) due to their inherent connections to certain structural properties of the underlying graph. However, most existing spectral attack methods focus on damaging the global graph structural properties and can hardly perform effective attacks on a target node. In this paper, we propose a novel targeted spectral attack method that can perform model-agnostic attacks effectively on the local structural properties of a target node. First, we define a novel node-specific metric—spectral density distance, which measures the difference of the local structural properties for the same target node between two different graph topologies. Then, we conduct attacks by maximizing the spectral density distance between the graphs before and after perturbation. Additionally, we also develop an effective strategy to improve attack efficiency by using the eigenvalue perturbation theory. Experimental results on three widely used datasets demonstrate the effectiveness of our proposed approach.

Keywords: Graph learning · Adversarial attack · Target attack

1 Introduction

Graphs have been widely used in many high-impact fields such as social media and infrastructure systems [9]. Graphs contain rich structural information and many powerful graph learning (GL) models [6,17] have been designed to exploit the graph structure for various predictive tasks (e.g., node classification and link prediction). However, recent works [18] have shown that GL models are

© The Author(s), under exclusive license to Springer Nature Singapore Pte Ltd. 2024
D.-N. Yang et al. (Eds.): PAKDD 2024, LNAI 14646, pp. 352–363, 2024.
https://doi.org/10.1007/978-981-97-2253-2_28

vulnerable to adversarial attacks. Specifically, malicious attackers can mislead GL models to make incorrect predictions by making carefully crafted perturbations on the input graph data. Among the existing attack methods on graph data, spectral attack [1, 2, 7] has recently attracted much attention because they are effective and need less information. Compared with conventional attacks, the spectral attack focuses on altering the spectral properties of the graph and does not need any information from victim models. It is usually implemented by changing the spectrum (eigenvalues) of graph topology-related matrices (e.g., Laplacian matrix) [12]. The graph spectrum implies a plethora of important graph structural properties (e.g., the number of connected components [12], algebraic connectivity [4], and degree distribution). Spectral attacks can effectively damage these structural properties by altering the spectrum of a graph. Moreover, the spectral attack is especially powerful in attacking GL models with spectral filters [11], such as Graph Convolution Networks (GCNs) [6].

Despite the effectiveness of spectral attacks, most studies [1, 2, 7] focus on jeopardizing the global structural properties. They design an attack loss function based on the eigenvalues to reflect the changes in structural properties and then flip edges to maximize the loss. However, the attack loss is designed for the whole graph and the attacker can only select from the same set of edges to flip even for different target nodes instead of making adjustments for different nodes. As a consequence, they lack explorations on attacking the local structural properties of a specific node, which limits their effectiveness. On the other hand, many studies have pointed out that eigenvectors can provide information on the local structural properties of nodes, e.g., node membership in clustering [4] and return probabilities of random walks starting from a specific node [3]. For example, the eigenvector of the second smallest eigenvalue, called the Fiedler vector [4], can be used to divide the nodes into two groups. Nodes with the same sign in the Fiedler vector are similar and can be split into the same cluster. Recent studies [3] have shown that the eigenvalues and eigenvectors of the normalized adjacency matrix can be used to calculate the return probabilities of random walks with different steps starting from a node. Various steps of return probabilities reflect interactions between a node and its neighbors with different resolutions. These pieces of evidence demonstrate that eigenvalues indicate the global structural properties of a graph, and eigenvectors can supplement by providing important local (node-level) information. Thus, spectral attacks can be further improved to disrupt the local structural properties of a specific node by perturbing both eigenvalues and eigenvectors.

In this paper, we study the novel problem of targeted spectral attacks on GL models. However, there remain two challenges: 1) The first challenge is to design a proper node-specific attack loss function to measure the attack damage w.r.t. a target node after graph perturbation. This loss function can provide guidance for altering the graph structure (e.g., selecting and flipping edges) to achieve targeted attacks. Nevertheless, it is difficult to bridge the gap between the eigenvalues and eigenvectors of the graph topology-related matrices (e.g., adjacency matrix and Laplacian matrix) and local structural properties (to attack).

2) The second challenge is to achieve such spectral attacks with high efficiency. With a node-specific attack loss function, a straightforward attack strategy is to calculate the attack loss after flipping each edge and then select the edges that lead to high losses. In this process, it is necessary to perform eigendecomposition for each edge flip, which is very time-consuming, as the time complexity is $O(n^3)$ in the worst case, where n is the number of nodes.

To address these challenges, we propose a novel targeted spectral attack method called Spectral Density Attack (**SD-Attack**). To design a node-specific attack loss function (the first challenge), we take advantage of a node-specific metric named Point-wise Density of States (PDOS) [3]. Specifically, the rationale of PDOS is to measure the local structural properties of a target node. For example, the return probabilities of random walks with any steps starting from the target node can be measured by PDOS [3]. We propose a novel node-specific metric named spectral density distance, which measures the Wasserstein distance of a target node's PDOS between the original graph and the perturbed one. We can achieve model-agnostic targeted attacks by flipping edges to maximize the spectral density distance. To avoid frequent computations of eigenvalues and eigenvectors (the second challenge), we leverage eigenvalue perturbation theory [15] to approximate the change of eigenvalues and eigenvectors. With such an approximation strategy, we only need to perform eigendecomposition *once* for the original graph, and the eigenvalues and eigenvectors of the perturbed graph can then be approximated for each edge flip. In experiments, our proposed method is evaluated on widely-used benchmarks (Cora [8], Citeseer [5], and Pubmed [14]) and also tested on two GL models (GCN [6] and GAT [17]). Empirical results show that our method can achieve effective and efficient attacks on target nodes. The main contributions of our work are summarized as follows.

- We study a novel problem: spectral attacks on the local structural properties of a target node. To the best of our knowledge, this is the first study on this important problem.
- We propose a novel node-specific metric spectral density distance to measure the attack effectiveness for any target node. Based on such a metric, we propose a SD-Attack to address the problem of targeted spectral attack.
- We evaluate the proposed method on three well-known datasets with extensive experiments. Experimental results indicate that SD-Attack can generate more effective attacks on a target node than state-of-the-art baselines.

2   Preliminaries

Let $G(V, E)$ represent the structure of an undirected graph, where $V = \{v_1, ..., v_n\}$ is the set of nodes. Let $A \in \{0, 1\}^{n \times n}$ be the adjacency matrix where $A_{ij} = A_{ji} = 1$ if there is an edge between node i and j. The normalized adjacency matrix is $\hat{A} = D^{-\frac{1}{2}} A D^{-\frac{1}{2}}$, where D is a diagonal matrix with $D_{ii} = d_i = \sum_{j=1}^{n} A_{ij}$.

2.1 Adversarial Attack on Graphs

Given a finite budget M, which is the maximum number of edges the attacker can flip, the targeted attack on node k can be formulated as the following problem:

$$\arg\max_{A'} \mathcal{D}_k(A, A'), \qquad (1)$$

$$\text{subject to } |\Delta A| \leq 2M.$$

Here $\mathcal{D}_k$ is defined as the attack loss measuring the attack damage on the node k. $\Delta A = A' - A$ contains the total flipped edges, where A and A' are the original and perturbed adjacency matrices, respectively. ΔA_{ij} is 1 if the attacker adds an edge between node i and node j, and -1 if the edge is deleted. The goal of the attack is to find the best-attacked graph A', which disrupts the local structural properties of node k most.

2.2 Density of States

A real symmetric matrix H can be diagonalized as $H = U \Lambda U^T$, where $\Lambda = diag(\lambda_1, ..., \lambda_n)$ is a diagonal matrix whose diagonal elements are eigenvalues, and $U = [q_1, ..., q_n]$ is a square matrix whose i-th column is the eigenvector q_i. The *density of states* (DOS) [3] induced by H is defined as a generalized function:

$$\mu(\lambda) = \frac{1}{n} \sum_{i=1}^{n} \delta(\lambda - \lambda_i), \qquad (2)$$

where δ is the Dirac delta function. $\mu(\lambda)$ is also called the spectral density, because it describes the distribution of eigenvalues that relate to the global structural properties of a graph. For a specific node k, the Point-wise Density of State (PDOS) is a weighted version of DOS:

$$\mu_k(\lambda) = \sum_{i=1}^{n} |e_k^T q_i|^2 \delta(\lambda - \lambda_i), \qquad (3)$$

where e_k is the k-th standard basis vector with the k-th entry equal to 1 and 0 for other entries. q_i is the i-th eigenvector. $\mu_k(\lambda)$ describes the local information of the node k.

The PDOS is related to the local structural properties of a specific node. For example, the PDOS encodes return probabilities of any steps starting from the target node. We propose to attack the target node by perturbing its PDOS in our proposed framework.

3 The Proposed Method SD-Attack

In this section, we propose a novel spectral attack strategy to disrupt the local structural properties of a target node. We leverage PDOS to establish a connection between the local structural properties and the eigenvalues and eigenvectors of the normalized adjacency matrix. Then, we design a node-specific metric

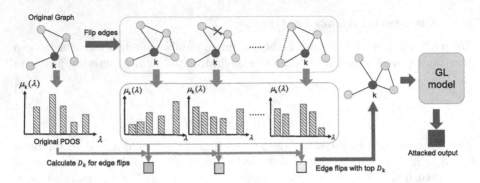

Fig. 1. The framework of Spectral Density Attack. SD-Attack aims to attack the local structural properties of the target node k by flipping edges to maximize the spectral density distance. The GL block can be any general GL models.

called spectral density distance to measure the damage of an attack on the target node. We flip edges that maximize the spectral density distance between the original graph and the perturbed version. To improve the efficiency, we employ eigenvalue perturbation theory [15] to efficiently compute the spectral density distance. The overall framework of our proposed method is shown in Fig. 1.

3.1   Spectral Density Distance

We first propose a novel metric—spectral density distance, to measure the distance between the original graph and the perturbed one w.r.t. their PDOS.

PDOS is a distribution of eigenvalues. One choice to evaluate the difference between two distributions is the Wasserstein distance (a.k.a. the Earth-mover distance). We use this metric to measure the distance between the original PDOS and the perturbed one. The spectral density distance between the original graph and the perturbed one on node k is defined as the W_1 distance (1-Wasserstein distance) of their PDOS:

$$\mathcal{D}_k(A, A') = W_1(\mu_k, \mu'_k) = \int_0^1 |F(q) - F'(q)| dq \qquad (4)$$

where μ_k and μ'_k are PDOS of the original graph and perturbed graph, respectively. $F(q)$ and $F'(q)$ are the quantile functions of μ_k and μ'_k (e.g., $F(0.5)$ is the median value of the distribution μ_k).

3.2   Why Wasserstein Distance

We use an example to further explain why we select the Wasserstein distance rather than other popular metrics. We first give an example and then compare how different metrics will behave in that example.

Example. Here, we use two simple PDOS as an example for illustration. We let the original PDOS be $\mu_k(\lambda) = \delta(\lambda - x)$ and the perturbed PDOS be $\mu'_k(\lambda) = \delta(\lambda - y)$. This means $\mu_k(\lambda)$ has a value of 1 when $\lambda = x$ and 0 when λ takes other values. $\mu'_k(\lambda)$ has a value of 1 when $\lambda = y$ and 0 when λ takes other values. Obviously, in this example, $\mu_k(\lambda)$ and $\mu'_k(\lambda)$ are close to each other if x and y are close to each other. Then we demonstrate how different metrics will behave.

Kullback-Leibler divergence (KL-divergence). The KL-divergence $D_{KL}(\mu'_k(\lambda)\|\mu_k(\lambda)) = +\infty$ if $x \neq y$ and $D_{KL}(\mu'_k(\lambda)\|\mu_k(\lambda)) = 0$ if $x = y$.

Jensen-Shannon divergence (JS-divergence). In our example, the JS $D_{JS}(\mu'_k(\lambda), \mu_k(\lambda)) = log2$ if $x \neq y$ and $D_{JS}(\mu'_k(\lambda), \mu_k(\lambda)) = 0$ if $x = y$.

Total Variation distance (VT-distance). In our example, the TV-distance $TV(\mu'_k(\lambda), \mu_k(\lambda)) = 1$ if $x \neq y$ and $TV(\mu'_k(\lambda), \mu_k(\lambda)) = 0$ if $x = y$.

Wasserstein distance. The Wasserstein distance is $W_1(\mu'_k(\lambda), \mu_k(\lambda)) = |x-y|$.

By comparing different distances and divergences in our example, the Wasserstein distance is the best metric to measure the distance between two PDOS. W_1 will become smaller when μ_k and μ'_k are closer and bigger when they are further away. However, other distances or divergences do not follow this trend and result in an infinite value or remain constant.

3.3   SD-Attack

With the spectral density distance defined in Eq. (4), we take advantage of it in our loss function for the attack. A straightforward approach to achieve the attack is to flip the edges to maximize the loss. However, such an approach requires frequent eigendecomposition operations, which can be prohibitively expensive when the graph is large. To efficiently calculate the spectral density distance, we propose to approximate the change in eigenvalues and eigenvectors with the eigenvalue perturbation theory [15].

Eigenvalues and Eigenvectors Approximation. To improve the efficiency, we approximate the perturbed eigenvalues and eigenvectors. For an edge flip ΔA_{ij}, we estimate the change in eigenvalues and eigenvectors.

Theorem 1. *Considering a generalized eigenvalue problem $Au_k = \lambda_k Du_k$ and a flip of a single edge $\Delta A_{ij} = \Delta A_{ji}$, the change in the k-th generalized eigenvalue $\Delta \lambda_k$ can be approximated as follows:*

$$\Delta \lambda_k \approx \Delta A_{ij}(2u_{ki}u_{kj} - \lambda_k(u_{ki}^2 + u_{kj}^2)). \tag{5}$$

We can also approximate the change of eigenvectors. For a target node v_c, the PDOS takes every c-th value from all eigenvectors and the change of it can be calculated as follows:

Theorem 2. *Considering a generalized eigenvalue problem $Au_k = \lambda_k Du_k$ and a flip of a single edge $\Delta A_{ij} = \Delta A_{ji} = 1 - 2A_{ij}$, the change of the c-th value of eigenvector u_k is:*

$$\Delta u_{kc} \approx \Delta A_{ij}((\lambda_k u_{ki} - u_{kj})(A - \lambda_k D)_{ci}^+ + (\lambda_k u_{kj} - u_{ki})(A - \lambda_k D)_{cj}^+), \tag{6}$$

where $(.)^+$ is the pseudo-inverse.

The Eq. (5) and Eq. (6) are proposed by [15]. The pseudo-inverse term in Eq. (6) is independent of any edge flip and can be precomputed. In conventional attack strategies that do eigendecomposition after each edge flip, the time complexity is $O(n^3)$ in the worst case. With our approximation method, the time complexity of calculating the spectral density distance after each edge flip can be reduced to $O(n)$.

The Overall Attack Algorithm. Given a target node k, candidate set C, and budget M, the problem is to flip M edges from the candidate set C so that the spectral density distance $D_k(A, A')$ in Eq. (4) is maximized. There are $\binom{|C|}{M}$ different numbers of different combinations of edges. Even though we can efficiently calculate the spectral density distance, it is still time-consuming to try every possible combination of edges and calculate the spectral density distance for each one. As a result, instead of exhausting every possible combination of edges, we calculate the spectral density distance for each single edge flip with Eq. (5) and Eq. (6), and then we can select the final perturbations based on the calculated spectral density distances. A straightforward way to conduct attacks is to use a greedy strategy to select these candidate edges. The greedy strategy in our setting is to calculate the spectral density distance for each edge flip, select the edge with the highest spectral density distance, and repeat this process until the budget is reached. However, the greedy strategy still leads to a high time complexity because we have to repeat the calculation of the spectral density distance for the remaining edges. As a result, we choose to directly select edges with the top-M spectral density distance on the original graph as the final perturbations. The rationale for such an approximation is that the perturbation budget M is often very small, therefore choosing the candidate edges on the original graph all at once and choosing the candidate edges greedily does not yield much difference.

Time Complexity. The overall time complexity of the algorithm is $O(n^3)$ in the worst case, where n is the number of nodes in the whole graph. The time complexity of the overall algorithm comes mainly from the eigendecomposition of the original normalized adjacency matrix. The time complexity for a single node attack is $O(n * |C|)$, where $|C|$ is the number of edges in the candidate set.

4   Experiments

In this section, we evaluate our proposed attack method SD-Attack with extensive experiments on three widely-used graph datasets [5,8,14]. Specifically, to validate the effectiveness of SD-Attack, we aim to answer the following two research questions:

– **RQ1:** How well can SD-Attack generate effective attacks for a target node compared to baselines?
– **RQ2:** How well does the spectral density distance help SD-Attack to attack a target node?

4.1   Setup

Datasets. We evaluate our proposed method on three well-known datasets: Cora [8], Citeseer [5], and Pubmed [14]. As it is time-consuming to perform targeted attacks on every unlabeled node, we randomly select 20% nodes as test nodes in Cora and Citeseer and 1.5% nodes as test nodes in Pubmed. We follow the data preprocessing setting in [2], where nodes' features are all normalized, and only the largest connected component is considered.

Baselines. Across all three datasets, we evaluate the proposed attack model SD-Attack against four baselines:

- **Degree** [16]: This method flips edges (inserting or removing) based on degree centrality (the sum of degrees at two ends of the edge). Edges with higher degree centralities are considered to be more important and flipped.
- **A_{class}** [1]: A_{class} focuses on attacking network representation learning models. A_{class} formulates the representation learning process as a matrix factorization process and flips edges to damage that process.
- **GF-attack** [2]: GF-Attack (Graph Filter Attack) attacks the graph filters. They formulate GNN models as an approximation process. They then attack filters by maximizing the loss function for the approximation.
- **SPAC** [7]: SPAC focuses on attacking the eigenvalues of the normalized graph Laplacian. It perturbs the filters of the GCN models by maximizing the spectral distance which is defined as the $L2$ distance of the eigenvalues.

Variants of SD-Attack. We conduct ablation study with following variants:

- **SD-DOS:** To demonstrate the importance of local information of the target node, SD-DOS replaces the PDOS (Eq. 3) with DOS (Eq. 2) which indicates the global structural properties of a graph [3].
- **SD-Eigenval:** SD-Eigenval only considers the change of eigenvalues.
- **SD-Eigenvec:** SD-Eigenvec only considers the change of eigenvectors.
- **SD-KL:** SD-KL replaces the Wasserstein distance with KL divergence.

Configurations. We use two widely-used graph learning models as victim models: GCN [6] and GAT [17]. In our experiments, we conduct experiments on different layers of GCN models (2,3, and 4 layers), and the output dimension of each layer is 32 (except the output layer). GAT has two layers, and the output dimension of the first layer is set to 32. In our setting, the attack is an evasion attack which means that attacks occur after the victim model is trained.

For a target node k, the candidate set is defined as: $C = \{(v, u)|v \in N_k \cup \{k\}, u \in V\}$, where N_k is the set of neighbors of node k. For each correctly predicted node in the test set, we take it as the target node and carry out the attack. We calculate the decrease in accuracy as the attack performance.

Attacker Knowledge and Capacity. In many real-world scenarios, it is hard for attackers to obtain all information (e.g., victim models and training data labels), and we assume that such information is not accessible. Attackers are only allowed to access the graph topology. For capacity, attackers can add or remove edges within a limited budget. Similarly to [18], since high-degree nodes are more difficult to attack, the budget is set as the degree of the target node.

Table 1. The change in classification accuracy after different attack methods (lower is better). Results are averaged for ten trials.

Model	Attack	Cora	Citeseer	Pubmed
GCN	Degree	-1.65 ± 0.12	-1.93 ± 0.29	-1.05 ± 0.03
	A_{class}	-4.59 ± 0.18	-4.95 ± 0.34	-9.12 ± 0.21
	GF-Attack	-4.63 ± 0.29	-3.42 ± 0.31	-5.57 ± 0.14
	SPAC	-5.13 ± 0.27	-4.62 ± 0.34	-4.73 ± 0.12
	SD-Attack	$\mathbf{-10.56 \pm 0.33}$	$\mathbf{-7.76 \pm 0.38}$	$\mathbf{-10.34 \pm 0.14}$
GAT	Degree	-5.73 ± 0.83	-3.44 ± 0.31	-2.09 ± 0.19
	A_{class}	-5.29 ± 0.19	-4.55 ± 0.24	-11.32 ± 0.42
	GF-Attack	-6.84 ± 0.76	-4.67 ± 0.57	-4.76 ± 0.23
	SPAC	-5.19 ± 1.02	-3.11 ± 0.40	-3.07 ± 0.21
	SD-Attack	$\mathbf{-9.24 \pm 0.71}$	$\mathbf{-6.20 \pm 0.22}$	$\mathbf{-13.72 \pm 0.41}$

4.2   Attack Performance

To answer **RQ1**, we evaluate our proposed attack strategy with other baselines. We first train the GCN [6] and GAT [17] models on the original graph. Then the models are fixed, and the attackers perform attacks for each target node. We repeat experiments for ten trials and get the average results. The averaged changes of accuracy and standard deviation are shown in Table 1.

In general, our attack strategy achieves the best performance on three datasets. Among the baselines, the method based on degree centrality [16] can hardly perform effective attacks. Its performance is worse than the baselines of the spectral attacks [1,2,7]. This indicates that flipping edges with a high degree of centrality would not be enough to mislead GL models. Spectral attacks have better results than the method based on degree centrality, which demonstrates the effectiveness of attacking graph spectra. However, they ignore the local structural properties of a target node, which limits the attack performance when the objective is to attack a target node. SD-Attack damages the local structural properties of the target node by maximizing the spectral density distance. Compared with other baselines, SD-Attack can generate more effective attacks on three well-known datasets.

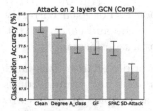 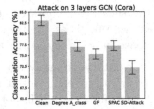 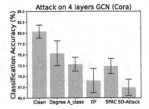

Fig. 2. Comparison of different attacks on different layers of GCN on Cora dataset.

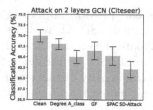 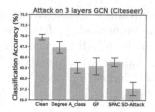 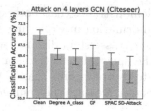

Fig. 3. Comparison of different attack strategies on different layers of GCN on Citeseer dataset.

To further evaluate the effectiveness of our approach, we compare its attack performance with baselines in victim GCN models with different numbers of layers. The experiment is repeated ten times and the averaged results with standard deviations are shown in Fig. 2 and Fig. 3. Compared to other baselines, the degree method [16] has little influence on the accuracy of the victim model. Other baselines [1, 2, 7] perform better and are more effective in attacking the victim model. In different cases, our method always outperforms other baselines, further indicating the effectiveness of SD-Attack.

Table 2. The change in classification accuracy under different variants (lower is better). Results are averaged for ten trials.

Model	Attack	Cora	Citeseer	Pubmed
GCN	SD-DOS	−6.52 ±0.36	−4.03 ±0.34	−8.14 ±0.17
	SD-Eigenval	−4.95 ±0.23	−4.55 ±0.21	−7.43 ±0.17
	SD-Eigenvec	−9.05 ±0.30	−7.43 ±0.34	−6.86 ±0.10
	SD-KL	−8.59 ±0.41	−5.90 ±0.65	−9.86 ±0.08
	SD-Attack	**−10.56** ±0.33	**−7.76** ±0.38	**−10.34** ±0.14
GAT	SD-DOS	−5.61 ±0.50	−3.21 ±0.11	−8.55 ±0.20
	SD-Eigenval	−5.92 ±0.43	−4.43 ±0.27	−7.97 ±0.25
	SD-Eigenvec	−8.29 ±0.61	−5.92 ±0.27	−6.08 ±0.35
	SD-KL	**−9.50** ±0.52	−6.14 ±0.28	**−15.24** ±0.50
	SD-Attack	−9.24 ±0.71	**−6.20** ±0.22	−13.72 ±0.41

4.3 Variants of SD-Attack

To answer **RQ2**, we compare different variants of our method and report the results in Table 2. By comparing SD-Attack with SD-DOS, we observe that the performance of attacks improves. The reason is that PDOS can provide more local and node-specific information about the structural properties of the target node [3]. SD-Attack has better results than SD-Eigenval and SD-Eigenvec, which demonstrates that both eigenvalues and eigenvectors can provide important information in exploiting the vulnerability of the target node. SD-KL replaces the Wasserstein distance with the KL divergence to validate the effectiveness of the spectral density distance in our attack method. Overall, SD-Attack generates better results than SD-KL, indicating that the spectral density distance can better measure the damage caused by attacks on the local structural properties.

5   Related Works

Spectral Adversarial Attack. Recently, several works focus on spectral attacks because most of them only need the network structure without access to the information about the victim models. Network representation learning can be viewed as solving a matrix factorization problem, and the learning process can be viewed as performing singular value decomposition (SVD) on certain graph topology-related matrices [13]. The loss function can be viewed as the square of the sum of small singular values. To damage the quality of representation, A_{class} [1] flips the edges and maximizes the loss function of the victim models. Inspired by the observation that output embeddings of GNN models have low-rank properties [10], GF-Attack [2] formulates GNN models as an approximation problem. GF-Attack then flips edges to maximize the loss function for the approximation problem. SPAC [7] focuses on attacking the graph filter of GCN models. SPAC proposes to maximize the changes on the eigenvalues of the graph Laplacian, so that the spectral filters and graph convolution can be influenced. SPAC defines the spectral distance as the $L2$ distance of eigenvalues of the Laplacian matrix and then flips the edges to maximize it. These methods ignore the local structural properties of a specific node. In this paper, we introduce a novel spectral attack strategy that attacks the local structural properties of a target node.

6   Conclusion

In this paper, we propose a novel method SD-Attack to attack the local structural properties of a target node. More specifically, we first leverage the Pointwise Density of State (PDOS) for the target node to connect the local structural properties of that node with the eigenvalues and eigenvectors of the normalized adjacency matrix. Based on PDOS, we define a node-specific metric called spectral density distance to measure the difference between two graphs on the target node. We then attack the local structural properties of the target node by flipping edges to maximize the spectral density distance between the original graph and the graph after perturbation. Furthermore, we improve the attack efficiency by approximating the perturbed eigenvalues and eigenvectors with the eigenvalue perturbation theory. We also conduct experiments with baselines and variants on widely used datasets. By comparing the results of our model and the baselines, we validate the effectiveness of our method in targeted attacks on different victim models.

References

1. Bojchevski, A., Günnemann, S.: Adversarial attacks on node embeddings via graph poisoning. In: International Conference on Machine Learning, pp. 695–704. PMLR (2019)
2. Chang, H., et al.: A restricted black-box adversarial framework towards attacking graph embedding models. In: Proceedings of the AAAI Conference on Artificial Intelligence, vol. 34, pp. 3389–3396 (2020)
3. Dong, K., Benson, A.R., Bindel, D.: Network density of states. In: Proceedings of the 25th ACM SIGKDD International Conference on Knowledge Discovery & Data Mining, pp. 1152–1161 (2019)
4. Fiedler, M.: Algebraic connectivity of graphs. Czechoslov. Math. J. **23**(2), 298–305 (1973)
5. Giles, C.L., Bollacker, K.D., Lawrence, S.: Citeseer: an automatic citation indexing system. In: Proceedings of the third ACM Conference on Digital Libraries, pp. 89–98 (1998)
6. Kipf, T.N., Welling, M.: Semi-supervised classification with graph convolutional networks. arXiv preprint arXiv:1609.02907 (2016)
7. Lin, L., Blaser, E., Wang, H.: Graph structural attack by perturbing spectral distance. In: Proceedings of the 28th ACM SIGKDD Conference on Knowledge Discovery and Data Mining, pp. 989–998 (2022)
8. McCallum, A.K., Nigam, K., Rennie, J., Seymore, K.: Automating the construction of internet portals with machine learning. Inf. Retrieval **3**(2), 127–163 (2000)
9. Milanović, J.V., Zhu, W.: Modeling of interconnected critical infrastructure systems using complex network theory. IEEE Trans. Smart Grid **9**(5), 4637–4648 (2017)
10. Nar, K., Ocal, O., Sastry, S.S., Ramchandran, K.: Cross-entropy loss and low-rank features have responsibility for adversarial examples. arXiv preprint arXiv:1901.08360 (2019)
11. Nt, H., Maehara, T.: Revisiting graph neural networks: all we have is low-pass filters. arXiv preprint arXiv:1905.09550 (2019)
12. Oellermann, O.R., Schwenk, A.J.: The Laplacian spectrum of graphs. Graph Theory, c, Appl. **2**, 871–898 (1991)
13. Qiu, J., Dong, Y., Ma, H., Li, J., Wang, K., Tang, J.: Network embedding as matrix factorization: unifying DeepWalk, LINE, PTE, and node2vec. In: Proceedings of the Eleventh ACM International Conference on Web Search and Data Mining, pp. 459–467 (2018)
14. Sen, P., Namata, G., Bilgic, M., Getoor, L., Galligher, B., Eliassi-Rad, T.: Collective classification in network data. AI Mag. **29**(3), 93–93 (2008)
15. Stewart, G., Sun, J.: Matrix Perturbation Theory. Elsevier Science, Computer Science and Scientific Computing (1990)
16. Tong, H.E.A.: Gelling, and melting, large graphs by edge manipulation. In: Proceedings of 21st ACM International Conference on Information and Knowledge Management, pp. 245–254 (2012)
17. Veličković, P., Cucurull, G., Casanova, A., Romero, A., Lio, P., Bengio, Y.: Graph attention networks. arXiv preprint arXiv:1710.10903 (2017)
18. Zügner, D., Akbarnejad, A., Günnemann, S.: Adversarial attacks on neural networks for graph data. In: Proceedings of the 24th ACM SIGKDD International Conference on Knowledge Discovery & Data Mining, pp. 2847–2856 (2018)

Improving Structural and Semantic Global Knowledge in Graph Contrastive Learning with Distillation

Mi Wen[1], Hongwei Wang[1(✉)], Yunsheng Xue[1], Yi Wu[2], and Hong Wen[3]

[1] Shanghai University of Electirc Power, Shanghai 201306, China
wanghongwei@mail.shiep.edu.cn
[2] State Grid Shanghai Electric Power Co., Ltd., Shanghai 200122, China
[3] University of Electronic Science and Technology of China, Chengdu 611731, China

Abstract. Graph contrastive learning has emerged as a pivotal task in the realm of graph representation learning, with the primary objective of maximizing mutual information between graph-augmented pairs exhibiting similar semantics. However, existing unsupervised graph contrastive learning approaches face a notable limitation in capturing both structural and semantic global information. This issues poses a substantial challenge, as nodes in close geographical proximity do not consistently possess similar features. To tackle this issue, this study introduces a simple framework for Distillation Node and Prototype Graph Contrastive Learning (DNPGCL). The framework enables contrastive learning by harnessing similar knowledge distillation to obtain more valuable structural and semantic global indications. Experimental results demonstrate that DNGCL outperforms existing unsupervised learning methods across a range of diverse graph datasets.

Keywords: Contrastive learning · Self-supervised learning · Graph representation learning · Graph neural network

1 Introduction

Graph contrastive learning (GCL) [17] stands as a potent pretext task within the realm of self-supervised learning(SSL) [8], designed to enhance representations by maximizing the agreement between two augmented views of a given graph through contrastive loss in latent space. Despite the current success of unsupervised graph contrastive learning, many existing GCL methods in this domain rely on partial domain information, largely due to the limited depth of traditional Graph Neural Network (GNN) [13] layers. However, in practical scenarios, as depicted in Fig. 1, it is evident that "nodes with similar characteristics are not always geographically close", emphasizing the need for algorithms to possess a global perspective. Achieving this goal poses a formidable challenge for existing GCL methods based on shallow GNNs. These models face inherent

© The Author(s), under exclusive license to Springer Nature Singapore Pte Ltd. 2024
D.-N. Yang et al. (Eds.): PAKDD 2024, LNAI 14646, pp. 364–375, 2024.
https://doi.org/10.1007/978-981-97-2253-2_29

limitations in capturing both structural global knowledge and semantic global knowledge. Specifically, at the structural level, due to the complexity of the graph data, capturing structural global knowledge requires long interactions between nodes. While deepening GNN layers may seem like a straightforward solution, it introduces a potential bottleneck in information propagation [18], potentially hindering the model's ability to effectively capture the global graph structure. At the seminal level, existing approaches focus on modelling instance-level structural similarities, but fail to discover the underlying global structure of the entire data distribution.

Fig. 1. Nodes with similar characteristics are not always geographically close: in a social network with similar interests, node A and node B may be from different cities or even different countries, but they still share common interests and communicate with each other.

In this paper, we propose a simple and effective graph contrastive learning method for Distillation Node and Prototype Graph Contrastive Learning (DNPGCL) to address the above limitations. The framework proposes two new contrastive learning algorithms. In order to obtain interactions between distant nodes, distillation node contrastive learning, which can be interpreted as a kind of unlabelled knowledge distillation. It passes input graphs with two different data augments to the student network and the teacher network, and improves the differentiation of nodes by directly predicting the output of the teacher network using the standard cross-entropy loss function. Meanwhile, the distillation prototype contrastive learning algorithm further enhances intra-cluster compactness and inter-cluster divisibility to model the underlying semantic structure of graph data by clustering semantically similar graphs into the same group for better consolidation of the semantic structure from a global perspective. By jointly optimizing the two contrasting learning losses, the pre-trained encoder network can learn representations suitable for a variety of downstream tasks without using any manually labeled labels. We summarize our contributions as follows:

- We develop a novel GCL framework, we call DNPGCL, which learns node representations by employing a simple, parameter-free encoder network in an unsupervised manner.

- We propose distillation node contrastive learning algorithms and distillation prototype contrastive learning algorithms that significantly acquire global structural and semantic information of the input graph.
- We conduct extensive experiments to demonstrate that our approach outperforms other state-of-the-art GCL methods on a variety of downstream tasks.

2   Related Work

Graph contrastive learning aims to maximize the Mutual Information (MI) between instances with similar semantic information. InfoGraph [14] emphasizes maximizing mutual information between graph-level representations and patch representations to learn, obtaining good performance graph representations and achieving competitive results on graph classification tasks. GraphCL [17] designs four types of graph data augmentation and proposes a graph contrastive learning framework for pre-training of GNNs, which allows learning of a variety of perturbation-independent graph structural data representation. GCA [19] designs a joint adaptive data augmentation model to provide multiple contexts for nodes at the topology level by removing edge and node attribute perspectives for feature masking at the topology level, respectively. SimGRACE [16] introduces perturbations at the encoder level, compares the semantic similarity between the views obtained after perturbing two encoders, and learns from the comparison using vector representations of semantically similar vectors in the approximate vector space. While these methods have focused on modeling instance-level feature similarity, they often ignores the feature that "nodes with similar features are not always geographically close" despite maintaining local smoothness within individual instances. Therefore, our proposed DNPGCL strives to address this limitation by incorporating feature similarity modeling at both the instance and prototype levels. This comprehensive approach enables our method to uncover the underlying semantic structure of the entire dataset.

3   Methodology

In this section, we elaborate our DNPGCL framework, as depicted in Fig. 2. Our proposed framework consists of three main components: (1) a simple (e.g., 1-layer) encoder network (2) a distillation node comparison learning algorithm (3) a distillation prototype comparison learning algorithm.

3.1   Distillation Node Contrastive Learning

Existing graph contrastive learning methods aim to maximize the MI between different augmented views to achieve node-wise discrimination. While they have achieved some accomplishments, they usually ignore global structural information. In this section, we propose the distillation node contrastive learning algorithm, which shares similarities with knowledge distillation [7]. We approach

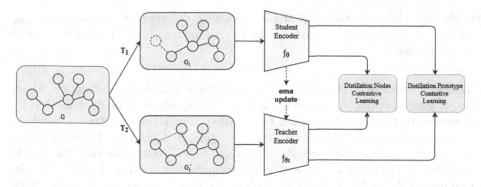

Fig. 2. Overview of DNPGCL. Two graph data augmentations T1 and T2 are applied to the input graph G. Subsequently, two views and are sent to the student-teacher network These two networks have the same architecture but different parameters; the parameters of the teacher network are updated with exponential moving average (ema) as the student network's parameters are updated. This process yields graph representations, z_i and z_i'.

distillation node graph contrastive learning from a different perspective. We illustrate the distillation node contrastive learning method in the left of Fig. 3.

Knowledge distillation is a learning paradigm where we train the student network $f_{\theta s}$ to match the output of the teacher network $f_{\theta t}$, parameterized by θs and θt, respectively. Given an input graph G, after undergoing two different data augmentation approaches, the outputs of both networks are represented by P_s and P_t. The probability P is obtained by normalizing the output of the network f_θ with a softmax function. More precisely,

$$P(x)^{(i)} = \frac{\exp\left(g_{\theta s}(x)^{(i)}/\tau\right)}{\sum_{k=1}^{K} \exp\left(g_{\theta s}(x)^{(k)}/\tau\right)} \tag{1}$$

where $\tau > 0$ is the temperature parameter controling the sharpness of the output distribution, and similar formulas apply to P_t with temperature τ_t. Given a fixed teacher network $f_{\theta t}$, we match these distributions by learning to minimize the cross-entropy loss. The parameters of the student network θ_S:

$$\mathcal{L}_{\text{node}} = \min_{\theta_s} H(P_t(x), P(x)), where \quad H(a,b) = -a\log b \tag{2}$$

In the following, we will provide a detailed explanation of how to adjust the self-supervised learning problem in Eq. 4. We construct different augmented views of the graph using a perturbation-based strategy [17]. This strategy includes node perturbation, edge perturbation, subgraph sampling. More precisely, from a given graph, we generate a set of different views V. This set contains two global views, G_1 and G_1'. We minimize this loss:

$$\min_{\theta_s} \sum_{x \in \{G_1, G_1'\}} H\left(P_t(x), P_s(x)\right) \tag{3}$$

Both networks share the same architecture with different sets of parameters $f_{\theta s}$ and $f_{\theta t}$. We learn the parameters θs by minimizing Eq. (6) using stochastic gradient descent.

Teacher Network. In contrast to traditional knowledge distillation where the teacher network $f_{\theta t}$ is pre-trained, our approach circumvents this step by constructing the teacher network from earlier versions of the student network. Notably, we employ Exponential Moving Average (EMA) on student weights, specifically a momentum encoder [5], which is well-suited for our framework. Originally introduced as an alternative to queues in contrast learning [5], the update rule $\theta_t \leftarrow \lambda\theta_t + (1 - \lambda)\theta_s$, utilizes a cosine schedule for λ ranging from 0.996 to 1 during training [3]. It is crucial to emphasize that, in our framework, the momentum encoder serves a distinct purpose without relying on a queue or a traditional contrast loss such as InfoNCE. Instead, it aligns more closely with the role of the average teacher used in self-supervised training [15]. Remarkably, we observe that this teacher network exhibits a form of model averaging with exponential decay, akin to Polyak-Ruppert averaging. Leveraging Polyak-Ruppert averaging mitigates oscillations during training, enhancing stability and generalization performance. Our findings indicate that this teacher network consistently outperforms the student network throughout training, guiding the training of student network by providing superior-quality target features.

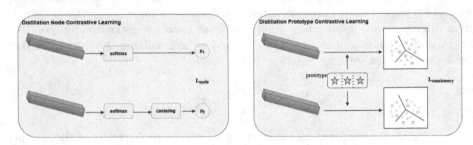

Fig. 3. In the distillation node contrastive learning, the output of the teacher networks is decentralized, then softmax normalized, and subsequently measured for their similarity using cross-entropy loss. In the distillation prototype contrastive learning, the representations are learned by encouraging clustering consistency between relevant views, where prototype vectors are also updated with the encoder parameters via backpropagation.

3.2   Distillation Prototype Contrastive Learning

The distillation node contrastive learning algorithm enhances node-level discrimination by utilizing global structural knowledge acquired from the student-teacher network. However, akin to existing graph contrastive learning methods, it faces a common limitation in explicitly describing the semantic structure of the input graph. To address this shortfall and further capture semantic global knowledge, we propose the distillation prototype contrastive learning algorithm. This novel algorithm aims to cluster graphs with semantic similarity and promote consistency in clustering between different augmented views, fostering cohesion among semantically similar nodes around their prototypes (cluster centroids). As illustrated in the right side of Fig. 3, this approach facilitates the formation of meaningful clusters.

Representing a graph neural network as $z_i = f_\theta(G_i)$, where f_θ maps the graph sample G_i to the representation vectors $z_i \in \mathbb{R}^D$, we aggregate these representations z_i to form K clusters. These clusters are represented by the trainable prototype vector collection in matrix $C \in \mathbb{R}^{K \times D}$, with column vectors $c_1, ..., c_K$ serving as the prototype vectors. Unlike traditional clustering methods such as K-means with fixed mean cluster centers, our approach employs prototype vectors as cluster centers. These vectors are the trainable weight matrices of the feedforward network, initialized with He initialization [6]. Given a graph G_i, we compute the similarity between the graph G_i and the K prototypes using Eq. (4):

$$p(y|z_i) = \text{softmax}(C \cdot f_\theta(G_i)) \tag{4}$$

Similarly, z_i' can be used to represent the prototypical assignment for G_i'. To encourage clustering consistency between different augmented views G_i and G_i' of the same graph, i.e., we would like the clustering assignment of G_i' to result in a z_i rather than the original z_i'. We define the clustering consistency objective by minimizing the average cross-entropy loss:

$$\ell(p_i, q_{i'}) = -\frac{1}{N} \sum_{y=1}^{N} q(y \mid z_i') \log p(y \mid z_i) \tag{5}$$

In Eq. (5), $q(y \mid z_i')$ represents the prototype assignment for view G_i'. The clustering consistency goal acts as a regularizer to encourage the clustering similarity between related views. If we exchange the positions of z_i and z_i' in Eq. (5), we can obtain another similar objective:

$$\mathcal{L}_{\text{consistency}} = \sum_{i=1}^{N} [\ell(p_i, q_{i'}) + \ell(p_{i'}, q_i)] \tag{6}$$

The consistency regularizer can be interpreted as comparing the clustering distributions of multiple views, rather than their specific representations to perform comparisons for inter-view comparison. However, in practice, clustering algorithms such as deep cluster [1] may produce degenerate solutions, where all data is assigned to a single cluster, leading to issues with the distribution q. To avoid this, we add the equipartition constraint to q, aiming to evenly distribute data points among the K clusters:

$$\min_{p,q} \mathcal{L}_{\text{consistency}} \text{s.t.} \forall y: q(y \mid z_i) \in [0,1] \text{and} \Sigma_{i=1}^{N} q(y \mid z_i) = \frac{N}{K} \tag{7}$$

These constraints can be understood as introducing a prior assumption, assuming a uniform label distribution when we do not know the true distribution of labels.

We denote that $P \in \mathbb{R}^{K \times N}$ and $Q \in \mathbb{R}^{K \times N}$, where $\mathrm{P} = \frac{1}{N}p(y \mid z_i)$ and $Q = \frac{1}{N}q(y \mid z_i)$. The multiplication by $1/N$ ensures that P and Q are probability matrices, i.e., $\sum_{p \in P} p = 1$. We define the feasible solution space T for Q, which satisfies the equipartition constraint :

$$T = \{Q \in \mathbb{R}_+^{K \times N} \mid Q \not\Vdash_N = r, Q^{\mathsf{T}} \not\Vdash_K = c\} \tag{8}$$

where $r = \frac{1}{K}\not\Vdash_K$, $c = \frac{1}{N}\not\Vdash_N$. The loss function in Eq. (7) can then be rewritten as:

$$\min_{p,q} \mathcal{L}_{\text{consistency}} = \min_{Q \in \mathbf{T}} \langle Q, -\log P \rangle - \log N \tag{9}$$

where $\langle \cdot \rangle$ denotes the Frobenius dot-product between two matrices, i.e., element-wise multiplication between matrices. By moving log N to the other side, we rephrase Eq. (9) as:

$$\min_{p,q} \mathcal{L}_{\text{consistency}} + \log N = \min_{Q \in \mathbf{T}} \langle Q, -\log P \rangle \tag{10}$$

For $\min_{Q \in \mathbf{T}} \langle Q, -\log P \rangle$, this is a standard optimal transport problem, where -log P serves as the cost matrix in the optimal transport problem, and during the optimization process is fixed. To make the problem easier to solve, we add the regular term $\mathrm{KL}(Q \parallel rc^{\mathsf{T}})$, which transforms the Wassertein distance into the Sinkhorn distance, which we solve iteratively by the Sinkhorn-Knopp algorithm [2]:

$$\min_{Q \in \mathbf{T}} \langle Q, -\log P \rangle + \frac{1}{\gamma}\mathrm{KL}(Q \parallel rc^{\mathsf{T}}) \tag{11}$$

where KL represents the Kullback-Leibler divergence, and rc^{T} can be viewed as a K $\times$ N probability matrix. For Eq. (11), the Sinkhorn-Knopp algorithm shows that its optimal solution satisfies the following form:

$$Q = \mathrm{Diag}(\alpha) P^{\eta} \mathrm{Diag}(\beta) \tag{12}$$

where α and β are two vectors of selected scaling coefficients. Exponentiation is meant element-wise vectors, α and β can be obtained by simple matrix scaling iterations as follows:

$$\forall y \colon \alpha_y \leftarrow [P^{\lambda}\beta]_y^{-1} \forall i \colon \beta_i \leftarrow [\alpha^{\mathsf{T}} P^{\lambda}]_i^{-1} \tag{13}$$

3.3   Model Learning

Overall Loss. In order to train our model and learn the encoder $f_\theta(\cdot)$ in an end-to-end manner, we jointly optimize the distillation node and distillation prototype contrastive learning loss. The overall objective function is defined as:

$$\mathcal{L} = \gamma \mathcal{L}_{kno} + (1 - \gamma)\mathcal{L}_{pro} \tag{14}$$

Here, our objective is to minimize $\mathcal{L}$ during training, where γ is a balancing parameter to control the weight of each contrastive learning loss. In our work, γ is a learnable parameter, which enables our model to progressively optimize its performance during training to approximate the optimal solution. Compared to a fixed parameter γ, our model exhibits strong adaptability and expressive power.

4   Experiments

This section is dedicated to demonstrating the effectiveness of the DNPGCL method. In our experiments, we compare with state-of-the-art competitors in the unsupervised graph classification task [14,17]. The results demonstrate our method outperform the baseline model in several different datasets to achieve optimal results.

4.1   Experimental Setup

Evalution Datasets and Compared Methods. We evaluate DNPGCL on eight publicly available benchmark datasets, including four bioinformatics datasets (MUTAG, PTC, PROTEINS, NCI1, NCI109) and one social network dataset (COLLAB). We compare with six unsupervised methods including Graph2Vec [11], InfoGraph [14], MVGRL [4], GCC [12], GraphCL [17] and PGCL [9].

4.2   Experimental Results

Baseline Comparisons. For unsupervised graph classification, we adopt the methodology outlined in [14,17] to access the performance of DNPGCL. The experimental results are presented in Table 1, demostrating the efficacy of our approach in graph-level representation for downstream graph classification tasks. Overall, from the table, it is evident that our method establishes state-of-the-art performance across all eight datasets when compared to other unsupervised models. DNPGCL consistently outperforms the unsupervised baseline, showcasing its superiority. Notably, on the MUTAG dataset, our method attains an accuracy of 92.8%, marking an absolute improvement of 1.7% over the previous state-of-the-art method (PGCL [9]). These results affirm the robustness and effectiveness of DNPGCL in unsupervised graph classification tasks.

Table 1. Graph classification accuracy (%) for supervised, kernel, and unsupervised methods. We report the average 10-fold cross-validation accuracy over 5 runs.

Datasets	MUTAG	PTC	PROTEINS	NCI1	NCI109	COLLAB
# graphs	188	344	1113	4110	4127	5000
Avg # nodes	17.9	14.3	39.1	29.9	29.7	74.5
Unsupervised						
GRAPH2VEC [11]	83.2 ± 9.3	60.2 ± 6.9	73.3 ± 2.1	73.2 ± 1.8	72.1 ± 0.4	–
INFOGRAPH [14]	89.0 ± 1.1	61.7 ± 1.7	74.4 ± 0.3	73.8 ± 0.7	71.3 ± 0.9	67.6 ± 1.2
MVGRL [4]	89.7 ± 1.1	62.5 ± 1.7	-	75.0 ± 0.7	74.5 ± 1.4	68.9 ± 1.9
GCC [12]	86.4 ± 0.5	58.4 ± 1.2	72.9 ± 0.5	66.9 ± 0.2	67.5 ± 0.3	75.2 ± 0.3
GRAPHCL [17]	86.8 ± 1.3	58.4 ± 1.7	74.4 ± 0.5	77.9 ± 0.4	77.1 ± 1.1	71.4 ± 1.2
PGCL [9]	91.1 ± 1.2	63.3 ± 1.3	75.7 ± 0.2	$\mathbf{78.8 \pm 0.8}$	76.4 ± 1.2	76.0 ± 0.3
DNPGCL(ours)	$\mathbf{92.8 \pm 1.3}$	$\mathbf{64.5 \pm 1.5}$	$\mathbf{76.0 \pm 0.6}$	78.2 ± 0.9	$\mathbf{78.4 \pm 1.7}$	$\mathbf{76.5 \pm 0.7}$

Ablation Studies. In order to access the efficacy of distillation node contrastive learning and distillation prototype contrastive learning within DNPGCL, we conducted ablation studies in MUTAG, NCI1 and PROTEINS datasets Specifically, we systematically removed one of the contrastive learning components from either variant of DNPGCL and examined the impact on performance in terms of node classification, as detailed in Table 2. The results reveal a noticeable decline in the performance of DNPGCL when either the distillation node contrast or distillation prototype contrast component is omitted. Notably, the comprehensive DNPGCL, incorporating both components, exhibits superior performance. This finding underscores the synergy between distillation node contrast and distillation prototype contrast, as their combined use proves most effective. Thus, our ablation studies provide compelling evidence of the individual effectiveness and mutual complementarity of each contrastive learning component within DNPGCL.

Table 2. Ablation study on contrastive components

Method	MUTAG	NCI1	PROTEINS
Our DNPGCL($\mathcal{L}_{node}$)	90.5 ± 0.5	78.0 ± 0.9	76.0 ± 0.6
Our DNPGCL($\mathcal{L}_{consistency}$)	91.1 ± 1.2	77.5 ± 0.8	78.3 ± 0.7
Our final DNPGCL($\mathcal{L}_{node} + \mathcal{L}_{consistency}$)	$\mathbf{92.8 \pm 1.3}$	$\mathbf{78.2 \pm 0.9}$	$\mathbf{79.1 \pm 0.5}$

Sensitivity to the Number of GNN Layers. In this section, we discuss the crucial consideration of selecting the number of GNN layers (L).Figure 4(a) presents the performance of DNPGCL across different number of layers ranging from 1 to 20, focusing on MUTAG and PTC datasets. Observations indicate that the optimal performance is achieved when the GNN employs 2 to 3 layers. Notably, while there is a slightly performance decrease with an increase in the number of layers, it avoids severe overfitting - a common concern in traditional GNN models, where an excessive number of layers may lead to a drastic decline in performance. Overall, our DNPGCL demonstrates adaptability and stability to variations in the number of GNN layers.

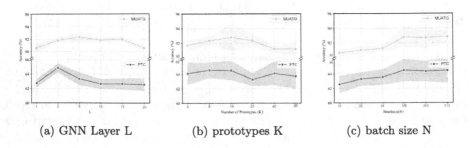

(a) GNN Layer L   (b) prototypes K   (c) batch size N

Fig. 4. Sensitivity analysis

Sensitivity to the Number of Prototypes. In this experiment, we evaluate the impact of prototypes count (K) on model performance. Figure 4(b) illustrates the classification accuracy of our DNPGCL on the MUTAG and PTC datasets across a range of prototype counts, varying from 6 to 80. The line plots depict a discernible trend, indicating that initially increasing the number of prototypes contributes to improve performance. However an excessive number of prototypes leads to a marginal decline in performance. In general, our DNPGCL exhibits robustness in response to variations in the choice of prototype count (K), showcasing its stability and adaptability in handling different prototype configurations.

Sensitivity to Batch Size. Figure 4(c) presents the classification accuracy of our model on the MUTAG and PTC datasets across different batch sizes ranging from 16 to 512. The line plots reveal a notable trend where larger batch sizes lead to better performance. In our experiments, for fairness and comparability with other competitors, we adapt a consistent setup, training the GNN encoder with a batch size of 128 and an epoch number of 20.

Visualization Results. In order to demonstrate the high-quality graph representations learned by DNPGCL, we use T-SNE [10] to visualize graph representations and prototype vectors for the number of clusters K = 10 on the three datasets. The visualization results, depicted in Fig. 5, represent each dot as a graph representation, with point color indicating its true label. model excels in uncovering the underlying linguistic structure of the entire data distribution. The visualization highlights the superior intra-class compactness and inter-class distinctiveness achieved by DNPGCL. These results underscore the model's capacity to generate meaningful representations, substantiating its effectiveness in capturing intricate graph relationships and semantic structures.

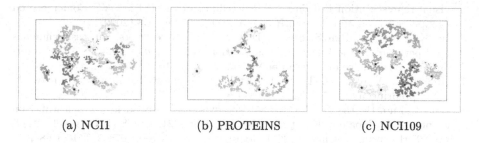

| (a) NCI1 | (b) PROTEINS | (c) NCI109 |

Fig. 5. Visualization of T-SNE for learning representations on three datasets, "⋆" denotes the prototype vector.

5   Conclusion

In this paper, we propose a novel GCL framework called DNPGCL. Our framework effectively captures global knowledge in unsupervised graph learning, addressing both structural and semantic levels. Through the synergistic optimization of distillation nodes and distillation prototypes contrastive learning loss, we effectively train an encoder network employing simple graph neural networks. This enables the learning of excellent representations without the need for manually labeled data. Our comprehensive experiments demonstrate the superiority of the DNPGCL method across multiple publicly available benchmark datasets. Notably, DNPGCL outperforms current state-of-the-art unsupervised GCL methods, showcasing its efficacy in advancing the field of unsupervised graph representation learning.

Acknowledgments. This work was supported by the National Natural Science Foundation of China under Grant No. U1936213, Program of Shanghai Academic Research Leader No. 21XD1421500, Shanghai Science and Technology Commission Project No. 20020500600.

References

1. Caron, M., Bojanowski, P., Joulin, A., Douze, M.: Deep clustering for unsupervised learning of visual features. In: Proceedings of the European Conference on Computer Vision (ECCV), pp. 132–149 (2018)
2. Cuturi, M.: Sinkhorn distances: lightspeed computation of optimal transport. In: Advances in Neural Information Processing Systems, vol. 26 (2013)
3. Grill, J.B., et al.: Bootstrap your own latent-a new approach to self-supervised learning. Adv. Neural. Inf. Process. Syst. **33**, 21271–21284 (2020)
4. Hassani, K., Khasahmadi, A.H.: Contrastive multi-view representation learning on graphs. In: International Conference on Machine Learning, pp. 4116–4126. PMLR (2020)
5. He, K., Fan, H., Wu, Y., Xie, S., Girshick, R.: Momentum contrast for unsupervised visual representation learning. In: Proceedings of the IEEE/CVF Conference on Computer Vision and Pattern Recognition, pp. 9729–9738 (2020)
6. He, K., Zhang, X., Ren, S., Sun, J.: Delving deep into rectifiers: Surpassing human-level performance on ImageNet classification. In: Proceedings of the IEEE International Conference on Computer Vision, pp. 1026–1034 (2015)
7. Hinton, G., Vinyals, O., Dean, J.: Distilling the knowledge in a neural network. arXiv preprint arXiv:1503.02531 (2015)
8. Lin, B., Luo, B., He, J., Gui, N.: Self-supervised adaptive aggregator learning on graph. In: Karlapalem, K., et al. (eds.) PAKDD 2021. LNCS (LNAI), vol. 12714, pp. 29–41. Springer, Cham (2021). https://doi.org/10.1007/978-3-030-75768-7_3
9. Lin, S., et al.: Prototypical graph contrastive learning. IEEE Trans. Neural Netw. Learn. Syst. **35**, 2747–2758 (2022)
10. Van der Maaten, L., Hinton, G.: Visualizing data using t-SNE. J. Mach. Learn. Res. **9**(11), 2579–2605 (2008)
11. Narayanan, A., Chandramohan, M., Venkatesan, R., Chen, L., Liu, Y., Jaiswal, S.: graph2vec: learning distributed representations of graphs. arXiv preprint arXiv:1707.05005 (2017)

12. Qiu, J., et al.: GCC: graph contrastive coding for graph neural network pre-training. In: Proceedings of the 26th ACM SIGKDD International Conference on Knowledge Discovery & Data Mining, pp. 1150–1160 (2020)

13. Scarselli, F., Gori, M., Tsoi, A.C., Hagenbuchner, M., Monfardini, G.: The graph neural network model. IEEE Trans. Neural Networks 20(1), 61–80 (2008)

14. Sun, F.Y., Hoffmann, J., Verma, V., Tang, J.: Infograph: unsupervised and semi-supervised graph-level representation learning via mutual information maximization. arXiv preprint arXiv:1908.01000 (2019)

15. Tarvainen, A., Valpola, H.: Mean teachers are better role models: weight-averaged consistency targets improve semi-supervised deep learning results. In: Advances in Neural Information Processing Systems, vol. 30 (2017)

16. Xia, J., Wu, L., Chen, J., Hu, B., Li, S.Z.: SimGRACE: a simple framework for graph contrastive learning without data augmentation. In: Proceedings of the ACM Web Conference 2022, pp. 1070–1079 (2022)

17. You, Y., Chen, T., Sui, Y., Chen, T., Wang, Z., Shen, Y.: Graph contrastive learning with augmentations. Adv. Neural. Inf. Process. Syst. 33, 5812–5823 (2020)

18. Zhang, W., et al.: Evaluating deep graph neural networks. arXiv preprint arXiv:2108.00955 (2021)

19. Zhu, Y., Xu, Y., Yu, F., Liu, Q., Wu, S., Wang, L.: Graph contrastive learning with adaptive augmentation. In: Proceedings of the Web Conference 2021, pp. 2069–2080 (2021)

DEGNN: Dual Experts Graph Neural Network Handling both Edge and Node Feature Noise

Tai Hasegawa[1,2], Sukwon Yun[3], Xin Liu[2(✉)], Yin Jun Phua[1], and Tsuyoshi Murata[1,2]

[1] Department of Computer Science, Tokyo Institute of Technology, Tokyo, Japan
hasegawa.t@net.c.titech.ac.jp, {phua,murata}@c.titech.ac.jp
[2] Artificial Intelligence Research Center, AIST, Tokyo, Japan
xin.liu@aist.go.jp
[3] Industrial and Systems Engineering, KAIST, Daejeon, Republic of Korea
swyun@kaist.ac.kr

Abstract. Graph Neural Networks (GNNs) have achieved notable success in various applications over graph data. However, recent research has revealed that real-world graphs often contain noise, and GNNs are susceptible to noise in the graph. To address this issue, several Graph Structure Learning (GSL) models have been introduced. While GSL models are tailored to enhance robustness against edge noise through edge reconstruction, a significant limitation surfaces: their high reliance on node features. This inherent dependence amplifies their susceptibility to noise within node features. Recognizing this vulnerability, we present DEGNN, a novel GNN model designed to adeptly mitigate noise in both edges and node features. The core idea of DEGNN is to design two separate experts: an edge expert and a node feature expert. These experts utilize self-supervised learning techniques to produce modified edges and node features. Leveraging these modified representations, DEGNN subsequently addresses downstream tasks, ensuring robustness against noise present in both edges and node features of real-world graphs. Notably, the modification process can be trained end-to-end, empowering DEGNN to adjust dynamically and achieves optimal edge and node representations for specific tasks. Comprehensive experiments demonstrate DEGNN's efficacy in managing noise, both in original real-world graphs and in graphs with synthetic noise.

Keywords: Graph Neural Networks · Graph Structure Learning · Graph Self-Supervised Learning

1 Introduction

Graphs are essential data structures for modeling a wide range of real-world phenomena, such as social networks, transportation networks, and chemical molecules. Graph Neural Networks (GNNs) have emerged as a powerful paradigm for modeling such graphs, primarily due to their message-passing

D.-N. Yang et al. (Eds.): PAKDD 2024, LNAI 14646, pp. 376–389, 2024.
https://doi.org/10.1007/978-981-97-2253-2_30

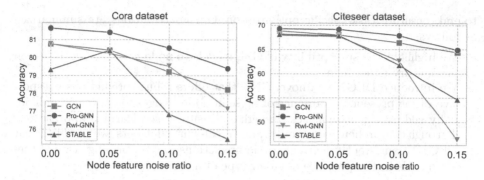

Fig. 1. A comparison of semi-supervised node classification between GCN and GSL models on Cora and Citeseer dataset when there is noise in the node features.

mechanism that aggregates node representations via edges. These GNNs can be applied to various tasks, including node classification [11, 24], link prediction [37], node ranking [3, 7], community detection [29], and graph classification [4].

Despite their success, it is well known that the quality of real-world graph data is often unreliable [8]. In other words, real-world graphs are known to contain noise. For instance, in citation networks, references to unrelated papers can introduce noise in the form of inaccurate edges. Recent studies on adversarial attacks and defenses have highlighted the susceptibility of GNNs to noise within graphs [9, 10]. In response, Graph Structure Learning (GSL) has been developed as a method to optimize graph structure, thus improving graph representations and ensuring more resilient predictions amidst edge noise.

Besides edge noise, node features can also contain noise. For instance, in social networks, users might provide inconsistent, overstated, or even false information about their interests or attributes, which introduces noise into the node features. GSL models often struggle with such node feature noises. The vulnerability of GSL models against node feature noise arises as these models disseminate noisy node features via message passing scheme. Additionally, their reliance on node features to rewire edges compounds the challenge. As illustrated in Fig. 1 (refer to Sect. 4 for the experiment details), recent GSL models such as Pro-GNN [18], Rwl-GNN [35], and STABLE [20] tend to underperform in prediction accuracy when compared to traditional GNN model, GCN [11] when the node feature noise increases. Although Pro-GNN marginally surpasses the performance of GCN, there remains ample scope for improvements.

In this paper, we introduce Dual Experts Graph Neural Network (DEGNN)[1], a novel GNN model crafted to offer robust predictions irrespective of the presence or absence of noise in edges, nodes, or both. At the heart of DEGNN is its distinctive architecture, which employs specialized branches, termed "experts", to individually learn and refine node features and edges. Using a self-supervised learning approach, these experts are seamlessly integrated and co-trained end-

---

[1] Codes are available at: https://github.com/TaiHasegawa/DEGNN.

to-end, ensuring task-specific optimization. Our contributions are summarized as follows:

- We highlight the susceptibility of GSL models to node feature noise based on preliminary experiments.
- We introduce DEGNN, a novel GNN that offers robust predictions irrespective of the presence or absence of noise in nodes, edges, or both, by individually addressing these noises through a self-supervised learning approach.
- Through comprehensive experiments on real-world datasets, we establish that DEGNN consistently delivers stable predictions, outperforming state-of-the-art models in the presence of either type of noise.

2   Related Work

2.1   Graph Neural Networks

GNNs have emerged as a powerful tool for learning from graph-structured data, effectively capturing the complex relationships and interdependencies between nodes [1]. Their successful development across numerous practical fields underscores their extensive applicability and effectiveness [16,17,26–28]. Generally, GNNs can be classified into two categories: spectral-based methods [6,11] and spatial-based methods [12,13]. Spectral-based GNNs hinge upon spectral graph theory [5] and employ spectral convolutional neural networks. To streamline the intricacies of spectral-based GNNs, various techniques have emerged, including ChebNet [6] and GCN [11].

On the other hand, spatial-based GNNs are engineered to tackle challenges related to efficiency, generality, and flexibility, as highlighted in [15]. They achieve graph convolution operation through neighborhood aggregation. For instance, GraphSAGE [12] selectively samples a subset of neighbors to grasp local features, while GAT [13] employs an attention mechanism for adaptive neighbor aggregation. However, these models are susceptible to edge noise.

2.2   Graph Structure Learning

The purpose of GSL is to improve prediction accuracy by modifying the given graph into an optimal structure. In this paper, we primarily focus on GNN-based graph structure learning models. LDS [14] jointly optimizes the probability for each node pair and the parameters of GNNs in a bilevel way. Pro-GNN [18] aims to learn the optimal graph structure by incorporating several regularizations, such as low-rank sparsity and feature smoothness. Gaug-M [19] directly computes the edge weights by taking the inner product of node embeddings. STABLE [20] utilizes self-supervised learning to acquire node embeddings and then modifies the graph structure based on their similarities. These obtained node embeddings and the modified graph structure are employed in downstream tasks. Each of these models demonstrates the ability to robustly predict against edge noise, as shown in their respective papers. However, as elucidated in Sect. 1, their pronounced reliance on node features during edge rewiring inherently exposes them to vulnerabilities in situations characterized by noisy node features.

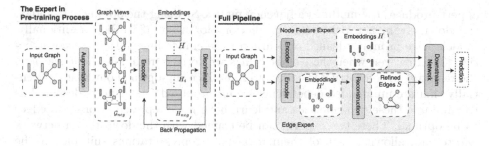

Fig. 2. The overview of DEGNN and the expert in its pre-training process.

3   The Proposed Model

In this section, we introduce our proposed approach, DEGNN. We begin by formulating the problem definition, followed by an overview of the model, and then provide a detailed description of its architecture and learning procedure.

3.1   Problem Definition

Let $\mathcal{G} = \{\mathcal{V}, \mathcal{E}, X\}$ represent an undirected graph, where $\mathcal{V} = \{v_i\}_{i=1}^N$ is the set of N nodes, $\mathcal{E}$ is the set of edges, $X = [x_1, ..., x_N] \in \mathbb{R}^{N \times D}$ indicates the node feature matrix and each $x_i \in \mathbb{R}^D$ is the feature vector of node v_i. The set of edges is represented by an adjacency matrix $A \in \{0,1\}^{N \times N}$, where A_{ij} denotes the connection between nodes v_i and v_j. Following the common semi-supervised node classification setting, only a small portion of nodes $\mathcal{V}_L = \{v_i\}_{i=1}^l$ are associated with the corresponding labels $\mathcal{Y}_L = \{y_i\}_{i=1}^l$ while the rest of the nodes $\mathcal{V}_U = \{v_i\}_{i=l+1}^N$ are unlabeled.

Given graph $\mathcal{G} = \{\mathcal{V}, \mathcal{E}, X\}$ and the available labels $\mathcal{Y}_L$, the goal of graph structure learning, aimed at refining node features and graph structure, is to learn optimal node embeddings $H \in \mathbb{R}^{N \times D'}$ with hidden dimension D', a modified adjacency matrix $S \in \mathbb{R}^{N \times N}$, and the GNN parameters θ in order to improve the predictive accuracy of $\hat{\mathcal{Y}}_L$. The objective function can be formulated as

$$\min_{\theta, S, H} \mathcal{L}(A, X, \mathcal{Y}_L) = \sum_{v_i \in V_L} \ell(f_\theta(H, S)_i, y_i), \tag{1}$$

where $f_\theta : \mathcal{V}_L \to \mathcal{Y}_L$ is a function learned by downstream GNNs that maps nodes to the set of labels, $f_\theta(H, S)_i$ is the prediction of node v_i, and ℓ is the loss measuring difference between prediction and true label such as cross-entropy.

3.2   Overview

The overview of DEGNN and the expert in its pre-training process are illustrated in Fig. 2. The input graph is passed through both the node feature expert and the edge expert. The node feature expert outputs node embeddings H, while the edge expert produces the modified adjacency matrix S. Using the obtained H and S, the downstream network makes predictions for a specific task. Traditionally, models have primarily leaned on learned node embeddings for edge reconstruction [20]. However, when these embeddings are derived from noisy node features, the resulting in reconstructed edges often produce sub-optimal outcomes, potentially that are unintended dependencies. Our proposed dual experts design aims to eliminate these dependencies and learn both node representations and edges to be optimal. These two experts can be trained with the downstream network end-to-end, allowing each of them to acquire representations suitable for the specific task. The details of the model are described in the following sections.

3.3   Node Feature Expert

A straightforward approach to predict robustly against node feature noise is to generate node embeddings H that capture node features more effectively than the input node features X. To obtain the node embeddings H, we utilize self-supervised learning, since it enables the model to achieve better performance, generalization, and robustness across a wide range of downstream tasks [32].

Graph Augmentation. Generating views is a key component of self-supervised learning methods. For instance, in the field of computer vision [21], views are created by rotating or cropping images. This allows for mitigating the impact of differences in angles and scales on the model's predictions. Similarly, for graph data, we presume that by generating views and exposing the model to modified edges and nodes, it can achieve enhanced generalizability, equipping it with a stronger capacity to manage noise emanating from both edges and nodes during predictions. Therefore, we generate graphs of positive view $\mathcal{G}_1 = (\tilde{A}, X)$ with noise added to the edges, $\mathcal{G}_2 = (A, \tilde{X})$ with noise added to the node features, and $\mathcal{G}_3 = (\tilde{A}, \tilde{X})$ with noise added to both edges and node features, where $\tilde{A} \in \{0,1\}^{N \times N}$ and $\tilde{X} \in \mathbb{R}^{N \times D}$ denote the noisy adjacency matrix and noisy node features, respectively.

Our proposed model augments edges by randomly rewiring them. Formally, we first create a mask matrix $P \in \{0,1\}^{N \times N}$ to rewire edges, where P is obtained from a Bernoulli distribution $P_{ij} \sim \mathcal{B}(p)$ with a hyper-parameter p that controls the rewiring probability. And then, $\tilde{A}$ is calculated as follows:

$$\tilde{A} = (1 - A) \odot P + A \odot (1 - P), \tag{2}$$

where $\odot$ is the Hadamard product. For node feature augmentation, unlike other studies that use node feature masking [22,23], in this paper, we shuffle elements in each row of X randomly with the probability q to replicate scenarios where there is noise in the node features.

Besides positive graphs, we generate a negative graph $\mathcal{G}_{neg} = (A_{neg}, X_{neg})$ that is entirely different from the original graph $\mathcal{G}$, where $A_{neg} \in \{0,1\}^{N \times N}$ and $X_{neg} \in \mathbb{R}^{N \times D}$ denote the negative adjacency matrix and negative node features, respectively. A_{neg} is given by $A_{neg} = (1-A) \odot P_{neg}$, where $P_{neg} \in \{0,1\}^{N \times N}$ is a mask matrix that is obtained from a Bernoulli distribution $P_{negij} \sim \mathcal{B}(\frac{\|\mathcal{E}\|}{N^2 - \|\mathcal{E}\|})$. X_{neg} is obtained through row-wise shuffling of X.

Encoder. The encoder f_ϕ parameterized by ϕ learns embeddings for each of the generated views. In this paper, we use a 1-layer GCN [11] as the encoder, which is formulated as follows:

$$f_\phi(X, A) = \sigma(\hat{D}^{-1/2}\hat{A}\hat{D}^{-1/2}XW), \qquad (3)$$

where $\hat{A} = A + I_N$, $\hat{D} = D + I_N$, D is the degree matrix of A, I_N is the identity matrix, σ is a non-linear activation function and W is weight matrix transforming raw features to embeddings. The embeddings for the original graph and each view $H, H_1, H_2, H_3, H_{neg} \in \mathbb{R}^{N \times D'}$ are obtained using the encoder as follows:

$$H_2 = f_\phi(\tilde{X}, A) \qquad (6)$$

$$H = f_\phi(X, A) \qquad (4)$$

$$H_3 = f_\phi(\tilde{X}, \tilde{A}) \qquad (7)$$

$$H_1 = f_\phi(X, \tilde{A}) \qquad (5)$$

$$H_{neg} = f_\phi(X_{neg}, A_{neg}) \qquad (8)$$

Objective Function. Following other graph contrastive learning methods [20, 31], we train the encoder to maximize the mutual information between the original graph $\mathcal{G}$ and positive graphs $\mathcal{G}_1$, $\mathcal{G}_2$ and $\mathcal{G}_3$ while minimizing agreement between original graph $\mathcal{G}$ and negative graph $\mathcal{G}_{neg}$. Formally, the objective function for node feature expert can be formulated via using binary cross-entropy loss between positive samples and negative samples as follows:

$$\mathcal{L}_N = -\frac{1}{2N}\sum_{i=1}^{N}(\frac{1}{3}\sum_{j=1}^{3}(log\mathcal{D}(h_i, h_i^j) + log(1 - \mathcal{D}(h_i, h_i^{neg})))), \qquad (9)$$

where h_i, h_i^j, h_i^{neg} are the embeddings of node v_i in $\mathcal{G}, \mathcal{G}_j$ and $\mathcal{G}_{neg}$, and $\mathcal{D}$ is the discriminator built upon an inner product, i.e., $\mathcal{D}(a, b) = ab^T$. In conclusion, by training the node feature expert, f_ϕ, we derive a noise-robust node embedding H that subsequently serves as input for the downstream network.

3.4   Edge Expert

To differentiate between the impacts of node feature noise and edge noise, and to adeptly address each situation, we further implement an edge expert. The edge expert learns node embeddings H' using self-supervised learning as with the node feature expert. More precisely, H' is obtained through an encoder f_ψ with different parameters from the node feature expert f_ϕ, and the objective function for edge expert using the binary cross-entropy loss L_E can be formulated as follows:

$$\mathcal{L}_E = -\frac{1}{2N} \sum_{i=1}^{N} \left(\frac{1}{3} \sum_{j=1}^{3} \left(log \mathcal{D}(h'_i, h_i^{j'}) + log(1 - \mathcal{D}(h'_i, h_i^{neg'})) \right) \right), \tag{10}$$

where h'_i, $h_i^{j'}$, $h_i^{neg'}$ are the embeddings of node v_i in $\mathcal{G}$, $\mathcal{G}_j$ and $\mathcal{G}_{neg}$ obtained with encoder f_ψ.

Reconstruction. Once the high-quality embedding H' from the edge expert is obtained, it is used to reconstruct the graph structure. This reconstruction process rewires the edges using the pairwise similarity in the embeddings H' under the homophily assumption [2], which posits that nodes with similar features are more likely to be connected.

In the beginning, we compute the cosine similarity matrix $B \in \mathbb{R}^{N \times N}$, where B_{ij} represents the cosine similarity between H'_i and H'_j. Next, we remove the $k\%$ (k is a hyper-parameter) of the edges with the smallest cosine similarity between pairs of nodes from the original edge set $\mathcal{E}$, resulting in a sparser adjacency matrix $\tilde{S} \in \{0, 1\}^{N \times N}$ as follows:

$$\tilde{S} = A \odot M_1, \tag{11}$$

where $M_1 \in \{0, 1\}^{N \times N}$ represents the mask matrix for edge deletion, where M_{1ij} is 1 if its cosine similarity ranks over the smallest $k\%$, and 0 otherwise.

Finally, to obtain the modified adjacency matrix S, we add the same number of edges that were removed i.e., $k * |\mathcal{E}|$ with the highest cosine similarity between pairs of nodes from the set of node pairs $\mathcal{E}' = V \times V \setminus \mathcal{E}$ that are not included in the edge set $\mathcal{E}$. Formally, S is obtained as follows:

$$S = \tilde{S} + B \odot M_2, \tag{12}$$

where $M_2 \in \{0, 1\}^{N \times N}$ is the mask matrix for edge addition, where M_{2ij} is 1 for the edges within top cosine similarity ($k * |\mathcal{E}|$), and 0 otherwise. Here, it is important that the encoder in the edge expert can be trained through backpropagation from the downstream network, so the reconstruction needs to be differentiable. Consequently, the value associated with newly introduced edges in S is not clamped and retains its cosine similarity. To sum up, through training edge expert, f_ψ, we obtain a modified adjacency matrix, S, which is then utilized as input for the downstream network.

3.5   Downstream Network

Once we obtain the node embeddings H and the modified adjacency matrix S, we now use them as input for the downstream network (i.e., GNNs). It is worth noting that our proposed method allows the use of any GNNs such as GCN [11] or GAT [13], and also can be applied to various tasks beyond node classification, including link prediction [37] and graph classification [4]. In this paper, we use a 2-layer GCN f_θ as the downstream network. To tackle node classification, the model is trained to minimize the cross-entropy loss:

$$\mathcal{L}_{GNN} = \sum_{v_i \in V_L} \ell(f_\theta(H, S)_i, y_i), \tag{13}$$

where $\ell(f_\theta(H, S)_i, y_i)$ is the cross-entropy between the prediction and the ground-truth label for node v_i.

3.6 Training Methodology

In this paper, we propose two variants of DEGNN with different training methods: (i) the pre-training and fine-tuning model (referred to as DEGNN-I), and (ii) the modular learning model (referred to as DEGNN-II). DEGNN-I first pre-trains the node feature expert and edge expert separately. Then, all components are jointly fine-tuned in an end-to-end manner. This is to enable each expert to obtain the optimal node embeddings and graph structure for downstream tasks. During the fine-tuning process, it is trained to minimize the following objective function:

$$\mathcal{L} = \mathcal{L}_{GNN} + \alpha \mathcal{L}_N + \beta \mathcal{L}_E, \tag{14}$$

where α and β are hyper-parameters to balance the contributions of node embedding generation and edge reconstruction, respectively.

In contrast, DEGNN-II follows a two-step approach: initially training the node feature expert and edge expert and subsequently freezing them during the training of the downstream network. This methodology empowers each expert to attain task-agnostic representations. This allows robust predictions in contexts with limited labels or large biases, given that the approach does not depend on the provided label during training-essentially, a self-supervised paradigm.

4   Experiments

In this section, we evaluate our proposed method, DEGNN, on a variety of noisy graphs in the context of the semi-supervised node classification task.

4.1 Experimental Setup

Datasets. We used four open datasets, including two citation networks (i.e., Cora [19], Citeseer [19]) and two co-purchasing networks (i.e., Photo [33], Computer [33]) Regarding the train/validation/test split, we prepared 20 training labels for each class in all datasets. For validation and test, we used 500 and 1000 nodes, respectively.

Noisy Graphs. To assess the robustness of our model across various graph settings with noise, we compared models on graphs that included the following types of noise:

- **Clean Graphs**: The original graphs of the datasets which may contain inherent node feature noise and edge noise.
- **Edge Noisy Graphs**: We randomly remove a certain number of edges and insert the same number of fake edges.
- **Node Feature Noisy Graphs**: As in [36], we added independent Gaussian noise to the node features. Specifically, we obtained the reference amplitude r by calculating the mean of the maximum value across each node's features. For each feature dimension of each node, we introduced independent Gaussian noise $\lambda \cdot r \cdot \epsilon$, where $\epsilon \sim N(0, 1)$, and λ represents the feature noise ratio.
- **Edge and Node Feature Noisy Graphs**: We introduced both the edge noise and node feature noise described above.

When adding these noises, we employed a poisoning attack, which initially prepares a graph with noise added, and used it for both training and evaluation.

Baselines. We compare the proposed DEGNN-I and DEGNN-II with two categories of baselines: classical GNN models (i.e., GCN [11], GAT [13] and RGCN [34]) and graph structure learning methods (i.e., Pro-GNN [18], Rwl-GNN [35] and STABLE [20]).

Implementation Details. All hyper-parameters are tuned on the clean graph. All models are trained using Adam optimizer with a default learning rate of 1e-2 and a weight decay of 5e-4 when not explicitly specified. GCN [11], GAT [13], and RGCN [34] have a fixed number of layers at 2. For GCN and RGCN, the hidden dimension is chosen from $\{16, 32, 64, 128\}$. GAT's number of heads and head dimensions are selected from $\{1, 2, 4, 8, 16\}$ and $\{8, 16, 32, 64, 128\}$, respectively, with a total hidden dimension ranging from 16 to 128. Other baselines follow hyper-parameter combinations specified in their respective papers. For DEGNN, α and β are tuned from $\{0, 0.1, 1.0, 10\}$, the hidden dimension D' is tuned from $\{128, 256, 512\}$. k is searched in $\{1, 5, 10, 15, 20, 25\}$, p and q are searched in $\{0.2, 0.4, 0.6\}$. The learning rate in the pre-training process is tuned from $\{1e-2, 5e-3, 1e-3\}$.

4.2   Semi-supervised Node Classification

Performance Comparison. In this section, we evaluate the proposed DEGNN on semi-supervised node classification on original graphs. The average accuracy and standard deviation of the model across 10 runs are summarized in Table 1. OOM indicates out of memory. Based on the experiment, our proposed approach outperformed other models in citation networks (Cora, Citeseer), demonstrating the highest predictive accuracy. This suggests that the experts successfully

Table 1. The results (accuracy(%) ± std) of semi-supervised node classification on clean graphs. The top two performance is highlighted in bold and underline.

Dataset	GCN	GAT	RGCN	Pro-GNN	Rwl-GNN	STABLE	DEGNN-I	DEGNN-II
Cora	80.8 ± 0.9	79.5 ±0.8	79.6 ±0.7	<u>81.7 ± 0.7</u>	80.8 ±0.7	79.3 ±1.1	**83.0 ±0.9**	81.6 ±1.0
Citeseer	68.8 ±0.7	66.7 ±1.7	65.7 ±1.5	69.3 ±0.6	68.0 ±0.6	68.2 ±0.5	<u>70.9 ± 1.9</u>	**71.6 ±1.4**
Photo	**91.8 ±0.2**	85.5 ±17.3	91.7 ±0.5	<u>91.8 ± 0.7</u>	86.4 ±2.9	OOM	91.3 ±0.4	91.6 ±0.6
Computers	**85.0 ±0.9**	77.2 ±22.7	81.6 ±2.6	OOM	74.6 ±1.4	OOM	<u>83.5 ± 0.9</u>	82.8 ±0.8

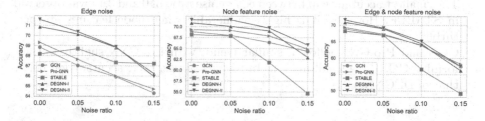

Fig. 3. Accuracy on Citeseer with noise added to either or both of edge and node features.

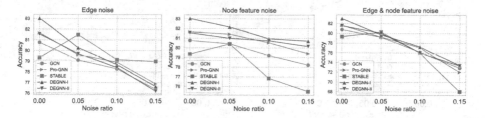

Fig. 4. Accuracy on Cora with noise added to either or both of edge and node features.

obtain superior representations for both edges and node features to eliminate latent noise within these graphs. However, in co-purchasing networks (Photo, Computer), our model achieved competitive results but was marginally surpassed by the traditional GCN. Considering that other GSL models also lagged behind GCN in accuracy and the fact that GCN's predictive accuracy in co-purchasing networks is significantly higher than in citation networks, it is conceivable that the co-purchasing networks may contain less inherent noise within the original graphs, and the GSL models might be unnecessarily altering the graph.

Robustness Evaluation. In this experiment, we evaluate the robustness of the models by comparing their performance on graphs with noise added to either or both of the edge and node features. The nature of these noise is described in Sect. 4.1. Specifically, we added edge noise with noise ratio 0.05, 0.1 and 0.15, while the parameter λ for node feature perturbation is also set to 0.05, 0.1, and 0.15. All experiments were conducted 10 times, and the average accuracy on

Citeseer and Cora dataset are shown in Figs. 3 and Fig. 4, respectively. From this experiment, we can obtain the following observation:

- When node feature noise is added, DEGNN-I or DEGNN-II demonstrated the highest accuracy among the compared models in all settings.
- When noise is introduced in only the edge and in both edge and node features, DEGNN demonstrated the best or competitive results compare to the baselines especially when the noise ratio is 0, 0.05 or 0.1.
- The smallest accuracy gap between edge noise ratio of 0 and 0.15 was observed in STABLE. However, it is highly vulnerable when node features are perturbed.
- GCN and Pro-GNN follow a similar trend, with Pro-GNN slightly outperforming GCN by a small margin. In many settings, their accuracy fell below that of DEGNN.

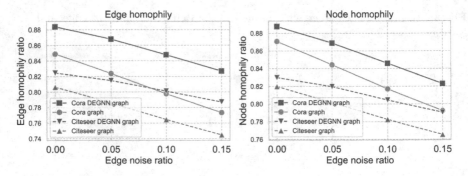

Fig. 5. Comparison of edge homophily ratio and node homophily ratio between noisy graphs and refined graphs by the edge expert on Cora and Citeseer.

4.3   Edge Expert Analysis

To evaluate the edge expert, we compared the edge homophily ratio [25] and node homophily ratio [30] between the noisy graph and the graphs refined by the edge expert (referred to as the DEGNN graph) on various edge noise added settings. In this experiment, we used DEGNN-II, and we set the hyper-parameter k, which controls the number of edge rewires, to 10%. Figure 5 shows the edge homophily ratio and node homophily ratio as edge noise is gradually added. The solid lines represent the results for the Cora dataset, while the dashed lines represent the results for the Citeseer dataset. This experiment reveals that the edge expert consistently promotes high edge homophily and node homophily in all scenarios.

4.4   Node Feature Expert Analysis

In this experiment, we compared the effectiveness of the node feature expert by comparing the models' accuracy when node features have noise added. Figure 6 shows the average of accuracy of 10 runs of GCN and DEGNN-I only having node feature expert (without edge expert) on Cora and Citeseer dataset with added node feature noise. Empirical results confirm that employing node embeddings derived from the node feature expert markedly elevates prediction accuracy across a majority of scenarios. This underscores both the indispensability and efficacy of the node feature expert.

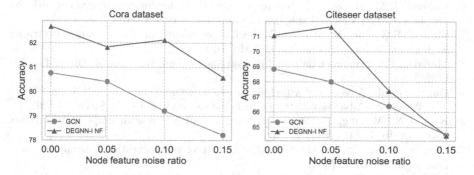

Fig. 6. Comparison of GCN and DEGNN-I without edge expert on Cora and Citeseer with added node feature noise.

5   Conclusion

In this paper, we identified the vulnerability of recent GSL methods to node feature noise and proposed a novel GNN model, DEGNN, to address this issue. DEGNN refines both edges and node features using two carefully designed experts via self-supervised learning, allowing it to robustly perform predictions in the presence of both edge and node feature noise. Extensive experiments verify the effectiveness of the experts in handling graphs with various noise scenarios.

In this paper, we employed GCN as both the encoder and downstream network. Additionally, simple methods were used for augmentation and edge reconstruction techniques. As future work, we plan to explore more specialized networks in the encoder and downstream network, as well as investigate different augmentation and edge reconstruction techniques.

Acknowledgements. This work is partly supported by JSPS Grant-in-Aid for Scientific Research (grant number 23H03451, 21K12042) and the New Energy and Industrial Technology Development Organization (Grant Number JPNP20017).

References

1. Zhou, J., et al.: Graph neural networks: a review of methods and applications. AI Open **1**, 57–81 (2020)
2. McPherson, M., Smith-Lovin, L., Cook, J.M.: Birds of a feather: homophily in social networks. Ann. Rev. Sociol. **27**(1), 415–444 (2001)
3. Maurya, S.K., Liu, X., Murata T.: Graph neural networks for fast node ranking approximation. In: TKDD (2021)
4. Zhang, M., et al.: An end-to-end deep learning architecture for graph classification. In: AAAI (2018)
5. Chung, F.R.K.: Spectral Graph Theory. number 92. American Mathematical Soc (1997)
6. Defferrard, M., Bresson, X., Vandergheynst, P.: Convolutional neural networks on graphs with fast localized spectral filtering. In: NeurIPS (2016)
7. Maurya, S.K., Liu, X., Murata, T.: Fast approximations of betweenness centrality with graph neural networks. In: CIKM (2019)
8. Marsden, P.V.: Network data and measurement. Ann. Rev. Sociol. **16**(1), 435–463 (1990)
9. Dai, H., et al.: Adversarial attack on graph structured data. In: ICML (2018)
10. Jin, W., et al.: Adversarial attacks and defenses on graphs. In: SIGKDD (2021)
11. Kipf, T.N., Welling, M.: Semi-supervised classification with graph convolutional networks. In: ICLR (2017)
12. Hamilton, W., Ying, Z., Leskovec, J.: Inductive representation learning on large graphs. In: NeurIPS (2017)
13. Veličković, P., Cucurull, G.: Arantxa Casanova. Pietro Lio, and Yoshua Bengio. Graph attention networks. In ICLR, Adriana Romero (2018)
14. Franceschi, L., Niepert, M., Pontil, M., He, X.: Learning discrete structures for graph neural networks. In: ICML (2019)
15. Wu, Z., Pan, S., Chen, F., Long, G., Zhang, C., Philip, S.Y.: A comprehensive survey on graph neural networks. IEEE Trans. Neural Netw. Learn. Syst. **32**(1), 4–24 (2020)
16. Jin, R., Xia, T., Liu, X., Murata, T.: Predicting emergency medical service demand with bipartite graph convolutional networks. IEEE Access **9**, 9903–9915 (2021)
17. Fan, W., et al.: Graph neural networks for social recommendation. In: WWW, pp. 417–426 (2019)
18. Jin, W., et al.: Graph structure learning for robust graph neural networks. In: SIGKDD (2020)
19. Zhao, T., et al.: Data augmentation for graph neural networks. In: AAAI (2021)
20. Li, K., et al.: Reliable representations make a stronger defender: unsupervised structure refinement for robust GNN. In: SIGKDD (2022)
21. Berthelot, D., et al.: MixMatch: a holistic approach to semi-supervised learning. In: NeurIPS (2019)
22. Zhu, Y., Xu, Y., Yu, F., Liu, Q., Wu, S., Wang, L.: Deep graph contrastive representation learning. In: ICML (2020)
23. You, Y., Chen, T., Sui, Y., Chen, T., Wang, Z., Shen, Y.: Graph contrastive learning with augmentations. In: NeurIPS (2020)
24. Maurya, S.K., Liu, X., Murata, T.: Simplifying approach to node classification in graph neural networks. J. Comput. Sci. **62**, 101695 (2022)
25. Zhu, J., Yan, Y., Zhao, L., Heimann, M., Akoglu, L., Koutra, D.: Beyond homophily in graph neural networks: current limitations and effective designs. In: NeurIPS (2020)

26. Marcheggiani, D., Titov, I.: Encoding sentences with graph convolutional networks for semantic role labeling. In: EMNLP, pp. 1506–1515 (2017)
27. Rakhimberdina, Z., Liu, X., Murata, T.: Population graph-based multi-model ensemble method for diagnosing autism spectrum disorder. Sensors **20**(21), 6001 (2020)
28. Djenouri, Y., Belhadi, A., Srivastava, G., Lin, J.C.: Hybrid graph convolution neural network and branch-and-bound optimization for traffic flow forecasting. Futur. Gener. Comput. Syst. **139**, 100–108 (2023)
29. Choong, J.J., Liu, X., Murata, T.: Learning community structure with variational autoencoder. In: ICDM, pp. 69–78 (2018)
30. Pei, H., Wei, B., Kevin, C.-C.C., Yu, L., Yang, B.: Geom-GCN: Geometric graph convolutional networks. In: ICLR (2020)
31. Suresh, S., Li, P., Hao, C., Neville, J.: Adversarial graph augmentation to improve graph contrastive learning. In: NeurIPS (2021)
32. Liu, Y., et al.: Graph self-supervised learning: a survey. IEEE Trans. Knowl. Data Eng. **35**(6), 5879–5900 (2022)
33. Shchur, O., Mumme, M., Bojchevski, A., Günnemann, S.: Pitfalls of graph neural network evaluation. In: NeurIPS Workshop (2018)
34. Zhu, D., Zhang, Z., Cui, P., Zhu, W.: Robust graph convolutional networks against adversarial attacks. In: SIGKDD (2019)
35. Runwal, B., Kumar, S.: Robust graph neural networks using weighted graph Laplacian (2022). arXiv preprint arXiv:2208.01853
36. Wu, T., Ren, H., Li, P., Leskovec, J.: Graph information bottleneck. In: NeurIPS (2020)
37. Zhang, M., Chen, Y.: Link prediction based on graph neural networks. In: NeurIPS (2018)

Alleviating Over-Smoothing
via Aggregation over Compact Manifolds

Dongzhuoran Zhou[1,2(✉)], Hui Yang[3], Bo Xiong[4], Yue Ma[3],
and Evgeny Kharlamov[1,2]

[1] Bosch Center for AI, Karlsruhe, Germany
{dongzhuoran.zhou,evgeny.kharlamov}@de.bosch.com
[2] University of Oslo, Oslo, Norway
[3] LISN, CNRS, Universite Paris-Saclay, Gif Sur Yvette, France
{yang,ma}@lisn.fr
[4] University of Stuttgart, Stuttgart, Germany
bo.xiong@ipvs.uni-stuttgart.de

Abstract. Graph neural networks (GNNs) have achieved significant success in various applications. Most GNNs learn the node features with information aggregation of its neighbors and feature transformation in each layer. However, the node features become indistinguishable after many layers, leading to performance deterioration: a significant limitation known as over-smoothing. Past work adopted various techniques for addressing this issue, such as normalization and skip-connection of layer-wise output. After the study, we found that the information aggregations in existing work are all contracted aggregations, with the intrinsic property that features will inevitably converge to the same single point after many layers. To this end, we propose the aggregation over compacted manifolds method (ACM) that replaces the existing information aggregation with aggregation over compact manifolds, a special type of manifold, which avoids contracted aggregations. In this work, we theoretically analyze contracted aggregation and its properties. We also provide an extensive empirical evaluation that shows ACM can effectively alleviate over-smoothing and outperforms the state-of-the-art. The code can be found in https://github.com/DongzhuoranZhou/ACM.git.

Keywords: Graph Neural Network · Over-smoothing · Manifold

1 Introduction

Graph neural networks (GNNs) [42] are potent tools for analyzing graph-structured data, including biochemical networks [38], social networks [13], and academic networks [9]. Most GNNs employ a message-passing mechanism for

---

D. Zhou and H. Yang—Equal contribution.

Supplementary Information The online version contains supplementary material available at https://doi.org/10.1007/978-981-97-2253-2_31.

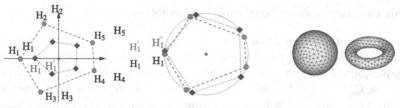

(a) Contracted   aggrega-(b) Aggregation on circle   (c) Compact Manifolds
tion

Fig. 1. (a) node features with *contracted aggregation* become closer and inevitably converge after many layers. (b) Our approach avoids contracted aggregation by aggregating on compact manifolds. (c) Compact manifolds: sphere and torus.

learning node features, involving information aggregation from neighbors and feature transformation in each layer [12]. This mechanism enables GNNs to effectively capture detailed information in graph data.

Though potential in various tasks, existing GNNs with many layers cause node representations to be highly indistinguishable, thus leading to significant performance deterioration in downstream tasks - a phenomenon known as over-smoothing [26]. Increasing efforts have been devoted to addressing the over-smoothing issue, such as batch norm [17], pair norm [41], group norm [43], drop edges or nodes [5,32], and regularize the nodes [15] or feature distance [18]. Most of existing studies focus on alleviating the over-smoothing based on enhancing the distinction in node dimension [17,41] or feature dimension [18] through normalization, adding regularization items [18], skip-connection [25,39], etc.

In this work, we show that past works on over-smoothing are confined to *contracted aggregation* on Euclidean spaces, where features inevitably converge to a single point after multiple layers of information aggregations. The intuition of *contracted aggregation* is illustrated in Fig. 1a, using the plane $\mathbb{R}^2$ as the embedding space and the mean of vectors as the aggregation function. The aggregation results (red points) become closer than the initial embedding (blue points), causing nodes to converge to a single point after several steps. In contrast, Fig. 1b shows that after *aggregation over compact manifold* (compact manifold see Fig. 1(c)), node features do not become closer, thus avoiding the *contracted aggregation* issue[1]. Even after numerous information aggregation steps, the aggregation results (red points) remain distinguishable and do not converge to a single point, unlike the Euclidean case.

In light of this, we perform an in-depth study on the over-smoothing in GNNs from the information aggregation perspective. We aim to bring the benefits of information aggregation on manifolds to alleviate over-smoothing. We develop a general theory to explain the reason for over-smoothing in Riemannian manifolds (including Euclidean spaces). Based on this theory, we propose *aggregation over compact manifolds* (ACM) to address the over-smoothing issue.

---

[1] In Fig. 1b, the unit circle is used as the embedding space. The aggregation of points is defined under *polar coordinates*. For example, the aggregation of two points in the unit circle with polar coordinate $(1, \theta), (1, \phi)$ is $(1, \frac{\theta+\phi}{2})$.

Our contributions can be summarized as follows:

- We propose the notion of *contracted aggregation* (Sect. 4) and prove that node features will converge to the same single point after many layers of contracted aggregation, leading to over-smoothing. We also prove that standard GNNs, e.g., SGC, GCN, GAT, are either *contracted aggregation* or mathematically equal. Our claim holds for embedding space of any Riemannian manifolds, including Euclidean Space.
- We propose *aggregation over compact manifold* (ACM) (Sect. 5) that can be integrated into each layer of GNNs models to prevent contracted aggregation, effectively mitigating over-smoothing across various GNN architectures.
- We provide extensive empirical evidence (Sect. 6) that shows the proposed ACM can alleviate over-smoothing and improve deeper GNNs with better performance compared to the state-of-the-art (SotA) methods. The improvement becomes more significant in a more complex experiment setting (i.e., missing feature) that requires massive layers for GNNs to achieve a good performance.

2 Related Work

Over-Smoothing in GNNs. Recent studies have shown that deep stacking of GNN layers can result in a significant performance deterioration, commonly attributed to the problem of over-smoothing [5,41]. The over-smoothing issue was initially highlighted in [26], where the authors show that node embeddings will converge to a single point or lose structural information after an infinite number of random walk iterations. Their result has been extended to more general cases considering transformation layers and different activate functions by [3,16,29].

To address the over-smoothing issue and enable deeper GNNs, various methods have been proposed [13,25,26,32]. One approach is to employ skip connections for multi-hop message passing, such as GraphSAGE [13] JKNet [26]. There are also several approaches developed based on the existing methods in other areas. For instance, the study of [25] leveraged concepts from ResNet [14] to incorporate both residual and dense connections in the training of deep GCNs; GCNII [6] utilizes initial residual and identity mapping. [32] proposed Dropedge that alleviates over-smoothing through a reduction in message passing by removing edges inspired by the use of Dropout [34]. Another mainstream approach is to normalize output features of each layer, such as Batch-Norm, PairNorm [41] and DGN [43]. DeCorr [18] introduced over-correlation as one reason for significant performance deterioration. APPNP [22] use Personalized PageRank and GPRGNN [7] use Generalized PageRank to mitigate over-smoothing. DAGNN [27] was introduced to create deep GNNs by decoupling graph convolutions and feature transformation. However, these works keep using the original information aggregation function and focus only on alleviating distinguishable after feature transformation.

GNNs on Manifolds. Recent attention in GNN research has been directed towards neural networks on *manifolds*, impacting various domains like knowledge graph embedding, computer vision, and natural language processing [2,11,19]. Many works focus on the *hyperbolic space*, a Riemannian manifold with constant negative curvature, with [4] introducing Hyperbolic Graph Neural Networks (HGNN). [1] proposes a general GNN on manifolds applicable to hyperbolic space and hyperspheres. This work concentrates on GNN over compact manifolds, encompassing hyperspheres but not hyperbolic spaces. We introduce a novel GNN framework over compact manifolds. Unlike existing methods, our model doesn't depend on special non-linear functions, such as the *exponential* and *logarithmic* maps [8] defined in hyperbolic space and hyperspheres, making it computationally more straightforward. Additionally, our model is more general, defined on more general kinds of manifold spaces, including hyperspheres.

3   Preliminaries

Compact Metric Space. We start with a brief introduction to basic topology (more details can be found in literature [28]). A *metric space* S is a set equipped with a distance function $d_S : S \times S \to \mathbb{R}_{\geq 0}$ (i.e., a function that is positive, symmetric, and satisfying the triangle inequality). A sequence $p_1, p_2, \ldots \in S$ *converge* if $\lim_{k \to \infty} \max\{d_S(p_i, p_j) \mid i, j \geq k\} = 0$. Metric space S is *closed* if for every convergent sequence $p_1, p_2, \ldots \in S$ there exits $p \in S$ such that $\lim_{i \to \infty} d_S(p_i, p) = 0$. Metric space S is *compact* if it is closed and *bounded*—that is, there exists a $r \in \mathbb{R}$ such that $d_S(p, q) < r$ for every $p, q \in S$. For example, the open interval $S_{\text{oint}} = (0, 1) \subseteq \mathbb{R}$ is a metric space associated with distance $d_{S_{\text{oint}}}(x, y) = |x - y|$. Then $2^{-1}, 2^{-2}, 2^{-3}, \ldots \in S_{\text{oint}}$ is a convergent sequence, but there is no point in S_{oint} that is the limit of this sequence. Therefore, S_{oint} is not closed. In contrast, a unit circle S_{circ} in $\mathbb{R}^2$ with $d_{S_{\text{circ}}}(x, y) = ||x - y||$ is closed, since $d_{S_{\text{circ}}}(p, q) < 3$ for all $p, q \in S_{\text{circ}}$, and compact.

Compact Manifolds. Next we provide a brief introduction to manifolds with the notions necessary for this paper (see further details in literature [24]). A d-dimensional *manifold* $\mathcal{M}$ is a hyper-surface in the Euclidean space $\mathbb{R}^n$ with $n \geq d$ such that each point has an (open) neighbourhood that is homeomorphic to an open subset of $\mathbb{R}^d$ (i.e., locally looks like $\mathbb{R}^d$). A *Riemannian manifold* $\mathcal{M}$ is a manifold along with a Riemannian metric, from which one can derive a distance function $d_{\mathcal{M}}(\mathbf{x}, \mathbf{y})$ for points $\mathbf{x}, \mathbf{y} \in \mathcal{M}$ thus making $\mathcal{M}$ a metric space (so, we can talk about closed and compact Riemannian manifolds). A *compact manifold* is a Riemannian manifold that is also being a compact metric space. Some examples of compact manifolds in $\mathbb{R}^3$ are shown in Fig. 1(c). A differentiable map $f : \mathcal{M} \to \mathcal{M}$ is *diffeomorphism* if it is bijective and its inverse f^{-1} is a differentiable map.

Graph Neural Networks. A (undirected) graph $G = (V, E)$ is a pair of a nodes set V and a edges set E. Let $\mathcal{N}(u) = \{v \mid \{u, v\} \in E\}$ and $\widehat{\mathcal{N}}(u) = \mathcal{N}(u) \cup \{u\}$, $\mathbf{D} = diag(d_1, \ldots, d_m)$ with $d_i = \sum_{j=1}^m \mathbf{A}_{i,j}$, where $\mathbf{A}_{i,j}$ denotes the elements of the adjacency matrix $\mathbf{A}$, as well as the augmented matrices $\widetilde{\mathbf{A}} = \mathbf{A} + \mathbf{I}$ and $\widetilde{\mathbf{D}} = \mathbf{D} + \mathbf{I}$, for $\mathbf{I} = diag(1, \ldots, 1)$. We denote the collection of all (undirected) graphs as $\mathcal{G}$.

Given a graph G and an n-dimensional node embedding $\mathbf{H}^{(0)}$ of G, A Graph Neural Network (GNN) $\mathcal{A}$ updates the embedding for ℓ layers as follows:

$$\mathbf{H}^{(k)} = Tran^{(k)}\left(Agg(\mathbf{H}^{(k-1)})\right); \tag{1}$$

The layer of vanilla-GCN [21] is defined by $Agg(\mathbf{H}) = \tilde{\mathbf{D}}^{-\frac{1}{2}}\tilde{\mathbf{A}}\tilde{\mathbf{D}}^{-\frac{1}{2}} \cdot \mathbf{H}$, $Tran^{(k)}(\mathbf{H}) = \sigma(\mathbf{H} \cdot \mathbf{W}^{(k)})$, where σ is activation function and $\mathbf{W}^{(k)} \in \mathbb{R}^{n \times n}$. GNN $\mathcal{A}$ is over manifold $\mathcal{M}$ in $\mathbb{R}^n$ if its aggregation and transformation functions are over $\mathcal{M}$.

In this work, we consider arbitrary aggregation functions that are universally defined over the embedding of all neighborhoods $\tilde{\mathcal{N}}(u_i)$ in any graph $G \in \mathcal{G}$. Therefore, all aggregate-transform GNNs with specific aggregations are included, such as GCN [21] and GAT [36], as well as SGC [37].

4   Contracted Aggregation Problem

This section is organized as follows. First, in Sect. 4.1, we identify a special property, called *contracted*, of aggregation functions. We show that aggregation functions in GCNs, GATs, and SGCs are contracted or mathematically equivalent. Then, in Sect. 4.2, we demonstrate that contracted aggregations cause over-smoothing by means of Theorem 1. Inspired by this, in Sect. 4.3, we develop a general approach for constructing non-contracted aggregations for alleviating over-smoothing using aggregation over compact Riemannian manifolds (Theorem 2). Finally, in Sect. 4.4, we motivate and introduce our aggregation function defined over a specific kinds of compact Riemannian manifolds denoted by $\mathcal{M}_{\mathbf{U}}$.

Next, we let $\mathcal{M} \subset \mathbb{R}^n$ be a d-dimensional closed Riemannian manifold, and let Agg be an aggregation function over $\mathcal{M}$. For each graph $G \in \mathcal{G}$ with nodes $\{u_1, \ldots, u_m\}$, we denote by $Agg^G : \mathcal{M}^m \to \mathcal{M}^m$ the *restriction of Agg on G* defined by restricting Agg over embeddings of G on $\mathcal{M}$ (i.e., $\mathbf{H} \in \mathcal{M}^m$).

4.1   Contracted Aggregation

The notion of contracted aggregation, which we introduce in this section, generalizes the usual mean (or avg) aggregation by extracting two center properties of the averaging function over Euclidean space $\mathbb{R}^n$: for all $\mathbf{x}, \mathbf{x}_1, \ldots, \mathbf{x}_h \in \mathbb{R}^n$ we have **(i)** $Mean(\mathbf{x}_1, \ldots, \mathbf{x}_h) = \mathbf{x}$ if $\mathbf{x}_1 = \cdots = \mathbf{x}_h = \mathbf{x}$; and **(ii)** $d_{\mathbb{R}^n}(\mathbf{x}, Mean(\mathbf{x}_1, \ldots, \mathbf{x}_h)) \leq \max(d_{\mathbb{R}^n}(\mathbf{x}, \mathbf{x}_1), \ldots, d_{\mathbb{R}^n}(\mathbf{x}, \mathbf{x}_h))$. Figure 2 illustrates the second property.

Definition 1 (Contracted aggregation). *For a graph G with nodes $\{u_1, \ldots, u_m\}$, Agg^G is contracted if for any $1 \leq i \leq m$, and for any embedding $\mathbf{H} \in \mathcal{M}^m$ of G with $\overline{\mathbf{H}} = Agg^G(\mathbf{H})$, the following results hold:*

1. If $\mathbf{H}_{j,:} = \mathbf{x}, \forall u_j \in \tilde{\mathcal{N}}(u_i)$, then $\overline{\mathbf{H}}_{i,:} = \mathbf{x}$.
2. For any $\mathbf{x} \in \mathcal{M}$, we have $d_{\mathcal{M}}(\mathbf{x}, \overline{\mathbf{H}}_{i,:}) \leq \max_{u_j \in \tilde{\mathcal{N}}(u_i)}\{d_{\mathcal{M}}(\mathbf{x}, \mathbf{H}_{j,:})\}$ and the equality holds if and only if Case 1 happens.

Moreover, we say Agg^G is equivalently contracted if there exists a diffeomorphism $g : \mathcal{M}^m \to \mathcal{M}^m$ s.t. $g^{-1} \circ Agg^G \circ g$ is contracted, where $\circ$ means the map composition, i.e., $f \circ g(x) = f(g(x))$. Finally, an aggregation Agg is (equivalently) contracted if Agg^G is (equivalently) contracted for any graph $G \in \mathcal{G}$.

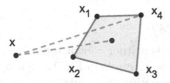

Fig. 2. Distance between $\mathbf{x}$ and the mean point (blue) is smaller than $d_{\mathbb{R}^2}(\mathbf{x}, \mathbf{x}_4)$ (Color figure online)

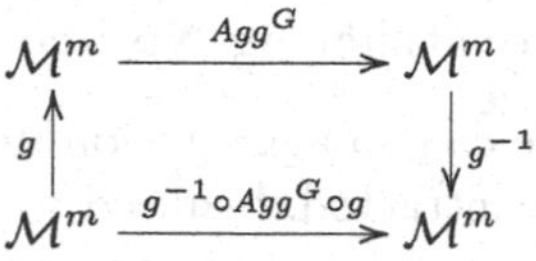

Fig. 3. Equivalently contracted aggregation

Intuitively, Agg^G is equivalently contracted if Agg^G is contracted after changing the "reference frame" by the equivalent transformation g on $\mathcal{M}^m$, which is the space of embedding matrix of G, as shown in Fig. 3. By the following result, we show that the standard aggregations used in SGC, GCN, and GAT are all contracted or equivalently contracted.

Proposition 1. *Let $\mathcal{M} = \mathbb{R}^n$ ($n > 0$), assume Agg is a aggregation function on $\mathcal{M}$ s.t. the restriction of Agg on a graph $G \in \mathcal{G}$ with node $\{u_1, \dots, u_n\}$ is defined by $Agg^G(\mathbf{H}) = \mathbf{L} \cdot \mathbf{H}$, $\forall \, \mathbf{H} \in \mathbb{R}^{m \times n}$, where $\mathbf{L} \in \mathbb{R}^{m \times m}$. Then we have the following results:*

1. *If $\mathbf{L} = (1 - \lambda)\mathbf{I} + \lambda\dot{\mathbf{D}}^{-1}\mathbf{A}$ for some $\lambda \in (0,1]$, then Agg is contracted;*
2. *If $\mathbf{L} = (1 - \lambda)\mathbf{I} + \lambda\tilde{\mathbf{D}}^{-\frac{1}{2}}\tilde{\mathbf{A}}\tilde{\mathbf{D}}^{-\frac{1}{2}}$ for some $\lambda \in (0,1]$, then Agg is equivalently contracted;*
3. *If $\mathbf{L} = Att(\mathbf{H})$ is defined by a function $Att : \mathbb{R}^{m \times n} \to \mathbb{R}^{n \times n}$ s.t. $\mathbf{L}_{i,j} > 0$ and $\sum_k \mathbf{L}_{i,k} = 1$ for any $\mathbf{H} \in \mathcal{M}^m$ and $1 \le i, j \le m$, then Agg is contracted.*

4.2 Over-Smoothing Due to Contracted Aggregations

Now, we show that contracted aggregations lead to over-smoothing by Theorem 1. This generalizes Theorem 1 of [26] to arbitrary equivalently contracted aggregations over Riemannian manifolds, from aggregations defined by *Laplace smoothing* (aggregations of Case 1 and 2 in Proposition 1) over Euclidean spaces.

Theorem 1. *Let $\{\mathbf{H}^{(k)}\}_{k \ge 0}$ be an infinite sequence of embeddings of G over $\mathcal{M}$ s.t. $\mathbf{H}^{(k+1)} = Agg^G\left(\mathbf{H}^{(k)}\right)$, for all $k \ge 0$. Assume G is connected. If Agg^G is equivalently contracted wrt a diffeomorphism $g : \mathcal{M}^m \to \mathcal{M}^m$ (i.e., $g^{-1} \circ Agg^G \circ g$ is contracted), then there exist $\mathbf{X}_0 = (\mathbf{x}_0, \dots, \mathbf{x}_0) \in \mathcal{M}^m$ s.t. $\lim_{k \to \infty} \mathbf{H}^{(k)} = g(\mathbf{X}_0)$.*

Note that Theorem 1 (proof in Appendix B.1) can be easily extended to non-connected graphs by considering their connected components. Theorem 1 shows that by repeatedly applying an equivalently contracted aggregation, the embedding points of nodes in the given graph converge to the same point (modulo the impact of the diffeomorphism g). That is, equivalently contracted aggregation leads to over-smoothing.

4.3   Constructing Non-contracted Aggregations

Next, we call an aggregation *non-contracted* if it is neither contracted nor equivalently contracted. Inspired by Theorem 1 and the example in Fig. 1, we propose to alleviate the over-smoothing problem by constructing non-contracted aggregations on manifolds. Indeed, by the following result, we show that any aggregation over a compact manifold is non-contracted. This result provides us with a general approach to constructing non-contracted aggregations.

Theorem 2. *Asuume $\mathcal{M} \subset R^n$ is a compact Riemannian manifold with dimension $d > 0$. If a graph G contains a node u_0 s.t. $|\widetilde{\mathcal{N}}(u_0)| > 2$, then for any aggregation Agg over $\mathcal{M}$, the restricted aggregation Agg^G is non-contracted.*

For instance, the unit circle is compact, thus, all aggregation over the unit circle is non-contracted. An example of non-contracted aggregation that could avoid over-smoothing problems is shown in Fig. 1. In contrast, Euclidean spaces $\mathbb{R}^n$ ($n > 0$) are not compact and thus can not benefit from the result of Theorem 2. The proof of Theorem 2 is in Appendix B.2.

4.4   Our Non-contracted Aggregation

To the best of our knowledge, there are mainly two kinds of aggregation defined over manifolds beyond Euclidean spaces. The *tangential aggregations* [4] and the *κ-Left-Matrix-Multiplication* [1]. However, the former aggregation is defined on *hyperbolic space*, which is not compact. In contrast, the second aggregation could be applied to *hyperspheres*, which are compact Riemannian manifolds. Their definition is based on the *gyromidpoint* introduced in [35] and uses *exponential* and *logarithmic* [8] maps defined on hyperspheres. In the following, we propose a simple way to construct aggregations over the specific family of compact manifolds denoted by $\mathcal{M}_{\mathbf{U}}$, which include hypersphere as a special case. Here, $\mathbf{U} \in \mathbb{R}^{n \times n}$ is a *positive-definite matrix*. That is, $\mathbf{x}\mathbf{U}\mathbf{x}^T > 0, \forall \mathbf{x} \in \mathbb{R}^n$.

Definition 2 (ACM). *Let $\mathbf{U} \in \mathbb{R}^{n \times n}$ be a positive-definite matrix. Consider the Riemannian manifold $\mathcal{M}_{\mathbf{U}} = \{\boldsymbol{x} = (x_1, \ldots, x_n) \in \mathbb{R}^n \mid \boldsymbol{x}\mathbf{U}\boldsymbol{x}^T = 1\} \subset \mathbb{R}^n$. ACM aggregation function Agg over $\mathcal{M}_{\mathbf{U}}$ is defined as below. For any graph $G \in \mathcal{G}$ with m nodes, the restricted aggregation Agg^G has the form:*

$$Agg^G(\mathbf{H}) = P_{\mathbf{U}}(\mathbf{L} \cdot \mathbf{H}), \quad \forall \, \mathbf{H} \in \mathcal{M}^m,$$

where $P_{\mathbf{U}}(\mathbf{x}) = \frac{\mathbf{x}}{\sqrt{\boldsymbol{x}\mathbf{U}\boldsymbol{x}^T}}, \quad \forall \mathbf{x} \in \mathbb{R}^n \setminus \{\boldsymbol{0}\}$ ($\boldsymbol{0}$ is the original point of $\mathbb{R}^n$), and $\mathbf{L} \in \mathbb{R}^{m \times m}$ takes one of the three forms as introduced in Proposition 1.

Since $\mathcal{M}_{\mathbf{U}}$ is compact, the above aggregation Agg is non-contracted (Theorem 2). For any graph $G \in \mathcal{G}$, Agg^G satisfies condition 1 of Definition 1 since $P_{\mathbf{U}}(k \cdot \mathbf{x}) = \mathbf{x}$ for any $k > 0, \mathbf{x} \in \mathcal{M} \subset \mathbb{R}_n$. Therefore, Agg can be regarded as a generalization of mean aggregations on compact manifold $\mathcal{M}_{\mathbf{U}}$. Moreover, $\mathcal{M}_{\mathbf{U}}$ is a hypersphere when $\mathbf{U}$ is the identity matrix. Thus, our aggregation also works for hyperspheres. The complexity of computing $P_{\mathbf{U}}(\mathbf{x})$ is $O(n^2)$.

5 Aggregation over Compact Manifolds

Here, we introduce ACM (**A**ggregation over **C**ompact **M**anifolds), which integrates our aggregation function over compact manifolds $\mathcal{M}_{\mathbf{U}}$ (Definition 2) into standard SGC, GCN and GAT. Let $G \in \mathcal{G}$ be a graph with nodes $\{u_1, \ldots, u_m\}$.

SGC. We integrate our aggregation into SGC by setting $\mathbf{L} = \tilde{\mathbf{D}}^{-\frac{1}{2}} \tilde{\mathbf{A}} \tilde{\mathbf{D}}^{-\frac{1}{2}}$. Then, the embedding is updated by $\mathbf{H}^{(k)} = P_{\mathbf{U}} \left(\tilde{\mathbf{D}}^{-\frac{1}{2}} \tilde{\mathbf{A}} \tilde{\mathbf{D}}^{-\frac{1}{2}} \cdot \mathbf{H}^{(k-1)} \right)$.

GCN. The adaptation of our aggregation on GCN is defined by Eqs. (2) below. The aggregation step of GCN is defined the same as above. Next, we show how to build a transformation function between the compact manifold $\mathcal{M}_{\mathbf{U}} \subset \mathbb{R}^n$.

Let $N_b = \{\mathbf{x} = (x_1, \ldots, x_n) \in \mathbb{R}^n \mid x_1 = b\}$ be a hyperplane in $\mathbb{R}^n$. Then, our transformation function from $\mathcal{M}_{\mathbf{U}}$ to $\mathcal{M}_{\mathbf{U}}$ takes three steps:

$$\mathcal{M}_{\mathbf{U}} \xrightarrow{\text{Step 1}} N_b \xrightarrow{\text{Step 2}} \mathbb{R}^n \xrightarrow{\text{Step 3}} \mathcal{M}_{\mathbf{U}}.$$

In Step 1, we map $\mathcal{M}_{\mathbf{U}}$ to the hyperplane N_b using *push forward* (*PF*) function as illustrated in Fig. 4; Then, at Step 2, we map N_b to $\mathbb{R}^n$ by the standard transformation functions (i.e., $\mathbf{x} \mapsto \sigma(\mathbf{x} \cdot \mathbf{W})$); In the last step, we map $\mathbb{R}^n$ to the manifold $\mathcal{M}_{\mathbf{U}}$ by the *push back* (*PB*) function as shown in Fig. 4. Finally, our GCN layer update $\mathbf{H}^{(k-1)}$ to a new embedding $\mathbf{H}^{(k)}$ as below.

$$\overline{\mathbf{H}}^{(k)} = P_{\mathbf{U}} \left(\tilde{\mathbf{D}}^{-\frac{1}{2}} \tilde{\mathbf{A}} \tilde{\mathbf{D}}^{-\frac{1}{2}} \cdot \mathbf{H}^{(k-1)} \right), \quad \mathbf{H}_u^{(k)} = PB \left(\sigma \left(PF(\overline{\mathbf{H}}^{(k)}) \cdot \mathbf{W}^{(k)} \right) \right), \tag{2}$$

where σ is an activate function, $\mathbf{W}^{(k)} \in \mathbb{R}^{n \times n}$. The functions PF and PB is a generalization of *Stereographic projection* and its inverse map from $\mathcal{M}_{\mathbf{U}}$ to N_b with center $\mathbf{x}_0 = (a_0, 0, \ldots, 0) \in \mathcal{M}_{\mathbf{U}}$, where $a_0 = \mathbf{U}_{11}^{-\frac{1}{2}}$, as in Fig. 4.

GAT Our GAT layer is defined similarly to our GCN layer but with the weighted $\mathbf{L} = Att(\mathbf{H}^{(k-1)})$, where Att is the attention function introduced in [36].

Finally, following the works of [14,26], for node classification task, we use *softmax* classifier in last layer of GNN to predict the labels of nodes in the given graph G.

6 Experiments

This section validates the following hypotheses: (1) ACM successfully mitigates over-smoothing in deep layers within GNNs, as evidenced by a reduced decline in performance; (2) ACM effectively alleviate over-smoothing in a more challenging context, specifically the scenario of missing features.

6.1   Experiments Setup

Datasets. We use four homophilic datasets (Cora, Citeseer, Pubmed [40], and CoauthorCS [33]) and five heterophilic datasets (Texas, Cornell, Wisconsin, Actor, and Chameleon [30]). In the heterophilic datasets, the assumption of neighboring feature similarity is not applicable. Our focus is on standard transductive node classification tasks across these datasets. Details can be found in Appendix C.1.

Experiment Settings. We conducted node classification experiments in two settings [41]: the standard scenario and the *missing feature scenario*. The latter aims to highlight deep Graph Neural Networks' performance by assuming that initial node embeddings in the test and validation sets are absent, replaced with zero vectors. This reflects real-world scenarios, such as in social networks, where new users may have limited connections [31], requiring more information aggregation steps for effective representation.

Implementation. We consider three standard GNN models, namely GCN [21], GAT [36], and SGC [37], and augment them with techniques alleviating over-smoothing: PairNorm [41], BatchNorm [17], DropEdge [32], DGN [43], and DeCorr [18]. Moreover, we incorporate recent SOTA models, including GPRGNN [7], GCNII [6], DAGNN [27], and APPNP [23]. To analyze the impact of different methods, we vary the number of layers (5/60/120 for SGC, and 2/15/30/45/60 for GCN and GAT). By convention, SGC typically has more layers than GCN and GAT due to its simpler design and fewer parameters, requiring more layers for best performance and to effectively capture data details. We hence evaluate deeper models. Hyperparameter settings for GNN models and optimizers follow previous works [18,41,43]. The hidden layer size is set to 16 units for GCN, GAT, and SGC. GAT has 1 attention head. Training involves a maximum of 1000 epochs with Adam optimizer [20] and early stopping. GNN weights are initialized using the Glorot algorithm [10]. Each experiment is repeated 5 times, and mean values are reported. For ACM, we employ *tanh* as the activation

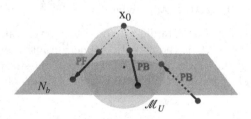

Fig. 4. Illustration of PB (orange) and PF (blue), points in $\mathcal{M}_U$ (resp. N_b) are in red (resp. black). The formal definitions can be found in the supply material. (Color figure online)

function in hidden layers. We introduce two distinct variants of the ACM, differentiated by the configuration of the parameter matrix U. (i) the first variant, still denoted as ACM, sets the matrix U to the identity matrix I; (ii) the second variant, denoted as ACM*, defines the matrix U as a positive definite diagonal matrix, where each diagonal element is a trainable parameter. Therefore, for ACM, the underlying manifold is the unit hypersphere, while ACM* offers a variable manifold (e.g., ellipsoid) that can be tailored through training to better suit the underlying data distribution.

6.2 Experiment Results

In this section, we compare the performance of ACM with previous methods that tackle the over-smoothing issue on SGC, GCN, and GAT. Refer to [18,43] for all the results of the methods other than ACM unless stated otherwise.

Mitigate the Performance Drop in Deep SGC. SGC is a proper model to show the effectiveness of ACM in relieving over-smoothing. As SGC has only one feature transformation step, the performance of SGC highly depends on the information aggregation steps.

We present the performances of SGCs with 5/60/120 layers. Table 1 summarizes the results of applying different over-smoothing alleviation approaches to SGC. "None" indicates the vanilla SGC without over-smoothing alleviation. As shown in Table 1, ACM outperforms the other over-smoothing alleviation methods in most of the cases. In particular, on the Citeseer dataset with 120 layers, the improvements over None, BatchNorm, PairNorm, and DGN achieved by ACM are by a margin of 57.5%, 19%, 19.8%, and 14.3%, respectively.

To show the detailed performance differences on each layer, we illustrate in Fig. 5(a,b) the test accuracy of SGC trained with different methods with layers from 1 to 120 on datasets Cora and CoauthorCS. We can see that ACM behaved the best in slowing down the performance drop on Cora and comparably with the best system DGN on CoauthorCS. The same conclusion holds for other datasets as on Cora (see Appendix D.1).

Table 1. Node classification accuracy (%) on SGC. The two best performing methods are highlighted in **blue** (First), **violet** (Second).

Datasets	Cora			Citeseer			Pubmed			CoauthorCS		
#Layers	5	60	120	5	60	120	5	60	120	5	60	120
None	75.8	29.4	25.1	**69.6**	**66.3**	9.4	71.5	34.2	18.0	89.8	10.2	5.8
BatchNorm	76.3	72.1	51.2	58.8	46.9	47.9	76.5	75.2	70.4	88.7	59.7	30.5
PairNorm	75.4	71.7	65.5	64.8	46.7	47.1	75.8	77.1	70.4	86.0	76.4	52.6
DGN	**77.9**	**77.8**	**73.7**	**69.5**	53.1	**52.6**	76.8	**77.4**	**76.9**	**90.2**	**81.3**	**60.8**
ACM	**78.5**	**80.0**	**77.0**	65.0	**67.4**	**66.9**	**76.9**	**78.3**	**78.6**	**90.9**	**80.1**	**57.4**

Mitigation of the Performance Drop in Deeper GCN, GAT. We integrated our ACM method into GCN and GAT and measured the performance of different GNNs with 30/45/60 layers in Table 2. The full table is in D.2. The results of None, BatchNorm, PairNorm, DGN at 45/60 layers were obtained by running the source code supplied by [43]. The results of Dropedge and Decorr at 45/60 layers were collected by running the source code from [18].

From Table 2, we can see that ACM can greatly improve the performance of deeper GNNs on these datasets. In particular, given the same number of layers, ACM consistently achieves the best performance for almost all cases, dramatically outperforms recent baselines on deep models (30/45/60 layers), and keeps the comparable performance in the lower-layers model. For instance, on the Cora dataset, at 60 layers of GCN setting, ACM improves the classification accuracy of None, BatchNorm, PairNorm, Dropedge, DGN and DeCorr by 56.5% 11.1%, 22.5%, 56.5%, 16.9%, 56.5%, respectively.

Table 2. Node classification accuracy (%) on GCN, GAT.

Datasets		Cora			Citeseer			Pubmed			CoauthorCS		
Methods	#Layers	L30	L45	L60	L30	L45	L60	L30	L45	L60	L30	L45	L60
GCN	None	13.1	13	13.0	9.4	7.7	7.7	18.0	18	18.0	3.3	3.3	3.3
	BN	67.2	60.2	58.4	47.9	38.7	36.5	70.4	72.9	67.1	84.7	80.1	79.1
	PN	64.3	54.5	47.0	47.1	43.1	37.1	70.4	63.4	60.5	64.5	70	66.5
	DropEdge	45.4	13	13.0	31.6	7.7	7.7	62.1	18	18.0	31.9	3.3	3.3
	DGN	73.2	67.8	52.6	52.6	45.8	40.5	**76.9**	73.4	72.8	84.4	83.7	**82.1**
	DeCorr	**73.4**	38.9	13.0	**67.3**	37.1	7.7	**77.3**	32.5	13.0	84.5	29	3.3
	ACM	**73.5**	**71.6**	69.5	55.1	**56.5**	**53.5**	75.7	**74.3**	74.4	**85.5**	**84.6**	82.4
	ACM*	72.4	**71.6**	**70.3**	56.5	**57.0**	53.4	74.4	**76.3**	**74.8**	84.3	**84.7**	70.7
GAT	None	13.0	13	13.0	7.7	7.7	7.7	18.0	18	18.0	3.3	3.3	3.3
	BN	25.0	21.6	16.2	21.4	21.1	18.1	46.6	45.3	29.4	16.7	4.2	2.6
	PN	30.2	28.8	19.3	33.3	30.6	27.3	58.2	58.8	58.1	48.1	30.4	26.6
	DropEdge	51.0	13	13.0	36.1	7.7	7.7	64.7	18	18.0	52.1	3.3	3.3
	DGN	51.3	44.2	38.0	45.6	32.8	27.5	73.3	53.7	60.1	75.5	20.9	44.8
	DeCorr	54.3	18.3	13.0	46.9	18.9	7.7	74.1	48.9	18.0	77.3	19.2	3.3
	ACM	**67.4**	53.5	48.5	49.5	38.8	38.4	75.4	72.0	68.4	84.8	79.8	74.2
	ACM*	**71.4**	**53.6**	**61.3**	**56.2**	48.3	**47.2**	**76.7**	**75.0**	68.5	**85.0**	**82.6**	**76.3**

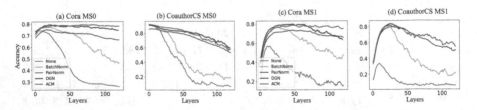

Fig. 5. Test accuracy of SGC with different methods and layers on Cora and CoauthorCS datasets under miss rate 0 (MS0) and miss rate 1 (MS1) scenario.

Table 3. Test accuracy (%) on missing feature setting.

Model	SGC				GCN				GAT			
Methods	Cora	Citeseer	Pubmed	CCS	Cora	Citeseer	Pubmed	CCS	Cora	Citeseer	Pubmed	CCS
None	63.4(5)	51.2(40)	63.7(5)	71(5)	57.3(3)	44(6)	36.4(4)	67.3(3)	50.1(2)	40.8(4)	38.5(4)	63.7(3)
BN	78.5(20)	50.4(20)	72.3(50)	84.4(20)	71.8(20)	45.1(25)	70.4(30)	82.7(30)	72.7(5)	48.7(5)	60.7(4)	80.5(6)
PN	73.4(50)	58(120)	75.2(30)	80.1(10)	65.6(20)	43.6(25)	63.1(30)	63.5(4)	68.8(8)	50.3(6)	63.2(20)	66.6(3)
DGN	80.2(5)	58.2(90)	76.2(90)	85.8(20)	67(6)	44.2(8)	69.3(6)	68.6(4)	67.2(6)	48.2(6)	67.2(6)	75.1(4)
ACM	80.7(90)	59.5(109)	76.3(91)	86.5(26)	76.3(20)	50.2(30)	72(30)	83.7(25)	75.8(8)	54.5(5)	72.3(20)	83.6(15)
ACM*	-	-	-	-	73.8(20)	49.1(30)	73.3(15)	84.3(20)	72.8(15)	46.5(6)	72.4(15)	83.7(15)

Table 4. Node classification accuracy (%) for eight datasets

Methods	Cora	Citeseer	Pubmed	Texas	Cornell	Wisconsin	Actor	Chameleon
GCN	82.3	71.7	78.6	67.2	57.4	56.0	26.7	43.3
GAT	81.2	69.5	77.4	71.6	63.6	64.1	27.8	48.1
DAGNN	84.4	**73.3**	**80.3**	78.6	68.1	71.3	32.4	47.9
APPNP	83.3	71.8	80.1	73.5	54.3	65.4	34.5	**54.3**
GPRGNN	83.0	71.3	71.5	**78.7**	76.7	71.2	37.1	48.8
GCNII	**84.5**	71.2	79.1	67.2	**80.8**	**72.1**	**33.4**	51.2
GPR+ACM	84.7	71.9	80.4	80.3	87.3	88.7	38.8	50.8

Enabling Deeper GNNs Under Missing Feature Setting. Table 1 and 2 show that deeper GNNs often perform worse than shallower ones. The "missing feature setting" experiments (see Sect. 6.1) executed the node classification task in more complex scenario. Such complex setting requires massive layers for GNNs to achieve good performance, for which ACM shows particular benefits since over-smoothing issue becomes severer with increasing layers.

In Table 3, Acc is the best test accuracy of a model with the optimal layer number #K. We observe that ACM outperforms the other over-smoothing alleviation methods in almost all cases. The average accuracy improvements achieved by ACM over None, BatchNorm, PairNorm, and DGN are 19.62%, 5.4%, 9.3%, and 1.18%, respectively. Moreover, the best performances were achieved with large layers. For instance, on the Cora dataset, ACM behaved the best with 90, 34, and 11 layers for SGC, GCN, and GAT, respectively, evidently higher than those widely-used shallow vanilla models, i.e., with 2 or 3 layers.

To better compare different methods with various layers, we run the experiments for all the layers from 1 to 120 on SGC. The result is illustrated by Fig. 5(c,d). We can see that ACM achieved best performance on deep SGCs on Cora and Citeseer. The same conclusion holds for other datasets (see Appendix D.1).

Combining with Other Deep GNN Methods. To ascertain whether ACM can function as a supplementary technique for alleviating over-smoothing, we choose to integrate it with one of the strongest baselines, GPRGNN. Our experiments on eight benchmark datasets, with the average accuracy reported for 10 random seeds in Table 4, shows that ACM consistently enhances GRPGNN across all datasets. For example, ACM enhance the GPRGNN by 10.6% on Cornell and 17.5% on Wisconsin, respectively. This supports the idea that ACM can enhance model performance.

7   Conclusion

This paper proposed a general theory for explaining over-smoothing in Riemannian manifolds, which attributes the over-smoothing to the contracted aggregation problem. Inspired by this theoretical result, we proposed a general framework ACM—by constructing non-contracted aggregation functions on compact Riemann manifold—to effectively mitigate the over-smoothing and encourage deeper GNNs to learn distinguishable node embeddings. The effectiveness of ACM has been demonstrated through the extensive experimental results.

In the future, we aim to extend our theoretical findings to encompass transformation functions. Additionally, we intend to employ the ACM framework in tasks requring deeper GNNs, such as multi-hop question answering.

References

1. Bachmann, G., Bécigneul, G., Ganea, O.: Constant curvature graph convolutional networks. In: ICML, pp. 486–496. PMLR (2020)
2. Balazevic, I., Allen, C., Hospedales, T.M.: Multi-relational poincaré graph embeddings. In: NeurIPS, pp. 4465–4475 (2019)
3. Cai, C., Wang, Y.: A note on over-smoothing for graph neural networks. CoRR abs/ arXiv: 2006.13318 (2020)
4. Chami, I., et al: Hyperbolic graph convolutional neural networks. In: NeurIPS, pp. 4869–4880 (2019)
5. Chen, D., Lin, Y., et al.: Measuring and relieving the over-smoothing problem for graph neural networks from the topological view. CoRR abs/ arXiv: 1909.03211 (2019)
6. Chen, M., Wei, Z., Huang, Z., Ding, B., Li, Y.: Simple and deep graph convolutional networks. In: ICML, pp. 1725–1735. PMLR (2020)
7. Chien, E., Peng, J., Li, P., Milenkovic, O.: Adaptive universal generalized pagerank graph neural network. In: ICLR (2021)
8. Ganea, O., Bécigneul, G., Hofmann, T.: Hyperbolic neural networks. In: NeurIPS, pp. 5350–5360 (2018)
9. Gao, H., Wang, Z., Ji, S.: Large-scale learnable graph convolutional networks. In: KDD, pp. 1416–1424. ACM (2018)
10. Glorot, X., Bengio, Y.: Understanding the difficulty of training deep feedforward neural networks. In: AISTATS, pp. 249–256 (2010)
11. Gülçehre, Ç., et al.: Hyperbolic attention networks. In: ICLR (2019)
12. Hamilton, W.L.: Graph Representation Learning. Synthesis Lect. Artifi. Intell. Mach. Learn. (2020)
13. Hamilton, W.L., et al.: Inductive representation learning on large graphs. In: NIPS, pp. 1024–1034 (2017)

14. He, K., et al.: Deep residual learning for image recognition. In: CVPR, pp. 770–778. IEEE Computer Society (2016)
15. Hou, Y., Zhang, J., et al.: Measuring and improving the use of graph information in graph neural networks. In: ICLR (2020)
16. Huang, W., et al.: Tackling over-smoothing for general graph convolutional networks. CoRR (2020)
17. Ioffe, S., Szegedy, C.: Batch normalization: accelerating deep network training by reducing internal covariate shift. In: ICML, pp. 448–456 (2015)
18. Jin, W., Et al.: Feature overcorrelation in deep graph neural networks: a new perspective. In: KDD, pp. 709–719. ACM (2022)
19. Khrulkov, V., Et al.: Hyperbolic image embeddings. In: CVPR, pp. 6417–6427. Computer Vision Foundation/IEEE (2020)
20. Kingma, D.P., Ba, J.: Adam: a method for stochastic optimization. In: ICLR (2015)
21. Kipf, T.N., Welling, M.: Semi-supervised classification with graph convolutional networks. CoRR abs/ arxiv: 1609.02907 (2016)
22. Klicpera, J., Et al.: Predict then propagate: graph neural networks meet personalized pagerank. In: ICLR (2019)
23. . Klicpera, J., et al.: Predict then propagate: graph neural networks meet personalized pagerank. In: ICLR (2019)
24. Lee, J.: Introduction to Smooth Manifolds. Graduate Texts in Mathematics
25. Li, G., Müller, M., et al.: Deepgcns: can gcns go as deep as cnns? In: ICCV, pp. 9266–9275. IEEE (2019)
26. Li, Q., Han, Z., Wu, X.M.: Deeper insights into graph convolutional networks for semi-supervised learning. In: AAAI, pp. 3538–3545 (2018)
27. Liu, M., Gao, H., Ji, S.: Towards deeper graph neural networks. In: KDD, pp. 338–348. ACM (2020)
28. Mendelson, B.: Introduction to topology (1990)
29. Oono, K., Suzuki, T.: Graph neural networks exponentially lose expressive power for node classification. In: ICLR (2020)
30. Pei, H., et al.: Geom-gcn: Geometric graph convolutional networks. In: ICLR (2020)
31. Rashid, A.M., Karypis, G., et al.: Learning preferences of new users in recommender systems: an information theoretic approach. SIGKDD Explor., 90–100 (2008)
32. Rong, Y., Huang, W., Xu, T., Huang, J.: Dropedge: towards deep graph convolutional networks on node classification. In: ICLR (2020)
33. Shchur, O., Mumme, M., Bojchevski, A., Günnemann, S.: Pitfalls of graph neural network evaluation. CoRR abs/ arXiv: 1811.05868 (2018)
34. Srivastava, N., et al.: Dropout: a simple way to prevent neural networks from overfitting. J. Mach. Learn. Res. **15**(1), 1929–1958 (2014)
35. Ungar, A.A.: Barycentric calculus in Euclidean and hyperbolic geometry: a comparative introduction (2010)
36. Velickovic, P., Cucurull, G., Casanova, A., Romero, A., Liò, P., Bengio, Y.: Graph attention networks. CoRR abs/ arXiv: 1710.10903 (2017)
37. Wu, F., et al.: Simplifying graph convolutional networks. In: ICML, vol. 97, pp. 6861–6871. PMLR (2019)
38. Xu, K., Hu, W., et al.: How powerful are graph neural networks? In: ICLR (2019)
39. Xu, K., Li, C., Tian, Y., et al.: Representation learning on graphs with jumping knowledge networks. In: ICML, pp. 5449–5458. PMLR (2018)
40. Yang, Z., et al.: Revisiting semi-supervised learning with graph embeddings. In: ICML. JMLR Workshop and Conference Proceedings, vol. 48, pp. 40–48 (2016)

41. Zhao, L., Akoglu, L.: Pairnorm: tackling oversmoothing in gnns. In: ICLR (2020)
42. Zhou, J., Cui, G., et al.: Graph neural networks: a review of methods and applications. AI Open **1**, 57–81 (2020)
43. Zhou, K., et al.: Towards deeper graph neural networks with differentiable group normalization. In: NeurIPS (2020)

Are Graph Embeddings the Panacea?
An Empirical Survey from the Data Fitness Perspective

Qiang Sun(✉) iD, Du Q. Huynh iD, Mark Reynolds iD, and Wei Liu(✉) iD

The University of Western Australia, 35 Stirling Hwy, Crawley, WA 6009, Australia
pascal.sun@research.uwa.edu.au,
{du.huynh,mark.reynolds,wei.liu}@uwa.edu.au

Abstract. Graph representation learning has emerged as a machine learning go-to technique, outperforming traditional tabular view of data across many domains. Current surveys on graph representation learning predominantly have an algorithmic focus with the primary goal of explaining foundational principles and comparing performances, yet the natural and practical question "Are graph embeddings the panacea?" has been so far neglected. In this paper, we propose to examine graph embedding algorithms from a data fitness perspective by offering a methodical analysis that aligns network characteristics of data with appropriate embedding algorithms. The overarching objective is to provide researchers and practitioners with comprehensive and methodical investigations, enabling them to confidently answer pivotal questions confronting node classification problems: 1) Is there a potential benefit of applying graph representation learning? 2) Is structural information alone sufficient? 3) Which embedding technique would best suit my dataset? Through 1400 experiments across 35 datasets, we have evaluated four network embedding algorithms – three popular GNN-based algorithms (GraphSage, GCN, GAE) and node2vec – over traditional classification methods, namely SVM, KNN, and Random Forest (RF). Our results indicate that the cohesiveness of the network, the representation of relation information, and the number of classes in a classification problem play significant roles in algorithm selection.

Keywords: Network Characteristics · Graph embedding

1 Introduction

Graphs are vital for representing non-IID (Independent and Identically Distributed) data, such as social, citation, and traffic networks. Graph embedding techniques effectively interpret these structures through low-dimensional representations, thus becoming indispensable in contemporary machine learning pipelines.

Existing surveys [2,3,5,12,15] on graph embedding techniques primarily examine the following key areas: theoretical foundations of embedding algorithms (including deep learning, matrix factorization, and edge reconstruction

© The Author(s), under exclusive license to Springer Nature Singapore Pte Ltd. 2024
D.-N. Yang et al. (Eds.): PAKDD 2024, LNAI 14646, pp. 405–417, 2024.
https://doi.org/10.1007/978-981-97-2253-2_32

approaches), preservation of graph properties (such as first or second proximity) in the embedding space, the types of outputs produced by these algorithms (be it node, edge, whole-graph or sub-graph embeddings), the downstream problems (e.g., classification and link prediction), and the application domains (e.g., social networks, citation networks). With the sole focus on comparing graph embedding techniques, these surveys, to some extent, overlooked an important question: *Are graph embeddings the "panacea" for network data?* Intuitively, while some network data may respond well to graph embedding techniques, others may not.

The study by Makaro et al. [12] can be considered as an exception. However, the work was on limited datasets, all for citation networks (Cora and Citeseer) and their focus was on performance comparison over different downstream tasks such as node clustering, link prediction, and node classification. To the best of our knowledge, there are no systematic investigations on embedding techniques from the dataset fitness perspective despite the urgent needs in answering the following real-world applicability questions:

1. Is there a potential benefit of applying graph representation learning?
2. Would structural information alone be sufficient?
3. Which embedding technique would best suit my dataset?

In this research, we selectively collected a diverse range of public datasets used extensively in literature [4,6,8,10,11] and propose to use network characteristics as proxies to measure dataset fitness, so as to quantify the applicability of graph representation learning algorithms. Through 1400 experiments across 35 datasets, we have evaluated four network embedding algorithms: three popular GNN-based algorithms (GraphSage [8], GCN [13], GAE [10]) and a widely-adopted random-walk based technique (node2vec [6]), over traditional classification methods, namely SVM, KNN, and Random Forest (RF).

Our systematic analysis confirmed that the effectiveness of graph embedding techniques for node classification tasks is highly dependent on the network characteristics. In general, *sparsely connected networks* that have longer average shortest paths, relative large numbers of classes and fewer edges, graph representation learning is not beneficial. On the contrary, *dense and cohesive networks* with moderate numbers of classes see significant benefits from graph representation learning. Graphs that effectively represent connection information can perform well in node classification tasks by utilizing node2vec [6] without incorporating node attribute information. Notably, node2vec offers good performance for networks that lack node features, while supervised Graph Neural Networks (GNNs) are preferred in feature-rich networks. The number of edges, diversity of class labels, and average shortest path length of networks are identified as pivotal determinants in the effectiveness of graph embeddings techniques.

The novel and core contribution of this research therefore lies in the proposed data fitness measurement framework. Based on network characteristics matching with performance profiles, the framework facilitates the quantification of how applicable a graph representation learning algorithm over a dataset is, which yields critical insights and interpretable decision trees.

2   Literature Review

Node classification in homogeneous graphs is a task that assigns labels to nodes based on the graph topology and node attributes [2]. Graph representation learning is essential for node classification, as it effectively transforms the complex structure of graphs into low-dimensional representations, thereby converting the challenge of classifying nodes in graph structures into a more tractable tabular data classification problem. Node-level graph embedding techniques for homogeneous graphs have evolved from matrix factorization to deep learning-based approaches, driven by advancements in deep learning and the increasing complexity of graphs [2,5,15]. A typical categorisation for node embedding techniques is illustrated in Fig. 1. In the context of deep learning-based graph embeddings, methods are primarily categorized into two groups: *feature-less random walk-based* and *neighbourhood feature-based* techniques. Within the *neighborhood feature-based* category, depending on the downstream task (*extrinsic tasks* such as node classification, or *intrinsic tasks* such as edge reconstruction or graph reconstruction), these methods can be further divided into three distinct subcategories [2]. Note *intrinsic* tasks are also known as *in-graph tasks*, can be handled by *unsupervised learning* techniques because no external labels are required.

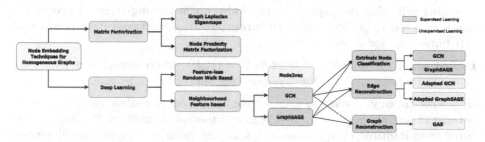

Fig. 1. Node Embedding Techniques for Homogeneous Graphs (Adapted from [2]).

Node2vec, GCN, and GraphSAGE listed in Fig. 1 are prominent deep learning models in node classification tasks. Based on random walks and inspired by SkipGram [7], node2vec [6] can effectively capture the structural context of nodes by balancing breadth-first (BFS) and depth-first searches (DFS) on the graph. *Graph Convolutional Networks* (GCNs) [11] use convolutional neural networks to generate node embeddings, aggregating features from the immediate neighbours of nodes. *Graph Sample and AggreGatE* (GraphSAGE) extends this approach by sampling a fixed number of neighbours and employing techniques like averaging, LSTM, or pooling to aggregate features [12]. *Graph Autoencoders* (GAEs) [10] use an encoder-decoder framework, with GCN or GraphSAGE being the encoding layers, to learn node embeddings that minimize the graph reconstruction loss.

If we treat each node as an independent observation, focus solely on node attributes, and ignore network topology, then the classification task becomes one that can be learned using a traditional method like Random Forest (RF), KNN, and SVM. In contrast, Node2vec primarily focuses on graph topology structure to learn node embeddings that can be used in the subsequent classification tasks. Meanwhile, GCN, GraphSAGE and GAE integrate both node attributes and structural topological information in their representation learning. GCN and GraphSAGE can directly perform node classification in a supervised manner, whereas GAE, similar to Node2vec, prioritize learning on node embeddings. GCN and GraphSAGE, on the other hand, possess the flexibility to adapt their loss functions to align with the approaches of Node2vec and GAE.

These models are ideal for our empirical study as they provide diversity for analyzing node features (attributes) and structure. We exclude matrix factorization methods due to scalability issues and the superior performance of deep learning models. We also limit our scope to node-level graph embedding techniques.

3   Network Characteristics of Datasets

Datasets with intrinsic connections, such as social networks which capture relationships between people, citation networks which map scholarly references, are typical examples of network datasets and can be described using the standard definition of graphs. There is also an increasing trend of processing non-networked data into graphs to capture relations between data points and attributes, e.g. for anomaly detection [14].

Definition 1. *A **graph** is defined as $\mathcal{G} = \langle V, E \rangle$, where a node $\nu \in V$ and an edge $e \in E$. Edge $e_{ij} \in E$ stands for an edge between node $\nu_i, \nu_j \in V$.*

In **graph theory**, diverse measures have been used to analyze structural characteristics and connectivity of networks and to capture network topological features from multiple perspectives. Let k be the random variable that describes the degrees of nodes in a graph. Then the *average node degree* ($\bar{k}$) represents the average number of edges for all the nodes. Other statistics of k that offer insights to the graph's density and degree distribution are also commonly used. The *distance* between two nodes in a graph is defined to be the number of edges in a shortest path connecting them. The *average shortest path length* (L) therefore quantifies the average node distance; in terms of information flow, this measure is important for measuring network efficiency. The *diameter* (D), which gives the longest distance between all node pairs, provides the upper bound on the reachability of nodes in a graph. The *betweenness centrality*, based on the number of shortest paths passing through a node, measures its criticality in forming sub-networks. The *average clustering coefficient* (C) measures how many nodes, on average, cluster together, while *transitivity* (T) looks at triangle connections among nodes. In scale-free networks [1], the distribution of node degrees follows a power law: $P(k) \sim k^{-\gamma}$, where the *degree exponent* γ quantifies whether the network has large hubs or not (i.e., having a few nodes with large degrees (for small γ) or many nodes with small degrees (for large γ)). The definitions of these measures are summarised in Table 1.

Table 1. Network characteristics: notations and definitions

$\|V\|$	Total number of nodes	$\|E\|$	Total number of edges
d	Dimension of node features	K	Number of classes
$\bar{k}$	Average degree of nodes ($= 2\|E\|/\|V\|$)	k_{var}	Second moment of degree distribution
k_{min}	Minimum node degree	k_{max}	Maximum node degree
L	Average shortest path between all node pairs	D	Diameter (The maximum distance between all possible pairs of nodes)
T	Transitivity (measuring likelihood of triangle formation)	C	Average clustering (quantifying the tendency of nodes to cluster together)
γ	Degree exponent of the power-law degree distribution, $P(k) \sim k^{-\gamma}$		

Table 2. Dataset Context, where each number inside the parentheses denotes the number of datasets for that given network type or dataset topic.

Network Type	Dataset topic	Description
Co-existence Network (7)	Actor (1)	Wikipedia co-occurrence of actors; Classify into categories.
	Amazon (3)	Product co-purchases, bag-of-words from reviews; Product/review categories.
	Coauthor (CS, Physics) (2)	Authorship network, paper keywords; Study fields classification.
	Tolokers (1)	Toloka worker data, shared tasks; Banned worker prediction.
Citation Network (15)	Cora, etc. (12)	Publications, citations, bag-of-words; Publication classes.
	WebKG (3)	University web pages, hyperlinks, bag-of-words; Page categories.
Social Network (8)	Twitch (6)	Streamers, mutual follows, game embeddings; Language use classification.
	BlogCatalog (1)	Blog platform users and friendships, TF-IDF from blogs; User categories.
	Github (1)	Developer relationships, locations, repos; Web/ML developer classification.
Knowledge Graph (2)	Wiki (1)	Wikidata graph, item relations, one-hot vectors; Item categories.
	Roman Empire (1)	Wikipedia articles, word connections, embeddings; Syntactic roles classification.
Social Knowledge Network (2)	Questions (1)	Yandex Q dataset, answered questions, user profile embeddings; Active user prediction.
	Flickr (1)	Users, images, metadata, text annotations; Tag-based classification.
Grid (1)	Minesweeper (1)	Grid cells, adjacent mines; Mine presence prediction.

3.1 Dataset Selection

The datasets commonly used in the four graph embedding techniques discussed in Sect. 2 are mostly citation networks, e.g. `Cora`, `CiteSeer`, and `PubMed`, as well as knowledge and social networks including `BlogCatalog`, `Wikipedia`, `NELL`, and `Reddit` [6,8,10,11], which are lack diversity. In contrast, our study adopts a **data first** perspective, examining a broader spectrum of public datasets.

After a comprehensive review of the literature, we compiled 35 datasets for node classification tasks, excluding larger ones like `Reddit` and focusing on homogeneous networks for ease of analysis. These datasets encompass various types and sizes of networks. These network types (see Table 2) include: (i) Co-existence networks (such as Actor and `Amazon`), where nodes are connected by shared contexts), (ii) Citation networks (including `CiteSeer` where connections are formed through citations, and `WebKG` which associates nodes via hyperlink references), (iii) Social networks e.g.`BlogCatalog`, `Github`, and `Twitch`, based on social interactions such as friendships or followings), (iv) Knowledge Graph networks (represented by `Roman Empire` and `Wiki` datasets, illustrating structured knowledge in graph form), and (v) a Grid network (`Minesweeper`, where nodes are connected based on grid adjacency).

3.2 Summary

The network characteristics for the 35 datasets are shown in Table 3. We can see that the characteristics span a wide range of values: $\|V\|$ (183 to 48,921), $\|E\|$ (298 to 519,000), d (1 to 12,047), K (2 to 70), $\bar{k}$ (2.52 to 88.28), and k_{var} (9.50 to

Table 3. Dataset Statistics, where the figures in bold under each column denote the smallest or largest number in that column (see Table 2 for the colour codes).

Dataset	$|V|$	$|E|$	d	K	$\bar{k}$	k_{var}	k_{min}	k_{max}	L	C	T	D	γ
Actor	7,600	30,019	932	5	7.90	400.56	1	1,304	4.11	0.05	0.04	12	2.81
AZ_COMPUTERS	13,752	245,861	767	10	35.76	6,221.40	0	2,992	3.38	0.34	0.10	10	2.83
AZ_PHOTO	7,650	119,081	745	8	31.13	3,204.10	0	1,434	4.05	**0.40**	0.17	11	2.92
AGD_BlogCatalog	5,196	171,743	8,189	6	66.11	7,376.53	5	769	2.51	0.12	0.08	4	4.06
AGD_CiteSeer	3,312	4,715	3,703	6	2.85	19.81	1	100	**1.08**	0.07	0.03	28	3.31
AGD_Cora	2,708	5,429	1,433	7	4.00	44.23	1	169	1.17	0.13	0.02	19	3.05
AGD_Flickr	7,575	239,738	**12,047**	9	63.30	21,303.96	1	1,881	2.41	0.33	0.10	4	2.76
AGD_Pubmed	19,717	44,338	500	3	4.50	75.51	1	171	6.34	0.03	0.01	18	4.20
AGD_Wiki	2,405	16,523	4,973	17	13.74	499.01	1	281	3.65	0.32	**0.44**	9	3.82
CF_CiteSeer	4,230	5,337	602	6	**2.52**	20.43	1	85	1.35	0.11	0.08	26	2.83
CF_Cora	19,793	63,421	8,710	70	6.41	118.33	1	297	1.16	0.26	0.13	23	3.38
CF_Cora_ML	19,793	63,421	8,710	70	6.41	118.33	1	297	1.16	0.26	0.13	23	3.38
CF_DBLP	17,716	52,867	1,639	4	5.96	123.03	1	339	1.16	0.13	0.10	34	3.23
CF_PubMed	19,717	44,324	500	3	4.49	75.43	1	171	6.34	0.06	0.05	18	4.17
CiteSeer	3,327	4,552	3,703	6	2.73	18.92	0	99	1.08	0.14	0.13	28	2.63
Coauthor_CS	18,333	81,894	6,805	15	8.93	162.75	1	136	5.42	0.34	0.18	24	5.18
Coauthor_Physics	34,493	247,962	8,415	5	14.38	449.23	1	382	5.16	0.37	0.18	17	4.88
Cora	2,708	5,278	1,433	7	3.89	42.53	1	168	1.17	0.24	0.09	19	2.98
Cora_Full	19,793	63,421	8,710	70	6.41	118.33	1	297	1.16	0.26	0.13	23	3.38
GitHub	37,700	289,003	128	2	15.33	6,761.61	1	**9,458**	3.25	0.17	0.01	11	2.54
HGD_AZ_Ratings	24,492	93,050	300	5	7.60	93.39	5	132	16.24	0.29	0.15	46	3.60
HGD_Minesweeper	10,000	39,402	**7**	2	7.88	62.44	3	**8**	46.67	0.22	0.33	99	2.79
HGD_Questions	**48,921**	153,540	301	2	6.28	774.50	1	1539	4.29	**0.02**	0.01	16	**1.83**
HGD_Roman_empire	22,662	32,927	300	18	2.91	**9.50**	2	14	**2331.56**	0.19	0.28	**6,824**	**6.34**
HGD_Tolokers	11,758	**519,000**	10	2	**88.28**	**33,978.74**	1	2,138	2.78	0.27	0.11	11	3.08
PubMed	19,717	44,324	500	3	4.50	75.43	1	171	6.34	0.06	0.05	18	4.18
TWITCH_DE	9,498	162,636	128	2	34.25	8,363.44	3	4,261	2.72	0.20	0.05	7	2.58
TWITCH_EN	7,126	42,450	128	2	11.91	634.28	3	722	3.68	0.13	0.04	10	2.79
TWITCH_ES	4,648	64,030	128	2	27.55	3,198.91	3	1,024	2.88	0.22	0.08	9	2.58
TWITCH_FR	6,551	119,217	128	2	36.40	7,328.28	2	2,042	2.68	0.22	0.05	7	2.61
TWITCH_PT	1,912	33,211	128	2	34.74	4,324.69	3	769	2.53	0.32	0.13	7	2.53
TWITCH_RU	4,385	41,689	128	2	19.01	2,098.57	3	1,231	3.02	0.17	0.05	9	2.58
WEBKB_Cornell	**183**	**298**	1,703	5	3.26	60.52	1	94	3.20	0.10	0.01	8	3.09
WEBKB_Texas	**183**	325	1,703	5	3.55	75.22	1	104	3.03	0.11	**0.01**	8	2.62
WEBKB_Wisconsin	251	515	1,703	5	4.10	81.85	1	122	3.26	0.11	0.01	8	2.50

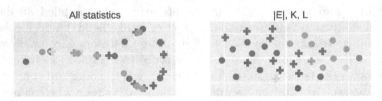

All statistics |E|, K, L

Fig. 2. t-SNE plots of network types (see Table 2) with 13 graph metrics and specific metrics (—E—, K, L). The "+" symbols indicate datasets that do not benefit from graph representation learning.

33,978.74). Among these, `HGD_Roman_Empire` has the lowest k_{var} but highest D, L, and γ. In contrast, `HGD_Questions` has the most number of nodes, yet fewest classes, smallest C, and lowest γ. The statistics of each dataset, represented as vectors and visualized through a t-SNE plot (Fig. 2), highlight the relationships and cluster groups of the network characteristics. Both co-existence networks and social networks have more similar network characteristics and separated from the clusters of Knowledge Graph and Citation Networks.

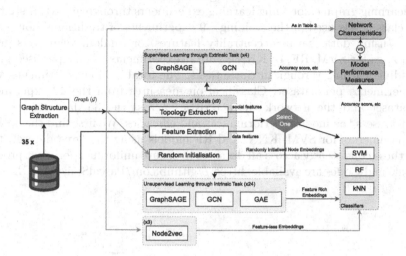

Fig. 3. Experiment Architecture

4   Experiments, Results and Discussions

4.1   Experiments

To answer the dataset fitness questions raised in the Abstract and Sect. 1, we designed a comprehensive set of experiments. Figure 3 shows that each dataset undergoes a Graph Structure Extraction phase, wherein the network characteristics (see Table 1) are computed. The Topology Extraction phase then extracts the *social features* [betweenness, degree] centrality for each node in the graph, while the Feature Extraction phase retrieves the d-dimensional attribute-based data features from the dataset. In addition, we initialize nodes within the graph using random vectors and labels, feed them into SVM, RF and KNN to establish the baseline for the classification task (**Random Guess**). The graph $\mathcal{G}$ that contains the node and edge information is combined with either the *data features* (node attributes) or the *social features*, [betweenness, degree]. These combined inputs are then passed to either a supervised graph embedding technique (**Supervised GCN/GraphSAGE**) or an unsupervised (**Node2vec; Unsupervised GCN/GraphSAGE/GAE**) graph embedding technique. The former techniques learn the embeddings specifically from the extrinsic task of node classifications, where the class labels are provided by the datasets. For the latter techniques, the learned node embeddings are used directly as inputs for the traditional classification models, namely, SVM, RF, and KNN. The data features (**Feature Only**) and social features [betweenness, degree] are also independently fed into these classification models.

Overall, our experiments are categorized into three main types: 4 supervised learning graph embedding learning experiments through an **extrinsic task** (with class labels used in the training), 9 experiments of traditional non-neural models (using data features, centrality features, or random vectors as graph embeddings for SVM, RF, or KNN), and 27 experiments of unsupervised graph embedding learning through intrinsic tasks. The total is 1400 experiments, with 40 experiments per dataset. Classification measures from the 40 experiments per dataset and the network characteristics across the 35 datasets are analyzed to assess the models' performances and efficacies. We first find the optimal hyperparameters for SVM, KNN, and RF models using the Cora dataset, then apply these consistently across all datasets to ensure uniform experiment process. Datasets and Codes are available: https://github.com/PascalSun/PAKDD-2024.

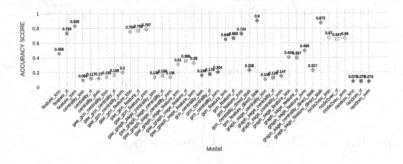

Fig. 4. The performance vector on accuracy for the Coauthor_CS dataset.

4.2   Results and Discussions

For each dataset, the models' performance measures are compiled into a 40-dimensional vector, giving $\mathbf{v}_{\mathrm{per}} \triangleq (\mathrm{per}_1, \cdots, \mathrm{per}_{40})$, where all the elements in $\mathbf{v}_{\mathrm{per}}$ are either F1 scores or accuracies. For example, Fig. 4 shows the performance vector for accuracy for the Coauthor_CS Dataset. In addition, the 13-dimensional vector, $\mathbf{v}_{\mathrm{net}} \triangleq (\mathrm{NC}_1, \cdots, \mathrm{NC}_{13})$, that captures the network characteristics (see Table 1) for each dataset is also extracted. Analyzing these vectors across all datasets provides insights into the interplay between network characteristics and the efficacy of graph embedding techniques.

Q1: Do all datasets benefit from graph structure representation learning?

To determine the benefits of graph structure in representation learning, we focus on comparing the components of $\mathbf{v}_{\mathrm{per}}$ for each dataset. We compared **Feature Only** models with those integrating node features and graph structure, noting that a superior performance by the latter would indicate a benefit from graph structures. Our study showed that 21 of 35 datasets, including Minesweeper,

Questions, Roman Empire, Tolokers, Twitch, WebKG, and Github, as well as most citation networks, benefited from this approach, with supervised learning models often outperforming others. In contrast, datasets such as Actor, Amazon Ratings, Wiki, BlogCatalog, Flickr, and PubMed, found the **Feature Only** models to be more effective. Specifically, the PubMed dataset, tasked with classifying diabetes-related literature into one of three classes, displayed the largest average shortest path (L) in the citation networks (a very sparse network), which may explain its unique status as an exception. In summary, our findings reveal that graph structure representation learning is not always advantageous; for certain datasets,"Feature Only" models suffice and outperform.

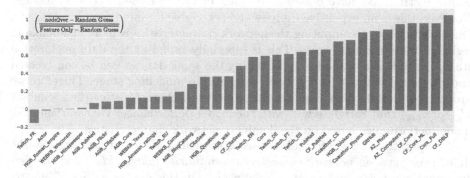

Fig. 5. Comparison of classification accuracies of the **node2vec** models versus the **Feature Only** models for all the datasets (the bar symbol represents the average accuracy of the model over the SVM, RF, and KNN classifiers).

Q2: Is structural information alone sufficient?

If we compare models that use only structural information (the graph $\mathcal{G}$ and/or [betweenness, degree], **Structure Only**) with those that integrate data features and structural information, the latter consistently outperforms the former, signifying the need for both feature and structural data. This finding aligns with our intuitive expectations. If we adopt a different perspective and compare **Feature Only** models with **Structure Only** models, we then found that for certain datasets, the performance of node2vec models, which solely rely on structural information, is comparable to that of feature-only models. To quantify this observation, we calculate the ratio of accuracy improvement of node2vec models over **Random Guess** compared to feature-only models against the same baseline, using the average accuracy from three classical classifiers (see Fig. 5). For Cora series, Amazon Photo, Amazon Computers, GitHub, the performances of **Feature Only** and node2vec models are comparable and both significantly outperform the **Random Guess**. However, for the datasets TWITCH_FR, Actor, Roman Empire, WebKG_Wisconsin and Minesweeper, **Structure Only** models show no improvement over **Random Guess**. While in the remaining datasets, **node2vec** models exhibit reasonable performance gains over **Random Guess**, but not to the extent of **Feature Only** models. Interestingly, `Amazon Ratings`, `Amazon Photos`, and `Amazon Computers`, all from the same co-existence network category, show varied diameters of 46, 10, and 11 respectively. The latter two networks, having shorter diameters, achieve notably better results for **Structure Only** models. Conclusively, our analysis shows that structural information alone is not sufficient for optimal model performance. However, the random walk-based node2vec approach can achieve comparable results to **Feature Only** models for dense networks. So, node2vec would be useful for dense but feature-less networks.

Q3: Which model(s) suit my dataset?

A classical approach to answer this question is to run various models on a given dataset and pick the best performing model according to a set of performance metrics. We have done this for each of the 35 datasets over 40 models. However, this is an expensive exercise and not viable for large complex datasets. On the other hand, computing the network characteristics is more attainable in terms of computational costs. This is especially useful at the data exploration stage, multiple graph models representing the same dataset can be constructed and compared without having to go through the modelling stage. Therefore we propose to find the best combination of network characteristics (i.e., a subset $\mathbf{v}'_{net}$ of $\mathbf{v}_{net}$) that correlate well with the model performance. We can then infer model performance based on the dominate network characteristics.

Computationally, we formulate this as searching for $\mathbf{v}'_{net}$ that minimise the differences between *network characteristic similarity matrix* M_{net} and the *performance similarity matrix* M_{per}, where M_{net} in this case is a 35×35 matrix calculated based on pairwise cosine similarity of the dataset network characteristics vectors $\{v_{net}\}$, and M_{per} is a pairwise cross dataset 35×35 similarity matrix based on 35 performance vectors, $\{v_{per}\}$, one for each of the 35 datasets. Figure 6 shows the heatmaps (M_{per}) of pairwise similarities across all 35 datasets on the 40-dimensional vectors of F1 score, accuracy, and one of the network characteristics similarity matrices (M_{net}) for the $\mathbf{v}'_{net} \triangleq (|E|, K, L)$ subset.

To identify a subset of 13 network characteristics, represented by the vector $\mathbf{v}'_{net}$ with a dimension n ($0 < n \leq 13$), that generates a 35×35 similarity matrix closely resembling the performance similarity matrix, we first need to measure the closeness between two heatmap matrices. Instead of going for numerical distance and similarity measures, the heatmaps' visual structural similarity is more important, we therefore choose to utilize the Structural Similarity Index Measure (SSIM) [9], a metric developed for image quality assessment. SSIM, by evaluating structural information of the heatmap as images, luminance, and contrast variations, is apt for this analysis as it prioritizes structural similarities over individual value discrepancies. An SSIM value close to 1 indicates higher similarity.

We conducted a search of all combinations of the 13 network characteristics, and found that the following combinations emerged as the most prominent: $(|E|, K, L)$ in Fig. 6 (right), with an SSIM of approximately 0.77. Similarly, the combination of $(|E|, K, D)$ also yields an SSIM score of 0.77. Comparing the

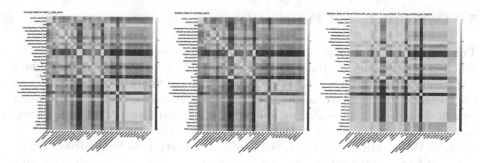

Fig. 6. Similarity matrices for: F1 score (left), Accuracy score (middle), Network Parameters ($|E|$, K, L) (right).

F1 score heatmap and the Accuracy heatmap, we get an SSIM of 0.95. Visually, the heatmaps in Fig. 6 (left) and (middle) do closely resemble each other.

To deepen our understanding of the previous two questions within the framework of ($|E|, K, L$), we classify the datasets into two categories respectively based on the observations in **Q1** and **Q2**. For **Q2**, datasets are categorized based on the threshold value of 0.5 in Fig. 5. This division distinguishes between datasets that achieve relatively good performance solely with structural information and those that do not. We then utilize the decision tree classifier, with ($|E|, K, L$) as inputs, for binary classification. All datasets were used in the training process. The findings from this analysis are also depicted in Fig. 7. For Q1, the accuracy score is 0.942, with feature importances 0.13, 0.45, and 0.41 for $|E|$, K, and L, respectively. We can see that K and L play a more crucial role when deciding whether a dataset benefits from graph representation learning or not. For Q2, the accuracy score is 0.88, with feature importances 0.895, 0, and 0.1 for $|E|$, K, and L. As shown in Fig. 7 (b), a graph with adequately represented connections alone can be sufficient for node classification.

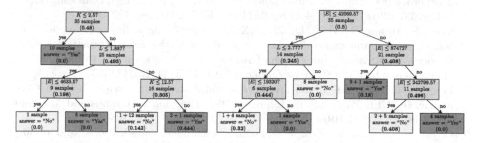

Fig. 7. Decision Trees for Q1 and Q2 based on $\mathbf{v}'_{net} = (|E|, K, L)$. The boldface numbers inside parentheses denote the Gini indices. Each leaf node is coloured in green (or pink) for the "Yes" (or "No") answer to the question. (Color figure online)

5   Conclusion and Future Work

This paper proposes to use network characteristics to quantify data fitness in graph representation learning algorithm selection. We have methodically and empirically identified a clear relationship between network characteristics and the performance of graph embedding techniques. We observe that not all datasets benefit from graph representation in node classification tasks. Additionally, even for datasets that do benefit, relying solely on structural information is insufficient, as the contribution of structural information tends to be relatively minor compared to node attribute information. In particular, sparse networks with long average shortest path lengths, large number of classes and limited number of edges are generally unsuitable for classification tasks via graph representation learning. Whereas for densely well-connected networks with rich edge information, random walk-based models (e.g. node2vec) demonstrate notable performance compared to attribute-only models. However, the most effective ones are neighbour-attribute-based supervised learning methods, which consistently outperform others by leveraging both structural and attribute information.

To proceed into the future, firstly, the dominance of citation networks urgently call for wider domains to improve **dataset diversity**. Secondly, the reliance on bag-of-words and one-hot encoding for textual attributes could benefit from semantic enriched content embeddings through Large Language Models (LLM). Last but not least, while random walk-based node2vec demonstrates impressive performance, our non-random walk-based models utilize only two layers. Further study on **deeper random-walk inspired networks** is needed.

References

1. Barabási, A.L.: Network Science by Albert-László Barabási. Cambridge University Press (2016). http://networksciencebook.com/
2. Cai, H., Zheng, V.W., Chang, K.C.C.: A comprehensive survey of graph embedding: problems, techniques, and applications. IEEE Trans. Knowl. Data Eng. **30**(9), 1616–1637 (2018) https://doi.org/10.1109/TKDE.2018.2807452
3. Chen, S., Huang, S., Yuan, D., Zhao, X.: A survey of algorithms and applications related with graph embedding. In: Proceedings of the 2020 International Conference on Cyberspace Innovation of Advanced Technologies, pp. 181–185. ACM, Guangzhou China (2020).https://doi.org/10.1145/3444370.3444568
4. Fey, M., Lenssen, J.E.: Fast graph representation learning with PyTorch Geometric. In ICLR Workshop on Representation Learning on Graphs and Manifold arXiv:1903.02428 (2019)
5. Goyal, P., Ferrara, E.: Graph embedding techniques, applications, and performance: a survey. Knowl.-Based Syst. **151**, 78–94 (2018). https://doi.org/10.1016/j.knosys.2018.03.022
6. Grover, A., Leskovec, J.: Node2vec: scalable feature learning for networks. In: Proceedings of the 22nd ACM SIGKDD International Conference on Knowledge Discovery and Data Mining, pp. 855–864. ACM (2016).https://doi.org/10.1145/2939672.2939754

7. Guthrie, D., Allison, B., Liu, W., Guthrie, L., Wilks, Y.: A closer look at skip-gram modelling. In: Proceedings of the Fifth International Conference on Language Resources and Evaluation (LREC'06) (2014)
8. Hamilton, W.L., Ying, R., Leskovec, J.: Inductive representation learning on large graphs (2018) arXiv:1706.02216 [cs, stat]
9. Hore, A., Ziou, D.: Image quality metrics: PSNR vs SSIM. In: 2010 20th International Conference on Pattern Recognition, pp. 2366–2369. IEEE (2010) https://doi.org/10.1109/ICPR.2010.579
10. Kipf, T.N., Welling, M.: Variational graph Auto-encoders (2016) https://doi.org/10.48550/arXiv.1611.07308
11. Kipf, T.N., Welling, M.: Semi-supervised classification with graph convolutional networks. In: ICLR (2017)
12. Makarov, I., Kiselev, D., Nikitinsky, N., Subelj, L.: Survey on graph embeddings and their applications to machine learning problems on graphs. PeerJ Comput. Sci. **7**, e357 (2021). https://doi.org/10.7717/peerj-cs.357
13. Pei, H., Wei, B., Chang, K.C.C., Lei, Y., Yang, B.: Geom-GCN: geometric graph convolutional networks (2020) https://doi.org/10.48550/arXiv.2002.05287
14. Senaratne, A., Christen, P., Williams, G.J., Omran, P.G.: Rule-based knowledge discovery via anomaly detection in tabular data. In: Make (2023) https://api.semanticscholar.org/CorpusID:260356626
15. Xu, M.: Understanding graph embedding methods and their applications. SIAM Rev. **63**(4), 825–853 (2021). https://doi.org/10.1137/20M1386062

Revisiting Link Prediction
with the Dowker Complex

Jae Won Choi[1]([✉]), Yuzhou Chen[2], José Frías[3], Joel Castillo[3], and Yulia Gel[4]

[1] Department of Computer Science, University of Texas at Dallas, Richardson, USA
jaewon.choi@utdallas.edu
[2] Department of Computer and Information Sciences, Temple University,
Philadelphia, USA
[3] Center for Research in Mathematics (CIMAT), Mexico, Mexico
[4] Department of Mathematical Sciences, University of Texas at Dallas, Richardson,
USA

Abstract. We propose a novel method to study properties of graph-structured data by means of a geometric construction called Dowker complex. We study this simplicial complex through the use of persistent homology, which has shown to be a prominent tool to uncover relevant geometric and topological information in data. A positively weighted graph induces a distance in its sets of vertices. A classical approach in persistent homology is to construct a filtered Vietoris-Rips complex with vertices on the vertices of the graph. However, when the size of the set of vertices of the graph is large, the obtained simplicial complex may be computationally hard to handle. A solution The Dowker complex is constructed on a sample in the set of vertices of the graph called landmarks. A way to guaranty sparsity and proximity of the set of landmarks to all the vertices of the graph is by considering ϵ-nets. We provide theoretical proofs of the stability of the Dowker construction and comparison with the Vietorips-Rips construction. We perform experiments showing that the Dowker complex based neural networks model performs good with respect to baseline methods in tasks such as link prediction and resilience to attacks.

Keywords: Graph Neural Networks · Link Prediction · Dowker Complex

1 Introduction

The emerging findings increasingly more often suggest that higher order interactions among multiple constituents of complex systems such as generators, buses, and substations in power grid networks or proteins in molecular reactions tend to play a fundamental role behind the system organization and response to adverse external events [26]. Such higher-order or polyadic relationships among nodes are ubiquitous and vital for the functioning of a wide range of complex systems. For example, in the ecosystem, a prey may react to a predator's presence

© The Author(s), under exclusive license to Springer Nature Singapore Pte Ltd. 2024
D.-N. Yang et al. (Eds.): PAKDD 2024, LNAI 14646, pp. 418–430, 2024.
https://doi.org/10.1007/978-981-97-2253-2_33

and decrease its foraging for other species, while human interference (or lack of there off) can further drastically alter the dynamics of these interactions, and as the recent results suggest, the higher order relationships among multiple species may be the key driving factor behind species coexistence and the ecosystem biodiversity. In turn, recent findings highlight the crucial role of social group interactions, like those among colleagues and family members, in improving contagion tracking and intervention strategies. This is not to mention the critical role of higher-order relationships in formation and evolution of criminal gangs. However, as the existing graph learning methods still predominantly focus only on pair-wise node relationships, our understanding of the key mechanisms of these higher-order relationship phenomena remains largely in its nascency.

One emerging direction to address this fundamental challenge is to describe the mechanism of message propagation on higher-order network substructures such as simplicial complexes. In the context of graph neural networks (GNNs), this idea advances the notion of convolution from the conventional combinatorial graph Laplacian to a higher-order Hodge-Laplacian, leveraging the concepts of the Hodge-de Raam theory. This direction allows for describing diffusion not only at a pairwise node level as the prevailing GNNs do, but also at multi-node level such as among edges via triangles and other higher-order graph substructures. This approach results in the notion of simplicial neural networks which are arguably one of the most recent techniques in geometric deep learning. The alternative, yet closely related approach is to capitalize on the ideas of persistent homology (PH), in particular, by extracting topological signatures from the observed graph and constructing various forms of topological layers. Despite their success in graph tasks like classification and forecasting, the utility of both approaches for link prediction is less explored. This is largely due to excessive computational costs of higher-order graph learning, particularly, PH, and albeit the intuitive premise that higher-order relationships may be one of the most critical factors behind the link formation.

In this paper we take the first step toward bridging these two directions for higher-order graph learning, capitalizing on their complementary strengths not only to better capture diffusion on graphs but also to enhance robustness of link prediction against perturbations. In particular, we introduce a notion of Dowker complexes to graph machine learning, allowing us to substantially improve computational costs, thereby addressing one of the key bottlenecks in the way of the wider applicability of PH for link prediction and, more generally, graph learning. We develop a fully trainable Dowker persistence representation layer and integrate it into a simplicial convolutional network, then design a topological convolutional neural network over topological features of different Dowker filtrations. Note that, for topological signature representation learning, we consider global shape descriptor and local shape descriptor of Dowker complexes respectively. We also prove the persistence stability guarantees for Dowker complexes and offer the in-depth theoretical exploration of the relationship between the Dowker complex and its currently most widely used but computationally costly counterpart, the Vietoris-Rips complex. Our extensive experiments suggest that the proposed DC-NETs leads to substantial gains both in link prediction performance and robustness.

Significance of our contributions can be summarized as follows:

- We propose a new summary of Dowker complex, namely, Dowker complex based topological summary. DC-NETs is the first approach introducing the concepts of Dowker complex to graph learning.
- We validate DC-NETs with link prediction on 5 datasets, showcasing superior performance over state-of-the-art baselines and robustness to noise.

2   Related Work

Link Prediction. Graph Convolutional Networks (GCNs) are recognized as powerful tools for addressing link prediction tasks. In particular, the Graph Autoencoder (GAE) [20] and its alteration version, Variational Graph Autoencoder (VGAE), are initially utilized to link prediction on citation networks. SEAL [30] adopts a distinctive approach by extracting local enclosing subgraphs surrounding target links and subsequently learning a function that maps subgraph patterns to link existence. Along with, the Hyperbolic Graph Convolutional Neural Networks (HGCN) [7] leverages both hyperbolic geometry and the GCN framework to derive node representations. Another recent strategy involves utilizing pairwise topological features with GCN to uncover latent representations of a graph's geometrical structure [29].

Simplicial Complexes in Machine Learning and Persistent Homology. Persistent homology is a key tool in topological data analysis to deliver invaluable and complementary information on the intrinsic properties of data that are inaccessible with conventional methods [9,18]. In the past decade, PH has become quite popular in various ML tasks, ranging from manifold learning to medical image analysis, and material science to finance. Modeling higher-order interactions on graphs is an emerging direction in graph representation learning. While the role of higher-order structures for graph learning has been documented for a number of years [1,13,19] and involves such diverse applications as graph signal processing in image recognition dynamics of disease transmission and biological network, integration of higher-order graph substructures into DL on graphs has emerged only in 2020. As shown by [4,11,26], higher-order network structures can be leveraged to boost graph learning performance. Indeed, several most recent approaches [10,12,17,25] propose to leverage simplicial information to perform neural networks on graphs.

3   Background

Along the present manuscript, all graphs are non directed and weighted. We start describing the metric spaces induced by weighted graphs. These spaces are very important given their strong combinatorial foundations and have demonstrated relevance in data analysis [3,22]. A positively weighted graph $\mathcal{G} = (\mathcal{V}, \mathcal{E}, \omega)$ is determined by set of vertices $\mathcal{V}$ ($|\mathcal{V}| = N$), an edge set $\mathcal{E} \subset \mathcal{V} \times \mathcal{V}$ ($|\mathcal{E}| = M$), and a weight function $\omega : \mathcal{E} \to \mathbb{R}_+$. The graph $\mathcal{G}$ is simple if it does not contain self-edges nor multiple edges. Given a path γ in a weighted graph $\mathcal{G}$, the length of γ is the sum of the weights of the edges in γ. A connected positively weighted graph $\mathcal{G}$ is then endowed with the geodesic distance $d_\mathcal{G} : \mathcal{V} \times \mathcal{V} \to \mathbb{R}_{\geq 0}$, defined on a pair of vertices $u, v \in \mathcal{V}$ as the minimum length among all the paths connecting u and v. The set $(\mathcal{V}, d_\mathcal{G})$ is a finite metric space In this case, the graph $\mathcal{G}$ is also referred as a metric graph. If we have a sequence of nested simplicial complexes $\mathcal{K}_0 \subseteq \mathcal{K}_1 \subseteq \cdots \subseteq \mathcal{K}_n = \mathcal{K}$, we say that the simplicial complex $\mathcal{K}$ is filtered. The aim of Persistent Homology is to track the geometric evolution of the simplicial complexes along the filtration when we increase the parameter α by means of the homology groups $H_p(\mathcal{K}_i)$, for every $i, p \geq 0$ [14]. A non-trivial element $\beta \in H_p(\mathcal{K}_i)$ is usually called a topological feature of dimension p and can be interpreted in dimension 0 as a connected component in the complex K_i, in dimension 1 is the number of loops in the complex. In general, a topological feature in dimension p represents a p-dimensional "hole" and has associated two levels of the filtration b_β and d_α, the filtration level at which the topological feature is born (when it appears) and dies (when it disappears or is merged with another topological feature that was born before), respectively. The history of all topological features along the filtration can be summarized in a persistence diagram $PD(\mathcal{K}) = \{(b_\beta, d_\beta) \mid \beta \in H_p(\mathcal{K})\}$ [14]. When we restrict to a particular dimension $n \geq 0$, we denote the persistence diagram as $PD^n(\mathcal{K})$.

One of the most widely used simplicial complexes in topological data analysis(TDA) is the Vietoris-Rips simplicial complex [8]. This complex has the important property of being a flag complex, namely, it is completely determined by its 1-skeleton, which makes it suitable for computations. To implement any computation in persistent homology, to construct the Vietorips-Rips complex, we only need to compute the distance between any pair of points.

Definition 1 (Vietoris-Rips Complex). *Let $\mathcal{G} = (\mathcal{V}, \mathcal{E}, \omega)$ be a weighted simple graph with induced geodesic distance $d_\mathcal{G}$. For $\alpha \in \mathbb{R}_{\geq 0}$, we define the Vietoris-Rips complex $VR_\alpha(\mathcal{G})$ as the abstract simplicial complex with vertices in $\mathcal{V}$ and, for $k \geq 2$, a k-simplex $\sigma = [x_0, x_1, \ldots, x_k] \in VR_\alpha(\mathcal{G})$ if and only if $d_\mathcal{G}(x_i, x_j) \leq \alpha$ for $0 \leq i \leq j \leq k$.*

Starting with a sequence of non-decreasing scale parameters, $0 \leq \alpha_0 \leq \alpha_1 \leq \cdots \leq \alpha_{n-1} \leq \alpha_n$, we can construct a filtration of Vietoris-Rips complexes constructed on a geometric graph $\mathcal{G}$ satisfying $VR_{\alpha_0}(\mathcal{G}) \subseteq VR_{\alpha_1}(\mathcal{G}) \subseteq \cdots \subseteq VR_{\alpha_n}(\mathcal{G})$, and then apply persistent homology. One problem with the Vietorips-Rips complex is that if the cardinality of $\mathcal{V}$ is considerable, the complex $VR_\alpha(\mathcal{G})$ may be excessively large to be analyzed using computational tools. A natural alternative to deal with the scalability problem is to construct another simplicial complex with set of vertices $L \subset \mathcal{V}$, where $|L| < |\mathcal{V}|$. One of such simplicial complexes is the witness complex [3,6,15,16].

Definition 2 (Witness Complex). *Let $\mathcal{G} = (\mathcal{V}, \mathcal{E}, \omega)$ be a weighted simple graph with induced geodesic distance $d_\mathcal{G}$, and take subsets $L, W \subset \mathcal{V}$. For $\alpha \in$*

$\mathbb{R}_{\geq 0}$, let $Wit_\alpha(W, L)$ be the abstract simplicial complex with set of vertices L, and a simplex $\sigma \in Wit_\alpha(W, L)$ if and only if for every $\tau \subseteq \sigma$ there exists $w \in W$ such that $d_\mathcal{G}(w, l) \leq d_\mathcal{G}(w, l') + \alpha$ for all $l \in \tau$ and $l' \in L \setminus \tau$. The complex $Wit_\alpha(W, L)$ is called the witness complex of $\mathcal{G}$ with set of witnesses W and landmarks L.

Witness complex has important geometric properties. For instance, in the case of a point cloud in an Euclidean space it is related to the Delaunay triangulation, namely, the sequence of scale values define a filtered simplicial complex. Note that if we select sets of landmarks and witnesses $L, W \subset \mathcal{V}$, there exists also a filtration of witness complexes.

4   Dowker Complex

The witness complex is used as an alternative to deal with the scalability problem of computations of Vietoris-Rips constructions. As mentioned before, the witness complex has important geometric implications for point clouds in Euclidean spaces. However, in the case of metric graphs, those geometric properties are not inherited and it is possible to restrict the constructions of witness complex to obtain the Dowker complex:

Definition 3 (Dowker Complex). Let $\mathcal{G} = (\mathcal{V}, \mathcal{E}, \omega)$ be a weighted simple graph with induced geodesic distance $d_\mathcal{G}$, and let $L, W \subset \mathcal{V}$. For $\alpha \in \mathbb{R}_{\geq 0}$, let $Dow_\alpha(W, L)$ be the abstract simplicial complex with set of vertices L and a simplex $\sigma \in Dow_\alpha(W, L)$ if and only if there exists $w \in W$ such that $d_\mathcal{G}(w, l) \leq \alpha$ for all $l \in \sigma$. The simplicial complex $Dow_\alpha(W, L)$ is called the Dowker complex of $\mathcal{G}$ with sets of witnesses and landmarks W and L, respectively.

Note that Dowker complexes construction strongly depend on the selection of the sets L and W. In [15], two algorithms to define the set of landmarks L are developed: random and maxmin algorithms. An interesting approach to define the a set of landmarks is via ϵ-nets [2,3]. This selection of landmarks satisfies two important properties: sparcity and proximity to the complete set of vertices, thereby reducing both, computational complexity and potential information loss.

Definition 4 (ϵ-Net). Let $(\mathcal{V}, d_\mathcal{G})$ be the finite metric space obtained from the weighted graph $\mathcal{G} = (\mathcal{V}, \mathcal{E}, \omega)$. Given $L = \{u_1, u_2, \dots, u_l\} \subset \mathcal{V}$ and $\epsilon \geq 0$ then:

(i) The set L is an ϵ-sample of $\mathcal{G}$ if the collection $\{\mathcal{N}(u_i)\}_{i=1}^l$ of closed ϵ-neighborhoods of points in L covers $\mathcal{V}$, i.e. for any $v \in \mathcal{V}$ there exists $u_j \in L$ such that $d_\mathcal{G}(v, u_j) \leq \epsilon$.
(ii) L is ϵ-sparse if for any two points $u_i, u_j \in L$, their distance $d_\mathcal{G}(u_i, u_j) > \epsilon$.
(iii) The set L is an ϵ-net of $\mathcal{G}$ if it is an ϵ-sample of $\mathcal{G}$ and is ϵ-sparse.

Following the definition of an ϵ-net, natural questions arise, for instance, given the metric graph $\mathcal{G}$ and ϵ: (i) how many points does it contain an ϵ-net of $\mathcal{G}$?, or (ii) what is the most suitable way to select the points in an ϵ-net? Concerning the later question, some algorithms were presented and compared in

[3], however the suitability of each algorithm will depend on the context in which they are compared. With respect to the number of points of an ϵ-net, an upper bound in terms of the diameter of the graph and ϵ is presented in Theorem 3 of [3]. As an example, consider the unweighted star graph S_n with central vertex v_0. It is clear that there exist two 1-nets, $L_1 = \{v_0\}$ and $L_2 = \mathcal{V}(S_n) \setminus \{v_0\}$, with cardinalities 1 and $N - 1$ (see Fig. 1). However, if $\epsilon \geq 2$, every ϵ-net has a single vertex, showing the difficulty of provide bounds for the cardinality of ϵ-nets for general graphs. We describe the construction of ϵ-nets with bounded cardinality for arbitrary unweighted graphs:

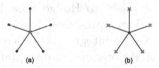

Fig. 1. Two 1-nets of an unweighted star graph. Marked points are the ϵ-net.

Lemma 1. *Let $\mathcal{G} = (\mathcal{V}, \mathcal{E})$ be a connected unweighted graph and $\epsilon > 0$. If $N = |\mathcal{V}|$, then there is an ϵ-net of $\mathcal{G}$ whose cardinality is bounded by $N/(\epsilon+1) + 2N/d$, where d is the diameter of the graph.*

Proof. Let T be a spanning tree of $\mathcal{G}$, which is rooted at one of its centers, say v_0 (remember that a tree has at most two centers). Given a vertex $v \in \mathcal{V}$, let $depth(v)$ be the depth of v with respect to v_0, and d_{max} the maximum depth among the vertices in $\mathcal{V}$. Since the graph $\mathcal{G}$ is unweighted (every edge has weight 1), we may assume that ϵ is a positive integer. Define the level sets $N_i = \{v \in \mathcal{V} \mid depth(v) = i\}$, and $M_j = \cup N_i$, where the union is taken on the indexes $i \equiv j$ (mod $\epsilon + 1$), for $j = 0, 1 \ldots, \epsilon$. Note that each set $M_j \cup \{v_0\}$ is an ϵ-sample of $\mathcal{G}$, and each of such sets contains at most $d_{max}/(\epsilon + 1) + 1$ of the level sets N_i. On the other hand, there exists at least one set M_k such that every level set contained in M_k has in average N/d_{max} points. Then the total number of points in M_k is upperly bounded by $(d_{max}/(\epsilon+1)+1)(N/d_{max}) = N/(\epsilon+1)+N/d_{max}$. Since $d \leq d_{max}$, the result follows.

5   Methodology

Recall that a k-simplex $\sigma_k = (v_0, \ldots, v_k)$ is a k-dimensional convex hull of nodes $v_0, \ldots, v_k$. Given the set of nodes $v_0, \ldots, v_k$ of σ_k, define an orientation on the simplex σ_k by choosing some (arbitrary) ordering of its nodes. Consider a real-valued vector space C^k with basis from the oriented k-simplices. Elements C^k are called k-chains. The *boundary* operator ∂_k is a homomorphism $\partial_k : C^k \to C^{k-1}$. The adjoint of ∂_k is the *co-boundary* operator $\partial_k^T : C^k \to C^{k+1}$. Let $\boldsymbol{B}_k$ and $\boldsymbol{B}_k^T$ be the matrix representations of ∂_k and ∂_k^T, respectively.

Definition 5. *Linear operator L_k over oriented k-simplices $L_k : C^k \to C^k$ is called the k-Hodge Laplacian, and its corresponding matrix representation is*

$$L_k = B_k^\top B_k + B_{k+1} B_{k+1}^\top, \tag{1}$$

where $B_k^\top B_k$ and $B_{k+1} B_{k+1}^\top$ are often referred to L_k^{down} and L_k^{up}, respectively.

That is, the standard graph Laplacian $L_0 = B_1 B_1^\top \in \mathbb{R}^{N \times N}$ is a special case of the above k-th combinatorial Hodge Laplacian and the matrix $L_1 \in \mathbb{R}^{M \times M}$ is the Hodge 1-Laplacian. While the Hodge theory allows us to systematically describe diffusion across higher-order graph substructures, or k-simplices of any k, all current studies are restricted solely to Hodge 1-Laplacian L_1 [25,26]. In this paper, we propose a new random walk-based block Hodge-Laplacian operator that enables us to simultaneously integrate knowledge on interactions among higher-order substructures of various orders into graph learning. The K-block Hodge-Laplacian $\mathfrak{L}_K$ is related to the Dirac operator in differential geometry. As such, $\mathfrak{L}_K$ has multiple implications for analysis of synchronization dynamics and coupling of various topological signals on graphs.

Furthermore, we propose the random walk-based block Hodge-Laplacian, i.e., r-th power of block Hodge-Laplacian representation (where $r \in \mathbb{Z}_{>0}$). Our random walk-based block Hodge-Laplacian is inspired by the recent success in random walk based graph embeddings and simplicial complexes but is designed to conduct informative *joint* random walks on higher-order Hodge Laplacians instead of limited powering Hodge 1-Laplacian [4,5]. Indeed, successfully travelling through higher-order topological features will provide us with additional feature information which is valuable for learning edge embeddings.

Adaptive Simplicial Complex Based Block. Our innovative method for understanding the complex topological structure of higher-order graphs is driven by a central inquiry: Is it possible to encapsulate the relational dynamics between k-simplices within a graph $\mathcal{G}$ that exhibit a separation beyond the immediate adjacency of orders $k-1$ and $k+1$? Although exploring this issue extends beyond the conventional scope of the Hodge-de Rham framework applied to simplicial Hodge Laplacians, we present a solution via the development of an Adaptive Simplicial Complex (ASC) block as follows

$$\mathbb{L}_2^{ASC} = \left[\begin{array}{c|c} L_{k_1} & g(\mathbb{P}_{d_2} L_{k_1}, L_{k_2}) \\ \hline g(\mathbb{P}_{d_2} L_{k_1}, L_{k_2})^\top & L_{k_2} \end{array} \right]. \tag{2}$$

Note that in general, the linear operator $\mathbb{L}_2^{ASC} \in \mathbb{R}^{(d_1+d_2) \times (d_1+d_2)}$ is no longer a Laplacian. For example, while $\mathbb{L}_2^B$ is symmetric (by construction), it may not satisfy the condition of positive semidefinitess. However, as shown below, ASC block opens multiple opportunities to better describe higher-order interactions on graphs. Armed with the ASC block, we can utilize the non-local message operation $g(\cdot, \cdot)$ to capture long-range relations and intrinsic higher-order connectivity among entities in the 2-block (reduced) Hodge-Laplacian $\mathfrak{L}_2$, i.e., $g(\mathbb{P}_{d_2} L_{k_1}, L_{k_2})$ placed in the off-diagonal. We now describe the choices of non-local message passing functions which can be used for $g(\cdot, \cdot)$ in the following.

Specifically, given two higher-order Hodge Laplacians $\boldsymbol{L}_{k_1}$ and $\boldsymbol{L}_{k_2}$, we define two types of the non-local message passing functions to capture the relations between $[\mathbb{P}_{d_2}\boldsymbol{L}_{k_1}]_i \in \mathbb{R}^{d_2}$ and $[\boldsymbol{L}_{k_2}]_j \in \mathbb{R}^{d_2}$ (i.e., topological embedding of the i-th graph substructure in $\mathbb{P}_{d_2}\boldsymbol{L}_{k_1}$ and topological embedding of the j-th graph substructure in $\boldsymbol{L}_{k_2}$ respectively) as $g([\mathbb{P}_{d_2}\boldsymbol{L}_{k_1}]_i, [\boldsymbol{L}_{k_2}]_j) = <[\mathbb{P}_{d_2}\boldsymbol{L}_{k_1}]_i, [\boldsymbol{L}_{k_2}]_j>$. Finally, the adaptive Hodge block can be formulated as $\tilde{\mathbb{L}}_2^B = \text{softmax}\left(\text{ReLU}\left(\mathbb{L}_2^B\right)\right)$, where the softmax function is employed to standardize the block operator $\mathbb{L}_2^B$, the ReLU activation function, defined as $\text{ReLU}(\cdot) = \max(0, \cdot)$, serves to filter out both the weak pairwise interactions among higher-dimensional simplices and weak connections. Consequently, the proposed ASC block $\tilde{\mathbb{L}}_2^{ASC}$ is adept at dynamically learning the interplay between simplices across varying dimensions. We now turn our attention to formulating the adaptive simplicial complex block convolutional layer (ASC-CL), which facilitates the learning of distances between nodes within the embedding space and can be formulated as $\boldsymbol{Z}^{(\ell+1)} = \left(\tilde{\mathbb{L}}_2^B \boldsymbol{Z}^{(\ell)} \boldsymbol{\Theta}_1^{(\ell)}\right) \boldsymbol{\Theta}_2^{(\ell)}$, where $\boldsymbol{Z}^{(0)} = \boldsymbol{X} \in \mathbb{R}^{n \times d}$ is node features matrix, $\boldsymbol{\Theta}_1^{(\ell)} \in \mathbb{R}^{d \times (d_1+d_2)}$ and $\boldsymbol{\Theta}_2^{(\ell)} \in \mathbb{R}^{(d_1+d_2) \times d_{\text{out}}}$ are two trainable weight matrices at layer ℓ, and d_{out} denotes the dimension of node embedding at the (ℓ)-th layer through the ASC-CL operation. For the link prediction task, we use the Fermi-Dirac decoder to compute the distance between the two nodes. Formally, $\text{dist}_{uv}^{\text{ASC-CL}} = (\boldsymbol{Z}_u^{(\ell+1)} - \boldsymbol{Z}_v^{(\ell+1)})^2$ where $\text{dist}^{\text{ASC-CL}}{}_{uv} \in \mathbb{R}^{1 \times d_{\text{out}}}$ is the distance between nodes u and v in a local spatial domain.

Dowker Complex Based Topological Layer. To learn critical information on the Dowker complex based topological features, we propose the Dowker complex based topological layer (DC-TL) denoted by Ψ. DC-TL is a Dowker complex based topological representation learning layer. Given Dowker complex based topological features $\boldsymbol{\Xi}$, the DC-TL operator will output the latent Dowker complex based topological representation $\tilde{\boldsymbol{Z}}$ with shape d_c as follows (d_c is the number of channels in output), i.e., $\boldsymbol{Z}_{\text{DC-TL}} = \Psi(\boldsymbol{\Xi}) = \phi_{\text{MAX}}\left(f_{\text{CNN}}^{(\ell)}(\boldsymbol{\Xi})\right)$ where $f_{\text{CNN}}^{(\ell)}$ is the convolutional neural network (CNN) in the ℓ-th layer, and ϕ_{MAX} denotes global max-pooling layer. Similar to the process of computing distances between learnable node embeddings with the adaptive simplicial complex block convolutional layer, we conduct convolution operation to evaluate the distance between the node embeddings of nodes u and v as

$$\boldsymbol{H}^{(\ell+1)} = [\tilde{\boldsymbol{L}}\boldsymbol{H}^{(\ell)}\boldsymbol{\Theta}_3^{(\ell)}, \boldsymbol{Z}_{\text{DC-TL}}], \quad \text{dist}_{uv}^{\text{DC-TL}} = (\boldsymbol{H}_u^{(\ell+1)} - \boldsymbol{H}_v^{(\ell+1)})^2, \quad (3)$$

where $[\cdot, \cdot]$ denotes the concatenation operation, $\tilde{\boldsymbol{L}} = \boldsymbol{D}_v^{-1/2}(\boldsymbol{A}+\boldsymbol{I})\boldsymbol{D}_v^{-1/2}$ (where $\boldsymbol{D}_v$ is the degree matrix of $\boldsymbol{A}+\boldsymbol{I}$, i.e., $[\boldsymbol{D}_v]_{ii} = \sum_j[\boldsymbol{A}+\boldsymbol{I}]_{ij}$), $\boldsymbol{H}^{(0)} = \boldsymbol{X} \in \mathbb{R}^{N \times d}$ is node features matrix, $\boldsymbol{\Theta}_3^{(\ell)} \in \mathbb{R}^{d \times d_{\text{out}'}}$ is the trainable weight matrix, $d_{\text{out}'}$ denotes the dimension of node embedding at the (ℓ)-th layer through the graph convolution operation, $\text{dist}_{uv}^{\text{DC-TL}} \in \mathbb{R}^{1 \times d_{\text{out}'}}$ is the distance between nodes u and v in global spatial domain. Finally, we wrap the concatenation of Hodge-style adaptive convolution and graph convolution operation outputs into the Multi-layer Percerpton (MLP) dist $=$

$\sigma(f_{\text{MLP}}([\alpha_{\text{DC-TL}} \times \text{dist}^{\text{DC-TL}}, \alpha_{\text{SC}} \times \text{dist}^{\text{ASC-CL}}]))$ where $[\cdot, \cdot]$ denotes the concatenation of the outputs from the ASC-CL operator and Dowker complex based topological layer, $\sigma(\cdot)$ denotes the activation function, f_{MLP} is an MLP layer that maps the concatenated embedding to a d_o-dimensional space, and $\alpha_{\text{DC-TL}}$ and $\alpha_{\text{ASC-CL}}$ are the hyperparameters representing the weight of each distance in the $(\ell+1)$-th layer. Lastly, the edges connection probability can be computed as $\mathcal{P}_{u,v} = [\exp((\text{dist}_{uv} - \delta)/\eta) + 1]^{-1}$, where δ and η are hyperparameters. Then, training via standard backpropagation is performed via binary cross-entropy loss function.

Table 1. ROC AUC and standard deviations for link prediction. Bold numbers denote the best results.

Model	Cora	Citeseer	Cornell	Texas	Wisconsin
GCN	89.14 ± 1.20	87.89 ± 1.48	79.14 ± 8.89	67.42 ± 9.39	72.77 ± 6.96
GAT	72.80 ± 0.20	74.80 ± 1.50	68.83 ± 4.34	70.42 ± 3.98	70.41 ± 4.03
GIC	91.42 ± 1.24	92.99 ± 1.14	57.26 ± 8.77	65.16 ± 7.87	75.24 ± 8.45
GAE	90.21 ± 0.98	88.42 ± 1.13	60.58 ± 3.71	68.67 ± 6.95	75.10 ± 8.69
GVAE	92.17 ± 0.72	90.24 ± 1.10	68.25 ± 4.21	74.61 ± 8.61	74.39 ± 8.39
SEAL	90.29 ± 1.89	88.12 ± 0.85	68.99 ± 3.77	71.68 ± 6.85	77.96 ± 10.37
TLC-GCN	94.22 ± 0.78	93.30 ± 0.60	63.49 ± 10.43	58.12 ± 5.67	58.57 ± 3.28
BScNets	95.09 ± 0.65	92.06 ± 0.66	72.50 ± 11.64	72.29 ± 14.49	84.83 ± 4.80
DC-NETs (ours)	$\mathbf{95.12 \pm 0.54}$	$\mathbf{93.87 \pm 0.83}$	$\mathbf{83.35 \pm 6.12}$	$\mathbf{82.35 \pm 6.99}$	$\mathbf{87.14 \pm 5.63}$

Table 2. Performance comparison between dowker complex and Vietoris-Rips complex.

Model	Cora	Cornell	Texas	Wisconsin
DC-NETs (ours)	$\mathbf{95.12 \pm 0.54}$	83.35 ± 6.12	$\mathbf{82.35 \pm 6.99}$	$\mathbf{87.14 \pm 5.63}$
RC-NETs	94.85 ± 0.39	$\mathbf{84.55 \pm 6.18}$	81.49 ± 2.25	86.23 ± 7.50

6   Experiments

Datasets and Baselines. We experiment on three types of networks (i) citation networks: Cora and Citeseer [27]; (ii) webpage datasets: Cornell, Texas, and Wisconsin [24]. We compare against eight state-of-the-art (SOA) baselines, including (i) Graph convolution network (GCN) [21] which multiplies the adjacency matrix to simplify the multi-layer graph convolutional networks; (ii) Graph Attention Networks (GAT) [28] which adopts attention mechanism to aggregate the information from neighbors, considering relative weights between two connected nodes; (iii) Graph InfoClust (GIC) [23] which enhances node representations by leveraging cluster-level content; (iv) Graph Autoencoder (GAE) and (v) its variational version, Variational Graph Autoencoder (GVAE) [20] which

Table 3. ROC AUC and standard deviations for link prediction.

Model	Cora	Cornell	Texas	Wisconsin
SEAL (0.05)	91.70 ± 0.73	76.02 ± 5.80	80.66 ± 5.54	67.91 ± 4.69
TLC-GCN (0.05)	92.24 ± 0.68	58.94 ± 7.77	56.47 ± 13.67	56.82 ± 5.77
BscNets (0.05)	**94.95 ± 0.64**	71.54 ± 10.08	72.74 ± 11.30	84.07 ± 5.88
DC-NETs (0.05)	94.10 ± 1.62	**79.15 ± 5.32**	**77.01 ± 4.48**	**87.04 ± 3.88**
SEAL (0.10)	90.91 ± 2.36	74.78 ± 4.678	79.83 ± 4.29	66.93 ± 4.22
TLC-GCN (0.10)	91.48 ± 2.13	54.96 ± 7.43	55.71 ± 8.83	53.62 ± 2.72
BscNets (0.10)	**94.52 ± 0.70**	69.63 ± 9.11	64.77 ± 13.01	83.71 ± 6.20
DC-NETs (0.10)	92.75 ± 3.56	**78.86 ± 7.26**	**76.94 ± 4.84**	**86.92 ± 4.43**
SEAL (0.15)	89.93 ± 0.46	73.78 ± 4.13	79.39 ± 4.53	66.28 ± 3.13
TLC-GCN (0.15)	88.98 ± 3.89	51.94 ± 4.50	54.90 ± 6.62	50.10 ± 5.86
BscNets (0.15)	**93.95 ± 1.03**	68.53 ± 12.66	71.55 ± 11.47	81.62 ± 7.86
DC-NETs (0.15)	92.24 ± 1.15	**76.55 ± 7.07**	**76.61 ± 4.26**	**86.63 ± 4.08**

is employing latent variables and GCN encoder to generate interpretable representations; (vi) SEAL [30] which predicts the relationship on graph-structure data based on graph convolution on local subgraph around each link; (vii) Topological Loop-Counting Graph Neural Network (TLC-GNN) [29] which injects the pairwise topological information into the latent representation of a GCN; (viii) BScNet [10] which revolutionize graph learning by incorporating multilevel interactions among higher-order graph structures using the Block Hodge-Laplacian.

Hyperparameter Settings and Training Details. DC-NETs consists of a one adaptive block Hodge convolution layer (*nhid1*) and two graph convolution layers (*nhid2*, *nhid3*), where $nhid1 \in \{1, 8, 16, 32, 64, 128\}$, $nhid2 \in \{16, 32, 64, 128\}$, and $nhid3 \in \{4, 16, 32\}$. In addition, we perform an extensive grid search for learning rate among $\{0.001, 0.005, 0.008, 0.01, 0.05\}$, the dropout rate among $\{0.1, 0.2, \ldots, 0.9\}$, power r for random walk among $\{2, 3, 4, 5\}$, and importance weights $\alpha_{DC\text{-}TL}$ among $\{0.01, 0.1, 1.0, 10.0, 100.0\}$ and $\alpha_{ASC\text{-}CL}$ among $\{1.0, 10.0\}$ respectively. The model is trained for 1,000 epochs with early stopping applied when the metric (i.e., validation loss) starts to drop.

6.1   Results

The evaluation results on 5 datasets are summarized in Tables 1 and 2. Table 1 shows the performance comparison among 8 baselines on 2 citation networks (i.e., Cora and Citeseer) and 3 webpage datasets (i.e., Cornell, Texas, and Wisconsin) where nodes indicate web pages, and edges show hyperlinks between them. On all 5 datasets, we find that our DC-NETs consistently perform far better than baseline models. In particular, the average relative gain of DC-NETs over the runner-ups is 3.812%, demonstrating the proposed DC-NETs' effectiveness on all common link prediction benchmarks. Regarding baseline methods, although GCN which is two spectral-based convolutional GNN can reflect the relationship between nodes, they fail to consider topological structure information. Furthermore, three spatial-based convolutional GNNs, such as GAT and SEAL, capture the weights of neighboring nodes, representing node-level information. This approach has limitations in expressing broader availability, making it more difficult to extract information from nodes located at a significant distance and to identify high-order spatial dependencies. Table 2 shows the performance comparison between the Dowker and Vietoris-Rips complex. We can observe that, for 3 out of 4 datasets, DC-NETs achieves an average of 0.24% higher performance than RC-NETs. The results of resilience analysis are summarized in Table 3. Table 3 shows performance comparisons among SEAL, TLC-GCN, BscNets, and DC-NETs with different perturbation rates from 0.05 to 0.15 on 1 citation network and 3 web page datasets. From Table 3, we can observe that (i) our proposed DC-NETs significantly outperform all baselines on 3 web page datasets across all 3 perturbation rates and (ii) although BScNets outperform DC-NETs on Cora dataset, our Dowker complex based topological layer improves the robustness with a slight performance degradation.

7   Conclusion

By introducing the concepts of Dowker complex to graph learning, we have addressed one of the primary bottlenecks in the way of wider adoption of topological methods – that is, computational complexity. We have developed DC-NETS, a novel approach for link prediction harnessing topological information on a graph in the form of the Dowker layer. Our numerical experiments have indicated that DC-NETS outperforms state-of-the-art competitors while requiring noticeably fewer computational resources than other topological approaches. This opens up new possibilities for future research in topological graph learning such as computationally efficient pooling, topological attacks, and topological defences.

Acknowledgements. This project has been supported in part by NASA AIST 21-AIST21_2-0059, NSF ECCS 2039701, TIP-2333703, and ONR N00014-21-1-2530 grants. The paper is based upon work supported by (while Y.R.G. was serving at) the NSF. The views expressed in the article do not necessarily represent the views of NSF, ONR or NASA.

References

1. Agarwal, S., Branson, K., Belongie, S.: Higher order learning with graphs. In: ICML, pp. 17–24 (2006)
2. Aksoy, S.G., et al.: Seven open problems in applied combinatorics. arXiv preprint arXiv:2303.11464 (2023)

3. Arafat, N.A., Basu, D., Bressan, S.: ϵ-net induced lazy witness complexes on graphs (2020)
4. Benson, A.R., Abebe, R., Schaub, M.T., Jadbabaie, A., Kleinberg, J.: Simplicial closure and higher-order link prediction. PNAS **115**(48), E11221–E11230 (2018)
5. Bodnar, C., et al.: Weisfeiler and Lehman go topological: message passing simplicial networks. In: ICML, pp. 1026–1037 (2021)
6. Boissonnat, J.D., Guibas, L., Oudot, S.: Manifold reconstruction in arbitrary dimensions using witness complexes. In: SoCG, pp. 194–203 (2007)
7. Chami, I., Ying, R., Ré, C., Leskovec, J.: Hyperbolic graph convolutional neural networks (2019)
8. Chazal, F., De Silva, V., Oudot, S.: Persistence stability for geometric complexes. Geom. Dedicata. **173**(1), 193–214 (2014)
9. Chazal, F., Michel, B.: An introduction to topological data analysis: fundamental and practical aspects for data scientists. Front. Artif. Intell **4** (2021)
10. Chen, Y., Gel, Y.R., Poor, H.V.: BScNets: block simplicial complex neural networks. Proc. AAAI Conf. Artif. Intell. **36**(6), 6333–6341 (2022)
11. Chen, Y., Gel, Y.R., Marathe, M.V., Poor, H.V.: A simplicial epidemic model for COVID-19 spread analysis. Proc. Natl. Acad. Sci. **121**(1), e2313171120 (2024)
12. Chen, Y., Jacob, R.A., Gel, Y.R., Zhang, J., Poor, H.V.: Learning power grid outages with higher-order topological neural networks. IEEE Trans. Power Syst. **39**(1), 720–732 (2024)
13. Chen, Y., Jiang, T., Gel, Y.R.: H^2-nets: hyper-Hdge convolutional neural networks for time-series forecasting. In: Koutra, D., Plant, C., Gomez Rodriguez, M., Baralis, E., Bonchi, F. (eds.) ECML PKDD 2023. LNCS, vol. 14713, pp. 271–289. Springer, Cham (2023). https://doi.org/10.1007/978-3-031-43424-2_17
14. Cohen-Steiner, D., Edelsbrunner, H., Harer, J.: Stability of persistence diagrams. In: SoCG, pp. 263–271 (2005)
15. De Silva, V., Carlsson, G.: Topological estimation using witness complexes. In: PBG, pp. 157–166 (2004)
16. Dey, T.K., Fan, F., Wang, Y.: Graph induced complex on point data. In: SoCG, pp. 107–116 (2013)
17. Ebli, S., Defferrard, M., Spreemann, G.: Simplicial neural networks. In: NeurIPS 2020 Workshop on TDA and Beyond (2020)
18. Hensel, F., Moor, M., Rieck, B.: A survey of topological machine learning methods. Front. Artif. Intell. **4**, 52 (2021)
19. Johnson, J.L., Goldring, T.: Discrete Hodge theory on graphs: a tutorial. Comput. Sci. Eng. **15**(5), 42–55 (2013)
20. Kipf, T.N., Welling, M.: Variational graph auto-encoders (2016)
21. Kipf, T.N., Welling, M.: Semi-supervised classification with gcns. In: ICLR (2017)
22. Liu, X., Feng, H., Wu, J., Xia, K.: Dowker complex based machine learning (DCML) models for protein-ligand binding affinity prediction. PLoS Comp. Biol. **18**(4), e1009943 (2022)
23. Mavromatis, C., Karypis, G.: Graph InfoClust: maximizing coarse-grain mutual information in graphs. In: Karlapalem, K., et al. (eds.) Advances in Knowledge Discovery and Data Mining. PAKDD 2021, Part I. LNCS, vol. 12712, pp. 541–553. Springer, Cham (2021). https://doi.org/10.1007/978-3-030-75762-5_43
24. Pei, H., Wei, B., Chang, K.C.C., Lei, Y., Yang, B.: Geom-GCN: geometric graph convolutional networks. In: ICLR (2019)
25. Roddenberry, T.M., Glaze, N., Segarra, S.: Principled simplicial neural networks for trajectory prediction. In: ICML, pp. 9020–9029 (2021)

26. Schaub, M.T., Benson, A.R., Horn, P., Lippner, G., Jadbabaie, A.: Random walks on simplicial complexes and the normalized Hodge 1-laplacian. SIAM Rev. **62**(2), 353–391 (2020)
27. Sen, P., Namata, G., Bilgic, M., Getoor, L., Galligher, B., Eliassi-Rad, T.: Collective classification in network data. AI Mag. **29**(3), 93–93 (2008)
28. Veličković, P., Cucurull, G., Casanova, A., Romero, A., Liò, P., Bengio, Y.: Graph attention networks. In: ICLR (2018)
29. Yan, Z., Ma, T., Gao, L., Tang, Z., Chen, C.: Link prediction with persistent homology: an interactive view (2021)
30. Zhang, M., Chen, Y.: Link prediction based on graph neural networks (2018)

GraphNILM: A Graph Neural Network for Energy Disaggregation

Rui Shang[1][(✉)], Siji Chen[1], Zhiqian Chen[2], and Chang-Tien Lu[1]

[1] Virginia Polytechnic Institute and State University Blacksburg, Blacksburg, VA 24061-0002, USA
{srui,sijic}@vt.edu, zchen@cse.msstate.edu
[2] Mississippi State University, 75 B.S. Hood Drive, Mississippi, MS 39762, USA
ctlu@vt.edu

Abstract. Non-Intrusive Load Monitoring (NILM) remains a critical issue in both commercial and residential energy management, with a key challenge being the requirement for individual appliance-specific deep learning models. These models often disregard the interconnected nature of loads and usage patterns, stemming from diverse user behavior. To address this, we introduce GraphNILM, an innovative end-to-end model that leverages graph neural networks to deliver appliance-level energy usage analysis for an entire home. In its initial phase, GraphNILM employs Gaussian random variables to depict the graph edges, later enhancing prediction accuracy by substituting these edges with observations of appliance interrelationships, stripping the individual load enery from the aggregated main energy all at one time, resulting in reduced memory usage, especially with more than three loads involved, thus presenting a time and space-efficient solution for real-world implementation. Comprehensive testing on popular NILM datasets confirms that our model outperforms existing benchmarks in both accuracy and memory consumption, suggesting its considerable promise for future deployment in edge devices.

Keywords: NILM · Energy Disaggregation · Graph Neural Network

1 Introduction

Energy conservation is a crucial research area in today's scientific world. Household and commercial electrical use accounts for approximately 60% of global energy consumption [7]. Real-time monitoring of power consumption is a useful approach for assisting homes, utilities, appliance manufacturers, and policymakers in making more informed decisions. However, obtaining individual appliance-level load data in real-time typically requires the installation of a sensor per load, which can be costly and impractical for older houses or office buildings. As a result, non-intrusive load monitoring (NILM) technology has gained popularity due to its low installation and maintenance costs, as well as its respect for privacy.

© The Author(s), under exclusive license to Springer Nature Singapore Pte Ltd. 2024
D.-N. Yang et al. (Eds.): PAKDD 2024, LNAI 14646, pp. 431–443, 2024.
https://doi.org/10.1007/978-981-97-2253-2_34

By gathering data from the main power measurements and computing the projected individual power consumption without additional measuring equipment, NILM provides a cost-effective solution for disaggregating energy consumption. Several surveys [1] have demonstrated the business case for NILM, revealing that energy savings outweigh installation costs. Moreover, research has shown that providing active energy data feedback to customers through NILM can reduce energy use by 5–20% [24]. However, NILM is inherently challenging due to the various load combinations in a given place and consumers' complex consumption preferences. Addressing this problem requires the development of innovative and efficient models that can accurately disaggregate energy consumption at the appliance level, which is the focus of this study.

Figure 1 depicts the NILM application scenario utilizing state-of-the-art disaggregation techniques versus our proposed GraphNILM method. In most houses or office buildings, the number of routinely used devices exceeds four. As the number of appliances increases, the amount of resources needed to estimate the instantaneous load power rises. GraphNILM, in contrast, utilizes roughly the same amount of

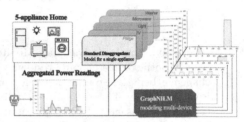

Fig. 1. NILM general explanations and comparisons of our proposed method with common state-of-art methods.

memory size to achieve comparable results, which seems more reasonable to be deployed on edge devices in the houses or office buildings.

There are three main challenges in the NILM field. **(1) A low rate of sampling.** The sampling rate in the NILM field, which is typically 1 Hz for common datasets, is significantly lower than the working frequency of the loads; and thus, the sampled data cannot be fully restored according to the Nyquist-Shannon sampling theorem. Adoption of classical algorithms, like hidden Markov models and their variants [17,22], yields restricted results under specific conditions, making widespread adoption challenging. **(2) Homogeneous data with restricted characteristics**. The majority of available data consists solely of aggregated power readings in a timely order, which makes it difficult for domain experts to quantify the dedicated load power numbers from readings only. Numerous studies therefore focus on the classification problem [8] by Convolutional Neural Network (CNN) and the Recurrent Neural Network (RNN) [5], rather than quantitative analysis, but even good classification results are of limited utility to consumers. Non-stable power consumption and the complex combinations of loads makes disaggregation a challenging problem. **(3) More memory resources required as more loads are introduced** In the NILM commercial sector, a regression model with instantaneous individual load power output appears more desirable than a categorization model. In recent years, non-linear regression models in NILM have emerged with respectable performance, but the vast majority of deep learning models have a high memory footprint and increased computational complexity [23]. To improve performance, the majority of regression models [13,26] repeat the same structure with different parameters for different loads. A residence with ten loads will necessitate the concurrent operation of ten times the proposed model, indicating the high cost of business implementations on edge devices. This research aims to develop an end-to-end model with reduced memory consumption and cutting-edge performance for future business deployments on edge devices. To solve the challenges highlighted in the NILM field, this paper designs the GraphNILM model and provides the following significant contributions:

- **Formulating a novel end-to-end framework for energy disaggregation.** The paper presents GraphNILM, a model that uses a modified convolutional neural network for initial disaggregation and a graph neural network for refinement, enabling simultaneous disaggregated power readings. Graph-NILM efficiently extracts power features from low-rate sequences, fine-tuning the results using load relationships before producing the final output.
- **Constructing an effective algorithm to characaterize load relations.** This paper categorizes relationships between distinct loads as synchronous and asynchronous. Supplementing our approach, we introduce a new algorithm to calculate the synchronous relations between the aggregated power readings and individual load by correlations, and asynchronous relations between loads by dynamic matching.
- **Designing a new structure for memory reduction.** In response to single-load targeted models, we construct a weighted graph for GraphNILM, transforming loads into nodes using pre-established relationships. This allows simultaneous power disaggregation, reducing memory and computational requirements by avoiding separate individual load trainings. We are the first to use dynamic time wrapping relationship structure in the NILM field.
- **Conducting extensive experimental performance evaluations.** The proposed GraphNILM network has been evaluated utilizing data from standard NILM datasets: REDD [16] and UK-DALE [14]. It often surpasses competing methods across different metrics, utilizing a fraction of memory compared to benchmarks. We also discuss the practical benefits of integrating GraphNILM into edge devices.

2   Related Work

The NILM field was pioneered by Hart [10] about three decades ago. In order to solve the NILM problem with low rate sampling data, Hidden Markov based Model (HMM, FHMM, etc.) [17] and its variants [22] were adopted in the early stage. Such methods were categorized as event-based methods, which usually contain three procedures: edge detection, feature extraction , and classification [9,18], and were broadened to handle NILM disaggregations problems in other fields [6]. With deep learning flourishing in most domains in recent years, deep neural networks and convolution neural networks have lowered the obstacles in the NILM field for researchers to extract power features without the help of domain experts [13]. Deep learning has lighted up a new direction for solving the NILM problem [19]. Long-short-term memory network (LSTM) [20] extracted dominate appliance usage from the aggregate power signals, which are collected at a low sampling rate. The widely-used Seq2Point model [26], showed great improvement on regression tasks in the NILM domain and received challenges all the time since its debut, but still need trainings per load introduced [4]. The introduction of graph signal processing (GSP) to the NILM field is a novel concept. Stankovic's research group [2,11,27] tracked this technique by segmenting aggregated energy sequences to do classification task in NILM. Similar work has been conducted by Bing and other groups [25], with all of them utilizing graph neural networks to perform load identification tasks which extracted information is insufficient to meet individual and commercial energy planning requirements.

Our aim is to create an easily implementable real-world model that can produce instantaneous disaggregated load energy with reduced memory consumption, while maintaining accuracy equivalent to the state-of-the-art techniques. Realized the ignorance of relations among loads in regression tasks and the difficulty of transforming homogeneity data to adapt to the multi-variate inputs needs for graph signal processing, we decide to design a new integrated framework for energy disaggregation by utilizing the advantages of the deep convolution neural network and the graph neural network. As a result, the proposed model can not only fine-tune the disaggregated power results, but also reduce total computations by incorporating the graph design.

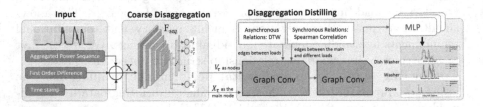

Fig. 2. GraphNILM Total Structure.

3   Proposed Model

3.1   Problem Setup: Disaggregation

Given the aggregated power meter readings, x_t, we wish to disaggregate the immediate power contribution of each individual load. The disaggregation of $x_t \in \mathbb{R}$ at time t is formulated as:

$$x_t = \sum_{n=1}^{N} y_t^n + \text{noise}, \qquad (1)$$

N is the number of loads which are to be monitored and y_t^n stands for the power of the n-th load at time t. The noise and unmonitored loads in the given dataset have been generalized to the niose. Our purpose is to get the individual load power together at the same time, as $Y_t = [y_t^1, y_t^2, ..., y_t^N]^T$ based on the measurement of the aggregated power x_t from our proposed model.

3.2   GraphNILM

The proposed GraphNILM model consists of two major components: coarse disaggregation and disaggregation distillation, see Fig. 2. It utilizes convolutional neural networks to extract the representations of power from the aggregated power sequence. The output of the coarse disaggregation component represents the power characteristics of various loads and will be used as the coarse disaggregation results for the disaggregation distilling's inputs. GCN components in our disaggregation distillation section will use these inputs along with the constructed relations as the graph edges to disaggregate the main power to the individual load power in the given house.

(1) Coarse Disaggregation

The coarse disaggregation part is for extracting power characteristics. The input $x_{t-W+1:t}$ is a length W time series representing the aggregated main power within W time stamps. Then we intentionally add the first order difference between two consecutive aggregated main power readings which brings approximately half sigma better performance in the later experiment. F_{seq} represents the convolutional layers in the proposed architecture, and it outputs

$$V_\tau = F_{seq}(x_{t-W+1:t}) = [v_\tau^1, v_\tau^2, ..., v_\tau^N]^T, \tag{2}$$

where $\tau = t - \frac{W-1}{2}$ representing the middle point of the window W time series, and V_τ stands for the extracted power characteristics of the input sequence $x_{t-W+1:t}$. The intuition behind the midpoint selection is based on the assumption that the model can learn the information of aggregated power before and after the midpoint [21]. These results will be further used as the nodes in the graph structure in the following Disaggregation Distilling part.

(2) Disaggregation Distilling

The introduction of GraphNILM's second component, graph structure, is a novel concept in the NILM field, as the accessible data in popular datasets include no relational information. A graph $\mathcal{G}$ usually consists of nodes set $\mathcal{V}$ and adjacency matrix A and is represented as $\mathcal{G} = \{\mathcal{V}, A\}$. $v_i \in \mathcal{V}$ denotes for the i-th node in $\mathcal{V}$, which is v_τ^i from equation (2). The adjacency matrix defines the edges $a_{ij} \in A$ and their weights in the graph. These weighted edges are mapping our expectations to utilize relations among loads in the design of our proposed model. Some loads have simultaneous direct relationships, while others may have asynchronous relations. Weighted edges in a graph can appropriately describe such relationships when they can be quantified.

To leverage the relational strengths of the graph model, we map both the aggregated power and the coarse disaggregated characteristics - obtained from the coarse disaggregation stage - onto the nodes in $\mathcal{G}$, signifying that $x_\tau, V_\tau \subseteq \mathcal{V}$. To avoid over-fitting, we add x_τ, the mid-point of the aggregated power sequence, as the central node in the graph.

Finding meaningful edges between nodes, i.e., interpreting the relationships between loads in NILM, is the core principle in our graph construction. According to our observations, there are primarily two sorts of relationships in given houses: synchronous and asynchronous. For example, in a given house, the owner prefers to watch television with her food prepared. Before turning on the television, she toasts a slice of bread and then boils some water in the kettle. These events occur sequentially and have strong relationships from an asynchronous perspective: the individual power reading peak for one load occurs close to the power reading peaks for other loads. On the other hand, it is straightforward to conclude that the aggregated power x_τ closely connected with each load at the same time, e.g. the aggregated power would rise at the same time she turns on a new load, a typical synchronous relationship. Though for each house, owners' habits may vary, the major loads for the house are still similar, which makes our pre-trained model transferrable.

For *synchronous* relations, spearman correlations can be used in this model since it measures the strength and direction of monotonic association between two variables. Thus we use it to denote weight edges from the aggregated main power meter to the disaggregaged individual load. For *asynchronous* relationships, dynamic time wrapping (DTW) is used to determine the correlations between each load. In both of the typical datasets with which we explored, missing values at different time for different loads hampered our ability to obtain relationships through simple correlation techniques. Therefore, using DTW is a good choice for asynchronous relations. For the calculation of the distance between two load sequences, we define the k-th load with p samples as s_o^k and the l-th load with q samples as s_o^l: $s_o^k = [y_1^k, ..., y_p^k]$ and $s_o^l = [y_1^l, .., y_q^l]$. However, since the power ranges of each load are different and DTW accumulates the absolute distance, meaning two asynchronous well-correlated loads with small power may get smaller results than two asynchronous uncorrelated loads with large power. Therefore, we must normalize the load sequence in order to have meaningful DTW results:

$$s^k = \frac{s_o^k}{E(s_o^{kON})}, \tag{3}$$

where $E(s_o^{kON})$ is the mean of the k-th load's active power when the load turned ON. Then, using DTW algorithm [11], we will obtain the final DTW distance as $\mathcal{D}(s^k, s^l)$. Apply the same rule to all loads and we will get DTW standard distances between loads. Next, we translate standard distances, ranging from 0 to ∞ to range 0 to 1, to ensure the asynchronous and synchronous relations lie in the same range, by

$$r_d = e^{-\mathcal{D}(s^k, s^l)/\alpha}, \tag{4}$$

where α is a scale factor chosen manually based on the average distances. The relations between the load and the aggregated main power, r_s, and the relations between the load and the other load, r_d, may thus be applied to the weighted edges in A in our graph $\mathcal{G}$. The entire process of how the adjacency matrix A is derived from loads and main power is shown in the Algorithm 1. The proposed method utilizes two weeks of known load data $S = [s^1, ..., s^N]$ concatenated from Eq. 3 in the given house to perform our algorithm. The standard distance between the k-the load and l-th load is the element positioned at the k-the row/column l-th column/row in A. By resampling the aggregated power sequence X to X^k, which has aligned samples with s^k in the given time, spearman correlation r_s^k can be performed simultaneously. Then the r_s^k could be placed in A's k-th row/column $N + 1$-th column/row. With two relations being calculated and put into the appropriate locations in A, the building of the adjacency matrix for the designed weighted graph is completed.

By mapping our design into standard GCN layer [15], our proposed method completes the initial distilling. With the repeated GCN layer nodes fully connected to the MLP layer at the output, the GraphNILM will return N results representing the disaggregated power readings for N separate loads.

Algorithm 1: Adjacency matrix from DTW and correlation

---
    input : X, S
    output: The adjacency matrix A
1 Start $\alpha = 1000$, $A = [\mathbf{1}]_{N+1, N+1}$
2 **while** *not all edges, r_s, r_d, in A have been computed* **do**
3     $\mathcal{D}(0,0) \leftarrow 0$ ▷ Initialize the start point
4     $s^k, s^l \sim S$ ▷ Sample s^k, s^l from S
5     $r_d = \exp(-\mathcal{D}(s^k, s^l)/\alpha)$ ▷ Equation (4)
6   X^k, X^l ▷ Resample X to match s^k, s^l
7   r_s^k, r_s^l ▷ X^k, s^k and X^l, s^l to perform correlation
8     $A(k, l) \leftarrow r_d, \quad A(l, k) \leftarrow r_d$
9     $A(N+1, k) \leftarrow r_s^k, \quad A(k, N+1) \leftarrow r_s^k$
10    $A(N+1, l) \leftarrow r_s^l, \quad A(l, N+1) \leftarrow r_s^l$
11 **end**

---

4 Experiment

This study involves the examination of two mainstream open-access datasets: REDD [16], UK-DALE [14]. All datasets have labeled appliance-level power consumption along with whole-house power consumption. We also use NILMTK [3] for data prepossessing and comparing results among benchmarking algorithms. All these algorithms are implemented in Python3 and run on NVIDIA QUADRO P5000 GPU. The model is implemented using Pytorch and Pyg. To ensure the consistency of our results, each experiment was performed 20 times with a fixed seed. The Adam optimizer was employed, with a learning rate 0.005 and the batch size is 1024. The training and testing split is 0.8 vs 0.2. The loss function applied in the proposed model is L2 loss. Standard normalization is used on both input and output with data from each appliance normalized separately. Any negative values post-denormalization are set to 0. An early stop was implemented after 49 non-improved validation losses. The reason for setting the threshold at 49 for monitoring validation loss is due to fluctuations observed in the validation loss. Initially, the model fits into the mean value of each appliance, causing the loss to decrease. However, once the mean is fitted, the model begins the disaggregation process, thereby inducing an increase in validation loss. The validation loss subsequently decreases once the disaggregation pattern is discerned. In addition to Seq2Point and GraphNILM model, Seq2MultiPoint model whose structure is largely similar to that of GraphNILM except the GCN layers is also tested for ablation study. GraphNILM* is for investigating the proposed model without first-order difference. This paper also evaluates the classical FHMM method for comparison.

4.1   Dataset

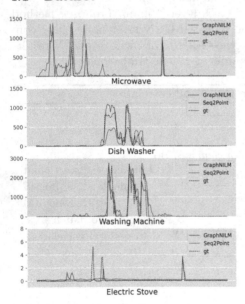

Fig. 3. REDD House 1 case study.

The **REDD dataset** contains 6 residential houses with 17 uniquely labeled appliances in Boston area from April 2011 to Jun 2011. Though it recorded the power consumption at very low rate, extending up to fifty seconds in some cases, we still choose the training data from 2011-04-20 to 2011-04-30 and test from 2011-05-01 to 2011-05-03 for classic dataset results comparison. **The UK-DALE dataset** contains 5 residential houses with 62 load-level unique labels in Southern England from November 2012 to January 2015. The experiment elected to utilize data from house 1 due to the superior number of appliances and data points collected every six seconds for each appliance therein. The training data is from 2014-02-01 to 2014-02-14 for the completeness of data during this period; the testing data is from 2014-02-15 to 2014-02-28. DTW relations are calculated from 2014-02-01 3 am to 2014-02-07 3am.

4.2   Metrics

For evaluating the performance, MAE and NEP metrics were chosen since they are the most encountered metrics to assess the disaggregated energy [12]. Mean

Absolute Error (MAE): MAE measures how accurately the disaggregated energy is compared to the true energy consumption. Normalized Error in Assigned Power (NEP) is an accuracy measures across different appliances.

$$\text{NEP} = \sqrt{\sum_{i=1}^{N}\sum_{t=1}^{T}(\hat{y}_t^i - y_t^i)/\sum_{i=1}^{N}\sum_{t=1}^{T}(y_t^i)}. \tag{5}$$

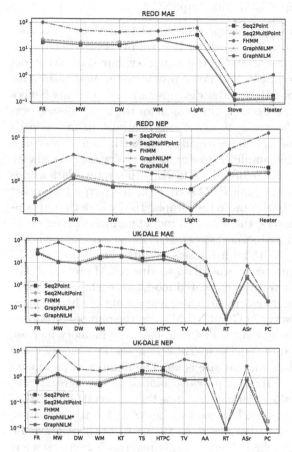

When the requirement is to compare performance across different appliances, the NEP provides a more effective measurement framework. The larger the MAE and NEP values are, the more error is produced by the model compared to the ground truth.

4.3 Results

Figure 3 shows the disaggregation results for 4 devices in REDD, revealing how Graph-NIML captures the patterns for different appliances as Seq2Point model does.

REDD. Follow the methodology discussed earlier, seven appliances in REDD dataset were chosen with the results in Fig. 4. Overall the proposed model produces comparable results compared to BM model when we only use 3.9% of memory compared to BM deep learning models. For classic method FHMM which uses least of memory resources, its performance is much worse than all the deep

Fig. 4. Comparison w.r.t. MAE and NEP (log y-axis). (left to right) Fridge, Microwave oven, Dish washer, Washing machine, Kettle, Toaster, Audio amplifier, Router, Active subwoofer, Computer.

learning models. The proposed model has 4 best performance appliances while BM model has 3 best performance appliances. Not surprisingly, Seq2Point model shows slightly better MAE and NEP on disaggregating fridge and microwave, whose pattern could be more easily to learn from separate single model. Graph-NILM performs closely to Seq2Point models on these three loads. With load relationship taken into consideration, washing machine, light, stove and heater results from the proposed model outperform results from the BM models. To understand the impact of the DTW and GCN layers in GraphNILM, it is compared to Seq2MultiPoint: GraphNILM is better in 6 out of 7 appliances in MAE and NEP, which means the GraphNILM solution is constantly providing extra information needed to diaggregate energy consumption. The first order difference introduced in the proposed design also contributes to the overall better performance by comparing GraphNILM and GraphNILM*.

UK-DALE. The same methodology is adopted here and 12 appliances are chosen in UK-DALE dataset. The result is shown in Fig. 4. In this experiment, the proposed method achieves the same level of performance compared to BM model while only uses 2.6% of memory. Still, Seq2Point produces the best results for fidge and microwave while GraphNILM better disaggregates all other appliances except the washing machine in this dataset. One reason for the better disaggregation in GraphNILM should be the strong asynchronous relationship among loads observed by DTW: the owner of the house usually uses kettle and toaster in a timely order, and computer, audio amplifier, router, active subwoofer are always functioning in nearby time slot. Therefore, these loads with strong relationships converted to graph edge weights in GraphNILM seem to help it outperform other models. When comparing Seq2MultiPoint to the GraphNILM where the only difference is the DTW and GCN layers, the proposed model performs better on 11 out 12 devices in both MAE and NEP, which stresses the importance of the DTW and GCN in the proposed design. With both datasets' results, the proposed model shows a good and reliable performance in general for solving NILM problems. The popular SOA solution Seq2Point trains a dedicated model for each chosen appliance separately hence the total memory size increases linearly along with the number of appliances increases. However in GraphNILM since the number of parameter increase is only nodes and edges in the distilling part, the memory increase is much less significant compared to Seq2Point. Table 1 shows the memory usage for Seq2Point and GraphNILM at window size is 99. The number of parameter in Seq2Point is calculated using NILMTK provided model. To disaggregate one device, Seq2Point requires 3.6 million parameters while the proposed GraphNILM model uses 832 thousands parameters. In a modern home, at least 5 appliances are presented to be disaggregate. In this way more than 80% of the energy consumption could be explained. GraphNILM model only requires 5.2% parameters compared to Seq2Point to disaggregate 5 devices. Besides, in UK-DALE experiment, the runtime for GraphNILM is 98.32 s while Seq2Point requires 254.74 s for

Table 1. Memory usage of Seq2Point and GraphNILM

Load amounts	1	4	8	12
Seq2Point	3.6 MB	14.4 MB	28.8 MB	43.2 MB
GraphNILM	832 KB	832 KB	832 KB	832 KB

12 loads training. Therefore, in terms of memory saving, efficiency and transfer implementation, GraphNILM shows competitive advantage.

Transferability. This paper also did a quick study on the transferability of our model by using our trained model from UK-DALE House 1 to predict UK-DALE House 5. The chosen houses have similar amounts and categories of loads, which is more like the office building usecase. Table 2 shows the proposed model is at least one sigma better than the Seq2MultiPoint model, stressing the DTW and GCN importance again in the proposed design.

Table 2. UK-DALE House 1 Model transferred to House 5 under metric 3*MAE

Model	FR	MW	DW	WM	KT	TS	HTPC	TV	AA	RT	ASr	PC	Overall
GraphNILM	**43.15**	30.04	22.73	**57.20**	**16.61**	5.64	**63.17**	24.20	**23.61**	**6.03**	4.03	**13.07**	**309.53 ± 13.99**
Seq2MultiPoint	44.37	**29.83**	**20.87**	66.56	18.68	**5.33**	70.22	**20.84**	25.10	6.03	**2.75**	13.08	323.72 ± 13.99

5 Conclusion

GraphNILM outperforms the benchmarks in terms of both the total memory saving, runtime efficiency and the overall MAE performance. Especially for loads with evident relationships, such as the TV, toaster, and kettle groups, the proposed method produces nearly all better results than the current state-of-art method. Even for an independent working device like a fridge or washer, Graph-NILM achieves comparable satisfactory results based on MAE and NEP. Given the proposed framework only consumes up to the reciprocal of the total load amount of the memory size in the benchmark, the computational cost of a house with typical loads is drastically reduced. Therefore, extensive experiments conducted on REDD and UK-DALE demonstrate the extraordinary competitiveness of less memory usage and better performance provided by GraphNILM. The deployment of the NILM technique in edge devices for commercial use seems to be around the corner.

References

1. Armel, C., Gupta, A., Shrimali, G., Albert, A.: Is disaggregation the holy grail of energy efficiency? the case of electricity. Energy Policy **52**, 213–234 (2013)
2. Batic, D., Tanoni, G., Stankovic, L., Stankovic, V., Principi, E.: Improving knowledge distillation for non-intrusive load monitoring through explainability guided learning. In: 2023 IEEE International Conference on Acoustics, Speech and Signal Processing (ICASSP) (ICASSP 2023), pp. 1–5 (2023)
3. Batra, N., et al.: Nilmtk: an open source toolkit for non-intrusive load monitoring (2014)
4. Çimen, H., Çetinkaya, N., Vasquez, J.C., Guerrero, J.M.: A microgrid energy management system based on non-intrusive load monitoring via multitask learning. IEEE Trans. Smart Grid **12**(2), 977–987 (2020)

5. De Baets, L., Ruyssinck, J., Develder, C., Dhaene, T., Deschrijver, D.: Appliance classification using VI trajectories and convolutional neural networks. Energy Build. **158**, 32–36 (2018). https://doi.org/10.1016/j.enbuild.2017.09.087
6. Dong, H., Wang, B., Lu, C.T.: Deep sparse coding based recursive disaggregation model for water conservation. In: Twenty-Third International Joint Conference on Artificial Intelligence (2013)
7. Faustine, A., Mvungi, N.H., Kaijage, S.F., Kisangiri, M.: A survey on non-intrusive load monitoring methodies and techniques for energy disaggregation problem. arXiv preprint arXiv:1703.00785 (2017)
8. Ferraz, F.C., Monteiro, R.V.A., Teixeira, R.F.S., Bretas, A.S.: A siamese CNN + KNN-based classification framework for non-intrusive load monitoring. J. Control Automat. Electric. Syst.
9. Ghosh, S., Chatterjee, A., Chatterjee, D.: Extraction of statistical features for type-2 fuzzy Nilm with IoT enabled control in a smart home. Exp. Syst. Appl. **212**, 118750 (2023)
10. Hart, G.W.: Prototype nonintrusive appliance load monitor. Tech. Rep. Progress Report 2, MIT Energy Laboratory (1985)
11. He, K., Stankovic, V., Stankovic, L.: Building a graph signal processing model using dynamic time warping for load disaggregation. Sensors **20**, 6628 (2020)
12. Huber, P., Calatroni, A., Rumsch, A., Paice, A.: Review on deep neural networks applied to low-frequency Nilm (2021)
13. Kelly, J., Knottenbelt, W.: Neural nilm: deep neural networks applied to energy disaggregation. In: Proceedings of the 2nd ACM International Conference on Embedded Systems for Energy-Efficient Built Environments, pp. 55–64 (2015)
14. Kelly, J., Knottenbelt, W.J.: The UK-DALE: a dataset recording UK domestic appliance-level electricity demand and whole-house demand. arXiv preprint arXiv:1404.0284 (2014)
15. Kipf, T.N., Welling, M.: Semi-supervised classification with graph convolutional networks (2016)
16. Kolter, J., Johnson, M.: Redd: a public data set for energy disaggregation research. Artif. Intell. **25** (2011)
17. Kolter, J., Jaakkola, T.: Approximate inference in additive factorial HMMs with application to energy disaggregation. In: International Conference on Artificial Intelligence and Statistics, pp. 1472–1482 (2012)
18. Liu, B., Zhang, J., Luan, W.: Load oscillation pattern detection for nilm based on scale space decomposition. In: 2023 8th Asia Conference on Power and Electrical Engineering (ACPEE), pp. 1590–1595 (2023)
19. Liu, Y., Xu, Q., Yang, Y., Zhang, W.: Detection of electric bicycle indoor charging for electrical safety: a NILM Approach. IEEE Trans. Smart Grid **14**(5), 3862–3875 (2023)
20. Mauch, L., Yang, B.: A New Approach for Supervised Power Disaggregation by Using a Deep Recurrent LSTM Network, pp. 63–67 (2015)
21. Oord, A.v.d., et al.: Wavenet: a generative model for raw audio (2016)
22. Parson, O., Ghosh, S., Weal, M., Rogers, A.: Non-intrusive load monitoring using prior models of general appliance types (2012)
23. Sykiotis, S., et al.: Performance-aware NILM model optimization for edge deployment. IEEE Trans. Green Commun. Netw. **7**(3), 1434–1446 (2023). https://doi.org/10.1109/TGCN.2023.3244278
24. Vine, D., Buys, L., Morris, P.: The effectiveness of energy feedback for conservation and peak demand: a literature review. Open J. Energy Efficiency **2**, 2013 (2013)

25. Zhang, B., Zhao, S., Shi, Q., Zhang, R.: Low-rate non-intrusive appliance load monitoring based on graph signal processing. In: 2019 International Conference on Security, Pattern Analysis, and Cybernetics (SPAC) (2019)
26. Zhang, C., Zhong, M., Wang, Z., Goddard, N., Sutton, C.: Sequence-to-point learning with neural networks for nonintrusive load monitoring (2016)
27. Zhao, B., Stankovic, L., Stankovic, V.: Blind non-intrusive appliance load monitoring using graph-based signal processing. In: 2015 IEEE Global Conference on Signal and Information Processing (GlobalSIP) (2015)

A Contraction Tree SAT Encoding
for Computing Twin-Width

Yinon Horev[1], Shiraz Shay[1], Sarel Cohen[1], Tobias Friedrich[2], Davis Issac[2],
Lior Kamma[1], Aikaterini Niklanovits[2], and Kirill Simonov[2(✉)]

[1] The Academic College of Tel Aviv-Yaffo, Tel Aviv-Yafo, Israel
{yinonho,shirazsh,sarelco,liorkm}@mta.ac.il
[2] Hasso Plattner Institute, University of Potsdam, Potsdam, Germany
{tobias.friedrich,davis.issac,aikaterini.niklanovits,
kirill.simonov}@hpi.de

Abstract. Twin-width is a structural width parameter and matrix invariant introduced by Bonnet et al. [FOCS 2020], that has been gaining attention due to its various fields of applications. In this paper, inspired by the SAT approach of Schidler and Szeider [ALENEX 2022], we provide a new SAT encoding for computing twin-width. The encoding aims to encode the contraction sequence as a binary tree. The asymptotic size of the formula under our encoding is smaller than in the state-of-the-art relative encoding of Schidler and Szeider. We also conduct an experimental study, comparing the performance of the new encoding and the relative encoding.

Keywords: Twin-width · SAT encoding

1 Introduction

Twin-width is a graph and matrix invariant recently introduced by Bonnet et al. [4,5,7], inspired by a width invariant defined on permutations by Guillemot and Marx [9]. Since its inception, twin-width received tremendous interest in the scientific community. From the algorithmic perspective, the benefits of twin-width are twofold. First, many diverse graph families are known to have bounded twin-width, for example graphs of bounded treewidth or clique-width, graphs excluding a fixed minor, planar graphs, posets of bounded width (in particular, unit interval graphs) [7]. Second, many NP-hard problems are solvable in polynomial time on graphs of bounded twin-width.

The latter property is formalized by Bonnet et al. [7] as follows: Given an n-vertex graph G, a witness that its twin-width is at most d, and a first-order sentence ϕ, it can be decided whether ϕ holds on G in $f(d, \phi)n$ time, where f is some computable function. It is worth noting, that this result does not give directly algorithms with practical running times since f is an extremely fast-growing function; this is a common drawback of algorithmic meta-theorems. However, several important NP-hard problems that are expressible by a first-order sentence

D.-N. Yang et al. (Eds.): PAKDD 2024, LNAI 14646, pp. 444–456, 2024.
https://doi.org/10.1007/978-981-97-2253-2_35

of bounded size, are also known to be solvable on graphs of bounded twin-width directly via dynamic programming with efficient running times; in particular, k-CLIQUE, k-DOMINATING SET, and k-ONES SAT for bounded k [5,8].

Given the above, the natural question is, whether the twin-width of a graph can be computed exactly. That is, given a graph G, can we find the smallest d such that d-contraction sequence exists? Unfortunately, in the general case this turns out to be intractable: even deciding whether a graph has twin-width 4 is NP-complete [2]. While for many graph classes twin-width is bounded, very few results are known for computing (or approximating) twin-width even when the given graph comes from a special graph class. For example, Král and Lamaison [11] showed that planar graphs have twin-width at most 8; however it is wide open whether we can compute the twin-width of a *given* planar graph (since it may also be smaller than 8). This motivates turning our attention to heuristic methods of computing twin-width. The high demand for such results is also illustrated by the 2023 edition of the PACE challenge[1], which focuses exclusively on computing twin-width. For the purpose of this work we only consider algorithms that yield a provably optimal contraction sequence; however, the running time is not necessarily bounded by a polynomial in n in general.

This line of research was pioneered by Schidler and Szeider [14], who devised a SAT encoding for computing twin-width. By supplying that encoding to a SAT solver, they were able to identify the exact value of twin-width for a variety of named graphs. In fact, they presented two different SAT encodings called *absolute* and *relative*. Interestingly, in all their tests the relative encoding vastly outperforms the absolute encoding, despite the fact that the formula in the relative encoding is larger: $\mathcal{O}(n^4)$ clauses versus $\mathcal{O}(n^3)$ clauses for the absolute encoding for an n-vertex graph. To the best of our knowledge, no other SAT encodings for computing twin-width were studied so far.

Our Contribution. In this work we propose an alternative SAT encoding for computing twin-width, which is conceptually different from the encodings of Schidler ans Szeider [14]. We prove formally the correctness of the encoding, and argue that the size of the formula in the encoding is only $\mathcal{O}(n^3)$ clauses, which is a asymptotically smaller than $\Omega(n^4)$ in the relative encoding of [14]. We also supply an implementation of the encoding, and conduct empirical tests comparing the performance of our encoding to that of the state-of-the-art relative encoding. The results show that there are cases where the new encoding allows to compute twin-width much faster, although the converse happens as well. We highlight that our main contribution is presenting a new SAT-encoding based on a binary contraction tree that is significantly different than the encoding presented so far. In the context of our theoretical analysis, we prove the correctness of our

---

[1] PACE stands for Parameterized Algorithms and Computational Experiments; the challenge is dedicated to bringing the gap between theoretical and practical parameterized algorithms. The official website of the challenge is https://pacechallenge.org/2023/.

SAT-encoding and analyze the big-O size of the formula, while the experimental part is mainly to empirically validate the correctness and feasibility of the encoding.

Related Work. Here we list some of the known results on twin-width. Graphs of twin-width 0 are exactly cographs, and can be recognized in poly-time [7]. Later it was shown that graphs of twin-width 1 can also be recognized in poly-time [6]. Jacob and Pilipczuk [10] show, among other results, that twin-width of a graph is at most $3 \cdot 2^{tw-1}$, where tw is the treewidth of the graph; it is also known that twin-width can be exponential in treewidth [3]. Balabán and Hliněný [1] show that twin-width is linear in the poset width, which implies that twin-width of unit interval graphs is at most two, and can be computed in poly-time. Twin-width of planar graphs is at most 8 [11], and can be as high as 7 [12].

2   Preliminaries

All graphs mentioned in this paper are simple, undirected and finite and we use standard graph-theoretic notations. In particular, given a graph G, we denote by $V(G)$ and $E(G)$ its set of vertices (or nodes) and edges respectively. Moreover, given a vertex set $S \subseteq V(G)$ we denote by $G[S]$ and $G - S$, the graph induced by the vertices of S and the graph induced by the vertices $V(G) - S$ respectively. When referring to the *open neighborhood* of a vertex $v \in V(G)$, i.e. the set of neighbors of v without v, we write $N_G(v)$, while we omit G when the graph we refer to is clear from the context. Similarly, we denote by $N_G[v]$ the *closed neighborhood* of v, i.e. $N_G(v) \cup v$. Two distinct vertices u, v are called *false twins* if $N(u) = N(v)$ and *true twins* if $N[u] = N[v]$. Given a pair $u, v \in V(G)$ we characterize the process of deleting those vertices and creating a new one with neighborhood $N(u) \cup N(v)$ as *contraction of u, v*. The graph that occurs from G after the contraction of u, v is denoted by $G/u, v$.

We now proceed in defining twin-width, following the definition of Bonnet et al. in [7]: A graph $G = (V, B, R)$ is a *trigraph* if B and R are two disjoint sets of edges on V (referred to as black and red respectively). Note that an ordinary graph is a trigraph where $B = E(G)$ and $R = \emptyset$. A trigraph (V, B, R) such that (V, R) has maximum degree d is called a *d-trigraph*. The neighborhood of a vertex v on a trigraph is all of its adjacent vertices, regardless of the color of the edge that connects them. Given a trigraph $G = (V, B, R)$ and a pair of distinct vertices $u, v \in V(G)$ we define the trigraph $G' = G/u, v$ such that, for the neighbors of the vertex w that occur from the contraction the following holds: A vertex $x \in N(w)$ is connected to w through a black edge if and only if it was connected through a black edge to both u, v in G. Otherwise (if at least one was already connected through a red edge or if x was not adjacent to both of the contracted vertices), the edge connecting x to w in G' is red.

A sequence of *d-contractions* for a trigraph is a sequence of d-trigraphs $G_0, G_1, \ldots, G_{n-1}$, where $G_0 = G$, $|V(G_{n-1})| = 1$ and G_i for $i \geq 1$ is obtained from G_{i-1} by a contraction (see Fig. 1 for an example). The twin-width of a graph is the smallest d for which a d-sequence exists and is denoted by $tww(G)$. Such a sequence is called a d-*contraction sequence*.

Since the main result of this paper is based on encoding a contraction sequence as a binary tree, we also define what the Conjunctive Normal Form (CNF) of a formula is. A literal is a (propositional) variable or the negation of it, and we call a clause a disjunction of literals. A formula ϕ is in Conjunctive Normal Form (CNF) if it is the conjunction of clauses.

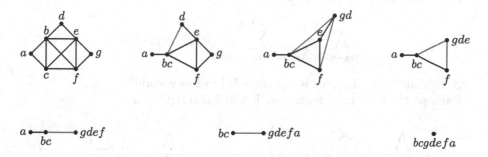

Fig. 1. A 2-contraction sequence of a graph.

3   Binary SAT Encoding

To obtain our binary SAT encoding of the d-contraction sequence problem we express all restrictions into propositional logic and then convert them to CNF. Throughout the description of our encoding we refer to $V(G)$ as vertices, and to $V(T)$ as nodes.

Let $G = (V, E)$ be the input graph and consider an initialization of E, denoting by $\neg edge_{i,j}$ the non-existing edges between vertices i, j, by $edge_{i,j}$ the existing ones, and by $\neg red_{i,j}$ the absence of red edges on the input graph.

$$\left(\bigwedge_{\substack{i,j \in [n], j > i \\ ij \notin E}} \neg edge_{i,j} \right) \wedge \left(\bigwedge_{\substack{i,j \in [n], j > i \\ ij \in E}} edge_{i,j} \right) \wedge \left(\bigwedge_{i,j \in [n], j > i} \neg red_{i,j} \right)$$

Observe that a contraction sequence can be represented by a rooted binary tree, where the leaves correspond to the vertices of G, internal nodes to subsequent contractions, and the root corresponds to the final contraction. Since twin-width is always at most d when the number of vertices does not exceed $d + 1$, we can avoid encoding the final $d + 1$ contractions in the sequence. In terms of the tree representation, this corresponds to removing the nodes of the tree representing the final $d + 1$ contractions, which results in a binary forest with at most $d+2$ trees. We refer to the binary forest of the sought-after optimal contraction sequence as T. We note that $|V(T)| \leq 2n - d - 2$ since the whole contraction tree contains $2n - 1$ nodes as a binary tree with $n = |V(G)|$ leaves, and the nodes corresponding to the final $d + 1$ contractions are not considered.

To encode T we define variables $lc_{i,j}$ and $rc_{i,j}$ for each pair of vertices, which are true if and only if j is the left child of i and j is the right child of i respectively. Using these variables we ensure that T is binary by encoding the following:

- Each node has at most one parent.

$$\bigwedge_{i \in [2n-d-3]} \left\langle \sum_{j=max(i+1,n+1)}^{2n-d-2} (lc_{j,i} + rc_{j,i}) \leq 1 \right\rangle$$

Note that the final vertex is not included in this encoding.
- Each parent node has exactly one left and one right child.

$$\bigwedge_{j \in [n+1, 2n-d-2]} \left\langle \sum_{i \in [j]} rc_{j,i} = 1 \right\rangle \qquad \bigwedge_{j \in [n+1, 2n-d-2]} \left\langle \sum_{i \in [j]} lc_{j,i} = 1 \right\rangle$$

Lemma 1. *The encoding above produces a binary forest on $2n - d - 2$ nodes.*

Proof. Assume that there is a node v_i that has at least two parents v_x, v_y, where $x, y > i$. Then the clause $\bigwedge_{i \in [2n-d-3]} \left\langle \sum_{j=max(i+1,n+1)}^{2n-d-2} (lc_{j,i} + rc_{j,i}) \leq 1 \right\rangle$ corresponding to v_i is false since $\sum_{j=max(i+1,n+1)}^{2n-d-2} (lc_{j,i} + rc_{j,i})$ includes $(lc_{x,i} + rc_{x,i}) + (lc_{y,i} + rc_{y,i})$ which is equal to 2, leading to contradiction. Note that we construct those clauses only considering values greater than n, since all the vertices of G correspond to leaf nodes.

Regarding the fact that each parent has exactly one left and one right child, we do not consider the leaves and hence we start from $n + 1$. Here it suffices to note that for each possible parent node we check all the possible right and left children to ensure that it has exactly one of each.

We proceed with encoding the already contracted vertices through the variable $vanish_{p,i}$. In particular, we define this variable to be true if i is contracted to any vertex with number at most p. Intuitively, this variable is true if and only if at the time when the node p is formed through some contraction, the node i doesn't exist any more. To implement the semantics we encode the following:

- Each leaf node is vanished at the moment its parent node is formed and hence, for each $i \in [n]$

$$vanish_{n+1,i} \iff lc_{n+1,i} \lor rc_{n+1,i}$$

- Iteratively defined, an internal node i is already vanished when p is formed, either if it is one of its children or if it has been vanished in some previous contraction, due to it being a child of some other contracted node. Hence, for each $p \in [n + 2, 2n - d - 2]$, $i \in [p - 1]$

$$vanish_{p,i} \iff lc_{p,i} \vee rc_{p,i} \vee vanish_{p-1,i}$$

– Lastly, in order to ensure that no node is considered to be "contracted to itself" at the time it is formed, we set for each $p \in [n + 1, 2n - d - 2]$

$$\neg vanish_{p,p}$$

We now encode the edges of G between the children of some node and the other nodes. In particular, the left adjacency of i, p, denoted $la_{i,p}$, means that there is an edge between the vertex represented by the left child of p and the one represented by i. Similarly the right adjacency of i, p is defined, and denoted by $ra_{i,p}$. In the list encoding these adjacencies below, the first argument refers to the child of p while the second denotes whether or not an edge (red or black) to vertex c exists. The minimum value is placed first when choosing the edge, to be consistent with how we initially encoded $E(G)$. For each $p \in [n + 1, 2n - d - 2]$, $c \in [p - 1]$, $i \in [p - 1]$ such that $i \neq c$, and similarly for their negation

$$lc_{p,c} \wedge edge_{min(i,c),max(i,c)} \Rightarrow la_{i,p}, \qquad rc_{p,c} \wedge edge_{min(i,c),max(i,c)} \Rightarrow ra_{i,p}$$
$$lc_{p,c} \wedge red_{min(i,c),max(i,c)} \Rightarrow lr_{i,p}, \qquad rc_{p,c} \wedge red_{min(i,c),max(i,c)} \Rightarrow rr_{i,p}$$
$$lc_{p,c} \wedge \neg edge_{min(i,c),max(i,c)} \Rightarrow \neg la_{i,p}, \quad rc_{p,c} \wedge \neg edge_{min(i,c),max(i,c)} \Rightarrow \neg ra_{i,p}$$
$$lc_{p,c} \wedge \neg red_{min(i,c),max(i,c)} \Rightarrow \neg lr_{i,p}, \quad rc_{p,c} \wedge \neg red_{min(i,c),max(i,c)} \Rightarrow \neg rr_{i,p}$$

Using the right and left adjacencies of each node, we are able to encode the edges created from the contractions. In particular, an edge connecting node i that occurs from some contraction to another node j exists, if any of the children of i (the ones that get contracted in order to create i) is adjacent in G to j, and j still exists at the moment i is created. Moreover, this edge is red if exactly one of those children is adjacent to j. Formally, this is encoded as follows:
For each $i \in [n + 1, 2n - d - 2]$, $j \in [i - 1]$

$$edge_{j,i} \iff (la_{j,i} \vee ra_{j,i}) \wedge \neg vanish_{i,j}$$
$$red_{j,i} \iff edge_{j,i} \wedge (lr_{j,i} \vee rr_{j,i} \vee (la_{j,i} \oplus ra_{j,i}))$$

The encoding of the edges occurring from the contraction we described above $edge_{j,i} \iff (la_{j,i} \vee ra_{j,i}) \wedge \neg vanish_{i,j}$ is converted to CNF as follows:

$$(\neg edge_{j,i} \vee la_{j,i} \vee ra_{j,i}) \wedge (\neg vanish_{i,j} \vee \neg edge_{j,i}) \wedge (\neg la_{j,i} \vee edge_{j,i} \vee vanish_{i,j})$$
$$\wedge (\neg ra_{j,i} \vee edge_{j,i} \vee vanish_{i,j}))$$

Similarly for the encoding of the red edges, $red_{j,i} \iff edge_{j,i} \wedge (lr_{j,i} \vee rr_{j,i} \vee (la_{j,i} \oplus ra_{j,i}))$ we have:

$$red_{j,i} \iff edge_{j,i} \wedge (lr_{j,i} \vee rr_{j,i} \vee ((la_{j,i} \vee ra_{j,i}) \wedge (\neg la_{j,i} \vee \neg ra_{j,i})))$$

$$red_{j,i} \iff (edge_{j,i}) \wedge (lr_{j,i} \vee rr_{j,i} \vee la_{j,i} \vee ra_{j,i})$$
$$\wedge(lr_{j,i} \vee rr_{j,i} \vee \neg la_{j,i} \vee \neg ra_{j,i})$$

$$\Rightarrow (\neg red_{j,i} \vee edge_{j,i}) \wedge (\neg red_{j,i} \vee lr_{j,i} \vee rr_{j,i} \vee la_{j,i} \vee ra_{j,i})$$
$$\wedge (\neg red_{j,i} \vee lr_{j,i} \vee rr_{j,i} \vee \neg la_{j,i} \vee \neg ra_{j,i})$$
$$\Leftarrow (red_{j,i} \vee \neg edge_{j,i} \vee \neg lr_{j,i}) \wedge (red_{j,i} \vee \neg edge_{j,i} \vee \neg rr_{j,i})$$
$$\wedge (red_{j,i} \vee \neg edge_{j,i} \vee \neg la_{j,i} \vee ra_{j,i}) \wedge (red_{j,i} \vee \neg edge_{j,i} \vee \neg ra_{j,i} \vee la_{j,i})$$

Now, in order to encode the red un-vanished edges at the moment of a contraction, to be able to restrict the maximum degree we introduce a new variable $reduv_{i,j,k}$. This is created at the moment node i is formed, and a red edge between the nodes j and k exists (and the nodes j and k of course still have not vanished). For each $i \in [n+1, 2n-d-2]$, for each $j \in [i]$, for each $k \in [j+1, i]$,

$$reduv_{i,j,k} \iff \neg vanish_{i,j} \wedge \neg vanish_{i,k} \wedge red_{j,k}$$

This is expressed in CNF as:

$$(\neg vanish_{i,j} \vee \neg reduv_{i,j,k}) \wedge (\neg vanish_{i,k} \vee \neg reduv_{i,j,k}) \wedge (red_{j,k} \vee \neg reduv_{i,j,k})$$
$$\wedge(vanish_{i,j} \vee vanish_{i,k} \vee \neg red_{j,k} \vee reduv_{i,j,k})$$

Lastly we need to ensure that the maximum red degree at any point is at most d which is expressed as

$$\bigwedge_{\substack{i \in [n+1, 2n-d-2], \\ j \in [i]}} \left\langle \sum_{k \in [i], j \neq k} reduv_{i, \min(j,k), \max(j,k)} \leq d \right\rangle$$

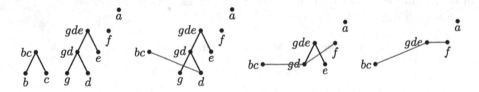

Fig. 2. The contraction forest and the sequence up to $2n-d-2$ for $d=2$ that occurs through the binary encoding for the 2-contraction sequence of Fig. 1. Observe that for $d=2$ we stop once 4 vertices remain since after the next contraction a graph on 3 vertices occurs which can have at most 2 red degree.

Theorem 1. *Given a graph G of order n and an integer d, we construct in polynomial time a CNF formula which is satisfiable if and only if $tww(G) \leq d$, and has $\mathcal{O}(n^3)$ clauses.*

Proof. We first prove that given an assignment of the variables that satisfies the SAT formula corresponding to a graph G, we are able to build a d-contraction sequence for G. First, we create the contraction tree following the parent-child relations occuring by the assignment values for the variables $lc_{i,j}$, $rc_{j,i}$. Then, we construct a contraction sequence based on the children of $n_1, \ldots, 2n - d - 2$. By the construction of our formula we ensure that all edges of G are encoded. Also, for every contraction node i, the assignment denoting whether an edge between its children and other nodes exists, and the red edges created though the contraction of a pair of nodes, are counted, as the clauses equivalent to $edge_{j,i} \iff (la_{j,i} \vee ra_{j,i}) \wedge \neg vanish_{i,j}$ are satisfied. Lastly, if during this contraction sequence the red degree becomes greater than d, the clauses equivalent to $\bigwedge_{i \in [n+1,2n-d-2], j \in [i]} \sum_{k \in [i], j \neq k} reduv_{i,min(j,k),max(j,k)} \leq d$ are not satisfied, leading to a contradiction. Hence, the sequence occurring from contracting the child nodes of $n + 1, \ldots, 2n - d - 2$ is indeed a d-contraction sequence of G.

Given a d-contraction sequence of G we construct an assignment that satisfies the formula corresponding to G as follows: For the ith-contraction pair, with the contracted vertices being x and y, where $i \leq 2n - d - 2$ we assign 1 to $lc_{x,n+i}, rc_{y,n+i}$ and 0 for the other values of i. Hence, we have "built" our contraction tree. We also initialize $edge_{i,j}$ to 1 for the adjacent pairs of vertices in G, and $red_{i,j}$ to 0 for all pairs of vertices, to encode G. Similarly for right and left adjacencies of the nodes of T. We then follow the changes each contraction causes on G, and assign truth values to the respective variables. For example, for each variable $vanish_{p,i}$ if the vertex i has been contracted when it is p's turn to participate in some contraction, we assign 1 to it. Similarly, we assign 1 to $reduv_{i,j,k}$ for the red edges that survive each contraction. Notice that because a d-contraction sequence of G is given, due to the construction of our formula, the assignment created so far also satisfies the clauses of $\bigwedge_{i \in [n+1,2n-d-2], j \in [i]} \sum_{k \in [i], j \neq k} reduv_{i,min(j,k),max(j,k)} \leq d$, satisfying the CNF corresponding to G.

We now analyze the size of the CNF formula created from a graph G and an integer d, by counting the clauses occurring from each information encoding. For initializing $E(G)$, we use one variable for each edge, one for each non-edge and one to denote the absence of red edges, hence $\mathcal{O}(n^2)$ in total. To encode the contraction tree we use, for the "one-parent property", for each $i \in [2n - d - 3]$ at least $2(n - d - 2)$ variables, meaning $\mathcal{O}(n^2)$, and for the "binary property", for each $j \in [n + 1, 2n - d - 2]$, j variables for each child, hence $\mathcal{O}(n^2)$ in total. When converted to CNF (the same holds for bounding the red degree by d, by [13]) we get $\mathcal{O}(n^2)$ clauses. For $vanish_{p,i}$ we get two literals for each leaf, at most $2n - d - 2$ for each of the internal nodes $2n - d - 2$ for the "self variables", meaning $\mathcal{O}(n^2)$ in total. To encode right and left adjacencies (edges between the vertices represented by the children of a node and vertices represented by other nodes) we have $\mathcal{O}(n^2)$ variables to choose which child to refer to, and $\mathcal{O}(n)$, meaning $\mathcal{O}(n^3)$ in total.

Table 1. Results for the PACE dataset, random graphs, and Paley graphs. The numbers under "binary" and "relative" are the running times of the encodings, in seconds.

name	n	m	tww	binary	relative	name	n	m	tww	binary	relative
EX_001	19	64	6	12.86	**9.23**	ER_0.1	30	44	3	304.23	**37.23**
EX_002	20	69	6	4.27	**1.15**	ER_0.2	30	100	6	**100.81**	933.68
EX_003	25	97	6	**48.64**	85.84	ER_0.3	30	120	7	**587.99**	703.2
EX_004	25	181	7	107.33	**68.72**	ER_0.4	–	–	–	–	–
EX_006	28	131	7	1390.23	**480.01**	ER_0.5	30	236	9	375.13	**319.4**
EX_007	28	205	7	**178.24**	181.24	ER_0.6	–	–	–	–	–
EX_008	28	210	10	26.83	**8.68**	ER_0.7	30	293	8	**409.17**	–
EX_009	28	228	7	98.99	**68.77**	ER_0.8	30	342	6	**502.77**	–
EX_010	28	235	6	735.61	**438.79**	ER_0.9	30	381	4	–	**1418.76**
EX_011	29	174	8	–	**1077.78**	P_013	13	39	6	0.25	**0.1**
EX_012	29	180	8	–	**1163.65**	P_017	17	68	8	0.83	**0.25**
EX_013	30	155	8	–	**1498.19**	P_029	29	203	14	14.27	**7.52**
EX_014	30	175	8	1026.46	**873.93**	P_037	37	333	18	85.19	**35.83**
EX_015	30	178	8	**414.99**	858.01	P_041	41	410	20	175.01	**58.7**
EX_016	30	195	8	–	**1527.49**	P_053	53	689	26	660.1	**352.25**
EX_017	30	207	8	–	**409.64**						
EX_018	31	52	3	648.12	**175.37**						
EX_019	32	90	5	**1578.88**	–						
EX_031	48	80	3	**974.02**	1206.32						
EX_034	51	240	4	484	**16.66**						
EX_035	52	53	2	253.07	**135.8**						

When encoding the existence of an edge as a relation of right and left adjacencies and the variable *vanish*, we get 4 clauses for each edge variable, and 4 for the existence of each red edge, so $\mathcal{O}(n^2)$ clauses in total. Similarly 4 clauses are created for each variable *reduv* that represents the existence of red edges at the moment of a contraction, making them $\mathcal{O}(n^3)$ in total. Lastly, for bounding the maximum red degree by d we need at most $\mathcal{O}(n^2)$ variables, which also produce $\mathcal{O}(n^2)$ clauses. Hence, the size of the formula produced by our encoding is $\mathcal{O}(n^3)$.

4 Experiments

We implemented the binary SAT encoding presented above and run it on several datasets. For the implementation, we used Python 3.10.12 with PySAT 3.1.0, and Cadical as the specific SAT solver. The tests were run on a PC with AMD Ryzen 7 6800HS CPU, 16 GB RAM, and Ubuntu 22.04, using only a single

Table 2. Comparison between binary encoding and *absolute* encoding on the tiny dataset of PACE.

name	n	m	tww	time, absolute	time, binary	factor
tiny001.gr	10	9	1	1.48	**0.09**	x16
tiny002.gr	10	10	2	3.7	**0.13**	x29
tiny005.gr	25	40	3	>10 h	**21.35**	>x2000
tiny007.gr	14	13	2	3131.77	**1.16**	x2689
tiny008.gr	10	15	4	65.80	**0.07**	x937
tiny009.gr	7	7	1	**0.01**	0.03	x0.33

thread. We also include an implementation of the relative encoding of [14], and compare the performance of our binary SAT encoding with the relative encoding. To make a fair comparison, both encodings were run using exactly the same settings and auxiliary code. We generally do not compare with the absolute encoding of [14] (except for Table 2), as it is hopelessly inefficient compared to both relative encoding and our binary encoding. We also implemented modular decomposition as the standard preprocessing step for computing twin-width; for detailed description of the preprocessing see [14]. The implementations and the testing data are provided in the supplementary material.

Next, we describe the datasets that were used for testing, and provide tables that compare performance of the binary encoding and the relative encoding. In all tests the computed twin-width value is the same for both encodings (as both are shown to output the optimal value of twin-width), so we only compare the time used by each encoding. In the results that we list, time is always measured in seconds. All tests were run with a time limit of 30 min (1800 s) and a memory limit of 10 gigabytes. Entries marked with "-" are used to denote that the solver exceeded time and/or memory limit on the corresponding instance.

Random Graphs. We first construct Erdős–Rényi graphs $G(n, p)$ where $n = 30$, and p ranges from 0.1 to 0.9 win an increment of 0.1. The graph $G(n, p)$ contains n vertices, and each edge is created with probability p. The results for Erdős–Rényi graphs are listed in Table 1 in the "ER" rows. It is interesting to note that while the relative encoding performs slightly better in the uniform setting ($p = 0.5$), and considerably better when the graph is very sparse ($p = 0.1$) or very dense ($p = 0.9$), the binary encoding vastly outperforms the relative encoding in the intermediate cases ($p \in \{0.2, 0.3, 0.7, 0.8\}$).

Moreover, the performance of both solvers is worse for larger p compared to smaller p; this implies that neither encoding is exploiting to the fullest the property that the twin-width of a graph is always equal to the twin-width of its complement. That is, computing the twin-width of $G(n, p)$ should be equally hard as computing the twin-width of $G(n, 1 - p)$, since in the complement of $G(n, p)$ each edge exists independently with probability $1 - p$.

PACE 2023 Challenge. We next compare the encodings on several instances from the PACE challenge. We first compare the binary encoding and the absolute encoding of [14] on the "tiny" sample set of the challenge[2]. The results are shown in Table 2, and highlight that the performance of the absolute encoding is much worse that of the binary encoding, despite the same asymptotic size of the encoding. For this reason, we do not include absolute encoding in the other tests.

Table 1 then shows the result of comparison between the binary and the relative encoding on the regular instances of the challenge, see rows starting with "EX". While the original dataset contains 200 graphs[3], we only list the instances where at least one of the two solvers was able to compute the twin-width under the time and memory limit of 30 min and 10GB. While the relative encoding is generally faster on this dataset, the binary encoding is still able to perform considerably better on several instances. This further shows that the binary encoding is conceptually different from the relative, and may be used to deal with the instances where the relative encoding is not efficient.

Paley Graphs. Finally, following the experiments of [14], we test the encodings on Paley graphs. We construct Paley graphs for several prime numbers; for a prime p with $p \equiv 1 (\text{mod } 4)$, Paley graph is a $\frac{p-1}{2}$-regular graph on p vertices. The vertices of the Paley graph are associated with the elements of the unique finite field of order p, and the edge between x and y appears if and only if $x - y$ is a square in the field; in the case $p \equiv 1 (\text{mod } 4)$, $x - y$ is a square if and only if $y - x$ is a square so the edges are symmetric. Last part of Table 1 lists the results. While the relative encoding is generally faster, it is interesting to observe that the memory usage of the binary encoding is lower on the Paley graphs. This may be attributed to the smaller size of the encoding, although this effect does not show on the other instances.

5   Conclusion

In this work, we introduced a novel SAT encoding for computing twin-width of a graph. Theoretically, we have shown that the encoding is sound, and that the size of the encoding is smaller than that of the state-of-the-art relative encoding of [14]. Further, we conducted experiments on several datasets, comparing the performance of the binary encoding, and the relative and absolute encoding of [14]. Similar to relative encoding, binary encoding vastly outperforms the absolute encoding. Comparing the binary and relative encoding, the results suggest that neither encoding dominates the other, while in many cases the binary encoding is much more efficient than the relative encoding (although the converse also happens). Therefore, the introduction of the binary encoding indeed improves the ability to compute twin-width in practice. This also motivates further work

---

[2] https://pacechallenge.org/2023/tiny-set.pdf.
[3] All instances are available at https://pacechallenge.org/2023/.

on comparing different twin-width encodings empirically, identifying families of instances where twin-width can be computed efficiently with either encoding.

References

1. Balabán, J., Hliněný, P.: Twin-width is linear in the Poset width. In: Golovach, P.A., Zehavi, M. (eds.) 16th International Symposium on Parameterized and Exact Computation (IPEC 2021). Leibniz International Proceedings in Informatics (LIPIcs), vol. 214, pp. 6:1–6:13. Schloss Dagstuhl – Leibniz-Zentrum für Informatik, Dagstuhl (2021). https://doi.org/10.4230/LIPIcs.IPEC.2021.6
2. Bergé, P., Bonnet, É., Déprés, H.: Deciding twin-width at most 4 is np-complete. In: Bojanczyk, M., Merelli, E., Woodruff, D.P. (eds.) 49th International Colloquium on Automata, Languages, and Programming, ICALP 2022, 4–8 July 2022, Paris. LIPIcs, vol. 229, pp. 18:1–18:20. Schloss Dagstuhl - Leibniz-Zentrum für Informatik (2022). https://doi.org/10.4230/LIPIcs.ICALP.2022.18
3. Bonnet, E., Déprés, H.: Twin-width can be exponential in treewidth. J. Comb. Theory Ser. B **161**(C), 1–14 (2023). https://doi.org/10.1016/j.jctb.2023.01.003
4. Bonnet, É., Geniet, C., Kim, E.J., Thomassé, S., Watrigant, R.: Twin-width II: small classes. In: Marx, D. (ed.) Proceedings of the 2021 ACM-SIAM Symposium on Discrete Algorithms, SODA 2021, Virtual Conference, 10–13 January 2021, pp. 1977–1996. SIAM (2021). https://doi.org/10.1137/1.9781611976465.118
5. Bonnet, É., Geniet, C., Kim, E.J., Thomassé, S., Watrigant, R.: Twin-width III: max independent set, min dominating set, and coloring. In: Bansal, N., Merelli, E., Worrell, J. (eds.) 48th International Colloquium on Automata, Languages, and Programming, ICALP 2021, 12–16 July 2021, Glasgow (Virtual Conference). LIPIcs, vol. 198, pp. 35:1–35:20. Schloss Dagstuhl - Leibniz-Zentrum für Informatik (2021). https://doi.org/10.4230/LIPIcs.ICALP.2021.35
6. Bonnet, E., Kim, E.J., Reinald, A., Thomassé, S., Watrigant, R.: Twin-width and polynomial kernels. In: Golovach, P.A., Zehavi, M. (eds.) 16th International Symposium on Parameterized and Exact Computation (IPEC 2021). Leibniz International Proceedings in Informatics (LIPIcs), vol. 214, pp. 10:1–10:16. Schloss Dagstuhl – Leibniz-Zentrum für Informatik, Dagstuhl (2021). https://doi.org/10.4230/LIPIcs.IPEC.2021.10
7. Bonnet, É., Kim, E.J., Thomassé, S., Watrigant, R.: Twin-width I: tractable FO model checking. J. ACM **69**(1), 3:1–3:46 (2022). https://doi.org/10.1145/3486655
8. Ganian, R., Pokrývka, F., Schidler, A., Simonov, K., Szeider, S.: Weighted model counting with twin-width. In: Meel, K.S., Strichman, O. (eds.) 25th International Conference on Theory and Applications of Satisfiability Testing, SAT 2022, 2–5 August 2022, Haifa. LIPIcs, vol. 236, pp. 15:1–15:17. Schloss Dagstuhl - Leibniz-Zentrum für Informatik (2022). https://doi.org/10.4230/LIPIcs.SAT.2022.15
9. Guillemot, S., Marx, D.: Finding small patterns in permutations in linear time. In: Chekuri, C. (ed.) Proceedings of the Twenty-Fifth Annual ACM-SIAM Symposium on Discrete Algorithms, SODA 2014, Portland, 5–7 January 2014, pp. 82–101. SIAM (2014). https://doi.org/10.1137/1.9781611973402.7
10. Jacob, H., Pilipczuk, M.: Bounding twin-width for bounded-treewidth graphs, planar graphs, and bipartite graphs. In: Bekos, M.A., Kaufmann, M. (eds.) Graph-Theoretic Concepts in Computer Science. WG 2022. LNCS, vol. 13453, pp. 287–299. Springer, Cham (2022). https://doi.org/10.1007/978-3-031-15914-5_21

11. Král, D., Lamaison, A.: Planar graph with twin-width seven. arXiv preprint arXiv:2209.11537 (2022)
12. Král', D., Lamaison, A.: Planar graph with twin-width seven. Eur. J. Combinator. 103749 (2023). https://doi.org/10.1016/j.ejc.2023.103749
13. Marques-Silva, J., Lynce, I.: Towards robust CNF encodings of cardinality constraints. In: Bessiere, C. (ed.) Principles and Practice of Constraint Programming. CP 2007. LNCS, vol. 4741, pp. 483–497. Springer, Cham (2007). https://doi.org/10.1007/978-3-540-74970-7_35
14. Schidler, A., Szeider, S.: A SAT approach to twin-width. In: Phillips, C.A., Speckmann, B. (eds.) Proceedings of the Symposium on Algorithm Engineering and Experiments, ALENEX 2022, Alexandria, 9–10 January 2022, pp. 67–77. SIAM (2022). https://doi.org/10.1137/1.9781611977042.6

Author Index

D.-N. Yang et al. (Eds.): PAKDD 2024, LNAI 14646, pp. 457–459, 2024.
https://doi.org/10.1007/978-981-97-2253-2

Printed in the United States
by Baker & Taylor Publisher Services

Printed in the United States
by Baker & Taylor Publisher Services